The Evolutionary Ecology of
Animal Migration

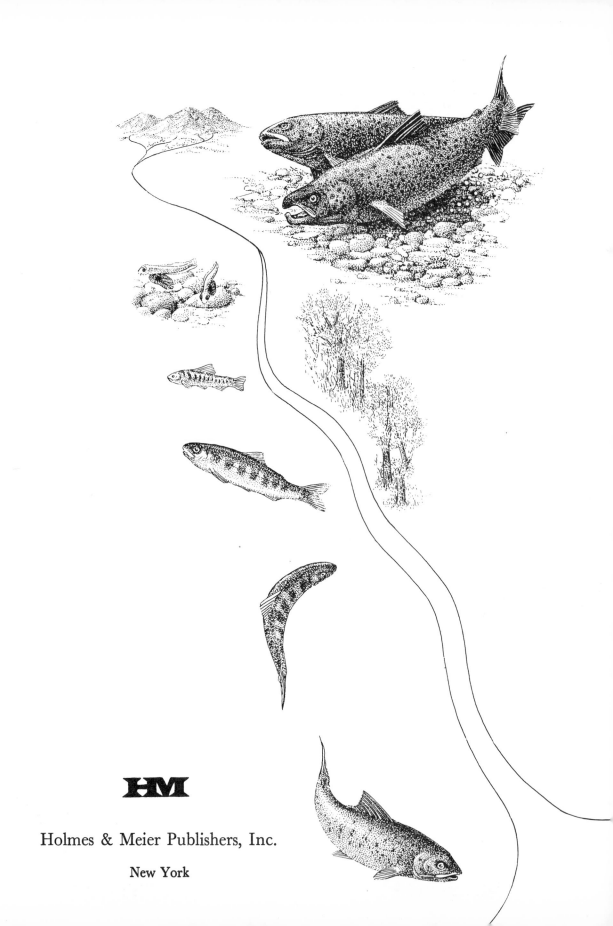

HM

Holmes & Meier Publishers, Inc.

New York

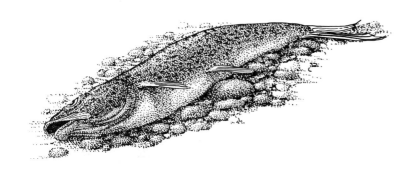

The Evolutionary Ecology of
Animal Migration

by

R R Baker

Department of Zoology,
University of Manchester

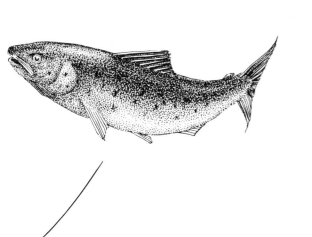

First published in the United States of America 1978 by
HOLMES & MEIER PUBLISHERS, INC.
30 Irving Place, New York, New York 10003

Library of Congress Cataloging in Publication Data

Baker, Reginald Robin, 1944–
  The evolutionary ecology of animal migration.

  Bibliography: p.
  Includes index.
  1. Animal migration.   I. Title.

QL754.B33 1978   591.5′2   78–34

ISBN 0–8419–0368–9

PRINTED IN GREAT BRITAIN

Preface and acknowledgements

It is now over forty-five years since Heape attempted to review, analyse and rationalise the ecological aspects of animal movement patterns. Since that time, students of different taxonomic groups have very much gone their own way with little or no reference to the advances being made with respect to other animals. The result has been a serious loss of perspective and increasing difficulty in initiating discussion between different factions. The situation has reached such a state that entomologists and ornithologists, students of the two animal groups that have received most attention with respect to movement patterns, at the present time each define the term 'migration' in such different ways that the movements of the other group are excluded. Thus, according to the entomological definition, birds do not migrate, whereas according to the ornithological definition insects do not migrate.

The problem of defining migration and the fact that different groups of zoologists define the term in different ways is not by itself, perhaps, particularly serious. What is serious, however, is the lack of contact between study-groups and the resulting loss of perspective that is reflected by the ornithological–entomological paradox. This situation is perhaps all the more surprising in view of the increasing value that zoologists place on a comparative approach to their subject. Moreover, during the past half-century there has been a rapid increase of interest in the movement patterns of such a wide range of animal groups that there is now no shortage of at least preliminary data on which a comparative approach could be based.

It is this background that effectively dictates the format of this book. A tripartite structure has been adopted, each part dealing with one of the major aims of the book. The first part represents an attempt to rationalise the use of the term migration. The second part represents an attempt to construct an evolutionary model of animal migration. Finally, the third and longest part represents an attempt to review the range of form of the lifetime track manifest by the Metazoa. In particular, the review attempts to place the lifetime tracks of conventional migrants (e.g. birds, whales, fish) into the perspective of the lifetime tracks of animals in general. Despite this, however, the review is not entirely an end in itself. Throughout Part III the choice of examples is geared to an evaluation of the theoretical model developed in Part II. As a result of this approach, many new interpretations of existing data and observations have suggested themselves. Some new, previously unpublished, field data are presented in Part III as also are many attempts at new collations and analyses of existing data and observations.

Although a definition and review of migration and a development and evaluation of an evolutionary model of metazoan lifetime tracks are the major aims of this book, there are other, secondary, aims. One of these, for example, is an integration, within the general framework of animal migration, of the lifetime tracks of humans, ranging from those of hunter-gatherers and nomadic pastoralists to those of modern industrialists. A second is an attempt to take a fresh look at the migration and navigation of birds and to consider whether the movements of these animals might not better be viewed from a comparative viewpoint rather than from the unique position that they are often accorded.

In the course of writing this book I have inevitably accumulated many debts of altruism with respect to my colleagues. Hopefully, I can reciprocate this altruism in full at some time in the future. In the meantime, perhaps I can make a small down-payment by acknowledging these debts in the next few paragraphs.

I am particularly indebted to Dr. Edward Broadhead of the Department of Pure and Applied Zoology at Leeds University, firstly for causing the book to be initiated at all and secondly as my Biological Editor for his patience and sympathetic encouragement each time I changed my mind over precisely what form the book should take.

I am also extremely grateful to Dr. John Croxall, the then Director of the Oiled Seabird Research Unit at Newcastle University but now with the British Antarctic Survey, who volunteered to read the entire manuscript in first draft form. Without his perceptive and constructive comments, the format of this book would have been quite different and much inferior.

A further debt is owed to Dr. Geoff. Parker of the Department of Zoology, University of Liverpool. The balance of our reciprocal altruism partnership has see-sawed frequently since we graduated together at Bristol in 1965. At the time of writing, however, there is no doubt that he is in credit. By reading the first drafts of Parts I and II of this book he saved me from committing a number of conceptual and mathematical errors. He did this despite the fact that during the $2\frac{1}{2}$ years that I was writing this book, during which time we exchanged a number of manuscripts, it became obvious to both of us that despite dealing with the superficially different problems of courtship and migration, we were nevertheless very often reaching the same conclusions. In order to avoid any confusion, however, I would like to echo Dr. Parker's statement (*in* Parker and Stuart 1976) that, except where otherwise acknowledged, we have reached these conclusions completely independently.

Finally, among those people that read parts of the manuscript, I would like to thank Dr. Judith E. Marlow (née King) of the University of New South Wales, Australia, for reading an early draft of the chapter on pinnipeds, and Dr. Peter Gabbutt of the Zoology Department, Manchester University, for reading my attempt to integrate pseudoscorpions with the rest of the animal kingdom. I am sure the accuracy of the book would have been much improved if all of the animals discussed had been subjected to the same specialist scrutiny as these two groups.

A number of other people made a considerable contribution to the production of this book. This contribution may not have been voluntary in the same sense as with the colleagues acknowledged above but was nevertheless real because of the extreme efficiency with which the people concerned carried out their everyday employment. I am particularly grateful to Mrs Lorna Appiah and her staff of the Inter-Library Loan section of the University Library at Newcastle upon Tyne. The efficient and cheerful, almost cavalier, way in which they processed a two-year long stream of requests for references allowed me to cover a much wider range of literature than I could ever have imagined possible when I first began. This acknowledgement must also be extended to that unseen entirety of the Inter-Library Loans Organisation, not only in Britain but also those sister organisations in Canada, the United States, South America, France, Germany, Russia, Japan and Australia, all of whom enabled me to include some piece of information or other that I could not otherwise have obtained.

Many of my colleagues have also contributed significantly to the range of animals and observations discussed in this book by drawing my attention to references that otherwise would certainly have escaped my attention. Among these I would like to mention Stuart Bailey, Gordon Blower, Lawrence Cook, John Dalingwater, Peter Gabbutt, Hugh Jones, and Derek Yalden of the Zoology Department, Manchester University. I would also like to thank all those colleagues and students that have allowed me to refer to their unpublished results. Due acknowledgement is given in the text and references for all such information.

Mr. J. F. Cameron of the Canadian Wildlife Service made a notable contribution to the accuracy of many sections of the text when he volunteered to send me a number of brochures in the Hinterland Who's Who series. I found these brochures an invaluable source of background biological information for a wide range of Canadian and North American animals. By putting me in touch with a number of Canadian wildlife photographers, Mr. Cameron, and Dr. John Kelsall, also of the Canadian Wildlife Service, made a further contribution, this time to any aesthetic merit that this book may be considered to possess. I would also like to thank all those photographers and agents that have submitted photographs for use in this book. They have shown considerable patience during what has been a long gestation period. Eric Hosking, Frank Lane and Bruce Coleman Ltd. all deserve special mention in this connection. Mallinson's freelancing organisation helped me to contact a number of photographers by allowing me to advertise free of charge in their journal. I also thank Gordon Howson (Newcastle) and Les Lockey (Manchester) for preparing prints for me from my slides and negatives.

A special mention is reserved for Y.S. who pointed out an alternative interpretation of Fig. 33.5.

It is also a pleasure to acknowledge the pleasant and patient way that Brian Steven, Susan Devlin, Jenny Rawlins, her predecessor as biology editor, Karen Wilby, and other members of the Hodder and Stoughton organisation, have guided and encouraged me through the tortuous process that has miraculously transformed a mountain of muddled, handwritten notes, sketches and photographs into a real, solid, ordered book. I am also grateful to Alan Whittle who copy-edited the original manuscript and read the proofs. Thanks to the efficiency with which he carried out a long and tedious task, many typographical, grammatical, and cross-reference errors that I missed during my own proof-reading did not find their way into this final version. My major thanks in this section, however, must go to Sharyn Coffey of the Design Department at Hodder and Stoughton who had sole responsibility for the paste-up of the book. When I opted during the writing of the book to use abundant, composite illustrations with lengthy legends, all of which for each figure was to appear on a double-page spread but in an appropriate position relative to the text, I had no idea of the real problems that I was creating for the paste-up stage. That these problems are no longer apparent is a tribute to the skilful and aesthetic way that Sharyn carried out her part of the publication process.

If the diagrams in this book are considered to have any aesthetic merit, it is due in large part to the illustrators concerned. In particular it is due to Josephine Martin who produced the beautiful animal drawings that accompany the majority of the figures but it is also due to Chartwell Illustrations who converted my pencil sketches of maps and graphs etc. into a form suitable for reproduction. Marriage of the diagrams and animal drawings was again the work of Sharyn Coffey who, for many of the figures, produced dynamic and interesting arrangements that were big improvements over my original suggestions.

Although a cliché, it is nevertheless true to say that by far the greatest contribution to the production and completion of this book has been made by my wife. Since the book was started in 1972 she has single-handedly raised our three sons, two of them from birth, and dealt with all the details and problems, both trivial and major, of everyday living, that normally I would have shared. It would not quite be true to say that she did so entirely without complaint, but nearly so, and it is certain that without such a buffer from the trials and tribulations of everyday family life I could never have written this book, at least not for many years.

My final acknowledgement, however, and I write this knowing that it will be taken in good part, is to various past and present members of the entomology section at Rothamsted Experimental Station. I often wonder whether, without their timely, but far from favourable, comments and publications relating to my work on insect migration, I would have found it possible to maintain the motivation and drive that is necessary to complete a work of this size.

R.R. BAKER
January 1976

Dedication:

To my family: past, present and future

Contents

Part IIIb Forms of the lifetime track

Part I The definition of migration

1

Four examples of animal migration

Some animals exhibit a pattern of movement that is likely to be termed a migration by the vast majority of zoologists and laymen. In this chapter, four examples are given that illustrate this extreme pattern of movement. All four of these migrations are described in greater detail in Part III of this book.

A North American icterid songbird, the bobolink, *Dolichonyx oryzivorus* (Fig. 1.1), nests in southern Canada, from British Columbia east to Nova Scotia, and south into northern United States, from northeastern California east to Pennsylvania. During winter in the northern hemisphere, however, *D. oryzivorus* is found in the southern hemisphere in eastern

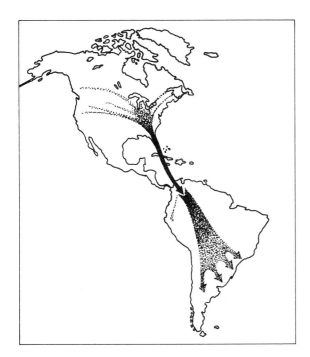

FIG. 1.1   The seasonal return migration of the bobolink, *Dolichonyx oryzivorus*

The bobolink is an icterid songbird about 14 to 19 cm in length. It lives in hay meadows and open fields and feeds on a variety of insects and seeds. The nest, a grass-lined

cup, is built among grass. The sexes are unlike, the photograph showing a male. During migration the species also frequents marshes and while in South America swamps are the preferred habitat-type.

[*Migration routes after Hamilton (re-drawn from Orr 1970). Biology from Peterson (1961) and Dorst (1962). Photo by G. R. Austing*]

FIG. 1.2 A typical harem of the northern fur seal, *Callorhinus ursinus*, consisting of an adult territorial male, several adult females, and recently born pups

[*Photo courtesy of National Marine Fisheries Service of the United States Department of Commerce*]

Bolivia, western Brazil, Paraguay, and northern Argentina. The maximum distance between the most northerly breeding ground and the most southerly wintering ground is approximately 10 000 km. *D. oryzivorus* provides but one example of a pattern of movement that is found in many species of birds in many parts of the world.

The humpback whale, *Megaptera novaeangliae*, has a similarly spectacular migration between summer feeding grounds and winter breeding grounds (Fig. 29.9). *M. novaeangliae* has northern and southern hemisphere populations between which there is little interchange. In both hemispheres the species feeds in polar regions during the polar summer. In the autumn, movement towards the equator occurs and during the polar winter breeding takes place in tropical and subtropical latitudes. In the spring, movement back to polar regions takes place. Such a migration pattern is typical for the mysticete whales (Chapter 29) and may involve movements of 7000 km or so.

The northern fur seal, *Callorhinus ursinus*, breeds in the summer on a small number of islands in the North Pacific (Fig. 28.15). The fur seals that breed on the Pribilof Islands are found in the vicinity of these islands from June to October during which time the young are born and mating and moulting takes place. At the end of October and continuing through to the beginning of December, the fur seals leave the breeding grounds. A small proportion move SW down the Japanese coast, but the majority move SE between the Aleutian Islands. The adult bulls winter only just south of the Aleutian chain but the young males and females and the adult females continue SE, staying within about 80 km from the shore, and from January to April may be found anywhere between Sitka, Alaska, and California up to a distance of some 5000 km from the Pribilof Islands. The northward movement begins in April. By May there are large numbers in the Gulf of Alaska and by June the majority are once more in the vicinity of the Pribilof Islands. Similar movements are performed by the *C. ursinus* that breed on Commander and Robben Islands in the western North Pacific except that the preferred autumn direction of these animals is SW, down the Japanese coast, with only a small proportion of the Commander Islands animals going E and SE down the American coast.

The sockeye salmon, *Oncorhynchus nerka*, spawns in rivers and lakes a considerable distance inland from the Pacific coast of North America (Fig. 31.10). The young fry, if they hatch in a river, hide in the river bed during the day and at night drift with the river current until they reach a lake where they remain and develop for a year or so. At the end of this time they have reached the stage of development, known as the smolt, at which they move to the lake outlet. As they move downstream and reach estuarine conditions, adaptations to marine life appear. The smolt then move out to sea and range widely over the North Pacific, though particularly in the Gulf of Alaska, for two or three years until they become mature. Movement back to the coast then occurs. The salmon finds and enters the mouth of the river from which it emerged a few years earlier. It then migrates upstream eventually finding its way back to the lake and the tributary in which it developed and hatched respectively. Spawning then takes place and the animal dies.

In introducing these four examples of geographical movement patterns, it was remarked that the majority of scientists and laymen would probably feel that migration was a suitable descriptive term. However, C. G. Johnson (1969, p. 7), in his book on insect migration, concluded that the adaptive change of breeding habitat was the primarily important fact of migration. Johnson thus produced what is probably the most useful concept to date for the discussion of insect migration. Yet, according to this concept, none of the four examples of animal movement described above can be considered to be migration, for in each case the movement results in the animals returning to the same breeding habitat. Nevertheless, for most people, certain conspicuous components of these movement patterns, such as the return to the original habitat, geographical directionality, periodicity, and the large distances involved, qualify these movements as migrations. However, when, as in the following pages, these movements are considered in the perspective of the movements of other animals, it is found that they are but extreme examples of a continuous range of movement patterns that at the other extreme is represented simply by movement from one point in space to another and which is performed at some stage during its life history by every animal species.

2

Variation in some components of migration

2.1 Seasonal change in geographical range

Probably the most widely known animal migrants, at least in temperate latitudes, are the many species of birds that are abundant and conspicuous for approximately one half of the year but absent during the other half. Many of these species have summer and winter ranges that are geographically distinct. Consequently, to many people, the existence of separate winter and summer ranges is an important component of the concept of migration. However, even among birds (Dorst 1962) there is a gradation between widely separate winter and summer ranges and considerable overlap. Such a spectrum is even more evident when other animals are also considered (Fig. 2.1).

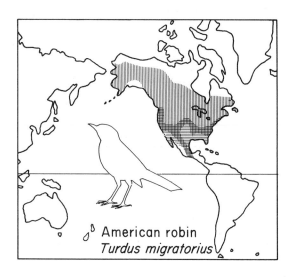

American robin
Turdus migratorius

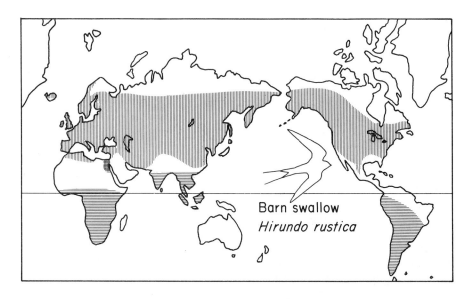

Barn swallow
Hirundo rustica

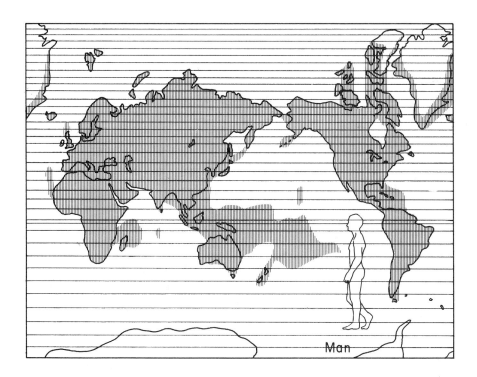

FIG. 2.1 Variation in the degree of overlap between the breeding and non-breeding range

The seasonal and/or ecological ranges (e.g. summer, winter, breeding, non-breeding) of a species may be geographically *separate*, as in the barn swallow or *overlap* slightly, as in the American robin, or considerably, as in man.

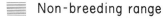

▦ Breeding range

▤ Non-breeding range

[*Drawn from diagrams and descriptions in Orr (1970a), Moreau (1972), and Long (1973)*]

2.2 Return movement

The existence of some degree of return movement is perhaps the prime component in most lay and all ornithocentric concepts of animal migration. Apart from C.B. Williams (1958), however, most recent entomologists do not consider that a movement need involve a return in order to qualify as a migration (C.G. Johnson 1969). A recent attempt to view animal migration comparatively over a wide range of animal groups, including insects, nevertheless decided in favour of the ornithocentric view of migration i.e. a periodic round trip (Orr 1970a, p. 2). However, spectacular though the return movement may be in the migration of animals such as the

four species described in Chapter 1, further consideration shows that virtually every gradation exists between a very precise return movement and a one-way migration.

The most extreme form of return movement is found in the adults of some birds in which movement occurs year after year from a precise nesting site to a precise overwintering site and back again. This is found, for example, in at least some individuals of the grey wagtail, *Motacilla cinerea*. More normally, however, the non-breeding site is a much larger, less precise, area than the breeding site. Hence, most sockeye salmon, *Oncorhynchus nerka*, and northern fur seals, *Callorhinus ursinus*, return to breed at exactly the same site that was vacated at some time, though not necessarily one year, earlier. After breeding, however, the animals and/or their offspring do not necessarily return to exactly the same feeding grounds as previously but range quite widely over certain areas of the North Pacific. Other animals do not have a precise return movement in either direction. For example, young birds cannot return, except in the case of colonially breeding species, to nest in exactly the same place as the previous year, for that site will normally be occupied by their parents. Hence, the return movement is not as exact for young as for adult birds. All that can be said is that movement takes place between two areas which may vary considerably in size from species to species and even at different stages of development within a single species.

The Atlantic eels, *Anguilla* spp., alternate between areas that are probably even less precise than those of young birds. Unlike salmon, which breed in the size-limited habitat of a river or river tributary, Atlantic eels spawn in a fairly large area of ocean, the Sargasso Sea. Further, it is most unlikely that the young eel will return to the same stream that was vacated by its parent, which dies after spawning. Thus the feeding and maturation area is also fairly imprecise, being represented by any of the streams opening into the sea along the coasts of the North Atlantic that are appropriate to the species. Some whales, such as the blue, *Balaenoptera musculus*, and finback, *B. physalus*, also alternate between fairly large imprecise areas. The leopard seal, *Hydrurga leptonyx*, is a species of solitary habits which breeds and moults around the outer fringes of the Antarctic pack ice between October and February. Between June and October, individuals range to most of the subantarctic islands and even as far north as the beaches of south Australia and New Zealand. It remains to be proved, but it is possible that there is no great precision of return in the movements of this animal, except insofar as it moves north for the winter and south for the spring and summer. Some other seals, such as the crabeater, *Lobodon carcinophagus*, in the southern hemisphere and the banded, *Histriophoca fasciata*, in the northern hemisphere, could be similar.

Although return movements are often carried out by the same individuals that left the area previously, this is not necessarily so, particularly as far as the feeding or overwintering areas are concerned. Among the birds that fly to the overwintering grounds each autumn are young birds that have never made the trip before. However, even if all the adults were to die off after reproduction, the return of the species to the overwintering grounds would still be considered a migration, even though it consists not of individuals that have made the trip previously but their offspring. This situation, which is hypothetical for birds, is nevertheless found in some fish. For instance, the eels, *Anguilla* spp., that return to European and American streams are not the same individuals that left previously, but their offspring. Similarly, the monarch butterflies, *Danaus plexippus*, that return to Canada in the spring and summer are possibly not the same individuals that flew south the previous autumn, but their offspring. The migration of *D. plexippus* is a return movement only in the sense that movement is to the north at one time of year and to the south at another.

Other northern hemisphere butterflies are similar to the monarch, *D. plexippus*, in that their mean flight directions are to the north at one time of year and to the south at another but differ in that seasonally consecutive movements are not in exactly opposite directions. Thus, if mean directions of movements are measured clockwise from south, the large white butterfly, *Pieris brassicae*, flies at 159° in the spring and at 333° in the autumn, so there is a measured difference of 174°, not 180°. The small white, *Pieris rapae*, flies at 158° in the spring and at 1° in the autumn, a difference of 157°. Finally, the small tortoiseshell, *Aglais urticae*, flies at 148° in the summer and at 19° in the autumn, a difference of 129°. Differences between spring and summer directions are even less. Hence, *A. urticae* flies at 167° in the spring and at 148° in the summer, a difference of 19°. All of these butterfly measurements were made in western Europe.

There is little difference between two consecutive movements, differing only by 19°, and movement that is one-way only, as in the case of most present-day migrations of man, *Homo sapiens*. In fact, in lay terms, an individual of *H. sapiens* is less likely to be considered a migrant if he later makes a return trip than if the movement is one-way only.

There is, therefore, a continuous gradation between one extreme, that of a precise return movement, as with the grey wagtail, *Motacilla cinerea*, and the other extreme, that of unidirectional movement, as with most modern migrations of man, *Homo sapiens*. As a last word on the continuous range of this feature of migration, the examples can be cited of the birds, the giant petrel, *Macronectes giganteus*, and the wandering albatross, *Diomedea exulans*, which apparently move forever eastward. Yet in so doing, the birds return eventually to their place of origin, having circled the Antarctic.

2.3 Geographical direction

It is probably true to say that most people, excluding entomologists, automatically think of migration as a movement oriented along a N–S axis. Movement along this axis is probably more common among extreme migratory species than movement along any other axis and is found not only in many birds, bats, whales, seals and terrestrial mammals, but also in many insects, including some locusts and butterflies. However, not all otherwise similar movements are oriented N–S. Some bird migrations, such as that of the bobolink, *Dolichonyx oryzivorus*, are along a NW–SE axis. Others, such as that of the European crane, *Grus grus*, are along a NE–SW axis. Yet others, such as that of the Alaskan race of the wheatear, *Oenanthe oenanthe*, are oriented more along a WSW–ENE axis. Finally, the sockeye salmon, *Oncorhynchus nerka*, the giant petrel, *Macronectes giganteus*, and the wandering albatross, *Diomedea exulans*, migrate along a more or less W–E axis.

2.4 Direction ratio

Another striking component of the four examples of clearly recognisable migration given in Chapter 1 is the small degree of scatter about the mean geographical direction of movement. I have previously (R.R. Baker 1969a) used the term direction ratio to describe the variation in the scatter of movement directions about a mean direction. Numerically, the direction ratio expresses the percentage of individuals within the population that move in four different compass sectors relative to the mean direction of movement ($a°$). In order of expression the figures that constitute the direction ratio are the percentages moving in the directions described by: $a° \pm 45°$; $(a° - 90°) \pm 45°$; $(a° + 90°) \pm 45°$; $(a° - 180°) \pm 45°$. This is illustrated in Fig. 2.2.

The scatter of geographical directions of individual bobolink, *Dolichonyx oryzivorus*, about the mean geographical direction must be very small, certainly less than 90° and probably less than about 40°. Similarly, the scatter of directions of humpback whales, *Megaptera novaeangliae*, about a mean N–S direction, is small and is mainly the result of diversion by land-masses. The European eel, *An-*

guilla anguilla, from its breeding concentration in the Sargasso Sea to the river estuaries along the coasts of Europe must have a slightly larger scatter of directions, as measured by straight lines from the Sargasso Sea to maturation places, of approximately 45° about the mean direction. *Dolichonyx oryzivorus*, *Megaptera novaeangliae* and *Anguilla anguilla* all, therefore, have direction ratios of 100:0:0:0.

The pink salmon, *Oncorhynchus gorbuscha*, spawns in the rivers of Asia from northern Korea to the Lena River in Siberia and in the rivers of North America from central California to the Northwest Territories of Canada. The scatter of geographical directions of *O. gorbuscha*, therefore, must be of the order of 180° from its feeding grounds in the North Pacific. No precise measurement of direction ratio is available, of course, but an estimate of 34:33:33:0 perhaps gives an adequate impression of the scatter of migration directions within the population. The northern fur seal, *Callorhinus ursinus*, is characterised by a marked mean movement direction, though the mean is different in different parts of the range. The majority of fur seals that breed on the Pribilof Islands move SE down the coast of America. A smaller proportion moves W and SW to the Commander Islands and down the Japanese coast. Finally, a very small proportion moves to the N (in summer) and is found around the Arctic coasts. Despite having a pronounced mean movement direction, therefore, the total scatter of directions for *C. ursinus* approaches 360°.

Students of seals are inclined to consider (e.g. J.E. King 1964) that individuals of species such as *C. ursinus* with a pronounced bias toward the mean direction of movement are 'truly' migratory whereas others, such as the young of the grey seal, *Halichoerus grypus*, which scatter in all directions, are not. The difference, however, is quantitative, not qualitative. Marking–release/recapture experiments on young *H. grypus* born on the Farne Islands off the northeast coast of England show that although individual seals may go in any direction, more go to the N or NE than in any other direction. As with *C. ursinus*, therefore, young *H. grypus*, although having a virtual 360° scatter of movement directions, nevertheless have a mean direction of movement. The difference between the two is the extent of bias in this direction which is much greater in the Pribilof *C. ursinus* than in the Farne *H. grypus*.

Little attempt has been made to measure the direction ratio of most vertebrate migrants. Fairly accurate measurements have been made, however, of the direction ratios of a number of western European butterflies that are generally considered, at least by entomologists, to be migratory, and can be added to the above list. Hence, the direction ratio of the painted lady butterfly, *Cynthia cardui*, is 54:18:22:6, of the large white, *Pieris brassicae*, is 45:21:21:13, of the peacock, *Inachis io*, is 36:22:25:17, and of the orange tip, *Anthocharis cardamines*, is 25:25:25:25. Finally, the list may be completed by the direction ratio of the migration of man, *Homo sapiens*, which despite certain temporal and geographic variation, such as a mean migration direction to the S in the British Isles over the past few decades, is likely also to be 25:25:25:25.

All of these direction ratios are summarised graphically in Fig. 2.3, which shows that again there is continuous gradation between the two extremes which in this instance are 100:0:0:0 and 25:25:25:25.

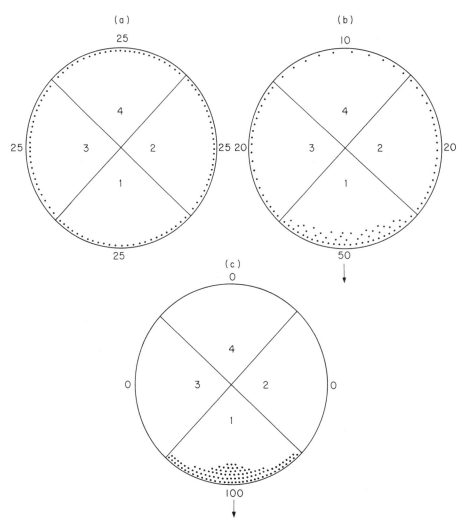

FIG. 2.2   Variation in direction ratio

Direction ratio is the percentage moving in each of four compass sectors relative to the mean movement direction. Arrow shows mean movement direction. Each dot represents the movement direction of one per cent of the group, race, species, etc. being studied. The numbers 1 to 4 indicate the order in which the per cent moving in the direction of that sector is expressed in the written direction ratio. Hence, the direction ratio of (a) is written as 25:25:25:25; of (b) as 50:20:20:10; and of (c) as 100:0:0:0.

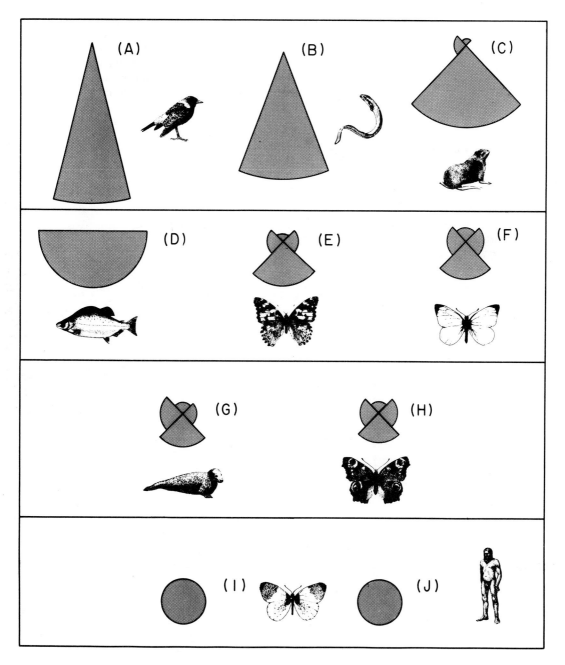

FIG. 2.3   Variation in the direction ratio of ten groups of animals

A, bobolink *Dolichonyx oryzivorus*; B, European eel, *Anguilla anguilla*; C, northern fur seal, *Callorhinus ursinus*, from the Pribilof Islands; D, pink salmon, *Oncorhynchus gorbuscha*; E, painted lady butterfly, *Cynthia cardui*, in Britain; F, large white butterfly, *Pieris brassicae*, in western Europe; G, Grey seal, *Halichoerus grypus*, young from Farne Islands, Britain; H, peacock butterfly, *Inachis io*, in western Europe; I, orange tip butterfly, *Anthocharis cardamines*, in Britain; J, man, *Homo sapiens*.

A, B, C, D, and J are estimated on the basis of observed movements of populations or demes. E, F, G, H, and I are based on field observation of movements of individuals. The direction ratio component of migration shows continuous variation between the extremes of 100:0:0:0 (A and B) and 25:25:25:25 (I and J).

2.5 Plane of movement

So far, discussion has been concerned with migration in relation to what may be regarded for convenience as a horizontal plane. In reality, of course, the plane of migration is not horizontal as the animals concerned are moving over the surface of a globe. Furthermore, it is occasionally important as in the case of the circumpolar migration of the wandering albatross, *Diomedea exulans*, to think in global rather than horizontal terms. For the most part, however, at least for the purpose of this present short discussion, it is convenient to term movement at right angles to the Earth's gravitational pull as 'horizontal' and movement in the same direction as the Earth's gravitational pull as 'vertical'.

Some species, although engaging in an essentially horizontal migration, nevertheless introduce a vertical component into the movement by settling preferentially at different altitudes at different times of year.

In the moose, *Alces alces*, the vertical component is more pronounced than the horizontal component. For instance, in Gray Park, British Columbia, *A. alces* spends the summer in subalpine forests at about 1500–2000 m. In October the animals start to move to lower altitudes and by December have reached the upper levels of lowland areas at 750–1500 m. Downward movement continues until March. The reverse movement begins in April, and by May the moose have returned to their summer grounds. The maximum distance between winter and summer grounds is about 60 km. In the elk, *Cervus canadensis*, some western North American aborigines, *Homo sapiens*, the mountain quail, *Orcortyx pictus*, and some accentors, the vertical component is perhaps even more pronounced. The direction of the vertical component is not always, as in these cases, to higher ground in summer and lower ground in winter. For instance, at the same time and in the same general area that the mountain quail, *Orcortyx pictus*, is moving toward lower ground, the blue grouse, *Dendragapus obscurus*, is moving toward higher ground. Similarly the insect, the convergent ladybird beetle, *Hippodamia convergens*, hibernates in the mountains of the southern United States. Finally, in the diurnal and seasonal vertical migration of plankton, the importance of the horizontal geographical component is at a minimum in relation to the vertical component. Once again, therefore, there is continuous gradation among animal movement patterns between the extremes of relative importance of the horizontal and vertical components.

2.6 Periodicity

As with the previous components of migration, the periodic component shows continuous variation between extremes which may be as different as once every 20 years and several times a day.

In considering the periodicity of animal movements it is necessary to distinguish between species and individuals. At the species level, most movements occur at least once a year, though there are exceptions. For example, in Washington and Oregon, USA, pink salmon, *Oncorhynchus gorbuscha*, migrate upstream to spawn every other year (in the odd years). In other *Oncorhynchus* spp. and the European eel, *Anguilla anguilla*, migrations occur annually in the sense that individuals of the species are seen to migrate at the same season at the same place every year. At the individual level, however, migration may occur only twice in a lifetime. Once when the animal moves from its hatching place to its feeding grounds, and once when it moves back to its hatching place to spawn. In the European eel, *Anguilla anguilla*, these two movements may be separated by an interval of 20 years, whereas in *Oncorhynchus gorbuscha* the interval may be only 2 years. In the sockeye salmon, *Oncorhynchus nerka*, each individual in effect performs three major migratory movements. The first movement is as a small fry from the hatching place in a small stream to a lake. A year or so later, as a smolt, the second movement takes place from the lake downriver to the open sea. The third movement takes place two or three years later when the animal moves from the open sea back to the stream of its birth in order to spawn and then die. The Atlantic salmon, *Salmo salar*, does not necessarily die after spawning and may move several times as an adult between breeding and feeding grounds.

Some migrant animals, therefore, although showing seasonal movements at the species level, are migrating at the individual level at intervals often considerably greater than one year. Some birds, especially the young of many species and all ages of species that nest only every other year, are similar. However, perhaps the most commonly recognised migratory movements, particularly among birds, are those that involve biannual movements at both the species and individual level. Many temperate birds, for example, show two movements per year that are so recognisably seasonal that they are normally described as spring and autumn migrations. It cannot be assumed, however, even for birds, that, when more is known about tropical species, a yearly migration cycle will be found to be an invariable fact. It is known for instance that nesting cycles do not always follow a twelve-month periodicity. The wide-awake tern, *Sterna fuscata*, for example, on Ascension Island, nests on average every ten lunar months (Dorst 1962).

Like most temperate migratory birds, the monarch butterfly, *Danaus plexippus*, in Canada, United States, and Mexico has a southward autumn movement and a northward spring movement. As with birds, the butterflies that fly north from the southern United States (though not necessarily the individuals that appear in Canada later in the summer) are the same individuals that were seen flying south the previous autumn. The situation is further complicated in *D. plexippus*, however, by the fact that the offspring of these spring individuals, at least those offspring produced in the southern United States, also apparently migrate northwards. Further, it is not this second generation but their offspring that change direction in the autumn and fly south once more. In the central United States, therefore, there are three migration seasons, though not all are equally conspicuous. Some individuals, those of the autumn generation, are relatively long-lived and migrate twice, once in the autumn and again in the spring. Other (summer) individuals have much shorter lives and migrate once only, albeit throughout their adult life (p. 428).

Many birds are similar to the monarch butterfly, *Danaus plexippus*, in that as a species there are three seasons at which migration, though not equally conspicuous migration, takes place. Apart from the autumn and spring migrations undertaken by all ages, the young of many species, particularly passerines, undergo considerable movement away from their birth site soon after fledging.

Some species of butterflies have three apparent migratory periods, as does the Monarch, *Danaus plexippus*, but differ in that all the generations are relatively short-lived adults. The small white butterfly, *Pieris rapae*, in Britain migrates at a more or less steady rate throughout its adult life, pausing during migration only to feed, court, oviposit, and to roost overnight and during inclement weather. Because adult life is only about 20 days, however, there are two or three recognisable peaks in observ-

able migratory periods that are simple reflections of the fluctuation in seasonal incidence. The hooded seal, *Cystophora cristata*, also has three major periods of movement per year, though for reasons different from those of the previous species. In the case of this seal the movements are: from the breeding grounds to the moulting grounds; from the moulting grounds to the main summer and autumn feeding grounds; and from the feeding grounds back to the breeding grounds.

The sunn pest, *Eurygaster integriceps*, in the Middle East also has three recognisable periods of major movement. At the beginning of summer, the bugs move up into the mountains to aestivate. In late autumn, the bugs move to hibernation sites which are at lower altitudes than the aestivation sites. Finally, in spring there is a movement onto the plains where the animals breed. Further, as this species is univoltine, there are three movements per year at both the individual and species level.

The fish, the grunion, *Leuresthes tenuis*, in the coastal waters of southern California and northwestern Baja California, Mexico, has a spawning migration that is functionally similar to that of salmon, *Oncorhynchus* spp., in that the animals move inshore. The movement differs from salmon, however, in that the fish do not enter rivers but instead spawn on the shore. The inshore movement of grunion, *L. tenuis*, is seasonal insofar as it takes place from late February to early September. Superimposed on this seasonal component, however, is a lunar monthly or semi-monthly component, and superimposed on the lunar monthly or semi-monthly component is a nightly component. Large masses of grunion move to the shore for the first three or four nights following either the full moon or the new moon.

Many Amphibia move from one habitat-type to another, (i.e. land to water, or water to land) in order to spawn, aestivate or hibernate. The periodicity of these movements varies from species to species and may involve migration between two and six times a year. However, many Amphibia perform movements that are identical with respect to distance, track, and orientation, and differ only in that the movement takes place at much shorter intervals. Fowler's toad, *Bufo fowleri*, for example, along the shores of Lake Michigan, performs a more or less weekly 'drinking' migration that with respect to distance and direction is perhaps identical to a spawning migration. This weekly migration, however, involves movement to the lake shore every fifth night or so and to the daytime retreat the following morning (p. 521).

The vertical migration of plankton has a daily as well as a seasonal component, the animals migrating to the surface at night and to deeper water during daylight.

If a wide variety of animals is considered, therefore, a range of movement periodicity can be found. At the individual level the range is from about once every twenty years to more than once a day. At the species level, the difference between the extremes of periodicity is only a little less great, between about once every two years to more than once a day.

2.7 Distance

Of all the components of migration under discussion in this chapter, that of distance is the most manifestly variable. Yet at the same time it is perhaps the distance component of some extreme examples of migration that most captures the imagination of zoologists and laymen alike. However, it is a simple matter to cite a number of examples of animal movement to show that the 16 000 km interpolar migration of the bird, the arctic tern, *Sterna paradisaea*, is but one extreme of a range of movement distances that pass through every gradation down to movements that are of a few centimetres only.

Humpback whales, *Megaptera novaeangliae*, migrate 7000 km from the Antarctic to tropical waters. Northern fur seals, *Callorhinus ursinus*, migrate nearly 5000 km from the Pribilof Islands to the coast of California. Some individuals of the monarch butterfly, *Danaus plexippus*, migrate 3000 km from Canada to Mexico. The barren-ground caribou, *Rangifer tarandus*, travels up to about 500 km from summer to winter areas. The small tortoiseshell butterfly, *Aglais urticae*, has been recaptured 150 km from its release point in West Germany. Moose, *Alces*

alces, may move 60 km between their winter and summer grounds. Blue grouse, *Dendragapus obscurus*, may move up to 45 km from their breeding area to the winter range. The blowfly, *Phormia regina*, has been known to move 45 km in less than thirteen days. The midge, *Culicoides tristriatulatus*, has been shown to move at least 8 km during adult life, and the white bass fish, *Roccus chrysops*, may move 2 km from open water to spawning sites on the north shore of Lake Mendota, Wisconsin, United States. The midge, *Culicoides furens*, has been caught in a trap 2–3 km upwind from its release point and presumably travels much further when it has a following wind. Many other small insects, small mammals, terrestrial reptiles, amphibians, and non-insect terrestrial arthropods may move only a kilometre or so away from their place of birth. Yet other, even less mobile animals, such as terrestrial gastropod molluscs, may move only a few metres away from their hatching place.

Although many vertical migrations involve movements over distances of many kilometres, as with the moose, *Alces alces*, the blue grouse, *Dendragapus obscurus*, and many insects, such as the convergent ladybird beetle, *Hippodamia convergens*, other vertical migrations may be measured in metres rather than kilometres. The vertical migration of some plankton, for instance, involves movement over a distance of only a few metres. The sycamore aphid, *Drepanosiphum platanoides*, flies in the summer the few metres from the lower to the higher branches of its host tree. Finally, the example may be cited of the hemipteran, the cabbage white-fly, *Aleyrodes brassicae*, which shows a movement identical in all respects except distance to that of *D. platanoides*. In the case of *A. brassicae*, however, the flight is over a distance of only a few centimetres from the lower, older, to the higher, younger, leaves of the cabbage, *Brassica oleracea*.

3
Migration and non-migration

In Chapter 1, four examples of animal movement patterns were given and it was suggested that certain components of these movements made it likely that most people would consider migration to be a suitable descriptive term. However, in Chapter 2 attention was drawn to the fact that, striking though each of these components may be, they are in effect just one extreme of a spectrum of variation within which there is every gradation between extremes. Hence, the movements of animals show almost every gradation between the spectacular periodic return movements between geographically widely separated areas and the movements over a few centimetres from one part of a plant to another. Over the range of movement patterns discussed so far there is no clear-cut qualitative break between any one type of movement and another that merits on the basis of anything other than pragmatism the drawing of a line between migration and non-migration. Nevertheless, the feeling exists among zoologists that migration and movement should not be synonymous, and various attempts have been made to define criteria, other than those already considered in Chapter 2, by which the qualitative distinction between migration and non-migration can be made. The criteria suggested can be divided for the purpose of the following discussion into two major categories, spatial and non-spatial.

3.1 Spatial criteria

The *Shorter Oxford English Dictionary* (1933) defines migration as the act of moving from one place of abode to another or, of some animals, the act of moving from one habitat to another.

At first sight the latter half of this definition, that

concerned with the habitat concept, appears to offer a solution to the problems encountered in attempting to distinguish between migration and non-migration. It could be said that a cabbage white-fly, *Aleyrodes brassicae*, flying from the bottom, older, to the upper, younger, leaves of a cabbage, is exchanging one habitat for another and is therefore migrating. On the other hand it could be said that a butterfly flying perhaps several metres from one flower to another is moving within a single habitat and is therefore not migrating. This distinction between inter-habitat movement (migration) and intra-habitat movement (non-migration) has been used by various authors for nearly 70 years, e.g. Pearson and Blakeman (1906), Heape (1931), Southwood (1962) and C. G. Johnson (1969). The term 'trivial movement' has been used to describe movement within a single habitat.

The above distinction between migration and non-migration rests on the concept of a habitat which unfortunately is itself an arbitrary and open-ended concept (Fig. 3.1) that is quite unsuitable for use in the making of qualitative distinctions. The habitat concept is arbitrary in that it is extremely difficult, if not impossible, to recognise the edge of a habitat unit by anything other than arbitrary criteria, and open-ended in that habitats are found within habitats which themselves exist within habitats (Fig. 3.1). The limit of the habitat of an individual aphid could be said to be the edge of the leaf on which it feeds and reproduces, or perhaps the tree on which the leaf is found, or the edge of the woodland in which the tree is found, or the valley in which the woodland is found, or the land mass on which the valley is found, or the continent of which the land is a part, or even the Earth, the Solar System, the galaxy, or the Universe. It is difficult to

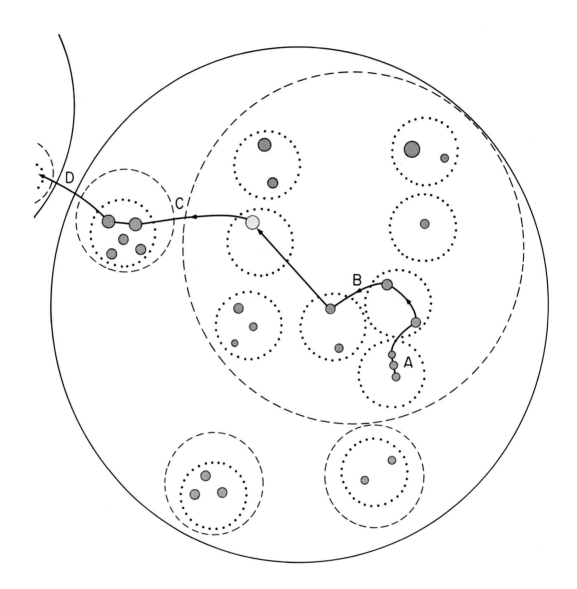

FIG. 3.1   The open-ended nature of the habitat concept and the shortcomings of the habitat concept as part of a definition of migration

Every single circle in the diagram could be termed a habitat. For example, the large circle could represent a valley; the dashed circles, woodland; the dotted circles, trees; and the hatched areas, branches. Habitats are found within habitats which themselves are found within habitats. Yet migration has been defined as movement from one habitat to another. Suppose an animal flies from branch to branch as indicated by the arrowed line. The animal, or different animals, may move from branch to branch on the same tree (A), or from a branch on one tree to a branch on another tree (B), or from a branch on a tree in one woodland to a branch on a tree in another woodland (C), or from a branch on a tree in a woodland in one valley to a branch on a tree in a woodland in another valley (D). The distinction between migration and non-migration can only be made in arbitrary terms.

see any qualitative justification for deciding that any one of these units represents a habitat whereas the others do not, though two, at first sight promising, criteria do perhaps suggest themselves.

First, a habitat could perhaps be defined as the minimum area within which an animal can obtain all of the ecological requirements necessary for normal life. However, the criteria by which it is possible to distinguish between necessary and unnecessary ecological requirements are as elusive and arbitrary as the spatial limits of the habitat we are seeking to qualify. The final conclusion can only be that all of the present range of movement of an animal is 'necessary'. For example, the north temperate United States clearly do not provide the ecological requirements necessary for the efficient overwintering of bobolinks, *Dolichonyx oryzivorus*. Nor is this species of bird adapted to a permanent life in the tropics. Hence it follows that the minimum range within which an individual *D. oryzivorus* can obtain all of the 'necessary' ecological requirements is the range within which the animal moves from winter to summer quarters at the present day.

Second, and perhaps more promisingly, a habitat, and consequently migration, could perhaps be recognised by reversing the normal procedure of defining trivial movement in terms of the habitat and instead to define trivial movement in terms of the sensory range of the animal and then to define a single habitat as that area within which the animal can move by the use of trivial movement. A trivial movement could be defined as a movement of an animal from one point to another in which the end point was within the sensory range of the animal before it vacated its starting point. Conversely, a migration could be defined as a movement of an animal from one point to another in which the end point was outside the sensory range of the animal before it vacated its starting point. Although, of course, practical difficulties present themselves in determining what is within an animal's sensory range and what is not, these definitions do at first appear to provide a qualitative distinction between trivial movement and migration. Hence, a butterfly flying from one flower to another within a flower patch is showing trivial movement, whereas a butterfly flying from a flower patch in one valley to a flower patch in another valley is migrating. A male moth flying 2 km upwind to a pheromone-producing female is showing trivial movement, whereas a moth flying 2 km from one woodland to another is migrating. However, even this apparently quali-

tative means of distinguishing between migration and trivial movement does not always lead to a satisfactory conclusion. For instance, the often spectacular upstream movements of salmon clearly require consideration in any discussion of animal migration. It seems likely, however, that this final stage of the return of salmon from the sea to the spawning ground is achieved by olfaction, by swimming up-current to a chemical source or sequence of sources. The fact that this phase of the migration may cover several hundred, if not thousand, kilometres is irrelevant and only seems remarkable to a human because of the latter's extremely optocentric view of the Universe. The movement is in fact equivalent to that of a man moving to a clearly visible topographical feature such as a hill or to that of a butterfly flying from flower to flower. Further, the movement, or at least part of the movement, is comparable in all respects except distance to that of a male moth flying a few or a few hundred metres upwind to a pheromone-producing female. Yet the movement of salmon upstream is generally considered to be a migration, whereas that of a male moth to a pheromone-producing female is not.

There is no doubt that both of the above criteria by which a habitat could be defined require consideration in any discussion of animal migration, and in Part II of this book both the spatial distribution of an animal's ecological requirements and the selective pressures that act when an animal loses sensory contact with its ecological requirements are argued to be important in the evolution of some aspects of animal migration. However, as far as distinguishing between migration and non-migration is concerned, the habitat concept appears to be of little qualitative use. Any definition of migration based on the habitat concept, apart from being difficult to apply in practice, has the disadvantage of excluding some of the movements, such as that of salmon, which one would normally wish to discuss in a consideration of migration. Johnson's suggestion (C. G. Johnson 1969) that the term migration should be restricted to the adaptive change of breeding habitat, suffers from the same shortcomings in that it excludes all of the spectacular return movements shown by many vertebrates.

3.2 Non-spatial criteria

The problems involved in trying to define migration in a purely spatial sense have been recognised on many occasions and from time to time the solution has been sought, particularly by entomologists, by defining migration by other criteria. These attempts have met with little success, however, and invariably zoologists have reverted to the use of spatial criteria. For example, during the 1950's and 1960's, entomologists used the criterion of undistracted flight suggested by Kennedy (1958, 1961) as the definitive characteristic of insect migration, and the suggestion still finds some favour even to this day (e.g. Dingle 1972). It was argued that migrants can be recognised by their lack of reaction to features of the environment that would normally inhibit flight and trigger off other non-migratory behaviour patterns such as feeding, courtship, or reproduction. Oddly, roosting is not often included in this list. A migrant could therefore be distinguished from a non-migrant by its transient accentuation of locomotor functions with depression of vegetative functions such that the animal now travels, by what method being a secondary matter (Kennedy and Stroyan 1959). After a period of quite general acceptance among entomologists, however, this view is now losing favour and is giving way once more to the use of spatial criteria in the definition of migration. For example, C. G. Johnson (1969, p. 6) in his major work on insect migration argued convincingly that definitions based only on behaviour can define migration only partially. More importantly, Johnson also pointed out that even among insects, migration was often the result of distracted flight. Locusts feed avidly during migration and many flies, butterflies, and dragonflies mate during migration. Many butterflies also feed and oviposit during migration and, of course, probably all animals roost at some stage during migration. Indeed, if animals are considered as a whole, migration by undistracted movement (in Kennedy's sense) is probably the exception rather than the rule. Many birds, such as swallows, terns, and shearwaters etc. feed throughout their migration. Most other birds migrate for only a small part of the day and feed and roost during the remainder before continuing on their flight the next suitable day (or night). Lemmings set up temporary quarters during migration, and most ungulates such as caribou, moose, reindeer and elk feed during migration. Most seals also feed during migration.

Undistracted movement is not, therefore, a definitive characteristic of migration, either for vertebrates or even for the insects for the migration of which the suggestion was originally made. Recently,

therefore, entomologists, such as Johnson (1969, pp. 8 to 9) have reverted to spatial criteria as the means by which migration can best be defined. Unfortunately, the spatial criteria used by Johnson (i.e. dispersal, and change of breeding habitat) are also unsuitable. Other entomologists, particularly those primarily interested in butterflies, such as C. B. Williams (1958) and Nielsen (1961) have felt that straightened-out flight under the control of the animal concerned was the primary characteristic of migration. Straightened out (rectilinear) flight was also an important facet of the 'undistracted movement' view of migration for without some straightening out of the movement track even undistracted movement could lead nowhere. For example, a gyrinid water beetle may swim continuously for long periods at the water surface without ever leaving its pond. Unlike the criterion of undistracted movement, however, there is no doubt that rectilinear movement is the commonest pattern by which animals migrate. However, even here there are exceptions. A flock of swallows, *Hirundo rustica*, or swifts, *Apus apus*, often moves slowly across country, feeding as it goes. Each individual performs a series of more or less circular movements with flight in one direction only slightly more frequent than flight in other directions. Honey bees, *Apis mellifera*, when moving in a swarm to set up a new colony, move in a similar way. Finally, there is the most extreme case of migration by random movement such as that performed by some nematodes. For example, *Trichostrongylus retortaeformis* is a nematode parasitic in sheep, *Ovis* spp. The eggs pass out of the sheep with the sheep's faeces in which subsequently the first two developmental stages are spent as a free-living larva. The third, infective, stage shifts from the faeces to the surrounding grass with which, when grazed, the larva once more gains access to its host. The movement pattern by which this shift is achieved is more or less random. Nevertheless, this 'random walk', which is habitually referred to by nematologists as a migration (Crofton 1948), is in both a spatial and ecological sense a migration quite comparable to that of other, more spectacular, animals.

The other aspect of the definitions proposed by Williams (1958) and Nielsen (1961), that of movement under the control of the animal concerned, now finds little favour, even among non-entomologists. Many insects are displaced from one place to another by wind currents and furthermore many of these insects have evolved complex behavioural mechanisms which ensure that the animal is displaced by the wind. Another group of terrestrial arthropods, the gossamer (aeronaut) spiders, have also evolved complex behaviour patterns in adaptation, presumably, to the same selective pressures. The first stage of the migration of many fish, such as salmon, is achieved by allowing themselves at night to be displaced downstream to a lake or to the sea, though throughout the major part of the displacement the animals may continue to face upstream. Green turtles, *Chelonia mydas*, are thought to be carried by oceanic currents from their breeding sites on the shores of Ascension Island in mid-Atlantic to their feeding sites on the Brazilian coast. Even some birds, such as the wandering albatross, *Diomedea exulans*, apparently migrate, though not rectilinearly, on the permanent westerly winds that blow round the Antarctic continent. Finally, animals as diverse as pseudoscorpions and man often migrate, not by their own locomotion, but by attaching themselves to some other motile object either living, such as a fly, a horse, or a camel, or non-living, such as a ship or an aeroplane.

In recent years a few entomologists have shown a tendency to overcompensate for the earlier view that only controlled movement was migration, uncontrolled movement being passive or accidental, by trying to show that almost all insect migration was the result of downwind displacement. However, this is clearly not the case, and even the migration of locusts, which was the prime example of a large strong-flying animal migrating by wind-displacement, has now been shown in an important paper by Waloff (1972) to require some reappraisal. While it is still true that locusts make use of the wind, they are not wind-displaced like a balloon or like weaker-flying insects such as aphids. Instead, the desert locust, *Schistocerca gregaria*, apparently orientates to some feature of its environment. Sometimes orientation is with the wind and at other times against the wind. Even when a swarm of locusts is migrating with the wind, the rate of movement is more or less under the control of the animals as a result of a large proportion of the swarm being on the ground at any one time (see pp. 454–61 for further details). In general, the impression gained is that locust swarms make use of the wind but that rate and direction of movement is very much under the animal's control. The rate and direction of migration of some other insects, notably butterflies and some moths, but probably also the larger flies, are also largely under the control of the animals concerned. The control is the result, as it is with birds, of the coordinated use of celestial and topographical features, although, again as with birds, adverse winds can inhibit migration.

Finally in this section on non-spatial criteria and the recognition of migration, there is the conception, most popular among laymen, that migration is a gregarious phenomenon. This is largely due to the fact that most conspicuous migrations are inevitably those that involve many individuals migrating as a cohesive swarm. Locusts and many birds are prime examples. A less superficial look at animal migrants, however, shows that swarming is not an essential component of migration. In fact, truly gregarious migrants are probably the exception rather than the rule, though because it often happens that many individuals start to migrate independently, but at the same time, the impression given is often that of swarming. Lemmings, *Lemmus lemmus*, migrate independently of one another. Many birds (e.g. most cuckoos), migrate singly. Some seals migrate in groups, the size of which varies considerably from species to species, whereas others, such as the leopard seal, *Hydrurga leptonyx*, migrate singly. Whales migrate in small groups (but see p. 767), particularly family groups, and so does man, *Homo sapiens*, though the latter also commonly migrates singly. Among insects, so far as is known, only locusts and some social insects show true swarm behaviour during migration.

4

The definition and use of the term migration

4.1 Some previous definitions of migration

Past definitions of migration, particularly those produced in relation to the movements of single taxonomic groups, have tended to be restrictive in order that discussion could be limited to certain specific movement patterns in which the author was particularly interested. For example, ornithologists interested in the seasonal long-distance changes of geographical range of many birds have defined migration in terms of seasonal to-and-fro movements on a geographical scale. Entomologists, who can find few examples of seasonal to-and-fro geographical movements among insects, have sought other definitions that permit a more wide-ranging discussion of insect movements. Even among entomologists, however, definitions have varied according to the group of insects or the aspect of migration in which the entomologist was most interested. Lepidopterists, for example, tend to choose either directional flight under the control of the animal as the criterion by which migration can be recognised (e.g. C.B. Williams 1958) or the existence of a peak geographical flight direction (R.R. Baker 1969a). Entomologists mainly interested in weaker-flying insects require a more general definition if they are to be able to discuss the migration of animals such as aphids. Hence, an ecological entomologist is likely to use a definition, such as that of C.G. Johnson (1969), which involves adaptive change of breeding habitat. A more physiological entomologist, on the other hand, is more likely to use a definition, such as that proposed by Kennedy (1961), which is based on aspects of behaviour such as persistent, undistracted,

rectilinear flight. A sociologist, on the other hand, is interested in the social consequences of human migration and so is likely to use a definition that involves a movement from one society to another (Eisenstadt 1955), usually a relatively permanent movement (Mangalam 1968).

Such a restrictive approach to the definition of migration is, of course, wholly justifiable in the vast majority of cases and in the past has given rise to a number of outstanding works on certain specialist groups of animals. For example, the books by Dorst (1962) on birds, C.B. Williams (1930) on butterflies, and Johnson (1969) on insects in general are all milestones in their own particular fields and would all have suffered if any attempt had been made to extend the definition of migration beyond the particular types of movement in which the respective authors were primarily interested. These authors accept, however, that their definitions are arbitrary and that migration, as they each define it, and non-migration, grade into one another.

4.2 A definition of migration

Chapters 2 and 3 have shown that there is no qualitative distinction between migration and non-migration. Most previous major works on migration have also accepted that migration and non-migration grade into one another and that distinction between the two can be made only on arbitrary grounds. Yet despite this, almost all previous definitions of migration have been restrictive. It is suggested here that a much safer and more realistic procedure is to legalise and emphasise the arbitrary nature of the term migration by the production and use of a definition that is in itself manifestly arbitrary. The definition proposed, therefore, is as follows:

Migration: the act of moving from one spatial unit to another.

This definition is close to, though even more general than, the dictionary definition (i.e. the act of moving from one place of abode to another). The term 'spatial unit', however, has no restrictive overtones unlike the terms 'place of abode' or 'habitat'. Hence, a zoologist interested in the animals on, say, a particular tree, could consider the confines of the tree to be a spatial unit. Movement within the confines of the tree could therefore be described as trivial movement, and movement from tree to tree could be described as migration. A zoologist interested in a particular woodland, on the other hand, would perhaps find it convenient to consider the edge of the wood as the boundary of a spatial unit. Movement within the wood could then be described as trivial movement whereas movement from wood to wood could be described as migration. Similar pragmatic criteria could be applied to, say, lakes, lake systems, countries, latitudes, or even planets, depending on the interests of the worker concerned. This, I believe, is in any case more or less the way the term migration has always been used. Problems have only arisen when workers have had to consider spatial units of different magnitudes. Hence, when an animal moves from tree to tree, from wood to wood, and from country to country, there has been a perhaps intuitive tendency on the part of the worker to feel that not all of the movements ought to be categorised as migration. The definition given here, however, would allow the term migration to be restricted to one or applied to all three of these movements, whichever use of the term is the most useful. In addition, the nature of the definition serves to emphasise the fact that these movements (i.e. from tree to tree, wood to wood, and country to country) are not distinct from one another but are part of a continuous gamut of movement. Consequently, it is automatically emphasised that no special significance can be attached to any one particular movement merely because it has been described as a migration.

The term migration is not only used to describe the movement of a single or several individual animals. Indeed, migration can be applied to the movement of inanimate objects, such as boulders,

buildings, etc. Also histologists frequently refer to the migration of cells within metazoan organs or tissues. Biogeographers and palaeontologists often refer to changes in the geographical range of a species as migration, as, for example, in describing the northward movement of species, both plant and animal, that took place following the retreat of the ice sheet at the end of the last ice age. All of these uses of the term migration are covered by the above definition.

4.3 The approach to migration used in this book

At first sight, it may appear that a definition of migration such as that given above will hinder any discussion of animal migration. Certainly, in the past the gradations, which are emphasised by the above definition, between migration and non-migration and between the different degrees and patterns of migration have been viewed as an inconvenience to be minimised by careful choice of definition. From the evolutionary viewpoint, however, such gradations are of primary importance in examining the processes that may have been involved in the evolution of any particular aspect of migration. In this book, therefore, it is in fact desirable not to limit by some restrictive arbitrary definition the range of phenomena to be discussed but rather to determine the range of phenomena evolutionarily relevant to the more extreme forms of migration and then to consider the range as a whole. To do otherwise is to risk missing some evolutionary sequence or some correlation that is important in our final understanding of the more extreme forms of migration.

Such an approach to the evolution of animal migration requires a quite different philosophy from that employed previously. It does not, however, necessarily require a description and consideration of all animal movement. Like previous authors that have considered animal migration, I too am most excited by and therefore most interested in the extreme forms of movement shown by each animal group. However, if required for the understanding of a particular extreme movement, I can see every reason to examine any other relevant movement, no matter how trivial it may seem by comparison. In the remainder of this book, therefore, examples are given some of which would not normally be considered as migration by specialists on those particular animal groups, at least according to criteria in popular use at the present day.

There is one further important consequence of the definition of migration given above. The usual approach to animal migration, and the question most asked, is: why does animal A migrate? More properly, the question should be: why does animal A migrate whereas animal B does not? As defined, however, migration is a characteristic of all animals. The difference between species A and species B is not, therefore, that species A migrates whereas B does not, but that A has evolved one pattern of movement whereas B has evolved some other pattern of movement. A more realistic approach, therefore, is to consider the factors that make one rate and pattern of movement optimum for A but another rate and pattern of movement optimum for B. For example, insect A may only ever fly from leaf to leaf within the confines of a single tree. Insect B, as

well as flying from leaf to leaf may also fly from tree to tree. Finally, insect C, as well as flying from leaf to leaf and from tree to tree, may also fly from woodland in one valley to woodland in another valley. In the past it is quite likely that the situation would have been discussed by considering why C migrates whereas A and B do not. A more satisfactory approach is to consider why the optimum movement for A is to migrate from leaf to leaf, for B to migrate from tree to tree, and for C to migrate from valley to valley. Apart from any other virtue, the latter approach emphasises the sequence by which the more extreme migration (i.e. either that of A or C) may have evolved. The migration/non-migration approach, on the other hand, suggests a qualitative break in an evolutionary sequence that simply does not exist.

Far from being a hindrance, therefore, the definition of migration given in this chapter is likely to assist a consideration of the evolution of animal migration.

5
Definition and interrelationships of terms used in this book

It is convenient in a book such as this if repeated lengthy phrases are to be avoided, to use a number of terms as a form of shorthand. Some of the terms used here are new and yet others, such as 'migration', it has been found necessary to re-define. This, of course, is normal practice and part of the licence of attempting a review and analysis of a phenomenon as complex as animal migration. The desirability of such practice, on the other hand, is questionable as it may so easily lead to confusion on the part of the reader. Such confusion, however, can perhaps be minimised by providing a central reference point from which the meaning and nuances of the terms used can be ascertained. This chapter is intended to be such a reference point. Many of the terms simply require definition. Others require more lengthy discussion.

5.1 Movement, locomotion and migration

Movement: change in position

Movement can only be defined in terms of change in position relative to some other object. In the universal sense of movement through space it is not possible to say what is moving and what is not, only that a body is changing position relative to some other body. In the terrestrial sense, however, movement is usually defined relative to the Earth's surface. Objects that are stationary in the terrestrial sense are nevertheless moving in the universal sense due to the movement of the Earth through space.

Locomotion (of an animal): movement of an animal achieved by force exerted by that animal.

Migration: the act of moving from one spatial unit to another.

Migration, unlike movement and locomotion, is a utilitarian rather than an absolute term in that the limits of the spatial unit are open to arbitrary definition by whatever criteria are convenient in the particular instance. If the spatial unit is defined as the position occupied by an entity at a moment in time, then migration is synonymous with movement. Any other definition of the limits of the spatial unit renders migration and movement as not synonymous. The migrating entity may be part of an animal, a single animal, a loose or cohesive group of animals, an animal species, a non-metazoan living organism, or a non-living organism.

As far as the interrelationships of movement, locomotion, and migration are concerned, it can be said that: movement may or may not involve locomotion and may or may not lead to migration; locomotion, by definition, involves movement but may or may not lead to migration; migration, by definition, involves movement but may or may not be the result of locomotion.

5.2 The subdivisions of migration

This book is concerned entirely with *individual migration* (i.e. the migratory movements performed by an individual, either independently or as a member of a loose or cohesive group, during the course of its lifetime). Little attention is paid to *species migration* (i.e. shift in the geographical range of a species over a period greater than a year) except where the latter provides data for the former, as it does, for example, in the spread of the evening grosbeak, *Hesperiphona vespertina*, across Canada from the Rockies in the west to Newfoundland in the east between 1823 and 1961 (Parks 1973), and as it does in the spread of the collared dove, *Streptopelia decaocto*, across Europe from Turkey to Britain between 1920 and 1952.

Two major categories of individual migration are recognised. These are accidental migration and non-accidental migration.

Accidental migration: migration that is initiated and continues due to a failure in the normal station-keeping mechanisms of an animal.

Non-accidental migration: migration that is initiated and continues by any mechanism other than the failure of a station-keeping mechanism.

These definitions would seem to warrant further discussion.

There is a temptation to contrast accidental migration with adaptive migration on the basis that the former is disadvantageous whereas the latter is advantageous. However, such a contrast would not be valid, for not only are accidental migrations not always disadvantageous, a large part of the terrestrial fauna of oceanic islands probably arriving as a result of what can only be described as accidental migration, but also some individuals, performing what can only be described as adaptive migration, may suffer considerable disadvantage as a direct result of that migration.

Accidental migration, therefore, cannot be defined in terms of disadvantageousness. Nor can it be defined solely on the basis of the mechanism by which migration is achieved. In the past, for example, 'accidental migration' was used to describe the migration of any animal that did not migrate by its own power but instead was carried 'passively' by environmental forces such as air and water currents (e.g. Williams 1958). It has since been realised, however, that the passivity involved in migrations such as those of an aphid being displaced by the wind or a fish being displaced by a water current is essentially similar to the passivity involved in migrations such as a pseudoscorpion being carried by a fly or a man being carried by an

aeroplane. In all four cases the initiation of the migration is as much under the control of the animal concerned as that of any other animal. Furthermore, in the case of an aphid, the animal can only remain airborne if it continues to beat its wings (Johnson 1969). Clearly, therefore, these 'passive' migrations are initiated and terminated just as actively and are no more accidental than other so-called 'active' migrations. It would seem, therefore, that a definition of accidental migration based only on the migration mechanism and which ignores the events involved in the initiation of migration is also unsuitable.

The final conclusion, therefore, is that accidental migration is best described, not only in terms of the migration mechanism, but also in terms of the conditions that influence the initiation of migration. In order to examine this conclusion, let us consider the following two groups of examples of migrations each of which results in the animal concerned going beyond its normal range of distribution.

Group 1 consists of: a planktonic larva of an inhabitant of continental rocky shores that is carried out to mid-ocean and happens to settle on the rocky shore of an island; a gossamer spider that is carried by wind out to sea and that lands on an oceanic island; a butterfly that flies out to sea and lands on an island; a lemming that swims out to sea and lands on an island; a bat that flies out to sea and lands on an island; a walrus on an ice-floe that enters an unusual current and makes a landfall outside of its normal range.

Group 2 consists of: a planktonic larva of a southern hemisphere barnacle that settles on a ship instead of on a shore and which happens to release its larvae in the northern hemisphere; a spider on a branch that breaks off a tree, is washed out to sea, and lands on an oceanic island; a caterpillar on a cabbage that is shipped from Europe to the United States; a rat that boards a ship to search for food and which disembarks on the opposite side of an ocean; a giant tortoise that walks onto a beach, is caught by the tide, and which floats to an oceanic island.

Both of these groups of examples could perhaps be said to illustrate accidental migration. Nevertheless, there appears to be a fundamental difference between them, even though in the field the distinction may not be clear, especially if only the end result is available for study. All of the group 1 migrations are performed by the normal migration mechanism of the species concerned and were initiated by the animals themselves presumably because, on average, conditions should have favoured migration (though not the migration actually performed). None of the second group of migrations, however, was initiated by the animal concerned. Instead, the initiation of migration was imposed upon the animals by some environmental force. Furthermore, group 2 migrations were all performed by some means other than the migration mechanism that is normal for the species. In my view, only the group 2 migrations should be described as accidental. The individuals in group 1 are all non-accidental migrants that happen to go over the edge of their normal range of distribution. It is perhaps almost unnecessary to point out that the animals of both groups, because they arrive at destinations previously unoccupied by the species, may have considerable evolutionary significance. Yet for neither group is there any reason to suppose that the mechanisms by which the migrations were performed evolved specifically in order that these distant, extra-range destinations could be reached.

In order to emphasise the differences between the two groups of examples, it helps to reconsider the same movements but this time on the assumption that the distances travelled are small, that no open oceans are crossed, and that the destinations are all within the normal range of distribution of the animals concerned. When this is done, it can be seen that the group 1 migrations are indistinguishable from the migrations that are being performed throughout the range of the species. The group 2 migrations, on the other hand, can still be seen to be accidental and to be different from the migrations being performed by the vast majority of individuals of that species.

It is suggested, therefore, that use of the term 'accidental migration' should be restricted to those situations in which both the initiation and the mechanisms of migration are beyond the control of the animal concerned, the initiation and continuation of the migration being imposed on the animal by some environmental force that overcomes the normal station-keeping mechanisms of the animal.

The examples of accidental migration given in group 2 are fairly homogeneous. When a large number of accidental migrations are considered, however, there is at least one important respect in which the category is heterogeneous. Some, perhaps most, accidental migrations, such as those in group 2, seem unlikely to be capable of accentuation by selection. Others, however, seem capable of being acted upon by selection to produce a non-accidental migration. As an example of the former type of accidental migration it would seem that the spider sitting on a branch that breaks off a tree and is washed out to sea is performing a migration that is unlikely to become established by selection as a characteristic of the species. By way of contrast, however, and as an example of an 'accidental' migration that may through selection become non-accidental, reference can be made to those pond or stream inhabitants that migrate from one pond or drainage system to another in the mud that adheres to the legs or bodies of birds or other animals (Maguire 1963, Proctor and Malone 1965). Presumably, migration between such habitats was initially the result of accidental attachment to the vectors, and in many cases the initial introduction of the ancestor into this habitat may also have been by this agency. Any advantage of migration between these habitats, however, would have been likely to impose selection on the aquatic organisms to aggregate or reproduce in those areas of the ponds or streams where such 'accidental' attachment was most likely to occur, as well, of course, as on mechanisms of attachment. On occasion, therefore, non-accidental migration may have evolved from a pre-adaptation of accidental migration. The vast majority of non-accidental migrations, however, seem likely to have evolved by modification of a previously existing non-accidental migration.

The whole of Parts II and III of this book is concerned with non-accidental migration, two categories of which are recognised (Fig. 5.1). These categories are non-calculated migration and calculated migration.

Non-calculated migration: migration to a destination about which, at the time of initiation of the migration, the animal has no information, either memorised, through direct perception, or through social communication.

Calculated migration: migration to a specific destination that is known to the animal at the time of initiation of the migration, either through direct perception, previous acquaintance, or social communication.

Two categories of non-calculated migration are recognised (Fig. 5.1). These are exploratory migration and removal migration.

Exploratory migration: a migration beyond the pre-migration limits of the familiar area during which the ability to return to that familiar area is retained, though not necessarily exploited.

Removal migration: a migration away from a spatial unit (U), usually ending in some other comparable spatial unit, which is not followed by a return to U, or at least not until the passage of a sufficiently long, but arbitrarily determined, interval of time.

Calculated migration can also be divided into two categories, one of which is removal migration, and the other of which is return migration.

Return migration: migration to a spatial unit that has been visited previously.

Return and removal migration should not be considered to be mutually exclusive categories, for although they may conveniently be used as such, they also interact in a number of ways. First, if it is to be biologically meaningful, an arbitrary time scale has to be attached to the definition of removal migration. The time scale involved in return migration, however, is fixed by the moments of departure from and return to the specified spatial unit. Thus it is possible to break down any given return migration (and, for that matter, exploratory migration) into a sequence of removal migrations, a facility that is extremely useful, theoretically, in the development of a return migration model. Secondly, whether or not two consecutive removal migrations are considered to be a return migration depends on the size of the specified spatial unit. Thus if a bird flies from tree T_1 in wood A to tree T_2 in wood B and then back, but to tree T_3, in wood A, the migration is a return migration with respect to wood A but not with respect to tree T_1. Indeed, the bird has performed a removal migration from tree T_1 to tree T_3.

Exploratory, removal, and return migrations can be described and analysed in terms of their various components. Many of these components have been discussed in Chapter 2 and here need only to be defined. Other components require a more extensive discussion. The subdivisions of migration and the components of individual migration are illustrated in Fig. 5.1.

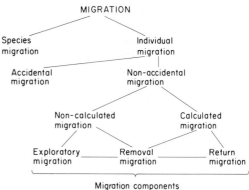

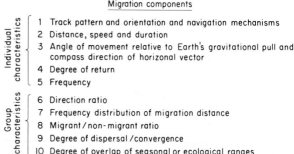

Fig. 5.1   The hierarchy of the sub-divisions of migration and the components of individual migration at the level of both the individual and the group

5.3 The components of individual migration

When an individual migrates from A to B it traces out a *track* in space. This track has a certain pattern that results partly from the *orientation and navigation mechanism(s)* employed by the individual and partly from the environmental forces (e.g. wind, water currents) to which it is subjected. The track has *distance* and the movement has *speed* and *duration*. The total movement from A to B subtends an angle to the gravitational pull of the Earth and any horizontal vector has a *compass direction*. When the same individual moves from B to C, the two movements (A to B, B to C) show a *degree of return*. Two or more movements have a *frequency*. These components are all characteristics of migration when analysed at the level of the individual.

When the migrations of more than one individual are observed, the group shows a *direction ratio* and a *frequency distribution of migration distance*. As a result of this frequency distribution, some individuals leave the relevant spatial unit whereas others do not.

There is thus a *migrant/non-migrant ratio*. The combination of these three components and the distribution of the group before migration is initiated results in there being a *degree of dispersal/convergence* to the group's migration. Depending on the interaction of all these components, the different seasonal and/or ecological ranges of the group show a *degree of overlap*. These components are all characteristics of migration when analysed at the level of the group.

5.3.1 Individual characteristics

5.3.1.1 Track pattern and orientation and navigation mechanisms

Most track patterns can be described simply by some term that conveys most nearly their spatial form. 'Zig-zag', 'spiral', and 'random' are three such terms, the last often being used to describe tracks for which no other pattern is readily apparent rather than only for tracks that are mathematically random. Perhaps the two commonest migration track patterns are those associated with rectilinear migration and nomadic migration.

Rectilinear migration: migration in a more or less straight line.

Nomadic migration: migration by an individual or group for which no fixed pattern of directions is apparent.

Heape (1931) proposed that the term 'nomadism' should describe the pattern of movement shown by many animal species which leave their original site and wander fairly randomly thereafter, returning only fortuitously to their original site. Heape thus contrasted nomadism with migration which he defined as a regular periodic movement in which an animal eventually returns to the site of its birth or hatching. However, as has already been shown in Chapter 2, animal movement patterns show every gradation between precise periodic return movements and one-way movement. On the other hand, many animals that wander more or less 'randomly' over wide areas during a large part of their lives nevertheless eventually return to their original breeding or feeding site. Among these animals may be included a number of seals which return at regular intervals and also various nomadic races of man. The family units of some tribes of Australian aborigines, *Homo sapiens*, for example, have a number of camp sites at each of which they stay for a short period before leaving for another. Eventually, however, and with some periodicity, they return to each camp site. Individuals of the South African fur seal, *Arctocephalus pusillus*, wander throughout their entire geographical range, yet each year the individual returns to the same site to breed.

No clear distinction is possible, therefore, between nomadism and migration, and it is probably best to regard nomadism as one pattern of migration.

As a result of the migrations performed by an animal during the course of its lifetime, the animal manifests a certain pattern of exploitation of space. A number of terms may be used in connection with the description of this pattern.

Lifetime track: the path traced out in space by an individual between birth and death.

Lifetime range: the total area (or volume) of space perceived by an individual between birth and death.

Linear range: a lifetime range that extends in more or less a single direction from the birth place and that is much longer than it is broad (see Fig. 18.1).

Limited area range: a lifetime range that occupies a restricted area within which the lifetime track frequently re-crosses itself.

Familiar area: any fraction of the lifetime range (plus for some animals an additional area or volume that is familiar to the animal through social communication) within which an animal can navigate from one point to any other point about which it has prior information, either as a result of direct perception, previous memorised sensory contact or as a result of social communication.

The familiar area is the commonest form of limited area range described in this book.

Home range: the total fraction of the familiar area physically visited by an animal in a given time interval.

These various aspects of the lifetime track are discussed in Chapter 18.

Various orientation and navigation mechanisms

have evolved in connection with the execution of the lifetime track. Here it is necessary to introduce only the terms orientation and navigation.

Orientation: the mechanism by which an animal moves in a given plane or compass direction. Animals using an orientation mechanism, when displaced laterally, continue to move in their original plane or compass direction.

Navigation: the mechanism by which an animal determines the position of a given point in space. Animals using a navigation mechanism to migrate to point *A*, when displaced laterally (within certain limits) can still migrate to point *A*.

Orientation and navigation mechanisms are discussed primarily in Chapters 19 and 33.

5.3.1.2 Distance, speed and duration

Long-distance migration: migration over relatively long distances.

In the case of the more spectacular vertebrate migrants, 'long-distance' may be replaced by a more descriptive term, such as transequatorial or interpolar.

Short-distance migration: migration over relatively short distances.

The description of the distance covered by migration as either long or short is, of course, completely arbitrary, but is sometimes useful and has been widely used in the past. Inherent in the use of the terms 'long-distance' and 'short-distance' is some indication of the capabilities of the species being discussed. Hence, a 50 km migration might be regarded as a short-distance migration for a bird but a long-distance migration for a small mammal.

The distance component of migration is, of course, readily expressed quantitatively. It is not, however, an absolutely single component but is the product of two separate components, speed of movement and duration of movement, both of which are open to independent, as well as dependent, variation. The three factors are, however, so interdependent that it is perhaps worth considering them together, rather than as three separate migration components.

5.3.1.3 Plane of movement and the compass direction of the horizontal component

Horizontal migration: migration in a plane at right angles to the gravitational pull of the Earth.

Vertical migration: migration in a line parallel to the gravitational pull of the Earth.

Migrations often have both horizontal and vertical components, but unless these are altitudinal, they are usually described in terms of their strongest component.

Altitudinal migration: a migration with both horizontal and vertical components that involves movement up or down major topographical features such as mountains or hills.

The horizontal component of migration can be further categorised in relation to geographical direction as north, south, east, west, etc. Often it is sufficient to describe directions qualitatively in these terms. On other occasions, a more precise description is warranted, in which case directions are described as angles measured clockwise from a fixed direction. The direction of species that orientate to the *Sun's azimuth* (the angle subtended to an observer between north and the point on the horizon vertically beneath the Sun when measured in a clockwise direction), but do not compensate for the movement of the Sun across the sky during the day, is expressed naturally relative to the Sun's azimuth. Consequently, in order to permit comparison with these species, and as the majority of examples refer to the Northern Hemisphere, the direction of most other species is expressed relative to south. This is contrary to the perhaps more normal practice of expressing direction relative to north as, for example, in the case of the Sun's azimuth. The expression of direction in angular terms is always potentially confusing as different authors become used to different conventions. I have tried to reduce this potential source of confusion to a minimum in the text by qualifying angular directions whenever these are used.

In all reference to movement direction it is important to bear in mind that movement direction has two components, course and track.

Course: the direction of orientation or heading of an animal and therefore the direction in which the animal would travel if no other forces were acting.

Track: the direction in which an animal travels.

Track is therefore the resultant of course and any other forces acting. An animal walking across a drifting ice field, swimming at an angle to a water current, or flying at an angle to an air current, has a track that differs from its course. All of the directional components of migration at both the individual and the group level can therefore relate either to course or track (e.g. mean course direction, mean track direction). Unless otherwise qualified as, for example, in discussing orientation mechanisms, direction, in this book, will be used in reference to track direction.

5.3.1.4 Degree of return

When an animal migrates from *A* to *B* and then from *B* to *C*, the degree of return can be expressed in a variety of ways. Perhaps the most useful is to express the distance between *A* and *C* as a proportion of the sum of the distances between *A* and *B* and *B* and *C*. The nearer this proportion comes to zero, the higher the degree of return. Thus a bird that performs an autumn migration from Britain to South Africa and a migration the following spring from South Africa to Poland has a higher degree of return than a bird that migrates from Britain to France to Poland. If the proportion exceeds about 0·7, the degree of return is so low that the migration would probably be described as a *one-way migration*.

Migrations with a high degree of return can be performed by executing any one of a variety of different track patterns. The best known track pattern for such migrations is probably that of a *to-and-fro migration* in which the animal returns by retracing more or less exactly its outward track. Some sea-birds and marine fish perform *loop* or *circular migrations* during which the animal does not meet its outward track until it arrives back at its starting point. Even ostensibly one-way migrations, such as those of the various pelagic Antarctic sea birds that move forever eastwards, may be circular, with a high degree of return, if the animal encircles the pole. Finally, a nomadic migration is a common track pattern associated with return migration.

If circular, to-and-fro, or nomadic migrations are found to involve the animal visiting a relatively constant succession of areas at regular intervals, it is often convenient to refer to that animal as performing a *migration circuit*. This normally replaces to-and-fro migration as the descriptive term only when more than two areas are involved. When a migration that was previously described as nomadic is found to involve a migration circuit, it should no longer strictly be described as nomadism, though the term may often be retained to imply a low constancy of direction during successive stages in the migration. Although of dubious semantic validity, therefore, it is occasionally convenient to refer to a nomadic migration circuit.

5.3.1.5 Frequency

Regular migration: migration that occurs at a more or less fixed periodicity.

Regular migration is usually qualified by the appropriate adjective (e.g. daily, monthly, seasonal, yearly). Even accidental migration can be regular if the stage of development or time of day or year at which an animal is prone to be accidentally displaced occurs at regular intervals.

Irregular migration: migration that occurs at irregular intervals.

Many authors use the term 'irruption' to describe irregular migrations, particularly when large numbers of animals are involved. Irruption may also be used even when the migration is regular if vastly different numbers of animals are involved on each occasion, thus giving the superficial impression of an irregular migration.

5.3.2 Group characteristics

5.3.2.1 Direction ratio

Direction ratio: the percentage of individuals within a species, subspecies, race, group, etc. that move in four different compass sectors relative to the mean direction of movement. If the mean direction of movement is $a°$, the direction ratio is expressed, in order, as the percentage of individuals moving in the directions described by: $a° \pm 45°$; $(a° - 90°) \pm 45°$; $(a° + 90°) \pm 45°$; $(a° - 180°) \pm 45°$.

Direction ratio is illustrated in Fig. 2.2. It can be seen that this component of migration can vary between the extremes of $25:25:25:25$ and $100:0:0:0$.

The frequency distribution of migration directions of a group can therefore be expressed by the mean direction of movement followed by the direction ratio. Hence, in Britain, the track directions of the small white butterfly, *Pieris rapae*, in spring and summer can be described as $158°/42:21:21:16$.

Obviously the frequency distribution of migration directions may be unimodal, bimodal, or multimodal, but perhaps most often in migration we are dealing with a unimodal distribution. The following formulae and statistics associated with unimodal direction data and the calculation and analysis of mean migration directions are taken from Batschelet (1965).

In statistical discussion, the mean direction of a series of directions $(e_j°)$ (relative to an arbitrarily determined direction from which all observed and calculated directions are measured) is referred to as the *mean vector*. The mean vector has both direction (the mean angle, $a°$) and length (r). The length of the mean vector measures the *bias* towards the mean direction and is the statistic used to determine whether the distribution of directions is significantly non-uniform. A uniform distribution (i.e. a direction ratio of $25:25:25:25$) has an r-value of zero. When all individuals migrate in precisely the same direction, $r = 1$.

The value of r for a given series of observations $(e_j°)$ for which n observations of direction are obtained is given by

$$r = \sqrt{\left[\left(\frac{\Sigma_j \cos e_j°}{n} \right)^2 + \left(\frac{\Sigma_j \sin e_j°}{n} \right)^2 \right]}$$

The direction $(a°)$ of the mean vector is given by

$$\cos a° = \left(\frac{\Sigma_j \cos e_j°}{n} \right) \Big/ r.$$

Note that because $\cos a°$ and $\cos - a°$ are the same, it is necessary to determine by inspection whether the direction of the mean vector is $a°$ or $-a°$.

It is necessary to correct for any grouping of data, such as when directions are observed only as the cardinal points, N, NE, E, etc. The mean vector statistic is valid only if the observed directions are normally distributed about the direction of the mean vector. The method of correction and the tests for normality are given by Batschelet, but need not be presented here.

Comparison of the observed distribution with a uniform distribution is made by means of Rayleigh's test. This test is valid only for a unimodal distribution and involves the calculation of the statistic z which is given by

$$z = \frac{(\Sigma_j \cos e_j°)^2 + (\Sigma_j \sin e_j°)^2}{n}$$

If $z > z_p$, a table for which is given by Batschelet, the distribution is unimodal as opposed to uniform. As an example, for $P = 0.05$, $z_p = 2.9957$ when n is infinity, 2.9579 when $n = 20$, and 2.9187 when $n = 10$.

If we assume a significant circular normal distribution, or at least a unimodal and symmetrical distribution, the calculated mean migration direction can be tested against a theoretical mean migration direction with respect to the null hypothesis that the two directions are the same. The test statistic is R, which is given by

$$R = \sqrt{[(\Sigma_j \cos e_j°)^2 + (\Sigma_j \sin e_j°)^2]}.$$

If $R > R_0$, the null hypothesis is rejected. R_0, for which a table must be consulted, is a function of X, where

$$X = R \cos (\text{observed mean direction} - \text{expected mean direction}).$$

Significance tests for the difference between two calculated mean directions are also given by Batschelet, but as these are not used in this book they need not be presented here.

Both the direction ratio and the length of the mean vector may be used to express the amount of bias shown by a given group of animals toward the mean direction. Of the two, the direction ratio tends to be used for descriptive purposes whereas the length of the mean vector tends to be used for statistical purposes.

5.3.2.2 The frequency distribution of migration distance

Although this distribution is of extreme theoretical importance, for descriptive purposes recourse is usually made to the migrant/non-migrant ratio described below. Otherwise, the usual terminology of frequency distribution is applied (e.g. normal, leptokurtic, unimodal).

5.3.2.3 Migrant/non-migrant ratio

Complete migration: the migration of 100 per cent of a group.

Complete migration refers to the original spatial unit occupied by the group, not to the distance covered once that spatial unit has been vacated. One hundred per cent exodus from a spatial unit warrants the term complete migration even if some individuals subsequently migrate much less far than others.

Partial migration: the migration of less than 100 per cent of a group.

Obviously complete and partial migrations are part of a continuous spectrum of variability and ideally should be considered as such. Their use is most justified when the frequency distribution of migration distance is clearly bimodal, as it appears to be for some birds.

5.3.2.4 Degree of dispersal/ convergence

To disperse: to increase the mean distance between the individual members of a group of organisms.

Perhaps the nearest opposite of 'to disperse' is 'to concentrate'. Most of the examples of concentration given in this book, however, are probably better described by the verb 'to converge' and its derivatives.

To converge: to come together within a particular spatial unit.

Migration can lead to both dispersal and convergence. In fact, for a great many animals, particularly those that have return migrations, it is inevitable that if one migration results in dispersal, the following migration results in convergence. Hence, the summer migration of many seals results in the dispersal of animals over wide feeding areas. The winter migration results in convergence at the much more localised breeding sites. The same applies, though to a lesser extent, to some whales, such as the blue, *Balaenoptera musculus*, and finback, *B. physalus*, except that in these animals it is the winter migration to the breeding latitudes that leads

to dispersal, the summer feeding grounds occupying much more localised areas. Similar migrations occur in some insects, particularly those such as the convergent ladybird, *Hippodamia convergens*, in the United States, which has a convergent migration to the mountain aestivation and hibernation sites but which is much more dispersed following the spring migration. The migrations of swarms of the desert locust, *Schistocerca gregaria*, in Africa, Arabia and India are from May to August more convergent than dispersive, bringing swarms from different areas to the vicinity of the inter-tropical convergence zone.

Not all migrations lead to marked convergence or dispersal. For instance, the migrations of the humpback whale, *Megaptera novaeangliae*, probably lead to little dispersal or convergence.

In some cases whether or not migration is considered to have a dispersive or convergent component or neither depends on the unit observed. In all species that have a more or less static geographic range, such as some seals, man, and many insects, the nett result of migration on the species is neither convergence nor dispersal. The distribution of the species remains the same, yet within that range migration has resulted in considerable redistribution of individuals (Fig. 5.2). Dispersal has occurred in the sense that groups born into one area have migrated to several areas. Conversely, the animals found in any one area at the end of migration have converged there by migration from several other areas.

In this latter example the same migration sequence leads to both convergence and dispersal or, if the whole range is considered, neither. In the previous examples, migration leads either to convergence, to dispersal, or to neither. All that can be said about the relationship between migration, dispersal, and convergence, therefore, is that dispersal and convergence can both result from migration, but that migration need not necessarily lead to either.

Attempts to justify the use of migration and dispersal as synonyms or as mutually exclusive terms are therefore untenable. The former view, that migration and dispersal are synonymous, has been proposed mainly by entomologists. Southwood (1962), for example, used definitions of migration and dispersal similar to those used here but suggested that as migration invariably led to dispersal, the two terms could to all intents and purposes be regarded as synonyms. C.G. Johnson (1969) took an even more extreme view and said that migration is

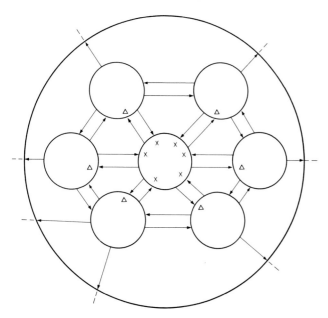

Fig. 5.2   An example of migration, dispersal, and convergence

The migration of some species leads to convergence and/or dispersal and/or neither, depending on the groups or areas considered. In the diagram, migration has led to the convergence of the individuals indicated by crosses and the dispersal of the individuals indicated by triangles. If the range is large relative to the distance moved by an individual and if individuals (the tracks of which are shown by dashed lines) that migrate beyond the edge of the geographical range (large circle) fail to reproduce, migration has led to negligible dispersal or to convergence of the species as a whole. Rather, migration has led to a redistribution of individuals within the geographical range of the species.

Each circle represents a habitat. Each arrow represents the migration of an individual.

adaptive dispersal. Johnson did not say what definition of dispersal he was using but in the same book he accepts the convergence of the convergent ladybird, *Hippodamia convergens*, as a migration.

The second view of the relationship between migration and dispersal, that the terms are mutually exclusive, is equally untenable. This is largely the view of ornithologists, though it is also the view I used in my previous attempt to define insect migration. Dorst (1962) refers to the post-juvenile movements of young birds, as well as to the winter movements of European herons, *Ardea cinerea*, gannets, *Sula bassana*, etc. as dispersal and therefore as something apart from migration. I suggested

(Baker 1969a) that animals with a direction ratio of 25:25:25:25 should be regarded as showing dispersal, whereas animals with a peak direction of movement should be regarded as migrating. However, as already shown (pp. 9–11), there is more or less continuous gradation between direction ratios of 25:25:25:25 and 100:0:0:0. Also all direction ratios can contain an element of convergence or dispersal, even a direction ratio of 100:0:0:0 if individuals fly different distances. There is, therefore, nothing sufficiently qualitatively unique about the absence of a peak movement direction to warrant a redefinition of dispersal only in terms of a direction ratio of 25:25:25:25. In addition to these arguments against my previous ornithocentric attempt to make the terms dispersal and migration mutually exclusive is the argument that to do so is to distinguish between phenomena at a group level that at an individual level can be indistinguishable. For instance, the orange tip butterfly, *Anthocharis cardamines*, performs a movement that is indistinguishable at the individual level from that of, say, individuals of the small white butterfly, *Pieris rapae*. Yet, according to my previous definition, because the track direction ratio of *A. cardamines* approximates to 25:25:25:25 and that of *P. rapae* (in Britain) is 42:21:21:16, *A. cardamines* would be said to disperse, not migrate, whereas *P. rapae* would be said to migrate, not disperse. In fact, of course, the difference between the two species, even in degree of dispersal, is small. At the individual level no difference can be seen at all. Clearly, therefore, nonsensical situations arise, particularly at the individual level, if migration is defined only at the group level. Dispersal, of course, must be defined at the group level.

In conclusion, therefore, it can perhaps be repeated that the only reasonable view of the interrelationships between migration, dispersal and convergence is an acceptance that dispersal and convergence can both result from migration but that migration need not necessarily lead to either. Consequently, any definition of migration in terms of dispersal or convergence must be untenable.

5.3.2.5 Degree of overlap of seasonal and/or ecological ranges

The geographical range of a species can be described as the area enclosed within a boundary outside of which no individuals of that species can be found

(Fig. 2.1). Many species, perhaps the majority, have a range that is more or less *static* (p. 7) and within the boundary of which all the ecological requirements of the species can be found at all times of the year. Migration can occur within the range but has no effect on the boundary. Since the colonisation of the western hemisphere and Australia, and up to the recent temporary extension of the non-breeding range to the Moon, man has been an example of a migratory species with a static range. Other species have ranges that shift seasonally or at various stages of development or of the reproductive or feeding cycle. More often than not, the ranges occupied at different seasons or stages *overlap* to a greater or lesser extent (pp. 6–7). At the opposite extreme to species with a static range are those species, of which many birds, seals, whales, some bats, and other vertebrates are examples, in which the ranges occupied at different times or stages are *separate* (p. 6).

6

Summary of conclusions

Some animals show periodic, to-and-fro, return movements between geographically separate, seasonal ranges. These movements are achieved by every individual within the population moving in a more or less straight line over long distances in a limited number of geographical directions. The majority of people are likely to have little hesitation in describing these movements as migrations. These migrations, however, are not singular phenomena but are the composite manifestation of a number of components (e.g. degree of return, degree of overlap of seasonal ranges, geographical direction, direction ratio, degree of dispersal/convergence, proportion of the population migrating, track pattern produced by an orientation mechanism, distance, and periodicity) each of which is shown by these examples in its extreme form. If each of these components is viewed separately in the perspective of animal movement as a whole, it is found that every gradation exists between the one extreme shown by clearly recognisable migrants and the opposite extreme. No qualitative break can be found in any of the components between one extreme and the other that is sufficient to distinguish between a migration and a non-migration.

Most attempts to distinguish between migration and non-migration have been based on spatial criteria. The failure of these attempts has led other authors, though with even less success, to seek the qualitative distinction between migration and non-migration in the terms of non-spatial criteria.

The failure of all attempts to produce an absolute definition of migration has led to the acceptance that any definition of migration must be arbitrary. In the past, therefore, definitions of migration have been framed in such a way that discussion could be limited to the movements in which each author was particularly interested.

Although previous definitions have been accepted as arbitrary, the actual definitions proposed have never been manifestly arbitrary. This has often led to the term migration assuming unwarranted absolute powers. A safer procedure is not only to accept the arbitrary nature of any definition of migration but also to frame the definition in such a way that it is manifestly arbitrary. '*Migration:* the act of moving from one spatial unit to another' is the definition proposed and used in this book. This definition is manifestly arbitrary, makes no qualitative statement about any of the components of which the more extreme forms of the phenomenon are composed, and furthermore has certain advantages in the consideration of the evolution of the more extreme forms of animal migration. Because the extreme forms of migration are composite phenomena, it is unrealistic to consider why one animal should migrate whereas another does not. A more realistic approach is to consider each of the components of migration separately and then to consider why the optimum for each component should differ from species to species or group to group within a species.

Migration can be divided into a number of categories in a hierarchical way that is illustrated in Fig. 5.1. This book is concerned primarily with the non-accidental initiation of migration by individuals, either independently or as a member of a cohesive group. The smallest categories of migration recognised in this book are exploratory, removal, and return migration. It is possible, however, to analyse exploratory and return migrations as a series of removal migrations. All of these migrations can be described by a number of components that are listed in Fig. 5.1. Some of these appear when the migrations are analysed at the level of the individual, but some only appear when the migrations are analysed at the level of the group.

Part II The development of a migration model

This part of the book represents an attempt to lay the foundations of a migration model by means of which the multitudinous observations of animal migration may be placed into some sort of total evolutionary context. Although further hypotheses are included in Part III, most have been extracted and are presented here. This has the advantage of providing a general synopsis of the total view of migration presented in this book and could be used as such by the reader who wishes to examine the gross form of what is to follow without being involved in much detail. Such extraction has the additional advantage of separating the theoretical (Part II) and natural history (Part III) sections of the book, though it should be stressed that Part III is presented specifically as an evaluation and further development of the model presented in Part II, rather than solely as a review of animal migration.

7
The migration equation

7.1 Removal migration and the lifetime track

During the interval from birth to death any animal must trace out a largely invisible track in space. This track may be termed the *lifetime track*. No matter how long an animal remains in each volume of space occupied by its body and no matter by what means movement or locomotion is achieved, the track must be continuous both in time and space. Despite this continuum, it seems to be a reasonable procedure to analyse the lifetime track by breaking it down into a number of migration units from one spatial unit to another. This procedure seems to remain justifiable no matter how tortuous or convoluted the track and no matter to what extent one migration unit retraces a previous migration unit of an individual's lifetime track.

Each migration unit can be chosen so as to be a removal migration (e.g. from A to B) rather than a return migration (e.g. from A to B to A) (see p. 26). Any return migration that may occur during the lifetime track can be analysed as two or more removal migrations (e.g. in the previous example, one migration from A to B followed by another from B to A). Examples of the way in which observed sections of lifetime tracks can be broken down into their component removal migrations are presented in Chapters 18 and 27. Here, the main point of interest is that if a model can be developed that can account for the observed characteristics of removal migration, then a major step has been taken toward an understanding of the observed characteristics of the lifetime track which, in effect, is the eventual aim of any consideration of migration.

Although it seems likely that the lifetime track may be analysed in terms of its component removal

migrations, it does not follow that other selective pressures do not come into force at other levels. For example, a common form for the lifetime track in vertebrates is that of the limited area range, which is usually a familiar area, and out of which crystallises a home range and/or territory (Chapter 18). It is possible to recognise selective pressures that act upon the optimum size and degree of overlap of the areas occupied by neighbouring individuals or groups. Nevertheless, although these selective pressures may be readily appreciated at the level of the familiar area, in many ways they may be even better understood through their influence on the component removal migrations by means of which an animal establishes and maintains a familiar area. As becomes clear from Part III of this book, there are, in fact, very few features of the lifetime track for which it is necessary to postulate factors acting at levels other than that of the component removal migrations.

The aim of this chapter, therefore, is to lay the foundations for the analysis of the lifetime track by developing an equation that identifies the advantage of removal migration.

7.2 An introduction to the development of a migration model

It should be stressed that the model developed in this book is not intended to represent a primarily mathematical exercise. Instead I am concerned mainly with attempting to identify the conditions under which a particular migration is advantageous or disadvantageous and on the basis of such identification with an attempt to predict the different strategies that will be optimum for different animals. My model, therefore, never reaches the elegant proportions of the transport equation produced by Rotenberg (1972) which describes animal movement through points in space, age, and time but which, at least as framed, does not lend itself to evolutionary consideration.

It is, of course, necessary before developing any evolutionary model to state the premises upon which the model is based. In the case of the model developed in this book the major premise is that metazoan migration has evolved in response to selection that acts at the level of the individual. I am, therefore, in agreement with Murray (1967) that some of the advantages often attributed to migration, such as the 'benefit to a population' of range extension (Howard 1960, Johnston 1961), interdeme gene exchange (Howard 1960, Johnston 1961) and regulation of population density (Lidicker 1962) are unselected consequences of selection at the level of the individual.

It is also necessary before a model can be developed to have an initial thesis that serves as the starting point from which later aspects of the model can be developed. In the present case, the thesis is that: *migration is an advantage when the realisation of potential reproductive success achieved on the way to and in the spatial unit to which an individual animal migrates is greater than the realisation of potential reproductive success that would have been achieved during the same period if the animal had remained in the spatial unit vacated.*

7.3 Discussion of parameters and development of a model

7.3.1 Discussion of parameters

7.3.1.1 Reproductive success

The reproductive success of an individual, A, is given by some quantitative measure of the surviving members of the lineage to which individual A gives rise. When the sole surviving (nth generation) member of the lineage founded by individual A finally dies without leaving offspring, the reproductive success of individual A becomes zero. Apart from a zero measure, however, there can be no final or absolute measure of reproductive success, for as long as a lineage has survivors, the reproductive success of the founder of the lineage can continue to change with time. At best, therefore, the reproductive success of a lineage founder can be measured by selecting some arbitrary moment in time and measuring the quantity of survivors of that lineage at that moment in time. Normally, number of survivors would be the measure used. However, due to the 'dilution' effect of cross-breeding in gonochoristic (separate males and females), non-parthenogenetic and in self-sterile hermaphroditic metazoa, number of survivors is not necessarily a good measure of reproductive success. A more valid measure would be the surviving quantity of hereditary material that has been replicated from the hereditary material of the lineage founder, or from a replica of the hereditary material of the lineage founder, or a replica of a replica, and so on. The measure of quantity can be the product of the number of surviving lineage members and the proportion of their hereditary material that has been replicated from that of the lineage founder as just described. Some estimate of this proportion can be made if the in-breeding coefficient and rate of chromosomal cross-over is known. Such a measure of reproductive success would take into account the effect of in-breeding, periods of parthenogenesis, etc. The problem remains, however, of deciding which moment in time is the most meaningful for the measure of reproductive success, given that the moment chosen, except that of lineage extinction, can only be selected on the basis of arbitrary or practical, but not absolute, criteria.

The procedure normally followed is to choose that measure of reproductive success that is most appropriate for the phenomenon being considered. Three examples may be given to illustrate this procedure, of which the most notable is perhaps that of L.C. Cole (1954) who was concerned with life-history phenomena. In this case the most appropriate measure of reproductive success is the so-called intrinsic rate of increase, the calculation of which was developed by Cole and which represents a milestone in the understanding of life-history phenomena. Ecologists interested in the optimum age for colonisation, however, such as Dingle (1972), use instead the related concept of reproductive value (i.e. the expected contribution of an individual of specified age to future population growth) first developed by R.A. Fisher (1930). Both intrinsic rate of increase and reproductive value avoid the problem of future fluctuations in survival and birthrate etc., and their effect on reproductive success by the postulation of unlimited resources. Finally, Parker (1974a) in a consideration of male time-investment strategy in a sexual selection context uses fertilisation rate as the measure of reproductive success (or 'fitness' as reproductive success is currently being termed).

None of these measures, however, lends itself to the development of a model based on the hypothesis suggested at the end of the previous section. The main problem is that many migrations take place that do not directly involve reproduction. For example, the seasonal return migrations performed by many male bats in temperate regions consist of: an autumn migration from a copulation site to a hibernation site; a winter migration between hibernation sites; a spring migration from a hibernation site to a feeding site or a succession of feeding sites; and finally in late summer a migration from a feeding site to a copulation site. In order to develop a single model which enables predictions concerning any type of migration, not only migrations between, say, reproduction sites, it is necessary to be able to express the advantage of all migrations in the same units. Reproductive success is the obvious measure to use. However, it is only useful if the measure adopted is sufficiently flexible for it to be possible, say, to express the advantage of migration from a hibernation site to a summer feeding site in terms of its effect on reproductive success. This problem is solved as far as this book is concerned by the development of the concept of potential reproductive success.

At the earliest relevant stage of ontogeny, which may be fertilisation or even the beginning of the development of the female gamete within the female parent of individual A, individual A can be considered to have a maximum potential reproductive success, S_{max}, i.e. the reproductive success as measured at some future time, t, after the death of individual A, that would be realised given ideal conditions for individual A and its offspring from that earliest stage of ontogeny until time t. At any stage of the ontogeny of individual A it can be postulated that there would be a certain state, i.e. size, state of nutrition, state of maturity, etc., that indicates that it is still possible for individual A to realise maximum potential reproductive success, S_{max}. Any deviation from this state indicates that it is no longer possible for individual A to do so. Instead,

individual A is capable only of realising its present potential reproductive success, S_p. In any natural situation, the value of the present potential reproductive success, S_p, steadily decreases. After the death of individual A, present potential reproductive success, S_p, of individual A is no longer a function of the conditions affecting individual A (unless A is eaten by its offspring or its corpse is utilised by the offspring in some other way) but will instead be a function of the conditions affecting the offspring of individual A.

During the immature and pre-reproductive stage of the ontogeny of individual A, present potential reproductive success, S_p, consists of only one component, i.e. action-dependent potential reproductive success, S_d. After the onset of reproduction, however, present potential reproductive success, S_p, has two components, i.e. action-dependent potential reproductive success, S_d, and action-independent potential reproductive success, S_i. Action-dependent potential reproductive success, S_d, may be defined as that component of the present potential reproductive success of an individual that can be influenced by the present and future actions of that individual. Conversely, action-independent potential reproductive success, S_i, may be defined as that component of the present potential reproductive success of an individual that cannot be influenced by the present and future actions of that individual, only by the present and future actions of the offspring of that individual. It has to be acknowledged, of course, that it is not possible to say that the reproductive success of any individual can ever be totally uninfluenced by the actions of any other individual, no matter how far apart the two individuals may be. Doubtless a case could be made out that the death of a spider in Australia will influence the reproductive success of a man in Norway. For the purposes of this book, however, it will be assumed that when parent and individual offspring lose sensory contact with each other for the last time, all further actions on the part of the parent fail to influence the potential reproductive success of the offspring and also therefore that portion of the potential reproductive success of the parent that is realised through that offspring.

The most useful aspect of this concept of potential reproductive success from the point of view of migration is that any action by an animal, including migration, can be considered to maintain or reduce its potential reproductive success, whether or not that action is directly concerned with reproduction.

All statements in this book concerning the advantage or disadvantage of a particular behaviour strategy are therefore based on the effect of that strategy on the potential reproductive success of an individual using that strategy.

7.3.1.2 Habitat suitability

The concept of the habitat as it relates to the definition of migration has already been discussed at some length (pp. 15–17). It was concluded that the current use of the term 'habitat' rendered it undesirable at the present time as part of a definition of migration. I am not advocating, however, that the term 'habitat' should be dropped from the biological dictionary. On the contrary, if popular usage of the term changes, it may be possible at some time in the future to make out a case for the re-definition of migration as movement 'from one habitat to another'. Certainly 'habitat' lends itself to many more less-cumbersome sentence constructions than does, for example, 'spatial unit'. For this reason, therefore, the term 'habitat' is used for the remainder of this book. However, it is hoped that definition of the term and the contexts in which it is used will together reduce the danger of ambiguity of interpretation to a minimum.

The following is the definition of habitat used in the remainder of this book.

Habitat: any spatial unit that can be occupied by an individual animal, no matter how briefly.

It is hoped that such a definition draws attention to the pragmatic function of the term and indicates that the limits of the spatial unit concerned can be constructed on practical and arbitrary grounds to suit the convenience of the instance under consideration. It is now possible, having defined the term 'habitat', to construct a measure of habitat suitability in such a way that it lends itself to the development of a model of the evolution of migration.

The concept of habitat suitability has been used in a general way by a wide range of ecologists. It is a more or less self-evident concept that nevertheless is extremely difficult to quantify. The following paragraphs by no means represent the first attempt to produce a measure of habitat suitability (e.g. Fretwell 1972) and presumably will not be the last. As with reproductive success, the main reason for the construction of different measures of the same concept by different authors seems to be that a measure

that is extremely useful for one type of investigation is much less useful for another. The most useful measure of habitat suitability, h, for present purposes seems to be one based on the effect of a particular habitat, H, on the potential reproductive success of an individual A that spends time T in that habitat.

As discussed in the previous section, an individual A at any particular moment can be considered to have a present potential reproductive success, S_p, that is the sum of action-independent potential reproductive success, S_i, and action-dependent potential reproductive success, S_d. Thus,

$$S_p = S_i + S_d.$$

During the immature and pre-reproductive stage of ontogeny of individual A, $S_i = 0$.

At the moment of arrival at habitat H, let the potential reproductive success, S_{arr}, of individual A be given by $S_{i(arr)} + S_{d(arr)}$, where the latter two terms are respectively the action-independent and action-dependent potential reproductive success of individual A at that moment. At the end of time T individual A leaves habitat H. Let the potential reproductive success, S_{lve}, of individual A at time T be given by $S_{i(lve)} + s_i + s_d$, where $S_{i(lve)}$ is the new value of that portion, i.e. $S_{i(arr)}$, of the potential reproductive success of individual A that was already action-independent at the time of arrival at habitat H and where s_i and s_d are respectively the values of the action-independent and action-dependent portions after time T of that portion of the potential reproductive success of individual A that remained action-dependent at the time of arrival at habitat H.

Any change in value during time T of that portion of the potential reproductive success of individual A that was already action-independent at the time of arrival of individual A at habitat H cannot, by definition, be attributed to the effect of habitat H on individual A. The effect of habitat H on the potential reproductive success of individual A is therefore given by the relationship between the potential reproductive success, S_{arr}, at the moment of arrival at habitat H minus that portion, $S_{i(arr)}$, that was already action-independent and the potential reproductive success, S_{lve}, at the moment of departure from habitat H minus the new value, $S_{i(lve)}$, of that portion that was already action-independent at the moment of arrival at habitat H. For example,

$$\frac{S_{lve} - S_{i(lve)}}{S_{arr} - S_{i(arr)}}.$$

This expression provides a good measure of the suitability, h, of habitat H in relation to individual A during time T.

Hence, habitat suitability, h, may be given by

$$h = \frac{S_{lve} - S_{i(lve)}}{S_{arr} - S_{i(arr)}} = \frac{s_i + s_d}{S_{d(arr)}}, \qquad (1)$$

or, in words, habitat suitability, h, equals the sum of the values after time T of the action-independent, s_i, and action-dependent, s_d, portions of the new value of that portion of the potential reproductive success of individual A that remained action-dependent at the time of arrival at habitat H divided by the original value of that portion at the time of arrival. The terms s_d and s_i require further description.

The action-dependent potential reproductive success, s_d, of individual A at the moment of departure from H can be realised through any offspring with which individual A has not permanently ceased sensory contact (including, in the case of a male, sensory contact with a female that carries the sperm or offspring of that male) and all future offspring that individual A still has the potential to produce. s_d is only zero, therefore, when individual A dies before leaving habitat H.

The action-independent potential reproductive success, s_i, is more complex and is not necessarily measured at the moment of departure of individual A from habitat H. The quantity s_i is realised through the offspring that become independent, i.e. permanently lose sensory contact with individual A, during the period that individual A is resident in habitat H. The value of s_i in relation to habitat suitability is given by the sum of the potential reproductive successes of these offspring (taking into account the generation dilution effect—p. 37) with the potential reproductive success of each offspring being measured at the moment of departure of that offspring from habitat H. The value of s_i is zero, therefore, only for a habitat in which individual A does not permanently cease sensory contact with any offspring while in that habitat, including, in the case of a male, sensory contact with a female that carries the sperm or offspring of the male.

The suitability, h, of an ideal habitat can be seen from eq. (1) to be 1·0, because in such a habitat an individual suffers no reduction in $S_{d(arr)}$ during the period between arrival at and departure from H and the transformation of $S_{d(arr)}$ into s_i and s_d. Conversely, the suitability is zero of a habitat in which an animal dies without, in the period from arrival to death, producing any offspring capable of an independent

existence. Assuming, that is, that the action-dependent portion, $S_{d(arr)}$, of the potential reproductive success of individual A was greater than zero at the moment of arrival.

Habitat suitability, h, of a habitat, H, changes with time in relation to individual A for two reasons. First, the conditions within the habitat may change. Secondly, the requirements of individual A may change. For example, whereas during a particular stage of ontogeny, season or time of day, etc., individual A may require food and shelter, at another time it may also require an abundance of mates and reproduction sites. It may well happen that whereas habitat H may have a high suitability during the former period, it may have a low suitability during the latter period.

It will be noticed that in keeping with the definition of habitat given, no mention has been made of the order of size to which 'habitat' is intended to refer. Size limits can only be set on practical grounds for the convenience of consideration of the particular movement to which the model is to be applied. For example, suppose the model is to be applied to the movements of a bird from its nesting site to some food source within its territory. In this case, the relevant 'habitats' are the nesting site and the feeding site and discussion concerning stay-times in each habitat centre upon optimum time-investment strategy for feeding and nesting. However, if the point of interest is the time spent by the bird in its territory before migrating to some other territory, then in this case 'habitat', as far as the model is concerned, is the entire territory.

One of the main aims in the following pages is to construct a model in such a way that it can be applied to any movement, no matter how short or long, between any two spatial units, no matter how small or large and no matter to what extent the spatial units provide the resources necessary for a given individual to live and reproduce.

7.3.2   Development of a model

Let us consider the situation where an individual A vacates a habitat H_1 at time T_{1vac} and migrates to another habitat H_2 arriving at time T_{2arr}. At time T_e, individual A has lived in habitat H_2 for a period of $T_e - T_{2arr}$. Let h_1 represent what would have been the suitability of habitat H_1 in relation to individual A during the period from T_{1vac}, the time of departure from habitat H_1, until time T_e if individual A had remained in habitat H_1. Let h_2 represent the suitability of habitat H_2 in relation to individual A during the period from $T_{2\text{arr}}$, the time of arrival at habitat H_2, until time T_e. Let S_1 represent what would have been the action-dependent portion of the potential reproductive success of individual A at $T_{2\text{arr}}$, the time of arrival at habitat H_2, if individual A had remained in habitat H_1, and let S_2 represent the action-dependent portion of the potential reproductive success of individual A at $T_{2\text{arr}}$, the time of arrival at habitat H_2. Finally, let $s_{i(1)}$ and $s_{d(1)}$ represent respectively what would have been the action-independent and action-dependent portions of the new value of S_1 at time T_e if individual A had stayed in habitat H_1, and let $s_{i(2)}$ and $s_{d(2)}$ represent respectively the action-independent and action-dependent portions of the new value of S_2 at time T_e.

We can now compare the two strategies available to individual A at time $T_{1\text{vac}}$, i.e. either to remain in habitat H_1 until time T_e or to migrate to habitat H_2 and remain there until time T_e. According to the hypothesis given in the introduction to this chapter, migration from habitat H_1 to habitat H_2 is an advantage if the realisation of potential reproductive success achieved by migration to, and residence in, habitat H_2 is greater than the realisation of potential reproductive success that would have been achieved during the same period in H_1. As that portion of the potential reproductive success of individual A that is already action-independent by time $T_{1\text{vac}}$ is unaffected by either strategy, it is only necessary to compare the effects of the two strategies on the action-dependent portion of potential reproductive success at time $T_{1\text{vac}}$ (i.e. $S_{1\text{vac}}$).

Migration from habitat H_1 to habitat H_2 is therefore an advantage when

$$\frac{s_{i(1)} + s_{d(1)}}{S_{1\text{vac}}} < \frac{s_{i(2)} + s_{d(2)}}{S_{1\text{vac}}},$$

i.e. when

$$s_{i(1)} + s_{d(1)} < s_{i(2)} + s_{d(2)}. \tag{2}$$

From eqs. (1) and (2), it follows that migration from habitat H_1 to H_2 is an advantage when

$$S_1 h_1 < S_2 h_2. \tag{3}$$

The next point to consider is the relationship between S_1, the action-dependent portion of the potential reproductive success that individual A would have possessed at $T_{2\text{arr}}$, the time of arrival at habitat H_2, if it had remained in habitat H_1, and S_2, the action-dependent portion of the potential repro-

ductive success of individual A at $T_{2\text{arr}}$, the time of arrival at habitat H_2. In this consideration it is assumed that during the time from $T_{1\text{vac}}$ to $T_{2\text{arr}}$ no portion of $S_{1\text{vac}}$, the action-dependent potential reproductive success of individual A at time $T_{1\text{vac}}$, becomes action-independent. This assumption is discussed further at the end of this section.

The relationship between S_2 and S_1 indicates the extent to which the action-dependent portion of the potential reproductive success of individual A has decreased as a result of migration from H_1 to H_2. This relationship can therefore be used as a measure of m, the migration cost. For example,

$$m = S_1 - S_2.$$

Let M, the migration factor, be given by:

$$M = S_2/S_1, \tag{4}$$

in which case

$$M = \frac{S_1 - m}{S_1} = 1 - \frac{m}{S_1}.$$

The numerical value of M, the migration factor, can be greater than $1\cdot0$, because the possibility exists that S_2, the action-dependent potential reproductive success of individual A at $T_{2\text{arr}}$, the time of arrival at habitat H_2, can be greater than it would have been at the same time, $T_{2\text{arr}}$, if individual A had remained in habitat H_1. This could arise for a number of reasons, but in particular it may arise in those few cases in which migration involves less energy expenditure than not migrating, such as perhaps the wandering albatross, *Diomedea exulans*, (Fig. 27.8) during circumpolar migration, and in those, possibly more numerous, cases in which predation is less during migration than when not migrating.

It is probably not possible, even if it is desirable, to generalise about the relative predation upon migrants and non-migrants. It could be argued that it is in the habitats in which a species is normally resident that the predators (and parasites) specialised to attack the species are most likely to occur and that therefore mortality as a result of such agencies would be greater within such a habitat than away from it. On the other hand, it could be argued that the species concerned has some morphological and/or behavioural protection against predators etc. while in its residential habitat which it does not have when away from that habitat, and that therefore mortality through predation would be less while in such a habitat. It can perhaps be said that a species

Fig. 7.1   There is probably no general rule for animals concerning the risk of predation and parasitisation while in a feeding (etc.) habitat relative to the risk while migrating between such habitats. Many arguments presented later (e.g. for birds, p. 635) are based on the thesis that migrants and non-migrants experience the same potential reproductive success. Thus, animals, such as aphids, may well suffer considerable mortality during migration-by-flight. They also suffer considerable mortality while in their feeding habitat from a wide range of parasites and predators. Among these are hymenopterous parasites, a cocoon of which can be seen beneath the dead shell of the aphid in the top picture, and the larvae of hoverflies, one of which is shown sucking out the contents of an aphid in the bottom picture.

[*Photos by R.R. Baker*]

which rarely migrates is better adapted to the physical and biological components of its residential habitat-type than to other habitat-types. Increase in the frequency of migration between such residential habitats, however, is likely to lead to improved adaptation to the physical and biological components away from the residential habitat-type. Perhaps relative mortality through predation when migrating and not migrating should be regarded as variable, i.e. more frequent when migrating than when not migrating for some migrations of some species and less frequent for other migrations of other species.

Having discussed the value of M, the migration factor, its position in a model of the evolution of migration can now be established. It follows from eqs. (3) and (4) that migration from habitat H_1 to habitat H_2 is an advantage as measured at time T_e when

$$h_1 S_1 < h_2 S_1 M,$$

i.e. when

$$h_1 < h_2 M. \qquad (5)$$

As measured at time T_e, therefore, migration from H_1 to H_2 is an advantage when h_1, the suitability of habitat H_1, is less than the product of h_2, the suitability of habitat H_2, and M, the migration factor.

In the following chapters eq. (5) will be termed the migration equation and forms the basis for further development of a model for the evolution of migration patterns. For most aspects of the further development of this model, therefore, it is necessary only to describe the advantage of migration in terms of habitat suitability and the migration factor, thus precluding further extensive use of the rather cumbersome concept of potential reproductive success. Nevertheless, because habitat suitability and the migration factor are measured respectively in terms of the effect of the habitat and migration on the potential reproductive success of the migrant, some uses of the model are permissible and some are not. The derivation of the migration equation has, therefore, constantly to be borne in mind in any further model development.

As an example of this it can perhaps be pointed out that the migration equation, although it can be applied to animals that pause to feed, drink, roost, etc. during migration, cannot be applied to animals the young of which become independent during migration. This is because it was decided in the derivation of the equation to assume that no part of $S_{1\text{vac}}$, the action-dependent portion of the potential reproductive success of individual A at $T_{1\text{vac}}$, the time of departure from habitat H_1, has become action-independent by $T_{2\text{arr}}$, the time of arrival at habitat H_2, at which time both S_1 and S_2 are measured. It should be stressed that it was not essential for this assumption to be made. Any portion of $S_{1\text{vac}}$ that did become action-independent between times $T_{1\text{vac}}$ and $T_{2\text{arr}}$ could easily have been described algebraically and accommodated within the migration equation. To have done so, however, would have resulted in the migration equation and further developments of the model including more terms than at present. It was considered that any gain through increase in completeness of the model as a result of this accommodation was more than outweighed by the increased difficulty of interpretation of the model. This was particularly thought to be so because the lack of application of the model to animals the young of which become independent during migration is not particularly serious. In practice, the problem can be overcome by assuming that a given removal migration has terminated as soon as part of the previously action-dependent potential reproductive success becomes action-independent. For example, if a male insect during migration successfully courts and inseminates a female, the area in which copulation takes place must be recognised as habitat H_2 and any further movement from the copulation area must be considered as a separate migration. On the other hand, if that insect pauses to feed, roost, or unsuccessfully to court a female, that behaviour and the associated effect on the potential reproductive success of that insect can be included as part of the cost of migration to habitat H_2 and need not, unless so desired, be considered to take place in habitat H_2.

8

The calculated and non-calculated initiation of migration

8.1 Introduction

The hierarchy of the categories of migration recognised in this book have been illustrated in Fig. 5.1 (p. 26). This chapter is concerned with the penultimate level of this hierarchy (i.e. calculated and non-calculated migration).

The distinction between calculated and non-calculated migration is concerned solely with the initiation of migration. Thus:

Calculated migration: migration to a specific destination that is known to the animal at the time of initiation of the migration, either through direct perception, previous acquaintance, or social communication.

Non-calculated migration: migration to a destination about which, at the time of initiation of the migration, the animal has no information, either memorised, through direct perception, or through social communication.

This chapter is concerned with identifying the point at which it becomes advantageous for an animal to initiate calculated or non-calculated migration from a particular spatial unit at a given moment in time.

8.2 The calculated initiation of migration

According to the migration equation (p. 43), migration is an advantage when $h_1 < h_2 M$ (i.e. when the suitability, h_1, of the habitat occupied, H_1, is less than the product of the suitability, h_2, of the habitat

to which the animal migrates, H_2, and the migration factor, M). From the definitions in the previous section it follows that in calculated migration the animal has direct information concerning all three parts of the migration equation (i.e. h_1, h_2, and M). In non-calculated migration, however, the animal has information concerning only h_1.

The hypothesis of this section may be developed as follows. Maximum advantage from migration would be gained if an individual initiated migration only when $h_1 < h_2 M$. However, the non-calculated initiation of migration results inevitably in some individuals performing a migration in which h_1 happens to be greater than $h_2 M$. Any given migration is more likely to be advantageous, therefore, if it is calculated rather than non-calculated. As long, that is, as the criteria by which h_1, h_2 and M are assessed are reliable and as long as the comparison by the individual of h_1 with $h_2 M$ using these criteria is accurate. Comparison of h_1 with $h_2 M$ is only possible if either: (1) H_2 and the migration route are both within the sensory range of the animal while still in H_1; (2) H_2 and the migration route, although outside the sensory range, are both familiar to the animal as a result of previous experience; or (3) H_2 and the migration route are both familiar to the animal as a result of socially communicated information.

If we accept that any given migration is more likely to be advantageous if it is calculated rather than non-calculated, it follows that selection is likely to favour: those individuals that assess h_1, h_2 and M on the basis of those proximate factors that correlate best with h_1, h_2 and M; those individuals with the greatest capacity for accurate comparison of h_1 with $h_2 M$ on the basis of these factors; and those individuals that employ criteria for habitat assessment and other strategies that increase the ratio of calculated to non-calculated migrations performed by those individuals.

Thirteen examples of the calculated initiation of migration are given in Chapter 14. On the basis of these examples, some evaluation of the above hypothesis is possible (pp. 152–6).

8.3 The non-calculated initiation of migration

It was suggested in the previous section that, given the evolution of accurate assessment of habitat suitability on the basis of reliable criteria, migration from H_1 to H_2 is more likely to be advantageous if it is initiated on a calculated rather than on a non-calculated basis. On this assumption it was then suggested that selection might be expected to favour the evolution of strategies that increased the ratio of calculated relative to non-calculated migration. Despite these strategies, however, it is obvious that for any animal there is a finite limit to the size of the area or volume within which calculated migration can take place. In which case, whenever it happens that migration beyond this limit is advantageous, non-calculated migration becomes the optimum strategy.

According to the migration equation, migration from H_1 to H_2 is an advantage when h_1 is less than h_2M. When the migration is initiated on a non-calculated basis, however, in contrast to the situation discussed in Section 8.2, an animal has no information concerning its likely destination at the moment of initiating migration from H_1.

Consider an animal A, at the moment of departure from H_1. A number of potential destinations are available to A in that they are within the possible range of movement of A. Let A settle in a destination, H_d, of suitability h_d, having experienced a migration factor M_d. Now let a number of different individuals leave H_1 in different directions and travel different distances such that individuals arrive in each of the potential destinations. Each of these individuals experiences a particular value for h_dM_d. Let $\overline{h_dM_d}$ represent the mean product of the suitabilities of the potential destinations and migration factors experienced by these individuals. It follows that this function is the *mean expectation of migration* for any individual that initiates migration from H_1. Let the mean expectation of migration be represented by $\bar{E}$,

so that

$$\bar{E} = \overline{h_dM_d}. \qquad (6)$$

$\overline{h_dM_d}$ is a function of the abundance, spatial distribution, and suitability of habitats within a particular area. Consequently, the mean expectation of migration, $\bar{E}$, can be taken to be a characteristic of a certain area at a certain time. It is inevitable, however, that $\bar{E}$ has different values in different parts of the range of a species. It is also inevitable that the value of $\bar{E}$ changes with time. As far as long-term changes are concerned, this is particularly likely to have been the case over the past few centuries during which time the spatial distribution of habitat-types has altered radically. The possibility has always to be born in mind, therefore, that the value of $\bar{E}$ currently experienced by an animal species may be quite different from the value of $\bar{E}$ at the time that the animal's behavioural repertoire was evolving. In particular this possibility should be considered whenever any apparently disadvantageous migration patterns are encountered. Short-term changes in $\bar{E}$ are also likely. For example, meterological and/or hydrological conditions i.e. the migration cost variables of Section 16.4.3 (p. 251), affect the value of $\bar{E}$. The result of this variation is argued (p. 260) to be the inclusion of migration-cost variables within the migration threshold.

One aspect of the concept of $\bar{E}$ requires discussion here, however, before going on in the next section to consider the place of the concept in a model for the evolution of migration. A few animals, such as the planktonic larvae of sessile marine or inter-tidal organisms, may conceivably have little or no control over the distance of migration. Upon release from the parent it is perhaps possible (though unlikely) that the larva has an equal probability of settling in any of the following situations: on the same rock as the parent; on a different rock nearby; on a different shore; in a different bay; and even on a different coastline. Perhaps in the same way (though again it is unlikely) an aphid that launches itself from a leaf on a tree has an equal probability of settling in each of the following destinations: different leaf on the same tree; different leaf on a different tree in the same wood; different leaf on a different tree in a different wood in the same valley; different leaf on a different tree in a different wood in a different valley; etc. If this were true, the mean expectation of migration for a newly released planktonic larva or an aphid that flies off from its leaf would have to be based on the mean suitability of all rocks or leaves available to the animal and the mean cost of migration to each of these leaves. Even in the case of planktonic larvae and aphids, however, it seems likely that the animal can control to some extent the spatial unit in which it remains and the spatial unit from which it departs. There is no doubt, for example, that the sycamore aphid, *Drepanosiphum platanoides*, when it flies from its natal leaf, can control to a large extent whether or not it remains within the confines of its natal tree and probably also its natal wood, or whether it becomes part of the aerial plankton and migrates, say, to a different valley (A.F.G. Dixon 1969). Certainly the majority of animals have considerable control over the spatial unit in which the animal remains or from which it departs. In this case, it is most important that application of the concept of the mean expectation of migration that is used in this chapter is based on a relevant spatial unit. To emphasise this point, let us return to the example of an aphid on a leaf on a tree in a wood in a valley. Clearly an aphid that flies off from its natal leaf but which 'intends' to remain within the confines of its natal tree has a relatively high mean expectation of migration. An aphid that 'intends' to leave the natal tree but to remain in the same wood has a slightly lower mean expectation of migration, but not as low as an aphid that 'intends' to fly to a different wood or a different valley. Consequently in applying the model developed in this chapter, it is most important to choose a sensible spatial unit for consideration. It would be quite worthless, for example, to attempt to account for the initiation of migration of an aphid from one leaf to another on the same tree in terms of the mean expectation of migration from one valley to another.

Finally, but in similar vein, it should be pointed out that the value of the mean expectation of migration, $\bar{E}$, for migration from, say, feeding habitat to feeding habitat may be different from the value for migration from, say, reproduction habitat to reproduction habitat (in species where feeding and reproduction take place in different habitats). Similarly, depending on differences in size and defence against predators, etc. which may affect migration cost, the value of $\bar{E}$ may be different for different ontogenetic or sex classes. All of these characteristics of $\bar{E}$ have to be borne in mind during application of the model that is developed in the following chapters.

On the basis of eqs. (5) and (6) the migration equation can now be re-written in a way that renders it applicable to non-calculated migration. Thus, the initiation of non-calculated migration from habitat H_1 is likely to be advantageous when

$$h_1 < \bar{E}. \tag{7}$$

It can be postulated, therefore, that selection acts upon an animal to recognise when conditions within H_1 are such that h_1 is less than $\bar{E}$ and to respond to these conditions by the initiation of migration.

When H_1 is the natal habitat and in the absence of preliminary exploratory migrations, an animal living in H_1 has no means by which it can assess directly the value of $\bar{E}$. Selection can only act, therefore, on perception of some variable present in H_1 that correlates with the relative value of h_1 and $\bar{E}$. The likely result of such selection is the evolution of a threshold

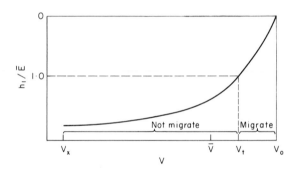

Fig. 8.1   Possible selection associated with the unpredictable variation of habitat suitability: the evolution of a migration threshold to habitat variables

The advantage or disadvantage experienced as a result of the initiation of non-calculated removal migration from habitat H_1 is given by $h_1/\bar{E}$, where h_1 is the suitability of the habitat H_1 and $\bar{E}$ is the mean expectation of migration from that habitat. As a result of the way in which h_1 and $\bar{E}$ have been derived, when $h_1/\bar{E}$ is less than $1\cdot0$, removal migration is an advantage, and, when greater than $1\cdot0$, a disadvantage.

The value, v, of the habitat variable, V, correlates with the suitability of the habitat. This habitat variable may be temperature, amount of food, number of roosts, number of mates, number of competitors, number of predators and/or parasites, etc. v_x represents the value of the variable at which the habitat is at maximum suitability; $\bar{v}$ is the mean value of the variable among the habitats available to the animal upon migration; v_t is the threshold value of the variable at which migration is neither advantageous nor disadvantageous; and v_0 is the value at which the habitat becomes of zero suitability. The difference between $\bar{v}$ and v_t is due to the cost of migration. If there is no migration cost, $\bar{v}$ and v_t are coincident.

Selection is likely to lead to a migration threshold to the habitat variable such that migration is initiated when h_1 becomes less than $\bar{E}$.

response to this variable such that migration is initiated when the value of the variable corresponds to the situation in which h_1 is just less than $\bar{E}$.

A number of variables come to mind, such as rate of encounter with conspecifics, mates or food of a certain quality, etc., as well as others such as temperature, humidity, and rainfall, all of which seem likely to correlate well with h_1. Time is another variable that for certain animals is likely to correlate well with h_1. For any given species it would be expected that selection would favour those individuals that use as a criterion for habitat suitability assessment that variable (or combination of variables) that has the best correlation with h_1.

The details of the hypothesis for the evolution of a threshold for the initiation of migration in response to some proximate factor in H_1 are given in the legend to Fig. 8.1. This threshold in relation to some variable (or variables) beyond which migration is initiated is referred to in the remainder of this book as the *migration threshold*. Those proximate factors that directly affect the reproductive success of the animal while in H_1 and of which, therefore, h_1 is a function, are referred to as *habitat variables*.

8.4 The migration threshold in non-calculated migration: facultative and obligatory initiation

For many years students of migration have recognised two forms of the type of migration that I now propose to call non-calculated migration. These two forms are distinguished on the basis of the different conditions apparent to an observer at the initiation of migration and, therefore, are particularly relevant to this chapter. Although I have used a different terminology in the past (Baker 1969a), in this book I shall refer to these two types as facultative and obligatory migrations. In facultative migration an animal initiates migration in response to a situation that an observer interprets as current adversity (e.g. overcrowding, absence of food, an abundance of predators, etc.). In obligatory migration an animal initiates migration without apparent reference to habitat suitability and often in a situation that an observer interprets as being in all respects favourable. Obligatory migration is usually

characterised by being performed at a fixed time of year or day or at a fixed stage of ontogeny.

In fact, there seems little doubt that both facultative and obligatory migrations are a response to current adversity (if current adversity is defined in terms of habitat suitability). Furthermore, most facultative migrations include some obligatory elements at the moment of initiation of migration and many obligatory migrations involve appreciation of environmental conditions. Nevertheless, it is convenient at this stage to retain the distinction between the two types, though some re-appraisal is eventually necessary (p. 264).

8.4.1 Facultative migration: some expected features of the migration threshold to habitat variables

According to the model developed in Section 8.3, selection favours those individuals of a species with a migration threshold that is that value, v_t, of that habitat variable, V, which is the best indicator of habitat suitability h_1. Furthermore, v_t should be that value of variable V that has corresponded during the evolution of the migration threshold to the situation in which $h_1 = \bar{E}$ (where $\bar{E}$ is the mean expectation of migration). Because the migration threshold is an evolved function of the value of $\bar{E}$, consistent ontogenetic, sexual and spatial differences in the value of $\bar{E}$ experienced by a species during the evolution of the migration threshold lead to consistent ontogenetic, sexual and spatial differences in the migration threshold. Ontogenetic and sexual differences in the migration threshold are discussed in Chapters 10 and 17. Spatial aspects of the migration threshold, however, are central to the theme of this chapter and require further discussion.

In Section 8.3 (p. 45), using the example of an aphid on a leaf on a tree in a wood in a valley, it was pointed out that the value of $\bar{E}$ would be greatest for migration from leaf to leaf on the same tree and least for migration from one valley to another. Consequently, it would be expected that there would be a lower migration threshold for migration from one leaf to another on the same tree than for migration from valley to valley. The fact that an aphid initiates migration from its natal leaf implies only that the threshold for migration from one leaf to another on the same tree has been exceeded. Only if subsequently the threshold for migration from one valley to another is exceeded should the longer-distance migration be initiated. The same applies to any animal that has the ability to exercise control over the spatial unit in which it should remain or from which it should depart. Indeed, the disadvantage of migrating (with a low $\bar{E}$) between large spatial units simply because the threshold for migration (with a high $\bar{E}$) between small spatial units has been exceeded is likely to have imposed stringent selection on animals to be able to control the spatial unit in which they should remain or from which they should depart. The likely result of this selection is a *hierarchy of migration thresholds*.

One of the problems confronting an animal that finds suddenly that its threshold for a given migration has been exceeded is assessment of how local or how temporary the situation may be. The extent of this problem is primarily a function of the degree of correlation between the habitat variable(s) and h_1. This problem is explored in Fig. 8.2. It is concluded that the migration threshold is not a simple response to the value v_t of variable V but rather the product of the amount by which v_t is exceeded and time of exposure.

In Section 8.4.3.2, seasonal and diel changes of habitat suitability are discussed in relation to the evolution of the obligatory initiation of removal migration. As indicated there in Fig. 8.10, seasonal or diel change in habitat suitability inevitably leads to seasonal or diel change in the value of $\bar{E}$. Seasonal or diel fluctuation in migration cost would similarly result in a seasonal or diel change in $\bar{E}$. It follows, therefore, that the value of V that corresponds to $h_1 = \bar{E}$ also varies with season and time of day. In species that become adapted to habitats that show a seasonal or diel variation in suitability, therefore, optimum migration strategy would include a migration threshold that also varies with season or time of day. Selection would favour those individuals with a migration threshold that varied seasonally or through the diem in the way most similar to the value of V that corresponds to $h_1 = \bar{E}$.

From the rather simple situation derived in Fig. 8.1, therefore, a rather more complex concept of migration threshold has emerged. As far as the action of selection is concerned, however, the situation remains relatively simple in that the optimum strategy is to initiate migration whenever h_1

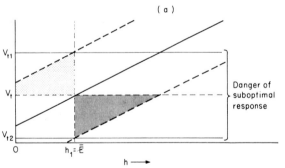
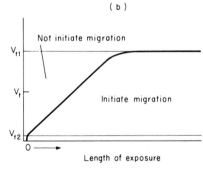
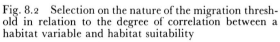

Fig. 8.2   Selection on the nature of the migration threshold in relation to the degree of correlation between a habitat variable and habitat suitability

The effect of the degree of correlation between habitat variable and habitat suitability.
In graph (a) a hypothetical habitat variable, V, is plotted against habitat suitability, h. The regression of habitat variable on habitat suitability is shown (solid diagonal line). In accordance with the hypothesis of this section, the value v_t corresponds to the situation in which $h = \bar{E}$ (where $\bar{E}$ is the mean expectation of migration). Dashed lines on either side of the regression line indicate the range of values of V that an animal might experience at any given moment (or period if V is a rate) of time for any given value of h. The lightly stippled area indicates situations in which an animal that responds only to the value of V does not migrate even though it is advantageous to do so. The darker area indicates situations in which a similar

animal migrates even though it is disadvantageous to do so. Over the range of values of V indicated, therefore, there is the danger of a suboptimal response between the values v_{t1} and v_{t2} if an animal has a migration threshold that is a simple response to the value of V.

Selection in favour of the use of length of exposure to different values of the habitat variable
In graph (b) the same hypothetical habitat variable is plotted against length of exposure of the animal to the variable. It is assumed that for any given value of h in graph (a) an animal is exposed for longer periods to the value of V given by the regression line and that the length of exposure to other values of V decreases on either side of the regression line. On this assumption, the optimum strategy for the animal is a migration threshold of the type shown in graph (b), i.e. a threshold that is a function both of the value of the habitat variable and the length of exposure to that variable.

becomes less than $\bar{E}$. As so far discussed in relation to facultative migration, the migration threshold that is the central component of such a strategy is likely to be a function of: the value of the habitat variable V that correlates well with habitat suitability h_1; the period of exposure to particular values of V; time of year; and time of day. The migration threshold is also likely to be a function of age and sex as well as varying in an optimum way from individual to individual. These latter aspects are discussed further in Chapters 10 and 17.

8.4.2 The migration threshold to migration-cost variables in both facultative and obligatory migration

So far, discussion concerning the migration threshold has been restricted to migration thresholds in relation to habitat variables that correlate with changes in h_1. Such a migration threshold is by definition peculiar to facultative migration, for obligatory migration is considered to have occurred only when migration is initiated without the involvement of any appreciation of habitat variables.

It seems likely, however, that the value of $\bar{E}$ could also be influenced in the short-term by environmental variables, in particular those that influence migration cost. Such environmental variables, referred to in the remainder of this book as *migration-cost variables*, are particularly those meteorological and/or hydrological conditions of wind, rain, temperature, cloud, fog, and water currents, etc. that are likely to influence the value of $\bar{E}$ through their effect on migration cost. Furthermore, and most importantly, information concerning the state of these variables at any particular time is usually available to an animal in its habitat before migration is initiated.

All migration-cost variables act on the value of $\bar{E}$ primarily through their effect on migration cost. Some migration-cost variables may in addition influence the value of $\bar{E}$ through their effect on $\bar{h}$ (mean habitat suitability). Rain, for example, in, say, seasonally arid or arid environments or in relation to pond or puddle dwelling organisms may influence $\bar{E}$ in a variety of ways. First, while rain is actually falling, migration cost (to, say, an insect) may be increased. Once rain has finished, however, migration cost may be much decreased due to the

reduction in mean inter-habitat distance that results from the sudden abundance of new habitats. Finally, if new habitats are more suitable than old habitats (Section 8.4.3.1, p. 54), the sudden abundance of new habitats increases the mean migration factor, $\bar{M}$, due to a decrease in inter-habitat distance, increases mean habitat suitability ($\bar{h}$) and hence increases $\bar{E}$. All of these effects are likely to impose selection on the migration threshold.

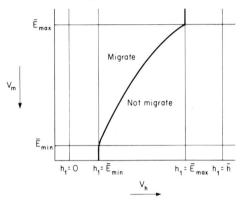

Fig. 8.3 The migration threshold for facultative migration as the combined result of migration-cost and habitat variables

The figure shows a possible form of the migration threshold (solid line) for facultative migration that is likely to result when an animal is exposed during evolution to a habitat variable V_h that correlates well with habitat suitability h_1, and a migration-cost variable V_m that correlates well with the mean expectation of migration, $\bar{E}$. Values of V_h that correspond to different values of h_1 are indicated by vertical lines. Note that the value of V_h that corresponds to $h_1 = \bar{E}_{max}$ (where $\bar{E}_{max}$ is the maximum value of $\bar{E}$ that an animal may experience due to the migration-cost variable being most favourable for migration) is less than the value that corresponds to $h_1 = \bar{h}$ (where $\bar{h}$ is the mean suitability of all habitats available to the animal) due to the effect of migration cost. The value of V_h that corresponds to $h_1 = \bar{E}_{min}$ (where $\bar{E}_{min}$ is the minimum value of $\bar{E}$ that an animal may experience due to the migration-cost variable being least favourable for migration) is even less and may conceivably be coincident with the value of V_h that corresponds to h_1 being zero. The values of the migration-cost variable V_m that correspond respectively to $\bar{E}_{max}$ and $\bar{E}_{min}$ are indicated by horizontal lines.

The action of selection should be to produce a curve of migration threshold that at all points corresponds to the situation at which $h_1 = \bar{E}$. Note that the figure illustrates the simplest possible situation, i.e. a migration threshold that is the combined result of absolute values of a single migration-cost variable and a single habitat variable. As discussed elsewhere (Figs 8.2 and 8.4), however, period of exposure to each variable is also involved, and in nature the migration threshold is likely to be the combined product of several variables of both types.

As far as facultative migration is concerned, the migration threshold is likely to consist of a combination of values of migration-cost and habitat variables. The type of effect that might be expected to evolve is illustrated in Fig. 8.3. The more environmental variables that are involved, the more complex is the migration threshold. The nett result of selection should always be, however, a *composite migration threshold* that corresponds to the situation at which $h_1 = \bar{E}$. This should remain true no matter how many migration-cost and habitat variables are involved in the composite migration threshold.

The arguments concerning the evolution of obligatory migration in the next section are based on the assumption that at some predictable moment (t_i) in time, h_1 is likely to be less than $\bar{E}$. If, however, at time t_i the migration-cost variables are such that h_1 is not less than $\bar{E}$, any migration that is initiated is disadvantageous. Selection should act in just the same way on obligatory migration as on facultative migration in this respect, i.e. to favour those individuals with a migration threshold that is a function of the state of these migration-cost variables.

The likely form of this migration threshold is explored in some detail in Fig. 8.4. It is concluded that the optimum strategy involves the total absence of migration until such time, t_0, that migration under the most favourable conditions of migration-cost variables likely to be experienced by the animal becomes more advantageous than migration can ever be under the least favourable conditions likely to be experienced by the animal. From this time onwards, however, a migration threshold to migration-cost variables should exist such that when conditions are more favourable than the threshold value migration should be initiated. This threshold value should decrease with time at a rate that is related to the existence of two further moments in time: (1) that moment, t_1, at which the initiation of migration is most advantageous, given optimum conditions; and (2) that moment, t_2, at which the initiation of migration is most advantageous even under the most unfavourable conditions likely to be experienced by the animal.

8.4.3 The obligatory initiation of non-calculated migration

In Section 8.4.1 it was argued that animals would evolve the habit of 'monitoring' one or a few habitat variables (and also migration-cost variables, but this is common to both facultative and obligatory migrants—see previous section) that correlate well with habitat suitability. Whenever this environmental variable exceeds a certain state for a certain length of time, the optimum strategy for the animal is the initiation of non-calculated migration. Selection should act to favour those individuals with a migration threshold that corresponds to the situation in which h_1 (habitat suitability) equals $\bar{E}$ (the mean expectation of migration).

There are two inherent disadvantages to this strategy. First, 'monitoring' habitat variables involves time and energy. Second, in many cases, the animal does not initiate migration until after disadvantageous conditions have been experienced for a period (Fig. 8.2). Despite these disadvantages, facultative removal migration is likely to be an optimum strategy in relation to unpredictable fluctuations in habitat suitability for which no other strategy is available.

Some animals, however, live in habitats the suitability of which fluctuates with predictable as well as unpredictable components. In these animals the disadvantages of facultative migration outlined above would favour those individuals that employed a strategy that took advantage of this predictability. It is this selection that in the following pages is argued to have led to the evolution of the obligatory initiation of non-calculated migration.

In the next three sections an attempt is made to identify and evaluate some of the predictable components of variation in habitat suitability and to suggest ways that these may lead to the evolution of obligatory migration. These components are suggested to fall into three major categories: time investment components (p. 57); seasonal or diel components (p. 56); and components associated with habitat ontogeny. These latter factors are perhaps the most varied and are discussed first.

8.4.3.1 Selection due to predictable decrease in habitat suitability

Predictable decrease in habitat suitability was first suggested to be a selective pressure on migration in a classic and stimulating paper by Southwood (1962). Although it is now possible to elaborate on Southwood's suggestions, his paper represents a milestone in the study of animal migration. Southwood concluded that the changing pattern of the environment has been the primary factor in the evolution of

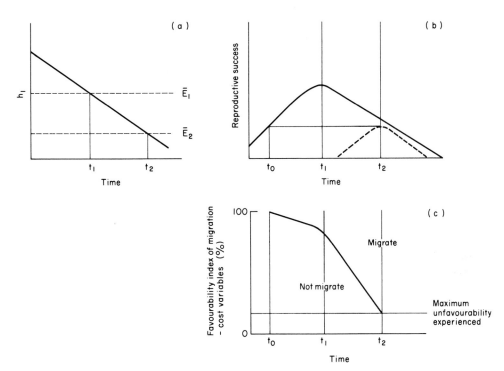

Fig. 8.4 Change with time in migration threshold to migration-cost variables as an optimum strategy for an individual due to migrate at a particular time

Diagram (a) shows the type of relationship between h_1 (habitat suitability) and $\bar{E}$ (mean expectation of migration) that is likely to lead to selection for obligatory migration. The most important component of the situation is that decrease in h_1 relative to $\bar{E}$ is a predictable function of time. $\bar{E}_1$ is the mean expectation of migration given favourable conditions of migration-cost variables. Unfavourable conditions increase migration cost and hence decrease the value of $\bar{E}$ from $\bar{E}_1$ to $\bar{E}_2$. For simplicity, only favourable and unfavourable conditions are recognised, though, of course, in practice there is continuous variation between these extremes.

Diagram (b) shows a possible relationship between time of initiation of migration and reproductive success under conditions of migration-cost variables that are favourable (solid line) and unfavourable (dashed line) for migration. The difference in optimum times for migration (t_1 and t_2) is due to the difference in the time that h_1 falls below $\bar{E}_1$ and $\bar{E}_2$ respectively as shown in diagram (a). The disadvantage of migrating before the optimum time (t_1 or t_2) is due to the disadvantage of leaving a habitat before $h_1 < \bar{E}$. The disadvantage of migrating after the optimum time is due to the disadvantage of spending time in a habitat for which $h_1 < \bar{E}$. The difference in peak reproductive success for migration under favourable and unfavourable conditions is given by the difference between $\bar{E}_1$ and $\bar{E}_2$. After time t_0, initiation of migration under favourable conditions is more advantageous than migration under unfavourable conditions at any time.

Diagram (c) shows the type of variation in migration threshold (in relation to migration-cost variables) that is likely to represent an optimum strategy for such a migrant. Favourability index for such variables can be roughly expressed as the mean expectation of migration for any given set of values of migration-cost variables as a percentage of the mean expectation of migration under those values of migration-cost variables that are optimum for migration. The gross form of optimum variation in the migration threshold is indicated by the heavy line. No matter how favourable the values of migration-cost variables may be, no migration should occur before time t_0. Animals living in areas with a high risk of unfavourable conditions at the appropriate time should show a gradual reduction in migration threshold after time t_0. Animals living in areas with a low risk of unfavourable conditions, on the other hand, may not show a reduction in migration threshold until nearer t_1. Given highly favourable conditions for migration, therefore, potential migrants are likely to initiate migration some time between t_0 and t_1. Towards t_1, migration threshold should show an ever-increasing rate of decrease which continues after t_1. By time t_2 an animal should be prepared to initiate migration no matter what the conditions. When $\bar{E}_2 = 0$, then t_2 is only reached when $h_1 = 0$. It is unlikely, however, that an obligatory migrant with the migration threshold pattern shown in diagram (c) would find itself in such a position. A facultative migrant, on the other hand, may from time to time find itself in such a position.

Conclusion. Optimum strategy for a migrant involves a migration threshold in relation to meteorological and/or other migration-cost variables. This migration threshold should first appear at time t_0 and decrease thereafter until time t_2, by which time the animal should initiate migration no matter what weather conditions etc. prevail.

migration such that part of the adaptation of a species to a temporary habitat is the evolution of obligatory migration. Southwood also points out that rates of appearance and disappearance of habitats are linked, like two sides of a coin. Hence a species exposed to a temporary habitat-type is also inevitably exposed to an environment in which there is a high rate of appearance of new habitats. This is not inevitably true, of course, because some habitat-types become extinct and yet others increase considerably in number. This must have been a general trend over the past 10 000 years in particular, since the advent of agriculture by man, but has been particularly pronounced in some parts of the world over the past few centuries. Nevertheless, as long as the possibility of exceptions is borne in mind, it can be accepted as generally true that the rates of appearance and disappearance of a particular habitat-type show a positive correlation. Having acknowledged the debt that the study of migration owes to Southwood's paper, however, I now propose to remove from its original context the central thesis of that paper—i.e. that obligatory migration enables an animal to keep pace with the changing pattern of its environment—and to integrate it into the general model developed in this book.

Southwood recognised many different facets of the changing pattern of the environment, and in a way it is unfortunate that he chose to discuss these facets in terms of temporariness and permanence. Choice of such terminology placed the emphasis upon habitat longevity, and it is this component on which subsequent authors have concentrated. Unfortunately, it now appears (pp. 53 and 55) that habitat longevity *per se* is perhaps one of the least important selective pressures involved in the evolution of obligatory migration.

In the following pages, the predictable change of habitat suitability, excluding seasonal and diel effects and change due to change in resource requirements, is divided into two components: habitat longevity and the ontogeny of habitat suitability. Eventually all habitats reach zero suitability. The entire Earth reaches zero suitability, at least as far as all present life-forms are concerned, in about 5000 million years when the Sun enters the red-giant phase of its ontogeny (Friedman 1969). The difference between habitats, therefore, is not that some are permanent and some are temporary but rather that some deteriorate to zero more rapidly (i.e. have a shorter longevity), than others. During the period from first appearance to attainment of zero suit-

ability (with respect to the required resource) habitats may differ markedly in their ontogeny of suitability, though it has to be admitted that very few examples that demonstrate such variability seem to be available. Perhaps the commonest form of ontogeny of habitat suitability is that shown in Fig. 8.6 in which suitability increases to a maximum soon after appearance and then gradually decreases to zero, though presumably any form of ontogeny may exist, including a major peak just before disappearance.

According to the modified version (p. 46) of the migration equation, the initiation of non-calculated migration is advantageous whenever h_1 (suitability of the occupied habitat) falls below $\bar{E}$ (the mean expectation of migration). Many animals pass through a stage of ontogeny during which a particular non-calculated migration either cannot occur or for which the migration cost would be so high (and hence $\bar{E}$ would be so low) that h_1 is unlikely to become less than $\bar{E}$. The pre-imago stages of insects fit into this category (as far as migrations requiring flight are concerned, excluding for the moment the inclusion of young larvae, etc. in the aerial plankton). So too do pre-fledgling birds and

Fig. 8.5 Eventually, all habitats reach zero suitability. The Earth reaches zero suitability, at least as far as all present life-forms are concerned, in about 5000 million years when the Sun enters the red-giant phase of its ontogeny.

[Photo courtesy of the NASA Johnson Space Center]

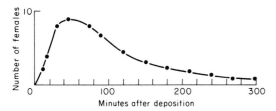

Fig. 8.6   An example of variation, with time, of habitat suitability: variation, with time, of the suitability of cow dung as a reproduction habitat for male yellow dung flies, *Scatophaga stercoraria*

The diagram shows the number of female dung flies around a cow dropping with time after deposition in August. This curve could be taken as a measure of the variation with time of the suitability of a cow dropping as a habitat to a male dung fly.

[*Re-drawn from Parker (1970c)*]

pre-independence young mammals (as far as independent migration from the parental home range is concerned). Suppose that the rate of decrease of habitat suitability is so high that by the time (ontogenetically) removal migration becomes a possibility, h_1 is always less than $\bar{E}$. In this situation, as soon as an animal reaches the appropriate stage of ontogeny, non-calculated migration is advantageous. Any time and energy spent searching the habitat in order to assess habitat suitability would be disadvantageous. The optimum strategy, therefore, would be the initiation of non-calculated migration without reference to the suitability of the immediate surroundings as soon as the appropriate stage of ontogeny is reached. Such an animal, experimentally released in a very favourable habitat (i.e. in which h_1 greatly exceeds $\bar{E}$), just before attaining the relevant stage of ontogeny is nevertheless likely to initiate non-calculated migration, thus manifesting the chief characteristic of obligatory migration. Other characteristics of this migration, such as the distance travelled before reference is made to the suitability of the habitats encountered, are discussed in later chapters (in particular, see p. 83).

Although the emphasis in the above paragraph was placed on the attainment of an ontogenetic stage preceded by a stage or stages during which migration was impossible, linkage of any time-correlated event with the initiation of migration could be favoured by selection. Biological clock mechanisms in particular come to mind. Obligatory migration may therefore be observed either at a particular stage of ontogeny or after a particular stay-time.

It remains, therefore, to attempt to identify the way that change in habitat suitability can act as a selective pressure that favours obligatory migration. The critical factor in this connection seems to be change in habitat suitability in relation to the generation interval of the animal concerned. The validity of this conclusion is based on the assumption that whereas many pre-adaptations (e.g. clock mechanisms, ontogenetic phases) exist upon which selection can act to produce the ability to measure time during an individual's own life-span, there are, with the exception of social communication, few if any pre-adaptations upon which selection can act to produce the ability to measure time in terms of generations. This section, therefore, is concerned with those species that are adapted to habitats with rapid change in suitability in relation to the generation interval. Such habitat-types are particularly those in the early stages of ecological succession which nowadays in most temperate regions are roadsides or waste ground but which previously would have been areas where small or large areas of ground were likely to be disturbed by fire, landslides, wind-blown trees, digging or grazing animals, etc. In addition, puddles and small ponds, dung, and, as far as parasites are concerned, hosts with a high mortality, are also habitats likely to show rapid changes in suitability. The ontogeny of suitability is, of course, not only a reflection of the gradual change in abundance of food and of suitable roosting sites and microclimatic conditions etc. but also of changes in deme density and in the abundance of predators and parasites.

Figure 8.7 illustrates the conditions that seem to be necessary to generate selection for obligatory migration. It is concluded that an ontogeny of habitat suitability that reaches a maximum soon after first appearance but which falls permanently below $\bar{E}$ before the habitat is older than the time interval between birth of an individual and the ontogenetic stage under consideration is the only type of habitat ontogeny that can generate such selection. Other more detailed conclusions are also reached. One of these conclusions is that the value of $\bar{E}$ for an individual of the appropriate ontogenetic stage is a function of the rate of appearance of new habitats. Independently of the rate of disappearance of older habitats at this time, an increase in the rate of appearance of new habitats leads to a greater value for $\bar{E}$. Animals subjected to a sudden appearance of new habitats at a particular season, therefore, such as locusts at the beginning of the wet

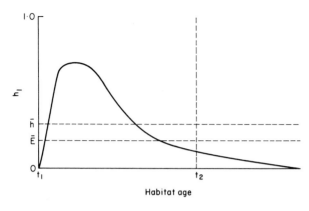

Fig. 8.7   Conditions necessary to generate selection for obligatory migration

The figure shows a possible relationship between habitat suitability (h_1) to a reproducing adult and habitat age in relation to the ontogeny of an animal adapted to reproduce in that habitat. The relationship shown is considered to be that most likely to generate selection for obligatory migration. Habitat suitability increases to a maximum soon after first formation of the habitat and subsequently decreases to zero. Mean habitat suitability for the species is indicated by $\bar{h}$. The mean expectation of migration, $\bar{E}$, is also indicated. Note that $\bar{E}$ is less than $\bar{h}$ because of the effect of migration cost. The time interval between t_1 and t_2 also corresponds to any time interval measurable by the animal but particularly the interval between birth and some time-correlated event, X, in ontogeny, e.g. adult emergence in insects or independence from parents in birds or mammals.

When, as shown, habitat suitability, h_1, is always less than $\bar{E}$ by habitat age t_2, it is always an advantage for the animal to initiate migration at event X and selection therefore favours obligatory migration coincident with event X.

These conclusions are valid only if there is no possibility that h_1 could be greater than $\bar{E}$ after habitat age t_2. Curves of habitat suitability with habitat age that do not result in h_1 being permanently less than $\bar{E}$ after t_2 are more likely to select for facultative migration than obligatory migration.

The difference between h_1 and $\bar{E}$ at t_2 depends partly on the value of $\bar{h}$ and partly on migration cost. Species with a high migration cost (and therefore low E) are less likely to evolve obligatory migration than species with a low migration cost.

Increase in $\bar{h}$ increases the probability of selection for obligatory migration. $\bar{h}$ is in part a function of the relative proportions of old and new habitats. Species that always attain event X coincidentally with a sudden increase in appearance of new habitats are exposed to a higher value of $\bar{h}$ and therefore more probability of selection for obligatory migration irrespective of the rate of habitat disappearance that is occurring at the same time.

season or pond- or puddle-dwelling beetles in Scandinavia during the spring thaw, are highly likely to have evolved a seasonally obligatory migration that may be ontogenetic if attainment of a particular ontogenetic stage coincides with a particular season. It should be noted that this obligatory migration is independent of the rate of disappearance of these habitats. The peak of disappearance of habitats that inevitably occurs later represents a selective pressure on a different migration, perhaps at a different ontogenetic phase or possibly by a different generation. Furthermore, this disappearance is much more likely to favour facultative migration than obligatory migration. Although there is some justification for assuming that rates of habitat appearance and disappearance are linked (p. 52), it does not follow, therefore, as proposed by Southwood (1962) that habitat appearance and disappearance do not act independently as selective pressures. On the contrary, it seems likely that they often act independently as in the above examples.

Habitat disappearance, where it acts as a selective pressure upon obligatory migration, does so only as an extreme case (Fig. 8.8) of the general conclusion that the critical condition is a value of h_1 that is invariably less than $\bar{E}$ after a certain stay-time or upon attainment of a particular ontogenetic stage. Even when habitat longevity is a constant, differences in the ontogeny of habitat suitability may be sufficient to generate selection for or against obligatory migration (Fig. 8.9).

The chief conclusion of this section, therefore, is that adaptation to a habitat the suitability of which reaches a maximum soon after habitat formation but which falls permanently below $\bar{E}$ before the habitat is older than the time interval between birth of an individual and the relevant ontogenetic stage is the only non-seasonal and non-time investment condition that can generate selection for obligatory migration at that stage. Given a curve of habitat ontogeny with a single peak soon after formation, any invariable coincidence between the ontogenetic stage under consideration and periods of maximum rate of *appearance* of new habitats also favours the evolution of obligatory migration. Rate of *disappearance* of habitats and habitat longevity are not necessarily involved in this selection except as an extreme of the general situation, and where they do occur in their own right are more likely to select for facultative than for obligatory migration.

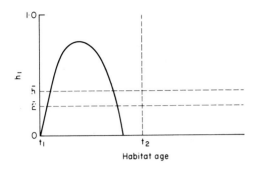

Fig. 8.8 Habitat longevity as a selective pressure for obligatory migration

The figure shows the situation in which habitat disappearance generates selection for obligatory migration. The curve shows variation in h_1, habitat suitability (in this case to a reproducing adult) with age of habitat after first formation (t_1). Habitat suitability increases to a maximum soon after first formation of the habitat and then rapidly decreases to zero. Mean habitat suitability for an animal species adapted to this habitat-type is indicated by $\bar{h}$. The mean expectation of migration, $\bar{E}$, is also indicated. Note that $\bar{E}$ is less than $\bar{h}$ because of the effect of migration cost. The time interval between t_1 and t_2 corresponds to the interval between birth of an animal of this species and some time-correlated event, X, in the animal's ontogeny. Because h_1 is zero and therefore less than $\bar{E}$ by habitat age t_2, it is always an advantage to initiate migration at event X and selection therefore favours obligatory migration at event X. Habitat longevity can, therefore, generate selection for obligatory migration but it can be seen to be only one extreme of the general selective pressure illustrated in Fig. 8.7 and scarcely merits consideration as an independent selective pressure.

Note that h_1 refers to habitat suitability in relation to a reproducing adult. Decrease of h_1 to zero before habitat age t_2 does not mean, therefore, that the habitat has ceased to exist as far as developing individuals are concerned. Only that if these individuals remain in the habitat to reproduce, their offspring will fail to reach reproductive age, for by that time the habitat will also have ceased to exist as far as developing individuals are concerned.

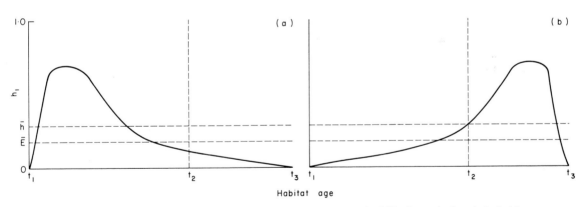

Fig. 8.9 The ontogeny of habitat suitability can generate selection for obligatory migration independently of habitat longevity

The migration equation, as modified for non-calculated removal migration, states that the initiation of migration will be advantageous if h_1, habitat suitability at the place of origin, is less than $\bar{E}$, the mean expectation of migration. Diagrams (a) and (b) suggest the possible influence of the ontogeny of habitat suitability on the values of h_1 and $\bar{E}$ in relation to the evolution of the obligatory initiation of migration. $\bar{h}$ refers to the mean habitat suitability experienced by an animal species adapted to this habitat-type.

In both diagrams, the longevity (i.e. time from appearance, t_1, to disappearance, t_3) of the habitat concerned is the same. The value of h_1 that corresponds to $\bar{E}$ is indicated by the dashed horizontal line. In diagram (a), maximum suitability is reached early in habitat ontogeny and in diagram (b) maximum suitability is reached late in habitat ontogeny. The dashed vertical line indicates a time, t_2. The time interval between t_1 and t_2 is the same as the time interval from birth of an animal of the species concerned to the ontogenetic stage X at which migration becomes possible.

The diagrams show that the ontogeny of habitat suitability exerts a selective pressure that is independent of habitat longevity. In the situation shown in (a), all individuals upon attaining the appropriate stage of ontogeny, X, are confronted by a habitat for which h_1 is less than $\bar{E}$. In situation (b), individuals may or may not be confronted with such a habitat, depending on the age of the habitat when oviposition or birth occurred. In (a), 100 per cent obligatory migration upon attainment of stage X is likely to evolve. In (b), partial obligatory migration and/or facultative migration is likely.

8.4.3.2 Selection due to seasonal and diel change of habitat suitability

Diel and seasonal cycles represent potential sources of predictable fluctuations in habitat suitability. First, let us consider habitat suitability in relation to the seasonal cycle. A particular type of habitat that has been favourable $(h_1 > \bar{E})$, say, all summer, by virtue of its exposure or food availability, may decrease in suitability in the autumn or winter. However, if all habitats are affected similarly by season, h_1 is unlikely to vary predictably in relation to $\bar{E}$, though the value of $\bar{E}$ inevitably changes. The critical factor is whether the suitabilities of different habitats are affected differentially, but predictably, by season. Figure 8.10 explores this possibility and indicates the way that seasonal change of habitat suitability can lead to the obligatory initiation of removal migration. Once again, at a certain time of year, h_1 may always be less than $\bar{E}$ for some species, and once again, therefore, time and energy spent in the assessment of habitat suitability is likely to be disadvantageous. The optimum strategy, therefore, should be initiation of removal migration at a certain season without proximate reference to habitat suitability.

Selection in favour of individuals with this strategy must include selection in favour of the ability to detect season. It seems likely, however, that the ability to detect season, which is so widespread amongst animals, could well have been an early metazoan adaptation, possibly in response to the advantage of seasonal variation in physiology. Phylogenetically, therefore, selection on a particular lineage in favour of the initiation of removal migration at a particular season, in the vast majority of cases is likely to have been acting on animals that already possessed the ability to appreciate season. In few cases does it seem likely that this ability had to evolve *de novo* before adaptation to selection on migration took place. In the majority of cases, therefore, selection seems likely to have favoured those individuals with optimum linkage between the initiation of migration and time of year.

Precisely similar arguments can be applied to animals adapted to habitats such that h_1 and $\bar{E}$ fluctuate with time of day. In this case selection seems likely to have favoured those individuals with optimum linkage between the initiation of migration and time of day.

Another type of seasonal effect, that resulting from seasonal peaks in formation of highly suitable new habitats, has been discussed in the previous section (p. 54).

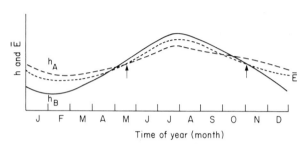

Fig. 8.10   Selection on the initiation of removal migration due to seasonal variation of habitat suitability

Suppose that two types of habitat (A and B) are available to an animal, and further suppose that the suitabilities (h_A, dashed line; and h_B, solid line) of these two types of habitat vary with season. The dotted line indicates the likely seasonal variation of $\bar{E}$, the mean expectation of migration. The difference between $\bar{E}$ and the more suitable of the two types of habitats is a function of the migration cost.

If the suitabilities of A and B vary identically (not shown) with season, the variation is unlikely to generate selection in favour of the obligatory initiation of migration at a certain season.

When, as in the diagram, the suitabilities of A and B, although both varying with season, do so to a differing degree, selection may favour obligatory migration at a certain season for the reasons described below. Time of day can be substituted for time of year in the diagram without in any way affecting the arguments.

In the diagram, in January and February habitats of type A in general have a suitability that is higher than habitats of type B. The mean expectation of migration, $\bar{E}$, is therefore the mean product of the probability of finding a habitat of type A and the migration factor which is a measure of the cost of finding such a habitat. As usual, an animal should initiate migration if $h_1 < \bar{E}$. In May, habitats of type B become more suitable than those of type A, even though both types of habitat have increased in suitability, a factor which by itself increases $\bar{E}$. Animals that may have spent the winter (JFM) in habitats of type A have a mean expectation of migration that is a function of the probability of finding a habitat of type B and the migration cost. Selection favours, therefore, those individuals that in spring initiate migration from a habitat of type A when (arrow) the suitability h_A of that habitat predictably becomes less than $\bar{E}$. In autumn, selection favours the initiation of migration from a habitat of type B when (arrow) the suitability, h_B, predictably becomes less than $\bar{E}$.

Note that seasonal variation in habitat suitability may be either a function of the habitat or a function of seasonal change in the animal's requirements.

8.4.3.3 Selection due to change of habitat suitability in relation to optimum time-investment strategy

It has been pointed out in the initial discussion of habitat suitability (Chapter 7, p. 40) and again in Fig. 8.10) that habitat suitability can change for one or both of two reasons. The most obvious reason for change in habitat suitability is some change of the habitat itself. A less obvious, but nevertheless equally important, reason is that the requirements of the animal change. As the animal's requirements are quite likely to change in a predictable way with ontogenetic stage or time of day or year, this source of variation in habitat suitability is very relevant to the consideration of obligatory migration. In this particular case, however, migration cannot be considered except as an integral part of a much wider phenomenon, that of the evolution of an optimum time-investment strategy.

Optimum time-investment strategy can be defined simply as the optimum allocation of times spent on given activities so as to achieve maximum reproductive success (Parker 1974a). One time-investment component increases, therefore, at the expense of the others. The problem of time investment has been explored in the paper by Parker (cited above); here I wish only to deal with those aspects of the problem that particularly concern migration. However, some degree of general introduction is necessary.

If an animal were surrounded by a constant environment, with an unlimited supply of food available to it without locomotion, and were capable of replicating itself without contact with other individuals and without hindering the feeding process, the optimum time-investment strategy would be to spend 100 per cent of the time feeding so as to maximise the offspring mass. The environment is such, however, that this is not possible. First, there are powerful reasons why metazoans have evolved the male/female phenomenon (Baker and Parker 1973, Parker et al. 1972). Some time has to be set aside, therefore, in connection with the process of bringing male and female gametes together. This process requires minimum time in animals in which selection against mobility has been so strong that the only available method for bringing male and female gametes together is to liberate them into some medium, particularly water, where either currents or the movement of the gametes themselves can bring them together. It is, moreover, in water that selection against mobility can be greatest because the movement of the environment itself brings to the animal at least certain types of food. In such sedentary spawning invertebrates a premium is likely to set upon feeding so as to maximise the gametic mass (Parker 1974a).

However, animals that are adapted to foods that are associated with selection for mobility on the part of the animal are automatically confronted with a number of other time-investment components. Firstly, and directly, migration between feeding sites becomes an important component. At the very least, therefore, optimum time-investment strategy would involve the optimum allocation of time between feeding and migration between feeding sites. However, selection for mobility invariably leads to the action of certain sexual selective pressures. As far as males are concerned, those individuals may be favoured which reduce feeding-time investment (at some cost in terms of gamete productivity) and increase time spent searching for and spawning as close as possible to females (Parker 1974a). This may increase reproductive success by increasing the chances that it is their sperm rather than that of their competitors which fertilise the limited number of ova. This sexual selective pressure could have been one of the major factors in the evolution of copulation (Parker 1970a). With the evolution of copulation it seems likely that an inevitable result would be a considerable reduction in male feeding-time investment, with a reduction in sperm production, with a consequent increase in time available for female-searching and other activities that increase the rate at which females are encountered and inseminated. Because in species that copulate, or spawn in pairs, it is the rate at which females are encountered that is so important to males, selection is likely to favour either extreme male mobility or the guarding of areas where receptive females are most likely to be found (Baker 1972a, Parker 1974a). In either case, the time investment of a male in staying with any one female will decrease the rate at which further females will be encountered. Extension of the period spent with each female is, therefore, only likely to evolve if the increase in number of offspring produced by that female that are sired by that male is greater than the increase in number of offspring sired by that male if the same time were spent searching or waiting for further females (Parker 1974a).

In females of many animals, time investment might be expected to remain heavily biased towards feeding (i.e. toward maintaining maximum production of gametes, eggs, or, in viviparous animals, young of optimum size). However, increased investment of time in the offspring (i.e. parental care), might in general be expected to reduce the feeding-time investment and hence the gamete production (of male as well as female if both sexes show parental care) and therefore only to evolve if the reproductive success is greater than if no time were spent in parental care. The many other evolutionary facets of parental investment have been considered by Trivers (1973).

Animals perform, therefore, a range of activities, e.g. feeding, drinking, sleeping, mating and associated behaviour, and parental care, and there is likely to exist an optimum allocation of time to each activity that leads to maximum reproductive success. One final point has to be considered, however, before the initiation of migration between the sites where each of these activities takes place most efficiently can be discussed. Not only is there likely to be an optimum time investment in each activity but environmental rhythms are likely to result in there being an optimum time of day, month, year, etc. for each of these activities to be performed. Hence, the feeding-time allocation tends to be concentrated into one time of the day or year, courtship and copulation into another time of day or year, sleeping or roosting into another time of day or year, and so on. These diel, seasonal, and ontogenetic cycles of time allocation are so widespread among animals and so well known that it is unnecessary in the present context to enter into further details. It should perhaps be remarked, however, that although it has been demonstrated that segments of behaviour patterns represent optimum time-investment strategies (e.g. Parker 1970c, 1974c, Krebs *et al.* 1974), a demonstration that the total time investment strategy of an animal is an evolved optimum is as yet lacking.

Despite this lack of experimental support, it is accepted in this book that there exists for any given animal an optimum daily, seasonal, and ontogenetic time-investment strategy, and that the reproductive success of any particular individual is a function of the time investment shown by that individual. An optimum time-investment strategy involves not only the length of time spent in the performance of a given activity but also the sequence in which given activities are performed and the synchronization of these sequences with diel and seasonal environmental rhythms. Recent papers by Schoener (1971), Trivers (1973) and Parker (1974a) have respectively elucidated some of the selective pressures that affect the optimum investment strategies of feeding, parental investment, and male courtship. There is still a great need, however, for these and other considerations to be brought together to produce a coherent model of time-investment strategy. It is not intended here, however, to enter further into the details of these other components of time investment except insofar as migration is concerned.

As far as the initiation of migration is concerned it is necessary only to take the simple viewpoint that migration from a spatial unit in which one activity is performed to a spatial unit in which another activity is performed is initiated at the end of the optimum period of investment in the original activity. In this context, migration need only be viewed as the means by which an animal moves from one type of habitat to another. The time invested in migration is an inevitable consequence of the pattern of distribution of the different types of habitat (i.e. feeding habitat, courting habitat, roosting habitat, etc.). Time spent

in migration is invariably disadvantageous, therefore, and an optimum time-investment strategy is one in which time spent in migration is reduced to the minimum that is compatible with the most efficient performance of other time-investment components. This conclusion is particularly relevant to any discussion of track pattern (Chapter 18), migration distance (Chapter 11), and orientation and navigation mechanisms (Chapter 33).

To avoid any ambiguity in relation to this conclusion, it should be stressed that it is not implied that migration itself is disadvantageous, only the time spent in migration. The advantage of migration is that the animal moves from one habitat to a more suitable habitat elsewhere. The disadvantage of time spent during migration has already been included in the migration equation (Chapter 7) in that it is a component of migration cost. For any given migration from H_1 to H_2, however, all other things being equal, such as energy expenditure and survival rate, selection for optimum time investment favours those individuals that migrate from H_1 to H_2 in the shortest possible time.

If we now return to the main point of this section, i.e. consideration of the evolution of obligatory migration, a certain clear conclusion has emerged. Remembering that the same arguments apply whether the habitats in which different activities are performed are a few millimetres or a hundred kilometres apart, migration in relation to time investment is likely to occur either at the end of the optimum period for the performance of a given activity or when the time of day or season is such that it is an advantage to change over to some other activity. In either case, migration is likely to be a response to some endogenous state rather than to any characteristic of the immediate surroundings and consequently to the experimentalist will appear as an obligatory migration. The endogenous trigger may either be based on a clock mechanism, or on a physiological state induced by photoperiod and/or temperature or the attainment of a particular state such as a full stomach, empty seminal vesicle, etc.

It should be noted that this section has been concerned with migration in relation to time investment and has, therefore, been concerned with migration between habitats in which *different* activities are performed. Section 8.4.3.1 (p. 50), on the other hand, was concerned with migration between habitats in which the *same* activities are performed. Section 8.4.3.2 (p. 56), was concerned with both types of migration.

8.4.4 Some expected physiological features of facultative and obligatory migration

In all of the arguments so far, use has been made of the concept of $\bar{E}$ (the mean expectation of migration). As far as the individual migrant is concerned, however, whether or not a particular migration has been advantageous depends on the value of $h_2 M$ (h_2 = suitability of destination; M = the migration factor) actually experienced and its relationship to h_1 (p. 43). The possible relationships between $h_2 M$ and $\bar{E}$ and the selective pressures that result from these relationships are central to several subsequent chapters, particularly those concerned with stay-time and migration distance (Chapter 11). One facet of the relationship between $h_2 M$ and $\bar{E}$ is important in relation to the initiation of migration, however, and is considered in this section.

A given animal increases the advantage gained from migration (or decreases the disadvantage) by increasing the value of $h_2 M$ actually experienced, and this remains true whether or not $h_2 M$ is less than $\bar{E}$ or even whether or not $h_2 M$ is less than h_1. It must happen, more or less often, depending on the species, that an individual, having initiated non-calculated migration when $h_1 < \bar{E}$, nevertheless experiences a zero value for $h_2 M$. One of the contributory factors to this could be the level of energy reserves at the moment of initiation of migration. All other things being equal, the chances of encountering a habitat of above-zero suitability are likely to be directly proportional to the area effectively searched during migration. Area searched, as well as being a function of track pattern (Chapter 18), is likely in turn to be directly proportional to the energy reserves at the moment of initiation of migration, at least as far as those individuals are concerned that fail to find a habitat of above-zero suitability or that find such a habitat just as their energy reserves become depleted. Selection is therefore likely to favour those individuals with greatest energy reserves (as long as accumulation of these does not conflict with optimum time investment on feeding before migration is initiated) at the moment of initiation of migration. Any physiological mechanism that leads to an increase in energy reserves in advance of migration is therefore likely to be advantageous.

Most facultative migrants (see below) are in something of an evolutionary impasse in this respect

if, as suggested, that moment in time when facultative migration is advantageous is unpredictable. In this situation, energy reserves cannot be increased in advance in any way that can be favoured by selection. When facultative migration does become advantageous, when h_1 is less than $\bar{E}$, the animal finds itself in just that type of habitat in which increase in energy reserves is least efficient. Furthermore, because increase in energy reserves requires time, any such increase must take place during a period when, because h_1 is less than $\bar{E}$, the individual suffers a disadvantage as a result of not initiating migration. The only available strategy would seem to be for the animal to maintain some energy reserves in case migration should at some time become advantageous. It can perhaps be assumed that for any particular animal there is an optimum energy-investment strategy that, at least in part, is a function of the optimum time-investment strategy (p. 57). Although there may be some selection in favour of maintaining energy reserves for a migration that may never take place, there is also likely to be selection against such energy reserves. The time and energy involved in building up, maintaining, and carrying reserves that may never be used could well be disadvantageous. The most likely evolutionary result of these opposing selective pressures seems to be that animals whose evolution has exposed them to selection in favour of facultative migration will most frequently invest relatively more time and energy in reserves that are specifically for the contingency of facultative migration than animals that have been exposed to such selection less frequently.

Although as argued here all facultative migration should be an adaptation to unpredictable variation in h_1, it has to be accepted that different degrees of unpredictability exist, some of which may generate selection for physiological preparation for migration in essentially facultative migrants. So far, discussion has assumed that it is the moment that h_1 falls below $\bar{E}$ that is unpredictable. It is possible, however, that in some habitats the falling of h_1 below $\bar{E}$ is only the end product of a predictable sequence of events, most of which take place while h_1 is greater than $\bar{E}$. In this situation the migration is still facultative if the initiation of this sequence of events is unpredictable, but, if the animal can appreciate when this sequence has begun, the probability of h_1 falling below $\bar{E}$ becomes predictable. In this situation, time is available in a habitat for which h_1 is still greater than $\bar{E}$, during which physiological preparation for migration could occur. Selection seems likely to favour such a strategy.

This situation approaches that experienced by obligatory migrants. Because obligatory migration, as suggested here, is an evolved response to a predictable decrease in habitat suitability, any advantage of a certain level of energy reserves at the moment of initiation of migration is likely to favour those individuals with physiological mechanisms that most nearly achieve this level at the appropriate time. On the basis that the time and energy involved in building up, maintaining and carrying such reserves represents a disadvantage at all times except just before migration, selection is likely to favour those individuals with the appropriate physiological mechanism which only begins to operate at the optimum time before migration becomes advantageous.

8.4.5 The relative incidence of facultative and obligatory non-calculated migration

It has been argued in the preceding sections that facultative migration is an evolved response to unpredictable fluctuation in habitat suitability. Obligatory migration, on the other hand, has been argued to be an evolved response to predictable fluctuation in habitat suitability. All habitats fluctuate unpredictably. Some habits also happen to fluctuate predictably. On this assessment, all animals are likely to have evolved the characteristics of facultative migrants. Some animals are also likely to have evolved the characteristics of obligatory migrants. Some animals, therefore, are likely to show both facultative and obligatory characteristics within their repertoire of migrations, reacting facultatively to unpredictable fluctuations in habitat suitability but obligatorily to predictable fluctuations in habitat suitability.

In the introduction to Section 8.4, facultative migration was portrayed as a response to current adversity, whereas obligatory migration occurred at a certain ontogenetic stage or time of day or year irrespective of the suitability of the immediate surroundings. If the arguments of this chapter are accepted, it follows that facultative and obligatory migrations are responses to equal adversity in that in both cases selection has acted to favour the initiation of migration, usually based on a migration threshold, whenever h_1 becomes less than $\bar{E}$. The major difference between the two types of migration is not, therefore, that one is a response to current adversity whereas the other is not, but that in one case response to the existence of adversity is linked with appreciation of the state of a habitat variable, whereas in the other case the response to the existence of adversity is linked with some time-correlated mechanism that may either be entirely endogenous or may involve appreciation of some *indirect variable* such as photoperiod or the light/dark cycle. In both cases the migration threshold is likely to involve migration-cost variables.

8.5 The migration threshold in calculated migration

It has been argued in the previous section that no matter how complex the situation and no matter how many habitat, migration cost, and indirect variables are involved in the composite migration threshold, selection should always favour non-calculated migration being initiated at the point at which h_1 becomes less than $\bar{E}$. The level of the composite migration threshold to variables perceived while in H_1 is set in this case by the value of $\bar{E}$ during the evolution of that threshold.

In calculated migration, the animal still responds to a given set of conditions perceived in H_1, the level of conditions at which migration is initiated still qualifying as a composite migration threshold. In this case, however, the composite migration threshold is set by conditions perceived by the animal in and between those habitats that surround H_1.

9

The initiation of removal, return, and exploratory migration

9.1 Introduction

The distinction between removal, return, and exploratory migrations, which together constitute the lowest level of the migration hierarchy illustrated in Fig. 5.1 (p. 26), is concerned not with the initiation of the migration but with the function of the migration. Thus:

Exploratory migration: migration beyond the pre-migration limits of the familiar area during which the ability to return to that familiar area is retained, though not necessarily exploited.

Removal migration: a migration away from a spatial unit (U), usually ending in some other comparable spatial unit, which is not followed by a return to U, or at least not until the passage of a sufficiently long, but arbitrarily determined, interval of time.

Return migration: migration to a spatial unit that has been visited previously.

Although the distinction between these three categories is not concerned with their initiation, it is obvious that, because they fulfil different functions, a given animal in a given situation is more likely to initiate one type of migration than either of the other two types. The aim of this chapter is to attempt to identify the conditions that are likely to favour the initiation of one type of migration rather than another. The approach adopted is first to identify the factors influencing the relative advantage of return and removal migration. Having done so, because of the usual sequential association between exploratory and removal migration (i.e. an animal first initiates exploratory migration and then performs a calculated removal migration, which is

inevitably a return migration, to one of the habitats encountered during exploratory migration), an attempt is then made to identify some of the characteristics of the exploratory-calculated migration sequence. Finally, an attempt is made to determine under what conditions selection can favour exploratory migration being followed by non-calculated removal migration.

In this discussion, two categories of $\bar{E}$, the mean expectation of migration, are recognised. These are $\bar{E}_n$, the mean expectation of non-calculated removal migration, and $\bar{E}_{ec}$, the mean expectation of a calculated removal migration preceded by a period of exploratory migration.

9.2 Selection on return and removal migration

Imagine an individual, A, that performs a removal migration from habitat H_1 to habitat H_2. Let us accept that the migration was advantageous (i.e. $h_1 < h_2 M$; p. 43). Suppose now that h_2, the suitability of habitat H_2, becomes less than $\bar{E}_n$ (p. 46). In this case, initiation of a further removal migration is likely to be advantageous. Two alternative strategies now present themselves to individual A. Either it can migrate without return, and thus arrive at a further habitat, H_3, or it can migrate back to H_1 (ignoring for the moment the navigational problems involved that are discussed in Chapter 33). We may now consider the factors that determine which migration is the optimum strategy.

From previous considerations (Chapter 8), it may be assumed that by migrating from H_1 to H_2 animal A experienced $\bar{E}_n$, the mean expectation of non-calculated removal migration. Similarly, if the animal were to migrate from H_2 to H_3, it should again experience $\bar{E}_n$. There is no contradiction in these two statements. Although the suitability of H_2 at the moment of arrival can be taken to be $\bar{h}$, mean habitat suitability, and the mean expectation of the suitability of H_3 at the moment of arrival is also $\bar{h}$, during the period that the animal has resided at H_2, h_2 has decreased from $\bar{h}$ to a level at which $h_2 < \overline{hM}$ (where $\overline{hM} = \bar{E}_n$). From the arguments of previous chapters, therefore, migration without return is advantageous if h_2 is less than $\bar{E}_n$, whereas return migration is an advantage if h_2 is less than $h_1 M_1$ (where M_1 is the migration factor for migration from H_2 to H_1). It follows, therefore, that return migration is optimum strategy if $\bar{E}_n$ is less than $h_1 M_1$, whereas if $h_1 M_1$ is less than $\bar{E}_n$, then a non-return removal migration is advantageous.

If there is no factor that has an inherent effect on $h_1 M_1$, then on average $h_1 M_1 = \bar{E}_n$, and the optimum strategy to the animal is one that takes no account of previous movements. However, there are a number of factors that are likely to influence the relative values of $h_1 M_1$ and $\bar{E}_n$. Three of these factors are that: the animal has previously spent some time in H_1; has initiated migration from H_1; and has previously travelled between H_1 and H_2. If this previous behaviour of the animal renders $h_1 M_1$ less than $\bar{E}_n$, non-return removal migration is optimum strategy, whereas if it renders $h_1 M_1$ greater than $\bar{E}_n$, return migration is the optimum strategy.

The first, and perhaps most important point to be made about the influence of previous behaviour is that when the animal initiated migration from H_1, the value of h_1 must, according to the migration equation, have been less than $\bar{E}_n$. If there is no change in h_1, therefore, then unless the value of M_1 for migration from H_2 to H_1 is much greater than the value of M_1 for migration from H_1 to H_2, return migration is unlikely to be advantageous. It follows, therefore, that return migration is most likely to be advantageous if the level of h_1 fluctuates about the level of $\bar{E}_n$. Likely sources of such fluctuation in the suitability of a habitat-type are diel or seasonal rhythms of change in the structure of the habitat (Fig. 8.10) and also those endogenous rhythms associated with the time-investment strategy of the animal (p. 57). Hence, a given habitat-type may be of suitability greater than E_n at one time of day or year or when an animal is engaged in one time-investment component, but of a suitability less than $\bar{E}_n$ at some other time of day or year when the animal switches to some other time-investment component. Fluctuation in suitability of a habitat-type, how-

ever, can lead only to removal migration to an example of that habitat-type (Fig. 8.10) and not necessarily to return migration to the specific habitat H_1 (but see Chapter 20, p. 480). In order for return migration to be favoured, therefore, previous residence in H_1 must render H_1 the most favourable destination of its habitat-type.

We have established, therefore, that the most likely situation in which return migration will be favoured is when: (1) H_1 and H_2 are of different habitat-types that fluctuate in relative suitability such that part of the time one habitat-type is the more suitable and the remainder of the time the other habitat-type is the more suitable; (2) after a period of residence in H_2 selection on migration favours movement to a habitat-type of which H_1 is an example; and (3) previous residence in the habitat renders H_1 the most favourable destination of its habitat-type to the individual being considered. Two situations seem capable of leading, either separately or in combination, to the existence of case (3). The first of these situations relates to the relative abundance and spatial distribution of the two habitat-types of which H_1 and H_2 are respective examples. This situation is illustrated in Fig. 9.1 and describes the selective pressures that on the one hand could lead to return migration and on the other hand could lead to non-return removal migration. The second of these situations relates to the relative suitability of H_1 and other examples of its habitat-type that can result from the previous presence of the animal in habitat H_1. During the period of previous residence, interaction with the structure of H_1 could occur that either decreases or increases the future level of h_1 relative to the suitability of other habitats. Hence, if during previous residence the animal depleted the habitat of the resources required by the animal to such an extent that the suitability of that habitat could not return to even average suitability for its type during the period that the animal is in some other habitat-type, selection acts against return migration at the end of that period. On the other hand, if during previous residence the animal made some form of learning, social or construction investment in the habitat such that the suitability of that habitat is still above the average suitability for its type at the end of the period that the animal spends in some other habitat-type, selection acts to favour return migration at the end of that period. As long, that is, as this advantage of return migration is not outweighed by a disadvantage of the type described in Fig. 9.1.

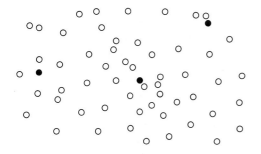

Fig. 9.1 Selection on second-order migration due to the relative abundance and spatial distribution of two different habitat-types when previous residence in a habitat does not influence its suitability

Each circle represents a 'habitat'. The open and solid circles represent respectively two different habitat-types. The suitabilities of the two types to an animal fluctuate out of phase with one another. Examples are one habitat-type favourable for roosting and the other habitat-type favourable for feeding, or alternatively, one habitat-type favourable for winter occupation and the other favourable for summer occupation, and so on.

Removal migration from one habitat-type to another becomes advantageous when the time-investment strategy of the animal and/or diel or seasonal fluctuation in the relative suitability of the two habitat-types reduces the suitability of the habitat occupied below the mean expectation of migration. The result of such migration is that the animal eventually settles in a habitat of the second type. After a period of residence in this habitat, removal migration again becomes advantageous and, due to further fluctuation in relative habitat suitability, migration results in eventual settlement in a habitat of the first type once more.

Consider an animal that starts in a habitat of the open-circle type and that performs removal migration to a habitat of the solid-circle type. When removal migration to an open-circle type next becomes advantageous, it is clear that on average the nearest open circle is one other than that originally occupied. Consequently, if all of the open-circle habitats are of equal suitability, the migration factor results in selection favouring migration to some habitat other than the one originally occupied. Selection should favour, therefore, those animals that settle in the first open-circle habitat encountered rather than those that navigate back to their original habitat. For an animal that starts in a solid-circle habitat and that settles in the first open-circle habitat encountered, the nearest solid-circle habitat is, on average, the one previously occupied. In this case, therefore, selection favours those individuals that navigate back to their original habitat rather than performing non-calculated removal migration and settling in the first solid-circle habitat encountered. As long, that is, as any extra migration cost that may result from the navigation process does not outweigh the decrease in migration cost that results from return migration being shorter than non-calculated removal migration. Return migration to the original solid-circle habitat is not favoured only if the animal visits so many open-circle

habitats between those periods of greater suitability of solid-circle habitat-types and is so mobile relative to the inter-habitat distance that its movement from open-circle to open-circle takes it such a distance that the solid-circle habitat originally occupied is no longer the nearest of its type.

In conclusion, therefore, when previous residence in a habitat does not influence its suitability, return migration is favoured by selection when the habitat originally occupied is the nearest example of its habitat-type when migration to such a habitat-type becomes advantageous. Non-return removal migration is favoured when this is not the case or when any extra migration cost involved in navigating back to the original habitat outweighs the decrease in migration cost that results from the original habitat being the nearest example of its habitat-type.

It remains, therefore, to consider the sequence of events by which return migration could evolve. The raw material upon which selection for return migration can act is undoubtedly those individuals that arrive by chance after two or more removal migrations at a habitat previously occupied. Although the concept of chance is useful in this connection to imply the low probability of the event, it follows nevertheless that these individuals arrive back at a habitat previously occupied because of their sensory-motor interactions with the environment during successive removal migrations. If these sensory-motor interactions have an inheritable component that influences (increases) the probability that the individuals' offspring will also perform return migration, then the raw material exists upon which selection can act to increase the incidence of return migration. If selection continues to act to increase the incidence and efficiency of return migration, the evolutionary result is a navigation mechanism of optimum efficiency for the animal concerned (Chapter 33).

It is perhaps worth pointing out that the possibility of return migration occurring by chance is a function of the distance of the previous removal migration. It follows that return migration is more likely to evolve from removal migration over short distances than over long distances. It is a more reasonable approach, therefore, to derive long-distance return migrations from short-distance return migrations than from long-distance removal migrations, a point that I had not fully thought out in an earlier note on the evolution of long-distance navigation in birds (Baker 1968c).

When selection favours the evolution of return migration such that optimum strategy for an animal that migrates from H_1 is to return to H_1 at some time in the future, the difference between the cost of return migration and the cost of removal migration is likely to lead to selection for slightly different migration thresholds. The mean expectation of return migration, $\bar{E}_R$, may be given by

$$\bar{E}_R = \bar{h}\overline{M_R^x}$$

where M_R^x is given by $M_{1-2}M_{2-1}$ (where M_R is the mean migration factor for the outward and return journeys, M_{1-2} is the migration factor for migration from H_1 to H_2, and M_{2-1} is the migration factor for the return migration from H_2 to H_1). When the migration factor is the same for both outward and return stages of the return migration, $x = 2$.

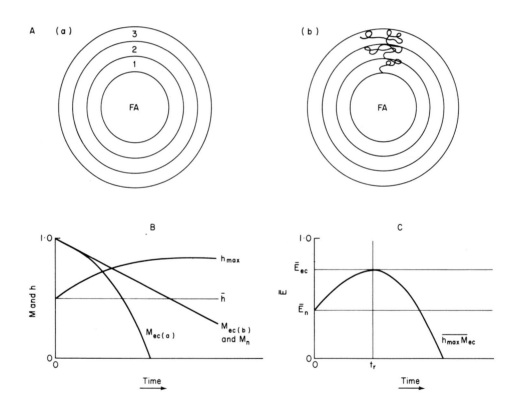

Fig. 9.2   Some features of the exploratory–calculated removal migration sequence

The diagrams assume that habitats are randomly distributed around the familiar area, both spatially and with respect to suitability.

In diagram A, three zones are recognised around the initial familiar area (FA). The diagnostic feature of exploratory migration is that it involves retention of the ability to return to FA. In most cases it also involves occasional or frequent return to FA (for feeding, roosting, territorial defence, etc.). Using diagram Aa, the situation can be considered in which from time to time during

exploratory migration an animal returns to FA. In this case selection is likely to favour thorough exploratory migration of zone 1 before exploratory migration is extended to zone 2, and so on. Zone 1 is, therefore, incorporated into the familiar area of the animal before zone 2, zone 2 before zone 3, and so on. Diagram Ab can be used to consider the situation in which the animal retains the ability to return to FA and to any other point visited during exploratory migration, but in which the animal either does not return to FA or at least not until exploratory migration is terminated without finding any habitat more suitable than FA. In this case, the animal is as likely, if not more so, to leave zone 1 in the direction of zone 2 as in the direction of *FA*, as indicated by the hypothetical track.

In diagram B, the probable variation with time is shown for M, the migration factor, and h, habitat suitability. In this case, M is M_{ec}, the migration factor for exploratory migration followed by calculated removal migration and h is the suitability of the eventual destination discovered as a result of exploratory migration. Of the two curves of M_{ec} shown, $M_{ec(a)}$ is based on the situation in diagram Aa. An animal has to travel further and thus invest more time and energy in exploratory migration of zone 2 in comparison with exploratory migration of zone 1. Consequently, as exploratory migration is performed in the nearest zones first, migration cost per unit time increases with time as further and further zones are explored. The variation of $M_{ec(a)}$ with time is therefore likely to be of the form shown. In the situation shown in diagram Ab, migration cost per unit time should remain more or less constant with time, in which case the decrease in migration factor with time should be a straight line as shown for $M_{ec(b)}$. This straight line may also describe change with time of the migration factor M_n of a non-calculated removal migrant. Exploratory migration is terminated when the animal performs calculated removal migration by settling in that habitat which is the habitat with the highest suitability of all those encountered during exploratory migration. As exploratory migration proceeds, from time to time a habitat should be encountered that is even more suitable than the previous most suitable habitat. Let h_{max} represent the suitability of the most suitable habitat so far encountered at any particular moment during exploratory migration. The value of h_{max} therefore increases as exploratory migration proceeds. If it is assumed, however, that habitats with a suitability greater than $\bar{h}$ (mean habitat suitability) are less abundant than habitats of suitability $\bar{h}$ (e.g. if habitats are normally distributed with respect to suitability), it follows that rate of increase in h_{max} is not linear but decreases with time as shown. In summary, therefore, the exploratory migration factor (M_{ec}) may either show a linear decrease with time or an increasing rate of decrease with time, depending on the optimum track pattern for the species concerned. Rate of increase of h_{max} is, however, always likely to show a decreasing rate of increase with time.

In diagram C, the conclusions based on diagram B are translated into mean expectation of migration, $\bar{E}$, where $\bar{E} = \bar{h}M$ (p. 45). An animal that initiates a series of exploratory migrations that will last a certain length of time has a mean expectation of $\bar{h}_{max}M_{ec}$ that is appropriate for that time. Variation in $\bar{h}_{max}M_{ec}$ with time is likely to be as shown. During the early part of exploratory migration, $\bar{h}_{max}M_{ec}$ increases. A peak value is reached after time t_r, and thereafter further exploratory migration becomes disadvantageous. Selection should therefore favour those individuals that terminate exploratory migration after time t_r. Given that an individual terminates exploratory migration at t_r, the value of $\bar{E}_{ec}$, the mean expectation of calculated removal migration preceded by exploratory migration, is given by the maximum value of $\bar{h}_{max}M_{ec}$ as shown. Finally, given the premises included in these diagrams, $\bar{E}_{ec}$ is always likely to be greater than $\bar{E}_n$, the mean expectation of non-calculated removal migration, which is given by $\bar{h}M_n$ (diagram B).

9.3 The exploratory-removal migration sequence

Some lineages will have been exposed to selection for adaptation to spatially separate habitat-types that fluctuate in relative suitability such that part of the time one habitat-type is the more suitable and the remainder of the time the other habitat-type is the more suitable. Some of these lineages will have consisted of animals that interact with their habitat in such a way that when next they require a habitat of a given type, the most suitable example of that type is the one previously occupied. It is in these lineages that, according to the previous section, return migration, and the associated mechanisms of navigation, make their major contribution to the gross form of the lifetime track.

In the lineages in which this is so, the animals invariably occupy a more limited lifetime range than those members of lineages in which return migration makes little contribution to the gross form of the lifetime track. It is these lineages, the members of which occupy a relatively limited range, that are most likely to be exposed to selection for strategies to increase the ratio of calculated to non-calculated migration (p. 44). Exploratory migration is one of these strategies (see also Chapter 15) and this section is concerned with an identification of some features of the exploratory-removal migration sequence.

Let the mean expectation of a calculated removal migration preceded by exploratory migration be $\bar{E}_{ec}$. If it is accepted for the moment that $\bar{E}_{ec}$ can be substituted for E in all of the arguments presented so far, it follows that exploratory migration is advantageous whenever h_1 becomes less than $\bar{E}_{ec}$. This selective pressure would be expected to lead to the evolution of exploratory migration thresholds, associated both with the facultative and obligatory initiation of exploratory migration (p. 47).

To this extent, and certainly as far as facultative exploratory migration is concerned, selection on the initiation of exploratory migration can be expected to be similar to selection on the initiation of non-calculated removal migration. One slight, but important, difference in the action of selection can be expected with respect to obligatory migration. It is concluded in Fig. 9.2 that there is an optimum time t_r between the initiation of exploratory migration and the initiation of calculated removal migration. Suppose that h_1 becomes less than $\bar{E}_{ec}$ at time t_{opt}. The advantage of a delay of t_r between the initiation of exploratory migration and the initiation of calculated removal migration imposes selection in favour of the obligatory initiation of exploratory migration some time in advance of t_{opt}, i.e. at $t_{opt} - t_r$.

In the legend to Fig. 9.2 it is also concluded that, given certain premises, the value of $\bar{E}_{ec}$ should always be greater than $\bar{E}_n$ for an animal that has evolved the use of a familiar area that is greater than the sensory range (animals for which selection has not favoured use of such a familiar area can by definition not perform exploratory migration). Evolved migration thresholds in relation to the obligatory and facultative initiation of exploratory migration, therefore, should always be lower than similar migration thresholds evolved in relation to non-calculated removal migration (i.e. animals should always first initiate exploratory migration).

Let us now consider a hypothetical animal living in a home range that is part of a larger familiar area. Let this home range be habitat H_1, though any other convenient spatial unit could be so defined without changing the validity of the arguments. This animal is faced by a decreasing suitability (h_1) of H_1. Suppose that the most suitable alternative home range within the existing familiar area is H_f of suitability h_f, and that the migration factor for migration from H_1 to H_f is M_f. If h_1 ever becomes less than $h_f M_f$, then ideally calculated removal migration to H_f is initiated—unless, that is, $h_f M_f$ happens to be less than $\bar{E}_{ec}$, the mean expectation of calculated removal migration preceded by exploratory migration. In this case, the threshold for exploratory migration is exceeded before calculated removal migration from H_1 to H_f is initiated. There is little problem, therefore, in identifying the different situations in which calculated removal migration should either be initiated directly or should be preceded by exploratory migration. If, after time t_r (Fig. 9.2), however, the most suitable habitat discovered during exploratory migration should happen to be less suitable than either H_f or H_1, the optimum strategy to the individual concerned should be to settle in whichever of these habitats is currently assessed to be most suitable.

There remains now the problem of identifying under what conditions non-calculated removal migration from a familiar area is more advantageous than exploratory migration. The central requirement for exploratory migration to be followed by non-calculated removal migration is that after time t_r of exploratory migration an animal still has not encountered a habitat the suitability of which is as great as $\bar{E}_n$. In this case, at time t_r the initiation of

non-calculated removal migration becomes advantageous. This event, and the situation that may give rise to it, is illustrated in Fig. 9.3. It is concluded that exploratory migration can be followed by non-calculated removal migration and that this is most likely to occur when an animal's familiar area is surrounded by a sufficiently large area in which all habitats are of low or zero suitability. The conclusion remains valid, however, that animals that make use of a familiar area larger than their sensory area should first initiate exploratory migration and then only initiate non-calculated removal migration after time t_r if no habitat has been encountered for which h is greater than $\bar{E}_n$.

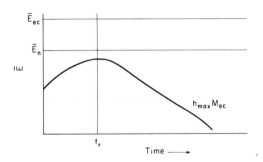

Fig. 9.3   A situation in which exploratory migration followed by non-calculated removal migration may be the optimum strategy

The diagram plots the value of $h_{max} M_{ec}$ against time since the initiation of exploratory migration in relation to two values, $\bar{E}_{ec}$ and $\bar{E}_n$, of mean expectation of migration. h_{max} = the suitability of the most suitable habitat so far encountered after a given period of exploratory migration; M_{ec} = the migration factor after a given period of exploratory migration; $h_{max} M_{ec}$ = the product of h_{max} and M_{ec} after a given period of exploratory migration; t_r = time after initiation of exploratory migration that $h_{max} M_{ec}$ is at a maximum; $\bar{E}_{ec}$ = mean expectation of calculated removal migration preceded by exploratory migration after time t_r for an animal adapted to habitats that are randomly distributed both spatially and with respect to suitability; $\bar{E}_n$ = mean expectation of non-calculated removal migration.

The diagram shows a possible relationship between $h_{max} M_{ec}$ and time for an individual that finds itself in such a position that habitats are not randomly distributed spatially or with respect to suitability but are instead distributed such that those habitats near to the place of origin are of very low suitability. In such a situation it is possible that after time t_r of exploratory migration the animal may still not have encountered a habitat the suitability of which is as great as $\bar{E}_n$. In this case, at time t_r, the initiation of non-calculated removal migration becomes advantageous if the suitability, h_1, of the habitat of origin, H_1, has also become less than $\bar{E}_n$.

10

Inter- and intra-specific variation in migration incidence

10.1 Introduction

The previous chapters have attempted to identify the conditions under which migrations with different functions are likely to become advantageous. Migration incidence, with which this chapter is concerned, is a function of the frequency with which these conditions occur.

It follows from Chapter 7–9 that non-calculated removal migration is initiated with a frequency that is a function of the frequency with which h_1 becomes less than $\bar{E}_n$ (p. 61). The critical frequency for exploratory migration is that with which h_1 becomes less than $\bar{E}_{ec}$ (p. 65) and for calculated removal migration that with which h_1 becomes less than $h_f M_f$ (p. 66). As far as removal migration incidence is concerned, therefore, inter-specific variation should be fairly readily explained in terms of migration cost and frequency of decrease of h_1 (relative to mean habitat suitability). The frequency and pattern of decrease of habitat suitability and its influence on the initiation of migration has been discussed in Chapter 8 (p. 44). The effect of migration cost is that the lower the cost, the higher the value of the migration factor, the higher the value of $\bar{E}$ (mean expectation of migration) or $h_f M_f$, and hence the greater the frequency with which removal migration becomes advantageous. One of the major components of migration cost is likely to be a function of inter-habitat distance. In general it seems likely that for any given group of animals of similar size that use a similar migration mechanism, migration cost is a direct function of inter-habitat distance. Thus, within a species, short distance migrations should be more frequent than longer distance migrations, and between species, those with a shorter inter-habitat distance should have a greater migration incidence than those with a longer inter-habitat distance.

Species that show a high removal-migration incidence should therefore be those adapted to habitats with a high rate of decrease of habitat suitability (relative to mean habitat suitability) but with a low migration cost, largely due to a short inter-habitat distance. Species that show a high return migration incidence, on the other hand, should be those adapted to spatially separate habitat-types (etc.— p. 61) that show the most rapid fluctuation in relative habitat suitability. The question of how animals can become adapted to spatially separate habitats in the first place is discussed elsewhere (p. 248).

Inter-specific variation in migration incidence may thus fairly readily be incorporated into the model developed here. Inter-sexual variation can be incorporated in exactly the same way. Ontogenetic variation in migration incidence, however, although using the same arguments, leads to conclusions that are similar for a wide variety of species and as a result merits rather more extensive consideration at the theoretical level than inter-specific or inter-sexual variation which can really only be discussed further in relation to specific examples. Ontogenetic variation is discussed in Section 10.3 (p. 73).

The other category of intra-specific variation in migration incidence, that concerned with variation between individuals of the same age and sex, involves a slightly different development of the model produced in Chapters 7–9, and hence requires a more extended discussion.

10.2 Variation in the probability of initiation of migration between individuals of the same age and sex

There are two major sources of variation between individuals of the same age and sex with regard to initiation of migration. These are individual differences in experience and individual differences in migration threshold.

10.2.1 Individual differences in experience

Individual differences in experience are relevant to individuals that initiate migration in response to some environmental variable, V (the value of which is given by v), when the value of v is v_t, where v_t is the threshold for removal migration which correlates with a habitat suitability of $h = \bar{E}$ (p. 47). Suppose V is deme density and v is therefore rate of contact with other individuals. Further, suppose that two animals living in the same area have the same migration threshold v_t. It is quite possible that, on occasion, because the position and movement of the two individuals are slightly different, even though they are in the 'same' habitat, that the rate of contact experienced by one individual may be above the migration threshold whereas that experienced by the other may be below the migration threshold. In this case, one individual would initiate removal migration whereas the other would not. Of course, suboptimal response such as this is selected against in a manner that has already been discussed (Fig. 8.2). Nevertheless, it seems unlikely that the ideal adaptation is ever totally achieved with the result that individual differences in experience always remain as a source of variation in migration response between individuals of the same age and sex.

10.2.2 Individual differences in migration threshold

The nature of individual differences in migration threshold depends to a large extent on the position on the territorial spectrum that is occupied by the

deme (= all of the conspecific individuals that occur in an arbitrarily determined geographical area) to which the individuals belong. The territorial spectrum is discussed fully in Chapter 18, but it is necessary here to introduce three aspects insofar as they relate to individual differences in migration threshold.

First, the individual members of any deme differ considerably from one another in their ability to defend resources in the face of competition from other individuals. Each individual may thus be said to have a specific *resource-holding power* (Parker 1974b). Individual differences in resource-holding power result not only from age differences and phenotypic differences in such characteristics as size, stature, and behaviour, but also from differential acquisition of disease organisms and the effect that these may also have on size, health and mobility.

At the *despotic* end of the territorial spectrum (see p. 402) the reproductive success of an individual is a direct function of the resource-holding power: the greater the resource-holding power, the greater the ability of the individual to gain access to limited resources. In such a situation, migration threshold is also a function of resource-holding power, individuals with a high resource holding power having a high migration threshold (i.e. less likely to migrate in any given situation). Further discussion of this type of individual variation in migration threshold may be delayed conveniently until Chapter 18.

The remainder of this chapter, therefore, is concerned with animals living at the *free* end of the territorial spectrum (see p. 402). Here, although individuals continue to differ in size and stature, etc., no individual is denied access to a resource as a result of the presence of another individual with a higher resource-holding power. The theoretical aspect of the free situation has been developed in relation to different aspects of migration on a number of occasions (Hamilton *et al.* 1967, Parker 1970c, Baker 1972a, Fretwell 1972) but it was Fretwell who proposed the term 'free' (and 'despotic') to describe the situation. The chief characteristic of the free situation is that the individual members of a deme distribute themselves in time and space in such a way that all individuals have the same potential reproductive success (at least as far as the latter is a function of the animal's position in time and space). Only when a deme manifests such a state due to evolved frequency distributions within the deme of migration thresholds, migration distances, and migration directions, should selection stabilise. This section is concerned with the evolution of a stable frequency distribution of migration thresholds.

10.2.2.1 Selection due to correlation between deme density and habitat suitability

The suitability of a habitat to an animal with respect to the required resource(s) can be divided into two components. One is the result of the physical and biological structure of the habitat excluding other individuals of the species concerned or excluding individuals of other species that compete with the species concerned for the supply of the required resource(s). The other component results from the presence of individuals of all of those species that are competing for some or all of the supply of the required resource(s). As already discussed in Chapter 8, therefore, the migration threshold is likely to be based on a number of habitat variables, one of which is rate of contact with other competing individuals. Hence, a given level of food availability in the habitat may well indicate a value of h that is greater than $\bar{E}$ if the individual concerned is the only one present but a value of h that is less than $\bar{E}$ if other individuals are present.

The vast majority of habitat variables that correlate with habitat suitability are to some extent a function of deme density. For example, food intake, ease of finding a mate, ease of finding a roosting site, are all affected by the number of competing individuals per unit area. Consider, then, the situation if all individuals responded by the initiation of migration to the same value of v_t (i.e. if all individuals had the same migration threshold). The result would be an empty habitat. An individual that did not migrate when the value of habitat variable V reached v_t could well therefore find itself left in a habitat that now had a value of h that was greater than $\bar{E}$. Selection would therefore favour some individuals that did not respond to v_t by the initiation of migration. The likely result of this situation is that at the free end of the territorial spectrum selection favours variation in v_t between individuals such that the number of individuals that initiate migration for any particular value of v is such that the potential reproductive success of those that migrate is the same as that of those that stay behind. The frequency distribution of migration thresholds within a deme stabilises when this situation has been achieved and all individuals experience potentially

equal reproductive success in relation to migration.

It should perhaps be stressed again that the stabilisation of selection when all individuals experience potentially equal reproductive success is possible only in a free situation. In a despotic situation this position is not reached, due to the advantage gained by some individuals as a result of preventing other individuals from gaining access to a particular resource (Fretwell 1972).

The reproductive success, S_m, of each individual that migrates from a habitat of suitability h_m is given by the reproductive success that results from movement (i.e. initial potential reproductive success, S, times mean expectation of migration, $\bar{E}$). That is,

$$S_m = S\bar{E}. \qquad (8)$$

The reproductive success S_r of each individual that does not migrate results from the increase in habitat suitability from h_m to h_r due to the migration of other individuals. Hence, using arguments similar to those of the previous paragraph, the reproductive success of an individual that does not migrate is given by

$$S_r = Sh_r. \qquad (9)$$

The suggestion made in this section is that selection acts to produce the situation in which

$$S_r = S_m,$$

i.e. from (8) and (9)

$$Sh_r = S\bar{E}.$$

Therefore, assuming that both migrants and nonmigrants have equal initial potential reproductive success, S, selection should act to produce the situation in which

$$h_r = \bar{E}. \qquad (10)$$

In other words, selection should stabilise at the situation in which, no matter what level of habitat suitability should arise, just sufficient individuals initiate migration for habitat suitability to revert to a level equivalent to the mean expectation of migration. This conclusion renders it possible to calculate the proportion of migrants that should be favoured by selection for any given level of habitat suitability.

Let h, the suitability of a habitat, consist of two components, h_p, the component of habitat suitability that results from the physical and biological construction of the habitat excluding all other competing individuals, and h_d, the component that results from the presence of competing individuals. Habitat suitability is therefore given by

$$h = h_p h_d. \qquad (11)$$

The critical factor in this section is the relationship between h_d and deme density and the effect, therefore, that deme density may have on the value of h. In many situations, if the individual being considered is the only individual present in a habitat, h_d is likely to equal $1 \cdot 0$ and h is given solely by h_p. However, in other situations there may be a deme size greater than one individual that is necessary for the maximum realisation of potential reproductive success. According to Allee's principle, for example (Allee *et al.* 1949) the suitability of a habitat increases to a peak with the arrival of the first n individuals but thereafter each increase in deme density leads to a decrease in habitat suitability.

Clearly, the form of the relationship between h_d and deme density must be evaluated for each specific situation. However, there are two points in this relationship that are of particular theoretical importance. The first of these points is given by N_E, which represents the deme density that gives a value of h_d that interacts with h_p to give a value of h equal to $\bar{E}$, the mean expectation of migration. The second of these two points is given by N_m, which represents the deme density that gives a value of h_d that interacts with h_p to give a value of h equal to h_m, the habitat suitability to which the migration being considered is a response.

If, as suggested (eq. (10)), selection should stabilise when for all values of habitat suitability a proportion of individuals initiate migration such that h_r, habitat suitability after their departure, equals $\bar{E}$, then for any given value of h_p (see eq. (11)), f_m, the migratory fraction of a deme, is given by

$$f_m = 1 - \frac{N_E}{N_m}, \qquad (12)$$

this fraction being just that necessary to restore habitat suitability to $\bar{E}$.

The distribution of migration thresholds that is optimum for any particular deme is a function of the way that N_E and N_m relate respectively to $\bar{E}$ and h_m, and also a function of the relationship between h_p and h_d. In any case, it is clear that the migration threshold must involve both habitat variables that correlate with h_p and habitat variables, such as rate of contact with conspecifics, that correlate with h_d, though some habitat variables, such as rate of energy uptake, correlate with both.

The above and following considerations are concerned with free situations in which there is an inverse relationship between deme density and habitat suitability, at least, that is, at levels of habitat suitability less than $\bar{E}$. Of course, if habitat suitability does not decrease with increase in deme density at habitat suitabilities less than $\bar{E}$, then clearly there is no longer a relationship between migratory fraction and deme density, for migration of some individuals in response to $h < \bar{E}$ has no effect on the fitness (a measure of potential reproductive success) of individuals that do not migrate, or may even decrease their fitness. For example, if there is a positive correlation between habitat suitability and deme density (remember that for migration we are only concerned with values of h that are less than $\bar{E}$) and if d is the deme density below which h is less than $\bar{E}$, then at all densities less than d, 100 per cent of the deme should initiate migration. Such habitats can only be colonised by the arrival at more or less the same time of enough individuals to give a density greater than d.

Having derived a model of variation in the initiation of migration, it is worth considering the range of situations for which the model is relevant. The arguments presented in this section should be applicable to any free situation in which habitat suitability is an inverse function of deme density, not just those situations in which selection favours the evolution of a migration threshold to habitat variables. Figure 8.10 (p. 56), for example, illustrates a situation in which two habitat-types, A and B, fluctuate in suitability with season or time of day in such a way that selection favours linkage between appreciation of indirect variables, such as the photoperiod/temperature regime or the light/dark cycle, and the initiation of migration. However, suppose that the suitability of these two habitat-types, as well as fluctuating seasonally (or with time of day) was also an inverse function of deme density. Suppose also that throughout the summer 100 per cent of the deme is found in the most suitable habitat-type (B in Fig. 8.10). In autumn, h_B, the suitability of habitat-type B, falls below $\bar{E}$ soon after h_A has risen above $\bar{E}$. Individuals leave habitats of type B and settle in habitats of type A. In doing so, however, they raise h_B and decrease h_A, both to the level of $\bar{E}$. As the seasonal deterioration of B continues, however, to a level less than $\bar{E}$, more individuals migrate. This process should continue either until 100 per cent of the deme has migrated to type A habitats or until the seasonal deterioration of type B habitats ceases, whichever is sooner. The reverse happens in the spring. As this seasonal variation in the relative suitability of type A and type B habitats is assumed to be predictable in Fig. 8.10, the result of this selection should be a particular frequency distribution of linkage within the deme between different values of the photoperiod/temperature regime (or light/dark cycle) and the initiation of migration. This distribution should be such that migration leads to a fluid distribution of deme density between the habitat-types such that either $h_A = h_B = \bar{E}$ throughout the year or at some times of year one habitat-type remains the more suitable despite supporting 100 per cent of the deme.

Similar arguments can be applied to selection for ontogenetically obligatory migration. Figure 8.7 (p. 54) illustrates the type of situation that is considered to generate selection for ontogenetically obligatory migration. Such selection seems likely when a habitat must always be of a suitability less than $\bar{E}$ by the time an animal attains a particular ontogenetic stage. However, as before, it is conceivable that the migration of some individuals raises the value of h, possibly to $\bar{E}$, in which case migration by other individuals would no longer be advantageous. The main advantage of obligatory migration, however, is suggested to be that time and energy need not be invested in monitoring habitat variables (p. 50). Obligatory migrants, therefore, have no information concerning deme density and habitat suitability. In this situation, selection is likely to stabilise at a point at which the proportion of the deme that performs obligatory migration is the maximum that on average increases habitat suitability to the level of $\bar{E}$ without exceeding this level. The remainder of the deme are likely to be facultative migrants with a distribution of migration thresholds to deme density of the form already described (p. 69).

So far in this discussion, attention has been concentrated on the distribution of migratory characteristics (e.g. migration thresholds and proportion of obligatory migrants), within a deme. It is perhaps also worthwhile to examine the other end of the spectrum of individual differences, particularly the proportion of obligatory non-migrants within a deme. Many animals, if migration is a possibly advantageous future event, need to develop and maintain certain physio-morphological mechanisms or structures. Examples are the flight apparatus of some insects and fat deposits of insects and birds. If migration is an unlikely advantageous future event,

reproductive success is likely to be maximised by not maintaining and/or developing the appropriate physio-morphological mechanisms or structures.

When fluctuation in habitat suitability is predictable, selection is likely to favour a proportion of obligatory non-migrants within the deme. If it is predictable that habitat suitability will fall to zero, then the optimum proportion of obligatory non-migrants will also be zero. Otherwise, remembering that this section is concerned with habitat suitability as an inverse function of deme density, selection should stabilise at a proportion of obligatory non-migrants that is the maximum that, on average, in the absence of other (migrated) individuals, gives a deme density that will not decrease h_1 below $\bar{E}$. In some cases, particularly species adapted to habitat-types with a negligible rate of formation of new sites, a relatively great habitat longevity, and a large inter-habitat distance, i.e. species with a low $\bar{E}$ and high and stable h, this maximum proportion could be 100 per cent.

In the case of predictable fluctuation of relative suitability of two habitat-types, A and B, with respect to season or time of day, it was pointed out that selection on the initiation of migration should stabilise if the distribution of deme density between habitat-types was such that $h_A = h_B = \bar{E}$ throughout the year, or if the habitat-type supporting 100 per cent of the deme remained the more suitable. In some situations, as described, this will result in a proportion of the deme spending all year in the same habitat-type. Selection is likely, therefore, to favour the situation in which this proportion does not possess the mechanisms necessary to be able to migrate from type A to type B habitats. This does not necessarily mean, however, depending on the situation, that this proportion will not possess the mechanisms necessary to be able to migrate between habitats of the same type (e.g. type A to type A), if these mechanisms are different from those necessary for migration from type A to type B.

10.2.2.2 Selection due to unpredictable variation in $\bar{E}$ about a mean value

In the previous section, discussion was concerned with the consequences of a correlation between habitat suitability and deme density. Because at least some variables with which an animal is in sensory contact from time to time are likely to correlate with habitat suitability, it can be assumed that the adaptations to the various selective pressures in the previous and present chapter constitute behavioral strategies in relation to fluctuations in habitat suitability that are as near optimum as possible for a biological situation. Except in relation to sensory contact with migration-cost variables (p. 49), however, fewer pre-adaptations exist upon which selection can act in relation to short-term unpredictable fluctuations in $\bar{E}$, the mean expectation of migration. This is because most of the factors that influence the value of $\bar{E}$ are incapable of being perceived by the animal (see also p. 82 concerning $\bar{E}_e$, the mean expectation of migration based on recent experience) before the initiation of migration.

One likely consequence of short-term unpredictable fluctuations in $\bar{E}$, however, seems to be the generation of further selection for individual differences in migration threshold, and the performance of obligatory migration. The form of this selection is examined in Fig. 10.1. It is concluded that as far as

individual variation in migration threshold to migration-cost and habitat variables is concerned, selection is likely to stabilise when the distribution of migration thresholds within the deme corresponds to the frequency with which $\bar{E}$ takes on values different distances from the mean value. As far as the performance of obligatory migration is concerned, selection is likely to stabilise at a proportion of obligatory migrants that is a function of the proportion of occasions that h_1 is less than $\bar{E}$ at the appropriate ontogenetic stage.

With regard to distribution of stay-times in habitats for which h_1 is less than $\bar{E}$ at the moment of arrival, variation in $\bar{E}$ leads to a different optimum stay-time in all habitats. Increase in $\bar{E}$ is likely to lead to a decrease in optimum stay-time in all habitats and a decrease in $\bar{E}$ leads to an increase in optimum stay-time in all habitats. Unpredictable fluctuations in $\bar{E}$ about the mean value, therefore, are likely to lead to a stabilisation of selection when the distribution of stay-times within the deme for

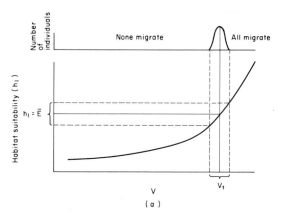

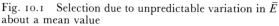

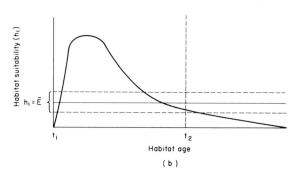

(a)

(b)

Fig. 10.1 Selection due to unpredictable variation in $\bar{E}$ about a mean value

The diagram reproduces two figures presented in Chapter 8 (Figs. 8.1 and 8.7) that demonstrated respectively selection for a migration threshold to habitat variables and for ontogenetically obligatory migration. These diagrams have been modified by including a range of variation in the value of $\bar{E}$ about a mean value.

In (a), h_1 is the suitability of the habitat occupied, $\bar{E}$ is the mean expectation of migration, v is the range of values of the habitat variable V, which in this case is not a function of deme density, the solid curve is that of h_1 on v, and v_t is the migration threshold to V for the species concerned. The solid horizontal line represents the value of h_1 that corresponds to the mean of $\bar{E}$. The dashed horizontal lines represent the limits of fluctuation of $\bar{E}$ about this mean value. The result of this selection is that on some occasions a lower v_t corresponds to $h_1 = \bar{E}$ than on others.

In this situation, selection is likely to stabilise as shown in the upper part of the figure when the distribution of migration thresholds within the deme corresponds to the frequency with which $\bar{E}$ takes on values different distances from the mean value. This figure is equally applicable when V is a migration-cost variable.

In (b), the same conventions are used. In this case, however, the curve shows habitat suitability as a function of habitat age. The time interval between t_1 and t_2 corresponds to any time interval measurable by the animal, but particularly the interval between birth and some time-correlated event, X, in ontogeny. The fluctuation in value of $\bar{E}$ about a mean value is such that on some occasions migration is advantageous at time X and on other occasions not. Depending on the age of habitat into which an animal is usually born, selection is likely to produce a proportion of obligatory migrants that is a function of the proportion of occasions that h_1 is less than $\bar{E}$ at the attainment of ontogenetic stage X.

each habitat suitability corresponds to the distribution of frequency of different values of $\bar{E}$.

The ramifications of models of selection on individual differences in the initiation of migration are clearly considerable. In this and the previous section, selection due to a correlation between deme density and habitat suitability and selection due to unpredictable variation in $\bar{E}$ about a mean value have been considered separately. The increase in complexity that occurs if these and possibly other relevant selective pressures are considered together is self-evident and no attempt at such an integration is made here.

10.3 Variation with age

10.3.1 Previous theories

One fact that I hope emerges during the course of this book is that there are almost as many patterns of migration as there are species. Consequently, any model developed to account for any component of migration should be capable of accounting for differences as well as for similarities. However, if there is any exception to this general diversity of behaviour it lies in the realm of ontogenetic variation in the probability of initiation of removal migration. Throughout the Metazoa there is an almost universal pre-reproductive peak in the incidence of removal migration.

Because this phenomenon of a pre-reproductive peak is so widespread it has come to the attention of a number of zoologists and, perhaps more than any other aspect of migration except orientation and navigation, has been discussed by students of a number of animal groups, but particularly students of insects and small mammals. Most such considerations, however, have been directed toward a single animal group. Yet the phenomenon is so widespread in the Metazoa that an understanding seems likely only to be reached by means of a comparative approach. Any evolutionary model must be as applicable to man as to the black bean aphid and as applicable to the blue whale as to the barnacle. Furthermore, such a model must account for the differences and similarities in ontogenetic variation between males and females as well as for differences and similarities between species. Before making a few preliminary suggestions concerning possible relevant selective pressures, however, it is necessary to examine some of the theories that have been advanced previously.

Margalef (1963) suggests that animals tend to spend their lives in more-mature ecosystems (maturity being a composite measure of diversity, primary production, and biomass) but to reproduce in less-mature ecosystems and to send larvae or reproductive elements into them. Lloyd (in Margalef 1963) suggests that less-mature ecosystems 'maintain themselves less efficiently' and that therefore energy for growth is more readily available and competition is less severe. Consequently, it is an advantage 'to send the young outside to less-mature ecosystems to gather energy and bring it back'. Margalef's thesis has recently been used by Jonkel and Cowan (1971) in their interpretation of the pre-reproductive migration of the black bear, *Ursus americanus*.

Even if we suppose that the concepts expressed in the above phrases represent real selective pressures, it is difficult to see how they can act in the manner suggested to produce a migration by young animals either into a 'less-mature ecosystem' or back to a 'more-mature ecosystem' or both. The only selective pressure suggested, that energy is more readily available and competition is less severe in 'less-mature ecosystems' would act just as much on adults as on young and would not by itself favour migration between 'more-mature' and 'less-mature' ecosystems at different stages of ontogeny. Margalef supports the thesis by pointing out that many insects produce young in aquatic habitats (which are considered to be less-mature ecosystems) but as adults are integrated into the (more-mature) terrestrial ecosystem. Such insects certainly perform a pre-reproductive migration but then so too do insects that are adapted to an aquatic habitat throughout larval and adult life, such as water beetles. A pre-reproductive peak of removal migration is also characteristic of many other insects, such as aphids, which occupy more or less identical feeding habitats as adults and larvae.

Cohen (1967) suggests that more young migrate than adults because the young in their natal area are in competition with adults with established breeding and feeding sites, whereas if they migrate they have an equal opportunity to establish themselves. This argument, however, assumes first that both adults and young migrate and secondly that it is only the migrants that compete for sites at the destination whereas in fact established adults are likely to be present both in the natal area and at the destination. To be a valid suggestion it must be postulated that competition from established adults is less at the destination than at the source, a suggestion that, if extended to include other factors that affect habitat suitability as well as factors that affect migration cost, would then be in agreement with the thesis of this book that migration is an advantage if h_1 is less than h_2M. Quite possibly, of course, in some cases competition with established adults may well be a contributory selective pressure towards a pre-reproductive peak of removal migration. As a general thesis, however, the suggestion is inadequate, for a pre-reproductive peak of migration is widespread even among animals, such as many insects, for which, because adults die before the young become adult, there is no question that there is competition between young and 'established' individuals.

Many insects, particularly females, exhibit a marked pre-reproductive (i.e. post-teneral) peak of migration by flight. The existence of this peak led C.G. Johnson (1969) to postulate a physiological relationship between migration by females and the relative states of development of the flight apparatus and the ovaries. This relationship is termed the oogenesis-flight syndrome (Johnson, p. 9). It is suggested that migration is favoured (physiologically) by a highly developed flight apparatus and undeveloped ovaries but is inhibited by the hormonal mechanisms involved in oogenesis. At times, e.g. in relation to dragonflies, Johnson goes to such lengths to argue that migrant yet gravid females may be in the process of terminating migration rather than migrating after bouts of oviposition, that the impression is given that oogenesis must inevitably inhibit migration rather than that the onset of oviposition and the reduction of migration are independent ontogenetic events that selection has often favoured to coincide and to be physiologically related. Yet Johnson himself presents sufficient examples (see also p. 351 *et seq.* of this book) of migration after the ontogenetic onset of oviposition for it to be clear that the view is not tenable that the inhibition of migration by oogenesis is inevitable.

Johnson (1969, pp. 607–611) stresses that his book is concerned primarily with the physiological mechanisms involved in insect migration rather than with any evolutionary consideration of the phenomenon, and as far as I can determine there is only one sentence in his whole book that attempts to ascribe a selective advantage to pre-reproductive migration. It is suggested (Johnson 1969, p. 174) that, as far as female insects are concerned, 'sexual immaturity during migration can be regarded as an ecological

adaptation ensuring that eggs are laid in another place, more suitable and possibly distant, and not in an old habitat that may become unfavourable'. However, as it stands, even this statement cannot be accepted as a description of selection for pre-reproductive migration. First, though of minor importance, nothing is added to the description of the postulated selective pressure by reference to the possibility that the old habitat may become unfavourable, the critical factor being covered by the suggestion that the new habitat may be more suitable than the old, by whatever sequence of events. Secondly, it is not sexual immaturity but migration that ensures that eggs are laid in another place, possibly distant, though neither phenomenon, but especially not sexual immaturity, can ensure that this place is more suitable than the old habitat. The only influence of sexual immaturity during removal migration is that *all* eggs are laid in another place. If any advantage of sexual immaturity during migration (or, more correctly, migration during sexual immaturity) is to be sought, therefore, it is likely to be related to the advantage of laying *all* of the eggs elsewhere, rather than just *some* of the eggs, a conclusion that once again brings us back to the relative values of h_1 and h_2M as the centrally important factor in the advantage of migration. Although Johnson's book considers the pre-reproductive migration of female insects at great length, therefore, it contains little in the way of an evolutionary evaluation of the phenomenon.

One of the problems experienced by Johnson, having formulated the concept of the oogenesis-flight syndrome and having stressed its central role in insect migration, was to account for the migration of males. Males manifestly do not have a spermatogenesis-flight syndrome, even though the majority may migrate more when young than when old. It has to be accepted, therefore, either that there are other selective pressures at work sufficiently powerful to lead to the evolution of migration quite independently of any gametogenesis-flight syndrome, or that male migration is explicable totally in relation to female migration. This latter view seems to be the one adopted by Johnson. He suggests (pp. 170 and 609) that if females are inseminated before migration, the males may not migrate. If, on the other hand, as in the desert locust, *Schistocerca gregaria*, the females are inseminated towards the end of migration, then the males accompany the females on migration. Unfortunately, as far as this line of reasoning is concerned, there are some species, such as the mottled umber moth, *Erannis defoliaria* (Williams 1958), and all the doryline army ants (Schneirla 1971) in which only the males migrate, the females being wingless. Johnson's suggestion (p. 609) that in such cases the males 'migrate rather purposelessly' is perhaps not totally acceptable.

It should be stressed, perhaps, that it is not the intention of this section to imply that the oogenesis-flight syndrome, to retain such a purely entomological phrase for just one more paragraph, is non-adaptive. Clearly some endocrinological relationship does exist between the state of development of the flight apparatus of many female insects and the initiation of oogenesis. Indeed, some female insects, such as termite reproductives (p. 240), migrate just once, after which the flight muscles atrophy, presumably utilising the energy thus released in the production of eggs, a physiological system that has clear advantages in an animal for which a once-only removal migration is the optimum strategy. However, my feeling is that entomologists in recent years have exaggerated the importance of the oogenesis-flight syndrome out of all proportion to its real place in the evolution of insect migration, and if the syndrome is viewed in the perspective of animal migration as a whole it can be seen to be a very small aspect of a much more general phenomenon. The view of the oogenesis-flight syndrome propounded here is that female insects, in the same way as male insects and all other animals, have been exposed to a number of selective pressures that have acted upon the ontogenetic curve of frequency of initiation of removal migration. One consequence of this selection is that for a great many animals, not only females and not only insects, there has evolved a peak of removal migration during the pre-reproductive phase of ontogeny. Superimposed upon this adaptation, selection has also favoured many insects utilising some method of locomotion other than flight when moving within their feeding and breeding habitat(s) with the consequence that flight has been retained only for the purpose of migration between these habitats. In such cases the evolutionary opportunity has presented itself for some physiological interaction between development of the flight apparatus and oogenesis such that there is an optimum utilisation of reserves. The problem with which this section is chiefly concerned, therefore, is that of elucidating the selective pressures that favour in so many animals a pre-reproductive peak of removal

migration. At the same time it is necessary to consider why many other animals, particularly males, should continue to migrate at intervals after the ontogenetic onset of reproduction. First, however, it is necessary to consider one other viewpoint on the existence of a pre-reproductive peak to the initiation of migration.

This view is based on the concept of reproductive value formulated by Fisher (1930) and has been developed particularly by students of the theory of island biogeography, such as MacArthur and Wilson (1967). Reproductive value tells which age a propagule (i.e. the minimum unit necessary for reproduction), is most likely to establish a successful colony. This age is that of maximum reproductive value, which usually approximates the age at which reproduction commences. Consequently, the most successful colonist is a pre-reproductive migrant, all other things being equal.

MacArthur and Wilson (1967) go further, however, and suggest that for the above reason evolution would be expected to favour migration (= dispersal, in their terms) at the age of maximum reproductive value. This extrapolation from success at colonising islands to optimum strategy on the mainland is, however, of doubtful validity. It seems unlikely, due to the lack of genetic feedback, that success at colonising islands will have been a major selective pressure on the *removal* migration of a mainland species, though clearly the situation may be different with respect to *return* migration. Further, it seems likely that although the migration from mainland to island may prove to be an advantage to an individual that successfully colonises the island, in the vast majority of cases migration away from the mainland is likely to be of such high migration cost that on average such migration is disadvantageous. Any characteristic of a migrant, such as a high reproductive value, that may be an advantage in the colonisation of an island, is unlikely, therefore, to spread through a population unless it is also an advantage to migrants on the mainland. Characteristics such as a high reproductive value that may be important in island colonisation are therefore irrelevant in relation to removal migration on the mainland unless it can be demonstrated that the factors that are important in island colonisation are also important in the 'colonisation' of mainland habitats.

Some authors, such as Dingle (1972), have assumed that mainland habitats can be considered to be islands surrounded by 'seas' of unsuitable habitats and that therefore the theory of island biogeography can be applied directly to a consideration of all migration, whether within the mainland or from mainland to islands. Although Dingle (1972) was concerned mainly with insects, particularly female insects, his arguments are applicable to most animals, and as his paper probably represents the most intensive attempt to apply theories of island biogeography to mainland migration it is perhaps worthwhile to consider his argument in some detail. In the following sentences, the phrases that I consider to be the critical parts of his argument are italicised. According to Dingle, insect migrants are *colonisers and not refugees*. *As a coloniser*, upon arriving in a new habitat, the insect must be able to reproduce and leave descendants. *In this situation*, evolution favours high productivity and in order to do this a migrant would be expected to increase fecundity, reproduce earlier, and to have a high reproductive value. Since pre-reproductive adults have their reproductive life ahead of them and at the same time have already survived the causes of juvenile mortality, their reproductive values are usually high relative to juveniles or older adults and *thus have a high colonisation potential*. Evolution would be expected, therefore, to favour migration *in most species* just prior to reproduction.

Promising though this argument appears at first sight, it has certain shortcomings as an evolutionary argument and is perhaps not totally acceptable in its present form. First, I cannot agree that migrants are *colonisers and not refugees*. Neither concept on its own adequately describes the advantage of migration. To use the often quoted example of migration of a chicken crossing a road, it can be argued that a chicken does not cross a road in order to get to (colonise) the other side, nor does it cross a road to escape from the side of origin. Rather, and the whole of this book is based on this argument, it seems likely that a chicken crosses a road because the other side is sufficiently more suitable than the side it is on for the risk involved in crossing the road to be worth taking. Secondly, it is not *as a coloniser* that an insect, or any other animal, must be able to reproduce and leave descendants. All animals, whether 'colonisers' or not, must be able to reproduce and leave descendants. *In any situation*, therefore, evolution favours high productivity, though, if we follow Lack (1954) (but see Wynne-Edwards 1962), maximum reproductive success need not necessarily be given by maximum fecundity nor by earlier reproduction, a point that is neglected by any argument based on the concept of the intrinsic rate of increase de-

veloped by Cole (1954) which requires unlimited resources in an unlimited universe. Far from always being associated with increased fecundity and earlier reproduction, therefore, migration is often associated with decreased fecundity and delayed reproduction. Indeed, the relationship between migration and delayed reproduction is one of the central facets of the oogenesis-flight syndrome of female insects (Johnson 1969).

The major part of Dingle's argument is unacceptable, therefore, and we are left with his final point that since pre-reproductive adults have their reproductive life ahead of them and at the same time have already survived the causes of juvenile mortality, their reproductive values are usually high relative to juveniles or older adults and *thus have a high colonisation potential*. Consequently selection would be expected to favour migration *in most species* just prior to reproduction. In fact, if the point is valid, evolution would be expected to favour migration in all species, males as well as females, just prior to reproduction. The question to be considered, however, is whether the point is valid as it stands.

The concept of high colonisation potential, although extremely useful to an island biogeographer, is of doubtful application to mainland migrants. MacArthur and Wilson (1967) themselves point out that there is a danger in assuming that mainland habitats can be treated as true islands. On true islands the immigration rate is low. The major selective pressures acting on the reproductive value of a colonising propagule, therefore, are those associated with the production and survival of sufficient numbers of offspring for individuals to be able to find mates. In this situation, successful propagules are much more likely to be those with a high reproductive value. On the mainland, however, the vast majority of migrants arrive in habitats that either are already occupied by members of the same species or will receive such members soon afterwards. The offspring, also, are free to migrate with a reasonable chance of such migration being advantageous. Mainland migrants are not, therefore, colonisers in the same sense as island immigrants. Indeed, an animal with a high colonisation potential is also an animal with a high reproductive potential in its natal habitat. In fact, reproductive value is higher if the animal remains in its natal habitat than if it migrates because it can reproduce earlier and is not exposed to migration cost. All other things being equal, therefore, if it is claimed that a high reproductive value is the important factor, it can only be used as an argument against the initiation of migration, the fact that a high reproductive value confers a high colonisation potential being irrelevant. This is only not true if there is some reason why a high reproductive value is more important in one habitat than in another.

We come back, therefore, to the conclusion that the critical situation for migration to be advantageous is that h_1 is less than $h_2 M$. Consequently, if reproductive value is at all involved in ontogenetic variation in the initiation of migration, it can only be because reproductive value influences the relationship between h_1 and $h_2 M$ to different extents at different ontogenetic stages, and is in no way related to colonisation potential in the sense the term is used by island biogeographers.

10.3.2 Development of a model of adaptive ontogenetic variation in the probability of initiation of removal migration

The thesis of this book is that migration is an advantage when h_1, the suitability of the habitat occupied, is less than $h_2 M$ (where h_2 is the suitability of the destination and M is the migration factor). The application of this thesis depends on whether the removal migration being considered is calculated or non-calculated. Calculated removal migration is advantageous when h_1 is less than $h_f M_f$ (where h_f is the suitability of the most suitable habitat with which the animal is familiar and M_f is the migration factor for migration to that habitat). Non-calculated removal migration is on average advantageous when h_1 is less than $\overline{hM}$ (the mean product of habitat suitability and migration factor for all potential destinations). In Chapter 8, the factor $\overline{hM}$ was represented by $\bar{E}$ (the mean expectation of migration).

The probability that a given individual will initiate removal migration is, in effect, the probability of that individual encountering environmental variables that exceed the migration threshold for that individual. The migration threshold to environmental variables in non-calculated removal migration is that value of the environmental variable that corresponds to the situation in which $h_1 = \overline{hM}$ (Chapter 8, p. 45). Similarly, the migration threshold to environmental variables in calculated removal migration is that value of en-

vironmental variables that correspond to the situation in which $h_1 = h_f M_f$ (Chapter 8, p. 66). The probability that a given individual will initiate removal migration is, in effect, therefore, the probability of that individual encountering environmental variables that correspond to situations in which either $\overline{hM}$ or $h_f M_f$ exceeds h_1. Alternatively, it is the probability of that individual encountering a situation in which either $\overline{hM}/h_1$ or $h_f M_f/h_1$ exceeds unity.

If we consider a single individual, migration should ideally only be initiated if that individual encounters a situation for which either $\overline{hM}/h_1$ or $h_f M_f/h_1$ is greater than unity, and even then initiation of migration depends on the migration threshold of that individual as part of the optimum frequency distribution of migration thresholds within the deme of which the individual is a member (as discussed in Section 10.2). Field measurement of the probability of the initiation of migration, however, is usually based on observed migration incidence, and migration incidence is invariably measured either as per cent migrants or as an average rate of initiation. If we are to develop a model of adaptive ontogenetic variation in migration threshold that is to be to some extent testable against field data, therefore, this group effect has to be taken into account. Incorporation of this group effect into the arguments of the previous paragraph, however, is relatively simple. It follows from these arguments that migration incidence is, in effect, a measure of the probability of encountering a situation in which either $\overline{hM}/h_1$ or $h_f M_f/h_1$ exceeds unity. If the mean suitability of the habitats occupied by the different individual members of the group being considered is represented by $\bar{h}_1$, then the proportion of individuals likely to initiate migration is a direct function of the value of $\overline{hM}/\bar{h}_1$ or $\overline{h_f M_f}/\bar{h}_1$. Clearly, the nearer $\overline{hM}/\bar{h}_1$ or $\overline{h_f M_f}/\bar{h}_1$ approaches unity from zero, the greater the probability that some individuals will encounter a situation for which $\overline{hM}/h_1$ or $h_f M_f/h_1$ exceeds unity. Similarly, the more that either $\overline{hM}/\bar{h}_1$ or $\overline{h_f M_f}/\bar{h}_1$ (which may be referred to as the initiation factor, i', for, respectively, non-calculated and calculated removal migration) exceeds unity, then the less the probability that some individuals encounter a situation for which $\overline{hM}/h_1$ or $h_f M_f/h_1$ is less than unity. It also follows from the arguments presented in Chapter 11 (p. 82) that the greater the numerical value of the initiation factor, the shorter is the optimum stay-time. Whether migration incidence is measured as per cent migrants or as a rate of initiation of migration, therefore, it would be expected from the arguments of this section that migration incidence should be a direct function of the value of i', the initiation factor. It follows, therefore, that if the value of the initiation factor varies with age or sex, then ideally migration incidence should similarly vary with age or sex. The statement that I, migration incidence, is proportional to i', the initiation factor $(\overline{hM}/\bar{h}_1$ in non-calculated removal migration; $\overline{h_f M_f}/\bar{h}_1$ in calculated removal migration) can be considered to be a model of adaptive ontogenetic variation in the probability of initiation of migration.

It is, of course, unlikely, because of the derivation of the initiation factor, that the relationship between i' and I is linear throughout their entire range. Moreover, for practical reasons, an additional source of error is introduced into this model for the remainder of this section and in the evaluation of the model in Chapter 17. The true value of i' is given by $\overline{hM}/\bar{h}_1$ or $\overline{h_f M_f}/\bar{h}_1$ as just demonstrated. The factor $\overline{hM}$ is the mean product of habitat suitability and migration factor for all potential destinations available to an animal upon departure from H_1 (see p. 45). Although there is a strong correlation between the two functions, however, the value of the function $\overline{hM}$ is not the same as the value of the function $\bar{h}\overline{M}$ (the product of $\bar{h}$, the mean suitability of all destinations available to an animal that leaves H_1, and $\overline{M}$ the mean migration factor for migration to each of these habitats). Nevertheless, in evaluating the above model in Chapter 17, it becomes much more convenient and meaningful to express the value of the initiation factor, not as i' (where $i' = \overline{hM}/\bar{h}_1$ or $\overline{h_f M_f}/\bar{h}_1$) but as i (where $i = \bar{h}\overline{M}/\bar{h}_1$ or $\bar{h}_f \overline{M}_f/\bar{h}_1$). The advantage of this is that the initiation factor can be broken down, not into $\overline{hM}$ and $\bar{h}_1$ but into $\bar{h}/\bar{h}_1$ and $\overline{M}$ (or $\bar{h}_f/\bar{h}_1$ and $\overline{M}_f$). The former term, $\bar{h}/\bar{h}_1$ or $\bar{h}_f/\bar{h}_1$, is referred to in the remainder of this book as the mean habitat quotient. This is an extremely useful and biologically meaningful term, the advantages of which are sufficiently great as to outweigh the dangers of the above procedure. In any case, because there is such a strong correlation between i' and i and because the weakest point of this model is the assumption of a linear correlation between either i' or i and I, the danger of drawing erroneous conclusions through using i rather than i' is not at all great.

Let us now consider briefly some of the sources of ontogenetic variation in the initiation factor. Details of such variation for a wide range of animals are presented in Chapter 17 and it is only necessary here to make a few general observations.

There are a number of factors that could influence the value of the mean habitat quotient $\bar{h}/\bar{h}_1$ or $\bar{h}_f/\bar{h}_1$ at different stages of ontogeny. Some insects, for example, that are adapted to habitats that are most suitable soon after they have formed (p. 49) often have a seasonal incidence that is such that adult emergence coincides with the sudden appearance of new habitats in the environment. Such insects are confronted as young adults with a mean habitat quotient that greatly exceeds unity. Depending on the rate of appearance of new habitats, however, once the insect has migrated (so that the habitat occupied is now a new habitat) or at least as the insect ages and the rate of appearance of new habitats decreases, the mean habitat quotient $(\bar{h}/\bar{h}_1)$ approaches unity. Exploratory migrants are likely to experience a mean habitat quotient $(\bar{h}_f/\bar{h}_1)$ that exceeds unity by an increasing amount as exploratory migration continues (p. 64). On the other hand, species that live in a home range or territory invariably make learning and other (e.g. construction) investments in their area. Assuming that $\bar{h}_1 = \bar{h}$ with respect to all other characteristics of the habitat (e.g. food availability, climate), it follows that the mean habitat quotient is invariably less than unity because the individual is far less efficient at exploiting the food available and taking shelter, etc. in a new home range than in the present home range. The availability of food, etc. in the other home range would have to be far superior to that in the present home range before the habitat quotient approached or exceeded unity for any particular individual. The greater the average total 'investment' in a home range, for example as an individual ages, then the further the mean habitat quotient falls short of unity. On the other hand, if a species, such as the beaver, *Castor canadensis*, not only makes an investment in its home range (in this case a construction investment as well as a learning investment) but at the same time exploits an effectively non-replaceable resource (in this case timber for construction and young wood for food), then the sheer presence of the animal eventually decreases $\bar{h}_1$ below $\bar{h}_f$, notwithstanding the effect of the investment. In such an animal, the mean habitat quotient is likely to show a characteristic cyclic variation above and below unity with progression of ontogeny.

The value of $\overline{M}$, mean migration factor, is also likely to show ontogenetic variation. In order to understand this variation, however, it is necessary to consider the derivation of the migration factor. The value of M is given by

$$M = S_2/S_1, \qquad \text{(p. 41)}$$

where S_2 is the value of S_d, action-dependent potential reproductive success, of an animal at the end of a migration (time t_{arr}) and S_1 is the value of the action-dependent potential reproductive success that the animal would have had at time t_{arr} if it had remained in its original habitat.

The difference between S_1 and S_2 is m, the migration cost (i.e. $m = S_1 - S_2$). The value of M is therefore given by

$$M = \frac{S_1 - m}{S_1} \quad \text{or} \quad 1 - \frac{m}{S_1}. \qquad (13)$$

Ontogenetic variaton in m, migration cost, is considered below.

If, for the moment, however, migration cost is considered to be a constant, the form of the relationship between M and S_1 according to this derivation can be illustrated as in Fig. 10.2. It can be seen that as action-dependent potential reproductive success decreases, so too does the value of M. This relationship has important repercussions throughout all the components of the model developed in this book that make use of the migration factor. For example, it means that a potential calculated removal migrant with a low action-dependent potential reproductive success requires the difference in suitability of h_f and h_1 (where h_f is the suitability of the most favourable habitat known to the animal) to be much greater for migration to be advantageous than an individual with a higher action-dependent potential reproductive success. Similarly, a potential non-calculated removal migrant with a high S_d, action-dependent potential reproductive success, has a higher value for $\bar{E}$, mean expectation of migration ($\bar{E} = hM$; p. 45) than individuals with a low S_d. This means that individuals with a high S_d are likely to have shorter stay-times (p. 82), lower migration thresholds (i.e. more probability of migration) to habitat variables, indirect variables, and migration-cost variables (p. 265), and a greater proportion of migrants over a wide range of habitat suitabilities (p. 69). As S_d for an individual starts at a high level at birth and thereafter decreases until it reaches zero at death, the relationship between S_d and M is clearly of considerable influence in the ontogenetic variation in the probability of initiation of removal migration.

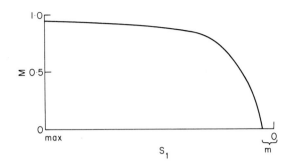

Fig. 10.2   General form of the relationship between the migration factor and the action-dependent portion of present potential reproductive success.

The action-dependent potential reproductive success, S_d, of an individual may be defined as that component of the present potential reproductive success that can be influenced by the present and future actions of that individual (see p. 38).

The migration factor, M, is given by S_2/S_1, where S_2 is the action-dependent potential reproductive success at time t_{arr}, the moment of termination of a migration, and S_1 is the action-dependent potential reproductive success that the animal would have had at time t_{arr} if it had remained in its original habitat. It is assumed—see p. 43—that no portion of S_d becomes action-independent during the migration.

The difference between S_2 and S_1 is the migration cost, m, such that $m = S_1 - S_2$. The factor M is therefore given by the formula $M = 1 - (m/S_1)$, the general form of which is illustrated in the diagram, assuming for present purposes that m is constant and very small. As the action-dependent portion of the potential reproductive success at the initiation of migration decreases, so too does the migration factor for the migration about to be performed.

It was assumed for convenience in the previous argument that given constant conditions of environmental variables, there would be no variation in migration cost with age. There are a number of factors, however, that seem likely also to influence migration cost in relation to ontogenetic stage. For example, mature males may often suffer a reduction in opportunity to copulate during migration whereas immature males are not subject to this migration cost. Also, many animals for which learning is an important strategy are likely gradually to improve with age in their ability, during migration, say, to avoid predation, to find potential destinations, or to transport and/or protect offspring, and are therefore likely to experience a gradual reduction in migration cost with age. Finally, in animals in which migration occurs by flight but other activities are performed by remaining sedentary or walking, a situation that is found primarily among some insects, such as aphids, variation of wing-loading with age can be important. The wing surface area of insects remains more or less constant throughout life. The body weight, however, may change with age. In particular, females increase weight gradually from a low soon after emergence (depending on the species) to a high when gravid, and then gradually to a low as energy reserves become spent. Males may show a similar variation but to a much lesser extent than females. For insects such as those that fly only during relatively long-distance migration, therefore, the cost of such migration in terms of energy expenditure (p. 41) may be relatively low in the post-teneral pre-reproductive phase, may increase toward the beginning of reproduction, and then perhaps gradually decrease again. This effect is much greater on the variation in the migration cost of females with age than on the migration cost of males. It is particularly important to stress that this selective pressure applies mainly to those species that fly only during such migration and within their habitat usually remain still or walk. If a species flies while within its habitat, as for example, do butterflies and many larger Diptera, then as much energy can be expended per unit time within the habitat as during migration and the effects of wing-loading on migration cost are much less.

Clearly, there are many sources, in addition to those mentioned here, of ontogenetic variation in migration cost. This variation in m along with that in S_d invariably produces ontogenetic variation in M, the migration factor. This ontogenetic variation in M along with that in mean habitat quotient ($\bar{h}/\bar{h}_1$ or $\bar{h}_f/\bar{h}_1$) in turn invariably produces ontogenetic variation in i, the initiation factor ($\bar{h}\bar{M}/\bar{h}_1$ or $\bar{h}_f\bar{M}_f/\bar{h}_1$). It remains, therefore, in Chapter 17, to evaluate for a wide range of migrations the extent to which this model can describe observed ontogenetic variation in migration incidence.

11

Migration rate and distance

11.1 Introduction

Chapters 7–10 have been concerned with identifying firstly the point at which the initiation of migrations of different types becomes advantageous and secondly the way in which the probability of initiation of a given migration is likely to vary between and within species. In this and the next chapter we are concerned for the first time with the selective pressures acting on the animal once it has initiated migration. In this chapter the component of migration to be considered is distance, along with the associated component of rate (i.e. the average rate of cross-country progression taking into account both periods of migration and periods spent within a suitable habitat).

As an animal migrates, it encounters a succession of spatial units or habitats, many of which are of zero suitability, others of a suitability greater than zero but less than $\bar{E}$ (in non-calculated migration) or $h_f M_f$ (in calculated migration) and some of a suitability greater than $\bar{E}$ or equal to $h_f M_f$. Clearly in non-calculated removal migration, the migration distance is to a large extent a function of the distance between H_1 and the first habitat for which h_1 is not less than $\bar{E}$, though for many species, even if h_1 is greater than $\bar{E}$ upon arrival, it is likely rapidly to become less than $\bar{E}$ once more. Even in the former situation, however, the identification of likely migration distance is not simple, for it seems probable that the value of $\bar{E}$ changes as migration proceeds.

The approach used in this chapter is first to consider the selective pressures acting on migrations for which the distance is not fixed at the initiation of migration, and secondly to consider the differences in the selective pressures elicited when the distance is fixed. These are all individual considerations. The chapter ends with a brief consideration of selection on the frequency distribution of migration distances within a deme.

11.2 Selection when migration is not to a fixed destination

According to the arguments presented in previous chapters, if an animal initiates migration from a habitat, H_1, for which h_1 is less than $\bar{E}$, then *on average* migration is advantageous. Many individuals that migrate in response to this situation, however, find themselves in a habitat, H_2, the suitability of which is such that $h_2 M_1$ (where M_1 is the migration factor for migration from H_1 to H_2) is not greater than h_1. To such individuals, the migration has proved to be a disadvantage. These individuals, arriving in a habitat the suitability of which is such that $h_2 M_1$ is not greater than h_1, have two strategies available to them. Either they can remain in H_2 and thus suffer a disadvantage from migration or they can initiate a further migration on the basis of the possibility that the next destination will be more suitable. A number of considerations are relevant here. First, if $h_3 M_1 M_2$ (where h_3 is the suitability of habitat H_3, M_1 is the migration factor for migration from H_1 to H_2 and M_2 is the migration factor for migration from H_2 to H_3) is greater than h_1, then the total migration from H_1 to H_3 will have been an advantage. On the other hand, if $h_3 M_2$ is greater than h_2 but $h_3 M_1 M_2$ is less than h_1, then the total migration from H_1 to H_3 will not have been an advantage but the second removal migration from H_2 to H_3 will have been more advantageous than having stayed in H_2. However, as far as migration from H_2 is concerned, the same considerations apply as before (pp. 45–50) and *on average* the initiation of migration is an advantage if h_2 is less than E.

It would be expected, therefore, that removal migration will be repeated by an animal until it arrives in a habitat for which h is not less than $\bar{E}$. This should be the best strategy for an individual that happens to experience encounters with a series of habitats for each of which h is less than $\bar{E}$, though the nett result of the strategy may not in every case be an advantage compared with having remained in the initial habitat, H_1 (e.g. if $h_3 M_1 M_2$ is less than h_1).

Other factors are also likely to be involved. For example, the individual under consideration could make no direct estimate of $\bar{E}$ in advance of migration from H_1, and therefore could only respond to some endogenous evolved standard for $\bar{E}$ as manifest by response to the threshold value, v_t, of some environmental variable. However, an animal that performs removal migration several times between a number of habitats is presented with some opportunity for assessing $\bar{E}$. The endogenous standard for $\bar{E}$ involved in the initiation of the first removal migration can only have evolved on the basis of the value of $\bar{E}$ to which the animal's ancestors have been exposed during the evolution of this endogenous standard. Even when the mean value of $\bar{E}$ remains relatively constant with time, fluctuation about this mean occurs. At any moment in time, therefore, the mean expectation of migration for a particular individual could be considerably different from the value of $\bar{E}$ in relation to which the endogenous standard has evolved. Hence an animal that repeatedly finds itself in habitats for which h is less than $\bar{E}$ could well be doing so because at that particular time of that particular year or in that particular area the actual $\bar{E}$ is much lower than the mean $\bar{E}$ in relation to which the endogenous standard has evolved. The optimum strategy for an animal, therefore, seems to be to initiate the first migration on the basis of an endogenous standard and then, as migration takes place and is initiated several times for the reasons already discussed, to assess the relationship between the endogenous standard and the actual $\bar{E}$ for that area at that time. In other words, the initiation of migration from the natal area should occur when h_1 is less than $\bar{E}_i$ (the innate mean expectation of migration) in response to an environmental variable the value of which is beyond v_{ti} (the innate threshold value). Subsequent initiations of migration, however, should take increasing account of $\bar{E}_e$ (the expectation of migration based on

the recent experience of the animal). Practically, what should be observed is an increase in the threshold for migration at each migration if the suitability of each habitat encountered is such that h is less than the current value of $\bar{E}$.

So far, therefore, it has been concluded that removal migration should be repeatedly initiated by an animal until it arrives in a habitat for which the suitability is not less than the mean expectation of migration but that with each migration that this is not realised the expectation of migration should be lowered. Now, let us consider the optimum stay-time in each of the habitats encountered for which h is less than the current value for $\bar{E}$.

During the period of migration from H_1 to H_d (where H_d is the first habitat encountered by the animal for which h is greater than $\bar{E}$) selection should favour those individuals that invest the minimum time possible in the intervening habitats. Ideally this should be the time necessary to migrate through or over such habitats. However, the process of assessment of habitat suitability may involve some finite time investment. It may be possible for some habitats to determine that h is less than $\bar{E}$ without the need for any change in migration speed or track. Other habitats, however, particularly those the suitability of which approaches $\bar{E}$, may require considerable time investment on the part of the animal before it can be determined that h is less than $\bar{E}$. For this reason, an animal migrating across country may at intervals pause and invest time in a given habitat before continuing the migration.

Similar behaviour will be shown, but for different reasons, by those animals adapted to habitats for which, even if h is greater than $\bar{E}$ upon arrival, habitat suitability is likely rapidly to become less than $\bar{E}$. Such rapid reduction in habitat suitability to a level below $\bar{E}$ may on occasion be due to the resource being rapidly consumable, but most often probably results from changes in an animal's resource requirements. For example, suppose that an animal initiates migration from a reproduction habitat, H_r, because h_r is less than $\bar{E}_r$, the mean expectation of migration between reproduction habitats. As migration proceeds, the animal uses up energy reserves until at some point feeding becomes the optimum time-investment strategy. The animal continues to migrate until a feeding habitat, H_F, is encountered for which h_F is greater than $\bar{E}_F$, the mean expectation of migration between feeding habitats. The animal stays in the habitat either until h_F becomes less than $\bar{E}_F$ because the food resource has been consumed (yet feeding remains the optimum time-investment strategy), or until the suitability of the feeding habitat *as a reproduction habitat* becomes less than $\bar{E}_{Fr}$, the mean expectation of migration between feeding and reproduction habitats, because reproduction has become the optimum time-investment strategy. This type of migration is found in animals adapted to spatially separate habitat-types.

It can be seen that in the above situation there is no direct selection on migration distance. An individual moves across country at a rate that is entirely a function of the suitability of the habitats encountered and the distance between habitats for which $h > \bar{E}$. The end-point of the migration sequence, and hence total migration distance, is determined by chance encounter with a habitat for which h is greater than $\bar{E}$, though because $\bar{E}$ often decreases with age and with recent experience, in most cases the chance of encountering such a habitat increases with time. A migration distance that results from this type of strategy may be described as *facultative*.

Although most instances of non-calculated removal migration seem likely to be based on this strategy, there seems to be an inherent disadvantage to such a strategy. This disadvantage is that time has to be spent monitoring and assessing the suitability of all habitats encountered. As long as the distance between H_1 and the nearest habitat for which h is greater than $\bar{E}$ is unpredictable, this disadvantage is likely to be more than outweighed by the danger of missing such a habitat as a result of not employing such a strategy. Suppose, however, that the distance between H_1 and the nearest suitable destination along the track taken by the migrant were to be predictable. Selection would then be likely to favour those individuals that took advantage of such predictability by not investing time in monitoring or assessing intermediate habitats until they had migrated a distance appropriate for arrival at such a suitable destination. It is convenient to describe such a migration distance as *obligatory*. It should be noted, however, that migration over an obligatory distance is likely to be followed by a period during which migration distance is facultative while the animal attempts to locate a habitat for which h is greater than $\bar{E}$.

Perhaps the clearest situation in which selection favours an obligatory migration distance is when selection favours the evolution of a migration that alternates between a high degree of convergence and a high degree of dispersal. When animals, such as bats, ladybirds, or moths, etc., converge in large numbers within a limited area, usually in order to hibernate or aestivate, and then later begin to disperse, it seems likely that until a certain distance has been travelled, deme density cannot have been reduced sufficiently for h to be greater than $\bar{E}$. Such situations inevitably also involve selection on the frequency distribution of migration distances (p. 84) and directions (p. 85).

11.3 Selection when migration is to a fixed destination

Calculated migrants have a migration distance that is fixed when they initiate migration. Some non-calculated migrants may also have a migration distance that is more or less fixed. Thus the facultative and obligatory migration distance strategies described above may in some lineages have led invariably to a given distance of cross-country migration. Examples are those inhabitants of seasonally arid regions for which the rainfall zone at a given time of year has invariably occupied a position a given distance in a given direction. The situation is similar for those inhabitants of temperate regions that have become adapted to pass the winter at lower latitudes where a given resource remains continuously available. In all of these cases additional selective pressures seem likely to be elicited that favour those animals that vacate H_1 at the optimum time, arrive at the destination at the optimum time, and migrate across country between the two in such a way as to minimise the cost of the total migration. Model development in this case was found to be easier in relation to a specific example, that of birds, and the reader is referred to pp. 678 *et seq.* for the development of this part of the model. Although in those pages the model is developed in relation to long-distance movement across country by a vertebrate, the same model is applicable whatever the animal and over no matter what distance for all movement between relatively fixed sources and destinations.

11.4 The frequency distribution of migration distances

The frequency distribution of migration distances, like the frequency distribution of migration thresholds and migration directions, is a function of the position of the deme concerned on the territorial spectrum (p. 68). In a despotic situation, individuals of highest resource-holding power migrate that distance in that direction that maximises h_2M.

Individuals with a lower resource-holding power migrate shorter or longer distances in such a way as to produce a value of h_2M that is the maximum possible for an individual with that particular level of resource-holding power. In a free situation, selection should stabilise when the combination of migration thresholds, migration distances, and migration directions lead to a distribution of the deme through space and time that is such that all individuals experience the same potential reproductive success. For an evaluation and application of this model see p. 802 (for fish) and p. 696 (for birds).

12

Migration direction and the direction ratio

12.1 Introduction

Selection for migration in a given direction occurs when there is an environmental gradient that is the resultant of habitat suitability and migration cost. An animal migrating in a plane and direction parallel to this gradient experiences, given a constant value for action-dependent potential reproductive success (see p. 38), a continuous change per unit distance in the mean expectation of further migration. In such a situation, selection favours those individuals that migrate in that plane and direction that gives the greatest increase in $\bar{E}$, the mean expectation of migration, per unit distance. Selection should further favour those individuals that continue to migrate in this direction at an optimum rate (Chapter 11) until $\bar{E}$ no longer continues to increase, whereupon the migration should end in a habitat for which h is greater than $\bar{E}$.

There is no difficulty, therefore, in extending the model developed in Chapters 7–11 to include selection on migration direction. Further consideration is necessary, however, in order to include selection on the direction ratio (= frequency distribution, within a deme, of migration directions). It is this consideration that occupies the remainder of this chapter. Two situations are discussed: return migration; and re-migration (= a return migration by the offspring of the outward migrants, but not necessarily by the outward migrants themselves).

12.2 The direction ratio in return migration

In return migration, each individual returns to its starting point. There is thus no nett shift in the centre of distribution of the deme during the course of a generation. The situation is consequently relatively simple and can be discussed solely in terms of the optimum distribution of the deme through space and time, excluding the pressures that arise as the result of individuals being lost over the edge of the geographical range. As before (pp. 69 and 84), the situation is a function of position on the territorial spectrum. In a despotic situation, individuals of the highest resource-holding power should migrate to destinations in the most advantageous directions and thus exclude individuals of lower resource-holding power that should then eventually come to occupy destinations in less advantageous directions. In a free situation, individuals are eventually distributed through space and time such that all individuals experience the same potential reproductive success (insofar as the latter is a function of position in space and time). For example, if too many individuals migrate in the most advantageous direction, the habitats lying in this direction come to support a deme density that is so high that migration in this direction is less advantageous than migration in slightly different directions. The result is selection in favour of individuals that migrate in these directions and selection against individuals that migrate in what was previously the most advantageous direction. The result is an increase in the proportion of the deme that migrates in the former directions and a decrease in the proportion that migrates in the latter direction. Until, that is, these proportions are such that all directions are once more equally advantageous.

The result is that more individuals migrate in the direction that, if the effects of deme density were to be excluded, would be the most advantageous. On either side of this direction the proportion of the deme that migrates in a given direction decreases. If there is no environmental gradient that is the resultant of habitat suitability and migration cost, selection should favour a direction ratio of 25:25:25:25. This model of direction ratio in return migration is developed further in the chapter on the return migration of birds (p. 654).

12.3 The direction ratio in re-migration

In the simplest form of re-migration, one generation of an animal performs a removal migration, pro-

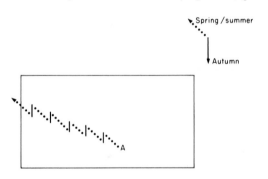

Fig. 12.1 Selection for and against a peak migration direction as a result of selection for re-migration

Re-migration is a return migration in which the return migration is performed by the offspring of the outward migrants, but not necessarily by the outward migrants themselves. It is often characterised by a low degree of return of the offspring with respect to the place of origin of their parents.

The arrows in the top right of the diagram reflect a typical re-migration situation for some insects in western Europe. The dotted arrow indicates the most advantageous migration direction for spring and summer generations relative to the solid arrow which indicates the most advantageous migration direction for the autumn generation. Selection should thus favour any lineage that shows a peak migration direction parallel to these arrows in the different generations at the different times of year.

The main part of the diagram shows the history of a lineage that starts at A and shows a direction ratio of 100:0:0:0 in the most advantageous direction. The large rectangle represents the geographical range of the species. Within a very few generations the lineage disappears over the edge of its geographical range.

Selection should favour a direction ratio that is the optimum compromise between 100:0:0:0 and 25:25:25:25.

duces offspring, and then dies out. Some time later the offspring also perform a removal migration. Often the most advantageous migration direction is in one direction during the first generation and in another direction, sometimes differing by nearly 180°, during the second generation. Usually, however, one generation migrates much further than the other, and often the two generations do not migrate in precisely opposite directions. If the direction ratio were 100:0:0:0 with respect to the most advantageous direction, therefore, the population would eventually disappear over the edge of the range (Fig. 12.1). On the other hand, a direction ratio of 25:25:25:25 would take no advantage of the optimum direction. The result seems likely to be selection for a direction ratio that is the best compromise between 100:0:0:0 and 25:25:25:25.

A computer model can be produced that compares the reproductive success of lineages with different direction ratios when in competition with lineages with other direction ratios. The method of computation has been described in detail elsewhere (R.R. Baker 1968d, 1969a) and need not be reproduced here. It should be noted, however, that only very few computational methodologies generate optimum direction ratios other than 25:25:25:25. As these methods successfully reproduced observed direction ratios in certain British butterflies, they were adopted in preference to those that did not, especially as the methods that were successful seemed to be biologically correct (Baker 1968d), an assertion that has, however, recently been questioned, though not dismissed (Gilbert and Singer 1975).

Figure 12.2 shows the computed effect on the optimum direction ratio of varying certain parameters. The chief conclusions are as follows. Unless the most advantageous direction is such that successive generations show a high degree of return, the optimum direction ratio is 25:25:25:25. The higher the degree of return and the steeper the environmental gradient of mean expectation of migration, the further the optimum direction ratio departs from 25:25:25:25. Only when the offspring return precisely to the place of origin of their parents can the optimum direction ratio be 100:0:0:0. In such a situation, however, the arguments of the preceding section would also take effect.

At first sight it appears that a peak migration direction, even if intermediate between 100:0:0:0 and 25:25:25:25, when not associated with a precise return migration, must eventually lead to an accumulation of the population along one edge of the range. Calculations have shown (Baker 1969a), however, that this need not occur. Any distribution of deme density can be maintained through an animal's geographical range by factors such as differential predation and parasitism, etc., without seriously altering the calculated optimum direction ratios shown in Fig. 12.2. Paradoxical though it may seem, a peak migration direction can persist within a population both in the absence of a precise return

migration and in the absence of an unusual spatial distribution of the species.

The action of selection on direction ratio seems to be to favour individuals that produce offspring that manifest the optimum direction ratio (Baker 1969a). The advantage gained by an individual does not, therefore, derive from its own migration direction, as long as it fits the optimum direction ratio for the deme (see p. 88), but from the direction ratio of its offspring.

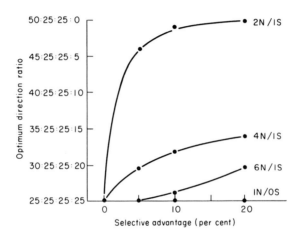

Fig. 12.2 Variation in optimum direction ratio

Selective advantage refers to the advantage of migrating in the most advantageous direction from emergence to death. It also refers to the disadvantage of migrating the same distance in the opposite direction. This advantage and disadvantage results from environmental gradients of $\bar{E}$, the mean expectation of migration.

2N/1S indicates that the most advantageous migration direction is in direction N for the first two generations but in the opposite direction S for the third generation. The calculation assumes that the diameter of the geographical range is 21 times the distance between the place of

emergence of an individual and the place of emergence of its average offspring.

In order to facilitate interpretation of variation in optimum direction ratio, comparison was artificially restricted during calculation to direction ratios the middle two figures of which were —:25:25:—.

The less steep the gradient of $\bar{E}$ and the more generations that migrate in direction N per generation that migrates in direction S, then the nearer does the optimum direction ratio approach 25:25:25:25.

[*Simplified from Baker (1968d, 1969a) to which reference should be made for full details of the method of calculation*]

13

General discussion

Part II of this book has been concerned with the development of the foundations of a migration model. Chapter 7 developed a migration equation that embodied the thesis of this book and identified the situation in which it would be advantageous for a given individual to initiate migration. Chapter 8 demonstrated the way that the selection associated with the advantage of migration would act on an animal to produce migration thresholds to habitat, migration cost, and indirect variables. These migration thresholds not only exist at the moment of initiation of migration but are part of an ordered sequence of primarily physiological events, fixed by selection, that are triggered, often long in advance of the initiation of the actual migration, sometimes by habitat variables, sometimes by indirect variables, and sometimes by an endogenous mechanism that is independent of environmental events. Migration cost thresholds appear at the approach of the optimum time for the initiation of the migration itself. Chapter 9 discussed the interrelationships of thresholds for migrations that fulfilled different functions (exploratory, removal, return) and Chapter 10 discussed the way that these thresholds differed between species and between different age and sex groups within a species.

Having considered many of the major features of the initiation of migration, Chapters 11 and 12 went on to consider the way that the behaviour of an animal during migration could be shaped by selection to produce a strategy that was optimum for the given situation. These chapters were concerned primarily with the rate, distance, and direction of migration.

Chapters 7–12 by no means contain all of the theoretical discussion in this book because in many cases it is easier to discuss particular aspects of the model in relation to specific examples. The majority of the points so far made are evaluated, along with further developments, in Part III by a process of example and analysis. In general, observed characteristics of migration can be readily accommodated within this theoretical framework.

Most of the points made in Chapters 7–12 are discussed further in later chapters in relation to specific examples, and need not be discussed further here. There is one point, however, that is perhaps sufficiently paradoxical to merit further discussion before proceeding to test its applicability to field observations. It has been argued (pp. 69, 84, and 85) that in a free situation a deme of animals shows characteristic frequency distributions of migration thresholds, migration distances, and migration directions. These frequency distributions are evolutionarily stable only when they are such that the deme distributes itself through space and time in such a way that all individuals experience the same potential reproductive success (insofar as reproductive success is a function of position in space and time). If this is so it means that an individual does not gain an advantage *relative to other individuals* as a result of its actions. The only advantage gained is in the sense that if the individual did not behave in the way dictated by its migration threshold, such that it migrates a certain distance in a certain direction, then it would *suffer a disadvantage* (as well as imparting a disadvantage to some members of the deme and an advantage to others). Only by fitting into the stable pattern for the deme as a whole can an individual achieve maximum reproductive success. This point is important on a number of occasions in the chapters that follow.

Part III The evaluation of the model: examples and analysis

Part IIIa   Removal|migration

14
The calculated initiation of migration

14.1 Introduction

A calculated migration is a migration to a specific destination that is known to the animal at the time of initiation of migration, either through direct perception, previous acquaintance, or social communication (p. 25). It is advantageous whenever h_1, the suitability of the habitat occupied, becomes less than $h_2 M$ (where h_2 is the suitability of the destination and M is the migration factor for migration from H_1 to that destination). The action of selection on calculated migration has been argued in Chapter 8 to favour those individuals that use that combination of criteria that gives the best indication of $h_2 M$. Or, to use statistical description, selection should favour those individuals that weight each environmental variable being used as a criterion for assessment in such a way as to give the maximum coefficient of determination of $h_2 M$. Selection should also favour those individuals that make the most accurate comparison of $h_2 M$ with h_1 such that migration is only initiated when h_1 becomes less than $h_2 M$.

This chapter presents a number of examples to illustrate the three major mechanisms (direct perception, previous acquaintance, social communication) by which an animal can possess or obtain information concerning h_1, h_2, and M before it initiates calculated removal migration from H_1. Particular attention is given, wherever possible, to the criteria for assessment adopted by the animals concerned.

14.2 Sensory comparison of habitat suitability while still in H_1

14.2.1 General comments

An animal that requires, say, food, as it depletes the food resource in H_1, might be expected to scan the surroundings with the appropriate sense(s) for other feeding sites. If an alternative feeding site is located by such scanning, there is a possibility that the suitability h_2 of this new feeding site, H_2, could be assessed while still in H_1. In a similar way, a male searching for a female in H_1 might be expected to alternate search in its immediate surroundings with scanning at a distance. If a female is located (at H_2) at a distance, the male then has the opportunity to assess whether the chances (h_1) of finding and successfully courting a female nearby (in H_1) during the time taken to move to the distant locality (H_2) and successfully court the female already sensed, renders $h_1 < h_2$. In both cases, not only have h_1 and h_2 to be assessed on the basis of such direct criteria as food or mates already sensed, but also the area in between must be assessed before h_1 can be compared with $h_2 M$. As far as assessment of M is concerned, the distance between H_1 and H_2, the likely ease of travel, the amount of cover, etc. would all seem to be reliable and available criteria if the animal can make sensory contact with H_2 while still in H_1.

This type of migration, because it is limited by sensory range, usually involves only short distances such as that of a butterfly flying from one flower to another. Other examples of direct sensory comparison of habitat suitability that may be given are male moths assembling to a pheromone-producing female and birds that are already feeding but which come to bread thrown out onto a nearby lawn. Numerous cases may also come to mind of animals apparently making an advance assessment of migration route suitability on the basis of direct sensory contact. The reluctance of rodents to cross open ground, the reluctance of many terrestrial animals to cross water, and the apparent ability of the potto, *Perodicticus potto*, to plan in advance its tortuous route through the vines and branches of tropical forests (Charles-Dominique 1971) are examples. It seems likely that assessment of habitat and route suitability and comparison of h_1 with h_2M is a major part of the everyday movements of a wide variety of animals. Direct sensory comparison of habitat suitability and assessment of migration cost probably describes a large proportion of very short-distance animal migration.

14.2.2 The group-to-group migration of the female gorilla

It is generally accepted (Crook 1970) for primate social groups that the efficiency of protection from predators, of finding food, and of reproducing, and hence the degree of realisation of potential reproductive success of each individual within a group, is in large part a function of group composition (i.e. the numbers and relative proportions of each age and sex class within a cohesive social group). On this assumption, departure from the group composition that is optimum for the species in the particular habitat being considered results in a decrease in the potential reproductive success of each member of the group. In particular, individuals of the age and sex class that is relatively too numerous are likely to suffer the greatest disadvantage.

Very often, different social groups of a species encounter one another within their respective familiar areas which, particularly in the case of the gorilla, *Gorilla gorilla* (Schaller 1963), may often overlap (p. 385). The opportunity then presents itself for members of the two groups to compare directly the composition of their own and neighbouring groups.

Schaller (1963, pp. 105–109) gives some detailed field notes on the changes in group composition of the gorilla, *G. gorilla*, in the Kabara region, north of Lake Tanganyika. The group-to-group migration of female gorillas, unlike that of males, is performed as a calculated migration when two groups are in contact and when, therefore, the opportunity exists for the relative suitabilities of the two groups to be assessed and when the migration cost is minimal ($M \simeq 1.0$). I have analysed (Tables 14.1 and 14.2) Schaller's observations of these movements according to the proximate criteria on the basis of which the females seemed likely to attempt to assess the relative suitability of the two groups.

Table 14.1 The reactions of females of the gorilla, *Gorilla gorilla*, to social groups of differing composition (compiled from field observations by Schaller 1963)

Attractiveness of group*	Number of individuals in group†				
	SBM	BBM	F	J	I
5	1	3	6	5	4
5	1	0	3	2	2
5	4	1	10	3	6
5	5	1	10	3	7
20	4	1	10	3	9
5	4	1	12	3	10
5	3	1	12	3	10
5	2	2	3	2	2
5	4	2	3	4	2
−10	1	1	9	2	7
5	1	1	8	2	7
5	1	1	8	2	8
5	1	2	6	4	5
5	1	2	6	4	6
10	1	2	6	4	7
5	1	2	7	4	7
5	1	2	8	3	7
5	1	1	3	0	2
−30	0	1	3	0	1
30	1	0	2	0	0
20	1	1	3	0	1
5	1	1	5	0	2
5	1	1	4	0	2
−30	0	1	3	0	2
30	1	0	4	0	1
40	1	1	3	0	2

*Attractiveness
−30   3 females leave the group
−10   1 female leaves the group
  5   no females leave the group despite contact with another group
 10   1 female joins the group
 20   2 females join the group
etc.

†Age and sex classes
SBM   silverback males (>10 years)
BBM   blackbacked males (6–10 years)
F   females (>6 years)
J   juveniles (3–6 years)
I   infants (0–3 years)

Fig. 14.1   Part of a social group of the gorilla, *Gorilla gorilla*

[*Photo by G.B. Schaller, courtesy of Bruce Coleman Ltd.*]

Table 14.2 The possible weighting of criteria for habitat assessment used by the female gorilla: step-wise multiple regression analysis of the data in Table 14.1

R^2	Group characteristic*	\multicolumn{9}{c}{t-values for multiple regression coefficients†}
0·425	$SBM:J$	4·21	6·39	5·10	4·27	3·79	4·13	4·44	4·47	4·23
0·216	$(F–Y):SBM$		−3·72	−4·67	−3·48	−2·73	−1·49	−0·16	−0·03	0·37
0·071	adults:Y			2·34	1·94	0·57	0·37	1·04	1·56	1·96
0·014	$(SBM+BBM):J$				−1·02	−1·69	−2·27	−2·71	−2·97	−3·00
0·023	SBM					−1·37	−1·81	−2·24	−2·31	−0·82
0·026	BBM						1·50	1·49	1·67	1·70
0·022	$(F–I):(SBM+BBM)$							−1·40	−1·84	−2·19
0·016	$SBM:Y$								−1·19	−1·66
0·014	$F:Y$									−1·14
Total 0·828		17·75	20·55	18·18	13·92	11·98	10·98	10·16	9·27	8·56
		$F_{1,24}$	$F_{2,23}$	$F_{3,22}$	$F_{4,21}$	$F_{5,20}$	$F_{6,19}$	$F_{7,18}$	$F_{8,17}$	$F_{9,16}$

* $SBM:J$ ratio of number of silverback males to number of juveniles
$(F–Y):SBM$ ratio of number of females minus the number of young to the number of silverback males
$(F–I):(SBM+BBM)$ ratio of females minus infants to the sum of silverback males and blackback males
etc.

† t-values underlined indicate that the regression coefficient is significantly different ($P<0.05$) from zero.

Interpretation of the step-wise multiple regression analysis presented in Table 14.2 is not particularly straightforward but the group characteristics that give the highest coefficient of determination of the migration behaviour of female gorillas and that at the same time are consistent with the variation in t-values for the regression coefficients as fresh variables are added are the number of silverback males relative to the number of females without young or, if they have young, that are not infants.

Schaller records several instances of gorillas, even silverback males, being killed by leopards, *Panthera pardus*, and notes that they are also killed for food by the indigenous humans. Silverback males fulfil a protective role within the group. Quite possibly it is degree of protection for themselves and their young that is being assessed by the female gorilla. This would seem to be the only criterion that could generate the multiple regression characteristics given in Table 14.2. The more juveniles and perhaps also blackbacked males there are per silverback male, the less efficient the protection received by each female and her offspring seems likely to be. It also seems possible that females without young and females with juveniles or blackbacked male offspring are likely soon to be entering into oestrus and to elicit above-average attention from the mature males, even though such attention may not be as conspicuous as in many other primates.

Whether or not protection for the female and her offspring is the major factor involved, the high ($R^2 = 0.828$) coefficient of determination for the multiple regression indicates that the major part of the observed features of calculated migration of female gorillas can be attributed to the relative compositions of the social group of which the female is a member and any other social group with which they come into contact. Group-to-group migration is most likely to occur when the female is a member of a group containing relatively few silverback males per individual of certain age and sex classes (i.e. post-weaning immatures, particularly juveniles, and adult females without offspring) and that group comes into contact with another group in which there are relatively many silverback males per individual of these age and sex classes.

14.2.3 The dung-to-dung migration of male dung flies

The yellow dung fly, *Scatophaga stercoraria*, of Europe, Asia and Africa, oviposits on fresh dung, particularly that of the larger ungulates. Males arrive at the dung within a few minutes of dung deposition and may attain quite high densities. All females, whether virgin or not, as they arrive at the dung to oviposit, are quickly seized by a male and copulation takes place. Both males and females apparently locate the dung by olfaction followed by flight up-wind in a series of long 'hops', each one followed by re-orientation (Parker 1970b,c). Information concerning the age of a dung pat can presumably be derived by olfaction from the chemical composition of the gases given off by the dung. Only fresh dung attracts females and hence only fresh dung offers a highly suitable habitat to male dung flies (but see Chapter 17). As the dung pat (H_1), on which the male is searching, ages, there is therefore an advantage in scanning the environment for newly-deposited dung (H_2). This advantage is presumed to have resulted in selection on the male to spend a proportion of its search-time in the surrounding grass a few centimetres up-wind of H_1 in which position the male is presumably not confused by the chemicals being given off by H_1. It is not known, however, to what extent migration from H_1 to H_2 occurs only in the presence of olfactory information concerning H_2 (Parker, *pers. comm.*).

Fig. 14.2   The yellow dung fly, *Scatophaga stercoraria*

[*Photo by G.A. Parker*]

14.2.4 The dry season pasture-to-pasture migration of wildebeest within their familiar area

Many of the ungulates of the great Serengeti-Mara Plains of Africa perform relatively long-distance migrations from grazing area to grazing area. In particular, gazelles, *Gazella* spp., zebra, *Equus quagga*, wildebeest, *Connochaetus taurinus*, and to a lesser extent eland, *Taurotragus oryx*, may move hundreds of kilometres from dry season to wet season grazing areas (Grzimek and Grzimek 1960, Talbot and Talbot 1963).

The wildebeest, *Connochaetus taurinus*, is a bovid artiodactyl that is adapted to feed on grass less than about 10 cm high, in particular freshly sprouting grass (Talbot and Talbot 1963). Wildebeest drink every day when water is readily available and in any case do not go more than about five days without water. During the dry season the great herds may move more than 80 km/day if food and water are widely separated. At this time of year, and particularly near the beginning of the wet season, the distribution of rain storms, with consequent stimulation of fresh grass growth, is of primary importance to the species. It is reported by Talbot and Talbot that at this time wildebeest react to the sound of thunder up to 25 km distant by movement in the direction of the storm. Migration is also initiated toward the characteristic dark cloud-capped storm columns that may be visible for distances of up to 80 km. Over half of the movements observed were across or down-wind. It appears, therefore, that sight and hearing are the senses primarily concerned with the detection of suitable distant habitats and that if olfaction is involved it is only used as an additional sense on those occasions that rain falls up-wind of the wildebeest. More factors are involved, however, than just the automatic initiation of migration from poor pasture toward distant rain. The familiar area (p. 27) of some herds, though not necessarily of every individual within those herds, on the Serengeti-Mara Plains may consist of many thousands of square kilometres (Talbot and Talbot 1963). A herd near the edge of its familiar area seems to have the ability to judge whether a distant storm is producing rain within the boundaries of the familiar area or outside. Only storms producing rain within the familiar area lead to the initiation of migration by the herd.

Fig. 14.3   Migrating wildebeest, *Connochaetus taurinus*

[*Photo by Norman Myers, courtesy of Bruce Coleman Ltd.*]

14.3 Calculated migrations within a memorised familiar area

14.3.1 General comments

Most workers on animals that make use of a familiar area, whether the animals are reptiles (e.g. Tinkle 1967b) or mammals (e.g. Burt 1940), have been impressed by the importance of the familiar area to the animal in terms of escape from predators. No matter where the animal may be disturbed within its familiar area, it usually gains direct access to shelter, without search, and usually by way of the shortest route. Many mammals, of course, such as the black-tailed prairie dog, *Cynomys ludovicianus*, a North American sciurid rodent (Fig. 18.26) have a number of burrows scattered through their familiar area and it is to the nearest of these that the animal retreats, in a more or less straight line, as soon as danger threatens, even though the burrow may not, at least at first, be visible. Of course, the route followed need not necessarily be navigated entirely by memory. Many animals with a familiar area, including the

Fig. 14.4   A coyote at the entrance to a den in a rock cavity   [*Photo by Pat Kirkpatrick, courtesy of Frank W. Lane*]

prairie dog, have a system of trails within their familiar area that have been produced either by continued use, as in the caribou, *Rangifer tarandus* (Banfield 1954), by marking, as in the mountain lion, *Felis concolor*, of America (Musgrave 1926), or by some degree of construction engineering, as in termites (Imms 1964) and man.

Trails, as well as memory, may also be involved in the second major function for calculated migration within a familiar area, migration from food supply to food supply. In many cases this involves movement between two places previously encountered by the animal and which offer different food resources. In some cases, however, especially among rodents but also among birds, the destination may be some habitat in which the animal has previously cached a supply of food, the position of which has been marked and/or memorised. In yet other cases, such as some species of ant and some groups of humans, the destination may be a habitat that is used for the purposes of agriculture or animal husbandry.

Calculated migrations within a familiar area, either as part of an optimum strategy of food acquisition or in the form of escape from predators, are so common, particularly among vertebrates, that I do not propose to single out a specific example, preferring instead to concentrate on some of the more complex examples of similar type such as those that are described in the remainder of this chapter.

14.3.2 The den-to-den migration of coyote families

The coyote, *Canis latrans*, a North American canid carnivore, frequents brush country and the edges of woods and open farmlands and feeds largely on rabbits, rats and mice (Schwartz and Schwartz 1959) and also, particularly in winter, on carrion (Murie 1940). A coyote family consists of a male, female and usually about 5 to 7 young of the year (Schwartz and Schwartz 1959). The young are born in May into a den which is usually located in unused fields, often close to timber, and which may be in a bank, under a hollow tree or log or in a rock cavity. Dens are only used from May through August during the early development of the young. For the rest of the year all coyotes merely sleep on the ground in some concealed and protected spot. Several dens are maintained by the parents within their familiar area but usually only one is used in any one year. Occasionally, however, the whole family

migrates from one den to another.

The criteria by which h_1 is compared with $h_2 M$ in this case are unknown. In some cases, however, the original den, H_1, is reported as harbouring a heavy infestation of ectoparasites (Schwartz and Schwartz 1959).

14.3.3 Thermoregulation and hibernation and the winter migrations of bats

Many species of temperate bats, such as the little brown myotis, *Myotis lucifugus*, Keen's myotis, *M. keenii*, the Indiana myotis, *M. sodalis*, and the eastern pipistrelle, *Pipistrellus subflavus*, in New England, United States (Folk 1940), the big brown bat, *Eptesicus fuscus*, in Indiana, United States (Mumford 1958), and the greater horseshoe bat, *Rhinolophus ferrumequinum*, in western Europe (Ransome 1968), show considerable movement within and between hibernation roosts during the winter months. However, before the significance of these movements can be discussed in relation to calculated migration within a familiar area, it is necessary to digress for a moment and consider in general terms some of the pressures to which hibernating bats are subjected.

Bats can be active over a wide range of ambient temperatures. For example, O'Farrell and Bradley (1970) record that the western pipistrelle, *Pipistrellus hesperus*, the California myotis, *Myotis californicus*, and the desert pallid bat, *Antrozous pallidus*, were active and feeding around a desert spring in South Nevada throughout the year and over a temperature range of from −8°C to 31°C.

The body temperature of active bats of all families lies within the range of $37 \pm 2°C$ (Lyman 1970). The larger, tropical Megachiroptera and the larger tropical Microchiroptera seem to be truly homeothermic and maintain their body temperature within this range throughout adult life. At low ambient temperatures the wings may be used as a blanket and an increase in metabolism and shivering serve as homeostatic mechanisms. At high temperatures the wings, ears and scrotum can be vasodilated and the bats may also pant. Some of the smaller tropical and subtropical megachiropteran bats are not truly homeothermic and at times allow a reduction in body temperature (Lyman 1970). However, these types seem unable to arouse themselves (i.e. raise their body temperature to the active level by shivering) from body temperatures of less

Fig. 14.5 A hibernating cluster of the Indiana myotis, *Myotis sodalis*, on the roof of a cave. Note the single individual awake near the bottom of the picture [*Photo by Roger W. Barbour*]

than about 16°C without a corresponding rise in ambient temperature. All smaller bats, when not active, have body temperatures that follow ambient temperature at least down to about 5°C.

Only three families of bats have colonised temperate regions. Of these only the vespertilionid and horseshoe bats (Vespertilionidae and Rhinolophidae) have produced species that overwinter by hibernation. The characteristics of bat hibernation, such as choice of hibernation site and the amount of winter activity and movement, can be viewed as representing the optimum strategy to overcome the problems of freezing, starvation and dehydration. The problems presented by all three of these potential mortality factors are overcome to a large extent by a suitable choice of hibernation site. As far as cave-roosting species are concerned it seems that one of the prime requirements for a cave to be chosen as a hibernation roost is that it contains water from which the bats may drink during periods of arousal (W.H. Davis 1970). The majority of temperate species hibernate in caves and those human artefacts, such as mines, wells, cellars, burial vaults and buildings, that share certain characteristics with caves. A few species, such as the tree-bats, *Lasiurus* spp., of North America and the noctule, *Nyctalus noctula*, and serotine, *Eptesicus serotinus*, of northwest Europe, regularly hibernate in hollow trees or under loose bark. Certain age and sex groups of other species may also hibernate in trees.

In Eurasia, and probably elsewhere, there seems to be some correlation between climate and the preferred degree of shelter in the hibernation site (Strelkov 1972). In colder climates the hibernation sites tend to be more sheltered, often more subterranean, than in milder climates. Despite the use of sheltered hibernation sites, however, bats do not always escape being subjected to freezing temperatures, and at such times other behavioural and physiological adaptations are involved in winter survival. Clusters of the noctule, *Nyctalus noctula*, in hollow trees can survive temperatures outside the tree of at least −14°C (Sluiter and Heerdt 1966) and hibernating bats as a whole may be exposed to a temperature range of from −17°C to >21°C, though those that hibernate in caves are subjected to much less temperature fluctuation (Davis 1970).

The length of time an individual is subjected to freezing temperatures depends partly on the rate at which body temperature decreases per unit decrease of ambient temperature and this rate in turn is likely to depend partly on the amount of clustering of the bats (Twente 1955) and partly on body size. In males of the little brown myotis, *Myotis lucifugus*, the rate at which body temperature drops is in part a function of body size such that larger males lose heat less quickly than smaller males (Stones and Wiebers 1966). Nearly all of the species that hibernate in trees or solitarily in caves are relatively large. Interaction of body size and degree of clustering can be seen by a consideration of the different temperature regimes selected for hibernation by the different species of bats within a given cave system. J.S. Hall (1962) has described the cluster formation and microclimates selected by different species of North American bats. For example, the Indiana myotis, *Myotis sodalis*, forms single-tiered, tight, compact clusters and selects a part of the cave that has a steady minimum temperature of from 4° to 6°C. The slightly larger cave myotis, *M. velifer*, forms less compact clusters and selects areas of the cave that have markedly fluctuating temperatures. The big brown bat, *Eptesicus fuscus*, occurs singly at temperatures of −2° to 5°C. Finally, the small eastern pipistrelle, *Pipistrellus subflavus*, occupies warm regions of the cave where the temperature is a constant 12° to 13°C.

Apart from behavioural adaptations to avoid freezing such as choice of hibernation site and degree of clustering, both of which may be a function of body size, bats show some other behavioural and also physiological adaptations to survive low temperatures. Some species can supercool to −5°C but cannot arouse themselves from this condition without a rise in the ambient temperature (Davis 1970). More frequently, however, hibernating bats respond to low temperatures either by arousing themselves and moving to a more suitable shelter, or by increasing their body metabolism sufficiently to maintain body temperature above 5°C. Frequent exposure to low temperatures, therefore, results in an increased energy expenditure. High temperatures during hibernation also lead to more rapid utilisation of body reserves. The little brown myotis, *Myotis lucifugus*, has five times the metabolic rate at 20°C than it has at 2°C (Hock 1951). At 10°C the metabolic rate is double that at 2°C. Below 2°C metabolic rate rises once more as a result of the thermoregulatory process. Hibernating bats live off their fat reserves and gradually lose weight throughout the winter (e.g. Hall 1962). This loss of fat reserves is critical and in some winters many individuals die of starvation (Ransome 1968). Young individuals in their first winter are particularly prone to this mortality due to a low reserve level at the onset of winter. There appears to be stringent selection, therefore, on young bats in temperate regions to accumulate as much reserve as possible in the short period between midsummer birth and the onset of hibernation. This selection also operates on adults, though perhaps to a lesser extent as a longer feeding period is available.

Let us now return to a consideration of winter movement between hibernation sites by summarising the relevant major selective pressures. Hibernating bats suffer increased energy expenditure which leads to increased depletion of body reserves which in turn leads to increased possibility of starvation in two quite different situations: (1) when ambient temperatures are near or below freezing, in which case increased metabolism is necessary to maintain the body temperature above a lethal level; and (2) when ambient temperatures are high, in which case there is increased energy expenditure because of the effect of temperature on metabolic rate. However, the strategies available to the bats in these two situations are quite different. Energy expended in the avoidance of freezing by increased metabolism or by arousal and movement cannot be offset by any energy gain. Increased energy expenditure due to temperatures in excess of about 10°C, however, can be offset to some extent because a few insects may be flying and some feeding is possible. Of course, intermittent feeding during winter raises problems of digestion and excretion and species that feed during hibernation need to arouse themselves occasionally in order to excrete and need a reasonably elevated temperature for digestion to proceed. Sometimes the excretion and roosting sites are separate, though usually within the same cave system, and the bats fly between the two areas. Movement between defaecation areas, which can be recognised by the pile of guano that accumulates, and the roosting area within the same cave system has been described for the Indiana myotis, *Myotis sodalis* (Hall 1962). The energy expended on these excretory movements, however, is unlikely to exceed the energy gained as a result of feeding.

The possibility of feeding at higher temperatures and the avoidance of excess energy expenditure at lower temperatures has led to complex winter strategies of movement in some species. For instance, it has been shown for the greater horseshoe bat, *Rhinolophus ferrumequinum*, in southwest England (Ransome 1968) that the preferred temperature for roosting changes throughout the winter. In October

and November, when insects are more abundant than at any other time during the winter, the bats select caves or parts of caves with temperatures of about 11°C. In February and during periods when no insects are flying, the bats select sites with temperatures of about 7°C. The decision as to whether to return, after foraging, to a high or low temperature site seems to depend on the conditions prevailing during foraging. At any time during the winter, if the temperature of the night is above 10°C, which is about the threshold for insect flight, the bats leave their roost and feed. The overwintering strategy of this species, therefore, involves a considerable amount of movement, either to take advantage of conditions favourable for feeding or to move from cave to cave in order to maintain the temperature regime that is optimum for that particular stage of winter. It has not been proved, but it seems likely that, although the distance between different roosts may be as great as 30 km, these winter movements are taking place within a familiar area and that the different characteristics of the different caves within the area are known to the bats.

14.4 Extension of the familiar area by exploratory migration

14.4.1 Two types of removal migration by the muskrat

The muskrat, *Ondatra zibethicus*, a microtine cricetid rodent of North American origin but which has also been introduced into Europe, has a body length of 23–36 cm and lives in marshes, streams, rivers, ponds and lakes, particularly still or slowly running water with vegetation in the water or along the shore (Schwartz and Schwartz 1959). The species feeds mainly on roots, stems, seeds and leaves but also on some animal food such as clams, snails, crayfish, fish and frogs. The roosting site is usually dug into a bank. Being adapted to such a moist habitat, muskrats are particularly susceptible to drought if it is sufficiently severe for the water to become too low or to dry up. On such occasions the muskrats move first to the limits of their familiar area, about 200 m distant, where a temporary roost is established. The familiar area is therefore enlarged before any further

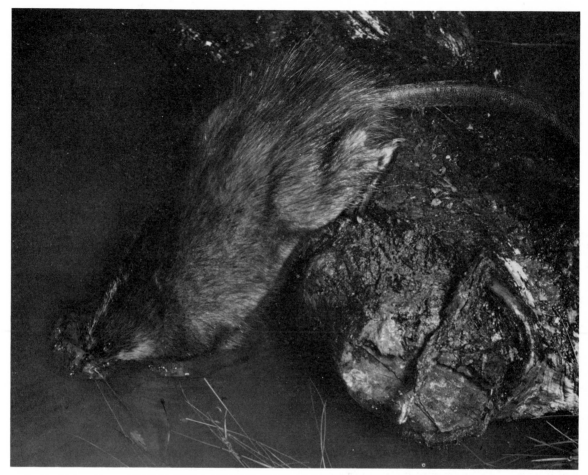

Fig. 14.6   The muskrat, *Ondatra zibethicus*, diving off a log   *[Photo by Leonard Lee Rue III, courtesy of Frank W. Lane]*

removal migration takes place. These temporary roosts may be established in the burrows of other species in cornfields and other 'terrestrial' localities. There is a return migration to water, if not to the original roost, when water is again available. How often extension of the familiar area is repeated during periods of drought is unknown but perhaps it occurs only once. Once an animal moves out of its familiar area it may cover a distance of up to 35 km. In the absence of drought a muskrat may remain in its home range throughout a year.

Young newly independent muskrats show similar familiar area extension as a preliminary to their first removal migration (Schwartz and Schwartz 1959). Weaning occurs at between 3 and 4 weeks of age. The young may then either remain in their natal roost or move 10–60 m away and establish a new roost. As in the response to drought, this initial migration probably takes the young muskrat to the edge of its existing familiar area, which is then extended before further removal migration takes place.

14.4.2 Establishment of a new distant territory by adult prairie dogs

The black-tailed prairie dog, *Cynomys ludovicianus*, is a burrowing North American sciurid rodent that lives in large 'prairie dog towns' which occupy an area of from 3 to 30 ha (J.A. King 1955). Within this area the town is subdivided by topographical features such as ridges, trees etc. into wards. A ward in turn consists of a number of territories (termed coteries by King) each with an area of about 2800 m² and each of which is occupied by a territorial unit of prairie dogs. This territorial unit consists of 1 or 2 ($\bar{x} = 1 \cdot 65$) adult males, 2 or 3 ($\bar{x} = 2 \cdot 45$) adult females, and 0–39 young. The species feeds on various parts, including the seeds, of grass and various dicotyledonous plants and practises the cutting of non-foodplants, thus encouraging the growth of foodplants and maintaining good visibility. There is no storage of food.

Removal migration of adult prairie dogs from their territory to a new territory occurs predominantly in June and July when the young of the year become independent. When the new territory is in a previously uninhabited area the adults move backward and forward between the old territory and the new for up to a month before removal migration takes place. Details of the process of the extension of the familiar area, modification of the new area and subsequent removal migration are given in Fig. 14.7.

In *C. ludovicianus* it is the adults that leave the old territory while the newly independent young remain, many of them to perform removal migration themselves a year later when about 15 months old and while still sexually immature. First copulation occurs when about 23 months old.

Removal migration of adults from their old territory to a new but already occupied territory again involves extension of the familiar area, this time into the occupied territory, as a preliminary to removal migration. In this case exploratory migration is followed by an attempt by the individual to establish itself in the new territory in the face of territorial interaction with the inhabitants. Usually this involves an initial occupation of a particular burrow in the new territory and an attempt to make use of the immediate surroundings. Successful establishment followed by gradual acceptance by the residents completes the removal migration. In the case of a male, successful removal migration may involve displacement of the male already in the territory.

The criteria by which h_1 is compared to $h_2 M$ are probably numerous. In the case of removal migration to an uninhabited area as in Fig. 14.7, food availability per individual is probably the criterion that receives the greatest weight in the comparison

The inset (bottom right) shows the seasonal behaviour of the species. Dotted line, copulation period; solid line, parturition period; dashed line, lactation period; waved line, period for which diagram is relevant.

The area delimited by a double line (bottom left), was uninhabited for the two years previous to the year considered. The four areas delimited by a single line represent the four nearest territories (one of which divided temporarily into two) at the original edge of the prairiedog 'town'. Inside each of these areas the number of adults that comprise the territorial unit are indicated along with the number of young that are produced during the spring of the year under consideration.

In early spring individuals of all age and sex groups visited the uninhabited area to feed but returned to their territories to roost or when danger threatened. Later in the spring and throughout May and June, burrows were dug and enlarged in the new area. These individuals were not from the nearest territories and could have been from any distance. There was a constant turnover of individuals. In June, adults from one of the four nearest territories started to dig burrows in the area and to travel back and forth between the new area and their old territory in which the young of the year were still permanently underground. By late July, turnover of individuals ceased, and the adults from the near territory spent more and more time in the new area, thus leaving their old territory entirely to the 31 young that had been produced that spring. Apart from these adults, the other migrants were yearling (one year old—pre-reproductive) individuals that came from other parts of the adjacent 'town' and some of which had probably crossed unsuitable areas of 100 m or so. Through July and August the males gradually restricted themselves to certain areas indicated by dashed lines and which therefore became incipient territories. The majority of the females, however, continued to move throughout the area, associating with all of the now-territorial males. By the end of August there was some indication of the beginning of the establishment of territorial units.

[*Based on diagrams, tables, and descriptions in J.A. King (1955)*]

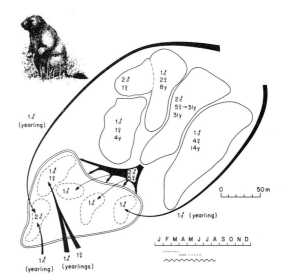

Fig. 14.7 The removal migration of a territorial mammal: the black-tailed prairie dog, *Cynomys ludovicianus*, a North American rodent, in the Black Hills of South Dakota

of h_1 and h_2. It may well be significant that in Fig. 14.7 all seven adults left the territory that contained 38 individuals whereas none of the adults left the territories that contained 6–19 individuals. Migration cost may well be assessed on the basis of distance between old and new territories, particularly in relation to the availability of burrows for shelter from predators. This is likely to be particularly important during the period of modification of the new area as many journeys are made during this time between old and new territories. In the later stages of the migration, the final destination of the males is likely to depend on topography and dietary suitability and on performance in territorial interaction with other males. Once males have established incipient territories, the final stages of assessment by females may perhaps involve some compromise between mate assessment and territory assessment.

The criteria involved in removal migration to a territory that is already occupied may involve group composition as in gorillas as well as topographic and dietary suitability of the territory and the chances of success in territorial interaction.

14.4.3 Calculated removal migration by the rabbit

The European wild rabbit, *Oryctolagus cuniculus*, was introduced by man into Britain, probably during the eleventh century, and more recently has been introduced into other parts of the world, notably to Australia and New Zealand (H.N. Southern 1964). The species forms complex burrow systems, lives in small social groups, and as an adult occupies a well-defined home range (Myers and Poole 1959).

The vast majority of movements take place within the home range but from time to time exploratory migrations beyond the limits of the home range are performed (Myers and Poole 1961). Exploratory migrations are carried out quickly, at a steady lope, with frequent stops for investigation of the new area. Each stop is accompanied by smelling and intense observation, often taking the form of standing against objects and craning in all directions. Such familiar area extensions usually commence late in the evening after the main period of feeding and are usually brief, lasting only 15–20 minutes or so, after which the rabbit concerned returns to its home range. Removal migration resulting in shifts of home range are invariably preceded by exploratory

migration and thus take place within the familiar area (Myers and Poole 1961).

Working in Australia, Myers and Poole observed groups of rabbits in enclosures about 0·8 ha in area. Within these enclosures, exploratory migration was mostly limited to the summer, in the non-breeding season, when food was inadequate either in quantity or quality. During most of the experimental period, adjacent enclosures were separate. Occasionally, however, access from one enclosure to another was made possible. Whenever this happened, removal migration was one-way, from the more-dense to the less-dense deme. On one occasion, 17 males and 17 females migrated from a deme of 115 individuals to a deme of 52 individuals resulting in demes of 81 and 86 individuals. Of these 34 migrants only two of the males and two of the females were adults. On another occasion six rabbits, four females and two males, migrated from a deme of 53 individuals to a deme of 39 individuals resulting in demes of 47 and 45 individuals.

Fig. 14.8   The rabbit, *Oryctolagus cuniculus*

[Photo by François Merlet, courtesy of Frank W. Lane]

In conclusion, therefore, rabbits live predominantly within a home range but from time to time brief exploratory migrations are performed. Exploratory migrations increase in frequency as nett rate of energy uptake per individual decreases. Removal migration occurs invariably to a destination within the familiar area as established by exploratory migrations. The correlation between migration direction and deme density suggests that deme density, food abundance, and/or nett rate of energy uptake per individual represent the criteria on the basis of which habitat suitability may be assessed.

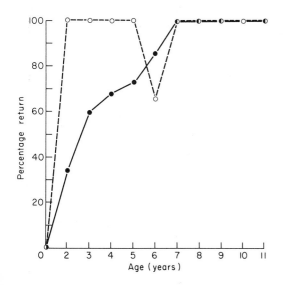

Fig. 14.9 Ontogenetic change in degree of return in males and females of the northern fur seal, *Callorhinus ursinus*

Percentage return refers to the percentage of individuals, born on the Pribilof Islands, that, when they return to the Pribilof Islands, return to the shore on which they were born. Open circles, females; solid circles, males.

Females are sexually mature at about three years and may produce their first young when four years old, though five or even older is more usual. Males are sexually mature at five or six years, begin to copulate at about eight years, but are not successful at obtaining a territory and harem until about twelve years. Individuals born on the Pribilof Islands that return to the Pribilof Islands to breed, return to the shore on which they were born.

[*Graph drawn from data in Kenyon and Wilke (1953). Photo courtesy of the National Marine Fisheries Service of the United States*]

14.4.4 The colony-to-colony migration of pinnipeds

The conspicuous and impressive feature of pinniped (i.e. seals, sea lions, fur seals and walruses) migration is the high degree of return. Individuals return to the same spot on the same shore that they occupied previously often after round journeys of 6000 km or more and often, as in the young grey seal, *Halichoerus grypus*, after wandering at sea for two years. Of the northern fur seals, *Callorhinus ursinus*, that were born on the Pribilof Islands and that return to these islands to breed, over 90% return two or more years

later to the breeding shore on which they were marked soon after birth (Fig. 14.9).

Impressive though these observations are, however, they should not be considered in isolation. To conclude that all pinnipeds return to the place of their birth is to neglect a most important aspect of pinniped migration: removal migration. The above observations of *C. ursinus*, for instance, have to be reconciled with the results reported by Chugunkov (1966) that over 3000 individuals in five years transferred from the Pribilof Islands to the Commander Islands (Fig. 14.10a). Although it may be true, therefore, that almost all individuals born on

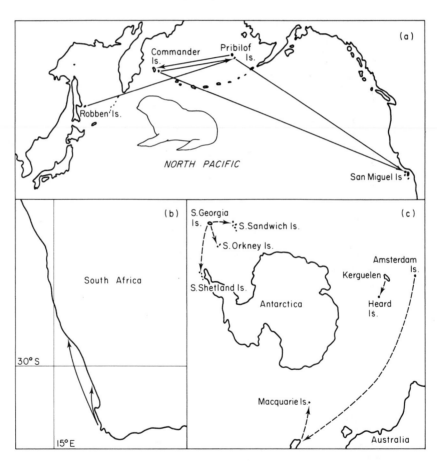

Fig. 14.10 Some known and suspected removal migrations of fur seals (Arctocephalinae)

Solid arrows, removal migrations as determined by marking–release/recapture experiments; dashed arrows, removal migrations suspected from observed changes in distribution, especially breeding distribution. All arrows join known or suspected place of origin with eventual

destination and are not intended to represent the track taken.

(a) Northern fur seal, *Callorhinus ursinus*.
(b) South African fur seal, *Arctocephalus pusillus*.
(c) Kerguelen fur seal, *Arctocephalus tropicalis*.

[*Compiled from Rand (1959), Csordas (1962), J.E. King (1964), Chugunkov (1966), Peterson et al. (1968) and Marine Mammal Biological Laboratory (1970)*]

the Pribilof Islands that subsequently start to breed on the Pribilof Islands do so on the breeding shore of their birth, it is not true that all individuals born on the Pribilof Islands return to breed on the Pribilof Islands. A significant proportion show long-distance removal migration.

The existence of removal migration is evident not only from marking–release/recapture experiments but also from the extension of breeding range that many species have undergone over the past few decades. The Kerguelen fur seal, *Arctocephalus tropicalis*, grey seal, *Halichoerus grypus*, southern elephant seal, *Mirounga leonina*, and northern elephant seal, *M. angustirostris*, for example, have recently started to breed on islands that for some time previously had not supported breeding colonies (Figs. 14.10 and 14.11).

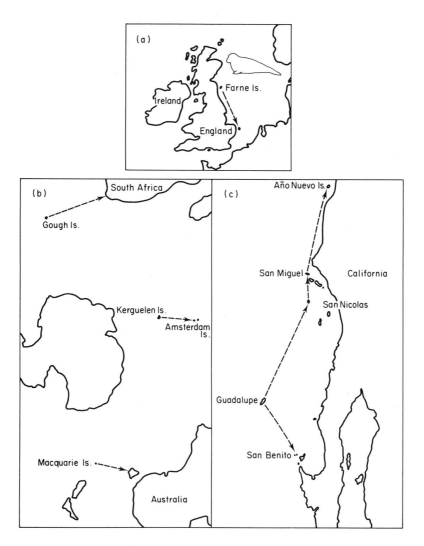

Fig. 14.11 Some suspected removal migrations of seals (Phocidae)

Dashed arrows, removal migrations suspected from observed changes in breeding distribution. Arrows join the new breeding site with the nearest previous breeding colony and are not intended to represent the track taken.

(a) Grey seal, *Halichoerus grypus*.
(b) Southern elephant seal, *Mirounga leonina*.
(c) Northern elephant seal, *M. angustirostris*.

[*Compiled from J.E. King (1964) and Radford et al. (1965)*]

The status of the initial colonising individuals unfortunately in most cases is unknown, largely because a colony has usually become established before it is first noticed. However, the observation that adults have a much greater degree of return than young pinnipeds and that young pre-reproductive individuals are much more likely than adults to haul-out on strange shores during migration suggests strongly that it is the individuals that have not yet reproduced that initially colonise new breeding sites. Of particular interest in this connection is the breeding colony of the northern fur seal, *Callorhinus ursinus*, that has recently been found on San Miguel Island, off Point Conception, California, United States (Peterson *et al.* 1968). When first described in 1968 the colony consisted of one adult male, about sixty females and about forty new-born young. Five tagged females showed that the colony consisted of individuals that had been born both on the Pribilof and Commander Islands (Fig. 14.10a). The average age of these five females in 1968 was about eight years. The first time that *C. ursinus* was seen on San Miguel Island was in 1965 and the first time that breeding appeared to have taken place was 1966. There is every indication, therefore, that many if not all of the initial female colonisers of San Miguel Island were about five years old or less, which is just about the age of first reproduction in this species.

The juveniles (i.e. young post-weaning pre-reproductive individuals) of the northern fur seal, *C. ursinus*, and of a number of other species, for example the South African fur seal, *Arctocephalus pusillus*, Kerguelen fur seal, *A. tropicalis*, and grey seal, *Halichoerus grypus*, migrate much greater distances than the adults and are nomadic throughout a much wider area. Yet at the same time it seems likely that these long-distance migrants retain the ability to return to their natal breeding site. It should be stressed, however, that this does not seem to have been proved. There seem to have been no instances of marked individuals being recaptured at great distances from their natal site and then re-captured back at their natal site. Nevertheless, as long-distance wandering seems to be a characteristic of more or less all juveniles of most of the species mentioned above, and as degree of return is so high, it seems reasonable to suppose that the ability to return is retained by all juveniles of these species of pinnipeds.

Fig. 14.12 Mother grey seal, *Halichoerus grypus*, suckling young on Farne Islands

[*Photo by C. Stephen Robbins*]

Almost immediately after weaning, many young grey seals begin a period of extensive exploratory migration which may last two years and which may take them up to 750 km from their natal site. It seems likely, however, that the ability to return to the natal site is retained throughout this period.

So, juveniles of the above and possibly other species, while retaining the ability to return to their natal colony, extend their nomadic range over a much wider area than any other age or sex group. It is also the juveniles, as in most animals (Chapter 17), that are most likely to perform removal migration from one breeding colony to another. This coincidence makes it difficult to avoid the conclusion that the increased nomadism shown by juveniles of some species of pinnipeds is in some way connected with the possibility of removal migration. The suggestion, therefore, is that juvenile pinnipeds develop a familiar area in excess of their later requirements as adults as part of a sampling process of the suitability of available feeding and breeding sites etc. Throughout this sampling process, or exploratory migration, the option to return to the natal site is retained. The option eventually taken (i.e. calculated removal

migration or return migration) presumably depends in part on the individual's threshold for removal migration (Chapter 17) and in part on the relative attributes of the natal site and the sites visited during the pre-reproductive nomadic phase.

The suggested relationship between removal migration and extension of the familiar area by juveniles can be evaluated to some extent by a comparison of the distance components of removal migration and return migration for those species for which information is available. A positive correlation between the distance components of removal and return migrations, particularly the return (= exploratory?) migration of juveniles, would support the hypothesis. Figure 14.13, which is based on Table 28.5 and Figs. 14.10 and 14.11, provides this support.

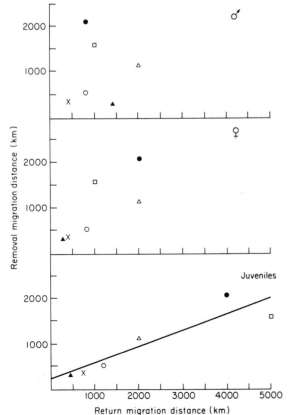

Fig. 14.13 Relationship between the distance component of removal migration and the distance component of return migration of the juveniles and adult males and females of six species of pinnipeds

Solid circle, northern fur seal, *Callorhinus ursinus*; open circle, South African fur seal, *Arctocephalus pusillus*; open square, Kerguelen fur seal, *A. tropicalis*; cross, grey seal, *Halichoerus grypus*; open triangle, southern elephant seal, *Mirounga leonina*; solid triangle, northern elephant seal, *M. angustirostris*.

Line, calculated significant ($P < 0.05$) regression line.

The distance component of return migration represents for adults the average maximum distance (straight line) away from the breeding site achieved during the longest return migration of the yearly cycle. For juveniles, figures refer to the average maximum distance from the birth site achieved during the period from birth to first reproduction back at the birth site. The distance component of removal migration is measured as the distance (straight line) between the destination and the known or suspected place of origin. In the graphs the figure represents the average distance of all known removal migrations > 100 km. Only in the case of *C. ursinus* can the figure be calculated on the basis of the proportion of individuals performing movements of different distances.

There is a positive correlation between the distance of removal and return migration for juveniles ($r = 0.913$), females ($r = 0.794$) and males ($r = 0.057$) but only for juveniles ($P < 0.02$) is the correlation significant (for females $P > 0.05$ and for males $P > 0.10$).

The available evidence is consistent with the suggestion that removal migration invariably occurs to a place visited during return or exploratory migration and is usually performed during the pre-reproductive phase of the life history. The greater correlation for adult females than for males is perhaps the result of the greater similarity between the return migrations of females and juveniles in some species than between males and juveniles.

The question now presents itself as to the criteria upon which a juvenile pinniped could base an assessment of the relative suitability of the natal and any other potential breeding site that it may encounter. Only one possible criterion, that of deme density, is considered here. It is not intended to suggest, however, that other possibilities, such as topography and climate, are not equally likely to be important in habitat assessment. As far as deme density and migration are concerned, the only species for which data have been obtained in any quantity is the northern fur seal, *Callorhinus ursinus*. Even for this species, however, data are insufficient to permit any firm conclusion. Nevertheless, presentation of the data is perhaps worthwhile, if only for its uniqueness.

C. ursinus performs its terrestrial activities, i.e. parturition, copulation and moulting, during the period from May to October (Fig. 28.15). Until 1965, these activities were restricted to three oceanic island groups in the North Pacific (Fig. 14.10a). These were Robben Island and the Commander Islands in the western North Pacific and the Pribilof Islands off the coast of Alaska. As already mentioned, however, in 1965 San Miguel Island off the coast of California became established as a breeding habitat.

Male *C. ursinus* haul out on land in advance of the females, establish territories, and later collect harems of up to 50 females (Fig. 28.15). Yearling fur seals do not arrive and haul out to moult until September when the harems have broken up and all reproductive activity for the year, except lactation, has ceased (Fig. 28.15). Sexually mature males which arrive in May but which fail to collect a harem form bachelor groups on the breeding beach. From November until May, all individuals are more or less permanently at sea and considerable segregation of age and sex classes takes place (Fig. 28.15).

Over the past few decades, American and Soviet scientists have carried out extensive marking programmes on the various breeding colonies of *C. ursinus* and a great deal of data has accumulated. The following figures have been taken from J.E. King (1964), Roppel *et al.* (1965), Chugunkov (1966), Peterson *et al.* (1968), Bureau of Commercial Fisheries (1969), North Pacific Fur Seal Commission (1969) and the Marine Mammal Biological Laboratory (1970).

In 1961, 490 000 fur seals were born on the Pribilof Islands. Of these, 71 011 died at the birth site. The following year, 245 943 returned to the Pribilof Islands. At about the same time, an average of 660 fur seals/year were known to show removal migration from the Pribilof Islands to the Commander Islands.

At about the same time, about 33 000 fur seals were born on the Commander Islands. If the survival of these up to the end of their first year is assumed to be the same as that of the Pribilof fur seals, 16 570 should subsequently return to the Commander Islands. The known yearly numbers of removal migrants from the Commander Islands to the Pribilof Islands average about 30.

The recently formed breeding colony on San Miguel Island off the coast of California consists of individuals both from the Pribilof Islands and from the Commander Islands. If the much greater distance to San Miguel Island from the Commander Islands and the effect that this may have on the chances of removal migration is ignored and if we consider only the relative sizes of the Commander and Pribilof breeding colonies, it might be expected that San Miguel Island would have been colonised by Pribilof-born fur seals and Commander-born fur seals in the ratio of about 15:1. In fact, of the five marked seals found on San Miguel Island, the Pribilof:Commander ratio was only 4:1.

The sample is too small, of course, to base any firm conclusion on the difference between the two ratios and, in any case, other factors, such as the proportion of Commander and Pribilof demes that are marked, would have to be considered before the significance of the observation could be assessed. Furthermore, it is doubtful whether the assessment by the fur seals of the relative suitability of, say, San Miguel Island and the Pribilof Islands can be analysed solely by consideration of deme size. Deme density (i.e. deme size divided by area of shore available for breeding) is only a little more relevant. In an animal such as the fur seal which forms compact groups (harems), space available per individual is unlikely to show a linear relationship with deme density as calculated above. As far as breeding suitability is concerned, therefore, an individual is unlikely to compare, say, San Miguel Island to the Pribilof Islands. If comparison is based on deme density at all it is much more likely to be based on the deployment of adults and young on a particular shore on San Miguel Island in relation to the deployment on a particular shore on the Pribilof Islands. There is, however, one respect in this case in which habitat suitability might be a function of the total number of individuals breeding on an island group rather than a function of the detailed situation on a particular shore. As Ashmole (1963) has pointed out for tropical oceanic birds, as colony size increases on an island or group of islands, beyond a certain level each increase in size results in each individual having to travel further to find sufficient food. Energy expenditure per unit energy uptake therefore increases and potential reproductive success of each individual member of the colony is likely to decrease. Assuming, that is, that increase in colony size is not accompanied by some advantage, such as increased protection from predators, which does not often seem to be the case on oceanic islands (Ashmole 1963, Lack 1966). In the case of the northern fur seal, *Callorhinus ursinus*, competition for food, mainly fish and cephalopods, may not be of primary importance to the male which does not leave land to feed for several weeks during the guarding of its harem. Competition for food may be very important to a female, however, for during the lactation period the female leaves the young on land for several days at a time while she goes to sea to feed and to replenish her mammary glands.

Only one criterion for habitat assessment, that of deme density, is being considered in this example, other possible criteria being neglected. Even as far as deme density is concerned, however, it would seem that assessment is unlikely to involve a single factor. Maximum reliability of assessment of the relative suitability of two potential breeding shores should involve a complex of criteria, including the deployment of males, females and young on the two shores as well as some measure of the total number of individuals on these and neighbouring shores insofar as the total number is likely to be relevant to feeding efficiency. In order to assess any of these factors, of course, it is necessary for the young pre-reproductive individuals to be in a position to visit a number of breeding shores while these shores are occupied by breeding colonies and not just, as throughout the winter, when the colonies are empty of all fur seals. It could be significant, therefore, that yearling *C. ursinus* do not arrive back at their natal islands until the breeding season is well advanced and lactation is nearly completed (Fig. 28.15). Perhaps, therefore, this absence of yearling fur seals from their natal breeding shores should not be viewed solely as a mechanism that enables them to avoid competition for food and space with the breeding adults but also as a period of assessment of the relative suitability of a number of other potential breeding habitats. Though whether individuals would be far enough

away from their natal site at this time for it to have been possible for them to have assessed an island as distant as San Miguel remains to be investigated.

The available evidence outlined earlier in this example shows, if nothing else, that the removal migration of the northern fur seal, *C. ursinus*, takes place in both directions between areas of high deme density, such as between the Pribilof and Commander Islands, and also from areas of high deme density to an area, such as San Miguel Island, which, initially at least, supported no individuals at all. Expansion of the breeding range of other species (Figs. 14.10 and 14.11) shows that removal migration to previously uninhabited areas may be of general occurrence among pinnipeds. On the other hand, the rapid increase in numbers of the northern fur seal, *C. ursinus*, which cannot be attributed solely to reproduction, once San Miguel Island was colonised (Fig. 14.14) and the similar logarithmic increase in the numbers of the northern elephant seal, *Mirounga angustirostris*, on Año Nuevo Island (Orr and Poulter 1965, Radford *et al.* 1965) suggests that removal migration is more likely if the destination already supports a breeding colony.

It is possible, of course, that from artefacts a *C. ursinus* could recognise even during the winter months when no fur seals are present that a certain shore was used for breeding by its own species the previous summer. Assessment would be much more reliable, however, if the shore was visited during the breeding season. The increased rate of removal migration once even a small colony was established on San Miguel Island (Fig. 14.14) tends to suggest that the shore was visited by further potential removal migrants during the period that the breeding colony was in position on the shore. The fact that yearling immature *C. ursinus* are at sea during this period and that juveniles are the most likely individuals to perform removal migration also provides some support, albeit circumstantial, for this suggestion.

To recapitulate, juveniles of a number of species of pinnipeds extend their familiar area beyond that necessary to meet their requirements when adult. Furthermore, there is some evidence that juveniles, like those of many other animals, show a greater incidence of removal migration than any other age or sex class. As far as the northern fur seal, *C. ursinus*, is concerned, the evidence concerning deme density at the natal site and at the destination when removal migration takes place is insufficient to determine whether deme density is used as a criterion in the assessment of relative habitat suitability. There is some evidence, however, that removal migration rate to a previously uninhabited site increases immediately after the initial colonisation of that site. A pinniped is most likely to observe that a shore has been colonised if it is at sea during the breeding season of the adults. In the northern fur seal, *C. ursinus*, this is the case with yearling individuals. It is suggested, therefore, that although the ability to return to the natal site is retained, young pinnipeds extend their familiar area as part of a procedure of assessment of the relative suitability of a number of potential breeding sites. As yet, the evidence supporting this suggestion is entirely circumstantial.

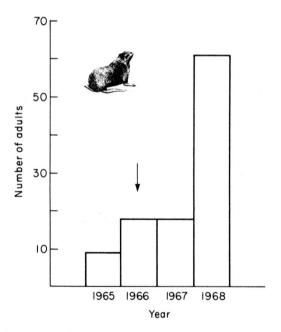

Fig. 14.14   Removal migration of the northern fur seal, *Callorhinus ursinus*, to San Miguel Island in relation to the presence or absence of other individuals

The histogram shows the number of adult fur seals present on the island for the years 1965 to 1968. The arrow indicates the year that newly born fur seals were first observed on the island.

For many years, San Miguel Island was uninhabited by fur seals. As soon as the island was colonised, however, further removal migration took place. This indicates that although a few individuals may show removal migration to a previously uninhabited site, the majority of individuals show removal migration only to a site that already supports a breeding colony.

[*From data in Peterson et al. (1968)*]

14.5 Extension of the familiar area by migration and social communication

14.5.1 The group-to-group removal migration of male–female human pairs

More data are available on the migrations of modern agricultural and industrial man than on the migrations of any other animal. Virtually every paper in journals such as *Human Biology* includes references to, or data concerning, migration. Add to this wealth of information the mass of data accumulated by the periodic censuses carried out by many nations and some appreciation can be obtained of the task awaiting any author attempting to gain some insight into human migration. Innumerable studies have of course been carried out by anthropologists, social scientists and economists and various theories are extant at the present time. To my knowledge, however, no zoological synthesis of human migration has been attempted. The various discussions of human migration that are scattered through this book represent my attempt at such a synthesis.

I hope that my colleagues in other scientific fields who are also involved in the study of human migration will excuse me if I start from first principles and develop my own hypothesis. In the final analysis it may well be that the hypothesis developed here and the theories developed by anthropologists, social scientists and economists are one and the same and that apparent differences are due merely to differences in phraseology and frames of reference. On the other hand it may well be that different scientific disciplines have different requirements of their theories. My own concern is with an understanding of the evolution of human migration patterns. Such an understanding is both aided by and improves an understanding of the evolution of animal migration patterns in general and is an integral part of any consideration of the evolution of animal migration. This should not be taken necessarily to imply, however, that there is an inherent reason why such a zoological approach should be of use to, or even interest, members of any other scientific discipline except zoology.

A zoological approach to the evolution of human migration must begin by a consideration of the periods of time over which different selective pressures have acted. As far as is known, the human lineage (Fig. 14.15) has consisted of hunter–gatherers ever since the hominids first invaded the African savanna somewhere between the lower and upper Miocene. Modern hunter–gatherer humans, therefore, have an unbroken history of hunting and gathering of between 10 and 20 million years. Not until about 10 000 years B.P., within the distribution of the hunter–gatherers of the North East, did agricultural practice first become established. Since that time, however, agricultural practice has spread and is still spreading through the human population, rapidly at first and only slowing down or halting at the edges of areas unsuitable for cultivation (Fig. 14.16). Cities have been in existence for some 5000 years (Barnett 1971), but only for the last 200 years or so, during and since the industrial revolution, have cities taken on their modern form and become the main destination of human migrants all over the world.

When we consider, therefore, the rural–urban and urban–urban migrations of modern industrial man in, for example, Western Europe, North America or Japan, we are dealing with the migrations of animals that have been subjected to hunter–gatherer selection (i.e. the selective pressures that act on hunter–gatherers) for perhaps more than half a million generations, to agricultural selection for 200–300 generations, and to industrial selection for 0–10 generations. The question to be borne in mind, therefore, in any consideration of the migrations of industrial man, is to what extent the migrations should be considered in terms of industrial man and to what extent they should be considered in terms of hunter–gatherer or agricultural man but in an industrial setting.

In this example I am concerned with the calculated removal migration of newly formed pairs of humans in relation to the migrations that take place before pairing. The general plan is first to develop a hypothesis to account for the migrations of hunter–gatherer human pairs and then to attempt to follow the course of this postulated primitive migration pattern through the various changes in human socio-ecology that have taken place over the past 10 000 years.

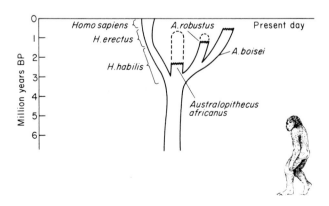

Fig. 14.15   A possible phylogeny of the hominid lineage over the past six million years   [After Tobias (1973)]

14.5.1.1 The evolution of some major features of the socio-ecology of human hunter–gatherers

The hunting and gathering economy: Although the timing of the first hominid colonisation of savanna is open to debate (Howell 1970), there seems little doubt that it is to this event in Africa that the origin of a hunter–gatherer economy can be traced. Hominid hunter–gatherers appear to have had a long history in Africa. All finds of undisputed *Australopithecus* remains have been there (Howell 1970). There is a suspicion, however, that before hunter–gatherer australopithecines first started to use tools to dismember and kill their prey, speciation had taken or was taking place into gracile (*A. africanus*) and robust (*A. robustus* and *A. boisei*) forms (Fig. 14.15). The gracile form remained an inhabitant of the savanna, continued to be a hunter–gatherer, gradually increased in size, and gradually developed the ability to use and manufacture tools. The robust form reinvaded more forested regions, became almost entirely a gatherer or possibly even a browser, was large from an early evolutionary stage, and may never have used or manufactured tools (Robinson 1961, Howell 1970) except to the extent, say, that modern chimpanzees use tools (Lawick-Goodall 1968).

The *Homo–Australopithecus* dichotomy, which can now be dated to about 3 million years B.P. (Fig. 14.15), undoubtedly took place within the population of savanna-dwelling, hunter–gatherer, tool-using australopithecines. At or soon after this dichotomy, hominids, in the form of hunter–gatherer *Homo* spp. spread out from Africa and gradually spread throughout continental Eurasia. During this spread, the now adaptable hunter–gatherer economy came into contact with a number of different habitat-types, from forest to tundra. The earliest evidence of *Homo sapiens* in Australia dates from 32 000 years B.P., but by 20 000 years B.P. man had occupied almost all principal ecological zones on the continent (R. Jones 1973). At about the same period, between about 30 000 and 15 000 years B.P. (Cruxent 1968), *Homo sapiens* seems likely to have arrived in North America. Beringia, the land mass that connected Siberia and Alaska and across which man almost certainly spread, was in existence for some 15 000 years at this time and was of large extent, perhaps 1000 km from north to south (J.B. Griffin 1968). During most of this time Beringia would have supported tundra and a variety of animals such as mammoth, mastodon, caribou and muskox, all of which are known to have been hunted.

Migration from Asia to Alaska seems to have been performed by two major groups of hunter–gatherers. Before 19 000 years B.P., at which date the most likely route into North America, east of the Rocky Mountains, was closed by ice (Griffin 1968), hunter–gatherers from northeast Siberia entered North America and reached Central America, perhaps by 20 000 years B.P. (Cruxent 1968). Northern South America may have been reached between 20 000 and 10 000 years B.P., either by gradual spread through Central America, by sea, or through the chain of Caribbean Islands, or perhaps all three (Cruxent 1968). Western and southern South America seem likely to have been reached between 14 000 and 8000 years B.P. (Fig. 14.16).

The second group to arrive were coastally specialised hunter–gatherers from the Pacific side of the Chukchi peninsula who probably spread round the southern coast of Beringia and reached southwest Alaska and the Aleutian Islands between about 10 000 and 8 000 years B.P. (Griffin 1968). These people, ancestors of present-day Eskimoes and Aleuts, upon the submergence of Beringia about 10 000 years B.P., spread around the coast and reached the eastern Canadian Arctic between about 5000 and 4000 years B.P. North Greenland was reached about 4000 years B.P.

Modern hunter–gatherers around the Arctic Circle are almost entirely hunters, subsisting on a high protein and fat but low carbohydrate diet derived almost entirely from animal sources. By way of contrast, those modern hunter–gatherers that have re-invaded the tropical forests show a reduced amount of animal protein in their diet, probably only about 33 per cent in the case of the pygmies of the African tropical rain forest (Fischer 1955), and subsist largely as gatherers. In seasonally wet and dry conditions hunter–gatherers such as the African Hadza subsist mainly as gatherers in the wet season and almost entirely as hunters in the dry season (Tomita 1966).

The symbiotic hunting relationship between man and dogs developed about 11 000 years B.P. (Barnett 1971), spread throughout Africa and Eurasia, and reached Australia between about 8000 and 7000 years B.P. (Tyndale-Biscoe 1973).

Sexual division of labour: Considerable differences are found in the migration patterns of male and female

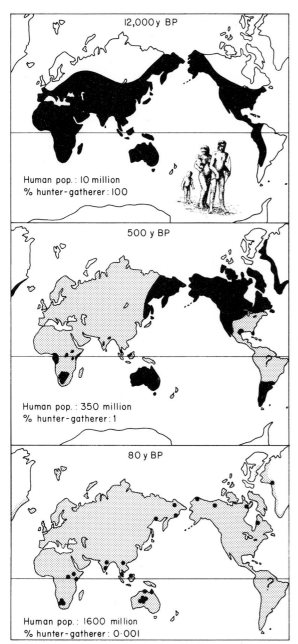

Fig. 14.16 The spread and decline of man as a hunter–gatherer

The solid shading shows the World distribution of hunter–gatherer humans at three points in time during the past 12 000 years. The stippled shading shows the World distribution of humans with an economy based on agriculture or animal husbandry. Population size and per cent hunter–gatherers are also indicated.

[Modified from Lee and DeVore (1968)]

humans. Interpretation of these differences is not possible without some discussion of the division of labour between the two sexes over and above those normally associated with reproductive differences.

Human hunter–gatherers show a marked sexual division of labour such that the males hunt and the females (and young) gather. The same trend is seen in the other two primate hunter–gatherers, chimpanzees (Lawick-Goodall 1968) and savanna baboons (Potter, in Dart 1964), though in the latter group old females may also play some part in hunting.

In general, there does not seem to be an inherent evolutionary reason why males rather than females should become hunters. On the contrary, in most animals, such as carnivorous eutherians, carnivorous marsupials, odontocetes, birds and sharks, for example, participation in hunting is shared equally between males and females. The only deviation from this state occurs in those terrestrial mammals the young of which are born into a den. In some of these species, such as the red fox, *Vulpes vulpes*, the female may stay in the den during the early periods of lactation while the male hunts and brings food to the female (H.N. Southern 1964).

Specialisation for hunting only by the male would therefore seem to be a function of the primate situation, and the evolutionary explanation for the phenomenon must be sought in primate preadaptation rather than in some universal facet of maleness and femaleness. The main difference between primate and other hunters seems to be that in other hunters selection for hunting seems likely to have acted on ancestors in which males and females lived independently, in which there was little sexual dimorphism and in which the young were either left in a nest or den or were capable of independent movement from an early age. In primates, however, selection for hunting did not act until after the evolution of social groups consisting of males and females, considerable sexual dimorphism, and in particular after the evolution of the habit of the young being carried by the mother. To this list of primate preadaptations can be added one other adaptation that applies particularly to hominids and which is likely further to accentuate sexual division of labour with respect to hunting. Selection for bipedalism is generally considered to have occurred from an early stage in hominid evolution and probably to have been the main selective force causing the emergence of the hominid line. Selection for bipedalism also involves selection on the female–

young relationship. In an upright position not only is the young primate likely to have greater difficulty in clinging to the fur of the female but the ventral position no longer has the protective advantages that it has in quadrupedal locomotion. Selection for bipedalism, therefore, also involves selection on the female to use an arm to support and protect the young offspring. This renders even more unlikely selection for the involvement of females in defence and hunting. These differences in preadaptation when selection for hunting began seem to be likely to result in the evolution of sexual division of labour in primates, particularly hominids, but not in other groups.

Sexual division of labour within a group consisting of males and females in relation to defence from predators is a much more general phenomenon, being found widely in ungulates as well as in primates. The strongest correlation with this behaviour seems to be marked sexual dimorphism, though the habit of primate females of carrying their young also seems likely to reduce selection for participation in group defence. It is tempting to suggest that when selection, presumably mainly sexual selection, acts on males more than females to increase size and the effectiveness of other characteristics instrumental in successful offence and defence, mainly against other males, then selection also acts to increase the role of males in group defence against predators. The reproductive success of all individuals in the group, including, therefore, that of the male(s), is increased by protection of the group by the individual(s) most capable of protecting the group. If to this is added the possibility that the reproductive success of subadult males is increased by gaining experience in group defence before becoming a reproductive member of the same or another group, then selection for sexual division of labour in defence can be seen not necessarily to involve selection for altruistic behaviour by males.

In summary, therefore, sexual division of labour in group defence can be traced back through the human lineage about 50–60 million years to a time soon after the emergence of primates as a group. Sexual division of labour for hunting can be traced back through the human lineage some 10–20 million years to the first hominid or pre-hominid invasion of the savanna. It seems reasonable to suggest that the gathering behaviour of male and female humans has been acted on by selection for some 60 million years or perhaps much longer if we accept an unbroken sequence of a diet of fruit, seeds and invertebrates

from modern man back to the early mammalian forms. The hunting behaviour of male humans, on the other hand, has been acted on by selection for only 10–20 million years.

Social communication concerning the familiar area: Direct communication by sound, scent or movement is widespread among animals. Communication by artefact is less widespread though the marking of territories by faeces and scratching falls into this category (Chapter 18). Efficient hunting by a group requires organisation of the group. As optimum group deployment varies from situation to situation, the most flexible mechanism of group organisation is one based on communication between individuals in advance as well as during each hunting expedition. There seems little doubt that communication systems that facilitate such flexible group organisation have evolved on many occasions. For example, A. Murie (1944) describes the way that wolves, *Canis lupus*, before setting off on a hunting expedition, come together in a group and spend a period of time interacting by uttering sounds and wagging tails, etc. It is difficult to avoid the conclusion that during this pre-hunt aggregation information is being exchanged and decisions reached of which use is later made during hunting. Certainly the wolves move in a direct line to the area in which the night's hunting is to take place (S.P. Young 1964). Similar pre-hunting communication is likely in all carnivores and primates, and probably ants as well, that hunt in coordinated groups. There is, however, a limit to the amount and accuracy of information that can be exchanged by sound and gesticulation even when these are as complex as in human language. Movement, as in the dancing of bees and humans, represents a relatively efficient method of conveying accurate information. So, too, does the employment of artefacts, as in drawing or the manipulation of objects. It is perhaps not surprising, therefore, that dancing and artefacts developed such an important role in the communication of hunter–gatherer man. By the time the Cro-Magnon hunter–gatherers of southern Europe were producing sophisticated cave drawings and small models and carvings some 30 000 years B.P. (Howell 1970), the genus *Homo* probably had a considerable history of communication by artefacts. Presumably early drawings, solely for the purposes of communication, would have been temporary and made in accessible sites where preservation was unlikely.

Gathering by a group also introduces selection for

social communication concerning the relative suitability and location of potential foraging sites. That this is so is manifest by the communication systems evolved by social insects, gatherers as well as hunters (p. 146). Less is known, however, about the ability of non-human primates to communicate information concerning the relative suitability and location of potential foraging sites. Nishida (1968) has suggested that a correlation may exist for social hominoids between a gatherer economy and an 'extensible' social system in which the maximum social group splits up from time to time into subunits which disperse over a wide area and announce vocally the discovery of a particularly abundant food source. Hunter–gatherer humans and chimpanzees, *Pan troglodytes* (Itani and Suzuki 1967), both seem to possess this type of social organisation. The gorilla, *Gorilla gorilla*, however, which is a browser, forms relatively small and closed social units (Schaller 1963). A similar correlation exists between economy and social organisation in non-hominoid primates (Eisenberg *et al.* 1972). If it is assumed, then, that those dryopithecines that are part of the pongid and hominid lineages were gatherers, it seems reasonable to assume that they too had an extensible social system. It would be difficult to accept that an extensible social organisation could evolve without at the same time the evolution of social communication of information concerning the relative suitability and location of potential foraging sites. The role of reciprocal altruism in such a social strategy is discussed elsewhere (p. 201).

In conclusion, therefore, it seems reasonable to suppose that when the hominid lineage first invaded the savanna some 10–20 million years B.P. the ability to communicate information concerning the location and suitability of different parts of the familiar area had already evolved. The level of sophistication of this social communication was probably on a par with or below that of modern forest-dwelling chimpanzees, *Pan troglodytes*, but perhaps below that of modern honey bees, *Apis mellifera*. The history of hominid hunter–gatherers over these 10–20 million years has probably been marked by a gradual sophistication of the ability to communicate information largely in response to the selection acting on the efficiency of hunting and gathering. This sophistication has involved the further evolution of various communication mechanisms including vocalisation, body movements, dancing, drawing and the manipulation of objects.

Variation in deme size and density: Modern hunter–gatherers live in maximum territorial units of, on average, about 350 individuals (Coon 1965), but often as few as 100 (Tobias 1964), 80 (Tomita 1966), or even 20 (Murdock 1959). Rarely does the number exceed 700 individuals (Coon 1965). By non-human primate standards these maximum social units are large, though, of course, they are small by agricultural and industrial human standards. Deme density, however, is low both by the standards of non-human primates (Aldrich-Blake 1970) and of human agriculturalists (Barnett 1971). The territory occupied by these maximum social units is large, about 2500–7000 km^2 (Coon 1965). Before the introduction of agriculture, between 6000 and 5000 years B.P., the whole of England supported only about 250 individuals (Barnett 1971).

When food is scarce, hunter–gatherer maximum territorial units break up into smaller units which disperse throughout the territory (Chapter 25). When food is abundant, the entire maximum unit converges on a single site at which a temporary or semi-permanent, possibly communal, shelter may be constructed e.g. the Yumbri of North Thailand (G. Young 1962). Most of the time, however, hunter–gatherers are more or less permanently nomadic within their territory and, at least in tropical regions, construct only temporary shelters which may vary from a mere windbreak to a structure that approximates to a hut (Murdock 1959, Bleakly 1961, Tobias 1964).

The subunits into which hunter–gatherer groups divide when food is scarce are usually either families (i.e. one reproductive male, one or more reproductive females and their young, and perhaps one or more adult but actually or effectively post-reproductive males and/or females) or some multiple of a family, depending on how scarce food may be. In this respect the social organisation of human hunter–gatherers is different from that of chimpanzees, *Pan troglodytes*, the subunits of which may consist of any combination of males, females and young (Nishida 1968). Some reports of the Yumbri human hunter–gatherers of the monsoon forest of North Thailand (G. Young 1962), however, indicate a social organisation somewhat intermediate between that of chimpanzees and the organisation normally associated with human hunter–gatherers. Nevertheless, in general the smallest social unit of human hunter–gatherers is the family. Many authors have derived the family unit from the type of situation found in chimpanzees (e.g. Itani and Suzuki 1967). Other authors (e.g. Weiner 1971) envisage such difficulty in this that they prefer to derive the human situation from an ancestral state similar to that of modern gibbons which on the whole are monogamous. Yet other authors (e.g. Imanishi 1961) prefer to derive the human family system from the one-male or age-graded male (Eisenberg *et al.* 1972) unit of the gorilla, *Gorilla gorilla*.

Nishida's observation (Nishida 1968) of the correlation between a gathering economy and an extensible social organisation, however, suggests that it is likely that the early hominids had an extensible social organisation. The explanation for the evolution of a family unit from the extensible system of a gatherer can perhaps be attributed to selection that acts upon males once exposed to selection for hunting. The advantage to a male of sharing a kill with other males that cooperated in the hunt is discussed elsewhere (p. 201). When food is scarce, however, and groups have to split up into smaller groups such that a male may hunt singly yet at the same time maintain frequent contact with one or a number of females in case they come into oestrus, a kill still probably provides more food than the male can consume before the food deteriorates or is eaten by scavengers. In such a situation the male could increase his reproductive success by sharing the excess food with his own offspring and with a female that is suckling and/or raising his offspring. Any behaviour that keeps male and offspring together is therefore an advantage to both as well as to the female parent(s) of these offspring. The most likely evolutionary result seems to be formation of a pair bond or bonds between the male and a female or females such that the young that accompany these females and with which therefore the male shares food have a high probability of being the male's offspring. The complex of behavioural strategies that are likely to have evolved in response to the selective pressures connected with extra-pair matings by both male and female and also connected with cuckoldry have been elegantly described by Trivers (1973).

I would, therefore, derive human pair-bonding, with all the implications this has for migration patterns, from selective pressures acting during periods of food scarcity on the extensible social organisation of a gatherer with incipient hunting by males. Once hunting by males has led to the evolution of a family unit, frequent periods of food scarcity are likely to have other far-reaching evol-utionary effects. The fact that males hunt whereas the females and young gather and that at times hunting provides the major means of subsistence means that, unlike most primates, the reproductive success of human hunter–gatherers depends to a large extent on the amount of food that can be captured by the males rather than on the amount of food available to all individuals. The ecology of hominid primates, therefore, shows more similarities to that of non-game birds than to the ecology of non-hominid primates. Conversely, the ecology of non-hominid primates at least once the young have been weaned perhaps shows more similarities to that of game birds than to the ecology of hominid primates. The 'optimum clutch-size' hypothesis of Lack (1954) (i.e. that non-game birds produce the clutch-size from which the maximum number of offspring can be raised rather than the maximum clutch-size physiologically possible) has far more applicability to human hunter–gatherers than to other primates. Modern human hunter–gatherers are invariably found in areas, such as deserts and tropical rain forests, where food, at least at ground level, is scarce but where agricultural practice has failed to pene-trate. General (Tindale 1953), seasonal (Tomita 1966) or irregular (Tindale 1953, Tobias 1964) food scarcity limits the number of young that can be raised. Virtual monogamy is more or less universal among modern hunter–gatherers (Murdock 1959, Cappieri 1970). Furthermore, delayed maturity, extended lactation and a reduction in sexual in-teractions (paired male and female Yumbri of North Thailand may meet for only three months per year—Young 1962) reduce the birth rate to about four per female lifetime (Carr-Saunders 1922). Of these four only two or three may survive to maturity. If we apply Lack's principle it would be expectd that males that show parental investment to more than one female plus offspring and also pairs that produce more than four births would in fact raise fewer young to maturity.

All of the arguments and observations so far suggest that food availability or, more precisely, nett rate of energy uptake, was a critical factor in the evolution of hominid hunter–gatherers. Several pieces of evidence suggest that this was, and still is, the case. First, a differential can be seen in the deme density of immediate post-pleistocene hunter–gatherers specialising upon different foods. Some 10 000–12 000 years ago, after the last glaciation, hunter–gatherer man existed in most places at deme densities of between 0·01 to 0·1 individuals/km^2. In certain places, however, such as along streams, the edges of lakes and sea coasts, where specialisation on the more reliable sources of animal food such as fish and other sea-foods took place, the density was more of the order of 1 individual/km^2. Secondly, the evolution of agriculture and animal husbandry led directly to a 10–20-fold increase in reproductive success (Barnett 1971). Not only are agriculture and, to a lesser extent, specialisation on fishing, likely to have affected gross energy uptake as a result of increased food availability, they are also likely to have affected gross energy expenditure. According to Doxiades (1970), human hunter–gatherers expend 50–60 per cent of muscular energy on movement, perhaps even as much as 80–90 per cent a million years ago before the sophistication of hunting techniques. As an agriculturalist, however, man expends only about 25 per cent of available muscular energy on movement. On these figures it seems reasonable to conclude that the reproductive success of human hunter–gatherers correlated well with nett rate of energy uptake per individual.

A similar relationship between food availability and reproductive success is evident among modern hunter–gatherers. Laughlin (1966) compares the polar eskimoes of North Greenland and the extinct Sadlermiut eskimoes of Southampton Island in Hudson Bay with the Aleuts and Koniag eskimoes of Kodiak Island. For nine months of the year the former two groups have or had to hunt along the edge of the ice or at punch holes. The food supply during this period is therefore entirely dependent on fit males. Reproductive success is relatively low, with a high death rate in particular among young and old individuals. Deme density is also low and there is a large inter-deme distance. On the other hand the Aleuts and Koniag eskimoes have an intertidal zone from which sea food can be collected all year. Freshwater lakes, streams and sheltered bays, etc. also provide an abundance of food that can be gathered by females and young and old individuals. Reproductive success is relatively high, deme density is also relatively high and there is a relatively short inter-deme distance. By way of contrast, the aborigines on the forested east coast of Australia have a deme density of 1·3 individuals/km^2, whereas on the central arid plains of the continent deme density is 0·005 individuals/km^2 (Bergamini 1965).

One final facet of deme size and density is important in the formulation of a hypothesis of the migration of hunter–gatherer male–female pairs. Higher deme densities are likely to have a higher incidence of disease. The age group most affected by a high incidence of disease consists of young individuals from birth until the full development of a spectrum of immunological response. As far as migration is concerned, therefore, disease acts as a selective pressure on male–female pairs with young offspring. High deme densities are probably more disadvantageous to this ontogenetic class than to any other.

The selective pressures acting on the deme size and density of human hunter–gatherers and thus on the migration of human hunter–gatherers in relation to deme size and density seem likely to have remained relatively unchanged since selection for invasion of the savanna led to the evolution of hunting by males and the evolution of the family unit. Hunter–gatherers have therefore been exposed to selection on migration in relation to deme size and density for some 10–20 million years. During this period actual density gradually increased, e.g. tenfold from the early manufacture of stone tools to modern hunter–gatherers (Howell 1970). It is unlikely, therefore, that any particular deme density became genetically fixed as an optimum. It seems more likely that the optimum strategy was to seek lower deme densities when the nett rate of energy uptake began to fall and higher deme densities when the rate began to rise, irrespective of the actual deme density being experienced at the moment. It should perhaps be pointed out here that variation in the abundance of water invariably poses slightly different selective pressures from variation in the abundance of food (Chapter 25).

Territorial interaction: The maximum territorial unit of modern human hunter–gatherers ranges from about 20–700 individuals and occupies a territory of from about 400 km^2 to about 100 000 km^2, depending on the availability of food. For reasons already discussed, territorial defence is undertaken solely by the males.

Evidence concerning territorial defence by hunter–gatherers is scarce (Gardner 1968). Territorial defence does occur, however, and such observations as are available suggest that, as with many other animals, the intensity of defence and offence increases when a resource becomes limiting and the intruding unit threatens that resource. Some of the territorial units of the hunter–gatherers of the Andaman Islands in the Bay of Bengal, for instance, which by their delayed maturity and reduced reproductive success show every sign of being limited by resources, are notorious for the intensity of their territorial resistance (Cappieri 1970). Observations on the bushmanoid hunter–gatherers of the Kalahari Desert of Africa also suggest that territorial resistance fluctuates, both within and between maximum territorial units, with fluctuations in resource abundance, the resources in this case being food and water (Tobias 1964).

Territorial resistance to intruders may therefore depend partly on the availability of particular resources. It may also depend on the number of attempted intruders that threaten those resources. Territorial resistance is greatest when a maximum territorial unit is confronted with an equivalent territorial unit which inevitably threatens all resources or when confronted with intruders that threaten a particularly scarce resource. Resistance is least, however, to individual intruders. Hence, the territorial units of most, if not all, hunter–gatherers (e.g. Australian aborigines (Tindale 1953), North Thailand Yumbri (G. Young 1962), and Kalahari bushmen (Tobias 1964)) have contained some individuals, males as well as females, that have migrated from some other territorial unit.

14.5.1.2 The group-to-group removal migration of male–female pairs of human hunter–gatherers: a hypothesis

The general plan of this section is first to present eight components of a hypothesis of the migrations of hunter–gatherers during that period of ontogeny from the time of first being capable of existing independently of the parents to the time of first production of offspring. Each of the eight components of the hypothesis is then discussed separately. In advance it should be pointed out that this hypothesis has been developed from a number of starting points. Wherever possible, the starting point has been observation of some facet of the migration of modern hunter–gatherers. These observations have then been interpreted from an evolutionary viewpoint. Where appropriate observations of modern hunter–gatherers have not been available, I have employed *a priori* evolutionary reasoning in predicting what is most likely to occur. The nett result is a hypothesis of hunter–gatherer migration which I shall then assume to have been manifest by man at the end of the Pleistocene period some 10 000 to 12 000 years B.P. This hypothesised hunter–gatherer migration pattern can then be followed

through the development and spread of agriculture, urbanisation and industrialisation.

It is suggested that the migrations of human hunter–gatherers up to the early reproductive phase of ontogeny should show the following components:

1 extension of the familiar area by sub-adult males and sub-adult females;
2 greater familiar area extension by sub-adult males than by sub-adult females;
3 familiar area extension by migration by sub-adults based largely on socially communicated information;
4 less correlation between the areas visited by sub-adults during familiar area extension and food availability and deme density than for any other ontogenetic class, though stay-time may be longer in areas of high deme density;
5 less territorial resistance to immigrants when food is abundant than when food is scarce;
6 removal migration of a pair some time after formation of a permanent or semi-permanent pair-bond;
7 the migration destination is within the combined familiar area of the pair and should be the destination that is assessed to be most suitable;
8 nett rate of energy uptake per individual and deme size and density are the criteria on the basis of which pairs are most likely to assess habitat suitability, with greater weight being given to the former.

Each of these components is discussed below.

1 *Extension of the familiar area by sub-adult males and sub-adult females.* Some qualification of this component is necessary because obviously all individuals, no matter what their age or sex, extend their familiar area continually, both by occasional movement beyond the edge of their existing familiar area and also by social communication with individuals from outside their familiar area. From the moment of birth individuals gradually establish a familiar area by social communication and by movement within the familiar area of their parents. This first component of the hypothesis, however, implies an increase in the rate of familiar area extension, both by social communication and by movement, from the time of incipient independence from the parents until the time of pairing at which time rate of extension should decrease.

There are several reasons for expecting selection to favour an increased rate of extension of familiar area by sub-adults (see also Chapter 17). The first selective pressure results from the advantage of calculated migration as opposed to non-calculated migration. This selective pressure is one of the main concerns of Chapter 15 and does not need separate discussion here. The second selective pressure results from the advantage of pairing with the most suitable available mate. Research on animals that pair for long periods, such as some birds (Coulson 1966), has demonstrated the importance to an individual of pairing with a mate such that the characteristics of the two individuals are complementary and combine to form an efficient unit for parental care. Presumably the same applies to humans. Certainly individual humans show fairly precise preferences during pair-bonding (Coon 1965). Such individual preferences may be inherent or learnt. Quite possibly the probability of formation of an optimum pair-bond and hence of maximisation of reproductive success increases with increase in the number of potential mates encountered during the sub-adult period.

Extension of the familiar area inevitably results in a decrease in the efficiency of food acquisition because some time is spent in an area with which the individual is not directly familiar. This imposes a stringent selective pressure on individuals with mates and offspring, for the reproductive success of these individuals should be a function of the food they acquire over and above their own physiological requirements. The potential reproductive success of sub-adults is maintained, however, as long as sufficient food for their own physiological requirements is obtained.

The nett result of all these selective pressures is likely to be an increased rate of extension of familiar area of sub-adults compared with the rate of extension of other independent ontogenetic classes. Observations on some African hunter–gatherers reported by Murdock (1959), which are described in component (2), provide for males some support for this prediction. I have found no observations, however, that confirm that female hunter–gatherers also increase their rate of familiar area extension during the sub-adult phase of their ontogeny.

2 *Greater familiar area extension by sub-adult males than by sub-adult females.* In the discussion of the previous component it was suggested that the probability of formation of an optimum pair-bond and hence of maximisation of reproductive success increases with the number of potential mates encountered during the sub-adult period. The reproductive success of males is also increased by an increase in the number of females with which copulation but no further parental investment takes place (Trivers 1973). This selective pressure does not, of course, apply to females and consequently is likely to render familiar area extension to be more favourable to males than to females. Furthermore, the evolution of sexual dimorphism in man for reasons already discussed (p. 116) is likely to be associated with and accentuate selection against familiar area extension by lone females. Sexual dimorphism of the human type in any animal (Parker 1974a) leads to an increase in the role of rape as part of the reproductive strategy of the male. Rape is a disadvantage to female humans, particularly unpaired sub-adults, because it is not usually associated with further parental investment on the part of the male (Trivers 1973). The lack of selection on the female for the evolution of a morphology suited to offence and defence also seems likely to render lone females more liable to predation than lone males.

Altogether, therefore, it seems likely that selection favours a greater extension of the familiar area by sub-adult males than by sub-adult females and there is some evidence that this is the case, at least for the hunter–gatherer bushmen of the Kalahari Desert of Africa (Murdock 1959). Among these people the sub-adult male extends his familiar area considerably, leaving his natal home range and possibly, though not necessarily, the territory of his natal territorial unit, and settles for short periods with other groups. The sub-adult female may also extend her familiar area but usually only within the range of movement of her natal group. What is not clear for any hunter–gatherer is whether the familiar area extension during the sub-adult stage of the female is any greater than during other ontogenetic phases.

It is, of course, almost as great an advantage to the parents of a sub-adult female as to the sub-adult female herself that she should form an optimum pair-bond, thus maximising her own and therefore also the parental reproductive success. It is also, therefore, an advantage to the parents that their sub-adult female offspring does not fall victim to rape or, of course, predation. Quite possibly, therefore, we might expect selection to favour parents that accompany their sub-adult female offspring during extension of her familiar area. The parental instincts that could have evolved in response to this selection may well account for the cultural development in some hunter–gatherers of the habit of parents of taking their sub-adult female offspring to be paired to a male of the same or a neighbouring group, such pairings having been arranged in advance by the respective parents (Bergamini 1965).

As well as extension of the familiar area in company with the parents, the disadvantage of familiar area extension by lone sub-adult females could perhaps lead to aggregation of sibling and other sub-adult females during the period of familiar area extension.

3 *Familiar area extension by migration by sub-adults based largely on socially communicated information.* Familiar area extension involves either exploration of areas totally unknown to the individual or acquisition by social communication of information concerning such areas. Social communication represents an efficient mechanism for rapid expansion of the familiar area of each individual within a group. As far as sub-adults are concerned, however, extension of the familiar area by social communication cannot completely replace extension by migration. Only the latter increases the rate of encounter with potential mates and probably only the latter permits accurate assessment of the suitability of an area as a suitable destination for removal migration after pairing. It was pointed out in relation to component (1), however, that one disadvantage of familiar area extension by migration was that some time was spent in areas previously unknown to the individual and in which, therefore, food acquisition was relatively less efficient. To some extent this disadvantage could be reduced by extending the familiar area by migration mainly within areas that are within the socially communicated familiar area of that individual. It might be expected, therefore, that selection favours those sub-adults that accumulate most information by social communication concerning areas previously unknown to them followed by migration to and from some of these areas. Unfortunately, I have no data concerning the frequency with which familiar area extension of hunter–gatherer sub-adults by migration takes place within the familiar area previously established by social communication.

4 *Less correlation between the areas visited by sub-adults during familiar area extension by migration and food availability and deme size and density than for any other ontogenetic class, though stay-time may be longer in areas of high deme density.* It was suggested in relation to component (1) that the advantage of familiar area

extension by sub-adults is firstly that it increases the number and range of potential mates encountered and secondly it increases the range of potential destinations available to the individual once pairing has taken place. If this suggestion is correct, it might be expected that area covered and number of potential mates encountered are the critical factors as far as sub-adults are concerned rather than any other characteristics of individual sites. It has already been pointed out that nett rate of energy uptake is less critical to a sub-adult than to a paired individual with offspring. Furthermore, sub-adults have developed a spectrum of immunological response but do not yet have offspring, which have not. Food availability and incidence of disease are therefore much less stringent selective pressures on sub-adults than on other ontogenetic classes. Less correlation might be expected, therefore, between the areas visited by sub-adults during familiar area extension and food availability and deme size and density. Availability of potential mates, however, is likely to be important, and as rate of encounter is likely to be greater in areas of high rather than low deme size and density, it might be expected that frequency of visits and stay-time might show some positive correlation with these factors. Unfortunately, again I do not have data for hunter–gatherers concerning this component of the hypothesis.

5 *Less territorial resistance to immigrants when food is abundant than when food is scarce.* It was concluded earlier (p. 120) that territorial resistance to all ontogenetic classes increases as food, or more precisely, nett rate of energy uptake per individual, decreases. There is some evidence to support this suggestion but there is an obvious need for a more quantitative analysis. Among Australian hunter–gatherers, for example, the percentage of pairings that involve the removal migration of an individual, usually the female, from one territory to another varies from 15 per cent in predominantly desert and scrub areas where deme density and presumably also nett rate of energy uptake is low, to 21 per cent in the more-forested regions of the northeast where deme density and presumably also nett rate of energy uptake is relatively high (Tindale 1953, Bergamini 1965). Also, inter-territory migration may be rare among bushmen in the interior of the Kalahari Desert but where food is abundant 60 per cent of pairings may be between individuals from separate territories (Tobias 1964).

It may be postulated, therefore, that when food is scarce, territorial resistance to intrusion by individuals from neighbouring territorial units is relatively great. Individuals have little chance to extend their familiar area beyond the territorial boundary of their natal territorial unit, and therefore little chance to gain information concerning neighbouring territories or to meet and pair with individuals from neighbouring territories. Calculated removal migration from one territory to another is therefore less likely. When food is abundant the converse applies and the chances of interterritorial removal migration are increased.

6 *Removal migration of a pair some time after formation of a permanent or semi-permanent pair-bond.* Pair formation, unless it involves siblings, inevitably involves removal migration of one or both members of the pair. If both members of the pair are from the same deme, the distance of removal migration may be small. If the two individuals originate from different demes, then the distance of removal migration is relatively large. Three possibilities are open to the pair. Either they can establish a centre of activity near to the previous centre of activity of the female, or near to that of the male, or they can shift their centre of activity away from the previous centres of activity of both male and female. The factors affecting these possibilities are considered further under component (7). The point of interest here, however, is that the process of pair formation must involve selection for a low threshold for removal migration in both male and female. Differential selection on the two sexes in this respect is discussed below.

7 *The migration destination is within the combined familiar area of the pair and should be the destination that is assessed to be most suitable.* The suggestion that the migration destination of the pair should be within their combined familiar area as established independently during the sub-adult phase of familiar area extension is based on the postulated advantage of calculated as opposed to non-calculated migration. This postulated advantage is one of the main topics of Chapter 15 and need not be discussed separately here. On the contrary, the main importance of this example in this chapter is the demonstration that human removal migration does take place within a familiar area. As far as hunter–gatherers are concerned there is considerable evidence that this is the case. However, in most cases it seems that hunter–

gatherer pairs show a limited range of migration destinations in that the vast majority settle within not only the familiar area but also the natal home range of the male. In the Kalahari bushmen, for example, the newly formed pair, although they usually remain for some time with the natal group of the female, soon after the birth of the first offspring migrate and usually join the natal group of the male (Murdock 1959). Pairs of Australian aborigines also invariably settle within the familiar area of the male (Tindale 1953). Consideration of the frequency with which pairs of human hunter–gatherers settle within the familiar area of the male requires discussion within the context of the sexual and ontogenetic differences in migration patterns shown by a wide variety of animals. This consideration is delayed, therefore, until Chapter 17. Anticipating the conclusions reached in that chapter, however, it would seem that male hunter–gatherers make a greater investment in their familiar area than do females, with the result that all else being equal the efficiency of the pair is greater in the familiar area of the male than of the female. Such invariable migration within the familiar area of the male, usually back to the natal group of the male, could be considered to offer evidence against the existence of assessment of habitat suitability and therefore against the advantage of familiar area extension for any purpose other than mate selection. Sufficient exceptions occur, however, to suggest that assessment of habitat suitability is also involved.

8 *Nett rate of energy uptake per individual and deme size and density are the criteria on the basis of which pairs are most likely to assess habitat suitability with greater weight being given to the former.* The data that support the suggestion that the reproductive success of human hunter–gatherers is a function of the nett rate of energy uptake have already been given (p. 118). Selection is therefore likely to favour those individuals that base their assessment of habitat suitability on the nett rate of energy uptake they experience while in that habitat. Pairs using this criterion to assess habitat suitability would certainly frequently assess the area occupied by the natal group of the male to be most suitable. This is because the male has established hunting relationships and strategies within that group over a much longer period than elsewhere and consequently is also more familiar with the area. These learning investments (Chapter 17) seem likely to be more important for hunting efficiency than for gathering efficiency.

Consequently the nett rate of energy uptake for the pair is likely to be greater among the natal group of the male than of the female. Hence the likelihood that a pair often assess the area occupied by the natal group of the male to be most suitable.

The possibility is always present, however, despite the disadvantage of discarding the investment made into the natal area and of having to re-invest elsewhere, that nett rate of energy uptake will be found somewhere during familiar area extension to be greater than in the natal area. There are two main factors that may produce this result. First, food may be much more abundant in this other area. Secondly, deme size and density may be greater. The arguments concerning these two points have already been given (p. 119). The conclusion reached was that if the nett rate of energy uptake per individual is high or increasing it is an advantage to join a larger, more dense, deme rather than a smaller, less dense, deme. It might be expected, therefore, that in areas and at times when food is abundant and nett rate of energy uptake per individual is high and/or increasing, selection favours those pairs that assess a larger, more dense, deme to be more suitable than a smaller, less dense, deme. On the other hand, in areas and at times when nett rate of energy uptake per individual is low or decreasing, especially if associated with an increased incidence of disease, it might be expected that selection favours those pairs that assess a smaller, less dense, deme to be more suitable. The reasons why selection would act such that pairs with young offspring should be the first to react to this situation have already been discussed (p. 119).

14.5.1.3 Some features of the origin and early evolution of agriculture and animal husbandry

Economy: Many animals that feed on seeds store surplus food in underground caches (Vaughan 1972). It is also not unknown for otherwise non-agricultural animals, such as the black-tailed prairie dog, *Cynomys ludivicianus*, of North America (J.A. King 1955), to weed out non-foodplants to encourage the growth of foodplants. Possibly both of these behaviour patterns were shown by pre-agricultural hunter–gatherers and could even have been major factors in the early development of agricultural practice. In any case, around 10 000 years B.P., recognisable agricultural practice became evident in the Near East, based on wheat

and barley, and independently in China, based on millet (Barnett 1971). As agricultural practice spread to other geographic and climatic regions, local foodplants became adopted for agricultural use. There is still debate whether the agricultural practices developed around pearl millet, *Pennisetum spicatum*, and hungry rice, *Digitaria exilis*, in western North Africa about 6500 years B.P. and around maize and potato in Central and South America about 3000 years B.P. were independent discoveries or were modifications of introduced practice (Murdock 1959, Barnett 1971, Weiner 1971). Where there were no suitable local foodplants, the spread of agriculture ceased and the local inhabitants retained a hunter–gatherer economy.

The parallel discovery of increased food supply as a result of domesticating, rather than hunting, certain of the animal prey species was made in the Near East at about the same time (9000–10 000 years B.P.) as cereal cultivation (Barnett 1971) some thousand years after the domestication of the dog. (For convenience I shall refer to the development of symbiotic relationships between man and other animals by the conventional anthropocentric term 'domestication'.) Cattle, sheep, goats and pigs were domesticated first for food, with the use of the horse, ass, ox and camel for transport and traction coming later (6000–5000 years B.P.) at about the time of the development of the plough and the wheel. At least two other aspects of animal husbandry, that of the use of sheep's wool for clothing (4000 years B.P.), and dairy farming (3000 years B.P.), seem to have originated in Mesopotamia.

It should be stressed that the adoption of agriculture and animal husbandry by a particular group of humans did not result in the immediate cessation of hunting and gathering. Large-scale gathering persisted long into the agricultural period (Renfrew 1973) and even in western Europe, at least until the last few decades, the gathering of wild fruits, seeds and nuts was a small but not insignificant part of the rural economy. Similarly, hunting has persisted until the present day. In largely urbanised regions such as the United States and western Europe, hunting has more or less disappeared except in connection with animal husbandry, particularly specially reared and protected game birds. In other parts of the world, however, hunting may still make a considerable contribution to an otherwise agricultural economy. The Bantu of mid-Zambia, for example, derive 40 per cent of their food from agriculture, 20 per cent from animal husbandry, and as much as 40 per cent from hunting and fishing (Murdock 1959). In other regions, however, such as highland New Guinea, where terrain is rugged, vegetation dense and game scarce, hunting makes a relatively minor contribution to the economy of the resident humans (Coon 1965).

Deme size and density: The spatial distribution of humans underwent considerable change during the advent and spread of agriculture. Perhaps the major change involved the rapid shrinking of territory size from the 2500–7000 km^2 of hunter–gatherers to the 26–90 km^2 found in some modern neolithic agriculturalists (Brookfield and Brown 1963, Gardner and Heider 1969). Yet at the same time the maximum territorial unit increased from the hunter–gatherer level of 20–700 individuals to about 2000–4000 individuals. To judge from modern neolithic agriculturalists, this maximum territorial unit would have been subdivided into roosting aggregations which would have been distributed throughout the territory. This large increase in deme size and density can be attributed to the increase in nett rate of energy uptake per individual and hence also of reproductive success that resulted from the adoption of the practices of agriculture and animal husbandry.

Stay-time: As a hunter–gatherer, man used and uses shelters that most of the time in most places were relatively temporary and often little more than windbreaks. Depending on the climate, however, and on the availability of food, rather more elaborate constructions are produced (Bleakly 1961). Hunter–gatherers, of course, having no pack animals, cannot carry their shelter with them as can, for instance, nomadic pastoralists (Chapter 25). When food is abundant, however, hunter–gatherers do build or make use of more elaborate shelters that are occupied for periods longer than the one or few days for which more temporary shelters are used. Hence, hunter–gatherers in seasonally wet environments, such as the Hadza of East Africa (Tomita 1966) and the Yumbri of the monsoon forests of North Thailand (G. Young 1962) may occupy relatively complex huts for periods of a few months during the wet season. Similarly, those post-pleistocene hunter–gatherers that lived along rivers or the shores of lakes or the sea and which specialised on fish or other sea foods also formed more or less permanent settlements (Clark 1962). Shallow pits in the ground covered with a roof of animal hides were being used

as shelters by European hunter–gatherers about 23 000 years B.P. (Howell 1970) and by 11 000 years B.P., before the advent of agriculture, substantial huts were being constructed (Barnett 1971).

The development of fixed settlements with the advent of agriculture emerged naturally, therefore, from behaviour patterns that had evolved during the 10–20 million years or so that man had been a hunter–gatherer. Presumably selection had acted in favour of the strategy: when food is scarce, spend little time and energy in constructing shelters and move frequently; when food is abundant, construct more elaborate shelters and increase the stay-time. Food became relatively abundant with the advent of agriculture and animal husbandry. The evolved hunter–gatherer response to this situation would have been to build more-permanent shelters and to increase stay-time. As it happens, this behaviour is well-suited to agricultural practice and animal husbandry (except in seasonally arid or severe conditions in which nomadic pastoralism evolves—Chapter 25). This fact, that movement and stay-time behaviour patterns that were optimum for hunter–gatherer ecology were also optimum for agricultural ecology, seems likely to have been one of the factors that contributed to the rapid spread of agriculture through the human population.

It is probably no accident that agriculture first developed in the valley of the 'fertile crescent' of Syria and Mesopotamia in areas that during the pleistocene received a much greater rainfall than at present and which coincided with the distribution of the wild ancestors of wheat and barley (Barnett 1971). As already described, hunter–gatherers tend to form relatively permanent settlements along rivers due to the relatively permanent supply of food in these places. Only in relatively permanent settlements is it likely to be advantageous to develop practices such as seed storage and weeding without which agricultural practice is unlikely to have originated and developed. Furthermore, once agricultural practice has developed, only in the alluvial deposits of a river valley where new soil is laid down by flood water year after year is it possible to grow crops in the same place for long periods (Weiner 1971). It was also along major river systems such as the Nile and Danube that agriculture first spread most rapidly.

Away from major rivers, soil rapidly loses what little fertility it has if single crops are grown for any period of time. Adoption of agricultural practice in such areas resulted, therefore, in the development of 'extensive' agriculture in which there is frequent use of new ground as the old loses its productivity. Extensive farming persists today in many tropical areas. In the New Guinea highlands, for example, cultivation involves land clearing and burning (Brookfield and Brown 1963). Seedlings are planted in holes made with sticks and the land is allowed to continue bearing until the yield decreases. No attempt is made to fertilise the soil artificially and the re-establishment of humus and mineral content depends entirely on the area being left fallow. Depending on the nature of the soil, an area is cultivated with short fallow for 4–10 years and then left fallow for 5–20 years, and necessarily involves movement of the old settlement as the distance from village to fields increases beyond a certain limit (Chapter 18). Some villages of relatively long standing in North Thailand, for example, may now be almost a day's walk from the nearest suitably fertile land. In this area the village inhabitants stay at the same site for 5–15 years and then the whole group migrates and establishes a new village elsewhere (G. Young 1962).

Territorial interaction between hunter–gatherers and agriculturalists: Since its first development in one or several centres, agricultural practice has spread throughout almost the entire human population. The final stages of this process are being witnessed at the present time (Tobias 1964). To some extent this spread has resulted from the adoption of agricultural practice by hunter–gatherers when agriculturalists are encountered. Spread has also been facilitated by the predominantly one-way migration of hunter–gatherers that is discussed in the next section. Probably more important than either of these, however, has been the displacement of hunter–gatherers by agriculturalists as a result of territorial interaction between the two. In some cases, of course, such as the colonisation of North and South America by agricultural European caucasoids at the expense of the predominantly hunter–gatherer Amerindians and the similar colonisation of Australia, there was considerable difference in the weapons available to the two groups. Throughout Eurasia and Africa, however, the two groups are likely to have had similar weapons. Nevertheless, the advantage in any territorial interaction may always have been with the agriculturalists because of the greater numerical size of their territorial units. In addition, agriculturalists have more muscular energy available for movements other than those

required for the acquisition of food (Doxiades 1970).

As far as smaller-scale territorial interactions are concerned, it might be expected from the arguments given previously (p. 120) that agriculturalists, because of their greater nett rate of energy uptake per individual, would show less group resistance to immigration by individual hunter–gatherers than hunter–gatherers would show to individual agriculturalists. This expectation is supported by the greater proportion of transient individuals found in groups of modern neolithic agriculturalists, such as the New Guinea highlanders (Brookfield and Brown 1963), when compared with modern hunter–gatherers, such as Australian aborigines (Tindale 1953). Coon (1965) has suggested that the rate of interchange between Pleistocene hunter–gatherer territories may have been as low as 1 per cent per generation. In support of this low level of inter-territorial migration and therefore high level of inbreeding, Coon cites the high incidence (75 per cent) of sacral anomalies and (56 per cent) of infant mortality during the first two years of life that has been found for hunter–gatherers that made use of a cave in northeast Morocco between about 12 000 and 10 500 years B.P. In making this suggestion, however, Coon ignores the observations (p. 120) on modern hunter–gatherers which indicate the relationship between food availability, territorial resistance, and frequency of inter-territorial migration.

14.5.1.4 The hunter–gatherer migration pattern in relation to the spread of agriculture and animal husbandry

At various points on the Earth's surface, ever since the early development of agriculture and animal husbandry in the Near East some 10 000 years B.P., hunter–gatherers have been in the process of encountering for the first time groups of agriculturalists. Quite possibly in parts of tropical South America and of South East Asia there still exist hunter–gatherer humans that have never encountered agriculturalist humans. In these places and in a few parts of Africa, Australia and the Arctic Circle, the vanguard of agricultural advance is still moving. However, by far the major proportion of humans are now agriculturalists (Fig. 14.16). The main concern of this section is to consider how individuals with the evolved pattern of hunter–

gatherer migration respond in terms of migration when they encounter an agricultural settlement. This is done by considering the eight components of the postulated pattern of hunter–gatherer migration in relation to this situation.

1–3 Extension of the familiar area by migration by sub-adults, particularly males, largely within the socially communicated familiar area. It is assumed that the first contact between a group of agriculturalists and a group of hunter–gatherers occurs most frequently when the agriculturalists settle within the territory of hunter–gatherers. The removal migration of the entire deme of agriculturalists, possibly as the result of the splitting up of a still larger deme (R.G. Ward 1961, Wada 1969), is also likely to occur within the combined familiar area of its members. Some contact between individuals of the two groups may therefore have occurred even before the agricultural deme has migrated. In any case, at some stage contact between the two groups involves some individuals extending their familiar area by migration beyond even their socially communicated familiar area.

Once first contact has been made, information concerning the agricultural group is likely to spread through the hunter–gatherer group by social communication. Although large-scale interaction between the maximum territorial units of both groups is likely to occur, at the individual level some individuals, presumably mainly sub-adult males, spend some time with the other group during familiar area extension.

4 Less correlation between the areas visited by sub-adults during familiar area extension and food availability and deme density than for any other ontogenetic class though stay-time may be longer in areas of high deme density. The postulated low correlation between the areas visited by sub-adults during familiar area extension and food availability and deme density is likely to result both in sub-adult agriculturalists spending some time with groups of hunter–gatherers and sub-adult hunter–gatherers spending some time with groups of agriculturalists. According to the hypothesis these individuals should be predominantly, if not entirely, males. During this sub-adult stage some copulation without further parental investment is likely to occur resulting in genetic interchange between neighbouring groups of agriculturalists and hunter–gatherers. The existence of past genetic interchange between neighbouring agriculturalists and hunter–

gatherers has been well demonstrated (e.g. Barnicot *et al.* 1972, Szathmary and Reed 1972).

Although familiar area extension should occur in both directions, the hypothesised hunter–gatherer migration pattern is likely to lead to hunter–gatherer sub-adult males spending more time amongst agriculturalists than agricultural sub-adult males spend amongst hunter–gatherers. This is because agriculturalists have larger, more dense, demes than hunter–gatherers and it has been argued that stay-time may be longer in areas of high deme density. I do not have data that test this prediction.

5 Less territorial resistance to immigrants when food is abundant than when food is scarce. If this suggested component of hunter–gatherer migration is correct it would be expected that when a group of hunter–gatherers comes into contact with a group of agriculturalists the latter group would show less territorial resistance to immigration by individuals of the former group than vice-versa. This is because food availability and nett rate of energy uptake per individual is likely to be greater for the agriculturalists than for the hunter–gatherers. This factor would tend to reinforce the effect described under component (4), and the nett result should be a greater proportion of hunter–gatherer sub-adults, largely males, spending time with groups of agriculturalists than agricultural sub-adults, again largely males, spending time with groups of hunter–gatherers.

6–8 Removal migration of a pair some time after pair-bond formation to that area within the combined familiar area of the pair that on the basis of the criteria of nett rate of energy uptake and deme size and density is assessed to be most suitable. When a sub-adult male agriculturalist pairs with a female hunter–gatherer there seems little doubt that, according to the hypothesised pattern of hunter–gatherer migration, the pair should migrate to rejoin the natal group of the male. Unless, that is, there is some reason, such as the male having been expelled by his natal group, for such a migration to be disadvantageous. Normally, however, by rejoining the natal group of the male the pair would be migrating to an area where nett rate of energy uptake and deme size and density are all greater. The situation is less clear, however, for pairs in which the male is a hunter–gatherer and the female is an agriculturalist. In this case the pair have within their combined familiar area one area in which the male has made the greatest investment and another area in which nett rate of energy uptake per individual and deme size and density are all high. Which area is assessed by the pair to be most suitable probably depends on the nett rate of energy uptake experienced by the male while residing with the agriculturalists as compared with the nett rate experienced when last residing with his natal group of hunter–gatherers. In any case, using the postulated criteria of nett rate of energy uptake and deme size and density, a sub-adult male hunter–gatherer seems more likely to settle away from his natal group when pairing takes place with an agricultural female than when pairing takes place with a hunter–gatherer female.

On the basis of the postulated pattern of hunter–gatherer migration, the following predictions can be made concerning the migration of male–female pairs when one partner is a hunter–gatherer and the other is an agriculturalist: more or less 100 per cent of agricultural male/hunter–gatherer female pairs migrate to join the natal group of the male; some hunter–gatherer male/agricultural female pairs join the natal group of the male, some the natal group of the female, but more join the natal group of the female than in the case of hunter–gatherer male/hunter–gatherer female pairs. In addition it can be predicted that some hunter–gatherer male/hunter–gatherer female pairs migrate to join agricultural groups if the male spent some time with such groups during familiar area extension as a sub-adult.

I have so far been unable to trace any quantitative evidence that permits evaluation of these predictions. In particular, it is not always clear whether genetic interchange, such as that between agricultural European caucasoids and some hunter–gatherer groups of North American Amerindians studied by Szathmary and Reed (1972) is the result of the migration of pairs or is the result of copulation without further parental investment. It is important to distinguish between these phenomena if the results are to be useful for the testing of these predictions. Similar problems arise when considering the genetic interchange between the hunter–gatherer Hadza of East Africa and their agricultural negroid neighbours studied by Barnicot et al. (1972). Nevertheless, some indirect evidence concerning these predictions is available and has been derived mainly from the review of African history by Murdock (1959).

First, let us consider the establishment of the zone of contact between the hunter–gatherer pygmoid

Africans and the agricultural negroid Africans in part of the tropical rain forest. When agricultural practice reached western North Africa from the Near East about 6000 years B.P. (or was alternatively discovered independently—p. 124), its progress ceased at the edge of the tropical rain forest. This was because, whatever its origin, the crops on which the agriculture was based were suited only to savanna conditions and none of the local cereals, vegetables and fruits could be adopted for cultivation under rain forest conditions. The rain forest remained, therefore, the domain of the hunter–gatherer African pygmies. Some 3500 years later, plants that had originated in the rain forests of Malaysia (e.g. rice, yams and bananas (*Musa* spp.)) arrived on the East African coast and spread westward, north of the tropical rain forest. Adoption of these plants some 2000 years B.P. removed the barrier to the spread of agriculture into the tropical rain forest. Throughout the region, therefore, pygmoid hunter–gatherers have come into contact with negroid agriculturalists. Contrary to the situation elsewhere in Africa, however, extinction of the hunter–gatherer economy has not occurred. Pockets of hunter–gatherers still exist. The point is, however, that these hunter–gatherers are not self-supporting. Everywhere hunter–gatherers and agriculturalists exist side by side in the tropical rain forest of Africa, they have entered into a symbiotic relationship. The pygmies hunt and gather in parts of the forest where the agriculturalists do not or cannot penetrate. The two groups then exchange products to their mutual benefit. In effect, therefore, hunter–gatherers and agriculturalists form a single social unit. In such a case, it may seem unjustified to claim that the situation provides support for the suggestion that hunter–gatherers have shown removal migration to the agricultural group rather than vice-versa. However, at the present time, whenever an individual does change life-style the migration is in the direction of from hunter–gatherers to agriculturalists. The continuing existence of hunter–gatherer 'groups' can be attributed to the difficulties facing agricultural practice in a tropical rain forest environment and hence the existence of areas in which a hunter–gatherer economy still permits optimum exploitation of the habitat.

Further perspective to the situation in the tropical rain forest and support for its implications concerning the direction of migration is found where hunter–gatherers and agriculturalists have contacted one another in the past elsewhere in Africa.

Over the rest of the continent expansion of agricultural practice has proceeded at the expense of another morphological group of hunter–gatherers, the bushmanoids. Until about 1200 years B.P. and the emergence of the negroid agriculturalists from the equatorial rain forest (Murdock 1959), the bushmanoids occupied all of Africa south of about 10°S (Tobias 1957, 1961). Many of the tribes of agriculturalists that now occupy eastern and southern Africa include among their social organisation a particular group of individuals which largely in-breed, which are concerned mainly with hunting, and which often show a number of bushmanoid characteristics. It is tempting to suggest that these individuals are the offspring of bushmanoid hunter–gatherer individuals that performed removal migration to a group of agriculturalists. Contrary to the situation in the tropical rain forest, however, virtually all of the land was suitable for cultivation, the hunter–gatherer economy disappeared, and migrant hunter–gatherers became integrated into the agriculturalists' society.

Finally, in Australasia, the inter-group migration of pairs since the arrival of agricultural European caucasoids has been almost entirely from the groups of hunter–gatherers to the groups of agriculturalists (Kirk, in Baker and Weiner 1966).

There is evidence, therefore, that the movement of pairs between groups of hunter–gatherers and agriculturalists has been predominantly from hunter–gatherers to agriculturalists. The evidence presented does not, of course, test the finer points of the predictions concerning differences between different types of pairings.

Some conclusions can now be drawn from this consideration of the possible interaction of the hypothesised hunter–gatherer migration pattern in relation to the spread of agriculture and animal husbandry. First, the extension of the familiar area that occurs during the sub-adult phase of ontogeny seems likely to have resulted in considerable diffusion of information among hunter–gatherers concerning agricultural practice. Second, hunter–gatherer sub-adults would be likely to spend more time among agriculturalists than among hunter–gatherers. Third, newly formed pairs would be likely, on the basis of hunter–gatherer criteria, to assess areas occupied by groups of agriculturalists to be more suitable than areas occupied by hunter–gatherers, though not all pairs would respond in the same way in this respect, depending in part on the cultural origin of male and female. This last aspect of

the migration pattern is likely to result in a marked flow of people, particularly newly paired females, into groups of agriculturalists. Altogether, therefore, the spread of agriculture at the expense of hunting and gathering is likely to have been greatly facilitated by the migration patterns that evolved in response to the selective pressures that had acted for the previous 10–20 million years. Furthermore, because the migration performed seems likely to be advantageous, the nature of selection on migration is unlikely to change during the spread of agriculture.

14.5.1.5 The hunter–gatherer migration pattern in relation to the group-to-group migration of early agriculturalists

Having concluded that selection on migration was unlikely to change during the spread of agriculture and animal husbandry, it would be expected that early agriculturalists would continue to show the hunter–gatherer migration pattern during group-to-group migration and to use hunter–gatherer criteria for habitat assessment. No evidence is available, of course, concerning the group-to-group migration of the early mesolithic or neolithic agriculturalists. It is possible, however, to examine the migrations of modern people, such as the New Guinea highlanders, that until very recently practised a neolithic type of agriculture. It has to be accepted, of course, that the migration patterns shown by these modern groups may not be representative of the migrations of early mesolithic and neolithic humans in general. This possibility would be most critical if such people were found to have a migration pattern that is grossly different from that hypothesised for hunter–gatherers. In which case, when we come to consider industrial humans, difficulty would be experienced in deciding whether the neolithic ancestors of industrial man were acted upon by selection similar to that experienced by their hunter–gatherer ancestors or by selection similar to that acting on modern neolithic humans. As it happens, this problem does not arise and the conclusion reached below is that not only do modern neolithic agriculturalists show the hunter–gatherer migration pattern but also the components of selection have remained the same.

Although the New Guinea highlanders employ, or rather employed at the time of study, a neolithic type of shifting agriculture, there are some respects in which they are probably atypical. In particular, animal husbandry is reduced to a minimum due to the unsuitability of the terrain, climate and vegetation (Coon 1965). Pigs only are reared. Even so, growth and reproduction of the livestock is so slow that individual pigs are killed and eaten at only irregular intervals. Large-scale killing and consumption of the majority of the herd occurs only every 5–10 years once the herd reaches a size of about 1·5 pigs/human individual (Brookfield and Brown 1973, Gardner and Heider 1969). Adjacent territorial units are usually separated by an uncultivated area upon which territorial interactions take place. Such areas are suitable for the grazing of pigs, and since adjacent territorial units, presumably not entirely by chance, alternate the killing of their pigs, the intervening land is in demand for grazing by only one unit at a time. The scarcity of game animals (Coon 1965) further means that

hunting, although important (Buchbinder and Clark 1971), makes only a small contribution to the diet as a whole. Consequently, subsistence depends almost entirely on cultivation of the sweet potato, *Ipomoea batatas*, which provides about 92 per cent of the calorific intake (Brookfield and Brown 1963).

Each maximum territorial unit of these agriculturalists of the New Guinea highlands occupies an area of 26–90 km^2, much smaller than that of comparable hunter–gatherers. Yet the maximum territorial unit may consist of 2000–4000 individuals (Brookfield and Brown 1963, Gardner and Heider 1969). In most of the New Guinea highlands the majority of the territorial interactions take the form of displays rather than wars and result in the death of only about 10–20 males per year from each territorial unit (Gardner and Heider 1969). Occasionally, however, there is wholesale failure of territorial

Fig. 14.17   Territorial interaction of neighbouring territorial units of New Guinea highlanders

[*Photo courtesy of Film Study Center, Harvard University*]

defence whereupon the whole or part of the losing territorial unit shows removal migration, usually to the territory of another unit which is recognised as sharing common descent (Brookfield and Brown 1963).

Most of the territorial interactions of New Guinea highlanders seem to be initiated in some way by conflict over pigs or occasionally females. Territorial units frequently organise small-scale raiding parties of males which make incursions into the neighbouring territory and attempt either: (1) to acquire pigs owned by the neighbouring unit; (2) to re-acquire pigs previously owned but which were lost to a similar raiding party organised by the neighbouring territorial unit; and/or (3) to restore the numerical balance when a fellow male has been killed in or by a raiding party. Larger-scale territorial interactions involving the majority of the males of the territorial unit tend to arise from a series of these smaller territorial interactions (Gardner and Heider 1969).

Territorial resistance is greatest, therefore, to intruding units, such as raiding parties, that threaten the scarce resource, pigs, and to maximum territorial units, which threaten all resources. Territorial resistance to individual intruders, however, is relatively small. Possibly this can be attributed to the fact that food in general i.e. mainly sweet potato, is not particularly scarce and that therefore an individual intruder does not constitute a major threat to any particular resource.

Having described some of the major characteristics of the socio-ecology of the New Guinea highlanders, such evidence as is available concerning group-to-group migration can be described under the headings of the hypothesised components of the group-to-group migrations of hunter–gatherers.

1–3 Extension of the familiar area by migration by sub-adults, particularly males, largely within the socially communicated familiar area. The maximum territorial unit of about 2000–4000 individuals of New Guinea highlanders is subdivided into roosting aggregations which are based largely on individuals of common patrilineal descent, except for the effects of removal migration. Within these roosting aggregations, further subdivisions of groups of 75–200 individuals can be recognised. The basic social unit is the family which is an age-graded male group (according to the terminology of Eisenberg *et al.* 1972) consisting of one adult reproductive male, one to several adult reproductive females and their young, and a varying number of adult but actually or effectively post-reproductive males and females (Brookfield and Brown 1963). However, even within a family group the sexes rarely roost together and aggregate daily only for the late afternoon meal. Males older than five years roost in a large communal hut. Each reproductive female has her own hut within the garden or grazing ground that is the responsibility of her mate. The female shares this roost with all her female offspring, male offspring younger than 5 years, any related post-reproductive females, and in some areas also the pigs. The distance between male and female roosts may be as much as 3 km (Fig. 18.11). Within the communal male roost, at any one time 11 per cent are transient sub-adults that were born into some other roosting aggregation, some of which may have been in a neighbouring territory. Of this 11 per cent, about 60 per cent have shown further removal migration after 16 months.

It seems clear, therefore, that New Guinea highlanders conform to the postulated pre-pairing behaviour of hunter–gatherers insofar as extension of the familiar area by sub-adult males takes place. Again I can present no evidence that confirms or contradicts the suggestion that the females also extend their familiar area during the sub-adult stage. There seems no doubt, however, that if such extension does take place, it is less pronounced than for males. Nor can I present evidence concerning the extent to which familiar area extension by migration by sub-adults takes place within the socially communicated familiar area.

4 Less correlation between the areas visited by sub-adults during familiar area extension and food availability and deme density than for any other ontogenetic class though stay-time may be longer in areas of high deme density. Data presented by Malcolm *et al.* (1971) suggests that a male may pair with a female from a wide range of deme sizes, which implies that a male includes a wide range of deme sizes within his familiar area. However, from the data presented, it is not possible to determine whether stay-time is greater in larger or smaller demes.

5 Less territorial resistance to immigrants when food is abundant than when food is scarce. At any one time only 79 per cent of males are found with the roosting aggregation into which they were born (Brookfield and Brown 1963). This figure is probably toward the lower limit of range of variation shown by hunter–gatherers. Certainly the proportion of inter-territorial pairings (i.e. about 50 per cent), is near the 60 per cent maximum reported for hunter–gatherers (Tobias 1964) and well in excess of the 21 per cent maximum reported for Australian hunter–gatherers (Tindale 1953). It would seem, therefore, that the greater nett rate of energy uptake per individual experienced by even neolithic agriculturalists when compared with hunter–gatherers elicits the suggested hunter–gatherer response of a decrease in territorial resistance to individuals.

6–8 *Removal migration of a pair some time after pair-bond formation to that area within the combined familiar area of the pair that on the basis of the criteria of nett rate of energy uptake and deme size and density is assessed to be most suitable, but with the male threshold for migration perhaps being higher than the female threshold.* It was suggested for hunter–gatherers that selection should favour habitat assessment mainly on the basis of the criterion of nett rate of energy uptake per individual but also on that of deme size and density, such that when nett rate of energy uptake per individual was high or increasing, demes of large size and high density should be assessed to be most suitable, though the threshold for migration of the male may be higher than that for the female. Pairs of New Guinea highlanders invariably settle within the natal group of the male. Of all the pairings censused by Malcolm *et al.* (1971) among the Bundi, 33·3 per cent involved the inter-roosting group migration of one or both partners. Of these, 5·1 per cent of the pairs settled within the natal group of the female, 8·4 per cent within the natal group of neither the male nor the female, and 86·5 per cent within the natal group of the male.

Although information concerning migration in relation to inter-deme variation in nett rate of energy uptake per individual is not available, Malcolm *et al.* (1971) do present some data concerning deme size and density. By evolved hunter–gatherer standards, the nett rate of energy uptake per individual experienced by New Guinea highlanders should be high. If the thesis concerning the criteria used by hunter–gatherers is correct, and if New Guinea highlanders still retain the migration pattern evolved during their hunter–gatherer ancestry, pairs, when they do not settle within the natal group of the male and as long as the nett rate of energy uptake per individual at the destination is high, should settle predominantly among the larger, more-dense demes. Of pairs formed in which the natal deme of the female consists of fewer than 200 individuals, 0 per cent settle within that deme. If the

natal deme of the female consists of 200–700 individuals, 0–21 per cent of pairs settle within that deme. If the natal deme of the female consists of more than 700 individuals, 11–44 per cent of pairs settle within that deme. Malcolm *et al.* (1971) discuss these figures in terms of exogamy, and without comparable figures for males, final interpretation of these figures is not possible and the extent to which they provide real support for the thesis of this section cannot be evaluated.

Agricultural practice in the New Guinea highlands is the responsibility of both males and females. Human sexual dimorphism (p. 116), however, has resulted in some division of labour such that the males perform heavy work such as clearing trees and digging irrigation channels, whereas the females plant, tend and collect crops (Gardner and Heider 1969). Care of the pigs is also the responsibility of the females and the young. Although there is this division of labour, sexual differences in agriculture would not seem to be sufficient for the nett rate of energy uptake necessarily to be greater for the pair among the natal group of the male than among the natal group of the female. However, some hunting is performed by the male. In addition, of course, males 'hunt' pigs within neighbouring territories (Gardner and Heider 1969). This hunting is also likely to be more efficient in country with which the male is familiar and in company with other males with which the male is familiar. As long as hunting, territorial interactions, and other activities that are performed more efficiently by males in cooperating groups persist and make a significant contribution to reproductive success, investment in the natal home range seems likely to continue to be greater for the male than for the female. This sexual difference in investment seems likely to be less, however, for agriculturalists than for hunter–gatherers. The continuing advantage of cooperative activities is also likely to maintain selection in favour of assessing larger, more-dense demes to be more suitable as long as the nett rate of energy uptake per individual remains high or increases.

Insofar as any conclusion can be drawn from this section, it seems that there is no reason to suppose that there is any evolved difference between the migrations of hunter–gatherers and the agriculturalists of the New Guinea highlands. On the contrary, what little evidence is available suggests that these agriculturalists are performing the migrations and using the criteria for assessment of habitat suitability that would be expected for hunter–

gatherers with a nett rate of energy uptake per individual of the level experienced by neolithic agriculturalists. Furthermore, selection seems likely to continue to act to favour individuals that show hunter–gatherer migration patterns using hunter–gatherer criteria for habitat assessment.

14.5.1.6 The evolution of some major features of the socio-ecology of human agriculturalists and industrialists

The agricultural, urban and industrial economy: the history of agriculture and animal husbandry over the past 10 000 years has been one of gradual sophistication and increase in efficiency. Irrigation was practised by 7000 years B.P. (Bobek 1962). The use of the horse, ass, ox and camel for transport and traction came later, between 6000 years and 5500 years B.P. at about the time that the plough and wheeled vehicles first appeared (Barnett 1971). Bronze was in use in Asia by about 5000 years B.P. and iron by about 3000 years B.P. The use of sheep's wool for clothing appeared in 4000 years B.P. and dairy farming was developed about 3000 years B.P., both in Mesopotamia. Water mills (2000 years B.P.) and crop rotation (1000 years B.P.) appeared later. The last century has seen the development of mechanised traction, improved techniques of soil fertilisation and pest control, and improved transport for the distribution of agricultural produce. Throughout the 10 000 years there has also been a gradual sophistication of utensils and buildings for storage.

Limited food storage must have been practised by Pleistocene hunter–gatherers (Howell 1970), but did not become a major aspect of human economy until about 7000 years B.P. (Barnett 1971). Some non-sexual division of labour in connection with the manufacture of tools and also defence and hunting etc. probably also existed among Pleistocene hunter–gatherers. The development of agriculture, however, and increase in food production and storage raised the possibility of some individuals specialising in certain activities and trading their activity or artefacts for food. It is a relatively short step from trading within a territorial unit to trading between territorial units, particularly when, as in this case, all factors favour such a step. For example, as artefacts increase in importance the requirements from the environment become more diverse. Yet at the same time, the adoption of agriculture is associated with a marked decrease in territory size (p. 125) with the result that the chances that the territory contains all necessary resources decreases. It has been argued earlier that increase in food availability, such as occurs with the development of agriculture, decreases territorial resistance and increases the incidence of the extension of the familiar area of individuals into neighbouring territories. All of these factors, therefore, favour the development of trading over wide areas.

Part of the increase in efficiency of the trading process was the development of monetary systems, thus overcoming the inherent inconvenience of direct exchange that results from the need to carry items that are often large and bulky to the place of exchange. The origins of trading can perhaps be traced back to the beginning of a sexual division of labour whereby the females gather and the males hunt and excess food of each type is shared and exchanged. The evolution of the trading instinct therefore seems likely to have taken place relatively early in hominid history. Once the trading instinct has evolved in relation to hunting and gathering, subsequent cultural evolution is likely to be a function of the frequency and nature of excess food and other resources produced by a particular group as already discussed. Once the trading principle is established, however, the transition to a money economy seems a relatively small cultural and psychological step. In this respect it is significant that all people, whether hunter–gatherers, neolithic or iron-tool agriculturalists, experience relatively little difficulty in transferring from a subsistence to a monetary economy (Kirk, in Baker and Weiner 1966, Gugler 1969).

In areas where particular resources are plentiful and access is relatively easy, aggregations of individuals specialising in trading resources are likely to occur. These individuals would not, themselves, need to be actively engaged in food production. It is this situation that probably led to the development of the first cities (Barnett 1971). Cities appeared first in the valleys of the Nile, Tigris and Euphrates (5000 years B.P.), Indus (4000 years B.P.) and perhaps independently in the valley of the Yangtse-kiang in China and in Central and South America (about 3000 years B.P.).

The development of nomadic pastoralists is considered in Chapter 25, but one phase of their development is relevant in the present connection. Not unreasonably, horse-riding first appeared among the nomadic pastoralists of the Eurasian steppe at about 4000 years B.P. (Bobek 1962) and for a time, before the practice spread, gave these groups considerable advantage in territorial interactions with settled farmers and city inhabitants. The advent of horse-riding almost certainly added new impetus to the development of the city as a defendable area. Thereafter, city development proceeded slowly and gradually. Not until the last two or three centuries in western Europe or the last hundred years elsewhere, with the arrival of the industrial revolution, did city development receive the major impetus of massive rural–urban migration.

Deme size and density: Before the adoption of horse-riding by settled agriculturalists, the size of territory was limited to the area that could be defended by walking individuals and through which information could be transmitted for efficient mobilisation of the territorial unit. The familiar area, of course, is likely to have been much larger than the territory. The familiar area of groups living next to bodies of water in particular may have been large, especially after the development of the sail about 6000 years B.P. (Barnett 1971). Further inland, however, it was the adoption of horse-riding by settled agriculturalists that was the first major step in the increase in size of their territory and familiar area. Similar use of the dromedary in desert regions had the same effect (Bobek 1962). Apart from an increasing use of horses for pulling wheeled vehicles, however, further major influences on the size of territory and familiar area as determined by migration probably did not arise until the last 200 years or so with the development of the steam engine, internal combustion engine, and finally air and space travel.

Social communication: The use of written numbers and alphabetic scripts dates from about 5000 years B.P. (Barnett 1971) and the role of communication by artefact has increased in importance ever since, particularly over the past 300 years or so with the development and spread of printing, postal services, and more recently of still and moving photography and television. Finally, as far as direct communication is concerned, the development of the telephone, radio, and communication satellites have had a dramatic effect on the potential size of the socially communicated familiar area of a modern industrial human.

14.5.1.7 The hunter–gatherer migration pattern in relation to rural–urban and urban–urban migration

It has been concluded so far that the migration pattern and criteria for assessment of habitat suitability that evolved in response to the selective pressures that act on hunter–gatherers persisted through the spread of agriculture and into neolithic agricultural societies. Furthermore, it has been concluded that the selective pressures that act on neolithic agriculturalists would tend to maintain hunter–gatherer migration patterns and continue to favour those individuals that assessed habitat suitability on the basis of hunter–gatherer criteria. If these conclusions are correct, and it has to be admitted that the evidence on which they are based is not exhaustive, then it might be expected that the migrations performed and the criteria for habitat assessment used by modern industrial humans are those characteristic of hunter–gatherers. A separate question is whether the selective pressures that act on modern industrial humans tend to maintain or to change this pattern of hunter–gatherer migration. The aim of these last two sections is twofold: first to review the evidence that the migration pattern of modern industrial humans is that of a hunter–gatherer; secondly to consider whether selection acts to maintain or change this pattern. This first section is once again organised in relation to the eight postulated components of hunter–gatherer migration. However, as ontogenetic and sexual variation in the migration incidence of modern industrial humans is discussed in detail in Chapter 17, only a minimum of data are presented here.

1 *Extension of the familiar area by sub-adult males and sub-adult females.* The following evidence, taken together, seems to support the suggestion that there is greater extension of the familiar area during the sub-adult phase of ontogeny than at any other time. In the United States, the first shift in centre of activity away from the parental home is a shorter distance than subsequent moves (Brunner 1957), a pattern that would be expected on the basis of familiar area extension. Studies in India (Anderson and Banerji 1962), Subsaharan Africa (Gugler 1969), Japan (Kasahara 1958), western Europe (L.B. Brown 1957), and the United States (Brunner 1957, Beshers and Nishiura 1961) all demonstrate that the greatest mobility is shown by individuals less than 30 years of

age. Furthermore, this trait can be traced back more than 100 years in the United States (Hotz 1955, Bauder 1959). Most of these studies, but particularly those in Subsaharan Africa, Japan and western Europe, indicate that most migrations are performed first by an individual while unpaired, though subsequent migration may occur after pairing. It is also evident from any census (e.g. General Register Office 1968) that this extension of the familiar area by unpaired sub-adults involves females as well as males.

2 *Greater familiar area extension by sub-adult males than sub-adult females.* This is a particularly difficult facet of migration to demonstrate quantitatively because of the conflicting effects of pre-and post-pairing behaviour on census data. Two types of evidence however, add support to the existence of this component in the migration of modern industrial humans. The first derives from the observation that among the mobile females of the under 30 years age-group, although some are unpaired, a greater proportion are paired than are males. This observation has been made in Subsaharan Africa (Gugler 1969), India (Anderson and Banerji 1962) and the United States (Allen *et al.* 1955). The second type of evidence derives from an analysis of pre-pairing migration distance (Fig. 14.18) and of long-distance

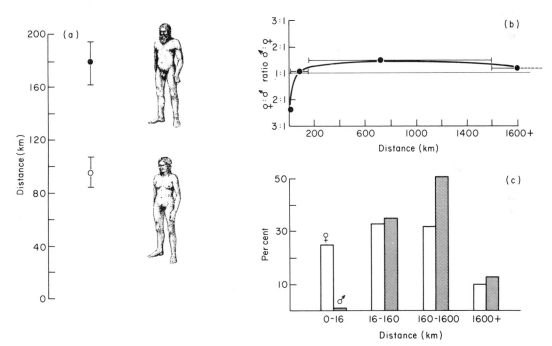

Fig. 14.18 Sexual difference in the removal migration of man, *Homo sapiens*: the pre-reproductive distance component within a predominantly urban ecosystem, 1900–1950, in the United States.

(a) Mean ± SE of the median distance between birth-place and place of mate-acquisition. Solid circle, males; open circle, females. Males migrate significantly ($t = 4.018$; $P < 0.001$) further than females. Between 1900 and 1950 there was a tendency for the mean median migration distance to increase for both males and females. The correlations between mean median migration distance and time ($r = 0.293$ for males and $r = 0.414$ for females; $n = 10$ in each case), however, is not significant ($P > 0.10$) for either sex.

(b) Change in number of male and female migrants over four distance categories. Solid circle, ratio of male:female

(when males outnumber females) and female:male (when females outnumber males) migrants with a given pre-reproductive distance component; bar, range of distances for which ratio calculated; horizontal line, position at which the number of male migrants equals the number of female migrants. Migrants moving less than about 20 km between birth and mate-acquisition site are predominantly female. Migrants moving further are predominantly male.

(c) Percentage of all males and all females with distance components in four different categories in 1950. Over 50 per cent of all males migrate further than 160 km between birth and mate-acquisition. Over 50 per cent of all females migrate less than 160 km.

[*Diagrams and analyses based on data obtained by Spuhler and Clark (1961) for Ann Arbor, Michigan, United States*]

migrants. The results of the analysis of migrants from Britain to New Zealand by L. B. Brown (1957) can be taken to be typical. It was found that the most common category of such long-distance migrants is unpaired males under the age of 30 years that have a history of frequent shorter-distance migration. Furthermore, the incidence of return in such international migration is high (Richmond 1969), thus indicating that such migration is better interpreted as extension of the familiar area rather than as a permanent migration.

The evidence is fairly convincing, therefore, that although familiar area extension is carried out by sub-adults of both sexes, there is greater extension by sub-adult males than by sub-adult females.

3 *Familiar area extension by migration by sub-adults based largely on socially communicated information.* Increase in the number and use of maps, books and communication media has resulted in the potential socially communicated familiar area of a modern industrial human consisting of virtually the whole surface of the Earth plus parts of the Moon. To this socially communicated familiar area could perhaps be added those other parts of the Universe that have been explored by sensory extensions such as telescopes and other instruments that have been operated both on Earth and extra-terrestrially.

Although the potential socially communicated familiar area is so vast, it does not necessarily follow, of course, even though it seems likely, that all familiar area extension by migration takes place within the socially communicated familiar area of the individual concerned. In any case, there are degrees of familiarity. Little interest seems to have been shown in this connection in relation to short-distance familiar area extensions by migration, though there is some evidence from Subsaharan Africa that short-distance, rural–urban familiar area extension by migration by sub-adults occurs to a large extent within the socially communicated familiar area (Gugler 1969). More extensive supporting evidence is found, however, when long-distance familiar area extensions are considered. L. B. Brown (1960) compared 100 unpaired males born in England that were about to migrate to New Zealand with 100 similar males that were not about to migrate. It was found that one of the main differences between the groups concerned the amount of socially communicated information concerning New Zealand to which the two groups had been exposed. Similarly, Richardson (1959) found

that migration from Britain to Australia was more likely to be performed by individuals with a history of social communication with an individual living in Australia. The 'chain' migration (p. 202) that results from the effects on familiar area extension of social communication has been a marked feature of the migration of southern Europeans to Australia (Price 1969).

4 *Less correlation between the areas visited by sub-adults during familiar area extension and food availability and deme density than for any other ontogenetic class though stay-time may be longer in areas of high deme density.* Lind (1969) provides data for a number of regions of England and Wales. The data consist of: deme density (inhabitants/acre); average weekly income per male; average weekly income per person; and nett migration (balance of immigration and emigration during one year) for various age and sex groups expressed as a percentage of that age group living in the region. The data are relevant to the early 1960's. I have carried out a number of partial correlations on this data, the results of which are given below. As far as the present component is concerned, this analysis is consistent with the postulated hunter–gatherer migration pattern in that there is less correlation between the areas visited by sub-adults during familiar area extension and income and deme density than for any other ontogenetic class. It can be assumed that food acquisition correlates positively with income (Mackenzie 1954).

Males in the 15–24 years age-group, the majority of which are unpaired (Fig. 17.8), show a partial correlation between nett migration and income per male with deme density partialled out of 0·448, which is not significant ($t = 1·227$; $df = 6$; $P > 0·1$). The partial correlation for this group between nett migration and deme density with income per male partialled out is 0·004, which is also not significant ($t = 0·010$; $df = 6$; $P > 0·1$).

Females in the 15–24 years age-group, which, however, show a greater proportion of paired individuals than males (Fig. 17.8), show a partial correlation between nett migration and income per person with deme density partialled out of 0·528 which is not significant ($t = 1·523$; $df = 6$; $P > 0·1$). The partial correlation for this group between nett migration and deme density with income per person partialled out is $-0·406$ which is also not significant ($t = 1·088$; $df = 6$; $P > 0·1$).

As far as comparison with older age-groups is concerned, these correlations are conservative in that a proportion of the individuals analysed, particularly the females, were no longer unpaired. Similarly, a proportion of the individuals in the older age-group, partial correlations for which are discussed below, are unpaired. Nevertheless, the 15–24 years age-group show no significant correlations with income or deme density whereas, as shown later, the 25–44 years age-group do show significant correlations with both.

The partial correlations given above for England and Wales in the mid-twentieth century support the suggestion of relatively low selectivity of destination during the sub-adult phase of ontogeny.

Since it has been shown that sub-adults have a relatively low selectivity of destination, it follows that the second part of this postulated component, that stay-time may be longer in areas of high deme density, will be difficult to detect without a large sample size. However, one feature of the above correlations may be relevant. To demonstrate this possible relevance, however, it is necessary to point out some features of the evolution of rural–urban migration. Virtual wholesale migration from the country to the towns was a predominantly nineteenth and early twentieth-century phenomenon in western Europe (Richmond 1969, J. Sutter 1969) and the United States (Hitt 1956). Elsewhere in the world rural–urban migration began later and in some areas is only just beginning. Confining our attention to western Europe, and in particular Britain, however, by the mid-twentieth century 60 per cent of all migrants were urban–urban and only 13 per cent rural–urban (Jansen 1969). In fact, if the few British counties with a deme density of less than 125 individuals/km^2 are ignored, there is now a correlation coefficient of -0.23 between deme density and nett migration gain (Lind 1969), albeit significant only at the 10 per cent level. The conclusion is, therefore, that although emigration from areas of relatively very low deme density, such as rural Scotland, continues, elsewhere in Britain a new trend has appeared in which continuing rural–urban migration is swamped by the migrations away from the more congested areas. This new migration trend is described further on p. 139. My own analyses for different age and sex groups support Lind's conclusion for all groups except one. Unlike all other age and sex groups which show a negative, though not always significant, partial correlation between nett migration and deme density with income partialled out, males of the 15–24 years age-group show a slight, though not significant, *positive* correlation, albeit of only 0.004. Straight correlation

between nett migration and deme density is also unique in my analysis in being positive, though again not significant, with a value of 0.140. These unique positive values could be taken to provide support for the hypothesis that although sub-adult males visit a wide variety of places during familiar area extension, stay-time is greater in areas of higher deme density. Females of the 15–24 years age-group, however, show the usual negative correlation, even though not significant, between nett migration and deme density. Whether this indicates that stay-time is greater in areas of low deme density or whether it is due to the relatively high proportion of paired individuals in this class cannot be evaluated from the data I have available to me.

5 *Less territorial resistance to immigrants when food is abundant than when food is scarce.* I can present little meaningful analysis in relation to this component except to point out that the increase in food availability that has resulted from the gradual sophistication of agriculture and the process of industrialisation seems to have been associated with an increase in the frequency of inter-deme migration. The greatest territorial resistance to immigrants at present still seems to be shown by modern hunter–gatherers and modern neolithic agriculturalists and by the inhabitants of those areas with a relatively low nett rate of energy uptake per individual. On the whole, territorial resistance probably has little influence on the frequency of inter-deme migration in modern industrial regions where nett rate of energy uptake per individual is relatively high.

6 *Removal migration of a pair some time after formation of a permanent or semi-permanent pair-bond.* Migration studies of modern industrial humans invariably seem to support the existence of this component of migration. Studies in southeast Brazil (Salzano 1971), the United States (Martin 1955, C. H. Brown, 1960, Walz 1960), England (Dickinson 1958, Jansen 1969), India (Anderson and Banerji 1962) and Subsaharan Africa (Gugler 1969) all suggest that the migration threshold for a pair is lower immediately after pairing than later.

7 *The migration destination is within the combined familiar area of the pair and should be the destination that is assessed to be most suitable.* The concept that the migration of industrial humans may take place within a familiar area first appeared when Boyce

et al. (1967) suggested that the probability of migration to a particular destination was a function of the degree of familiarity with that destination or, in their own terminology, a function of neighbourhood knowledge. Boyce and later Swedlund (1972) developed and then tested their hypothesis as follows.

Individuals are most familiar with areas near to their home and least familiar with areas distant from their home. If migration is a function of familiarity, therefore, removal migration frequency should decrease with distance. Using marriage distance (i.e. distance between natal sites of marriage partners) as a measure of removal migration distance, this prediction is tested for two nineteenth-century villages, one in England and one in the United States, by plotting marriage distance against frequency. The curves obtained, which are leptokurtic (i.e. in comparison to a normal distribution have a greater proportion of short and long distance movements) are taken to provide evidence in support of the hypothesis, though it is pointed out that number of previous visits to a particular site would be a better measure of familiarity than distance. Unfortunately, the leptokurtic curves obtained in these two studies, although they do not contradict the hypothesis, cannot be regarded as providing convincing supporting evidence. Similar leptokurtic curves are obtained for a variety of organisms (Bateman 1950), some of which are most unlikely to be migrating within a familiar area—such as pollen!

A convincing demonstration that the migration destination of newly formed pairs is a function of degree of familiarity does not as yet, therefore, seem to exist. A full analysis of the situation ought to take into account, according to my hypothesis, habitat assessment and sexual differences in investment (see Chapter 17) as well as degree of familiarity. Furthermore, it would be necessary to include the socially communicated familiar area as well as the area that is familiar as a result of migration. Although undoubtedly a large proportion of the migrations of pairs of modern industrial humans takes place within the area that is familiar through migration, a significant proportion probably takes place to destinations within the socially communicated familiar area but which have not previously been visited.

Discussion of the extent to which pairs migrate to the destination assessed to be most suitable is given below. There is, however, convincing evidence that most often the natal home range investment of the male is assessed by the pair to be higher than that of the female. As far as non-industrial human societies are concerned, Murdock and Wilson (1972) have reviewed the demographic information available for numerous demes throughout the world. From the data given concerning the destination of newly formed pairs it seems that by far the commonest behaviour is for the pair to settle among the natal deme of the male. Only in one type of deme, that with incipient agriculture, is it common for the pair to settle among the natal deme of the female. As this type of cultural organisation is just that in which areas other than that occupied by the natal deme of the male may well offer the pair a greater nett rate of energy uptake, the world variation observed is perhaps consistent with the hypothesis of habitat assessment that includes sexual differences in home-range investment. As far as modern industrial humans are concerned, the higher male than female migration threshold after pairing seems to be well documented. Studies in Peru (Lasker and Kaplan 1964), United States (Allen *et al.* 1955, Brunner 1957, Walz 1960) and India (Anderson and Banerji 1962) all indicate that newly formed pairs are more likely to settle in the natal deme or at least the familiar area of the male than of the female. For a discussion of the nature of the migration threshold in this type of situation see pp. 347–8.

8 *Nett rate of energy uptake per individual and deme size and density are the criteria on the basis of which pairs are most likely to assess habitat suitability, with greater weight being given to the former.* The prediction made when considering the migration of hunter–gatherer pairs was that nett rate of energy uptake per individual was the criterion likely to be given most weight in habitat assessment. In addition, when nett rate of energy uptake per individual was high or rising, areas of high deme size and density should be assessed to be most suitable. Conversely, when nett rate of energy uptake per individual was low or falling, areas of low deme size and density should be assessed to be most suitable.

Throughout the nineteenth and early twentieth century, there seems likely to have been a gradual increase in the nett rate of energy uptake per individual in western Europe. At the same time, industrialisation offered much greater and more diverse opportunity than previously of a money income in an urban environment. On both counts, therefore, of nett rate of energy uptake and deme size and density the evolved hunter–gatherer response seems likely to have been to assess urban areas to be more suitable than rural areas. However, not only was the nineteenth and early twentieth century in western Europe marked by massive rural–urban migration, it was also marked by large-scale migrations from Europe to the relatively low deme-density areas of America and Australia (Beijer 1969). To determine whether these migrations also fit with the hypothesised hunter–gatherer criteria for habitat assessment I have analysed figures provided by Cowan (1961) for migration from Ireland to the United States in the mid-nineteenth century.

In Ireland in the mid-nineteenth century the staple food was the potato, *Solanum tuberosum*. The ease of cultivation of this crop and the relatively high food value had resulted by 1840 in the island of Ireland having the highest deme density of man then known in the temperate zone (Cowan 1961). Peaks of removal migration to the United States at this time invariably coincided with a partial failure of the potato crop. In the autumn of 1846 a fungal disease, the potato blight, *Phytophthora infestans*, which was first recorded in Europe about 1840 (Brooks 1953) resulted in a total failure of the potato crop. This failure was followed by five years of intense removal migration during which some villages lost nearly 70 per cent of their occupants through migration (Cowan 1961). The destination of most of these migrants was the United States, though some went to Scotland and England.

Cowan (1961) presents data for deme density and percentage migrants during these years and suggests that there is a correlation between the two. Partial correlation, however, (Fig. 14.19) suggests that any apparent correlation between deme density and proportion of migrants is an artefact of the correlation between deme density and time after crop failure. In fact, proportion of migrants correlates better with time after crop failure than it does with deme density. Doubtless various interpretations of these correlations are possible but the one suggested here is as follows. Crop failure leads to an immediate decrease in nett rate of energy uptake per individual, whereupon areas within the familiar area with a relatively high nett rate of energy uptake per individual and with a relatively low deme size and density are assessed to be most suitable. Thereafter, nett rate of energy uptake per individual is affected much more by the recovery of the potato crop, which is a function of time, than by deme density. Eventual increase in the nett rate of energy uptake per individual leads to areas of high deme size and density (i.e. within Ireland), once again being

assessed to be most suitable.

My own interpretation of nineteenth and early twentieth-century migration patterns in western Europe for present purposes is therefore as follows. Pairs that experienced a relatively high or rising nett rate of energy uptake assessed areas with a high deme size and density to be most suitable and became part of the rural-to-urban migration stream. Pairs that experienced a relatively low or falling nett rate of energy uptake assessed areas with a low deme size and density to be most suitable and became part of the international migration stream. In all cases, however, migration was always to an area with a higher nett rate of energy uptake, and when this criterion was contrary to the criterion of deme density, less weight was given to the latter in fixing the destination.

Evidence that the use of these same criteria continues to the present day is presented in Fig. 14.20, which is an analysis of inter-regional migration in Great Britain (excluding Northern Ireland) in the last decade (circa 1970). The chief conclusion of this analysis is that modern industrial humans continue to assess habitat suitability on the basis of nett rate of energy uptake per individual (measured in the analysis as weight of protein per person per week) and deme density. When measured at a regional level, modern industrial humans, which may be considered to experience a relatively high nett rate of energy uptake per individual, assess regions of higher deme density to be more suitable than regions of lower deme density. Much less weight is given, however, to deme density than to protein uptake in the assessment of habitat suitability.

At the regional level, therefore, the migration of modern industrial humans is entirely consistent with the pattern that is suggested to have evolved during their 10–20 million years of hunter–gatherer ancestry. Until recently, this conclusion would also have applied to migration between smaller geographical units. In the last few decades in Britain a new pattern has emerged (Osborne 1956, Lind 1969). The data presented by Lind (1969) for England and Wales in the early 1960's show that there is a growing movement of pairs away from areas of high deme size and density. To a large extent, though not entirely, this migration has been associated with an increase in the distance of daily return migration (Lind 1969). According to Doxiades (1970), as an inhabitant of a city, until the past 10–20 decades, a human travelled up to 1 km from

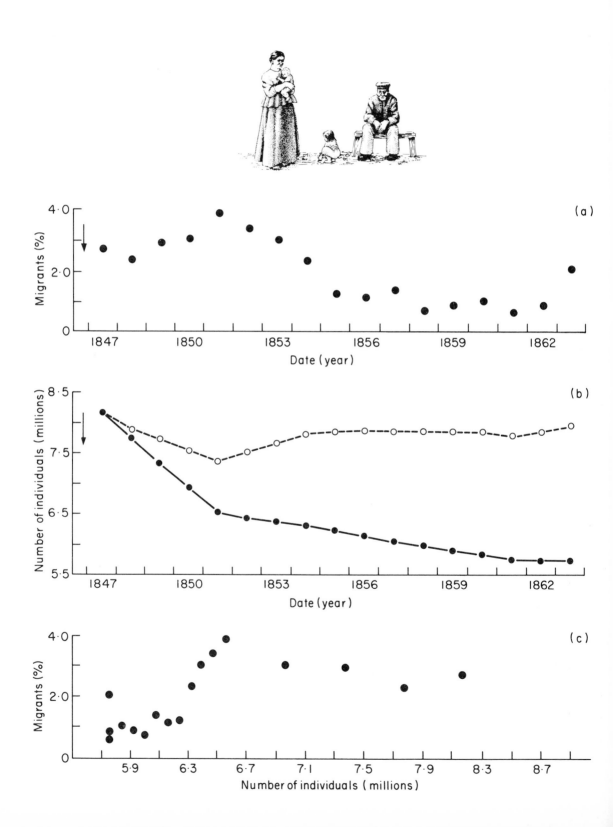

Fig. 14.19 Calculated removal migration of a group of man, *Homo sapiens*, with a predominantly agricultural economy: migrant/non-migrant ratio in relation to food shortage and population density in Ireland, 1846 to 1863

In Ireland in the mid-nineteenth century the staple food of *H. sapiens* was the potato, *Solanum tuberosum*. The ease of cultivation of this crop and the relatively high food value had resulted by 1840 in the island of Ireland having the lowest ratio of land to the population unit of *H. sapiens* then known in the temperate zone. In the autumn of 1846 a fungal disease, the potato blight, *Phytophthora infestans* (Mont.), which was first recorded in Europe about 1840 (Brooks 1953), resulted in the total failure of the potato crop. This failure had a marked effect on the population density and migrant/non-migrant ratio. The data in (a) and (c) refer to long-distance removal migration (i.e. from Ireland to England, Scotland and in particular to the United States), and not to removal migrations within Ireland.

(a) Change in migrant/non-migrant ratio with time after crop failure. Arrow, time of crop failure; solid dot, number of removal migrants during the year as a percentage of the number of individuals in Ireland at the beginning of the year. The migrant/non-migrant ratio reached a peak five years after the crop failure and thereafter gradually decreased with time. There is a significant $(P < 0.001)$ negative correlation $(r = -0.750)$ between time after crop failure and percentage migrants. A calculated partial correlation coefficient, with population density partialled out, is also significantly negative $(r_p = -0.641; P < 0.01)$.

(b) Change in population density with time after crop failure. Arrow, time of crop failure; solid dots, number of individuals present in Ireland; open dots, number of individuals present in Ireland plus the total number of emigrants from Ireland since 1846. Death-rate continues to exceed birth-rate for five years after the initial crop failure. Thereafter, birth-rate exceeds death-rate and the decrease in population density that continues for a further nine years is entirely the result of migration.

(c) Change in migrant/non-migrant ratio with population density. Solid dots, number of removal migrants that left Ireland in one year as a percentage of the number of individuals present at the beginning of the year. There is a significant $(P < 0.02)$ positive correlation $(r = 0.590)$ between percentage migrants and population density. However, this seems to be an artefact due to the correlation between population density and time after crop-failure that is evident from (b). Calculation of a partial correlation coefficient reveals that when time after crop failure is partialled out, the correlation between population density and percentage migrants is no longer significant $(P > 0.10)$ and is, in fact, negative $(r_p = -0.342)$.

In summary, in an agricultural economy, sudden and almost total crop failure followed by gradual recovery results in an initial increase in the proportion of long-distance removal migrants of man, *H. sapiens*, followed by a gradual decrease that is more a function of time than of population density.

[*Diagrams and analyses based on data in a British Parliamentary Paper of 1864 which is tabulated in Cowan (1961)*]

home during daily return migration. In the United States in the late 1960's, the equivalent average distance was 9 km and was increasing at the rate of about 120 m per year.

Increase in commuting distance cannot entirely account for the migration, however, for migration as analysed by workplace shows the same trend as migration as analysed by residence (Lind 1969). That this migration is performed mainly at the time of pairing is dramatically illustrated by the nett migration balance for 1960 to 1961 (expressed as a percentage of the particular age and sex group) for the high deme density area of London and the southeast of England as given by Lind. The 15–24 years age-group shows a nett migration balance of $+0.12$ for males and $+0.20$ for females. The 0–14 years age-group (i.e. young children migrating with their parents), however, shows a nett migration balance of -0.78 for males and -0.76 for females. Figures for the 25–44 years age-group are -0.86 and -0.82 for males and females respectively. The inescapable conclusion, therefore, is that whereas immigration of sub-adults into areas of very high deme size and density continues during familiar area extension (see also p. 137), as soon as pairing and reproduction commences other parts of the familiar area are assessed to be more suitable.

The meaning of this changed reaction to deme density over the past decade or so in terms of the thesis being presented here is not clear. It could be taken to reflect that the nett rate of energy uptake per individual in the large conurbations, such as London, has decreased over the last one or two decades, at least in comparison with less densely populated areas. In which case the hunter–gatherer response would be to assess as more suitable those areas with an equally high or higher nett rate of energy uptake per individual but with a lower deme size and density. An alternative but perhaps less likely possibility is discussed in the next section.

14.5.1.8 Selection acting on the migration of modern industrial humans

With the possible exception of the movement away from the most densely populated areas that began in Britain some time during the past few decades, I feel some justification in claiming that the migration pattern shown by modern industrial humans is that evolved during the 10–20 million years of hunting–gathering. All that remains now is to consider

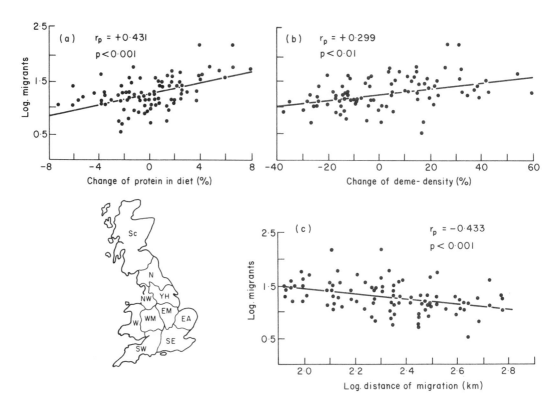

Fig. 14.20 Analysis of some components of the inter-regional migration pattern of newly formed male–female pairs of humans in England, Wales, and Scotland from 1965 to 1971

Inter-regional migrants. In this analysis a migrant is an individual that moves from one region of England, Wales, or Scotland to another. Ten regions are recognised as shown on the map. These are: Sc, Scotland; N, the north; YH, Yorkshire and Humberside; NW, the northwest; W, Wales; WM, west midlands; EM, east midlands; EA, East Anglia; SW, the southwest; and SE, the southeast.

The measure of migrants. In order to segregate data for the migration of male–female pairs during the years following pairing, census data for individuals from 1–14 years of age have been used. As pairs migrate more frequently with young than with older offspring, the sample is strongly biased toward the migration of male–female pairs with young offspring. The number of migrants (1–14 years) from one region to another during the period from April 1965 to April 1966 was then expressed as a percentage of the number of individuals between 5 and 14 years of age in the region of origin in 1971. This figure was then multiplied by 100. Finally, a logarithmic transformation was applied to the data. From each region of origin some individuals migrated to each of the other nine regions, giving a sample size of 90.

The criteria for habitat assessment (graphs (a) and (b)). The proportion of migrants was plotted against the criteria for habitat assessment, the use of which has been postulated to have evolved as a result of selection that acted on the hunter–gatherer ancestors of modern 'industrial' humans: nett rate of energy uptake per individual and deme density. Regression lines were then calculated.

In graph (a), proportion of migrants is plotted against percentage change in proportion of protein in diet that results from the migration. The change of proportion of protein in the diet was measured as follows. The total consumption (by weight) of food/person/week (excluding tea, coffee and other similar beverages) in each region was calculated from published data on the assumption that each egg weighs 56.7 g and that each pint of milk weighs 0.62 kg. The percentage (by weight) of this weekly consumption of food that consists of milk, cheese, butter, margarine, fat, eggs, meat and fish was then calculated for each region. The proportion of these protein-rich foods in the diet was thought to be a better measure of nett rate of energy uptake per individual than more gross measurements such as total food uptake per person per week or total calorie uptake per person per week in that it measures quality as well as quantity of food and yet at the same time is less affected by regional variation in number of children per household and other age reproductive differences. Analysis shows a highly significant ($P < 0.01$) positive correlation ($r = 0.849$), between average income per household for a region and the per cent (by weight) of protein-rich foods in the diet as measured above. Change of protein in diet as a result of migration is given by the per cent of protein-rich foods in the diet at the destination minus the per cent of these foods at the region of origin expressed as a percentage of the per cent of these foods at the region of origin.

In graph (b), proportion of migrants is plotted against the change of deme density experienced by the migrant. Deme density (where deme = occupants of a region), measured as individuals/km², was transformed to logarithms. Change of deme density is given by the logarithm of deme density at the destination minus the logarithm of deme density at the region of origin expressed as a percentage of the logarithm of deme density at the region of origin. Partial correlation coefficients (r_p) were calculated for migrants against each of these two variables, i.e. diet and deme density, for which the other variable and distance (see graph (c)) were partialled out. In each case there is a significant ($P < 0.05$) positive partial correlation which, however, is stronger for diet than for deme density. These partial correlations are consistent with the suggestion that modern 'industrial' humans are using hunter–gatherer criteria for habitat assessment.

Migration within the familiar area (graph (c)). In graph (c) the proportion of migrants from one region to another is plotted against the logarithm of the distance (km) between the centres of the two regions. The partial correlation coefficient (r_p) between these two variables has been calculated with changes in diet and deme density that result from the migration partialled out. This partial correlation coefficient is negative and highly significant ($P < 0.001$). This correlation could be interpreted as suggesting that migration distance is limited by the size of the familiar area established by exploratory migration. Although this is probably true, analysis of distance alone is insufficient to prove the point, for many non-calculated migrant organisms show a similar correlation.

Conclusion. The inter-regional migration pattern of newly formed male–female pairs of humans in England, Wales, and Scotland in circa 1970 is consistent with the continuing use of hunter–gatherer criteria for habitat assessment in that those areas in which protein uptake per person and deme density are relatively high are assessed to be most suitable, with greater weight being given to the former criterion (according to step-wise multiple regression of the data, 24.8 per cent of the variance of migration incidence can be attributed to migration to an area of higher protein intake but only a further 5.8 per cent can be attributed to migration to an area of higher deme density).

[Raw data from *Central Statistical Office* (1972a,b, 1973). Photo by R.R. Baker]

whether the selection that is acting at the present time on modern industrial humans is tending to perpetuate or to change the hunter–gatherer pattern.

Let us consider first the selective pressures acting on sub-adults during familiar area extension. The hypothesised selective pressures acting during this phase result from the advantage of: (a) finding and pairing with an optimum mate; (b) in the case of a male, copulation with a number of different females but without further parental investment; (c) exploration of a variety of potential sites to which migration could occur after pairing; and (d) survival. The migration pattern shown results in extension of the familiar area to include a variety of places with the possibility of a greater stay-time in areas of high deme density. There seems to be no reason to suppose that selective pressures (a), (b) and (c) will act to change the hunter–gatherer migration pattern. However, it is well documented that survival is less in areas of high than in areas of low deme density (e.g. Registrar General 1894). It is possible, therefore, that for some time, perhaps 300 years, selection has been acting against sub-adults having a longer stay-time in areas of high deme density. Multitudinous data from all over the world is available that would permit evaluation of this point. However, as the greatest selection has probably acted as a result of rural–urban migration and as this migration was most marked toward the end of the nineteenth century, I have taken data from the 1892 census of England and Wales (Registrar General 1894) to analyse this possibility. Correlation of the deaths per 1000 living in relation to deme density by county for the 15–24 years age-group gives a correlation coefficient of -0.172. Although, surprisingly, the correlation is negative, it is not significant even at the 10 per cent level ($t = 1.145$; $df = 43$). The surprising conclusion, therefore, is that unlike all other ontogenetic classes the survival of sub-adults is either unaffected or may even be improved by an increase in deme density. It seems, therefore, that there is or has been no selection acting on sub-adults to change the hunter–gatherer migration pattern.

Finally, let us consider the selective pressures acting on newly formed pairs of modern industrial humans. The hypothesised selective pressures acting during this phase result from the advantage of: (a) producing and successfully rearing offspring; and (b) survival. The selective pressures acting on males and females in relation to the performance of extra-pair copulations (Trivers 1973) are probably of minor importance in the present connection. The migration pattern shown results in removal migration of one or both members of the pair soon after pairing to a site within their combined familiar area that is assessed to be most suitable. The criteria on which habitat assessment is based are nett rate of energy uptake and deme size and density. Habitats in which nett rate of energy uptake per individual is high are assessed to be most suitable. In addition, when nett rate of energy uptake per individual is high or increasing, areas with a large deme size and high density are assessed to be most suitable. Conversely, when nett rate of energy uptake per individual is low or decreasing, areas with a small

deme size and low density are assessed to be most suitable.

Selective pressures (a) and (b) seem likely to have acted on the criteria on which habitat assessment is based. It has been argued that culturally and psychologically food and money are readily interchangeable (p. 133). Furthermore, analysis of data presented by the Central Statistical Office (1973) for regions of England, Wales and Scotland for the early 1970's gives a correlation coefficient of $+0.622$ ($t = 2.247$; $df = 8$; $P < 0.05$) between income ($\pounds$/week/male older than 21 years) and consumption of meat and fish (oz/person/week). This conclusion, that there is a relationship between income and nett rate of energy uptake per individual, is in agreement with a similar conclusion reached earlier by Mackenzie (1954). Given this relationship, the question presents itself, therefore, of whether number of offspring correlates with income. On the surface, this is a perfectly straightforward question. Yet it has presented one of the main challenges to modern sociologists. However, all of the most recent analyses (e.g. Easterlin 1969, D.V. Glass 1969, Kuznets 1969) seem to be agreed that although there is no correlation between income and number of offspring produced when different countries and different division-of-labour groups are compared, there is a clear positive correlation when analysis is confined to a single division-of-labour group within a single country. Apparently, therefore, the individual's experience and habitat between birth and pairing has a more marked effect than income on number of offspring produced. For any given set of ontogenetic experiences, however, an increase in income results in an increase in number of offspring. As far as the individual is concerned, therefore, migration to a place where that individual receives a higher income is likely to result in that individual producing more offspring. Selection is therefore likely to perpetuate the habit of assessing habitat suitability on the basis of the money income and/or nett rate of energy uptake experienced such that places with a high income or nett energy uptake are assessed to be most suitable.

There remains, therefore, the question of whether migration to areas of large deme size and high density when nett rate of energy uptake is high, as in rural–urban migration, continues to be advantageous in relation to selective pressures (a) and (b). Before considering this question in detail, however, a general introduction is necessary.

Throughout the world, the process of industrialisation, urbanisation and 'modernisation' has been accompanied by an apparently fixed sequence of change in birth and death rates that has shown little, if any, geographical or historical variation. The main stages have been summarised by Bourgeois-Pichat (1951). Two centuries or more ago a rough equilibrium existed between a very high birth rate (i.e. birth rate = number of births per female lifetime = 'fertility' in sociological terminology) and a high death rate. The first stage of modernisation is accompanied by an initial decline in death rate without a corresponding change in birth rate, with the result that deme size and density increases rapidly. Scandinavia and France entered this first stage as early as the mid-eighteenth century. The second stage is marked by a decline in the birth rate. In most of western Europe this stage was reached about 1880, but France began on this stage before 1800. This decline in the birth rate is universally greater in urban than in rural areas. The third stage consists of an even more pronounced decline in the birth rate until, as in some western European countries at present, the birth rate is less than the death rate.

Everywhere that modernisation has entered the second stage, the rural birth rate exceeds the urban birth rate. In Korea, for example, which entered the second stage during the 1950's, the urban birth rate fell twice as fast after 1960 as the rural birth rate (Kirk 1969). In Europe between 1955 and 1962 the rural:urban ratio of birth rates (i.e. number of children less than 5 years of age per female between 15 and 49 years) ranged from 1·61 in Portugal to 1·05 in Scotland (Glass 1969). There seems little doubt, therefore, that rural–urban migration results in occupation of an area in which reproductive success is likely to be reduced unless there is a compensatory reduction in death rate, both of offspring and of parents. As already indicated, this is not the case. My analysis by county of England and Wales of the data presented by the Registrar General (1894) shows that in 1892 there was a significant ($t = 3·231$; $df = 43$; $P < 0·01$) correlation coefficient of 0·442 between deme density and death rate of humans less than one year old. In addition to decreased survival of offspring there is a significant ($t = 3·140$; $df = 43$; $P < 0·01$) correlation coefficient of 0·432 between deme density and death rate of humans in the 25–44 years age-group.

The only conclusion possible, therefore, is that assessment of habitats with a large deme size and high density as most suitable when nett rate of energy uptake per individual is high or increasing is no longer advantageous in an industrial society. Assessment of habitats with a low deme size and density as most suitable when nett rate of energy uptake per individual is low or decreasing, however, continues to be advantageous. It might be expected, therefore, that, with time, individuals that assess as most suitable habitats that have a small deme size and low deme density even when the nett rate of energy uptake per individual is high or increasing will come to predominate in the human population. It could even be suggested that the negative partial correlation between nett migration and deme density that has become apparent in Britain over the past two decades or so represents the first manifestation of adaptation to this selection. However, selection has been acting as described for 10 generations at the most in Britain. Over large parts of the country such selection has been acting for only one or two generations. It seems unlikely, therefore, even in Britain with its relatively long industrial history, that the change in selection that takes place with industrialisation could have resulted in such a major change in migration pattern. The interpretation given on p. 141 in the previous section still seems to be more acceptable.

In conclusion, therefore, modern industrial humans seem to show the pattern of migration that is suggested to have evolved during the long hunter–gatherer period of their history. Furthermore, in most respects selection continues to act to perpetuate this pattern. Only in one respect does selection appear to have acted or be acting to change the pattern. Thus, although it seems to remain advantageous to assess habitat suitability on the basis of money income and/or nett rate of energy uptake per individual, it seems that given two habitats in which this criterion is identical, over the past 100–200 years selection has favoured assessment of the area with the lower deme size and density to be more suitable. This is in contrast to pre-industrial times when given prior experience of a high or rising nett rate of energy uptake it was likely to have been advantageous to assess the area with the higher deme size and density to be most suitable, at least, that is, in areas in which activities that influenced reproductive success were performed more efficiently by large groups of cooperating individuals.

14.5.1.9 Summary

The human lineage subsisted on a hunter–gatherer economy for a period of 10–20 million years. Some modern humans have subsisted on an agricultural/animal husbandry economy for 10 000 years and on an industrial economy for at most 300 years.

The selective pressures acting on the migration pattern of human hunter–gatherers have been discussed and on the basis of the conclusions reached the following hypothesis of hunter–gatherer migration has been developed.

The sub-adult period of ontogeny should be characterised by an increase in the rate of familiar area extension by migration. Rate of extension should decrease once more after pairing (p. 293). Sub-adult males should extend their familiar area more than sub-adult females. Familiar area extension by migration occurs to a large extent within the socially communicated familiar area. There should be low selectivity of destination during familiar area extension, places being visited that show a wide variety of food abundance and deme size and density. Stay-time may be longer in larger, more dense demes. Demes experiencing a relatively high nett rate of energy uptake per individual should show less territorial resistance to intrusion by individuals than should demes experiencing a relatively low nett rate of energy uptake per individual. The migration threshold of both male and female should be particularly low just after pairing. Most often the threshold of the male should be higher than that of the female, due to different learning and other investments by the two sexes, with the result that the pair return to the natal deme of the male. Habitat assessment involves criteria other than previous investment, however, and if a habitat within the combined familiar area of the pair is assessed to be more suitable than the natal habitat of the male, the pair will migrate to the most suitable habitat. The criterion likely to be given most weight in habitat assessment is nett rate of energy uptake per individual. Of lesser weight, but also likely to be used in habitat assessment is deme size and density. When the nett rate of energy uptake is high or increasing, larger, more dense demes should be assessed to be most suitable. When the nett rate of energy uptake is low or decreasing, smaller less-dense demes should be assessed to be most suitable.

Evidence is presented that indicates that this hunter–gatherer migration pattern has persisted through human cultural evolution and is the pattern still shown by industrial humans at the present time.

The great migration streams of human history: from groups of hunter–gatherers to groups of agriculturalists; rural–urban migration; international migration; and urban–urban migration; can all be attributed to the evolved hunter–gatherer migration pattern.

Evidence is also presented that throughout most of human cultural evolution selection has acted to perpetuate rather than to change the hunter–gatherer migration pattern. However, over the past 100–200 years in industrial societies there seems likely to have been a disadvantage in rural–urban migration. Although rural–urban migration is quite consistent with hunter–gatherer criteria for habitat assessment, pairs performing the migration seem likely to have suffered a reduction in reproductive success in comparison with urban–rural and rural–rural migrant pairs. This disadvantage seems to have applied only to pairs, not to sub-adults.

14.5.2 The calculated removal migration of social units of social insects

14.5.2.1 Communication systems in social insects

In most social insects so far investigated, including various ants and termites (E.O. Wilson 1965), a perennial tropical wasp (Lindauer 1967b), honeybees (Lindauer 1967a,b, Frisch 1967) and various meliponine stingless bees (Lindauer 1967a,b), some form of communication system has been discovered, usually in relation to foraging or colony migration behaviour. Despite their close taxonomic relationship to the honeybee and stingless bees, however, no such communication system has been found in the bumblebees (Bombini) of temperate regions (Lindauer 1967b).

Ant and termite foragers, on encountering a food source, lay a scent trail during their return to the colony (Wilson 1965). In ants the secretion used is that of Dufour's gland which opens near the tip of the abdomen. Other potential foragers, upon encountering the trail, which persists for only a few minutes, depending upon the characteristics of the surface on which it is laid, follow the trail until they reach the food source. As long as the food source remains abundant, each new recruit reinforces the trail as it returns to the nest. Most of the communication of ants and termites, not only in relation to foraging but also in relation to alarm, nest repair, etc. seems to be based on scent production and olfaction (Wilson 1965, Schneirla 1971).

Communication among stingless bees and, as far as is known, wasps (Lindauer 1967b) consists firstly of the forager alerting potential recruits by movements, sounds and/or scent production. In most cases, the only information conveyed to the recruits while in the nest is the nature of the food source, as a result, at least in bees, of the scent of the food source being carried on the hairs of the body. In some species this may be the only information exchanged after other individuals have been alerted. In which case the novices leave the nest and search in all directions for this scent. In other species, however, such as primitive Meliponini and the tropical wasp, *Polybia scutellaris*, the forager may lead a group of recruited novices from the nest to the food source. In addition, the foragers of some species, such as *Trigona postica* in Brazil, mark out, by stopping at 2–3 m intervals, an olfactory trail between the food source and the nest. In none of the stingless species so far investigated, however, is information concerning direction and distance conveyed during the process of alerting other individuals within the nest.

On the other hand all four species of the genus *Apis*, the honeybees, which includes the European (but which originated in India and southeast Asia) honeybee, *A. mellifera*, convey both distance and direction to new recruits by means of the well-known bee dance. Points of similarity, which may also imply a common origin, between the communication system of honeybees and stingless bees include alerting potential recruits by the process of movement and sound production, both of which probably also convey the quality of the food source, and the conveyance of the nature of the food source

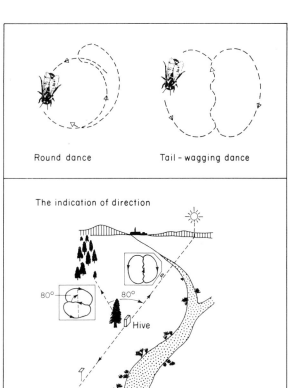

Round dance   Tail - wagging dance

The indication of direction

80°   80°

Hive

Fig. 14.21   The round and tail-wagging dances of the honeybee as mechanisms for the communication of the distance and direction of a resource

The round dance is used to announce resources discovered near to the nest. Different subspecies of *Apis mellifera* change from a round to a waggle dance at resource distances of 10, 30, 60, and 80 m.

The waggle dance consists of the execution of a figure-of-eight with the abdomen being waggled during the oblique part of the figure.

Distance is indicated by the tempo of the tail-wagging dance. With increasing distance to the resource the number of circuits per unit time decreases and the duration of the individual circuits increases. The correlation between dance tempo and distance exists for the range from about 80 m (or the nearest resource distance for which a tail-wagging rather than a round dance is performed) up to about 11 km.

Direction is indicated by the direction of the wagging run. When the bee is dancing on a horizontal surface in front of the hive the wagging run points directly toward the resource, the dancer maintaining the same angle relative to the Sun as she held previously during her flight from hive to resource. On the vertical surface of a comb inside the hive the dancer transposes the solar angle into a gravitational angle. As shown by the insets in the bottom figure, when the resource is in the same direction as the Sun the wagging run is directed vertically upwards and when in the opposite direction from the Sun it is directed vertically downwards. The direction of resources that deviate clockwise or anticlockwise from the Sun's direction are transposed by the same angle clockwise or anticlockwise from vertically upwards on the comb surface.

[*Re-drawn from Frisch (1954, 1967)*]

as a result of the flower or other scent carried on the body. In addition, the food source may be marked olfactorally by a secretion from Nassanoff's gland on the head. Unlike the alerting movements of stingless bees, however, the alerting movements or 'dance' of the honeybee have a precise form that conveys distance and direction.

When the food source is relatively near to the nest, the returned forager performs a round dance (Fig. 14.21). When the food source is further away the returned forager performs a 'waggle' dance (i.e. a figure-of-eight with the abdomen being 'waggled' during execution of the oblique part of the figure). Different subspecies of *A. mellifera* change from a round to a waggle dance at food-source distances of 10, 30, 60 and 80 m (Lindauer 1967a,b). The Indian honeybee, *A. indica*, changes over at only 2 m. The forager appears to assess distance on the basis of the energy expended during the outward flight rather than on distance as such or on the time taken. The precise way in which distance is communicated during the waggle dance remains as yet unknown (Lindauer 1967a) though the tempo of the dance (Fig. 14.21) seems most likely to be the critical component.

Direction is communicated during the waggle dance by the angle that the bee makes to the vertical, or to the Sun if the dance is performed on the hive entrance platform. Aspects of the mechanisms involved in the orientation and navigation of honeybees are discussed in Chapter 33 (p. 873).

Groups of experienced foragers that all forage in the same area aggregate in a localised part of the hive to exchange information about the area (Lindauer 1967a). Foragers working more distant areas use sites within the hive that are more distant from the hive entrance. Inexperienced novices may gather first at those dancing sites where the nearer food sources will be announced.

14.5.2.2 The exploratory–calculated sequence in the removal migration of social units

Social insects show a number of different types of removal migration, and in any discussion such as this it is important to specify which type of removal migration is under consideration. The simplest type of removal migration is probably that shown by individual, winged male and female reproductives of termites and ants. In this type of removal migration the unmated reproductive individual performs removal migration, mates, and eventually establishes another colony some distance away from the parent colony. The reproductive females of some bees and wasps, particularly in temperate regions,

Fig. 14.22   Two foraging workers of the honeybee, *Apis mellifera*, collecting water

[*Photo by Axel Doering, courtesy of Frank W. Lane*]

also perform removal migration unaccompanied by workers. In this case, however, mating usually occurs, although possibly at some distance from the parent colony, during the period that the female still roosts in the natal nest. In temperate regions these mated females may be the only individuals to survive the winter, forming a new colony the following spring some distance from the site of their natal colony. These removal migrations, particularly those of ant and termite reproductives, are almost certainly non-calculated and are discussed in Chapter 16.

All four major groups of social insects, however, contain species in which from time to time removal migration is also performed by social units consisting of one or several reproductives and a number of workers. It is this type of removal migration that is discussed in this section in relation to extension of the familiar area by migration and social communication. Three examples are given. The first concerns the removal migration of entire colonies of the fire ant, *Solenopsis saevissima*. The second concerns the bivouac-to-bivouac removal migrations of entire colonies of doryline army ants. The third concerns the removal migration of the colony-establishing swarms of the honeybee, *Apis mellifera*.

Worker individuals of the fire ant, *S. saevissima*, are continually extending their familiar area (E.O. Wilson 1962a,b). Furthermore, if a new and therefore previously unexplored area is created within the previous familiar area of an ant colony by the introduction of a relatively large object, the first individual to encounter the object recruits other individuals to explore the object by laying a scent trail between the object and the nest. The trail is laid by this and other individuals even though no immediate 'reward' in the form of food has been obtained from the object. Recruitment of further individuals by trail laying continues until the object has been thoroughly explored.

If, during such exploratory migration, workers discover a nest site that is more suitable than the present nest site, scent trails are laid on the homeward journey. Other workers recruited by this trail investigate the new site and may or may not lay trails themselves depending on the suitability of the site. If removal migration of the colony is initiated, trails are laid by individuals migrating back to the old site to collect and transport more brood. Newly emerged adult workers, pre-reproductive but fertile adult males and females, and the functional reproductive female of the colony migrate under their own power.

The criteria by which h_1 and h_2M are compared in this case do not seem to have been analysed. Presumably the migration cost is fairly high in that foraging activity must be reduced and in that the danger of predation on brood and reproductives is likely to be greater than if removal migration did not occur. The value of M is therefore likely to be relatively low. In this case the difference in suitability between the two nest sites must be large before removal migration is likely to be advantageous.

The predatory doryline army ants of the genus *Eciton*, studied in Panama by Schneirla (1971), establish daily bivouacs during the nomadic phase of their movement cycle (see Fig. 17.30). A bivouac is a temporary base, usually in a hollow log or some similar protected site, in which the queen and brood roost for the day in association with a cluster of workers. The composition of the cluster is constantly changing as individuals leave the cluster, move along a chemical raiding trail laid down by the continuous ant traffic between the bivouac and the raiding area, catch some item of food, and move back to the bivouac to feed their prey to the larvae. These raiding trails may stretch for hundreds of metres from the bivouac yet consist of an unbroken column of ants moving between the bivouac and the raiding area.

Daily removal migration from one bivouac site to another is characteristic of the nomadic phase of army ants. The nomadic phase coincides with the later stages of development of the current batch of larvae. When these larvae pupate a statary (see p. 345) or non-migratory phase is initiated during which a single bivouac is used. The statary phase persists throughout the pupal stage of the current brood. During this time the queen lays the next batch of eggs and the young larvae pass through the early stages of development. The next nomadic phase is initiated when all of the pupae have produced workers.

The distance between the bivouacs of consecutive days during the nomadic phase ranges from 100 to 450 m for *Eciton* spp. Removal migration from the current bivouac site to the next is a largely nocturnal phenomenon and takes place along one of the raiding trails to some site that has been encountered by workers using that trail during the previous day's raiding. The removal migration is therefore initiated on a calculated basis. Several raiding trails may radiate from a single bivouac and the mechanism by which the colony, which may consist of between 150 000 and 2 000 000 individuals, reaches a decision as to the most suitable site for the next bivouac does not seem to be known as it is for honeybees. Toward the end of the nomadic phase when the current batch of larvae are nearly full grown, a number of caches of food are made during the day along the raiding trail and these caches are then used to feed the larvae during the removal migration. The workers form the vanguard of the removal migration and start to assemble in the new bivouac site while contact with the old bivouac site which still contains the queen is maintained by workers moving backward and forward between the two sites to collect and carry the brood. Only once a large proportion of the brood has been removed does the queen leave the old bivouac site and migrate along the trail to the new site accompanied by an entourage of workers. Once the queen has left the bivouac the remaining brood is collected. The rearguard of workers then pass along the trail to the new bivouac site and only then does the 'living trail' of ants between the two sites disappear.

The criteria by which a new site is assessed are unknown, though possibly, apart from appropriate conditions of shelter and microclimate at the bivouac site itself, the relative abundance of invertebrate and other prey in the surrounding area is given particular weighting in habitat assessment.

Much more is known about the establishment of new colonies of the honeybee, *Apis mellifera*, largely due to the extremely elegant work of Lindauer(1955). At some time in the late spring, when nectar is abundant, the worker force of the bee colony has built up in numbers, and there is insufficient space in the existing nest for the bees and for storing food (Lindauer 1967b), the probability that removal migration will occur is at its greatest (Chapter 17). When removal migration does occur, the old queen and about half of the workers vacate the colony leaving behind the other half of the worker force and a new queen which may not yet have emerged from the pupa. Before leaving, the workers appropriate large quantities of honey from the colony store. The migrant swarm will feed on this appropriated honey over the next few days until the new colony has been established and workers are available to forage for further nectar and pollen. Soon after leaving the parent site, the migrant swarm settles nearby as a hanging cluster. This cluster is not a random aggregation of individuals, however, but has a well-organised structure (Meyer 1955). The interior of the cluster consists of loosely

branching chains of bees with space between for other bees to pass. On the outside, older bees form a compact cover of about three layers in which the bees are close together. Only one gap, which is used as a flight entrance, is left in this living cover. The cluster remains in position for a period of hours or days and does not depart until a decision has been made concerning the most suitable destination.

Within fifteen minutes of the cluster having formed, scout bees start to dance on the surface of the cluster. In so doing, the scouts are continuing a process that was begun up to three days before the swarm departed from the parent nest. This is the time that scout bees first become active and start to visit potential nesting sites around the site of the parent colony. It seems likely that considerable exploratory migration may be performed by these scout bees at this time. After discovering a potential site the scout bee returns to the parental nest and starts to dance, communicating information concerning the site discovered to other individuals. This dance is similar to that performed by returning foragers which communicate information concerning a food source. There are differences, however, such that fellow workers are unlikely to confuse the information being given. In particular scout bees do not distribute nectar to their fellows but solicit food instead. Furthermore, instead of dancing for one or two minutes as would a forager, a scout may dance non-stop for 15–30 minutes and, with interruptions, for many hours. Presumably this marathon dancing is important not only in conveying information concerning a particular site but also in the recruitment of individuals to join the swarm. By the time that the hanging cluster is formed, therefore, a number of potential destinations have been discovered and the process of deciding to which site the swarm should migrate begins.

The information content of the dance is the same as for foraging. In particular the 'intensity' of the dance conveys the suitability of the site. The criteria by which suitability is assessed are considered below. First, however, let us consider the process by which a decision is reached concerning the site to which the swarm should migrate. At first, different scouts announce different potential nesting sites (e.g. a hollow tree, hole in a wall, empty beehive, hole in the ground, etc.). Each dancing scout recruits fresh bees to examine the site being announced. At intervals the scout returns to the site and, it is presumed, attracts to the right spot other scouts that have gone to the general area by means of the pheromone produced by Nassanoff's gland. The number of fresh bees recruited to examine and in turn dance in support depends in part on the intensity of the dance performed by the original scout and in part on the assessment of the site made by the various recruits. The intensity of the dance, relating to the suitability of the site, seems to be based on an innate rather than learned scale of values. Direct comparison of sites has to take place, however, before a decision can be reached. A bee dancing in support of a moderately suitable site, if it sees another bee dancing more intensely than itself, will run after that bee and go and inspect the site being announced. If that site is assessed to be more suitable than the site the individual was previously announcing, on returning to the cluster the individual dances in support of the new site. Eventually, this process results in all scout bees dancing in support of the same site. Not until this situation is reached does the cluster move. Figure 14.23 provides an example of the fluctuations

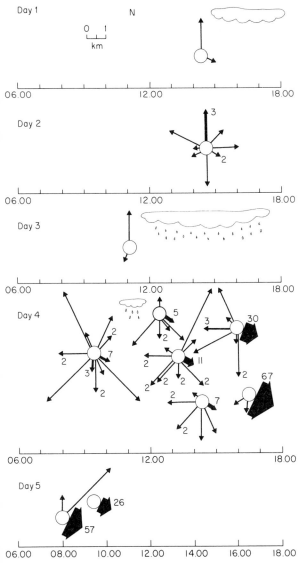

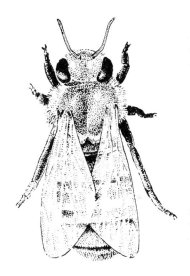

Fig. 14.23 Variation in support for different potential nesting sites during the period between emergence of a honeybee swarm from the hive and migration to a new nesting site

The diagram shows the number of new recruits dancing on the surface of a honeybee cluster in support of different potential nesting sites. The length of each arrow shows the distance from the cluster of the potential nesting site being advertised by the bee. The width of the arrow and the number printed at the arrow head indicates the number of new recruits since the previous observation that are dancing in support of that site. Where no number is printed, only a single individual is involved.

This particular swarm of bees required five days (26–30 June) to reach a decision as to the most suitable destination. Much of this time the weather was cloudy or raining as indicated. Early on day 4, strong support appeared for a site 300 m to the ESE and although support for other sites continued for most of the day, and to some extent also on the morning of day 5, by 09.00 hours on day 5 this site was the only one receiving support. At 09.40 hours the swarm initiated migration to this site.

[Modified from Lindauer (1955)]

in support for different sites that may occur before agreement is finally reached. Once the scouts have reached agreement they perform a series of buzzing runs over the surface of the cluster. Within a few minutes the whole cluster is alerted and takes off. The swarm cloud proceeds gradually, but within the swarm a few hundred bees, which are presumed to be the scouts, shoot ahead through the swarm in the direction of the selected destination. Once they have cleared the swarm they fly back slowly before repeating the process. In this way direction is conveyed to the entire swarm.

Once the decision-making process had been elucidated, the way was open for Lindauer to experiment with different potential sites and thus evaluate the criteria on which the bees were assessing site suitability. Experimental procedure involved changing the characteristics of a site being announced by the bees and noting the effect of this manipulation on the support being given for that site. A wind-break erected next to an exposed site being poorly supported by dancers on the cluster increased recruitment. On the other hand a drop of oil of balm (*Melissa*) within a site that was well supported produced a marked drop in support and recruitment. Two experimental nesting sites, judged by Lindauer to be identical, one of which was 30 m from the parent site and one of which was 250 m, showed that greater support is given for a nesting site that is not too close to the parent site. A site that was discovered when the sky was cloudy was at first well supported. When the sun came out, however, and the temperature inside the site rose to 40°C, support fell. An experimentally produced shade cover produced renewed support for the site. Clearly, therefore, bees take into account a number of factors in the assessment of habitat suitability, including shelter from wind and sun, distance from parent site, and odour characteristics. At first the scouts translate these characteristics against an innate scale of suitability, but at some stage in the reaching of agreement at least some individuals, if not the majority, have had to make a direct comparison of two or more sites. Not until all scouts are in agreement does the swarm complete its migration.

One striking aspect of this migration is that the swarm initiates migration before the final destination is known. To this extent, therefore, the migration is different from all other examples of calculated removal migration given in this chapter. On the other hand, scouts have started to announce possible sites before the swarm leaves the parent nest. Although the actual destination is not known when migration is initiated, therefore, the fact that suitable sites are available has been established. The decision-making process during clustering is in effect concerned only with deciding which is the most suitable site, not with establishing that sites are available. The initiation of migration still qualifies, therefore, as calculated rather than non-calculated.

14.6 Discussion

14.6.1 Criteria for habitat assessment

According to the hypothesis developed in Chapter 8, selection is likely to favour those individuals that assess h_1 (the suitability of the habitat occupied), h_2 (the suitability of the potential destination) and M (the migration factor) on the basis of those proximate factors that correlate best with h_1, h_2 and M. Selection is also likely to favour those individuals with the greatest capacity for accurate comparison of h_1 with $h_2 M$ on the basis of these factors. Finally, it was suggested that selection should favour those individuals that used criteria for habitat assessment that increased the ratio of calculated to non-calculated migrations performed by those individuals.

It might be expected that the proximate factor(s) that would correlate best with h_1 and h_2 would be some quantitative measure(s) of the resource(s) that the animal requires from H_1 and H_2. In four of the six examples in which the criteria being used for habitat assessment are known or presumed on the basis of at least some circumstantial evidence, this expectation is realised. Hence, escape and feeding movements within a familiar area (p. 96) seem likely to be based on an assessment of the destination on the basis of its suitability either as a place of refuge or as a feeding site. In the group-to-group migration of human pairs (p. 142) nett rate of energy uptake per individual was the criterion presumed on the basis of at least some circumstantial evidence to be given most weight except perhaps at times and in places with a money economy. Honeybees (this page) apparently base their assessment of the suitability of a nest site on direct criteria such as shelter, temperature, smell and distance.

In most of the examples given, any assessment of migration cost (a component of the migration factor—p. 41) is based on indirect criteria, such as amount of shelter (p. 92), rather than on the basis of direct criteria, such as the number of predators. Dall's sheep, *Ovis dalli*, however, in Alaska, appear on occasion to assess migration cost on a direct criterion. Whenever Dall's sheep move down to low altitudes they are particularly vulnerable to predation by wolves, *Canis lupus*. During seasonal return migration from summer to winter feeding grounds in some areas it is necessary to descend from one set of mountain slopes, cross a valley bottom, and ascend the other side to new feeding grounds. Before crossing a valley bottom on this migration, the sheep are reported (A. Murie 1944) to spend hours or even days at some vantage point overlooking the valley and it is assumed by Murie that during this period the sheep are observing the absence or movements of wolves.

In some of the examples given, therefore, assessment of h_1, h_2 and M is based directly on assessment of the critical resource. In other of the examples given, however, this is not the case and indirect criteria are employed. Hence, male dung flies, *Scatophaga stercoraria* (p. 94), searching for a habitat in which females are abundant, do not base their assessment of habitat suitability on the basis of number of females but, it is assumed, on the chemical composition of the odours given off by the habitat. Wildebeest, *Connochaetus taurinus*, on the Serengeti Plains of Tanzania (p. 95), searching for a habitat containing food, do not base their assessment of habitat suitability on the basis of observation of food but on the location of rain storms. A similar criterion is used by the bedouin of Kuwait, a deme of human nomadic pastoralists. Migration is initiated by day toward the direction of lightning observed the previous night (Dickson 1949). Male–female human pairs (p. 119), searching for a habitat with a high nett rate of energy uptake per individual, as well possibly as basing their assessment of habitat suitability directly on this criterion, may also use the indirect criteria of availability of money and of deme size and density. Deme size and density are indirect criteria in that, when nett rate of energy uptake per individual is high or increasing, assessment of a larger, more dense deme as more suitable is advantageous not because it is immediately so but because any future variations in health will be buffered within a larger deme (assuming the existence of reciprocal altruism) and because any future territorial interactions will have a greater likelihood of success. When nett rate of energy uptake per

individual is low or decreasing, assessment of a smaller, less dense deme as more suitable is again advantageous, not as a result of immediate effects (these are negated by giving greatest weight to the direct criterion of nett rate of energy uptake per individual) but because future death rate is lower in such a deme. In other examples it is problematical whether the criteria used are direct or indirect, largely because of lack of information. Hence, female gorillas, *Gorilla gorilla* (p. 92), using the criterion of the ratio of silverback males to certain other age and sex classes, are using a direct criterion if individual attention to the female and offspring from a male is advantageous, but in the absence of relevant data, this cannot be evaluated. Greater horseshoe bats, *Rhinolophus ferrumequinum* (p. 101), may be using the direct criterion of cave temperature, gained through present or past experience, in assessing the cave most suitable for roosting at a particular time. On the other hand, some indirect criteria such as cave size and structure may be used.

The relatively high incidence of the use of indirect criteria for habitat assessment in the examples given, even taking into account that these examples are not a random selection, could be argued to provide evidence against the hypothesis developed in Chapter 8. Two arguments can be presented, however, that this is not so. First, although the criteria used are indirect, they nevertheless are likely to show a high correlation with habitat suitability. Hence, chemical composition of dung, in that it indicates age of the dung, is likely to show a high correlation with the presence of female dung flies. At the beginning of the wet season on African plains and Arabian deserts, rain is likely to show a high correlation with the abundance of food and water for young-grass feeders such as wildebeest and the ungulate symbionts of human nomadic pastoralists. Cave size and structure, if such criteria are used, are likely to correlate well with winter temperature and the suitability of a cave as a hibernation site for bats. Deme size and density when used as a secondary criterion, correlates well (p. 119) with the future suitability of a deme to a male–female human pair. Finally, as far as migration cost is concerned, indirect criteria such as the amount of shelter, are likely to correlate well with survival during migration.

The second argument is that not only are the criteria used likely to correlate well with h_1, h_2 and M, but they also increase the ratio of possible calculated to non-calculated migration. In which case the use of these indirect criteria provides supporting evidence rather than evidence against the general thesis. Hence, use of airborne chemicals rather than direct observation of females by male dung flies greatly increases the distance over which calculated migration can be performed. Similarly, observation of distant rain clouds rather than direct observation or olfactory sensing of food and/or water greatly increases the area over which man or wildebeest can perform calculated migration. Finally, although the criteria by which cave bats assess a suitable hibernation site has not been demonstrated, use of indirect criteria such as size, structure and exposure of the cave rather than the direct criterion of temperature, would permit calculated migration in advance of the critical winter period. J.S. Hall (1962) has suggested that one advantage to bats of returning to the same hibernation site year after year is that if they did not do so they would be faced with the problem of selecting a suitable hibernation site a few months in advance of the critical period. Although this is probably true, assessment of cave suitability on the basis of indirect criteria that nevertheless correlate well with temperature during the critical period would also solve this problem. Selection, therefore, is likely to have favoured individuals that both assess cave suitability on the basis of indirect criteria and return to the same cave in subsequent years should assessment prove to have been accurate.

Use of the indirect criterion of deme size and density by male–female human pairs represents another facet of selection on criteria for habitat assessment. In effect, use of the direct factor of nett rate of energy uptake per individual as the primary criterion for habitat assessment leads to the migration being an immediate advantage. As long, that is, as assessment is accurate. Use of the indirect factor of deme size and density as a secondary criterion for habitat assessment on the other hand is likely to result in the migration continuing to be an advantage longer into the future.

I have not distinguished in the above discussion between innate and learned responses to criteria for habitat assessment. Although an attempt at such a distinction may be of interest for its own sake, it is not particularly important as far as the arguments of this section are concerned. At one extreme, as for example in the case of the male dung fly (p. 94), it seems clear that selection has acted to favour those males that show an innate response to the smell of fresh dung by migrating up-wind even though they

are already in the vicinity of a dung-pat. At the other extreme the response to distant storms of bedouin humans (p. 153) and wildebeest (p. 95) seems much more likely to be a learned response that has been passed on by social communication from one generation to the next. Nevertheless, individuals of these species that are capable of learning this response, either by direct experience or by social communication, are at an advantage relative to those individuals that are less capable. Selection on the criteria for habitat assessment is still involved, except that in these two examples it acts on learning process differentials. Whatever the proximate mechanism involved, whether it is a learned or an innate response, selection still favours those individuals that respond to those proximate factors that correlate best with habitat suitability. This remains true with those intermediate examples in which it is not clear whether the response is learned or innate. It is problematical, for example, whether the response of female gorillas to group composition (p. 92) or the response of sub-adult male and female humans (p. 137) and the different response of newly formed male–female human pairs (p. 143) to deme size and density are learned or innate or a combination of both.

To sum up, therefore, the criteria for habitat assessment in the examples given fit the general thesis presented in Chapter 8 in that they seem likely to correlate well with habitat suitability and migration cost and at the same time permit the maximum ratio of calculated to non-calculated migration by the animals concerned.

A separate question is whether comparison of h_1 with $h_2 M$ on the basis of the various criteria is accurate. In order to answer this question it would be necessary to determine the frequency with which calculated migration is advantageous. How often do male dung-flies vacate one dung-pat for another on the basis of air-borne chemicals only to arrive at a dung-pat at which the female quota is less than it would have been at the original pat? How often do wildebeest migrate across the African plains toward a distant cloud only to find that rain has not fallen and grass is not growing? How often does a bat respond by hibernation to the size, structure and exposure of a cave only to be killed in mid-winter by freezing temperatures? How often do human pairs migrate from one place to another in response to an assessed difference in nett rate of energy uptake per individual only to experience a reduced nett rate of energy uptake per individual at the destination?

How often does a swarm of honeybees settle in a new nesting site in response to the amount of shelter from the Sun offered by the site only to die out some time later through overheating? As far as I know the information is not available that would permit evaluation of the accuracy with which animals can compare h_1 with $h_2 M$ on the basis of the criteria normally used. In any case, in such an evaluation it would be necessary to make the often very difficult distinction between accuracy of comparison and how well the criteria used correlate with habitat suitability or migration cost. Often it is not possible to separate these two phenomena completely. Wildebeest migration, for example, could be disadvantageous either because the animals cannot accurately distinguish between rain and non-rain clouds or, alternatively, because the presence of a storm cloud over a certain area may not always be associated with later growth of grass. On the whole, however, it seems a reasonable assumption that selection has produced animals that can accurately compare h_1 with $h_2 M$, even though as yet supporting evidence does not seem to exist.

14.6.2 Strategies that increase the ratio of calculated to non-calculated migration

One type of adaptation has already been described that leads to an increase in the ratio of calculated to non-calculated migration. That is evolution of the use of indirect criteria that correlate with habitat suitability or migration cost approximately as well as direct criteria but which increase the area or period over which calculated migration may take place. As well as adoption of the best criteria for habitat assessment, extension of the area or volume within which calculated migration can take place can also be achieved by certain behavioural strategies. From the examples given it seems that selection has favoured the evolution of these strategies on a number of occasions.

Probably the commonest strategy for extension of the area or volume within which calculated migration can take place is extension of the familiar area by migration (= exploratory migration). Hence muskrats, *Ondatra zibethicus* (p. 101), when exposed to drought conditions, do not immediately perform non-calculated migration but instead establish a centre of activity at the edge of their previous familiar area from which, presumably, exploratory migrations are performed. As a result of such exploration it is presumed that a place suitable for temporary settlement is discovered or, failing that, non-calculated migration is performed. Adult black-tailed prairie-dogs, *Cynomys ludovicianus* (p. 102), having discovered a suitable destination during exploration, move backward and forward between old and new habitats for some time until the new habitat has been modified sufficiently for colonisation. The juveniles of some species of pinnipeds (p. 105) seem to establish a much larger familiar area than they will use as adults, and although return to the natal site is usual, removal migration to another site, probably within the familiar area, occurs frequently. Sub-adult humans, particularly males (p. 134), extend their familiar area more during this than any other ontogenetic phase. When pairing takes place it seems likely that the pair migrate to that destination within their combined familiar area that is assessed to be most suitable.

A possibly less common strategy than exploratory migration for extension of the area or volume within which calculated migration can take place is extension of the familiar area by social communication. Hence, not all members of a herd of wildebeest, *Connochaetus taurinus* (p. 95), have the same familiar area. Some individuals are young and have not established as large a familiar area as older individuals. Herds are often the result of coalescence of smaller herds which previously had different but overlapping familiar areas. It is only necessary, however, for rain to fall in a distant part of the familiar area of some individuals in the herd for migration to occur. Migration toward a distant storm is initiated by some individuals which are then followed by others (Talbot and Talbot 1963). Furthermore, trails are laid by the pedal glands during migration and these scent trails may be followed by other herds. Possibly the individuals that initiate the migration are those within the familiar area of which the rain is falling. The fact of directed migration by these individuals seems to be sufficient to communicate the suitability of the distant habitat to other members of the herd and to other herds that later come upon the scent trail. Of course, the possibility that other forms of communication may also be involved, particularly during the pre-migration 'milling' that occurs cannot be ruled out entirely. In man (p. 136), it seems likely that extension of the familiar area during the sub-adult phase of ontogeny takes place to a large extent within the socially communicated familiar area.

Furthermore, although the majority of migrations of newly formed male–female pairs probably occur within the combined familiar area (by migration) of the pair, a significant proportion probably take place within the combined socially communicated familiar area. This is particularly likely to be so in modern industrial humans in which the potential socially communicated familiar area is so vast. Migration of social groups of insects (p. 148) is likely to take place within the combined familiar area of the group (i.e. within the familiar area (by migration) of some individuals, the familiar area (by social communication) of other individuals, and beyond the familiar area of the remainder).

14.7 Summary of conclusions concerning calculated removal migration

1. The subject of this chapter is the calculated initiation of migration in which the animal can make a direct assessment of h_1 and h_2M before initiating removal migration from H_1. Migration from H_1 to H_2 is only initiated, therefore, when the animal assesses h_1 to be less than h_2M.

2. Three broad categories of calculated migration may be recognised, any combination of which may be shown by any animal. The first category consists of migrations initiated from a situation in which the individual concerned, while still within H_1, can establish sensory contact with H_2. The second category consists of migrations performed entirely within the limits of a familiar area that has been established by a series of exploratory migrations. The third category consists of migrations initiated as a result of assessments based on socially communicated information. Examples of all three categories are given.

3. Selection that acts on calculated migration is likely to favour: (a) those individuals that assess h_1, h_2 and M on the basis of those proximate factors that correlate best with h_1, h_2 and M; and (b) those individuals with the greatest capacity for accurate comparison of h_1 with h_2M on the basis of these factors. It is argued that as a result of this selection any given migration is more likely to be advantageous if it is calculated rather than non-calculated. On this assumption it follows that selection also favours those individuals that employ criteria for habitat assessment and other strategies that increase the ratio of calculated to non-calculated migrations performed by those individuals.

4. It might be expected that the proximate factor(s) that would correlate best with h_1, h_2 and M would be some quantitative measure of the resource(s) that the animal requires from H_1 and H_2. Although this is often the case, some examples are given in which habitat suitability is assessed on the basis of indirect criteria. However, whenever indirect criteria are used, not only are they likely to correlate well with h_1, h_2 and M, but they also increase the distance or time over which assessment is possible. In all the examples given in which the criteria for habitat assessment are either known or assumed with some certainty, the thesis of this chapter is supported in that the criteria are likely to correlate well with habitat suitability and migration cost and at the same time permit the optimum ratio of calculated and non-calculated migrations by the animals concerned.

5. No data are available that permit evaluation of the accuracy with which animals are capable of comparing h_1 with h_2M.

15

The possible incidence in the Metazoa of strategies to increase the ratio of calculated to non-calculated removal migration

15.1 Introduction

A number of examples were presented in the previous chapter that demonstrated the range of movements that qualified for consideration as calculated removal migrations. These migrations were then discussed in terms of the criteria for habitat and migration-cost assessment adopted by the animals concerned. Finally, it was pointed out that associated with these migrations were other behavioural strategies that could only reasonably be interpreted as mechanisms for increasing the ratio of calculated to non-calculated removal migration.

Broadly, these strategies may be divided into three categories: (1) adoption of criteria for habitat assessment that can be perceived over relatively long distances; (2) exploratory migration; and (3) extension of the familiar area by social communication. If, as suggested in Chapter 8, it is so much more advantageous for any given removal migration to be initiated on a calculated rather than on a non-calculated basis, it might be expected that calculated migration would be much more common among the Metazoa than it appears to be. In this chapter the possibility is considered that calculated migration is a much more common phenomenon than is usually recognised by students of migration. Before doing so, however, it is necessary to stress that this chapter should not be interpreted to imply that non-calculated migration is disadvantageous. This is obviously not so and Chapters 8 and 16 consider at some length the selection that acts on non-calculated migration. The hypothesis of this chapter states only that migration from H_1 to H_2 is more likely to be advantageous if it is a calculated rather than a non-calculated migration.

In considering the possible incidence of calculated migration in the Metazoa I am not concerned so much with the removal migrations themselves. Rather, I am concerned with pointing out some behaviour patterns one of the functions of which could be *extension* of the familiar area, thereby *increasing* the ratio of calculated to non-calculated migration. In this consideration I am therefore concerned primarily with the incidence of exploratory migration and socially communicated extension of the familiar area among the Metazoa.

In drawing attention to the various behavioural phenomena described below I have undoubtedly on occasion made far more of a particular situation than is warranted by the information available. To some extent this is intentional and represents an attempt to compensate for the scant attention that the possibility of calculated migration has received from students of migration. Quite probably many of the behaviour patterns discussed below will subsequently prove to have no connection with calculated migration. However, if inclusion of these phenomena in this chapter provokes research designed to test the possibility of calculated migration, then I shall feel that such inclusion has been justified even if the phenomena subsequently prove to have no relation to calculated migration.

15.2 Bats

Bats perform daily and seasonal return migrations (Chapter 26). The familiar area of a bat therefore consists of at least the total area visited by the bat during performance of an annual migration circuit (R. Davis 1966). In addition to these return migrations, bats also perform removal migration in that one year an individual bat may use, say, a summer roosting site that is different from the roosting site used the previous summer. Removal migration may be performed at any time during the animal's life but seems to be most frequent during the juvenile phase of ontogeny (Chapter 17). The possibility presents itself that most such removal migrations take place within the familiar area of the individual bat. In this case the possibility also presents itself that behavioural strategies have evolved to increase the ratio of calculated to non-calculated migration. Two behaviour patterns in particular come to mind, each of which could possibly represent such a strategy.

First, it is well documented that both the young and adults of many temperate bat species become nomadic in late summer and autumn after the young of the year have been weaned. Often, as in the case of the American free-tailed bat, *Tadarida brasiliensis*, in the southern United States and Schreiber's bat, *Miniopterus schreibersii*, in southeastern Australia, this post-weaning migration results in a considerable dispersal of the members of a single maternity colony, particularly of the newly weaned young. This dispersal could be interpreted either as non-calculated removal migration or as exploratory migration. The critical factor to determine is whether or not these post-weaning migrants retain the ability to return to the original maternity site. If the ability is not retained the bats are performing non-calculated removal migration (Chapter 9). If the ability to return is retained, the most likely explanation is that the bats are extending their familiar area and that the following summer they will return to the site within that familiar area that is assessed to be the most suitable, which may or may not be the original maternity site. If it is not the original maternity site that is assessed to be most suitable then the summer roost to summer roost removal migration that has been performed from one summer to the next is a calculated migration. Critical evidence relevant to this situation does not as yet seem to exist (Chapter 33). Homing experiments are the obvious approach to the problem but

are difficult to interpret. The critical factor is the *ability* to return. However, most homing experiments test mainly the *motivation* to return. Post-weaning migrants in particular are likely to have a low motivation to return to the original maternity site until the following summer and only then if a more suitable site has not been discovered in the meantime. Despite problems of interpretation of results, however, homing experiments on post-weaning migrants offer the only method by which the above two alternatives can be distinguished. Speed of post-weaning migration is often so high that it gives the impression of being non-calculated rather than exploratory. Schreiber's bat, *Miniopterus schreibersii*, for example, is a wide-ranging species (Fig. 26.17) which, at least in east continental Australia, occurs in distinct demes each based on one or a few maternity colonies (Dwyer 1966). Parturition occurs at the beginning of December (Fig. 26.17). Juvenile bats leave the maternity colony during March and are on the move from one transient roost to another until May. By mid-April they can be up to 150 km

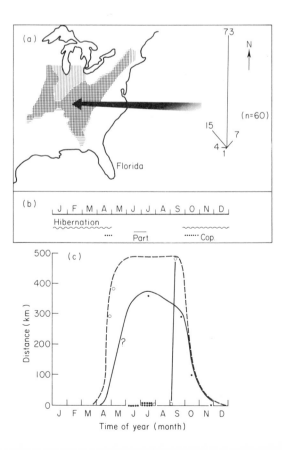

Fig. 15.1 Seasonal return migration of the Indiana myotis, *Myotis sodalis*

(a) Distribution and direction ratio. Map shows southeastern North America. Horizontal lines, winter distribution; vertical lines, summer distribution. Direction ratio is shown for a group (males and females summated) that hibernates in Kentucky as shown by the arrow. Length of line and number indicates the percentage of individuals that was recaptured in the summer in each of eight directions from the hibernation site in which the bats were marked. As indicated by winter and summer distribution, *M. sodalis* in the centre of its range has a peak post-hibernation direction to the north. The same is unlikely to be true in the eastern and western parts of the range.

(b) Seasonal behaviour. Solid line, parturition period; waved line, hibernation period; dotted line, copulation period. Cavernous limestone caves associated with river-ways are used as hibernation roosts by aggregations of < 500 to 100 000 individuals (photo). Tight, compact, single-tiered clusters are formed in regions of a cave that maintain a steady 4–6°C during winter. Females in summer occur singly or perhaps form maternity colonies at day-roosts that are possibly under bridges, under the bark of trees and in buildings. Males are thought not to have permanent day roosts during summer but to move frequently. All males and many females visit their hibernation roost once during the summer for 1–2 days.

(c) Distance component of seasonal return migration. A solid dot (male) and open dot (female) refers to the distance an individual has been recaptured from the hibernation site in which it was marked. The maximum distance limits are indicated for males (solid line) and females (dashed line). Straight line shows the movement of a female that was marked during the summer visit to the hibernation site. This bat moved 480 km in 8 days.

In summary, seasonal return migration of *M. sodalis* consists of: (1) a post-hibernation movement away from the hibernation site during which, at least in the central part of the range, both males and females show a peak direction to the north; (2) a summer return to the hibernation site for one to two days; (3) a pre-copulatory summer movement away from the hibernation site; and (4) an autumn return to the hibernation site at which most copulation takes place in the autumn and again in the spring.

[*Compiled from data in Hall (1962) and Barbour and Davis (1969). Photo by Roger W. Barbour*]

from the maternity roost, the maximum rate recorded being 216 km in six days. Such a high speed of migration might be supposed to indicate non-calculated removal migration rather than familiar area extension. It is quite conceivable, however, that a bat following some major linear topographical feature, such as a drainage system or range of hills or mountains, could cover this distance in such a time without losing the ability to return (Chapter 33). Such topographical features are known to be important in the movements of Schreiber's bat (Dwyer 1966), and individuals often perform return migrations from such distances (Fig. 26.17).

A second behaviour pattern that may also represent a strategy for an increase in the ratio of calculated to non-calculated migration in bats is that of 'swarming' at potential hibernation sites during the summer and autumn. In the Indiana myotis, *Myotis sodalis* (Fig. 15.1), groups of bats visit during the summer the guano pile (p. 101) in the cave system that also contains the hibernation roost. These groups consist of 90 per cent males and stay in the cave for just one to two days. Hall (1962) has shown that virtually all the males that hibernate in the area make this visit at least once during the summer. The little brown myotis, *M. lucifugus* (Fig. 26.7) also returns briefly to its hibernation site during the summer (Fenton 1969). The visit starts in August at which time the visitors are predominantly (62 per cent) young of the year. By September, when the adults are beginning to copulate, only 26 per cent of the visitors are young of the year. The average length of the visit is 61 ± 21 min. Each individual usually visits the cave only once. If the cave is visited a second time, this does not occur until September. Of individuals banded during hibernation, 3 per cent were caught while visiting the cave in the summer. Of individuals banded during their summer visit, 15 per cent were subsequently caught at that cave during hibernation. It is tempting to interpret these summer visits to be part of a process of roost assessment whereby the roost assessed to be most suitable for hibernation is the roost to which the bats migrate later in the autumn. The fact that the majority of bats return to the same hibernation site year after year does not conflict with this interpretation. As far as a yearling or adult is concerned, the fact of survival the previous winter renders it likely to be advantageous to assess the hibernation site of the previous winter to be most suitable. Only if, on whatever criteria are optimum for habitat assessment in these species, some other

hibernation site is assessed to be even more suitable should removal migration occur. As long as the energy expended is not too great, therefore, the possibility that a more suitable hibernation site may exist is likely to be sufficient to render advantageous summer visits to a variety of potential hibernation sites. As far as young of the year are concerned, summer visits to potential hibernation sites could be interpreted as part of the process of familiar area extension which again could increase the ratio of calculated to non-calculated migration. One final possibility also deserves mention. Summer visits to hibernation roosts are performed by large numbers of individuals. Hence the term 'swarming' by which these visits are described by chiropterologists. Furthermore, in the little brown myotis, *M. lucifugus*, adults and young, which may or may not represent parents and offspring, visit the site together. In view of the rapidly accumulating list of examples of socially communicated information in animals, it would perhaps be unwise to dismiss prematurely the possibility that information concerning the suitability of various hibernation sites is being exchanged during these visits. If this were so it would represent yet another example of behaviour that increases the ratio of calculated to non-calculated migration. In view of the seasonal timing of the behaviour, it seems unlikely that 'swarming' represents a display of the type envisaged by Wynne-Edwards (1962) whereby an animal can assess the size of the breeding population and adjust its fecundity accordingly. I have not seen correlations between size of breeding population and size of population the previous summer and autumn. As with many temperate birds (Lack 1966), however, because bat mortality is largely a function of winter conditions (Ransome 1968), it seems unlikely that breeding population will correlate well with size of the population the previous summer. A display for assessment of the size of the breeding population would, therefore, be expected in spring, not autumn. Even apart from other theoretical objections, therefore, the summer and autumn 'swarming' of bats is unlikely to be such a display.

To conclude, therefore, relatively little is known about the removal migrations of bats and it is by no means clear to what extent such migrations are initiated on a calculated or non-calculated basis. If calculated migration does occur, it is possible that the autumn post-weaning dispersal of maternity colonies in temperate regions represents a strategy that has evolved in part in response to the advantage of increasing the ratio of calculated to non-calculated migration. The 'swarming' of adults and juveniles at potential hibernation roosts that occurs at this time or of males that can occur at any time throughout the summer may well represent a facet of the habitat assessment process that would be an integral part of such a strategy.

15.3 Rodents and lagomorphs

The majority of rodents and lagomorphs live within a familiar area, though it has been suggested that some Arctic rodents, such as the brown lemming, *Lemmus trimucronatus*, and the varying lemming, *Dicrostonyx groenlandicus*, particularly the males, are wide-ranging and continually on the move (C.J. Krebs 1964). The varying lemming, for example, has been reported to move up to 20 km over a few days even in winter (Degerbøl and Freuchen 1935). An alternative explanation could be that the familiar area of these species is extremely large, as seems to be the case for other Arctic rodents. The Barrow ground squirrel, *Citellus parryi*, for example, at Point Barrow in Alaska in July has been seen to travel over the tundra for up to 1 km from its burrows (Mayer 1953). It can be assumed, however, that the majority of rodents live within a familiar area. Only part of the familiar area is used at any one period and this part is usually known as the home range. There may be some seasonal shift of home range and in species that live more than one year this seasonal shift probably takes place within the familiar area. The first ontogenetic shift of home range could occur as described for the muskrat, *Ondatra zibethicus* (p. 101), the black-tailed prairie-dog, *Cynomys ludovicianus* (p.102), and the rabbit, *Oryctolagus cuniculus* (p. 104) by initial exploratory migrations that extend the familiar area followed by calculated migration. This appears to be the case for the woodchuck, *Marmota monax*, a North American vegetarian ground-burrowing squirrel, the young of which leave their natal burrow in midsummer when about three months old (Schwartz and Schwartz 1959). After leaving their natal burrow the young woodchuck dig temporary burrows within the parental home range before moving further and establishing a permanent burrow. Whether this final removal migration is calculated, as suggested by the preliminary behaviour, or non-calculated is not known.

There is some further evidence of exploratory migration in rodents. For example, wild Norway rats, *Rattus norvegicus*, although living in a relatively small home range may travel as far as 1 km from the roost in order to obtain a particularly attractive food (Steiniger 1950). In order to find such a feeding place in the first instance periodic exploratory migrations must be made beyond the normal range, unless social communication or long-distance olfaction is involved.

Although strategies that increase the ratio of calculated to non-calculated migration seem to have evolved quite frequently in rodents, and may, when investigated further, prove to be even more common, there seems little doubt that many removal migrations by rodents are also non-calculated. This is particularly likely to be true of those longer-distance migrations that seem to be performed by many rodents and lagomorphs, particularly as juveniles, but also when adult. For example, males of the eastern fox squirrel, *Sciurus niger*, of the United States, which may migrate 2–22 km from their previous home range with a known maximum of 64 km (Schwartz and Schwartz 1959) are unlikely to be performing calculated migration or exploratory migration.

15.4 Primates

Primates invariably live within a home range within the confines of which the majority of movements take place. Some prosimians are solitary within their home range (Jolly 1966) as also is the orang utan, *Pongo pygmaeus* (Rodman 1973) but, except for certain ontogenetic and sex classes, the vast majority of anthropoids form social groups the members of which have a common home range. Most often the members of a social group move as a unit but in some species, particularly among gatherers and hunter–gatherers, the group is extensible (Nishida 1968), continually dispersing through the home range as smaller but inconstant groups and then converging to form a large-sized group once more. The impression gained from the literature on primate movements is that exploratory migration by the entire group is rare or non-existent, with the result that the familiar area and home range are the same. Except, that is, insofar as areas outside the home range are nevertheless familiar as a result of being within sensory range. This impression may result in part from the relatively short period of time over which most studies of primate movements are carried out and in part from the difficulty of keeping track of primate movements, particularly in more forested areas. Radio-tracking, as used by Moor and Steffens (1972) for the vervet monkey, *Cercopithecus aethiops*, in the deciduous gallery forest of north-eastern South Africa, offers some hope that in the near future studies of primate movements can be more extensive than previously.

In the examples of the group-to-group migrations of female gorillas, *Gorilla gorilla* (p. 92) and of newly formed pairs of humans (p. 123), it was argued that the migrations performed took place within a familiar area. The occasional meeting of separate groups of gorillas could be interpreted as a strategy for increase in the ratio of calculated to non-calculated migrations. Of particular interest in this respect is the observation by Schaller (1963) that when the only silverback male of a gorilla group dies or is killed by a leopard, *Panthera pardus*, or by man, the remaining females and blackbacked (sub-adult) males perform rapid migration as a group and soon join with another group, usually a group that previously had a relatively high silverback male-to-female ratio. It is tempting to suggest that as a result of previous meetings between the two groups the composition and approximate whereabouts of the group was already known and that the migration is therefore calculated. Large-sized groups of chimpanzees, *Pan troglodytes*, also sometimes approach one another and mingle without antagonism, particularly in forests in which the deme density is relatively high (i.e. about 4 individuals/km²) (Sugiyama 1968). Migration of adult and young males from one group to another during such mingling has been observed, though it was not certain that the two groups between which migration occurred were from separate large-sized groups. In the woodland savanna of western Tanzania, however, definite large-sized group to large-sized group removal migration by four sub-adult females was observed by Nishida (1968). In such areas, however, where deme density is low (about 0·5 individuals/km²) it seems that large-sized groups usually avoid one another, even where their nomadic ranges overlap (Fig. 18.23) (Izawa 1970).

So far this discussion of primate migration has centred on removal migration between groups that have had some contact, and it seems fairly clear that such migrations will prove to be initiated on a calculated basis. It seems less clear, however, whether the removal migration of individuals, particularly sub-adult males, that involves the individual living solitarily or joining a single-sex group for a period, will prove to be initiated on a calculated basis, being preceded by exploratory migration as described for humans (p. 121), or on a non-calculated basis. Once again the critical question is whether or not the ability to return to the natal group, or any other group encountered during exploration, is retained. The list of species in which the existence of a solitary phase in male ontogeny has been demonstrated or inferred is growing (Aldrich-Blake 1970, Eisenberg *et al.* 1972), but the species for which most information is available is the Japanese macaque, *Macaca fuscata* (Nishida 1966). In areas in which the species has a relatively continuous distribution, males rarely return to their natal group. In other areas, however, where for some historical or topographical reason a group may be isolated from conspecific groups, males may frequently return to their natal group. In addition, an individual male becomes solitary only gradually. During the first few months of solitarisation the male returns to his natal group frequently but with decreasing regularity. Both of these observations could be interpreted as suggesting that the solitary phase is a period of familiar area extension as described for humans. Solitary male Japanese macaques associate for varying periods with bisexual groups, particularly during the annual four-month copulation period. During this period the male may consort and copulate with females and then later become solitary once more. Alternatively, the male may settle permanently with an already formed group or may collect females from such a group, thus forming a new group (Itani 1963, Nishida 1966).

Male gorillas, *Gorilla gorilla*, behave similarly (Schaller 1963) as perhaps do the males of the vervet monkey, *Cercopithecus aethiops* (Gartlan and Brain 1968), rhesus macaque, *Macaca mulatta* (A.P. Wilson 1968), and chimpanzee, *Pan troglodytes* (Itani 1966, 1967, Itani and Suzuki 1967). I have analysed (Tables 15.1 and 15.2) Schaller's observations (Schaller 1963) of the changes in group composition of the gorilla in relation to the proximate factors that may influence males to join or leave a particular group. The conclusion reached (Table 15.2) is that a silverback male is most likely to become solitary if the group of which he is a member has a high silverback male:receptive female ratio. A solitary male, on the other hand, is most likely to join a group, either temporarily or permanently, if the group has a low silverback male:receptive female ratio. Solitary male gorillas may appear 30 km or so

Fig. 15.2 A solitary silverback male gorilla, *Gorilla gorilla*

[*Photo by George B. Schaller, courtesy Bruce Coleman Ltd*]

from known gorilla habitats (Schaller 1963). Analysis of other data presented by Schaller suggests that for about six years the males live more or less solitarily or in temporary all-male associations of two or so individuals. During these six years the males are likely to associate with, on average, two bisexual groups. Such association may involve copulation with females within the group. The male may depart from the group even if copulation has occurred. Solitary migration seems to terminate once a male becomes a dominant member of a group. Dominance may be achieved by forming a new group by attracting females away from an existing group or by taking over an existing group by superseding another male. Bisexual groups have also been observed 30 km or so from the nearest known gorilla areas. These groups may be performing non-calculated removal migration. Alternatively, the male, having formed or obtained a group in one part of his familiar area, could well be leading the group to some other part of his familiar area encountered during exploratory migration and which has been assessed to be more suitable. In such migrations possible criteria for habitat assessment could be deme density of conspecifics, the abundance of secondary forest vegetation, and/or the density of predators.

If the migrations of sub-adult male Japanese macaques and gorillas have evolved in response to the advantages of: (1) copulating without further parental investment with as many females as possible; (2) pairing with the optimum mates and/or of becoming the dominant member of a group; and (3) increasing the ratio of calculated to non-calculated migration, then such migrations show considerable similarity to the migrations of sub-adult male humans (p. 121). If this proves to be the case it could be taken to indicate that the exploratory migrations of sub-adult male humans have a considerable evolutionary history, extending back perhaps some 50 million years to the common ancestors of modern anthropoids.

The aggregation of sub-adult and unpaired males into all-male groups that is found in some primates (Aldrich-Blake 1970) has perhaps evolved on several occasions from a situation in which such males are solitary. Even in the Japanese macaque and gorilla, solitary males may associate with each other for brief periods. All-male groups are usually nomadic over a wide area such that they move through the home ranges of several bisexual groups. From time to time the all-male group of the common Indian langur,

Table 15.1 The reactions of solitary males of the gorilla, *Gorilla gorilla*, to social groups of differing composition (compiled from field observations by Schaller 1963)

Attractiveness of group*	Number of individuals in group†				
	SBM	BBM	F	J	I
0	4	1	10	3	6
10	5	1	10	3	6
5	5	1	10	3	6
10	5	1	10	3	6
5	5	1	10	3	7
5	4	1	10	3	7
5	4	1	12	3	9
0	3	1	12	3	10
10	3	2	3	4	2
10	3	2	3	4	2
0	2	1	8	2	7
0	2	1	8	2	7
0	2	1	8	2	8
5	1	1	3	0	2
10	1	1	5	0	1
10	1	1	7	0	3
−5	2	3	6	5	4
−5	3	2	3	2	2
−5	2	2	6	4	5
−5	2	2	8	3	7

*Attractiveness
10   male joins group
5   male leaves group
0   male associates with group at periphery but does not join
−5   solitary males not associate with group

†Age and sex classes
SBM   silverback males (>10 years) (number includes the individual concerned)
BBM   blackbacked males (6–10 years)
F   females (>6 years)
J   juveniles (3–6 years)
I   infants (0–3 years)

Presbytis entellus (Yoshiba 1968), succeeds in displacing the single adult male from one of the bisexual groups within their familiar area. The males then fight amongst themselves until one of them establishes himself as leader of the bisexual group. The new dominant male then kills all the infants in the group and the females come into oestrus.

Young males that join all-male groups not only extend their familiar area by migration, as do solitary males, but also extend their familiar area by social communication. How big an advantage this may be remains to be determined but it can certainly be added to the other advantages of associating with other males rather than living solitarily. Among these advantages may be included protection against predators (Crook 1970) and, at least in the langur, the possibility of greater success in the displacement of males from bisexual groups. Against these advantages of joining an all-male group must be balanced the following disadvantages: (1) having displaced a dominant male, the males must then

Table 15.2 The possible weighting of criteria for habitat assessment used by solitary male gorillas: step-wise multiple regression analysis of the data in Table 15.1

R^2	Group characteristic*	t-values for multiple regression coefficients†				
0.336	$SBM:(F-J)$	−3.019	−3.472	−6.055	−5.618	−5.752
0.130	I		−2.030	−5.924	−6.389	−5.821
0.357	$SBM:BBM$			5.677	5.949	6.675
0.028	$F:SBM$				1.674	2.746
0.034	$SBM:(F-BBM)$					2.046
Total 0.886		9.115	7.409	24.753	21.356	21.552
		$F_{1,18}$	$F_{2,17}$	$F_{3,16}$	$F_{4,15}$	$F_{5,14}$

* $SBM:(F-J)$ ratio of the number of silverback males to the difference between the number of females and the number of juveniles

I number of infants

$SBM:BBM$ ratio of number of silverback males to number of blackbacked (sub-adult) males

etc.

† t-values underlined indicate that the regression coefficient is significantly different $(P<0.05)$ from zero

Conclusion.
Solitary males are more likely to invest time in a given social group if there is a low silverback male to female ratio. Furthermore, as a high value for $F-J$ and a low value for I and BBM could all indicate that more females were about to come into oestrous, the critical factor in habitat assessment could well be a low silverback male to receptive female ratio. The high value $(R^2=0.886)$ for the coefficient of determination for the multiple regression suggests that most of the social behaviour of solitary male gorillas is explicable in terms of a search for a group in which reproductive success is likely to be high.

compete amongst themselves for the vacant position; (2) in comparison with a solitary male a member of an all-male group is likely to have less chance of copulation with a female without further parental investment; and (3) a member of an all-male group almost certainly reduces his chances of attracting females away from an already formed group to establish a new group. Evidently in the langur, the balance of these selective pressures favours those males that join an all-male group. The extent to which this balance is affected by the advantage of increasing the familiar area by social communication is unknown, though some possible facets of this advantage come to mind. Increase in the familiar area results in an increase in the available information concerning the status of more bisexual groups and hence an increase in the possibility of calculated migration to the home range of the group the male of which is most likely to be displaced. Once a male has become established within a bisexual group, a large familiar area presents the opportunity of assessing the area in which displacement by the all-male group is least likely.

In conclusion, therefore, there is a strong possibility that a number of primate behaviour patterns, such as occasional contact between adjacent social groups, male solitarisation, and the formation of all-male groups, are in fact strategies that *in part* function to increase the ratio of calculated to non-calculated migration. The extent to which this is true and to which primates perform calculated rather than non-calculated migration remains to be evaluated.

15.5 Ungulates

Ungulates often live in social groups the inter-specific variety of which rivals, if not surpasses, that of primates. Some species, such as most members of the 'horse' group of the Equidae, form one-male bisexual groups, the remaining males living either solitarily or within an all-male group (Antonius 1937, Klingel 1969). Other species, such as most members of the 'ass' group of the Equidae, have sexes that remain separate for the greater part of the year. The females and juveniles form groups which are led by an experienced female with a large familiar area, while the males are solitary or form smaller groups (Antonius 1937). In the breeding season the males approach the female groups and each attempts to mate with as many oestrous females as possible. The banteng, *Bos javanicus*, a Javan bovid, also seems to show this behaviour pattern (Hoogerwerf 1970). Others, such as the muskox, *Ovibos moschatus*, on the Canadian tundra, form age-graded or multi-male groups of 4–100 individuals (Tener 1954).

Some species, such as the barren-ground caribou, *Rangifer tarandus*, in the Canadian Arctic and sub-arctic (Banfield 1954) and the African wildebeest, *Connochaetus taurinus* (Talbot and Talbot 1963) have vast familiar areas of thousands of square kilometres or so within which seasonal shifts of home range occur, whereas others, such as the vicuña, *Vicugna vicugna*, of the semi-arid rolling grasslands of the Andean highlands of south Peru, remain in a territory of 8–40 ha throughout the year (Koford 1957). Smaller groups may aggregate to form great herds in which the basic social units may either retain their integrity as in the plains zebra, *Equus quagga* (Klingel 1969), or not, as in the wildebeest, *Connochaetus taurinus* (Talbot and Talbot 1963). If group integrity is lost, the sexes may either mix within the herd, as in the wildebeest, or may segregate, as at times in the barren-ground caribou, *Rangifer tarandus*, on the Canadian tundra (Banfield 1954).

Migration from one basic social unit to another, where such migration has been investigated, seems to be very similar to that of primates. In the vicuña, *Vicugna vicugna*, for example, the basic social units of which, like primates, remain within a relatively small home range, sub-adult males when about a year old leave the bisexual group (1 male plus 1–18 females plus 0–9 juveniles) into which they were born and join all-male groups (Koford 1957). All-male groups, many of which consist of more than 30 males and often as many as 75, consist pre-dominantly (90 per cent) of males between one and three years of age. All-male groups are temporary aggregations that males join and leave freely and their composition may change several times a day. During this phase of ontogeny a male ranges widely and establishes a large familiar area, at least in comparison to the small home range that will be occupied once a harem has been gathered. Females of any age, but particularly sub-adults, may migrate

from the territory of one male to that of another. Unlike the gorilla, *Gorilla gorilla*, however, which lives in forest and which, apart from auditory reception, must mingle with other groups before their suitability can be assessed, the vicuña lives on semi-arid rolling grasslands over which the visual range is considerable. Even females, therefore, have a familiar area that is considerably greater than their home range, with the result that the majority of removal migrations are likely to be calculated rather than non-calculated. With the exception of the difference in sensory range, therefore, the situation found in the vicuña is, as far as migration strategy is concerned, apparently very similar to that already described for some primates and it can be assumed that similar selective pressures are acting.

Another animal for which the example of the gorilla may be used to illustrate similar movement patterns between different mammalian groups is the Indian elephant, *Elephas maximus*, which has been observed in Malaya (Mohamed Khan bin Momin 1969) to perform a migration that seems to be similar to that described for groups of gorillas in which the silverback male has died (p. 162). In Malaya the Indian elephant lives for the most part within discrete home ranges which overlap to some extent such that the occupant herds meet from time to time. These herds consist of 3–25 individuals and consist of 1–8 males, 1–10 females, and 1–7 young. When a herd becomes reduced to 3 individuals or less, however, the survivors migrate to join another herd. It seems likely that this migration is calculated and is based on prior assessment of the relative suitability of neighbouring herds.

The plains zebra, *Equus quagga*, shows a situation (Fig. 17.20) similar to that found in the vicuña, with the exception that all individuals of the zebra have a relatively large familiar area as a result of the seasonal return migrations that take place (Chapter 25). In the zebra, therefore, basic social units are in continuous visual contact with large numbers of other social units within the herd and must frequent-ly contact new social units. Removal migration of both males and females from the natal group does not occur until the zebra are one year old, and may not occur until the age of four (Klingel 1969). By the time removal migration from the natal group takes place, therefore, an annual migration cycle has been performed and the individual has established a large basic familiar area. Removal migration by the female involves leaving one bisexual group (1 male, 1–6 females, and 0–9 juveniles) for another when two groups or the bisexual group and a solitary male come into contact. Removal migration by the female is therefore likely to be calculated and possibly could be based on the criterion of the male : female ratio in the two groups as well as on the relative attributes of the two males. Once again, therefore, the situation is similar to that described for primates. Removal migration of males from one group to another also involves departure from one bisexual group and eventual membership of another. Several years, however, may intervene between these two events, during which time the male lives either solitarily or as a member of an all-male group of up to nine individuals. In view of the time interval between these two events it is doubtful whether the initiation of group-to-group removal migration is calculated. Although changes in the composition of zebra groups are infrequent (Klingel 1969), they occur with sufficient frequency for the natal group of a male zebra to have more or less totally lost its integrity by the time the male is becoming established as a dominant male. Further, and in contrast to the gorilla, as the natal group did not occupy a discrete home range as such that is likely still to be occupied by even the remnants of the natal group, it is doubtful if there is any meaningful way in which

Fig. 15.3 Part of a herd of plains' zebra, *Equus quagga*

Although the herd moves as a unit, the primary social groups (consisting of a single adult male plus one to several females and their offspring) retain their identity. In such a situation, group-to-group removal migration is not the same as area-to-area removal migration.

[Photo by H. Klingel]

the initiation of group-to-group removal migration by a young male zebra can be said to be calculated. Initiation of group-to-group removal migration by male zebra, therefore, must be non-calculated. Once a male has established a harem, however, it seems likely that its subsequent movements may well be based on habitat assessments within the familiar area established as a solitary male. The movements of solitary males may not, therefore, represent a strategy to increase the ratio of calculated to non-calculated group-to-group migration but may still represent such a strategy in relation to area-to-area migration. In all of the mammals considered so far, group-to-group removal migration has been inseparable from area-to-area removal migration be-cause the groups involved also inhabited a discrete area. In the plains' zebra and many ungulates in which large herds perform long-distance seasonal return migrations this is not necessarily the case. The remainder of this consideration of the possible existence of strategies that increase the ratio of calculated to non-calculated migration in ungulates is concerned with area-to-area removal migration.

In some ungulate species, such as the white-tailed deer, *Odocoileus virginianus*, which occupies predominantly brush-type habitats in the United States (A. Kramer 1971), individuals occupy a relatively small home range at any particular season. Seasonal return migration between winter and summer home ranges, however, often over distances of ten or more

Fig. 15.4   Females and male of the white-tailed deer, *Odocoileus virginianus*

[*Photos by Leonard Lee Rue III (courtesy of Frank W. Lane) and C.G. Hampson*]
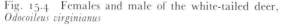

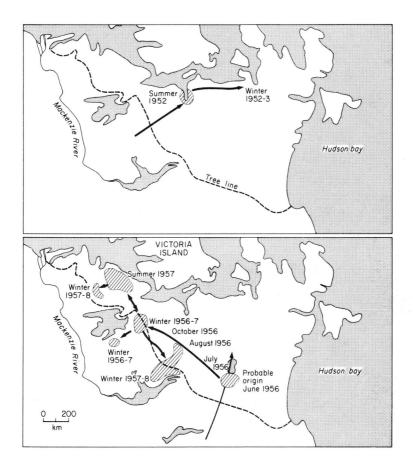

Fig. 15.6 Long-distance removal migration by large herds of barren-ground caribou, *Rangifer tarandus*, on the Canadian tundra

The 'normal' seasonal return migration axis of caribou in the region shown is perpendicular to the tree line as indicated by the first arrow in the top diagram and by the thin arrow in the bottom diagram (see also Fig. 25.6). From time to time, however, long-distance removal migrations are performed in some quite different direction. In the top diagram, instead of returning to the tree line for the winter, the caribou herd continued ENE to coastal wintering grounds. In the bottom diagram, instead of continuing their seasonal migration to the NNE, after parturition the herd, which at times involved 20 000 to 40 000 animals, migrated to wintering grounds several hundred kilometres to the NW. The following year part of the herd continued the migration, part returned some distance to the SE, and part remained in the same general area. All eventually established 'normal' seasonal return migration circuits.

The eastward migration in the top diagram occurred at a time when deme density was much greater in the western part of the Canadian tundra. The westward migration in the bottom diagram occurred four years later at a time when deme density was much greater in the eastern part of the range. Assuming a causal relationship, how could the caribou have determined the direction of the area of low deme density? The migration in the top diagram led to the recolonisation of areas devoid of herds for 15 years. However, this is well within the lifetime of at least some individuals. Quite possibly, the removal migration is a calculated migration to part of the familiar area of the older members of the herd. In this case upon initiation, perhaps in response to a subthreshold nett rate of energy uptake per individual, younger individuals are likely to delegate direction determination to older individuals. If the migration is calculated and within the familiar area of older individuals, it is necessary to postulate that caribou are capable of developing a familiar area that consists of a very large proportion of the Canadian tundra.

Even though as a result of these removal migrations the modal deme density may oscillate between the western and eastern parts of the range with a period of perhaps a decade or so, some individuals are found in most parts of the range at all times. Presumably, therefore, there is a frequency distribution of migration thresholds of the type discussed in Chapter 17.

[*Compiled from descriptions and diagrams in Kelsall (1968). Photo of caribou by J.P. Kelsall (courtesy of Canadian Wildlife Service)*]

directly to an area 80–160 km away (Banfield 1954). Whether this is a calculated or a non-calculated migration is not known. If it is a calculated migration the question presents itself of how the migration was possible. Conceivably the new winter area was encountered during exploratory migration by all or part of the herd the previous winter. Alternatively, the destination may be within the familiar area of older members of the herd that had used the area several years previously. The latter of these two alternatives is considered further in relation to the even more spectacular examples of removal migration shown in Fig. 15.6.

To conclude, therefore, ungulates seem to show a number of behaviour patterns that increase the ratio of calculated to non-calculated migration. As with other mammals, particularly primates, these behaviour patterns consist of contact between adjacent social units, exploratory migration, and possibly also social communication. Contrary to the situation when considering most other terrestrial mammals, however, it is necessary to distinguish between group-to-group and area-to-area removal migration. Migration patterns that may well be part of non-calculated group-to-group removal migration may at the same time be exploratory with respect to calculated area-to-area removal migration.

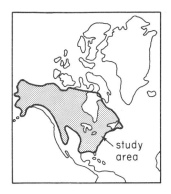

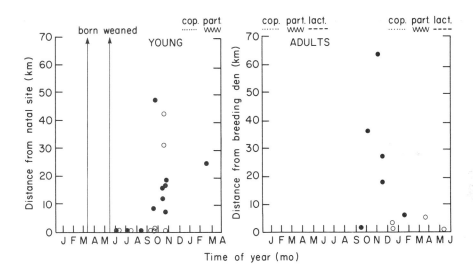

Fig. 15.7   Migration distance of the young and adults of the red fox, *Vulpes vulpes*, in New York State

The stippled area in the map shows the geographical range of the red fox in North America. Seasonal behaviour is indicated by the dotted line (pairing and copulation season), zig-zag line (parturition season), and dashed line (lactation season). Large dots (○, females; ●, males) show the distance from the spring and summer marking site that foxes marked as young and adults were recaptured.

Young foxes move little from their natal den during the summer. In the late autumn, however, at least some individuals of both males and females migrate some distance from their natal den. Whether this migration is exploratory, the ability being retained to return to the natal area or any other area encountered during the migration, or whether the migration is a non-calculated removal migration, the ability to return not being retained, is unknown. Certainly the movement leads to

removal migration, either calculated or non-calculated, at least one male breeding some distance from its natal site.

Adult foxes remain within a few kilometres of their breeding den during spring and summer. Although the home range of the female may expand slightly during winter, little migration beyond the breeding home range occurs. Males, however, may migrate long-distances in late autumn and early winter. This migration could either be exploratory and followed by calculated removal migration in late winter to the area assessed to be most suitable, or it could be a non-calculated removal migration. Data are insufficient at present to distinguish between these two alternatives.

Similar behaviour is shown by the red fox in Ireland, though there may be less long-distance migration by adult males.

[*North American range and seasonal behaviour from descriptions and diagrams in Churcher (1973). Marking–release/recapture data from Sheldon (1950, 1953). Irish study by Fairley (1969, 1970). Photo of red fox by Wilford L. Miller (courtesy of Frank W. Lane)*]

15.6 Carnivora

Carnivorous mammals have a relatively large familiar area. Among North American mustelids, for example (Schwartz and Schwartz 1959) the long-tailed weasel, *Mustela frenata*, which has a body length of about 25 cm, has a home range of about 160 ha. In some parts of their range males may have a hunting circuit of some 3 km in length and 2 km in width. The more aquatically inclined mink, *M. vison*, which is intermediate in habits between a weasel and an otter and which has a body length of about 36 cm, has, at least in the male, a large home range of up to 8 km in diameter and takes approximately 2 weeks to cover the entire circuit. The home range of a female mink is only about 8 ha. The mountain lion or cougar, *Felis concolor*, of America, which has a body length of 2·1–2·5 m, has a home range with a radius of 100 km or so in Arizona (Musgrave 1926). The wolf, *Canis lupus*, in the United States in the non-breeding season performs a hunting circuit of some 200–300 km (S.P. Young 1964) which is completed about every 10 days or so (Thompson 1952). Finally, the African hunting dog, *Lycaon pictus*, which lives in groups, defends a territory of about 1 km^2 around its burrows against conspecifics or any other species of carnivore. Round this territory there exists a hunting range of 100–200 km^2 which may overlap the ranges of other groups (Kühme 1965). Later in the dry season, when the burrows are vacated and the dogs become nomadic, the hunting range may increase to 2000 km^2. Because of the size of the home range of carnivores and the even greater size of the familiar area, it is difficult to interpret the results of marking–release/recapture experiments, particularly when recapture invariably involves the death of the individual concerned so that only one recapture of that individual is possible.

Probably the majority of carnivorous mammals live solitarily within their home range. This is particularly true of the felids except for the lion, *Panthera leo*, though in other groups there may be some cooperation by male–female pairs during the parturition and lactation period. Yet other species live in complex social units, usually either one-male or age-graded male groups as in the lion (Stevenson–Hamilton 1947) or multi-male groups as in the African hunting dog and, in more northern parts of its range during the winter, the wolf. Unlike some ungulates, however, group-to-group removal migration is probably also area-to-area removal migration.

The American red fox, *Vulpes fulva*, a North American canid that is conspecific with the European red fox, *V. vulpes* (H.N. Southern 1964, Churcher 1973), feeds mainly on rabbits, rats, and mice and prefers the borders of forested areas and adjacent open lands (Schwartz and Schwartz 1959). While the young are still being fed in the den, from April to June, the home range is about 3 km in diameter or even less (Ables 1969). From September to March foxes sleep in any sheltered place on the ground, and the home range is increased to an area 8–16 km in diameter (Schwartz and Schwartz 1959). In September the young become independent but for a month or so stay within about a kilometre of the natal den. Late in the autumn removal migration from the natal home range takes place (Fig. 15.7). Adult males may also show removal migration in the autumn and some may migrate as far as 65 km from their former home range.

At first sight such long-distance migration suggests a clear example of non-calculated removal migration. The critical question, however, of whether the ability to return to the natal or previous familiar area is retained remains unanswered. For example, Musgrave (1926), on the basis of observation of the mountain lion in Arizona, claims that an adult male may travel more than 30 km in a single night and that it returns by the same route even when it has travelled 100 km or so. Furthermore, as part of movement within the home range such individuals may often move from one mountain range to another across deserts up to 80 km wide. Unless the situation is very different in Idaho, therefore, recapture of two young females at distances of 65 km and 110 km from the natal site (Hornocker 1970) need not necessarily prove the existence of non-calculated removal migration. Neither, perhaps, does the recapture of two young males at distances of nearly 200 km from their natal site. Two juvenile wolves, *Canis lupus*, that were marked in the Rocky Mountains of West Alberta, Canada, have been recaptured at distances of 25 and 250 km from the marking site after intervals of 4 years and 2 years respectively. The latter individual, a young male when marked, had crossed seven rivers during his migration (Banfield 1953). From such marking results, however, it is still not possible to determine whether the animal arrived at its destination as a result of non-calculated migration or whether the destination was assessed to be the most suitable habitat within an extended familiar

Fig. 15.8 Mountain lion, *Felis concolor*, in southwest Colorado

The lake in the background has been produced by the activity of beavers, *Castor canadensis*.

[*Photo by Wilford L. Miller (courtesy Frank W. Lane)*]

area that included both the original home range and the new home range.

Unfortunately most of the information relevant to this consideration for the Carnivora as a taxonomic group comes not from carnivorous animals in the ecological sense but from omnivores such as bears and raccoons. Perhaps not surprisingly the movements of these animals seem to parallel those of rodents and primates rather than other Carnivora. The black bear, *Ursus americanus*, of America, for example, feeds on green vegetation in the spring, berries in the autumn, and insects and other animals whenever possible (Hatler 1972). The home range is small by carnivore standards, measuring, along its longest axis, only 2·5 km for females and 6 km for males (Jonkel and Cowan 1971). Yet occasionally in

the spring some individuals migrate 16 km or so beyond the boundary of their home range to visit a rubbish dump. Similar movements may be made to a salmon stream at the appropriate time of year. The implication is, therefore, that as for rats, *Rattus norvegicus* (p. 161), the familiar area is considerably greater than the home range and that at some time exploratory migration must have occurred. It is possible, of course, that both the rubbish dump (or salmon stream) and eventual destination were discovered during non-calculated migration and that only the route between the two was learnt. It seems unlikely, however, that the individual should have learnt only that part of its route during migration, especially as the position of the eventual home range would not have been known on this interpretation when the rubbish dump or stream was discovered. The only reasonable alternative to discovery during exploratory migration, therefore, is some form of long-distance sensory perception of the existence of a rubbish dump or salmon stream after the individual has arrived at the home range by non-calculated migration. The postulated existence of exploratory migration need not, of course, exclude the possibility of non-calculated removal migration unless it is postulated that exploratory migration is continuous from more or less the moment of birth. Non-calculated migration could be postulated, for example, to be the type of migration by which the bear leaves its natal site and arrives at its eventual home range. Exploratory migration from this base could then result in discovery of a rubbish dump or salmon stream well outside the normal home range. The reasonable alternatives in this case, therefore, are: (1) exploratory migration based on the natal home range followed by calculated removal migration to the habitat assessed to be most suitable but with continued use of any other sites discovered during exploration even though these may be well outside the new home range; (2) non-calculated migration from the natal home range followed by exploratory migration based on the new home range; and (3) non-calculated or calculated migration from the natal to the new home range followed by long-distance sensory perception of food sources well outside the new home range. The data accumulated by Jonkel and Cowan (1971) in northwest Montana do not really permit a decision to be made between these alternatives. Young black bears are weaned at 7 months and leave their mother at 1·5 years at about the time that an adult male joins the mother for copulation. The young bears often stay in the mother's home range for a time after this and two females were observed to remain in the mother's home range permanently, though whether they performed exploratory migrations beyond this is not known. It cannot be decided, therefore, whether these two females had a high threshold for non-calculated removal migration or whether, having extended their familiar area, the mother's home range was assessed to be the most suitable habitat within that familiar area. One male juvenile about 2·5 years old was recaptured 48 km from its natal home range. The maximum distance recorded seems to be 140 km, also for a bear 2·5 years old. Bearing in

mind the 16 km foraging migrations recorded for black bears and their ability to home after displacements of up to 80 km (Kolenosky 1970), 48 km does not seem unreasonable for exploratory migration. The maximum record of 140 km, however, is perhaps more likely to be a non-calculated removal migration. However, black bears are not sexually mature until 5 or 6 years old and the crucial question of whether juvenile migrants, even if they travel 100 km or more, retain until the onset of sexual maturity the ability to return to any area visited since leaving the natal home range remains unanswered.

Fig. 15.9   Kodiak bear, *Ursus arctos*, catching salmon

Like the black bear, *U. americanus*, the different forms of the holarctic brown bear, *U. arctos*, such as the grizzly and Kodiak bears, may migrate some distance to salmon streams during the peak of the salmon run.

Data for perhaps the most carnivorous of bears, the polar bear, *Ursus maritimus*, of the Arctic Ocean, collected by Perry (1966) and T. Larsen (1968) raise more problems than are solved. On land, polar bears feed on grass and berries in the spring after denning. Thereafter, seals form the main part of the diet until summer when grass, berries, and birds etc. are consumed on land. In the autumn virtually any-thing edible is eaten, including seals by those bears hunting on the pack-ice. While hunting for seals on the pack-ice at any time of year, polar bears drift on the ice with the current. The pack-ice drifts clock-wise around the Arctic Ocean but there are so many local currents and eddies that it is doubtful if any polar bear ever circumnavigates the North Pole. Nevertheless, polar bears probably travel consider-

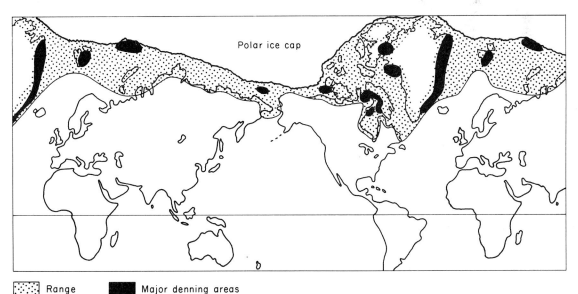

Range Major denning areas

Fig. 15.10   The circumpolar distribution of the polar bear, *Ursus maritimus*

Fossils of the polar bear are absent in deposits pre-dating the last glaciation. Anatomical and fossil evidence in-dicates that both the polar bear, *Ursus maritimus*, and the brown bear, *Ursus arctos*, stem from a common ancestor, the Pleistocene bear, *Ursus etruscus*, the polar bear repre-senting a fairly recent, carnivorous offshoot that is specialised for Arctic survival. [*Distribution and evolution from Harington (1970). Photos by Steve and Dolores McCutcheon (courtesy Frank W. Lane) and J. Twiselton*]

able distances, perhaps many hundreds or thousands of kilometres. Sometimes in summer polar bears may be carried to land on the pack-ice which then, with a change in wind direction, may be carried so far off shore that it is beyond the swimming capabilities of even the polar bear. Such individuals, therefore, may be stranded on land for the entire summer. In such circumstances, particularly in Greenland, polar bears are reputed to show well-marked navigational powers, being able to return often considerable distances overland to their original starting point. On the one hand, therefore, polar bears are reputed to be complete nomads, drifting with the ice for considerable distances and totally unattached to any particular home range. Yet, on the other hand, the species is reputed to form demes (Larsen 1968) and to show marked navigational powers (Perry 1966). Clearly much more objective rather than anecdotal data are necessary before polar bear migration can be interpreted with any degree of confidence. In this connection recent attempts at long-range radio-tracking of polar bears (Baldwin 1972) raise some hopes that data may be available in the near future.

To conclude, therefore, the large home ranges of most carnivores make the interpretation of marking–release/recapture experiments difficult. Many of the migrations reported could well have been, or could have been preceded by, exploratory migrations. On the other hand, other migrations involve such long distances that it is difficult to believe that they can have been initiated on anything other than a non-calculated removal basis. There seems little hope of understanding the removal migrations of carnivores as long as the methodology of investigation depends upon the death of the individual for recapture. The increasing use of radio-tracking, however, offers some hope that suitable data will become available in the near future.

15.7 Amphibious mammals

Amphibious mammals usually either roost on land or ice and feed in the water, as in pinnipeds, otters, amphibious rodents, and the duck-billed platypus, *Ornithorhynchus platypus*, or they may roost in the water and feed on land as in the hippopotamus, *Hippopotamus amphibius*. Marine mammals vary from being solitary except for mating, as in some pinnipeds that produce their young on ice, through

being intensely gregarious during the breeding season but more or less solitary at other times as in the northern fur seal, *Callorhinus ursinus*, to being more or less gregarious both on land and at sea as in the southern sea lion, *Otaria byronia* (Chapter 28). Among the freshwater species the hippopotamus is more or less gregarious when in the water, particularly during the dry season, but more or less solitary on land at night. Freshwater otters live in family groups for a large part of the year as may some amphibious rodents, such as the beaver, *Castor canadensis*. Other amphibious rodents may live more or less solitarily, as also does the platypus.

The familiar area may be extremely large, particularly in some pinnipeds. An individual sea otter, *Enhydra lutris*, may make use of 17 km of coastline in the female or even more in the male and may include several hauling-out places (Kenyon 1969). The river otter, *Lutra canadensis*, of North America also has a large familiar area which may include a shoreline of 80–160 km (Schwartz and Schwartz 1959) though at any one season a family may have a home range that includes only 5–16 km of stream. By way of comparison, the home range of the beaver, *Castor canadensis*, which feeds mainly on bark, extends only to within 1 km of the roost. Finally, the platypus, which feeds predominantly on small aquatic invertebrates such as worms, crustaceans, insect larvae and molluscs, has a home range that includes up to 5 km of stream (Burrell 1927).

Extension of the familiar area in a number of pinnipeds, particularly the northern fur seal, *Callorhinus ursinus*, has already been discussed (p. 105) and it does not seem too unreasonable to suggest that the initiation of removal migration in pinnipeds is often calculated and preceded by exploratory migration. Possible exceptions are those pelagic pagophilic (ice-loving) species such as the crabeater seal, *Lobodon carcinophagus*, leopard seal, *Hydrurga leptonyx*, and Ross seal, *Ommatophoca rossi*, about which little is known and which may or may not make use of a familiar area.

Extension of the familiar area has also been described in the muskrat, *Ondatra zibethicus* (p. 101). In view of the linearity of their habitat, animals that live in the banks of streams and rivers should experience little difficulty in retaining their ability to return to any area visited during familiar area extension, including the natal home range, and it might be expected that many such animals when performing removal migration along a stream precede such migration by extension of the familiar area

Fig. 15.11   Crabeater seal, *Lobodon carcinophagus*

[*Photo courtesy of British Antarctic Survey*]

and habitat assessment. For example, in the beaver, *Castor canadensis*, male and female often pair for life (Schwartz and Schwartz 1959). During the summer, however, the male vacates the home range occupied by the female, their two yearling offspring and their two young of the year. The main advantage of this behaviour seems likely to be to reduce the foraging distance for the female and young as a result of a reduced utilisation of resources near to the roost. The possibility cannot be totally excluded, however, that the male is also performing an exploratory migration. From time to time, particularly in autumn soon after the male returns, beaver families perform removal migration to a new home range, usually when the suitable food within 100–200 m of the water's edge in the old home range has been utilised (Schwartz and Schwartz 1959). Quite possibly the destination is an area that has been visited by the male during the summer.

The hippopotamus, *Hippopotamus amphibius*, is found in most lakes and rivers of Africa, south of the Sahara Desert, though it is commoner in the east than in the west (J.R. Clarke 1953). The species is diurnally aquatic and nocturnally terrestrial and shows a daily return migration from daytime aquatic roosting sites to nocturnal terrestrial grazing sites. Groups of hippo at the roosting site consist predominantly of females and young. Lactating females prefer shallow water where the young offspring does not need to swim. Consequently the central roosting site is often a submerged sandbank (Field 1970).

Around this central area the males have individual roosting sites (Verheyen 1954) though some may also roost with the females and young. In the Queen Elizabeth National Park in Uganda this central group consists on average of 19, though on occasion as many as 107, individuals (Laws and Clough 1966). The home range toward the end of the dry season may extend up to 7 km from water which at this time is usually part of a permanent lake or river system (Field 1970). During the wet season, however, there is considerable movement along waterways (Clarke 1953) which could be interpreted as exploratory. There certainly seems little doubt that the familiar area is considerably larger than the dry season home range. However, exploratory migration cannot be confined to waterways, for during the wet season the roosting aggregations disperse and make use of temporary inland roosts (wallows). Males seem to show a preference for wallows whereas females continue to show a preference for lakes (Field 1970) which could indicate a larger familiar area for the males than the females. Certainly tracks have been followed for up to 30 km overland from one river to another (Clarke 1953) though without knowing the history of movements of the individual concerned it is not possible to be sure whether such a movement was exploratory or was initiated on a calculated or non-calculated removal basis.

The duck-billed platypus, *Ornithorhynchus platypus*, is found in Australia east of about 138°E and south to Tasmania. It shows a preference for high places but is not confined to these areas. It is found both in rapid running water and lakes but prefers clean, clear water (Burrell 1927). Except during mating, individuals are solitary. A relatively simple roosting burrow or nest is built individually by both sexes and is used at all times of year. Depending on food abundance, several such roosting burrows are maintained within the home range. During the breeding season the female leaves that part of her familiar area that is most suitable for feeding and in a suitable bank builds a complex nesting burrow in which 1–3 eggs will be laid. Nesting burrows are found at densities of about one burrow to 5 km of river. Not only does each individual extend its familiar area along the river, however, but it seems likely that exploratory migrations away from the river over land also occur. Platypus released up to 2 km from water are reported to head directly back toward the water, thus suggesting familiarity with the area, though without controlled experiments the possibility of long-distance sensory perception of the water cannot entirely be ruled out. Whether or not migrations of up to 8 km that occur occasionally from one river to another across intervening high land are, or have been preceded by, exploratory migrations remains unknown.

To conclude, except for the muskrat, *Ondatra zibethicus*, the existence of exploratory migration as a preliminary to removal migration has not definitely been established for any amphibious mammal. Such observations that are available, however, suggest that a large proportion of the removal migrations performed by amphibious mammals are initiated on a calculated rather than on a non-calculated basis and have been preceded by exploratory migration.

15.8 Aquatic mammals

The relatively small amount of information that seems to be available concerning the movements of dugong, *Dugong dugong*, in the western North Pacific (Nishiwaki 1967) and manatee, *Trichechus manatus*, in Florida, United States (Layne 1965) in effect means that the only group of aquatic mammals for which information is available is the Cetacea. Even here, problems exist that limit considerably any attempt to consider the possible existence of strategies that increase the ratio of calculated to non-calculated removal migration. First, except for the sperm whale, *Physeter catodon*, and to a lesser extent the killer whale, *Orcinus orca*, odontocete cetaceans are of relatively little interest to the whaling industry and hence have received relatively little attention. The main body of data, therefore, refers to the mysticete or baleen cetaceans. Even here, however, because of their considerable mobility, extreme difficulty exists in recognising individual home ranges or familiar areas. Many species perform seasonal migrations from polar to equatorial regions (Fig. 15.12) and often in these species the northern and southern hemisphere demes are more or less distinct (Chapter 29). So too are the Atlantic and Pacific demes. Any individual that migrated between southern and northern polar regions or between the Atlantic and the Pacific could therefore be considered to be an inter-deme removal migrant. As far as I have been able to determine, however, marking–release/recapture experiments have so far failed to demonstrate such a migration. Other species, such as Bryde's whale, *Balaenoptera edeni*, are equatorial in distribution (Fig. 15.13). Presumably

in these cases even movements between the northern and southern hemispheres could be movements within the familiar area and need not necessarily indicate removal migration.

Fig. 15.12 Seasonal return migration of the sei whale, *Balaenoptera borealis*

General ecology. The sei whale is a mysticete that feeds on copepods (✳✳✳), euphausids (✳✳), cephalopods (✳✳), and fish (✳✳). It employs both skimming and swallowing types of feeding behaviour (p. 746) and seems to be adapted to exploit areas in which euphausids occur at only moderate density. Feeding occurs throughout the day but less at dusk and dawn. The species is fairly solitary, occurring either singly or in groups of 1–6. Pairs are usually both males. The seasonal incidence of copulation (· · · ·) in the northern and southern hemispheres is shown, respectively, above and below diagram B. The seasonal incidence of parturition (———) and lactation (– – –) in the southern hemisphere is shown below diagram B. The gestation period is 10–12 months. Preferred surface temperature in winter (April to October) off South Africa is 18°C.

Sexual maturation occurs at a length of 13·3 m in males and 14.0 m in females. Average maximum length is 15·0 m for males and 16·5 m for females. For females there is an interval of 2 years between successive births.

Seasonal return migration
A. Geographical distribution in August (△) and February (○)

Each dot shows a sighting. The arrows indicate possible migration routes.

B. Seasonal shift of latitude

The dots indicate the latitude at which sei whales were sighted in different oceans for each month of the year. The lines join the latitudes and times of release and recapture of marked individuals.

There is a clear indication of a polar migration in summer and tropical migration in winter. Off South Africa, sei whales are swimming to the north until midwinter (July) and south thereafter. There is no clear separation between northern and southern hemisphere demes, especially in the Indian Ocean and Pacific. However, this species is so often confused with the tropical Bryde's whale, *Balaenoptera edeni* (Fig. 15.13) that not too much confidence can be placed in tropical sightings.

Analysis of size and sex distribution in the southern hemisphere in summer shows that the proportion of males increases towards the pole and that the mean length of individuals also increases. Thus at latitudes lower than 50°S, mean body length is 14·5 m for males and 14·7 m for females. At higher latitudes, mean body length is 14·9 m for males and 15·3 m for females.

[*Compiled from data in Nemoto (1959), Slijper (1962), Omura and Ohsumi (1964), Slijper et al. (1964), Bannister and Gambell (1965), Jonsgård (1966), Nasu (1966), Nishiwaki (1966, 1967), Best (1967), Gaskin (1968b), Neave and Wright (1968), and Nasu and Masaki (1970)*]

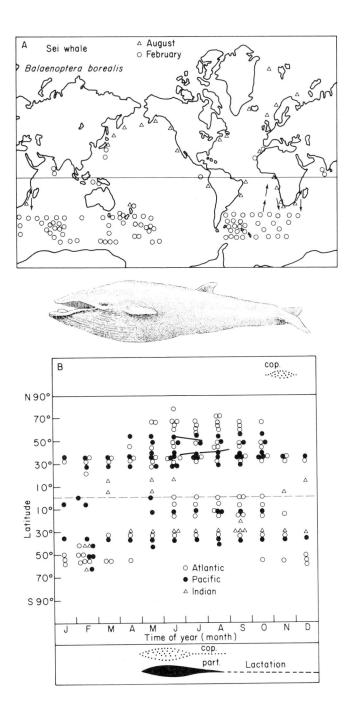

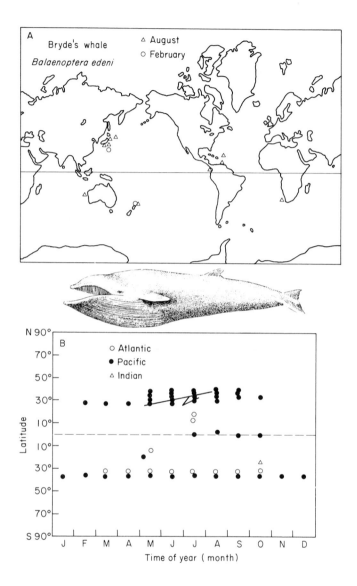

Fig. 15.13   Geographical distribution of Bryde's whale, *Balaenoptera edeni*

General ecology. Bryde's whale is a mysticete but feeds predominantly on fish (✳✳✳✳), although it also takes euphausids (✳✳) and copepods (✳). It employs a swallowing type of feeding behaviour (p. 746). Both copulation and parturition take place throughout the year, though each deme may have its own peak season. Sexual maturity is reached at lengths of 11·9 m for males and 12·2 m for females. The species is difficult to distinguish from the sei whale, *Balaenoptera borealis*, and has only recently been generally accepted as a separate species.

Seasonal return migration
A. Geographical distribution in August (△) and February (○)

B. Latitudinal distribution throughout the year

The dots indicate the latitudes at which Bryde's whales have been sighted in different oceans during each month of the year. The lines join the latitudes and times of release and recapture of marked individuals.

The distribution is tropical and shows little seasonal latitudinal shift. Presumably each deme within the population has its own pattern of seasonal change of home range. There is some indication that the deme occupying the northwest Pacific migrates to the north in summer and to the south in winter.

[*Compiled from data in Omura (1959), Slijper (1962), Omura and Ohsumi (1964), Clarke and Aguayo (1965), Jonsgård (1966), Nishiwaki (1966, 1967), Best (1967), and Gaskin (1968b)*]

It is generally supposed, however, that whales live in smaller demes than say 'North Pacific' or 'Indian Ocean' demes and in one case at least this supposition seems to be well substantiated. In the southeast Indian Ocean and southwest Pacific, the humpback whale, *Megaptera novaeangliae*, breeds in three more or less distinct areas, these being: (1) the west coast of Australia; (2) the east coast of Australia; and (3) New Zealand and parts of Oceania (Fig. 29.9). The humpbacks that breed in these three areas seem to constitute three distinct demes. Dawbin (1959) has even suggested that within the demes that breed along the east coast of Australia and the southwest Pacific Islands some segregation takes place into smaller demes such that individuals return each year to the same coast or island group to breed. Chittleborough (1965), however, has analysed recapture data in this region and concluded that although this may be true in some years, in other years individuals seem to return at random to the different breeding sites within the same general area.

These humpback whales remain in their breeding area during the southern winter, during which time parturition and copulation take place (Fig. 29.9). In the southern spring the seasonal return to the rich plankton of Antarctic waters takes place. Analysis of marking–release/recapture results by Dawbin (1964) and Chittleborough (1959, 1965) has shown that the three breeding demes described above remain more or less distinct during the period that they are in Antarctic waters. We can accept, therefore, that at any time during seasonal return migration any movement of individuals from one of the areas shown in Fig. 29.9 to another constitutes removal migration.

An individual that does not perform removal migration is likely to perform the same migration between Antarctic and tropical waters many times and thus to establish a familiar area some 4000 km or so in length. Migration between winter and summer grounds is not a continuous direct process and it seems clear that for shorter or longer periods en route individuals settle down in a particular area. One individual humpback with certain recognisable morphological characteristics was observed by Schevill and Backus (1960) to traverse repeatedly on a daily basis an area of about 50 km² off the coast of Portland, Maine, eastern United States. This individual remained within this area for a period of at least 10 days from 29 October through 7 November during the period of southward migration.

Given, then, that no matter what orientation and navigation mechanisms may be used (Chapter 33), humpback whales establish a familiar area that includes winter and summer grounds and the route in between, the question presents itself as to the nature of the initiation of removal migration. A number of observations seem to be relevant in this connection. First, migration has been found to take place from one breeding ground to another during a single winter. This has been reported for an adolescent female that was marked in New Zealand waters and which was recaptured later that same summer off the east coast of Australia. It remains unknown whether such a migrant would be engaged in exploratory migration or in calculated or noncalculated removal migration. As far as can be determined from marking–release/recapture experiments (Chittleborough 1959, 1965, Dawbin 1964), however, it seems that most removal migration occurs during the summer while the whales are in Antarctic waters. Furthermore, most removal migration occurs when there is greatest overlap between the demes during this period. The surprising aspect of this is that such overlap, which could be interpreted to be a consequence of more familiar area extension in some years than others, leads to some individuals taking on the seasonal migration route of the neighbouring deme. This could, of course, be interpreted as resulting from such individuals being incapable of returning to their original familiar area. North–south orientation may then bring them to suitable breeding grounds or alternatively they may associate with and follow other members of the deme they have just joined. However, not all such removal migrants are adolescent (Chapter 17). Many are adult and are likely to have been members of their original deme for many years. Suggestions that such individuals could be incapable of returning to a familiar area 4000 km in length, therefore, must be treated with some reservation. An equally, if not more, acceptable interpretation is that although these removal migrants retain the ability to return to their original seasonal migration route, the new seasonal migration route is assessed to be more suitable. This is consistent with the observation that such removal migrations are more frequent some summers than others. Possibly familiar area extension is more likely when food is scarce. Discovery of a more abundant food source during such exploratory migration could well lead to calculated removal migration between feeding grounds. It does not explain, however, why seasonal migration now takes place to the breeding grounds of the new deme rather than return to the original seasonal migration route. Such a return, of course, would increase the distance of seasonal return migration. Quite possibly the disadvantage of such increase in distance has resulted in selection favouring individuals, having performed removal migration at one stage in the seasonal return migration, adopting the new seasonal return migration route in its entirety. Alternatively, social communication concerning the relative suitability of the different breeding grounds could take place while on the feeding grounds. Acoustic communication of relatively complex information is known for cetaceans (Bastian 1967), perhaps over extremely long distances (p. 883).

To conclude, therefore, consideration of the existence of behaviour patterns that increase the ratio of calculated to non-calculated removal migration in aquatic mammals is hindered by lack of data and the difficulty of deciding when removal migration has occurred. Only in one species of mysticete cetacean, the humpback whale, has a beginning been made on an analysis of the type of data that could permit such a consideration. Even in this case it is still not possible to distinguish between exploratory migrations and calculated and non-calculated removal migrations. The possibility of the existence of exploratory migration followed by calculated removal migration seems quite high, however. There is also the possibility that some social communication of information concerning habitat suitability may occur.

15.9 Birds

Despite the wealth of information that has been accumulated on the seasonal return migrations of birds, it is probably true to say that as far as the strategies involved in removal migration are concerned little more is known for these than for any other group of animals.

First, let us consider sea-birds and in particular those such as albatrosses, penguins, auks, shearwaters, petrels, boobies, and many gulls, etc. that nest on land but feed at sea. These species show considerable similarity to pinnipeds in their migrations. Both groups tend to converge to breed and to disperse to feed. Both groups show an extremely high degree of return to the same breeding site each year (i.e. show an extremely low rate of removal migration). When removal migration does occur it is invariably carried out by juveniles or sub-adults and it is also this ontogenetic class that wanders most while still, apparently, retaining the ability to return to the natal site. It is tempting, therefore, to come to the same conclusion for sea-birds as for pinnipeds: that the nomadism of juveniles in part represents exploratory migration the function of which is to increase the ratio of calculated to non-calculated

Fig. 15.14 Two Antarctic sea birds, the giant petrel, *Macronectes giganteus*, and the adelie penguin, *Pygoscelis adeliae*

In both species the young, newly fledged birds seem to establish a huge familiar area in the Southern Ocean out of which crystallises a much smaller adult home range.

[*Photos by D.G. Bone*]

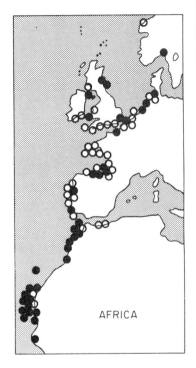

Fig. 15.15   Winter recoveries (November to February) of young and adult gannets, *Sula bassana*, banded in Great Britain

o, recoveries of adults; ●, recoveries of first-year birds.
  Note that on average the young birds travel much further than do the adults.

[*Modified from Dorst (1962) after A.L. Thomson. Photo by Eric Hosking*]

migration. A particularly striking example is the giant petrel, *Macronectes giganteus* (Dorst 1962), in which the adults move relatively little from their breeding grounds, whereas the young seem likely to circumnavigate the Southern Ocean several times before starting to reproduce. Doubtless the advantage of increase in food availability to both adults and young may also have resulted in selection in favour of such segregation. Which aspect of such ontogenetic differences in migration pattern is most advantageous, exploratory migration or feeding segregation, remains to be determined. The adélie penguin, *Pygoscelis adeliae*, nests all round the shores of Antarctica and in mid-February moves over the ice to the north. The adults travel the minimum distance, between 160 and 650 km, necessary to provide open sea for feeding on krill (large planktonic crustacea) during the southern winter (Orr 1970a). The young depart long before the last adults leave, probably travel much further, and do not return until two years old. The northern gannet, *Sula bassana*, nests in the North Atlantic from the Gulf of

St. Lawrence to the British Isles. The adults remain to feed during the winter relatively near to the breeding sites. The young, on the other hand, which do not breed until four years of age, may be found up to 7000 km away, often as far south in winter as the tropical and subtropical waters of North Africa and the Gulf of Mexico (Orr 1970a). Many similar examples could be given in which the young of pelagic sea birds migrate through a much greater area than the adults.

Most pelagic sea-birds do not reach sexual maturity for several years after fledging (Lack 1966). Even then, young birds tend to be less successful at rearing young than more experienced birds. Many factors may be involved, but one possibility in particular is relevant here. Success in rearing young seems likely to be in part a function of the efficiency with which food is obtained, particularly just before and during the breeding season. Pelagic birds that feed on fish, squid, crustaceans etc., are presented with a scattered food source that is constantly changing in distribution. Presumably, however,

Fig. 15.16   Canada geese, *Branta canadensis*

The parents and young-of-the-year of geese and swans remain together for the first autumn and spring migrations of the young which therefore establish an initial familiar area encompassing areas suitable for summer, winter, and transient home ranges by social communication. The following summer and autumn the young (now yearlings) perform independent exploratory migrations. In at least one species of swan (Fig. 27.40), however, parents and yearling offspring are capable of recognising one another if they meet during the following winter.

[*Top photo by D. Muir. Lower photo by Mary Hampson*]

under certain conditions certain areas are more likely to contain food than others. Considerable learning is likely to be involved, therefore, not only because the necessary foraging area is large but because the most suitable feeding areas are likely to vary according to the condition of the wind, tide and time of day and year. In this situation establishment of the familiar area is likely to require a relatively long period of time that extends over more than one annual cycle. As establishment of the familiar area inevitably involves some exploratory migration, an individual's potential efficiency as a parent is likely to increase only slowly with the result that selection for delayed maturity is likely.

An individual that becomes sexually mature only after several years is likely, during establishment of the familiar area, to encounter breeding colonies other than its natal colony. The possibility of comparison of habitat suitabilities therefore presents itself. In this case the advantage of encountering as many potential breeding sites as possible could well result in selection for exploratory migration to increase the ratio of calculated to non-calculated removal migration from one breeding site to another. Even if at the time a particular breeding site is encountered during exploratory migration it is not assessed to be more suitable than the natal site, the information obtained may well function to permit calculated removal migration at some time in the future should conditions at the natal site deteriorate.

The evidence for habitat assessment during exploratory migration, if the migration of juveniles and sub-adults is assumed to be partly exploratory, is meagre. However, Coulson and Horobin (1972) have presented evidence for the fulmar, *Fulmaris glacialis*, that prospective breeders may visit a number of colonies during winter and spring. Large established colonies are probably visited first, thus suggesting that high deme size and density is one of the criteria by which breeding habitats are assessed to be most suitable. If the birds fail to establish themselves within such a colony, new cliffs are visited and the birds eventually attach themselves to recently formed colonies. In the kittiwake, *Rissa tridactyla*, birds may breed in a colony up to 160 km from their natal colony (Coulson 1966). New colonies are formed entirely by young birds (Coulson 1971).

Among anatid waterfowl, the male, female, and offspring of geese and swans stay together in a family unit for nearly a year (Hochbaum 1955). The young

therefore acquire a basic familiar area of breeding ground, winter ground and migration route by social communication from their parents. Exploratory migration and/or removal migration does not therefore occur until the young have acquired the migration circuit of their parents which may or may not also represent the total familiar area of the parents. In the ducks, on the other hand, the male leaves the female before the young hatch and forms a moulting aggregation with other males. In river ducks, such as the mallard, *Anas platyrhynchos*, the female stays with the young until they can fly and then leaves them to moult her flight feathers. In diving ducks, such as the canvasback, *Aythya valisneria*, however, the female abandons the young days or weeks before they fly. After abandonment by the mother, young ducks gradually establish a familiar area within the breeding range. That this post-fledging migration is exploratory rather than a removal migration is suggested, though not proved, by the observation that the following spring some individuals do return to their natal site (Sowls 1955). As demonstrated for sub-adult humans (p. 136) exploratory migration is characterised by a low selectivity of temporary destination, even when performed within a socially communicated familiar area. Low selectivity also characterises the post-fledging migration of ducks (Hochbaum 1955) which visit small and large water bodies alike.

Fig. 15.17 The canvasback, *Aythya valisneria*, a North American diving duck

See also Fig. 27.6. [*Photo by Eric Hosking*]

Toward autumn, however, the more suitable habitats are selected and the juveniles aggregate with the older birds before beginning the southward seasonal migration. The following spring these juveniles, now yearlings, are likely to return, many of them to breed, to some part of the familiar area established the previous summer and autumn. It is tempting to suggest that the destination the following spring is the habitat assessed the previous autumn to be most suitable. This habitat may or may not be the natal site.

In ducks, familiar area extension is not confined to the period spent in the breeding range. Marking–release/recapture experiments on ducks captured in autumn at sites on the seasonal migration route and in the winter area show that subsequent movements may occur in any direction, including the 'reverse' direction (i.e. northwards in autumn) (Hochbaum 1955). Again juveniles are more likely to show this migration than adults. One possible interpretation is that whereas the young ducks may well learn a migration route by association with adults as suggested by Hochbaum (1955), at the same time, using each temporary stopping place on the seasonal migration route as a base, a familiar area may be established on either side of the route. Quite possibly as a result of such exploratory migration during their first seasonal migration, some young ducks may utilise slightly different stopping sites from the adults that they accompany during the first year. The same may apply to the winter area. Exploratory migration may well be an important strategy in ducks, therefore, that increases the ratio of calculated to non-calculated removal migration, not only between breeding sites but also between winter sites and between seasonal transient sites. Such a picture of duck migration is much more flexible than that envisaged by Hochbaum and would allow appreciation of the way in which ducks, while still using the same migration circuits year after year, at the same time come to include new bodies of water within these circuits or alternatively establish new circuits to include these new bodies of water.

Temporary, perhaps long-distance, reversal of the direction of seasonal migration is an important characteristic of the migration of many species of birds, not only ducks. P.R. Evans (1972) has investigated this problem in relation to autumn night-migrant passerines in northeast England. In two autumns, reversed migration was observed on 17 occasions and on 5 occasions it was observed two nights in succession. On all of these nights the pre-dominant migration direction was NNW instead of the usual SSE. Analysis of the weather records for these nights showed that the movement was not always downwind nor was reversal associated with atypical temperatures. Similarly there was no absolute correlation with cloud cover though reversal was marginally more common when the sky was less than half-clouded than when there was total overcast. The only significant factor to emerge from meteorological analysis was that on all occasions windspeed was less than 15 km/h and was always blowing from between SE and SW. That reversal is not a direct response to wind direction, however, is shown by the fact that there was SSE migration on 52 per cent of nights with light opposed winds in August and September and on 36 per cent of such nights in October and November. Rabøl (1970) has suggested that reversal is a reflection of the fact that the seasonal migration of birds is a time-programmed movement such that on any date a migrant should be at a particular locality. If on any day, therefore, a bird 'overshoots' its destination there should be a reversal the next day to regain the intermediate goal. As argued by Evans (1972), however, it is unlikely that all individuals would overshoot the appropriate destination. Evans suggests instead that the birds concerned, which are almost entirely juveniles, respond to change in photoperiod such that with an increasing photoperiod the spring migration direction is shown whereas with a decreasing photoperiod the autumn migration direction is shown. Hence reversal may occur when, as a result of cloud conditions at dawn and dusk, the following day in autumn has a longer not a shorter photoperiod than the previous day. Previously, I had made a similar suggestion to account for the similar reversal that is sometimes shown by the small white butterfly, *Pieris rapae* (Baker 1968a). In butterflies, however, such response to day-to-day fluctuations in photoperiod can be adaptive (Chapter 19). By contrast, the hypotheses of Rabøl (1970) and Evans (1972) for bird reversal imply that migrant birds are frequently 'victims' of their migration mechanisms. I can see no reason why this should be so and would prefer a more flexible and adaptive interpretation.

Most often bird reversal involves a 180° change in direction and results in the birds, predominantly juveniles, retracing their previous migration track. The birds are by no means disorientated (Evans 1972). These birds are, therefore, moving between one area with which they have had recent direct experience and another area through or over which they have recently passed. Whether the destination of reversal migration is any point on the migration track or a previous stopping place does not seem to be known. A movement between areas with both of which an animal has some degree of familiarity has all the ingredients of a calculated migration. It is possible, therefore, that having arrived at one destination the birds assess it to be less suitable than an area previously visited or perceived. The importance of feeding during migration to the energy reserves of migrants has been well shown for fringillid finches (Newton 1972) and has probably been the major selective pressure to favour nocturnal migration. That habitat assessment does occur during seasonal migration is suggested, again by data for European fringillids, by the observation that when such birds in spring do not return to their nesting site of the previous year they most frequently stop at some other site on the seasonal migration route (Newton 1972).

Let us assume for the moment (but see p. 617), therefore, that reversal is a calculated removal migration from one transient area to another that takes place during seasonal migration. On this assumption it is possible to make some predictions concerning likely characteristics of the situation. According to the migration equation, migration is likely to be advantageous when $h_1 < h_2 M$. In this case, h_1 is the suitability of H_1, the most recent destination before reversal, h_2 is the suitability of H_2, an area visited or perceived previously and which is also the target for reversal migration, and M is the migration factor for migration from H_1 back to H_2. The migration cost is not only a function of the energy expended in migrating from H_1 to H_2 and then later back to H_1 but also of the effect of reversal in delaying eventual arrival at the wintering grounds. Comparison of h_1 with h_2 is most important for birds in need of replenishing their fat reserves. Dolnik and Blyumental (1967) have made out a case and presented supporting evidence that the 'waves' of migrant chaffinches, *Fringilla coelebs*, in Europe are the result of the synchronising effect of the sudden onset of conditions suitable for migration followed by the depletion of fat reserves. Migratory chaffinches carry about 2 g of fat reserves and during stops during seasonal migration can replenish their fat at the rate of about 0·5–1·0 g/day. Lean chaffinches stop more often and for longer than fat ones. Let us assume for the moment that on the basis of food availability a bird has assessed h_1 to be less than h_2. Reversal may still not occur unless h_1 is also

assessed to be less than $h_2 M$ (on the assumption that selection has acted on birds to make accurate comparison of h_1 with $h_2 M$ using ideal criteria). The higher the value of M (i.e. the lower the migration cost), the more likely is reversal to occur. Reversal is likely to occur, therefore, when energy expenditure is low and the delay in reaching the overwintering grounds is also relatively unimportant. Energy expenditure is likely to be low when skies are clear (Hamilton 1966) and the wind is favourable. The delay is less likely to be disadvantageous when temperatures are not considerably below the seasonal norm.

The following predictions can therefore be made concerning reversals on the assumption that they are calculated removal migrations: (1) they should be most common from areas that are less suitable than areas further back along the seasonal migration track; (2) they should most commonly involve juveniles which have no experience of conditions further along the migration track and which, therefore, have no means of comparing the relative advantage of continuing in the appropriate direction and of reversing; (3) they should be most common when birds have a low fat content and consequently should be performed by the majority of individuals at the same time for the same reasons that migrations occur in 'waves' as argued by Dolnik and Blyumental (1967); (4) they should be most common but not confined to nights when the sky is not overcast so that celestial orientation is possible as well as there being sufficient light for efficient navigation by landmarks; (5) they should be most common when the wind is favourable; and (6) they should not occur when the temperature is markedly below average. The results obtained by Evans (1972) in northeast England are consistent with this hypothesis. However, without evidence in support of the basic premise that reversal is a calculated migration, no matter how much other evidence may be consistent with the above predictions such evidence can only be circumstantial.

Although I have discussed the reversal of seasonal migration in this section I do not, of course, intend to imply that seasonal migration has evolved as a strategy to increase the ratio of calculated to non-calculated removal migrations. The evolution of the seasonal migration circuits of birds is discussed elsewhere (p. 604). While seasonal migratory behaviour was evolving, however, other selective pressures related to the advantage of calculated removal migration could well have acted to produce behaviour patterns such as those described above.

The same argument applies to removal migration from one breeding site to another on the seasonal return migration route as already described for some fringillids. The migration is likely to be calculated but is not preceded by any behaviour pattern, other than habitat assessment, that has evolved specifically to increase the frequency of calculated breeding site to breeding site removal migration. This may not be the case, however, in the breeding site to breeding site removal migrations of other birds, particularly juveniles. The common pattern among land birds is for the young to undergo a post-fledging migration that may result in them subsequently breeding at distances from the natal site of between a few metres to hundreds of kilometres. In relatively sedentary species, such as the blue tit, *Parus caeruleus*, and great tit, *P. major*, in Britain (Lack 1966) such post-fledging migration may be the only long-distance migration performed by the species (Goodbody 1952). In temperate species that also perform seasonal migrations, however, post fledging migration precedes the autumn seasonal migration in the same way as described for ducks. The following spring, the young may return either to the natal site or elsewhere. Many juvenile passerines appear to return to the area occupied immediately prior to their autumn migration rather than to their natal area unless the natal area was the area so occupied (Bellrose 1972). Collared flycatchers, *Muscicapa albicollis*, were found to be imprinted with the cues that enabled them to return to the area occupied only during the two-week period prior to autumn departure from Germany (Löhrl 1959). The question is whether the birds arrive in the area in which imprinting takes place as a result of non-calculated migration or whether post-fledging migration is really a period of exploratory migration and habitat assessment. If the latter, it must be argued that fledglings retain in the short-term some navigation mechanism that enables them to return that same summer or autumn to the habitat that is assessed to be the most suitable of all those visited. The bird then returns to that habitat during its first autumn, whereupon the imprinting takes place that permits return to the area following seasonal migration (see Fig. 33.28).

There is little difficulty in accepting that the short-distance post-fledging migration of such species as the song sparrow, *Melospiza melodia*, of temperate North America (Nice 1937, Johnston 1961) consists of exploratory migration, habitat assessment and

calculated removal migration. It also seems possible that an exploratory migration–habitat assessment–seasonal migration–exploratory migration–habitat assessment–seasonal migration–calculated removal migration sequence could account for the removal migration of white storks, *Ciconia ciconia*, from natal site to breeding site in Germany. Young storks migrate from Europe to Africa during their first autumn (Fig. 27.3) usually in company with adults. The following spring these yearlings often remain for the summer in Africa or return only as far as the Mediterranean (Lack 1966). This behaviour seems likely to be primarily an evolved response to the advantage of segregation from the adults and a shorter seasonal return migration, rather than to be connected with future removal migration. The following spring the young return to Europe and the vicinity of their natal area. The migration that then occurs could be interpreted as exploratory and to involve habitat assessment. These two-year old sub-adults occupy nests but do not breed. If the area is suitable, breeding occurs there the following year

Fig. 15.18   White storks, *Ciconia ciconia*, and nestlings See also Fig. 27.3.

[*Photo by Eric Hosking*]

after seasonal return migration. If the area is unsuitable, however, further migration (exploration + habitat assessment?) occurs before autumn and the following year the individual is likely to be found on average 20 km or so away breeding in a more suitable area. Preferred breeding areas are low-lying lands. Of individuals born in the Rhine Valley, 70 per cent eventually breed within 50 km of the natal site. In a declining population, however, of individuals born in more hilly country, only 17 per cent of those that survive return to within 50 km of the natal site. Such figures are consistent with an exploratory migration–habitat assessment–calculated removal migration hypothesis, but also do not contradict a facultative non-calculated removal migration hypothesis.

The year-to-year breeding site to breeding site removal migration of adult passerines that occurs, particularly among females, though to a much lesser extent than for juveniles, could also be interpreted as being within the familiar area established by that adult either as a juvenile or during the previous breeding season. Alternatively, such removal migrations could be interpreted as non-calculated migrations or even, in the case of long-distance seasonal return migrants, as simply representing a failure of the navigation mechanism. Of these three alter-

natives, the first is perhaps most likely. However, until some technique for following the movements of an individual bird through more than one season can be developed there seems little hope of establishing beyond doubt the nature of the year-to-year breeding site to breeding site removal migration of birds. Telemetry has been used successfully to follow an individual bird during part of its seasonal return migration (Cochran 1972) and it is to be hoped that in the near future such a technique may be applied to the study of the post-fledging migration.

The postulation of familiar area extension, habitat assessment and calculated removal migration would explain some anomalies that can be found in bird studies. The removal migration of the starling, *Sturnus vulgaris*, for example, has been studied in Europe (Kluijver 1935). It was found that the median distance of year-to-year breeding site to breeding site removal migration for the species was close to 400 m and the mean over 500 m. It should be stressed, of course, that studies concentrated at a particular centre always run the risk of under-estimating the distance of removal migration. If, for the moment, however, we accept Kluijver's conclusion that the starling shows removal migration over only relatively short distances, a problem presents itself. Sixty starlings were released in New

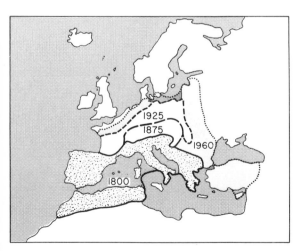

Fig. 15.19 The spread of the serin, *Serinus serinus*, in Europe

Over a century ago, the serin was confined to Mediterranean woodland. Following the adoption of parks and gardens with scattered trees as a breeding habitat, however, a habitat-type similar in appearance to its original habitat-type, the serin spread north into much of central

Europe during the nineteenth century, and further in the present century. In most of this range the species performs a seasonal return migration to the N in spring and to the S in autumn. Breeding first occurred in Britain in 1967 following several years of a steady increase in the occurrence of non-breeding (exploratory?) migrants.

[*Modified from Newton (1972), partly after Mayr*]

York City, North America, in the year 1890 (Orr 1970a). By the year 1900, individuals had reached Connecticut and New Jersey and forty years later the species was breeding in California. It is difficult to reconcile these two observations unless it is assumed that during post-fledging dispersal and seasonal return migration habitat assessment occurs. In Europe, where the places visited during these migrations are all occupied by the species, habitats near to the natal site may well be assessed to be most suitable whereas in North America during the spread of the species the areas encountered were unoccupied and hence more likely to be assessed to be most suitable. A similar behaviour pattern would account for the observation that the spread of the serin, *Serinus serinus*, a European fringillid, from Continental Europe to Britain (Fig. 15.19) was preceded by the irregular arrival of birds that did not breed (Ferguson-Lees 1968). These birds could well have been individuals engaged in exploratory migration that at first did not assess Britain to be more suitable than other parts of their familiar area. Only when deme density had increased on neighbouring parts of the continent was Britain assessed to be suitable.

The postulated existence for any animal of a free distribution (Fretwell 1972) in which each individual settles in an equally suitable habitat (p. 402) inevitably involves exploratory migration, habitat assessment and calculated removal migration. The same is not necessarily true, however, for a despotic distribution. The red grouse in Scotland provides a fine example of such a distribution,

Fig. 15.20   A female red grouse, *Lagopus lagopus*

[Photo by Eric Hosking]

though the population biology of the species has been the subject of a continuing controversy. The red grouse is probably a subspecies of the circumpolar willow grouse or willow ptarmigan, *Lagopus lagopus*, of subarctic willow and birch forest and woody or shrubby tundra (Lack 1966). The British deme probably also lived in this type of habitat before tree clearing, sheep grazing and regular burning produced heather moors. At present, the red grouse feeds primarily on heather shoots. According to the terminology of Fretwell (1972), the red grouse on a heather moor should be described as showing a despotic distribution in which individuals in the more favourable habitats have a greater reproductive success than individuals in the less favourable habitats (N.B. 'favourable' in this sense is not synonymous with 'suitable' as used in this book: suitability is defined in terms of reproductive success (p. 39) whereas favourability may be used to describe the relative attributes of a habitat but without taking into account deme density).

One period of territorial interaction takes place in autumn but during periods of snow feeding flocks may be formed. With the advent of milder weather, males may spend the whole day in territories where pairing takes place. Non-territorial males are found predominantly (95 per cent) within 5 km of the breeding area, chiefly around the edge of the moors. I find it difficult to accept the concept of Jenkins *et al.* (1963) that most 'surplus' non-territorial birds die soon after failing to obtain a territory. There are two reasons for this. First, if this were true after the autumn period of territorial interaction, there would be no late winter or spring territorial interactions between territorial and non-territorial birds unless the territories were different sizes at the two seasons. Some birds that fail to obtain a territory in autumn are likely at least to survive the period of autumn territorial interaction for long enough to become members of the winter feeding flocks. Second, whenever during the year a territory holder is killed he is immediately replaced by another bird.

An alternative interpretation to that of Jenkins *et al.* (1963) is that birds that fail to obtain a territory at any time of the year perform virtually continuous exploratory migration within the confines of the moor and occasionally beyond. During such exploratory migration the birds feed and search for territories in which the previous owner has died or been killed. As a result of exploratory migration such birds establish a relatively large familiar area over the moor. Such an interpretation of the behaviour of non-territorial birds would account for many of the observations that appear anomalous to the suggestion that surplus birds die soon after failing to obtain a territory. For example, if surplus grouse spend most of their time in exploratory migration within the confines of the moor, it would account for the fact that comparatively few grouse seem to have died on marginal ground or outside the moor, the majority being found on the moor between established territories (Jenkins *et al.* 1963). It would also account for the observations that 'surplus' birds appear on the territories in early winter, join winter feeding flocks, and immediately find and establish themselves in any territory that becomes vacant. It can be assumed that territorial birds not only live in sites in which shelter from predators is better but also, living in a small area, are very familiar with the location of such shelter (p. 96). It can also be assumed that non-territorial birds, as a result of living within such a large familiar area and as a result of performing frequent exploratory migration to sites not previously visited on the moor, are less familiar with the location of suitable shelter that is accessible (in the sense that it is outside the territory of any other male). It is to be expected, therefore, that non-territorial birds will have a higher death rate as a result of predation than territorial birds. That this is so is clearly demonstrated by figures presented by Jenkins *et al.* (1963). In addition, of course, territorial males have considerably more access than non-territorial males to females during the mating season. There is, therefore, little doubt that grouse that fail to obtain a territory have a lower reproductive success than grouse that do obtain a territory (i.e. the distribution is despotic (Chapter 18)). What is open to doubt, however, is the view that 'surplus' birds leave and die soon after failing to obtain a territory. Until clear evidence that this is so can be presented, the hypothesis that seems most consistent with the data is that grouse that fail initially to obtain a territory perform exploratory migration and establish a large familiar area within the confines of the moor. Within this familiar area calculated removal migration is performed as soon as a territory becomes vacant.

To recapitulate, therefore, there is no clear evidence for any species of bird that behavioural strategies have evolved that increase the ratio of calculated to non-calculated migration. There seems a strong possibility, however, that the extensive migrations of juvenile sea birds, the post-fledging migrations of terrestrial birds, the dispersal of ducks from transient stopping sites during seasonal migration, and the migrations of non-territorial individuals of territorial demes all represent such a strategy. Seasonal return migration may also result in an increase in the ratio of calculated to non-calculated migration, both in relation to 'reversal' during migration and in relation to breeding site to breeding site removal migration. The possible evolutionary relationship between seasonal return migration and familiar area extension is discussed in Chapter 27.

15.10 Reptiles

Undoubtedly the most spectacular migrants among the reptiles (except for birds!) are the sea turtles. Thanks mainly to the pioneer work of A.F. Carr, many details of the migrations of these animals, particularly of the green turtle, *Chelonia mydas*, have recently been uncovered (Fig. 30.1). Although there are still some gaps, the general pattern of return migration, at least of those demes of green turtles that breed along the coasts of Costa Rica and Ascension Island (Carr 1972) and probably also those that breed along the coasts and islands of the Indian Ocean (Hirth and Carr 1970) seems to be as follows. Eggs are buried above the high-tide line in sandy shores of tropical and subtropical mainland and island coasts. As soon as the young hatch they migrate across the sand to the sea whereupon they effectively disappear. Nothing is known about where they go or on what they feed during their first year of life, though it is possible that they live among the floating rafts of Sargasso weed. Adult green turtles feed predominantly on turtle grass, largely off continental coasts. Feeding and breeding grounds may be 2000 km or so apart. Every two or three years females leave the grazing grounds, migrate to the breeding grounds, lay from three to seven batches of 100 or so eggs at 12 day intervals, and then return to the grazing grounds. Degree of return of adult females to the breeding grounds is high. Whether the same degree of return is shown to the feeding grounds seems to be unknown. Males are also found around the breeding grounds, copulation taking place offshore. These copulations fertilise the eggs that will be laid in two or three years time. The periodicity of male migrations between feeding and breeding grounds is unknown.

It seems incredible to suppose that large numbers of a species that both feeds and breeds on coasts, as does the green turtle, could possibly find an island,

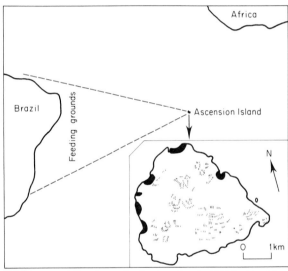

Fig. 15.21   The six egg-laying beaches (solid black) of the green turtle, *Chelonia mydas*, on Ascension Island in mid-Atlantic

[Map modified from Carr (1972). Photo by Paul Turner]

such as Ascension Island, which is 2200 km from the feeding ground, as a result of exploratory migration as a juvenile or sub-adult. It seems equally unlikely that the young, newly hatched turtles have the opportunity to assess directly the suitability of their hatching area. The most likely conclusion is that reached by Carr (1972) that, as with migratory salmon, some olfactory learning occurs during the first day(s) or week(s) of life which in association with down-current drift followed later by olfaction and up-current orientation results somehow in navigation back to the hatching area (Chapter 33).

Whether exploratory migration, habitat assessment, and then calculated removal migration occurs later, once the turtle has returned as an adult to its hatching site, is not known. The tagging results obtained by Carr, however, have so far failed to demonstrate egg-laying by females more than a few kilometres from their original marking site. The only possible example of exploratory migration and habitat assessment leading to calculated removal migration in sea turtles so far seems to be the beach-to-beach removal migration of females between successive egg batches. On Ascension Island, for example, where there are six beaches suitable for egg-laying (Fig. 15.21), 17 per cent of females return to a different beach from one oviposition to the next during a single laying season (Carr and Hirth 1962). This removal migration could be interpreted as failure to navigate back to the original beach but it is possible that it is a calculated migration that was preceded by exploratory migration and habitat assessment.

A number of species of the order Chelonia are amphibious, alternating between land and ponds, streams and/or smaller water bodies. According to F.R. Cagle (in Oliver 1955) some genera of terrapins, such as *Chrysemys* and *Pseudemys*, have a home range that probably includes part or all of several water bodies. Migration across land from one water body to another which occurs from time to time is almost certainly, therefore, to be initiated on a calculated basis. Establishment of such a familiar area, which in times of drought and at certain seasons increases the ratio of calculated to non-calculated migration, must have involved exploratory migration and possibly also habitat assessment. Similar overland movements seem to be performed by the Mississippi alligator, *Alligator mississippiensis* (Oliver 1955), and the Egyptian crocodile, *Crocodilus niloticus* (Stevenson-Hamilton 1947), and a similar role for exploratory migration and habitat assessment may be postulated.

Many terrestrial Chelonia, or tortoises, may also move within a familiar area. One example could be the seasonal movement between distinct winter quarters and summer feeding grounds in the desert tortoise, *Gopherus agassizii*, in the southwestern United States (Carr 1952), though other interpretations are possible. Possibly the altitudinal return migration of the giant tortoise, *Testudo gigantea*, on the Galapagos Islands in which the adult females migrate from the cooler, moist highlands to the coastal lowlands to lay eggs in the soil or sand

(Orr 1970a) could also be interpreted as a calculated migration within a familiar area. Such a migration is possible only if the young, upon hatching, establish a familiar area while at the same time assessing habitat suitability within that area. The same species on Aldabra in the Indian Ocean moves primarily between feeding sites, shade, wallows, and water bodies (Gaymer 1968). It is not clear, however, whether these movements are performed within a familiar area or whether the tortoises wander at random but respond to sensory cues from a distance. There is a stronger suggestion, however, that exploratory migration occurs in the common box turtle, *Terrapene carolina*, of the eastern United States. This species, whose habitat-type is open woodland near streams or ponds, is mainly terrestrial, predatory upon slow-moving invertebrates when young, but herbivorous when adult. The home range has a diameter of about 250 m (Carr 1952) but the species shows good homing from distances of up to 1200 m and can choose the shortest route from point to point. Other explanations can be given (Chapter 33) but the possibility that the more distant points have previously been visited during exploratory migration cannot be entirely ruled out.

Some migration patterns shown by snakes are known (Oliver 1955) but in insufficient detail for it to be worth considering at any length the possibility of the existence of strategies to increase the ratio of calculated to non-calculated migration. In the Sundanese rock python, *Python bivittatus*, however, the newly hatched young return each night to the empty shells of their eggs and some may crawl into them (Delsman 1922). The early movements of this species, therefore, involve exploratory migration and it is possible that later migration of the young individual is a calculated removal migration to the place within the explored area that has been assessed to be most suitable.

Similar movements may be performed by some lizards, most of which live their adult lives in a specific home range. Homing from even short distances such as 100 m is not always well developed (Tinkle 1967a,b), which may suggest that the home range and familiar area are more or less the same. In the species concerned, however, the side-blotched lizard, *Uta stansburiana*, in Texas the distance from hatching site to the geometric centre of the adult home range varies from 2 to 80 m in males and from 4 to 55 m in females. As the homing experiment was performed from outside the area within which the individual migrated, it cannot be used to assess whether removal migration from hatching site to adult home range was calculated or non-calculated. One possible example of a behaviour pattern that increases the ratio of calculated to non-calculated migration is provided by the rainbow lizard, *Agama agama*, in Nigeria (V.A. Harris 1964). This insectivorous species, which feeds mainly on ants, forms male–female pairs which live in a territory defended by the male. Also in the territory are male and female young and juveniles, some or many of which are the offspring of the pair, though removal migration of the young between territories may take place. The individuals within the group, excluding those younger than 2–4 months, are gregarious. When juveniles perform removal migration at about 4 months it is preceded by movement between territories which could be interpreted as exploratory and to involve habitat assessment. Females that mature (15–18 months) during the dry, non-breeding season may leave their original group and search for an unpaired territorial male. During this search, which may have a large non-calculated removal component, the female may join other groups for periods but eventually settles with an unpaired male. The movements of the female while temporarily settled with a group are unknown, but it is possible that this territory is used as a base from which exploratory migration takes place. In this case, the habit of settling temporarily within the territory of an already paired male could be viewed as part of a strategy that increases the ratio of calculated to non-calculated migration.

In conclusion, studies of reptile movements have so far provided few examples of behaviour patterns that could be interpreted as having evolved in response to the advantage of calculated as opposed to non-calculated migration. Only for the beach-to-beach migrations of female sea turtles during a single laying season, the territory-to-territory migrations of some lizards, and the hatching site to adult home range migration of a snake is there circumstantial evidence of such a behaviour pattern.

15.11 Amphibians

Most anurans seem to live in a home range that includes either one water body, as in most *Rana* spp. and the American toad, *Bufo terrestris*, in northeastern United States, or several water bodies, as with male *B. terrestris* in Oklahoma and male green frogs,

Rana clamitans, in south Michigan (Oliver 1955). In general, the year's home range of most frogs seems to be from < 1–3 km in diameter. Similarly, the migrations of urodeles from terrestrial feeding to aquatic breeding sites rarely seem to involve movements greater than 1 km.

Most amphibia seem to show a strong attachment to their spawning site even when, as has been done in the case of the Pacific tree frog, *Hyla regilla*, in west Oregon (Jameson 1957), individuals are released in another pond that contains a breeding chorus. Some salamanders also, such as the spotted salamander, *Ambystoma maculatum*, on Rhode Island, eastern United States (Whitford and Vinegar 1966) show a high degree of return to their breeding site. The western red-bellied newt, *Taricha rivularis*, of the redwood belt of northwest California, which breeds in streams and ponds, shows good return to the 'home' segment of stream from distances of up to 4 km even when released in a different stream (Twitty *et al.* 1964). Twitty *et al.* consider it is improbable that such release points are within the familiar area of the individuals concerned.

In conclusion, most studies of amphibian migration to date have served to emphasise the high degree of return that exists in relation to a specific breeding site. Removal migrations, such as from breeding site to breeding site or natal site to adult breeding site have received little attention. Such evidence as is available implies that exploratory migration, habitat assessment, and calculated removal migration is relatively unimportant (but see Chapter 33).

15.12 Fish

There is some evidence that stream fishes occupy a home range and/or a territory in much the same way as do terrestrial vertebrates (Gerking 1959). It seems reasonable, therefore, to suggest that exploratory migration (see Kleerekoper *et al.* 1974, for the goldfish, *Carassius auratus*) and habitat assessment may well be found to precede such removal migrations as those from natal site to adult home range or those from home range to home range. Such removal migrations are particularly noticeable when they result in migration from one drainage system to another by way of the sea as has occurred, for example, in the striped bass, *Roccus saxatilis*, on

the Pacific coast of the United States (Calhoun 1952). The striped bass was introduced to the Sacramento River from the Atlantic coast in the year 1879 and again in 1882. Twenty years later it had spread north into Oregon and in 1906 was being caught 1000 km to the north in the mouth of the Columbia River (Fig. 31.5). Occasional individuals, usually juveniles, are also caught hundreds of kilometres to the south of the Sacramento. The question that remains unanswered is whether or not such individuals retain the ability to return to any river mouth visited during the migration, including that of the natal river. If the ability is retained such migrations could be interpreted as exploratory. If not, it must be assumed that such individuals are performing non-calculated removal migration.

Among anadromous fish, such as salmon, shad, and lamprey, that migrate between marine feeding grounds and freshwater spawning grounds there is little opportunity for the young fry to perform exploratory migration with respect to potential breeding sites during the period from hatching and reaching the sea (Chapter 31). There is presumably some opportunity for exploratory migration of streams and creeks by adults returning to spawn. However, selection on the females to spawn before being preyed upon and sexual selection on the males is likely to impose stringent selection for a high degree of return. Experimentally determined degree of return for such fish has been high (> 90 per cent for the sockeye salmon, *Oncorhynchus nerka*, and the pink salmon, *O. gorbuscha*, in the streams of the Pacific coast of North America, the steelhead trout, *Salmo gairdneri*, in California, and the Atlantic salmon, *Salmo salar*) but some removal migration from the natal stream to some other spawning stream has been found to occur (Harden–Jones 1968). Whether such removal migrations are calculated or non-calculated remains unknown. Catadromous fish, such as freshwater eels, *Anguilla* spp., have slightly more opportunity for removal migration from stream to stream in that they occur in such habitats during the relatively long feeding period of perhaps 20 years (Orr 1970a) and in that they may migrate overland (Bertin 1956). Ecologically, therefore, freshwater eels show certain similarities to amphibious mammals, reptiles, and amphibians. The arguments that have been presented for these animals in that case probably also apply to freshwater eels during this phase of ontogeny.

Little is known with certainty concerning the removal migrations of oceanodromous fish that spend their entire life history in the sea. Let us consider some movements of the spring-spawning Norwegian deme of the Atlantic herring, *Clupea harengus*, as examples of the types of removal migration that may take place (Harden Jones 1968). The young herring spends the first year of its life in fjords or coastal waters at a distance from the spawning site that is a function of drift during the first few months. For the next two to five years or so the juveniles perform relatively local seasonal return migrations that are to the north and toward the coast in summer and to the south and away from the coast in late autumn. Once the fish have reached a certain critical length they initiate an ontogenetic return migration that at first is directed away from the coast and is not completed until the fish are ready to spawn one or two years later. Spawning occurs first between the ages of four and seven years. Most herring leave the Norwegian coast for the Norwegian Sea where they feed on zooplankton. The limit of summer feeding is from Spitsbergen through Jan Mayen to Iceland (Fig. 31.12). Ripening herring winter to the north of the Faeroes and in December and January pre-spawning concentrations move toward the Norwegian coast. There is some evidence of a return to the same general spawning area or grounds but no evidence of a very high degree of return resulting in return to the natal spawning bed. There is perhaps some spawning area to spawning area removal migration from Norway to Iceland and Norway to East Anglia. Such removal migration is not confined to juveniles and may occur even after an individual has spawned at its natal area. Some of these destinations, such as Iceland, are within the area covered during ontogenetic and seasonal return migrations, in which case the possibility of calculated removal migration cannot be ruled out. On the other hand, other destinations, such as East Anglia, would appear to be outside the range of ontogenetic and seasonal migration and therefore to be the result of non-calculated migration. As the herring is a shoaling species, the possibility that spawning ground to spawning ground migration is a result of shoal-to-shoal migration cannot entirely be ruled out.

In conclusion, little is known about the removal migration of fish and although exploratory migration and habitat assessment may occur in, say, the overland migration of eels, the stream-to-stream migration of anadromous fish, and the spawning ground to spawning ground migration of oceanodromous fish, no supporting evidence is available.

15.13 Adult insects

There seems little doubt that the vast majority of insect migrations over all but short distances are non-calculated and few strategies seem to have evolved to increase the ratio of calculated to non-calculated migration. The evolution of this situation can perhaps be attributed to the absence of selection for a memorised familiar area (Chapter 18) in most species. Major exceptions, of course, are provided by social insects which show well-marked behaviour patterns, such as exploratory migration and social communication that increase the ratio of calculated to non-calculated migration (p. 148). Even among social insects, however, the removal migration of solitary reproductives as in ants and termites is non-calculated.

Apart from social insects and those solitary Hymenoptera known collectively as 'hunting' wasps (Chapter 33), there are few examples of insects that live even temporarily within a memorised familiar area. One example is provided, however, by the territorial males of the peacock butterfly, *Inachis io*, a Eurasian nymphalid. The territorial behaviour of this and the related nymphalid species, the small tortoiseshell, *Aglais urticae*, has been studied in southern England (R.R. Baker 1972a). Territorial butterflies seem to provide an example of a free distribution, though confirmatory measurements are not available. In southern England males of the peacock butterfly establish territories in spring along the edge of a wood. The territory, about 50 m² in size, is defended against other males from midday until either the Sun becomes permanently obscured by cloud or the temperature drops below a critical level. The following morning, after roosting, the male initiates non-calculated removal migration and at midday occupies a territory usually at least 0·5 km from the previous territory and probably often much further. Females perform non-calculated removal migration across country from one oviposition site to another, the oviposition site being a patch of nettles, *Urtica dioica*, preferably alongside trees. Females during migration, as do males, maintain a constant angle to the Sun's azimuth (Chapter 19) but are also deflected along topographical edges such as hedges, walls, rows of trees, etc. The result is that a male with a territory along such an edge is likely to have a greater female quota than a male with a territory away from such an edge. The greatest female quota, however, is likely to be experienced by a male with a corner territory and there is some evidence that male *Inachis io* settle preferentially in a corner territory. On a fine spring afternoon a male with a corner territory may often be adjacent to males with an edge territory. Such 'edge' males from time to time make flights beyond the confines of their territory, whereupon territorial interactions involving chasing and spiralling with the 'corner' male may occur. Invariably after such flights and interactions the males return to their respective territories. If the 'corner' territory becomes vacant at any time, either as the result of experimental removal of the territorial male or as a result of the male leaving his territory in pursuit of a female (copulation does not occur until evening and may occur some distance, perhaps even kilometres, from the male's territory), the adjacent 'edge' male, when next he flies beyond his own territory, immediately settles in the 'corner' territory. It seems a reasonable interpretation, therefore, that the occasional flights beyond the territory are exploratory migrations and the fact that the 'corner' territory is immediately occupied when found to be vacant indicates that habitat assessment is involved. Removal migration of *I. io* from one territory to another during a single afternoon seems, therefore, to be calculated and to be preceded by behaviour patterns that increase the ratio of calculated to non-calculated migration. Day-to-day removal migration from one territory to another, however, seems almost certain to be non-calculated (Chapter 16). The distances involved in exploratory migration depend on the spacing of territories which in turn are a function of topography. The distance from the territory that an individual may travel during exploratory migration may range from 5 up to about 100 m. It now seems also (Turner 1971) that some *Heliconius* butterflies may establish permanent familiar areas, presumably as a result of the exploratory–calculated migration process described in Chapter 18.

The second example in which the possibility of exploratory migration, habitat assessment and calculated removal migration are perhaps worth considering concerns the branch-to-branch migration within the same tree of the sycamore aphid, *Drepanosiphum platanoides*, studied by A.F.G. Dixon (1969). Unlike many species of aphid, adults of the sycamore aphid, except for the oviparous female which is produced in late autumn, are winged and retain the ability to fly throughout their adult life. Like most aphids, however, throughout the spring and summer the females produce young parthenogenetically and

viviparously. A conspicuous behaviour pattern of the species is the hovering flight that occurs within the canopy of the tree. This hovering flight has been shown by Dixon to be correlated with a change in the vertical distribution of the aphid within the canopy of the tree and to result in movement from areas of high to low density. It is tempting to suggest that the hovering flight is an exploratory migration during which habitat assessment occurs with the result that calculated removal migration takes place. Final acceptance of this suggestion must await investigation, which will be difficult, of whether an individual, having departed from the natal area, is capable of settling in the most suitable part of the tree, including back at the natal area if that should prove to be the most suitable part. Other interpretations are, of course, equally, if not more probable. For example, departure from the natal area within the canopy could be initiated on a non-calculated basis, followed by vertical movement in the direction appropriate for the time of year to take the aphid to an area of lower density, followed by settling on the first branch or leaf in which the deme density is below a certain threshold.

In the Odonata (dragonflies) the immature stages are aquatic. The adults upon emergence are generally considered to migrate away from the water, though the distance involved may vary from a few metres to about 3 km (P.S. Corbet 1962). While away from the water the insects feed and become sexually mature, a process that may take up to 4 weeks. Once sexually mature, the adults return to the vicinity of water for feeding, copulation and oviposition. What is not known is the extent to which the ability to return to the natal water body is retained during the period of maturation. If the ability to return is retained, the migration of immature imagines could be considered to be partly exploratory and to involve assessment of the suitability of neighbouring water bodies, particularly if removal migration is also shown to occur. On the other hand, the migrations could be entirely non-calculated removal migrations of the type described for other large insects in Chapter 19. The nature of dragonfly migration cannot be determined until some estimate of degree of return has been made. Jacobs (1955) in a study of the territorial behaviour of the North American dragonfly, *Plathemis lydia*, marked 129 newly emerged males and 209 newly emerged females. Of these, 7 per cent of the males returned after a period of 8–14 days and 3 per cent of the females after 13–24 days. However, without knowing what proportion of surviving individuals these returns represent and without knowing the distance from the natal water body travelled by these individuals before initiating return, it is not possible to reach any conclusion concerning degree of return.

In the cockchafer beetle, *Melolontha melolontha*, the females of which migrate from the natal field to a neighbouring woodland maturation site, a regular and high degree of return to the natal site for oviposition has been established (C.G. Johnson 1969). In this case, however, the frequency of removal migration has not been established. Whether exploratory migration occurs from the woodland maturation sites to possible oviposition sites other than the natal site remains unknown.

In summary, therefore, except for social species, the possibility that calculated removal migration may occur in insects remains largely uninvestigated. There seems little doubt that the majority of removal migrations performed by non-social insects are non-calculated. A few short-distance migrations such as territory-to-territory migrations may be preceded by exploratory migration and habitat assessment. Longer-distance exploratory migrations may occur in dragonflies and some other large insects but relevant data are at present not available.

15.14 Other invertebrates

The possibility cannot entirely be ruled out at present that a great many terrestrial and inter-tidal invertebrates establish a familiar area in much the same way as terrestrial vertebrates. This seems particularly likely for such animals as spiders (see p. 878), scorpions, isopods, millipedes, centipedes, molluscs (see p. 878), echinoderms, decapod crustaceans (see p. 879), etc. Certainly some of the larger inter-tidal molluscs, such as the owl limpet, *Lottia gigantea* (Stimson 1973), of the coast of California and Mexico, live in well-defined territories from within which the animal is capable of return to the smooth, clean, slightly depressed area on the rock surface that is used between tides as a roosting place. Such limpets seem to retain the ability to 'home' by virtue of memory of the tactile reception of variations in the rock surface (Galbraith 1965). Examples of removal migration from such territories and the movements preceding such migration do not seem to have been documented.

Terrestrial invertebrates such as spiders, scorpions, and other arachnids are reputed to remain in a relatively restricted area (Savory 1964) which may or may not be a familiar area and from which exploratory migrations may or may not be performed. Many colonial insect larvae, particularly those that construct extensive silk tents, tend to defoliate the area around the tent. The area is a familiar area insofar as the larvae perform return migrations between the tent and available food, though possibly the movement takes place entirely within sensory range. Removal migration to a new tent site presumably also takes place within sensory range (p. 91) though as far as I know the possibility of preliminary exploratory migration by some individuals has not been investigated.

Davenport and Dimock (in S.M. Evans 1971) cite the case of the polychaete, *Arctonoë*, which is a commensal with echinoderms. This polychaete, when tested in a Y-tube olfactometer, shows a preference for the host from which it has been removed. After a few weeks in a different host there is a preference to return to the new host, again when tested in a Y-tube olfactometer. What this means in terms of migration behaviour is not known, but possibly it indicates that the polychaete leaves its host from time to time on (short-distance) feeding or exploratory migrations during which the ability to return to the host is retained.

Most long-distance migrations performed by invertebrates other than insects involve the use of mechanisms that prohibit the possibility of strategies to increase the ratio of calculated to non-calculated migration. Hence, many marine and inter-tidal animals produce young that perform removal migration by swimming into such a position that displacement by water currents takes place. Pseudoscorpions and mites may cling to the leg of an insect and some spiders produce long gossamer threads that cause them to be displaced by the wind (Savory 1964). The final stages of such migrations, however, may well involve some exploratory behaviour and habitat assessment. The larvae of the serpulid polychaete, *Spirorbis borealis*, for example, when ready to settle, tests various substrata, and if a suitable substratum is not found, attachment, settlement, and metamorphosis are all delayed (Knight–Jones 1951). The extent to which a larva retains the ability to return to a substratum visited previously, however, does not seem to be known. It is, of course, dangerous to attempt to interpret the potential capability of any animal, particularly if such interpretation is likely to be biased by anthropomorphism. However, it seems unlikely that an animal with the neural simplicity of a polychaete larva is capable of storing sufficient information to permit exploratory migration, habitat assessment and then calculated removal migration. A much more likely explanation is that the migration is non-calculated throughout, that the larva settles as soon as it encounters a suitable substratum, and that the threshold requirements for settling change with time (p. 82).

The migration patterns of large, mobile invertebrates such as cephalopods are relatively unknown though in some species, such as the short-finned squid, *Illex illecebrosus* (Squires 1957), they seem to be similar to some fish. Presumably, therefore, the discussion that was relevant for oceanodromous fish (p. 197) may also be relevant to these invertebrates.

In conclusion, little is known about removal migration in invertebrates other than insects, and where data are available, particularly for longer-distance removal migration, only non-calculated removal migration seems to be implicated. Some possibility remains, however, for animals as diverse as commensal polychaetes, inter-tidal molluscs, colonial larval insects, and perhaps also spiders and scorpions, that exploratory migration and habitat assessment may precede removal migration.

15.15 Discussion

15.15.1 The necessity for the distinction between non-calculated and calculated migration

The possibility of the existence of calculated migration has received scant attention from students of migratory phenomena. Apart from anything that may be gained from a study of calculated migration for its own sake, failure to recognise the existence of the phenomenon and the selective pressures associated with it can have important repercussions in the study of other facets of migratory phenomena. This is particularly so in the study of removal migration distance and orientation and navigation mechanisms.

As far as removal migration distance is concerned, it may on occasion only be possible to reconcile apparently contradictory results by appreciation of

the possibility that the removal migration may be calculated rather than non-calculated. For example, study of a species in one area may well lead to the conclusion that it is highly sedentary or, if it is a return migrant, that it has a high degree of return because individuals are found to breed in successive breeding seasons in the same restricted area. When the species is studied elsewhere, however, or alternatively introduced into a new country, it may well be found that far from being sedentary it has a high rate of spread. Such results are difficult though not impossible to reconcile on the basis of non-calculated removal migration. On the basis of exploratory migration–habitat assessment– calculated removal migration, however, such results are easily reconciled. In the first area the distribution may have been in an equilibrium state (p. 402), typical of a free situation, with the result that despite exploratory migration the animals return to breed to their natal site (p. 325). In another area, on the other hand, particularly if the species has been introduced into virgin territory, the area covered during exploratory migration may well contain a number of more suitable sites with the result that although some individuals settle at the natal site others may settle up to the limits of the area covered during exploratory migration. In considering the potential rate of spread of a species about to be introduced, therefore, attention should not be restricted to the breeding site to breeding site removal migration distance that is 'normal' for the species in the area from which individuals are to be collected. If the species is a calculated removal migrant it is the exploratory migration distance that is the critical factor.

The necessity to determine whether an animal normally performs non-calculated or calculated removal migration is especially important in the design and interpretation of experiments on orientation and navigation mechanisms. This problem is discussed at length in Chapter 33 and need not be discussed here beyond the presentation of a single example.

The homing pigeon, *Columba livia*, is a domesticated version of the rock dove. It is frequently pointed out that the rock dove is a 'non-migratory' species, by which is meant that it does not perform seasonal return migration (but see p. 895). Because the wild ancestor was a 'non-migratory' species, it is therefore assumed that the navigational abilities now observed in the homing pigeon are the product of selection during domestication (Bellrose 1972).

Although there has undoubtedly been selection on homing ability as a result of domestication, such statements neglect the possible importance to the wild ancestor of strategies to increase the ratio of calculated to non-calculated removal migration. In fact, it is highly probable that rock doves, particularly during the juvenile phase of ontogeny, in common with other 'sedentary' species, perform exploratory migrations in relation to habitat assessment (p. 187). Such migrations, if they evolve in response to the advantage of increase in the possibility of calculated migration, must be associated with the evolution of the ability to return to the natal site. Basic ability to navigate and/or orientate to the home site is therefore highly likely to be present in wild rock doves. What may not be present in the rock dove, or any other 'sedentary' bird, is the motivation to return to an area in which the bird has been captured, handled, ringed, and transported. Such a bird may well no longer consider the home site to be the most suitable known to it and may, therefore, navigate to some other site. Quite possibly the process of domestication has selected for birds that have a high threshold for removal migration, are more tolerant of handling, and that therefore may continue to assess the home site to be the most suitable habitat despite the frequent handling and displacement to which they are subjected. On such an interpretation the observed differences in homing behaviour of rock doves and other 'sedentary' birds (Matthews 1968) and the homing pigeon may well indicate differences in the criteria for habitat assessment and in the threshold for removal migration rather than differences in navigational ability. This situation is discussed further in Chapter 33.

15.15.2 The socially communicated familiar area and altruism

Using the concept of potential reproductive success defined earlier (p. 38) altruistic behaviour may be defined as behaviour that decreases the potential reproductive success of the individual showing the behaviour relative to the potential reproductive success of the individual(s) toward which the behaviour is directed.

It is necessary to distinguish between altruistic and other behaviour that may appear similar but which possibly does not decrease the potential reproductive success of the individual showing the behaviour relative to that of the individuals toward which the behaviour is directed. For example, if we accept that there is an advantage to individual wildebeest (p. 95) in living in a large herd in terms of protection from predators and that this advantage is greater than the disadvantage of sharing areas of grass with large numbers of individuals, the habit of communicating the direction of distant food by movement and the laying of scent trails is likely to be advantageous even to the individual communicating the information. Similarly, 'donation' of information by an individual that is a member of a cooperating group (of hunters, for example) may not be disadvantageous, even if no information is received in return. On the assumption, that is, that each individual increases its nett rate of energy uptake by hunting in a group such that there is no advantage to an individual becoming solitary and exploiting sources known to it rather than imparting information. In this situation it is advantageous to every individual that the group should hunt at the most suitable site known to the group. This cannot be achieved without an information exchange and decision-making mechanism. Exchange of information concerning the familiar area between the male and female of a reproducing pair or between closely related kin, is similarly not altruistic for, up to a point, in maintaining the potential reproductive success of any of these categories the individual (or gene) is also maintaining its own reproductive success.

The 'donation' of information concerning an ephemeral but superabundant resource may also not always qualify as altruism in that if the exploitation of such a resource by the 'donor' is not affected in any way by sharing it, then any advantage gained by association with another individual may lead to a nett advantage. True altruism (in the short term) comes, however, in the situation in which information is donated to an individual that is not a member of a cooperative group nor a mate or a potential mate nor a close kin and is not concerned with an ephemeral yet superabundant resource. In such situations, the habit of 'donation' of information is likely to evolve only if there is a degree of reciprocity, either immediately or at some time in the future (i.e. altruism in relation to the familiar area evolves only if it elicits similar altruism from other individuals and particularly, though not necessarily, if receipt of information is more advantageous than donation of the same information is disadvantageous). For example, donation of information concerning an abundant resource, even if the donor suffers some disadvantage through sharing, is likely to be less of a disadvantage than receipt of information concerning a different resource is an advantage. As long as a unit of altruistic behaviour elicits a similar or greater unit of altruistic behaviour from other individuals such that over a sufficient period of time the behaviour is advantageous to all individuals concerned, then altruistic behaviour will evolve. Most additional selection in this situation seems to be concerned with 'cheating' (Trivers 1971) (i.e. soliciting information but then imparting no or false information in exchange). Selection has presumably acted on individuals to recognise potential 'cheats', to avoid cooperation with such individuals, and if possible to give retribution when false information has been received. One advantage to a male human of returning to the natal group after pairing is likely to be that the male has learned which individuals are 'reliable'. The advantage of associating with known individuals, the 'cheating' potentials of which have been assessed, also seems likely to divorce altruism from the chain migration phenomenon (p. 136). When one individual of a cooperating group has performed removal migration, it is an advantage to that individual to recruit other individuals from the group to perform the same removal migration. If removal migration is in any case an advantage to another individual, the advantage may be even greater if migration is to the same destination as the earlier migrant. Both individuals therefore gain by social communication concerning their respective familiar areas.

Recently, as reviewed by Lin and Michener (1972), evidence has accumulated that workers may make a much more significant contribution to the production of males within a colony of social insects than was formerly realised. If there is some possibility that some time in the future any individual worker may become reproductive, it is an advantage to every individual, both workers and reproductives, that the colony moves to an optimum site and that the colony amasses food reserves as efficiently as possible. Upon discovery of a suitable place for colonisation or of an abundant supply of food it is an advantage to the individual concerned to communicate this information to other individuals and to recruit them to exploit this food source. In the latter case, sharing an ephemerally superabundant food source, such as nectar, with other individuals involves the selective pressures already discussed. All individuals gain,

including the informant, by efficient collection and storage of this food source for later use. Full consideration of the ramifications of this situation in social insects, however, must await demonstration of the factors that cause some worker individuals but not others to become reproductive.

In conclusion, therefore, many of the examples of social communication concerning the familiar area may not be altruistic, even in the short term, because the informant may not directly reduce its potential reproductive success relative to other individuals. In those examples where the behaviour is altruistic, however, the behaviour is only likely to evolve if it elicits similar behaviour from other individuals such that over a period of time all individuals gain as a result of the behaviour. Where altruism occurs, so too does selection in relation to the possibility of 'cheating'. This selection may well have been involved in the evolution of some facets of calculated migration such as return to the natal group and removal migration to an area occupied by an individual the 'cheating' potential of which has already been assessed. Full interpretation of the social communication of social insects in relation to altruism must await demonstration of the factors that result in some worker individuals but not others becoming reproductive.

15.16 Summary

1. Examples are given which support the hypothesis that in animals which perform calculated migration, strategies have evolved that increase the ratio of calculated to non-calculated migration. In addition to adoption of appropriate criteria for habitat assessment, these strategies may be divided into two broad categories: (a) exploratory migration; and (b) extension of the familiar area by social communication.

2. One special type of exploratory migration involves contact between neighbouring social groups. All animals that live in a social group occasionally contact or even mingle with neighbouring groups and from time to time removal migration of individuals occurs from one group to another. In such species it is tempting to interpret occasional contact or mingling as a strategy to increase the ratio of calculated to non-calculated migration. Clear examples of such behaviour are provided by some primates and ungulates and it is probable that other mammals that live in social groups as well, possibly, as animals such as shoaling fish may show similar behaviour.

3. Exploratory migration in the more conventional sense (i.e. a migration beyond the limit of the pre-migration

familiar area for the purpose of suitability assessment of an area rather than a group, but during which the ability to return to any point visited is retained) seems to be widespread in the Metazoa, but particularly among vertebrates. The vast majority of vertebrates live within a familiar area which is established by exploratory migration and/or social communication. Once a familiar area has been established, animals usually restrict their movements over a limited period of time to only a part, known as the home range, of this area. Between birth and the onset of reproduction a vertebrate performs a series of migrations which often result in removal migration occurring between the natal home range and the (eventual) adult home range. In very few cases, however, are the data available that can evaluate whether the migrations that are involved in this removal migration are non-calculated or exploratory, in which latter case the removal migration is calculated. Hence, the post-fledging migration of bats and birds, the migration of the juveniles of most rodents, carnivores and reptiles, the 'solitarisation' of the males of some primates and ungulates, and the migration of immature imagines of dragonflies are all examples of migrations that could be either exploratory or non-calculated removal migrations and which require more data before evaluation is possible. Only for some rodents, some pinnipeds and man is there strong, though circumstantial, evidence that the migrations of juveniles and sub-adults are exploratory and that subsequent removal migration is calculated. Adults of vertebrates and some invertebrates may perform removal migration from one home range and/or social group to another. Again exploratory migration may or may not precede such a migration. Hence, the autumn dispersal migration of some adult bats, birds, and rodents, the overland migrations of some amphibious mammals and reptiles, and the occasional movements of some whales beyond the boundary of the area occupied by their deme, are all examples of migrations that could be exploratory and

that could have evolved in response to the postulated advantage of increasing the ratio of calculated to non-calculated migration but which require more data before evaluation is possible. Only for the adults of some rodents, some social insects, and some butterflies is there convincing evidence that such exploratory migrations occur. In some species, particularly birds, removal migration from one home range to another of the same type (e.g. breeding site to breeding site), takes place within the familiar area established as a result of seasonal return migration.

4. In many species the familiar area is established initially by social communication. Most often, as in many mammals, some birds (particularly waterfowl), and some social insects, the type of social communication involved is association of experienced and inexperienced individuals during migration or by the laying of trails. In other cases, however, such as man and honeybees, communication of information concerning a familiar area is achieved by some form of language. The possibility cannot be totally excluded that the 'swarming' of bats, the all-male groups of primates and perhaps also ungulates, have evolved in part in response to the advantage of familiar area extension by social communication. Both exploratory migration and calculated removal migration may take place within the socially communicated familiar area.

5. Many of the examples of social communication concerning a familiar area do not qualify as altruistic because the informant gains as much as the recipient as a result of information donation. Other examples of social communication concerning a familiar area, although qualifying in the short term as altruistic, in the long term seem likely to confer an advantage on all individuals concerned.

6. Distinction between calculated and non-calculated migration is necessary, not only for its own sake, but also in relation to the study of the distance component of migration and of navigation mechanisms.

16

The non-calculated initiation of migration: exploratory migration and non-calculated removal migration

16.1 Introduction

The two previous chapters have been concerned with the calculated initiation of migration and with the strategies that may have evolved *in part* in response to selection for an increase in the ratio of calculated to non-calculated removal migrations. Paradoxically, it is the need for such strategies that, at some point, in association with selection for longer-distance migration, renders non-calculated removal migration more advantageous than calculated removal migration (Chapter 8). In order to perform a calculated migration an animal has first to obtain information concerning a range of potential destinations. The acquisition of such information involves the expenditure of time and energy. The level of this expenditure is a function of distance between H_1 and the potential destinations. Beyond a certain distance, the disadvantage of the time and energy expended in strategies to permit calculated migration is likely to become greater than the advantage of calculated relative to non-calculated removal migration. In many animals for which there is selection against return migration (p. 63), the distance at which calculated removal migration becomes less advantageous than non-calculated removal migration is likely to be set by the limit of the sensory range. Such animals (e.g. many insects), spend little if any time in a memorised familiar area. Other animals (e.g. probably all vertebrates and a variety of invertebrates) have been exposed to selection for return migration, and in these the distance at which calculated removal migration

becomes less advantageous than non-calculated removal migration is further than the limit of the sensory range. Such animals spend a large part of their lives within a memorised familiar area. The distance at which calculated removal migration ceases to be advantageous (i.e. the optimum size of the memorised familiar area) for such animals is discussed in Chapter 18. Chapter 16 is concerned with the factors involved in the initiation of migration when non-calculated removal migration is more advantageous than calculated removal migration.

Removal migration is not the only form of non-calculated migration. Although usually a preliminary to calculated migration, exploratory migration is also invariably non-calculated (p. 25). This is because it is a migration beyond the limit of the pre-migration familiar area, albeit during which the ability to return to this area is retained. Even though exploratory and non-calculated removal migrations perform quite different functions, therefore, because they are both initiated on a non-calculated basis the factors involved in their initiation are broadly similar. The theory concerning selection on the non-calculated initiation of migration has been presented in Chapters 8 and 9. This present chapter is an attempt to evaluate the theses that have been presented. During the main part of this chapter, the distinction between the facultative and obligatory initiation of migration (p. 47) is retained. At the end, however, the distinction is discussed in the light of the examples presented. The summary examines the degree of support provided by this chapter for the theses presented in Chapters 8 and 9.

16.2 Exploratory migration

16.2.1 The facultative initiation of exploratory migration

Enough examples were given in Chapter 15 to illustrate the difficulty at the present time of distinguishing in the field between exploratory migration and calculated and non-calculated removal migrations. It has to be accepted, therefore, that the examples given in this and subsequent sections of this chapter cannot be guaranteed to be the type of migration under consideration. Wherever possible, supporting evidence is given, but in other cases the decision has been made according to subjective criteria that may or may not prove subsequently to be sound. For example, long-distance migration by a lemming or other rodent is assumed to be non-calculated removal migration rather than exploratory. A similar migration by a snowy owl, *Nyctea scandiaca*, however, is assumed to be exploratory on the basis that such an animal would be more likely to retain the ability to return to its natal or some other area. The subjective nature of many of these decisions on my part should be borne in mind by the reader during the remainder of this chapter. However, in the present state of knowledge of animal migration, such subjectivity seems unavoidable.

The model presented in Chapter 8 suggests that adaptation to unpredictable variation in habitat suitability involves the evolution of a threshold response to that habitat variable (or those habitat variables) that correlates best with habitat suitability. In many cases this variable (or these variables) is some facet of nett rate of energy uptake per individual (e.g. rate of contact with suitable food or rate of contact with conspecifics). Rate of contact with conspecifics can, of course, also correlate with suitability for other reasons, such as the influence of deme density on mortality risk due to predation or on competition for mates, etc. For whatever reason, however, rate of contact with suitable food or rate of contact with conspecifics seem to be two of the habitat variables that correlate best with unpredictable fluctuations in habitat suitability, and it is perhaps no coincidence that the literature abounds with anecdotal information concerning migrations initiated in response to these factors. In few cases, however, has it been proved that a threshold re-

sponse to these factors exists or even that the migration being described was indeed facultative and a response to the factors suggested. Only a few of the better authenticated observations are described here.

Exploratory migration by the European wild rabbit, *Oryctolagus cuniculus*, has already been described (p. 104) and the calculated removal migrations were shown to equilibrate deme densities in neighbouring areas. Myers and Poole (1961) working on the species in Australia in enclosures about 0·8 ha in area found that the initiation of exploratory migration was mostly limited to the summer (the non-breeding season) and seemed to correlate with periods when food was inadequate, either in quantity or quality. When migration between adjacent enclosures was made possible, however, only individuals from the more densely populated enclosure migrated. It seems, therefore, that exploratory migration was initiated when rate of energy uptake per individual fell below a certain threshold but that removal migration did not occur unless a more suitable habitat was encountered.

The snowy owl, *Nyctea scandiaca*, in Canada preys extensively on the various species of rodents to be found within its range but particularly on lemmings (some species of cricetid rodents). From time to time, however, snowy owls move to the south in noticeable numbers and appear in New England. Some evidence is available (Fig. 16.1) that the appearance of snowy owls in New England is associated with a decline in the lemming population at a time when the snowy owl population is high in numbers following a peak in abundance of lemmings. The evidence, however, is far from conclusive. Final acceptance of a relationship between lemming numbers and the initiation of exploratory migration must await the measurement of the migrant/non-migrant ratio of owls in years with different lemming and owl densities. It is possible, for example, that a constant proportion of the owl population initiates migration and that 'invasion' years in New England are simple reflections of a large owl population at source. It should not be forgotten, either, that although lemmings may be an important constituent of the snowy owl diet, other prey species, both mammals and birds, are also taken. Furthermore, the numbers of some of these species, such as hares, also fluctuate considerably and not necessarily synchronously with lemming numbers.

That the extra-breeding range migrations of snowy owls are exploratory rather than removal migrations is supported by observations of the recent

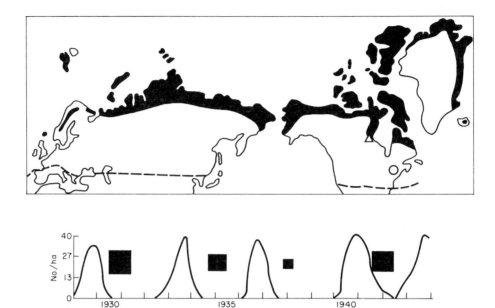

Year

Fig. 16.1   Exploratory migration of the snowy owl, *Nyctea scandiaca*, as a possible response to the availability of food

N. scandiaca preys to a large extent on the various species of small rodents, particularly lemmings, that occur within its breeding range (shaded area). The graph shows the density of lemmings estimated by their number to the hectare in the Churchill area of Canada (triangle on map). The size of the square shown on the graph is proportional to the number of owls that migrate beyond the limits of their breeding range into New England (open circle on map).

Each decrease in the lemming population is followed by a large movement of snowy owls to the south. The evidence suggests that the owls are responding to a shortage of food by initiating removal migration. Final acceptance of this conclusion must await the measurement of the migrant/non-migrant ratio of owls in years with different lemming and owl densities. The approximate limit of the exploratory range is indicated by the dashed line.

[Compiled from Shelford (in Dorst 1962), and British Ornithologists' Union (1971). Photos by Eric Hosking]

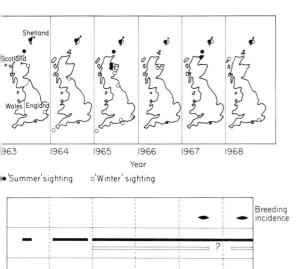

Fig. 16.2 Calculated removal migration preceded by exploratory migration in the snowy owl, *Nyctea scandiaca*

In the upper diagram, a dot indicates a reported sighting of a snowy owl in the British Isles. A solid dot refers to a 'summer' sighting (April through September) and an open dot refers to a 'winter' sighting (January through March; October through December) for the year indicated. It is assumed that the sightings refer to wild individuals though there is a possibility that some may be feral.

In the lower diagram the horizontal bars refer to the presence of snowy owls in the Shetland Islands. Sightings have been collated such that the diagram postulates the least number of individual birds necessary to account for all sightings. Each horizontal bar refers to the continued presence of a single individual: solid bar, male; open bar, female. A gap in a bar implies that the individual was not present on the Shetland Isles during that period. The uppermost section of the diagram indicates the incidence of reproductive activity. The middle section refers to the presence of two individuals that are assumed to have constituted the breeding pair in 1967.

Until 1967, the British Isles had been outside the recent breeding range of snowy owls. Throughout the nineteenth century, snowy owls were frequent visitors to Scotland (especially the Shetland Isles and to a lesser extent the Orkney Islands and Hebrides) and Ireland (especially the north and northwest) and occasionally also to England and Wales. During the first half of the twentieth century such visitors were less frequent and few were reported anywhere in the British Isles between 1950 and 1962. From 1963 onwards, however, there has been a marked increase in visits of snowy owls to Britain, an increase that culminated in the establishment of a breeding pair on Fetlar of the Shetland Islands from 1967 onwards. This diagram illustrates the events that preceded this removal migration of two individuals beyond the previous breeding range of the species.

Snowy owls breed both in Iceland and above the tree-line in Scandinavia (Fig. 16.1) and visitors to the British Isles are perhaps most likely to originate from these areas. It is perhaps significant that from 1960 onwards, just before the increase in number of visitors to Britain, there appears to have been an increase in numbers and an expansion of the breeding range in Scandinavia. Between 1963 and 1966 several snowy owls were noted, mainly in winter, throughout the British Isles as far south as southern England. If the initial migration is exploratory, the fact that the British Isles experienced very severe weather during the first months of 1963 could well have influenced the assessment of the suitability of Britain by any individual that initiated exploratory migration that year. This could well account for the subsequent persistence of individuals throughout the year in the east Scottish highlands and the Shetland Isles and for the reappearance of individuals in England during 1964 and 1965.

In 1963 at least one male was present in Shetland in June and July, the normal incubation and young-raising period for the species, and what may have been the same individual re-appeared in February 1964 and remained until November. This male may well have been joined by other individuals (exploratory migrants from Scandinavia and/or Iceland?) in 1965 for snowy owls were seen on the Shetland Islands every month of that year and at least three males and one female were involved. Two males and one female (the same individuals?) were present all year during 1966 and in June of 1967 a pair (the same individuals?) were found to have nested on Fetlar. The main prey species were a mammal (the rabbit, *Oryctolagus cuniculus*) and two birds (oystercatchers, *Haematopus ostralegus*, and Arctic skuas, *Stercorarius parasiticus*). Although the male remained more or less permanently within the area of the Shetland Islands, the female may have left for several months. Nevertheless what was assumed to be the same pair nested on Fetlar in May 1968 and have continued to do so each year since (to 1971).

The interpretation offered here is that in 1962/3 a male exploratory migrant arrived in Britain. Winter was spent in more southerly areas and summer in more northerly areas. Unlike previous and subsequent exploratory migrants this male assessed the area to be the most suitable part of his familiar area, an assessment that may have been influenced by the deme density experienced in the Scandinavian (?) part of his familiar area and the winter conditions that year in the British part of his familiar area. Consequent upon this assessment, a breeding territory was established on the Shetland Islands in 1964. A female exploratory migrant, encountering the male either in his territory or elsewhere (in December 1964?) subsequently assessed the male plus his territory to be the most suitable part of her familiar area and subsequently settled in the area (1965) and commenced breeding (1967). If this interpretation is accepted, it is only necessary for it to be demonstrated that migrating snowy owls retain the ability to return to their place of origin and snowy owl visitors to Britain can be considered to be exploratory migrants.

[*Diagrams compiled from data in Tulloch (1968, 1969), British Ornithologists' Union (1971) and the reports of rare birds in the British Isles published in the journal* British Birds *(1964–1969), Vols. 57 to 62.*]

history of the species in the British Isles (Fig. 16.2) as well as by analogy to other 'irruptive' species (p. 210).

Lack (1966) has argued that the feeding strategy of most birds that feed on rodents such as voles and lemmings, the demes of which show marked fluctuations in density, can be attributed to the lack of inter-deme synchrony of density fluctuations. The short-eared owl, *Asio flammeus*, for example, feeds in Britain on the short-tailed vole, *Microtus agrestis*, and further north in Scandinavia on the lemming, *Lemmus lemmus*. Both of these rodents show marked fluctuations in deme density. The owl, therefore, although markedly territorial at any one time, is liable to initiate migration from the territory at any time of year, winter or summer, whether breeding or not, whenever (it is assumed) food availability falls below a certain threshold level. A new territory is then presumably established in an area where the rodents are in a more numerous phase of their density cycle. Unless the short-eared owl can predict in which part of its familiar area rodents are likely to be most common at any particular time, the only reasonable strategies are either non-calculated removal migration or exploratory migration. Of the two the latter would seem initially to be most advantageous, though it remains to be demonstrated that it actually occurs. Lack (1966) suggests that the pomarine jaeger (or skua), *Stercorarius pomarinus*, and the snowy owl, *Nyctea scandiaca*, both predators of tundra rodents, show a similar strategy to the short-eared owl, a suggestion that is in accord with the interpretation of snowy owl movements given in Figs. 16.1 and 16.2. Although the facultative initiation of exploratory migration may be the normal way of life for the snowy owl, the possibility cannot be ruled out completely on present evidence that some of the individuals that visit areas such as New England (Fig. 16.1) and the British Isles (Fig. 16.2) are juvenile or sub-adult obligatory exploratory migrants of the type described in Chapter 17.

Not only the avian predators of 'cyclic' rodents seem to show facultative exploratory migration when the rodents decrease rapidly in number. Behaviour that appears to be similar has also been reported for mammalian predators of these animals. For example, Sutton and Hamilton (1932) in a discussion of the eight species of mammals that live on Southampton Island in Hudson Bay, Canada, suggests that foxes and weasels, as well as owls and skuas, initiate (exploratory?) migration whenever lemmings become scarce.

Species of birds and mammals that prey upon 'cyclic' rodents are not the only animals that exploit a food source of highly variable abundance that fluctuates synchronously over large areas but not over the whole range. As argued most convincingly by Newton (1972), birds that feed on the seeds of trees are faced with a similarly variable food source. The abundance of tree seeds is far less predictable than that of the seeds of herbaceous plants. The size of the seed crop of a tree is a function partly of the endogenous rhythm of the tree and partly of the weather. Most trees require more than one year to accumulate the reserves necessary to fruit, the necessary period being longer at higher latitudes. Spruce, for example, tend to fruit every two to three years on the mountains of central Europe, every three to four years in the southern part of the boreal forest, and every four to five years further north. As well as this endogenous rhythm, a good seed crop is dependent on fine and warm weather when the fruit buds form, usually in autumn, and again when the flowers set, usually in spring. In the absence of suitable weather at the critical time, seed abundance is delayed for a further year. As all the trees in any one area are subjected to the same weather they are likely to produce seed in synchrony. Each year, therefore, there is a patchy distribution of crops with each productive area extending over hundreds or thousands of square kilometres and being separated from the next productive area by an area in which the trees are devoid of seeds. Furthermore, the areas of greatest productivity differ from year to year. Presumably such a description of spatial distribution could be applied equally well to 'cyclic' rodents, though the causal mechanisms are less well understood.

Measurements of food availability in areas that have a reasonable probability of being the area of origin have been made for at least some of the birds that feed on the seeds of trees. These measurements lead to the same conclusion as for the snowy owl (Fig. 16.1). Thus the appearance of Clark's nutcracker, *Nucifraga columbiana*, in the Californian lowlands correlates with the size of the cone crop of ponderosa, sugar and Jeffrey pines (*Pinus ponderosa*, *P. lambertiana*, and *P. jeffreyi*) and the white fir (*Abies concolor*) on the west slope of the Sierra Nevada (Davis and Williams 1957, 1964). A similar correlation has been reported between the appearance in eastern Europe of another species of nutcracker, *Nucifraga caryocatactes*, and the abundance in Siberia of seeds of the Siberian arolla, *Pinus cembra* (Formozov in Dorst 1962). The nature of the correlation is such that migration is most likely to be initiated in years when a light seed crop follows one, or more often two, years when the seed crop has been heavy. Although there has been some discussion (Davis and Williams 1957, 1964) over whether the threshold variable is food abundance or deme density, the factor that seems likely to correlate best with habitat suitability is food availability per individual. However, a threshold that is affected both by rate of

contact with food and rate of contact with competitors for food seems to be more advantageous than a threshold based only on nett rate of energy uptake per individual. Both of these nutcracker species cache seeds in the autumn for use over winter and an abundance of seeds per individual in autumn is assumed to lead to a high winter survival and a high deme density the following autumn. Migration is initiated in California in September and October at the time that the winter caches are normally assembled and it is possible that rate of caching may be one of the threshold variables for migration in these species.

Many other birds fit into a similar category to the nutcrackers, such as, for example, those 'irruptive' species of European finches (Fringillidae) described by Newton (1972) as well as similar North American species such as the grosbeak, *Hesperiphona vespertina*. Further reference is made to these species in Chapters 17 and 27 in discussions of degree of return and the evolutionary relationship between exploratory migration and seasonal return migration in birds. There is one other species group that adds to the present discussion, however, and that merits further description here. This species group is the crossbills, *Loxia* spp., which are also fringillids. Of the three species discussed by Newton (1972) only the common crossbill, *L. curvirostra*, is considered here.

The crossbill breeds primarily in the coniferous woods of Europe, Asia and North America (Fig. 16.3) and, at least in Europe, specialises on the seeds of spruce. From time to time the species appears in some numbers outside of its breeding range. In southwest Europe, for example, such 'invasions' have taken place on 67 occasions between the years 1800 and 1965. On some of these occasions the birds

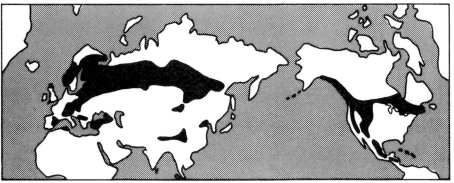

Fig. 16.3   The breeding range of the common crossbill, *Loxia curvirostra*

[*Redrawn from Newton (1972). Photo by Eric Hosking*]

have come mainly from Fenno-Scandia and on others from further east. The main point of interest concerning the crossbill, however, in the present discussion, is that it appears in southwest Europe in summer rather than autumn or winter as do most migrants from further north. Furthermore, the migration occurs just once a year.

According to Newton (1972), the seeds of spruce in northern Europe are formed, like those of most conifers, in late June. The seeds then remain in the cones over winter and are shed the following May or June. Crossbills can feed on these seeds at any stage but cannot take them off the ground. During the period, which may be as long as two months, between the falling of one seed crop and the formation of the next, crossbills feed chiefly from cones of pine. It is when the new crop of spruce seeds is forming in June that crossbills are most likely to initiate (exploratory?) migration. The birds even-

tually settle in an area where there has been a heavy set of seeds. In most years the removal migration distance is probably of the order of 100 or so kilometres. In other years it may be as much as 4000 km. Having found a suitable area (large number of seeds set/individual bird?) the crossbills settle down, breed the following spring, and may initiate further (exploratory?) migration the following May or June (if the habitat variable in the new habitat falls below the threshold value?). Unlike most seasonal return migrants, therefore, and also unlike most facultative exploratory migrants such as the nutcrackers and other 'irruptive' fringillids, the crossbill initiates migration at only one season (i.e. mid-summer).

In the case of the crossbill, there is some evidence to support the suggestion that the migration is exploratory (i.e. a migration beyond the familiar area during which habitat assessment takes place

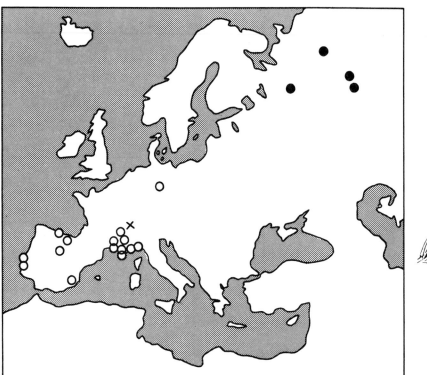

Fig. 16.4   Long-distance migrants of the crossbill, *Loxia curvirostra*, as exploratory migrants

The diagram shows the subsequent history of immigrant crossbills to Switzerland in the 'invasions' of 1959 and 1963. Such birds are emigrants from the boreal forests to the north and east. Birds marked in Switzerland (×) were

recaptured mostly within a year of ringing (○) but four, all males, were recaptured in later years (●). There is a clear indication that although the birds may spend a year in southwest Europe, during which time breeding may often occur, the ability to return to the boreal forest 4000 km or so away is retained.

[*Modified from Newton (1972)*]

and throughout which the ability to return at least to some of the areas visited, including the original familiar area, is retained (p. 26)). First, there is no doubt that the migration takes the animal beyond its previous familiar area, at least as far as juveniles are concerned. Secondly, the existence of habitat assessment can be inferred from the observation (Newton 1972) that the birds leave an area where the seed-set is poor and settle where the seed-set is good. Third, recapture of ringed individuals (Fig. 16.4) indicates that even some long-distance migrants in some subsequent year may return, at least to their natal latitudes, even after breeding once or more in the area occupied as a result of the initial migration.

In all of the examples given so far, habitat suitability has been unpredictable to the individual because of the nature of its variation with time. There is another type of unpredictability to which all animals are exposed early on in their ontogeny. Whether or not the natal habitat for an individual is suitable depends to a large extent on the choice of habitats available to the parents and in particular to the relative success of the parents in any territorial interaction. This situation is considered further in Chapter 17 in relation to the ontogenetic variation in migration threshold. For the moment, however, it is sufficient that the suitability of the habitat into which an individual is born is often unpredictable and that a migration threshold is likely to evolve that is functional at an early stage of ontogeny. Two examples of such thresholds for exploratory migration may be given, both concerning birds.

The case for the pre-reproductive migrations of the white stork, *Ciconia ciconia*, in Germany being exploratory has already been given (p. 190). This species feeds its young on small vertebrates and large invertebrates collected from damp ground. Hence, low-lying land is the preferred type of breeding habitat. The distribution could well be free (Chapter 18), however, and at high deme densities many individuals nest in hilly country. In recent years the German deme of white storks has been declining (Lack 1966) with the result that nesting sites have become vacant in both low and high ground areas. In favourable low-lying areas, such as the Rhine valley, 70 per cent of birds breed less than 50 km from their natal site, whereas birds born in hilly country show only 17 per cent nesting within 50 km of their natal site. A problem of interpretation should, however, be pointed out here. These results could well indicate that birds born in hilly country are more likely to initiate exploratory migration

than birds born in low-lying country. They could equally well indicate that birds of both areas initiate exploratory migrations of similar distance with similar frequency and that all birds settle within the most suitable habitat encountered during exploratory migration.

A similar problem of interpretation is involved with the second species, the great tit, *Parus major*. The great tit shows a free distribution in much of its range (Fretwell 1972), though in some places it may be (debatedly) despotic (J.R. Krebs 1971). In either event, the most favoured type of habitat is broad-leaved forest, particularly oak (*Quercus* spp.), though the species also occurs but at a lower density in pine woods. A greater proportion of post-fledgling young birds disappear from pine than from mixed forests (Kluijver 1951), an observation that could be interpreted in a variety of ways but which could indicate that the availability of food is more likely to exceed the migration threshold in pine forest than in mixed forest.

So far, although the existence of an exploratory migration threshold relative to some habitat variable that correlates well with habitat suitability has been implicit in the observations reported, no actual measurements of the threshold have been given. Indeed, measurement of such thresholds seems to be rare in migration literature. However, one example can be given of a measured threshold for migration, though no supporting evidence can be presented that the migration concerned is exploratory rather than calculated or non-calculated removal migration.

The barren-ground caribou, a subspecies of the circumpolar *Rangifer tarandus*, spends the summer months on the Canadian tundra. In winter, however, the majority of individuals are found south of the tree-line in the taiga (Fig. 25.6). Caribou (and reindeer, another subspecies of *R. tarandus*) are exclusive grazers, not feeding on twigs or stems (Kelsall 1968). Despite popular belief, however, lichens are not the sole food source and even during the winter when their importance as a food source is at a maximum, lichens constitute only up to 30–55 per cent of caribou stomach contents. During the winter, when a large proportion of their food species are under snow, caribou dig feeding craters to reach their food. Feeding craters are dug only twice in any given unit of snow, however, after which the snow is so hard that undisturbed snow has to be sought before further feeding can take place. Furthermore, not only does the nature of the snow cover affect the availability of food to the caribou, but it also correlates with the likely danger of predation by wolves, *Canis lupus* (Kelsall 1968).

The characteristics of the snow cover are, therefore, likely to correlate well with habitat suitability. Furthermore, because snow fall is unpredictable, it might be expected that caribou will have evolved a migration threshold with respect to snow. Pruitt (1959) analysed the distribution of caribou during the winter months and concluded that such a threshold does exist. In fact, there seem to be at least three characteristics of snow cover (i.e. depth, hardness and density), each of which contributes to a composite migration threshold. In forest, the migration threshold for snow depth seems to be 60 cm; for hardness, 50 g/cm^2; and for density, 0·19 to 0·20. This last threshold is particularly interesting for, as

pointed out by Kelsall (1968), at snow densities less than about 0·21, wolves sink into the snow up to their chest.

Thus, caribou prefer areas where the snow is quite soft, light and thin, and when the snow exceeds any of the above threshold values, migration (exploratory?) is initiated. However, unless snow of a density less than 0·21 is available elsewhere, a probability that will vary with time of year, initiation of migration at snow densities greater than 0·21 is unlikely to be advantageous. A lot depends, therefore, on the mean expectation of migration ($\bar{E}$) at a particular time of year, and as already argued (p. 48), if the value of $\bar{E}$ changes with time of year so too should the migration threshold. A habitat-type the suitability of which is a function of snow-cover might be expected to provide a fine example of a

Fig. 16.5   A herd of caribou, *Rangifer tarandus*

Thousands of caribou resting or moving up Ghost Lake, District of Mackenzie, during spring migration. The dot in

the middle of a large open space surrounded by caribou is a wolf, *Canis lupus*.

[*Photo by Frank Banfield (courtesy of the Canadian Wildlife Service)*]

habitat-type for which $\bar{E}$ is likely to vary with time of year. For instance, in late autumn an animal finding itself in a habitat with a thin snow covering has a high chance of finding a habitat with no snow if it initiates migration. In mid-winter, however, an animal finding itself in a habitat with a thin snow covering and initiating migration is only likely to find itself in a habitat with a deeper snow covering. If, as therefore seems likely, the mean expectation of migration, $\bar{E}$, changes as winter progresses, it follows (p. 48) that the migration threshold should also change as winter progresses. As far as snow depth is concerned, for example, the migration threshold should increase as winter progresses. Such an increase in migration threshold does seem to occur (Pruitt 1959).

So far, attention has been confined to situations in which in effect nett rate of energy uptake per individual has been the critical facet of habitat suitability that varies unpredictably. Rate of contact with suitable food and rate of contact with conspecifics and other competitors for food are both likely to be habitat variables that correlate well with habitat suitability in this situation, and so far no attempt has been made to distinguish separate migration thresholds for these variables. In the vast majority of migrations initiated in response to overcrowding (e.g. Clarke (1953) for the hippopotamus, *Hippopotamus amphibius*), data simply are not available by which such a distinction can be made.

As far as likely exploratory migrations are concerned there are one or two examples, however, which suggest that rate of contact with conspecifics may be involved in the migration threshold as a habitat variable that is independent of nett rate of energy uptake per individual. For example, the social unit of the warthog, *Phacochoerus aethiopicus*, an African suid ungulate, consists of two or more mature females and their young and sometimes also an adult male (Frädrich 1965). In the Nairobi National Park, each group has its own home range, which includes the sleeping holes, but the ranges overlap considerably. The female leaves the group when about to give birth to a litter and may or may not return to the same group.

The possible significance of this observation, which could perhaps be repeated for many other ungulate species, may be made more apparent by drawing attention to a second situation which has been described in Chapter 14 (p. 102). This is the initiation of exploratory migration followed by

calculated removal migration to new territories by a number of adults of the black-tailed prairie-dog, *Cynomys ludovicianus*. It was shown in Fig. 14.7 that one of the territorial units observed by J.A. King (1955), a unit that consisted of two males and five females, experienced a highly successful breeding season in the year in question and produced 31 surviving young, more than twice the number produced by any of the other territorial units. In June, in the latter half of the lactation period, but before the young appeared above ground, all seven adults of this territorial unit but apparently none of the adults of the other territorial units initiated exploratory migration that led them to an unoccupied area 50–150 m away from the nearest part of their territory. During the next month or so, as the young were weaned and began to come above ground, the adults travelled back and forth between their original territory and the new area in which burrows were dug and vegetation cleared. Eventually the adults settled permanently in the new area, leaving the old territory to the 31 young.

In these cases, it seems that something other than a direct response to nett rate of energy uptake being experienced by the individual is involved in the initiation of exploratory migration. It is perhaps only worth discussing the situation in relation to the better-documented prairie-dog, but similar arguments could well apply to the female warthog or,

Fig. 16.6   A pair of warthogs, *Phacochoerus aethiopicus*

[*Photo by John Karmali (courtesy of Frank W. Lane)*]

indeed, to any case in which the female initiates exploratory and/or removal migration soon after giving birth to offspring that survive. As far as the prairie-dog is concerned, exploratory migration was initiated before the young were making demands on the resources of the territory, except possibly for any increased feeding needed to be carried out by the lactating females. There is a suggestion, therefore, that there may be in the prairie-dog a threshold level of breeding success in any one season above which exploratory migration is initiated. A threshold response to breeding success rather than a threshold response to nett rate of energy uptake per individual would seem to confer one major advantage to any prairie-dog individual possessing such a threshold. On the assumption, that is, that breeding success correlates with future nett rate of energy uptake per individual. This advantage is that exploratory migration can be initiated earlier and that the optimum time $(t_r$—Fig. 9.2) between the initiation of exploratory migration and the initiation of removal migration can be spent based on a home range that is more suitable than it would be if exploratory migration was initiated later. This time-difference seems likely to be particularly important to a prairie-dog for which t_r is not only a function of search-time but is also a function of the time required to modify the new habitat by digging burrows and cropping vegetation. Against this advantage of the early initiation of exploratory migration, of course, must be set the disadvantage that results from the in-creased energy expenditure and increased danger of predation due to exploratory migration at a time when the young are not yet weaned and not yet capable of living an independent existence.

Although emphasis has been placed in the above paragraph on an exploratory migration threshold in relation to breeding success it seems likely that the habitat variable that correlates best with habitat suitability would be some combination of number of adults and number of young, as well possibly as present nett rate of energy uptake per individual in the habitat. In the case of the warthog, for example, whether or not the initiation of migration is likely to be advantageous following the successful production of a litter is likely to be a function both of the number of other adults and young in the social group and of the availability of food and sleeping sites, etc. in the home range.

The probable existence of such an appreciation of group size and availability of food in relation to the initiation of other behaviour patterns has recently been demonstrated in a superb film of African hunting dogs, *Lycaon pictus*, made by Baron Hugo van Lawick. During the three month or so parturition and lactation period at the beginning of the dry season, the social group, consisting of males, females, and sub-adults, establish central roosting burrows into which the young are born and from which the group performs hunting return migrations. While the group is hunting, an individual of either sex is usually left behind to guard the young

Fig. 16.7   African hunting dogs, *Lycaon pictus* | *[Photo by Leonard Lee Rue III (courtesy of Bruce Coleman Ltd.)]*

(Kühme 1965). The guard and, after weaning, the young are fed on meat carried back to the burrows and regurgitated by all other members of the group that form a cooperative hunting unit. The critical factor in the reproductive success of the reproductive members of the group appears to be whether or not the young are provided with sufficient meat for them to attain a sufficient size and strength to perform the often long-distance migrations necessary to maintain contact throughout the year with suitable prey species. As the dry season progresses, these prey species move away from the area in which the dog burrows are located.

Male and female pairs leave the group for copulation and may not re-assemble with other members of the group but instead raise their young independently. Such pairs may well be the originators of a separate group. In other cases, however, pairs rejoin the group after copulation. Large hunting groups are capable of killing large prey species, such as zebra, whereas pairs probably subsist on smaller prey species.

Hugo van Lawick's film relates the sequence of behaviour patterns shown by a female member, A, of a group of hunting dogs toward another female member, B, of the same group. Female A produced her litter at the appropriate time. Female B, however, had conceived late, and as shown by subsequent events there was no possibility of her offspring being capable of performing the seasonal migration circuit that the group had established. As parturition time for female B drew near, she was constantly harassed by female A until eventually female B initiated what seemed likely at first to be removal migration from the group but, as subsequent events proved, the ability to return to the group was retained. Unlike another female, C, that had left the group earlier, however, female B was not accompanied by her mate. Whether or not this was the critical factor, as seems likely, female B rejoined the group rather than perform removal migration from the group as had female C. Female B, having rejoined the group, subsequently produced a litter. Over a period of time, however, female A succeeded in removing from their burrow, one by one, the offspring of female B, and killed them. During this time, also, other members of the group refused to regurgitate meat to female B. Only one, the largest and therefore the most likely to attain sufficient size to survive migration, of the offspring of female B was not killed, even though female A had ample opportunity to do so. Eventually, this young dog began to

receive meat brought back by the pack and, although considerably smaller than the offspring of female A, survived to begin and complete part of the migration circuit, during which time it received protection from predators, from its mother and other members of the group. It failed, however, to complete the journey and was eventually abandoned by the group.

The selective advantage of her own behaviour to female A is clear in that by removing competitors for a limiting resource, meat, she increased her reproductive success by increasing the chances of survival of her offspring. If all of the offspring of female B had lived to consume part of the meat brought back to the burrows by the hunting group then quite possibly some of the offspring of female A might have failed to attain sufficient size and strength by the optimum time for the initiation of seasonal migration. The main point of interest for the present discussion, however, is that if this interpretation is accepted, the behaviour of female A is only explicable if it is also accepted that female A was capable of assessing future group size in relation to the meat-gathering capacity of the group in that particular habitat. This assessment may, of course, be based either on a system of innate thresholds or on previous experience or on a combination of both. Subsequent events proved, furthermore, that female A had assessed the situation with remarkable accuracy. Permitting one of the offspring of female B to survive appeared to have no adverse affect on the survival of her own offspring. On the other hand, even the largest of the offspring of female B could not have survived as a member of the group with its particular seasonal migration circuit (Chapter 25) and it must be considered doubtful whether any of the offspring of female B would have survived even if female B had left the group accompanied by her mate. Possibly this explains the failure of her mate to accompany her when female B initiated exploratory migration from the group. Certainly the failure of her mate to accompany her seems likely to explain why female B did not initiate removal migration from the group.

In other years, if there had been more rain or game was more abundant, it is perhaps possible that such a late conception would not have been disadvantageous. This and many other of the questions raised by this sequence of events can only be answered by more extensive long-term studies of a number of hunting dog groups. For the moment, however, Hugo van Lawick's remarkable film strongly suggests that zoologists must be prepared to accept that animals other than man are capable of making, by whatever mechanism, far more subtle assessments of their surroundings than perhaps has previously been anticipated. Certainly, by comparison with the hunting dog situation, it seems quite reasonable to accept that prairie-dogs, warthogs, and many other mammals have evolved a migration threshold in response to deme density, breeding success, and food availability rather than simply nett rate of energy uptake per individual. Unless, that is, nett rate of energy uptake per individual correlates better with present and, where relevant, future habitat suitability than any combination of the separate habitat variables.

It has been suggested that primate social groups may split in response to an adverse age and sex composition (Itani *et al.* 1963) such that the group becomes 'unstable', and it is perhaps worthwhile to consider this suggestion in the present context. Long-term studies of primate groups of sufficient detail to investigate this suggestion are rare. The study with which Itani *et al.* are particularly concerned, however, is that of the Japanese macaque, *Macaca fuscata*, and provides a splendid example of the type of detailed long-term investigation that will probably be necessary for other species before zoologists are in a position to answer many of the questions posed by primate social dynamics. In 1952, Japanese zoologists started to provision a social group of macaques consisting of about 220 individuals (Sugiyama 1960). The main advantage of such provisioning was that the group appeared at a fixed point in their home range that was suitable for observation with sufficient regularity for individuals to be marked and recognised and for changes in group composition and behaviour to be noted. Provisioning also had the effect of increasing reproductive success and deme size and by 1962 the deme consisted of about 740 individuals (Itani *et al.* 1963). The original single group, however, by this time consisted of three social groups, each with their own home range but each of which included within their home range the provisioning area. The question that presents itself, therefore, is whether the subdivision of the original group was a result of an imbalance in the composition of the group or whether there is some other more likely explanation.

In 1954, two years after provisioning began, the socio-economic sex ratio (adult males as a percentage of adult females) was 36·7. In 1959, when the first group division took place, the socio-economic sex ratio was 61·2. The main effect of

provisioning, therefore, seems likely either to have been an increase in male survival or, more probably, an inhibition of migration by males. Male migration in this species seems likely to be exploratory followed by calculated removal migration (p. 162). Inhibition of migration may either, therefore, be non-initiation of exploratory migration or more frequent assessment of the natal home range as most suitable. Both effects could be attributed to provisioning which could either prevent habitat suitability falling beneath the exploratory migration threshold or increase the suitability of the natal home range relative to that of surrounding habitats.

The first group division involved sub-adult and adult males and females associating with a nucleus of adult males. This new group, *B*, gradually developed a home range of its own, separate from that of the larger group *A* (Sugiyama 1960). Three years after the division, the two groups, *A* and *B*, had distinct but overlapping home ranges, both of which included the provisioning area (Fig. 16.8). During the development of these home ranges some individuals of both sexes alternated between groups. The second group division involved sub-adult and adult males and females of group *A* associating with a solitary male that had contacted the group. Once again, this new group, *Y*, gradually developed a home range of its own, separate from that of the larger group but including the provisioning area (Itani *et al.* 1963). The socio-economic sex ratio of group *A* at this time, just before the division, was 52·0.

It could possibly be suggested, therefore, that at socio-economic sex ratios greater than about 50, social groups of the Japanese macaque become unstable and divide to form separate groups. However, group division involves the migration of individuals from one group to the nucleus of another group. This nucleus may be either an immigrant solitary male or a male or males from within the parent group. There is no evidence to suggest that the migrants are driven out of the parent group. If, consequently, the migration is a response to an unfavourable socio-economic sex ratio, it would be expected that the migrants would migrate only to a group in which the socio-economic sex ratio is more favourable. Yet the socio-economic sex ratio of group *B* in 1959 after the first major phase of group-to-group migration was 92·3, well above the threshold value of about 50 inferred from the situation at which division commenced. Similarly, the socio-economic sex ratio of group *Y* in 1962 was 111·8.

Furthermore, by this time the socio-economic sex ratio of group *B* had increased to 156·3. Group *B* in 1962, in addition, consisted of 150 individuals, nearly as many as in the original group in 1952 when provisioning began. An exploratory migration threshold in relation to socio-economic sex ratio followed by calculated removal migration to the group encountered with the most favourable socio-economic sex ratio is not, therefore, an interpretation that is supported by the available evidence.

Consideration of the most likely threshold variable is perhaps facilitated by stressing some of the clearest results to emerge from this study. First, both group divisions took place when the size of the parent deme was between 500 and 600 individuals. By 1962, after the formation of group *Y*, group *A* consisted of 517 individuals, group *B* consisted of 150 individuals, and group *Y* consisted of 73 individuals (Itani *et al.* 1963). Secondly, the provisioning area was retained within the home range of all groups.

Thirdly, separate home ranges were established as a result of the newly formed groups initiating a series of exploratory migrations beyond the limits of the home range of the parent group. The interpretation that best fits the facts, therefore, is that some resource of the home range other than that found in the provisioning area interacted with deme size such that the exploratory migration threshold of some individuals was exceeded. The high socio-economic sex-ratio of the newly formed groups could on this interpretation be attributed to a much lower exploratory migration threshold for males than females. This possibility has already been considered for primates in Chapter 15 and is discussed in much greater detail in Chapter 17. Which resource in the home range could have interacted with deme size in this way must remain unknown for the present. Perhaps the most likely candidates are those food items that are not available in the provisioning area, though other resources, such as roosting sites, could also be involved.

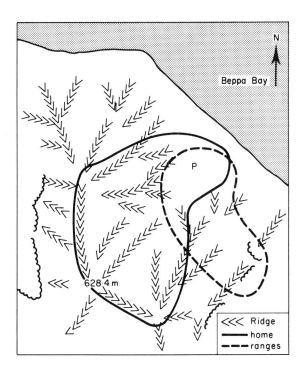

Fig. 16.8 Formation of a new home range by exploratory migration by some individuals from a social group of the Japanese macaque, *Macaca fuscata*

The Japanese macaque lives in multi-male social groups. In 1952, Japanese zoologists started to provision a social

group of macaques consisting of 220 individuals. By 1959, the group consisted of over 500 individuals. In this year, the group began to divide into two. The nucleus of the new group was a group of four adult males with which other individuals of both sexes began to associate. The new group gradually began to forage separately from the 'parent' group and to perform short exploratory migrations beyond the previous home range. During the early stages of division there was frequent interchange of individuals between the two groups. By 1962, however, the two groups had consolidated and the exploratory migrations by the new group had led to the establishment of a home range (outlined by dashed lines in the diagram) that was separate from the home range of the parent group (outlined by a solid line). Note especially that the two home ranges continue to overlap in the region of the provisioning area (P).

[Modified from Itani et al. (1963)]

The Japanese macaque, in common with the vast majority of primates as well as most ungulates, many cetaceans, some carnivores and a number of other mammals and birds, lives in a relatively large social group. Many more vertebrates, however, including probably the majority of mammals, live more or less solitarily when adult. It should be noted that use of the term 'solitary' to denote an animal with a home range that overlaps either not at all or minimally with that of other individuals, is retained solely for the convenience of description. It is accepted, however, that, as pointed out by Leyhausen (1964), even the most solitary (by this definition) of animals nevertheless have a real and complex social life. Schaller's (1967) study of the tiger, *Panthera tigris*, clearly illustrates this point. It is convenient in the present context, however, to distinguish between species the individuals of which exclude (by whatever means—Chapter 18) other individuals from their home range and species the individuals of which share their home range with other individuals. In the present discussion the terms 'solitary' and 'gregarious' are used to make this distinction.

In 'gregarious' animals, such as the Japanese macaque, the migration threshold to deme size can be any number from two upwards. The 500+ threshold deduced for the Japanese macaque is almost certainly exceptional and should only be viewed within the context of a provisioned group. That this is so is implied by the marked increase in the socio-economic sex ratio that followed the beginning of provisioning. This observation suggests strongly that the exploratory migration threshold of males was less likely to be exceeded once provisioning began, a conclusion that also emphasises the interaction of deme size with other habitat variables in the production of the migration threshold. Only about 50 per cent of males up to the age of 11 years initiated (exploratory?) migration from the provisioned group, whereas by this age up to 90 per cent of males of other, non-provisioned, groups had initiated migration (Nishida 1966).

'Solitary' species, on the other hand, are likely to have a migration threshold to a deme size of one. The fact that the individuals of a species are 'solitary' implies that the optimum size of the social unit in these species is one adult or one adult female plus dependent offspring and that if two individuals beyond a certain age attempt to share a home range for any length of time both individuals suffer a reduction in reproductive success (Chapter 18). It

follows, therefore, that by the time an individual reaches a certain age it is advantageous both for it and for its parent(s) that parent(s) and offspring should occupy separate home ranges. Early in ontogeny, therefore, as long as the parent(s) survives, it becomes advantageous for either the parent or the offspring to initiate migration. Usually it is the young that initiate migration. The reasons for this, however, can be described much more elegantly in relation to ontogenetic variation in migration threshold. Further consideration to this situation is given, therefore, in Chapter 17 rather than here.

On the whole, this section has been concerned with exploratory migration thresholds that in some way involve deme size and/or density. Such thresholds undoubtedly present the greatest scope for discussion (see Chapter 17). Nevertheless, there must be innumerable habitat variables of less general relevance in response to which exploratory migration thresholds have evolved. Most reports of such thresholds are anecdotal and difficult to quantify and there is probably little to be gained by presenting many such instances. The European mole, *Talpa europaea*, for example, initiates a series of migrations whenever the soil in which it is living becomes either too dry or too wet (Godfrey and Crowcroft 1960). These (exploratory?) migrations often result in a return to the original home range when the physical condition of the soil improves. The majority of animals seem likely to have evolved migration thresholds to parallel physico-climatic conditions. In addition, most species probably manifest rather esoteric migration thresholds specific to the peculiar niche occupied by the species. For example, the female polar bear, *Ursus maritimus*, spends the Arctic winter in a substantial den into which the young are born (Perry 1966). This den may be a cave or, if the female is out on the pack ice, it may be a hole in the snow and ice. In this latter case, if the iceberg above the den starts to tip during the winter, the mother and cubs initiate (exploratory?) migration. Presumably there is some angle of tip that constitutes a migration threshold.

16.2.2 The obligatory initiation of exploratory migration

In the previous section, attention was confined to migration initiated in response to unpredictable variation in habitat suitability. An attempt was made to present some evidence that animals exposed to such fluctuations have evolved a migration threshold relative to some proximate habitat variable (or combination of variables) that correlates well with habitat suitability. Occasionally, as in the prairie-dog, *Cynomys ludovicianus*, the habitat variable correlates well with *future* suitability (see discussion, p. 153). Although the migration threshold may vary with time (e.g. seasonally as in the caribou, *Rangifer tarandus*), there has been no question of the migrations under consideration being obligatory. At some times migration from *A* may occur, at other times not, depending on whether or not the exploratory migration threshold is exceeded at *A* at that particular time.

Following the arguments presented in Chapter 8, this section is concerned with migrations that are a response to predictable variation in habitat suitability relative to $\bar{E}_{ec}$ (the mean expectation of calculated removal migration preceded by time t_r of exploratory migration). Such predictable variation is most likely to be due to: a relatively high rate of change of habitat suitability relative to the generation interval; environmental cycles of habitat suitability; or optimum time-investment strategy. Such migrations should, therefore, be characterised by being initiated either at a fixed ontogenetic stage, or a fixed time of day or year, or at regular intervals.

In Chapter 8, which considered the obligatory initiation of migration from a theoretical viewpoint, it was suggested that rate of change in habitat suitability relative to the generation interval may well be a critical selective pressure. At that stage in the argument it was not necessary to distinguish between the initiation of non-calculated removal migration and the initiation of exploratory migration. Now, however, certain correlates of exploratory migration require to be taken into consideration.

Exploratory migration, because in the present context it is related to the use of a familiar area larger than the sensory range (Chapter 18), is found predominantly among animals characterised by a relatively large body size, a relatively complex central nervous system, and a relatively long generation interval. As already indicated in the survey in Chapter 15, therefore, exploratory migration is found primarily among vertebrates. When it is found in invertebrates, it is either associated with social behaviour or else the familiar area is small and/or temporary (p. 374). Furthermore, use of a familiar area is likely to evolve only in relation to habitats (equivalent in size to the familiar area)

that have a relatively low rate of decrease of suitability. The arguments relating to this assertion are given in Chapter 18. For the moment, however, it is sufficient to accept that vertebrate evolution has involved adaptation to relatively permanent habitats whereas invertebrate evolution has involved adaptation to habitats with all degrees of rate of decrease of habitat suitability relative to the generation interval.

As far as exploratory migration is concerned, therefore, because in contrast to non-calculated removal migration the animals involved are primarily vertebrates, predictable rate of decrease of habitat suitability is rarely likely to be such that selection favours ontogenetically obligatory initiation.

The behaviour that comes nearest to qualifying as an ontogenetically obligatory exploratory migration is probably that of the post-weaning, post-fledging, post-hatching or post-metamorphosis migration of 'solitary' vertebrates described at the end of the previous section and in Chapter 17. In the case of the post-weaning migration of 'solitary' mammals (exploratory?) migration is probably always initiated at a certain age if the parent still survives in the natal home range. The migration is likely to be facultative with a migration threshold to deme size of one adult. However, as the young is unlikely to reach this ontogenetic stage unless the parent also survives, it invariably happens that migration is advantageous. It is possible, therefore, that selection has favoured obligatory migration. However, as the time and energy required to 'monitor' the continued survival of the parent during the period of parental care is minimal, there is unlikely in this case to have been much selection against facultative migration. Consequently, as facultative migration is more advantageous than obligatory migration in that migration may not be initiated on those few occasions that the parent does die at the critical time, selection seems likely to have favoured facultative rather than obligatory migration. Unfortunately, observations of whether or not migration occurs when the parent has died at the critical time are insufficient to evaluate this conclusion.

For various reasons, therefore, it appears that ontogenetically obligatory exploratory migrations have evolved only rarely. A similar scarcity of examples is encountered when seasonally obligatory exploratory migrations are considered, and once again it seems likely that there are good reasons why this should be so.

When behaviour patterns are found that could be described as seasonally obligatory exploratory migrations, they seem to be more properly described as seasonal return migrations. This lack of a clear distinction between the two types of migration is considered further in Chapters 21–33 and is suggested to imply a close evolutionary relationship between exploratory and seasonal return migration. In part this suggested relationship is based on Fig. 8.10, which considered the factors necessary for selection to favour the seasonally obligatory initiation of migration.

The most important factor suggested in Fig. 8.10 was that seasonally obligatory migration would evolve only if the suitability of a given habitat-type A always became less than $\bar{E}$ (the mean expectation of migration) at a particular season. Clearly this is only possible if the suitability of A also becomes greater than $\bar{E}$ at some other time of year (though not necessarily a predictable time of year). Such consistent fluctuations in the suitability of A and the value of $\bar{E}$ are likely only if the suitability of habitat-type A fluctuates consistently relative to other habitat-types which were categorised as B in Fig. 8.10. This in turn is likely only if habitat-types A and B show consistent differences either in composition or in location (i.e. topographical and/or geographical). Automatically, therefore, selection for obligatory seasonal initiation of migration is associated with selection favouring individuals that spend different seasons in different habitat-types or in different topographical or geographical situations. In this situation, individuals of species likely to perform exploratory migration (i.e. mainly vertebrates), are likely to be exposed to selection that favours a relatively high degree of return (Chapter 18).

Perhaps the nearest we can get to an example of a seasonally obligatory exploratory migration that is not necessarily associated with a very high degree of return are those rodents such as the short-tailed vole, *Microtus agrestis*, which in the autumn migrates from open fields to forest borders, and the house mouse, *Mus musculus*, which, also in autumn, tends to migrate to drier habitats such as hayricks or houses (L.E. Brown 1962). It is by no means certain, however, that such removal migrations are preceded by exploratory migrations, nor that such exploratory migrations, if they exist, are seasonally obligatory rather than facultative.

It appears, therefore, as though selection for the initiation of exploratory migration leads either to

facultative migration or to the evolution of obligatory seasonal return migration rather than to the evolution of ontogenetically or seasonally obligatory exploratory migration. This can be attributed to the fact (Chapter 18) that exploratory migration beyond the limits of a familiar area that is larger than the sensory range is predominantly a vertebrate phenomenon, and vertebrates are also characterised by the following relative attributes: large body size; complex central nervous system; great adult life-span; and adaptation to permanent habitats (size of habitat = size of familiar area). Consequently, predictable rate of decrease of habitat suitability is unlikely to be great relative to the generation interval, and any selection for a seasonally obligatory migration is likely to lead to selection for a high degree of return with the results noted above.

The only selective pressures that seem likely to generate obligatory exploratory migration, therefore, are those associated with optimum time-investment strategy. In Chapter 9 (p. 64), the main conclusion reached concerning exploratory migration was that selection would favour those individuals that spent time t_r in exploratory migration before initiating calculated removal migration. In the present context, the main point for consideration is the optimum method of integration of time t_r within the total time-investment pattern for the species.

In order to illustrate the integration of exploratory migration into a total time-investment strategy, the example has been chosen of the European rabbit, *Oryctolagus cuniculus*, in Australia, studied by Myers and Poole (1961). The relevant information concerning this species has already been presented (p. 104) and the behaviour described has also been used as an example of the facultative initiation of exploratory migration (p. 205). It is hoped that the use of this example in relation both to obligatory and to facultative exploratory migration will illustrate the way that a single migration sequence may contain both facultative and obligatory components, and that, therefore, the only useful distinction between the two migration types is made at the sequence level (p. 265).

In their study enclosures, Myers and Poole (1961) noted that exploratory migration was mostly limited to the summer, in the non-breeding season, when food was inadequate either in quantity or quality. In the diel, exploratory migration usually commences late in the evening after the main period of feeding and is usually brief, lasting only 15–20 minutes or so, after which the rabbit concerned returns to its home range. Even at a time of year when the only major time investment activities are feeding and resting, therefore, the rabbit invests no more than 15–20 minutes per day (i.e. approximately $1 \cdot 0 – 1 \cdot 4$ per cent of its time) in exploratory migration. However, the fact that the behaviour shows such a marked diel incidence suggests that rather more is involved than simple proportional time investment. Possibly, in the absence of direct evidence, the diel incidence of predation risk and other migration-cost variables is such that late evening is the most favourable time of day for exploratory migration to be performed. Although this selective pressure may account for the diel incidence, however, it probably does not, by itself, account for the restriction of exploratory migration to a single 15–20 minute time investment each day. For this other factors must be sought.

It has been argued (p. 58) that selection for optimum time investment would reduce time invested in migration to the minimum that was still consistent with the dictates of the migration equation (p. 43). Furthermore, from the diel incidence of exploratory migration in the rabbit it seems that time is invested in the behaviour at the expense, probably, of time investment in resting (i.e. efficient digestion, wound repair, etc., and maximum protection from predators). There is, therefore, likely to be strong selection to reduce the time spent in exploratory migration even at a time of day that may be least disadvantageous for it. Yet at the same time it has been argued (Fig. 9.2) that calculated removal migration should not be performed until exploratory migration has been performed for a finite time, t_r. Presumably, the shorter each bout of exploratory migration, the more bouts must be performed and the longer the absolute time between the initiation of exploratory migration and the initiation of calculated removal migration.

Unfortunately, data are not available, either for the rabbit or for any other animal as far as I am aware, that would enable further evaluation of the way that t_r is integrated into the total time investment pattern for the species. Indeed, perhaps this would be a suitable moment to stress that the existence of an optimum time, t_r, spent on exploratory migration before initiating removal migration is at present only a hypothetical abstraction (Fig. 9.2). So far, I have been unable to find in the literature any evidence that either supports or contradicts the suggestion that such an optimum time investment exists. Arguments, such as those given above, that

make use of this concept, should therefore be interpreted accordingly.

One point that does emerge from the example of the rabbit, however, is the way that both facultative and obligatory components may be integrated within any particular migration sequence. It has been suggested (p. 205) that the rabbit initiates exploratory migration when some habitat variable, possibly nett rate of energy uptake by the individual, falls below the exploratory migration threshold. However, as just described, exploratory migration occurs primarily late in the evening after the main feeding period and lasts just 15–20 minutes. One interpretation is that the nett energy uptake for the day by the end of the evening feeding period is the threshold variable. In this case, each day that the individual experiences a sub-threshold nett energy uptake by the end of the evening feeding period, 15–20 minutes of exploratory migration is immediately initiated. An alternative suggestion is that the threshold variable is some combination of nett rate of energy uptake and time (in days) as argued in Fig. 8.2. Once this threshold is exceeded an exploratory migration time investment of t_r is initiated such that each evening after feeding, no matter what rate of energy uptake is experienced on a particular day, 15–20 minutes are invested in exploratory migration. This investment would then be repeated at nightly intervals (unless the migration-cost variables are below the threshold of favourability—Fig. 8.3) until time t_r had been invested, whereupon either removal migration occurs or the rabbit remains in its original home range and exploratory migration ceases. The suggestion is, therefore, that the facultative response to the appropriate habitat variable when the latter exceeds the exploratory migration threshold is the attainment of an exploratory migration physiology. Part of this exploratory migration physiology is a biological clock or clocks that: (1) trigger exploratory migration at the appropriate time of day (late evening in the rabbit); (2) terminate each bout of exploratory migration after the optimum time investment (15–20 minutes in the rabbit); and (3) repeat the bouts at appropriate intervals until the optimum total time in exploratory migration has been invested (t_r according to Fig. 9.2). Attainment of the exploratory migration state is therefore facultative, but the actual initiation of each exploratory migration may be obligatory. The total sequence of events, however, seems to be described best as facultative (p. 265).

16.3 The fixed sequence of initiation of exploratory migration and non-calculated removal migration

It was concluded in Chapter 9 that because the mean expectation of exploratory migration followed by claculated removal migration, $\bar{E}_{ec}$, must always be greater than the mean expectation of non-calculated removal migration, $\bar{E}_n$, the suitability, h_1, of the habitat occupied, H_1, must always become less than $\bar{E}_{ec}$ before becoming less than $\bar{E}_n$. Consequently, animals should always initiate exploratory migration before initiating non-calculated removal migration. However (Fig. 9.3, p. 66), if after the optimum total time investment, t_r, in exploratory migration even the most suitable habitat, H_{max}, discovered as a result of exploratory migration has a suitability that is such that $h_{max}M_{max}$ (where M_{max} is the migration factor for migration from H_1 to H_{max}) is less than $\bar{E}_n$, then if by time t_r, or soon after, h_1 has also become less than $\bar{E}_n$, non-calculated removal migration becomes the optimum response. It was suggested that this situation was most likely to arise when an individual's familiar area is surrounded by a sufficiently large area in which all habitats are of low to zero suitability. A few examples can now be presented that support the main aspects of this hypothesis.

The sea otter, *Enhydra lutris*, the only marine representative of the lutrid carnivores, is found in the North Pacific in waters up to about 60 m depth (Kenyon 1969). The species feeds on a number of benthic invertebrates and fish and occurs up to 1 km, or in shallow water up to 16 km, from shore. The home range of a female includes up to 17 km of shoreline and includes several hauling-out places. Within this home range the female occurs more or less solitarily or with her young of up to a year old, but groups of up to fifteen individuals do occur, and home ranges of individuals overlap considerably. Much of the time there is considerable segregation of the sexes and except when searching for a female the home range of a male rarely exceeds more than a few hundred metres radius from the hauling-out site which may be used by up to 100 individuals at a time. When searching for a female the male migrates through several female areas as part of a much larger home range than that used for feeding. After copulation, the male guards the female against the attentions of other males for about three days.

Because the sea otter was reduced nearly to extinction during the past 200 years, there exist within its geographical range (Fig. 16.9) many unoccupied but otherwise favourable habitats. When such an area is colonised the usual course of events appears to be a gradual rise in deme density to a figure as high as about 16 individuals/km²,

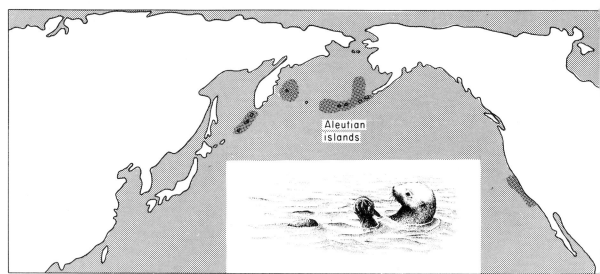

Fig. 16.9 Geographical distribution of the sea otter, *Enhydra lutris*, shown by dark areas

[*Modified from Nishiwaki (1967)*]

refer to a mean dung pat with a surface area of 500 cm²) $\bar{E}$ is given by $\overline{h_2 M}$, the mean product of h_2, the suitability of grass as a copulation site and M, the value of the migration factor for migration from dung to grass. In this case the value of M can be assumed to be approximately 1·0 as the distance involved is so small and requires little time. Hence, migration may be expected to occur when v_t is of the value that corresponds to the situation in which $h_1 = \bar{h_2}$. The suitability of the dung surface (h_1) and of the surrounding grass $(\bar{h_2})$ as habitats for copulation is a function, as far as males are concerned, of the fertilisation rates achieved by males in the two habitats.

A. The effect of temperature on the value of h_1 and h_2. In the left-hand graph the solid dots show the temperature on the dung surface in relation to the temperature in the surrounding grass. The dashed line shows the relationship if these two temperatures were equal. The temperature of the dung surface is higher than the temperature of the surrounding grass and this affects the copulation time in the two habitats as shown in the right-hand graph. Here, solid dots show the duration of copula on the dung surface and open dots show the duration of copula in the surrounding grass. Copulation takes longer in the surrounding grass. During copulation other males may take over (i.e. displace an already copulating male), and in the dung fly it is the last male to mate that fertilises the majority (80 per cent) of the next batch of eggs. Hence, a longer copulation time leads to a greater probability of take-over and consequently a reduced fertilisation rate. Longer copulation time also reduces the fertilisation rate by increasing the cycle (i.e. searching for, mating with, guarding, and then leaving a female and searching for another) duration.

B. The effect of male density on the value of h_1 and h_2. The diagram shows the rate at which copulating pairs are

encountered by a searching male in relation to distance from the dung pat and male density on the dung surface. Figures are shown for male densities of 10, 20 and 30 males/500 cm² of dung. The probability of take-over is proportional to the number of encounters a copulating pair has with searching males. Encounter rate decreases with distance from the edge of the dung.

C. Observed and calculated thresholds for removal migration of copulating pairs of dung flies from dung to grass in response to male density. The left-hand graph shows the percentage of pairs that migrate from the dung surface to the surrounding grass in relation to the density of searching males on the dung surface. If the threshold for migration is taken to be the point when 50 per cent of pairs migrate, the observed threshold occurs when there are 5 males/500 cm² of dung. The right-hand graph shows the calculated fertilisation rates at different male densities achieved by males when the pair stays on the dung surface (dashed line) and when the pair migrates to the surrounding grass (solid line). Calculated fertilisation rates are based on the measures shown in A and B as well as other measures such as probability of take over, percentage of the eggs laid in the next and subsequent batches by the female partner that are fertilised by the male as a result of the copulation, mean longevity, etc. According to these calculations, a greater fertilisation rate is achieved by staying on the dung to copulate when male density is less than five but at greater densities a greater fertilisation rate is achieved by migrating into the grass. In the case of the removal migration of copulating pairs of dung flies from a dung to a grass habitat, it can be calculated that $h_1 = \bar{E}$ when $V = 5$ males/500 cm² of dung, a calculation that agrees well with the observed migration threshold.

It should be stressed that these calculations are concerned with the suitability to the male of the two copulation habitats. A calculation of v_t for the female gives a figure of 27·5 males/500 cm². However, it is the male that performs the migration when the migration is performed by flight, the female being carried passively as the male flies her to the surrounding grass. It also seems likely that when the migration to the grass occurs by walking, the initiation of migration is still under the control of the male. Hence, although it is the copulating pair that performs the migration, the migration is in effect a male migration and occurs at a value of V that seems to be optimum for the male but sub-optimum for the female.

In conclusion, a male yellow dung fly, *Scatophaga stercoraria*, upon beginning to copulate, may either cause the copulating pair to remain on the dung surface or to migrate to the surrounding grass. Migration to the surrounding grass occurs as a function of male density on the dung surface and has a threshold of 5 males/500 cm² above which density migration takes place. The available evidence supports the theory that this threshold is an evolved optimum and corresponds to the situation where the suitability of the habitat occupied equals the mean expectation of migration. As the migration either occurs or does not occur immediately upon the commencement of copulation, the male response must be to experience of male density prior to contact with the female.

[*The calculations of fertilisation rate are given in greater detail in Parker (1971) from which paper this figure was compiled. Photo by G.A. Parker*]

aphids caught in a suction trap (trap opening 1·4 m above ground level and situated 7 m south and 22 m west of the nearest sycamore trees) during the spring migration is divided by the number (B) of fourth instar individuals at their peak deme density preceding the spring migration, the result can be compared with the total number (T) of nymphs present on the leaves. Over three years values for A/B of 1·1, 2·4, and 18·0 were associated with values for T of 65, 130 and 161, respectively. There is some indication, therefore, that when nymphs are more abundant in spring, the resulting adults are more likely to initiate migration.

These first-generation adults have a high reproduction rate and are usually all dead by the end of June. If survival of offspring is high, the resultant adults are often present in considerable numbers during July. Neither migration within the canopy of the tree nor mortality from any cause(s) can account for the observed numerical fluctuations of these adults. However, a regression coefficient relating the numbers of adults remaining after the decline in numbers (late July to mid-August) to the peak numbers (late July) of adults is significantly less than one ($b = 0·605$; $P < 0·01$). Furthermore, at the time of the rapid decline of aphid numbers there is an increase in the number taken in the suction trap. Over three years, the number of adults disappearing from samples of forty leaves was 228, 715, and 844. These numbers were associated with total suction trap catches of 775, 1104, and 2236, respectively. There is a strong indication, therefore, that in summer as well as in spring an individual sycamore aphid is more likely to initiate migration when exposed to higher deme densities. There is a similar indication for the autumn migration flight.

Dixon's work is notable in that it provides information concerning the initiation of migration by aphids subjected to the minimum of experimental interference. Further support for the conclusions reached in the field was obtained in the laboratory, though the techniques used could not identify a migration threshold nor could they identify the relative involvement of contact with conspecifics and energy uptake. Other studies on aphids, however, have at least succeeded in the second respect.

The sycamore aphid produces entirely winged adults during spring and summer and only in late autumn are wingless adults produced, these being the sexual females that mate and lay the overwintering eggs. Many other aphid species, however, particularly those that live in relatively temporary habitats, such as on herbaceous plants, are polymorphic, producing both wingless (apterous) and winged (alate) viviparous females. This polymorphism is extremely convenient from an experimentalist's viewpoint. As winged individuals of these polymorphic species are usually, though not always, more likely than wingless individuals to initiate migration to another habitat, then the entomologist is provided with an immediate and obvious manifestation of whether or not an individual aphid's migration threshold has been exceeded during early development.

It seems likely that, as suggested by Shull (1938) for the potato aphid, *Macrosiphum euphorbiae*, and Johnson and Birks (1960) for the cowpea aphid, *Aphis craccivora*, that all parthenogenetic females are potential alatae during their embryonic or first instar phases of ontogeny. The course of development of these individuals thus seems to be dependent on the conditions prevailing during these early stages of development, perhaps even as late as the second instar.

Parthenogenetic wingless adult females of the vetch aphid, *Megoura viciae*, have been reared singly on plants and then confined for twenty-four hours in empty tubes, some singly, others in groups of ten (Lees 1961). They were then released singly onto host-plants. Any effect of food uptake on migration threshold can therefore be excluded. The females that had been kept in groups of ten produced some

Fig. 16.11 Wing polymorphism in the rose aphid, *Macrosiphum rosae*

Both of the aphids in the picture are mature females. One is winged, the other is not.

[Photo by R.R. Baker]

alatae amongst their offspring but those that had been kept singly produced none.

Not all polymorphic aphids are influenced by the conditions experienced by their mothers. Crowding adults of the alfalfa aphid, *Therioaphis maculata*, did not alter the proportion of alatae produced (Toba *et al.* 1967). However, alatae are produced if the young larvae are crowded for more than 12 hours in the first 18 hours after their birth. The effect seems to be the result of physical contact between living larvae. Pheromones do not seem to be involved. Although the migration threshold for this species is exceeded only after several hours of 'crowding', parthenogenetic adults of the cowpea aphid, *Aphis craccivora*, produce alatae among their offspring if they are in physical contact for only two minutes (B. Johnson 1965).

The studies of aphids seem to come the nearest to supporting the suggestion (Fig. 8.2) that migration thresholds will be found to be the resultant of a given value of habitat variable and the length of time of exposure to that value. The dynamic relationship between these two components of a simple migration threshold in any single species remains to be fully evaluated.

It will be noted that so far the initiation of migration in response to deme density has been discussed solely in relation to the initiation of migration by flight from the natal leaf, stem or branch. In Chapter 8 it was suggested (p. 48) that the value of $\bar{E}$, the mean expectation of migration, on which selection should act in relation to the migration threshold is that value of $\bar{E}$ relevant to the particular migration. An aphid that initiates migration from its natal leaf or branch can either migrate: (a) to another leaf or branch on the same tree; or (b) to another leaf or branch on another tree but in the same group of trees; or (c) to another leaf or branch on another tree in a different group of trees but in the same valley; and so on. The value of $\bar{E}$ is different for each of these migrations, and of those described is likely to be least for migration (c) and greatest for migration (a). Let us consider this variation in relation to the migration threshold to deme density. The threshold density is likely to be least for migration (a) and greatest for migration (c), according to the arguments given previously. However, a developing aphid experiences only the deme density that exists on and around the natal leaf, twig or branch. When migration by flight is initiated from the natal leaf, etc. in response to the lowest migration threshold, it should ideally be a migration of type (a). Migrations of type (b) should be initiated only in response to a much greater deme density and, furthermore, a deme density that is relevant to the natal tree, not simply the natal leaf. Similarly, migrations of type (c) should be initiated only in response to a still greater deme density and, furthermore, a deme density that is relevant to the natal group of trees, not simply the natal tree or leaf.

The question presents itself, therefore, of what behavioural strategy is likely to evolve in response to these postulated selective pressures. Clearly, optimum strategy is for aphids first to initiate migration by flight from the natal twig or branch when the migration (a) threshold is exceeded. If the deme density on the natal tree does not exceed the migration (b) threshold, then the aphid should settle elsewhere on the natal tree. If the deme density on the natal tree does exceed the migration (b) threshold, then the aphid should initiate migration to other trees in the natal group of trees. If, however, the deme density in the natal group of trees exceeds the migration (c) threshold, then the aphid should initiate migration away from the natal group of trees to another group of trees.

Dixon (1969) in his field study on the sycamore aphid, *Drepanosiphum platanoides* (p. 225), captured, by means of a suction trap, aphids migrating away from (and towards?) the study group of trees. In addition sticky traps were mounted 3 m from the ground level at the edge of the canopy of two trees within the study group of trees. By taking the ratio of suction trap to sticky trap captures of aphids, some measure is possible of the incidence of migration within the canopy of the trees and, presumably, between trees (migrations (a) and (b) above) relative to the incidence of migration between groups of trees (migration (c) above).

The situation is complicated by fluctuations that can probably be attributed to variation in migration threshold with adult age and also variation in relation to migration-cost variables (wind, rain, temperature, etc.). In addition, the migration threshold seems to show a gradual increase from spring to autumn. Nevertheless, some clear trends emerge that on the whole support the suggestion that the sycamore aphid at least has evolved the strategy derived above. First, migration between groups of trees seems to be preceded by a period of migration within the group of trees. This migration within the canopy of a single tree and between adjacent trees leads to a redistribution of aphids within the canopy. Not until redistribution has taken place is there a marked increase in migration between groups of trees. Furthermore, migrations of types (a) and (b) seem to give way to migration of type (c) more rapidly at higher deme densities than at lower deme densities.

These findings can be extrapolated, with some reservations, in an attempt to consider the behaviour of an individual sycamore aphid. The observed trends would be consistent with an aphid initiating migration from the natal branch when the migration (a) threshold of deme density is exceeded. The individual then searches (see p. 198) the canopy of the natal tree for a leaf or branch with a sub-threshold deme density. If such branches are available, then redistribution occurs within the natal tree. Quite possibly, the migration threshold for migrations of types (b) and (c) not only involve deme density as for migration (a) but also involve search time (p. 82). Hence an aphid, having initiated type (a) migration, but finding no branches with a sub-threshold density, beyond a certain threshold search time initiates type (b) migration and beyond a longer threshold search time initiates type (c) migration. However, the time and energy involved in an aphid sampling even just one tree could be considerable, and any individual that senses environmental clues that correlate well with deme density on the tree or group of trees as a whole and that shorten the search time would be likely to be favoured by selection. Such clues would, therefore, be likely to be incorporated into the type (b) and type (c) migration thresholds and/or the search strategy.

One possibility is that at particular seasons some levels of the canopy are more likely to possess sub-threshold deme densities than other levels (it should be noted from Dixon's work, however, that the threshold density seems to differ at different canopy levels and it seems likely that the migration threshold does in fact consist of a number of interacting habitat variables, not only deme density). Quite possibly, therefore, an aphid can direct its search to that canopy level that at that season is most likely to have a sub-threshold density of aphids. A second possibility is that an aphid can assess the probability of being able to find a sub-threshold deme density on a particular tree or group of trees without actively searching the trees themselves. It seems quite likely that the chances of finding a branch with an aphid deme of sub-threshold density correlates with the density of aphids in flight, either within the canopy of a single tree or between trees. In this case,

selection is likely to favour those individuals with a migration threshold, for migration of type (b) and type (c), that includes or perhaps consists entirely of rate of sensory contact with other flying aphids. It should be noted that although, if this migration threshold exists, the behaviour of the aphids would be 'epideictic' of the type envisaged by Wynne-Edwards (1962), it is divorced from any connotations of group selection.

The sycamore aphid has in the past been considered to be 'non-migratory' in comparison with those 'migratory' aphids adapted to more temporary host plants (Southwood 1962). Although this distinction can no longer be accepted (p. 246), it is worth considering the question of whether aphids adapted to more temporary habitats have also evolved the hierarchy of migration thresholds corresponding to different values of $\bar{E}$ as suggested for the sycamore aphid. It should be stressed, however, that we are here only concerned with aphids migrating between similar habitat-types (e.g. from summer habitat-type to summer habitat-type). Many aphid species, particularly those that during the summer months feed on herbaceous plants, nevertheless pass the winter months on woody plants, usually as an egg (p. 242). The arguments concerning a hierarchy of migration thresholds do not apply to those individuals (which are obligatory migrants—p. 241) that initiate migration from the summer host to the winter host or *vice-versa*. As far as these individuals are concerned, no part of the natal site, neither the natal plant nor the natal group of plants, is a potential destination and there is much more likely to be a single mode, rather than several, to the distribution of values for hM (h = habitat suitability; M = migration factor). Consequently a single value for $\bar{E}$ ($\bar{E} = \bar{hM}$) is a reasonably acceptable measure of the expectation of migration. This single value is a function of the mean distance between summer-type habitats and winter-type habitats.

Consideration of the evidence that aphids adapted to herbaceous plants also have a hierarchy of migration thresholds in summer should perhaps start with a consideration of the habitat variables involved in the migration threshold for the initiation of migration from the natal leaf or plant. One of the disadvantages involved in wing polymorphism as an adaptation to facultative migration is that for mechanical reasons the development of wings has to be initiated early in life some time before migration can take place. Yet the fluctuation of habitat suitability in response to which facultative migration is an adaptation is, as postulated, unpredictable. It must frequently happen, therefore, that at the time wing development must be initiated, h_1 (habitat suitability) is less than $\bar{E}$ (the mean expectation of migration) but that by the time a winged adult capable of migration has developed, h_1 is greater than $\bar{E}$. Clearly the optimum strategy is for the facultatively produced winged aphid to be a facultative migrant, not an obligatory migrant. Evidence that this strategy has evolved in at least two species has been produced by M.J.P. Shaw (1970a,b) for the black bean aphid, *Aphis fabae*, and by Dixon *et al.* (1968) for the vetch aphid, *Megoura viciae*.

Wingless parthenogenetic females of the vetch aphid were induced to give birth to winged offspring by crowding them in small vials for 24 hours. These offspring were then reared in different ways until just before they became adult. Some were raised in isolation and some were raised in crowded conditions. Just before becoming adult both groups were then either isolated or each individual was crowded with 30 third-instar aphids. The proportion of each that subsequently initiated migration was then recorded. It was found that regardless of nymphal experience, crowding acting directly on the adult resulted in three times more winged individuals flying before they reproduced than when they were kept in isolation.

The black bean aphid shows a similar response but to conditions that exist somewhat earlier than the adult stage. Winged adult parthenogenetic female black bean aphids show a range of behaviour. Some initiate migration before depositing nymphs, others deposit some nymphs before initiating migration, and others never initiate migration. When early third-instar alate aphids are isolated, 60 per cent subsequently fail to initiate migration. When late third-instar alate aphids are isolated, 47 per cent fail to initiate migration. However, when late fourth-instar alate aphids are isolated, only 9 per cent fail to initiate migration.

There is some evidence that even those parthenogenetic black bean aphids that do initiate migration, in the first instance initiate only the type of migration that involves movement from one plant to another in the same patch or in the same localised area (e.g. garden to neighbouring garden). This can be inferred from the observation that parthenogenetic black bean aphids migrating from one plant to another of the summer host species, in this case broad bean, *Vicia fabae*, are prepared to settle and examine a leaf presented to them after a flight of only

10–30 s (B. Johnson 1954, Kennedy and Booth 1963a). The probability that the aphid will settle and reproduce on the leaf is a function of the duration of the migration flight (Johnson 1958, Kennedy and Booth 1963a) such that the shorter the duration of migration (and therefore search-time?) the less likely is the aphid to settle permanently on the first leaf encountered (see p. 82). It is also a function of the suitability of the leaf. Johnson (1958) examined the response of aphids to hosts of different suitabilities after migration flights of 30 s. Aphids alighting on paper remained on average only 1 min 30 s before initiating further migration. Aphids alighting on ivy, *Hedera helix*, a non-host species, remained on average only 2 min 9 s before initiating further migration. 100 per cent of aphids alighting on mature leaves of the host plant initiated migration within an hour. 88 per cent initiated migration within an hour of alighting on a young shoot of the host species, but only 3 per cent initiated migration within an hour of alighting on a seedling less than 6 cm in height. Furthermore, only 10 per cent of those alighting on a young shoot stayed overnight and larviposited but 82 per cent of those that alighted on the seedling did so. Kennedy and Booth (1963b) repeated Johnson's investigation using a rather more sophisticated technique and effectively confirmed his findings. However, the critical experiment of the presentation of the *most* favourable habitat was not carried out in this later work. Young folded bean leaves were not presented to the aphids for the unusual reason that such leaves are 'powerfully arresting'.

The termination of migration by parthenogenetic females of the summer generation of black bean aphids is therefore a function of the search-time and the suitability of the available habitats, a finding that also provides strong support for the thesis developed (p. 82) in relation to optimum stay-time. For present purposes, however, the critical fact is that aphids that initiate migration are prepared to examine potential destinations after migration times of only 10–30 s. H.J. Müller (1953) noted that black bean aphids in the field alight and take off again frequently. Davidson (1927) and Moericke (1941, 1955), however, noted that such flights near the natal plant or group of plants are not continued indefinitely and that migration away from the area may soon be initiated.

All of these observations taken together and in association with the sycamore aphid data are consistent with the suggestion that facultative migrant aphids have a hierarchy of migration thresholds, each corresponding to a different mean expectation of migration. These thresholds are such that initiation of migration from the natal plant or twig, etc., is in the first instance directed to nearby plants or twigs. Only when higher migration thresholds are exceeded, thresholds that are possibly based on rate of sensory contact with other flying aphids as well as search-time, are longer and longer-distance migrations initiated. The time interval between initiation of short-distance and longer-distance migration by an individual need not be long. Selection presumably favours those individuals that initiate longer-distance migration as quickly as is consistent with assessment of the probability of not finding a suitable habitat near to the place of origin.

Although deme density and some other habitat variables seem to be implicated in the facultative initiation of migration of many insects (reviewed by C.G. Johnson 1969), there do not seem to be any other examples that permit as extensive an evaluation of the situation as those described above for aphids.

16.4.2 The obligatory initiation of non-calculated removal migration

The main aim of this section is to provide a few examples of animals that initiate non-calculated removal migration at a particular stage of ontogeny, time of day or year, or in relation to optimum time-investment strategy, without apparent reference to habitat variables (but with reference to migration-cost variables—p. 49—and in some cases also to indirect variables—p. 265). It is inevitable that this section should consist entirely of examples of invertebrates. Most vertebrates live within a familiar area that is greater than their sensory-range area (Chapter 18). For reasons already presented (p. 64), such animals should always initiate exploratory migration before non-calculated removal migration, the latter behaviour being initiated only if exploratory migration fails to find a habitat of suitability greater than $\bar{E}_n$, the mean expectation of non-calculated removal migration. In most vertebrates, therefore, although the initiation of exploratory migration may be obligatory, the initiation of removal migration is likely to be facultative. In invertebrates, for many of which the familiar area is likely to be coincident with their sensory-range area,

the initiation of non-calculated removal migration may be obligatory.

As with most cases of undoubted non-calculated removal migration, the majority of examples of the ontogenetically obligatory initiation of migration are drawn from insects. Furthermore, most of these examples are concerned with migration by flight. Before presenting a few examples of these migrations, however, it seems worthwhile to point out that essentially similar, though in terms of numbers and distance much less spectacular, migrations are performed by walking. Such migrations are often facultative when, for example, large numbers of caterpillars, having denuded suitable vegetation in their natal habitat, migrate en masse and may or may not find an alternative suitable habitat, thus giving rise to popular accounts of 'plagues' of caterpillars (Fryer 1937). Ontogenetically obligatory migrations are also performed, however, by the larvae of insects and other invertebrates. The first example, however, concerns a migration by walking by an adult insect.

This example, which concerns the copulation site to copulation site migration of a fly, the sepsid, Sepsis cynipsea, is presented, not because it has been analysed in great detail, but because, viewed in conjunction with the example of the yellow dung-fly presented in the previous section (p. 226), it illustrates the way that slightly different conditions can make the difference between selection for facultative migration in one case and selection for obligatory migration in the other. This example is also taken from the work of Parker (1972a,b). Sepsis cynipsea is one of the commonest flies around cattle droppings in the Palearctic region. It is a small shiny black fly (about 2·5–3·0 mm) which has the black apical wing spots (one on each wing) typical of the genus. S. cynipsea in Britain takes over from the yellow dung-fly during the summer as one of the most dominant dung insects and shows many similarities in its reproductive ecology. Males arrive at the female's oviposition site (fresh dung) first and in large numbers and seize the females as they arrive to oviposit. In contrast to the dung-fly, however, the pair enter a passive phase first while the female lays her eggs.

Migration from the dung surface to the surrounding grass, which takes place after oviposition, also differs from the dung-fly migration in that it is obligatory, not facultative, and is initiated by the female, not the male. Also unlike the dung-fly, avoidance of 'take-overs' does not seem to be the critical selective pressure. Analysis similar to that for the dung-fly shows that in *S. cynipsea*, remaining on the dung surface would be likely to delay the termination of copulation, as a result of take-overs, by about 2·7 min. Migrating to the surrounding grass, however, gives a mean expectation of delay of the termination of copulation of 9·5 min. The nett result of migration to the female would be a time loss of at least 6·8 min. Although it may be a contributory factor, therefore, avoidance of time-loss due to take-over cannot alone account for the obligatory migration by the female from dung to grass before permitting copulation.

Parker considers, therefore, that the main selective pressure involved on this relatively small insect is avoidance of predators. Pairs and individuals of *S. cynipsea* on the dung surface have been observed to fall prey to staphylinid beetles (probably *Philonthus* spp.), and also occasionally to vespine wasps (*Vespula* sp.) and to the yellow dung-fly, *Scatophaga stercoraria*. In contrast, no examples were observed of predation upon individuals or pairs some distance from the dropping and it seems likely that the advantage of migration from dung to grass in avoiding predation would outweigh the disadvantage of a longer *copula* duration. If this is always so, then once oviposition ends, h_1 is always less than $\bar{E}$ (in relation to this migration) for *S. cynipsea* and the

situation is just that suggested to be necessary for selection to favour obligatory migration (p. 54).

The next example is an ontogenetically obligatory migration performed by the larva of the small white butterfly, *Pieris rapae* (see p. 243). This species lays its eggs singly on the underside of the leaves of a variety of cruciferous plants, but primarily (Richards 1940) on the various types of *Brassica oleracea* (i.e. cabbage, broccoli, cauliflower, kale, etc.). When the young larva hatches, it first eats its eggshell and then feeds on the leaf in the immediate vicinity. The first, second and first half of the third instar are normally spent within a few centimetres of the place where the individual was first deposited as an egg. Unless, that is, the leaf is very old or in other ways unsuitable. Towards the end of the third instar, however, the larva initiates a migration which takes it down the petiole of its natal leaf to the main stem of the plant and then upwards. In many 'open' varieties of *B. oleracea* this migration takes the larva to the young, not yet unfurled, leaves that form the crown of the plant. In other varieties, however, which have 'hearts', such as cabbages, the migration takes the larva into the heart (Fig. 16.13). In both cases, therefore, the moult to the fourth instar takes place in a relatively sheltered position. On a cabbage the remainder of larval development takes place in the heart where the larva is subjected to much less

Fig. 16.12 Copulating and ovipositing pairs of the sepsid fly, *Sepsis cynipsea*   [*Photos courtesy of Dr. G.A. Parker and the editors of* Behaviour.]

predation by birds (Baker 1970). On 'open' plants, however, the remainder of larval development takes place on the upper surface of the youngest but unfurled leaves. This migration by third-instar larvae of *Pieris rapae* seems to be obligatory and to be performed by more or less 100 per cent, at least of the British deme. On the other hand the migration is totally absent in both the closely related green-veined white, *P. napi*, and the large white, *P. brassicae*.

The larvae of all three of these species, however, in common with many other species of Lepidoptera, show the second type of ontogenetically obligatory removal migration manifested by larvae of *P. rapae*. At the end of the fifth instar, when the larva is full grown and feeding has been terminated, migration

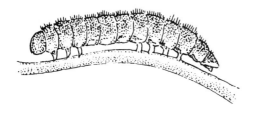

Fig. 16.13 The migration route of the third-instar larva of the small white butterfly, *Pieris rapae*

Female small whites lay their eggs singly on the underside of an outside leaf of a cabbage. When the resulting larva enters its third instar, it performs an obligatory migration to the cabbage 'heart' as indicated by the white arrow.

[*Photo by R.R. Baker*]

to a pupation site is initiated. In most individuals, this migration involves crawling down the main stem of the natal plant, crawling across the ground in a more or less straight line for a period that also seems likely to be obligatory and to be controlled by a clock mechanism, and then, at the end of this time, crawling up the nearest suitable structure (e.g. wall, fence, hedge), upon which pupation occurs. Once again, the migration seems to be obligatory for the majority of individuals. There are, however, a number of larvae that pupate on their natal plant. In Britain, the migrant/non-migrant ratio varies with the season, more larvae pupating on their natal plant in summer than at any other time of year. In the late autumn nearly 100 per cent of larvae migrate. It seems likely that the initiation of this pre-pupation migration from the natal plant is associated with the appreciation of indirect variables, such as the photoperiod/temperature regime. However, the possibility that habitat variables are also involved and that the migration, for a proportion of individuals, is facultative cannot at present be ruled out entirely.

Occasionally, in the autumn, third- or fourth-instar larvae of *P. rapae* may be observed to leave their natal plant and perform what can only be interpreted as a pre-pupation migration of the type just described for fifth-instar larvae. Such individuals are found to be parasitised by the solitary larva of the braconid hymenopteran, *Apanteles rubecola*. When the young *Pieris* larva arrives at what would, for a fully grown fifth-instar larva, be a pupation site, the *Apanteles* larva completes its development and kills its host through the body wall of which it bores to reach the outside world. It then spins a cocoon which it attaches to the wall, stick or other substratum selected by the *Pieris* larva, and pupates. Soon after, the now shrivelled body of the *Pieris* larva that transported the parasite to its pupation site falls to the ground. It seems likely that the *A. rubecola* larva, having reached the appropriate stage of development, in some way changes the physiology of its host, possibly by influencing its hormonal state, such that the behaviour patterns are initiated that are normally characteristic of a much later stage of the host's larval development.

Such ontogenetically obligatory migrations are not confined to insect larvae. Nematodes of the genus *Trichostrongylus* which are parasites of the gut of sheep, *Ovis* spp., produce eggs which pass out with the host's faeces. The first two larval instars are free-living. Third-instar larvae, however, initiate a migration that takes them from the faeces to a position on the surrounding grass from which a new host is likely to be infected (Crofton 1948). This migration seems to be ontogenetically obligatory.

Perhaps the most spectacular obligatory migrations that do not involve flight, however, are those of various species of marine polychaetes that spend most of their lives on the sea bottom but that perform a migration to the upper levels of the sea as a preliminary to reproduction. In a number of species males and females congregate at the surface of the sea to spawn and in some cases spectacular swarms are formed. In the Atlantic palolo worm, *Eunice fucata*, in the West Indies the worm backs out of its burrow at about 03.00 to 04.00 hours of the morning of the migration. The posterior portion (epitoke) then breaks off and migrates to the surface, the gametes being released when the Sun rises (Clark and Hess 1940). These obligatory pre-spawning migrations from sea-bottom to sea-surface are both ontogenetic and seasonal and in some species, particularly the palolo worms, are famous for their precise timing and lunar periodicity. *Eunice fucata*, for example, swarms on two or three consecutive nights close to the third lunar quarter in July, though if the third quarter falls late in the month, spawning takes place at the preceding first quarter (Clark and Hess 1940). *Eunice viridis*, on the other hand, swarms in October or November but never on more than three consecutive nights (Caspers 1961). The range of synchronous spawning behaviour shown by polychaetes has been reviewed by S.M. Evans (1971).

Before describing a few of the many observations of aerial migrations which involve flying insects, there is one other noteworthy aerial migration that is performed by arthropods. This migration is performed by various species of spiders and the migrants are known alternatively as gossamer spiders or as aeronaut spiders. The migration is initiated when the spider mounts the top of the vegetation and raises its abdomen as high in the air as possible. A drop of silk is squeezed out which is then pulled out by the wind into a strand. When the spider releases its foothold the wind then carries it away (Bristowe 1939). This migration seems likely to be ontogenetically obligatory when performed by young spiders soon after hatching and possibly also when performed by adults just before reproduction. Presumably, however, the migration mechanism can also be employed for facultative migrations.

The larvae of some insects are also occasionally found among the aerial plankton (C. G. Johnson 1969). However, the young larvae of many insects, particularly Lepidoptera, more or less continuously produce silk which they attach to the surface of the leaves and stems etc. of the plant on which they are feeding. When the larva is dislodged this strand of silk serves as a life-line up which the larva can climb and re-attain its original position on the plant. It is by no means certain, therefore, that all of the larvae that appear in the aerial plankton are non-accidental migrants of the spider type. It is possible that many happen to have a strand of silk attached as a result of their normal station-keeping mechanism and are therefore effectively accidental migrants (Chapter 5). When such larvae belong to species with wingless or brachypterous females, however, and/or the larvae are extremely hairy, thus facilitating wind displacement, there would seem to be a strong possibility that the migration is non-accidental. More work is required before the nature of the migration can be assessed with any confidence.

As far as insects are concerned, ontogenetically obligatory migrations have, in the previous few paragraphs, been shown to occur at various stages during early development. The most conspicuous ontogenetic transition experienced by the majority of insects, however, is that from an animal incapable of flight to that of an animal capable of flight. Suddenly, migrations become possible that were previously impossible by virtue of distance. The onset of the ability to fly is often associated with the obligatory initiation of removal migration. Migration, whether facultative or obligatory, at this stage of ontogeny will be referred to as post-teneral migration.

Entomologists frequently use the term 'teneral' to describe the period that immediately follows adult emergence and during which the cuticle hardens and flight becomes mechanically possible. As argued by C.G. Johnson (1969, p. 51), however, the teneral period cannot be precisely defined. Use of the term represents convenience of description rather than an attempt at great precision. In this book I therefore propose to follow Johnson and describe as a post-teneral migration any exodus made relatively soon after emergence on the first flight or on one of the very early flights. The length of the teneral period varies considerably from insect to insect. It ranges from a few seconds in many dipteran flies, particularly those with aquatic larvae, through a few hours in most butterflies, to several days in locusts.

The selective pressures acting on the length of the teneral period are likely to be diverse but are likely to include selective pressures connected with the optimum time for migration. However, migration by flight cannot be earlier than post-teneral according to the definition adopted here and the question of interest in the present discussion is how soon after flight becomes mechanically possible is a particular type of migration initiated. Very often the first flight made by an insect is a relatively long-distance removal migration, and in these instances the migration is often likely to be obligatory, though perhaps equally often the migration physiology may have been triggered by above-threshold values of habitat variables during larval development.

In the black bean aphid, *Aphis fabae*, for example, at least some of the winged females of the parthenogenetic summer generation are 'obligatory' migrants, initiating migration irrespective of the suitability of the habitat with which they are experimentally put into contact (B. Johnson 1958). Not until flight of 10–30 s has occurred, which takes the female away from its natal plant, are other plants searched and tested for suitability. However, the proportion of a particular generation that 'obligatorily' initiates migration depends in a facultative way on the sequence of deme densities encountered during larval development up to and including the fourth instar (p. 231).

The occurrence of post-teneral migration in insects has been reviewed by C.G. Johnson (1969) and only a few examples are presented here. There seems little doubt, however, that where *ontogenetically* obligatory migration (by flight) occurs in insects it is most commonly post-teneral. This observation is discussed further in Chapter 17.

Adults of the dragonfly, *Hemigomphus heteroclitus*, have been observed to crawl out of their exuviae and within ten minutes to initiate migration away from their natal water body to a nearby forest (Tillyard 1917). There is a possibility, however, that this migration is exploratory rather than a non-calculated removal migration (p. 198).

Newly moulted adults of the desert locust, *Schistocerca gregaria*, are soft-bodied and do not fly, or fly only feebly (Kennedy 1951). Older adults of phase *gregaria* are harder-bodied, though still relatively soft, and can fly readily but only for short periods and the swarm does not initiate migration. Groups of locusts may 'stream' but do not leave the area of the swarm and readily settle if wind-speed increases. As ageing and cuticular hardening continues, flight becomes more vigorous, the swarm 'surges', and begins to migrate, for short distances at first and then further and further. The period between emergence and full migratory behaviour is relatively long and extends over several days. Nevertheless, the sequence seems to be obligatory.

The female reproductives of all social insects (p. 148) are likely to perform removal migration from the natal colony, though in the honey bee, *Apis mellifera*, this removal migration is neither non-calculated nor necessarily post-teneral (p. 332). In the majority of social insects, however, but particularly ants and termites, the migration is markedly post-teneral and involves males and females. Male and female winged reproductives accumulate in the colony and show no sexual behaviour. There seems to be an abrupt migration threshold to migration-cost variables which results in not only many individuals from the same colony but also from different colonies initiating migration at the same time. Ants often initiate copulation in flight, land on the ground and then the female sheds her wings before attempting to establish a new colony (Talbot 1964). Termites also shed their wings when they alight on the ground which, unlike many ants, is also where mating takes place (Harris and Sands 1965).

'Thousands' of individuals of the painted lady butterfly, *Cynthia cardui*, were observed by Skertchly (1879) emerging from pupae near the western edge of the Red Sea. Within an hour of emergence the butterflies took off more or less simultaneously and flew with a peak direction between (toward) E and NE. The butterfly, *Catopsilia pyranthe*, in Sri Lanka, is reputed to initiate migration soon after adult emergence in no matter what part of the island emergence happens to occur (Manders 1904).

Newly emerged but nectar-fed adults of the salt-marsh mosquito, *Aedes taeniorhynchus*, fly straight upwards during the initiation of migration. The flight levels off at just above tree-top level (Fig. 16.14).

It should not be concluded from the above examples that post-teneral migration is the only ontogenetically obligatory migration that is found among adult insects. There are at least two factors, however, that have led in the past (e.g. C.G. Johnson 1969) to this type of migration receiving particular attention from entomologists. First, the majority of insects, in common with the vast majority of animals, are more likely to perform removal migration pre-reproductively than after reproduction has begun (Chapter 17). Consequently, as in most

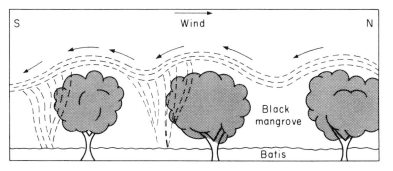

Fig. 16.14 Obligatory migration by the salt-marsh mosquito, *Aedes taeniorhynchus*

The individuals involved in the migration are newly emerged but have fed on nectar before migrating. Males and females are represented in more or less equal numbers and the migration is accompanied by a high-pitched humming of millions of wingbeats. The humming ceases suddenly when most individuals leave the immediate locality.

[*After Haeger (1960)*]

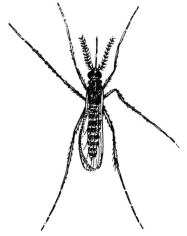

insects reproduction begins relatively soon after the teneral period, this inevitably means that more migration is post-teneral than post-reproductive. Secondly, the end of the teneral period, even though it is not an easily defined stage of ontogeny (p. 235) is much easier than other stages to recognise in the field. Yet insects do pass through a number of ontogenetic phases, though these may not be clearly demarcated, particularly in males for which the end of the teneral phase and perhaps the ontogenetic (as opposed to diel) onset of the search for females are more or less the only landmarks. Copulation occurs too unpredictably for each copulation to be considered an ontogenetic stage. Females, on the other hand, particularly those of species that lay egg batches at regular intervals, may have a more recognisable sequence of ontogenetic phases (i.e. teneral phase, pre-copulation phase, pre-oviposition phase, and then an alternating sequence of oviposition/inter-oviposition phases). Each of these phases may be associated with characteristic migration behaviour. Where this occurs, the migrations are obligatory in that they are part of optimum time-investment strategy. It should also be noted here, though it is discussed in more detail later (p. 249), that obligatory post-teneral migration is also very

often the result of selection connected with the change in the animal's resource requirements that occur at this stage rather than selection due to predictable changes in habitat suitability that result from changes in the habitat itself. It is most important to distinguish between these two types of selection in any consideration of the adaptive nature of a particular migration sequence.

The following two examples, the peacock butterfly, *Inachis io*, and the small tortoiseshell butterfly, *Aglais urticae*, will serve to illustrate the variety of obligatory migrations, other than post-teneral, performed by a species in the course of its lifetime. These examples (Figs. 16.15 and 16.16) are based on continuing research on the species in southern England, results of part of which have been published (R.R. Baker 1972a). They show the way that a sequence of obligatory migrations associated with time of day, season and optimum time-investment strategy combine to produce the cross-country removal migration of an insect species. The main emphasis in this section, however, is on the initiation of each migration. Other components of the migration, such as distance and direction, are discussed elsewhere (pp. 422 and 435).

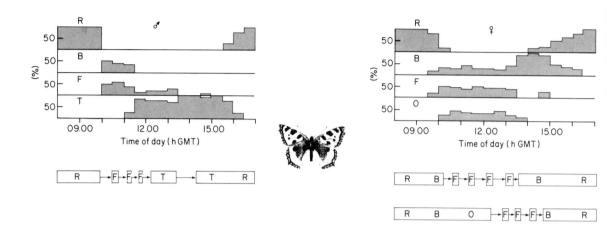

Fig. 16.15   The diel pattern of time investment in spring by males and females of the small tortoiseshell butterfly, *Aglais urticae*, in southern England

The histograms show the percentage of individuals counted at various times of day (30 min units) engaged in different behaviour patterns. R = roosting; B = basking (thermoregulation); F = feeding; T = territorial behaviour; O = ovipositing. The horizontal bars that form the lower part of the figures indicate diagrammatically the diel sequence of behaviour (assuming continuous sunshine) in 'average' individuals. Each arrow indicates a migration between two habitats (boxes). The letter inside the box refers to the predominant behaviour or behaviour sequence performed in each habitat. Note that the number of feeding habitat to feeding habitat migrations is likely to be highly variable and to be a function of the suitability of the habitats visited. In the diagram for females, the upper sequence refers to females on days between ovipositions and the lower sequence refers to females on days that oviposition occurs.

In the morning, as soon as migration-cost variables are above threshold, males immediately initiate migration from the roosting habitat. The roosting site in spring is usually a patch of nettles, *Urtica dioica*, though other plant species, such as docks, *Rumex* spp., may also be used. This first diel migration, which may cover tens, hundreds or thousands of metres, terminates in a suitable feeding habitat. After a certain stay-time in each feeding habitat, further migration is initiated. Migration from feeding habitat to feeding habitat continues throughout the morning until, from mid-day onwards, migrations terminate not at a feeding habitat but at a suitable territory. However, if a male *Aglais urticae* fails to associate with a female in a particular territory, a further migration is initiated. The critical stay-time in a fruitless territory seems to be about 90 min.

Some females (upper female sequence in diagram) behave as males when they emerge from roost and initiate migration as soon as migration-cost variables are above threshold. It is assumed that such females also terminate this first diel migration when a feeding habitat is encountered. Other females (lower female sequence in diagram), however, fail to initiate migration and instead remain at their nettle roosting habitat. When this happens, oviposition commences in middle or late morning. Females of this species lay a batch of eggs of a hundred or more on the underside of a terminal leaf of a young plant of nettle. Oviposition may last for 20 to 90 min, depending on temperature and number of eggs laid. If oviposition is interrupted, the female flies a few metres away and then after a few minutes returns to the original leaf and recommences oviposition. Topographical learning rather than scent marking seems to be involved, for if the plant is moved a metre or so while the female is away the female returns first to the original place and only after failing to find her eggs does she search around the area. It seems, therefore, that until an entire batch of eggs has been laid, the female's migration threshold is such that migration is unlikely to be initiated. As soon as the batch has been laid, however, the female immediately initiates migration from the roosting/oviposition habitat. When it has been possible to follow such migrants, sometimes over several hundred metres under favourable (for the author!) conditions, the migration has terminated at the first feeding habitat encountered but from which further migration has been initiated after a certain stay-time. Females of *A. urticae* in spring arrive at roosting habitats early in the afternoon, from 13.00 hours GMT onwards. Once the female has encountered a suitable roosting habitat after about 13.00 hours GMT, no further migration is initiated that day, the remainder of which is passed in basking (thermoregulation). Some feeding may occur if flowers are present, but the presence of adults' foodplants does not seem to be an essential part of habitat suitability, most feeding taking place in the morning and early afternoon.

Males establish territories at potential roosting habitats and intercept the females as they arrive. Once a female has been encountered, male courtship strategy involves maintaining close contact with the female for the remainder of the afternoon and attempting to prevent other males from doing the same. If the male is successful, copulation takes place immediately after the female goes to roost and lasts all night.

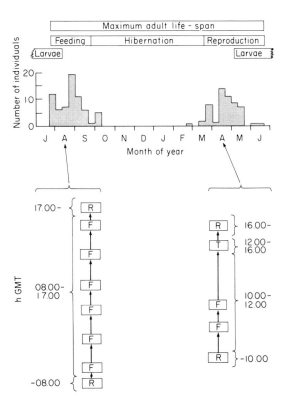

Fig. 16.16 Seasonal incidence and seasonal variation in diel patterns of time investment in the peacock butterfly, *Inachis io*, in southern England

The histogram shows the seasonal incidence of *I. io* in southern England. Number of individuals refers to the number performing migration counted by the author in 10-day units in the period 1966–1970. The horizontal bars above the histogram indicate the maximum adult life-span, the predominant behaviour during three major phases of adult life, and the time of year that larvae are to be found. The lower two parts of the figure indicate the diel migration sequence (assuming continuous sunshine) for males and females in autumn (August) and for males in spring (April). Information concerning the sequence for females in spring is not yet available. Each arrow indicates a migration between habitats (boxes). The letter inside each box indicates the habitat-type. R = roosting habitat; F = feeding habitat for adult; T = territory. Approximate times of day occupied by the different units of the behaviour/migration sequence are indicated.

In England, *I. io* is univoltine. Adults emerge from the pupa from mid-July onwards but do not become sexually mature before entering hibernation. Time investment during the period from adult emergence to hibernation is divided amongst roosting, feeding, thermoregulation, migration, and the examination of potential roosting and hibernation sites. At night and when the Sun is obscured by cloud, time investment involves roosting only. At this time of year (autumn), the roosting site is similar, if not identical, to the type of site that will be used for hibernation (e.g. a hole in a tree, any type of building providing a dark background, a barrel, or an upturned wheel-barrow). The diel behaviour sequence soon after emergence is as follows. When migration-cost variables are above threshold, a migration from the roosting site is initiated that terminates when a suitable feeding site is reached. Behaviour at the feeding site includes thermo-regulation, by spreading and closing the wings according to temperature, and the examination of potential roosting sites, as well as, of course, as feeding. After a certain stay-time (Chapter 11) at the first feeding site, a second migration is initiated, which terminates at the next feeding site to be encountered which may be a few metres or a few kilometres away. This behaviour is repeated throughout

the day but as the afternoon progresses stay-time in each feeding habitat increases and the emphasis is placed less on food abundance and more on the availability of potential roosting sites. Finally, depending on time of day and conditions of sunshine and temperature, the butterfly enters a roosting site. This diel sequence is repeated each day. Although more data are needed, marking experiments and direct observation in a group of rural gardens (area approximately 1 ha) containing abundant nectar (mainly *Buddleia* bushes) and roosting sites but surrounded by open dairy fields containing neither, suggest that stay-time in the area ranged (in actual time) from a few minutes to about three days with a mean (in sunshine hours) of about one hour. Stay-time seems to increase as both the day and the season progress. In England the species enters hibernation early in comparison with other closely related nymphalid butterflies. Few individuals are seen on the wing after mid-September.

I. io emerges from hibernation on more or less the first fine (warm and sunny) day from early March onwards. Diel obligatory migration then shows the following pattern until the individuals die some time in June. Once again, at night and other times the Sun is obscured, time is invested only in roosting behaviour. During spring and summer, however, the roosting site is on dry ground amongst fairly thick but broad-leaved vegetation. Dry ground, covered with ivy, *Hedera helix*, at the edge of a wood or row of trees is a favoured roosting site. Given a fine day, migration from the roosting site is initiated between 09.00 and 10.00 hours GMT. Males spend the morning migrating across country from one feeding site to another, stay-time at each habitat possibly being a function of the suitability of the habitat as a feeding area. At about 12.00 hours GMT a male ceases to settle at feeding sites but instead settles in the first unoccupied territory encountered. A territory is a sunlit area of low vegetation along the edge of a wood or row of trees where potential roosting sites are available. A territory need not necessarily contain adult foodplants (nectar-producing flowers) or larval foodplants (nettles, *Urtica dioica*). The male remains in the territory until either: (1) he leaves to follow a female (facultative); (2) he 'loses' a territorial interaction with another male, where-upon the male initiates (facultative) migration until a further unoccupied territory is encountered; or (3) either time of day (obligatory) or weather conditions (facul-tative) favour the male going to roost. The same sequence is repeated on the next fine day with the result that a male rarely occupies the same territory two days in succession.

[*Photo by Roger Gaymer*]

Each individual butterfly when it initiates mi-gration across habitats of low suitability that are larger than its sensory range always takes up the same direction relative to the Sun's azimuth (Chap-ter 19). The nett result, therefore, of the migration sequences illustrated in Figs. 16.15 and 16.16 is that the butterfly travels across country, often, during its life-time, covering a total distance of hundreds or even thousands of kilometres (Chapter 19). The above examples show, however, the way that this total migration consists of a series of shorter mi-grations, each one of which is initiated either when the habitat occupied by the animal decreases in suitability (below $\bar{E}$, the mean expectation of mi-gration?) or after a certain stay-time. Some of these migrations are initiated obligatorily on a diel basis as part, presumably, of diel variation in optimum time-investment components. Hence, roosting inevitably ceases to be optimum behaviour beyond a certain time of day. Instead, depending on sex and physi-ological state, as well as season, feeding or ovi-position becomes optimum behaviour. Yet, because for the above species the most suitable roosting habitats are spatially separated from the most suitable feeding habitats (see p. 249 for discussion), this diel change in optimum behaviour inevitably involves the obligatory initiation of migration. Simi-lar arguments apply to variation in optimum behaviour with season if seasonally optimum habi-tats are spatially separate. Migration between habitats of different types (e.g. roosting to feeding habitats, etc.), is therefore initiated when the animal's resource requirements change. Migration between habitats of the same type, on the other

hand, may be initiated either when h_1 becomes less than $\bar{E}$ or when a certain stay-time has elapsed (p. 82).

Some insects undoubtedly perform a single migra-tion, probably usually post-teneral, after which no further migration is initiated. The clearest examples of such a single migration are the female repro-ductives of various species of ants and termites. For instance, the winged reproductive individuals of the desert damp-wood termite, *Paraneotermes simplicornis*, accumulate in the termitarium for about three weeks before migration is initiated (Nutting 1966). They remain quiescent and aggregated until migration-cost variables, particularly temperature, are above threshold whereupon they become active, emerge from the nest, and initiate migration. In common with the majority of termites (Imms 1964), females lose their wings after the single migratory flight, autolyse their flight muscles, mate, become gravid, and found a new colony.

In the majority of insects, as with most animals, the total migration consists of a series of migrations divided up into diel and seasonal units. In many, perhaps the majority of insects, particularly females, the migration threshold also varies ontogenetically, most and longer-distance migration units being in-itiated pre-reproductively (Chapter 17). In many other species, however, as in the two butterfly species described in Fig. 16.15 and Fig. 16.16, migration continues throughout adult life. Females of the Moroccan locust, *Dociostaurus maroccanus*, phase *gregaria*, in Iraq initiate migration, joining in passing swarms, after each of the two to five bouts of oviposition (Bodenheimer 1944). The desert locust, *Schistocerca gregaria*, behaves similarly (Rainey 1963, Michel 1969, Waloff 1972). The black bean aphid, *Aphis fabae*, may initiate migration after each of the first two or three bouts of larviposition, as well as post-tenerally (C.G. Johnson 1969), though even-tually no further migrations are initiated and the flight-muscles autolyse. Females of the army cut-worm moth, *Chorizagrotis auxiliaris*, of North America, collected in the field after the immediate post-hibernation migration, initiated migration (as deduced by performance on a flight-mill) between successive bouts of oviposition (Koerwitz and Pruess 1964). Although more difficult to observe, houseflies and blowflies (muscids and calliphorids) seem to migrate across country between roosting, feeding, mating and oviposition sites in much the same way as described above for the butterflies, *Aglais urticae* and *Inachis io*. The frit fly, *Oscinella frit*, also continues to migrate between bouts of oviposition (Southwood *et al.* 1961).

The final part (before attempting some degree of analysis) of this presentation of examples of obligatory migration is concerned with the seasonal aspect of the obligatory initiation of migration. In the examples of the small tortoiseshell butterfly, *Aglais urticae* (Fig. 16.15) and peacock butterfly, *Inachis io* (Fig. 16.16), the seasonal aspect of migration is obscured to some extent in that seasonal effects are integrated into the diel pattern. Nevertheless, the *total* autumn migration sequence could be said to take the butterflies from their sites of larval development to their hibernation sites. Similarly, the *total* spring migration sequence could be said to take the butterflies from their hibernation sites to their death sites, reproduction taking place *en route*. The well-known migration (Urquhart 1960) of the monarch butterfly, *Danaus plexippus*, in North America is also often described in these terms, though the diel and seasonal sequence seems to be similar to that for *Aglais urticae* and *Inachis io*. In other species, however, particularly those for which flight is much less involved in attaining feeding, mating or roosting sites or for which such sites are contained within more-or-less a single habitat, description of the migrations in such simple seasonal terms may be reasonably adequate. The sunn pest, *Eurygaster integriceps*, a scutellerid bug, is a widespread pest of wheat in parts of Russia, the Middle East, and North Africa (Arnol'di 1955, E.S. Brown 1962, 1965). The species is univoltine (one generation/year) and each individual performs three seasonally obligatory migrations. The extent, if at all, to which each of these migrations is divided into diel or daily units remains unknown. Basically, the individuals become adult in the summer. They feed for nearly two weeks on the plains on which they developed and then initiate migration to aestivation sites. Near the tropics, these aestivation sites are on neighbouring mountains at altitudes of 2000 m or more. Later, there is a second, presumably obligatory, migration to a slightly lower altitude (1500 to 2000 m) at which hibernation occurs under fallen leaves and other litter. In more northerly latitudes the insects hibernate either on hills or on the plains at altitudes similar to those at which they breed. In the spring, first the males and then, slightly later, the females initiate migration to the plains on which copulation and oviposition occur.

The black bean aphid, *Aphis fabae*, in common with other host-alternating aphid species, also per-

forms seasonally obligatory migrations. The life-history of *A. fabae* is illustrated in Fig. 16.17. The initiation of migration between secondary host plants by parthenogenetic viviparous females of the summer generations has already been discussed in some detail (p. 230). In the autumn, however, winged obligatory (presumably) migrants are produced (i.e. males and parthenogenetic viviparous females), that settle on a plant of the primary host species, the males mating with the wingless oviparous females that are the offspring of the female migrants. In the spring the aphids that hatch from the overwintered eggs are wingless, parthenogenetic, viviparous females. After two or so generations, however, winged parthenogenetic viviparous females are produced that (presumably) are obligatory migrants and that settle on plants of the secondary hosts.

Other insects may also be similar to host-alternating aphid species in that the spring and autumn migrations are obligatory whereas the summer migrations are facultative. For example, E.C. Young (1963), in his study of the migration of corixid bugs (Heteroptera) considered that the spring migration by pre-reproductive individuals was obligatory but that summer migration by individuals at any stage of ontogeny was a response to extreme environmental conditions.

Most examples of seasonally obligatory migration in temperate regions are similar to the black bean aphid in that they involve migration between a habitat-type suitable for breeding or feeding and a habitat-type suitable for hibernation. In tropical regions, seasonally obligatory migration may involve migration between a habitat-type suitable for the dry season and a habitat-type suitable for the wet season. Seasonally obligatory migration may therefore occur either at the beginning of the dry season or at the beginning of the wet season, or both. The red locust, *Nomadacris septemfasciata*, for example, is found in regions of equatorial Africa where there is a long dry season. The species is univoltine, producing just one generation during the year, and breeds only during the wet season. Nevertheless, the species is adult for the major part of the year and passes the dry season in a reproductive diapause. It seems to be usual that the red locust does not initiate migration from the natal habitat for 3–6 months after becoming adult (Rainey *et al.* 1957). This period of reproductive diapause seems to be spent among long grass and total movement is short distance (1–2 km) only. At the end of the dry season, however, the

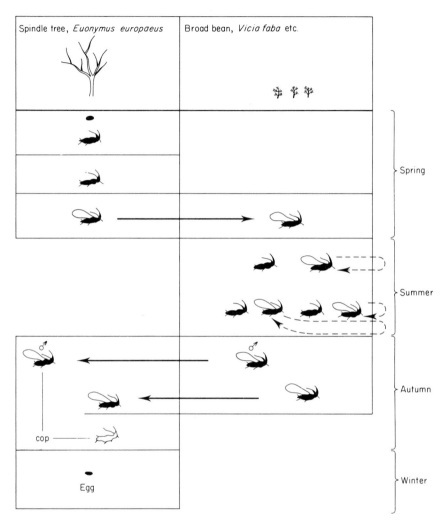

Fig. 16.17   The life-history of a host-alternating aphid species: the black bean aphid, *Aphis fabae*

The left-hand column indicates the part of the life-history spent on the primary host, the spindle tree, *Euonymus europaeus*, and the right-hand column indicates the part of the life-history spent on secondary host plants such as the broad bean, *Vicia faba*. The black oval shape at the top and bottom of the spindle tree column represents a fertilised egg. All aphids shown are females except for the two indicated. The presence or absence of wings in the adults is shown for each generation but during the summer the proportion of winged to wingless forms varies considerably and is not necessarily 1:1 as shown. Females shown in black are parthenogenetic and viviparous. The generation of females shown in white are oviparous and copulate with males as shown. Horizontal lines separate generations except during the summer when the number of generations is variable. Continuous arrows indicate obli-

gatory migrations. Dashed arrows indicate facultative migrations.

Aphis fabae overwinters as a fertilised egg on the spindle tree. In spring wingless viviparous females hatch from these eggs and in one or two generations produce winged viviparous females which are obligatory migrants and which larviposit on secondary host plants. A variable number of generations of viviparous females are produced during the summer, some individuals of which are winged and some of which are wingless. Winged individuals are produced facultatively and depending on conditions up to their fourth instar may or may not initiate migration when adult. In autumn, winged males are produced which are obligatory migrants and which settle on spindle trees. The same generation includes winged viviparous females which are also obligatory migrants. Having settled on spindle trees these migrant females produce a generation of wingless females that copulate with the males and then produce the eggs that overwinter. With minor modifications this seasonal sequence is applicable to many host-alternating aphid species.

reproductive diapause is broken and the locusts initiate migration first into the surrounding bush and then often tens, hundreds or thousands of kilometres to an area where rain has fallen and reproduction commences.

It is perhaps unnecessary to point out that the above examples of seasonally obligatory migrations should not be confused with other, conspicuously seasonal, migrations which are merely reflections of curves of seasonal incidence of the adult insect. For example, an observer looking for a migration of the orange tip butterfly, *Anthocharis cardamines*, in England would note that migration only occurs during April, May and June. However, the species is only adult during April, May and June. The seasonal peak of migration is, therefore, merely a reflection of the seasonal peak of incidence of the adult. A similar explanation undoubtedly accounts also for the peaks of migration of the large white butterfly, *Pieris brassicae*, reported by C.B. Williams (1958) and for many other examples.

The first part of this section on the obligatory initiation of non-calculated removal migration has been concerned primarily with a presentation of examples of some of the wide variety of migrations that seem to be obligatory. Little attention has so far been paid to the selective pressures that may have produced the different migrations. It is a consideration of some of these selective pressures that forms the second part of this section.

Perhaps the largest single class of selective pressures on obligatory migration are those associated with the spatial distribution of the habitat-types that are most suitable for different time-investment components. The simplest example of this was the sepsid fly, *Sepis cynipsea* (p. 232). The optimum habitat-type for oviposition for this species is fresh dung. The optimum habitat-type for copulation, however, seems to be grass. Yet optimum time investment appears to involve oviposition followed immediately by copulation. The result is selection for the obligatory initiation of migration from the dung surface to the surrounding grass as soon as oviposition is terminated. Similar considerations undoubtedly apply to the pre-spawning migration by some polychaetes, such as the palolo worms, *Eunice* spp. (p. 235). Here it is assumed that the sea-bottom is the most suitable habitat for all activities except spawning but that the sea-surface is the most suitable habitat for this latter activity. Furthermore, it is assumed that a particular time of year is optimum for spawning and that reproductive success of all

individuals is greatest when the individual performs the migration at the same time as most other individuals. Given these selective pressures, none of which have been demonstrated, a seasonal *cum* ontogenetic obligatory migration of the type observed would evolve.

Some other situations offer rather more hope that experimental evidence can be obtained to demonstrate the selective pressures that may have been involved in adaptation to spatially separate habitat-types. Such work is in progress on the migration (p. 233) of third-instar larvae of the small white butterfly, *Pieris rapae*, from the outer leaves to the heart of the larval foodplant, *Brassica oleracea*. The oviposition site chosen by the female *P. rapae* (i.e. near the edge of a leaf more or less as far as possible from the main stem), seems to be optimum for at least two reasons. First, it decreases considerably the risk of predation by house sparrows, *Passer domesticus*, and similar birds, the structure and size of which is such that they can only take eggs or larvae while they (the birds) are standing either on the ground or on the stalk of the cabbage leaf. From such positions most *Pieris rapae* eggs laid in their normal position are out of reach. Secondly, a position on a leaf away from the stalk necessitates any ground-dwelling arthropods that are potential predators of the egg searching a larger area than would an oviposition site near to the main stem. It has also been demonstrated (R.R. Baker 1970) that larval development in the heart of the plant is more advantageous than development on the outer leaves. The main advantage seems to be much less predation risk by birds. This advantage more than outweighs the disadvantage that results from a greater risk of succumbing to disease, both viral and bacterial, in the heart than on the outer leaves. There is some evidence, therefore, that there are selective pressures favouring oviposition on one part (H_1) of the plant and larval development on another part (H_2) of the plant. Such selective pressures must inevitably generate selection for migration as long as h_1, as well as being less than $\bar{h}_2$ (the mean suitability of H_2), is also less than $\bar{h}_2 M$ (i.e. as long as mean migration cost does not outweigh the difference in suitability between H_1 and H_2 as a site for larval development). So far, however, it has not been demonstrated that selection always favours migration being initiated during the third instar. This possibility is being studied by measuring the cost of performing the migration at different larval stages, and the effect of migration at different stages on the risk of predation and disease.

Most obligatory migrations that are associated with the spatial distribution of the habitat-types most suitable for different time-investment components are more complex than the above and unfortunately it seems likely to be some time before experimental evaluation of the selective pressures involved will be possible. The following few paragraphs, therefore, must remain speculative and non-quantitative. However, the principle seems likely to be the same whether the migration is simple or complex, namely that one habitat-type is most suitable for one time-investment component but another habitat-type, spatially separate from the first, is most suitable for another time-investment component.

Selective pressures generated by diel and seasonal environmental cycles seem often to have led to the association of different time-investment components with different times of day or year. Roosting, for example, depending on the species, may be the optimum time-investment component during daylight, or during dark, etc., and also during winter (hibernation) or summer (aestivation). Feeding and/or reproduction, however, are likely to be optimum time-investment components at times of day and/or year quite different from those times that roosting is optimum. If roosting, hibernation or aestivation habitats are spatially separate from feeding habitats and if all are spatially separate from reproduction habitats then seasonal, diel, and ontogenetic obligatory migrations of the types described are inevitable.

Many insects have larvae and adults that are adapted to quite different habitat-types. The clearest examples are those, such as dragonflies, in which the larvae are aquatic whereas the adults are not. Very often in such cases the immediate vicinity of the natal water body does not offer the optimum conditions for adult feeding. Less obvious, but equally real differences in adult and larval requirements are found among other insects, especially endopterygotes, for which group such a dichotomy of requirements seems to be primitive, any species for which adult and larval requirements are similar, as in water beetles, representing secondary evolution. Hence, many Lepidoptera, including the vast majority of butterflies, as adults require abundant nectar but as larvae require abundant foliage of an appropriate plant species. For many Lepidoptera the larval foodplants grow in habitat-types that are spatially separate from habitat-types providing the most abundant nectar. Such a situation can only

have evolved if selection has favoured the linkage of the obligatory initiation of migration by females with certain physiological states or times of day or year such that migration is initiated as optimum periods (diel, seasonal, ontogenetic) for time investment in feeding give way to optimum periods for oviposition and vice versa. The results of such selection are presumably exemplified by the migration patterns shown by females of the small tortoiseshell butterfly, *Aglais urticae* (Fig. 16.15). As far as males are concerned, habitat-types in which receptive females are most likely to be found may be spatially separate from habitat-types in which food is most abundant or from the habitat-type in which the male developed as a larva. In this case the action of selection would be similar to that on the female. The migration patterns shown by males of the peacock butterfly, *Inachis io* (Fig. 16.16), and the small tortoiseshell butterfly, *Aglais urticae* (Fig. 16.15), could again be presumed to have resulted from such selection.

A relatively large amount of space has been devoted here to a consideration of removal migrations that have evolved in response to selection on animals to perform different time-investment components in spatially separate habitat-types. There are several reasons for this attention here. First, the importance of such migration has not previously been recognised by students of insect migration (see, for example, Kennedy 1961, Southwood 1962, and C.G. Johnson 1969). Second, the analysis of this type of migration requires a quite different approach to that required for the analysis of obligatory migration between habitats of a single type. Third, and finally, a thorough consideration of removal migration in relation to time-investment components and the spatial separation of habitat-types facilitates the consideration of various essentially similar forms of return migration (Part IIIb).

For the moment, however, let us leave the consideration of migration in relation to time-investment strategy and instead consider the evidence for the involvement of other selective pressures in the evolution of obligatory migration, particularly those selective pressures of the 'rate of appearance of new habitats:pattern of habitat ontogeny:habitat longevity' complex discussed in Chapter 8 (p. 50). The accepted method for investigating the action of this selective pressure complex is that developed by Southwood (1962) and involves a comparison of the initiation of migration by species adapted to habitats that show different degrees of this selective pressure complex. For example, habitat-types such as rivers, lakes, perennial plants including trees of climax vegetation such as woodlands, salt marshes, heath lands, and marshes fringing lakes and rivers are likely to have a relatively great longevity and a relatively low rate of formation of new sites. Habitat-types such as dung, carrion, fungi, plant debris (such as logs and straw), annual and perennial plants of seral communities, and some ponds, on the other hand, are likely to have a relatively short longevity and a relatively high rate of formation of new sites.

In order to increase the relevance of analysis to the migration-type under consideration and in order not to include the effects of obligatory migration due to the spatial separation of different habitat-types, it is necessary to restrict analysis in the first instance to groups that have similar requirements as adults and larvae and that as adults feed, roost and reproduce within the same habitat-type. This considerably restricts the scope of analysis. Furthermore, even when such a group has been identified, it is impossible to sustain even moderate confidence that the results indicate cause and effect in the evolution of *obligatory* migration. As an example, let us consider the analysis of water beetles carried out by Southwood (1962). This group lends itself as much as, if not more than, any other to this type of analysis. Not only does it satisfy the requirements just described but also the quantification of migration can be based on evidence that is rather more reliable than reports of field behaviour that have not been thoroughly investigated. This is because, in this group, flight seems only to be used for migration from one water body to another. Within the group, however, the proportion of individuals that are able to fly varies considerably from species to species. Although not definitely proven, it seems reasonable to assume that the incidence of the initiation of migration by a species correlates with the proportion of individuals of that species that are able to fly.

Figure 16.18 demonstrates clearly a relationship between the rate of formation of new habitats and/or habitat longevity and the ability of water beetles to fly. It cannot be automatically assumed, however, that these selective pressures have acted to produce the variation in initiation of *obligatory* migration that is the main concern of this section. The work on aphids (p. 230) has shown quite clearly that apparent obligatory post-teneral migration by winged parthenogenetic females of the summer generations (alienicolae) is in fact dependent for its occurrence

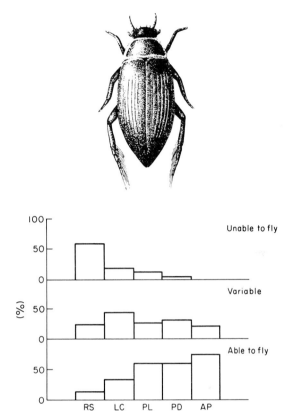

Fig. 16.18 Variation in the ability of water beetles to fly in relation to the rate of formation of new habitats and/or habitat longevity

The figure divides 159 species of west European water beetles into five categories according to the type of habitat to which they are adapted and three categories according to their ability to fly. Categories of habitat-type are: RS = rivers and springs; LC = lakes, tarns and canals; PL = brackish water, peat pools, and bog ponds; PD = ponds and ditches; and AP = artificial ponds, gravel pits and cattle troughs, etc. The arrow indicates an increasing rate of formation of new sites and a decreasing habitat longevity.

A clear relationship emerges such that the greater the rate of formation of new sites and/or the less the habitat longevity the more likely is a species to be able to fly. It is not clear, however, whether the figure is analysing obligatory migration as intended or whether the arrow also indicates an increasing probability that conditions will favour facultative migration. In the latter case, selection would be likely to be such that the greater the possibility of facultative migration being advantageous the greater the advantage (or less the disadvantage) to an individual of developing and maintaining flight apparatus.

[Figure based on an analysis by Southwood (1962)]

on above-threshold values of certain habitat variables. Development and use of the flight apparatus is therefore a facultative effect. E.C. Young (1963) has suggested that water body to water body migrations of corixid bugs in summer are facultative. Furthermore, those species of water beetles adapted to habitats with a high rate of formation of new sites and/or short longevity in Fig. 16.18 are likely to be those species with the greatest probability of exposure to migration-favourable above-threshold values of habitat variables. Artificial ponds, gravel pits, cattle troughs, ponds, ditches, peat pools, and bog ponds are just those habitats that are most likely to become too hot, too cold, too deep, too shallow, too crowded or too deserted, etc. Retention of the ability to fly may therefore have evolved in response to this high probability of the advantageousness of facultative migration rather than in response to selection for obligatory migration. The variation shown in Fig. 16.18, therefore, does not necessarily prove that a high rate of formation of new sites and/or a short habitat longevity has played a functional part in the evolution of obligatory migration by water beetles, even though it may well be found subsequently that this is the case. What is required is experimental analysis of the proportion of water beetles initiating migration under a variety of different conditions.

However, some positive support for the existence of an ultimate relationship between rate of formation of new sites (as a selective pressure independent of habitat longevity) and obligatory water beetle migration is provided by the results obtained by Landin and Stark (1973) for the water beetle, *Helophorus brevipalpis*, in Sweden In this species, peaks of migration occur in spring and again in July and August. Both of these peaks coincide with peaks of formation of new sites rather than with times at which habitat variables are likely to become unfavourable. In spring, numerous ponds are refilled by the thaw and by precipitation. There is also increased precipitation from July onwards. Pajunen and Jansson (1969) concluded a similar relationship between the appearance of new habitats and the initiation of migration by rock pool corixids.

A similar selective pressure seems to be the only reasonable explanation for the evolution of obligatory migration in locusts. These insects live in generally or seasonally arid or semi-arid parts of the world. The most suitable oviposition sites in such regions are habitats in which rain has recently fallen. Yet the locusts live in areas where the spatial distribution of rainfall is unpredictable. Hence, although there is a high rate of formation of new habitats, the probability is not particularly high that such habitats will be formed in the natal area. The situation is very much the type in which h_1 is almost always likely to be less than $\bar{E}$ and in which, therefore, selection should favour the evolution of obligatory migration.

Quite possibly, rate of formation of new habitats could also account for the evolution of the interspecific variation in initiation of aerial migration by spiders (Fig. 16.19). In this case, however, the same problem is encountered as for the cross-species analysis of water beetles (Fig. 16.18). Without more-detailed knowledge of the factors that influence, or fail to influence, the initiation of migration by individuals of each species, it is not possible to be sure that we are not analysing the incidence of facultative migration in different species. The same problem is encountered in attempting to interpret the observations of Vachon (1947) for pseudoscorpions. These small arachnids initiate aerial migration by grasping the leg of a fly or other insect. The migration is terminated when the pseudoscorpion releases its hold as the fly lands in a suitable habitat. This migration mechanism is known as phoresy. Pseudoscorpions that are commonly phoretic, such as *Lamprochernes nodosus*, live on vegetable detritus, manure heaps and tide-line debris, etc. In contrast, those species that live on tree bark or are soil dwellers are not phoretic.

Aphids illustrate well the problems that are involved in interpreting this type of correlation. Unlike water beetles, for which it seems that rate of formation of new habitats may well have been a critical selective pressure, the more that is learned about aphids the less the rate of formation of new habitats and/or habitat longevity appears to be a critical factor, at least in relation to the *present* function of migration. When Southwood (1962) first analysed the group, there seemed to be a clear distinction between tree-dwelling aphids, which performed only 'trivial' flights, and the 'migratory' aphids which during the summer months fed on annuals or other plants associated with 'temporary' habitats. It is now clear from the work of Dixon (1969) on the sycamore aphid, *Drepanosiphum platanoides*, that tree-dwelling species may be migratory in the same sense as species adapted to more temporary foodplants, such as the black bean aphid. *Aphis fabae*. Furthermore, it seems that the mechanisms are just the same in both of these species, individuals being more likely to initiate migration if they are subjected during development to a high deme density, a low nett rate of energy uptake or, presumably, a combination of these and/or other above-threshold habitat variables. Both species also seem to have evolved a hierarchy of migration thresholds, each associated with a migration with a different value for $\bar{E}$, the mean expectation of

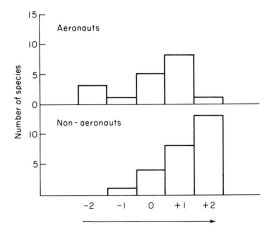

Fig. 16.19 Distribution of the occurrence or non-occurrence of aerial migration by British spiders according to habitat-type

The figure divides 44 species of British spiders into two categories according to whether or not the spiders are considered to be aerial migrants (aeronauts) and into five categories according to a subjective estimate of the relative rate of formation of new habitats and/or habitat longevity of the habitat-type to which they are adapted. Categories of habitat-type are achieved by assigning different numerical values to different habitat-types: woods, heather and soil $= +2$; bushes and grass $= +1$; everywhere $= 0$; low plants and fallen leaves $= -1$; detritus and straw etc. $= -2$. When a species is adapted to two of these habitat-types the appropriate categorisation is achieved by summing the two values. Hence, a spider adapted to detritus in woods is assigned to a category of $(+2) + (-2) = 0$. The arrow indicates a decreasing rate of formation of new sites and an increasing habitat longevity.

The only clear difference in incidence of aerial migration is found in spiders adapted to woods, heather and soil, etc. $(+2)$, the great majority (93 per cent) of which are non-aeronauts. There may also be found subsequently to be a significantly higher proportion of aerial migrants among species adapted to detritus and straw, etc. (-2). All other habitat-types seem to have little, if any, effect on the incidence of aerial migration. Whether the migrations concerned are obligatory or facultative is not known. The arrow could also indicate a decreasing probability of facultative migration becoming advantageous.

[Figure, but not conclusions, based on an analysis by Southwood (1962)]

migration. Both species also are found in the aerial plankton (Johnson 1969, pp. 40 and 588). Finally, consecutive generations of both species are migrating from spring through autumn. The only clear difference between the two species is that, once in the spring and again in the autumn, an entire generation of *Aphis fabae* consists of obligatory migrants that settle, not on the same species of plant on which they were born but on a different species (Fig. 16.17).

The difference between the sycamore aphid, *Drepanosiphum platanoides*, and the black bean aphid, *Aphis fabae*, therefore, is not that the former is non-migratory whereas the latter is migratory but that

1 mm

the former is predominantly, if not entirely, a facultative migrant whereas the latter, as well as being a facultative migrant, also includes two phases of obligatory migration within its year's behavioural repertoire. Even the existence or non-existence of obligatory migration does not correlate with habitat-type for, although in autumn the black bean aphid obligatorily initiates migration from a 'temporary' host (e.g. broad bean), in the spring the species obligatorily initiates migration from a 'permanent' host (i.e. the spindle tree). The only conclusion, therefore, is that both the tree-dwelling aphid and the host-alternating aphid are migratory, but that associated with the evolution of obligatory host-alternation is the evolution of obligatory migration, an association that is inevitable and that tells us little about the evolution of either behaviour.

A demonstration of a correlation between obligatory host-alternation and obligatory migration is as meaningless as the demonstration of obligatory migration by any other species obligatorily adapted to different and spatially separate habitat-types suitable for different time-investment components. The crucial consideration for all such situations is the sequence of selective pressures and adaptations that led in the first place to an aphid or other animal becoming adapted to spatially separate habitat-types. It is a consideration of this question that occupies the remainder of this section.

We have concluded so far that for at least some species, such as the water beetle, *Helophorus brevipalpis*, various rock pool corixids, and locusts, it seems likely that the present advantage of obligatory migration results from the high rate of formation of new habitats. It seems reasonable to assume for these species that this selective pressure has also, in the past, been the critical selective pressure that led to the present form of initiation of migration. This selective pressure could also be the one that accounts for the inter-specific variation in incidence of the initiation of migration observed within groups, such as spiders (Fig. 16.19), pseudoscorpions (p. 246), water beetles (Fig. 16.18) and various other insect groups (T. Lewis 1961, Southwood 1962), for which adult and larval requirements are similar and the different adult time-investment components can all be performed most efficiently in a single habitat-type. In most of these examples of cross-species correlation within taxons, however, it is possible that the causal factor(s) is not rate of formation of new habitats but rather the probability of facultative migration becoming advantageous. If, as seems

likely (p. 246), probability of facultative migration is greatest in habitats with a high rate of formation of new sites and/or short habitat longevity, it will not be possible to reach any conclusion concerning the critical factors until much more is known about the initiation of migration in the individual species, as for example, *Helophorus brevipalpis*, for water beetles (p. 246). For present purposes, however, it is sufficient to accept for species with similar adult and larval requirements and with adults adapted to a single habitat-type, that a correlation exists between rate of formation of new habitats and/or habitat longevity and the probability of the initiation of migration. The causal factor, however, may be proximate, as for facultative migration, or ultimate, as for obligatory migration, though even with facultative migration there may be some ultimate effect, such as the level of the ability to fly (p. 246).

Clearly, no modern animal can be adapted to perform different time-investment components in spatially separate habitat-types unless its ancestor was sufficiently migratory to encounter such habitat-types in the first place. Furthermore, the proportion of individuals that initiate migration, for whatever reason, seems likely to be a critical factor in any evolutionary sequence concerning adaptation to an additional habitat-type. Consider two species, *a* and *b*, adapted respectively to habitat-types *A* and *B*. Suppose that both species contain individuals that initiate migration, either facultatively or obligatorily, but that a much greater proportion of *a*, because of the characteristics of habitat *A*, initiate migration than *b*. Suppose further that during migration individuals of both species encounter habitat-type *C* and that individuals that invest some time in *C* happen to be at an advantage to individuals that do not, the advantage gained being the same for an individual of species *a* as for an individual of species *b*. The result will be adaptation to habitat-type *C*. However, because a greater proportion of *a* than *b* encounter habitat-type *C* in each generation, rate of spread of any characteristic adaptive in habitat-type *C*, including optimum time investment, through the population of species *a* will be greater than the rate of spread through the population of species *b*. Quite possibly both species may eventually become adapted to *C*, but if any competitive exclusion occurs it is likely to be *a* rather than *b* that becomes adapted to *C*. The result of this sequence is likely to be a modified version of the original species *a* that is now adapted to invest time in spatially separate habitat-types *A* and *C* and a less modified version of the original species *b* that still remains adapted only to habitat-type *B* and may actively avoid investing time in habitat-type *C*. Further selection on *a* will act to favour obligatory migration between *A* and *C* at the optimum times.

If this argument is valid, a correlation would be expected between adaptation to spatially separate habitat-types (which must involve migration) and a high incidence of the initiation of migration by the ancestor. As the highest incidence of initiation of migration, for whatever reason, has been shown to occur in species adapted to (single) habitat-types with a high rate of formation of new sites and/or a high probability of facultative migration, it might be expected that species adapted, say, during the larval stage to habitat-types with a high rate of formation of new sites, etc., should also be the species that are adapted to different and spatially separate habitat-types.

It is significant, therefore, that Southwood (1962) found a correlation between habitat 'temporariness' and the probability of migration for all arthropod groups that he analysed, not just those adapted to a single habitat-type. For example, of the British species of anisopteran dragonflies, 0 per cent (0/6) of those adapted to rivers, streams and runnels, 43 per cent (6/14) of those adapted to broads, lakes and canals, 42 per cent (5/12) of those adapted to peat pools and rushy ponds, and 73 per cent (8/11) of those adapted to ponds, gravel pits and other relatively temporary bodies of water, are recognised as migrants. Yet there is little doubt that these migrations are initiated at the present time because the most suitable adult feeding sites are spatially separate from the larval habitat. Indeed, there is even a suspicion that these migrations, at initiation, may be exploratory rather than removal migrations (p. 198).

A similar analysis for British Macrolepidoptera, the migrations of which are also, by and large, the result of adaptation to different and spatially separate habitat-types, leads to a similar conclusion (Southwood 1962). Hence, 43 per cent (3/7) of those adapted as larvae to a wide variety of plants, 29 per cent (17/58) of those adapted as larvae to plants of arable land, 19 per cent (22/116) of those adapted as larvae to non-climax plants of hedgerows and gravel pits, etc., 0.4 per cent (1/284) of those adapted as larvae to bushes and perennial climax plants of moors, woods, saltmarshes, marshes, chalk downs, and heaths, and 0 per cent (0/310) of those adapted as larvae to woodlands, including lichens and mosses

growing on trees as well as the trees themselves, were recognised as migratory by Southwood. Furthermore, 13 of the 310 woodland species are characterised by brachypterous females incapable of flight. Subsequently, I carried out a rather more detailed analysis of British butterflies (Baker 1969a) in which inter-habitat distance, another major component of $\bar{E}$, was included, and came effectively to the same conclusion as Southwood (1962), that is that species adapted as larvae to more temporary (and in this case also, closely spaced) habitat-types are 'more migratory' than species adapted to more permanent habitats.

Despite this correlation, species that show a relatively high frequency of initiation of migration do so almost entirely in relation to movement between different habitat-types rather than in relation to movement from one breeding site to another. Furthermore, migration between breeding habitats cannot account for the pronounced autumn migrations of adults in reproductive diapause such as the peacock butterfly, *Inachis io* (Fig. 16.16) and the well-known monarch butterfly, *Danaus plexippus*, in North America (p. 426). In addition, migration between oviposition habitats cannot often account for the migrations of males. Finally, it is now known that the estimate of 0 per cent for migratory species of Lepidoptera that feed as larvae on trees is incorrect. The brimstone butterfly, *Gonepteryx rhamni*, the holly blue butterfly, *Celastrina argiolus*, and white-letter hairstreak, *Strymonidia w-album* (Baker 1969a) as well as the camberwell beauty, *Nymphalis antiopa* (Roer 1970) are frequent and undoubtedly obligatory migrants. In all of these cases, which must be considered exceptions to the correlation, as well as the much larger number that are not exceptions, the migration is associated with adaptation to spatially separate habitat-types. The apparent correlation, which includes exceptions, between the rate of formation of habitats and incidence of migration can therefore be re-interpreted as a correlation, with fewer if any exceptions (in statistical terms an exception in this sense would be a species outside of the confidence limits set by the variance of the regression), between rate of formation of habitats and the probability of becoming adapted to spatially separate habitat-types (p. 248). Even tree-dwelling species can become adapted to spatially separate habitat-types, given a sufficiently high incidence of facultative migration in the ancestor such as is found at present in, say, the sycamore aphid, *Drepanosiphum plata-*

Fig. 16.20   Three butterflies that as larvae feed on trees yet still show a relatively high migration incidence as adults

Top right, brimstone, *Gonepteryx rhamni*; top left, holly blue, *Celastrina argiolus*; bottom, white-letter hairstreak, *Strymonidia w-album*

[*Photos by R.R. Baker*]

noides, but such species are less likely to become adapted to spatially separate habitat-types than those species adapted to habitats with a high rate of formation of new sites and/or short longevity for which migration incidence is likely to be greater.

In conclusion, therefore, non-calculated removal migrants can be divided broadly into two classes: those in which all adult requirements and juvenile requirements are present within a single habitat-type; and those in which the different habitat-types in which different time-investment components are performed are spatially separate. The present function of migration for the first class of migrants is that it *maximises individual reproductive success by allowing the individual to exploit the changing pattern of habitat suitability in space and time* (the italicised phrase seems to be more accurate than Southwood's thesis that 'migration enables a species to keep pace with the changes in the location of its habitat'—Southwood 1962). This function is fulfilled by both facultative and obligatory migration, the type that evolves depending on the predictability with which h_1 falls below $\bar{E}$. When the critical selective pressure is the predictable rate of appearance of new habitats (increase in $\bar{E}$) as in the water beetle, *Helophorus brevipalpis*, some rock pool corixids, and locusts (p. 53), obligatory migration represents optimum strategy. When the critical selective pressure is the decrease in suitability of the present habitat (decrease in h_1), optimum strategy will be obligatory migration when the decrease in suitability is predictable (no examples) and facultative migration when the decrease is unpredictable, as for the sycamore aphid, *Drepanosiphum platanoides*, and for the summer generations of the black bean aphid, *Aphis fabae*. There seems to be a general correlation for this first class of migrants between the incidence of the initiation of migration (facultative and/or obligatory) and the rate of formation of new habitats and/or habitat longevity. As exemplified by the aphids, however, the distinction is not between migratory and non-migratory species but rather a graded response of probability that conditions will favour the initiation of migration.

The *present* function of migration for the second class of migrants, however, is that it *maximises individual reproductive success by allowing the individual to perform each time-investment component in the habitat-type most suitable for the performance of that component.* In such species, migration is inevitably obligatory and is an integral part of time-investment strategy. Migration may be initiated on a diel basis, as in many butterflies (Figs. 16.15 and 16.16) or on a seasonal basis, as in host-alternating aphids, or both. It is quite meaningless in evolutionary terms to demonstrate for such species a correlation between the spatial separation of habitat-types and migration between these habitat-types. The critical consideration is the sequence of events that led the ancestors of such species to become adapted to spatially separate habitat-types. It is concluded that species with a higher probability of initiation of migration, whether facultative or obligatory, are more likely to become adapted to spatially separate habitat-types than species with a lower probability of initiation of migration. It is this differential probability, acting on the ancestors of modern forms, that is suggested to account for the observed correlation between adaptation to spatially separate habitat-types and larval adaptation to habitats with a high rate of formation of new sites and/or short habitat longevity. Whether or not adaptation to spatially separate habitat-types occurs in any particular lineage depends not only on incidence of migration but also on competition from other species for the use of a particular habitat-type. Even species with a relatively low incidence of initiation of migration could become adapted to spatially separate habitat-types if there was little or no competition from other species with a higher incidence of initiation of migration. Possibly this accounts for the high incidence of obligatory migration in some species adapted as larvae to habitats with a low rate of formation of new sites, such as some species of butterflies, the larvae of which feed on trees.

16.4.3 Migration thresholds in relation to some migration-cost variables

Once again, insects are the only animal group to have received experimental attention in relation to exploratory or removal migration thresholds to migration-cost variables. All the examples in this and the next section are therefore taken from this group.

16.4.3.1 Temperature

The temperature range over which an insect flies is usually, if not always, less than the range of ambient temperatures to which the insect is subjected in its natural environment. The range of flight-muscle temperatures over which the flight muscles are functional is usually even narrower than the range of ambient temperatures over which the insect flies. Although this phenomenon has been investigated physiologically, it has received negligible attention from an evolutionary viewpoint. An evolutionary consideration of any length is outside the scope of this book, though a few of the more obvious deductions that can be made may possibly facilitate consideration of temperature as a migration-cost variable.

It can perhaps be assumed that there is an advantage to be gained from the maintenance and use of only a part of the enzyme systems and other physiological mechanisms that would be necessary for flight to take place over the complete range of temperatures to which a particular insect species is subjected. If the energy that is saved through not maintaining systems and mechanisms that function at other temperatures represents an advantage that is greater than the disadvantage of being behaviourally inactive at these temperatures, then selection will favour individuals that are capable of activity over only a limited range of temperatures. Presumably the systems and mechanisms that will be selected against most strongly will be those that are functional at temperatures to which an insect of a certain size and diel incidence occupying a certain niche will be subjected least often. Large insects, by virtue of their surface area to volume ratio, not only have proportionately greater difficulty in gaining heat from the environment, but also, if heat is generated within the body during vigorous activity, have proportionately greater difficulty in losing heat to the environment. The most frequent temperatures to which large insects in flight are likely to be subjected, therefore, are above ambient. In large moths, for example, heat loss approaches heat production when the thoracic temperature minus the ambient temperature approaches 20°C (Dorsett 1962). Consequently, selection is likely to favour large insects specialising in enzyme systems and other physiological mechanisms that function most efficiently at relatively high temperatures. Unless different systems and mechanisms are to be used for take-off from those used for flight, therefore, a disadvantageous situation that would require energy to be expended to maintain and use two mechanisms instead of one, some behavioural and/or physiological mechanism is necessary to raise the temperature of the flight muscles to that at which the evolved systems function most efficiently. Whether the energy involved in the maintenance of physiological systems that function most efficiently at different temperatures coupled with frequency of exposure of a species to different internal temperatures is a sufficient selective pressure to account for the variation in insect behaviour with temperature must await further consideration. Quite probably, many other selective pressures are also involved. For example, physiological systems that function most efficiently at higher temperatures than other systems are also likely to be inherently more powerful. The power required to raise the weight of an insect from the ground increases as the 3·5th power of linear dimension whereas wing area only increases as the square (Maynard Smith 1954). Consequently, large insects need more power per unit weight than small ones for take-off, an effect that could also result in selection for larger size having been associated with selection for flight muscles that work most efficiently at higher temperatures. It is perhaps significant in this connection that a relationship has been demonstrated (Dorsett 1962) between the temperature ($T°C$) at which flight occurs and wing loading (L mg/cm^2) for sphingid hawk moths in Nigeria. This relationship is described by the regression line $T = 0·11L + 27·9$. Clearly any evolutionary explanation of variation in flight-muscle efficiency with temperature will be complex, but the above paragraph, though inevitably naive, is perhaps sufficient to serve as background to a consideration of temperature as a migration-cost variable.

If energy expenditure per unit distance is less at certain ambient temperatures than at others, then ambient temperature is inevitably a migration-cost variable and it is to be expected according to the

arguments (p. 49) of Chapter 8 that there is a threshold value of ambient temperature that corresponds to the situation in which $h_1 = \bar{E}$. No experimental verification of this is available that is comparable to the dung-fly verification for a habitat variable (p. 226) but there are several measurements that demonstrate the existence of migration thresholds in relation to ambient temperature.

The black bean aphid, *Aphis fabae*, and the water beetle, *Helophorus brevipalpis*, provide clear examples of a migration threshold to ambient temperature (Fig. 16.21). The desert locust, *Schistocerca gregaria*, also provides a fairly clear example of a migration threshold to ambient temperature in that the initiation of migration by a swarm, often described as 'stream-away' was associated consistently with air temperatures of between 17°C and 23°C (Gunn *et al.* 1945). In fact, the locust situation is especially relevant to the present consideration in that it provides support for the suggestion that temperature is a migration-cost variable in relation to which a migration threshold has evolved (p. 49) rather than that the migration threshold is simply the result of a physiological limitation. Desert locusts, like other large insects, have evolved behavioural and physiological mechanisms by which the temperature of the flight muscles can be raised to their functional level independently of the level of ambient temperature. Gunn *et al.* showed that before sunrise the thoracic temperature of settled desert locusts approximates to that of the air. When the Sun begins to shine, however, the locusts orientate themselves to make maximum use of insolation, whereupon the body temperature rises, sometimes to in excess of 11–16°C above air temperature, well above the 20°C or so at which spontaneous flight becomes possible. In fact, the initiation of stream-away was more consistently correlated with the temperature of the air than with body temperature.

Despite the involvement of body temperature in the possibility of flight, other factors such as the maintenance of body temperature, the efficiency of flight, and hence energy expenditure, are a function of the cooling effect of the air. Migration cost is therefore a function of ambient temperature and no matter what behavioural mechanisms may be involved in achieving an above-threshold body temperature, a migration threshold to ambient temperature would still be expected to evolve.

Most butterflies also seem likely to show a migration threshold to ambient temperature as well as a migration threshold to thoracic temperature, and not until both thresholds are exceeded will migration be initiated. The red admiral, *Vanessa atalanta*, for example, flies at all ambient temperatures above 20°C, but usually first warms up the thoracic muscles to 26–35°C by vibrating the wings (Krogh and Zeuthen 1941). Most butterflies also gain body heat from insolation by basking in the Sun with wings outstretched before take-off. Indeed it is rare for butterflies to initiate migration when the Sun is not shining.

Take-off by large moths, such as hawk-moths (Sphingidae) is independent of ambient temperature which merely affects the duration of the wing-vibration warming-up period to the thoracic threshold of 34° to 45·5°C (Dorsett 1962). This work, however, has not been related to migration. Medium-sized moths, such as noctuids, show a clear migration threshold to ambient temperature (L.R. Taylor 1963), and quite possibly the same is true for sphingids. Moths tend to migrate in the warmest available layer of air (C.G. Johnson 1969) which, as under conditions of temperature inversion in which temperature increases with height above ground up

to a certain height and thereafter decreases with height, is not necessarily the layer next to the ground. Unlike most diurnal insects, therefore, nocturnal insects such as most moths do not, before take-off, have information available to them concerning the ambient temperature at which migration can take place. Such information can only be obtained by vertical flight until the warmest available layer is identified. Such flight, of course, cannot be initiated until the thoracic temperature threshold for take-off has been exceeded. Optimum strategy for an insect such as a medium to large moth, therefore, may be to warm up to the thoracic temperature threshold, fly vertically until the warmest air layer is identified, and then, if this layer is above the migration threshold to ambient temperature, to initiate migration. If the temperature of this layer is not above the migration threshold to ambient temperature, optimum behaviour could be to return to ground level, either to roost or to perform other behaviour patterns such as courting or feeding etc. which may have lower ambient temperature thresholds. This interpretation would reconcile the results obtained by Dorsett (1962) in the laboratory in relation to take-off and Taylor (1963) in relation to migration at heights of 2·7–17 m in the field as well as explaining the spectacular dusk ascent recently discovered by the use of radar (Chapter 19). However, it is always possible, of course, that

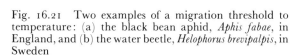

Fig. 16.21 Two examples of a migration threshold to temperature: (a) the black bean aphid, *Aphis fabae*, in England, and (b) the water beetle, *Helophorus brevipalpis*, in Sweden

In (a), migration was monitored by a suction trap at crop height, 0·3–1·2 m, and temperature was monitored at the same height. In (b), migration was monitored by a ground-level glass trap which simulated a water surface and which captured beetles migrating between ponds, and temperature was monitored at a height of 1·5 m. Samples at times of day that light intensity was above the migration

threshold were divided into time units: 30 min units in (a); and 60 min units in (b). The presence or absence of the species was noted for each time unit and the catch occurrence (percentage of time units that a positive catch of any number occurred) was calculated for time units characterised by a particular temperature. Note that the maximum temperature during the time unit is used in (a) but the minimum temperature is used in (b).

Taking into account the methods employed, both species show a marked migration threshold to temperature.

[*Modified from Taylor (1963) and Landin and Stark (1973)*]

Fig. 16.22   An example of an upper and lower flight threshold to temperature: the soldier beetle, *Rhagonycha fulva*, in England

Flight was monitored by a suction trap at crop height, 0·3–1·2 m, and temperature was monitored at the same height. Samples at times of day that light intensity was above the migration threshold were divided into 30 min units. The presence or absence of soldier beetles was noted for each time unit and the catch occurrence (percentage of time units that a positive catch of any number occurred) was calculated for time units characterised by a particular maximum temperature.

A clear upper and lower flight threshold is apparent.

[*After Taylor (1963). Photo by R.R. Baker*]

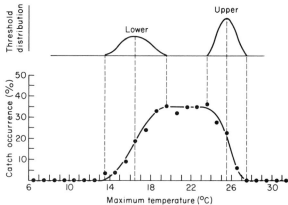

sphingids and noctuids, possibly in consequence of their difference in size, behave differently in this respect and that sphingids do not have a migration threshold to ambient temperature.

It has been argued (p. 51) that migration thresholds to migration-cost variables should evolve to include time of exposure to the variable, such that the migration threshold becomes lower with time. To some extent this suggestion is supported, with respect to temperature, by the work of Gunn *et al.* (1945) on the desert locust, *Schistocerca gregaria*. As stated earlier (p. 252) the initiation of stream-away was associated with a migration threshold to ambient temperature of between 17°C and 23°C. It was noted, however, that the higher temperatures operated on bright, sunny days and the lower ones on damp days. Hence, the migration threshold seems to start high early in the day but if ambient temperature remains low into the day due to damp conditions then the migration threshold lowers to 17°C.

So far, discussion of migration thresholds to

temperature has been concerned only with low temperatures. It seems likely, however, that flight efficiency also begins to decrease, and hence migration cost begins to increase, with a rise in temperature beyond a certain level. On the same arguments as previously, therefore, it might be expected that there will be an upper as well as a lower migration threshold to temperature. In the absence of knowledge of either the flight efficiency of different species at different temperatures or the presence or absence of behaviour patterns to continue migration even when measured temperatures (at, say, the trap) exceed those at which flight is thought to become inefficient, there is no possibility of predicting which species should show upper threshold temperatures and which should not. Taylor (1963) found no upper threshold temperature for the black bean aphid, *Aphis fabae*, but such a threshold was detected for the beaded chestnut moth, *Agrochola lychnidis*, and mouse moth, *Amphipyra tragopogonis*, as well as for the soldier beetle, *Rhagonycha fulva* (Fig. 16.22).

16.4.3.2 Wind

As a migration-cost variable, wind influences energy expenditure, the ability to settle in a perceived destination, and probably also survival. An insect maintaining a migration course against the wind (Chapter 19) must expend more energy per unit distance relative to the ground when the wind is strong than when the wind is light. Having perceived a suitable destination, an insect may need to expend more energy to settle in that habitat when the wind is strong than when the wind is light. There is also the danger that the insect may be unable to find or may even be physically incapable of returning to a habitat perceived while being subjected to a

Fig. 16.23   A female small white butterfly, *Pieris rapae*, trapped in a spider's web

Insects that cannot control their track as a result of wind speed while flying below tree height probably suffer an increased mortality risk through increased predation by spiders and vertebrates. Birds may often take insects while the latter are trapped in a spider's web.

[Photo by R.R. Baker]

strong wind. Finally, an insect's avoidance of environmental obstacles, such as water, as well as predators, particularly predators such as web-weaving spiders, is likely to be a function of the speed of approach. There can be few field entomologists that have not witnessed even large insects, such as butterflies, falling victim to avian predators as a result of a gust of wind temporarily blowing the insect into a bush or tree or perhaps into a spider's web, whereas normally their sensory–motor avoidance system would have operated. As wind speed increases, the danger of drowning and predation seems likely also to increase for all insects flying below tree-top level and presumably this is one of the main selective pressures that has led insects either to avoid flight at above-threshold wind speeds or to fly above tree-top height. Flight above tree-top height in strong winds, however, raises problems of settling in perceived potential destinations in that by the time an insect has reduced height the wind has taken the insect beyond the perceived location.

There is absolutely no doubt that many observed facets of insect migration at the population level are attributable to wind displacement. The involvement of wind displacement in the migration strategy of an individual insect, however, is far less clear and the point cannot be adequately discussed until the advantages and disadvantages of different migration distances have been discussed. The problem is, therefore, considered in Chapter 19. For present purposes, however, it is suggested that for large and small insects alike, wind is a migration-cost variable. As wind speed increases so too does migration cost and, in consequence, $\bar{E}$, the mean expectation of migration, decreases. Selection is therefore likely to favour a migration threshold that includes a wind-speed component.

Landin and Stark (1973) used glass-pane traps, which simulate water's surface, to sample water insects migrating to and from a pond on the island of Öland, Sweden. A special study was made of the migration thresholds to temperature and wind speed for the small (length about 2·5 mm) water beetle, *Helophorus brevipalpis*. A migration threshold of about 12°C to temperature was found (Fig. 16.21). When hourly catches below this threshold are excluded there is a clear relationship between catch occurrence and mean wind velocity (Fig. 16.24). A 50 per cent catch occurrence was found at mean wind velocities of about 10 km/h (measured 1·5 m above ground-level) and no beetles were captured at mean wind velocities above about 22 km/h.

Haine (1955) exposed individuals of various aphid species to continuous winds of known velocity in a wind tunnel. His results strongly support the postulated existence of a migration threshold that results from an optimum interaction between a migration-cost variable of a given value and period of exposure to that variable (p. 51). Aphids readily initiate migration in still air but if exposed to a continuous wind of nearly 5 km/h the black bean aphid, *Aphis fabae*, and the cabbage aphid, *Brevicoryne brassicae*, delayed initiation for several hours. If exposed to a continuous wind of 8 km/h these aphids did not initiate migration until after a period of 24 h or more. This type of interaction between wind speed and period of exposure is precisely what one would expect (Fig. 8.4) in an insect for which wind is a migration-cost variable, but not necessarily what one would expect for an insect adapted solely to migration by aeolian transport.

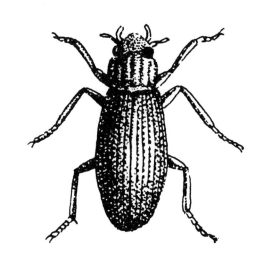

Fig. 16.24 An example of a migration threshold to wind velocity: the water beetle, *Helophorus brevipalpis*, in Sweden Migration was monitored by a ground-level glass trap which simulated a water surface and which captured beetles migrating between ponds. Wind velocity was measured at a height of 1·5 m as the total wind run during one hour, from which the mean wind velocity was calculated. Samples at times of day that both light intensity and temperature were above the migration threshold were divided into time units of 1 h. The presence or absence of the species was noted for each time unit and the catch occurrence (percentage of time units that a positive catch of any number occurred) was calculated for time units characterised by a particular mean wind velocity.

The migration threshold to wind velocity, which appears to be between 1 and 6 m/s, is probably also a function of length of exposure to a particular wind velocity.

[*After Landin and Stark (1973)*]

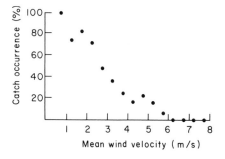

Spontaneous flights by the non-swarming *transiens* phase of the red locust, *Nomadacris septemfasciata*, of Africa do not occur at wind speeds greater than about 10·5 km/h (Rainey *et al.* 1957) and similar thresholds to wind speed seem to exist in other relatively large strong-flying non-swarming insects. For example, Williams (1961, 1962) examined over several years the relationship between wind and other meteorological factors on catches of moths, and other insects, in light and suction traps in the Highlands of Scotland. Williams demonstrated statistically what over the years has been the common experience of countless numbers of amateur moth-catchers, that is that an increase in wind speed decreases the number of moths captured. It seems extremely likely that this represents the failure of moths of different species to initiate migration at wind speeds above their respective thresholds.

My own observations of the initiation of butterfly migration further support the existence of a migration threshold to wind as well as indicating some of the complexities of the situation if wind speed is measured by an anemometer at a fixed height in a fixed site. As an example, let us consider the small white butterfly, *Pieris rapae*. There is a strong diel flight periodicity as well as clear, but as yet unmeasured, thresholds to light intensity, temperature and humidity. Migration is obligatory and is associated with the marked spatial separation of the habitat-types most suitable for the various adult time-investment components (p. 249). Migration continues at a more or less constant rate throughout adult life in both males and females (but see p. 334). In this discussion, the initiation of migration is considered to be the initiation of a non-return movement across an area which contains neither larval nor adult foodplants and which is wider than the presumed sensory range of the species.

All wind speeds quoted in the following paragraph are subjective assessments according to normal Meteorological Office convention (Meteorological Office 1939, p. 14) and refer to wind speed at a height of 10 m. These speeds are not, therefore, the wind speeds experienced by the butterfly which are less. Small white butterflies initiate migration at all wind velocities of up to about 12 km/h. Butterflies continue to initiate take-off at wind speeds of up to 34–45 km/h, but as wind speed increases to these speeds, increasing use is made of topographical features such as hedges, trees and buildings that provide shelter from the wind. Migration may continue to be initiated at all wind speeds up to 34 km/h in areas where topographical features provide the opportunity for continuous shelter from the wind during migration (Chapter 19, p. 416). In any sheltered spot, feeding, oviposition and courtship etc. are continued beyond wind speeds at which migration away from the area is no longer initiated. In open country, however, neither migration nor any other behaviour that involves take-off may be initiated at wind speeds above about 12 km/h.

As far as the butterfly is concerned, therefore, there is probably a well-marked threshold to wind for immediate initiation of migration (all other threshold variables also being favourable) of about 8–12 km/h at about 1 m above ground level. However, even on days with wind velocities in excess of 30 km/h, as far as a meteorological station or fixed anemometer at a height of 10 m is concerned, some migration is likely to be initiated by individuals in areas with such topography that by appropriate behaviour the butterflies can migrate in wind of sub-threshold velocity.

A much more detailed analysis of the association between migration routes and wind-breaks for a wide variety of insects is presented by Johnson (1969).

16.4.3.3 Light intensity

Whether or not light intensity by itself can influence migration cost is an open question. Depending on the sensory repertoire of the insect concerned, different levels of light intensity could well influence the ability of an individual to perceive a potential destination or to detect and avoid areas in which death-risk is high. Whether or not this is so, light intensity is likely to correlate well with time of day which, by virtue of diel variation in many of the biological and meteorological factors that affect flight efficiency and death-risk, is likely to influence migration cost and hence $\bar{E}$, the mean expectation of migration. Light intensity, insofar as it fluctuates with the appearance and disappearance of the sun behind clouds, is also likely to correlate well both with temperature and rain risk, both of which are likely to influence migration cost. Either as a migration-cost variable in its own right, therefore, or because it is a rapidly and easily detectable clue that correlates well with other migration-cost variables, it might be expected that many insects would have evolved a migration threshold to light intensity.

Many, if not the majority, of the larger diurnal insects such as butterflies, locusts and dragonflies do not initiate migration or any other behaviour involving flight if the Sun is obscured by cloud (C.B.

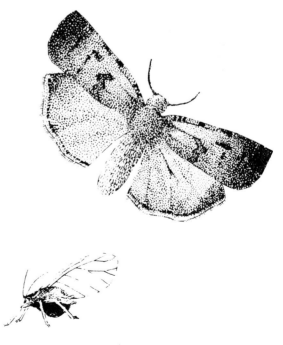

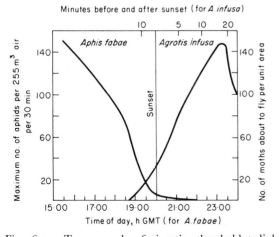

Fig. 16.25 Two examples of migration thresholds to light intensity and/or time of day: the black bean aphid, *Aphis fabae*, in England and the bogong moth, *Agrotis infusa*, in Australia

In *Aphis fabae*, migration ceases as light intensity diminishes at sunset but in *Agrotis infusa* migration is initiated as light intensity diminishes at sunset.

[*Simplified from C.G. Johnson (1969), after L.R. Taylor (1957) and Common (1954)*]

Williams 1930, Waloff and Rainey 1951, Pajunen 1962). In these cases it seems likely that even though light intensity may be involved as a proximate factor, the ultimate factor is much more likely to be either temperature or rain-risk. Quite possibly it is the correlation between an ultimate factor, such as temperature, or perhaps time of day, and a proximate factor, light intensity, that has led to the evolution of migration thresholds to light intensity such as that shown by the fruit fly, *Drosophila pseudoobscura*, which fails to initiate flight when light intensity exceeds 1000 lx (Lewis and Taylor 1964).

Most information concerning migration thresholds to light intensity also relates to time of day and once again it is difficult to decide the extent to which light intensity is the proximate or ultimate factor. Two examples, one a diurnal migrant, the black bean aphid, *Aphis fabae*, and one a nocturnal migrant, the bogong moth, *Agrotis infusa*, are given in Fig. 16.25. Particularly interesting in this respect are those species of locusts, such as the desert locust, *Schistocerca gregaria*, migratory locust, *Locusta migratoria*, and red locust, *Nomadacris septemfasciata*, that migrate by day as coherent swarms when in phase *gregaria* and by night as solitary individuals when in phase *solitaria* or *transiens* (Davey 1959, Waloff 1962, 1963). In brief, locusts that experience crowded conditions during development become adults of a particular colour and morphology and qualify as being referable to phase *gregaria* whereas those that develop more solitarily become adults of phase *solitaria*. Phase *transiens* adults are intermediate and indeed there is continuous variation between extremes. The effect of crowding may be cumulative over several generations. Physiological mechanisms, including pheromones, have evolved which increase synchronisation of development of crowded individuals. In general, adults of phase *gregaria* have a colour that tends to render them less conspicuous against backgrounds consisting of more substratum and less vegetation, whereas adults of phase *solitaria* have a colour that tends to render them less conspicuous against backgrounds consisting of more vegetation and less substratum. Morphologically, phase *gregaria* adults, when compared to phase *solitaria*, have longer wings relative to body size, less wing loading, a greater supply of energy reserves available as 'fuel' for flight, and less gamete production.

Although interesting, it is not so surprising that locusts of phases *gregaria* and *solitaria* should differ so markedly in diel periodicity of migration and/or

migration threshold to light intensity. In effect, the two phases are quite different animals exposed to quite different environmental pressures. Phase *gregaria* have developed in an area with many conspecifics per unit area and where food and shelter is limited. Phase *solitaria*, on the other hand, have developed in an area with few conspecifics per unit area and where food and shelter are relatively abundant. It is quite possible, therefore, that phase *gregaria* adults would be subject to greatest ground predation by storks, etc. by day whereas phase *solitaria* would be subject to greatest ground predation by night. This selective pressure by itself would be quite sufficient for such markedly different diel periodicities of migration to evolve.

Phase *gregaria* adults of the desert locust, *Schistocerca gregaria*, also provide the most likely example of a migration threshold to light intensity as a migration-cost variable in its own right. Waloff and Rainey (1951) conclude that the marked effect of sunshine on flying activity is not confined to the contribution of solar radiation to the thermal balance of the flying locust. The migration of phase *gregaria* locusts is dependent for its adaptiveness to a considerable extent on the cohesiveness of the swarm (Chapter 19). It is an advantage to each individual locust that it can perceive the presence and behaviour of other members of the swarm and also the presence or absence of other individuals both above and below it, often at considerable heights. Quite possibly, the efficiency of perception is a function of light intensity. In this case light intensity becomes a migration-cost variable to each individual and a migration threshold to light intensity would be expected.

16.4.3.4 Rain

The combination of rain, wind, and low temperatures can prove fatal to insects migrating across water. Williams (1958) cites various spectacular examples of large numbers of insects having been drowned as a result of meeting a storm while migrating across water. However, insects migrating over land may not be at such a disadvantage in rain. Nevertheless, few insects initiate or continue migration during rain, though some exceptions have been noted. The painted lady butterfly, *Cynthia cardui*, has been observed migrating in the United States in cloudy drizzle (Tilden 1962) and the great southern white butterfly, *Ascia monuste*, has been recorded in Florida to migrate in light but not heavy rain (Nielsen 1961). It is also a common experience that some moths will enter a light trap even during quite heavy rain. As far as butterflies are concerned, however, such observations are infrequent and are far outnumbered by observations that flight ceases during rain. This is generally true even for the two species mentioned above. Locusts also resume flight after they have stopped flying in the rain, though it is difficult to dissociate this effect from the disappearance and re-appearance of the Sun and possible cooling effects due to the rain wetting the thorax (Waloff and Rainey 1951). In any event, observations of insects migrating in the rain are not critical. What is required is a set of measurements of flight efficiency or survival rates in rain compared with no rain. In the meantime it may be suspected, but not proven, that rain is a migration-cost variable.

It is difficult also to demonstrate the existence of a migration threshold to rain intensity. This is particularly true for diurnal insects, for which it is difficult to dissociate rain thresholds from temperature or light intensity thresholds. Indeed, light intensity thresholds of the type associated with the appearance and disappearance of the Sun behind clouds (p. 257) may well have evolved in response to the advantage of searching for a favourable roosting habitat in advance of the onset of rain. However, there is at least one reliable example, that of males of the European pine shoot moth, *Rhyacionia buoliana* (Fig. 16.26), of flight behaviour that involves a threshold to rain intensity.

Possibly, exceptional observations of insects migrating in the rain are concerned with individuals that have received little food or opportunity to oviposit or copulate during the preceding few days. Starvation is known to have a marked effect on the migration threshold to migration-cost variables (p. 261).

16.4.3.5 Atmospheric pressure

It seems most unlikely that atmospheric pressure could, in its own right, be a migration-cost variable. Any reliable demonstration of a migration threshold to atmospheric pressure is therefore likely to reflect the predictive value (i.e. correlation) of the variable with future incidence of rain, and changes in wind speed and temperature. It will also correlate with altitude for those insects transported to great heights by convective air currents. In the same way as suggested earlier (p. 257) for many examples of migration thresholds to light intensity, atmospheric pressure is likely to be a proximate factor that is related to some other ultimate factor. With this in mind it is perhaps possible to reconcile observed migration thresholds to atmospheric pressure to the evolutionary thesis of this chapter.

The two series of experiments that can be related most reasonably to the flight behaviour of insects in response to the passage of weather systems are those

Fig. 16.26 The influence of rain intensity on the flight activity of male European pine-shoot moths, *Rhyacionia buoliana*

The histogram indicates the number of male moths observed flying at 10 min intervals with time of day and rain intensity on 5 July 1958. The solid line indicates light intensity and the number of rain drops indicates rain intensity.

The number of males in flight varied inversely with changes in rain intensity.

[*Modified from Green (1962)*]

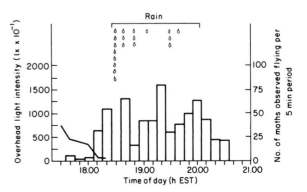

of Broadbent (1949) on aphids and Haufe (1954) on a species of mosquito. The cabbage aphid, *Brevicoryne brassicae*, showed the same frequency of initiation of take-off inside covered dishes at all pressures when these were kept constant. A drop in pressure of about 1·3 mm of mercury, however, and fluctuating pressure were associated with an increase in take-off rate. In the mosquito, *Aedes aegypti*, it was found, when number of take-offs per unit time was used as an index, that flight activity was more or less the same at all static pressures between 550 and 800 mm of mercury. Above 735 mm, a moderate decrease in pressure produced about twice as many take-offs as a similar increase in pressure. Below 735 mm, however, the reverse was the case. Insofar as a threshold is relevant in this situation, maximum take-offs in a pressure decreasing at about 1 mm/min occurred at 780 mm, and in a similarly rising pressure at 735 mm.

In general, therefore, insects, in the same way as human barometer owners, pay rather less attention to the level of atmospheric pressure than to whether it is rising or falling. A decrease in pressure usually correlates with future rain and, depending on rate of decrease, strong winds, whereas an increase in pressure usually correlates with the future absence of rain and lighter winds. Change in pressure also correlates with future change in temperature. In summer in temperate regions a sudden increase in pressure usually coincides with the passage of a cold front and the arrival of a period of lower temperatures. Subsequent increase in pressure, however, usually correlates with rising temperatures. On a world-wide and seasonal scale, however, the situation is complex.

Let us consider an insect approaching the optimum time for the initiation of migration (Fig. 8.4, p. 51). Atmospheric pressure, starting from a high position but beginning to fall indicates future adverse migration-cost variables of temperature, wind, rain and possibly light intensity that are likely to combine to give a low mean expectation of migration and hence a high migration threshold. Depending on how near the individual may be to the optimum time for the initiation of migration, optimum strategy may be the initiation of migration in advance of this time, thus suffering a below maximum advantage, but before the arrival of a weather system that may delay the initiation of migration, possibly even to time t_2 (Fig. 8.4), thus causing the individual to gain even less advantage from the migration. Increasing atmospheric pressure, start-

ing from a high position, suggests that migration-cost variables are likely to exceed threshold values at the optimum time for the initiation of migration and that therefore maximum advantage is gained by delaying the initiation of migration until this time. In an experimental situation these conflicting considerations would result in more individuals initiating migration with a decreasing pressure than with an increasing pressure if the starting pressure was relatively high. Starting from a relatively low pressure, however, a decreasing pressure is likely to correlate with imminent adverse migration-cost variables whereas an increase in pressure is likely to correlate with future improvement in migration-cost variables such that if present variables are favourable (above threshold) there is likely to be no disadvantage in immediate initiation of migration.

It must be stressed that this interpretation is highly speculative and should be considered to be no more than a preliminary suggestion. Until the behaviour of insect species with known responses to experimental fluctuations in pressure is observed in relation to the large-scale weather systems referred to above, speculation is all that is possible. Nevertheless, the experiments on insect behaviour in relation to pressure, if interpreted as above, are consistent both with the thesis of Fig. 8.4 and with what little is known concerning other migration-cost variables as discussed in the preceding pages.

16.4.4 The composite migration threshold as the combined result of several habitat variables and/or migration-cost variables

In other sections of this chapter it has been shown that facultative migration is a response to a number of habitat variables. The main effect of these habitat variables is that they cause the animal to develop a migration physiology, a major aspect of which is some clock mechanism that results in the initiation of migration at the optimum stage of ontogeny or time of day or year. When the migration is not facultative, the animal initiates migration independently of the value of any habitat variable. Even obligatory migration is only initiated, however, if migration-cost variables are favourable (above threshold). In facultative migration, therefore, the migration threshold consists of a combination of

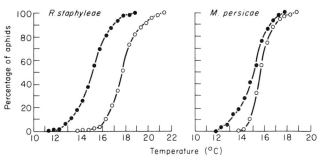

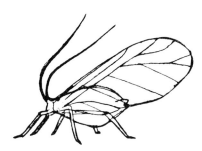

Fig. 16.27 Two examples of a migration threshold that is the combined result of a habitat variable and a migration-cost variable: the initiation of migration by the aphids *Rhopalosiphoninus staphyleae* and *Myzus persicae* in relation to the availability of food and temperature

The diagram shows the accumulative percentage of aphids that flew from two surfaces as the temperature rose above 10°C. Each curve represents at least 100 aphids. Solid dots, aphids taking off from a glass surface; open dots, aphids taking off from a sugar-beet plant.

When habitat suitability is high, migration is initiated at a higher temperature than when habitat suitability is low. The curves are consistent with the suggestion that selection has acted to produce a migration threshold to a number of interacting variables such that the migration threshold always corresponds to the situation at which h_1 (suitability of the occupied habitat) $= \bar{E}$ (the mean expectation of migration).

[*Figure after Heathcote and Cockbain (1966)*]

habitat and migration-cost variables, even though these different variables may be registered by the animal at quite different times, but in obligatory migration only migration-cost variables are involved in the migration threshold, though in many cases the sequence of events leading up to migration is triggered by indirect variables. The aim of this section is to present a few examples of the way in which different variables combine to produce the migration threshold.

Individuals of the mosquito, *Aedes aegypti*, were kept, after emergence, at 22°–24°C, 100 per cent relative humidity, and in darkness. Under such conditions the flies were immobile. Experiment involved exposing the insects to light of increasing intensity, as at day-break, and the cumulative take-off frequencies were measured in relation to time (Haufe 1962). Take-off was initiated once the light intensity had reached a threshold value but was also influenced by other factors, especially the hygro-thermal environment. For fed insects, light intensity served as a limiting factor and flight did not occur at below-threshold light intensities. Above threshold

light intensities, temperature and relative humidity combine to produce upper and lower flight thresholds. Thus initiation of sustained flight occurs at ambient temperatures above 14°C and below 33°C and above water vapour pressures of approximately 10 mm. Starvation, however, radically alters all flight thresholds so that even in the dark some starved mosquitoes initiated flight. The migration threshold for an individual *A. aegypti* is the result of the combination of at least one habitat variable, energy uptake, and at least three migration-cost variables or their correlates—light intensity, temperature and water vapour pressure.

Figure 16.27 shows the accumulative percentage of aphids of two species that initiated migration from a more suitable habitat, a sugar-beet plant, and a less suitable habitat, a glass surface, in relation to temperature. There is a clear interaction between the habitat variable, food availability, and the migration-cost variable, temperature, in the production of the migration threshold. The variation is consistent with the suggestion that selection has favoured individuals that initiate migration in response to that combined value of threshold variables that corresponds to the situation at which $h_1 = \bar{E}$. The variation is also, therefore, consistent with the suggestion that temperature is a migration-cost variable. However, the figure is not consistent with the alternative suggestion that the temperature threshold merely indicates a physiological inability to fly at lower temperatures and is therefore unconnected with the adaptive nature of migration.

It was suggested earlier that the spring and autumn generations of host-alternating species of aphids are likely to be obligatory migrants and that therefore the migration threshold would consist of migration-cost variables and probably indirect variables. Because habitat variables are not involved (i.e. any value of habitat variable exceeds the migration threshold), it might be expected that the obligatorily migrant spring and autumn generations may have lower thresholds to migration-cost variables than the facultatively migrant summer generations. Such a situation was found by Haine (1955) in experiments on the influence of wind on migration threshold.

16.5 Physiological preparation for exploratory or removal migration

Most work on physiological preparation for migration has been concentrated on the return migrations of vertebrates rather than on removal or exploratory migration. One or two examples may be presented, however, that will serve to illustrate the similarity that exists between physiological preparation for the initiation of exploratory, removal, and return migrations.

In Chapter 8 it was suggested that many facultative migrants could make no physiological preparation for migration, whereas if decrease in habitat suitability was in any way predictable selection would favour those individuals with physio-morphological characteristics that increased their potential migration distance. This conclusion still applies to facultative migrants if clues are available that indicate in advance that h_1 will become less than $\bar{E}$. This situation has already been described in the example of the degree of development of the flight apparatus of the black bean aphid, *Aphis fabae* (p. 231), in relation to deme density during early development. A similar situation also seems likely to exist in a species of bird, the crossbill, *Loxia curvirostra*, in relation to facultative exploratory migration (p. 210). Evidently, clues concerning the size of the next spruce seed crop in the area occupied are available before habitat suitability becomes a function of that crop. This is implied by the observation (Newton 1972) that the birds deposit fat in advance of the initiation of long-distance exploratory migration. It is not known, however, whether such fat deposits are laid down every year and clearly the situation requires much more detailed investigation. If nothing else, however, the observation emphasises that the migration is not the death-wandering of starving birds as was once thought (e.g. Dorst 1962) but is an adaptive behavioural strategy.

Both of the above examples involve variation in habitat suitability that correlates well with some advance value of a habitat variable. When habitat suitability varies predictably with season the variable that correlates best with habitat suitability is neither a habitat variable nor a migration-cost variable. It is, instead, likely to be an indirect variable, the photoperiod/temperature regime. The combination of photoperiod and temperature provides a reliable indicator of season and latitude and it is to be expected that any seasonally predictable aspect of migration is likely to have produced selection on animals for the optimum linkage between appreciation of the photoperiod/temperature regime and the appropriate migration mechanism. However, although such a linkage can be demonstrated for the initiation and direction of return migration in birds (Chapter 27) and for the direction of removal migration in a butterfly (Chapter 19), it has not so far been demonstrated either for the initiation of removal migration or for the initiation of physiological preparation for migration. M.J. Norris (1959, 1962, 1964) has shown that ovaries of the red locust, *Nomadacris septemfasciata*, of equatorial Africa, mature when the photoperiod is increased from 12 to 13 hours, a change that corresponds to the natural day-length in the late dry season. As the initiation of migration in this species is associated with ovarian development (p. 241) it seems likely that photoperiod is involved in the initiation of migration. The point has not been proved, however, nor has any linkage been demonstrated between appreciation of photoperiod and physiological preparation for migration.

Experiments on the black bean aphid, *Aphis fabae* (Fig. 16.28), have shown that the flight duration is a function of the initial quantity of stored fat. The aphid stores most of its fat in the fat-body of the abdomen and among the flight muscles but may also use fat from other parts of the body (Cockbain 1961). Of post-teneral aphids flown to apparent exhaustion and then starved, only 28 per cent were able to fly the next day. Of similar individuals placed on young bean plants on which they fed, 77 per cent were able to fly the next day, though only for two hours or so. 97 per cent of unfed female desert locusts, *Schistocerca gregaria*, fly for 4 h on the first day in a flight mill, but only 70 per cent fly on the second day and with lessened speed and duration (Weis-Fogh 1952). On the third day, after feeding, the original performance was more or less restored. Popov (1954) has recorded a swarm of the species descending to feed only 3 km from the morning departure point, despite apparently favourable weather conditions Further migration was not initiated until 2 h had been invested in feeding. Gunn *et al.* (1948) suggested that the time of initiation of migration on one day may well be a function of the time invested in feeding the previous day.

Problems of flight duration in relation to fuel consumption and feeding during migration are not, however, of major concern in this section, the main requirement for which is an example that demonstrates that species or individuals that obligatorily initiate migration possess greater energy

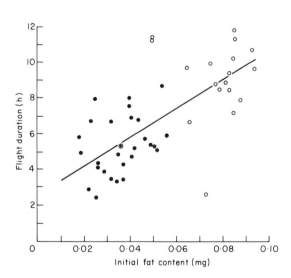

Fig. 16.28 Flight duration of the black bean aphid, *Aphis fabae*, in relation to initial fat content

Solid dots indicate results obtained with aphids from a culture. Open dots indicate results obtained with aphids from the field. The aphids used were parthenogenetic viviparous females of the summer generations.

There is a clear positive correlation between flight duration and initial fat content in this species. However, although this indicates that fat reserves can be used as fuel during migration, the first hour of flight uses glycogen. Quite possibly the majority of migrants do not need to use their fat reserves for migration.

[*After Cockbain* (*1961*)]

reserves than species or individuals that migrate only facultatively. It might be expected, for example, that insect species that perform obligatorily post-teneral migration accumulate relatively more energy reserves during larval development than species that do not obligatorily initiate such a migration. Similarly, in species in which some individuals are obligatory migrants and others are not, the obligatory migrants might be expected to possess more energy reserves than other individuals. The situation is complicated in aphids with wing polymorphism in that the energy budget has also to accommodate the development and maintenance of the flight apparatus in migrants. Information concerning stored energy by migrant and non-migrant aphids is therefore extremely difficult to interpret in the present context. Phase *gregaria* locusts have a greater supply of 'fuel' than phase *solitaria* locusts (Kennedy 1956) though whether this supports the above expectation, as it may appear at first sight, cannot be evaluated until the initiation of migration by *solitaria* locusts (Davey 1959) is better understood in relation to predictable (to the locust) and unpredictable variations in habitat suitability.

Although relevant information may exist, therefore, that I have been unable to locate, I can present no conclusive evidence that the selective pressures associated with predictable variation in habitat suitability have resulted in obligatory removal or exploratory migrants making physiological preparation for migration in the same way as has been demonstrated for return migrants (Part IIIb). The only exception involves the rather specialised situation of the development of the flight apparatus in wing-polymorphic aphids. Such aphids also provide examples of linkage between appreciation of photoperiod and preparation for seasonally obligatory migration (Lees 1966), though the effect has been monitored by the production of sexual individuals that should migrate to the winter host rather than by a direct observation of the initiation of migration.

16.6 Discussion

Most of the discussion of this chapter has been presented in the main body of the chapter as the various points were raised. In addition, the degree of support available for the different aspects of the hypothesis presented in chapters 8 and 9 is reviewed in the chapter summary. There is, however, one aspect of the initiation of non-calculated removal migration and exploratory migration that requires a concluding discussion. That is the distinction between facultative and obligatory migration and the associated distinction between indirect, migration-cost, and habitat variables.

16.6.1 The distinction between facultative and obligatory migration

It is now possible to distinguish between facultative and obligatory migration in a reasonably useful way:

facultative migration is initiated as a result of the attainment of an above-threshold value of at least one habitat variable in the course of some sequence of events;

obligatory migration is initiated as a result of a sequence of events but is not due to the attainment of above-threshold value(s) of habitat variable(s).

The main facultative effect in any sequence of events leading to migration is a triggering of a 'migration physiology' which includes the setting of a number of clock mechanisms and/or responses to various migration-cost variables in relation to which migration is subsequently initiated. Hence, the migration physiology of the rabbit, *Oryctolagus cuniculus*, seems to be triggered by exposure to a sub-threshold level of energy uptake by the individuals and/or deme density. The migration physiology includes a clock mechanism and/or a response to light intensity (a migration-cost variable) that is such that exploratory migration is initiated each evening after the main feeding period and is terminated after a period of 15–20 minutes (p. 220). Similarly, the migration physiology of the summer generations of many species of aphids, which for many species involves the development of a flight apparatus, is a response to extra-threshold conditions of deme density and/or nett rate of energy uptake by the individual. If extra-threshold conditions exist from early development through to adulthood, the final manifestation of a migration physiology is developed which is a response to extra-threshold values of various migration-cost variables (temperature, wind-velocity, light intensity and/or time of day, etc.) by the initiation of migration.

In obligatory migration, the migration physiology develops either as an inevitable part of ontogeny or alternatively is triggered at a particular season and/or time of day, either by endogenous rhythms or by such rhythms in association with indirect variables such as the photoperiod/temperature regime or the light/dark or lunar cycles, or by indirect variables alone. The migration physiology includes a response to extra-threshold values of various migration-cost variables. In the palolo worms, *Eunice* spp., for example (p. 235), the first stage in the obligatory migration seems to be an ontogenetic physiological change that involves the animal becoming receptive to indirect variables such as day-length and the way this is influenced by the Moon. Certain day/lunar length values trigger an endogenous clock-mechanism that after a certain time leads to the initiation of migration at a certain stage of the light/dark cycle which may involve a separate clock mechanism. The migration is therefore obligatory, being performed independently of the value of habitat variables.

Having drawn this broad distinction, it is possible to list the types of environmental variables involved in the different stages of the sequence of events leading up to the initiation of migration. Variables involved at the moment of initiation of migration are invariably (except in experimental situations) migration-cost variables such as temperature, wind-velocity, rain, light intensity, etc. all of which are likely to vary considerably from moment to moment and to influence migration cost and hence $\bar{E}$ (the mean expectation of migration).

Environmental variables that are likely to be involved in the initiation of the sequence of events of which the initiation of migration is the end-result, however, can be divided into two broad categories corresponding respectively to obligatory and facultative migration. The variables of the first category may be described as indirect variables and are such factors as the photoperiod/temperature regime, the lunar cycle, and the light/dark cycle. These indirect variables may correlate either with fluctuations in h_1 or $\bar{E}$ or both, but are characterised by being predictable and hence correlate with predictable variations in the relationship between h_1 and $\bar{E}$. The second category, on the other hand, may be described as habitat variables and are such factors as deme density, availability of food, temperature, etc. Habitat variables correlate with fluctuation in h_1 and are characterised by being unpredictable. However, the falling of h_1 below $\bar{E}$ may be only the end-result of a sequence of habitat changes that, once initiated, follows a predictable course. Hence, whether or not the set of spruce seeds will be light or heavy is unpredictable from year to year but beyond the stage of seed formation, whether the habitat will be suitable later in the year for a crossbill, *Loxia curvirostra*, is likely to be relatively predictable.

It will be noticed that some environmental variables fit into all three categories. Unpredictable extremes of temperature, for example, constitute a habitat variable. Relatively predictable seasonal fluctuation in temperature in association with photoperiod constitutes an indirect variable. Finally, the temperature at the moment of initiation of migration may constitute a migration-cost variable. It may not always be easy, or even necessary, to distinguish the nature of temperature as an environmental variable in any particular instance, but it seems desirable to bear in mind this possible triple involvement of a single variable in the initiation of migration.

16.7 Summary of conclusions concerning the non-calculated initiation of migration

16.7.1 Exploratory migration and non-calculated removal migration

1. Two major types of non-accidental migration beyond the pre-migration familiar area can be recognised: exploratory migration and non-calculated removal migration (p. 26).
2. In exploratory migration the intermediate destinations are unknown to the animal when migration is initiated. However, the ability is retained throughout the migration to return to the original familiar area which is often also the eventual destination.
3. In non-calculated removal migration the destination is beyond the pre-migration familiar area, the ability to return to which is not retained during the migration.

16.7.2 Facultative and obligatory migration

4. Both exploratory migration and non-calculated re-

moval migration can be further categorised into facultative and obligatory migrations according to the factors involved in the initiation of migration.
5. In facultative migration a sequence of mainly physiological events, the end-result of which is the initiation of migration, is initiated in response to a certain value of a habitat variable or variables (p. 264).
6. In obligatory migration a sequence of mainly physiological events, the end-result of which is the initiation of migration, is initiated without reference to habitat variables. The initiation of this sequence of events may either be an invariable part of ontogeny or it may involve reference to indirect variables (p. 264).

16.7.3 Indirect, habitat, and migration-cost variables

7. Environmental variables involved in the initiation of migration can be broadly divided into three classes: indirect variables, habitat variables, and migration-cost variables (p. 264).
8. Indirect variables are factors such as the photoperiod/temperature regime, the lunar cycle, and the light/dark cycle that correlate with predictable fluctuations in an animal's environment (p. 60).
9. Habitat variables are factors such as deme density, availability of food, microclimate, etc. that correlate with unpredictable fluctuations in the suitability of an animal's habitat (p. 47).
10. Migration-cost variables are factors such as temperature, wind velocity, rainfall, light intensity, etc. that correlate with migration cost (p. 49).

16.7.4 Evaluation of the hypothesis (Chapters 8 and 9)

Basic hypothesis

The main thesis of this chapter is that selection should favour the initiation of non-accidental migration beyond the pre-migration familiar area when h_1 (the suitability of the occupied habitat) is less than $\bar{E}$ (the mean expectation of such migration). Facultative migration evolves when fluctuations in h_1 are unpredictable. Obligatory migration evolves when fluctuations in h_1 are predictable.

Facultative exploratory and non-calculated removal migration

1. *Thesis:* In facultative migration selection is likely to favour the evolution of a migration threshold to a habitat variable that corresponds to the situation in which $h_1 = \bar{E}$ (p. 47).

Support: Strong support can be derived from an analysis of the copulation site to copulation site migration by copulating pairs of the yellow dung-fly, *Scatophaga stercoraria*, in relation to male density in the original habitat (Fig. 16.10).

2. *Thesis:* When the distribution of values of hM (where h = habitat suitability, and M = the migration factor for migration to that habitat) has a single mode there is some justification in accepting a single value for $\bar{E}$ (where $\bar{E} = \overline{hM}$). When the distribution of hM shows a number of modes and as long as the animal can exert some control over which spatial unit it leaves and in which spatial unit it remains, several values for $\bar{E}$ corresponding to these modes should be recognised. In such situations a hierarchy of migration thresholds would be expected to evolve, each one corresponding to a different value of $\bar{E}$ (p. 48).

Support: Some support can be obtained for the existence of a hierarchy of migration thresholds in two species of aphids, the sycamore aphid, *Drepanosiphum platanoides* (p. 229) and the black bean aphid, *Aphis fabae* (p. 231). However, there are no data relating migration thresholds to $\bar{E}$.

3. *Thesis:* The migration threshold should be a function both of the value of the habitat variable and the length of exposure to that variable (p. 48).

Support: No data.

4. *Thesis:* The migration threshold is likely to be composite and to consist of a number of interacting variables, both habitat variables and migration-cost variables, though the action of these two categories of variables may be separated in time. The action of selection should be to produce a curve of migration threshold that at all points corresponds to the situation at which $h_1 = \bar{E}$ (p. 50).

Support: In two aphid species the migration threshold to the migration-cost variable of temperature was higher for individuals on a relatively favourable habitat than for individuals on a relatively unfavourable habitat (p. 261), an observation that is consistent with the thesis.

5. *Thesis:* When subjected to above threshold values of migration-cost variables that inhibit migration, the migration threshold should lower with time until eventually migration is initiated irrespective of the value of migration-cost variables (Fig. 8.4).

Support: Aphids subjected to continuous above threshold wind-velocities eventually initiate migration (p. 255), though the initiation is much delayed.

6. *Thesis:* When habitat suitability fluctuates seasonally, the value of $\bar{E}$ also fluctuates seasonally. In such cases, the migration threshold should also fluctuate seasonally (p. 48).

Support: The barren-ground caribou, *Rangifer tarandus*, in Canada in winter shows at least three separate migration thresholds to the habitat variable

of snow cover (p. 212). The migration threshold changes as winter progresses.

7. *Thesis:* Animals adapted to habitats such that there is a high probability that facultative migration may become immediately advantageous at some time in their lives should show a higher level of constant physio-morphological preparedness for migration than other animals (p. 59).

Support: Water beetles adapted to habitats with a short longevity show a greater ability to fly than water beetles adapted to habitats with a greater longevity (Fig. 16.18). Whether this correlation supports this thesis or thesis 10 for obligatory migration remains to be demonstrated.

8. *Thesis:* Animals adapted to habitats such that fluctuations in suitability are unpredictable but for which the moment at which facultative migration becomes advantageous is predictable beyond a certain point are likely to make physiological preparation for a facultative migration (p. 262).

Support: The crossbill, *Loxia curvirostra*, seems to be able to determine the size of the seed crop of spruce before it becomes dependent on that crop (p. 210). Fat is laid down prior to migration. It is not known, however, whether or not fat deposition occurs in years when facultative migration does not occur.

9. *Thesis:* Optimum stay-time in habitats encountered for which h is less than $\bar{E}$ should be an inverse function of the extent to which $\bar{E}$ exceeds h (p. 82). As migration continues, however, and more and more habitats are encountered for which h is less than $\bar{E}$, the migration threshold to migration-cost or habitat variables also increases (p. 82).

Support: Strong support from experiments on the settling responses after migration of the black bean aphid, *Aphis fabae* (p. 231).

Obligatory exploratory and non-calculated removal migration

10. *Thesis:* Migration is ontogenetically obligatory when h_1 is always less than $\bar{E}$ at a certain stage of ontogeny. When a species is adapted to a habitat with an ontogeny of suitability such that newly formed sites are more suitable than older sites and when this curve of suitability interacts with the generation interval in a certain way (Fig. 8.7), the selective pressure of rate of appearance of new habitats should lead to the evolution of obligatory migration.

Support: A water beetle, rock pool corixids, and locusts emerge from hibernation or become adult coincidentally with a period of high rate of appearance of new habitats. These insects seem to be obligatory migrants. A general correlation for water beetles between ability to fly and rate of appearance of new habitats (Fig. 16.18) may or may not add further support to this thesis.

11. *Thesis:* When ontogenetic stage, season, or time of day coincides with a change in an animal's resource requirements, and when these requirements occur in spatially separate habitats, migration is ontogenetically, seasonally, or diurnally obligatory. It is suggested that the *present* function of migration for such species is that it maximises individual reproductive success by allowing the individual to perform each time-investment component in the habitat-type most suitable for the performance of that component (p. 251).

   Support: Although many examples were given (pp. 243–4) of animals adapted to perform different time-investment components in spatially separate habitat-types, and although in some it seemed likely that the habitat-types used were those most suitable for the performance of the relevant time-investment component, in no case has this been demonstrated conclusively.

12. *Thesis:* It is quite meaningless in evolutionary terms to demonstrate a correlation between the spatial separation of habitat-types and migration between these habitat-types. The critical consideration is the sequence of events that led the ancestors of such species to become adapted to spatially separate habitat-types. It is suggested that ancestral species with a higher probability of initiation of migration, whether facultative or obligatory, were more likely to become adapted to spatially separate habitat-types than ancestral species with a lower probability of initiation of migration (p. 248).

   Support: A correlation exists for many insects between adaptation to spatially separate habitat-types and larval adaptation to habitats with a high rate of formation of new sites and/or short habitat longevity (p. 249).

13. *Thesis:* The sequence of events that is likely to precede the seasonally obligatory initiation of migration is likely to be triggered by indirect variables (p. 60).

   Support: The pre-spawning migration of palolo worms, *Eunice* spp., is triggered by a combination of lunar cycle and day-length (p. 235). In host-alternating aphids (Fig. 16.17) the production of sexual individuals, which are often obligatory migrants, is a function of the photoperiod/temperature regime (p. 263).

14. *Thesis:* Obligatory removal migrants are likely to make physiological preparations in advance of initiating migration such that potential migration distance is increased (p. 59).

   Support: Phase *gregaria* locusts have more fuel for flight than phase *solitaria* locusts (p. 263). Until more is known about the migration of phase *solitaria* locusts, however, the degree of support for the thesis offered by this observation is difficult to evaluate.

Exploratory migration

15. *Thesis:* There is an optimum time interval between the initiation of exploratory migration and the initiation of calculated removal migration (Fig. 9.2).

    Support: No data.

16. *Thesis:* In animals in which exploratory migration and non-calculated removal migration are alternative strategies, exploratory migration should always be initiated first, either facultatively or obligatorily. Only if no habitat is found that is more suitable than $\bar{E}_n$ (the mean expectation of non-calculated removal migration) after the optimum period of exploratory migration, and only if by that time h_1 has decreased below $\bar{E}_n$, should non-calculated removal migration be initiated (p. 66).

    Support: This sequence of events seems to occur in the sea otter, *Enhydra lutris*, and the muskrat, *Ondatra zibethicus* (pp. 223 and 101). However, no measurements of migration thresholds or $\bar{E}_n$ are available.

16.7.5 Conclusion

Although there is general support for many aspects of the hypothesis presented in Chapters 8 and 9, in no case, with the exception of the copulation site to copulation site migration of the yellow dung fly, *Scatophaga stercoraria*, are the data fully quantitative.

17

Intra-specific variation in removal migration incidence in a free situation

17.1 Introduction

Chapters 14–16 have been concerned with the initiation of migration, particularly with the behavioural and physiological strategies that increase the chances of a given migration being advantageous to the individual concerned. Chapter 16 also examined some instances of inter-specific variation in removal migration incidence. This present chapter, which is the last to be concerned solely with the initiation of migration, examines intra-specific variation in migration incidence. Both the variation between individuals of the same age and sex and the variation with age and sex are discussed. The theoretical aspects of this situation have been discussed in Chapter 10 and the extent to which the theses presented in that chapter are supported by the available evidence is evaluated in the summary at the end of this chapter (p. 356). It should be noted that this chapter is concerned primarily with intra-specific variation in migration incidence in a free situation. Variation in a despotic situation is discussed in more detail in Chapter 18.

17.2 Variation between individuals of the same age and sex

In the theoretical discussion (Chapter 10) of the evolution of individual differences in the initiation of migration, the topic was discussed in relation to two categories of selective pressures. The two categories were first, selection due to correlation between deme

density and habitat suitability, and second, selection due to unpredictable variation in $\bar{E}$, the mean expectation of migration, about a mean value. Let us consider first some of the evidence that relates to the main suggestions made in connection with selection due to correlation between deme density and habitat suitability.

Tanner (1966) examined the results of a number of laboratory and field studies on deme growth rates and calculated correlation coefficients for deme growth rates against deme density. On Tanner's assumptions it would follow that a negative correlation coefficient would indicate that habitat suitability decreases as deme density increases. Out of a total of 64 species consisting of 4 invertebrates other than insects, 15 insects, 7 fish, 19 birds, and 19 mammals, only two species, a fish and a mammal, showed a positive correlation. Furthermore, only one of these positive correlations, that for the mammal, which was man, was significantly different from 0·0. If we accept Tanner's analysis, therefore, it follows that it is of fairly widespread occurrence that habitat suitability decreases as deme density increases. More recently, however, Eberhardt (1970) has pointed out that even using random numbers a negative correlation of −0·5 can sometimes be generated using Tanner's methodology. If this is so then for the purpose of relating habitat suitability to deme density, correlation coefficients should be compared, not with 0·0, but with some negative figure. Although it seems likely, therefore, that habitat suitability is often an inverse function of deme density, the relationship may not be quite as widespread as was once thought.

The thesis on which the calculations of optimum distribution of migration thresholds within a deme are based (p. 69) is that, given a free (as opposed to despotic) situation and given an inverse relationship between habitat suitability and deme density for levels of habitat suitability less than $\bar{E}$, selection will stabilise when the frequency distribution of migration thresholds is such that no matter what the habitat suitability, the reproductive success of each individual that initiates migration is equal to the reproductive success of each individual that does not initiate migration. Evidence that supports this hypothesis is again provided by the work of Parker (1970c) on the common yellow dung-fly, *Scatophaga stercoraria* (see also pp. 94 and 226), this time in relation to the dung pat to dung pat migration of males. This evidence is presented in Fig. 17.1.

Figure 17.1 shows clearly that, as far as reproductive success as a function of migration threshold is concerned, the frequency distribution of migration thresholds within the deme is such that all individuals achieve the same reproductive success, thus supporting the hypothesis.

As well as permitting some evaluation of the theoretical basis of selection for the distribution of migration thresholds within a deme, the dung-fly work also permits an evaluation of the model developed in Section 10.2.2.1 (p. 69) for the calculation of the distribution of migration thresholds. Before this can be done, however, it is necessary to discuss the determination of the values of the various components of this model. This involves discussion of the values of h_p, h_d, h, and $\bar{E}$.

Variation in the value of h_p, the component of habitat suitability that results from the physical and biological construction of the habitat excluding all other competing individuals is shown in effect in Fig. 17.1(a). As the field data have been corrected to refer to a constant ambient temperature and constant dung pat size on a single date it is possible to represent h_p as far as a male dung-fly is concerned solely by rate of arrival of receptive females. h_p is also, of course, a function of degree of insolation, pat size, time of year, and abundance of predators, etc. and appropriate migration thresholds seem to have evolved (Parker 1970b). As far as the present analysis is concerned, however, it is reasonable to take female arrival rate as a measure of h_p. It can be seen from Fig. 17.1(a) that there is a female arrival rate of about 0·22 females per min at about 10 min after deposition of the pat. For convenience, a value of 0·22 females per min can be taken to represent $h_p = 1·0$, though by extrapolation it seems that maximum habitat suitability occurs even sooner

after pat deposition. By 300 min after pat deposition the value of h_p has decreased to about 0·076.

Variation in the value of h_d, the component of habitat suitability that results from the presence of competing individuals (see p. 70), is shown in effect in Fig. 17.1(b). If only a single male were to be present at the dung pat it could not possibly manage to copulate with all the arriving females unless it modified the normal copulation strategy, for under normal competitive conditions, copulation plus female guarding lasts on average about 50 min. There is, therefore, a range of male densities from one male per 500 cm² of dung surface up to some greater number throughout which each male achieves the maximum possible fertilisation rate. Normally, however, males outnumber females so much, by at least 3 : 1, that the reproductive success of an individual male is likely to be an inverse function of male deme density. In this case, the value of h_d is given by

$$h_d = 1/N,$$

where N is the number of males around the pat. It can be seen from Fig. 17.1(b), therefore, that h_d reaches a low of about 0·033 approximately 30 min after pat deposition, and thereafter rises until by 300 min after deposition, $h_d = 0·367$.

The value of h, habitat suitability, is given by $h_p h_d$ (p. 70). Variation in the value of h is shown in effect in Fig. 17.1(c). A value for h of 1·0 would be given by a female quota of 0·22 females/male/min. A female quota of 0·006 females/male/min, therefore, represents a value for h of 0·027. It can be seen from Fig. 17.1(c) that $h = 0·031$ about ten minutes after pat deposition, but that from about 200 min after deposition onward it has reached a fairly stable low of $h = 0·020$.

The value of $\bar{E}$, the mean expectation of migration, is estimated in this case by the product of $\bar{h}$, the mean suitability of the habitat at which the individual is likely to arrive and $\bar{M}$, the mean migration factor (see p. 78). The value of $\bar{M}$ can be estimated (see p. 41) by dividing the nett female quota (Fig. 17.1(d)), which is 0·00485 females/male/min, by the average female quota while around the pat, which is 0·00527 females/male/min (Parker 1970c, p. 214). This gives a value for $\bar{M}$ of 0·92. The value of $\bar{h}$ may be taken to be the value of h 4 min after pat deposition, the average time of male arrival. This cannot be measured directly from Fig. 17.1(c), but by extrapolation would seem to be about 0·032. From a value for $\bar{h}$ of 0·032 and a value for $\bar{M}$ of 0·92, a value of 0·0295 is obtained for $\bar{E}$, the mean expectation of migration. A value for $\bar{E}$ of 0·0295 corresponds to a female quota of about 0·0065 females/male/min. The curve of habitat suitability (Fig. 17.1(c)) falls below this value at about 15 min after pat deposition. According to the migration equation, therefore, no migration should be initiated until about 15 min after pat deposition. This agrees extremely well with the situation observed in the field (Parker 1970c, p. 217). Further support for the theoretical basis of the model for the calculation of optimum distribution of migration thresholds is given in Fig. 17.2.

It should be noted that in both Figs 17.1 and 17.2, stay-times have been substituted for migration thresholds. It is assumed here, however, that the dung-flies are responding, not to an endogenous clock, but to migration thresholds to habitat variables such as the state of the dung surface and the deme density of competitors. It is also assumed that in general there is a strong correlation between time

after pat deposition and the value of threshold habitat variables. Any calculation of the individuals that should initiate migration during a particular period after pat deposition, therefore, is in effect a calculation of the individuals the migration thresholds of which are likely to be exceeded during that period. The advantage of using stay-times, however, is that they constitute a much more readily quantifiable parameter than most threshold habitat variables, as long, that is, as compensation is made for such variables as ambient temperature and degree of insolation.

The dung-fly situation, of course, lends itself extremely well to this type of analysis in that reasonable measures of habitat suitability, reproductive success, mean expectation of migration, deme density, etc. are more or less readily available. Such situations are few and far between, however, and it is not surprising that comparable examples

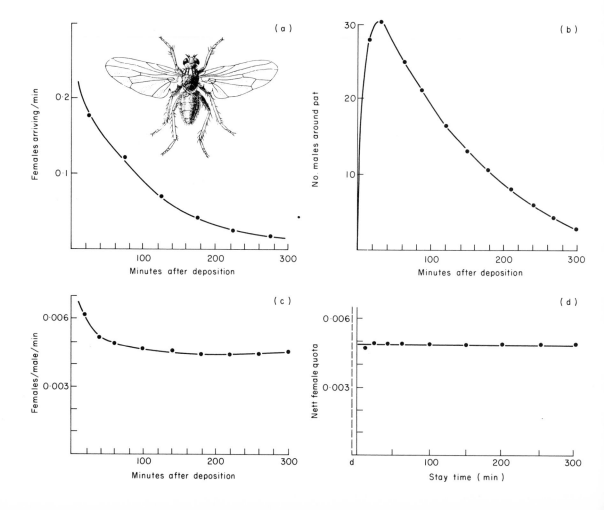

Fig. 17.1 A test of the hypothesis that selection on the frequency distribution of migration thresholds within a deme should stabilise when for all habitat suitabilities the reproductive success of each individual that initiates migration is equal to the reproductive success of each individual that does not initiate migration: the dung pat to dung pat migration of males of the common yellow dung-fly, *Scatophaga stercoraria*, in western England

Female dung-flies oviposit in the freshly deposited dung of a large number of animals but in particular that of cows, *Bos* spp. Male dung-flies aggregate on dung pats and divide their search time for females between the dung surface and the surrounding grass. Females show a level of promiscuity that is unusual among female insects and copulate every time they arrive at the dung to oviposit, even though such matings are not necessary to ensure 100 per cent fertility of the eggs. During copulation the male flushes out 80 per cent of the sperm already present in the female's reproductive tract from previous matings with other males. After copulation, while the female oviposits in the dung, the pair remain in a 'passive phase' (i.e. genital contact has ceased but the male remains in position on the female's back). During the passive phase the male fends off other males with his legs. From time to time attacking males succeed in 'taking-over' by displacing the original male. Take-overs may occur during copulation though they occur rather more often during the passive phase. Females allow full copulation by males that have taken over from other males. During this copulation the new male again displaces 80 per cent of the sperm of other males from previous matings.

After a certain stay-time searching for a female on a single dung-pat a male initiates migration to another pat. The factors involved in the initiation of this migration have not been fully evaluated. The migration threshold seems likely to involve not only deme density but also the thermal and chemical state of the dung surface, both of which probably correlate with pat age. Furthermore, while searching for females, males spend some time in the grass on the upwind side of the dung pat from which position olfactory information concerning other freshly deposited pats may be received (p. 94). It is assumed, therefore, that stay-time is primarily a function of response to exogenous rather than endogenous events. There is, however, a frequency distribution of stay-times within the deme studied that indicates a similar frequency distribution of migration thresholds to habitat variables. The critical question in relation to the hypothesis being examined is whether the distribution of stay-times is such that all individuals experience the same potential reproductive success.

In the following figures the reproductive success of a male dung-fly is measured as the 'female quota' for that male where female quota is the number of females inseminated per minute by that male. All data given refer to an ambient temperature of 20°C and a pat surface area of 500 cm^2 in late August 1966.

Observation and calculation have produced the following data. The average female quota for a male dung-fly while around a dung pat is 0·00527 females per minute. The average time taken for a male to find, mate and 'guard' a female during oviposition is 190 min while

around the dung. Of these 190 min, 140 are spent searching for the female.

(a) The average arrival rate of female dung-flies with time after deposition of the dung

The maximum rate of arrival occurs soon after deposition. This rate falls off steeply during the first 2 hours but after 5 hours females are still arriving, though at only about 10 per cent of their initial rate.

(b) The average incidence of male dung-flies around a pat with time after deposition of the dung

(c) The average female quota experienced by males at different times after deposition of the dung

This figure is obtained by dividing the rate of arrival of the females given in (a) by the total number of males present at the same time, given in (b). If there was no migration cost, selection would be favouring those males with a short stay-time that arrived at a pat as soon as possible after deposition, exploited the initially high female quota, and then migrated to another freshly deposited pat. During the time invested in migration to another pat, however, males that do not initiate migration are experiencing a positive female quota. Furthermore, any increase in males with a short stay-time decreases the competition experienced by other males for later-arriving females, thus increasing the female quota of those males with longer stay-times. If it is assumed that all males arrive at the pat at the same time, 4 minutes after pat deposition (where 4 minutes is the average arrival time after deposition) this figure also shows the average female quota for males of different stay-times.

(d) The average *nett* female quota experienced by males of different stay-times if the time invested in migration to a new pat equals the average time of arrival after deposition of a pat

d is the time of deposition of the dropping. Nett female quota is female quota taking time invested in migration from pat to pat into account. The figure is derived from (c) on the assumption that it takes a male 4 minutes to locate and arrive at a new pat after initiating migration from the original pat. The cost of this time invested during migration is greater when the original pat is still fresh and the average female quota is still high. Consequently, in comparison with (c), migration cost reduces the nett female quota of males with shorter stay-times more than that of males with longer stay-times. It can be seen from the figure that males staying around the pat for any duration longer than about 12 min all experience an equal nett female quota.

The distribution of male stay-times at a dung pat in the deme investigated is such that no males stay less than about 15 min. The data in (d) support the suggestion that selection on the frequency distribution of migration thresholds within a deme should stabilise when for all habitat suitabilities the reproductive success of each individual that initiates migration is equal to the reproductive success of each individual that does not initiate migration.

[*Compiled from Parker (1970c) to which paper reference should be made for details of the calculations involved*]

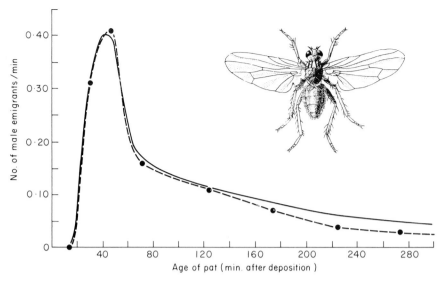

Fig. 17.2 Evaluation of a model for the calculation of the equilibrium distribution of migration thresholds within a deme: the dung pat to dung pat migration of males of the common yellow dung-fly, *Scatophaga stercoraria*, in relation to pat age.

The diagram shows the calculated and observed curves of rate of initiation of migration from one pat to another by males of the dung-fly, *Scatophaga stercoraria*. The close similarity between the two curves supports the validity of the theoretical basis of the model by which the calculated curve was obtained.

Dashed line, calculated curve; solid line, observed curve. All figures refer to an ambient temperature of 20°C and a dung pat with a surface area of 500 cm² in late August 1966 in western England.

Theoretical basis

The theoretical basis of the following calculation is that selection will have stabilised when the distribution of migration thresholds is such that for any habitat suitability between $h_1 = \bar{E}$ (where h_1 is the suitability of the habitat occupied and $\bar{E}$ is the mean expectation of migration) and $h_1 = 0$, the reproductive success of each individual that initiates migration is equal to the reproductive success of each individual that does not initiate migration.

The value of h_1 is given by $h_1 = h_p h_d$ (where h_p is the component of habitat suitability that results from the physical and biological construction of the habitat excluding all other competing individuals, and h_d is the component of habitat suitability that results from the presence of competing individuals). In this case, h_p is a function of x, the rate of arrival at the dung pat of receptive females (see Fig. 17.1(a) such that $h_p = x/0.22$. The value of h_d is a function of N, the number of males/500 cm² of dung surface such that $h_d = 1/N$. It follows, therefore, that h_1, the suitability of the habitat occupied is given by $h_1 = x/0.22N$. When $h_1 = \bar{E}$, $N = N_E$. The number (N_E) of males/500 cm² that result in h_1 being equal to $\bar{E}$ is therefore given by $N_E = x/0.22\bar{E}$. The value of $\bar{E}$ is a constant that for the pat-to-pat migration of males has been calculated (p. 270) to be 0.0295. In this case, therefore, N_E is related to x, the rate of arrival at the dung pat of receptive females, by the formula $N_E = x/0.0065$. The proportion, f_m, of a deme that should migrate over a particular period is given by $f_m = 1 - N_E/N_m$ if the reproductive success of both migrants

and non-migrants is to be equal (where N_m is the number of males/500 cm² of dung-surface at the beginning of the period, and N_E is the value of N_E that is appropriate to the value of x at the end of the period).

Specimen calculation

The value of N_E exceeds the value of N_m until 15 min after pat deposition (i.e. h_1 is greater than $\bar{E}$ during this period). Consequently, no migration is initiated. From 15 min after deposition onwards, however, N_E is less than N_m (i.e. h_1 is less than $\bar{E}$) and some individuals initiate migration to another pat. Let us determine the expected rate of initiation of migration during the period from 15–45 min after pat deposition. The value of N_m is given by the average total number of males that arrive at the pat during the first 45 min after deposition. This may be taken to be about 33 individuals (Parker 1970c, Figs. 1 and 12). The value of N_E after 45 min can be calculated from the above formula where $x = 0.144$ females/min and N_E is therefore 22.2 males. The proportion of individuals, f_m, that should initiate migration between 15 and 45 min after pat deposition is given by $f_m = 1 - 22.2/33.0$, i.e. $f_m = 0.327$. However, some of the males that arrive at the pat succeed in pairing with females. It may be assumed that the migration threshold (for pat-to-pat migration) of males paired to females is considerably higher than that of unpaired males such that paired males do not initiate migration at least until copulation and female guarding have been completed and only then if their unpaired threshold to pat age has been exceeded. On average, about 4.5 males are paired to females during the first 45 min after pat deposition (Parker 1970c, Figs. 4 and 5). Of these 4.5 males, it follows that 32.7 per cent, i.e. 1.47 would have initiated migration between 15 and 45 min after pat deposition if they had not paired with a female. The number of males that would be expected to initiate migration from the pat between 15 and 45 min after pat deposition, therefore, is given by $(33 \times 0.327) - 1.47$, i.e. 9.33. Mean rate of initiation of migration during the period is therefore given by 9.33/30, i.e. 0.311 males per min. This is in close agreement with the rate of 0.326 males per min that is observed (Parker 1970c, Fig. 12).

[*In the production of the above diagram, calculations have been carried out for the periods 15–45, 45–50, 50–100, 100–150, 150–200, 200–250, and 250–300 min after pat deposition. The curve of observed rates of initiation of migration are calculated from Parker (1970c, Figs. 1 and 12)*]

are in short supply in the literature, if they exist at all. In most cases, it is only possible to evaluate the support offered in relation to theoretical considerations in a much more qualitative way.

The partial migration from summer to winter sites by the alfalfa weevil, *Hypera postica*, has already been described (p. 225). It seems from the work of Prokopy and Gyrisco (1965) that only 46 per cent of

Fig. 17.3 Dung-flies, *Scatophaga stercoraria*, showing attempted take-over during copulation (bottom) and passive phase (top)

the deme studied were obligatory migrants, the remaining individuals possessing migration thresholds to the quality of the food, moisture of the soil, degree of protection from high temperatures, and deme density. Possibly 3 per cent of the weevils are obligatory non-migrants. In the absence of detailed information concerning reproductive success in different habitats under different conditions over a number of years, however, it is difficult to decide whether these proportions of obligatory migrants and non-migrants and the distribution of migration thresholds among the facultative migrants represents a response to selection due to a correlation between deme density and habitat suitability (p. 69) or a response to selection due to unpredictable variation of $\bar{E}$ about a mean value (p. 71).

Rather more information is available concerning obligatory non-migrants than obligatory migrants as the former can often be identified morphologically. This is particularly the case for those species that are unable to fly yet that are members of groups such as insects and birds for which the ability to fly is normal. Even here, however, no analysis is possible comparable to that for the dung-flies. Some of the problems involved in the analysis of the degree of manifestation of the ability to fly have been discussed in the previous chapter (pp. 245 and 266) in relation to the flight ability of water beetles. It seems clear, however, that the species with the maximum proportion of individuals unable to fly are those adapted to habitats, such as rivers and springs, for which $\bar{E}$, the mean expectation of migration (by flight) is likely to be relatively low, and for which h_1, suitability of a habitat occupied, is likely to be relatively high and stable. In such habitats the frequency with which h_1 is likely to fall below $\bar{E}$ is likely to be minimal, perhaps even zero over long periods in time. This is just the situation in which selection is suggested to lead to a deme of 100 per cent non-migrants. Note, however, that $\bar{E}$ in this case refers only to migration by flight from one river or spring to another. The value of $\bar{E}$ for migration by some other mechanism, such as walking or transport by water currents, between different parts of the same river may be much higher and may well fall within the range of variation of h_1. In this case, at least some of the deme are likely to be obligatory or facultative migrants between one part of the river and another. If this is so, then the proportion of the deme that performs such a migration may well be a function of the frequency with which h_1 is less than $\bar{E}$.

Insects and birds living on small, remote islands and insects living on mountain peaks are in habitats for which the value of $\bar{E}$ for migration over the level of distance for which flight is necessary is likely to be extremely low (due to a large inter-habitat distance) and it must be rare for h_1 to become less than $\bar{E}$. Lack of selection for the maintenance of flight mechanisms for migration is likely to be frequent and it is well known (e.g. Udvardy 1969) that in such habitats winglessness is often manifest. One family of birds in particular, the rails (Rallidae), show a marked tendency to become flightless on islands. Not only does a low value of $\bar{E}$ cause selection for migration by flight to decrease but also, if wind-induced accidental migration (Chapter 5) is at all probable, a low value leads to selection against any characteristic, such as flight, that may increase the chances of such accidental displacement.

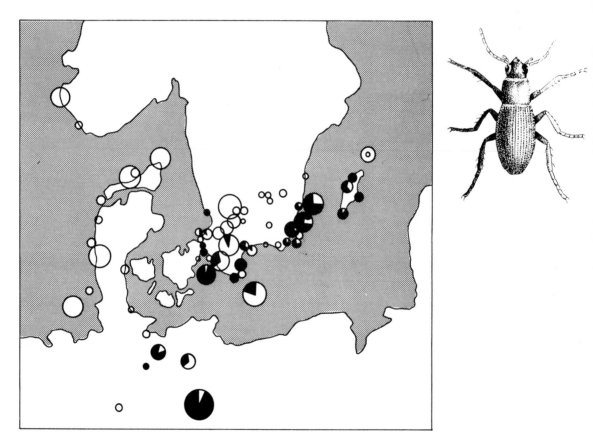

Fig. 17.4 Geographical distribution of obligatory non-migrants: the distribution of wing-forms of the dimorphic carabid beetle, *Calathus mollis*, in southern Scandinavia

Black circles and sectors, brachypterous individals that cannot fly; Open circles and sectors, macropterous individuals some or all of which can fly. The area of the circle is proportional to the number of individuals in the sample.

Macropterous individuals are recessive homozygotes and migrate further than brachypterous individuals. In general, demes furthest from the main areas of distribution of the species i.e. central and eastern Europe, consist entirely of macropterous individuals. The distribution is consistent with the suggestion that as the ice from the last glaciation has receded, new areas have been colonised first by macropterous individuals. The absence of brachypterous individuals in these areas, therefore, may not be explicable in terms of variation in habitat suitability in relation to mean expectation of migration, only in terms of the recent history of colonisation.

[*Simplified from Lindroth (1949)*]

Selection due to a low value of $\bar{E}$ cannot, by itself, however, account for all examples of flightlessness in birds and insects. In birds for example, flight is also a mechanism for escape from ground predators. If such predators are absent and if there is no selection for migration, either return or removal, over any distance requiring flight, then flightlessness may still evolve. This is the generally accepted explanation (Udvardy 1969) for the evolution of the flightless rail, *Notornis*, of New Zealand where mammalian ground predators were absent before the arrival of man some 600 years B.P.

Interpretation of the geographical distribution of flightlessness is often confused by founder effects in which the genetic constitution of the colonising individual(s) has a profound effect on the phenotype of subsequent generations. This is particularly the case when the species concerned is polymorphic for the ability to fly but where the most mobile phenotype is genotypically a recessive homozygote. In such a situation it seems reasonable that evolutionary interpretation of distribution should place more emphasis on relatively recent history than on adaptive significance, particularly at the edges of a species distribution range. A fine example of this type of situation is provided by Lindroth (1949) in relation to the ground beetle, *Calathus mollis*, in Scandinavia (Fig. 17.4). This beetle is dimorphic for wing development, existing either as a macropterous or brachypterous individual. Brachyptery is a dominant genetic trait in this species with the result that macropterous individuals are recessive homozygotes. Macropters, however, perform much longer-distance migrations than brachypters. The more-distant areas to have been colonised (or recolonised) during the short (7000–10 000 years) postglacial period, therefore, support entirely macropterous demes. Whether the relative distribution of the two forms is explicable in adaptive terms as well as historical terms must await investigation of the reproductive success of brachypterous individuals in areas that at present support only macropterous individuals.

Although a historical explanation for geographical variation in removal migration seems attractive for animals such as dimorphic ground beetles, similar explanations are far less attractive for such animals as birds, particularly those with a flight apparatus sufficiently well developed to perform long-distance return migrations. Nevertheless, such an explanation has been advanced by Haartman (1960) and promoted by Udvardy (1969) for the female pied flycatcher, *Ficedula hypoleuca*. Haartman (1949, 1960) found that mature female pied flycatchers in southern Finland and Sweden showed a much lower degree of return (=longer-distance removal migration; =less philopatry, in Udvardy's terms) during their seasonal return migration than females of the same species in Germany, the Netherlands, and Great Britain (Fig. 17.5). The suggested explanation was that during the last glaciation, the flycatcher was confined to areas in southern Europe. In postglacial times, however, it was individuals with a low degree of return that formed the vanguard of the advance of the species northwards. Scandinavia, being the area that has most recently shed its ice cover and become wooded, contains the greatest proportion of individuals with a low degree of return. There are, however, certain problems in the acceptance of this explanation. In particular, it is the males of this species, in common with the males of many similar passerine species, that establish the breeding territories in the spring, the females arriving slightly later. If either sex, as adults, is to be mainly responsible for the extension of the breeding range, therefore, it must be the male. Yet males show a uniformly high degree of return in all areas. Breeding range extension, therefore, is likely to be largely a function of removal migration (calculated, following exploratory migration the previous autumn?—p. 190) by the juveniles, both sexes of which show a relatively low degree of return (Fig. 17.26). The only age–sex class that seems unlikely, therefore, to be involved in the initial stages of breeding range extension following retreat of the ice after the last glaciation is that of adult females! There is, therefore, a major obstacle to the acceptance of Haartman's explanation. Finding an alternative explanation, however, is not easy. Nevertheless, one possibility derives from the suggestion made in Chapter 15 (p. 200) that differences in removal migration incidence between individuals that perform calculated removal migration after a period of exploratory migration need not necessarily have a genetic basis. The difference may be solely one of the relative suitability of the habitat occupied and habitats encountered during exploratory and/or, in this case, seasonal return migration.

Ontogenetic and sexual differences in migration incidence are considered later (Fig. 17.26) and it is only necessary here to consider the geographical variation in migration incidence shown by adult females. One of the major components of habitat suitability to a female pied flycatcher is likely to be

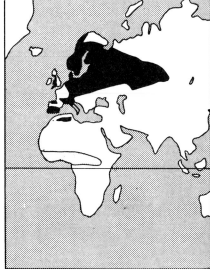

Fig. 17.5 Seasonal return migration of the pied fly-catcher, *Ficedula hypoleuca*

The pied flycatcher breeds (solid shading) in both broad-leaved and coniferous forests. The species winters (stippled shading) north of the equator in Africa in the canopies of evergreen forest and woodland savanna as well as in forest clearings and savanna. The autumn migration track has a strong southwesterly bias, birds from as far east as Poland and Moscow being recovered in western Iberia as well as in northern Italy.

In western Europe, juveniles show a low degree of return in their first spring. Adult males (bottom photo), on the other hand, show a high degree of return. Adult females (top) that have bred in Finland and Sweden show in general a lower degree of return than adult females that have bred in Great Britain, the Netherlands, and Germany.

[*Diagram modified from Moreau (1972). Description from Haartman (1960) and Moreau (1972). Photos by Eric Hosking*]

the presence of a competent and complementary male in a favourable habitat. It is reasonable, therefore, as a first consideration, to look for geographical differences in the abundance of such males. Fortunately, sufficient information is available for the pied flycatcher to permit at least a preliminary consideration of this suggestion.

The studies of various demes of pied flycatchers in Europe have been collected together by Lack (1966) from which the following data have been taken. One of the most notable features of the breeding biology of the pied flycatcher is the incidence of bigamy. Males establish several nesting holes in the spring, after their return migration (Fig. 17.5), and often these nesting holes are separated by a nesting hole maintained by another male. Presumably later-arriving females, if they are unable to find an unpaired male, occupy a male's second nesting hole rather than not breed at all. This is unlikely, however, to be a 'free' situation with more-competent males having more mates such that all females experience an equal reproductive success, because the male only feeds the offspring of the female with which he first establishes a pair bond. Females that, as a 'second' female, share a male seem likely, therefore, to be at a disadvantage relative to a female that either has a male to herself or shares a male but as a 'first' female (but not relative to an unpaired female). More females, therefore, experience a low value for h_1, the suitability of habitat occupied, in areas where bigamy is common than in areas where bigamy is less common.

Where information is available, there is a marked geographical variation in incidence of bigamy. In southwest Finland, 67 per cent of males have two mates or more. In central Europe, however, the range is from only 3 per cent near Berlin to 10 per cent near Dresden. On this limited information, therefore, a higher incidence of removal migration by females is found in areas in which females have a higher probability of sharing a male. Geographical variation in incidence of bigamy, therefore, which presumably is a reflection of geographical variation in differential survival by males and females, could well be the environmental variable that produces the geographical variation in removal migration incidence of adult females. The mechanism by which this variation could be produced, however, must await a better understanding of the process of removal migration in birds (see Chapter 15). According to the arguments of Chapter 8, a low h_1 would increase the probability of the migration

threshold (exploratory or non-calculated?) being exceeded. If removal migration is achieved by an exploratory–calculated sequence (Chapter 15), however, not only must the exploratory migration threshold be exceeded but also, if removal migration is to occur, H_f, the most suitable habitat encountered during exploratory migration, must be sufficiently suitable for h_1 to be less than $h_f M_f$, a probability that is also reduced in areas where bigamy is widespread. However, until more is known concerning: (1) the individual histories of the females that perform removal migration; (2) whether migrants go from a bigamous to a monogamous situation; (3) whether the migration is calculated or non-calculated; and (4) when, if the migration is calculated, habitat assessment by adult females takes place (i.e. autumn or spring, or both); there is little point in developing this hypothesis further.

17.3 Variation with age and sex

The format used in this section is that of a detailed consideration of migrations involving a wide range of animal species. For each migration, the consideration begins with an application of the model of the initiation of removal migration developed in Chapters 8–10. Using this model an attempt is made to generate a curve of variation in the probability of initiation of migration with age or time in a relative way for each of the species concerned. Where possible, quantitative techniques have been used, though unfortunately field data are scarce in most instances.

Ontogenetic curves of i, initiation factor ($\bar{h}\bar{M}/\bar{h}_1$ or $\bar{h}_f \bar{M}_f/\bar{h}_1$), should, according to the arguments of Chapter 10, be the same shape as ontogenetic curves of migration incidence observed in the field. Any agreement between these two sets of curves therefore provides support for these arguments and for the model of adaptive ontogenetic variation developed from them. Ideally, all analyses in this section should begin with a presentation of ontogenetic curves of $\bar{S}_d$, mean action-dependent potential reproductive success, and $\bar{m}$, mean migration cost, that have been obtained by quantitative measurement in the field. The ontogenetic curve of $\bar{M}$, mean migration factor, could then be calculated according to the formula $\bar{M} = 1 - \bar{m}/\bar{S}_d$. Then, again ideally, this ontogenetic curve for $\bar{M}$ should be modified by an ontogenetic curve of habitat quotient ($\bar{h}/\bar{h}_1$ or $\bar{h}_f/\bar{h}_1$), also measured in the field, to give an ontogenetic curve for i, initiation factor. The goodness of fit of the shape of this curve should then be tested statistically against that of observed ontogenetic variation in migration incidence. In such an ideal situation the model would have true predictive quality and the extent to which it could describe observed onto-genetic and sexual variation in probability of in-itiation of migration could then be evaluated in a totally objective manner.

Unfortunately, there are very few animals for which it is possible to construct a curve of $\bar{S}_d$ from quantitative field data, and in absolutely none have all components of $\bar{m}$, $\bar{h}_1$, $\bar{h}$ or $\bar{h}_f$ been measured. In no case, therefore, is it possible at present to evaluate the model in a totally quantitative and objective manner.

17.3.1 Ontogenetic variation in the incidence of group-to-group (type-GH) migration in non-avian vertebrates

In most of the animals discussed in this section, all that is possible is to evaluate the extent to which a curve of i, initiation factor, that fits the observed ontogenetic curve of migration incidence, and ac-ceptable curves of action-dependent potential repro-ductive success, migration cost, and habitat quotient can be integrated to form an 'adaptive package' that is consistent with the formula that $i = \bar{h}\bar{M}/\bar{h}_1$, where $\bar{M} = 1 - \bar{m}/\bar{S}_d$. Occasionally, if enough information is available, the development of such a 'package' for one sex or one type of migration enables certain predictions to be made for the other sex or another type of migration.

First, the procedures involved are described in detail for the group-to-group migration of a primate, the Japanese macaque, *Macaca fuscata*. Then a number of 'packages' are presented for a variety of mammalian species but with a minimum of procedural description. All of this first batch of examples are concerned with ontogenetic variation in animals that live for a number of years. Super-imposed on this long-term variation there are often short-term, particularly seasonal, variations. These are not discussed, however, until after the longer-term variations have been analysed.

Fig. 17.6 Ontogenetic variation in the incidence of short-distance deme-to-deme migration by male and female Japanese macaques, *Macaca fuscata*: application of a model of adaptive ontogenetic variation in the prob-ability of initiation of migration

A. Fitting a curve of ontogenetic variation in initiation factor to observed ontogenetic variation in migration incidence

The histograms show the proportion of males and females of different age-groups that performed removal migration to a neighbouring deme (=social group; =troop) in the early stages of deme division (formation of Y-troop; Itani *et al.* 1963). The process of deme division as a migration has been discussed in Chapter 16 (p. 216). The data for individuals older than about 17 years is unreliable, due to a shortage of individuals that were available to initiate migration.

The dashed curve shows the ontogenetic variation in i, the initiation factor, that, drawn by eye, is an acceptable fit to the data, bearing in mind the shortage of information for older individuals. The vertical axes were matched on the basis of the estimate that the time of maximum migration incidence for males was a time when the habitat quotient (see B(a)) was about 0·6.

B. Adaptive packages for the incidence of short-distance deme to deme migration

The ontogenetic variation of the initiation factor, i, shown in A, should, according to theory, be directly proportional to observed migration incidence. The value of i is given by the formula: $i = \bar{h}_f \bar{M}_f/\bar{h}_1$

(a) Ontogenetic curves of $\bar{h}_f/\bar{h}_1$, the habitat quotient, and $\bar{M}_f$, the mean migration factor

The curves of mean migration factor are calculated from the data in (b) according to the formula $\bar{M}_f = 1 - \bar{m}/\bar{S}_d$. In the case of short-distance migration the value of $\bar{M}_f$ is likely to be 1·0 throughout life for both males and females.

The curves of habitat quotient must, therefore, accord-ing to the above formulae, be identical to the curve of initiation factor shown in A. The curves produced are acceptable as curves of habitat quotient for the Japanese macaque and would be consistent with the following interpretation. Until the age at which weaning occurs (marked by the vertical arrow, W) all habitats other than that occupied by the mother are of zero suitability and the habitat quotient is, therefore, zero. After weaning, the suitability of habitats other than the one occupied by the mother become increasingly of above-zero suitability and the habitat quotient rises. However, the main component of habitat suitability during early years is the success enjoyed in competition with other members of the deme in obtaining the resources of the habitat (i.e. the positions reached in the feeding and other social hierarchies). As this success is considerably influenced by the activities of the mother, it is unlikely that the habitat quotient will reach unity. If, as seems likely, the main advantage of the hierarchical system is that of learning to avoid competition with individuals against which there is little chance of success, then once a position in a hierarchy has been established the individual makes the most efficient invest-ments of time and energy that is possible in that particular

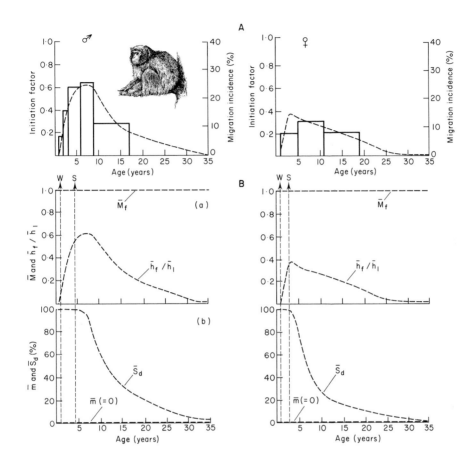

deme. If the individual migrates to another deme, however, until that individual has become integrated into the hierarchy, a process which in itself involves time, energy, and wound risk, then that individual will be making sub-optimal investments of time and energy. It seems likely, therefore, that the habitat quotient is always, on average, less than 1·0 for both males and females of the Japanese macaque. The onset of sexual maturity is indicated by the vertical arrow, S. The production of offspring begins more or less immediately after S in females but only after some delay in males. From the beginning of offspring production, however, the number of offspring present in the habitat seems likely to become the major component of habitat suitability. As more, younger offspring are likely to be present in the deme occupied by the parent than in any other deme, and as more offspring are produced, then it seems likely that the habitat quotient will begin to fall. Furthermore, as the presence of offspring is likely to be a larger component of habitat suitability for females than for males, then the habitat quotient is always likely to be less for females than for males. However, as at least some offspring are likely to migrate to another deme, and as Japanese macaque females are capable of recognising their offspring for some time after they have migrated (Yamada 1963), it seems likely that there will always be

some habitat other than the one occupied that, by virtue of the presence of an offspring, is of above-zero suitability. The habitat quotient is unlikely, therefore, ever to decrease entirely to zero, even for females.

(b) Ontogenetic curves of $\bar{S}_d$, mean action-dependent potential reproductive success, and $\bar{m}$, mean migration cost

It is assumed that there is no migration cost in moving from one deme to a neighbouring deme while the two are in sensory contact and that, therefore, the value of $\bar{m}$ is zero throughout life for both males and females. Any disadvantage that results from the migration is accounted for by the habitat quotient.

The curves of $\bar{S}_d$ for males and females have been drawn on the basis of what seems likely to be the reproductive history for a polygynous *cum* promiscuous primate with an age of onset of sexual maturity and longevity as shown by the Japanese macaque. It is assumed that the peak reproductive performance by females comes much sooner after sexual maturity than that by males. In the present case, however, because migration cost is assumed to be zero, the ontogenetic curve of initiation factor (in A) is unaffected by the curve of $\bar{S}_d$.

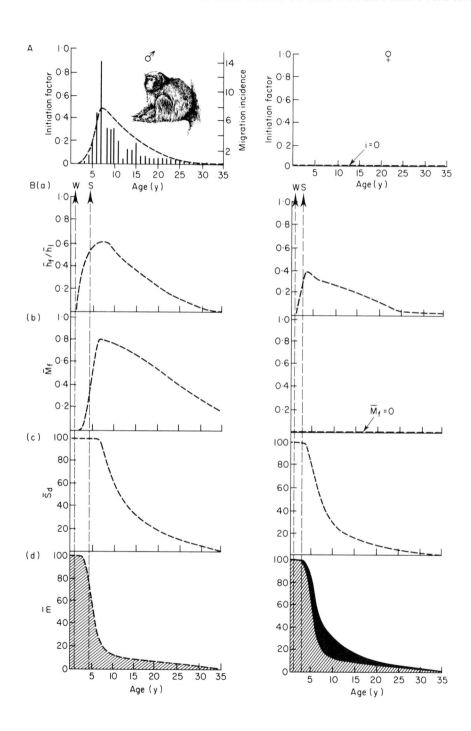

Fig. 17.7   Ontogenetic variation in the incidence of long-distance deme-to-deme migration by male and female Japanese macaques, *Macaca fuscata*: application of a model of adaptive ontogenetic variation in the probability of initiation of migration

A. Fitting a curve of ontogenetic variation in initiation factor to observed ontogenetic variation in migration incidence

The vertical lines show the number of males of different ages that have been observed to initiate solitary migration (=solitarisation) from their original deme (=social group; = troop) after a residence of some years as recorded by Nishida (1966). Such males seem likely initially to be performing exploratory migration (Chapter 15, p. 162) and may travel up to 23 km from their original deme. During the peak four months of copulation these solitary males associate with another deme and may copulate with some of the female members. These males may then either remain with their new deme or may initiate a further migration. The histogram does not distinguish between first and further migrations and therefore measures probability of initiation of migration at different ages after residence in a deme for some years. Note that it has not been possible to express the number of males observed to initiate migration at each age in the most meaningful way as a percentage of the total number of males of each year-class that were observed.

Females have not been observed to initiate long-distance migration.

The dashed curve shows the ontogenetic variation in i, the initiation factor, that best satisfies two criteria: (1) it gives an acceptable fit to the field data, bearing in mind that the field data are not expressed as percentages; and (2) permits description in B by curves of $\bar{h}_f/\bar{h}_1$, $\bar{m}$, and $\bar{S}_d$. In matching the vertical axes as shown, the limited data presented by Itani *et al.* (1963) for the initiation of long-distance solitary migration by males from the same deme to which the data for short-distance migration (Fig. 17.6A) refer were useful in fixing the extent to which the initiation factors in Fig. 17.6 and here should differ. The difference shown seems to represent the *minimum* difference in initiation factor for long-and short-distance migration.

B. Adaptive packages for the incidence of long-distance deme to deme migration by males and females

The ontogenetic variation of i, the initiation factor, shown in A, should, according to theory, be directly proportional to observed migration incidence. The value of i is given by the formula: $i = \bar{h}_f \bar{M}_f / \bar{h}_1$.

(a) Ontogenetic curves of $\bar{h}_f/\bar{h}_1$, the habitat quotient

It has been assumed that ontogenetic variation in habitat quotient is the same for long-distance migration as for short-distance migration. The curves shown are therefore the same as in Fig. 17.6B(a). It seems possible, however, that because long-distance migrants become familiar with more demes than short-distance migrants, the habitat quotient would be slightly greater for a long-distance than for a short-distance migrant.

(b) Ontogenetic curves of $\bar{M}_f$, the mean migration factor

The curves of mean migration factor are calculated from A and B(a) according to the formula: $\bar{M}_f = i\bar{h}_1/\bar{h}_f$.

(c) Ontogenetic curves of $\bar{S}_d$, the mean action-dependent potential reproductive success

The curves shown are the same as those in Fig. 17.6B(b).

(d) Ontogenetic curves of $\bar{m}$, mean migration cost

The curves of mean migration cost have been calculated from (b) and (c) according to the formula: $\bar{m} = \bar{S}_d(1 - \bar{M}_f)$. The form of the curve for the male and female suggests that the most likely interpretation is that the major component of migration cost is an increase in death-risk due to increased predation and possibly also decreased ability to find food as a result of spending long periods away from the social group. The details of the interpretation for males would then be that there was a high death-risk when young, which then decreases due to increasing experience at avoiding predation and finding food. After the age of about 7 years, when reproduction commences, absolute migration cost (as a percentage of $\bar{S}_d$ at birth) due to death-risk inevitably decreases. However, migration cost as a proportion of $\bar{S}_d$ increases as evidenced by the decrease in $\bar{M}_f$ in (b). This could either imply that death-risk increases after the age of 7 years or that death-risk remains constant at up to 20 per cent probability per migration, or perhaps decreases while some other, but non-proportional, component of migration cost, such as copulation loss, remains constant or increases. The latter interpretation seems the most likely.

The form of the curve for females shows an increased migration cost for this sex over males as indicated by the solid segment. Two interpretations seem possible. Either females are physically less able to avoid predation and find food when away from a social group, or they are more prone to predation and inability to find sufficient food because they are encumbered by young. As the contribution to total migration cost of this additional component for females appears at sexual maturity and continues to increase until the age of about 6 years, it appears that the latter interpretation is the more likely.

The deme-to-deme (=social group to social group; = troop to troop) migrations of the Japanese macaque, *M. fuscata*, have already been considered in two contexts, exploratory migration (p. 162) and exploratory migration thresholds to habitat variables (p. 216). Briefly, it has been concluded that female macaques initiate deme-to-deme removal migration only when the two demes are in sensory contact. Male macaques, on the other hand, may also initiate exploratory migration from their deme of origin without the migration being directed toward an adjacent deme. Short-distance exploratory migrations, scarcely beyond sensory contact of the original deme, may be performed by groups of males (=peripheral males). Longer-distance (up to 23 km) exploratory migrations (=solitarisation), which often develop from shorter-distance exploratory migrations, are usually performed by solitary males. Longer-distance exploratory males associate and copulate with females of other demes during the peak copulation season. Both long- and short-distance exploratory male migrants may form the nucleus of a new deme, depending on the state of the habitat variables to which the deme is exposed from which the male(s) recruit other individuals to form a new deme (see p. 216).

Data for ontogenetic variation in the probability of the initiation of migration are available for long-distance deme-to-deme male migrants (=solitary males) from observation of the age of such migrants (Nishida 1966) and for short-distance deme-to-deme male and female migrants from observations of troop division (Itani *et al.* 1963). These field data are shown in Figs. 17.6 and 17.7, which also present the 'packages' of ontogenetic curves of habitat quotient, migration factor, migration cost, and action-dependent potential reproductive success. All of these packages are consistent with the formula $i = \bar{h}_f \bar{M}_f / \bar{h}_1$, where $\bar{M}_f = 1 - \bar{m}/\bar{S}_d$, and with each other. The procedure followed in the development of these packages and the conclusions that it is possible to draw from their inter-consistency is described in the legends to the respective diagrams.

Before presenting any further packages for the migrations of other species it is perhaps necessary to discuss some of the conventions that have been used in the graphical presentation of the components of these packages, particularly in relation to $\bar{S}_d$ and $\bar{m}$. The first of these, $\bar{S}_d$, mean action-dependent potential reproductive success, has been expressed throughout as a percentage of its value at birth. The value of S_d must decrease from birth onwards. In the diagrams, however, S_d has been represented as remaining at 100 per cent until the onset of reproduction, some time after the attainment of sexual maturity. This convention, which made the figures far easier to construct and, it is hoped, easier to digest is unlikely to invalidate any of the conclusions.

The graphical representation of $\bar{m}$, mean migration cost, raised a number of problems. Eventually, it was decided to express $\bar{m}$ as a percentage of the value of $\bar{S}_d$ at birth. Even this convention, however, fails to solve some problems. These result mainly from the fact that migration cost consists of a number of components with different characteristics. Some of these components, such as energy expenditure, copulation loss, and death-risk to offspring, represent an absolute loss equivalent to x

units of reproductive success and consequently can easily be expressed as a percentage of $\bar{S}_d$ at birth. Migration cost due to such migration-cost components can, it is hoped, readily be appreciated from the figures presented. Other components of migration cost, however, but in particular death-risk to the individual itself represent a proportion not of S_d at birth but of *present* potential reproductive success. For example, suppose that death-risk during migration is 100 per cent (i.e. all individuals die during migration); a pre-reproductive migrant has a migration cost of 100 per cent as expressed in the figures. However, an individual that does not migrate until its S_d is only 10 per cent of its S_d at birth only loses this 10 per cent. Proportional components of migration cost, therefore, appear in the diagrams to decrease as an animal ages, even though the level of risk remains the same. An additional problem can be seen if we now suppose that death-risk during migration is 10 per cent (i.e. one in ten individuals die during migration). In the diagrams this will appear as one-tenth of the present value of $\bar{S}_d$ and, as before, will appear to decrease as $\bar{S}_d$ decreases. However, because the component is death-risk and is thus a proportion, no individual animal will actually experience the value of migration cost shown in the diagram. Either an individual survives migration, in which case this component of migration cost is zero, or an individual dies, in which case S_d is reduced to zero for that individual. This component of migration cost can only act as a selective pressure on an individual by virtue of its influence on succeeding generations of offspring of that individual.

Another problem of presentation and interpretation relates to the interaction of $\bar{S}_d$ and $\bar{m}$. It will be noticed that if migration cost for one unit of time is subtracted from the $\bar{S}_d$ for that unit of time, the difference does not correspond to the value of $\bar{S}_d$ illustrated for the next unit of time. Such subtraction, however, would not be a valid manipulation of the figures because it makes no allowance for two factors, one of which is the advantage gained from migration. Although a migrant must inevitably, because of migration cost, reduce its S_d below the $\bar{S}_d$ shown in the diagrams, subsequent realisation of remaining S_d must, if the migration is advantageous, be greater than that shown in the diagrams. The value of $\bar{S}_d$ illustrated for any particular age, therefore, should be taken to represent the mean value for that age within the deme concerned, many individuals of which may not have migrated by that age. The second factor results from the fact that $\bar{S}_d$ refers only to individuals that are surviving at that age and does not include individuals with zero S_d as a result of succumbing to proportional components of migration cost such as death-risk. On the other hand, the value of $\bar{m}$ for a particular age refers only to the mean migration cost experienced by individuals that migrate at that age (see also p. 79).

Although the conventions adopted present these problems of visual interpretation, they nevertheless seem to solve more problems than the other possibilities that were considered. It is, of course, essential to the calculation of initiation factor that $\bar{S}_d$ and $\bar{m}$ should be expressed in the same units. Expression as per cent of $\bar{S}_d$ at birth has the additional advantage that, where quantitative data on usual number of offspring are not readily available, graphical representation is still possible.

The next few pages examine the deme-to-deme migration of man. The data presented refer to two geographical demes, Britain and the United States, that culturally are relatively homogeneous, being essentially industrialist, even in the most agricultural areas. Despite this it seems likely that the conclusions reached are applicable to a wide range of human demes for, as pointed out in Chapter 14 (p. 146), the ontogenetic and sexual variation in migration pattern is similar over a wide range of cultures from hunter–gatherers to industrialists and from the United States and Japan to India and Sub-Saharan Africa. There is reason to believe, therefore (see also p. 294) that ontogenetic variation in migration threshold is a physiological human trait that evolved in response to the selective pressures that acted during man's long evolutionary period of hunting and gathering. The possible evolution of the migration pattern of industrial man from the likely pattern shown by his hunter–gatherer ancestors has already been discussed in Chapter 14 (p. 112).

Figures 17.8–17.11 show the calculation of the variation of initiation factor with age for males and females with an 'average' reproduction history (i.e. dependence on the parents until the onset of sexual maturity and the beginnings of independence at around 15 years; 'trial' pair-bond formations of varying degrees of permanence until the onset of the production of children at about 21 years for females and 24 years for males; continuing production of children until the ages of about 45 years for females and 48 years for males; and finally care of later-born children and assistance in the upbringing of grandchildren). The curves of variation of initiation

factor with age for these 'average' males and females are then compared with observed ontogenetic variation in migration incidence in the United States and in Great Britain (Fig. 17.11).

It is also informative to examine the ontogenetic variation in initiation factor for individuals with an unusual reproduction history. Figures 17.12 and 17.13 permit comparison of calculated curves of initiation factor with observed curves of migration incidence for male and female humans that never marry and the large majority of which, therefore, never reproduce.

The agreement found between the ontogenetic variation in initiation factor and the ontogenetic curves of migration incidence in Figs 17.11 and 17.13 provides some support for the suggestion that ontogenetic variation in the probability of initiation of deme-to-deme migration can be described to a large extent by a model of adaptive variation in initiation factor. In which case, the factors involved in the calculation of initiation factor can be used as an aid in the interpretation of the observed variation in migration incidence.

The migration threshold for deme-to-deme migration independently of the parents begins to decrease (low migration threshold ≡ high initiation factor ≡ high probability of initiation of migration) from the age of about 15 years onwards (Fig. 17.14), reaching its lowest point in individuals with an 'average' reproduction history at the age of about 22 years in females and about 26 years in males. Most migrations during this period of decrease in migration threshold are considered to be exploratory (Chapter 14, p. 120) and are characterised by a low selectivity of destination (p. 136).

The decrease in migration threshold during this period of exploratory migration is almost entirely a function of decrease in migration cost (Fig. 17.9(b)) due to increase in physical stature and the ability to live an independent existence. Males between the ages of the onset of sexual maturity and the formation of a long-term pair bond have the opportunity to increase their reproductive success by copulation without further parental investment. Females, on the other hand, do not have this opportunity. In fact, females that copulate with males that will not make any further parental investment are faced with the necessity of making an above-average investment in any child that results, a factor that is likely to reduce the potential reproductive success of that female. This difference in selection on the two sexes during this period of ontogeny has probably not only resulted in differences in courtship strategy (Trivers 1973) but has probably also led to males performing a higher proportion of solitary migrations than females. This is conspicuously the case for hunter–gatherers and modern neolithic agriculturalists and is probably also the case for modern industrial humans. Quite possibly the degree of solitariness during migration is also based on thresholds to migration-cost variables

which may constitute an equilibrium situation similar to that suggested for thresholds for the initiation of migration (p. 272). In this case, it may be expected that males and females experience the same migration cost as a result of their differing aggregation strategies. If this were so it would be expected that the differences between the two sexes in degree of solitariness during migration would be greatest when the difference in migration cost to a

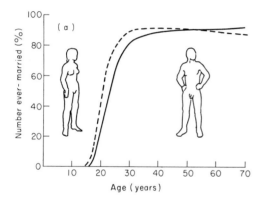 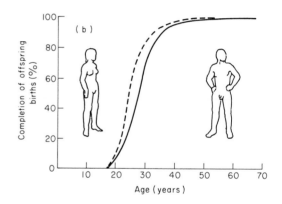

Fig. 17.8 The relationship between age and percentage completion of average total reproductive performance during life for man, *Homo sapiens*

Dashed line, females; solid line, males.

(a) Variation with age in percentage of males and females that have ever been married (= ever-married)

Figures are based on the 1971 census of England and Wales. On average, females are married by the age of 20 years and males are married by the age of 23 years. Biologically, marriage is likely to post-date long-term pair-bond formation by more than a year but is perhaps an effective measure of the onset of reproduction.

(b) Variation with age in the percentage completion of average total reproduction performance during life

The curve for females is calculated from the number of live births per married female in Great Britain after 5, 10, 15, 20, 25, and 30 years of marriage (Central Statistical Office 1972b) and is the average calculated from four cohorts, those that married in 1925, 1940, 1950 and 1960. Where cohorts had completed less than 30 years of marriage by the 1971 census the following procedure was adopted as exemplified for the 1960 cohort. From the 1925, 1940, and

1950 cohorts, an average completion of average total reproductive performance of 81·2 per cent after 10 years of marriage was calculated. The number of live-births to the 1960 cohort by 1970 (i.e. 2·16), was taken to represent 81·2 per cent of average total reproductive performance during life. On this basis, the 1·47 live births to the 1960 cohort after 5 years of marriage represent a percentage completion of average total reproductive performance during life of 55·3 per cent.

The curve for males is calculated from the number of live births to males of different age-groups married to females of any age presented by J.Z. Young (1971). Rate of live-births per year per married male was calculated by dividing the number of live births by the product of the number of males of that age-group that are married and the number of years spanned by that age-group. By multiplying this figure for each year of age by the proportion of males of that age that are married it is then possible to calculate the average total number of offspring born to a male of a particular age and to express this as a percentage of the total number born to a male of 70 years as shown in the figure.

On average, females have completed 50 per cent of their total reproductive performance (in terms of live births) by 25 years of age and males by 28 years of age.

solitary male and a solitary female is greatest as, say, in hunter–gatherers and modern neolithic agriculturalists, and least when the difference in migration cost is least, as, say, in modern industrialists. This interpretation seems, in the absence of any quantitative analysis, to fit observed cross-cultural differences in degree of solitariness during migration. If accepted, it follows that despite differences in physical stature, male and female sub-adults experience

the same migration cost during migration, and that the relative shapes of their curves of migration cost with age become solely a function of the development of the ability to live independently of the parents. The migration cost to females is therefore likely to decrease some three years in advance of that to males (Fig. 17.9(b)). Consequently, the migration threshold of females also decreases some three years in advance of that of males.

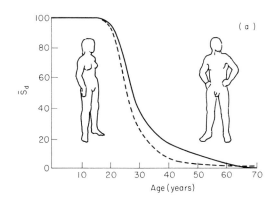

Fig. 17.9 Ontogenetic variation in action-dependent potential reproductive success and migration cost in male and female humans, *Homo sapiens*

Dashed line, females; solid line, males.

(a) Probable form of ontogenetic variation in $\bar{S}_d$, mean action-dependent potential reproductive success, in males and females

The value of $\bar{S}_d$ at each age is expressed as a percentage of the value of $\bar{S}_d$ at birth. The curve of $\bar{S}_d$ is obtained from the curve, with age, of percentage completion of average total reproductive performance during life shown in Fig. 17.8(b), but with the following assumptions. First, it is assumed that both the reproductive effort by both male and female that goes into copulation, conception, the survival and welfare of the female during pregnancy, parturition, and the survival and welfare of the offspring up to the age of about 2 years represents about 90 per cent of parental investment in that individual offspring (based on probable risk of: failure of implantation; miscarriage; pre- and perinatal mortality; and infant mortality). Consequently, over most of the curve there is little if any lag between the curves shown and those shown in Fig. 17.8(b). By the age of 45 years in females, and 50 years in

males, however, by which ages more or less 100 per cent of offspring have been born (Fig. 17.8(b)), it is assumed that about 5 per cent and 9 per cent respectively of $\bar{S}_d$ remain which will be realised by rearing the later-born children to independence and by assistance with grandchildren. It is assumed that males of 50 years have an $\bar{S}_d$ that is higher than females of 45 years due to the possibility of siring further offspring and due to the possibly slighter greater role in the provision of meat, protection from predators, education of male offspring, and protection of female offspring from other males intent on copulation without further parental investment. From the age of about 60 years onwards, however, it is assumed that the female's role in the carrying and care of grandchildren represents a greater value for $\bar{S}_d$ than the male's role which by this time is likely to have been taken over by the male offspring or by the mates of the female offspring. Hence, also, the selection for greater longevity in females.

Although the curve shown has as its quantitative base the reproductive performance of mid-twentieth century humans in England and Wales, it is assumed, because it has been converted to percentages, that it can be used as an expression of the average curve of $\bar{S}_d$ with age that evolved in response to selection on hunter–gatherer ancestors. In some areas, of course, marked deviations occur. Hence, female Andamanese hunter–gatherers do not start to reproduce until 28 years of age (Cappieri 1970). Similarly, male Australian hunter–gatherers, particularly in arid regions, do not start to reproduce until 25–30 years of age or later (Yengoyan 1968). In general, however, although there are marked regional differences in *total* number of offspring produced during the life-time, curves of offspring produced at different ages in different parts of the world (see Young 1971) suggest a curve of $\bar{S}_d$ with age that approximates to that shown above which is based on the abundant and accurate data available for humans in England and Wales. That the curve shown in (a) is acceptable as a general curve of $\bar{S}_d$ with age for the species *Homo sapiens* is particularly important in view of the conclusion of Chapter 14 (p. 146) that the migration of modern industrial humans is essentially still that of a hunter–gatherer rather than a migration adaptive to modern selection.

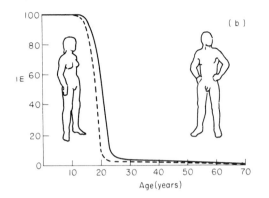

(b) Possible form of ontogenetic variation in $\bar{m}$, the mean migration cost

Mean migration cost, $\bar{m}$, for migration, by an individual of a particular age, to H_f, the most favourable habitat known to the individual other than H_1, the habitat occupied, is given by $\bar{S}_1$ minus $\bar{S}_2$, where $\bar{S}_1$ is the mean value, for an individual of that age, of S_d that would have obtained at the same point in time that migration terminates had migration from H_1 to H_f not occurred, and where $\bar{S}_2$ is the mean value, for an individual at that age, at the termination of the migration. It is assumed that the distance between H_1 and H_f and the state of all other factors to which the animal would be subjected during migration remain constant. We are concerned here, therefore, with ontogenetic variation in $\bar{m}$ that results from ontogenetic change in the animal.

The following reasoning is used in the production of the curves shown in (b). First, the value of $\bar{m}$ shown refers to removal migration unaccompanied by the parents and is based on a hunter–gatherer situation, being relevant to a local group to local group migration (i.e. within a territory). It is considered that due to physical stature and inexperience, young individuals of both sexes are likely to suffer a high value for $\bar{m}$ which up to the age of 15 years may approach $\bar{S}_1$. Consequently, if, as here, $\bar{m}$ is expressed as a percentage of the value of $\bar{S}_d$ at birth, then, up to the age of about 15 years, $\bar{m}$ approaches 100 per cent.

Males between the ages of 15 years and 25 years are assumed, as a result of increase in physical stature and experience, to suffer a rapidly decreasing value of $\bar{m}$ (i.e. decreasing death-risk, increasing ability to acquire food independently) down, eventually, to a minimal, though still decreasing, level.

Female hunter–gatherers have fewer techniques to learn than males before being capable of fulfilling the adult role (p. 115). Hence, presumably, the selection that seems to have existed for the earlier onset of sexual and physical maturity in females than in males. Any increased migration cost likely to be experienced by a sub-adult or young mature female relative to that experienced by a similar male due to their relative physical strengths and the disadvantage to a female of conceiving without previous formation of a stable pair-bond, is likely to be more than offset by the adaptation of rarely performing migrations solitarily but rather in the company of siblings, other females, or a male with which a short- or long-term pair bond has been established. It is suggested that the nett result of earlier maturity and less performance of solitary migration is likely to result in the migration cost of young females decreasing to the minimum level some three years in advance of that of young males.

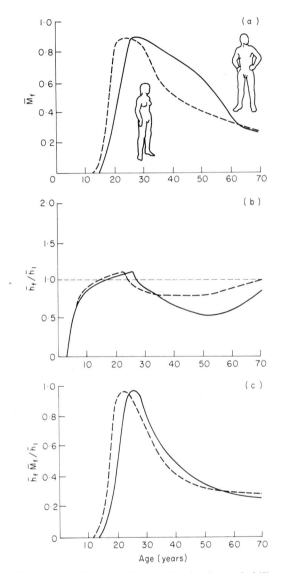

Fig. 17.10 Ontogenetic variation in the probability of the initiation of deme to deme removal migration by man, *Homo sapiens*: application of an adaptive model of the probability of initiation of migration

Dashed line, females; solid line, males.

In the migration being considered a habitat is a deme of a size equivalent to a local group, village, town, or city, depending on the culture.

(a) Ontogenetic variation in $\bar{M}_f$, the mean migration factor

The value of $\bar{M}_f$, the mean migration factor for migration to the most suitable habitat with which the animal is familiar other than the habitat occupied, may be calculated from the curves of $\bar{S}_d$, mean action-dependent

potential reproductive success, and $\bar{m}$, mean migration cost, shown in Fig. 17.9. The value of $\bar{M}_f$ is given by the formula $\bar{M}_f = 1 - \bar{m}/\bar{S}_d$.

(b) Ontogenetic variation in the habitat quotient

The ontogenetic variation shown for the habitat quotient, $\bar{h}_f/\bar{h}_1$ (where, for a number of individuals, $\bar{h}_f$ is the mean of the suitabilities of the most suitable habitat with which each individual is familiar other than the habitat occupied, and $\bar{h}_1$ is the mean of the suitabilities of the habitats occupied), has been derived according to the following reasoning. Up to the age of 2–3 years, all demes other than that occupied by the parents approach zero suitability. Thereafter, other habitats increase in relative suitability and from the age of about 15 years in females and 18 years in males, exploratory migration is likely to have led to encounters with demes of suitability equal to if not slightly greater than that of which the parents are a member. When the main phase of exploratory migration terminates with calculated removal migration to the most suitable habitat encountered, which now becomes the habitat occupied, the habitat quotient falls to 1·0 (dashed horizontal line) or below. By this time most individuals are paired and have started to produce offspring, whereupon the major components of habitat suitability are likely to be those factors that affect the health, growth and survival of the offspring. Among such factors may be included the efficient acquisition of sufficient food of sufficient quality, protection from climate, protection from predators, and protection from the results of territorial interaction. All of these factors are facilitated by both members of the pair making learning and other (e.g. social, construction) investments in the habitat occupied. For hunter–gatherers, agriculturalists, pastoralists, and debatedly even for industrialists, the male makes a larger investment than the female. As the investment in the habitat occupied increases, then this habitat becomes on average relatively more suitable than other habitats and the habitat quotient decreases. As the reproductive roles change from the ages of 45 years onwards in females and about 50 years onwards in males to that of providing assistance in the rearing of grandchildren, it is the presence of first and second filial generation offspring that is the major component of habitat suitability. As these are increasingly likely to have migrated from their parental deme, the probability gradually increases with age over 45 years for females and about 50 years for males that some other habitat may be as, if not more, suitable than the habitat occupied. Even so, more-efficient assistance in the rearing of grandchildren can probably be given in the habitat already occupied than in some other habitat with which the individuals (now grandparents) are less familiar. The habitat quotient is therefore likely to remain less than 1·0 as long as some children produce offspring in the deme of which their parents are a member.

(c) Ontogenetic variation in the initiation factor

According to the arguments in this chapter, the ontogenetic curve of i, the initiation factor, where $i = \bar{h}_f \bar{M}_f/\bar{h}_1$, should be directly proportional to the probability of the initiation of removal migration and hence to observed migration incidence. The ontogenetic curve of i can be calculated directly from figures (a) and (b).

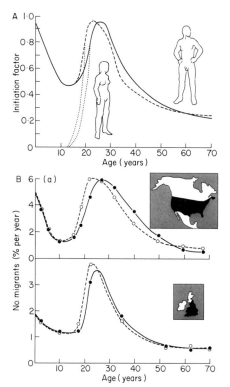

Fig. 17.11   Comparison of expected and observed curves of ontogenetic and sexual variation in the probability of the initiation of deme-to-deme removal migration by man, *Homo sapiens*

Dashed line, females; solid line, males.

A. Ontogenetic variation in initiation factor

The curves of ontogenetic variation in i, the initiation factor, have been derived from Fig. 17.10(c). The dotted lines plus the dashed line from 20 years of age onwards and the solid line from 23 years of age onwards represent the initiation factor for removal migrations of individuals unaccompanied by parents. In order to compare these curves with observed curves of migration incidence as normally expressed by census data, it is necessary to modify the curves shown in Fig. 17.10(c) to include the migration of young individuals due to the migration of their parents. This is done by making the assumption, from Fig. 17.8(b), that on average an offspring is born to male and female parents aged, respectively, 28 and 25 years. The value of i for an individual of 0 years is, therefore, the average of that for a male of 28 years and a female of 25 years. Similarly, the value of i for an individual of 10 years is the average of that for a male of 38 years and that for a female of 35 years. From the age of 13 years onwards, a measure of i is best given by taking a proportion of the sum of the values of i for parents and offspring, such that the proportion taken gives increasing weight with age to the value of i calculated for an independent offspring. The final results are the dashed and solid curves shown in the figure. According to the

arguments of this chapter, the ontogenetic curve of i should be directly proportional to the probability of the initiation of removal migration and hence to observed migration incidence.

B. Observed ontogenetic and sexual variation in migration incidence

Curves (a) and (b) show ontogenetic variation in migration incidence in, respectively, the United States and England and Wales. The curves for the United States are for rural–urban migration and are based on data in Thomas (1959). The curves for England and Wales are for inter-regional migration (see Fig. 14.20) in the year 1965–66 and are based on data from the General Register Office (1967, 1968).

Reservations

In evaluating the support provided for the validity of the theoretical basis of the model from the agreement between the curves shown in A and B the following factors have to be borne in mind. In particular, the curve shown in A is the product of three sets of curves, one set of which (Fig. 17.9(a) for $\bar{S}_d$ with age) is based almost entirely on quantitative data and the other two of which (Fig. 17.9(b) for $\bar{m}$ with age, and Fig. 17.10(b) for the habitat quotient) are based entirely on qualitative reasoning. It follows, therefore, that the more the curve is a function of the curve of $\bar{S}_d$, the greater the confidence that can be attached to any support provided. The relative importance of $\bar{S}_d$, $\bar{m}$, and $\bar{h}_f/\bar{h}_1$ in generating the shape of the curve in A can be assessed. First, that part of the curve for ages from 25 years to 50 years is largely a function of $\bar{S}_d$. The shape of the curve from 50 years onwards cannot reasonably be calculated from the curves in Fig. 17.9 and has been produced by extrapolation from the previous portion of the curve. The shape of the curve from 0 years to 13 years is based on the curve from 25 years to 38 years for females and from 28 years to 41 years for males and, therefore, given a constant very high migration cost for independent migration by individuals less than 13 years of age, is also mainly a function of the curve of $\bar{S}_d$. The main part of the curve generated by the curve of $\bar{m}$ is, therefore, that between the ages of 13 years and 25 years. The confidence that can be attached to the shape of this part of the curve should be a function of the confidence that can be attached to the suggestion that migration cost falls from a very high level at age 13 years to a very low level at age 20 years to 25 years and that, for whatever reason, the migration cost for females decreases about three years in advance of that for males. Comparison of the curves in Fig. 17.10(a) and Fig. 17.10(c) shows that the habitat quotient plays a relatively small, though not insignificant, part in determining the total form of the curves of initiation factor.

Conclusion

In view of the major dependence of the curves in A on the curves of $\bar{S}_d$ and the relatively simple and perhaps acceptable requirements of the curve of migration cost to generate that portion of the curve for ages from 13 years to 25 years, then the general agreement between the shapes of the curves shown in A and B is taken to provide some support for the model of ontogenetic variation in the probability of the initiation of migration developed in this chapter.

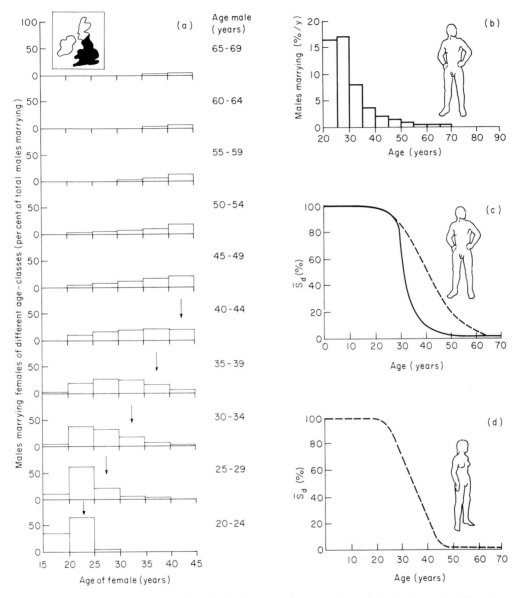

Fig. 17.12   The calculation of ontogenetic variation in action-dependent potential reproductive success for never-married males and females of man, *Homo sapiens*

(a) Variation for males of ten different age-classes in the probability of the age of female that a previously never-married male will marry

The histogram shows the percentage of males that are marrying for the first time that will marry females of six different age-classes. The arrows indicate males marrying females in the same age-class as themselves. Data refer to England and Wales, 1969. Percentages calculated from raw data in Office of Population Censuses and Surveys (1971).

(b) Ontogenetic variation in the probability that a never-married male will marry during the coming year

The histogram shows the average per cent per year of previously never-married males that marry during eight different ontogenetic periods. Data refer to England and Wales, 1969. Percentages calculated from raw data in Office of Population Census and Surveys (1971).

(c) Ontogenetic variation in $\bar{S}_d$, mean action-dependent potential reproductive success, for never-married males

The dashed curve shows reproductive success for males that marry at different ages. The curve is obtained from (a) and (d) by calculation for a male of a given age by summing the products of the per cent of males that marry females of different age groups (from (a)) and the per cent of maximum potential reproductive success that is capable of being realised by a female of that age-group that marries for the first time (from (d)). This calculation assumes that the realisation of potential reproductive success by a female is unaffected by the age of the male, an assumption that is more or less justifiable (J.Z. Young 1971). However, (b) shows that the probability of marriage decreases with age. When this is taken into account, the solid curve is produced which, although it does not take into account the potential reproductive success that can be realised by never-married males through copulation with females to whom they are not married, is probably an adequate description of ontogenetic variation in $\bar{S}_d$ for males that never marry. The curve is shown as just above zero even after the age of 50 years to allow for any action-dependent potential reproductive success that may be realised through assistance to kin.

(d) Ontogenetic variation in $\bar{S}_d$, mean action-dependent

potential reproductive success, for never-married females
The value of $\bar{S}_d$ is expressed as a percentage of the value of $\bar{S}_d$ at birth. The curve shows the calculated change in $\bar{S}_d$ with age for never-married (='single' in census data) females, and is calculated from the total number of live-born offspring produced during life by females marrying for the first time at different ages as presented by D.V. Glass (1969). Data refer to females marrying in England and Wales in 1945. In general, females that delay marriage up to the age of 20 years suffer no reduction in the total number of live-born offspring they produce during life. Thereafter, each year's delay in marriage results in a reduction in the total number of live-born offspring produced until, after the age of 45 years, the total number of live-born offspring produced during life approaches zero. The curve is shown as just above zero even after the age of 45 years to allow for any action-dependent potential reproductive success that may be realised through offspring that may have been born to never-married females and/or through assistance to kin.

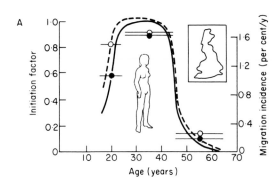
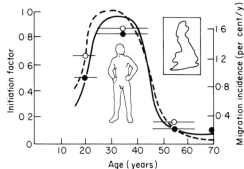

Fig. 17.13   Calculated and observed ontogenetic variation in the probability of the initiation of removal migration by never-married males and females of man, *Homo sapiens*

A. A comparison of observed and calculated curves of probability of initiation of migration by never-married males and females

Data for the number of 'single' male and female overseas emigrants of different age-groups from Great Britain in 1969 expressed as a percentage of the total number of such males and females in each group present in Great Britain on 30 June 1969 were compiled from Office of Population Censuses and Surveys (1971) and Registrar General for Scotland (1971). Such data are available for age-groups 15–24 years, 25–44 years, 45–64 years, and older than 65 years. Consequently, averages for only these large age-groups can be calculated. These averages are shown as solid dots, the age-group to which they refer being indicated by a horizontal bar. The best curve to describe these averages was then fitted by eye and is shown by the solid line.

The curve of initiation factor, i, was calculated from B according to the formula $i = \bar{h}_f \bar{M}_f / \bar{h}_1$, and is shown by the

dashed line. Averages for the same age-groups for which the observed migration incidence is available were calculated from the curve and are shown as open dots. The vertical scales were matched so that the maximum value for i was approximately level with the maximum value of per cent emigrants from Great Britain.

B. The calculation of ontogenetic variation in i, the initiation factor.

(a) Ontogenetic variation in $\bar{h}_f / \bar{h}_1$, the habitat quotient

Up to the age of about 18 years in females or 21 years in males the habitat quotient is assumed to be the same as for males and females that marry at the ages, respectively, of 20 years and 23 years as derived in Fig. 17.10(b). Unlike ever-married males and females, however, the major component of habitat suitability for never-married males and females seems likely to be the probability of finding a mate. Failure to find a mate after a certain stay-time in each habitat occupied probably means that the habitat quotient is greater than unity (horizontal line). As the most favourable habitats within the familiar area and those encountered during further exploratory migrations are visited without producing a mate, however, the habitat quotient is likely to decrease towards unity, but

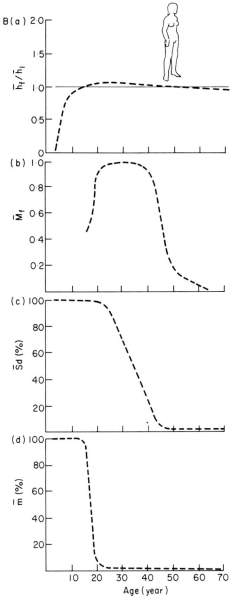

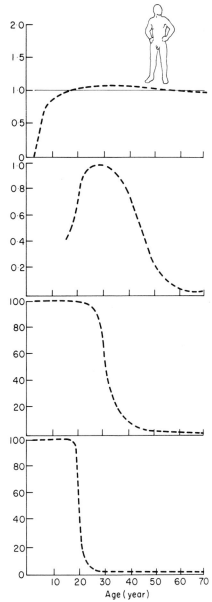

H_1, the habitat occupied, seems unlikely to be more suitable than other habitats as long as the search for mates continues. After menopause in females, however, and about the age of 50 years in males, if it is assumed that optimum behavioural strategy for never-married individuals is the assistance of kin, the habitat occupied can be marginally more suitable than other habitats if it contains the nearest kin. In this case the habitat quotient can be less than unity.

(b) Ontogenetic variation in $\overline{M}_f$, the mean migration factor for migration to the most favourable habitat with which an animal is familiar

The value of $\overline{M}_f$ can be calculated from (c) and (d) according to the formula $\overline{M}_f = 1 - \overline{m}/\overline{S}_d$.

(c) Ontogenetic variation in $\overline{S}_d$, the mean action-dependent potential reproductive success

The curves shown are taken from Fig. 17.12(c) and Fig. 17.12(d).

(d) Ontogenetic variation in $\overline{m}$, mean migration cost

The curve of migration cost is taken to be identical to that shown in Fig. 17.9(b) for males and females that marry.

Conclusion

The best agreement in A between the calculated curves of i and the observed curves of migration incidence occurs from the age of 20 years onwards when the subjectively estimated values of $\overline{m}$ and $\overline{h}_f/\overline{h}_1$ are more or less constant and the shape of the curve is largely a function of the quantitatively based curve of $\overline{S}_d$. Even so, the decrease in migration cost from a high to a low value which in B(d) is shown to begin at about 15 years and terminate at about 20 years in females, and three years later in males, needs only to be delayed by two years for there to be almost total agreement between the two curves.

It is concluded that A provides some support for the model of the probability of the initiation of migration developed in this book.

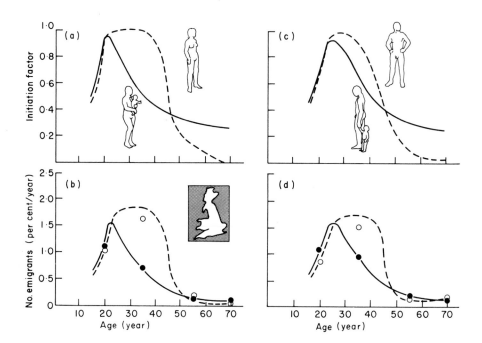

Fig. 17.14 Ontogenetic variation in the probability of initiation of removal migration by married and never-married males and females of man, *Homo sapiens*

(a) Calculated curves of ontogenetic variation in the initiation factor for married and never-married females

Dashed line, curve for a never-married female (from Fig. 17.13A); solid line, curve for a female that marries at age 20 years (from Fig. 17.11A).

(b) Observed curves of ontogenetic variation in migration incidence for married and never-married females

Data for the number of never-married female overseas emigrants of different age-groups from Great Britain in 1969 expressed as a percentage of the total number of such females in each age-group present in Great Britain on 30 June 1969 were obtained from: Office of Population Censuses and Surveys (1971); and Registrar General for Scotland (1971). Such data are available for age-groups 15–24 years, 25–44 years, 45–64 years, and older than 65 years. Consequently, averages for only these large age-groups can be calculated. These averages are shown for never-married females as open dots. The best curve to describe these averages was then fitted by eye and is shown by the dashed line. A similar curve and set of averages for married females are shown by a solid line and dots.

(c) Calculated curves of ontogenetic variation in the initiation factor for married and never-married males

Dashed line, curve for a never-married male (from Fig. 17.13A); solid line, curve for a male that marries at age 23 years (from Fig. 17.11A).

(d) Observed curves of ontogenetic variation in migration incidence for married and never-married males

The conventions used for males in this figure are the same as those used for females in (b) (i.e. dashed line, never-married emigrants; solid line, married emigrants). Data are also from sources as in (b).

Conclusion
The similarity between (a) and (b) and between (c) and (d) supports the suggestion that ontogenetic variation in probability of initiation of migration and the variation that occurs between males and females with different reproductive histories can be described by a model of adaptive variation in initiation factor as defined and calculated in this book. The model is least accurate for males and females older than about 50 years. This can be attributed to the speculative nature of the value of $\bar{S}_d$, mean action-dependent potential reproductive success, from this age onwards compared with the quantitative measure possible for all ages less than about 45 years.

Even if individuals do form a long-term pair bond and begin to reproduce between the ages of 15 years and 22 years in females and 15 years and 26 years in males, the decrease in S_d, action-dependent potential reproductive success, that occurs is likely to be small relative to the decrease in migration cost (Fig. 17.9). Consequently, little difference is to be expected, nor is it observed (Fig. 17.14(b), (d)), between the probability of initiation of migration by married and never-married individuals of these age-groups. Decrease in migration threshold during this ontogenetic period is, therefore, predominantly interrelated to the increase in physical maturity and experience that takes place and it seems likely that selection has favoured decrease in migration threshold and increasing physical maturity to be part of the same physiological syndrome.

Once migration cost has reached a fairly stable minimum level, however, at the age of about 22 years in females and about 26 years in males, changes in migration threshold are largely a function of S_d, changes in which are dependent on reproductive history (Figs. 17.9 and 17.12). Individuals that have formed a pair-bond and have started to reproduce have a rapidly decreasing S_d after this time, which results in a rapidly increasing migration threshold. Individuals that have failed to form a pair-bond of sufficient permanence to commence the production of offspring, however, have initially a less rapidly decreasing S_d (Fig. 17.12). In such individuals, therefore, migration threshold remains at a relatively low level for a few more years until S_d reaches a low level at, in females, about 45 years of age, whereupon there is a rapid increase in migration threshold (Fig. 17.13). Because of this association between migration threshold and reproductive history, it seems likely that the procedure of producing and raising children induces physiological changes in both males and females. These changes in physiology are likely to result not only in those changes in ovulation and implantation rate,

Fig. 17.15 In humans, migration threshold decreases throughout the period of maturation, reaching its lowest point at about the age of 22 years in females and about 25 years in males. The period of ontogeny normally associated with temporary and later permanent pair-bond formation is characterised by an increasing performance of exploratory migration, particularly by males.

[*Photo by Elizabeth Gabriel*]

Fig. 17.16 After the age of about 22 years in females and about 25 years in males, the production of and interaction with children seems to trigger a series of physiological changes, part of which is an increase in migration threshold. Individuals that do not produce children exhibit a low migration threshold until the age of about 40 years in males and until menopause in females.

[*New Guinea photo by Film Study Center, Harvard University*]

gametogenesis, and intra-pair sexual interest etc., that lead to the characteristic curve of variation in rate of production of offspring at different ages, but also in changes in migration threshold. Once again, because, according to the model, selection is likely to favour such a close adaptive relationship between migration threshold and reproductive history, all of these physiological changes are likely to be part of the same syndrome, probably under the same hormonal control. Individuals not directly involved in the production and rearing of children should not have these physiological changes induced and should retain a relatively low migration threshold until, in females, the age of about 45 years (Fig. 17.13), whereupon migration threshold should increase to a very high level. Once again, because of the close adaptive relationship between migration threshold and remaining reproductive potential it seems likely that this rapid increase in migration threshold should be part of the same physiological syndrome that governs reproductive physiology and in never-married females is likely to be as much a part of the menopause syndrome as is, say, the cessation of ovulation and menstruation.

Figure 9.2 (p. 64) outlined the likely relationship between the cessation of exploratory migration and the value of M, the migration factor. From the curves given in Figs. 9.2 and 17.10, it follows that continuance of exploratory migration is more likely if M is high and increasing (in Fig. 17.10) than if M is low and decreasing. It follows from Figs. 9.2 and 17.10, therefore, that the major ontogenetic phase of exploratory migration is likely to terminate soon after the age at which M starts to decrease (i.e. 22 years and 26 years respectively for married females and males (Fig. 17.10) and about 40 years for both never-married males and females (Fig. 17.13)). As the majority of both males and females have married two or more years earlier than the ages of 26 years or 22 years respectively, however, it follows that selection is likely to favour exploratory migration continuing a little beyond the first production of offspring. This is particularly so for individuals that marry early. Beyond these ages, however, the major ontogenetic phase of exploratory migration should cease and calculated removal migration should take place to the habitat encountered during exploratory migration that is assessed to be most suitable. An analysis has already been presented (p. 142) that shows that after the age of about 25 years migrations are characterised by a much greater selectivity of destination.

Before summarising the conclusions reached in this section concerning ontogenetic variation in the probability of initiation of deme-to-deme migration by man there is one point to which it is perhaps worth drawing attention, even though it is implicit in all of the arguments presented and also is dealt with more explicitly later. This point is that the model indicates that two individuals, living in the same habitat H_1, equally acquainted with the more suitable habitat H_2, and likely to be subjected to exactly the same migration cost, may nevertheless react quite differently if they are of different ages and hence have different values of M, the migration factor, due to differences in S_d, action-dependent potential reproductive success $(M = 1 - m/S_d)$. For one individual h_1 may well be less than $h_2 M$ whereas for the other it may not. In this case, ideally, only the first individual would initiate migration from H_1 to H_2.

In conclusion, the following hypothesis of deme-to-deme human migration can be presented. Young humans up to the age of about 12 years are unlikely to initiate removal migration independently of their parents. Thereafter, part of the physiological syndrome that transforms the individual to physical maturity is a gradually decreasing migration threshold that reaches a minimum level at the age of about 22 years in females and about 26 years in males. This decrease in migration threshold is effectively independent of the history of courtship and reproduction during these years, and is scarcely affected by the intervention of marriage and the production of children. The migrations initiated are largely exploratory and are characterised by a low selectivity of destination.

Females older than 22 years and males older than 26 years that produce and rear offspring are induced to undergo a number of physiological changes, one of which is an increase in migration threshold that continues at a steady but ever-decreasing rate until reaching a fairly stable level at the age of 55–60 years. Females older than 22 years and males older than 26 years that fail to produce offspring on the other hand, have a migration threshold that remains at a relatively low level until the age of about 45 years in females and about 40 years in males, after which there is a rapid increase, particularly in females. In such females, this increase in migration threshold is likely to be a part of the physiological syndrome of menopause.

In the next few pages a number of 'packages' (of ontogenetic curves of initiation factor, i, mean habitat quotient, $\bar{h}_q$, mean migration factor, $\bar{M}$, mean migration cost, $\bar{m}$, and mean action-dependent potential reproductive success, $\bar{S}_d$, the latter three of which can be related by the formula $\bar{M} = 1 - \bar{m}/\bar{S}_d$) are presented for the group-to-group migrations of a variety of other non-avian vertebrates. In each case the procedure involved in the production of the package is similar to that for the Japanese macaque.

Consequently, to reduce the necessity for tedious repetition of procedure, description of the methods involved has been reduced to a minimum. In all of these 'packages' the important point is that they are capable of being constructed in such a way that they can not only describe observed sexual and onto-genetic variation in migration incidence but at the same time can be acceptable descriptions of the

Table 17.1 Some field observations of ontogenetic and sexual variation in incidence of group-to-group removal migration

| Species | Migra-tion[†] | Place | Relative incidence[‡] before and after onset of reproduction | | | | Reference |
| | | | Male | | Female | | |
			pre-	post	pre-	post	
MAMMALS							
Primates							
titi monkey							
Callicebus moloch	GH	S. America	***?	o?	***?	o?	Mason 1966
common Indian langur							
Presbytis entellus	GH	India	***	*	?	?	Yoshiba 1968
lutong							
Presbytis cristatus	GH	Malaysia	***?	*?	?	?	Bernstein 1968
vervet monkey							
Cercopithecus aethiops	GH	E. Africa	***?	**?	?	?	Gartlan and Brain 1968
black mangabey							
Cercocebus albigena	GH	E. Africa	***	*	**	o?	Chalmers 1968
Japanese macaque							
Macaca fuscata	GH	Japan	***	**	**	*	Fig. 17.6
olive baboon							
Papio cynocephalus anubis	GH	E. Africa	?	*	?	?	DeVore and Hall 1965
man							
Homo sapiens	HG	Earth	**	*	**	*	Fig. 17.11
chimpanzee							Itani and Suzuki 1967
Pan troglodytes	GH	E. Africa	**	*	**	?	Sugiyama 1968
							Nishida 1968
gorilla							
Gorilla gorilla	GH	E. Africa	***	*	*?	*?	Fig. 17.18
white-handed gibbon							
Hylobates lar	GH	Malaysia	***	o?	***	o?	Fig. 17.17
Carnivores							
lion							Carr 1962
Panthera leo	GH	Africa	***	**	**	*	Schaller 1972
Ungulates							
plains' zebra							
Equus quagga	G	E. Africa	***	o	***	o	Fig. 17.20
vicuna							
Vicugna vicugna	GH	Andes	***	o	***	*	Koford 1957
Cetaceans							
sperm whale							
Physeter catodon	G	Pacific	***	*?	?	?	Fig. 17.19
REPTILES							
rainbow lizard							
Agama agama	GH	Nigeria	***	o?	**	o?	Harris 1964

† Types of group-to-group migration
  G = habitat assessment almost entirely concerned with group characteristics
GH = group characteristics probably given more weight than area characteristics in habitat assessment
HG = area characteristics probably given more weight than group characteristics in habitat assessment
‡ *** to * = decreasing migration incidence
    o = minimal, but not zero, incidence

ecology and reproductive and social biology of the species concerned. The packages should not, however, be interpreted as having been predictive except where indicated.

Table 17.1 summarises the results of some field studies that have provided data concerning ontogenetic and sexual variation in the group-to-group migration of a variety of animals that live in recognisable social groups. For descriptive convenience, group-to-group migration will often be referred to as type-GH migration. This category includes the type-G and type-HG migrations listed in the table. Table 17.1 is by no means an exhaustive list and represents a fairly random sample of the literature. It will be noticed that the list is marked by a liberal sprinkling of question marks that reflect the absence of quantitative data even where relevant observations have been made. The difficulties of obtaining data on type-GH migrations by individuals of an animal such as the sperm whale,

Physeter catodon, are self-evident. However, the difficulties are equally real with animals such as primates. Even when the species lives in fairly open vegetation where frequent observation is possible, a long-term study is required before all individuals in a social group can be recognised. Yet, if type-GH migration is to be studied, and particularly if removal migration is to be distinguishable from mortality, all individuals in a number of social groups over a relatively large area must be recognisable. Even then, to determine incidence of removal migration at different ages it would be necessary to continue observation over a period of several years. It is not surprising, therefore, that such studies are rare. Apart from man, only the Japanese macaque, *Macaca fuscata*, has been studied in any detail. Even here, however, detailed study was only possible by implementing the device of provisioning which, as already discussed (p. 218), by altering habitat suitability, inevitably altered migration incidence.

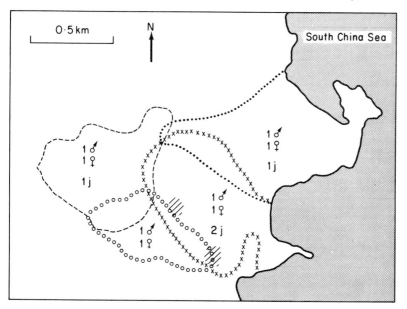

Fig. 17.17   Territory to territory removal migration by the common white-handed gibbon, *Hylobates lar*

Dotted, dashed, ringed and crossed lines show the edges of four territories on the Malay peninsula. The territorial unit is a family group as indicated. Territorial defence consists of vocal and behavioural demonstrations and on about one per cent of occasions by fighting between adult males. Territorial interactions are most common in the areas of territory overlap. As the juveniles (j) mature and become sub-adult they gradually spend their time more and more on the periphery of the parental territory. The hatched areas were occupied by single males. As the adult

sex ratio in a territorial unit is 1 : 1, it must be assumed that sub-adult females also become peripheral or, at least, perform territory-to-territory removal migration. Presumably sub-adult individuals of either sex do not remain permanently on the periphery of the parental territory. It seems likely, though it has not yet been observed, that if, after a certain period, an acceptable 'vacancy' does not appear in one of the adjacent territories, a sub-adult shows exploratory or removal migration until a mate and/or territory are encountered that together are above threshold suitability.

[*Based on map and observations of Ellefson* (1968)]

In most cases, therefore, data concerning incidence of type-GH removal migration derives from observations that only permit certain tentative deductions to be made. For example, the South American titi monkey, *Callicebus moloch*, has a territorial system similar to that illustrated for the white-handed gibbon, *Hylobates lar*, in Fig. 17.17, the territorial unit consisting of one adult male and one adult female plus their juvenile and early sub-adult offspring (Mason 1966). Although adult titis were always to be found in their territories, sub-adults were observed consistently to disappear from the forest area being studied which contained nine territories. Furthermore, individuals have been reported crossing the savanna between forests. It seems likely, therefore, that there is an ontogenetic peak of exploratory/removal migration during the sub-adult phase, with migration incidence falling to a low level after reproduction begins. In this case, because of the adult sex ratio in the territorial units, it seems likely that there is little sexual difference in migration incidence.

The circumstantial evidence available for other species, however, is not always as convincing as this. For example, Bernstein (1968), working in Malaysia, evaluated the social structure of five social groups of the silvered leaf monkey, or lutong, *Presbytis cristatus*. Each of these groups occupied a home range, the size of which averaged about 20 ha, and consisted on average of a single male to sixteen females. Yet within the study area of about 100 ha, no single males nor all-male bands were seen. Furthermore, some sub-adult males disappeared during the 1332 hours of observation. Although other explanations, such as differential mortality, are possible, it again seems likely that males have an ontogenetic peak of exploratory/removal migration during the sub-adult phase.

For other species the evidence is not so much circumstantial as limited. The basic social group of chimpanzees, *Pan troglodytes*, is a 'large-sized group' (Itani and Suzuki 1967) consisting of 20–40 individuals. A large-sized group consists of several adult males, adult females, sub-adults, juveniles and infants and in this respect is similar to the multi-male social units of many other primate species (Aldrich-Blake 1970). The large-sized group of the chimpanzee is characterised, however, by an extreme of 'extensibility' (Nishida 1968) (i.e. formation of subgroups of fluid composition that continually separate, disperse, converge, split and join). The subgroups of this extensible large-sized group may

consist of almost any sexual/ontogenetic combination. This extreme of extensibility makes it extremely difficult to observe removal migration from one large-sized group to another for, until the observer is intimately familiar with the total constitution of a given group, it is not possible to be absolutely sure that a movement of an individual from one group of individuals to another is not a movement between sub-groups. Both movements, of course, qualify as migration in the sense adopted in this book, but it seems likely that it is migration between large-sized groups of chimpanzees that is equivalent to the type-GH removal migration of other monkeys, man and apes.

Despite the difficulties involved, a few observations are available that suggest the form of ontogenetic and sexual variation in large-sized group to large-sized group migration by chimpanzees. First, sub-adult males are conspicuously absent from large-sized groups (Itani and Suzuki 1967), thus suggesting for males an ontogenetic peak in migration incidence during the sub-adult stage. As already described for the gorilla, *Gorilla gorilla* (p. 92) and the Japanese macaque, *Macaca fuscata* (Fig. 17.7), female chimpanzees do not seem to migrate except when the two groups between which migration occurs are in sensory contact. For such short-distance group-to-group migration, however, there is some evidence, though again limited, that there is an ontogenetic peak in migration incidence just before the ontogenetic onset of reproduction. For example, Nishida (1968) observed the simultaneous migration of four sub-adult females from one large-sized group to another in the woodland savanna of western Tanzania.

There are a number of species for which quantitative data on ontogenetic and sexual variation in type-GH migration incidence are available. Two of these species, man and the Japanese macaque, have already been considered in some detail. The ontogenetic variation in type-GH migration incidence of the remaining species, the male gorilla, *Gorilla gorilla*, the male sperm whale, *Physeter catodon*, and male and female plains' zebra, *Equus quagga*, are illustrated respectively in Figs. 17.18–17.20. In all five of these species the ontogenetic pattern is essentially the same with the peak of removal migration incidence occurring just before the onset of reproduction and thereafter, with the attainment of physical maturity and the early stages of reproduction, decreasing. Given these relatively well-authenticated examples of ontogenetic variation in

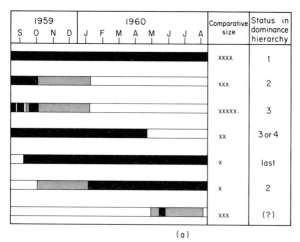

(a)

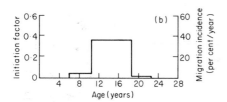

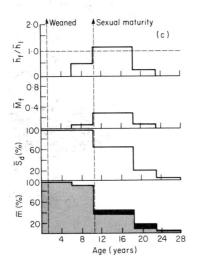

Fig. 17.18 Ontogenetic variation in the group-to-group migration of the gorilla, *Gorilla gorilla*, in the Kabara region, north of Lake Tanganyika, Africa: application of a model of adaptive ontogenetic variation in the probability of initiation of migration

Migration by males

Male gorillas may initiate migration from the social group of which they are a member at any time and live solitarily for a period. The migration seems likely to be exploratory at initiation. From time to time migrants associate temporarily with other heterosexual social groups and may even copulate with the female members. The male may initiate further migration even if copulation has occurred. Analysis has shown (see p. 164) that the habitat variable the perception of which (or some correlate of which) is the most important component of the male's migration threshold is the ratio of silverback males to receptive females in the group. Migration seems to terminate, however, if the male becomes the dominant member of a group, either by forming a new group by attracting females away from an existing group or by taking over an existing group by superseding another male.

(a) Migration of six silverback (> 10 years) males to and from a single heterosexual group of gorillas

Each horizontal bar refers to one individual. Solid black, with group; open, not with group; hatched, status not known. Except for the permanence of the dominant male, removal migration seems to be independent of size and position in dominance hierarchy.

(b) Ontogenetic variation in initiation factor and migration incidence

The histogram shows the annual percentage of males of different age-classes that initiate migration independently of their parents. To increase the number of usable observations, the figure was calculated for each age-class by the formula $X = 100(J+L)/2T$ (where X is the migration incidence shown in the histogram, J is the number of males of the age-class that join the group during a year, L is the number of males of the age-class that leave the group during a year, and T is the number of males of the age-class that were present in the group at the beginning of the year). The vertical axes for initiation factor and migration incidence were matched according to the assumption that when 100 per cent of males initiate migration, the initiation factor is likely to have a minimum value of 1·0.

(c) The adaptive package for the group-to-group migration of male gorillas: ontogenetic curves of $\bar{h}_f/\bar{h}_1$ (the habitat quotient), $\bar{M}_f$ (the mean migration factor), $\bar{S}_d$ (the mean action-dependent potential reproductive success), and $\bar{m}$ (mean migration cost)

The ages at weaning and the onset of sexual maturity are indicated by vertical dashed arrows. The two likely components of migration cost are indicated by shading. Solid shading, copulation loss; stippled shading, death-risk through predation. Gorillas, including adult males, are killed and eaten by two common indigenous predators: leopards, *Panthera pardus*, and man, *Homo sapiens*.

The initiation factor, i, shown in (b) is derived from the formula $i = \bar{h}_f \bar{M}_f / \bar{h}_1$, where $\bar{M}_f$ is given by $1 - \bar{m}/\bar{S}_d$.

[*Field data from Schaller (1963, 1965). Diagram A redrawn from Schaller (1963). Photo by G.B. Schaller (courtesy of Bruce Coleman Ltd)*]

migration incidence, it appears from the circum-stantial and limited data for the other species shown in Table 17.1 that a pre-reproductive peak of removal migration is a general rule for type-GH migration, no matter whether the species concerned is a reptile, as in the rainbow lizard, *Agama agama* (see p. 195), or a primate. There are only two instances in Table 17.1 where the evidence available does not strongly indicate such a pre-reproductive peak. One of these is the female gorilla, for which the limited data available (Schaller 1963), when ana-lysed as for the male (Fig. 17.18), could be taken to indicate a more or less constant migration incidence throughout life with the pre-reproductive peak, if it exists at all, being slight. The second is the male vervet monkey, *Cercopithecus aethiops*. This African monkey is typically an animal of riverine vegetation and of forest fringes, especially where these adjoin

grassland, and in effect represents an intermediate between an arboreal species and a fully terrestrial species such as the patas monkey, *Erythrocebus patas*. A typical group in a fairly rich forest habitat-type consists of eleven individuals of which two are adult males, four are adult females, and five are young. The males, however, seem to have an extremely high rate of type-GH removal migration. One group, for example, that was observed for thirteen months by Gartlan and Brain (1968) had a total of five males associate with it during that time. Only one of these, however, was present continuously, the others in-itiating removal migration, one of them to a neigh-bouring group. It could be that these observations indicate a high but constant rate of removal mi-gration by males throughout life. However, the situation is similar to that of the male gorilla (Fig. 17.18(a)), and it is possible that, if stay-times were

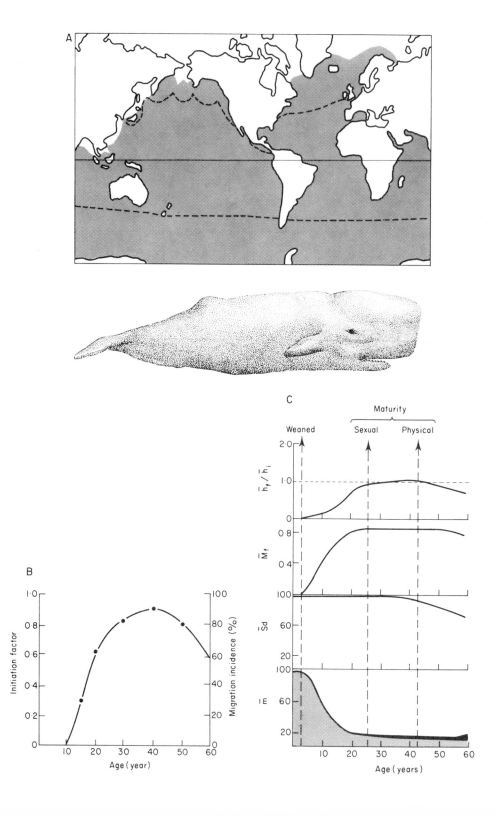

Fig. 17.19 Ontogenetic variation in the heterosexual group to heterosexual group migration by the male of the sperm whale, *Physeter catodon*: application of a model of adaptive ontogenetic variation in the probability of initiation of migration

General ecology
The basic social unit of the sperm whale, *Physeter catodon*, is a heterosexual group (=school) consisting of one or more adult males and 20–50 adult females and their young. When food is locally abundant, several heterosexual groups may aggregate temporarily to form a herd which may consist of a thousand or more individuals. Sperm whales feed in water more than about 100 m in depth, mainly on mid- but deep-water squid, as well as on octopus, cuttlefish and fish. Large numbers are consumed, up to 28 000 squid with an average length of 1 m having been found in a single stomach. Squid up to 10 m in length and sharks up to 3 m in length have been known to be taken. Young sperm whales may be preyed upon by sharks and young and larger individuals by killer whales, *Orcinus orca*, but, until the last few generations, adult males were probably subject to little predation in the open sea.

A. Approximate geographical distribution of males
The stippled area shows the maximum geographical range of solitary male sperm whales. Females, and hence males that are members of heterosexual groups, do not range as far north and south as solitary males, the approximate limits of female distribution being indicated by a dashed line.

Migration by males
The migration considered here involves a male leaving one heterosexual group and eventually becoming a member of another. Many years may intervene between these two events. Until about the time of attainment of sexual maturity, such migrant males often form all-male groups (=pods) of up to 40 individuals which, although the individuals may spread over an area of 50 km², appear to be in sensory contact and to move together. From the age of attainment of sexual maturity (19–26 years), males tend to become solitary and to attempt to become a member of a heterosexual group, presumably either by displacing another male from an already existing group or by attracting females away from their original group and thus forming a new group.

B. Ontogenetic variation in initiation factor and migration incidence
The graph shows the proportion of males of different ages that are in the process of migration (i.e. are not a member of a heterosexual group) (data from Ohsumi 1966). The curve shown can also be taken to represent ontogenetic variation in *i*, the initiation factor. The matching of the axes for initiation factor and migration incidence is based on the assumption that if 100 per cent of individuals are migrating, the initiation factor is likely to have a minimum value of 1·0.

C. The adaptive package for the heterosexual group to heterosexual group migration of male sperm whales: ontogenetic curves of $\bar{h}_f/\bar{h}_1$ (the habitat quotient), $\bar{M}_f$ (the mean migration factor), $\bar{S}_d$ (the mean action-dependent

potential reproductive success) and $\bar{m}$ (mean migration cost)

The ages at weaning and the attainment of sexual and physical maturity are indicated by vertical dashed arrows. The two likely components of migration cost are indicated by shading. Solid shading, copulation loss; stippled shading, death-risk through predation.

The initiation factor, *i*, shown in B, is derived from the formula $i = \bar{h}_f \bar{M}_f / \bar{h}_1$, where $\bar{M}_f$ is given by $1 - \bar{m}/\bar{S}_d$.

[*Data from: Slijper (1962), Caldwell et al. (1966), Nishiwaki (1967), Ohsumi (1965, 1966), Bannister (1968), Clarke et al. (1968), Gaskin (1968a), H.S.J. Roe (1969), Best (1969a,b, 1970a)*]

to be analysed, a pre-reproductive peak of migration incidence would be found for the male vervet monkey also.

Figures 17.18–17.20 present adaptive packages for type-GH migration by the males of three species, the gorilla, sperm whale, and plains' zebra, and for the female of the plains' zebra. All three of these species live in social groups in which adult females outnumber adult males. Only in the case of the zebra, however, is there always only one adult male per social group. Let us now consider, from a comparative viewpoint, the components of the adaptive packages for these three species together with those presented for man (Figs. 17.10 and 17.13) and the Japanese macaque (Figs. 17.6 and 17.7) in relation to the more general data presented in Table 17.1.

The illustrated habitat quotients are of the same general pattern for both sexes of all species from birth to the cessation of parturition in females and of virtual cessation of copulation in males and for the most part can probably be attributed to the same selective pressures. During the period of dependence on the parent(s), not only for the acquisition of food but also for protection and for the acquisition of those techniques and strategies necessary for maximum fitness as an adult, it may be assumed that the habitat quotient approaches zero for type-GH migration independent of the parents. As the young animal becomes increasingly capable of acquiring its own food, protecting itself, and generally performing adult functions, and perhaps also as the parent increasingly directs its attention toward younger offspring, the habitat quotient is likely to increase and the parental deme need be no longer inevitably the most suitable habitat with which the animal is familiar. This is particularly the case for young females which are likely to receive equal, if not greater, protection from the adult male member(s) of any group to which they migrate. The

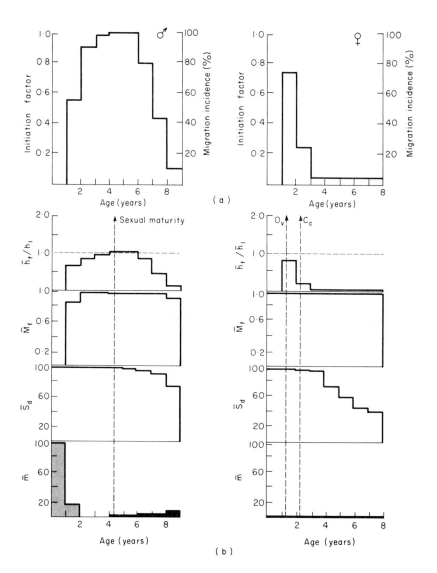

Fig. 17.20 Ontogenetic variation in the heterosexual group to heterosexual group migration of the plains' zebra, *Equus quagga*, in the Ngorongoro crater, Tanzania: application of a model of adaptive ontogenetic variation in the probability of initiation of migration

The basic social unit within the herd of the plains' zebra, *Equus quagga*, (=Burchell's zebra, *Equus burchelli*) is a heterosexual group consisting of one adult male, 1 to 6 ($\bar{x}$ =3) adult females, and 0 to 9 ($\bar{x}$=3) young. This figure is concerned with the migration of males and females from one heterosexual group to another.

Migration by males
Migration by males involves leaving one heterosexual group and eventually becoming a member of another. Several years may intervene between these two events, however, during which time the male alternates between a 'solitary' existence (though still remaining a member of the herd) and being a member of an all-male group of up to nine individuals. Removal migration by males from the group into which they were born is initiated soon after the age of 1 year. All males have initiated removal migration by the age of 4.5 years. The initiation of removal migration by the young males is not related to aggression on the part of the adult male and could be obligatory (but see p. 219). At the age of 5–6 years, males attempt to become a member of a heterosexual group either by displacing another male from an already existing group or by attracting females away from their original group and

thus forming a new group. During this stage the male is likely to take over and form several groups only to be displaced subsequently by some other male. With time, however, the male is less and less likely to be displaced.

Migration by females

Removal migration by females involves leaving one heterosexual group for another and usually occurs in response to a new male that may or may not already be a member of a heterosexual group. Any male that tries to attract or drive a female of any age from her present group is opposed by the adult male of that group. It seems likely that the removal migration of females is calculated and is dependent in part on the female's threshold and in part on the relative attributes of the original and new males and/or groups. Removal migration of females is most likely to occur between first ovulation and first conception.

(a) Ontogenetic variation in initiation factor, i, and migration incidence

The histogram for males shows the proportion of individuals of different ages that are in the process of migration (i.e. are not a member of a heterosexual group).

The histogram for females shows the proportion of individuals of different ages that initiate migration.

The matching of the axes for initiation factor and migration incidence is based on the assumption that if 100 per cent of individuals migrate, the initiation factor is likely to have a minimum value of 1·0.

(b) The adaptive package for the heterosexual group to heterosexual group migration of male and female zebra: ontogenetic curves of $\bar{h}_f/\bar{h}_1$ (the habitat quotient), $\bar{M}_f$ (the mean migration factor), $\bar{S}_d$ (the mean action-dependent potential reproductive success) and $\bar{m}$ (mean migration cost)

The ages of attainment of sexual maturity by males and of first ovulation (Ov) and first conception (Cc) by females are indicated by vertical dashed arrows. The two likely components of migration cost for males are indicated by shading. Solid shading, copulation loss; stippled shading, death-risk through predation.

The initiation factor, i, shown in (a), is derived from the formula $i = \bar{h}_f \bar{M}_f/\bar{h}_1$, where $\bar{M}_f$ is given by $1 - \bar{m}/\bar{S}_d$.

[*Compiled, with some extrapolation, from data in Klingel (1969). Photo by H. Klingel*].

adaptive package for the female zebra (Fig. 17.20) is particularly instructive in this context and not only emphasises the pressures that could produce a peak of group-to-group migration incidence in females at about the time of onset of reproduction but also provides further support for the arguments of Trivers (1973) concerning cuckoldry. The detailed field work of Klingel (1969) indicates that female zebra have a peak of type-GH removal migration at about the time of first ovulation but that, as the probability that conception has occurred increases, migration incidence falls off to a low and probably constant level. An adult male zebra makes a large parental investment in each of the adult female members of his group, not only by protecting them and their offspring from predators but also by protecting them from the frequent attention of other males. Such parental investment is only advantageous, however, if the offspring produced by that female are sired by that male. Females about to ovulate for the first time, therefore, present to a male an advantageous subject for investment. If the female has ovulated, however, and there is a chance that she has already been fertilised by some other male, optimum strategy is for the male to withhold parental investment until such time that the probability is sufficiently great that any offspring produced by the female were not sired by any other male. The greater the possibility that the female has been fertilised by another male, the less likely is the female to elicit parental investment from a male. Such behaviour by male zebra has not yet been investigated. If the arguments are accepted, however, it follows that the habitat quotient for female zebra should reach a peak at about the time of first ovulation and then rapidly decrease to a low level.

This peak of ability to attract parental investment from an adult male at about the time of first ovulation seems likely to be a major factor in the variation of habitat quotient in relation to migration by females between one-male social groups, including monogamous pairs, in which the danger of cuckoldry imposes considerable selection on the male. Other species, however, live in social groups the optimum structure of which (see Crook 1970) inevitably raises problems of cuckoldry to any male attempting to make considerable parental investment in a particular female and her offspring. Such species are particularly those that live in multi-male heterosexual groups such as the Japanese macaque, gorilla and sperm whale. Presumably, because of this danger of cuckoldry, males of such species tend

to make rather less parental investment in individual females and their offspring. It would probably be a mistake, however, to consider that such males make no parental investment other than that involved in courtship and copulation. As the group is likely to contain a male's offspring, any investment that the male makes in increasing the fitness of the group as a whole may be considered to be parental investment. It would probably be true to say that there is a broad correlation between the probability that a group contains a given male's offspring and the amount of investment (protection etc.) that the male makes in the group's welfare. Indeed, it is only by such a view of the behaviour of males of species that live in multi-male heterosexual groups that sense can be made of the apparent weight given by female gorillas to the silverback male:female and young ratio in their assessment of habitat suitability (p. 92). However, because the males of such species direct their investment to the group as a whole, any reduction in individual investment experienced by a female once she has passed the age of first ovulation is likely to be much less severe than in such species as the zebra. It seems likely, therefore, that the less individual the parental investment a female receives from the male members of the social group in relation to which migration is being considered, then the less marked should be the peak of habitat quotient that occurs at the age of first ovulation, or just before the age of first conception if this is likely to be different as it is in man. Note that this conclusion is as applicable to female humans as to the females of the other species considered. The multi-male heterosexual groups of man between which migration has been considered in Figs. 17.9–17.14 contain primarily monogamous sub-groups, the evolution of which was discussed in Chapter 14 (p. 118). It is the single male of this sub-group from which the female receives individual parental investment. A marked peak in habitat quotient for male-to-male migration would therefore be expected before the age of first conception, an expectation that is in accord with the peak of 'trial' pair-bond formations before first conception followed by infrequent male-to-male migration thereafter that is such a feature of human reproductive biology from hunter–gatherers to industrialists. In the deme-to-deme migration of man that has been considered in detail in the previous section, however, the female does not receive individual parental investment from the male members of the group and consequently does not have such a marked peak of habitat quotient.

Degree of individual parental investment is unlikely to be the only component of habitat quotient, however, even for females. Another component seems likely to exist before the onset of reproduction and to affect both males and females, though not necessarily to the same extent, and results from the learning, social, construction, and other investments that an individual makes in the habitat occupied. These investments commence soon after birth and also immediately after each removal migration. Habitat quotient is an inverse function of the size of such investment and consequently is likely to decrease with time after birth and after each removal migration. However, not only is the contribution of this component to the total habitat quotient a function of time but it is probably also a function of the demands for efficiency made on the individual. These demands seem likely to be greatest on males and females of monogamous pairs and adult males of single-male heterosexual groups and least on subadults of either sex and on adult males of species that form multi-male heterosexual groups.

One final major component of the habitat quotient seems likely to appear once an individual of either sex begins to produce offspring and concerns the distribution of the offspring produced. It has already been concluded that adult males of species that form multi-male groups (but without the formation of monogamous or polygamous subgroups as in man and the hamadryas baboon) influence their reproductive success by their protective and other investments in the welfare of the group most likely to contain their offspring. If it is assumed that a male by migrating from such a group reduces the fitness of each member of that group, at least for a short period, then it follows that a male achieves greatest reproductive success, all else being equal, by remaining with the group that has the greatest probability of containing his offspring. The greater the probability that a group contains an offspring of a given male or the more of that male's offspring the group is likely to contain, then the less is the habitat quotient for a male living with that group. As the male ages, therefore, and the probability increases that he has produced offspring, the less becomes the habitat quotient.

The critical factor in this situation seems to be that such males, if they initiate type-GH removal migration, are unlikely to be accompanied by the majority of their offspring. It should be noted, however, that there are observations for the gorilla (Schaller 1963) of adult males being accompanied

on their group-to-group migration by juveniles (3–6 years old), particularly males. Without knowing the filiation of the two individuals, however, and without knowing their recent histories, such associations are difficult to interpret.

The situation is slightly different, however, for males of species such as man that form monogamous pairs and in which the male, female and offspring perform type-GH migration as a unit. Such males, and, of course, females of virtually all species that perform type-GH migration, are invariably accompanied by their dependent offspring. As offspring become less and less dependent, however, it becomes less and less inevitable that the parents, if they migrate, will be accompanied by their young. The distribution of offspring as a component of habitat quotient results in a decrease in habitat quotient after the onset of reproduction in these males and all females as it did in adult males of multi-male heterosexual groups. In these males and all females, however, the decrease due to this component does not occur until a little later, that is not when the offspring are conceived but when they have been weaned and achieved a certain degree of independence.

To sum up this discussion of ontogenetic variation in habitat quotient, therefore, in relation to group-to-group migration, in all cases the habitat quotient is likely to begin at a low level during dependence on the parents and to increase after weaning to reach a maximum level just before the age of first conception in females and just before the age of first fertilisation of (as opposed to copulation with) a female in males. Thereafter, all components of habitat quotient act to produce a decrease in value. How rapid this decrease is likely to be, however, depends on the social structure of the species concerned. As far as females are concerned, decrease in habitat quotient is likely to be greatest in those species that form single-male heterosexual groups, including those that form monogamous pairs if it is a male-to-male and/or a male territory to male territory migration that is being considered, and that produce offspring that achieve independence rapidly relative to the potential length of reproductive life when adult. The decrease is likely to be least for females of species that form multi-male heterosexual groups and that, as individuals, do not elicit significant parental investment from any male member of the group, and also for females of species that produce offspring that achieve independence only slowly relative to the potential length of reproductive life when adult. As far as males are concerned, habitat quotient is likely to decrease most rapidly after the beginning of successful fertilisation of females in species that form single-male heterosexual groups and in which the male would not be accompanied by the offspring if it did migrate to another group. Decrease is likely to be least rapid in males of species that form multi-male heterosexual groups and that are likely to be accompanied by their offspring if they did migrate.

It is suggested by the various adaptive packages that have been presented that the habitat quotient makes an important contribution to ontogenetic variation in migration incidence. If the above conclusions are valid, therefore, they should be reflected by a more-pronounced pre-reproductive peak in species that form single-male heterosexual groups than in species that either form multi-male groups or that, as in the case of the lion, *Panthera leo*, have females that receive little individual parental investment from males other than that involved in

Table 17.2 Possible relationship between social structure and the extent of the pre-reproductive peak of group-to-group migration incidence

Species	Social group	Migration type†	Relative incidence‡ after onset of reproduction	
			Males	Females
titi monkey *Callicebus moloch*	monogamous pair + offspring	m	o?	o?
man *Homo sapiens*	,,	m	o	o
white-handed gibbon *Hylobates lar*	,,	m	o?	o?
rainbow lizard *Agama agama*	,,	m	o	o
common Indian langur *Presbytis entellus*	single-male group	g	*	?
lutong *Presbytis cristatus*	,,	g	*?	?
plains' zebra *Equus quagga*	,,	g	o	o
vicuña *Vicugna vicugna*	,,	g	o	*
vervet monkey *Cercopithecus aethiops*	multi-male group	g	**?	?
black mangabey *Cercocebus albigena*	,,	g	*	o?
Japanese macaque *Macaca fuscata*	,,	g	**	**
olive baboon *Papio cynocephalus anubis*	,,	g	*	?
man *Homo sapiens*	,,	g	**	**
chimpanzee *Pan troglodytes*	,,	g	*	?
gorilla *Gorilla gorilla*	,,	g	*	***?
lion *Panthera leo*	,,	g	**	**
sperm whale *Physeter catodon*	,,	g	*	?

Pronounced peak (bracket grouping titi monkey through vicuña); Less pronounced peak (bracket grouping vervet monkey through sperm whale)

† Types of group-to-group migration
m = mate-to-mate and/or mate's home range to mate's home range or, in many males, home range to home range migration
g = group-to-group and/or group's home range to group's home range migration
‡ Incidence expressed relative to that (standardised as ***) before onset of reproduction, where more asterisks indicate greater incidence and o indicates minimal, but not zero, incidence (i.e. *** indicates no pre-reproductive peak; o indicates pronounced pre-reproductive peak of migration incidence)

copulation. Table 17.2 compares these two group-types with respect to their migration incidence after the ontogenetic onset of reproduction expressed in relation to the level of pre-reproductive migration incidence. A difference does seem to exist between the two groups in the direction that would be expected from the above arguments concerning habitat quotient (i.e. males and females of species that form single-male groups, whether the reproductive unit is monogamous or polygamous, in general have a more-pronounced pre-reproductive peak of migration incidence than males and females of species that form multi-male groups).

Ontogenetic variation in S_d, mean action-dependent potential reproductive success, seems likely to be straightforward and to follow more or less the same pattern described for humans with an 'average' reproduction history in Fig. 17.9. In all of the species for which adaptive packages for type-GH migration are illustrated and for all the species listed in Table 17.1, including the rainbow lizard, *Agama agama*, males achieve reproductive maturity later than females. In all of these species, therefore, the ontogenetic curve for males starts to decrease later than that for females.

Migration cost seems likely to be another composite variable that again shows consistent interspecific features. Of particular importance, however, is the suggestion (Fig. 17.20) that when migration occurs by an individual moving from one group to another while the two groups are in sensory contact (= short distance migration, for present purposes), then migration cost is likely to approach zero at any age and whether or not the migration is advantageous is entirely a function of the relative suitability of the two groups (= habitat quotient). If migration cost is zero at any age, then the migration factor is 1·0 at any age, and any ontogenetic variation in migration incidence must, according to the present model, be entirely due to ontogenetic variation in habitat quotient. When it is possible, as with the Japanese macaque, to compare ontogenetic variation in migration incidence for short-distance and long-distance migration (i.e. in the present context a migration initiated when the group that is the eventual destination is not within the sensory range of the migrant), any difference can be attributed to the migration factor, which is a function of S_d and migration cost ($M = 1 - m/S_d$). When we find, therefore, as for the Japanese macaque, *Macaca fuscata* (Figs. 17.6 and 17.7) and zebra, *Equus quagga* (Fig. 17.20) as well as for the gorilla,

Gorilla gorilla (Schaller 1963), chimpanzee, *Pan troglodytes* (Nishida 1968), black mangabey, *Cercocebus albigena* (Chalmers 1968), and vicuña, *Vicugna vicugna* (Koford 1957), and probably for the vast majority of polygamous and promiscuous species that live in social groups, that females perform only short-distance migrations, it can perhaps be assumed that for the females of these species migration cost for long-distance migration approaches S_d, a situation that implies a death-risk of 100 per cent. The example of the Japanese macaque is particularly instructive in this situation because it is possible to compare males and females for both short- and long-distance migration (Figs. 17.6 and 17.7). Because the short-distance migration fixes the curve of habitat quotient, the model takes on a degree of objectivity for the curves of migration cost for long-distance migration. The curve of migration cost for immature females for long-distance migration is the same as that for a male during the same age period, but thereafter a difference appears. The difference increases to reach a maximum around the age that a female would for the first time be maximally encumbered by offspring. This suggests, therefore, that during the evolution of the species, ground predators, among which man should probably be included, were sufficiently active for a female encumbered by young to have little chance of escaping predation if she spent time away from a social group. Solitary males, on the other hand, unencumbered by young, experienced a moderate death-risk during migration that, however, was sufficiently less than 100 per cent for migration on average to be advantageous, depending of course on the suitability of the habitat previously occupied (p. 43).

Death-risk to self would seem to be the major component of migration cost for long-distance migration for most of the species considered. Especially during that period of ontogeny that migration cost has the greatest effect on migration incidence. This period of ontogeny is that from birth to the attainment of physical maturity, a state that also implies experience at escaping predation and finding food. For all of the species illustrated, therefore, migration cost due to death-risk to self starts at a high level during dependence on the parents and thereafter decreases. In the male zebra, *Equus quagga* (Fig. 17.20), the decrease is assumed to be rapid. Beyond a certain age, particularly when younger animals have been born into the group, the individual protection that a male zebra receives from either its mother or from the adult male of its natal group is likely to be minimal. Yet even a 'solitary' male zebra remains a member of the herd and gains some advantage from being surrounded by other individuals. Death-risk during removal migration is therefore likely to be minimal for a male zebra once it ceases to receive individual assistance and protection from its parent(s). In the male gorilla (Fig. 17.18), however, death-risk is assumed to decrease more gradually. Increase in stature from weaning to physical maturity seems likely to lead to a decrease in death-risk but even physically mature males, during the evolution of the species, will, to judge from the modern situation, have been subject to considerable predation by leopards, *Panthera pardus*, and man. Some decrease in death-risk with age after physical maturity seems likely to occur, however, due to increased experience.

A second component of migration cost that seems likely to be involved in male migration is copulation loss due to lack of access to receptive females. This component should appear at sexual maturity but is unlikely to make an important contribution to total migration cost until the male reaches the age at which normally he would begin to be successful at fertilising (as opposed to just courting or copulating with) females. As the male Japanese macaque approaches this age but before the migration threshold has risen sufficiently to cause the individual to settle permanently with a group, seasonal increase in the importance of copulation loss due to the existence of a peak copulation season (Nishida 1966) seems to be sufficient to raise the migration threshold enough during this season for the male to associate with a group temporarily. Presumably it is the copulations that occur during these temporary associations, acting as described through the habitat quotient (p. 305), that eventually raise the migration threshold sufficiently for a male to stay with a group for a longer period.

Other less proportional components of migration cost, such as death-risk to offspring and reduced nett rate of energy uptake, are largely self-explanatory but may well in some circumstances make an important contribution to total migration cost. These components are particularly likely to be important to females and monogamous pairs migrating with dependent young and could well contribute to the absence of long-distance migration by females of species that, because the optimum social group is polygamous or promiscuous, would not be accompanied on migration by males.

17.3.2 Ontogenetic variation in the incidence of home range to home range (type-H) and seasonal migration circuit to seasonal migration circuit (type-C) migration in non-avian vertebrates

Having reached certain conclusions concerning sexual and ontogenetic variation in group-to-group (type-GH) migration, the same approach can be used to consider sexual and ontogenetic variation in area-to-area migration. Table 17.3 summarises the results of some field studies that have provided data concerning ontogenetic and sexual variation in the area-to-area migration of a number of non-avian vertebrates. As with Table 17.1, this is by no means an exhaustive list and represents a fairly random sample of the literature. Although the northern fur seal, *Callorhinus ursinus*, and sea otter, *Enhydra lutris*, have, for taxonomic reasons, been included in this list, their removal migrations, which have already been described (pp. 105 and 222), are much more easily discussed in connection with that of sea-birds and are therefore excluded from the following discussion.

On the whole far more information is available for area-to-area than group-to-group migration and perhaps more confidence can be felt in the reality of the pre-reproductive peaks of migration incidence indicated in Table 17.3 than for those indicated in Table 17.1. Most of the species listed live solitarily within a home range that has more or less the same boundaries throughout the year. Removal migration (type-H migration in Table 17.3) involves the establishment by the individual of a new home range, the boundaries of which do not overlap those of the original home range. Investigation of these species involves plotting the home range over a period of time for a number of individuals on the basis of marking and trapping or, for larger animals, observation of recognisable individuals. During the course of such investigations migrant individuals, often called transients or nomads, are trapped or observed just once or twice within the home range of another individual. Combined data for: age of departure of young from the home range of their mother; the age distribution and gender frequency of the migrants relative to that of the 'resident'

Table 17.3 Some field observations of ontogenetic and sexual variation in incidence of area-to-area removal migration in non-avian vertebrates

Species	Migra- tion[†]	Place	Relative incidence[‡] before and after onset of reproduction				Reference
			Male		Female		
			pre-	post	pre-	post	
FISH							
California striped bass							
Roccus saxatilis	C	California	**?	?	**?	?	Calhoun 1952
REPTILES							
rusty lizard							
Sceloporus olivaceus	H	Texas	***	o	***	o	Blair 1960
side-blotched lizard							
Uta stansburiana	H	Colorado	***	*	***	*	Tinkle 1967a,b
MAMMALS							
Marsupials							
brush-tailed possum							Tyndale-
Trichosurus vulpecula	H	E. Australia	***	o	***	o	Biscoe 1973
quokka							
Setonix brachyurus	H	W. Australia	o	o	o	o	Dunnet 1962
Insectivores							
masked shrew							
Sorex cinereus	H	Manitoba	***	**	***	**	Buckner 1966
Arctic shrew							
Sorex arcticus	H	Manitoba	***	**	***	**	Buckner 1966
short-tailed shrew							
Blarina brevicauda	H	Manitoba	***	**	***	**	Buckner 1966
European mole							⎧ Godfrey 1957
Talpa europaea	H	Britain	***	*	***	*	⎨ Godfrey and
							⎩ Crowcroft 1960
Bats							
eastern pipistrelle							
Pipistrellus subflavus	C	E. USA	**?	o?	**?	*?	Fig. 17.21
Schreiber's bat							
Miniopterus schreibersii	C	E. Australia	*	o	**	*	Dwyer 1966
Pangolins							
giant pangolin							
Manis gigantea	H	Gabon	***	o	***	o	Pages 1970
Tree shrews							
Siamese tree shrew							
Tupaia belangeri	H	Thailand	***	*?	***	*?	Martin 1968
Primates							
orang utan							
Pongo pygmaeus	H	Indonesia	***	o?	***	o?	Rodman 1973
Rodents							
black-tailed prairie-dog							
Cynomys ludovicianus	H	S. Dakota	***	*	***	*	Fig. 17.27
13-lined ground squirrel							
Citellus tridecemlineatus	H	Texas	***	*?	**	o?	McCarley 1966
chipmunk							
Tamias striatus	H	New York State	***	**	***	*	Yerger 1953

† types of area-to-area migration
  H = total home range to total home range
  C = seasonal home range to seasonal home range
‡ convention for relative migration incidence
*** to * = decreasing migration incidence
  o = minimal, but not zero, migration incidence

Table 17.3, *cont'd.*

Species	Migra-tion[†]	Place	Relative incidence[‡] before and after onset of reproduction				Reference
			Male		Female		
			pre-	post	pre-	post	
woodchuck							Schwartz and
Marmota monax	H	E. USA	***	o?	***	o?	Schwartz 1959
eastern fox squirrel							
Sciurus niger	H	E. USA	***	**	***	**	Jordan 1971
grey squirrel							Schwartz and
Sciurus carolinensis	H	E. USA	***	**	***	**	Schwartz 1959
beaver							
Castor canadensis	H	Montana USA	***	**	***	**	Townsend 1953
muskrat							⎧Beer and
Ondatra zibethicus	H	Bavaria	***	**	***	*	⎨ Meyer 1951
							⎩Mallach 1971
Scandinavian lemming							⎧Kock and
Lemmus lemmus	H	Scandinavia	***	*	***	*	⎨ Robinson 1966
							⎩Bergström 1967
brown lemming							
Lemmus trimucronatus	H	Arctic Canada	***	**?	***	*?	Krebs 1964
varying lemming							
Dicrostonyx groenlandicus	H	Arctic Canada	***	**?	***	*?	Krebs 1964
prairie deermouse							⎧Blair 1958
Peromyscus maniculatus	H	USA	**	*	**	*	⎨Healy 1967
white-footed mouse							
Peromyscus leucopus	H	USA	***	**	**	*	Burt 1940
rice rat							
Oryzomys capito	H	Panama	***	o	***	o	Fleming 1971
rice rat							Negus *et al.*
Oryzomys palustris	H	Louisiana	***	*	***	*?	1961
spiny pocket mouse							
Liomys adspersus	H	Panama	***	o	***	o	Fleming 1971
spiny rat							
Proechimys semispinosus	H	Panama	***	o	***	o	Fleming 1971
Norway rat							
Rattus norvegicus	H	USA	***	*	**	*	Calhoun 1963
wood mouse							Flowerdew 1974
Apodemus sylvaticus	H	Britain	***	**	***	*	Southern 1964
European rabbit							Myers and
Oryctolagus cuniculus	H	Australia	***	*	***	*	Poole 1961
Carnivores							
weasel							
Mustela nivalis	H	Scotland	***	**	***	**	Lockie 1966
stoat							
Mustela erminea	H	Scotland	***	**	***	**	Lockie 1966
tiger							
Panthera tigris	H	India	***	*	***	**?	Schaller 1967
leopard							
Panthera pardus	H	Africa	***	*?	***	*?	Schaller 1972
mountain lion							
Felis concolor	H	Idaho	***	*?	***	*?	Hornocker 1970

† types of area-to-area migration
  H = total home range to total home range
  C = seasonal home range to seasonal home range
‡ convention for relative migration incidence
*** to * = decreasing migration incidence
    o = minimal, but not zero, migration incidence

Table 17.3, *cont'd.*

Species	Migra-tion[†]	Place	Relative incidence[‡] before and after onset of reproduction				Reference
			Male		Female		
			pre-	post	pre-	post	
cheetah							
Acinonyx jubatus	H	Africa	***	*	***	*	Schaller 1972
timber wolf							⎰ Murie 1944
Canis lupus	H	Canada	***	o?	**	o?	⎱ Banfield 1953
American red fox							
Vulpes fulva	H	E. USA	***	**	***	*	Sheldon 1950
sea otter							
Enhydra lutris	?	Aleutian Is.	**	*	**	*	Kenyon 1969
black bear							
Ursus americanus	H	Montana	***	o	**	o	Fig. 17.23
northern fur seal							
Callorhinus ursinus	?	N. Pacific	*	o	*	o	p. 105
Protoungulates							
aardvark							
Orycteropus afer	H	Gabon	***	o	***	o	Pages 1970
Ungulates							
white-tailed deer							⎧ Hawkins and
Odocoileus virginianus	H/C	E. USA	**	o	*	o	⎪ Klimstra 1970
							⎨ Downing *et al.*
							⎪ 1969
							⎩ Sweeney *et al.* 1971
Dall's sheep							
Ovis dalli	C	Canada	**	*	?	?	Geist 1971
bighorn sheep							
Ovis canadensis	C	Canada	**	*	?	?	Geist 1971
Cetaceans							
humpback whale							
Megaptera novaeangliae	C	SW Pacific	**	*	**	*	Fig. 17.25

† types of area-to-area migration
  H = total home range to total home range
  C = seasonal home range to seasonal home range

‡ convention for relative migration incidence
*** to * = decreasing migration incidence
    o = minimal, but not zero, migration incidence

---

individuals; and the frequency with which 'residents' shift their home range; usually show conclusively that there is a pre-reproductive peak of removal migration, the incidence of which falls off either gradually or rapidly after the onset of reproduction.

Rather more problems are involved in demonstrating removal migration from one seasonal migration circuit to another (type-C migration in Table 17.3). Each individual of such species invariably has a number of home ranges, each one of which is occupied at the appropriate season (see Chapter 18). Obtaining data relevant to removal migration in such species usually involves rather

longer-term field programmes than for the type-H removal migrants. One method of obtaining suitable data is described in the legend to Fig. 17.25 in relation to type-C migration by the humpback whale, *Megaptera novaeangliae*. More commonly, however, the research programme involves becoming acquainted with the home-ranges normally occupied by a given individual at a particular season. If an individual then appears one year in a different home range at the appropriate season or, in the case of a newly independent individual, in a different home range from that occupied the previous year with its parents, then such an individual can be considered to be a removal migrant. In view

of the difficulties involved in such a study it is not surprising that there are few type-C migrations for which detailed information is available. The work of Geist (1971) on American mountain sheep represents one of the most comprehensive of such field-studies as also does that of Dwyer (1966) on Schreiber's bat, *Miniopterus schreibersii*, in south-eastern Australia. The validity of considering type-C shifts in home range to be removal migrations is supported by the data obtained by Hanak *et al.* (1962) for the lesser horseshoe bat, *Rhinolophus hipposideros*, in Czechoslovakia. Several individual case-histories for marked bats recaptured several times in different parts of their seasonal migration circuit show clearly that when type-C removal migration takes place the individual adheres to the new migration circuit just as, before type-C migration, it adhered to the old migration circuit.

Occasionally, data for the type-C migration can be obtained by concentrating observation in one seasonal home-range type. In the case of the northern fur seal, *Callorhinus ursinus*, for example, described in Chapter 14 (p. 105) information is only readily accessible for breeding site to breeding site removal migration. It is rare for an individual to be captured more than once at other points on its migration circuit. Nevertheless, as long as conclusions are based as in the northern fur seal on information obtained through recaptures at the appropriate season but away from the original home range (breeding site in this case) reasonably reliable information concerning ontogenetic variation in incidence of type-C migration can be obtained. The problem comes, however, when conclusions are based only on the proportion of individuals that return to the season's home range of the previous year, for it is then not possible to distinguish between ontogenetic variation in removal migration and ontogenetic variation in survival rate. Perhaps this danger can be best illustrated by Fig. 17.21, in which data on degree of return of the eastern pipistrelle bat, *Pipistrellus subflavus*, to its hibernation site is presented that originally was interpreted solely in terms of ontogenetic variation in survival rate.

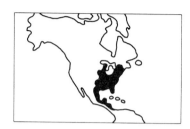

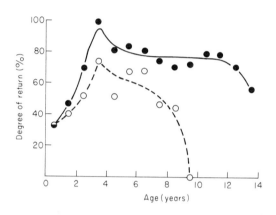

Fig. 17.21   Variation in survival and/or rate of removal migration with age in the eastern pipistrelle, *Pipistrellus subflavus*

A solid dot (male) or open dot (female) shows the degree of return (i.e. per cent return of individuals present the previous year but modified as described below), of bats of different ages to a hibernation roost in a West Virginia cave. As the experiment was intended to evaluate survival rate, the figures have been modified such that individuals that were found after more than one year were also assumed to have returned to the cave in the intervening years. In this way it could be argued that the figures obtained represent variation in survival rate with age. The curve could be taken to show that survival is low during the first two years of life, rises to a peak in about the fourth year, and thereafter decreases, though more rapidly in the female than the male (W.H. Davis 1966). The calculations involved in the production of this diagram, however, take no account of the very real possibility that the frequency of removal migration might also be a function of age. For instance, the figure could equally well be taken to show, if it were assumed that survival does not vary with age, that removal migration is most frequent during the first two years of life, is at a minimum in about the fourth year, and thereafter gradually increases with age, but being more frequent in females than males. Until such time that it is possible to distinguish between mortality and removal migration, the real meaning of this figure cannot be evaluated.

[*Diagram modified from Davis (1966)*]

Of the 50 species listed in Table 17.3, relatively convincing evidence is available for 49 that there is a peak of removal migration incidence during the pre-reproductive period. For only one species, the quokka, *Setonix brachyurus*, a small (height 25–30 cm; weight 2·5–5·0 kg), mainly nocturnal, non-burrowing, non-social, West Australian marsupial, did the author (Dunnet 1962) express the opinion that a pre-reproductive peak of area-to-area migration was absent. However, the techniques used in this investigation of quokka ecology were different from those described above for type-H migration, and although they could have demonstrated relatively long-distance removal migration, if it had occurred, they could not have demonstrated home range to home range removal migration. Home ranges for individuals were not studied, and as the species has a large familiar area, some individuals travelling up to 2 km to water and to rubbish dumps, it was impossible to distinguish between a shift in daily home range and animals captured during these longer-distance return migrations. In fact, the existence of such a large familiar area relative to the size of home range, as already argued for black and grizzly bears, *Ursus americanus* and *U. arctos horribilis* (p. 175), strongly implicates the existence of at least pre-productive exploratory migration and probably also, therefore, pre-reproductive removal migration. In view of the uncertainty that must still therefore remain concerning the reality of the absence of a pre-reproductive peak of type-H removal migration for the quokka, no attempt will be made in this section to account for such an absence.

Apart from the case of the quokka, the overwhelming implication of Table 17.3 is that it is general, at least for mammals, that ontogenetic variation in the incidence of both type-H and type-C area-to-area removal migration is characterised by a pre-reproductive peak in both males and females. Sexual differences seem to be slight. Of the 42 type-H migrations listed in Table 17.3, no sexual difference is indicated during the pre-reproductive stage for 36 (86 per cent). In the remaining six (14 per cent), greater migration incidence by males is indicated. Once reproduction has commenced, no sexual difference is indicated for 33 (79 per cent) of the species. Of the remaining nine species, eight (19 per cent) seem to show greater migration incidence by males and one (2 per cent) seem to show greater migration incidence by females. This latter species is the tiger, *Panthera tigris*, in which the adult females seem to initiate migration from the home range

around the time of ovulation (Schaller 1967). Whether this migration can be considered to be a type-H removal migration remains to be evaluated. Individually recognisable females have been observed to leave an area for several months and then to return. It is possible, therefore, that the female has a much larger home range during the optimum period for copulation than at other times. It is a reasonably well-documented phenomenon (p. 388) that the males of some species have a much larger home range during the breeding season than at other times. As discussed in Chapter 18, the optimum size of the home range is subject to selective pressures different from those on the optimum incidence of removal migration. The observation for tigers (Schaller 1967) is that adult transients are more often female than male and that such females are usually in oestrus. However, two interpretations are possible. One is that adult female tigers have a higher incidence of type-H removal migration than males and that migration threshold decreases at each ovulation, increasing again after conception. The alternative interpretation is that the majority of transient adult females are neighbouring individuals that are temporarily expanding their home range while in oestrus. In order to be conservative, the former interpretation has been adopted in Table 17.3. The true interpretation, however, remains to be elucidated. In general, therefore, the incidence of

Fig. 17.22 Tiger, *Panthera tigris*

[*Photo by G. B. Schaller (courtesy of Bruce Coleman Ltd)*]

Fig. 17.23   Ontogenetic variation in the home range to home range migration of the black bear, *Ursus americanus*, in northwest Montana: application of a model of adaptive ontogenetic variation in the probability of initiation of migration

General ecology
Black bears live mainly in heavily wooded areas and dense bushland, particularly mixed coniferous–deciduous forest. The basic social unit is a solitary adult or an adult female with one to four cubs ($\bar{x} = 2$). The home range of a female is 2·6 km along its longest axis and that of a male 6·2 km, but the familiar area is much larger, including occasional feeding sites that may be 16 km from their home range. There is minimal overlap in home range between individuals of the same sex, but the home range of a male overlaps with the home ranges of several females. Food is mainly green vegetation in spring, mainly berries in autumn, and fish, small mammals, birds, insects and carrion whenever possible. Bears also drink frequently. Food availability and acquisition is critical, the probability of successful conception and parturition possibly being a function of the berry crop the previous autumn. Death-risk through parasites, particularly cestodes and nematodes, may also be a function of the success at finding sufficient food of sufficient nutritional value. The main predators of black bears are grizzly bears, *Ursus arctos horribilis*, and wolves, *Canis lupus*, in parts of the range, and man, *Homo sapiens*, throughout the range (for 3000 generations at low intensity and approximately 10 generations at high intensity).

A. Geographical distribution

The asterisk indicates the study area for migration.
    The migration considered here is that from one year's home range (i.e. the total area in which a year's movements have been performed) to another, non-overlapping home range.

B. Ontogenetic variation in initiation factor, i, and migration incidence

The dashed histograms show the proportion of males and females of different age-classes that moved out of the study area of Jonkel and Cowan (1971). The migration by the one to two year age-class corresponds to 100 per cent migration by males and 80 per cent migration by females from the home range of the previous year which, in this case, is the home range of the mother. An estimate of the proportion migrating from their home range in subsequent years is made by assuming a constant relationship between this proportion and the proportion leaving the study area.
    The matching of the axes for initiation factor and migration incidence is based on the assumption that if 100 per cent of individuals migrate, the minimum likely value for the initiation factor is unity.

C. The adaptive package for the home range to home range migration of male and female black bears: ontogenetic curves of $\bar{h}_f/\bar{h}_1$ (the habitat quotient), $\bar{M}_f$ (the mean migration factor), $\bar{S}_d$ (the mean action-dependent potential reproductive success), and $\bar{m}$ (mean migration cost)

The ages at weaning and the attainment of sexual and physical maturity are indicated by vertical dashed arrows.
    The initiation factor, i, shown in B, is derived from the formula $i = \bar{h}_f \bar{M}_f / \bar{h}_1$, where $\bar{M}_f$ is given by $1 - \bar{m}/\bar{S}_d$.

[*General information from Kolenosky (1970). Feeding details from Hatler (1972). Home range, survival, and migration data from Jonkel and Cowan (1971)*]

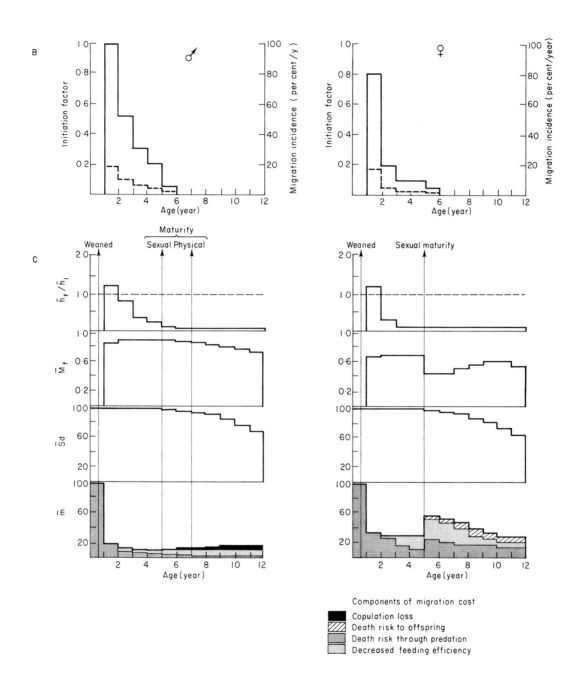

Components of migration cost

■ Copulation loss
▨ Death risk to offspring
▤ Death risk through predation
░ Decreased feeding efficiency

type-H removal migration by adults shows little sexual difference but when such a difference does occur it is invariably the male that shows the higher migration incidence.

Three type-C migrations for which estimates are available of the migration incidence of both sexes are listed in Table 17.3, one of which is the eastern pipistrelle bat, *Pipistrellus subflavus*, for which little confidence can be placed in the results obtained (Fig. 17.21). Of the two species remaining, there is no indication for one of them that there is any sexual difference in migration incidence. The remaining species, Schreiber's bat, *Miniopterus schreibersii*, shows some indication of a slightly greater migration incidence for females, an indication that incidentally is in accord with what would be concluded from Fig. 17.21 for the eastern pipistrelle if it was assumed that the results obtained were more relevant to incidence of removal migration than to survival rate.

Adaptive packages have been developed for three of the area-to-area migrations listed in Table 17.3. One of these migrations, that of the black-tailed prairie-dog, *Cynomys ludovicianus*, is considered later (Fig. 17.27). For the moment, discussion is concerned mainly with the other two migrations. One of these, that of the American black bear, *Ursus americanus* (Fig. 17.23) is a type-H migration. The other, that of the humpback whale, *Megaptera novaeangliae* (Fig. 17.25), is a type-C migration.

The components of habitat quotient that seem likely to be important in the area-to-area migration of vertebrates are similar to those that seem likely to be important in group-to-group migration, though inevitably some differences in emphasis must be postulated. Let us first consider the habitat quotient of type-H area-to-area migrants. Most of the species listed in Table 17.3 live solitarily, either in an exclusive home range or at least in such a way that their activities are relatively independent of other individuals the home ranges of which overlap with their own. If, as with the prairie-dog, *Cynomys ludovicianus*, the species does live in social groups, then the migration being considered, to 'qualify' as a type-H migration (rather than a type-GH migration—Table 17.1), must be a response to habitat variables other than those concerned with the age and sex structure of the group. This automatically decreases the importance of those components of habitat quotient, such as the parental investment received by females from males and such as the spatial distribution of offspring, that were considered to be important in type-GH migration. Even so, a number of components remain of those suggested for type-GH migration that seem likely also to be important in type-H migration.

It should perhaps be stressed at this point that the categorisation of removal migration as type-GH, type-H, and type-C has been carried out solely for descriptive convenience. In reality there is every gradation between clear group-to-group migration as in the zebra, *Equus quagga* (Fig. 17.20), through

Fig. 17.24   Adult and young of the grey squirrel, *Sciurus carolinensis*

In the vast majority of mammals, young pre-reproductive individuals have a lower threshold for the initiation of home range to home range migration than adults. If the two squirrels illustrated were sharing a home range, the suitability of which began to deteriorate, it would be the young squirrel in the right-hand picture that would initiate exploratory and then removal migration first. This would be so whether or not there was any aggressive interaction between them.

[*Photos by Paul Turner and R. R. Baker*]

intermediate situations of group-and-area to group-and-area migration as in the deme-to-deme migration of man, *Homo sapiens* (Fig. 17.11), to clear area-to-area migrations as in the black bear, *Ursus americanus* (Fig. 17.23). Similar gradation occurs beween type-H and type-C migration and between type-GH and type-C migration. The same components of habitat quotient and migration cost are likely to be involved in all types of removal migration. It is the relative contribution of the different components to the total variance of their wholes that is likely to differ from migration to migration and species to species.

In both type-GH and type-H migrations the habitat quotient is likely to approach zero during the period of dependence on the parent(s) and to increase as the now-juvenile individuals become increasingly capable of living an independent existence. Habitat quotient is particularly likely to increase in mammals as the next brood of offspring is produced by the parent(s) and the protection and attention received from the parent(s) by the juveniles in their natal home range decreases to a minimum level. Unlike the situation in type-GH migration, however, not only is the habitat quotient for type-H migration likely to increase with increasing independence but for juvenile or now sub-adult individuals it is likely to increase to a value greater than 1·0. If, as seems likely, the optimum size of the social unit for 'solitary' type-H migrant species is a single adult, then, as already argued (p. 219), as the young approach adulthood it becomes an advantage to both parent and offspring for only one adult to remain in their previously joint home range. The habitat quotient increases, therefore, for both parent and offspring as the latter approach adulthood. Other components of habitat quotient, however, make it seem likely that the habitat quotient is always much greater for sub-adults (i.e. individuals that are sexually mature in that gametogenesis has begun but which are not yet likely to produce offspring for other physical or social reasons) or near-adults (i.e. individuals that are near to sexual maturity of species for which sexual maturity and the production of offspring more or less coincide) than for their parent(s). In particular, the component of habitat quotient that results from learning, social (in the sense of Leyhausen (1964) that 'solitary' animals have social interactions with their neighbours), and construction investments seem likely to be important to type-H migrants.

Home range investments characteristically lead to a decrease in habitat quotient with time after arrival in the home range. This decrease is likely to be particularly pronounced in 'solitary' species because the investment(s) in the home range have all been made by a single individual whereas with animals that live in groups the investment and accuracy of investment per individual is probably less. Learning involves such factors as acquaintance with the position of optimum roosting, feeding and escape sites in different parts of the home range. In species with large home ranges and a longevity that spans several years, learning investment invariably also involves acquaintance with the way in which the positions of optimum sites vary at different times of year. Social investments involve the establishment of relationships with neighbouring individuals so that interactions are as efficient as possible. This involves, therefore, learning the limits of neighbours' home ranges and determining with which neighbours the individual can successfully compete during any dispute and which neighbours to avoid. For males, also, social investment involves learning which parts of his home range overlap or adjoin the home range of a female and also maintaining acquaintance with her epigamic state (i.e. receptivity to courtship). This latter investment involves knowing which females are rearing offspring, which females have lost offspring, which females are unhealthy, and in general maintaining familiarity with any clue that indicates whether or not a female is likely to be capable of being fertilised some time in the near future. Construction investments are probably made by most animals that live in a home range in the form of worn tracks, scent trails, roosts, escape shelters, burrows, and tunnels etc.

In summary, therefore, ontogenetically the habitat quotient is likely for most type-H migrants to start at a low level during the period just after birth, to rise as the animal develops toward first independence and then maturity, to reach a peak that in 'solitary' animals should exceed unity, and then to decrease once more as the investments made by the animal in its adopted home range gradually increase. The peak of habitat quotient should come at the age at which on average the presence of n developing young plus the parent(s) in the natal home range causes h, the suitability of that range to fall below the value of either $h_f M_f$ or $\bar{E}_{ec}$ (see p. 66). Although the situation is complex it may be expected for any one species that if h_p, the component of habitat suitability due to the physical and biological construction of, in this case, the home range apart from the presence of competing individuals (p. 70), is constant, the more offspring of a litter and succeeding litters that are produced and survive the earlier the peak will occur. This suggestion is consistent with the observation for many species, such as the vicuña, *Vicugna vicugna* (Koford 1957), that the age at which young individuals migrate is in part a function of whether or not the mother successfully produces her next litter.

The ontogenetic curve of $\bar{S}_d$, mean action-dependent potential reproductive success, for type-H migrants is likely to follow the usual form except that, in many species, the curve is likely to be similar for both males and females. The components of migration cost are also likely to be similar in both

type-H and type-GH migrants and are illustrated in Fig. 17.23 for the black bear. As for type-GH migrants, death-risk to self is likely to be high soon after birth and then to decrease with increase in stature and experience. Death-risk as a component of migration cost for type-H migrants, however, is always likely to be greater than zero because during migration there is no immediate access to escape shelters such as are available in a home range. Similarly, decrease in feeding efficiency and hence nett rate of energy uptake is likely to be greater away from a home range. Type-H migrant females, however, are less likely than type-GH migrant females to experience a migration cost of sufficient magnitude to negate any advantage of solitary long-

Fig. 17.25 Ontogenetic variation in the seasonal migration circuit to seasonal migration circuit removal migration by the humpback whale, *Megaptera novaeangliae*, in the southwest Pacific and Indian Oceans: application of a model of adaptive ontogenetic variation in the probability of initiation of migration

General ecology
The humpback is a baleen whale that filters its food from the plankton, euphausiid crustaceans and, to a lesser extent, fish forming the bulk of its diet. It lives either solitarily or in groups of two or three individuals. Pairs may consist of any sexual and ontogenetic combination but trios consist usually of two adults and one young. The possibility of widely distributed, loosely organised groups maintained by long-distance communication cannot be totally ruled out. The main predators, until the last few generations, particularly on young individuals, are sharks and killer whales, *Orcinus orca*.

A. Approximate geographical distribution and the nature of the removal migration considered

Although humpbacks occur throughout the major part of the Earth's oceans, it seems likely that the population is organised into a number of smaller demes. The three best-documented demes are those that inhabit the three areas indicated by hatching in the diagram. An individual marked in any one of these areas is unlikely to be recaptured in any other area. Within each area seasonal return migrations take place between summer feeding grounds in polar waters and winter copulation and parturition grounds near the tropics. Within each area, however, there is more or less random mixing of individuals. Each of these demes, therefore, is an assemblage of individuals that share more or less the same seasonal migration circuit. From time to time, however, an individual performs removal migration from one deme to another. It is the ontogenetic variation in the probability of initiation of this removal migration with which this figure is concerned.

B. Ontogenetic variation in initiation factor, i, and migration incidence

In the literature, recoveries of marked whales are recorded to give dates and positions of marking and recapture, sex, and body length at recapture. From the body length at recapture, an estimate of age at recapture and therefore, also, at marking can be made from the growth curve (male humpbacks are slightly smaller than females from the age of about 2 years onwards). The total number of marked and subsequently recaptured whales that have been at liberty during each age-year can thus be calculated. The number of removal migrations by each whale between marking and recapture is given by assuming just one removal migration if the whale is recaptured in the deme adjacent to that in which it was marked, but two removal migrations if it has migrated between the West Australian breeding deme and the New Zealand/Oceania breeding deme. If the number of removal migrations (usually zero) is divided by the number of years between marking and recapture, an estimate of the average rate of removal migration is obtained for each intermediate age-year. If these are summed for each age-year from all recaptures and divided by the number of marked and subsequently recaptured whales at liberty during that age-year, then the result is an estimate of the proportion of whales that initiate removal migration during that age-year. The results of these calculations are shown in the histogram. The matching of the axes for initiation factor and migration incidence is based on the assumption that if 100 per cent of individuals were to migrate, the minimum likely value for the initiation factor is unity.

C. The adaptive package for the seasonal migration circuit to seasonal migration circuit removal migration of humpback whales: ontogenetic curves of $\bar{h}_f/\bar{h}_1$ (the habitat quotient), $\bar{M}_f$ (the mean migration factor) $\bar{S}_d$ (the mean action-dependent potential reproductive success), and $\bar{m}$ (mean migration cost)

The ages at weaning and the attainment of sexual maturity are indicated by vertical dashed arrows.

The initiation factor, i, shown in B, is derived from the formula $i = \bar{h}_f \bar{M}_f / \bar{h}_1$, where $\bar{M}_f$ is given by $1 - \bar{m}/\bar{S}_d$.

[*Field data from Dawbin (1964) and Chittleborough (1959, 1965)*]

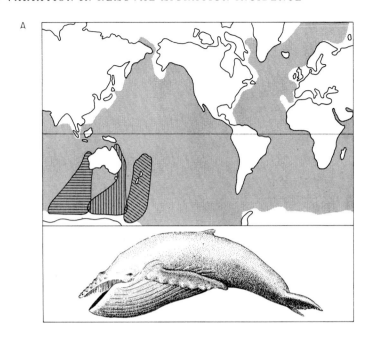

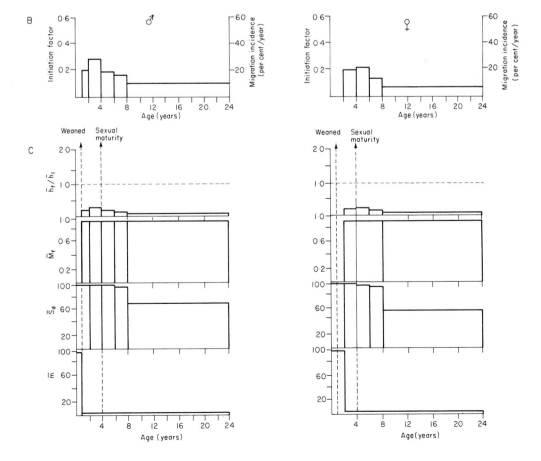

distance migration because, unlike the latter, even while in their home range they gain no extra protection from predators nor extra efficiency at food-finding as a result of being a member of a group. It is perhaps significant, therefore, that those species the females of which did perform long-distance type-GH migration were members of the smallest social groups (monogamous pairs—Table 17.2) and/or members of those demes for which migration cost is likely to be least.

Having considered the relative importance of the various components of an adaptive package for type-H migrants, let us consider briefly in a comparative way the components of an adaptive package for type-C migrants such as that presented in Fig. 17.25 for the humpback whale, *Megaptera novaeangliae*. On the whole, the same arguments can be applied to both. The difference in emphasis that can be postulated, however, results from one main point of difference in the biology of the two groups of species. On the whole, type-C migrants live far less solitarily than type-H migrants. Rather they tend to live in demes that share a migration circuit or at least part of a migration circuit. The members of this deme may or may not form social groups at different points on their migration circuit. For example, male and female mountain sheep, *Ovis dalli* and *Ovis canadensis*, form sexually segregated social groups along many points of the migration circuit but at other points, such as in the rutting range, form mixed-sex social groups (Geist 1971). The majority of temperate bat species hibernate in small or large clusters, often consisting of both sexes. In addition, the females of many bat species form so-called maternity or nursery groups in the home range used during the parturition period. On the other hand, as far as is known, members of demes of the humpback whale

(Fig. 17.25) form social groups of only two or three individuals (but see p. 767).

Unlike the situation for most type-H migrants, therefore, it is by no means inevitable that the habitat quotient for young type-C migrants will rise above 1·0 as they approach adulthood (compare Figs. 17.23 and 17.25C). In fact, habitat quotient is likely generally to be much lower for type-C than type-H migrants. According to the model, therefore, as long as there is no inevitable compensating influence from migration factor, migration incidence should generally be less for type-C than type-H migration. On the other hand, type-C migrants probably make equally marked investments in each of their seasonal home-ranges as type-H migrants make in their total home range, perhaps even a greater investment if the greater social investments and greater learning investments necessary to navigate successfully between successive seasonal home ranges are taken into account. The proportional decrease in habitat quotient is likely, therefore, at least to be equal in type-H and type-C migrants with the result that proportionately the pre-reproductive peak of type-C migration incidence should be at least as pronounced as that of type-H migration incidence, even though the general level of migration incidence should be less.

To some extent, this conclusion is supported by Table 17.4 which is based on the data in Tables 17.1 and 17.3. If we score three, two and one asterisks respectively as 3·0, 2·0 and 1·0 and score a 0-rating as 0·5, sum the scores respectively for type-H and type-C migrants, and divide by *n* to obtain the mean score, we obtain the following results, higher mean scores indicating higher migration incidence. The mean pre-reproductive migration score respectively for type-H and type-C migrants is 2·87 and 1·75 for

Table 17.4   Comparison of level of migration incidence and relative size of the pre-reproductive peak of migration incidence in four migration categories of non-avian vertebrates

Migration-type	Number of Examples		Mean score* for migration incidence before and after the onset of reproduction				Relative size (pre-/post) of pre-reproductive peak	
			Males		Females			
	Male	Female	pre-	post	pre-	post	Males	Females
GH one-male groups	7	5	3·00	0·64	2·80	0·60	4·68	4·67
multi-male groups	9	5	2·75	1·33	1·83	0·90	2·06	2·04
H	42	42	2·87	1·20	2·76	1·05	2·40	2·53
C	6	3	1·75	0·71	2·00	0·80	2·45	2·50

* Mean score = mean number of asterisks in Tables 17.1 and 17.3 with a 0-rating being scored as 0·5 asterisks

males and 2·76 and 2·00 for females. For individuals past the age of onset of reproduction the corresponding scores are 1·20 and 0·71 for males and 1·05 and 0·80 for females. At all ages for both sexes, therefore, type-H migrants have a higher migration incidence than type-C migrants. If we divide the score for pre-reproductive migration incidence by the score for migration incidence after the onset of reproduction in order to obtain a measure of the relative size of the pre-reproductive peak, we obtain values for type-H and type-C migrants respectively of 2·40 and 2·45 for males and 2·53 and 2·50 for females. In other words, the pre-reproductive peak is equally pronounced for both sexes in both type-H and type-C migration. Note that these figures are similar to those of 2·06 for males and 2·04 for females that can be obtained for type-GH migrants between multi-male heterosexual groups but are rather lower than the 4·68 for males and 4·67 for females obtained for type-GH migrants between single-male heterosexual groups. The available field data suggest, therefore, that the most pronounced pre-reproductive peak of migration incidence in non-avian vertebrates is found in type-GH migrants of both sexes between single-male heterosexual groups, an observation that is quite consistent with the arguments presented in this chapter (pp. 304 and 317) concerning ontogenetic variation in habitat quotient, given that habitat quotient makes an important contribution to ontogenetic variation in migration incidence.

17.3.3 Ontogenetic variation in the incidence of breeding site to breeding site migration in birds and amphibious marine mammals

Having considered removal migration in non-avian vertebrates in some detail, it is only necessary briefly to consider removal migrations by birds and amphibious marine mammals, for, with very little modification, the same arguments are applicable. Some introduction, however, is necessary. In particular, the categorisation of removal migrations into type-GH, type-H and type-C which was so useful in considering other vertebrates is inappropriate to the data available for birds. Some birds, particularly wildfowl as well as some sea birds (Chapter 27), perform definable migration circuits that are sufficiently clear for the removal migrations to be conveniently considered in type-C terms. An individual duck, goose or swan, for example, may use one water-body for breeding, another for aggregation in advance of seasonal migration, another for moulting, another or several as stopovers during migration and another or several for wintering, and the same water-bodies may be visited in the same sequence by that individual each year (Hochbaum 1955). Use of a different water-body for any one of these purposes could therefore be considered a type-C removal migration. Yet other birds, such as many individuals of the blackbird, *Turdus merula*, in England may occupy the same restricted home range, winter and summer, for several years (Lack 1966). Any removal migration by such a bird could therefore be considered a type-H migration. The vast majority of birds, however, are so mobile and their individual movements are so difficult to follow that it is simply not known to what extent the same individual performs the same migration circuit each year (but see p. 620). Nor, if the total home range is not a relatively circumscribed area, is it usually possible to define an individual's home range, especially if home ranges show considerable overlap (but see Odum and Kuenzler (1955) for the techniques that may be used in some cases). For most birds, therefore, all that is known is that they nest in one place one year and some other place x units of distance away (where the units may be centimetres or kilometres) the next year. The movements between these two events are relatively unknown except in the most general of terms for the species or deme as a whole. The recapture of ringed birds, although providing considerable information concerning many aspects of migration, does not lend itself to tracking the movement life-histories of individuals. Given the nature of the information available for most birds, therefore, the only reasonable approach to removal migration for this group is to measure the phenomenon in terms of the distance apart of nesting sites in successive years (i.e. the degree of return to the nesting site). Precisely similar arguments may be applied to amphibious marine mammals.

Table 17.5 lists some field observations of ontogenetic and sexual variation in the incidence of year-to-year nesting site to nesting site removal migration in birds. For pre-reproductive individuals the migration referred to is that from birth site to first nesting site. Asterisks in this case indicate in a very

Table 17.5 Some field observations of ontogenetic and sexual variation in mean distance of year-to-year nesting site to nesting site removal migration in birds

Species	Migration category†	Place	Relative incidence‡ before and after onset of reproduction				Reference
			Male		Female		
			pre-	post	pre-	post	
Impennae							
yellow-eyed penguin		New					
Megadyptes antipodes	c	Zealand	**	*	**	*	Lack 1966
Procellariiformes							
Laysan albatross							Bellrose
Diomedea immutabilis	c	N. Pacific	*	o	*	o	1972
black-footed albatross							Bellrose
Diomedea nigripes	c	N. Pacific	*	o	*	o	1972
slender-billed shearwater							Bellrose
Puffinus tenuirostris	c	Tasmania	*	o	*	o	1972
Manx shearwater							
Puffinus puffinus	c	Irish Sea	*?	o?	*?	o?	Harris 1972
Ciconiiformes							
white stork							
Ciconia ciconia	a	Germany	***	o	***	o	Lack 1966
Anseriformes							
mallard							
Anas platyrhynchos	a	Manitoba	****	***	***	o	Sowls 1955
pintail							
Anas acuta	a	Manitoba	****	***	***	o	Sowls 1955
gadwall							
Anas strepera	a	Manitoba	****	***	***	o	Sowls 1955
blue-winged teal							
Anas discors	a	Manitoba	****	***	***	*	Sowls 1955
shoveler							
Spatula clypeata	a	Manitoba	****	***	***	o	Sowls 1955
Charadriiformes							
common tern		Eastern					Bellrose
Sterna hirundo	c	USA	*	o	*	o	1972
kittiwake		NE					Coulson
Rissa tridactyla	c	Britain	**	o	**	o	1971
Columbiformes							
mourning dove							Tomlinson
Zenaidura macroura	a	Missouri	****	o	****	*	*et al.* 1960
Passeriformes							
cliff swallow							Bellrose
Petrochelidon pyrrhonota	a	USA	****	*	****	*	1972
tree swallow							Bellrose
Iridoprocne bicolor	a	USA	****	*	****	*	1972
bank swallow							Bellrose
Riparia riparia	a	USA	****	**	****	**	1972
kirtland warbler							Bellrose
Dendroica kirtlandii	a	Michigan	****	*	****	**	1972
ovenbird							Bellrose
Seiurus aurocapillus	a	Michigan	****	*	****	*	1972
house wren							Kendeigh
Troglodytes aedon	a	Illinois	***	o	****	*	1941
great tit							Kluijver
Parus major	a	Netherlands	***	*	***	*	1951
blue tit							Goodbody
Parus caeruleus	a	Britain	***	*	***	*	1952
pied flycatcher							
Ficedula hypoleuca	a	Germany	****	*	****	**	Lack 1966

† migration categories refer to columns (a), (b), and (c) in Fig. 17.26
‡ relative migration incidence
   *** to * = decreasing migration incidence
     o = minimal, but not zero, migration incidence

Table 17.5, *cont'd.*

Species	Migration category[†]	Place	Relative incidence[‡] before and after onset of reproduction				Reference
			Male		Female		
			pre-	post	pre-	post	
song sparrow *Melospiza melodia*	a	Ohio	*	o	*	*	Nice 1937
N. American grosbeak *Hesperiphona vespertina*	b	N. America	****	***	****	***	Newton 1972
siskin *Carduelis spinus*	b	Europe	****	***	****	***	Newton 1972
greenfinch *Carduelis chloris*	a	Britain	***	*	***	*	Newton 1972
redpoll *Acanthis flammea*	b	Europe	****	***	****	***	Newton 1972
brambling *Fringilla montifringilla*	b	Europe	****	***	****	***	Newton 1972
chaffinch *Fringilla coelebs*	a	Britain	***	*	***	**	Newton 1972
blackbird *Turdus merula*	a	England	*	o?	**	**?	Greenwood and Harvey 1976
song-thrush *Turdus philomelos*	a	England	**	o?	**?	?	Werth 1947

† migration categories refer to columns (a), (b), and (c) in Fig. 17.26
‡ relative migration incidence
   **** to * = decreasing migration incidence
     o = minimal, but not zero, migration incidence

general way the mean year-to-year migration distance. This table represents only a very small proportion of the information available for birds. It is hoped, however, that it indicates the existence of the three major categories of avian removal migration illustrated in Fig. 17.26. Representative examples of these three categories may now be indicated or described.

The first category involves perhaps the vast majority of land birds including many species of passerines and is of the pattern described for the pied flycatcher, *Ficedula hypoleuca* (p. 275). This type of removal migration is characterised by a moderately low degree of return by pre-reproductives and a high but not absolute degree of return by adults. In addition in many species, such as the pied flycatcher, the adult female shows a slightly lower degree of return than the male.

The second category involves various species of birds and is of the pattern described earlier for the short-eared owl, *Asio flammeus* (p. 208), and crossbill, *Loxia curvirostra* (p. 209). This type of removal migration is characterised by a low degree of return throughout life, though probably still with a pre-reproductive peak. Other examples are provided by Newton (1972) from among the European fringillid finches. All of the species that show a low degree of

return even when adult, such as the siskin, *Carduelus spinus*, redpoll, *Acanthis flammea*, pine grosbeak, *Pinicola enucleator*, brambling, *Fringilla montifringilla*, and the north European deme of the bullfinch, *Pyrrhula pyrrhula*, feed to a considerable extent on the seeds of trees, at least during the winter months. Perhaps it is significant, therefore, that much of the information for these species concerns winter range to winter range removal migration. Siskins and bramblings, for example, captured in Britain in one winter have been recaptured in subsequent winters on the European mainland. Other individuals of these two species banded in winter in Belgium have been recaptured in later winters in Turkey and the Balkans. No winter-to-winter recaptures of siskins and bramblings have been in the same place in two separate years and one siskin banded one winter in Germany was recaptured in a later winter 2200 km to the east. In North America the grosbeak, *Hesperiphona vespertina*, breeds in the coniferous forests, feeds on the large, hard fruits of trees, and migrates to the south or southeast in autumn. Degree of return is such that of 499 recaptures only 48 (9·6 per cent) were recaptured in a subsequent year near to the previous capture site. The remaining 451 (90·4 per cent) were found in subsequent years scattered among 17 American States and 4 Canadian

Provinces. Breeding site to breeding site recaptures also indicate a low degree of return. Summer-to-summer recaptures in Fenno-Scandia demonstrated a distance of 120 km between breeding sites for a siskin, *Carduelis spinus*, and 280 km and 550 km for two mealy redpolls, *Acanthis flammea*.

In areas where rainfall is erratic in distribution birds are often extremely nomadic with a low degree of return to all areas. Thus in Australia the honey-eaters (Meliphagidae), a predominantly nectari-vorous Australo-Papuan group which, with 69 species on the continent, is the largest Australian bird family, are characterised, particularly in areas where rainfall is limited and erratic, by a low degree of return that seems comparable to the situation described for tree and seed-feeding European finches. In the case of the honeyeaters, however, it is the flowering of the major nectar-bearing trees and shrubs that is the all-important factor (Keast 1968a). In any area the blossoming of a relatively few plant species accounts for most of the seasonal movements. Most movements in southern Australia are outside of the breeding season and there may be, therefore, a relatively high degree of return to the breeding sites in this area. In the interior and north, however, movements are less regular, particularly in dry years when breeding is inhibited. The erratic nature of many movements can be attributed to irregular blossoming, varying nectar flows from year to year, and to the major nectar-bearing trees flowering at different times in different places. According to Keast (1968a), 25 species (36·5 per cent) of honeyeaters in Australia have 'nomadic habits' that are at least moderately well developed. Indeed, some 23 per cent of all Australian bird species would seem to fit into the second category of removal migrants suggested here with only about 8 per cent of the avifauna having an annual north–south movement, constant in amplitude and timing that would presumably place them, along with most of the 'resident' species, into the first category of removal migrants suggested here. This high pro-portion of second-category species can be attributed to the year-to-year unpredictability of the rainfall over much of the Australian continent.

The third and final category involves various species of colonially nesting birds and in particular most, if not all, such species of penguins and procellariiform and charadriiform sea birds. A pos-sibly typical example is that of the kittiwake, *Rissa tridactyla*, on the northeast coast of England. The pre-reproductives of this species, the females of which breed for the first time when 3–4 years of age and males of which breed at 4–5 years of age (Coulson 1966, 1971), show a low to moderate incidence of removal migration, 25 per cent breed-ing in a colony other than the one into which they were born. Some of these young may start to breed in a colony nearly 200 km from their natal colony. When adult, however, removal migration incidence falls to a minimum level, only two adults having been observed to migrate in fifteen years from the colony in which they commenced breeding. The kittiwake was introduced as 'possibly typical' be-cause 25 per cent colony-to-colony removal migra-tion by pre-reproductives is rather higher than the estimated incidence for most similar birds (Bellrose 1972). The usual pattern, however, in the field study of degree of return is for the first estimates to indicate a 100 per cent return and then, as more and more-extensive studies are carried out, for this estimate to fall gradually to the sort of level shown by the kittiwake. A similar sequence seems to be usual in the investigation of colony-to-colony migration by pinnipeds (p. 105). There are many points of similarity in the return and removal migrations of pinnipeds and sea-birds, one of which is the level and pattern of ontogenetic variation in migration in-cidence, as can be seen from the rating for the northern fur seal, *Callorhinus ursinus*, and also, though not a pinniped, the sea otter, *Enhydra lutris*, in Table 17.3, and for the sea-birds in Table 17.5. The removal migration incidence of amphibious, ter-restrially colonial, marine mammals may therefore conveniently be included in this consideration of the removal migration of sea-birds.

Adaptive packages for these three categories of avian removal migration are presented in Fig. 17.26. Discussion of the components of these packages can be restricted to pointing out the similarities and differences in the nature of the components for birds and other vertebrates. The packages have been deliberately presented as histograms in yearly blocks in order that ontogenetic variation can be discussed apart from the influence of seasonal effects. A discussion of seasonal effects, which are inevitably more complex, follows in the next section.

Habitat quotient again seems to make an im-portant contribution to the form of ontogenetic variation in migration incidence. In all three bird-migration categories it is suggested that the habitat quotient is at a maximum during the pre-reproductive stage. The level of the maximum value, however, is likely to differ between the categories. In the first category the only inherent advantage to a young bird in returning to the area immediately surrounding its fledging site is that the very fact that the individual's parents succeeded in raising it to independence implies that the area is at least of minimum suitability. This inherent advantage of the natal area seems likely to be of considerable importance in the evolution of some forms of return migration (Part IIIb). Young birds, however, are extremely mobile and furthermore have sufficient time to be capable of visiting a large area around their natal site before the next breeding season. If other environmental clues, therefore, are also available to a young bird that correlate with future habitat suitability at the next breeding season, there seems to be no reason why habitats over a large area should not have equal probability of a suitability level similar to that of the natal site. The habitat quotient for such birds, therefore, is likely to approach unity. This is unlikely to be the case, however, for species of the second migration category which, on the whole, are those that exploit food sources such as tree seeds and cyclic rodents or, in irregularly arid areas, almost anything. As already argued in Chapter 16 (p. 208), it is highly likely that the most suitable habitats for these species are some distance away from the region of most suitable habitats of the previous year. The habitat quotient for the young of these species, therefore, is likely to be well above unity and has been given a nominal value of 2·0 in Fig. 17.26.

Juveniles of the third category of avian removal migrants, mainly colonially nesting sea-birds, are likely to be capable of visiting many other colonies as well as unoccupied potential nesting sites before they themselves reach maturity (p. 185). Most sea-birds, in common with most pinnipeds, are not accompanied by their parents once they become capable of living an independent existence. On the surface, therefore, there appears to be no reason for the habitat quotient for pre-reproductives to be anything other than 1·0. On the other hand, the apparently low incidence of removal migration by these animals and the observation, such as for the northern fur seal, *Callorhinus ursinus* (p. 105), that if the young return to breed on their natal islands then they are highly likely to return to their natal colony rather than to neighbouring colonies, implies that some behaviour is performed during the period from birth or hatching to independence from the parents that constitutes an investment in the natal colony sufficient to lower the habitat quotient below unity.

As this seems unlikely to be a construction investment or a learning investment into the topography of the nesting area, it can only be assumed that it is some form of social investment that, made when young, endures into reproductive life. This is discussed further in relation to adults.

In all three categories of avian migrants the habitat quotient seems likely to decrease after an early pre-reproductive peak. As it seems unlikely that the presence of independent offspring is an important component of habitat suitability, the arguments presented for type-H and type-C non-avian migrants seem to be most relevant. We may look, therefore, for increase with time in learning, social and construction investment in the breeding area to represent the main component leading to a decrease in habitat quotient from an early pre-reproductive peak. This possibility is supported for category-a migrants by an apparent correlation between greater investment in the breeding home range and a higher degree of return. Thus in many passerines, such as the pied flycatcher, *Ficedula hypoleuca* (p. 275), and the house wren, *Troglodytes aedon* (Kendeigh 1941), the males arrive first and establish the breeding territory. Females arrive a little later, play little part in territorial defence, and, at least during courtship, are fed by the male. In such species the adult and often the juvenile males tend to have a higher degree of return than the corresponding females. In many ducks, however, the male makes a relatively small investment in the breeding home range relative to the female and also has a lower degree of return. Both sexes of duck tend to arrive at the breeding site at about the same time but, although the male may play an active part in defending the territory during mating and early incubation, he leaves the nesting area even before the eggs hatch and joins other males at the moulting site (Hochbaum 1955). The female, on the other hand, stays with the young in the nesting area until just before or just after they can fly, the females of some species remaining in the nesting home range for up to five months (Sowls 1955). Such ducks are characterised by a much higher degree of return among the females than among the males. This low degree of return is associated in this adaptive complex with the phenomenon, particularly in river ducks such as the mallard, *Anas platyrhynchos*, of the male pairing with a female while on the winter range and then accompanying her on the spring migration, so closely in fact that it is the pair and not the individual that makes up the unit of a migrating

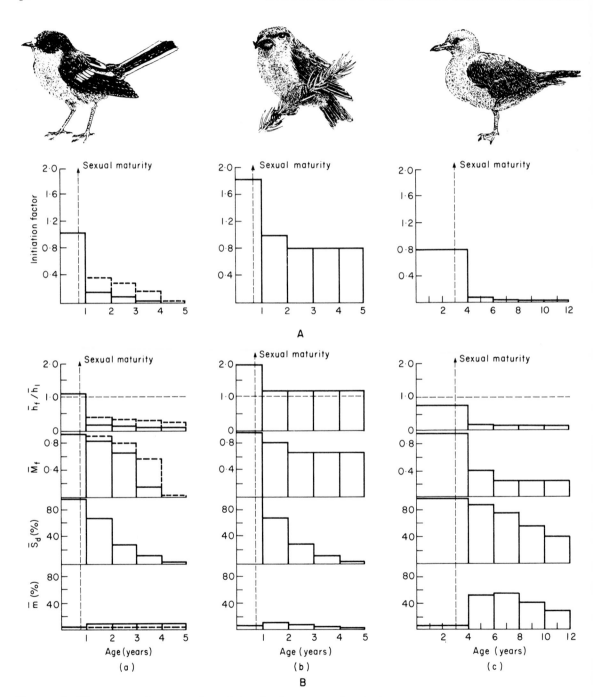

Fig. 17.26 Three types of ontogenetic variation in the breeding site to breeding site removal migration by birds: application of a model of adaptive ontogenetic variation in the probability of initiation of migration

At least three major types of ontogenetic variation in breeding site to breeding site removal migration can be recognised for birds and these are represented in columns (a), (b) and (c). In all these cases, removal migration may be achieved either by: (1) direct migration by a bird for which the home range for breeding is also the home range for the rest of the year; (2) using different parts of a familiar area for breeding in consecutive years; or (3) showing a low degree of return during a seasonal return migration.

Column (a) represents many species of birds, the individuals of which show a fairly high incidence of removal migration when juvenile but when adult show a fairly low incidence of removal migration. Among passerines adult females often have a higher incidence of removal migration than adult males. These species may feed on a wide variety of foods (except those listed under (b)) but are usually characterised by males that establish themselves in a nesting area before females select their mates. As well as being mainly responsible for territorial defence, the males also tend to feed the female to some extent before and during the incubation period.

Column (b) represents various species of passerines and birds of prey that have a high incidence of removal migration throughout their lives, though probably still with a maximum incidence when juvenile. These species seem to be characterised by an ecology based on a food source: (1) the abundance of which is synchronised over large areas; (2) with areas of abundance separated by areas of scarcity; and (3) the abundance of which in any one place is unpredictable but is such that a year of relative abundance is usually followed by a year or so of relative scarcity. Such food sources in temperate and sub-polar regions consist particularly of cyclic rodents and the seeds of trees, and in irregularly arid areas of almost anything.

Column (c) represents many species of colonially nesting birds that have a low to medium incidence of removal migration up to the commencement of breeding but thereafter have an extremely low incidence of removal migration. These species are usually characterised by the formation of long-term pair-bonds that may last for life.

A. Ontogenetic variation in initiation factor and migration incidence

The histograms illustrate the ontogenetic variation in calculated variation in initiation factor that seems to fit the general findings for migration incidence in relation to the onset of sexual maturity (vertical dashed arrow) in the three groups of birds. In column (a) the solid and dashed lines indicate a sexual difference (see p. 325). The sexes are not distinguished in columns (b) and (c).

B. Adaptive packages for three major types of breeding site to breeding site removal migration by birds: ontogenetic curves of $\bar{h}_f/\bar{h}_1$ (the habitat quotient), $\bar{M}_f$ (the mean migration factor), $\bar{S}_d$ (the mean action-dependent potential reproductive success), and $\bar{m}$ (mean-migration cost)

Approximate ages at the onset of sexual maturity are marked by vertical dashed arrows.

The initiation factor, i, shown in A, is derived from the formula $i = \bar{h}_f \bar{M}_f / \bar{h}_1$, where $\bar{M}_f$ is given by $1 - \bar{m}/\bar{S}_d$.

flock (Hochbaum 1955). Clearly such behaviour, the advantage of which is readily explicable in terms of sperm competition (Parker 1974a) and parental investment (Trivers 1973) can only evolve if there is also selection for a low degree of return by males or at least if any disadvantage of a low degree of return by males is less than the advantage gained from female-guarding, because the male is unlikely to be able to determine while on the wintering ground which females will be returning to which nesting areas. The most important part of this adaptive complex from which, perhaps, the other parts stem, evolutionarily, is that life on the water lends itself to an early independence of young in much the same way as does life on the ground. In such a situation male reproductive success becomes more a function of the fertilisation of eggs than a function of any parental investment directed toward the offspring. Hence the evolution of less male parental investment in the offspring and, associated with this, less investment by the male than the female in the nesting home range.

The adults of category-b birds (tree seed and cyclic rodent feeders and inhabitants of irregularly arid regions) are, like the juveniles, likely to experience a habitat quotient greater than unity. Yet, at the same time, the adults of even these species presumably make some investment in their nesting home range that lowers the habitat quotient relative to that of juveniles. Because the critical factor is the distribution of abundance of food, we are in effect dealing with a habitat quotient based only on that component, h_p, of habitat suitability that results from the physical and biological construction of the habitat other than that resulting from the presence of competing individuals (p. 70). Although the presence of abundant food (high h_p) one year invariably, for these species, means a low h_p in the succeeding year, h_p is unlikely to be reduced to zero. Any individual that does not migrate, therefore, experiences a high total suitability, h, because of the absence of competing individuals. This situation is identical to that considered in Section 17.2 (p. 268). Following the arguments presented there we would expect a frequency distribution of migration thresholds to evolve such that all individuals experience the same habitat suitability. More adults than juveniles, however, should have a high migration threshold because of the postulated effect of learning investment. It is possible, therefore, to envisage a population of category-b migrants, the individuals of which have a migration threshold that starts, ontogenetically, at a low level but that increases with age and as breeding or wintering home ranges are established. The level to which increase occurs, however, is different from individual to individual such that some move very little or have a high degree of return while others, in association with the year's juveniles, migrate some distance or have a low degree of return. The nett frequency distribution of migration thresholds for all ages, however, should be such that all individuals experience approximately the same potential reproductive success.

Category-c migrants (colonially nesting birds, particularly sea-birds, and amphibious marine mammals) seem also likely to be subjected to a decreasing habitat quotient once they become established in a breeding colony and, as with the other two categories, the most important factor seems likely to be the investment made in the nesting or breeding area. It was concluded earlier that the young of category-c migrants make some social investment in their natal colony that endures into adult life. Although there is no evidence as to what this investment may be for young individuals, there seems little doubt that reproductives make considerable social investment in the colony in which they breed. This investment seems to be sufficient to account for the decrease in habitat quotient to the low level shown for adults in Fig. 17.26(c). Adult male pinnipeds return to the same positions on the same beaches year after year and considerable time and energy must be saved as a result of the resource-holding power assessments of neighbouring individuals being carried over from year to year. Sub-adult males, also, perhaps increase their chances of obtaining females and, later, territories by becoming familiar with the relative resource-holding powers of the various territorial males and their distribution along the beach. Females, faced with the problems of ineffective or non-copulation and high offspring mortality if they join a large harem but copulation with a genetically inferior male if they join a small harem, probably also experience increased reproductive success by returning to the breeding colony in which they are familiar with the characteristics of the males and their harems.

Both pinnipeds and sea-birds breed in despotic situations, some regions of the breeding colony (e.g. the centre in many sea-birds) offering greater breeding success but being barred to many individuals by the presence of larger and/or stronger and/or older individuals, depending on the species. Establishment of the best position in the colony available to each individual, given its level of resource-holding power, involves time, energy, and injury risk. Selection should favour, therefore, individuals learning which positions are worth competing for given their resource-holding power, a process that is more rapid for individuals that return to the same colony at which breeding occurred the previous year. The argument remains valid even if it is not the same individuals that return to the same colony each year, though clearly, according to this argument, the advantage of return to the colony to one individual is even greater if colony composition and spatial distribution remains relatively constant from year to year so that the characteristics of the individuals can be learnt. The minimum requirement for return to the colony to be advantageous is only likely to be that approximately the same number of individuals with approximately the same frequency distribution of resource-holding power make up the colony each year. In this case, an individual with a relatively constant resource-holding power from one year to the next is likely to be able to attain the same type of position in the colony each year. However, once selection favours a high degree of return by adults then the colony composition tends to remain constant from year to year and this minimum requirement for selection to favour a high degree of return is soon exceeded.

Although habitat quotient seems to be the component that makes the largest contribution to the form of the adaptive packages shown in Fig. 17.26, it seems likely that migration factor also makes an important contribution. To a large extent, both components are influenced by an artefact that results from the difficulty of isolating ontogenetic and seasonal effects. The reasons for this difficulty are discussed in more detail in relation to Fig. 17.27 and are applicable to any type-C migrant in relation to breeding home range or even to any type-H or type-GH migrant that migrates just before or during the breeding season. In birds the effect results largely from the influence on migration cost of the possible delay that may be experienced in commencing to breed in a given breeding season as a result of an individual trying to establish itself in an unfamiliar area or colony. It is a matter of debate whether this factor should be included in migration cost or habitat quotient or both. In Fig. 17.26 it has been included in both.

17.3.4 The interaction of onto-genetic and seasonal variation in migration incidence

So far, we have been concerned strictly with onto-genetic variation in migration incidence in ver-tebrates. Yet for most of the species discussed there is also a very strong seasonal influence on migration incidence. In order to examine ontogenetic vari-ation independently of seasonal variation we have, in effect, taken the time of year of maximum migration incidence for each year of life and con-sidered the graph thus produced, ignoring the variation in migration incidence during the re-mainder of the year. Thus males of the Japanese macaque, *Macaca fuscata*, are most likely to initiate migration just after the four months of peak copu-latory activity (Nishida 1966). Young black bears, *Ursus americanus*, are most likely to initiate migration in their second summer (Jonkel and Cowan 1971). Humpback whales, *Megaptera novaeangliae*, of any age seem most likely to initiate removal migration during the summer while they are in polar waters (Chittleborough 1959, 1965). Most young tem-perate passerines and other birds that breed when one year old seem to decide during their first autumn how far from their natal site they will breed, even though they may then perform a seasonal return migration of many thousands of kilometres before actually settling to breed (p. 190).

Obviously, therefore, if the phenomenon of onto-genetic variation in removal migration is to be fully understood, some consideration must be given to the inter-relationship of these two sources of variation. For relative simplicity the example has been chosen of the black-tailed prairie-dog, *Cynomys ludovicianus* (Fig. 17.27), which is a type-H migrant. Neverthe-less, the principles that emerge are probably re-levant to most vertebrates and also many in-vertebrates.

If only the yearly peaks (June/July) of migration incidence are examined, the prairie-dog can be seen to follow the general pattern described earlier for type-H migrants. Migration incidence increases during immaturity to reach a peak just before the onset of sexual maturity, thereafter decreasing fol-lowing the establishment of a home range. As before, a large part of this ontogenetic variation can be attributed to variation in the habitat quotient and for the same reasons (p. 317). There is only one unusual feature in the *ontogenetic* variation in mi-gration incidence and this can be attributed to a

seasonal effect. That is that the migration threshold of first-year individuals seems to be higher than that of the parents with the result that, if habitat suitability decreases, the parental migration threshold is ex-ceeded first and they initiate migration, leaving their first-year offspring in their territory (see p. 213). This slightly unusual phenomenon, which argues against the inevitability of a pre-reproductive peak of migration incidence, can be attributed partly to the lengthy preparations that for maximum fitness a prairie-dog must make in its potential destination before finally initiating removal migra-tion, and partly to the influence of the seasonal availability of food on the timing of parturition. The seasonal breeding cycle would seem to have been subjected to selection for the appearance of lactating females and newly weaned young to coincide with the maximum abundance of nutrients. Other obser-vations (J. A. King 1955) suggest that if migrants are to modify the new habitat sufficiently to encourage the growth of appropriate food-plants and are to attain a suitable physiological state before the onset of winter conditions, then any removal migration to a new, previously unoccupied, territory must be completed by the end of July. In order to migrate by July it is necessary to begin exploratory migration and to start preparing the new habitat by May. The nett result of all these selective pressures is that it is most unlikely that a prairie-dog can perform an advantageous migration to a previously unoccupied territory during its first summer—unless, of course, the natal home range becomes of very low suit-ability. This is not so, however, for adults, with the result that selection seems to have favoured a lower migration threshold when adult than when newly weaned. By the second spring and summer of life, however, still immature prairie-dogs are even more likely than adults to be able to perform an advan-tageous migration because they have invested in the home range for a shorter period (a habitat-quotient effect) and also, because they are not involved in rearing offspring, they can spend more time, and from an earlier date, preparing a potential de-stination (a migration-cost effect). Perhaps a similar seasonal effect coupled with breeding when one year old can account for the possibly unusual ontogenetic variation in migration incidence in the water vole, *Arvicola terrestris*, in Scotland (Stoddart 1970). Here removal migration may be performed only by adult females. Stoddart feels, however, that removal mi-gration by juveniles has not yet been satisfactorily shown to be absent.

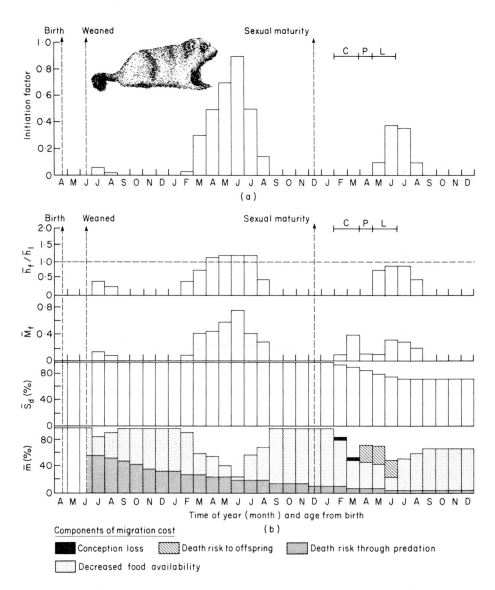

Fig. 17.27 Ontogenetic and seasonal variation in the territory-to-territory migration of the black-tailed prairie-dog, *Cynomys ludovicianus*, in the Black Hills of South Dakota: application of an adaptive model of the probability of the initiation of migration.

General ecology
The black-tailed prairie-dog is a North American rodent that grazes on grass and forbs, including seeds, and which is active all winter. Prairie-dogs aggregate in 'towns' which cover an area of 4–30 ha. Natural topographical features, such as ridges and trees, etc., divide the town up into 'wards'. The basic social unit, however, is a territorial unit consisting of 1–2 adult males ($\bar{x}=1\cdot65$), 2–5 adult females ($\bar{x}=2\cdot4$) and 0–39 young ($\bar{x}=5\cdot93$). Both males and females participate in territorial defence. A territory occupies an area of about 0·28 ha. Prairie-dogs establish a

complex system of runs and burrows within their territory and a notable feature of their behaviour is the way that they crop and modify non-food plants to improve visibility and to encourage the growth of food plants. Approximately 57 per cent of the active day is spent in feeding.

Migration
From time to time individuals leave the territory in which they have been living and establish themselves in a new habitat. Whether the new habitat is already occupied or whether it was previously uninhabited, the migration is usually preceded by a period of exploratory migration. The details of the process of territory-to-territory migration are described elsewhere (Fig. 14.7, p. 103). The incidence of territory-to-territory migration shows marked ontogenetic and seasonal incidence.

(a) Ontogenetic and seasonal variation in initiation factor and migration incidence

The histogram shown has been constructed from the descriptions given by J.A. King (1955) but does not have a quantitative basis. The vertical axis for initiation factor is based on the assumption that when almost 100 per cent of individuals migrate, as in the June of their second year, the likely minimum value of the initiation factor is 1·0. The vertical dashed arrows show the timing of birth, weaning, and sexual maturity, C, P, and L indicate respectively the seasons of copulation, parturition and lactation.

(b) An adaptive package for the territory-to-territory removal migration by black-tailed prairie-dogs: ontogenetic curves of $\bar{h}_f/\bar{h}_1$ (the habitat quotient), $\bar{M}_f$ (the mean migration factor), $\bar{S}_d$ (the mean action-dependent potential reproductive success), and $\bar{m}$ (mean-migration cost)

The likely components of migration cost are shaded as indicated.

The initiation factor, i, shown in (a), is derived from the formula $i = \bar{h}_f \bar{M}_f/\bar{h}_1$, where $\bar{M}_f$ is given by $1 - \bar{m}/\bar{S}_d$.

[Based on data in King (1955)]

The prairie-dog situation illustrates well the way that habitat quotient fluctuates seasonally as well as ontogenetically, in this case due to seasonal variation in abundance of foodplants and the extent to which previously uninhabited areas are amenable to modification. In many species, habitat quotient shows a change that is both seasonal and ontogenetic due to selection for seasonal synchronisation of the breeding cycle. Hence, the habitat quotient of black bears, *Ursus americanus*, always starts to increase during their second summer. Maximum food abundance for black bears seems to be in the late summer and autumn when berries are plentiful (Jonkel and Cowan 1971) and the whole breeding cycle seems to be geared to this period of abundance. Parturition occurs in winter after the pregnant female has built up a large fat deposit during the period of berry abundance. 1–4 cubs are born ($\bar{x}=2$) which become independent of the mother during their second summer at a time when it is presumably difficult for, on average, 2 now relatively large bears plus their mother plus a male consort to obtain sufficient food if they all stay and move together. The habitat quotient therefore rises rapidly at this time to surpass unity, not only for the ontogenetic reasons already described (p. 317) but because autumn food abundance is imminent when, presumably, even an inexperienced bear can maintain itself more or less equally efficiently whether in or away from its natal home range. In this case, therefore, it is unnecessary

to discriminate seasonal and ontogenetic effects in the ontogenetic rise of habitat quotient. Thereafter, ontogenetic and seasonal variation in habitat quotient can be readily discerned, for superimposed on the steady ontogenetic decrease in habitat quotient that begins as soon as the individual starts to establish a new home range is a seasonal oscillation from a low value in early spring to a high value in autumn.

In many ways consideration of seasonal aspects of migration incidence is simpler for type-C migrants than for type-H migrants because in the former type we are usually concerned with removal migration at one season only (e.g. breeding site to breeding site migration). Even so, complications can arise. There is a suspicion, for example, that nest-site to nest-site type-C migration in many birds, particularly juveniles of species that breed when one year old, is in effect performed not during the nesting season but during the previous autumn (p. 190). Having found a suitable nesting site or sites the bird may then perform the seasonal return migration of the species and not settle in the selected site until the following spring. The selective advantage of this behaviour in juveniles seems to be as follows. First both ontogenetic and seasonal effects, in the same way as just described for the black bear, probably combine for most species to produce a maximum value for the habitat quotient in early autumn. Furthermore, not only does food abundance help to produce a habitat quotient that approaches unity but it also decreases migration cost. Finally, locating a suitable nesting area in autumn reduces the time required to find an area and to start breeding the following spring, again therefore reducing migration cost.

Seasonal differences in migration cost can also help to explain seasonal differences in the incidence of type-C migration on occasions when there are no apparent seasonal differences in habitat quotient. The type-C removal migration of the humpback whale, *Megaptera novaeangliae*, for example, seems always to involve a shift in the entire seasonal migration circuit rather than just one season's home range (Fig. 17.25 and p. 183). Habitat quotient for such a migration seems unlikely to show any seasonal variation as a whole year's circuit is involved. Nevertheless, it seems that removal migration usually takes place during the summer months while the animals are feeding in polar waters. Migration cost, however, is likely to show a seasonal variation. During summer the only component of migration cost seems likely to be a reduced nett rate of energy uptake due to passage through waters less rich in plankton. The total migration cost, however, seems likely to be small. In the polar winter, however, while the animals are in their tropical breeding grounds, migration distance is greater and energy loss correspondingly more at a time when energy uptake is reduced and time and energy invested by males in courting females and by females in either protecting and feeding their young offspring or in courtship is at a premium. Migration cost, therefore, is likely to be greater in winter than summer. The combination of ontogenetic variation in habitat quotient and seasonal variation in migration cost therefore seems adequate to account for a large proportion of the variation in migration incidence shown by the humpback whale.

The discussion of sexual and temporal variation in migration incidence in relation to a home range or a seasonal succession of home ranges has so far been concerned entirely with vertebrates because it is for these animals that most information concerning the home range phenomenon is available. There is nothing in the arguments presented, however, that inevitably restricts the application of the model to vertebrates. Any animal that lives in a home range should be subject to similar selective pressures. To illustrate this point the 'nesting' site to 'nesting' site migration by a swarm of honey bees, *Apis mellifera* (p. 150), can be considered in a similar way to the type-H migration to a previously unoccupied area by the black-tailed prairie-dog, *Cynomys ludovicianus*, described in Fig. 17.27. It is necessary, however, to distinguish between the worker individuals of the honey bee which have a relatively short life-span of, during the summer, a few weeks and the queen with a life-span covering several years. The worker individuals that undertake the migration are predominantly young bees (pre-reproductive being an irrelevant concept relative to these sterile females) that have made relatively little learning investment in the original home range, thus conforming to the pattern that seems likely to be general to type-H migrants. As far as the queens are concerned, it is the old mature queen that migrates, leaving behind a young pre-reproductive adult queen which usually has not even become an imago. The critical pressure on a honey bee swarm is to amass sufficient food reserves for the new colony thus formed to survive the coming winter, with the result that the rapid onset of brood production and food gathering and storage is at a premium. The habitat variables that constitute the migration threshold for honey

bees are not well understood, but it seems likely that one of the most important is shortage of space for honey storage and brood production, a condition that is probably unpredictable early in the year and not likely to become predictable until late spring when there has been time for colony size and food stores to build up from their low level at the end of the winter. By this date, any swarm that is produced is already, depending on weather conditions during the coming months, short of time to make adequate provision for the coming winter. In southern England, swarms produced after mid-July rarely survive the coming winter. The situation is comparable to the time required by a prairie-dog to induce the growth of appropriate foodplants. Furthermore, for maximum fitness, honeybees have, in the same way as prairie-dogs, to make preparation for the migration in the form of exploratory migrations and decisions concerning which is the most suitable destination. In addition, prairie-dogs cannot complete their removal migration until the young are weaned and independent and honeybees cannot migrate until production of a second queen has been initiated. For both species, therefore, time is a critical factor. In the honeybee a few days extra delay in performing migration would result from waiting for the new queen to develop and emerge and to be ready to migrate. She would first need to become sufficiently flight-worthy and perhaps also sufficiently familiar with her home range to perform her mating flight so that she would be able immediately to commence oviposition once migration has taken place. Such a delay could be critical in determining whether the swarm as well as the parent colony survive the following winter. Migration by the mature honeybee queen rather than by the young pre-reproductive queen produced by the colony, because it avoids this delay, is therefore likely to be favoured in the same way as migration by adult rather than newly weaned prairie-dogs (Fig. 17.27).

In most of the consideration of variation in the incidence of removal migration, emphasis has been placed on the variation of the habitat quotient because for the species considered so far this single component of the adaptive package can probably account for more of the observed variation, particularly ontogenetic and sexual variation, than the remaining components put together. For maximum determination of the pattern of variation, however, it is necessary to consider all three (h_q, $\bar{S}_d$, and $\bar{m}$) of the illustrated components of the adaptive package, particularly as shown in Fig. 17.27 and in the four examples above, when seasonal variation is also taken into account. The necessity to consider all components of the adaptive package becomes even more apparent when the removal migrations are taken into account of animals, such as many invertebrates, that make relatively little investment in a home range and thus show a much less predictable ontogenetic change in habitat quotient.

17.3.5 Ontogenetic variation in the incidence of breeding and/or feeding habitat to breeding and/or feeding habitat migration by flight in adult insects

Animals that live in a home range, as do most vertebrates, inevitably induce ontogenetic variation in habitat quotient as a result of their investments in that home range and as a result of any offspring that may also be produced into that home range. Many animals, however, particularly insects but also other invertebrates have a qualitatively different life-style and do not inhabit an area that could meaningfully be termed a home range (Chapter 18). Such animals rarely, therefore, have been subjected during their evolutionary history to consistent ontogenetic variation in habitat quotient. Instead the habitat quotient most often seems likely to approach unity throughout adult life, all other habitats on average being of equal suitability to the habitat occupied. Occasionally, as shown in Fig. 17.28 and for the reasons presented in the legend, the habitat quotient seems likely to be slightly greater than unity throughout adult life. Such species are therefore likely to show a greater migration incidence than other species. Occasionally, also, in species such as locusts (p. 246) for which new habitats are more suitable than old habitats (p. 53) and other species for which not only is this the case but which have been subjected to selection for synchronisation of seasonal incidence and the seasonal peak of appearance of new habitats (p. 337), the habitat quotient may always have been higher early in ontogeny than later. In such cases habitat quotient may have contributed to selection for a pre-reproductive peak of migration incidence. On the

whole, however, habitat quotient can make little contribution to ontogenetic variation in migration incidence, though a difference in level of habitat quotient experienced by males and females could well contribute to a general difference in level of migration incidence shown by the two sexes. Consequently, according to the model developed in this chapter, ontogenetic variation in migration incidence in these insects and other invertebrates that do not live in a home range must be largely the result of the combined effects of $\bar{m}$, mean migration cost, and $\bar{S}_d$, mean action-dependent potential reproductive success.

Figure 17.28 presents adaptive packages for two species of butterflies for which field data on ontogenetic variation in migration incidence are available. The adaptive package for the female of the large white butterfly, *Pieris brassicae*, seems to come as near as is yet possible to providing quantitative support for the validity of the model developed in this chapter. On the assumption that habitat quotient remains constant, variation in mean migration factor, $\bar{M}$, can be calculated from the curve of observed migration incidence. Ontogenetic variation in $\bar{S}_d$, mean action-dependent potential reproductive success, can be calculated from the counts of number of eggs laid per day obtained by David and Gardiner (1962). From the curves of $\bar{M}$ and $\bar{S}_d$, ontogenetic variation in $\bar{m}$, mean migration cost (per day) can be calculated. The absolute value of $\bar{m}$ shown in Fig. 17.28 is, of course, a function of the matching of the axes for initiation factor and observed migration incidence. The variation in $\bar{m}$, however, within broad limits is more or less entirely a function of two sets of independently produced data and the model developed in this chapter. This variation can therefore meaningfully be analysed.

Table 17.6 permits examination of the extent to which the variance of migration cost can be attributed to two aspects of female physiology and behaviour. Female time-investment strategy in butterflies, if not most insects, during the egg-laying period of ontogeny is heavily biased toward maximising nett rate of energy uptake (i.e. uptake minus expenditure per unit time) and/or toward maximising efficiency of conversion of energy into egg production. In the large white butterfly the behaviour involved is not only searching for the adults' foodplants, feeding, searching for the foodplants of the larvae, and oviposition, but also thermoregulation. In any one day, therefore, females of this species not only lay the optimum number of eggs (in

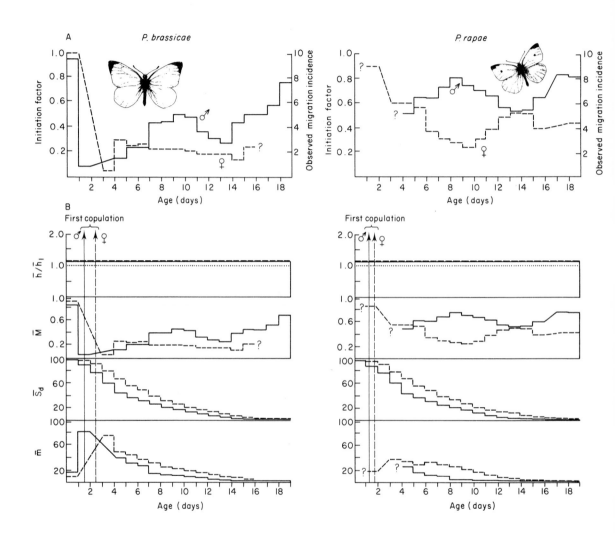

Fig. 17.28 Ontogenetic variation in breeding or feeding habitat to breeding or feeding habitat migration by two species of pierine butterflies, the large white butterfly, *Pieris brassicae*, and the small white butterfly, *Pieris rapae*, in southern England: application of a model of adaptive variation in the probability of initiation of migration

General ecology
Both the large and small whites feed as larvae on the leaves of a variety of cruciferous plants but in particular on the leaves of cabbages and other cultivated species of *Brassica oleracea*. As adults, both species feed on the nectar produced by a wide variety of flowering plants. Roosting at night and during cloudy periods during the day occurs under a leaf of almost any broad-leaved herb and also occasionally of shrubs and trees.

Females tend to oviposit in the mornings and to spend the afternoons feeding and thermoregulating, thus maturing eggs that will be laid the next day. Males search for females in the feeding habitats of both adults and larvae and also court any individual encountered during migration. No territories are established, many males searching for females in the same area at the same time. Females are prepared to copulate when two to three days old. In caged conditions, females may re-mate at intervals of 6–9 days (David and Gardiner 1961) but under natural conditions it seems likely that females are much less prepared to mate a second time. Consequently, males tend to achieve maximum realisation of potential reproductive success during the first few days as an adult when virgin females are most abundant.

Migration

The migration under consideration is the non-return movement of an individual across an area which contains the foodplants of neither the larva nor the adult and which is wider than the presumed sensory range of an individual of the species (100 m for present purposes).

A. Ontogenetic variation in migration incidence and initiation factor

The vast majority of individuals migrate every day that migration-cost variables are favourable, and the main source of ontogenetic variation in migration incidence is the variation in stay-time (Baker 1969a, Fig. 3). This variation was measured in the field in 1965 and 1966 as follows. The average age of the deme (counts were made in Wiltshire, England) on each day was calculated from observations of emergence of large numbers (100+) of pupae collected as larvae in a standardised way during the previous generation and maintained under field conditions. The average age derived in this way agreed well with calculations based on curves of seasonal incidence (Baker 1968a, Fig. 12) assuming an average physiological longevity of 21 days and a constant mortality rate at all ages. For each fine day that a count was made, the number of males and females of both species that crossed a field (approximately 1 ha) per hour between 11.00 and 13.00 GMT was determined. The mean number of individuals present at any one time in a nearby garden (described in Baker 1970) was also determined during the same two-hour period on the basis of a minimum of ten counts. The ratio of number migrating to number in the garden provides a relative measure of stay-time and hence also migration incidence. The method assumes that the garden retains a constant attractiveness during the experimental period (July–September 1965; April–October 1966). Although this assumption cannot be totally justified, the variation that must have occurred was considered not to be prohibitive.

The histograms show the migration incidence as the ratio of number crossing the field per hour to the mean number present in the garden at any one time during the same period. The matching of the axes for initiation factor and migration incidence is based on the arbitrary assumption that when the stay-time is at a minimum, the minimum likely value for the initiation factor is 1·0.

Dashed-line histogram, females; solid-line histogram, males. Question marks indicate an absence of data. Oblique lines joining histograms indicate that no data were obtained for the intervening age(s).

B. Adaptive package for the breeding or feeding habitat to breeding or feeding habitat migration by the large white butterfly and small white butterfly: ontogenetic curves of $\bar{h}/\bar{h}_1$ (the habitat quotient), $\bar{M}$ (the mean migration factor), $\bar{S}_d$ (the mean action-dependent potential reproductive success), and $\bar{m}$ (mean migration cost)

The solid and dashed vertical arrows indicate the age at first copulation of, respectively, males and females. Dashed-line histogram, females; solid-line histograms, males.

Because both species are adapted to spatially separate breeding, feeding and roosting habitats (Chapter 16), because presence in one habitat depletes the resources of that habitat, and because new resources are appearing frequently (e.g. different flower species open and produce nectar at different times of day—Beutler 1930, Kleber 1935), both sexes of both species are faced with a habitat quotient that is likely to be constantly slightly greater than 1·0 (dotted horizontal line).

Ontogenetic change in $\bar{S}_d$ for females is based on data for *P. brassicae* in David and Gardiner (1962). The same figure is assumed for both species. Males are assumed to achieve maximum rate of realisation of reproductive success at an earlier age than for females.

Ontogenetically, migration cost shows a low on the first day after adult emergence followed by a peak that is earlier for males than females, and thereafter shows a gradual decline. The variation is consistent with the suggestion that the main component of migration cost for males is copulation loss and for females is reduction in the rate of maturation of eggs, possibly due to decreased energy uptake, increased energy expenditure, and perhaps also less efficient thermoregulation.

The initiation factor, i, shown in A, is derived from the formula $i = \bar{h}\bar{M}/\bar{h}_1$, where $\bar{M}$ is given by $1 - \bar{m}/\bar{S}_d$.

Table 17.6 Data for stepwise multiple regression analysis of the effect on migration cost per day of the number of eggs being laid per day and the number of eggs being laid on the following day for the female of the large white butterfly, *Pieris brassicae*

age (day z)	m migration cost (%)	L no. eggs laid on day z	D no. eggs laid on day z+1
1	10	0	11
2	?	11	43
3	?	43	76
4	72	76	80
5	45	80	84
6	41	84	63
7	37	63	64
8	31	64	44
9	22	44	42
10	21	42	28
11	19	28	30
12	15	30	29
13	12	29	10
14	10	10	36
15	7	36	?

Data for column m from Fig. 17.28 and for columns L and D from David and Gardiner (1962)

this case optimum probably approaches the maximum possible given the state of the female's energy reserves, energy uptake, and climatic conditions) but also mature the optimum (again optimum probably approaches the maximum possible) number of eggs to be laid the next day. Migration is performed at the expense of these time-investment strategies and the components of migration cost, such as decreased energy uptake, increased energy expenditure, and perhaps also less efficient thermoregulation are likely to be manifest by a reduction in number of eggs laid and matured. Table 17.6 shows the daily migration cost as calculated in Fig. 17.28 for the first 15 days after adult emergence for the female large white butterfly. The other two columns (L and D) in Table 17.6 show the number (L) of eggs found by David and Gardiner (1962) to be laid on each of the first 15 days after emergence and the number (D) of eggs laid on day $z + 1$. The component of migration cost due to loss of oviposition time and associated factors can be assumed to be proportional to L, and the component of migration cost due to reduced rate and/or efficiency of egg maturation and associated factors can be assumed to be proportional to D. Missing values for one or other of the variables for three days reduces the number of observations to twelve. Stepwise multiple linear regression analysis of Table 17.6 gives a good fit ($F_{2,9} = 20.04$; $P < 0.002$) for the multiple regression line $m = 0.454D + 0.218L - 1.764$ to the data. There is surprisingly high predictability of m using just these two variables, the coefficient of determination for the multiple regression being 0.817. D and L do not, however, contribute equally to the variance of migration cost, their respective contributions being 78.9 per cent and 2.8 per cent. From this analysis, therefore, it would seem that those components of migration cost that result from the influence of migration on rate and/or efficiency of egg maturation are much more important than those that result from the influence of migration on the time available for oviposition.

If we assume that the female small white butterfly, *Pieris rapae*, shows similar ontogenetic variation in the rate of maturation of eggs, a similar interpretation of the variation in migration cost for this species also seems acceptable. Unfortunately, however, quantitative data for the mean reproductive performance of a male of these species is not available. It is assumed, however, on the basis of field experience, that the major part of male reproductive success is realised through virgin females.

Maximum copulation rate, on average, is therefore achieved early in adult life, when virgin females are most numerous, and thereafter decreases. On this assumption the ontogenetic variation in migration cost of the males of both species would fit the suggestion that its major component is decrease in female quota (females/male/min). Hence migration cost is low before the male is capable of copulation, rises rapidly as soon as copulation begins and virgin females are at their most abundant, and thereafter gradually decreases as the majority of females become non-virgin and cease to be receptive.

Further support for the importance of these components of migration cost may be derived from Fig. 17.29 in relation to the milkweed bug, *Oncopeltus fasciatus*.

The unexplained variance in migration cost in Fig. 17.28 for both males and females, albeit only 18.3 per cent for female large white butterflies, could well be due to a large extent to the other likely variable for insects suggested in Section 10.3.2 (p. 80), namely that of ontogenetic variation in wing-loading. Although, to the best of my knowledge, ontogenetic curves of wing-loading for the large and small white butterflies are not available, casual observation suggests that maximum wing-loading in females occurs at about day 4 and in males at about day 2 or 3. Thereafter, as fat reserves are used up, the insect loses weight until, just before death, it weighs only a fraction of its maximum. Yet wing area remains more or less constant, apart from the wear and tear that inevitably occurs. However, although wing-loading could well decrease the unexplained variance in migration cost in Table 17.6, it is likely to make far less contribution to migration cost variance in butterflies and also other insects such as moths, dragonflies, and many diptera etc. than in other insects such as aphids and beetles. This is because all of the former group are 'full-time' fliers, using flight as an integral part of almost all other behaviours except roosting. Consequently, wing-loading influences energy expenditure not only while migrating but also while within feeding and breeding habitats. Migration cost, as measured here (p. 41) is therefore much less subject to the influence of wing-loading, even though it must inevitably make some contribution. A number of moths, for example, such as, in North America, the spruce budworm, *Choristoneura fumiferana*, phantom hemlock looper, *Nepytia phantasmaria*, oak looper, *Lambdina somniaria*, and the pine shoot moth, *Rhyacionia buoliana*, all seem to have a higher migration threshold when gravid (Henson 1962, D.K. Edwards 1962, Green 1962).

Many other insects, however, are only 'part-time' fliers, using flight almost solely for the type of migration described in the legend to Fig. 17.28, with most other behaviours being performed by walking or jumping. The distinction between 'full-time' and 'part-time' fliers was in effect made by Southwood (1962) and seems likely to be particularly important in relation to the contribution of wing-loading to migration-cost variance. Part-time fliers should be particularly sensitive to variation in wing-loading and for such species this factor should make a much larger contribution to migration-cost variance than for full-time fliers. Wing-loading, however, in both full-time and part-time fliers should be much more important in females than males because the former show a much greater range of variation in wing-loading in insects than do the latter.

Wing-loading probably introduces a general ontogenetic pattern to migration-cost in that from the time of first maximum maturation of gametes there is probably a gradual decrease in migration cost for the majority of species, a decrease that is much more marked for females than males. Where females have oviposition cycles (i.e. batches of eggs laid at intervals separated by several days of egg maturation), wing-loading also introduces a cyclic element to migration cost with peaks coinciding with maximum female weight just before each oviposition. Species differ considerably, however, in the length of the period of sexual immaturity between adult emergence and the onset of copulation and oviposition. In the monarch butterfly, *Danaus plexippus*, in North America (Urquhart 1960) and the peacock butterfly, *Inachis io*, and small tortoiseshell butterfly, *Aglais urticae*, in Europe (Figs. 16.16 and 16.15), for example, this ontogenetic period lasts many months in individuals that emerge in late summer and autumn. In the large and small white butterflies (Fig. 17.28), however, the period lasts only one or two days. In yet other species, males are capable of copulating immediately the teneral period is over. The females of some species, such as the spruce budworm moth, *Choristoneura fumiferana*, emerge gravid from the pupa.

To recapitulate so far, it is suggested here that in insects that do not live in a home range the major part of the ontogenetic variation of initiation factor is a function of the migration factor, with habitat quotient only contributing when not only are new habitats more suitable than old but also selection has favoured synchronisation of either seasonal incidence or the onset of reproduction to coincide with the peak appearance of new habitats. In this case, habitat quotient tends to increase the initiation factor soon after adult emergence or soon after or just before the onset of reproduction, but thereafter makes little ontogenetic contribution. Ontogenetic variation in migration factor is a function of the interaction of $\bar{S}_d$, mean action-dependent potential reproductive success, and migration cost. Ontogenetic variation in migration cost for insects is suggested to be largely explicable in terms of ontogenetic variation in wing-loading in both sexes, but particularly in females and particularly in 'part-time' fliers, and reduced female quota in males and reduced rate of egg maturation in females, migration cost being greatest when wing-loading and copulation rate in males and wing-loading and egg maturation rate in females are at their ontogenetic peak.

Let us now consider to what extent these factors have a general effect on ontogenetic variation in initiation factor in relation to insect migration by flight. When a period of sexual immaturity exists between the end of the teneral period and the onset of copulation in males and onset of egg maturation in females, the value of $\bar{S}_d$ remains high at a time when reduced female quota and reduced rate of egg maturation are adding a minimal component to migration cost. Wing-loading is also likely to start at a lower level and then to increase gradually toward the onset of copulation in males and the onset of egg maturation in females, because in most such species this period is one of a gradual accumulation of energy reserves. At the ontogenetic onset of copulation in males and egg maturation in females, $\bar{S}_d$ begins to decrease and migration cost shows a marked increase. We may conclude, therefore, that where a period of sexual immaturity is found we would expect also to find that the initiation factor is at a maximum at this time, only beginning to decrease in males when receptive females are present and in females when egg maturation begins. On the other hand, if receptive females are present as soon as a male can fly and if females are maturing eggs at a maximum rate from the time of adult emergence or even earlier, then a peak of initiation factor immediately after the teneral period is no longer inevitable. Examples of the latter type of life-history are not common for species for which migration details are also available. Table 17.7 lists a number of insect species taken from examples given in this

Table 17.7 Some observations of the presence or absence of a post-teneral peak to the incidence of migration by flight in adult insects in relation to the presence or absence of immediate post-teneral reproductive activity.

Species	Males			Females		
	Relative migration† incidence		Reproductive§ activity	Relative migration incidence		Reproductive§ activity
	post-teneral	later‡		post-teneral	later‡	
Odonata						
Libellula quadrimaculata	***	*	×	***	o	×
Aeshna mixta	***	*	×	***	o	×
Gomphus vulgatissimus	***	*	×	***	o	×
Sympetrum fonscolombii	***	*	×	***	o	×
Orthoptera						
Schistocerca gregaria	***	**	×	***	*	×
Locusta migratoria	***	**	×	***	*	×
Nomadacris septemfasciata	***	*	×	***	*	×
Phaulacridium vittatum	?	?	?	***	*	×
Isoptera						
Paraneotermes simplicornis	***	—	×	***	—	×
Hemiptera: Homoptera						
Aphis fabae (alienicolae)				***	*	×
Myzus persicae (alienicolae)				***	**	×
Drepanosiphum platanoides				***	**	×
Circulifer tenellus				***	*	×
Aleyrodes brassicae				***	o	×
Hemiptera: Heteroptera						
Eurygaster integriceps	***	**	×	***	**?	×
Oncopeltus fasciatus	***	*	×	***	o	×
Lepidoptera						
Pieris brassicae	***	**	×	***	**	×
Ascia monuste	***	***	✓	***	o?	×
Danaus plexippus	***	**	×	***	**	×
Rhyacionia buoliana	o	o	✓	***	**	✓
Choristoneura fumiferana				o	**	✓
Coleoptera						
Semiadalia undecimnotata	***	*	×	***	*	×
Leptinotarsa decemlineata	?	?	?	***	*	×
Diptera						
Oscinella frit	?	?	?	***	**?	×
Aedes taeniorhynchus	*	*	✓	***	*	×
Hymenoptera						
Formica obscuriventris	***	*?	×	***	—	×
Eciton hamatum	***	—	×	—	—	×

† *** to   * = decreasing migration incidence
     o = minimal, but not zero, migration incidence
     — = no migration by flight
‡ later = stage of ontogeny immediately following the peak of reproductive activity, but not including the period from then to death
§ Reproductive activity
   ✓ = immediate onset of reproductive activity when sustained flight becomes physically possible
   × = a period of adult immaturity intervenes between onset of flight and onset of reproductive activity

book and in C.G. Johnson (1969) for which information is available concerning both adult life-history and ontogenetic variation in migration incidence. Despite a shortage of examples of species, particularly for females, that commence maximum reproductive effort immediately after completion of the teneral period, the comparison seems to be fairly convincing. Males and females with a pre-

reproductive period, in the sense used here, have a peak of migration incidence immediately following the teneral period. Males and females that commence maximum reproductive effort immediately following the teneral period may or may not, but usually not, have a peak of migration incidence at this time. Note that the peak referred to here is that occurring between the teneral period and the time of maximum probability of copulation in males and maximum rate of egg production (onset of oviposition probably being the critical ontogenetic event if wing-loading makes a major contribution to migration cost) in females. Events after this time are discussed below. Also the precise timing of the peak of initiation factor probably involves other components of migration cost, such as the acquisition of optimum energy reserves, that, although in general may have a lower coefficient of determination of migration-cost variance, could in some species have a marked effect during this period. The sunn pest, *Eurygaster integriceps*, in the Middle East, for example, does not fly immediately after adult emergence but can spend up to two weeks feeding on wheat, or weeds near wheat, until the fat body is filled, carbohydrates and proteins fill the mid-intestine, and the anterior part of the hind-gut is filled with food (E.S. Brown 1963). The chief selective pressure on this species at this time seems to be the danger of the food drying up before sufficient energy reserves can be acquired for the individuals to survive first aestivation and then hibernation (p. 241), both of which normally take place at higher altitudes than feeding and reproduction. In such a situation there is an obvious premium on feeding as a time-investment strategy and consequently a marked contribution to migration cost until optimum energy reserves have been accumulated. (In this case optimum is probably near maximum despite the effect on wing-loading during migration to aestivation sites.) It seems unlikely, however, that this and other factors with a low coefficient of determination of migration-cost variance have a consistent effect on ontogenetic variation in initiation factor, even though in some species with a particular ecology, such as the sunn pest, the effect may be important.

Inspection of Fig. 17.28 for the period after the time of maximum reproductive effort reveals that no generalisation is possible concerning the way that the interaction of changes in $\bar{S}_d$ and migration cost is reflected in changes in initiation factor. Relatively slight differences in the relationship between $\bar{S}_d$ and $\bar{m}$ can mean the difference between an increasing initiation factor with age or a constant or decreasing initiation factor. In both sexes of the large and small white butterflies, both $\bar{S}_d$ and migration cost decrease with age. Yet in the females of both species and the males of *Pieris rapae* from day 3 onwards the migration incidence remains more or less constant with age. In the males of the large white, *P. brassicae*,

however, from day 2 onwards migration incidence as presented in Fig. 17.28 shows a significant ($t = 2.80$; $df = 12$; $P < 0.02$) increase with age ($y = 0.21x + 1.96$, where y is migration incidence, as measured, and x is age in days after adult emergence; $F_{1,12} = 7.8$; $P < 0.05$) as analysed by linear regression. In such a situation, where minor differences in the relationship between $\bar{S}_d$ and $\bar{m}$ can have such a marked effect, the only confident generalisation that can be made is that after the first ontogenetic peak of reproductive effort, insects are likely to show highly variable curves of initiation factor. Having stressed this point, however, certain qualifications seem to be justified.

In particular, variation in $\bar{S}_d$ and migration cost cannot be totally independent when, as suggested here, those factors with the greatest coefficient of determination of migration-cost variance are intimately associated with the reproductive activity which must inevitably affect $\bar{S}_d$. If, therefore, insects show only a limited number of ontogenetic curves of reproductive effort, far less variation in initiation factor occurs than would otherwise be the case. Remembering that now we are concerned only with that ontogenetic period that comes after the first ontogenetic peak of reproductive effort, female insects in general, as far as is known, show a fairly constant pattern involving a gradual decrease from this first peak until death. Superimposed on this gradual decrease may be a cyclic variation in females of species that lay eggs in batches at intervals of several days with the result already described. Although not significantly different from zero, the regression coefficients for females of the large and small white butterflies after day 4 (Fig. 17.28) are both negative and it seems likely that in general interaction of $\bar{S}_d$ and $\bar{m}$ for female insects, according to this model, should produce either a constant or gradually decreasing level of initiation factor. The ontogenetic curve of reproductive effort for males, however, is far more variable.

In general, once a male becomes capable of copulation and once receptive females are available, a male's female quota reaches an early maximum. In the vast majority of insects, however, due to a variety of sexual selective pressures (Parker 1970a), female insects are not promiscuous. After usually a single copulation, female insects become non-receptive and reject any future attempts to copulate. Maximum realisation of potential reproductive success by most male insects, therefore, is usually achieved through virgin females. Virgin females, however,

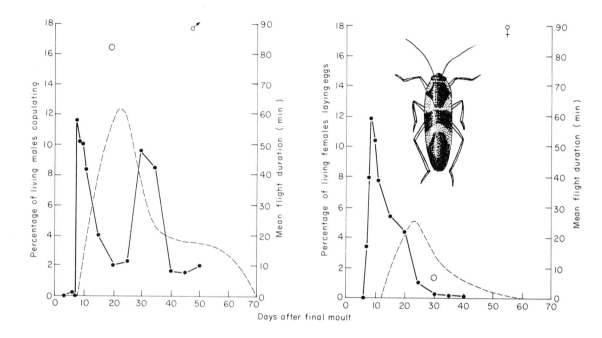

Days after final moult

Fig. 17.29   Ontogenetic variation in migration response in relation to reproductive activity in the milkweed bug, *Oncopeltus fasciatus*

General ecology
The large milkweed bug, *O. fasciatus*, is common over a large portion of the western hemisphere from southern Canada to Brazil and feeds almost exclusively during all developmental stages on milkweed plants (Asclepiadaceae). Mating occurs among patches of milkweed and eggs are laid, usually close to a developing seed pod.

Determination of migration threshold
Studies of various aspects of the migration of the species have been made in the laboratory by attaching a stick to the pronotum of the insect with wax and then using the sum of the durations of the first five flights as a measure of flight activity. Reservations must be attached to any interpretation of the duration of tethered flight in the laboratory in terms of migration. Bearing these reservations in mind, however, it is assumed here that tethered flight duration in the laboratory is directly proportional to migration incidence in the field. On this assumption, flight duration in this case should also be inversely proportional to migration threshold.

Relation between reproductive activity and migration threshold
Ontogenetic variation in flight duration in males and females, copulation frequency in males, and oviposition in females have all been determined under laboratory conditions (Dingle 1965) with results as shown. In the diagrams the dots joined by a solid line indicate flight duration and the dashed line indicates reproductive activity. The large open circles for males aged 20 days and

females aged 30 days show the mean tethered flight duration of virgin individuals that had been reared out of contact with the opposite sex.

Males show a migration threshold that varies directly with copulation frequency. At the end of the teneral period, migration threshold drops to a .low level (low migration threshold ≡ long flight duration ≡ high migration incidence) and thereafter increases as copulation frequency increases, maximum migration threshold (≡ minimum flight duration) coinciding with maximum copulation frequency. As copulation frequency decreases once more, the migration threshold also decreases only to rise again as copulation frequency levels off. An age effect as well as a response to copulation frequency is therefore indicated. Virgin males at 20 days show a much lower migration threshold than males that have had free access to females.

Females similarly show a minimum migration threshold immediately after the teneral period. Once egg maturation begins, however, the migration threshold increases once more to reach a constant maximum level at about the time of maximum rate of egg production. Virgin females which, however, mature eggs, show a migration threshold that is more or less the same as that of non-virgin females of the same age.

Support for the suggestion that it is migration threshold that varies with age and reproductive activity is provided by the observation that ovipositing females still initiate migration if subjected to a sufficiently low nett rate of energy uptake (p. 225).

[*Figures re-compiled from figures in Dingle (1965)*]

normally disappear rapidly from the deme concerned with the result that after an initial peak a male's female quota is also likely to decrease fairly rapidly. In the case of the male large white butterfly, *Pieris brassicae*, this decrease in female quota with age and hence decrease in migration cost interacts with $\bar{S}_d$ to produce an increase in initiation factor which can account for the observed significant increase in migration incidence. Although not significantly different from zero, the linear regression coefficient for migration incidence on age is also positive for the male small white butterfly, *Pieris rapae*. The males of the milkweed bug, *Oncopeltus fasciatus* (Fig. 17.29), also show a secondary increase in migration incidence that is associated with a decrease in female quota, and it could well be that future investigations will show a secondary increase in male migration incidence to be common in species characterised by non-promiscuous females.

The females of some species, presumably due to selective pressures associated with the problems involved in maintaining stored sperm sufficiently to ensure 100 per cent viability (Parker 1970a) become secondarily receptive to copulation towards the end of their life. If such secondary receptivity on the part of the females is sufficiently concentrated with respect to male ontogeny for males to experience a secondary peak in female quota, the result could well be a secondary increase in migration cost and hence a secondary decrease in initiation factor. This could well be the explanation for the secondary decrease in migration incidence found in the males of the milkweed bug, *Oncopeltus fasciatus*, after about 40 days (Fig. 17.29).

Finally, in a few insect species, such as the dung-fly, *Scatophaga stercoraria*, the females are promiscuous, being receptive to copulation at any stage of adult life. Such species seem invariably to be those for which the time and energy that would be involved in rejecting the copulation attempts of males seems to be greater than that involved in permitting copulation (Parker 1970a). The males of such species, once capable of copulating and once females first become receptive, should experience a more or less constant female quota throughout adult life. In this case migration cost might be expected to remain relatively constant which, in interaction with the decreasing level of $\bar{S}_d$, should produce a gradual decrease in initiation factor throughout adult life, though, without careful measurement, migration incidence may appear to be constant throughout life, though perhaps with a clear decrease toward death (Fig. 10.2). Such a situation may well occur in the male dung-fly, *Scatophaga stercoraria*, in relation to dung pat to dung pat migration.

To sum up this consideration of ontogenetic variation in incidence of migration by flight in insects, therefore, it has been concluded that for both sexes the major part of observed variation can be attributed to the interaction of $\bar{S}_d$ and migration cost, where the factors with the greatest coefficients of determination of migration-cost variance are those connected with ontogenetic variation in reproductive effort and associated changes in wing-loading. When the teneral period is not followed immediately by the ontogenetic peak of reproductive effort, the period that intervenes between these two events is likely to be the period of maximum initiation factor, though other variables, such as reduced rate of energy acquisition, may in some species influence the precise age at which peak initiation factor occurs. As reproductive effort increases, initiation factor decreases to reach a minimum level at the time of maximum reproductive effort. From this time onwards, which may coincide with the end of the teneral period in other species, greater variability can be expected. In females, however, the most likely pattern is either a constant or a continued gradual decrease in initiation factor with age. Superimposed on this may be a cyclic variation in level in species that lay a batch of eggs at intervals of several days with the periodic minimum levels of initiation factor coming just before each oviposition (but see p. 353). In males the commonest pattern seems likely to be a secondary increase in initiation factor as the number of virgin females decreases, with perhaps a secondary decrease in initiation factor in males of species the females of which show a second peak of receptivity. Males of species with promiscuous females, however, are likely to show a constant or gradually decreasing initiation factor.

17.3.6 Sexual and ontogenetic, seasonal and other temporal variation in migration incidence in invertebrates other than winged adult insects

In a sense it is unfortunate that most of the information available concerning ontogenetic variation in migration incidence of invertebrates comes from studies of adult insects. This is because in most cases, especially where the migration of adults occurs by flight, the migrations performed by juveniles and by adults cannot be compared in any very meaningful way. With the attainment of adulthood, migrations that were previously impossible became possible and conversely migrations that were possible for a larva became for an adult either impossible or insignificant. For example, migration by three different ontogenetic stages of the small white butterfly, *Pieris rapae*, have now been described. These are: the migration from the outer leaves of a cabbage plant to the heart by the third-instar larva (p. 233); the migration from larval feeding site to the pupation site by the post-feeding fifth-instar larva (p. 234); and the migration by flight from feeding and/or breeding habitats to feeding and/or breeding habitats by the adult insect. Obviously the only migration thresholds that are relevant to all three of these migrations are those to migration-cost variables such as rainfall and temperature (Chapter 16). Yet the migrations are so different that any ontogenetic variation in migration threshold to, say, temperature cannot meaningfully be analysed in terms of the ontogeny of a single individual. Inevitably, therefore, ontogenetic variation in migration threshold in insects is considered best by treating larvae and adults as different individuals.

This is not the case, however, for those other invertebrates, including some insects with wingless adults, that have a life-style similar to that of their larval offspring. Unfortunately, for most non-insect invertebrates the available information concerning removal migration does not lend itself to the type of consideration with which this chapter is concerned and in many cases also the behavioural ecology is known insufficiently to provide an adaptive model of the type presented for insects. However, a number of comments are possible.

Invertebrates other than insects that do not, as far as is known, live in a home range, are likely to experience similar components to migration cost as

insects. In particular it seems likely that a marked coefficient of determination between level of reproductive activity and migration cost could be of general application. This would automatically suggest, for the same reasons as for insects, that peaks of migration incidence occur between hatching from the egg and the onset of reproductive activity. At just which point during this period the peak is likely to occur cannot really be predicted without knowing some details of the behavioural ecology of the species concerned. Where migration is performed by locomotion, the migration cost due to energy expenditure of a given migration is likely gradually to decrease with age as locomotory ability increases. Hence migration from the outer leaves to the heart of a cabbage plant would represent a major migration with considerable migration cost to a first-instar larva, but represent a relatively minor migration with relatively little migration cost to a fifth-instar larva. Not only does locomotory ability in many species increase with age, but the energy reserves that can be carried also increase with age, with consequent decrease in the death-risk component of migration cost. The combined effect of decreasing migration cost with age when immature and then association between migration cost and level of reproductive activity when mature seems once again to place the peak of migration incidence at the largest most mobile yet immature stage which, presumably in most species, occurs just before the onset of reproductive activity.

In species with unusual migration mechanisms, however, such as aeronaut spiders (p. 235) and phoretic pseudoscorpions (p. 246) decrease in migration cost with age during immaturity may not be inevitable. In the first case, migration cost is likely to increase with age as proportionately more silk is required to transport the animal. Furthermore many spiders lay eggs in large batches, whereupon selection for dispersal is likely to occur (Chapter 11). The combination of these two factors, within the framework of the model developed in Chapter 10, could well account for the peak of migratory activity soon after hatching from the egg that seems to be general among aeronaut spiders (Southwood 1962). In the smaller species, however, migration may continue throughout the animal's life. In phoretic pseudoscorpions, the nature of ontogenetic change in migration cost during immaturity cannot easily be predicted. On the one hand, larger individuals are likely to slow down the insect being used as transportation and thus increase the probability of predation. On the other hand, ability to remain in position on the insect may be a critical factor, in which case younger individuals may be less capable of resisting the efforts of a fly to remove them than older individuals. Clearly we cannot construct an adaptive package for phoretic pseudoscorpions without more information. It appears that migration is most commonly performed by immature adults, particularly females (Southwood 1962) but the possibility cannot be ruled out that younger individuals and males use different hosts, less likely to be noticed by the casual observer. Most reports in the past have concerned pseudoscorpions attached to the legs of house flies.

Sexual differences in migration incidence in aeronaut spiders and phoretic pseudoscorpions, if they are real, seem most likely to result from differences in level of habitat quotient and migration cost. Ovipositing females seem most likely to maximise fitness by ovipositing in newly formed habitats (pp. 52 and 246) which automatically places the level of habitat quotient at greater than unity. For mature males, on the other hand, the major component of habitat suitability is likely to be the presence of receptive females which are likely to be more abundant in previously occupied than in new habitats, depending on the age at which females first become receptive. Furthermore, migration cost to males through reduced female quota is likely to be greater for animals using these non-locomotory migration mechanisms than migration cost to females due to reduced rate of maturation of eggs.

The migrations of the planktonic larvae of marine and inter-tidal animals seem likely to be mainly a response to the selective pressures associated with dispersal (Chapter 11).

Invertebrates, such as many molluscs and perhaps spiders and some of the larger members of the inter-tidal invertebrate fauna that establish a home range in which learning, social, and construction investments are made, are likely to be subjected to the same pressures described in relation to type-H migration for vertebrates. Thus the peak of migration incidence is likely to occur soon after 'independence' followed by a decrease once investments start to be made.

The larger myriapods would seem to be likely candidates for a lifetime track of the limited area type (p. 371). Whether or not this is so, there seems little doubt that some species, such as the relatively large (up to 5 cm) British snake millipede, *Tachypodoiulus niger*, show a high incidence of (exploratory or removal?) migration (Blower and Fairhurst 1968). At which position on the linear range/limited area range spectrum (p. 370) the lifetime track of such species lies is unknown. As distance from the breeding habitat increases, the males of *T. niger* captured in pitfall traps become progressively older (Fairhurst 1974), though whether this indicates a linear range or, for example, a larger home range or a longer-distance exploratory migration by older and more-mobile males is unknown. 'Activity', as measured by incidence of capture in pitfall traps, seems to increase with age, though this may reflect increased speed (distance per unit time) of movement rather than increased migration incidence and may not, therefore, be relevant here. *T. niger* seems to show seasonal variation in migration incidence, occurring most often away from recognised breeding habitats in spring, with sometimes a second peak in autumn (Blower and Fairhurst 1968). However, because of lack of data, it is not yet possible to analyse this seasonal variation in terms of the model developed here.

Finally, one further example of the application of the model can be presented that illustrates the range of types of temporal variation in migration incidence that can be accommodated. This example concerns the variation in incidence of bivouac-to-bivouac migration by colonies of the South American doryline army ant, *Eciton hamatum*, in relation to the brood cycle, and is presented in Fig. 17.30.

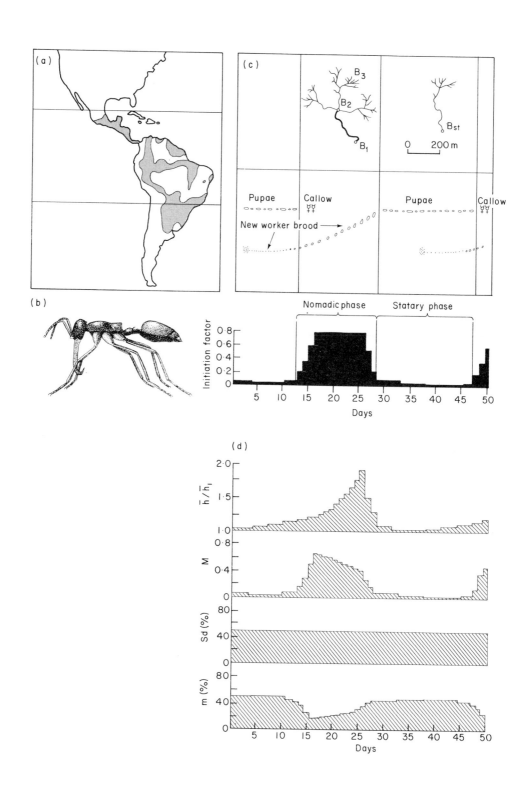

Fig. 17.30 The brood migration cycle of the South American doryline army ant, *Eciton hamatum*: application of a model of adaptive variation in the probability of initiation of migration

General ecology

(a) The known geographical distribution of the genus *Eciton*

Stippled area, known distribution. The gaps within the limits of the stippled area may be artefacts due to less collection.

Eciton hamatum, unlike many army ants, is a surface-raiding species that moves over the top of the ground and climbs trees, etc. Raids for the primarily arthropod food are carried out during the day and removal migration, when it occurs, takes place at night.

(b) A worker individual

The colony of *E. hamatum* consists of 150 000–500 000 worker individuals, which are sterile females, and a single reproductive female (=queen). All females are wingless, new colonies being formed by colony division, not by the flight of a solitary queen. Only males are winged. The colonies also contain the developing brood. Each brood is either all-worker, in which case it consists of about 80 000 larvae, or all-sexual, in which case it consists of about 1500 much larger (20×) larvae of which all but up to six become males.

The central social unit of the colony is a large cluster that forms a bivouac under or in a hollow log or some other protected site. The cluster consists of the queen, the brood, and attendant workers. Raiding trails radiate from the bivouac to which captured food is returned and fed to the brood. Raiding and migrating ants fall prey to other ants and termites, the colonies of which they raid, and also spiders, some birds, and a variety of mammals and amphibians.

Migration

The migration considered here is the bivouac-to-bivouac migration performed by the entire colony. There is a marked migration cycle, migratory (=nomadic, in the terminology of Schneirla 1971) and non-migratory (=statary) phases being clearly recognisable.

(c) Raising patterns, brood cycle, and variation in initiation factor and migration incidence during a migration cycle

During the nomadic phase, several raiding trails (thin lines) radiate from the central bivouac (top diagram). B_1, B_2 and B_3 mark the positions of the previous, present, and next bivouacs, respectively. Bivouac-to-bivouac migration takes place, at night, along a raiding trail of the previous day (wide line). Migration occurs nightly during the mid-nomadic phase. In the late nomadic phase, caches of food are assembled at intervals along the raiding trail along which migration is to take place. These provide food for the nearly full-grown larvae during the migration which, from initiation to completion may take 6–9 hours, but longer during the late nomadic than the early nomadic phase. When young, the brood are carried in 'packets' by the workers during migration, but when full-

grown the larvae have to be carried individually. The queen migrates by her own locomotion but with maximum protection from an 'entourage' of workers. During the nomadic phase the queen's body is contracted and she is incapable of laying eggs. By mid-statary phase the queen's body is fully distended and oviposition occurs. Raiding is much less intensive during the statary phase (B_{st} =statary bivouac), only a single raiding trail being formed and some days none at all.

The brood and migration cycles form an adaptive complex. The nomadic phase begins (middle diagram) with the emergence of a new batch of workers (callow =newly emerged with cuticle light-coloured and un-hardened) and ends with the spinning of cocoons by the mature larvae of the next batch. Oviposition occurs mid-way through the statary phase. The cycle shown is that of an all-worker brood. The larvae of sexual broods mature more quickly by about a week.

The illustrated variation in initiation factor is based on descriptions of migration incidence given by Schneirla (1971) but does not have a quantitative base. The absolute value of initiation factor is therefore arbitrary but the relative values for different parts of the cycle seem to fit Schneirla's descriptions. Although, given extreme disturbance, bivouac-to-bivouac migration takes place during the statary phase, it usually only involves migration over a few metres. During a single nomadic phase, however, the total shift in bivouac position may be as great as 3 km. One colony shifted its bivouac a total distance of 8 km during three migration cycles.

(d) Adaptive package for the bivouac-to-bivouac migration of *E. hamatum* during the maturation of an all-worker brood: ontogenetic curves of $\bar{h}_f/\bar{h}_1$ (habitat quotient), $\bar{M}_f$ (the mean migration factor), $\bar{S}_d$ (the mean action-dependent potential reproductive success), and $\bar{m}$ (mean migration cost)

The variation of habitat quotient is consistent with the relative depletion of food resources around each successive bivouac as the food required by the developing brood changes.

As each queen lives, on average, about 3 years, it is considered that $\bar{S}_d$ is relatively unchanged during the time interval shown.

Migration cost is consistent with the time and energy that would be required to complete migration at different stages of brood development. Also, the worker force probably gradually decreases from the emergence of one batch of workers to the emergence of the next. The variation is also consistent with death-risk to the queen as body size changes.

The initiation factor, i, shown in (c), is derived from the formula $i = \bar{h}_f \bar{M}_f / \bar{h}_1$ where $\bar{M}_f$ is given by $1 - \bar{m}/\bar{S}_d$.

[*Based on diagrams, tables and descriptions in Schneirla (1971) from which (a), (b) and the top two diagrams of (c) have been modified and/or re-drawn*]

17.3.7 Temporal and sexual variation in migration threshold

The migration threshold is that value, v_t, of either a habitat variable, V_h, migration-cost variable, V_m, or indirect variable, V_i, above which migration is initiated but below which migration is not initiated. It was argued in Chapter 8 that migration thresholds evolve when there is a good correlation between v, the value of the variable, and either h, habitat suitability, or m, migration cost. Habitat, migration cost, and indirect variables are all features of the environment such as, respectively, deme density, wind speed, and photoperiod, and it is quite probable that many of these at any one moment in time for an animal of a given age, sex, and reproductive history show a high correlation with either habitat suitability or migration cost. We have seen, however, in the previous sections of this consideration of variation in probability of initiation of migration that temporal and sexual variation in habitat suitability and migration cost can occur solely as a result of changes in the animal. Thus habitat suitability to many animals changes in relation to the level of learning investment made in a particular habitat. Migration cost to many animals changes as the physical stature of the animal changes or as the rate of egg maturation changes. In neither case need there be any change in the animal's external environment. It would be impossible, therefore, for the same value of V_h, V_m or V_i (relative values of V_h in H_1 and H_f in the case of calculated migration) to correspond (p. 46) to the situation in which $h_1 = h_2 M$ throughout an animal's ontogeny. In such a situation the ideal adaptation can occur only if the value of v_t varies not only during ontogeny and with time, generally, but also, and for the same reasons, between the sexes. In other words, when habitat suitability and/or migration factor vary with time and between the sexes as a result of variation in the animal rather than in the environment, selection is likely to favour variation in migration threshold between the sexes and with time. Similar reasoning may be applied to non-calculated migrants when $\bar{E}$ changes with time without any perceivable related change in h_1, such as when there is variation in rate of appearance of new habitats.

The situation is perhaps described best in the statistical terms normally applied to multiple regression analysis. Habitat suitability has, in this book, been defined in terms of the efficiency of realisation of potential reproductive success by an individual and cannot therefore be considered to exist except in relation to an animal. Let us first consider non-calculated migration and a situation in which habitat suitability is measured for a large number of individuals of a single sex of a single species in a wide variety of habitats, and in which at the same time measures are taken of a large number of habitat variables and various characteristics of the animals including age. The measures of habitat suitability become the Y-values in the multiple regression analysis and the measures of the habitat variables and the measures of age of the different individuals become the values of the different X-variables, X_1, X_2, X_3, etc. For the sake of argument all measures will be assumed to meet the assumptions of straight-line multiple regression analysis. First, from the arguments of Chapter 8, we would expect selection to favour individuals that monitor that combination of habitat variables that give maximum coefficient of determination of the variance of Y with a minimum of time and energy expenditure in the performance of the monitoring process. Selection only favours the animals monitoring a further variable if the advantage gained through increasing the coefficient of determination of Y (and hence greater likelihood of initiating migration at the optimum moment) is greater than the disadvantage that results from the increased expenditure of time and energy that is necessary to monitor the additional variable. In most of the animals discussed in this chapter, however, even if all habitat variables were taken into account, the coefficient of determination for the multiple regression would still not approach unity, due to the effect of the age of the animal on the variance of Y. Note that if we assume for the moment that $\bar{E}$, the mean expectation of migration, is a constant, there is in the ideal situation only one value of habitat suitability at which migration takes place. Because that value of habitat suitability is a function of age as well as of a number of habitat variables, any variation in age implies variation in the value of the various habitat variables for the continued determination of that value of habitat suitability. The relative amount of change involved is proportional to the relative coefficients of determination of the different variables.

In the field, a zoologist is likely to measure the level of a particular habitat variable at which an animal initiates migration. What is likely to be observed, therefore, is that as the animal ages, the level of habitat variable changes at which migration is initiated. In other words, the migration threshold to that variable changes with age.

Similar arguments can be applied to calculated removal migration. In this case, however, the Y-value in our regression analysis model is not h_1 but the ratio of h_f to h_1. Similarly the X-values concerned with habitat variables are not those values measured in H_1 but the values measured in H_f (the most favourable habitat with which the animal is familiar other than H_1) divided by the values measured in H_1. Translating this into the field, therefore, what is likely to be observed is that, as the animal ages, there is a change in the ratio of the levels of a habitat variable in H_f and H_1 at which migration is initiated. In calculated migration, the migration threshold to habitat variables is therefore the ratio of the values of a habitat variable in H_f and H_1 above which migration is initiated. In Chapter 14, when calculated removal migration was first introduced, animals were described as 'assessing the relative suitabilities of H_2 and H_1' and 'migrating if $h_2 M_2$ was assessed to be greater than h_1'. Such language is commonplace in behavioural ecology at the present time and considerably facilitates description. In the very near future it will no longer be necessary to justify the use of such language, for soon the majority of zoologists will accept that such terminology is describing biological phenomena just as precisely as the conventional, though much more verbose, biological terminology. In the present case, 'assessment' can be seen to describe the sequence of events during which an animal perceives some environmental feature in two separate habitats and then compares their relative value against an internal threshold by some psycho-physiological mechanism as yet unknown, the result of this comparison then being manifest as an item of behaviour.

So far we have been concerned with migration thresholds to habitat variables in both non-calculated and calculated migration. Similar arguments can be presented concerning migration thresholds to migration-cost variables. In both non-calculated and calculated migration the Y-variable in the multiple regression model is migration cost as measured for a large number of individuals. In non-calculated removal migration the X-variables, other than age, are the levels of those different migration-cost variables experienced while in H_1 just before the migration was initiated. In calculated removal migration some of the X-variables are the values of migration-cost variables perceived by the animal on those previous occasions that some part of the migration route eventually taken between H_f and H_1 was visited. Perhaps, if the migration route had been taken on a previous occasion during exploratory migration, one of the X values would be some measure of the mean migration cost experienced on previous occasions. Other X-variables will be, as for non-calculated migration, those different migration-cost variables experienced while in H_1 just before the migration was initiated. The same conclusion is reached as before which, translated into what is likely to be observed in the field situation, is that as the animal ages, the level changes of a particular migration-cost variable at which an animal initiates migration. In other words, the migration threshold to a migration-cost variable changes with age.

In these multiple regression models it has been assumed that either the mean expectation of migration, the migration factor, or the habitat quotient, as appropriate, remains constant. However, as argued in Chapter 10, the level of habitat suitability at which migration is advantageous is a function of migration factor (i.e. a function of action-dependent potential reproductive success and migration cost). Similarly, the level of migration cost at which migration is advantageous is a function of habitat quotient and, again, action-dependent potential reproductive success. All of these factors, therefore, would be expected, according to the model developed in this book, to influence both the level of habitat suitability and the level of migration cost at which migration is initiated and as all three of habitat quotient, action-dependent potential reproductive success, and migration cost are influenced by changes in the animal itself, almost any measured migration threshold is liable to vary with time (e.g. age, season, brood cycle) even without observable change in the environment. Furthermore, because gender is a characteristic of the animal, not the environment, and because gender influences habitat quotient, action-dependent potential reproductive success, and migration cost, measured migration thresholds are also liable to show male–female differences even where both sexes occupy identical habitat-types.

Migration threshold is measured as a value of some variable or combination of variables. At the physiological level, however, the sequence of events, stated naïvely, must be as follows. First, the level of the environmental variable is perceived by the animal. The sensory information is then processed, most often, perhaps, within the central nervous system. In some way, as yet unknown, the processed information is compared with some internal threshold that must have a physiological basis. Depending on the result of this comparison, migration is either initiated or not. Sexual or ontogenetic and other temporal differences in the measured migration threshold to habitat or migration-cost variables must reflect differences in the physiological state that constitutes the internal threshold. It is instructive to consider, as in the next section, the sensory and physiological mechanisms that could account for the changes in migration threshold that would be expected to occur on the basis of the model developed in this chapter, bearing in mind that changes in initiation factor and migration incidence only reflect changes in migration threshold to *all* environmental variables when either the critical variation stems from a change in the animal rather than the perceivable environment or from a change in the non-perceivable environment, such as change in $\bar{h}_2$ in non-calculated removal migration. It is clear from the multiple regression model, however, that, as already indicated in Chapter 8 (p. 49) a change in the level being experienced of any X-variable influences the threshold values of all other X-variables. Given, that is, that selection acts to favour the initiation of migration at only one value of the Y-variable (where $Y = \bar{E}$ in non-calculated migration (p. 46) and $Y = h_f M_f$ in calculated migration (p. 66)), changes in the animal and in the non-perceivable environment representing only special cases of this general phenomenon.

17.3.8 The physiological basis of ontogenetic change in migration threshold

The general conclusions reached from field observations of ontogenetic variation in migration incidence and from interpretation of the presented adaptive packages permit further speculation on the physiological basis of the migration threshold in a way similar to that already attempted for man (p. 293). The convention used is that a decrease in migration threshold refers to an increased probability of initiation of migration.

As far as group-to-group (type-GH) migration is concerned, the first conclusion that emerged from earlier discussion was that ontogenetic decrease in migration threshold is favoured by selective pressures intimately associated with the process of maturation. If this conclusion is valid, it follows that selection is likely to have favoured the physiological processes involved in the decrease in migration threshold being part of the physiological syndrome of maturation. The second conclusion was that subsequent ontogenetic increase in migration threshold is favoured by selective pressures intimately associated with the ontogenetic period of reproduction. In this case, therefore, as for male and female humans, it seems reasonable to postulate that it is some feedback from the reproductive process that induces change in the physiological processes that are the real basis of what is observed as a migration threshold. In the case of the female zebra (Fig. 17.20) we can perhaps be even more specific and suggest that the observed sudden decrease in migration threshold is part of the physiological syndrome of which the first ovulation is also a part and that the observed sudden increase in migration threshold is initiated, perhaps irreversibly, by the physiological changes that result from first conception. This sudden and perhaps irreversible increase in migration threshold following first conception is likely to be advantageous, for reasons described in relation to the habitat quotient (p. 304), in relation to any migration performed by a female in relation to a male from which the female receives parental investment as an individual (e.g. group-to-group migration by female zebra and male-to-male migration by female humans (p. 304) and rainbow lizards, *Agama agama* (p. 195)). Where selection favours a more gradual increase in migration threshold, different feedback mechanisms seem likely to have been favoured. In the deme-to-deme migration threshold for male and female humans it was suggested (p. 293) that the feedback resulted from visual, tactile and other stimuli received as a result of association with the offspring during their upbringing, and such stimuli would also seem to be optimum for all females except in relation to the migrations just mentioned. In particular, this feedback mechanism seems to be optimum for females of relatively promiscuous species that form multi-male heterosexual groups. Even this feedback mechanism, however, is inappropriate for the male members of these multi-male groups. If it is assumed, as seems likely, that males of such species have a low probability of association with their own offspring, except insofar as both may be members of the same group (p. 304), then the only reasonable feedback stimulus that could be favoured by selection is the incidence of copulation with females at or near the time of ovulation.

Type-H (total home range to total home range) and type-C (seasonal home range to seasonal home range or seasonal migration circuit to seasonal migration circuit) non-avian vertebrate migrants are likely to be similar to type-GH migrants during early ontogeny, insofar as the optimum physiological tie-up seems to be for the decrease in migration threshold that is favoured during this period to be an integral part of the maturation syndrome. Unlike the situation suggested for type-GH migrants, however, the minimum migration threshold should be reached some time in advance of the onset of sexual maturation and, optimally, should increase once more as the final stage of reproductive maturation takes place. In type-H and type-C migrants, therefore, we would postulate that ideally the immediate post-weaning stages of the maturation syndrome up to, and perhaps including, the onset of gametogenesis should provide the physiological basis for a decreasing migration threshold. The physiological changes that take place from just before or just after the onset of gametogenesis, depending on the species, up to the attainment of reproductive maturity, however, should include or provide the physiological basis for an increase in migration threshold. Although increase in migration threshold should be part of the final stages of reproductive maturation, this physiological syndrome should ideally only be associated with a slight increase in migration threshold. Major decrease in migration threshold of the extent characteristic of type-H and type-C migrants should ideally be the result of physiological changes induced by feedback from the environment. Unlike type-GH migrants, however, for which it was postulated that the distribution of offspring was likely to be the important component of habitat quotient, the feedback mechanism favoured by selection for type-H and type-C migrants is unlikely to be an aspect of reproduction. As it has been argued that the important component of habitat quotient for type-H and type-C migrants is the extent of the learning, social and construction investments made in a home range, it might be expected that selection would favour a feedback mechanism associated with the amount of investment in a home range the suitability of which is below the migration threshold. It can be postulated, therefore, that the process of establishment of a total home range for a type-H migrant or a season's home range or perhaps a migration circuit for a

type-C migrant induces physiological changes in an ontogenetically mature or nearly mature animal that are the basis of the marked increase in migration threshold that ideally should occur.

For once, a few field observations are available that provide at least some support for these postulated physiological syndromes. As far as the association of a decreasing migration threshold with the maturation syndrome is concerned, Howard (1949), working in Michigan on the prairie deermouse, *Peromyscus maniculatus*, found that those individuals that initiated type-H removal migration did so at the onset of sexual maturity. This association between gonad maturation and the initiation of type-H removal migration applied whether or not the animals reached maturity in the year of their birth or several months later in spring. Noting this association, Howard suggested that it was the physiological processes associated with gonad maturation that stimulated migration. Watts (1970), on the other hand, working in Britain on the bank vole, *Clethrionomys glareolus*, and the wood mouse, *Apodemus sylvaticus*, pointed out that in these species, both of which seem to show a pre-reproductive peak of type-H migration incidence, although maturation may stimulate migration in spring-born and summer-born individuals it does not do so in autumn-born individuals that do not reach puberty until spring.

Both of the above authors in the discussion of their results invoked the concept of migration being stimulated by gonad maturation. In this case their results indicate the presence of different physiological mechanisms in the species studied. It is probably important, however, to distinguish this concept of stimulation of migration from that invoked here of a decrease in migration threshold to environmental variables being a part of the maturation syndrome. Decrease in migration threshold is not the same as the stimulation of migration and can occur without the initiation of migration if either the suitability of the habitat occupied is sufficiently high or the migration factor is sufficiently low (Chapter 8). Suppose, therefore, that in the demes studied by Watts (1970) spring conditions were such that the migration threshold was not exceeded, even at its lowest point at about the age of sexual maturation, but that in the deme studied by Howard (1949) spring conditions were such that the migration thresholds of some individuals were exceeded. Given this, the observations of Howard and Watts would be consistent both with each other and with the suggestion made here that part of the maturation syndrome in type-H migrants is a decrease in migration threshold.

There is some support also for the suggestion of feedback from investment in a home range influencing an animal's physiology. Lockie (1966) working on the weasel, *Mustela nivalis*, and stoat, *M. erminea*, in Scotland noted that the testes of 'transient' males (type-H migrants) were small to medium in size by comparison with those of 'resident' males. Excluding ontogenetic pre-reproductives (i.e. individuals less than six months old) the age distribution of these transient males was similar to that of the residents. The decreased size of the testes, therefore, seemed likely to be correlated with the continuation of migration. Yet when these transient males encountered a home range that had become vacant due to the death of the previous occupant, they were indistinguishable from other 'resident' males in their ability to establish and maintain themselves within that home range. Furthermore, soon after establishing themselves in a home range, the size of the testes increased and they showed no more inclination to perform further removal migration than the previously resident males. This series of observations by Lockie seems to provide good support for the arguments of this section concerning the nature of the feedback mechanisms on the physiological basis of the migration threshold in type-H migrants.

It seems likely that the physiological basis of the migration threshold and its interaction with the environment through various feedback mechanisms is similar for birds to that for other vertebrates. It is possible that some birds that form social groups, such as colonially nesting species, may be subject to the selective pressures described for type-GH migrants. On the whole, however, it seems likely for most birds that the selective pressures described for type-H and type-C non-avian vertebrate migrants are more applicable. Ideally, after fledging, decrease in removal migration threshold should be associated with the maturation syndrome. The physiological events that take place from just before the onset of gametogenesis onwards, however, should form the basis of a subsequent gradual increase in migration threshold. Maximum increase in migration threshold should not occur, however, until the bird has begun to establish itself in a breeding home range. Selection should therefore favour some aspect of the establishment process inducing the physiological events that form the basis of what is observed as a migration threshold. It should be noted, of course, that these physiological events that form the basis of onto-

genetic variation in removal migration threshold must be, in type-C migrants of any vertebrate, avian or non-avian, separate from but integrated with those physiological events that form the basis of the seasonal return migration thresholds. This integration is discussed further in Chapter 27 once the pressures influencing the seasonal return migration thresholds have been discussed.

For reasons already discussed at length, changes in the animal that influence optimum type-GH, type-H or type-C migration incidence act to a large extent through changes in habitat quotient. In those adult insects and other animals that do not live within a home range, however, changes in the animal that influence optimum migration incidence act predominantly through the migration factor. In male adult insects, in relation to migration by flight, the ontogenetic absence of gametogenesis invariably relates to a period of high migration factor and hence low migration threshold. Selection is unlikely, however, to favour an increase in migration threshold with the production of sperm with the result that the physiological basis of the migration threshold should ideally be unaffected by spermatogenesis. The ideal adaptation seems to be a migration threshold that was influenced by the abundance of receptive females. Once spermatogenesis has begun, therefore, physiological feedback from rate of perception of receptive females such that the greater the rate of perception the higher the migration threshold, seems to be the most efficient mechanism available, given that some physiological influence of the ageing syndrome could provide optimum modification in accordance with the decrease in $\bar{S}_d$. We may predict, therefore, the following physiological basis for the migration threshold of male insects. The physiological syndrome of which the absence of gametogenesis is a part will provide the basis for a low migration threshold. The physiological changes associated with the onset of spermatogenesis will not by themselves influence the migration threshold but will provide a physiological environment sensitive to sensory input concerning rate of contact with receptive females. The direction of the sensitivity will be such that increased rate of contact results in physiological changes that produce an increase in migration threshold (=less probability of migration) and decreased rate of contact results in physiological changes that produce a decrease in migration threshold. The internal physiological environment in which these changes in migration threshold to external environmental variables other than rate of contact with receptive females take place will, however, be influenced by physiological changes involved in ageing such that a given rate of contact with receptive females induces an increasingly lower migration threshold with age.

To some extent these predictions for males are supported by the results obtained by Dingle (1965) for the milkweed bug, *Oncopeltus fasciatus* (Fig. 17.29). Comparison of flight duration of tethered males, some of which had almost continuous exposure to receptive females and some of which had been denied any access to females, showed a significantly greater flight duration by the latter group.

In females, as in males, the ontogenetic absence of gametogenesis relates to a period of high migration factor and hence low migration threshold. Unlike males, however, because it is rate of gametogenesis that is the critical factor for females, selection is likely to have favoured a direct association between the physiological basis for egg maturation and the physiological basis for an increase in the migration threshold, perhaps even to the extent of the same physiological mechanisms serving both functions. From what is known of the changes in insect physiology that occur at this time it is possible that migration threshold at this ontogenetic stage is a function of the combined activity of two endocrine organs, the prothoracic gland and the corpus allatum. Perhaps, for optimum adaptation, some sensory feedback concerning wing-loading from the wing base and/or wing muscles would also be necessary, though conceivably wing-loading is so intimately associated with internal physiology such as the state of the fat body and ovaries that such sensory feedback is unnecessary. Once the age of maximum egg production has passed, optimum migration threshold is related to decrease in $\bar{S}_d$ in relation to variation in migration cost. In the most likely situation described above, the internal physiological environment in which further changes in migration threshold takes place will, as suggested for males, be influenced by physiological changes involved in ageing such that, if there is no other source of variation, migration threshold will continue gradually to increase with age. In this case, for optimum adaptation, rate of egg maturation should have the opposite effect such that the physiological events that result in or from a reduced rate of egg maturation act to decrease the migration threshold. The nett result could be a migration threshold that increases, remains constant, or decreases depending on what is favoured by selection in that particular

instance. On the other hand, as rate of egg maturation is also invariably a function of age, rather than favour two physiological processes acting in opposition, selection may well have favoured the appropriate direction of change in migration threshold being based solely on the physiological processes associated with ageing. Alternatively, if selection favours a migration threshold that remains constant at the low level reached at the age of maximum rate of egg production, optimum adaptation could be a physiological basis for migration threshold that, having been produced by the physiological events involved in the first major flush of egg production, is unaffected by future changes in its physiological environment. Finally, in females that produce egg batches at intervals of several days and for which cyclic variation in migration threshold is optimum, it might be expected that the physiological basis of the migration threshold will either be intimately associated with, or the same as, the physiological mechanisms involved in the cycles of egg maturation and oviposition and/or will be sensitive to sensory feedback concerning wing-loading from the base of the wings or from the wing muscles. In some situations in which feeding and oviposition occur in the same habitat-type, the physiological basis should be such that egg maturation and increase in wing-loading increase the migration threshold, whereas oviposition and decrease in wing-loading decrease the migration threshold. In situations in which feeding and oviposition occur in different and spatially separate habitat-types, however, quite the opposite effect of egg maturation and wing-loading on migration threshold could be optimum (p. 353). In either case, this cyclic variation in migration threshold is likely to take place within a physiological environment influenced by physiological changes involved in ageing.

Credit for the first realisation of an association between the pre-oviposition period in female insects and migration incidence should, according to C.G. Johnson (1969), be given to Tower (1906). Credit for the expansion of this thesis to include most female insects and for placing it on a physiological basis, however, must be given to Johnson (1969). In essence the above account of the physiological basis of migration threshold for *female* insects is similar to that propounded by Johnson. There are differences in the models on which the physiological conclusions are based, however. In particular, Johnson's model of ontogenetic variation in migration incidence of female insects is framed in such a way that it cannot accommodate migration that occurs after oviposition. Thus species such as the frit fly, *Oscinella frit* (p. 240), the West African firefly, *Luciola discicollis* (p. 353), various moths (p. 336) and some aphids, some females of which clearly migrate when gravid as well as during the pre-reproductive adult phase, where such a phase exists, are considered to be 'exceptions to the general rule' (Johnson 1969, pp. 175 and 612). This inflexibility of the model also favours attempts to explain away other observations of migration by gravid and post-reproductive females (e.g. many dragonflies, the army cutworm moth, *Chorizagrotis auxiliaris*, and many butterflies— Johnson 1969, pp. 37, 182 and 181 respectively). For example, it is suggested in relation to butterflies observed to oviposit during migration that 'migrants recorded as sexually mature are probably at the end of the migratory flight'. In effect, therefore, Johnson's model is based on 'the association of migration with the immaturity of ovaries' (Johnson 1969, p. 215). By contrast, the model developed in this chapter is based on the migration threshold as an adaptive variable, ontogenetic variation in which is the evolutionary result of ontogenetic variation in the action of three categories of selective pressures, each category of which may be made up of any number of components that depend on the ecology of the species. Such a model is much more flexible than one based simply on an association between migration and immaturity of ovaries, and as shown does not only account for a pre-reproductive peak of removal migration in females where this occurs but also accounts for migration patterns in males, females after the onset of reproduction, and also unusual situations such as that of the honeybee, *Apis mellifera*, in which migration threshold reaches its lowest point much later in ontogeny than the onset of reproduction (p. 332) and also the West African firefly, *Luciola discicollis*, and yellow dung-fly, *Scatophaga stercoraria*, in which migration threshold seems to reach its lowest point at each oviposition (p. 353). Furthermore, the model is applicable to animals other than insects.

This section has been concerned with the physiological basis of ontogenetic change in migration threshold. A notable feature, however, has been the absence of measurements showing that migration threshold to a particular environmental variable does, in fact, change with age. As already described in Chapter 16, measurements of migration threshold of any kind are few and far between. However, some measurements of variation with age in migration

threshold to migration-cost variables are available for insects. Thus wind has been shown to decrease the flight activity of the black fly, *Simulium ornatum*, to a lesser extent when young than when older (L. Davies 1957) and in the mosquito, *Aedes cataphylla*, strong wind does not inhibit take-off after emergence as it does later in life (Klassen and Hocking 1964).

17.3.9 Temporal variation in migration threshold due to time-investment strategy and adaptation to spatially separate habitat-types

In Chapter 16 the situation was considered in which animals become adapted to perform different behavioural activities in spatially separate habitat-types. Thus the two butterfly species described in the legend to Fig. 17.28 feed in a habitat containing nectar-producing flowers, oviposit in a habitat containing a variety of *Brassica oleracea*, roost in a habitat containing plants such as nettles, *Urtica dioica*, or some other suitable broad-leaved species, and search for females in any of the other habitats. Furthermore, because different flower species produce nectar at different times of day (Beutler 1930, Kleber 1935) the habitat-type in which feeding occurs also varies during the day. These species have distinct diel cycles of activity of different behaviours (R.R. Baker 1969a for the small white, *Pieris rapae*) and undoubtedly there is also a distinct ontogenetic sequence of time investment in the different behaviours such as has been demonstrated for other insects (Caldwell and Rankin in Dingle 1972). We can assume for present purposes that these variations in behaviour represent optimum time-investment strategy.

The combination of adaptation to spatially separate habitat-types and diel and ontogenetic variation in behaviour seems likely to involve a migration threshold that varies in accordance with optimum time-investment strategy. Hence, an animal of an age and at a time of day for which, say, feeding is optimum behaviour is unlikely to initiate migration from a habitat containing abundant food. However, as the end approaches of the optimum time-investment period in feeding, the animal's migration thresholds to migration-cost variables and to habitat variables that correlate with food abundance are likely to decrease while the migration threshold to habitat variables that correlate with some other feature necessary for some other behaviour is likely to increase. Hence, female dung-flies, *Scatophaga stercoraria*, after adult emergence should have a very low migration threshold to dung but a high migration threshold to patches of vegetation containing insect prey species. As feeding commences and the female becomes gravid, the migration thresholds to migration-cost variables and to habitat variables that correlate with the abundance of food decreases. At the same time the migration threshold to the habitat variables that correlate with the suitability of oviposition sites, such as those habitat variables that correlate with the age of a dung pat, must increase. Presumably, this variation in migration threshold represents the physiological basis of the behaviour that is observed as migration away from the site of larval development upon adult emergence, settling in an area or succession of areas where food is abundant, and then, upon becoming gravid, migrating from the last feeding area, across other feeding areas in which previously the female would have settled, and arriving at a dung pat of below-threshold suitability. Note that in the dung-fly optimum time-investment strategy in an animal adapted to feed and oviposit in spatially separate habitat-types involves a peak of migration by a *gravid* female, a phenomenon that cannot be accommodated within the oogenesis-flight syndrome model of ontogenetic variation in migration developed by Johnson (1969) but which will probably be found to be relatively common among insects. The West African firefly, *Luciola discicollis*, a lampyrid beetle, provides another example and shows many similarities in its breeding biology to the dung-fly. The females are promiscuous, copulation being repeated throughout life but particularly at each oviposition (T. Kaufmann 1965a,b). Migration by the females is usually initiated on calm evenings and takes place at a height of 2·5–12 m above the ground, up to several hundred metres being travelled with the female flashing in a characteristic slow and regular manner. The females, which migrate mostly when gravid, terminate a migration in an area suitable for oviposition in which males are signalling.

In the model developed in this chapter the effects of time-investment strategy and adaptation to spatially separate habitat-types are manifest through variation in habitat quotient, and in the migrations just described must outweigh the effects of

the peak of migration cost that occurs when a female is gravid due to wing-loading. During a period of a single type of behaviour, such as feeding, an animal in a habitat-type suitable for feeding is experiencing a habitat quotient (h_q) of unity or below $(h_q = h_2/h_1)$. As the optimum time-investment period for this behaviour draws to a close, however, the habitat quotient rises above unity and continues to rise until migration is initiated and the animal arrives at a habitat of below-threshold suitability for the next behaviour, whereupon h_q again decreases to unity or below.

17.4 Discussion

This chapter has gone to some lengths to attempt to account for a number of forms of intra-specific variation in the initiation of migration, at least two of which may be felt by some zoologists not to require explanation. Hence, the individual variation discussed in Section 17.2 is often still dismissed in its entirety as inevitable biological variation that does not require an adaptive explanation. Similarly, the pre-reproductive peak of removal migration that appears to be so widespread through the Metazoa could also be considered to be inevitable and again, therefore, not to require an adaptive explanation. Jonkel and Cowan (1971), for example, suggest that the migration of young animals from any *stable* deme (the italics are mine), can be expected, though they do not elaborate on why it can be expected. It is quite possible, however, for a stable deme to be maintained by some system other than a pre-reproductive peak of migration, unless, of course, a stable deme is defined as a deme from which the young, but not the adults, migrate. For example, if older adults migrated from an area at the same rate as young animals matured into adults within the area, then a stable deme could be maintained by a peak of removal migration after some reproduction has occurred. Evidence that such a system can be adaptive is provided by the example of the honey-bee, *Apis mellifera* (p. 332) and to some extent also the black-tailed prairie-dog, *Cynomys ludovicianus* (p. 330). Similarly, if both young and adults showed removal migration from a deme at the same rate as other young and adults arrived in the deme, a stable deme could be maintained by a constant rate of removal migration throughout life, a situation that may yet prove to be found in the female gorilla, *Gorilla gorilla* (p. 299). Finally, if mortality and recruitment through reproduction exactly balance, it is possible to maintain a stable deme with the total absence of removal migration, a situation that must be approached by the vast majority of remote oceanic island demes of terrestrial species such as the deme of the hoary bat, *Lasiurus cinereus*, in the mid-Pacific. If, as suggested here, therefore, a stable deme can be maintained by the removal migration of animals of any age, then even if an individual were to gain an advantage from the maintenance of a stable deme, this advantage cannot impose selection for migration by younger rather than older individuals. A stable deme may be a characteristic of species the young of which migrate, but cannot account for the evolution of such a pattern. The only valid question seems, therefore, to be: at what age is it most likely to be advantageous for an individual of a given sex of a given species to perform a given removal migration? The extent to which this question has been answered in this chapter can now be discussed.

The first point to be stressed, which has already been made, is that we are far from being in a position for any animal to provide the quantitative data necessary to evaluate in an objective and quantitative way the validity of the model developed in this chapter. The nearest we can approach to such an evaluation seems to be for the female of the large white butterfly, *Pieris brassicae*, for which, given certain assumptions concerning habitat quotient, the variation in migration cost calculated from the model can be analysed in terms of its possible relationship to egg production. Even for man, the quantitative measurements of habitat quotient and migration cost necessary for a direct application of the model do not seem to be available. Such an absence of data removes any possibility at present of the model being used in a totally objective way. It was for this reason that the concept of the adaptive package was introduced (p. 278) as a means of evaluating to some degree the validity of the model in the absence of appropriate quantitative data.

In the preceding pages adaptive packages have been presented for a wide variety of types of sexual and temporal variation in migration incidence. It is hoped that these illustrate at least the possibility that a model based on habitat quotient, action-dependent potential reproductive success, and migration cost could account for 100 per cent of sexual and temporal variation in migration incidence. All three, however, are composite, consisting of a number of components. Most of the discussion of Section 17.3, therefore, has not been concerned with the validity of the model, support for which is provided simply by the production of adaptive packages, but with a consideration of which factors make the largest contribution to each of the three major variables. It is important to realise that there must be numerous factors that contribute to the variance of, say, habitat quotient, and that until all such factors have been identified it will not be possible to account for 100 per cent of the variance of habitat quotient in a given case nor, therefore, for 100 per cent of the variance of migration incidence. No attempt has been made here to identify all of the components of each of the three major variables. Instead the approach used has been to attempt to describe those components that in each situation seemed likely to make the largest contributions to the variance of each one. Hence, the major components of habitat quotient in type-GH migration were considered to be ability to live efficiently independently of the parents, distribution of offspring, the extent of the learning, social and construction investments in the home range occupied, and for females the ability to elicit individual parental investment from the adult male member(s) of the social group in relation to which migration was being performed. The major components of migration cost in type-H migration by males were considered to be death-risk, feeding efficiency, and copulation loss. Finally, the major component of action-dependent potential reproductive success, and in fact the only component described, was considered to be the proportion of the normal maximum number of offspring for the species that have been born or laid by a given age.

The above components and those others described for other migrations are unlikely, however, to account for 100 per cent of observed variation in each of the three variables that constitute the model, even though, because they are considered in each case to be the major components of the variable, they should account for a large proportion of observed variation. The model is such, however, that any other factor that is found to make a significant contribution to the variance of any one of the major variables can simply be added to the list of significant components. Undergraduates are inevitably suspicious of any model to which factors can be added if unexplained variation is encountered. The technique is a perfectly valid scientific procedure, however, and indeed forms the basis of multiple regression analysis.

To sum up this consideration of temporal and sexual variation in migration incidence, therefore, a model has been produced which it is suggested can account for observed variation. The model consists of three variables, habitat quotient, action-dependent potential reproductive success, and migration cost, and the suggestion is that these three variables can account for 100 per cent of observed sexual and temporal variation in migration incidence. Each of the variables, however, consists of a number of components. An attempt has been made for each variable to describe those components that make a major contribution to the variance of that variable. It is not suggested, however, that enough components have been described to explain 100 per cent of the variance of each variable, though it is hoped that a large proportion of the variance has been explained. In any subsequent detailed study of a particular migration, addition of further components to those described should decrease the unexplained variance. Although in the absence of quantitative data a high level of support for the model cannot be provided, it is felt that the production of adaptive packages and the accompanying discussion at least suggest that when quantitative data are available a large proportion of temporal and sexual variation in migration incidence can be explained by the present model. At the same time, because of the relationship (p. 77) between migration incidence and measured migration threshold, a large proportion of temporal and sexual variation in migration threshold can also be explained.

So far in this chapter variation between individuals of the same age and sex and temporal and sexual variation in migration incidence have been considered separately. In any real situation, however, both sources of variation will be found at the same time. Thus although all individuals should show the same pattern of increase and decrease in migration threshold and incidence with time and sex, the level of migration threshold is likely to vary from individual to individual in a way that in a free situation is likely to result in equal potential reproductive success for all individuals.

In addition to these combined sources of variation, in any particular study in which habitat variables and migration-cost variables are not controlled, changes with time in the level of these variables from below to above threshold and back again and vice versa could lead to a wide range of variation in observed behaviour. This could be the explanation of the wide variety of migration behaviour observed in the parthenogenetic females of the black bean aphid, *Aphis fabae* (Shaw 1970a,b). This species is likely to be exposed to rapidly fluctuating habitat variables due to the rapid increases and decreases in deme density that can occur through the high rate of reproduction and high incidence of predation by animals that once they arrive at a group of aphids tend to eat the majority of that group. The condition of a plant that is supporting aphids is also likely to change rapidly. The migration by flight of these females involves the development of a flight apparatus and physiology in response to above-threshold habitat variables early in larval life (p. 228). Amelioration of these habitat variables to a below-threshold level at any time up to the early fourth instar can lead to the development of the flight apparatus being incomplete. Amelioration in habitat variables after completion of the flight apparatus can still lead to migration physiology not being completed such that the adult does not migrate even though fully winged. There is considerable variation in individual response at each of these stages that could well fit the model developed in Section 17.2. Failure to develop a migration physiology in an individual with a fully developed flight apparatus results in that individual failing to show a pre-reproductive migration. This does not necessarily mean, however, that the migration threshold of such an individual is not at its lowest level during this ontogenetic phase, increasing as usual once larviposition has begun. Nevertheless, deterioration of habitat variables at a rate faster than ontogenetic increase in migration threshold could well lead to an above-threshold situation after larviposition has commenced. Certainly some individuals fail to initiate migration until after some larviposition has occurred. In general in this species, however, once larviposition has been proceeding for several days and hence action-dependent potential reproductive success has markedly fallen, the migration threshold seems likely to be so high that it is unlikely to be exceeded by the level of migration-cost and habitat variables. It is in such a situation that selection for flight-muscle autolysis seems most likely to occur, a phenomenon found in this species.

Although a detailed study of the initiation of migration such as that furnished by the study of the black bean aphid illustrates the complexity that can be encountered in a field situation, none of the facts that have emerged so far seem to be incapable of being accommodated by the two models developed in this chapter when applied in combination.

17.5 Summary of conclusions concerning intra-specific variation in the initiation of migration

1. Not all individual members of a deme migrate at the same time. For any single age and sex group within the deme there is a characteristic frequency distribution of migration thresholds to environmental variables. Migration threshold also varies with sex and time, particularly age and season.

Variation in migration response by individuals of the same age and sex

2. Variation in migration response can occur when two individuals with the same migration threshold in the same habitat nevertheless experience different habitat suitability. This could be due to the differences in the ability of the two individuals to exploit the habitat. Also it could be due to differential encounter with environmental variables that results from 'sampling error' produced by the two individuals not always being in the same place at the same time. Selection should favour adaptations that reduce such variation in migration response due to sampling error but it is unlikely that the ideal adaptation can ever totally be achieved (p. 68).

3. Variation in migration response can also reflect evolved innate individual differences in migration threshold. Items 4–9 of the summary are concerned with the selective pressures that act on the frequency distribution of migration thresholds in a single age and sex class of a deme. Only animals the spatial distribution of which is at the free end of the territorial spectrum are considered (p. 402). Items 4–8 of the summary are concerned only with animals for which habitat suitability, when less than the mean expectation of migration, is a negative function of deme density.

4. *Hypothesis:* Selection stabilises when the frequency distribution of migration thresholds within a deme is such that all individuals experience potentially equal reproductive success as a result of their migration response (p. 69).
   Support: Analysis of the frequency distribution of dung pat to dung pat migration thresholds in a deme of the common yellow dung-fly, *Scatophaga stercoraria*, by Parker (1970c) as reflected by stay-time on the pat provides strong support for the hypothesis. The frequency distribution of stay-times is such that on average all males achieve the same fertilisation rate irrespective of their stay-time (Fig. 17.1).

5. *Hypothesis:* Selection stabilises when for all values of habitat suitability a proportion of individuals initiate migration such that habitat suitability after their departure is equal to the mean expectation of migration (p. 69).
   Support: Variation with time after deposition of a dung pat of the proportion of males of the common yellow dung-fly, *Scatophaga stercoraria*, that, if they initiated pat-to-pat migration, would restore habitat suitability to the mean expectation of migration was calculated and compared with observed variation. The agreement between the two provides strong support for the hypothesis (Fig. 17.2).

6. *Hypothesis:* When a deme of an animal is adapted to two or more habitat-types that show a predictable seasonal fluctuation in that component of their relative suitability that is independent of deme density (p. 70), selection should act to favour a particular frequency distribution within the deme of linkage between different values of the photoperiod/temperature regime and the initiation of migration. This distribution should be such that migration leads to a fluid distribution of deme density between the habitat-types such that either the suitabilities of the habitat-types are equal throughout the year or at some times of year one habitat-type remains the more suitable despite supporting 100 per cent of the deme (p. 71).
   Support: No data.

7. *Hypothesis:* Selection stabilises when the proportion of a deme that performs ontogenetically obligatory removal migration is the maximum that on average increases habitat suitability to the level of the mean expectation of migration without exceeding this level. The remainder of the deme are likely to be facultative migrants (p. 71).
   Support: Insufficient data for evaluation (p. 273).

8. *Hypothesis:* Selection should stabilise when the proportion of obligatory non-migrants within a deme is the maximum that on average in the absence of other (migrated) individuals gives a deme density that will not decrease habitat suitability below the mean expectation of migration (p. 71).
   Support: Some support from the observed spatial distribution of wingless insects and birds but insufficient data for quantitative evaluation (p. 274).

9. *Hypothesis:* Selection due to short-term unpredictable variation in the mean expectation of migration about a mean value should stabilise when the frequency distribution of migration thresholds and/or stay-times within a deme corresponds to the frequency distribution of the mean expectation of migration. As far as obligatory migrants are concerned the stable situation should be when the proportion of ontogenetically obligatory migrants is a function of the proportion of occasions that habitat suitability is less than the mean expectation of migration at the appropriate ontogenetic stage (p. 72).
   Support: No data.

Variation in migration response with age and sex

10. Various theories of ontogenetic variation in migration response have been advanced previously. These are discussed (pp. 73–7) and found to be unsatisfactory.

11. A model is developed (pp. 77–80) based on the interaction of three variables. These are mean habitat quotient, $\bar{h}_q$, mean action-dependent potential reproductive success $(\bar{S}_d)$, and mean migration cost $(\bar{m})$. These interact to produce the initiation factor (i) according to the formula

$$i = h_q(1 - \bar{m}/\bar{S}_d).$$

12. *Hypothesis:* A model based on the interaction of habitat quotient, action-dependent potential reproductive success, and migration cost as expressed by the initiation factor, is capable of explaining 100 per cent of observed variation in migration incidence (p. 354).

    Support: Some quantitative support for the female large white butterfly, *Pieris brassicae* (p. 333), and to a lesser extent for man. Otherwise the support for the hypothesis rests on the production of adaptive packages (p. 278) that not only account for observed variation in migration incidence but that also provide acceptable descriptions of the ecology and social and reproductive biology of the species concerned. The fact that it is possible to produce adaptive packages for a wide range of temporal and sexual variation in migration incidence is taken to provide some support for the hypothesis.

13. Habitat quotient, action-dependent potential reproductive success, and migration cost are all composite variables. Although it is suggested that, as expressed through the initiation factor, they can account for 100 per cent of variation in migration incidence, it is not suggested that the described components of each variable can explain 100 per cent of the variation of each variable. Even so, because it is considered that the described components are those that make the largest contributions to the variation of the variable, it is thought that a large part of the variation has been explained (p. 354).

14. *Hypothesis:* The major components of habitat quotient are: learning, social, construction and other investments in the habitat occupied; parental presence when young; the spatial distribution of offspring; predictable temporal variation in the appearance of new and/or more suitable habitats; and, for females, the ability to elicit from a male or males parental investment other than that involved in copulation. The relative contribution of each of these components to the total variance of habitat quotient varies according to the type of migration and the species, sex and age of the animal being considered (pp. 317–21). The contribution of habitat quotient to variation in mi-

gration incidence is greatest when an animal lives in a home range and migrates unaccompanied by mate and offspring.

    Support: Some support, mainly from adaptive packages (Figs. 17.6, 17.11, 17.13, 17.18–17.20, 17.25–17.28, 17.30) and for many vertebrates from agreement between predicted and observed variation in the extent of the pre-reproductive peak of migration incidence.

15. *Hypothesis:* Action-dependent potential reproductive success is influenced by virtually every action of an animal. Perhaps the most important single component, however, and the only one considered in this chapter, is the production of offspring. Action-dependent potential reproductive success contributes most to variation in migration incidence when migration cost is high and/or individuals migrate accompanied by mate and offspring.

    Support: Some support, mainly from adaptive packages.

16. *Hypothesis:* The major components of migration cost are: death-risk to self; decreased nett rate of energy uptake; copulation loss; death-risk to offspring; and decreased rate of gametogenesis. The relative contribution of each of these components to the total variation of migration cost varies according to the migration and the species, sex and age of the animal being considered (pp. 307–41). The contribution of migration cost to variation in migration incidence is greatest in an animal that does not live in a home range and in animals, particularly females, that migrate accompanied by their offspring but not by their mate and that live within a home range but as a member of a large social group.

    Support: Some support, mainly from adaptive packages but also from analysis of migration cost for the female large white butterfly, *Pieris brassicae*, in relation to egg maturation and oviposition (Table 17.6).

17. *Hypothesis:* Variation in initiation factor can be categorised according to whether the variation results from: (1) a change in the environment that is detectable either directly or indirectly by the animal; (2) a change in the environment that is undetectable by the animal but that occurs at a predictable stage of ontogeny; and (3) a change in the animal. Optimum adaptation to category (1) changes involves migration thresholds to environmental variables such that the thresholds change as the level of any other environmental variable changes but which do not change in a consistent way with time (p. 346). Optimum adaptation to category (2) and category (3) changes involves migration thresholds that change in a consistent way with time (p. 348).

    Support: Some support, particularly from experiments on insect migration thresholds to migration-variables (p. 352).

18. *Hypothesis*: The migration threshold must have a physiological basis that at some stage must involve comparison of information received by sense organs with an internal threshold. When migration threshold varies with time, therefore, it must reflect a change in the internal threshold. The physiological events and feedback mechanisms that are likely to be involved in and to influence the physiological basis of this internal threshold can be postulated for a variety of animals and variety of types of migration on the basis of the conclusions reached from the adaptive packages. Different physiological events and different feedback mechanisms seem likely to be involved in the variation in internal threshold of different animals (p. 348).

Support: Some support regarding the physiological results of the beginning of investment in a home range in mustelids (p. 350). Otherwise most support is circumstantial and derives from observed coincidence of physiological events and changes in migration incidence.

19. Temporal variation in migration threshold due to time-investment strategy and adaptation to spatially separate habitat-types (p. 352) is considered to be a special case of category (3) of item 17 of this summary.

Part IIIb   Forms of the lifetime track

Chapters 14–17 and their theoretical counterparts, Chapters 7–10, have been concerned with identifying the range of situations in which the initiation of removal migration is likely to be advantageous. The behavioural and physiological mechanisms that are likely to evolve to increase the probability that an individual will initiate removal migration at the most advantageous time have also been discussed as have the factors involved in inter- and intra-specific variation in migration incidence in a free situation. Finally, towards the end of Chapter 18 (p. 403) some consideration is given to the factors involved in the initiation of migration in a despotic situation.

In Chapter 7 it was suggested that because the lifetime track of an animal could be broken down into a succession of removal migrations, the first and major step in an understanding of the lifetime track, which is the eventual aim of any study of migration, is an understanding of the factors involved in the initiation of removal migration. It is hoped that Chapters 7–10 and Part IIIa have gone a little way towards achieving such an understanding. The next step is to use the conclusions reached in these preceding chapters to attempt to analyse the various forms of lifetime track manifest by different members of the animal kingdom. Chapter 18 is concerned primarily with the selective pressures that act to produce the two major forms of the lifetime track, the linear range and the limited-area range. Chapter 19 examines the different mechanisms involved in the execution of a linear range by a variety of animals and at the same time examines the phenomenon of re-migration which emerges for some animals when the linear ranges of successive generations are studied. Finally, Chapters 20 onward examine perhaps the best known form of lifetime track, that in which the animals perform a regular, seasonally or ontogenetically synchronised, return migration.

18

Two basic forms of the lifetime track: the linear range and the limited-area range

18.1 Introduction

In Part IIIa of this book, consideration was restricted to the initiation of a single removal migration (i.e. from A to B). In a consideration of the lifetime track of an animal, however, we are concerned with a succession of removal migrations (i.e. from A to B to ? to ? to ?, etc.). Clearly the form of the lifetime track is to a large extent a function of the frequency with which selection acts for or against a high degree of return during any given removal migration. The theoretical aspects of this situation have been discussed in Chapter 9 and it is possible from the considerations of that chapter to recognise three categories of selection on any two consecutive removal migrations.

Once again, of course, it should be stressed that these three categories are constructed solely for descriptive convenience and in reality are three broad bands on a continuous spectrum of selection on successive removal migrations. Which part of the spectrum is associated with which evolutionary sequence depends partly on the pre-adaptations of the animals concerned and partly on the nature and spatial distribution of the habitat-types in relation to which speciation or other evolutionary change is taking place (Fig. 9.1). At one extreme of the spectrum can be recognised a category of selection that favours individuals that perform return migration (albeit with a particular timing and frequency, etc.). In the middle part of the spectrum can be recognised a category that favours removal migration taking place without regard to previous movements. Such selection occurs when the advantages and disadvantages of return and further removal migration cancel out such that both patterns are equally advantageous. Finally, at the other extreme a category can be recognised in which selection acts against individuals that perform return migration.

All bands of this spectrum of selection on removal migration are likely to act during the evolution of any given track pattern. The result is that the lifetime track of almost any animal is likely to consist of a series of return and non-return removal migrations. The way that these are integrated, however, can result in marked species differences in the gross form of the lifetime track. In this chapter we are interested particularly in two, perhaps basic, forms of lifetime track. One of these, the linear range, involves a pattern of exploitation of space that has a strong linear component and is manifest by animals such as butterflies and locusts. The second of these forms of the lifetime track is the limited-area range which invariably gives rise to the familiar area/home range/territory phenomenon (henceforth described as the familiar-area phenomenon) that is manifest by some invertebrates and the vast majority of vertebrates. Both of these forms of lifetime track have given rise secondarily to long-distance movements involving seasonal or ontogenetic changes of direction. The latter of the two track forms, the limited-area range as represented by the familiar-area phenomenon, is argued in subsequent chapters to have given rise to the long-distance seasonal and ontogenetic return migrations found in many vertebrates. The former, the linear range, is argued in Chapters 19 and 20 to have given rise to the seasonal major change of direction of insects such as butterflies and locusts. This chapter, however, is concerned only with the evolution of the basic form of the two patterns of spatial exploitation, excluding seasonal and the majority of ontogenetic effects.

18.2 The linear range

Figure 18.1 shows the track pattern of the small white butterfly, *Pieris rapae*. This track pattern may be divided into two parts. The first and most conspicuous part is the rectilinear movement across areas of low suitability. The second part is the movement characterised by a high turning rate that is performed within areas of higher suitability. This second part is considered in more detail in relation to another butterfly species in Fig. 18.4. Let us first consider the rectilinear part of the track pattern in relation to the conclusions already reached in previous chapters concerning removal migration.

Consideration of optimum stay-time (p. 82) suggested that optimum strategy in areas that provided none of the required resources (apart from respiratory and thermal requirements, etc.) was to stay the minimum time necessary to assess habitat suitability. For a small white butterfly that finds itself in an area of concrete and/or grass devoid of nectar-producing flowers and *Brassica oleracea* or other crucifers at a time of day that roosting is not optimum behaviour, this stay-time is that necessary to traverse the area. Having traversed the area, return is disadvantageous, for the area is unlikely to produce such plants in the time interval that the area

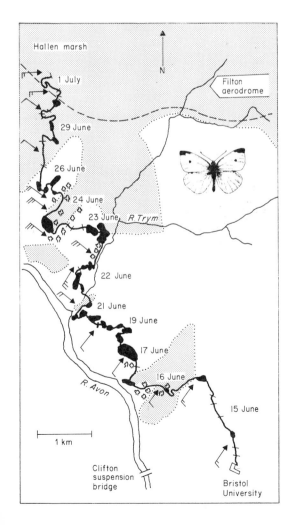

Fig. 18.1 Track of the small white butterfly, *Pieris rapae*: an example of the 'linear range' form of the lifetime track

The general ecology of this species is described in the legend to Fig. 17.28.

Track pattern

⊢——⊣, track of an individual butterfly from first sighting to the place at which it was lost from view. Upon such loss of an individual, observation was maintained at the place of disappearance until a second individual appeared following the same track as the previous individual. The process was then repeated. No distinction was made between males and females. Unless individuals are considered to behave differently when unobserved, the figure can be assumed to approximate to the track of a hypothetical single individual.

Solid black areas along the track represent areas in which flight ceased to be rectilinear and in which courting, oviposition, roosting, and feeding occurred. Such areas were largely private gardens containing flowers and/or cabbages, though in some places such areas were also patches of waste ground supporting wild crucifers.

Unshaded areas are largely urban and consist of streets, shops, and houses and little vegetation except that in gardens. Stippled areas are expanses of flat open grassland. Both habitat-types, except where the urban-type contained gardens, represent areas of low suitability for all activities of *P. rapae*. Some areas of woodland, also, in the cases concerned, representing relatively low suitability, are also indicated.

Wind speed and direction is indicated by arrows: one short feather = one unit; and one long feather = two units on the Beaufort scale. The dashed line indicates the boundary of the city and old county of Bristol.

The main point of interest for present purposes is the way rectilinear flight across areas of low suitability gives way to a track pattern characterised by an increased rate of turning in areas of high suitability. After a certain stay-time in an area of higher suitability, however, rectilinear flight is once again initiated and in more or less the same direction as previously. The result is that the individual moves gradually across country, spending shorter or longer periods in areas of higher suitability *en route*.

This figure is analysed further in Chapter 19 in relation to migration distance and orientation mechanisms.

is the nearest of its habitat-type available to the butterfly (p. 63). Selection therefore favours a track pattern that does not involve return to an area of low suitability already visited. Instead, selection favours a track pattern that maximises the search area. As shown in Fig. 18.2, such selection favours a track pattern involving a much reduced rate of turning. In the absence of any other selection, simplicity and efficiency of orientation is likely to favour an approximately rectilinear track pattern (Chapter 19). As long, therefore, as a butterfly, or any other animal, is searching for a relatively immobile resource in an area of low suitability, selection is likely to favour an approximately rectilinear track pattern.

The advantage of a rectilinear track pattern applies particularly to a single search sequence (i.e. a single removal migration). Consider now the situation in which an animal searches for a resource using a rectilinear track and succeeds in finding the resource. After a period exploiting the resource, migration away from the resource becomes advantageous, either because optimum stay time has been completed, exploitation of the resource has caused

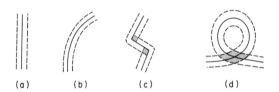

(a)   (b)   (c)   (d)

Fig. 18.2   Relative efficiency of four track patterns as searching strategies for an 'immobile' resource

An immobile resource for present purposes is one that is unlikely to appear, disappear, or move during the time that it is the nearest resource to a moving animal. Almost all areas of any size that constitute a habitat-type come into this category, as do most plants. Most non-sessile animals, on the other hand, are examples of mobile resources except to the extent that they are most likely to be found in a particular habitat-type that is relatively immobile.

The solid line represents the track of an animal. The dashed lines represent the limits of its sensory range. Selection on track pattern during search for a relatively immobile area or object should act to maximise the area (or volume) that comes within the animal's sensory range. In general, the lower the rate of turning, as in the rectilinear track shown in (a) or the gradual curve shown in (b), the greater the area searched per unit distance of movement. High rates of turn as in (c) and tracks that cross themselves as in (d) lead to a decrease in the area searched per unit distance of movement. The area marked by hatching in (c) and (d) is an area that is searched longer than is optimum.

habitat suitability to fall below the mean expectation of migration (p. 46), or optimum time-investment strategy favours switching to a different behaviour sequence for which the habitat occupied is below the mean expectation of migration (p. 243). Only in the latter case may it not be disadvantageous to return to part of the track followed during the first search and only then if the spatial distribution of habitats is such as to favour return (Fig. 9.1). Otherwise it is still disadvantageous to return to any part of the track followed previously.

This selection is further reinforced by the fact that butterflies and many other invertebrates are animals for which, in the light of Fig. 9.1, selection is least likely to favour return migration, either to previous tracks or to areas of higher suitability. This is because roosting sites, feeding sites, and oviposition sites etc. are for most species so abundant relative to the animal's mobility and so unaffected by the animal's previous visit that sites visited previously are unlikely to be the nearest or the most suitable available to the animal when next that particular resource is required. Hence, the roosting site, such as a nettle patch, of the small white butterfly, *Pieris rapae* (Fig. 18.1), used one night is unlikely to be the nearest or the most suitable roosting site available to the butterfly the following night. Selection may not act against return migration in this case, except through distance (Fig. 9.1) but nor will it act in favour of return migration. These butterflies, and similar insects, therefore, are acted upon by selection against recrossing previous tracks over areas of low suitability. Furthermore, there is absence of selection for return to previously occupied habitats of higher suitability. All tracks other than those that cause the animal to recross an earlier track are, however, likely to be equally suitable. The track least likely to result in the animal recrossing an earlier track is that in the same direction the animal was heading when it arrived at the habitat. Selection is unlikely to favour a high degree of precision of orientation, however, as long as on average the same direction is adopted. The validity of this argument is supported by Figs. 18.1 and 18.4. Individuals of both the small white butterfly, *P. rapae*, and meadow brown butterfly, *Maniola jurtina*, and indeed of many other butterflies for which similar observations have been made (Baker 1968a, 1969a), upon leaving an area of higher suitability, tend to take up more or less the same compass bearing that they were following when they arrived at the area. The precision of orientation, although not great (Chapter 19, p. 416) is sufficient to represent efficient adaptation to the selective pressures outlined above.

Figure 18.3 provides a further example of an animal, the desert locust, *Schistocerca gregaria* phase *gregaria*, that performs rectilinear migration across areas of low suitability and maintains a more or less constant direction.

When crossing areas of low suitability, migration cost is likely to be relatively low for further migration (i.e. the cost of migration is little greater than the cost of spending time in a habitat of low suitability —p. 41). In addition, the animal is likely to be searching for an example of a particular habitat-type. Such a habitat-type is likely to be recognised first through gross bio-topographical features and as such is likely to be relatively immobile. This combination of factors has just been argued to favour a rectilinear track pattern. When the animal arrives at an appropriate habitat-type, however, some or all of these selective pressures are likely to disappear.

Figure 18.4 shows the track pattern of a male and female meadow brown butterfly, *Maniola jurtina*, within an area of relatively high suitability surrounded by an area of low suitability. Let us first attempt to interpret the movements of the female. In late afternoon, the main resource requirements of the female meadow brown seem to be an area suitable for oviposition and thermoregulation in close proximity to a roosting site and perhaps also a feeding site. Having arrived at a superficially appropriate habitat-type as a result of rectilinear migration, the optimum track pattern for a butterfly is one that maintains the animal in the confines of the habitat and yet permits 'assessment' (see p. 347) of the suitability of the habitat. Optimum track pattern for exploratory migration is discussed on p. 376 and is not considered further here. Having perceived a suitable roosting site in close proximity to a suitable site for thermoregulation the female stays in that area until roosting becomes optimum behaviour, whereupon calculated removal migration within the sensory range (p. 91) takes place from a thermoregulation site to a roosting site. Shift of thermoregulation site seemed to result from the butterfly being increasingly shaded in one site by vegetation as the Sun's azimuth changed and consequent migration to a new site where the shading was less. Even these short-distance (a few metres) migrations therefore fit within the model developed in Part IIIa. The following morning, the first migrations were between roosting, basking and feeding sites and were again probably of sufficiently

short-distance to be calculated removal migrations within the sensory range, each migration becoming advantageous in connection with optimum time-investment strategy (p. 57). As oviposition became optimum behaviour, however, the situation changed. Let us accept that the array of parasites, predators and physical death-risk factors that are relevant to the niche occupied by *Maniola jurtina* are such that selection favours those females that distribute their eggs in space in batches of 1–4 at least 8 m apart. Deposition of a batch of eggs at one point in space, therefore, immediately renders disadvantageous an area within an 8 m radius of that point. Consequently, selection also favours those females with a track pattern that does not return the individual to an area previously visited. In this case, however, in contrast to the situation when crossing areas of low suitability, selection is against return to a previously suitable habitat rather than against recrossing a previous track. On the other hand, if we assume that a meadow is never homogeneously suitable for oviposition over its entire area, a female meadow brown is searching for a relatively immobile resource even within the meadow. A track pattern with a low rate of turn, such as rectilinear movement, again confers some advantage and for the same reasons as before (p. 362). Yet a rectilinear track pattern is just the one to take the animal most quickly out of a limited area if the area is not itself more or less linear. Rectilinear movement in a constant direction gives, therefore, the optimum track pattern until the butterfly arrives at or perceives the edge of the area of high suitability. If migration away from the entire habitat is not advantageous, optimum track pattern upon encountering the habitat edge involves a turn of between 90° and 180° followed by a further period of rectilinear movement in a constant direction. Alternatively, a gradual circle can be performed once the edge of the habitat is perceived, a pattern that retains many of the advantages of rectilinear movement without taking the animal away from the area of higher suitability.

Yet another set of selective pressures may be evoked when we consider the male meadow brown butterfly illustrated in Fig. 18.4. As with the small white butterfly, *Pieris rapae* (Fig. 18.1), the desert locust, *Schistocerca gregaria* (Fig. 18.3), as well as the female of its species (Fig. 18.4), a male meadow brown crossing an area of low suitability is searching for a relatively immobile resource, viz: an area of the type that offers a high probability of encountering a

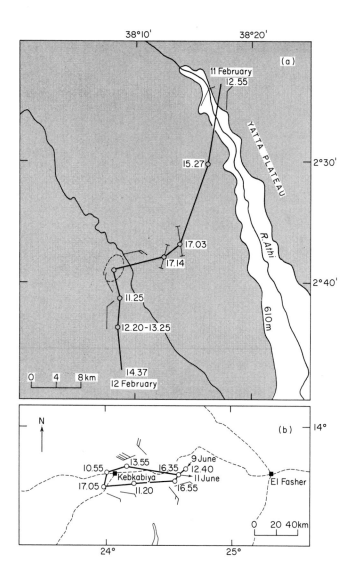

Fig. 18.3 Track patterns of swarms of the desert locust, *Schistocerca gregaria*, in areas of low and high suitability for oviposition

General ecology

The general ecology of this species is described in Chapter 19 (p. 452). Areas of high suitability for oviposition are those where rain has recently fallen. Any area containing vegetation, however, seems to be of moderate suitability for feeding and roosting. Locust swarms feed during migration, the swarm rolling along with feeding locusts taking off as the swarm passes overhead while other locusts at the front of the swarm settle and start to feed (p. 455).

(a) Track pattern in an area of low suitability for oviposition

The top figure shows the track pattern in Kenya of a swarm of sexually immature desert locusts across an area

in which rain had not recently fallen. The swarm when first located on 11 February measured 8 km^2 and individuals were flying at heights of up to 800 m above the ground. The open circle shows the position of the swarm centre at the time of day indicated and the cross-bar indicates the length of the swarm. The swarm settled for the night at about 19.00 hours, soon after sunset. The size of the roosting area is indicated by the dashed line. The swarm moved off again at about 09.00 hours the following morning. The track pattern shows a low rate of turn and takes the locusts across the area of low suitability for oviposition. At this time of year the main area of rainfall along the Inter-Tropical Convergence Zone (p. 452) lies to the south of the area shown.

(b) Track pattern in an area of high suitability for oviposition

The lower figure shows the track pattern in Darfur

province, Sudan, of another swarm of sexually immature desert locusts in the vicinity of the Inter-Tropical Convergence Zone. All three mornings of the period illustrated were fine to fair. Each afternoon, however, cumulonimbus developed with 3 mm of rain recorded nearby on the afternoon of the 10 June and 10 mm overnight on the 11/12 at El Fasher. Upon encountering such conditions, locusts develop to sexual maturity, copulate, and oviposit. The swarm when first encountered measured 20 km²

and contained individuals flying up to 900 m above the ground. At 11.20 on 10 June the swarm was just rising from its overnight roost with the topmost locusts at only 150 m. The track pattern shows a much higher rate of turn which maintains the swarm within the vicinity of the Inter-Tropical Convergence Zone, and takes it toward imminent rain (p. 457).

[Compiled from Rainey (1963). Photo courtesy of C. Ashall and COPR]

receptive female. Consequently, a rectilinear track pattern in a constant direction when crossing such low-suitability areas is likely to have been favoured by selection (Fig. 18.2). Once inside an area of higher suitability for encountering receptive females, however, the resource sought by the male is no longer relatively immobile. Even though a receptive female may not be present at a particular point in space one moment, there may be one present from any moment in time thereafter. Selection against return to a place previously visited is not therefore as great on the male once inside an area of higher suitability as it is on a female and also on both male and female when outside such an area. Nevertheless, the probability of encountering a receptive female is likely to remain greater in areas that the male has not previously searched than in areas that

have just been searched. This difference in probability is, however, likely to decrease with time after first visit as the opportunity for a female to arrive and settle in that space increases. As the area is unlikely to be homogeneously suitable over its entire area, a track pattern, such as one with a low rate of turn, that maximises the area searched, is likely to be favoured. Too low a rate of turn, however, rapidly takes the animal out of the area. A turn of 180° or two consecutive turns of 90°, however, whenever the animal finds itself entering an area of low suitability, leads to return to the area of higher suitability. This behaviour, which may be termed the 'edge reaction', is a well known feature of animal behaviour and is well shown by the male meadow brown in Fig. 18.4. The presence of an edge to the suitable area means that search has to be intensive within

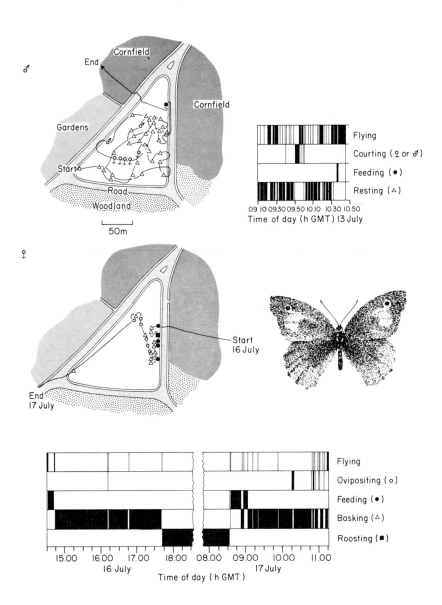

Fig. 18.4 Track pattern of a male and a female of the meadow brown butterfly, *Maniola jurtina*, in an area of moderately high suitability for finding a mate and for oviposition.

General ecology

The meadow brown is a satyrid butterfly that is extremely abundant in Britain and other parts of western Europe. Adults are abundant on the wing throughout mid-summer and early autumn with a tendency in southern Britain toward a bimodal curve of seasonal incidence. Females mate within a few days of adult emergence and thereafter are non-receptive to courting males. Eggs are laid very near to the soil, preferably among short grass (<5 cm) that has been grazed or cut. The larvae feed on a variety of species of meadow grasses and hibernate on or in the soil when quite small. They feed up in the spring and pupate in early summer. Males have a time-investment strategy that contains a large female-search component. Searching for females involves alternating between flight, during which females are sought both on the ground and in the air, and 'resting' on the ground, during which the male is prepared to intercept any butterfly that flies overhead within sensory range. Any brown butterfly encountered, whether male or female, is 'investigated' by a male and each such interaction is recorded as a courtship bout in the upper figure. An area of relatively high suitability for female-search and oviposition seems to be constituted by almost any sufficiently large area of grazed or cut grass of the appropriate grass species. Such an area is shown in the diagram and is sited on the northern outskirts of Bristol, western England. It consists of a triangular meadow of approximately 1 ha that had recently been used as pasture for cows. At the perimeter of the meadow was a hedgerow and bank that supported a number of nectar-producing flowers, such as bramble, *Rubus fruticosus*, and thistles, *Cirsium* spp., from which the meadow browns fed. The hedge also provided appropriate roosting sites. As a subjective estimate the habitat studied was probably of high suitability for oviposition, of average to low suitability for feeding and roosting, and of below average suitability for finding virgin females. The meadow was bordered on all sides by roads and beyond these by gardens, cornfields, and woodlands that were of low suitability with respect to all resource requirements except perhaps roosting sites and across which all meadow browns performed rectilinear movement. The area illustrated, therefore, consisted in effect of a habitat of moderate to high suitability surrounded by a strip over half a kilometre in width of very low suitability.

Track pattern and behaviour in an area of moderate to high suitability

As with the small white butterfly, *Pieris rapae*, in Fig. 18.1, meadow browns perform rectilinear movement over areas of low suitability. The geographical direction of movement is approximately the same before and after time spent in a habitat of moderate to high suitability (note directions of arrival and departure in the lower diagram). Within an area of moderate to high suitability, however, rate of turning increases considerably, particularly in males.

Males: The male illustrated was already present in the habitat when first observed and it is not known whether it had roosted there or had just arrived from elsewhere. For nearly 90 min after first observation the male indulged solely in a search for females. Periods of from 30 s to 3 min on the ground ('resting') were alternated with usually shorter periods of flight. Flight was slow and involved frequent close investigation of the ground. Each flight bout (= removal migration) had a low rate of turn and was often almost rectilinear. When flight track did depart from rectilinearity, it often took the form of a gradual curve. However, unlike movement across areas of low suitability, there was often a large angular difference in direction of the tracks of consecutive flight bouts. When an edge, such as the hedgerow, was crossed between areas of high and low suitability there was an immediate 90° turn followed by a second 90° turn in the same direction that brought the animal back into the area of high suitability, though a reasonable distance from the point where it left. During the stay in the habitat, the male encountered two unreceptive females and two males. No feeding had occurred during the first 90 min of observed behaviour despite an occasional encounter with appropriate nectar-producing flowers. The male's stay in the meadow was terminated by a sudden change in behaviour that involved first feeding and then rectilinear flight across and away from the meadow. The initiation of migration away from the habitat could be attributed to any combination of the factors discussed in Chapter 16. Possibly optimum stay-time in a female-search habitat of a suitability less than the mean expectation of migration had just been completed (p. 82). Alternatively, optimum time-investment strategy may have involved termination of the search for females and initiation of a period of feeding (p. 57), for which habitat suitability may again have been less than the mean expectation of migration.

Females: The female illustrated was first observed through binoculars approaching the meadow from across the cornfield on the afternoon of 16 July. After an initial period of feeding, the behaviour was strongly biased toward resting and, presumably, thermoregulation (= basking), though three eggs were laid in a single bout of oviposition. The female eventually went to roost in the hedge and was still in position the following morning. After feeding and basking the female described a more or less rectilinear track across the field, ovipositing at seven separate sites along this track. In all, twenty eggs had been laid in the field in eight bouts of oviposition, an average of 2·5 eggs per bout (range 1–4). Female behaviour also changed suddenly. The rectilinear track that took the female out of the meadow and away was at right angles to the oviposition track but at more or less the same angle to the Sun's azimuth as the rectilinear track by which the female had arrived at the meadow the previous afternoon. Migration away from the meadow may have become advantageous because the optimum stay-time or optimum amount of oviposition in a habitat of that suitability had just been completed. Alternatively, the optimum time-investment strategy for the female may well have involved cessation of oviposition and initiation of some other behaviour, such as feeding, for which habitat suitability was less than the mean expectation of migration.

boundaries and a track pattern that maximises area of search per unit of movement distance but without recrossing previous tracks would be difficult to execute. Given the advantage of the edge reaction and of the maximisation of the area searched per unit of movement distance, and the gradually decreasing disadvantage, with time, of recrossing a previous track, the track pattern illustrated in Fig. 18.4 for the male meadow brown seem to be an appropriate adaptation. This track pattern involves a gradual rate of turn coupled with an edge reaction such that there is minimum probability of immediate return to an area just searched but without such return being prohibited after a sufficient length of time.

Note that the hierarchy of migration thresholds discussed in Chapter 16 (p. 229) is applicable when discussing migration within and away from areas of higher suitability. Male time investment in Fig. 18.4 consisted of resting on the ground for periods of up to 3 min, apparently lying in wait for any female that flew overhead, followed by 30 s to 2 min of migration to a similar site elsewhere. During this resting site to resting site migration the male examined places on the ground that could either have contained basking females or that were a potential new resting site for the male. It may be assumed that three minutes represents the maximum advantageous stay time in any one resting site and that if, during that time, a female has not been intercepted, the suitability of the site is less than the mean expectation of migration. Just because resting site to resting site migration is advantageous, however, it does not follow that migration from the entire meadow is also advantageous. Presumably one stay time (approximately 3 min) is optimum for a resting site but another stay time (greater than 90 min) is optimum for the meadow. Until the optimum stay time for the meadow is completed, therefore, the edge-reaction is triggered whenever the male encounters an area of low suitability. After the stay time for the meadow is completed, rectilinear migration continues even when an area of low suitability is encountered.

In all of the examples described so far, selection has not favoured a return migration at any point along the lifetime track, not even within an area of relatively high suitability. The nearest selection has come to favouring a return migration has been for the male meadow brown just described where the advantage of maximisation of the area searched within a limited area and the decrease with time in the disadvantage of re-visiting areas previously visited has led to some recrossing of previous tracks while within an area of relatively high suitability. In some species with a linear range, however, selection favours at least some return movements within the areas of higher suitability. An example is given in Fig. 18.5.

To sum up, therefore, the main characteristic of animals with a linear range is that whenever they initiate removal migration across areas of low suitability they show a rectilinear track pattern in a direction that is more or less a constant for each individual. The result is that from birth to death the individual moves across country in a more or less straight line, pausing for shorter or longer periods in areas of higher suitability en route. Track pattern while within these areas of higher suitability depends on the size of the area relative to the animal's mobility and optimum stay-time and on the nature and distribution of the resources within the area. Great variation in evolved track pattern is therefore likely, ranging from a virtually rectilinear track in some species, as in the female meadow brown (Fig. 18.4), through a track with a higher rate of turn, as in the male meadow brown (Fig. 18.4), to a track consisting of a series of return migrations to a 'home' base in others (Fig. 18.5).

In all of the species illustrated there is a high migration incidence throughout life, even across quite wide areas of low suitability. Ontogenetic variation in migration incidence (Chapter 17), however, though not affecting the linearity of the lifetime track, may well mean that the major part of the linear track is performed when young rather than later. In this case, the animal spends more of its adult life in areas of relatively high suitability with the result that the track patterns performed in such areas are more 'normal' to the animal than the rectilinear movements performed when younger (i.e. an observer is more likely to see the species showing a high rate of turn in a limited area than performing rectilinear migration). The pattern shown by the small white, *Pieris rapae* (Figs. 17.28 and 18.1), in which linear movement is important throughout life, approaches one extreme for the linear range form of the lifetime track. At the other extreme are animals such as termite and ant reproductives that spend their early life in one area of relatively high suitability, the natal colony, initiate a single rectilinear movement across an area of low suitability (see Chapter 16, p. 240) and then spend the remainder of their life within a second area of relatively high suitability. Remove the rectilinear or

Fig. 18.5 Track pattern of a male peacock butterfly, *Inachis io*, in relation to an area of high suitability for intercepting females

General ecology
The general ecology of this species is described in the legend to Fig. 16.16. Male *I. io* have a linear range similar to that illustrated in Fig. 18.1. An area of high suitability to a male *I. io* on a fine spring afternoon after hibernation is an area in which they are most likely to intercept receptive females. Such an area contains an expanse of flat vegetation or bare soil open to the Sun but bordered on at least one side away from the Sun by a line of trees. Proximity to the foodplants used by the species as adults or larvae does not seem to be important but proximity to roosting sites is preferred. A corner site is also preferred, though non-corner sites are also used. Unless ousted by another male or leaving to follow a receptive female on her cross-country migration (copulation does not occur until the pair go to roost) the male remains in the same territory from soon after midday until going to roost. The following morning, cross-country migration is initiated with pauses in areas suitable for feeding.

Track pattern in relation to an area of high suitability for intercepting females
The diagram shows a plan view of the track pattern of a male that occupied a corner territory in a field at Failand, Bristol, on 7 April 1969. The triangle shows the starting position and the open circle shows the main vantage point taken up while in the territory. The major behavioural component during occupation of a territory involves resting on the ground and performing what appears to be thermoregulatory behaviour. The result seems to be the maintenance of a state of readiness to intercept and investigate passing insects. The male investigates almost any large insect that comes within sensory range. Individuals of other species are followed on average for about 3 s. After each such investigation the male returns to its territory. The solid line in the diagram indicates an observed track. The dashed line is a surmised track for a period that the butterfly was behind a stand of trees and hence unobserved. Female *I. io* that cross the territory are investigated and followed, often for relatively long distances as illustrated. Non-receptive females are abandoned and the male returns to his territory. Other male *I. io* that come within sensory range are similarly investigated. The two males then engage in a territorial interaction involving in effect a comparison of aerial manoeuvrability and during which the territorial male leads or chases the intruder away from the territory. On most occasions only the territorial male returns to the territory, the intruder continuing across country. The maximum distance that a territorial male has been observed to chase another individual away from his territory and still return is 200 m. The male the track patterns of which are illustrated ousted a previous male from the territory at 13.20 hours GMT on 7 April. It remained in possession of the territory until going to roost at 16.00 hours. Only a few of its sorties from the territory are illustrated. In all, 35 sorties were executed during the period the territory was occupied. Of these, 3 were directed at female *I. io*, 10 were directed at male *I. io*, 19 were directed at butterflies of other species, and one was directed at a queen bumble bee, *Bombus* sp.

[*Descriptions partly from Baker (1972a)*]

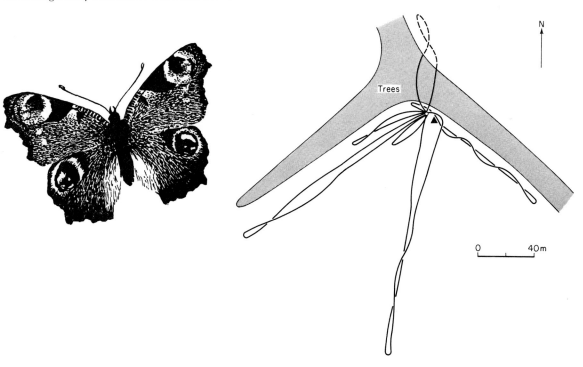

Trees

N

0   40m

similar non-calculated removal migration completely and/or replace it by an exploratory migration and the linear range pattern of lifetime track grades into the limited-area pattern. The gradation between these two types of patterns of lifetime track is illustrated in Fig. 18.6 and emphasises once again the absence of qualitative breaks in the range of migration patterns found in the Metazoa.

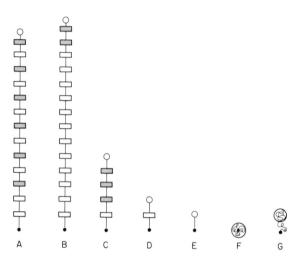

Fig. 18.6   Part of the range of variation in form of the lifetime track shown by the Metazoa

- • = area in which the animal is born
- ▭ = area in which the animal roosts and/or feeds
- ▭ = area in which the animal roosts and/or feeds and/or breeds
- ○ = area in which the animal roosts and/or feeds and/or breeds and/or dies

In addition, the straight line joining these areas indicates rectilinear migration across areas of low suitability for the required resource. The waved line indicates a track pattern with a high rate of turn. Such a track pattern also occurs in all of the areas indicated by rectangles or circles above.

A and F represent the two extremes of the lifetime track which may be described respectively as the linear range and the limited area range. Note that A, in effect, consists of a whole series of limited-area ranges joined together by rectilinear tracks. Gradation between A and F, therefore, involves a reduction in the number and distance of the bouts of rectilinear migration, eventually to zero, as illustrated. Another sequence can also be envisaged, however, in which there is an increase in the angular difference in direction between successive rectilinear bouts with directions following one another in an increasingly random sequence. Similarly, a decrease in rectilinearity when crossing areas of low suitability would also lead to a linear range becoming a limited-area range. Intermediate

examples of the latter two gradations are, however, difficult to find, perhaps because such behaviour rarely can represent an optimum strategy. Consequently, the first gradation sequence is the only one illustrated.

A. Linear range of the type shown by the small white butterfly, *Pieris rapae*, in Fig. 18.1. The direction of rectilinear migration is in a more or less constant geographical direction. This takes the animal from its site of larval development to first a succession of areas suitable for feeding and roosting and then later to a succession of areas suitable for each or any combination of feeding, roosting, and reproductive activity. Note that in many species this pattern includes a seasonal major change of geographical direction that gives rise to a return migration (Chapter 19) albeit usually with a very low degree of return (p. 29). It is also possible that in some species selection has not favoured the rectilinear track being in a constant direction and yet has not favoured the evolution of the limited-area phenomenon (F and G). Such animals would be considered to just wander, a pattern that is often suspected of animals such as snails and polar bears, *Ursus maritimus*, but which has yet to be demonstrated. No matter what the constancy of direction of the rectilinear track, each of the limited areas represented by rectangles and circles may constitute either a temporary familiar area, as in Figs. 18.5, 18.7, and 18.8 or, more simply, an area in which rate of turning is high but in which no marking behaviour or a spatial memory is employed, as perhaps in Fig. 18.4.

B. Linear range of the type shown by the desert locust, *Schistocerca gregaria*, phase *gregaria* (Fig. 18.3), and also the monarch butterfly, *Danaus plexippus* (Fig. 19.5), peacock butterfly, *Inachis io* (Fig. 16.16), and sunn pest, *Eurygaster integriceps* (p. 241). As with A except that ontogenetic and/or seasonal effects result in reproduction being concentrated toward the end of the linear range. In the desert locust, the reproductive part of the range often coincides with a decrease in directional constancy of successive rectilinear tracks (Fig. 18.3). In the other three species, reproductive activity is preceded by a major change in direction of the rectilinear track (Chapter 19).

C. Linear range similar to those in A and B but increased stay-time in each area of high suitability results in a shortening of the cross-country distance travelled. The meadow brown butterfly, *Maniola jurtina* (Fig. 18.4) and other butterflies with a similar ecology (Chapter 16) probably fit this category. Although the spectrum described here concerns a range of movements from hundreds of kilometres in A and B to a few metres in F, absolute distance is not an important part of the spectrum and the gradation is equally applicable to animals that move only metres, centimetres, or millimetres. It is the gradation in sequence of alternation between periods of rectilinear movement with periods with a higher rate of turn in a limited area that is important.

D. Linear range in which the animal performs a migration from the birth site, spends some time in an intermediate area for feeding, and then settles finally in an area suitable for reproduction. Those inter-tidal and benthic marine and freshwater animals that release their young into the plankton may be included in this category.

E. Linear range in which the animal performs rectilinear migration from the birth site and settles permanently in the first, or soon after (Chapter 11), habitat encountered that is above the threshold suitability for all resources. The clearest examples of species with this type of lifetime track are those social insects, such as ants and termites, in which the reproductives make a single rectilinear flight from their birth site, settling permanently in their first, or soon after, alighting area.

F and G. The limited-area range. Species not subjected to selection for rectilinear migration away from their birth site to a non-overlapping area. Such species are subjected to the same selection as species of A to E while in areas of high suitability but because of their longer stay-time are even more likely to evolve exploratory migration and a familiar area. There is no clear distinction between F and G and whether the exploratory migration–calculated migration sequence produces a final familiar area that includes or excludes the birth site depends on the distribution of resources and other individuals. F and G may be characteristic of a wide range of terrestrial, aquatic, and amphibious vertebrates and perhaps also many invertebrates, such as some insects, molluscs, decapod crustaceans, arachnids, and myriapods (Chapter 15).

18.3 The limited-area range

18.3.1 Selection for marking behaviour and/or a spatial memory in connection with the execution of an efficient track pattern within a limited area

All of the animals illustrated in Figs. 18.1 and 18.3–18.5 appear to have been subjected to selection for a high incidence of initiation of migration across relatively large areas of low suitability. Many other similar species, however, do not seem to have been exposed to selection for such a high incidence of migration, and for some it is very rare indeed for such a migration to be initiated. The possible reasons for this variation in selection has been discussed in Chapters 16 (p. 245) and 17 (p. 320).

The remainder of this present chapter is concerned with the selective pressures acting on those species that rarely initiate migration across relatively large areas of low suitability but which instead spend most of their lives in limited areas of relatively high suitability. Such animals, therefore, are continuously subject to the type of selection

described in relation to the meadow brown butterfly, *Maniola jurtina* (Fig. 18.4), and peacock butterfly, *Inachis io* (Fig. 18.5) while within areas of high suitability. Except, that is, that the gross area of high suitability occupied does not fall in suitability below the mean expectation of migration from that area, even though the suitability of sub-areas within the gross area may fall below the level of the mean expectation of migration from sub-area to sub-area within the gross area (p. 368).

The result of such selection is likely to be that such animals spend their entire lives within a circumscribed area performing some combination of the movements illustrated in Figs. 18.4 and 18.5. Such a situation, however, serves as a pre-adaptation upon which selection acts to increase the efficiency of the track pattern in a way that may not arise when stay-time is shorter. The problems encountered by the meadow brown, *Maniola jurtina*, and peacock, *Inachis io*, were those of avoiding areas visited previously, revisiting areas visited previously after an appropriate time interval, and returning to an area previously occupied by as efficient a track as possible. The selection associated with these problems would obviously favour some form of marking system and/or some degree of spatial memory, but is likely to be counteracted by the energy expenditure involved in both adaptations. When optimum stay-time in a given area is short, the decreased efficiency of a track pattern based solely on rate of turning (based either on an endogenous mechanism or on such a mechanism in relation to some environmental clue for orientation) may well nevertheless represent less of a disadvantage than the energy that would need to be expended to leave a scent trail or the energy expenditure or decreased neural capacity for other functions that would be needed to maintain a functional spatial memory. As optimum stay-time increases, however, a track pattern based solely on rate of turning must become increasingly less efficient and is likely at some point to represent a disadvantage that exceeds the disadvantage of marking and/or spatial learning.

There is no indication so far that butterfly species, such as the meadow brown, *Maniola jurtina*, or small white, *Pieris rapae*, that either have a short stay-time in areas of higher suitability or else, as with the female meadow brown, are relatively immobile in such areas, make use of any marking behaviour or spatial memory. There are, of course, some well-documented examples of invertebrates that do make use of a spatial memory during their movements

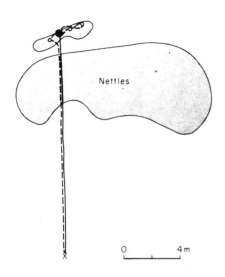

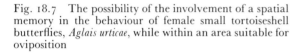

Nettles

0   4 m

Fig. 18.7 The possibility of the involvement of a spatial memory in the behaviour of female small tortoiseshell butterflies, *Aglais urticae*, while within an area suitable for oviposition

Female *A. urticae* lay their eggs in a batch on the underside of a near-terminal leaf of nettle, *Urtica dioica*, as shown in the photograph. In spring, a small, young plant at the edge of a nettle patch is usually chosen. Selection seems to have favoured those females of this species that lay their eggs in large batches rather than dispersed in ones and twos. Oviposition may last between 20 and 90 min depending on temperature and number of eggs being laid. Females may often be disturbed during the period of oviposition, particularly by birds but presumably also, though not observed, by small mammals. Selection for eggs to be laid in a large batch also favours a disturbed female to return and complete oviposition on the same leaf. This imposes selection for either a spatial memory, an orientation mechanism, or a marking mechanism. Some combination of the former two mechanisms seems to have evolved in this species, but scent-marking does not.

The large solid dot in the diagram indicates the oviposition site of a female that was disturbed experimentally by the close approach of the observer after the female had been ovipositing for 10 min. The stippled area represents the limits of the nettle patch. The solid line indicates the track taken by the female when disturbed and the cross represents the site at which she settled. After 150 s at this site she returned by the dashed line, settled on plants indicated by small open circles near to the oviposition site, and after a search of 3 min in this vicinity relocated the leaf on which previous oviposition occurred. Oviposition lasted a further 20 min. Upon completion, the female initiated rectilinear migration away from the area.

If a female is disturbed while ovipositing as described and if, while she is away, the leaf or plant containing the eggs is moved 4 m or more from the original site, the female, upon return, still concentrates her search in the vicinity of the original site. This remains the case even if the eggs are shifted up-wind.

The mechanism by which return is normally achieved has not yet conclusively been demonstrated. Reversal of orientation to some distant clue, such as the Sun or some major topographical feature, could account for the major part of the track. Such reversal of orientation could be combined either with a spatial memory or with a judgement of distance based on an internal clock or some other system such as energy expenditure. Either combination would be sufficient to take the female to the general vicinity of the oviposition site. The plant-removal experiment seems to rule out the involvement of a long-distance pheromone.

[*Photo by R.R. Baker*]

Fig. 18.8 The possibility of the involvement of a spatial memory in the behaviour of the male of the small tortoiseshell butterfly, *Aglais urticae*, in an area suitable for the courtship of receptive females

Soon after mid-day in spring male *A. urticae* establish territories at female roosting sites, particularly patches of nettle, *Urtica dioica*, the larval foodplant. In the deme studied, however, the territorial system tends to break down by mid-afternoon (see p. 408). Females fly in to the roosting area soon after mid-day and spend the remainder of the afternoon thermoregulating with the minimum of flight activity. A receptive female allows a territorial male to take up a position on the ground immediately behind her, and from time to time the male taps the female's wings with his antennae. Once a male has taken up such a position he abandons territorial behaviour for the remainder of the day. Males from neighbouring territories or incoming migrant males searching for a female may discover the pair.

The diagrams and photographs illustrate an interaction observed on 11 May at a large nettle patch at Whitley Bay,

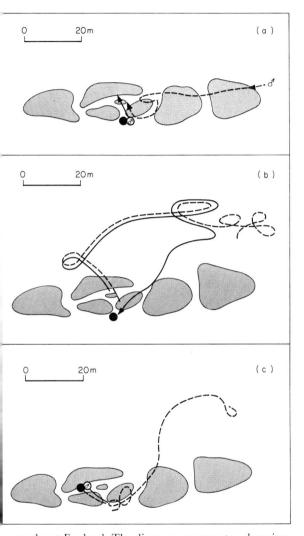

northeast England. The diagrams represent a plan view. The stippled areas indicate nettles.

In the top diagram an incoming migrant male (dashed line) finds a pair by the edge of a nettle patch. In the photograph the male has landed behind the pair but usually the first male flies up before this can occur. Males and females are visually identical and any individual flying up out of vegetation could be either a male or a female. The incoming male (dashed line) follows the first male (solid line) when it takes off. The female remains in the position indicated by a solid dot.

In the middle diagram the first male (solid line) leads the incoming male away from the female and then by a series of sharp turns attempts to disengage, behaviour similar to that shown by non-receptive, and occasionally receptive, females when first found by a male. When the first male has escaped the attention of the intruder the first male returns, from distances as great as 100 m and apparently from any compass bearing, more or less directly to the female that has remained in her original

position, leaving the intruder searching the area where it 'lost' the first male.

In the bottom diagram, as soon as the male settles behind the female he taps her with his antennae and she immediately takes off. The male follows and the pair fly some 5–10 m to a new location. The intruder eventually also returns to the site at which he first encountered the pair but by this time the pair are usually elsewhere. Except for interruptions from other males the pair stay in the position shown in the bottom photograph, shifting only if the Sun's movement causes the pair to become shaded by vegetation. Copulation occurs immediately the female goes to roost.

More experimental work is needed, but the ability of the male to return more or less directly from some distance and apparently any direction to a site previously occupied, often after a tortuous outward track, can only reasonably be interpreted on the basis of some form of spatial memory. If, while the two males are away interacting, the observer disturbs the female and forces her to fly away the males still return to the original site, not to the new site occupied by the female.

[Photos by R.R. Baker]

within a limited area. The most notable of these examples are the hunting wasp, *Ammophila pubescens*, (Thorpe 1950) and the honeybee, *Apis mellifera*, both of which are discussed in Chapter 33. Both of these species, however, seem to live in a more or less permanent familiar area and any removal migration beyond the confines of the previously existing familiar area seems to be preceded by exploratory migration, as already described for the honeybee (p. 150). These species, therefore, probably fall into categories F and/or G in Fig. 18.6. It now seems that some butterflies (e.g. *Heliconius*) also fit into this category (Turner 1971). There is, therefore, an absence of convincing evidence that any species with a linear range makes use of a spatial memory during the period that it is within an area of high suitability. The apparently greater ability of territorial males rather than incoming migrants of the peacock butterfly, *Inachis io* (Fig. 18.5), to return to the territory during their first two territorial interactions and the observation that this differential is a function of the length of the interaction (Baker 1972a) could be taken to indicate the involvement of a spatial memory. Further indication of the existence of a spatial memory in another species with a linear range is given in Figs. 18.7 and 18.8. None of the evidence so far, however, is absolutely conclusive although it is indicated that even species with an extreme linear range (type-A in Fig. 18.6) may make use of a spatial memory and thus form what are, in effect, temporary familiar areas.

If selection can favour the evolution of a spatial memory in connection with the execution of an optimum track pattern in animals that live in a type-A linear range, then it seems inevitable that when selection acts to favour an animal spending almost its entire life within a circumscribed area, it will also favour, given the appropriate pre-adaptations, the use of a spatial memory (and/or marking behaviour which can serve a similar function). Such adaptations themselves serve as pre-adaptations from which complex navigational mechanisms may evolve.

18.3.2 Selection for exploratory migration

Let us consider an animal for which selection on time investment favours alternation between habitat-types A and B. For example, A could be a habitat-type suitable for sheltering and B could

Fig. 18.9   Possible establishment of a familiar area by an individual of the grey field slug, *Agriolimax reticulatus*

General ecology

The grey field slug is perhaps the most abundant British slug, being found in great numbers in gardens, fields, and woodlands. It feeds on a wide variety of fresh and decaying vegetation. The animal is hermaphrodite and grows to a length of 3–4 cm when extended. Eggs are laid throughout the year. Young slugs hatch within a few weeks in spring and summer but only after up to three months in winter. The animals are largely nocturnal, though they also appear during the day when conditions are sufficiently moist. Roosting occurs either in a hole in the ground or under a leaf or stone etc. The young migrate from their hatching site but it is thought that when adult they may establish a familiar area.

Possible establishment of a familiar area

The diagrams show the movements and behaviour of an individual on five consecutive nights in an experimental arena containing topsoil and kept inside a greenhouse. The arena measured 91×68 cm and movements were recorded by time-lapse photography, one frame being exposed every 15 s using high-speed flash illumination.

In the diagrams the solid line indicates the track of the animal on the night in question. The dotted lines indicate the tracks of the animal on previous nights. Diagonally hatched oblongs show the positions of slices of carrot provided for the animal to feed. Open circles show the positions of holes in which roosting could occur.

The inset in each diagram shows the incidence and timing of four different behaviour patterns: cop, copulation; cwl, crawling; feed, feeding; and rest, resting or roosting. Solid shading refers to behaviour on the soil surface. Stippling indicates the period that the animal was below ground. It is assumed that the slug was roosting during these periods though the time spent below ground during night 1 could indicate a period of oviposition. When the slug rested on the soil surface, solid shading is used. The spatial position of an on-surface rest is indicated by a cross in the movement diagram. Similarly, the spatial position of a copulation is indicated by a solid circle. Vertical lines with diagonal edging in the insets indicate the times of sunset and sunrise.

When the slug crawls it secretes a clear, colourless mucus which forms a relatively durable trail. This trail could provide information to the animal concerning its movements over several previous nights via the senses of olfaction, touch, and/or taste. Various movements can be recognised. Directed movements toward co-copulants, food, and roosting holes are evident (Newell 1966) and may be based either on direct perception or on learned locations, or on both. Exploratory migrations are also in evidence and are characterised by a low turning rate and infrequent crossing of an earlier track, thus maximising the area searched (see also Fig. 18.10). A roosting hole to roosting hole removal migration took place on night 4 which may have been a calculated migration as it was to a hole encountered during exploratory migration on night 2. Unfortunately, the final stages of this removal migration were not filmed and the line indicated by a question mark is not a real track. The period with a high turning rate just

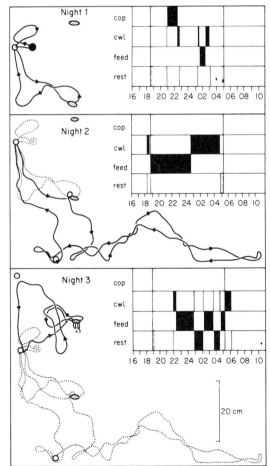

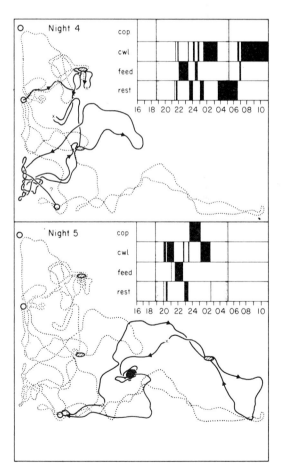

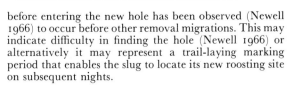

before entering the new hole has been observed (Newell 1966) to occur before other removal migrations. This may indicate difficulty in finding the hole (Newell 1966) or alternatively it may represent a trail-laying marking period that enables the slug to locate its new roosting site on subsequent nights.

[*Figure compiled with modifications from Newell (1966) from which some, but not all, of the interpretations have also been taken as indicated. Photo by R.R. Baker*]

be a habitat-type suitable for feeding. Suppose also that selection favours return migration to examples of both habitat-types for the reasons given on p. 63. For a period, therefore, an individual may alternate between habitat A_1, an example of habitat-type A, and habitat B_1, an example of habitat-type B. Suppose that after a time, B_1 falls below $\bar{E}_B$, the mean expectation of migration spent searching for a habitat of type-B. Optimum strategy for the animal, therefore, is to initiate migration from B_1 to search for another type-B habitat. Suppose, however, that, on average, having found another type-B habitat, return migration to A_1 is still advantageous. Selection would favour an individual that took a track from B_1 that not only maximised the efficiency of search for a type-B habitat but that also kept to a minimum the cost of migration from the discovered type-B habitat (B_2) back to A_1. The same argument applies whether the migration searching for a new type-B habitat is initiated directly from B_1 or after a period spent in A_1. The adaptation that results from selection for non-calculated removal migration (e.g. from B_1 to B_2) of such a form that a calculated (return) migration (e.g. to A_1) still occurs, has been termed an exploratory migration in this book (Chapter 5, p. 26). It follows from the above that selection on track pattern during exploratory migration acts to increase the efficiency of search for one habitat while increasing or maintaining the efficiency of return to another.

Figure 18.9 shows the observed track pattern of an individual of the grey field slug, *Agriolimax reticulatus*, over a number of consecutive days. The experimental design, particularly the use of an arena which limited the movements available to the animal, makes it difficult to extrapolate from these observations to a field situation. There is every indication, however, that each night the animal's time is apportioned between on the one hand feeding, resting, oviposition, and perhaps at intervals, copulation, and on the other hand, exploratory migration. The track patterns associated with feeding and copulation tend to be rectilinear when the destination appears to be within sensory range and to follow previous trails when not. It seems likely that a great deal of exploratory migration is a response to habitat suitability in relation to presence or absence of co-copulants. Whether it happened that alternative feeding and roosting sites were encountered during such exploratory migration or whether exploratory migration was also in part a response to the low suitability of the first roosting site

and/or feeding site cannot be determined. The roosting site to roosting site migration that took place, however, certainly gave every impression of being a calculated removal migration to a destination first discovered during exploratory migration. Finally, the general impression given by Fig. 18.9 is of an animal building up a series of tracks in an area such that movement to one point in the area, such as the roosting site, can be accomplished from any other site in the area. In this case, although the system is perhaps based on trail laying rather than on a spatial memory, the area developed by the slug may still be described as a familiar area.

Duval (1972) has questioned the conclusion that *Agriolimax* lives in a familiar area and suggests instead that the animal wanders at night, albeit circuitously, and at dawn moves in a straight line to the nearest shelter within sensory range. However, Duval measured track pattern by having the slugs crawl on a black plastic sheet. Assuming that such a surface would be assessed by a slug to be an area of low suitability, such treatment could well elicit the removal migration response of the slugs rather than return migration to a roosting site that borders on such an unsuitable area. Once again, therefore, the experimental design raises problems of interpretation and for the moment, especially in the light of observations (Chapter 22) of the seasonal movements of snails (Edelstam and Palmer 1950, Blinn 1963), the view that *Agriolimax* lives in a familiar area based on a system of marking seems the most attractive.

Accepting this for the moment, it is possible to consider the optimum form of the track pattern for exploratory migration when the search track is marked by a trail as in the slug. The area searched is maximised by not researching areas previously searched (Fig. 18.2). Re-tracing the outward track is therefore disadvantageous. Yet at the same time the animal has to regain contact with its earlier trail in order to retain contact with its familiar area. Optimum track pattern seems to be initially rectilinear followed by a rate of turn sufficiently gradual to maximise the search area (Fig. 18.2) yet sufficiently marked inevitably to lead the animal back to its outward trail. Upon encountering the trail the search area is maximised by crossing the trail at right angles and then initiating a second turn in the opposite direction to the first until the outward trail is encountered yet again. The process is then repeated until the previous familiar area is regained. The envisaged form of the optimum track for exploratory migration in a trail-producing animal is indicated in Fig. 18.10 and can be seen to approximate well to the track pattern of the slug in Fig. 18.9 during those periods that it is in an area that it has not previously visited.

Animals using navigation mechanisms other than trail-laying are likely to have different optimum track patterns during exploratory migration that are best discussed in relation to the navigation mechanisms themselves (Chapter 33).

Fig. 18.10   Optimum track for exploratory migration in a trail-laying animal

FA = the pre-exploratory migration familiar area. Solid line with arrows = outward and return tracks during exploratory migration. The width of the track represents the extent of the animal's sensory range.

Exploratory migration is a series of non-calculated migrations followed by a return either to the starting point (as here) or to one of the habitats visited earlier during the exploratory migration. Selection during exploratory migration acts to maximise the area searched while at the same time keeping the migration cost of the return migration to a minimum.

In a trail-laying animal the return migration has to maintain contact with the outward trail without re-searching the same area traversed on the outward journey. The track shown, consisting of a near-rectilinear outward track followed by a return track with sufficient rate of turn in first one direction and then the other to cross the outward track at infrequent intervals, seems to be optimum for a trail-laying animal, though other configurations based loosely on this pattern seem to be equally advantageous.

These arguments apply to any trail-laying animal, not only slugs.

[*Photo by R.R. Baker*]

18.3.3 The familiar area

The familiar area could be defined as the total area ever visited by an animal during its lifetime. In this case it would be equivalent to the lifetime range postulated by Jewell (1966). However, this would imply that the animal remained equally familiar with every area it ever visited, irrespective of the time interval involved. Although the term 'lifetime range' is descriptively useful, therefore, it is necessary to make a distinction between it and the familiar area. For present purposes it is convenient to define the familiar area as that portion of the lifetime range from any point in which an animal is capable of finding its way to any other point. Note that there is also a difference between lifetime track and lifetime range. The lifetime track is the track in space made by an animal's body during the period from birth to death. The lifetime range is most usefully defined as the area that comes within an animal's sensory range during the execution of the lifetime track. This definition of lifetime range is particularly necessary if the familiar area is to be considered as part of the lifetime range.

It is usually extremely convenient to think of the familiar area as an area in the form of a circle or ellipse with discrete boundaries, with one side of the boundary outside the familiar area and the other side inside the familiar area. As pointed out by Ewer (1968), however, in relation to territories, this is rarely the case. The reader has only to think of his own familiar area to realise that in reality the area consists of a number of reasonably discrete areas that are connected to one another by relatively narrow tracks. The areas between these regions are just as unfamiliar to the reader as any of the regions outside of a circle containing all of these smaller familiar areas and tracks.

It is instructive to examine the way that an animal builds up a familiar area, using the arguments presented in Chapters 14–16. Unfortunately, there are very few examples of the way in which this occurs in an individual followed from birth. Figure 18.9 provides some indication of the processes that are likely to be involved for the grey field slug, *Agriolimax reticulatus*. Drabble (1973) also provides an account for a female stoat, *Mustela erminea*, albeit under highly artificial surroundings. In both cases the sequence starts with the animal in a roosting *cum* sheltering site from which movements are made first within what seems likely to be the sensory range from this site. After scent-marking or trail-laying in this area, exploratory migrations are performed, during

which ability to return to the roost/shelter appears to be at a premium. During the course of these exploratory migrations a range of potential sheltering and feeding sites are encountered. Should any of these sites exceed the threshold for calculated removal migration (p. 347) then the new site is used rather than the old, a migration that appears to occur in the slug sequence in Fig. 18.9.

Unfortunately, however, neither of these observations was made in the field and in neither case was an animal observed from birth. We may surmise that the same procedure is involved, though it would be satisfying to have this confirmed in the field. In mammals, of course, such as described for lions, *Panthera leo*, by Schenkel (1966), and many birds, such as described for geese and swans by Hochbaum (1955) and for the reed warbler, *Acrocephalus scirpaceus*, by Catchpole (1974), the young animals establish a basic familiar area through accompanying their parents or other older individuals (see also Chapter 15). In the absence of detailed field observations, however, we may conclude that in those other vertebrates and invertebrates that do not show parental care, do not initiate immediate non-calculated removal migration, and are not subjected to selection for dispersal (Chapter 11), the following sequence of events is normal. The animal hatches from the egg and immediately searches for a suitable shelter, though in many species the eggs occupy a suitable shelter, as in slugs, or the empty egg may continue to serve as a shelter, as in the Sundanese rock python, *Python bivittatus* (p. 195). From this shelter, the animal moves to food where this is within sensory range or performs exploratory migration where it is not. The animal then alternates between feeding habitat and roosting habitat until the suitability of one or the other falls below the exploratory migration threshold for that habitat-type, perhaps due to increased size of the animal, or to increased competition, etc. Of course, most animals are adapted to a wide variety of habitat-types. Hence, there is no single feeding habitat-type but rather one habitat-type that provides one type of food, another type another type of food, and so on. Similarly, there are habitat-types that provide shelter suitable at one time of day but not at another. The young animal is therefore likely to continue to perform exploratory migration until it has encountered at least one example of each habitat type, to which the species is adapted, of a suitability greater than the exploratory migration threshold for that habitat-type. Exploratory migration can then cease until the suitability of one or other of the habitat-types again falls below the exploratory migration threshold. When selection for dispersal is involved, those exploratory migrations are favoured that are in a particular direction (e.g. Chapter 27, p. 654).

Lack of data concerning exploratory migration is largely the result of the virtual impossibility of monitoring this particular behaviour for most animals in the field. Evidence is mounting, however, that the *wheel running behaviour* of small mammals in cages, instead of being a general reflection of activity level, is in fact a specific reflection of reduced exploratory migration threshold (Baker *et al.* 1977). The phenomenon seems to be comparable to that of Zugunruhe and the seasonal migration threshold of birds (p. 628). In which case, the way is now open for laboratory studies to identify the way that the incidence of exploratory migration varies with time of day, age, sex, season and other environmental events.

Research into exploratory migration, as measured by wheel activity, has so far been confined to the golden hamster, *Mesocricetus auratus*, and mongolian gerbil, *Meriones unguiculatus*. Not only has a clear diel, ontogenetic, and sexual pattern emerged, but also it has been found that different environmental events (e.g. variation in food availability; appearance and disappearance of other individuals and/or their artefacts) influence differentially the time invested in wheel activity at different times in the diem. In both species it appears that time invested in exploratory migration during a clearly demarcated part of the night is concerned primarily with monitoring the presence, absence or movements of conspecifics whereas time invested during the rest of the night is concerned with the location of potential feeding and nesting sites. Ontogenetically, total daily wheel activity first appears at about the time of weaning, increases to a peak at about the age the animal would be establishing its own home range, and thereafter decreases until about the time of sexual maturity. Thereafter, the time invested stays at a low-level (zero or near-zero in male gerbils) if the individuals are allowed contact with members of the opposite sex but steadily rises again if not allowed such contact. After copulation, female hamsters for a few days show an increase in wheel activity at the time of night that seems to be sensitive to environmental events associated with the 'nest' site. This increase could indicate an increase after copulation in exploration for potential nest-sites suitable for parturition. A few days before parturition wheel activity drops to near-zero at all times of night, not normally increasing again until the young approach weaning.

One facet of exploratory migration and the establishment of a familiar area has not so far been mentioned but attains considerable importance later in this chapter. This is the optimum strategy for an animal in relation to all those examples of a particular habitat-type that are encountered during exploratory migration but which are not immediately exploited due to a suitability less than the currently most suitable habitat of that type with which the animal is familiar. All the examples of a particular habitat-type encountered by an individual during exploratory migration could be ranked by the animal according to the product of their suitability and the migration factor for the migrations necessary to exploit the habitat (for ease of description in the following paragraphs both of these factors will be assumed to be included in the term suitability). Time could then be invested by the individual in the most suitable of these habitats until its suitability fell below that of the second most-suitable habitat, whereupon the animal could from then on invest time in the previously second most suitable habitat. This process could then be repeated until none of the habitats of that type known to the animal were of a suitability above the exploratory migration threshold for that type. This process is referred to for shorthand in the remainder of this chapter as 'ranking strategy' or as the 'ranking system'. Whenever selection has favoured such a strategy, however, it is also likely to have imposed further selection on the exploratory migration process. Habitats change in suitability with time. The accuracy of an animal's ranking of a series of examples of a habitat-type encountered during different bouts of exploratory migration is likely to decrease with time after first assessment. Ranking may be sufficiently advantageous, therefore, in that it enables the most efficient exploitation of a range of habitats of a particular type, to impose selection for each habitat to be re-visited from time to time solely for the purpose of re-assessment (though some exploitation during the visit may be advantageous if it offsets to some extent the cost of the re-assessment migration). The optimum frequency for such visits would be a function of the average rate of change of suitability. Presumably, also, the higher-ranking habitats should be re-visited more frequently than the lower-ranking habitats, for the former are more likely to be required for exploitation than the latter.

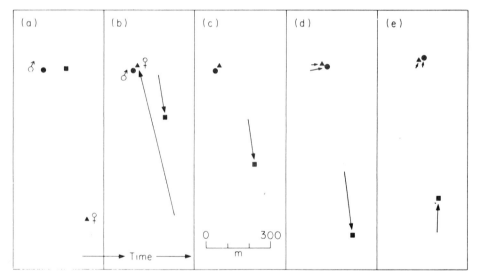

Fig. 18.11 Mode of function of the habitat ranking system in a field situation: calculated migration of man with respect to male and female roosting sites and food-growing sites in a deme in highland New Guinea

Solid square = centre of agriculture; solid triangle = roosting site of the adult and young female humans, male young less than 5 years old, and pigs; solid circle = communal roosting site of male humans older than 5 years.

The diagram shows the successive changes in position of these habitat-types within the familiar area of the deme studied. The top photographs show a female roosting area (or 'village') and the process of felling trees. The bare areas on the hillside in the distance are probably old centres of agriculture (or 'gardens') that have been abandoned. Smoke can be seen arising from an area that is probably being cleared to become a new garden. The lower two photographs show a male building a hut and a female planting sweet potato.

New Guinea highlanders practise 'extensive' or 'slash-and-burn' agriculture. Cultivation involves land clearing and burning. Seedlings are planted in holes made with sticks and the land is allowed to continue bearing until the yield decreases. No attempt is made to fertilise the soil artificially and the re-establishment of humus and mineral content depends entirely on the area being left fallow. Depending on the nature of the soil, an area is cultivated with short fallow for 4–10 years and then left fallow for 5–20 years and necessarily involves movement of the old settlement as the distance from roost to gardens increases beyond a certain limit (diagrams a to b).

As far as shifts in gardens are concerned, it is possible to envisage that within the familiar area around the current roosting site, areas potentially suitable for cultivation are ranked according to $h_g M_{r-g}$ (where h_g is the suitability of the area for clearing and cultivation and M_{r-g} is the migration factor for daily return migration between roosting site and garden). All of these areas will be re-visited from time to time for the purpose of re-assessment. As $h_c M_c$, the product of the suitability and migration factor for the area currently being cultivated, falls below the highest value of $h_g M_{r-g}$ for another area, the old area will be abandoned and the new area cleared and cultivated. This process will continue (b to d) for as long as areas are present for which $h_{r_1} h_g M_{r_1-g}$ (where h_{r_1} is the suitability of the present roosting site and M_{r_1-g} is the migration factor between the current roosting site and garden) is greater than $h_{r_2} M_{r_1-r_2}/h_{r_1}$ (where h_{r_2} is the suitability of the next most suitable roosting site and $M_{r_1-r_2}$ is the migration factor for shifting the roost from r_1 to r_2). When this ceases to be the case the roosting site will be shifted instead of or as well as the cultivation site (a to b). The nature of the complex of migration thresholds which form the basis of the mechanism by which such calculated migrations are initiated has been discussed in Chapters 14–17.

The existence of such a ranking system of habitat suitability is clear for humans and is probably the major system involved in calculated migration in all animals, whether vertebrate or invertebrate, that live in a familiar area.

[Photos by courtesy of the Trustees of the British Museum. Diagram based on figures and data in Brookfield (1968)]

The way in which the ranking system is envisaged to work and its involvement in an observed migration sequence is described in Fig. 18.11.

Re-visiting an area previously encountered during exploratory migration solely for the purpose of re-assessment provides a ready explanation for some animal movements that otherwise would be difficult to interpret. The gorilla, *Gorilla gorilla*, for example (Fig. 18.13), lives in a familiar area characterised by food that is so abundant, or at least appears so to an observer (Schaller 1963), that there is no attempt to exclude other gorilla groups from access. Most observers have remarked that gorillas move far more than seems to be necessary to obtain required resources (Schaller 1963). The chief advantage of a large familiar area to a gorilla seems to be that it permits optimum contact with other groups the composition of which is known (p. 162). Perhaps the movements of gorillas within their familiar area are adaptive to enabling the group to re-assess the suitability of other groups in the light of potential future changes in group composition. In this case the frequency of return to a particular area of forest (Fig. 18.13) could well reflect the rate at which gorilla groups are likely to change in composition through predation, births, or group-to-group removal migration.

Before leaving this discussion of exploratory migration and the familiar area, there is one further point that has already been encountered but which can now be considered further. It has been demonstrated for at least one species (p. 136) that exploratory migration is characterised by a low selectivity of the suitability of areas visited, though stay-time may be greater in some areas than in others. A low selectivity of areas visited is, of course, inevitable in any non-calculated migration, and exploratory migration can be no different in this respect from removal migration. Only in calculated migration can there be a high selectivity of destination and it was with the comparison of calculated and exploratory migration that we were concerned in Chapters 14 and 15. However, although selectivity of areas visited must be low, it is to be expected that stay-time in each area along the exploratory migration track will be subject to the same selective pressures already described for non-calculated removal migration (p. 82). The observed characteristics of exploratory migration appear to be quite consistent, therefore, with the model developed in this book.

Not only exploratory migration, however, but

also re-visits to an area for re-assessment of suitability as part of ranking strategy must also be characterised in comparison with calculated removal migration by a relatively low selectivity of destination. However, to some extent this is offset when compared with exploratory migration by the likelihood that the more suitable habitats will be re-visited more frequently than the less suitable habitats. Such a system makes sense when analysed in terms of the familiar area. Suppose, however, that we break the visits for re-assessment down into their removal migration components. We then have a situation in which an animal performs a calculated removal migration from an area of higher suitability (the habitat currently being exploited) to an area of lower suitability (a habitat lower in the rank-order of suitability). This migration, furthermore, may be

supposed to have a migration factor of less than unity. Such a migration is an exception to the migration equation as formulated in Chapter 7, and can only be accommodated by adding to the suitability of the habitat visited a value that takes into account the advantage of the visit at the familiar area level as part of the ranking system. Although exploratory migration, therefore, can be broken down into removal migration units all of which obey the migration equation, this is not the case for calculated migrations for re-assessment. Here, therefore, is an example of a migration that has evolved as a result of selection on exploratory, removal and return migrations, but which, upon producing a familiar area, has initiated feedback selection on migration at the level of its removal migration components.

Fig. 18.12 Some examples of exploratory and re-assessment migrations in man

A variety of human movements can be interpreted in terms of exploratory migrations and in terms of re-visiting for the purposes of re-assessment and ranking. Exploratory migrations are characterised by a low selectivity of destination (p. 136). Only the areas that are ranked high in suitability with respect to a specific resource, however, are likely to be re-visited for exploitation or re-assessment at relatively frequent intervals.

[Photos by J. Twiselton and the NASA Johnson Space Center]

18.3.4 Home range

A picture emerged in the previous section of an animal adapted to a particular range of habitat-types such that selection favoured those individuals that lived within a relatively circumscribed area. Such an animal contrasts with those for which selection favours individuals with a linear range. Selection that favours movement within a relatively circumscribed area invariably favours the evolution of trail-laying and/or a spatial memory and consequently a sequence of exploratory migrations and calculated migrations that results in the establishment of a familiar area. A further picture has emerged of this familiar area as a collection of habitats, a proportion of the total ever encountered. Some of these habitats are being exploited. Others are not being exploited but are re-visited from time to time for the purpose of re-assessment. Yet others are neither being exploited nor re-visited though their spatial position and characteristics are stored with varying degrees of detail in the animal's memory. Finally, there are those habitats which are part of the lifetime range (Jewell 1966) but about which the animal no longer retains information within its memory concerning either spatial position or characteristics.

According to this picture, an animal visits only a proportion of the habitats within its familiar area during any particular limited period of time. The collection of habitats visited during a stated period is referred to here as the home range. Other zoologists may or may not feel that this view of the home range concept has advantages over those propounded by other workers. It is suggested here because it facilitates some of the descriptions that follow and forms a useful link between the familiar area concept and the territory concept. Hence, those habitats within the familiar area visited during the course of a single day may be referred to as a day's home range. Those visited during the course of a season may be termed a season's home range, and those visited during the course of a year a year's home range, and so on. Note that the habitats visited by an animal such as a small white butterfly, *Pieris rapae* (Fig. 18.1), or a locust (Fig. 18.3) during the course of a day cannot strictly be termed a day's home range because, as far as is known, they are not part of a familiar area. In this book, the home range is an inseparable part of the familiar-area concept. If a term is required for the habitats visited by an animal such as the above in the course of, say, a day then it is suggested that the word 'home' is dropped and the area referred to as the

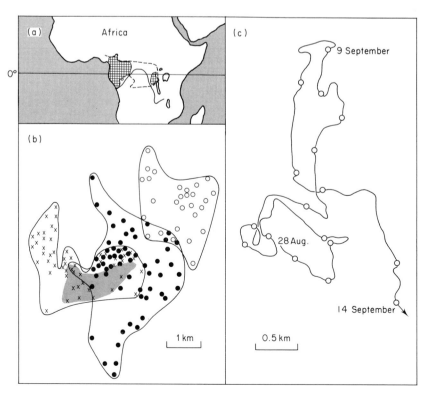

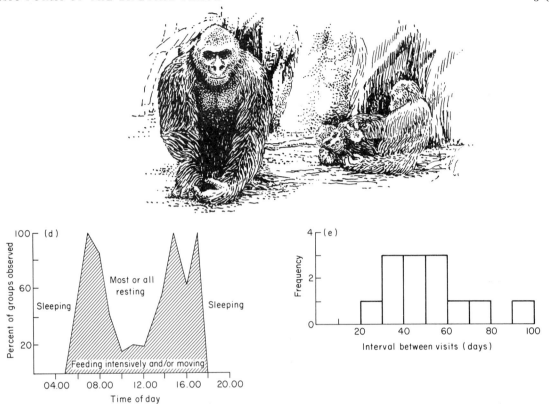

Fig. 18.13   The home range of the gorilla, *Gorilla gorilla*

(a) *Present geographical distribution*
Hatched area, distribution; dashed line, approximate
limit of rain forest. The majority of gorillas throughout the
range are found in lowland equatorial rain forest at
altitudes of 500–1700 m. Gorillas do not favour the mature
forest but prefer the secondary forest where herbs, shrubs,
and vines are plentiful. The food is leaves, stems, bark,
pith, and roots, and some fruit.

(b) *Home range and group composition*
Home range of three social groups near Kabara, north of
Lake Tanganyika. Each symbol (open circle, solid circle,
or cross) represents one encounter with, or definite sign of,
a particular group. Home range of a single group may be
25–40 km². The home ranges of different groups may
overlap only slightly or considerably. Up to six groups
may use one area of forest. Groups with overlapping
ranges may meet and mingle briefly and without aggression from time to time but most often part without changing composition. Group size in the region varies from
5 to 27 ($\bar{x}=16.9$) individuals consisting of: 1–4 ($\bar{x}=1.7$)
silverbacked (more than 10 years old) males; 0–3 ($\bar{x}=1.5$)
blackbacked (6–10 years old) males; 2–10 ($\bar{x}=6.2$) females (more than 6 years old); 1–5 ($\bar{x}=2.9$) juveniles (3–6
years old); and 1–7 ($\bar{x}=4.6$) infants (0–3 years old).
Females reach sexual maturity at 6–7 years of age and
males at 9–10. Females give birth about every 3–4 years.
Young are born throughout the year with a possible peak
in July to September. Longevity is generally about 25–30
years for those that reach maturity. Gorillas are killed and

eaten by two main indigenous predators, the leopard,
Panthera pardus, and man, *Homosapiens*. Stippled area,
see (c).

(c) *Track pattern*
Track pattern of a gorilla group on the slopes of Mt.
Mikeno behind Kabara, 28 August to 14 September,
within the stippled area shown in (b). Each circle
represents one night's roosting site. The gorilla group is
cohesive, group diameter rarely exceeding 70 m. Daily
movement in the region ranges from 100 to 2000 m
($\bar{x}=536$ m) at a mean walking speed of 0.16 km/h. Groups
moving, without feeding or pausing, from one part of the
home range to another move at a speed of 3–5 km/h.

(d) *Diurnal variation in movement*
Gorillas sleep from about 18.00 to 06.00 in constructed
roosts. Peak feeding and movement occurs in mid-
morning and late afternoon.

(e) *Frequency of return*
Frequency of intervals between arrival dates at a particular area of forest on the slopes behind Kabara. Time
interval between successive dates of arrival of a group
varied for three groups from 24 to 92 days. Having arrived
in the area a group may remain for from 2 to 26 days.
There was no indication of any seasonal variation of the
components of movement in the area. The periodicity of
gorilla return migration is irregular and fluctuates about a
mean of about 52 days.

[*Compiled from Schaller (1963, 1965)*]

day's range. Note also that non-calculated removal migration and exploratory migration always take the animal beyond the limits of both the previously existing home range and familiar area. Calculated migration, however, although it can never be beyond the familiar area, may or may not be beyond the home range for a particular period.

Track pattern within the home range is usually of similar form to that described for animals with a linear range while within an area of high suitability. Indeed, as execution of the track may in some species involve a spatial memory (Figs. 18.7 and 18.8) the limited area occupied by such species during periods between successive rectilinear migrations may quite justifiably be described as a familiar area and home range. Optimum track within a home range is likely to be a function of the nature of the resources sought and their spatial distribution. Apart from the presence of mechanisms for navigation and the higher proportion of rectilinear calculated migrations to known destinations, therefore, little difference is to be expected between track pattern within a home

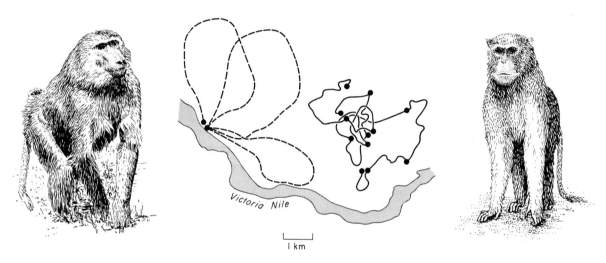

Fig. 18.14 Track pattern of the patas monkey, *Erythrocebus patas*, and the savanna baboon, *Papio cynocephalus*, in relation to areas suitable for roosting and feeding.

The diagram shows the daily track patterns of a social group of the patas monkey (solid line) and another of the savanna baboon (dashed lines) at Chobi, Murchison Falls National Park, Uganda. Solid dots indicate overnight roosting sites.

Both track patterns illustrate adaptation to minimum re-searching of areas already visited while searching for food in areas of relatively low suitability. Differences in track pattern may be attributed to the spatial distribution of the most suitable roosting sites.

The patas monkey is adapted both physically and socially to a full-time existence in a grass and woodland savanna habitat-type and has a continuous distribution over the African savanna north of the equator. The animal is capable of running extremely fast and is half the size of a baboon, the male weighing 13 kg and the female 6·5 kg. The food consists mainly of plant material but is supplemented by insects and lizards. Heterosexual-group size ranges from 9 to 31 ($\bar{x} = 15$) individuals, containing only a single adult male. Females determine the time and direction of movement, the male serving as a look-out on the periphery of the group. A day's movement ranges from

0·5 to 12 km and the home range may be as large as 52 km^2 for a group of 30 individuals. A shortage of very suitable roosts, mainly trees, seems to have resulted in the absence of selection for return to a previously occupied roosting site. The group disperses through a number of trees to roost.

The baboon occupies the whole of Africa south of the Sahara and is less specialised to savanna conditions than the patas monkey. As well as savanna it also inhabits veldt, woodland, and sometimes forests, though not equatorial rain forest. Adaptation to the high predation pressure of the savanna has centred more on defence than speed and has resulted in large multi-male heterosexual social groups, often consisting of up to 200 individuals but with a mean group size of 25–30. The food is again mainly plant material but is supplemented by a wide variety of animal material, often including relatively large mammals which are hunted by cooperating groups. Baboon groups are regular visitors to standing water and sometimes rivers, and where particularly suitable tree roosting sites are present in the home range, which averages about 8·4 km^2, the group may return to the same roost on several successive nights. A day's movement ranges from 0·5 to 12 km.

[*Modified from K.R.L. Hall (1965)*]

range and track pattern within a range. Hence, the track patterns of the gorilla group in Fig. 18.13 and that of the patas monkey, *Erythrocebus patas*, in Fig. 18.14 show many similarities to that of the male meadow brown in Fig. 18.4. In all three cases the required resources are more or less homogeneously scattered through the area and selection against return to a place just visited decreases with time after first visit. When one resource is limited in abundance and spatial distribution, however, periodic return to the limited resource is likely to be favoured as in the peacock butterfly, *Inachis io* (Fig. 18.5) and the group of savanna baboons, *Papio cynocephalus*, shown in Fig. 18.14.

The home-range concept is descriptively very useful and is made use of by many zoologists, particularly mammalogists, herpetologists, and, to some extent also, ornithologists. Most recent workers, however (e.g. L.E. Brown 1962, Jewell 1966, Vaughan 1972) have stressed the problems involved in the application of the concept. Most of the academic problems result from the fact that home range cannot be divorced from time, though in practice even greater problems are generated by shortcomings in the various methods of measuring and calculating home range. If we restrict ourselves here to the more academic problems, it can be seen that as an animal visits and re-visits more of the habitats within its familiar area the measured home range is likely to increase with time. On the other hand, if a restricted-period area (e.g. a day's home range) is measured over a longer period, the home range is likely to shift as exploratory migration and calculated removal migration take place (Fig. 18.15).

Fig. 18.15   Shift in home range with time: the history of a female bank vole, *Cleithrionomys glareolus*, on Skomer Island, Wales

General ecology
The bank vole is an inhabitant of the deciduous woodland zones of the western Palaearctic, though it extends also into the taiga zone. It is found north to the Arctic circle. It occupies a variety of habitat-types but occurs especially where there is a well-developed shrub layer. It feeds largely on bark, bulbs, and roots in winter and green vegetation in spring. The rest of the year the diet consists mainly of fruits, seeds, buds, and insects. It breeds from April to October and produces several litters of 3–6 young. The main predators seem to be mustelid carnivores and owls.

Home range
The figure shows the trap-revealed range of a female vole from July 1962 to August 1963. The female was already a breeding adult in the summer of 1962 and survived to breed again the following summer. It is tempting to suggest that the area occupied from July 1962 to April 1963 represented more or less the limits of the then familiar area of this individual, and that although shifts in home range took place during this period they were restricted to the familiar area as it existed by July 1962. With the advent of the breeding season in May, however, removal migration took place, presumably to an area encountered during exploratory migration in advance of the breeding season. During the course of the breeding season, the exploratory migration–calculated migration sequence appears to have continued, resulting in a gradual shift in home range which by August was several metres outside the limits of the familiar area as it appears to have existed only four months earlier.

[*Simplified from Jewell (1966). General ecology from Corbet (1966)*]

In view of the fact that the home range of an individual animal is a function of time, great care has to be exercised in comparing the home ranges of different individuals, different sexes, and different species. It is necessary also to bear in mind that although the home range conceptualised as an area having just length and width may be of considerable analytical convenience to a zoologist, unfortunately many animals also have a marked vertical dimension to their home range. Theoretically, comparisons should really involve volumes, not areas.

Fortunately, most size variation in home range between sex and species is sufficiently pronounced to be demonstrable despite all of these problems. Basically, seven variables can be recognised that have a major influence on the size of the home range: sex, season, availability of resources, body size, feeding strategy, optimum size of the social group, and deme density. This last variable is considered in detail in the next section in relation to territory rather than here.

Where males and females occupy separate home ranges, it is invariably the male that has the larger range. The selective pressures acting on a female with respect to size of home range are almost entirely those concerned with access to sufficient of the resources (food of sufficient quantity, quality, and diversity; water; roosting sites; etc.) to which the species has become adapted to ensure maximum fitness for herself and her offspring. The male also is subjected to these selective pressures, though because the male often is not accompanied by offspring these pressures are unlikely to generate a larger home range in the male than the female, unless this results from selection associated with body size described below. In addition to these pressures, however, the male is exposed to sexual selective pressures that do not act upon the female. Male reproductive success is a function of the number of females inseminated. The larger the male's home range the greater his female catchment area and the greater, therefore, his reproductive success.

Selection for increase in size of home range is, of course, counteracted by the associated increased cost of moving within and maintaining familiarity with a larger area. The evolved size is likely, therefore, to represent the optimum compromise between these selective pressures. Nevertheless, the

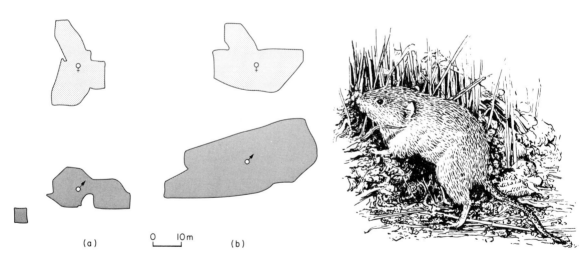

(a)   0   10m   (b)

Fig. 18.16 Increase in male home range during the breeding season: the history of a male and female bank vole, *Cleithrionomys glareolus*, on Skomer Island, Wales

General ecology
See Fig. 18.15

Home range
The figure shows the trap-revealed range of a male and female vole from September 1962 through August 1963. Both animals were immature adults in the autumn of 1962 and did not begin to breed until May 1963.

(a) During the non-breeding season from September to April, male and female voles have home ranges of similar size.

(b) During the breeding season, May to October, however, the male home range increases in size much more than that of the female. The result is that during the breeding season the male home range overlaps the home ranges of several females. The voles illustrated in (b) were trapped from May through August.

[*Simplified from Jewell (1966)*]

extra component for males to the array of selection for increased size of home range seems likely to be sufficient to generate different optima for size of home range in males and females. In species with a marked breeding season, the extra component acts on males only in relation to the breeding season home range. In many species, however, it may well happen that the territorial considerations discussed in the next section compel the male to maintain the home range at its breeding season size throughout the year. In other species, however, as in Fig. 18.16, this is evidently not the case and very often there is a pronounced increase in the size of the male home range at the approach of the breeding season.

The influence of body size and feeding strategy on the size of home range can be considered together and were identified by an analysis carried out by McNab (1963). In general, larger animals occupy a larger home range and 'hunters' occupy a larger home range than 'croppers' (Fig. 18.17). More recent authors (e.g. Jewell 1966) consider McNab's analysis as exploratory and indeed, as described below in another connection, McNab's formula cannot account for the observed home-range size of many mammals, particularly the larger carnivores. Nevertheless, at least until contrary evidence is forthcoming, McNab's general conclusions as phrased above can perhaps be accepted even though the details do not withstand close examination.

One of McNab's conclusions (Fig. 18.17) was that home-range size was a function of basal metabolic rate. From this conclusion it follows that any environmental event, such as for many mammals a period of low ambient temperature, that increases basic metabolic rate, also increases the optimum size of the home range during that period. As McNab pointed out, however, so many other factors are likely to influence optimum home-range size at different seasons that the contribution of change in metabolic rate would be difficult to detect.

It is clear that for many animals size of home range shows a marked fluctuation with season. Seasonal influences, however, are likely to be varied and simple measurement of changes in home-range size out of context of the familiar area rarely succeed in identifying the adaptive nature of the change. Indeed, very often superficially similar changes in home-range size could result from quite different environmental pressures. Suppose that rates of decrease in habitat suitability vary with season. It may be that at some times of year animals have to exploit and re-visit a greater proportion of habitats within

their familiar area than at other times. On the other hand, the mean expectation of exploratory migration may be greater at some times of year than at others (e.g. autumn for black bears, *Ursus americanus*, and many birds—pp. 331–2) thus causing a seasonal change in the exploratory migration threshold and hence also in exploratory migration incidence. Both of these seasonal fluctuations in environmental pressures would be reflected by seasonal change in size of home range as determined by conventional

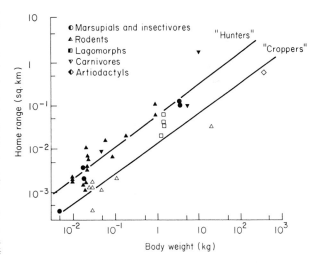

Fig. 18.17 The relationship of body size and basal metabolism to size of home range in some North American mammals

Open symbols, croppers; closed symbols, hunters.

The relationship between body weight and size of home range is significantly different ($P < 0.05$) between mammals with a hunting subsistence (i.e. feeding on seeds, fruits, and insects and larger animals) and mammals with a cropping subsistence (i.e. feeding on grass and other leaves and stems etc.) such that for a given body size a hunting animal has a larger home range than a cropping animal.

The relationship between body weight and size of home range is given by $R = 5.06 \, W^{0.71}$ for hunters and $R = 1.21 \, W^{0.69}$ for croppers (where R = area of home range in hectares, and W = body weight in kilograms). The values of 0.71 and 0.69 are not significantly different ($P > 0.05$) either from each other or from the power of 0.75 by which body weight can be related to basal metabolism according to the equation $M = 70 \, W^{0.75}$ (where M = basal metabolism in kcal/day). Substituting, therefore, a relationship is also obtained between size of home range and basal metabolism such that $R = 0.08 \, M$ for hunters and $R = 0.02 \, M$ for croppers.

[*Modified from McNab (1963)*]

measurements that cannot distinguish movements within and outside the familiar area. Yet on the one hand in the above situations it is harsh conditions, either as a result of giving greater necessity to exploit and re-visit more of the existing familiar area or as a result of exceeding the exploratory migration threshold, that leads to increased size of home range. On the other hand, it is good conditions that, as a result of decreasing the exploratory migration threshold, similarly leads to a measured increase in size of home range.

In the above case, quite different situations were postulated to lead to the same change in size of home range. The converse can also apply. Consider, for example, the effect of seasonal change in the abundance of food. In perhaps the more usual situation, such as that shown in Fig. 18.18, a seasonal decrease in food abundance leads to an increase in size of home range. This change is perhaps typical for the dry season in seasonally arid areas and for the winter season in temperate and sub-polar areas. The wolf, *Canis lupus*, for example, has a much larger range in winter than in summer (Thompson 1952). Similarly, the closely related North American carnivore, the coyote, *C. latrans*, has a home range that in summer, from May through August, may be as small as 5–7 km in diameter but which for the rest of the year may be as great as 40–50 km in diameter. In such species it may be assumed that the increased migration cost that results from extension of the size of the home range is more than compensated by the gain in energy uptake that results from exploiting more habitats. The balance point must exist, however, and it is not surprising that in some other species the increased migration cost appears to outweigh the gain in energy uptake such that the home range is smaller in winter than in summer. An adult female North American badger, *Taxidea taxus*, for example, that was tracked in east central Minnesota from a radio transmitter on a shoulder harness was found to occupy a home range of 750 ha during the period from 29 July to 27 September (Sargeant and Warner 1972). From 28 September to 30 November, however, the badger restricted her movements to an area of 52 ha. Finally, in winter, from 2 December to 9 January, the home range was only 2 ha. The habit of hibernation exhibited by many temperate mammals and reptiles can perhaps be viewed as an extreme manifestation of selection for a decrease in size of home range in winter. The same can perhaps be said of aestivation in seasonally hot and arid areas.

It follows from McNab's analysis that the more individuals that occupy the same home range, the larger the home range must be to offer an optimum supply of resources. Animals that live in a social group should occupy a home range such that the larger the social group (perhaps measured in terms of kcal/day for the total group rather than in terms of number of individuals, though the latter is a more accessible measure) the larger the optimum home range. The larger the home range, however, the greater the cost to the individual of migration within that home range. Selection on group size, therefore, consists of two major conflicting components. Selection for increased group size due to greater protection from predation and possibly also greater efficiency of resource exploitation (e.g. more efficient location of resource; better defence of resource against competitors) is counteracted by selection due to increased migration cost per individual. The group size and the size of the home range occupied should represent the optimum compromise between these selective pressures. Group composition is also likely to be sensitive to these selective pressures because males and females of different ages are likely to have different energy requirements and different predator detecting and deflecting abilities. For a more detailed discussion of the optimum composition of social groups the reader is referred to Crook (1970). The critical point for present purposes is that group size and composition and the size of the home range occupied should represent the optimum response to these major selective pressures. Any change in any one of these three variables, however, or in the supply of the different resources in the home range, shifts the optimum value of the other variables. The result should be a composite migration threshold to the different variables with an ontogenetic variation and frequency distribution within the deme as described in Chapters 14–17. Selection should stabilise when these distributions of migration thresholds are such that the response of individuals within the group to any change in any variable results in a change in group size, composition, or size of home range such that an optimum situation is restored. However, this need not necessarily result in a free situation as discussed in Section 18.3.5.

Having discussed some of the selective pressures that influence the optimum size of the home range, it is perhaps worthwhile to consider the possibility of feedback selection from the size of the home range upon the succession of removal migrations from

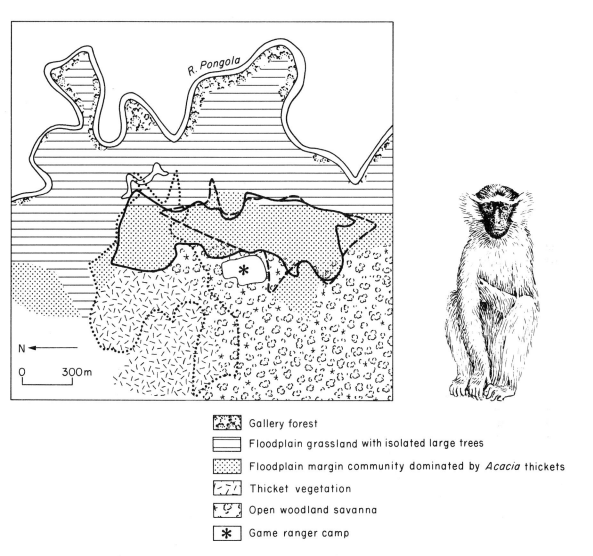

Gallery forest

Floodplain grassland with isolated large trees

Floodplain margin community dominated by *Acacia* thickets

Thicket vegetation

Open woodland savanna

* Game ranger camp

Fig. 18.18 Seasonal change in size and degree of overlap of home range: the vervet monkey, *Cercopithecus aethiops*

General ecology

The African vervet monkey is typically an animal of riverine vegetation and of forest fringes, especially where these adjoin grassland. The species represents an animal intermediate between on the one hand typical arboreal species and on the other hand more terrestrial species. A typical social group in a forest habitat consists of 11 individuals: 2 adult males; 4 adult females; and 5 young. The males, however, exhibit a high incidence of group-to-group migration (Chapter 17).

Home range

The home range of the vervet monkey has been studied in the Ndumu Game Reserve on the northeastern border of South Africa by means of a radio transmitter attached to one or two members of a number of social groups. In ecologically rich areas each social group (22–37 individuals) has a well-defined home range of from 9·4–27·4 ha, which overlaps little with the home ranges of neighbouring groups.

The diagram shows the home ranges of two groups that were living in a less favourable area. The dotted outline encompasses a year's home range of group A. The solid outline encompasses the dry season home range of group B. The wet season home range for group B is encompassed by the dashed line. A year's home range in this area was much larger (53·1 ha and 55·8 ha) than for groups in the more favourable area. Furthermore, during the dry season, when food was less abundant, there was a 25 per cent overlap in home range between the two groups. There was very little overlap, however, during the wet season when food was more abundant.

[*Re-drawn, with modification, from Moor and Steffens (1972)*]

which the home range crystallises. The hypothesis is as follows. The relationship between size of familiar area and home range is to some extent an inevitable part of the process by which the two are formed. The part of the familiar area outside of the home range is proportional to the number of habitats of a suitability below the exploratory migration threshold that are encountered before a habitat is encountered that is above the threshold. It is also likely, however, that there is selection on the home range/familiar area size ratio, the optimum being some compromise between the efficiency of the ranking system and the number of habitats held in reserve for possible future use. Accepting this, it follows that selection for a larger home range is likely to involve selection for a larger familiar area. As selection invariably seems to have acted to delay the onset of reproduction until

an animal is established in a home range (Chapter 17), the nett result of selection for an increase in size of home range should be further selection for an increase in the ontogenetic period of pre-reproductive exploratory migration.

Table 18.1 and Fig. 18.19 analyse the relationship between size of home range as calculated from McNab's formula (Fig. 18.17) for a number of insectivores, rodents, and carnivores for which information is available concerning length of the pre-reproductive period of exploratory migration. The period taken is that from the initiation of migration from the parental home range to the approximate age of first copulation. The real length of the period is somewhat less than this, for as discussed at length in Chapter 17 (p. 349), exploratory migration usually ceases some time before the onset of

Table 18.1 Data* for regression analysis of the association between body weight of a mammal and the age at the initiation of exploratory migration, age at first copulation, and the size of the home range

Species		Weight (kg)	Age (weeks) at		Size of home range (ha)†
			initiation of exploratory migration	first copulation	
Marsupials					
possum					
Didelphis marsupialis		3·630	13	43	12·441
Insectivores					
short-tailed shrew					
Blarina brevicauda		0·018	3	13	0·303
masked shrew					
Sorex cinereus		0·004	3	43	0·106
Rodents					
muskrat					
Ondatra zibethicus		1·200	17	48	1·375‡
beaver					
Castor canadensis		20·410	78	91	9·993‡
harvest mouse					
Micromys minutus		0·007	3	6	0·156
Lagomorphs					
European rabbit					
Oryctolagus cuniculus		1·750	5	15	1·790‡
Carnivores					
wolf	M	36·290	52	104	62·341
Canis lupus	F	36·290	52	156	62·341
red fox					
Vulpes fulva		5·450	26	52	16·535
raccoon					
Procyon lotor		10·890	26	52	26·844
American mink	M	0·900	17	52	4·687
Mustela vison	F	0·725	17	52	4·029
fisher	M	4·080	30	104	13·502
Martes pennanti	F	2·180	30	52	8·707
Canadian lynx	M	7·710	22	52	21·079
Lynx canadensis	F	7·710	22	52	21·079

* Data mainly from Peterson (1966) and Corbet (1966).
† Home range calculated from body weight according to the formula of McNab (1963)—see Fig. 18.17.
‡ Animal classed as a 'cropper'.

reproduction. Regression analysis of length of exploratory migration on size of home range gives a straight-line fit to the data that is significant at the 5 per cent level ($F_{1,15} = 12 \cdot 9$) but which has only a moderate coefficient of determination ($R^2 = 0 \cdot 463$). There are also significant fits at the same level for exploratory migration on body weight ($F = 7 \cdot 7$; $R^2 = 0 \cdot 339$), age at first copulation on size of home range ($F = 22 \cdot 8$; $R^2 = 0 \cdot 603$), age at first copulation on body weight ($F = 29 \cdot 4$; $R^2 = 0 \cdot 662$) and of the period of exploratory migration on the age at first copulation ($F = 44 \cdot 9$; $R^2 = 0 \cdot 749$). Finally, a multiple-regression analysis of the length of the period of exploratory migration on body weight, age at first copulation, and size of home range gives a significant fit to the data ($F_{3,13} = 33 \cdot 2$) with a fairly high coefficient of determination ($R^2 = 0 \cdot 885$).

Unfortunately, it was found during the course of this analysis that McNab's formula was unsatisfactory for the calculation of home-range size for the larger carnivores, giving a calculated range that is consistently much lower than that observed in the field. In retrospect, this characteristic of the formula is apparent from McNab's own analysis (see Fig. 18.17), the point for the raccoon, *Procyon lotor*, being considerably above the regression line. As two further examples, McNab's formula gives a home-range size for the wolf, *Canis lupus*, and the red fox, *Vulpes fulva*, of 62·3 ha and 16·5 ha respectively (Table 18.1). The figure of 62·3 ha for the wolf cannot begin to compare with the summer range of 38 400 ha observed by Thompson (1952) nor with the generally observed winter hunting circuit of up to 200 km in circumference (Peterson 1966). The figures are still not comparable even if the formation of a social group by this species is taken into account. Indeed, one may wonder how it would be possible in any case to make realistic comparisons between areas for an animal that at one time of year is based on a central roosting site from which it radiates and at another time of year performs a hunting circuit. The calculated home range of 16·5 ha for the red fox is similarly less than even the smallest range of 57 ha found for a fox in a radio-telemetry study by Ables (1969) in an area of marked ecological diversity in south-central Wisconsin. The home range for one of six other foxes studied in the same area was 160 ha, an order of magnitude larger than the calculated home range. Finally, another individual, the home range of which covered farmland, moved within an

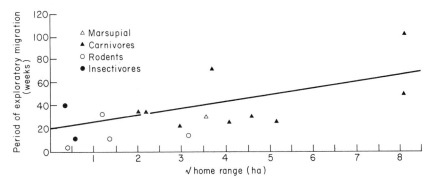

Fig. 18.19   The relationship for some mammals between the length of the pre-reproductive period of exploratory migration and size of home range when home range is calculated according to the formula of McNab (1963)

In the diagram, size of home range, as calculated from body weight according to the formula of McNab (see Fig. 18.17) is represented by the square root of the area of the home range expressed in hectares. This separates the points near to the origin.

The regression of *EM*, the length of the pre-reproductive period of exploratory migration in weeks, on *HR*, the area of the home range in hectares, according to the equation $EM = 0 \cdot 866 HR + 19 \cdot 96$ gives a significant fit to the data ($F_{1,15} = 12 \cdot 9$, $P < 0 \cdot 05$) and both the regression coefficient ($0 \cdot 866 \pm 0 \cdot 241$) and the intercept ($19 \cdot 96 \pm 5 \cdot 90$) are significantly ($P < 0 \cdot 05$) different from zero

($t = 3 \cdot 600$ and $3 \cdot 381$ respectively). The coefficient of determination for the regression is $0 \cdot 463$.

As discussed in the text, however (this p.), the virtual impossibility at present of comparison of home-range size in widely different animal species and the inaccuracy of McNab's formula inevitably result in an analysis such as this being extremely provisional. It may tentatively be concluded that increase in size of home range is associated with increase in the length of the period of pre-reproductive exploratory migration, a conclusion that is supported by the slightly more reliable analysis for primates presented in Fig. 18.20. Attempts to quantify the relationship, however, must await comparable measures of home-range size for a sufficient variety of animals.

[*The figure is based on the data in Table 18.1*]

area of 5100 ha. This study of the fox by Ables not only highlights the shortcomings of McNab's formula but also emphasises the difficulty of comparing home-range size between species. When the range of areas found for individuals of a single species can cover two orders of magnitude, such as from 57 ha to 5100 ha, one wonders how such measures can meaningfully be entered into an analysis of the type carried out by McNab. Nevertheless, although we may not yet be at the stage of being able to calculate the home range of a mammal given its body weight and feeding habits, the two main effects noted by McNab, that of increase in home-range size with increased body size and of larger home-range size in hunters than croppers, seem likely to be real enough, and it is perhaps not too surprising that they are strong enough to show through the analysis in Fig. 18.17 despite the difficulties involved. Furthermore, it must be added that McNab's formula seems to have a much higher predictability for smaller

mammals such as rodents and insectivores which perhaps also do not show such extreme within-species variation in size of home range.

Insofar as information is available it seems also that the home-range size of primates does not show within-species variation to the extent found in the two carnivores discussed above. Primates, therefore, perhaps lend themselves more to an analysis of association between length of exploratory migration and size of home range than do the larger non-primate terrestrial mammals. Table 18.2 and Fig. 18.20 present respectively the data for such an analysis and the results of the analysis itself. Here, observed rather than calculated size of home range is used for most species. The same conclusions are reached from this analysis as from that for other mammals in Fig. 18.19. Increase in size of home range is associated with increase in length of exploratory migration (measured as the period from the age an animal initiates its first migration inde-

Table 18.2 Data* for regression analysis of the association between body size of a primate and the age at the initiation of exploratory migration, age at first production of offspring, and the size of the home range

Species		Length of head and body (cm)	App. age (weeks) at		Size of home range (km^2)
			initiation of exploratory migration	first pro- duction of offspring	
Tree shrews Siamese tree shrew *Tupaia belangeri*		18·5	13	17	0·039†
Prosimians dwarf bush baby *Microcebus murinus*		13·0	17	37	0·020†
lemur *Lemur macaca*		38·0	30	78	0·057
Old world monkeys Japanese macaque *Macaca fuscata*	M	57·1	182	364	8·000
rhesus macaque *Macaca mulatta*	M	56·0	208	232	8·000
black mangabey *Cercocebus albigena*	M	54·3	104	156	0·180
Hominids man *Homo sapiens*	M	81·0	936	1250	56·000
	F	77·5	830	1040	56·000
Apes white-handed gibbon *Hylobates lar*	M	52·0	416	468	0·260
	F	51·5	416	468	0·260
gorilla *Gorilla gorilla*	M	95·0	624	830	32·000
chimpanzee *Pan troglodytes*	M	85·0	442	520	14·500

* Most data on body length and age at first production of offspring from Napier and Napier (1967), Jolly (1966), and Martin (1968). Size of home range and approximate age of initiation of exploratory migration also from these sources and from figures elsewhere in this book.
† Home range calculated from body weight according to the formula of McNab (1963) for hunters—see Fig. 18.17.

pendently of its natal group to the estimated age at which it produces its first offspring—see Chapter 17). This time, however, predictability is much higher with a coefficient of determination of 0·781. There are also significant correlations at the one per cent level between: body length[3] and age of first production of offspring ($r = 0·772$; $df = 10$); body length[3] and size of home range ($r = 0·721$); and highly significant correlations at the 0·1 per cent level between: size of home range and age of first production of offspring ($r = 0·923$); and age of first production of offspring and length of the period of exploratory migration ($r = 0·895$).

Despite the shortcomings of McNab's formula, therefore, the similar evidence for primates as for the other terrestrial mammals considered suggests that for most terrestrial mammals the variables considered form an adaptive complex. Any selection for change in any one of: body size; weight; size of home range; age at first reproduction; or length of the pre-reproductive period of exploratory migration, automatically imposes selection on each of the others. Furthermore, all of the variables are positively correlated, increase in one leading to increase in the others. An analysis of this type cannot, of course, identify cause and effect, only association. We may suspect, perhaps, but cannot prove, that most often during the adaptive radiation of terrestrial mammals selection initially has acted on this adaptive complex through selection on body size. Selection for increase in body size, for example, would then lead to selection for increase in size of home range. Increase in optimum size of home range would then impose selection for an increase in the pre-reproductive period of exploratory migration and, in consequence, a delay in the age of first reproduction. Such a sequence of cause and effect seems perhaps the most likely, but there is no inherent

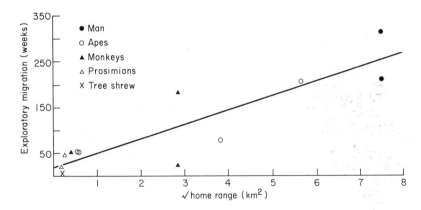

Fig. 18.20 The relationship between size of home range and length of the pre-reproductive period of exploratory migration in primates

In the diagram, size of home range is represented as the square root of the area of the home range. This separates the points near to the origin. The regression line is $EM = 30·57 HR^{0·5} + 21·69$, which gives a significant fit to the data ($F_{1,10} = 35·33$; $P < 0·02$), with a coefficient of determination for the regression of 0·779. A slightly higher coefficient of determination of 0·781 is obtained, however, for the straight line regression of EM on HR (where EM is the length of the period of exploratory migration in weeks and HR is the area of the home range in square kilometres). The calculated regression is $EM = 4·06 HR + 44·15$, which also gives a good fit to the data ($F_{1,10} = 35·66$; $P < 0·02$). The regression coefficient is significantly different from zero ($t = 5·972$; $P < 0·001$) and so too, though only just, is the intercept ($t = 2·573$; $P < 0·05$) though when the period of exploratory migration is regressed on the square root of the home range, the intercept is not significantly different from zero ($t = 1·104$; $P > 0·05$).

The data used are presented in Table 18.2. The size of the home range for the tree shrew, *Tupaia belangeri*, and the dwarf bush baby, *Microcebus murinus*, were calculated from McNab's formula for 'hunters' (see Fig. 18.17). The home-range sizes for the other species are taken directly from the literature except that for man which is a compromise between the home range for a family group of a hunter-gatherer (Fig. 25.15) and the area containing the roosting site and gardens of a highland village of New Guinea. Both males and females are included in the regression for man and the white-handed gibbon, *Hylobates lar*. In the other apes and monkeys only the male performs a solitary pre-reproductive exploratory migration.

It may be concluded that a relationship exists between size of adult home range and the length of the pre-reproductive period of exploratory migration such that an increase in size of home range is associated with an increase in the length of the pre-reproductive period of exploratory migration.

reason why selection should not enter the adaptive complex at any point.

18.3.5 Home range and territory

It was shown in Fig. 18.13 that the home ranges of neighbouring groups of the gorilla, *Gorilla gorilla*, overlap so much that as many as six gorilla groups may exploit the resources of a particular area of forest. The gorilla is surrounded, or so it seems to an observer, by an abundant supply of food and a gorilla group is unlikely to find itself short of food or any other resource even if it arrives at an area of forest that has just been vacated by some other group. The explanation for these six or so groups retaining their identity instead of coalescing to form a single group six times as large seems to lie in the direction of the selection on individual migration thresholds described in Chapters 14, 15, and 17. Each female that comes into oestrus increases the chance that the silverback males that are present in the group will remain and that transient males will associate with the group (p. 162). Increase in the number of silverback males relative to the number of females, blackback males, and juveniles increases the probability that females, non-receptive with young as well as receptive, will join the group from a neighbouring group (p. 92). As females become pregnant, infants are born, and the number of non-receptive females relative to the number of silverback males increases; however, silverback males other than those that had most copulatory access to the females when fertilisation was most likely are likely to leave the group and search for other groups with a more favourable composition. As the silverback male to juvenile and to female ratio decreases, so the group becomes less suitable to adult females and these become more likely to leave to join a neighbouring group, and so on. The result is that group size and composition tends to fluctuate about a mean. These fluctuations are not frequent because the physiological events upon which they are based are not frequent. Nevertheless, such fluctuations evidently do occur (Schaller 1963) though whether the observer is impressed by the fact that they occur at all, or by the fact that they occur so infrequently, depends entirely on the stand from which they are viewed.

Availability of environmental resources other than other individuals seems to impose little selection on the degree of overlap of home range in the gorilla, at least in the area studied by Schaller. In this respect, however, the gorilla is unusual, if not unique. As far as most animals that live in a home range are concerned, availability of resources is a critical factor that imposes considerable selection on the degree of overlap of the home ranges of neighbouring individuals or groups.

The ranking system for the suitability of habitats encountered during exploratory migration (p. 379) is likely to be a most important facet of the lifestyle of animals that live in a familiar area. Once selection on exploratory migration (p. 374) has led to the evolution of the behaviour that is pre-adaptive for the evolution of the ranking system, there is likely to be immediate selection for increased efficiency of the behaviour. Part of the adaptation to this selection is likely to be the re-visiting-for-re-assessment phenomenon already described. Although the advantages that result from an efficient ranking system could not account for the evolution of the familiar area, once an efficient ranking system evolved through the sequence already indicated, the advantages that result from it are likely in most modern animals to make a major contribution to the advantage of living within a familiar area. Any behaviour that tends to decrease the efficiency of the ranking system is likely to be acted against by selection.

Overlap of home range is likely considerably to decrease the efficiency of the ranking system. Calculated migration to an area ranked previously at a high level is much more likely to be advantageous if in the meantime the area has not been exposed to exploitation by other individuals. The result is that an inevitable consequence of selection for a ranking system is selection against home-range overlap. Often, however, there may be selection acting in favour of home-range overlap, such as when selection favours male and female cooperating in the raising of offspring, or favours offspring maintaining contact with their parent(s), or favours individuals remaining in sensory range to reduce predation incidence. Such selection may be greater than selection acting through the ranking system against home-range overlap. On such occasions, however, the nett result of selection for a ranking system and for home-range overlap is likely to be for the individuals subjected to such selection to travel together and/or to communicate with one another concerning habitat suitability, thus retaining the advantages both of a ranking system and of home-range overlap. In such cases the optimum number of individuals the home ranges of which overlap is set by the selection on group size and composition that has already been indicated and by such other selection as perhaps that on males to exclude other males from a particular group. Selection then acts on optimum degree of home-range overlap of neighbouring groups, again through selection on the efficiency of the ranking system.

Whether selection acts to favour a single individual, a mother and offspring, or a monogamous, polygamous, or promiscuous reproductive unit plus offspring, as the optimum social unit, it also acts through the ranking system on the optimum degree of overlap of the home ranges of neighbouring social units. The critical factor is the effect of exploitation by another unit upon the accuracy of the rank order

of habitat suitability previously determined. Thus, if a habitat provides an abundant or an abundant but ephemeral resource, then up to a point exploitation by other individuals is unlikely to affect the efficiency of an individual's exploitation of its familiar area. This explanation has already been advanced to account for the high degree of overlap of the home ranges of groups of gorillas. It also accounts for the fact that the home ranges of many animals overlap only in areas of an abundant resource. Hence, the home ranges of individuals of the same sex of black bears, *Ursus americanus*, and grizzly bears, *U. arctos horribilis*, overlap only in areas such as salmon streams, rubbish tips, and other sites of abundant resources (Jonkel and Cowan 1971). A similar overlap of home ranges in the vicinity of large fruit trees has been reported by McKinnon (1974) for the

Fig. 18.21 Overlap of home range

The available evidence suggests that slugs, such as the *Arion ater* illustrated, live in a familiar area. Like many other invertebrates and vertebrates, slugs seem to live in a

free situation and show considerable home-range overlap (see Fig. 18.25, position C).

[*Photo by R.R. Baker*]

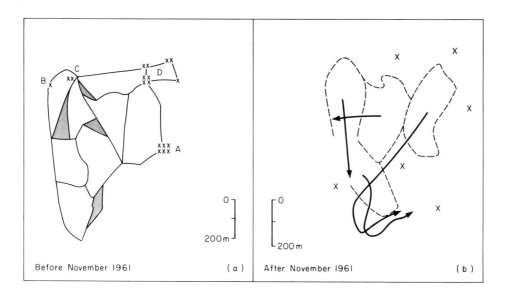

Before November 1961   (a)   After November 1961   (b)

Fig. 18.22   The effect of deme density on the territorial system of the weasel, *Mustela nivalis*, in Scotland

General ecology

The weasel is the smallest (body length: male, 210–230 mm, female, 160–190 mm; weight: male, 60–130 g, female, 45–60 g) mustelid carnivore, and indeed is the smallest of all the Carnivora. It occurs, wherever there is sufficient cover and food, throughout Europe (except Ireland and Iceland), Asia, parts of North Africa, and perhaps also North America. It feeds primarily on small rodents and to a lesser extent on shrews. The breeding season extends from April through August during which two litters of 4–8 young are produced. Some young of the first litter may breed the same year. The main predators appear to be birds of prey, particularly owls.

Territorial system

(a) Territories of male weasels in a young forest plantation in the Carron Valley, Stirlingshire, during the winter (November through March) non-breeding season

Heavy lines outline the territories. Hatching indicates either areas of overlap or areas where the edge of the territory could not be determined. The crosses indicate the history of a male yearling weasel. From November 1960 to January 1961 the male was caught only at *A*, within the territory of another male. Within the period of a few days in early January 1961, however, the male appeared first at *B* and *C* and then finally at *D*, where it took over a territory that had just become vacant due to the death of the occupant. Whether the migrant returned to *A* between visits to *B*, *C* and *D* is unknown.

(b)   Movements of surviving territory occupants following a sudden decrease in deme density

The situation illustrated in (a) persisted from November 1960 to October 1961. At this time, however, several territory holders disappeared, largely through predation and accident. The territories that became vacant are indicated by a large cross. The territories (dashed lines) of the surviving males did not expand to fill the area. Instead, the home ranges of these males gradually shifted with time (heavy arrows) thus indicating that the home-range size was the maximum beyond which further increase would have been disadvantageous (Fig. 18.25, position A). Shifts in home range were largely in the direction of a neighbouring unoccupied area. Incoming migrants similarly showed a shift in home range such that no more than two males were ever present at any one moment in time. This mobile situation persisted throughout 1962 and 1963. The indication is, therefore, that in the weasel the exploratory migration–calculated migration sequence leads to shift in home range at low deme densities if at least one side of the range is not in contact with the home range of a neighbouring individual. Only when deme density in an area is sufficiently high for the whole area to be occupied does a non-mobile territorial system develop.

[*Re-drawn, with modification, from Lockie (1966). General ecology from Corbet (1966)*]

orang utan, *Pongo pygmaeus*, and many such observations for other species are available (e.g. Fig. 18.21 for slugs and Keast 1968a for birds).

It is not, however, necessary for habitats to provide a resource in such spectacular abundance for selection not to act totally against home-range overlap. The general thesis is that the greater the effect of exploitation of habitats by one individual or group on the efficiency of the ranking system of another individual or group, then the less is the optimum degree of overlap. At one extreme, therefore, we have the case of the gorilla, with considerable overlap of home range (Fig. 18.13) and at the other we have situations such as illustrated in Fig. 18.22(a) for the weasel, *Mustela nivalis*, and in Fig. 18.26 for the black-tailed prairie-dog, *Cynomys ludovicianus*, in which there is minimum, if any, overlap of home range. The remainder of this section is concerned with the mechanisms involved in the maintenance of the optimum degree of home-range overlap and the way these are involved in home range to home range removal migration.

The disadvantage of home-range overlap is likely to be twofold. First, a resource is being exploited by more than one individual or group, and second, an unpredictable element is introduced (i.e. that any given habitat may or may not be within the familiar area of another individual) that reduces the efficiency of the ranking system. Adaptation to this disadvantage is likely to consist of one of the following alternative strategies: either both animals or groups can exploit the resource at the site of overlap; or only one of them can do so. That is, either an animal can no longer exploit the site, or it can exploit the site, or it can exploit the site and in addition invest time and energy preventing other individuals from exploiting the site. Which of these strategies is likely to be optimum depends on the relative abundance of the resource in parts of the familiar area other than the site of overlap. We will return later to the third possibility, that of investing time and energy to prevent other individuals from exploiting the resource. The first possibility, that of vacating the site without being forced to do so has been illustrated in Fig. 18.22(b). For the moment, let us consider the situation in which the required resource is not so abundant elsewhere that an animal would gain by abandoning the site of overlap completely, and yet is not in fact in such short supply that an advantage could be gained by investing time and energy preventing other individuals from exploiting the resource in the site of overlap. It is in this

intermediate situation that selection due to the inefficiency of the ranking system is likely to attain its maximum manifestation.

The main disadvantage of home-range overlap in this situation is not so much that the resource is being exploited by other individuals but that the suitability of the habitat at any one moment in time is unpredictable relative to that of other habitats within the familiar area. Suppose a habitat is given a certain ranking by an animal. If the habitat is not in an area of overlap, this ranking is likely to remain relatively accurate for a given period and only brief and infrequent visits for re-assessment should be necessary for the ranking system to function efficiently. If the habitat is within an area of overlap, its suitability is likely to change due to exploitation by the other animal(s). Nevertheless, if there is any consistency to the pattern of exploitation of a habitat of given suitability by a species, the change in habitat suitability that occurs should still be fairly predictable, though less so than for a habitat not in an area of overlap. Selection on the efficiency of the ranking system, therefore, imposes selection on an animal to be able to recognise whether or not a given habitat is within the home range of individuals other than itself.

Suppose that such detection were possible through visual and olfactory clues, no matter how subtle. Selection, apart from resulting in sophistication of the detection mechanism, would also be likely to favour individuals that ranked habitats in areas of overlap at a lower level than habitats exploited solely by that individual. The first strategy above, that of abandoning habitats in areas of overlap, is an extreme part of this adaptive process. Habitats ranked lower should probably ideally be re-visited for re-assessment less frequently and inevitably would not be visited for exploitation until later than habitats given a higher ranking. Once clues associated with the previous presence of other individuals became involved in the ranking process and the behavioural syndrome begins to spread through the population, selection is elicited that favours those individuals that increase the conspicuousness of the clues that indicate their previous presence in the area. Such increase in conspicuousness decreases the rate of visit of other individuals to habitats in the area of overlap. The marker therefore gains in that habitats within the area of overlap remain at a higher suitability than would otherwise have been the case and the responder gains in that the loss that results from decreased exploitation of

these habitats is less than the gain due to increased
efficiency of its own ranking system. The result is
selection for the production of marking/advertising
and detection systems that are advantageous to both
marker and detector. The raw material for such
marking systems are those artefacts, such as urine,
faeces, glandular secretions, scratch marks, broken
vegetation, etc. that would in any case have resulted
from an animal's visit to an area. The action of
selection is to increase the conspicuousness of the
artefacts or physical features and the efficiency of the
sensory apparatus used for detection. An excellent
review of marking systems in mammals is given by
Ewer (1968).

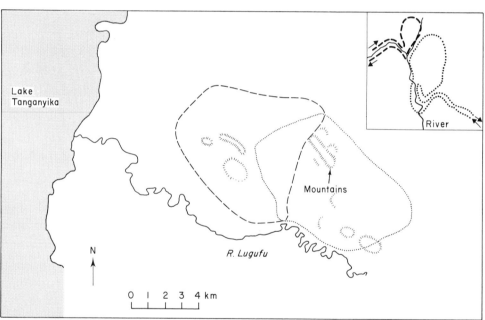

Fig. 18.23   Degree of overlap of home ranges and behaviour of two neighbouring large-sized groups of the
chimpanzee, *Pan troglodytes*, in the savanna woodland of
Western Tanzania

Social structure
A large-sized group of chimpanzees consists of 20–40
individuals containing several adult males, females, juveniles, and infants. The group spends a great deal of time
dispersed in sexual or nursery sub-groups and solitary
individuals. These sub-groups converge frequently when
food is abundant but less frequently when food is scarce.
The sub-groups maintain contact with one another, when
dispersed, by vocalisation. The individual constitution of
each sub-group is not constant and changes after each
congregation. Apart from large-sized groups, the other
relatively durable social unit of the chimpanzee is a small-
sized group, consisting of one or two adult males with one
or a few females and their young. Solitary individuals,

which are usually males, also occur. These social units do
not have independent home ranges comparable to the
large-sized groups but wander widely through the ranges
of several large-sized groups. The movements of these
other social units may be exploratory.

Home range
Although the home ranges of neighbouring large-sized
groups overlap as illustrated, the groups usually avoid
each other. The inset shows the tracks of the two large-
sized groups, the home ranges of which are illustrated, on
one occasion when they approached one another. On
other occasions, however, neighbouring large-sized
groups or sub-groups from neighbouring large-sized
groups may meet and mingle for a short period. Group-to-
group removal migration may occur on such occasions.

[*Re-drawn with some modification from Izawa (1970)*]

When a resource is extremely abundant in an area of overlap there is little disadvantage in neighbouring individuals visiting the habitat at the same time. Such communal feeding in animals that occupy separate but overlapping home ranges is a common occurrence and is well known for animals such as black bears at salmon streams, orang utans at large fruit trees, and rats at rubbish tips. When a resource is less abundant, however, individuals or groups usually avoid one another even in areas of home-range overlap (Fig. 18.23), presumably because both units cannot efficiently exploit the contained resources at the same time. Note, however, that some

mingling of social groups occupying overlapping home ranges does occur from time to time, a process that has been interpreted as exploratory in relation to group-to-group migration (Chapter 15, p. 162).

It has already been pointed out that on the whole, habitats in an area of overlap are likely to be given a lower ranking than habitats in a part of the home range exploited solely by a single individual or group. Concomitant with a low ranking should be, ideally, a lower rate of re-visiting for re-assessment and a lower probability of exploitation. The result is likely to be that an animal or group spends far more of its time in the area to which it has sole access than in the part of the range that it shares with other animals. A part of a home range that is used far more than the remainder of the home range is usually termed a core area (Fig. 18.24). It is common experience that the core area is that part of the home range that does not overlap that of other individuals or groups (Jewell 1966), a finding that provides support for the above interpretation.

So far, this discussion of degree of overlap has only indirectly taken into account deme density. However, this is clearly a most important environmental variable in this connection and has to

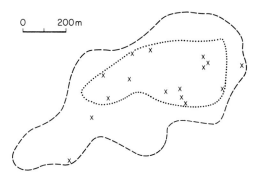

Fig. 18.24   Home range, core area, and roosting sites of a social group of the coati, *Nasua narica*, on Barro Colorado Island, Panama

Dashed line, limits of home range; dotted line, limits of core area; crosses, tree roosts.

The diagram shows the home range and core area of a maximum social unit of the coati, which consists of a number of adult females and their young less than two years of age. During the mating season each maximum social unit is joined by a single adult male which drives away other males. As the time for parturition approaches, the pregnant females become solitary and build a tree nest. After parturition the females remain solitary, except for looking after their young, for five weeks. During this

period, non-breeding females and juveniles remain together. At the end of five weeks when the young are capable of maintaining the rate of movement of the adults, the maximum social unit re-assembles, though the family unit within the group retains its integrity.

The core area is used for about 80 per cent of the time. The home range in this case may also represent the limits of the familiar area. Home ranges of different maximum social units overlap but the core areas do not.

Males live solitarily for most of the year and also have a home range, though this is larger than that of the females with which it overlaps. A single tree roost may be used for several days at a time, but usually there is almost daily movement between different tree roosts.

[*From data and figures in Kaufmann (1962)*]

be considered before going on to discuss the influence of home range configuration on home range to home range removal migration. Figure 18.25 presents a thesis for the influence of change in deme density on the size and degree of overlap of neighbouring home ranges. The chief conclusion is that the shape of the curve is unlikely to be simple and that a straight-line regression of home-range size on deme density is only to be expected over a relatively small range of deme densities.

We are now in a position to consider the influence of home range configuration on removal migrants. In this section it has been found convenient to leave a discussion of the relationship between home range and territory until after this consideration.

In a recent book, Fretwell (1972) made a distinction between animals exposed to free and despotic situations, and the reader is referred to this source for the original formulation of this distinction. The following description, however, although based on the free–despotic distinction has been tailored to suit present requirements. The selective pressures acting on an animal that produce a situation at some point on the free to despotic spectrum are given in effect in Fig. 18.25 in relation to the effect of deme density on the optimum size and degree of overlap of home ranges.

Consider an exploratory migrant from an area of low suitability that encounters two areas, A and B, of higher suitability. Let that component of habitat suitability attributable to the physical and biological structure of these two areas excluding the presence of competing individuals (p. 70) be given by h_a and h_b for A and B respectively. These suitabilities are such that $h_a > h_b$. Now suppose that there are four competitors already exploiting A but only one competitor already exploiting B. Suppose further that the total suitability of A ($=h_A$) to the individuals already present is $h_a/4$ and of B ($=h_B$) is $h_b/1$. As far as the exploratory migrant is concerned, therefore, $h_A = h_a/(4+1)$ and $h_B = h_b/(1+1)$. Optimum strategy for the migrant now depends on whether the species lives in a free or despotic environment.

Let us consider the free situation first and suppose that in the above situation $h_a/5 > h_b/2$. In a free situation, optimum strategy for the animal is to settle in A. Suppose, however, that $h_a/6 < h_b/2$. In this case, the next exploratory migrant to investigate these two areas will settle in B. The result of the free situation, therefore, is that deme density tends to be distributed in such a way that all habitats capable of

supporting at least one individual are of equal suitability, a concept that has already been discussed in relation to the frequency distribution of migration thresholds (Chapter 17, p. 269) and which is probably of common occurrence among the Metazoa.

In the despotic situation, however, which is also widespread, whether or not $h_A > h_B$ in the terms just described is not the only factor influencing in which area the animal will settle. Let us suppose as before that $h_a/5 > h_b/2$ and that therefore, all else being equal, optimum strategy is to settle in A. Suppose now, however, that in order to settle in A the migrant has to expend time and energy and expose itself to a certain wound- and/or death-risk to overcome resistance from the individuals already present in A. In the despotic situation, therefore, the component of habitat suitability attributable to competing individuals (h_d on p. 70) is not simply a function of deme density as in the free situation but a multiple function of deme density and the resource-holding power of each of the individuals already present relative to that of the migrant.

Resource-holding power is an extremely useful concept that has been developed in a recent paper by Parker (1974b). It refers to the ability of an individual or other territorial unit to defend and exploit a resource in the face of competition from other territorial units. The resource-holding power of any particular individual is obviously a function of the size, strength, experience, age, and general physical fitness of the individual. Parker has made out a convincing case that most selection in despotic situations is concerned firstly with efficiency of assessment of an individual's own resource-holding power relative to that of other individuals and secondly with efficiency of response once an assessment has been made. Parker suggests that when escalation of interaction is optimum strategy, as when two individuals are of similar resource-holding power, selection on fighting strategy acts through the probability that a particular action inflicts a loss of resource-holding power on the opponent. The pattern of territorial interaction evolves, therefore, in the form of a sequence of bouts, each bout consisting of an attempt by each interacting territorial unit to decrease the resource-holding power of the other unit. After each bout each territorial unit re-assesses its own resource-holding power relative to that of its opponent in the light of the bout just completed.

Parker's (1974b) paper on territorial interaction

in a despotic situation, along with the papers of Maynard Smith and Price (1973) and Maynard Smith (1974) on optimum strategy when two territorial units are evenly matched, forms a useful basis for a consideration of optimum strategy for migrants in a despotic situation. It is instructive, however, to consider the variation in optimum strategy at different points along the free to despotic spectrum. In all of the examples that follow, we are concerned with animals in which, all else being equal, selection favours a home range that does not overlap with that of another individual. We are not here concerned with animals for which selection favours either the formation of a social group or convergence, both of which are discussed elsewhere (pp. 390 and 405 respectively).

Consider the situation in which deme density is so low that there are large numbers of unoccupied habitats (i.e. the extreme left end of the curve in Fig. 18.25). Migrant individuals are confronted with a high mean expectation of migration ($\bar{E}$) and the sheer presence of any other individual should be sufficient to lower habitat suitability below $\bar{E}$ and for migration to be initiated or continued. Migrants, however, are likely to have lower migration thresholds than individuals that have already made some investment in the area (Chapter 17). Contact with another individual or its artefacts (p. 399) should be sufficient, therefore, for an individual to continue to migrate. Such a situation seems to have existed for part of the period of observation (Fig. 18.22(b)) in the deme of the weasel, *Mustela nivalis*, investigated by Lockie (1966) and also in the peacock butterfly, *Inachis io* (Baker 1972a). In the latter species, on 24 of 27 occasions that a territorial male (see Fig. 18.5) made its presence known to an intruder by intercepting it before the latter could pitch in the territory, the latter left without attempting to settle. On 15 occasions that the territorial male was experimentally removed, however, the first male to arrive established itself in the territory. In such species and situations as these, advertisement of presence seems to be sufficient defence of the area.

With increase in deme density, however, there is a corresponding decrease in mean expectation of migration. Beyond a certain point, it seems likely that for a migrant to detect the presence of another individual or group in an area is not sufficient for the suitability of that habitat necessarily to be less than $\bar{E}$. This is the point also at which overlap of territories becomes advantageous (Fig. 18.25). Over a range of deme density, however, bearing in mind

the discussion of ranking efficiency (p. 399), time and energy spent defending resources is unlikely to be advantageous. Over this range, marking mechanisms and other means of advertisement serve more to increase the efficiency of the ranking systems of all individuals concerned. Further increase in deme density, however, especially if coupled with a decreased efficiency of the ranking system, begins to impose selection for resource defence in an aggressive physical sense and further decreases the value of $\bar{E}$. At some point along this part of the spectrum, selection no longer favours a migrant continuing its migration simply because another individual is present but instead favours a stay-time of sufficient length to assess the resource-holding power of the occupant. To some extent, of course, some artefacts correlate with resource-holding power, quantity of urine, faeces, and glandular secretion and height of scratch marks, etc. all representing variables suitable for assessment. Comparison of resource-holding powers need not necessarily, therefore, involve the two individuals coming into physical or sensory contact with one another. At this stage, h_d, the component of habitat suitability due to competing individuals, is not given by deme density but by the relationship between the resource-holding power of the occupant and the migrant. Let $h_d = h_{rhp}$, where h_{rhp} is that component of habitat suitability due to the relationship between the resource-holding powers of the individuals concerned. As far as the migrant is concerned, h_{rhp} is 1·0 when the migrant's resource-holding power is so much greater than that of the present occupant that the migrant would suffer no reduction in reproductive success if it were to enter into a territorial interaction with the occupant. On the other hand, h_{rhp} is zero when the migrant's resource-holding power is so much less than that of the occupant that the migrant would die if it were to enter into a territorial interaction with the occupant.

On previous arguments (Chapter 17) the suitability, h, to a migrant of an encountered but occupied territory is given by $h = h_p h_{rhp}$. Migration rather than attempting to take over the territory is advantageous when $h < \bar{E}$. It follows, therefore, that continuation of migration is advantageous whenever $h_p h_{rhp}$ is less than $\bar{E}$. In other words, the more suitable the habitat due to its physical and biological construction (excluding the influence of the occupant), the higher the resource-holding power of the migrant, the lower the resource-holding power of the occupant, and the lower the mean expectation of migration, then the more likely is optimum strategy to a migrant to be to attempt to oust the occupant from the territory. It should be noted, however, that it is not implied that a straight-line or polynomial regression will fit the observed incidence of migration in a despotic situation any more than in a free situation. A migration threshold should still be found, in this case at the point at which $h_p h_{rhp} = \bar{E}$. Note, however, that the value of h_{rhp} is in part a characteristic of the animal rather than of its environment. According to the arguments of Chapter 17 (p. 348), therefore, an animal's measured response to habitat and migration-cost variables should ideally be influenced by its resource-

holding power. It can be postulated, therefore, that selection will have favoured linkage between an animal's resource-holding power and the physiological environment (=internal threshold) against which perceived environmental variables are compared (p. 347). Furthermore, any change in resource-holding power with time should similarly influence change in migration threshold with time.

Migration in a despotic environment can readily, therefore, be accommodated within the model developed in Chapters 7–10. The only modification necessary is that h_d becomes heavily, if not totally, biased toward relative resource-holding power rather than toward deme density.

Fig. 18.25   A hypothesis of the effect of increase in deme density on the optimum size and degree of overlap of home range

This figure is concerned solely with the influence of deme density on optimum adaption to selection against two or more individuals exploiting the same site of a resource. The figure is not concerned with situations such as that of the gorilla (p. 396) in relation to food, in which apparently there is minimal if any selection against exploitation of a resource site by more than one social unit. Nor does the figure take into account selection for aggregation due to predation. Food availability per unit area is also considered to be a constant.

The bottom diagram is concerned with the influence of deme density on size of home range (solid line) and core area (dashed line). The upper diagram is concerned with the degree of overlap of the home ranges of neighbouring individuals or groups. Discussion begins at the left-hand part of the curves and considers in turn positions A to E indicated by vertical lines.

A. Deme density is low such that there are areas of unoccupied habitats that are of greater suitability for colonisation by migrants than occupied habitats. There should be minimal overlap of home range. This virtual absence of overlap results from two individuals, whenever they find themselves exploiting the same sites, responding by each ranking these shared sites at low suitability, re-visiting them infrequently, and perhaps extending their respective home ranges into other, unoccupied, areas by a process of exploratory and calculated migration. Home-range size should be the optimum for the species and limited by the migration cost involved in further expan-

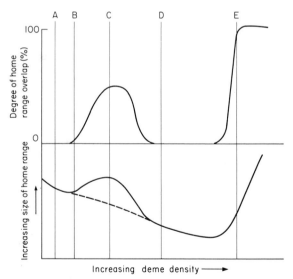

sion outweighing the gain from resources that results from such expansion. Increased deme density at this point probably causes a slight decrease in size of home range and appearance of a core area as peripheral habitats all become ranked at lower suitability than more central habitats. Incoming migrants rarely dispute areas but, upon finding them occupied, continue migration until they encounter an unoccupied habitat. The studied demes of the peacock butterfly, *Inachis io* (p. 403), and weasel, *Mustela nivalis* (Fig. 18.22—after November 1961), probably were at this point on the curve.

B. Deme density is still low as in A but unoccupied habitats are fewer and most invididuals are more or less surrounded by occupied habitats. Some overlap of home range and a large core area seem likely. Incoming migrants probably still continue to migrate until they encounter an unoccupied area. For reasons given in both A and C, change in deme density, whether increase or decrease, seems likely to lead to an increase in home-range size, but only increase in deme density leads to an increase in the degree of overlap and a decrease in the size of the core area. The variety of studies of insectivores and small rodents discussed by Stickel (1960) for which decrease in deme density led to an increase in the size of the home range may have occupied either this part of the curve or position D.

C. Deme density is higher than in B. All habitats are occupied of both higher and lower suitabilities (h_p— p. 70) but at different densities such that habitat suitability ($h = h_p h_d$—p. 70) is approximately the same in all areas that can support a minimum of one social unit. Core areas are a marked phenomenon and there is considerable degree of overlap. Because the resources in many habitats are being shared, home ranges may be larger than in B due to lower uptake from each site in the area of overlap. It is not yet, however, advantageous to invest time and energy in the defence of a resource in the area of overlap except perhaps when two individuals or groups happen to arrive soon after one another at the same site whereupon defence by the first to arrive and retreat by the second may be advantageous to both. Despotic effects are not yet clearly discernible but appear with only slight increases in deme density. Incoming migrants probably assess deme density in different areas (p. 402) but migration threshold may well involve also an assessment of relative resource-holding power. In years of low competition this point on the curve may be occupied by the blue tit, *Parus caeruleus* (Kluijver and Tinbergen 1953, Fretwell 1972) but increase in deme density leads directly to a despotic situation as in D (J.R. Krebs 1971). The coati, *Nasua narica*, may occupy this part of the curve (Fig. 18.24) as also may the males of the side-blotched lizard, *Uta stansburiana* (Tinkle 1967a,b).

D. The part of the curve between C and D covers an area of increasing despotism (i.e. increasing difference in reproductive success between different members of a deme due to the exclusion of some individuals from a resource as a result of the activities of other individuals), decreasing size of home range, and decreasing overlap until at D the core-area phenomenon has once again disappeared. Indeed, territorial defence may involve the animal spending proportionately more time around the edge of the range than elsewhere. All of these effects may be attributed to the decrease in efficiency of defence and increased migration cost with increase in distance between the centre of the range and the edge. This part of the curve is the only part in which size of territory may show a straight-line regression on deme density. Deme density is such that for many individuals with a low resource-holding power, optimum strategy is to subsist in areas of low suitability but to make frequent exploratory migrations or migrations for re-assessment into areas of higher suitability, especially to those habitats ranked most likely to become unoccupied. The migration threshold for both territory holders and incoming migrants consists to a large extent of relative resource-holding powers (p. 403). The red grouse, *Lagopus lagopus*, in Scotland (p. 192 and Watson and Miller 1971) seems to occupy this part of the curve. Here, increase in deme density leads to a decrease in size of territory and individuals with low resource-holding power subsist around and between occupied areas and readily take over any territory that falls vacant. The black-tailed prairie-dog, *Cynomys ludovicianus* (Fig. 18.26), also probably occupies this position as may the weasel, *Mustela nivalis* (Fig. 18.22(a)), and the females of the side-blotched lizard, *Uta stansburiana* (Tinkle 1967a,b).

E. Beyond a critical deme density the time and energy that would have to be expended to defend an area large enough to support an animal becomes so great that it outweighs the advantage gained as a result of defence. Despotism breaks down and individuals move freely within one another's home range, defending only the resource being exploited at the moment. The result is a sudden switch to almost 100 per cent home range overlap and a sudden increase in size of home range due to each area being exploited by several individuals. Deme density and habitat suitability (h_p—p. 70) probably make the largest contributions to the migration threshold. This situation develops regularly each afternoon in spring in the small tortoiseshell butterfly, *Aglais urticae* (p. 408). The transition from position A to position E in animals such as butterflies that are not pre-adapted for aggression occurs suddenly without really passing through positions B to D because these are to some extent dependent on one individual being able physically to prevent another individual from entering an area. This probably also accounts for the behaviour of most butterflies when many individuals utilise a single area: that is, for no attempt to be made to exclude other individuals from the area. When despotism disappears due to high deme density in an animal that lives in a familiar area, the ranking system becomes inefficient. The result is selection for individuals to travel together but to compete for the resource at source. Such selection can perhaps make a large contribution to the evolution of bird flocks, most of which probably occupy this position on the curve. However, the flocks then become a territorial unit and the curve is repeated over again with respect to size and overlap of the home range of neighbouring flocks.

Note that the curve is an optimum response curve that applies both within and between species, the threshold variable for the response being deme density. An individual of any species, experiencing a certain deme density, performs the optimum response. Hence, as deme density for a single species changes, so too does the optimum size of home range and optimum degree of overlap. The deme (Fig. 18.22) of the weasel, *Mustela nivalis*, studied by Lockie (1966), for example, changed from mid-way between positions C and D on the curve to position A in the space of a month. The blue tit, *Parus caeruleus*, and great tit, *P. major*, occupy a position that varies between C and D in different places in different years during the breeding season but in winter occupy position E and form mixed feeding flocks.

It should be noted, of course, that even though in a despotic situation some individuals seem to suffer very badly, they nevertheless perform the optimum strategy for themselves given their particular resource-holding power. As long as an individual does not die or is not rendered sterile by its behaviour, there is always the possibility that at some time in the future circumstances may change and some portion of potential reproductive success may be realised. In all interactions between residents and migrants, whether in a free or a despotic situation, both individuals are suggested to behave in the way most advantageous to themselves as defined in Chapter 17 and here. By definition, however, in a free situation both individuals achieve the same potential reproductive success as a result of their actions, whereas in a despotic situation the territory holder achieves a higher reproductive success than the non-territory holder.

Although the papers referred to in the preceding paragraphs have advanced considerably our qualitative understanding of territorial phenomena, there is still a need for quantitative data to evaluate some of the ideas put forward. This need is as real for the territorial phenomena themselves as for the migration phenomena associated with them. Apart from the observations of King (1955) for the black-tailed prairie-dog, *Cynomys ludovicianus* (Fig. 18.26), for example, I am not aware of the existence, apart from anecdotal accounts, of any detailed description of the procedure by which an exploratory migrant sets about firstly assessing h_p, then gaining a foothold in an already occupied territory, comparing its resource-holding power with that of the occupant, and then finally ousting the original occupant. Similarly, there is a need for measurements of migration thresholds of different components of resource-holding power in relation to the suitability of the required resource or habitat and the mean expectation of migration.

To a limited extent such measurements have been made for the peacock butterfly, *Inachis io* (R. R. Baker 1972a). In this species, which in the deme studied seemed to be existing in a free situation, migrant males sometimes entered into territorial interactions with territorial males. This escalation apparently took place, not because the mean expectation of migration was low, but because when the migrant arrived it received no clue that the territory was already occupied. This invariably resulted from the territorial male being temporarily absent as a result of following a female or interacting with

another male. On such occasions, when the territorial male returned, the pair engaged in territorial interaction. The territorial male was more likely to displace the migrant during the first two interactions. The success of the male was a function of the length of the interaction and hence the distance from the territory travelled by the pair (Fig. 18.5). Success during the first two interactions could therefore tentatively be attributed to the greater ability of the territorial male to return to the territory. Familiarity with the area surrounding the territory, therefore, seemed to increase resource-holding power. From the third interaction onwards, however, if the migrant succeeded in finding its way back to the territory after the previous interactions, the migrant and territorial male had an equal probability of success that was no longer a function of length of interaction. Instead, migration threshold rested on the result of a comparison of aerial manoeuvrability. During the territorial interaction each male attempted to attain and maintain a position above and just behind the other male. When one male achieved this position after a series of spiralling 'leap-frog' manoeuvres during which the pair climbed, often up to a height of about 10 m, the other male initiated a series of dives and climbs, often involving further spiralling, until the pair disengaged and one or both returned to the territory. If the male that gained the ascendancy during spiralling is referred to as the winner and the other as the loser then, on average, a male has only to lose two consecutive territorial interactions for it to initiate migration, whether or not it was previously the territorial male or the migrant.

If both males suffer a disadvantage as a result of sharing a territory, it is an advantage to *both* males (given a sufficiently high value for $\bar{E}$) for *one* of them to initiate migration. The male that suffers least disadvantage as a result of initiating migration is the male least likely to win any future competition between the two males for any receptive female that may enter the territory. As there is a large aerial component to the courtship of this species, this male is likely to be the one that aerially is the least manoeuvrable. These observations were made in a highly suitable corner territory with a large female catchment area. Furthermore, the males studied belonged to a deme for which the mean expectation of migration was likely to be high. We may conclude, therefore, that in such a habitat in such a deme the migration threshold difference in resource-holding power, once familiarity with the area is no

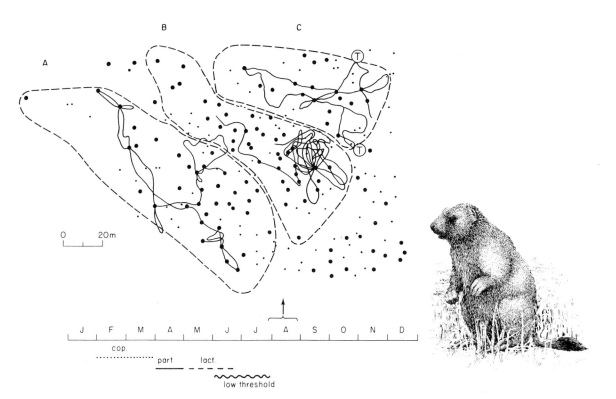

Fig. 18.26　Track pattern of the black-tailed prairie-dog, *Cynomys ludovicianus*

Seasonal behaviour as observed in the Black Hills of South Dakota is shown at the bottom of the diagram. Dotted line, copulation period; solid line, parturition period; dashed line, lactation period; waved line, period of low threshold for removal migration by the adults (see Fig. 17.27).

The main diagram shows the track pattern of three individuals within their territories as observed in August, after the period of greatest probability of removal migration by adults.

A. Track pattern of an adult female during a 286 min period. The territorial unit of which this female was a member consisted of 1 adult male, 4 adult females, and 14 young of the year.

B. Track pattern of an immature male during a 430 min period. The territorial unit of which this male was a member consisted of 31 young of the year.

C. Track pattern of an adult male during a 180 min period. The territorial unit of which this male was a member consisted of 1 adult male, 2 adult females, and 8 young of the year. The encircled letter T indicates where territorial interactions took place.

The large and small dots indicate respectively burrows with and without a crater and mound.

The distance travelled by a single individual during the course of a day ranges from 975 to 4300 m ($\bar{x} = 1690$ m). On the whole, feeding movements are relatively short, the longest movements being those concerned with territorial defence which involve the animal moving rapidly to the edge of its territory to intercept intruders. The track pattern that gives maximum coverage of the territory is one that takes the animal from one burrow to another with relatively infrequent retracing of steps. Such a track is shown by both the adult female in A and the adult male in C. The immature male in B was a young of the year in a territory from which the parents had recently performed removal migration (Fig. 14.7) and the intensive track based on a single burrow could be interpreted as exploratory migration, part of the process of the establishment of a familiar area.

In the absence of immediate territorial opposition (note the female in A) individuals may perform exploratory migration into a neighbouring or even more-distant territory. On occasion such exploratory migrants attempt to establish themselves in the already occupied territory. In the first place this usually involves an attempt at exploitation of the food resources of the territory followed by an attempt at establishment in one of the burrows. If during successive territorial interactions the immigrant succeeds in maintaining and then expanding its position it may eventually succeed in becoming a member of that territorial unit or, if a male, in ousting the resident male from that unit.

[*Compiled from diagrams, tables, photographs, and observations in J.A. King (1955)*]

longer a component of resource-holding power, is that sufficiently great for a male to lose two territorial interactions in succession.

In a deme of the small tortoiseshell butterfly, *Aglais urticae*, studied in the same way (Baker 1972a), deme density was much higher and it may be assumed that the value of $\bar{E}$ was much less. The critical factor, however, seemed to be $\bar{E}_e$, the mean expectation of migration based on recent experience (p. 82) for, although early in the afternoon migrant males, finding an area already occupied, initiated further migration in much the same way as *Inachis io*, as the afternoon wore on the migration threshold increased. The result was that, as butterflies are physically incapable of inflicting injury on one another, incoming males became less and less likely to initiate migration so that by mid-afternoon several males were likely to be sharing the more suitable territories and the territorial system had in effect broken down. Once a male contacts a receptive female in this species the pair remain in the territory, which is also the oviposition and roosting site, instead of the male following the female across country as in *I. io*. From then on until going to roost and copulating the male guards only the female (Fig. 18.8), not the entire territory.

A consideration of the use of the terms home range and territory has been left until the end of this section so that the range of relevant phenomena could be described in advance. So many people have now defined and discussed territory and the territory concept that consideration of the most useful application of the term has become highly personalised, and I do not claim that the use of the term in this section is inherently any more useful, except perhaps in the present context, than other uses of the term. Most authors include the concept of defence within their definition of territory and territorial behaviour and usually some indication of area is also involved. However, very often the area defended is not stationary but mobile, as for example in the case of the wildebeest, *Connochaetus taurinus*, that during the mating season defends the ground occupied by a moving group of females within an almost continuously moving herd (Talbot and Talbot 1963). A more general use of the term territorial behaviour is obtained, therefore, by defining it as defence of a resource rather than simply defence of an area,

though in most cases of course there is little distinction between the two. On this definition a territory becomes an area occupied by a defended resource or resources. A resource may be defined as some part of the environment without access to which an animal suffers a reduction in potential reproductive success. This concept of a territory evolved during discussions with Dr. G.A. Parker and, at least at the time of writing, represents a joint view of the phenomenon.

The generality of the above concept of territory and territorial behaviour depends, of course, on what is considered to be defence. The view adopted here is that any behaviour that reduces the probability of another individual exploiting the resource may be considered to be defence. Any form of advertisement (i.e. singing, marking, or simply manifestation of presence), because it decreases the probability of a migrant settling in the area concerned and exploiting the enclosed resources, constitutes just as real a defence in this sense as other directly more aggressive behaviour such as chasing or fighting. Because for the reasons already given, in almost all cases, even in such species with home-range overlap as the chimpanzee, *Pan troglodytes* (Fig. 18.23), the presence of one territorial unit influences the behaviour of another as far as exploitation of resources is concerned, there seems little point in distinguishing between a home range and a territory. Instead the descriptive convention that has been adopted in this book has been to use the terms interchangeably, with a tendency to use 'home range' when describing an area from the point of view of an animal's movements and to use territory when describing an area from the point of view of relative exploitation of resources by neighbouring individuals or groups. This approach will not suit many mammalogists but is the one found most useful here and indeed is the only approach that seems justifiable unless a case is made out for distinguishing between defence by advertisement (e.g. singing, marking, and manifestation) and defence by more-overt aggression (e.g. chasing and fighting). However, as advertisement and more-overt aggression often represent different stages of escalation in relation to a single area, such a distinction would be difficult to justify.

18.4 Summary of conclusions concerning the basic forms of the lifetime track

1. The lifetime track is the path through space traced out by all animals from birth to death (p. 27).

2. Although continuous, the lifetime track can be broken down for analysis into removal migration units from one point in space to another. For maximum usefulness the track of such a migration should not cross or touch itself, though it may cross or touch the track of earlier removal migrations (p. 36).

3. The selective pressures acting on removal migrations and the characteristics of removal migrations that are the result of these pressures have been the subject of Chapters 14–17. The form of the lifetime track can therefore be generated by application of the conclusions of these chapters to the removal migrations of which the lifetime track is composed (p. 359).

4. When two or more consecutive removal migrations are considered they form a migration pattern which may constitute either a return or a non-return removal migration (p. 61).

5. Return migration can only be advantageous to an animal adapted to spatially separate habitat-types that fluctuate in suitability out of phase with one another. Return migration is advantageous when, after a period in a second habitat-type the example of the first habitat-type to which migration is most advantageous is the example of that habitat-type that was previously occupied. Migration cost during return migration includes the added component that results from the cost of navigation (p. 63).

6. When neither of the situations mentioned in the previous item apply, non-return removal migration is the optimum migration pattern (p. 63).

7. The form of the lifetime track is the product of the way that return and non-return removal migrations follow one another. The two extremes of the lifetime track may be described as the linear range and the limited-area range. These forms represent the extremes of three separate spectra (Fig. 18.6), one of which, however, seems to be more common than the other two.

8. The linear range is characterised by a number of temporary limited area ranges, which may or may not be familiar areas, connected by rectilinear tracks across areas of low suitability, successive rectilinear tracks being in more or less the same direction. The result is that the individual travels across country with shorter or longer stays in areas of higher suitability en route. The observed track patterns within and between areas of higher suitability can be argued to be

adaptations that give optimum search strategy, optimum degree and frequency of return, and optimum stay time. To a large extent, also, the degree of manifestation of the linear range results from the hierarchical system of migration thresholds and the frequency with which the migration thresholds at different levels of the hierarchy are exceeded (p. 361).

9. The limited-area range most often results when only the 'lower' levels of the migration-threshold hierarchy are exceeded with any frequency. The result is that an animal spends its life in a relatively circumscribed area (p. 371).

10. Selection for time to be spent in a limited area invokes further selection for the evolution of either marking behaviour or a spatial memory that increases the efficiency of the track pattern (p. 371).

11. The migration sequence that results when selection favours non-return removal migration from one example of a given habitat-type to another example of that habitat-type but for return to the same example of another habitat type is an exploratory migration. Selection on the track pattern during exploratory migration acts to maximise the area searched for examples of one habitat-type while minimising the migration cost for eventual return to a specific example of the same or another habitat-type (p. 374).

12. The combination of exploratory migration and a spatial memory and/or trail laying results in a familiar area. A familiar area is an area from any one point in which an animal is capable of finding its way to any other point. Because this ability is likely to decrease with time after the previous visit to a point, the familiar area is somewhat less than the lifetime range (p. 378).

13. During the course of successive exploratory migrations, an animal encounters a large number of examples of each habitat-type that contains a resource for that animal. Optimum strategy with respect to these habitats is suggested to be to rank them in order of suitability (p. 379) so that when the suitability of the previously most suitable habitat decreases below that of the previously second most suitable habitat, calculated removal migration to this latter habitat takes place. This behavioural strategy has been termed the ranking strategy or ranking system (p. 379).

14. Because the accuracy of the ranking system is likely to decrease with time after ranking, the advantage of the system is likely to impose selection for animals to re-visit ranked habitats at intervals principally to re-assess their suitability (p. 379).

15. Although the advantage of the ranking system cannot account for the initial evolution of the familiar area, the advantage that most modern animals derive from their familiar area is likely to be largely a function of the advantage gained from the ranking system (p. 396).

16. The home range is considered to be that part of the familiar area utilised during a specified period. Home range size is, therefore, a function of time. It is also a function of sex, season, resource abundance, body size, feeding strategy, size of the social group occupying the range, and deme density (p. 384).

17. Body size, size of home range, age at first reproduction, and length of the pre-reproductive period of exploratory migration form an adaptive complex, at least for terrestrial mammals. Selection on any one variable also imposes selection on the other variables. It is suggested that most often during the adaptive radiation of terrestrial mammals selection has entered this adaptive complex through selection on body size. Selection for change in body size imposes selection for change in size of home range which in turn imposes selection for change in the length of the period of exploratory migration and change in the age of first reproduction (p. 395).

18. Selection on the efficiency of the ranking system imposes selection against overlap of neighbouring home ranges. Optimum size and degree of overlap of home range is a function of availability of resources and deme density. The relationship is unlikely to be simple, however, and over only a limited range of deme density can size of home range be expected to show an approximately linear regression on deme density (Fig. 18.25). Most of the adaptive responses of size, structure, and degree of overlap of home ranges can be attributed to limited resources or to selection against an inefficient ranking system (p. 399).

19. Territorial behaviour is considered to be the defence of a resource, which may or may not be stationary. A territory is the area occupied by the defended resource or resources. No distinction is made between a home range and a territory except that the term 'home range' tends to be used when discussing an area with respect to an animal's movements and the term 'territory' tends to be used when discussing exploitation of a resource by more than one individual or group (p. 408). Instead of distinguishing the two terms a spectrum of territorial interactions are noted ranging from free interactions at low and very high deme densities to despotic interactions at intermediate deme densities. In a free situation all individuals should have the same potential reproductive success. In a despotic situation an individual's potential reproductive success is a function of its resource-holding power and selection acts to favour individuals that perform the behaviour patterns that are optimum for their particular resource-holding power (p. 406).

20. At low deme densities, manifestation of presence is sufficient to cause incoming migrants to continue migration, and habitat suitability excluding the influence of resource-holding power is of major importance. At higher deme densities, deme density relative to habitat suitability excluding the effect of relative resource-holding power makes an important contribution to migration threshold. At yet higher deme densities, when a despotic situation prevails, deme density gives way to relative resource-holding power as the variable making the major contribution to migration threshold. At still higher deme densities, when the despotic situation has broken down, deme density relative to habitat suitability excluding the effect of resource-holding power again becomes most important (p. 405).

21. The characteristics of migration in a despotic situation can be accommodated within the model developed in Chapters 7–10 by adding the variable of relative resource-holding power (p. 403).

19

Aspects of the linear range: the mechanics of rectilinear migration and the phenomenon of re-migration

19.1 Introduction

The major portion of Part IIIb of this book is concerned with the limited-area range. This chapter and the next, however, are concerned with various aspects of the linear-range form of the lifetime track. Both chapters are concerned to some extent with the migration mechanisms (e.g. orientation mechanisms and interaction with environmental forces such as air and water currents) by which rectilinear (straight-line) migration is achieved. Both chapters are also concerned, however, with the selective pressures that have moulded the various spatial components of the migrations, particularly distance, direction, and direction ratio.

This first chapter is concerned primarily with re-migration, an aspect of the linear range that appears when the lifetime tracks of more than one generation of an animal are studied. In re-migration, the members of a given species move in one direction at one time and in another more or less opposite direction at another time. The different directions are usually taken up with some regularity and usually, but not always (e.g. lemmings), seasonally. Unlike return migration, however, the individuals involved in these different movements are not the same but belong to different generations.

The major part of this chapter concerns insects. Most vertebrates have a generation interval of sufficient duration for most selection for regular movements in opposite directions to favour return migration. Only in the case of the Scandinavian lemming, *Lemmus lemmus*, are there data available concerning the migration mechanisms involved in a vertebrate re-migration. Insects, however, of all sizes show a wide range of re-migration behaviour, much of which is still the subject of considerable controversy.

19.2 The Scandinavian lemming, *Lemmus lemmus*

Nearly all lemmings, not only *L. lemmus*, perform seasonal return migrations. These are described in Chapter 25 (p. 525). In the case of the Scandinavian lemming, which is primarily a montane species, this movement is altitudinal.

Superimposed upon this seasonal pattern there is a pattern of removal migration and re-migration that is less a function of season and more a function of deme density. In a flat, open area this removal migration would probably take place in all directions as seems to be the case for most other lemming species, but because the Scandinavian lemming occupies a habitat-type characterised by strong altitudinal zonation, a more definite pattern is recognisable. The preferred habitat-type of the species is the alpine zone of Scandinavian mountains (i.e. above the tree-line), and it seems likely that for a constant deme density, habitat suitability decreases with decreasing altitude through the birch wood and pine wood zones to the valley below (because $h = h_p h_d$—p. 70). Consequently it seems likely that the deme density at which the removal migration threshold is exceeded decreases with decreasing altitude.

Fig. 19.1   The Scandinavian lemming, *Lemmus lemmus*

[*Photo by Arthur Christiansen* (*courtesy of Frank W. Lane*)]

Lemming removal migrants are likely to begin as exploratory migrants (p. 64) but at some point, given the appropriate situation (p. 66), to become either calculated or non-calculated removal migrants. In the latter case, they are likely to adopt a straight-line course (p. 362). The direction of this straight-line course, to be an ideal adaptation, is perhaps likely to be a function of habitat variables encountered during the early exploratory migration sequence. An animal encountering low deme densities in the pine or birch wood zone should ideally perhaps orient exploratory migration entirely up the slope. Animals that encounter high deme densities at whatever altitude should ideally initiate exploratory migration and then later, perhaps, non-calculated removal migration, down the slope.

We may postulate, therefore, that a lemming for which the exploratory migration threshold has been exceeded, and these will usually, but not always, be young pre-reproductives (Chapter 17), initiates migration from its previous or natal home range. Following a period of exploratory migration, the lemming then either establishes a new home range following calculated removal migration or initiates a non-calculated removal migration up or down the slope as appropriate, attempting to establish a home range in the first suitable habitat encountered. Whether the situation is free or despotic (p. 402) cannot as yet be determined, though the latter seems likely from the amount of aggression that those animals usually exhibit. In this case it would be expected that the longer-distance migrants have a lower resource-holding power and achieve a lower reproductive success than those that move only short distances. However, because of ontogenetic variation in migration threshold (Chapter 17), it is to be expected that migrations consist primarily of pre-reproductives. At their peak, most observed lemming migrations consist of up to 80 per cent young individuals (Marsden 1964, Bergstrom 1967).

Curry-Lindahl (1962) described a sequence of events in 1960 that is quite consistent with this thesis. Throughout the summer, lemmings occupied their preferred positions on the Scandinavian fells, though some (exploratory?) migrants were always present. Gradually, however, as deme size and density increased, lemmings moved down into the willow scrub and birch regions and gradually from there down to the pine forest. The first wave of lemmings invaded the pine forests in May. A second increase in number of migrants appeared in June/July, consisting partly of the offspring of the lemmings that had settled and bred in the pine forests and partly of downward migrants from higher altitudes. A third major appearance of (exploratory and non-calculated removal?) migrants appeared in the pine forests from August to October.

Such increase in deme density does not necessarily result in the next generation migrating into the valley in large numbers. The critical factor, however, seems to be the deme density (relative to resource availability) reached in the pine forest. It should not be imagined, however, that increase in deme density occurs at a similar rate on all mountain slopes within a given area. Very often there is a considerable increase in deme density in a lemming deme on the slopes on one side of a valley and much less of an increase or even a decline on the other side (Marsden 1964). If habitat suitability for a given deme density decreases with decrease in altitude, there is likely to be a given deme density and food availability in the low altitude pine forest into which a given lemming is born, beyond which search for a suitable habitat in that forest is less likely to be advantageous than long-distance migration and search for a mountain slope that can perhaps be occupied at a higher altitude. This is yet another example of a migration-threshold hierarchy (p. 48), the highest migration threshold corresponding to the longest-distance migration (p. 229). It is now known that lemmings migrate up mountains as well as down (Marsden 1964, Bergstrom 1967). Lemmings migrating in opposite directions along a track evidently register one another as they pass (Bergstrom 1967) and quite possibly the decision whether or not to climb the opposite mountain slope is a function of the number of lemmings encountered migrating in the opposite direction. Quite possibly when large numbers are encountered, most lemmings from mountain slopes on both sides of the valley eventually turn and migrate down the valley instead of climbing the opposing slopes. Perhaps migration down the valley continues until lemmings migrating down from the local mountain slopes are encountered in sufficiently low numbers (the migration threshold rising as migration proceeds— p. 82) whereupon the lemmings may turn and migrate up the mountain slope until habitat suitability no longer increases.

Migrating lemmings retain their predominantly nocturnal activity pattern, though a few continue to migrate during the day (Marsden 1964). Unlike exploratory and short-distance non-calculated removal migrants at higher altitudes, long-distance non-calculated removal migrants take the straightest possible track across unsuitable habitats such as fields and bodies of water, thus suggesting that for these individuals the expenditure of time and energy makes a larger contribution to migration cost than does predation risk. Because the straightest track across country remains relatively constant from individual to individual, tracks become worn and are followed by later migrants, presumably as the line of least resistance (Bergstrom 1967). Reindeer and human tracks are also followed (Marsden 1964). Nevertheless, lemmings migrate individually with no evidence of orientation to other individuals. Although long-distance migrants migrate straight across fields, they take cover in the presence of predators.

When large water bodies are perceived at a distance lower down in the valley, lemmings adjust their orientation in an attempt to avoid having to cross the widest part of the water body (Bergstrom 1967). Any migration the advantage of which is crossing from one mountain slope to another must also involve the ability to cross a variety of water barriers ranging in size from streams to lakes. Upon encountering a water barrier, however, the lemmings do not immediately start to cross unless they can see the opposite bank and even then not unless the water is calm. In fog and wind, therefore, lemmings do not start to cross larger water bodies, such as lakes, and lemmings that accumulate on a lakeside by night migrate back into the forest to seek shelter during the day and delay attempting to cross the lake until conditions are favourable. However, time contributes to the migration threshold (p. 51). So too does deme density. If the delay is sufficiently long, therefore, and/or if large numbers accumulate at the water's edge, such as can happen particularly on a peninsula, many may start to swim even though conditions are not ideal. Once in the water, lemmings orientate to topographical features on the other side, if these are visible, or to islands. As long as the cumulative advantage of crossing water bodies throughout the species range is greater than the disadvantage that results from the increased mortality risk, selection will not favour the evolution of a total water barrier avoidance reaction (Baker 1971). Those individuals that die at the edge of the range as a result of entering the sea are the unfortunate few for whom a behaviour pattern that is advantageous elsewhere just happens to be disadvantageous at the unrecognisable edge of the species range.

When crossing ice, lemmings move at a rate of about 5–6 km/h in spring and 4 km/h in autumn (Marsden 1964). Unlike on land, no two individual tracks are superimposed, even for a short distance. Once lemmings set off over ice they move on a direct, but not necessarily the shortest, route to the opposite shore, apparently orientating to the silhouettes of hills across the lake. Some lemmings may not orientate directly to the opposite shore but may instead orientate to the highest mountain. At the same time, while aiming for the peaks, the lemmings avoid crossing wide stretches of open ice. These observations, coupled with the apparent difficulty of orientating while crossing rough water that obscures the opposite shore, suggests that lemmings do not make use of a compass, either celestial or magnetic, but instead orient entirely to topography. Orientation down the mountain slope at the start of long-distance migration, towards the opposite mountain slope while crossing water bodies if few lemmings have been encountered moving in the opposite direction, but down the valley if many individuals have been encountered, followed by orientation up a mountain slope when appropriate, seems to be the ideal orientation strategy if long-distance migration in lemmings has evolved for the reasons presented here.

The hierarchy of migration thresholds and the observed migration mechanism seems to be highly adaptive for an animal liable to violent fluctuations in deme density in an environment showing marked altitudinal zonation. When deme density declines, assuming that it does so first at lower altitudes, then as suggested above, young and other exploratory migrants should respond to habitat suitability by orienting their exploratory migrations up the mountain slope. In this way, as the deme declines, succeeding generations should rejoin the lemming demes in areas higher and higher up the mountain slope. The lemming generation interval is so short (p. 524), however, and ontogenetic variation in migration threshold is such (Chapter 17), that few of these migrants are likely to be the same individuals that migrated down the slope earlier. The migration is thus a re-migration rather than a return migration. The re-migration back up the slope is inevitably less often observed as it involves fewer individuals during periods of low deme density. Nevertheless, sufficient observations are available (Bergstrom 1967) for it to be accepted that re-migration back up the mountain slope does occur.

19.3 Butterflies
19.3.1 Migration distance and the migration mechanism

Cross-country migration by butterflies is achieved by an alternation between periods of residence in restricted areas during which the flight track shows a frequent change of direction (Fig. 18.4), and periods of rectilinear migration across areas that do not provide the required resource of the moment (Fig. 18.1). The rectilinear track is considered to represent an optimum search strategy for relatively immobile resources (Fig. 18.2). Inter- and intra-specific variation in migration threshold occurs and is discussed in Chapters 16 and 17. Thus individuals of some species never leave the restricted area into which they emerged as adults, whereas other individuals of other species rarely spend more than an hour or so, except overnight, in any given restricted area. Butterflies have migration thresholds, not only to habitat variables but also to the migration-cost variables of temperature, wind speed, and sunshine (Chapter 16). Perhaps as much as 99 per cent of butterfly migration takes place while the Sun is shining, at least in temperate regions. Consequently rate of migration in butterflies can only meaningfully be expressed in terms of distance per hour of sunshine.

Two broad categories may be recognised for the migration mechanism of butterflies, depending on whether selection has favoured migration *distance* being facultative or obligatory (p. 83). The facultative determination of migration distance has received most attention and is probably the more primitive of the two mechanisms. The associated migration mechanism is therefore considered first (Fig. 19.2).

Like some other strong-flying diurnal animals (e.g. bees—p. 875; birds—p. 857) butterflies take their primary orientation during cross-country migration from the Sun's azimuth but also make use of leading lines. Unlike these other animals, however, butterflies that use the mechanism shown in Fig. 19.2 do not compensate for the diel shift in the Sun's azimuth. It has been argued (Baker 1969a) that compensation for the Sun's movement is disadvantageous when distance is facultative. Compensation for the Sun's movement would increase the straight-line distance between the places of emergence and death but would not increase the area searched (compare the two left-hand tracks in Fig. 18.2).

Under conditions of selection for facultative distance, therefore, compensation is not advantageous. In this case, because of the extra mechanisms and energy that must be involved, first in learning the Sun's movement (p. 875) and then compensating for it, the process of compensation seems likely to be a disadvantage.

The altitude at which butterflies migrate seems to be determined by three major selective pressures. The first results from the height at which suitable habitats may be perceived (by vision or olfaction—Nielsen 1961) and at the same time permit the butterfly to exploit the habitat. A high-flying butterfly migrating with the wind may be unable to reduce height sufficiently quickly to locate and exploit the habitat. There is no advantage in perceiving a resource if it cannot efficiently be exploited. The second selective pressure concerns the size of the search area. Presumably a butterfly has a larger visual search area, but not necessarily a larger useful olfactory search area, if it flies higher. Optimum height is therefore given by that height that gives a maximum perceptual area and yet is not too high for a resource to be perceived. Thus a butterfly searching for solitary dispersed plants has a lower optimum migration height than a butterfly searching for plants or other resources that occupy larger or otherwise more conspicuous areas. Similarly, the optimum height for migration is higher when crossing areas in which resources are unlikely to be located (e.g. over water, or over grassland if the butterfly is a woodland species, etc.) than when crossing areas in which resources are highly likely to be located. The final selective pressure results from the effect of wind displacement. In general, wind speed, due to friction with the ground, increases with increase in height above ground level. Any advantage of wind displacement, therefore, tends to raise the optimum height for migration and any disadvantage tends to decrease it.

In all butterflies so far examined for which migration distance seems to be facultative, wind displacement seems to be disadvantageous, irrespective of wind speed and direction. Nielsen (1961) has shown that the great southern white, *Ascia monuste*, in Florida, maintains a rate of cross-country migration of 10–15 km/h, as measured over tens of kilometres, over a wide range of speeds of head and tail winds. Table 19.1 analyses marking–release/recapture data obtained by Roer (1968) for the small tortoiseshell, *Aglais urticae*. Again there is no association between wind speed and direction and migration speed.

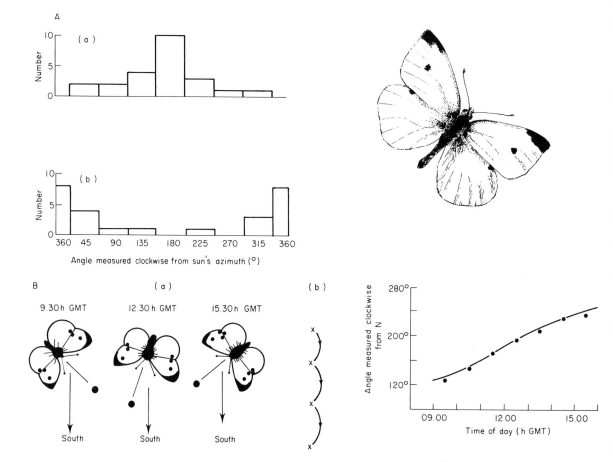

Fig. 19.2   The migration mechanism of butterflies when selection has favoured a facultative determination of migration distance

When inter-habitat distance is unpredictable, selection favours individuals that during migration maximise the area searched with maximum efficiency per unit distance and that settle in the first potentially suitable $(h > \bar{E})$ habitat encountered. Discovery that the habitat is unsuitable $(h < \bar{E})$, a change in the individual's resource requirements, or rapid decrease in the required resource, are all factors that can lead to a high migration incidence. In such a situation there is no selection on the straight-line distance of the lifetime track and migration distance is determined entirely facultatively by the conditions experienced by the individual during the execution of its lifetime track.

A. The reality of an orientation mechanism during rectilinear migration

Individuals of the small white butterfly, *Pieris rapae*, were captured during rectilinear migration across a field, displaced to a different field with different topographical features, and released. Individuals that were flying away from the Sun's azimuth at capture (a) continued to fly at that angle upon release. Individuals that were flying toward the Sun's azimuth at capture (b) continued to fly at that angle upon release. Thus there appears to exist an individual-specific preferred angle of orientation during rectilinear migration.

B. Primary orientation to the Sun's azimuth

Butterflies that seem to have been exposed to selection for the facultative determination of migration distance have, wherever critical data are available, been found to orient to the Sun's azimuth during rectilinear migration. Examination of the mean flight direction (●) of *P. rapae* in late autumn (graph) shows that it is directed toward the Sun's azimuth (solid line) at all times of day, there being no compensation for the diel shift of the Sun's azimuth. It follows, therefore, from A and B that each individual has a preferred angle of orientation to the Sun's azimuth such that the compass direction of the individual shifts during the day as shown in (a). An individual using this migration mechanism migrates across country in a series of semicircular loops as in (b). At other times of year and in other species the mean flight direction is at some angle to the Sun's azimuth other than 0° (p. 435).

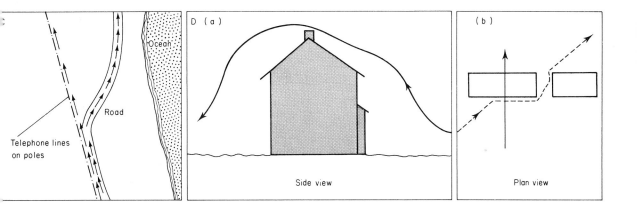

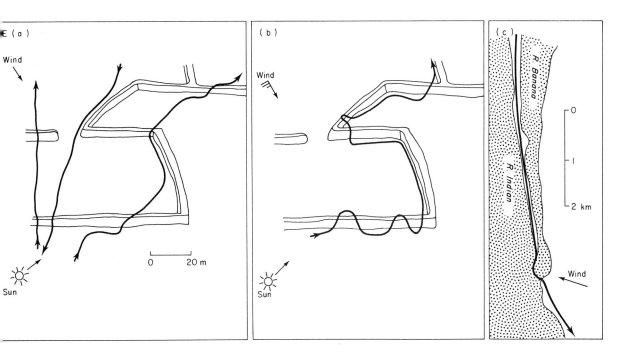

C. Secondary orientation to leading lines

When a leading line coincides with the preferred Sun orientation direction, butterflies delegate the maintenance of migration direction to orientation to the leading line. This secondary orientation ceases, however, whenever the deviation between the leading line and the preferred direction becomes too large.

D. Reaction to obstacles

Migrating butterflies fly over any obstacle encountered at right angles (solid line). When the obstacle makes a relatively small angle to the preferred direction, the edge of the obstacle functions as a leading line (dashed line).

E. Reaction to wind

In light winds, butterflies fly straight across open spaces (a). In strong winds, whether head winds or tail winds, butterflies migrate in the shelter of leading lines (b). Over water (c), the butterflies orient to a distant landmark in the preferred direction if such a landmark is available. Otherwise they either orient to the Sun, but do not compensate for wind drift, or simply orient downwind.

[*Compiled from Baker (1968a, 1969a, b) and Nielsen (1961) with some new material*]

Table 19.1   The absence of ranking correlation between the rate of migration of *Aglais urticae* over distances > 16 km as measured by marking–release/recapture experiments (Roer 1968) and wind speed and direction during the period (sunshine h) from release to recapture

No. (Roer 1968 tab. 2)	Rate of migration (km/h of sunshine)	Rank	Mean wind speed (km/h of sunshine)	Rank	$d°$ = angular difference between track and downwind displacement direction	
					$d°$	rank
1	0·50	47·5	2·40	73	107°	65·5
2	0·57	42	5·86	71	68°	55
3	1·26	13	6·64	65	127°	67
4	0·53	45·5	8·72	57	35°	38
5	0·88	25	11·47	38	31°	34
9	1·25	14	16·27	5	30°	32·5
10	4·17	1	18·90	2	7°	11
11	1·57	9	19·86	1	20°	21
12	1·00	22	16·53	4	52°	51
13	0·37	55	9·22	52	41°	45·5
14	0·53	45·5	11·86	33	136°	72
15	0·76	31	10·37	45·5	60°	53·5
16	1·12	16·5	10·37	45·5	128°	68·5
19	1·29	12	11·50	37	16°	18
20	2·42	3	8·58	59	34°	35·5
21	0·74	32	10·72	41·5	5°	7
22	1·15	15	12·83	18	3°	4
23	1·12	16·5	10·72	41·5	41°	45·5
24	0·28	66·5	9·97	48	60°	53·5
25	0·20	69	8·85	54·5	23°	26
26	0·89	24	8·85	54·5	40°	43·5
27	0·79	29	9·41	51	77°	59
28	2·55	2	10·82	40	19°	20
29	1·70	8	13·07	15	73°	57
30	1·79	6	13·47	12	71°	56
31	0·40	52·5	11·62	35	101°	64
32	1·97	4	13·47	12	50°	50
33	0·99	23	10·38	44	30°	32·5
34	0·62	38	9·10	53	6°	8·5
35	0·09	73	8·61	58	45°	48·5
37	0·50	47·5	6·99	62·5	13°	15
38	0·66	35	10·56	43	37°	40
39	0·40	52·5	6·99	62·5	25°	29
40	1·78	7	9·50	50	1°	2
41	0·70	33	6·88	64	24°	28
42	1·32	10·5	9·86	49	28°	30
43	1·91	5	4·69	72	160°	73
44	0·31	63	8·83	56	130°	70
45	0·85	27	8·16	60	84°	62
47	0·56	43	6·37	69	133°	71
48	1·01	21	6·38	68	107°	65·5
49	0·59	39	6·46	66·5	4°	5·5
50	1·11	18	6·22	70	91°	63
51	0·44	50·5	6·46	66·5	79°	60
52	0·86	26	10·11	47	128°	68·5
53	0·11	72	16·18	6	83°	61
54	0·13	71	17·38	3	76°	58
56	0·35	57	13·89	10	10°	14
57	0·34	59	13·14	14	0°	1
59	0·78	30	12·13	29·5	4°	5·5
60	0·58	40·5	14·61	7·5	17°	19
61	0·35	57	12·13	29·5	22°	23·5
62	0·29	65	12·78	19	34°	35·5
64	0·48	49	12·16	26·5	56°	52
65	0·28	66·5	13·04	16	44°	47
66	1·32	10·5	14·61	7·5	39°	41·5
67	0·58	40·5	12·16	26·5	45°	48·5
68	0·23	68	12·51	20	29°	31
69	0·63	36·5	12·16	26·5	40°	43·5
70	0·33	60·5	13·47	12	35°	38
71	0·63	36·5	13·94	9	7°	11

Table 19.1 *Continued*

No. (Roer 1968 tab. 2)	Rate of migration (km/h of sunshine)	Rank	Mean wind speed (km/h of sunshine)	Rank	$d° =$ angular difference between track and downwind displacement direction	
					$d°$	rank
72	0·17	70	11·95	31·5	8°	13
73	0·31	63	12·16	26·5	6°	8·5
74	0·31	63	11·60	36	39°	41·5
75	0·80	28	11·95	31·5	22°	23·5
76	1·04	20	8·00	61	23°	26
77	0·33	60·5	12·27	21·5	7°	11
78	0·54	44	11·20	39	2°	3
79	0·44	50·5	12·26	23·5	15°	17
80	0·67	34	12·93	17	21°	22
81	1·09	19	12·26	23·5	23°	26
82	0·35	57	12·27	21·5	35°	38
83	0·39	54	11·74	34	14°	16

Spearman's ranking correlation coefficient (ρ):

rate of migration/wind speed

$$\rho = -0·051; \quad t = 0·432; \quad P > 0·10;$$

rate of migration/wind direction

$$\rho = -0·023; \quad t = 0·194; \quad P > 0·10.$$

The justification for the use of mean wind over the total sunshine hours between release and recapture derives from Fig. 19.3. The justification for analysing wind speed and rate of migration independently of wind direction derives from Table 19.2 and Nielsen (1961).

The absence of a relationship between wind speed and rate of migration in this analysis may perhaps be attributed to most individuals being able to migrate in air the speed of which was below threshold (N.B. the absolute wind speed recorded was measured at a height of 10 m above ground level and is much faster than that experienced by the butterflies which fly much lower and are often able to migrate in the shelter of features such as hedgerows, etc.—Fig. 19.2).

Table 19.2 The effect of changes in wind speed and direction on the rate of migration of the small white, *Pieris rapae*

		Change in hourly number crossing open fields		χ^2	P
		increase	decrease		
Change in wind speed between one day and any of the 5 subsequent days	increase	7	23		
	decrease	23	11	14·41	<0·001
	same	12	16	0·57	>0·05
Change in wind direction between one day and any of the 5 subsequent days	head wind to tail wind	8	5		
	tail wind to head wind	5	17	3·74	>0·05
	same	23	27	0·32	>0·05

The justification for using hourly number crossing open fields as a measure of migration rate derives from Figs. 17.28 and 18.1. By restricting comparison of migration rate on any one day to the rate observed on the 5 subsequent days, the effect of seasonal incidence is reduced to a minimum, especially as observation continued throughout several generations.

Wind speed was assessed visually using the usual convention (Meteorological Office 1939) and was considered to be the same, to increase, or to decrease over any two days on the basis of comparison of their respective mean Beaufort numbers (converted to the nearest integer) during the period of observation on each day.

Categorisation of wind as a head wind or tail wind on any given occasion is relative to the mean migration direction for that time of year (i.e. to the NNW through spring and summer until 26 August and S from 27 August onwards). Hence in spring and summer a head wind was a wind from the W, NW, N, or NE and a tail wind was from the E, SE, S, or SW. In autumn a head wind was from the WSW, SW, S, SE, or ESE and a tail wind was from the WNW, NW, N, NE, or ENE. Days in spring and summer with cross-winds from the WSW or ENE or in autumn from W or E are excluded from the lower part of the table, hence the difference in total number of observations in the two halves of the table.

Analysis of the number of small whites, *Pieris rapae*, performing rectilinear migration in eight compass directions (N, NE, E, etc.) across a field in southern England, shows that the number flying in each direction is unaffected by variation in wind direction but is affected by change in wind speed (Table 19.2). Presumably this implies that wind speed but not wind direction contributes to the composite migration threshold. A single example will serve to illustrate the behavioural basis of this phenomenon.

A male small white, *P. rapae*, being followed on 21 June in connection with the experiment illustrated in Fig. 18.1, was exposed to a force 4 (Beaufort scale) NW wind. On two occasions the male attempted but failed to cross an open space approximately 100 m across. This individual stayed on the south side of the open space for 1·7 hours. During this time the flight track showed frequent change of direction, was confined to gardens, and was interspersed with basking, feeding, and (unsuccessful) courting. The open space was not crossed until the male encountered a sheltered track oriented approximately N–S along which there flew a steady stream of butterflies in both directions, though more to the N. Although the open-space barrier was eventually by-passed, there was no doubt that the continuation of the migration of this individual was delayed compared with the situation that would have obtained if the space had been encountered on a day with a lighter wind. This male had been restricted to the south side of the open space by a force 4 head wind. At the same time it was conspicuous that there were no individuals crossing the space from north to south with the wind. Not until observation was made at the sheltered track were south-flying individuals encountered. Observations were then made for a time at the northern edge of the space. Several south-flying individuals were observed to begin to cross the space but immediately on being caught by the wind they dropped on every occasion to within a few centimetres of the ground and flew back northward to the sheltered gardens that had just been left.

We may conclude from the above analyses and observations that the migration threshold to a tail wind occurs at about the same wind speed as the migration threshold to a head wind. It follows, therefore, that the migration strategy of the butterflies being considered here is based on an attempt to migrate across country in the individual-specific preferred compass direction at a given ground speed. The migration threshold to wind has been fixed by selection at that wind speed that is either too energy consuming (if a head wind) or at which the preferred ground speed is impossible (if a tail wind). Similar but not identical migration strategies are encountered in birds (p. 684) and fish (p. 795), that are also monitoring their environment during migration.

The apparent importance of ground speed rather than air speed (see also p. 415) seems to imply that visual scanning of the environment is of primary importance during butterfly migration. This is consistent with the view that the rectilinear track has evolved as part of a search strategy for resources (p. 362). Consider a butterfly migrating with a tail wind. As wind speed increases, the butterfly is carried over the ground faster. Visual perception of resources becomes more difficult. Even having perceived a resource the butterfly is liable to be carried beyond the resource before it can lose height with the result that it is forced either not to exploit that resource or to fly upwind to re-locate it. In addition, the faster the butterfly is carried across the ground, the less control it has in the avoidance of dangerous situations such as spiders' webs (Fig. 16.23) and stronger-flying aerial predators such as birds. As wind speed increases, a relatively constant ground speed can be maintained by the butterfly decreasing its air speed. Until, that is, the butterfly approaches stalling speed or wind speed becomes greater than the preferred ground speed. The neurophysiological basis of this behaviour is described by the oculomotor model of Kennedy (1951) but need not concern us here. Now consider a butterfly flying against the wind. As wind speed increases, the butterfly has to increase its air speed to maintain a constant ground speed, and thus expend more energy per unit cross-country distance.

It is clear, therefore, that given selection to maintain a relatively constant ground speed there will be a migration threshold to wind speed for both a head wind and a tail wind. On the basis of the above factors alone, however, it is unlikely that this threshold speed is the same for all wind directions. Furthermore, because individuals migrating downwind can maintain their ground speed for less energy expenditure per unit distance than individuals migrating upwind, we would expect preferential migration with the wind. Yet neither of these effects are found in the butterflies being considered. Some other factors must be involved. The only possible explanation is that energy expenditure per unit distance is not the only major factor contributing to migration cost. If some other factor were to make an equal contribution to migration cost but in the opposite direction (i.e. less migration cost per unit distance when migrating against the wind than when migrating with the wind), then selection would be expected to produce the situation observed (i.e. wind direction makes no contribution to the composite migration threshold). We cannot yet say what this opposing factor may be. It seems likely, however, that it is a function of air speed. A butterfly that maintains a constant ground speed has a much faster air speed in a strong head wind than in a strong tail wind. If a faster air speed reduces some component of migration cost to the same extent that it increases the energy expenditure component, the above situation evolves. One possibility is that predation risk, particularly by birds and perhaps also by dragonflies, is an inverse function of the butterfly's air speed. Migrating butterflies are preyed upon by a wide range of birds such as corvids (Williams 1958), flycatchers, tits, sparrows, swallows, and swifts. There are no data, however, relating the efficiency of predation to the air speed of their prey.

Returning momentarily to the optimum height for migration above ground level, it follows from the above arguments that optimum strategy is to fly at a height that permits the largest perceptual range to be scanned for resources at the optimum speed. Optimum height therefore decreases as wind speed increases. This behaviour has been reported on many occasions (e.g. Williams 1958, Nielsen 1961) and is familiar to anyone that has observed butterflies in the field, particularly in temperate regions.

So far we have been concerned with butterflies migrating over land. Butterflies also initiate migration across water when the water/land interface intersects at right angles the individual's preferred track as determined by Sun orientation. Otherwise the water/land interface acts as a leading line (Fig. 19.2). Field observation suggests that terrain may act as a migration-cost variable in the same way as for birds (p. 633) and that the migration threshold to other variables is higher for the initiation of migration over water if land cannot be perceived on the other side than it is for the initiation of migration over land. However, the threshold is not so high that migration across large water bodies is never initiated and on those few sunny days when sea breezes have not developed and wind speed is below threshold, butterflies may be observed in coastal regions initiating a migration over a sea that is so wide that they

are unlikely to reach land. This clearly disadvantageous behaviour is identical to that sometimes shown by lemmings and, as argued previously (Baker 1971), the evolutionary basis of the behaviour is likely to be the same in both groups (see p. 414).

Usually it is difficult to maintain an individual butterfly under surveillance for distances greater than 100–200 m. On a number of occasions, given exceptionally favourable circumstances, I have succeeded in following individuals of various species for distances of 1–2 km. A more profitable, and less exhausting, technique that gives results that are only slightly less useful is that illustrated in Fig. 18.1, in which a migration track is constructed based on the observation of a succession of individuals. The species involved in Fig. 18.1 is the small white, *Pieris rapae*. The migration illustrated, measured as a straight line from starting point to finishing point, is 7.7 km, and the total butterfly observation time involved was 9.85 hours of sunshine. This gives a rate of cross-country migration of 0.786 km per hour of sunshine. Roer (1961b) gives the results of a marking–release/recapture experiment on the large white, *Pieris brassicae*. From individuals recaptured between 11 and 95 km from the point of release it can be calculated that the average rate of migration was 1.094 ± 0.32 (SE) km per hour of sunshine. Despite the fact that the large white might be expected to have a slightly faster rate of cross-country migration, being a stronger flyer than the small white, this rate is not significantly different ($P > 0.05$) from the rate obtained for the small white by continuous observation.

These figures are entirely consistent with the suggestion that migration over long distances in these butterflies is achieved by a continuation of the behaviour observed over short distances. They do nothing to support the assertion of C.G. Johnson (1971, p. 34) that the displacement of individuals observed over short distances has no relevance to displacement over distances of 100 km or more. An even more convincing demonstration that long-distance migration is achieved by observed behaviour is provided by Nielsen (1961) for the great southern white, *Ascia monuste*, in Florida. By observing marked individuals several times at different points along their migration track it was shown that the migration of individuals over intermediate and long (up to 110 km) distances was a continuation of the direction and speed that was measured over very short distances. Furthermore, it was found that the butterflies performed cross-country migration at a relatively constant ground speed of 10–15 km/h over a wide range of wind speeds and directions. There was no question, therefore, that the long-distance movements were the results of a fundamental change in behaviour such as flight at high altitude or wind displacement. The only possible interpretation is that migration takes place along an oriented route with the butterflies showing every ability to compensate for changes in wind speed and direction.

It will be noted that the rate of migration of the great southern white measured by Nielsen (10–15 km/h) is considerably higher than that recorded for the small and large whites (approximately 1 km per hour of sunshine). The explanation for this difference emphasises the facultative determination of migration distance in these species in the areas concerned. When migration distance is determined facultatively (p. 83), the rate of cross-country progression by a given individual in a given area is very much a function of the rate of contact with areas containing required resources. Thus the rates of migration of the small and large whites were measured in suburban and rural areas where suitable habitats (Fig. 18.1) are common, inter-habitat distance often being as short as 100 m or so and rarely being greater than a few kilometres. The rate of migration of the great southern white, however, was measured in an area in which the distance between habitats containing larval foodplants (which along the coasts of Florida is *Batis maritima*) is often as great as 50–100 km. Once the early morning feeding period was completed, therefore, the great southern whites could migrate tens of kilometres before encountering a potentially suitable habitat. The small and large whites, however, would spend a large proportion of their time within suitable habitats (Fig. 18.1), often migrating no further than 50–100 m before investing 20–30 mins or so in the next cabbage-containing garden encountered. A few preliminary measurements of small and large whites over agricultural land and moorland shows that in such areas individuals that fail to encounter gardens containing cabbages or similar suitable habitats may indeed achieve a rate of migration of 8–9 km per hour of sunshine (as measured over only 1–2 km by the technique employed in Fig. 18.1). The slow rate recorded above, therefore, can be attributed entirely to frequent encounter with required resources.

Variation in the rate of cross-country migration, and presumably, therefore, total straight-line distance between the places of emergence and death, as a function of inter-habitat distance, emphasises that the distance of cross-country migration is not in itself an important component of migration by butterflies of the *Pieris rapae*, *P. brassicae*, and (in Florida) *Ascia monuste* type.

An inevitable result of the influence of encounter with suitable habitats on the rate of cross-country migration is that in the same way that the great southern white and small and large whites show different rates, different individuals of the same species will show different rates, depending on their particular experiences. An individual that encounters a succession of suitable habitats shows a slow rate of cross-country migration, whereas an individual that by chance encounters suitable habitats at only infrequent intervals shows a much faster rate of cross-country migration. This effect is exaggerated by the frequency distribution of migration thresholds that in any case is almost certain to characterise any butterfly deme (Chapter 17). The nett result of these two factors is that a given deme of butterflies in a given area is likely to be characterised by a given frequency distribution of migration rates. An example of this is given in Fig. 19.3 for the small tortoiseshell, *Aglais urticae*.

Fig. 19.3   Rate of migration of the small tortoiseshell, *Aglais urticae*, as calculated from the marking-release/recapture data of Roer (1968)

(a) Frequency distribution of rates of migration within the deme studied as assessed from all recaptures, 1965

Rate of migration in km per hour of sunshine is obtained for each individual by dividing the distance between release and recapture points by the total number of sunshine hours after release. The vertical arrow indicates the maximum rate ever recorded (individual no. 10, Table 19.1).

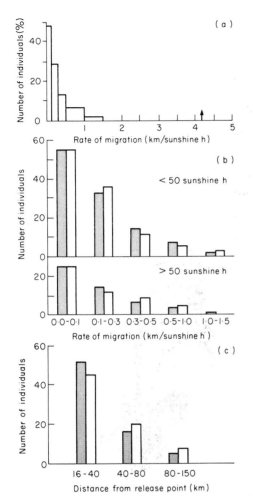

(b) Test of the constancy of rate of migration with time after release, 1965

The frequency distribution given in (a) is used to predict the frequency distribution of migration rates in two groups of individuals: those recaptured less than 50 sunshine hours after release and those recaptured more than 50 sunshine hours after release. Shaded columns, expected number; open columns, observed number. $X^2 = 2 \cdot 15$; $df = 2$; $P > 0 \cdot 05$.

(c) Frequency distribution of recapture distance (unshaded) relative to expected frequency distribution (shaded) if the distribution shown in (a) is a characteristic of *A. urticae* in the experimental area that remains constant from release group to release group and from year to year.

Based on all individuals recaptured at distances greater than 16 km (Roer 1968, Table 2) before entering hibernation. $X^2 = 3 \cdot 25$; $df = 1$; $P > 0 \cdot 05$.

The available evidence supports the suggestion that the frequency distribution of rate of migration is a characteristic of the population. There is in addition no support for the suggestion that migration in this species is concentrated within one or two days in the life of the individual.

In the example illustrated in Fig. 19.3, the frequency distribution of rate of migration remained relatively constant with time, at least for the duration of the experiment. This is unlikely always to be so, however, for as shown in Chapter 17, seasonal and ontogenetic variation in migration threshold is likely to occur, and will in turn influence the rate of migration. Figure 19.4 shows an analysis of data presented by Urquhart (1960) for the monarch butterfly, *Danaus plexippus*. This figure indicates that the mean rate of migration varies with the season.

It seems inevitable that most butterflies for which migration distance is facultatively determined will show a wide range of lengths to their lifetime tracks (measured as straight-line distance from place of emergence to place of death). Direct data concerning this range of total migration distance are unavailable for most species. A combination of observations of the type illustrated in Fig. 18.1 and marking–release/recapture data obtained by Roer (1961a,b, 1962, 1968, 1969, 1970) suggests that even for species such as the small and large whites, *Pieris rapae* and *P. brassicae*, which have a relatively high migration incidence throughout their lives (Fig. 17.28), the vast majority of individuals will die within 300 land km (plus the distance of any oversea crossings) from their place of emergence. Nielsen (1961) succeeded in following marked great southern whites, *Ascia monuste*, in Florida, throughout the major part of their lifetime track. This species seems to be characterised by a migration threshold that is high relative to the small and large whites, even during the post-teneral period. The migrant/non-migrant ratio is relatively low and 20–40 per cent of a given deme invariably fail to initiate migration from their natal habitat. Intra-deme variation in migration threshold perhaps reflects a major contribution of deme density to the composite migration threshold (Chapter 17). Those that do initiate migration settle in more or less the first breeding habitat perceived. This usually results in a total migration of 30–60 km that is achieved in 2–5 hours in a single day. When deme density is already high in the first breeding habitat encountered, however, or when large numbers of individuals are searching, the migration may continue beyond the first breeding habitat encountered. The migration involving the largest numbers of butterflies observed by Nielsen involved many individuals migrating 135 km in a single day before settling to roost in a breeding habitat. The following day many individuals continued their migration for a further 20–25 km. In this species, however, the males of which seem only to live 5 days and the females 10 days, there appears to be a rapid increase in migration threshold about 30 hours after emergence. Even in this species in Florida, therefore, the total migration distance falls within the range suggested above to be usual for temperate butterflies.

The small and large whites, the great southern white, and the small tortoiseshell are all species with a relatively low migration threshold and a relatively high migration incidence in comparison with butterflies as a whole. Many other species, as discussed in Chapters 16 and 18, have a much lower migration incidence. Such species have relatively short lifetime tracks and some (e.g. species listed in group IV,

Baker 1969a) may die within a few tens of metres from their place of emergence, having spent their entire lives within a limited area that is probably a familiar area (Turner 1971, Gilbert and Singer 1975).

So far we have considered the migration mechanism and range of lengths of the lifetime track associated with butterflies subjected to selection for the facultative determination of migration distance. Let us now consider whether any butterfly lineages are likely to have been subjected to selection for the obligatory initiation of migration distance as described in Chapter 11. Suppose that a given lineage encountered an area in which habitats suitable for, say, reproduction occupied an area that was always

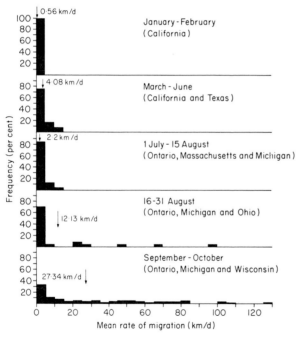

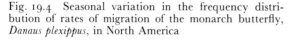

Fig. 19.4 Seasonal variation in the frequency distribution of rates of migration of the monarch butterfly, *Danaus plexippus*, in North America

Each histogram refers to butterflies both mark-released and recaptured in the time period indicated. Marking-release took place primarily in the areas listed. The vertical arrow and the associated rate indicate the mean rate of cross-country displacement at that time of year.

Movement away from the overwintering clusters is minimal throughout January and February and does not really begin until about 1 March. Butterflies of the spring and summer generations throughout North America show a low rate of cross-country migration of about 4 km/day

and not exceeding an average of 15 km/day even for the fastest individuals. Not until non-reproductive adults begin to appear in mid-August do individuals begin to appear with a high rate of migration. Even so, a proportion of the deme maintains a slow rate of cross-country migration.

It seems likely that slower rates of cross-country migration indicate that distance is facultative for the individuals concerned. The faster rates, however, particularly in autumn, perhaps indicate that, for these individuals, distance is obligatory.

[Analysis based on data presented by Urquhart (1960)]

a certain minimum distance in a certain direction from the place of emergence. Not only would selection favour those individuals the migration threshold of which produced a lifetime track that led them to cover this distance, it would also favour those individuals that covered that distance with the minimum migration cost. These individuals would be those with, among other things, the straightest track. It would also favour individuals that delayed reproductive activity and the search for reproduction habitats until they had migrated the appropriate distance. As a result, ground speed would no longer be of major importance (except during those times that feeding or roosting was the optimum behaviour). A strategy would become available, therefore, that is not available when distance is facultative, whereby individuals might maintain an optimum air speed (p. 683) in the appropriate direction. In this situation, ground range would be maximised by flying low against a head wind and high with a tail wind.

We may conclude, therefore, that lineages for which selection has favoured an obligatory determination of migration distance (in a given direction) should manifest compensation for the diel shift in the Sun's azimuth (on the assumption that they continue to orient to the Sun), thus straightening the lifetime track, and should fly high and with the wind when wind direction is favourable and low and against the wind when wind direction is unfavourable. Let us now consider whether any butterfly lineages have been exposed to such selection. The obvious candidates are those that live in the seasonally arid regions of the tropics (and that as larvae are adapted to some habitat-type the suitability of which is intimately associated with rainfall a certain time previously) and those temperate species that for some reason migrate to lower latitudes for the winter.

When adults of the former species emerge, the habitat occupied is usually no longer the highly suitable location that it was when their parents were ovipositing (p. 54). Rainfall in any single location may not be highly predictable, but the seasonal movements of zones of maximum rainfall probability are relatively constant from year to year. During the evolution of the components of migration that are characteristic of such butterflies, an individual emerging at a given time of year emerges into an environment in which the zone of maximum habitat suitability is a certain distance in a certain direction. The selective pressures thus engendered

are just those that have been described above and that are considered further in relation to locusts (p. 451) and the avian inhabitants of seasonally arid areas (p. 676).

Unfortunately, really systematic observations on tropical butterflies have not been made but such data as are available are in agreement with the above model. Thus, Hayward (1953) made a series of observations of the great southern white, *Ascia monuste*, form *automate*, in the grasslands of northeastern Argentina. It was found that when the direction of migration was upwind, the butterflies flew close to the ground. In sunny conditions with a favourable wind, however, the butterflies rose to heights of 90, 900, or even 1500 m above the ground and migrated with the wind. Parman (1926) observed that on a calm day the long-beaked butterfly or 'snout', *Libythea bachmani*, in the United States migrated 'as high as the eye could see' but as soon as the wind ceased to have a component to the E or SE the butterflies migrated near to the ground. As far as the orientation mechanism is concerned, systematic observations of migration directions at different times of day do not seem to have been made. However, a student of mine, K. M. Adams, who made preliminary observations on the brown-veined white, *Belenois aurota*, in Botswana, in January 1975, reports that migration was strongly to the E at all times of day. If these butterflies were orientating to the Sun, therefore, they were compensating for its diel shift. This is in contrast to butterflies for which migration distance is facultative (Fig. 19.2B), but is not unknown amongst insects living in the tropics. Thus honeybees, *Apis mellifera*, everywhere learn to compensate for the local pattern of diel shift in the Sun's azimuth (p. 875). In the tropics they can determine compass direction at all times except for those few minutes for a few days each year that the Sun is 2° or less from the zenith position (Frisch 1967).

The problems raised by those butterflies that seem to fit into the second category for which selection has been suggested to favour an obligatory determination of migration distance (i.e. those that migrate to lower latitudes for the winter) are much greater. Only three demes are considered here. These are the monarch, *Danaus plexippus*, in North America, and the red admiral, *Vanessa atalanta*, and painted lady, *Cynthia cardui*, in western Europe. These latter two species also occur in North America. The painted lady also occurs in the seasonally arid regions of the world where it may behave in a way similar to the species described in the previous paragraph (C.B. Williams 1930, Owen 1971). The migrations of these species have been the subject of considerable controversy (e.g. Beall 1946, Williams 1958, Urquhart 1960, Roer 1961a, French 1964, Warnecke and Harz 1964, Johnson 1969, Baker 1972b) and the matter is still not resolved. The view presented here is the one that seems to encompass the major portion of the available evidence (Figs. 19.5–19.7).

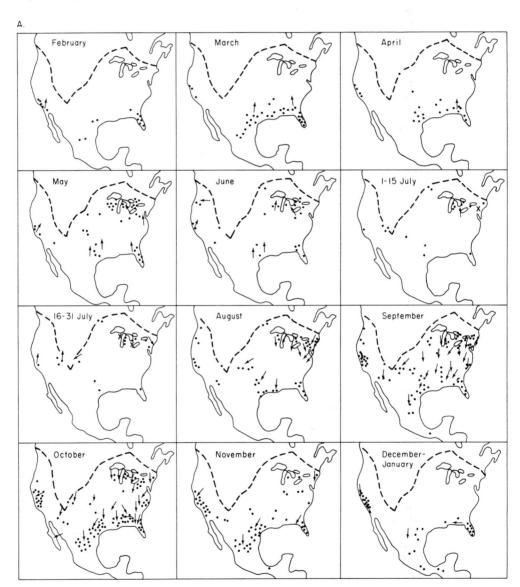

Fig. 19.5   The seasonal return and re-migration of the monarch butterfly, *Danaus plexippus*, in North America

The monarch butterfly is found throughout North and South America and in the last two centuries has spread to Hawaii, New Zealand, Australia, and the Canary Islands. Eggs are laid singly on various species of milkweed, *Asclepias* spp. The species is aposematic and distasteful, both as a larva and as an adult, but is eaten by a few birds such as the catbird, *Dumetella carolinensis*, black phoebe, *Sayornis nigricans*, and blackbilled cuckoo, *Coccyzus erythropthalmus*. The incidence of parasitisation by Hymenoptera varies from up to about 75 per cent in the south of its range in North America to about 2 per cent in the north. Males seem to adopt temporary territories (see p. 369) as part of their courtship strategy.

A. Seasonal distribution and migration

In the map series a dot indicates that adults have been observed in that month. An arrow indicates a reported flight direction. The dashed line on each map indicates the approximate northern limit of the species range. Reports of adult butterflies produce a range that shrinks away from this maximum range in winter and expands to fill it once more by late spring (May).

Overwintering: Monarchs overwinter as free-flying reproducing adults in Mexico, southern Florida, and southern Texas. At the northern limit of the range shown for December to February, which coincides with the edge of the polar front and thus represents regions that from time to time are subjected to outbreaks of cold polar air, the butterflies overwinter communally in clusters on trees. The best known of such sites occur along the length of the Californian coast and in Florida. Such communal roosts may also be formed in Louisiana. The occupants of these roosts make occasional feeding sorties in fine weather and begin to copulate in January. Butterflies do not begin to leave the roosts in any numbers, however, until about 1 March. Oviposition begins soon after. North of the polar front free-flying or exposed adults are not observed. Monarchs have, however, been reported in hibernation in tree crevices and critical evidence that they do not survive such hibernation does not exist.

Wing lengths: The summer generation of monarchs in the Great Lakes region of Ontario has a mean wing length ($\female = 51\cdot3$ mm; $\male = 52\cdot4$ mm) that is extremely similar to the wing length of the butterflies that are on the wing in the same region in autumn ($\female = 51\cdot9$; $\male = 52\cdot3$). These wing lengths, however, are greater than the mean wing lengths of butterflies that in winter are found in Mexico ($\female = 49\cdot3$; $\male = 50\cdot6$), Louisiana ($\female = 50\cdot2$; $\male = 50\cdot6$), and Florida ($\female = 50\cdot9$; $\male = 51\cdot4$). If a cline of wing lengths exists across North America as has been shown for some butterflies in Europe such that wing length is greater in the north, these data are consistent with the suggestion that winter butterflies in Mexico, Louisiana, and Florida are derived from all latitudes but that spring and early summer butterflies in Ontario have hibernated relatively locally.

Migration direction: Throughout the range, migration direction is to the N from late winter until mid-summer and to the S from late summer until mid-winter. Major change of peak migration direction seems to occur from N to S at the beginning of August, though perhaps earlier in the north, and from S to N in the south of the range some time in February.

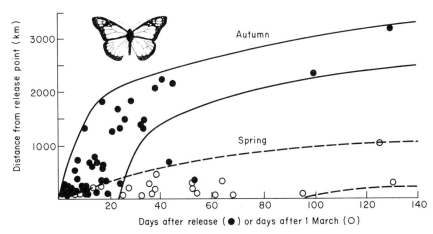

B

B. Migration rate

The graph shows the distance from their release point that marked monarchs have been recaptured at varying intervals of time after release in California in spring (o) and in the Great Lakes region in autumn (•). In both cases, migration distance increases with time, thus indicating that migration is a continual process (see Figs. 17.28 and 18.1). Migration rate varies, however, both with season (Fig. 19.4) and with time after release.

In spring, some individuals may migrate only relatively short distances from their overwintering site, at least by the end of May. The impression gained for these butterflies, which undoubtedly make considerable time investment in reproductive activity along the spring section of their lifetime track, is that of a facultatively determined migration distance. There is no indication that the initial migration away from the overwintering site occurs at a faster rate than subsequent migration.

In autumn, individuals may delay the initiation of migration for up to 20 days after marking. Once migration is initiated, however, it proceeds at a mean rate of about 120 km/day. Given continuous sunshine, monarchs are on the wing from 07.30 to 17.00 hours. Individual monarchs probably do not migrate throughout this period. It is probably necessary to postulate that each individual migrates for a minimum average of 4 hours/day to achieve the observed rate. Once the butterflies have travelled approximately 1500 km, the rate of migration slows down and approximates to that observed in spring. It is tempting to conclude that during the initial phase of autumn migration distance is obligatory but that thereafter it is facultative. Note, however, that some autumn individuals seem always to show a facultative distance.

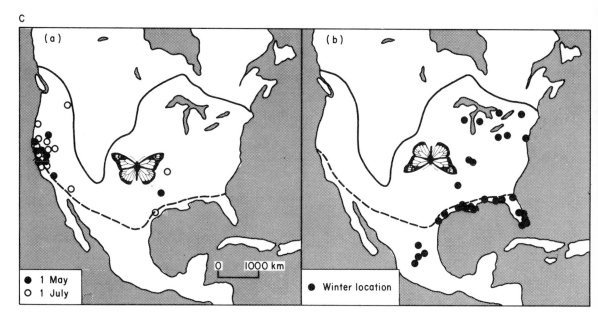

C. Migration distance

Using the curves in B, the diagram extrapolates the results of marking–release/recapture data for all monarchs recaptured after 10 days or more. This procedure permits some prediction of the location of the butterflies concerned on certain critical dates in (a) spring/summer, and (b) autumn/winter. In (a) the predicted locations are shown for 1 May and 1 July. In (b) the location was predicted for 1 November. If this location was south of latitude 34°N, the location was then predicted for mid-December. The dashed line indicates the limit of free-flying or clustered individuals in winter. The solid line shows the maximum extent of the monarch's range. The available evidence is against the suggestion that the butterflies that appear in the northern parts of the range in May are immigrants from the south that overwintered as free-flying individuals, even though by July or August some of these butterflies, some of which are still alive at least until July, or their offspring, may arrive. In autumn, however, at least 70 per cent of butterflies from the Great Lakes region migrate to areas in which they can overwinter as free-flying or communally roosting individuals. In Texas and Mexico, free-flying and perhaps reproducing long-distance immigrants are still alive in December and January. Note, however, that perhaps as many as 30 per cent do not make this journey and seem, after hibernation, to be the most likely source of the spring and early summer butterflies that appear in the northern parts of the range.

[*Analyses and diagrams based on data presented by Urquhart (1960) from which supplementary information was also derived. Wing lengths from Beall and Williams (1945) and Beall (1946)*]

Fig. 19.6 Seasonal return and re-migration of the red admiral butterfly, *Vanessa atalanta*, in Western Europe and North Africa

The red admiral is found in North Africa and through Europe and Asia Minor to Persia. It is found in the Azores and Canary Islands and also occurs in North America, south to Guatemala, Haiti, and New Zealand (where it is probably introduced). The eggs are laid singly on nettle, *Urtica dioica*, and the larva lives in a shelter made by drawing together the two sides of a leaf with silk. In autumn, adults feed on fallen fruit as well as on nectar. Males have a territorial system similar to that described on p. 369.

A. Seasonal distribution and migration direction

Each solid dot indicates a report of one or more free-flying adults. Each open dot on the January map indicates a report of one or more adults in hibernation. The dotted line on the January–April and October–December maps marks the approximate position of the 4.5°C isotherm. This is included solely to facilitate appreciation of the

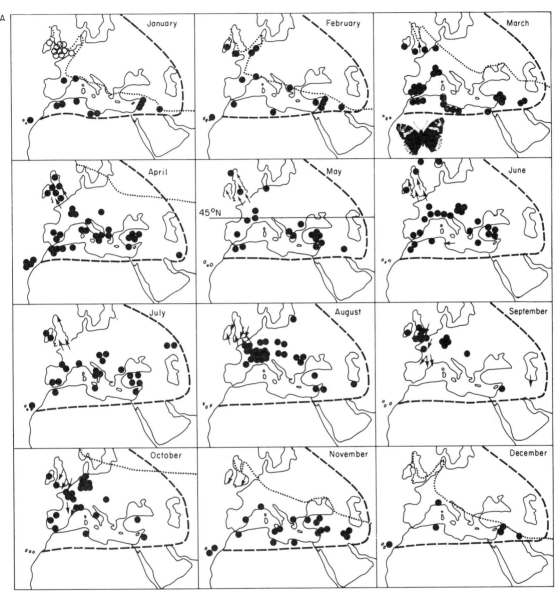

temperature regime over the area and is not intended to have any special biological significance. The dashed line shows the approximate limit of the range of the species in the area.

Reports of free-flying red admirals indicate a range that in autumn shrinks away from the northeastern edge of the maximum range and each spring expands to fill this range once more.

Overwintering: In North Africa, red admirals are free-flying and reproductive throughout the winter. In southern Europe they overwinter as non-reproductive adults that roost during cold weather but appear to feed during fine weather. Further north, individuals enter

hibernation. The only critical experiment, in Britain, which permitted the butterflies to select their own hibernation sites, demonstrated that, at least in Britain, red admirals could survive such hibernation. Similar critical experiments do not seem to have been performed further to the north and east. Although hibernation occurs in a variety of sites, the preferred site seems to be high up on a tree, either in a crevice, under ivy, or exposed on the bark.

Migration direction: Each arrow indicates a region for which sufficient observations are available to demonstrate a peak migration direction. Where data are available, peak direction is to the N or NNW in spring and summer. From mid-August until at least November, peak migration direction is to the S or SSW throughout the range.

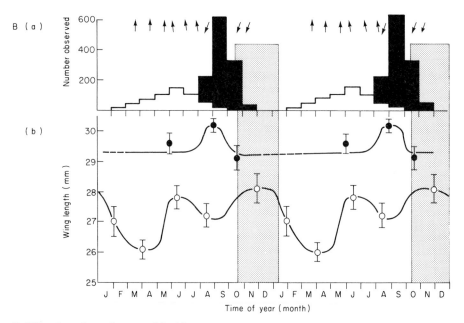

B. Wing lengths and seasonal incidence

The histogram in diagram (a) shows the seasonal incidence (expressed over two yearly cycles to facilitate interpretation) of the adults of the red admiral in Britain. Open shading indicates the butterflies are reproductive, solid shading that they are non-reproductive. The arrows show the peak migration direction in Britain and indicate a change from NNW to S on about 17 August. Diagram (b) shows the variation in wing length (middle of wing base to wing tip) $\pm$ SE of butterflies to the north ($\bullet$) and south ($\circ$) of latitude 45°N (see map for May). A detailed analysis of wing lengths has shown that the individuals that appear in Britain in the spring are most likely to have hibernated locally and are most unlikely to have originated from either northwest Africa, Spain, southern Europe, or the Middle East. The stippled columns indicate the only time of year that wing-lengths are sufficiently similar throughout Europe and North Africa for long-distance migration of a reasonable magnitude to have occurred. This situation occurs about 60 days after the first red admirals from Britain have begun to fly S and coincides with a marked decrease in numbers in Britain at a time when the species is still free-flying.

Circumstantial evidence indicates, therefore, that as with the monarch butterfly, *Danaus plexippus*, in North America (Fig. 19.5), northern spring butterflies have probably hibernated. Some of the offspring of these butterflies in autumn may migrate 2000–3000 km to southern Europe or North Africa where they reproduce respectively either early the following spring or during the winter. Some of the offspring of these butterflies may begin to arrive in northern Europe during the summer, though not before June/July/early August, where they rejoin and possibly interbreed with the individuals (or their offspring) that overwintered in the region.

[Compiled from many sources, particularly Baker (1969a, 1972b) and with some new material. Range information from Higgins and Riley (1973)]

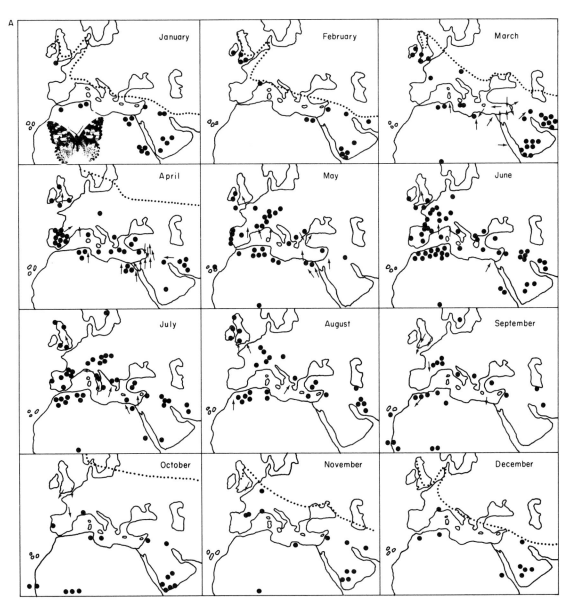

Fig. 19.7   Seasonal return and re-migration of the painted lady butterfly, *Cynthia cardui*, in Western Europe and North Africa.

The painted lady is found throughout the world except perhaps for South America. The eggs are laid singly, usually on thistle, *Cirsium* spp. (in Europe). Males have a territorial system similar to that described on p. 369. The tops of small hills seem to be preferred territories and territorial behaviour involves considerable 'patrolling' of the area concerned.

The conventions used in this diagram are the same as for the red admiral in Fig. 19.6 and only the differences between the two species are considered here.

Painted ladies are free-flying throughout the northern winter at all latitudes to the south of the Mediterranean. There are no critical data concerning the hibernation of painted ladies in northern Europe, though circumstantial evidence suggests that in Britain and Eire at least, hibernation does occur, probably high up on the bark of trees.

In the Middle East, migration is to the S until April and thereafter is to the N, at least until September. In North Africa, migration to the N begins in March. Migration is also to the N or NNW throughout spring and summer in Western Europe. Migration to the S or SSW begins in Britain in autumn at the end of August, and during September becomes established throughout Europe, the

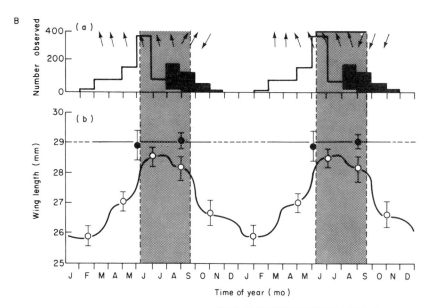

Mediterranean and North Africa. This peak migration direction persists in Europe until flight ceases for the winter.

Although the spring painted ladies in Britain cannot be attributed to immigration, the sudden increase in numbers in June coincides with the beginning of a period during which butterflies are of similar size throughout Europe and North Africa. This period ends in late September. As painted ladies do not begin to fly S in Britain until the end of August, long-distance migrants from the north, if they exist at all, can only be considered to contribute in any numbers to demes to the south of latitude 45°N during a period of about a month. Immediate flights to the S by individuals emerging from the

end of August onwards could account for the relatively small numbers of painted ladies recorded in Britain in the autumn. If such long-distance migrants are produced, it seems that those that escape the hibernation zone produce three generations before some of their offspring can rejoin hibernated individuals or their offspring in the north to produce the June peak of abundance. During this time the possibility has to be entertained that at least some of the members of these generations migrate even further to the south in Africa (e.g. down the west coast).

[Compiled from many sources, particularly Baker (1969a, 1972b) with some new material. Photo courtesy of Sdeuard C. Bisserot and Bruce Coleman]

It is concluded that each autumn some individuals enter hibernation throughout the breeding range of each species. Other individuals migrate 2000–3000 km to lower latitudes to an area where the winter can be passed with at least occasional access to food and often with access to habitats suitable for reproduction. Those that enter hibernation and those that have only limited access to food do not become reproductively mature until the following spring. Those that reach areas suitable for reproduction do so and probably die during the winter. The following spring, adults in the northern part of the range are hibernated individuals. Those in the south of the range have passed the winter torpid but with occasional feeding sorties, and those still further south are the offspring of indigenous and migrant individuals. During spring and summer these and the members of subsequent generations migrate, many of them to the N, but reproduce along the length of their lifetime track. Migration distance is probably facultative but a small proportion of the second or third generation offspring of the long-distance autumn migrants may arrive back in the northern part of the range and interbreed with the individuals, or the offspring of individuals, that hibernated. Approximately 70 per cent of monarchs from the Great Lakes region of North America seem to be long-distance migrants for which migration distance is obligatory. We may presume, however, that as with birds (p. 636) there is a cline of migrant/non-migrant ratio from south to north, being higher in the north. Possibly the same applies to the red admiral and painted lady in western Europe and northern Africa, but here it should be noted that as yet there is no direct evidence that long-distance migrants exist in these species. The only indication that they may exist derives from analogy with the monarch and the wing-length analyses shown in Figs. 19.6 and 19.7.

Returning now to the migration mechanism, we would expect some individuals of these three species in autumn to compensate for the shift in the Sun's azimuth and to take advantage of favourable winds by flying relatively high above ground level. Other autumn individuals, however, those for which migration distance is facultative, and all individuals in spring and summer, should not compensate for the shift in the Sun's azimuth and should have a composite migration threshold that receives a contribution from wind speed but not from wind direction.

As far as the orientation mechanism is concerned,

the evidence is equivocal. Critical observations for spring and summer individuals are not available. An analysis of unsatisfactory data (Baker 1968b) suggests that both the red admiral and painted lady in Britain in spring and summer (18 February to 17 August) orientate to the Sun's azimuth without compensating for its movement during the day as expected above. Analysis for these species in autumn (27 August to 25 November) using data from the same source, similarly suggests a lack of compensation for the Sun's movement. For both species, however, the suggestion is less convincing than for the spring and summer individuals. Urquhart (1960, p. 274) remarks that monarchs released in autumn were prone to fly SE in the morning and SW in the afternoon, though he did not indicate the proportion of individuals for which this was so. In all three species, therefore, there is an indication that orientation to the Sun but without compensation occurs throughout the year, though in the red admiral and painted lady it may be less marked in autumn than in spring. However, because even in autumn not all individuals have a migration distance that is obligatory, this situation might be expected. Since the above analysis was published, I have made a number of field observations of migrating red admirals in autumn in Britain. These observations are analysed in Fig. 19.8 and could well be taken to indicate that in autumn the British deme of red admirals does indeed consist partly of individuals that do not compensate for the Sun's movement and partly of individuals that do.

Observations of the height above ground level of the migration of the monarch in North America in relation to wind direction support the suggestion that in autumn the migration distance of some individuals is obligatory. Williams et al. (1942) describe a record from New York City of monarchs flying SW against the wind at low heights (less than 6 m) above the ground and another from New Jersey of butterflies heading to the S for two days following the course of a river and flying against a steady headwind. In Arkansas in October the butterflies normally fly SSE near the ground but when a north wind sets in they rise and go with it to the S, often very high in the clouds. Urquhart (1960, p. 270) similarly quoted an example of monarchs flying at about 30 m above the ground like a 'gigantic, brown carpet' and riding a Texas 'norther' coming down at 24 km/h. Once a cold front overtakes the butterflies, however, and the temperature falls, further migration is not initiated until the return of warmer weather (p. 251).

Let us accept for the moment that the monarch, red admiral, and painted lady are unusual amongst temperate butterflies in that they produce in autumn individuals for which migration distance is an obligatory 2000–3000 km. One immediately wonders what it is about this generation of these species that during their evolution has been associated with selection for such an obligatory migration distance. It could be significant that these three species have a phenology (=seasonal incidence) that is relatively unusual among temperate butterflies in that they overwinter as adults. In autumn, before entering hibernation, these butterflies are in a state of reproductive diapause that is not broken until either the following spring or perhaps until they have arrived at a sufficiently low latitude. Furthermore, it is among this generation of

reproductively diapausing adults, and only among this generation, that obligatory long-distance migrants are produced (Figs. 19.4 and 19.5). Such adults, of course, make no time investment in reproductive activities and except when feeding, or searching for hibernation and roosting sites, the only components of time investment available are basking/roosting and migration (p. 239). The next question, and one that cannot yet be answered, is whether those other species that similarly produce reproductively diapausing adults in autumn (e.g. in western Europe, the small tortoiseshell, *Aglais urticae*, peacock, *Inachis io*, comma, *Polygonia c-album*, camberwell beauty, *Nymphalis antiopa*, and brimstone, *Gonepteryx rhamni*) also produce obligatory long-distance migrants, and if not, why not?

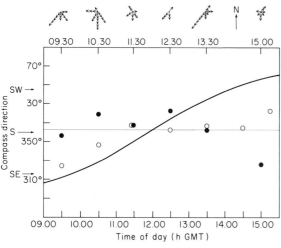

Fig. 19.8   Variation in flight direction with time of day in the autumn generation of the red admiral, *Vanessa atalanta*, in Britain

The mean flight direction of the red admiral in autumn has been found previously to be 3°W of S. If the butterflies orient to the Sun's azimuth but do not compensate for its diel shift, their mean flight direction should change during the day as indicated by the thick line. If they orient to the Sun but compensate for its diel shift, or if they orient to some other compass, such as the geomagnetic field, their mean flight direction should remain more or less constant as indicated by the horizontal line. Finally, if, as suggested here, the British deme of red admirals in autumn consists partly of individuals that orient to the Sun without compensating for its movement and partly of individuals that orient to the Sun but do compensate for its movement, observed diel change in mean flight direction should fall on a line intermediate between the two shown.

The solid dots indicate the mean flight directions of red admirals observed during rectilinear migration in

southern England in early September over several years and are the mean vectors of the directions presented in rose form in the top part of the diagram (1 arrow head = 1 butterfly). The open dots were calculated from the accumulated records of the Insect Immigration Committee of the South-East Union of Scientific Societies and indicate the mean vector of flight directions collected over many years from throughout Britain between 18 August and 25 November.

The open dots suggest a deme that over the autumn as a whole has a mixed orientation mechanism as postulated. However, even if all red admirals failed to compensate for the diel shift in the Sun's azimuth, the method of data collection would tend to displace the mean vectors toward the horizontal line (Baker 1968b). The solid dots suggest a constant compass direction throughout the day in early September but data are too few to dismiss the possibility of some relationship with the diel shift of the Sun's azimuth. The mean vector for 15.00 hours is not significantly different at the 5 per cent level from the position of the thick line, though those for 09.30 and 10.30 hours are significantly different at this level from the position of the thick line but not the horizontal line.

The data are consistent with, but do not prove, the thesis that: (1) in autumn the British deme of red admirals consists partly of individuals that compensate for the diel shift of the Sun and partly of individuals that do not; and (2) the proportion of the two forms varies during the autumn, the former being in the majority in early autumn but in the minority in late autumn.

[Modified from Baker (1968b) with some new material]

19.3.2 Migration direction and direction ratio

The interaction between migration distance and selection on migration direction is discussed in detail for birds (see p. 654). Briefly, latitudinal environmental gradients of habitat suitability exist and over a long enough distance emerge from the background 'noise' produced by local variations in habitat suitability. Animals that migrate only short distances experience only the local 'noise' selection and unless the local juxtaposition of resources imposes marked gradients of habitat suitability over short distances such as those produced by altitudinal gradients, a direction ratio of 25:25:25:25 is favoured. Animals that evolve longer-distance migration, however, are likely also to experience latitudinal environmental gradients. It is tentatively suggested in relation to birds that at migration distances above about 200 km, latitudinal gradients of selection may begin to act, and it seems reasonable to retain a similar guideline for butterflies. Some justification for this may be derived from an earlier demonstration in Britain (Baker 1968a) that latitudinal gradients of temperature over such a distance can be detected in the field through their effects on

Table 19.3 Comparison of the mean migration directions of demes of five species of butterflies with the directions expected on the basis of sun orientation and wind displacement

| Species | Physiological generation | marking–release/recapture experiments of Roer | | All angles measured clockwise from south mean directions expected due to | |
		release date(s)	calculated mean direction	sun orientation*	downwind displacement†
Pieris brassicae	spring and summer	5–6.54‡	169°	159°	45°
	autumn, after expected date of minor change of direction	8–9.54	8°	0°	70°
Vanessa atalanta	autumn (immature)	9.57–60	24°	15°	?
Inachis io	summer (mature) after expected date of major change of direction	21.7.66	317°	334°	270°
	autumn (immature)	17.9.59	355°	3°	360°
Aglais urticae	spring§	1–15.9.56–61 18–21.9.65	175°	167°	190°
	summer (mature) before expected date of major change of direction	6.60–61	175°	154°	202°
	summer (mature) after expected date of major change of direction	8.57–61 30.8–11.9.65	325°	334°	225°
	autumn (immature)	16–30.9.56–61 10.59–61 11.9.63	343°	14°	175°
Nymphalis antiopa	summer (immature) before expected date of major change of direction	6–7.62 6.68	195°	160°	213°

* Sun orientation direction is based on observed tracks of rectilinear migrants over distances of 100 m or so (Baker 1969a, Tables 1, 12, 13, 16).
† Mean direction expected if all individuals were wind displaced (p. 436).
‡ Dates are given in abbreviated form—for example:
  5–6.54 denotes May/June 1954.
  1–15.9.56–61 denotes September 1 to September 15 for each year from 1956 to 1961.
§ See p. 439.

the larval size of small and large white butterflies. If we accept this guideline it follows that we may expect a number of butterfly lineages to have been exposed to latitudinal gradients of habitat suitability.

Table 19.3 shows the peak flight directions of various species at different times of year in western Europe as determined on the one hand by direct observation of migrating individuals over short distances (approximately 100 m) and on the other hand by marking–release/recapture experiments. Mean vectors calculated from the two sources are in complete agreement, though the directions obtained by Roer for the small tortoiseshell, *Aglais urticae*, and peacock, *Inachis io*, require further discussion (p. 439).

In view of the agreement between mean direction as determined by direct observation over short distances (50–100 m) and marking–release/recapture data for migrations over distances of 16–100 km or so, it would be expected from the conclusions reached on p. 419 that the direction of migration as determined over long distances would be independent of wind speed and direction. Without critically analysing his results, however, Roer (1962, 1968, 1969, 1970) has concluded that the track directions of his marked and recaptured butterflies were the result of downwind displacement. Roer did not, however, take into account the possibility that the track directions could be the result of oriented migration in a direction that from time to time happened to coincide with the downwind direction. This latter possibility can readily be examined.

Using data for Köln presented in the Daily Weather Report of the British Meteorological Office (Meteorological Office 1954–1967), I have calculated by vector analysis the downwind displacement direction for every individual butterfly for which Roer (1961a,b, 1962, 1968, 1969, 1970) presents data. Similar successful calculations were also made for 13 balloons released and recovered by Roer. For butterflies the assumption was made that migration: (1) was evenly distributed about 12.00 hours GMT (Baker 1969a); (2) was proportional on any given day to the number of sunshine hours between 06.00 hours and 18.00 hours GMT (p. 415); and (3) proceeded at a constant rate per hour of sunshine from the date of release to the date of recapture (Fig. 19.3).

Figures 19.9 and 19.10 and Table 19.3 present the results of this analysis and demonstrate clearly that

Fig. 19.9 Migration direction, as determined by marking–release/recapture, as a function of preferred compass and downwind displacement directions in four species of western European butterflies

Observation of the direction of the rectilinear flight of butterflies over short distances (100 m or so) has clearly demonstrated an individual-specific preferred compass direction (Fig. 19.2). The direction ratio for these preferred compass directions produces for many species a peak migration direction. The direction of the mean vector for the direction ratio is species- and season-specific. Recently it has been claimed that butterfly behaviour over such short distances has no relevance to cross-country displacement over many tens or hundreds of kilometres (Johnson 1971). The only data that give credence to this claim are the results of marking–release/recapture experiments carried out by Roer in West Germany. These data have never been critically analysed in terms of the relative contribution of downwind displacement and preferred compass direction to the migration track of the butterflies concerned. In the analysis presented here, the convention adopted by Roer is followed and only those butterflies recaptured at distances greater than 16 km (but before entering hibernation) are considered. All angles are measured clockwise from south.

The left-hand graphs compare the observed migration direction of each recaptured individual with the mean compass direction that has either been identified or (in the case of the Camberwell beauty) predicted for the members of the appropriate physiological generation of that species. Note that the scatter about this mean direction is due to the direction ratio and not to a shortcoming in the preferred compass direction model.

The right-hand graphs compare the observed migration direction of each recaptured individual with the vectored downwind displacement direction during the sunshine hours between release and recapture (see left). Note that any deviation of migration direction from the downwind direction is attributable partly to error in the calculation of the downwind vector from the wind data available and partly to the shortcomings of the downwind displacement model.

The solid line on each graph shows the regression line that is to be expected if the model concerned is valid. Dashed lines are calculated regression lines but are included only if the calculated line is a significant fit to the data at the 5 per cent level and also if it is not significantly different at the same level from the solid regression line. In the left-hand graphs a partial correlation coefficient (r_p) is presented only when the partial correlation coefficient between observed and preferred compass directions (with downwind displacement direction partialled out) is significant at the 5 per cent level. In the right-hand graphs the same applies with respect to the partial correlation coefficient between observed migration direction and downwind displacement direction (with the preferred compass direction partialled out).

In the Camberwell beauty, *Nymphalis antiopa*, partial correlation and regression analyses cannot be performed. However, the observed migration direction is described more completely by the preferred compass direction model than by downwind displacement. In all of the other

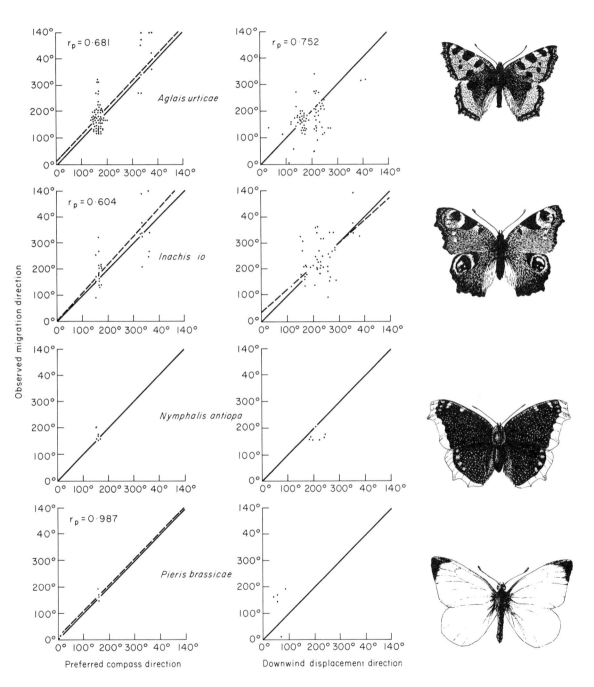

Preferred compass direction   Downwind displacement direction

three species, both r_p and the calculated regression line are significant with respect to the preferred compass direction model but for no species are both significant with respect to the wind displacement model.

We may conclude that the migration tracks of butterflies over distances of 16–100 km are the product of individual-specific preferred compass directions and are not produced by downwind displacement. Any similarity between migration track and wind direction is due to a coincidence between the latter and the preferred compass direction of the individual concerned.

[Analysis based on marking–release/recapture data presented by Roer (1961a,b, 1962, 1968, 1969, 1970). Wind data are also taken from these sources when presented, otherwise from the Daily Weather Report of the British Meteorological Office (Meteorological Office 1954 to 1967) using data for Köln. Preferred compass direction data from Table 19.3]

butterfly track direction is unaffected by wind even over long distances and presumably is the result of the mechanisms already described. We may now turn to a consideration of the environmental gradients of selective pressure that could have acted to favour observed peak flight directions.

In a previous paper (Baker 1969a) an analysis of the mean flight directions of British butterflies in spring and summer showed that the only clear correlation was with the angle made by isotherms across western Europe during the period of development of their larval offspring. This correlation is shown in Fig. 19.11. An attempt to identify the critical selective pressure(s) has been made (Baker 1968a, 1969a) and a summary of conclusions is also presented in Fig. 19.11.

If we now turn to a consideration of autumn migration direction we find (Fig. 19.12) that it is

necessary to distinguish between those species, such as the small white, *Pieris rapae*, and clouded yellow, *Colias croceus*, that overwinter as larvae or pupae and those such as the red admiral, *Vanessa atalanta*, and painted lady, *Cynthia cardui*, that overwinter as adults.

In all of the butterflies discussed so far there is a major change of peak flight direction at some time during the year. In western Europe this change is from a flight to the NNW in spring and summer to a flight to the SE, S, or SW, depending on the species, later in the year. In most species, this major change in peak flight direction takes place in late summer/early autumn (August/September). In the large white, *Pieris brassicae*, it occurs in mid-summer (mid-July) (Baker 1968a, 1969a).

Figures 19.5–19.7 and 19.12 show the geographical and seasonal distribution of peak migration

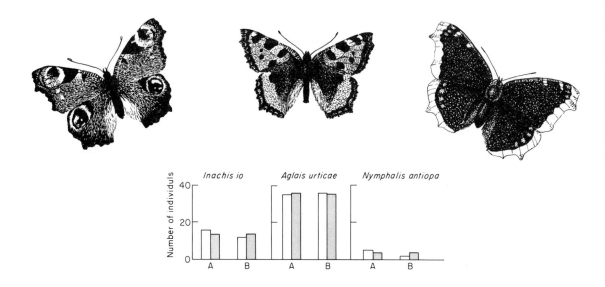

Fig. 19.10   Test for the occurrence of wind displacement from a track primarily determined by orientation to the Sun in three species of butterflies

It was concluded on the basis of Fig. 19.9 that the migration tracks of butterflies over distances of 16–100 km are the product of individual-specific preferred compass directions and are not produced by downwind displacement. Where known, orientation in the preferred compass direction is achieved by reference to a Sun compass, either with or without compensation for the Sun's diel movement. The possibility remains, however, that although the migration track is primarily determined by Sun orientation, the butterfly may be displaced from this track by the wind.

Columns represent the frequency with which the difference between the observed track direction (as determined by marking–release/recapture over distances greater than 16 km) and the preferred track direction is in the direction that: A, can be attributed to wind displacement; and B, cannot be attributed to wind displacement. Stippled columns, frequency expected on the null hypothesis that wind has no effect on the direction of the migration track; unshaded columns, observed frequency.

In no case is the null hypothesis disproved at the 5 per cent level. There is no indication that butterflies are displaced by wind from a preferred track direction that is primarily determined by orientation to the Sun.

[Sources as for Fig. 19.9]

directions of various species in North America and Europe. In all cases it seems that, as would be expected (Fig. 19.12), the major change of peak flight direction occurs earlier in the north than in the south. Previously (Baker 1968a, 1969a), I have attempted to calculate the rate and timing of the southward progression of the optimum date of change of flight direction for the small white, *Pieris rapae*, and other species and Fig. 19.12 has shown the way that this date progresses southward across western Europe in autumn and also shows a few observations of flight direction that suggest that the model used is not wildly inaccurate. Fig. 19.13 illustrates another attempt to monitor and analyse the southward progression of this zone.

As far as the small white is concerned, the date of change of direction cuts across a generation and could only be achieved by the same individual butterflies flying in different directions before and after this date. Figure 19.13 presents an analysis of the factors involved and shows clearly that the butterflies respond to the seasonal change in the photoperiod/temperature regime by flying in a different direction. Whether the evolved photoperiod/temperature threshold permits an optimum response anywhere in Europe or whether it is necessary to postulate a latitudinal cline cannot as yet be determined. In view of the mobility of the species, however, it seems likely that a latitudinal cline, if it exists, will be relatively slight.

Although in the small white it is the adult stage that has a photoperiod/temperature threshold with respect to flight direction, this cannot be the case for all species. In the large white, *Pieris brassicae*, for example, no combination of photoperiod and temperature as experienced by the adults could be used to permit May/June/early July butterflies in Britain to fly NNW and late July/August butterflies to fly SSE. In this species, therefore, it must be the photoperiod/temperature regime experienced by the larva that determines the migration direction when adult, larvae developing at high temperatures and long days producing adults that fly SSE and larvae developing at low temperatures and short days producing adults that fly NNW after hibernation as a pupa. Perhaps larvae developing at high temperatures and short days produce adults that fly NNW. In the small white, *P. rapae*, the effect of photoperiod/temperature on flight direction seems to be reversible (Baker 1968a and Fig. 19.13).

Because of the influence of photoperiod and temperature on migration direction, great care has to be taken in the handling of butterflies prior to experiments on migration direction. A fine example of this seems to be provided by the marking–release/recapture results of Roer in West Germany. In his early experiments, the butterflies released by Roer (1959, 1961a,b) in autumn flew to the S in the same direction as their free-living conspecifics. In later experiments (Roer 1968, 1969, 1970), however, the butterflies released in autumn flew not to the S but to the N in the direction that is typical for the species in spring. This sudden change, which cannot be attributed to factors such as wind direction (Figs 19.9 and 19.10), apparently coincided with the adoption of the experimental practice of keeping the butterflies in a coldroom at 9°C, at unspecified photoperiod, for intervals of up to a fortnight before releasing them (Roer 1968, p. 315). The butterflies concerned, the small tortoiseshell, *Aglais urticae*, peacock, *Inachis io*, and Camberwell beauty, *Nymphalis antiopa*, are all species that fly S in autumn, hibernate as adults, and then fly N in the spring. Presumably, therefore, it is conditions experienced during hibernation, such as low temperature and short days, that not only induce the non-reversible physiological change from non-reproductive to reproductive but also induce the change in preferred migration direction from S to N. It appears, therefore, that inadvertently Roer converted his butterflies, at least as far as flight direction was concerned, from 'autumn' butterflies to 'spring' butterflies.

Bowden and Johnson (in Johnson 1971, p. 33) have claimed that analysis of my published data has shown that the seasonal reversal of direction of the small white, *Pieris rapae*, could be equally well explained by the effect of prevailing wind strength and direction. The analysis on which this claim is based has never been published, but Dr. J. Bowden of Rothamsted has been good enough to let me examine his data. These consist of a demonstrated correlation between wind direction and migration direction of *P. rapae* on 12–16 August 1966 (a temporary reversal of direction took place in the Bristol region on 14 and 15 August—Baker 1968a). However, if the analysis is extended just a few days either side of this period it is found that the correlation with wind direction breaks entirely. Figure 19.14 presents a complete analysis of the situation and shows that the major change of direction of *P. rapae* in autumn is the result of a real change in compass orientation by the butterflies and cannot in any sense be attributed to the wind.

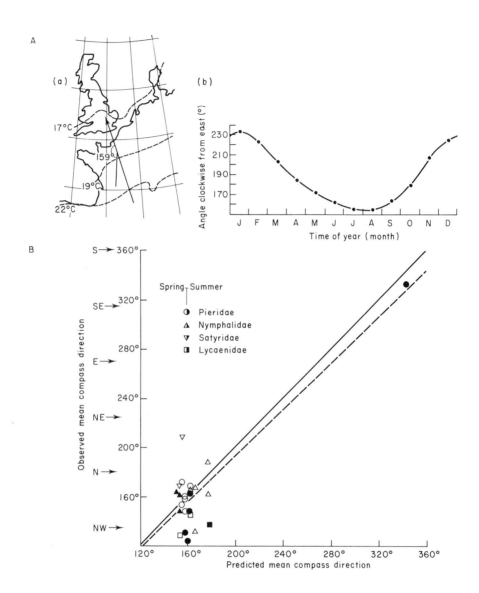

Fig. 19.11   The mean migration direction of the spring and summer generations of temperate butterflies as an adaptation for the maximisation of the body size of the offspring

A. Seasonal variation in the isotherm angle across western Europe

The isotherm angle is the angle subtended to S by a line drawn perpendicular to isotherms (corrected to sea level). (a) shows the situation in western Europe in July. At this time, the isotherm angle is 159°. The isotherm angle changes with the season as shown in (b). Across western Europe, the isotherms run from NE to SW in summer and from NW to SE in winter (see Fig. 19.6).

B. The mean migration direction of the spring and summer

generations of British butterflies as a function of the isotherm angle

The mean migration direction of the spring and summer generations of all British butterflies for which data are available have been analysed with respect to a variety of environmental latitudinal gradients that seemed to be potential selective pressures on migration direction. The best correlation obtained is that illustrated between mean migration direction and the isotherm angle during the period of development of the larval offspring of the butterflies concerned. The correlation coefficient is 0·867 ($P < 0$·001). The solid regression line indicates the line expected on the basis of the 'temperature during larval development' model. The dashed line is the calculated regression line. There is no significant difference at the 5 per cent level between the two lines.

C. The paradox of the large white, *Pieris brassicae*

The highly significant agreement between migration direction and the 'temperature during larval development' model stands only if the large white in summer is entered into the model as migrating towards higher temperatures whereas the large white in spring and all other species in both spring and summer are entered into the model as migrating towards lower temperatures. The rose diagram compares the observed flight directions of the small and large whites during the period from 9 July to 27 August. During this period the small white has a mean migration direction (white arrow) to the NNW and the large white has a mean migration direction to the SSE.

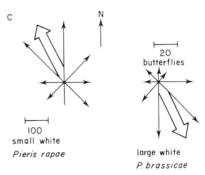

C

N

20 butterflies

100
small white
Pieris rapae

large white
P. brassicae

9 July – 27 August

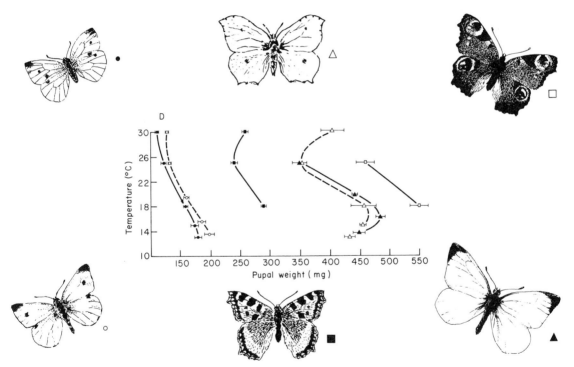

D

D. The effect of temperature during development on body size

The graph shows the effect of temperature during the larval stage on the weight (± SE) of the resulting pupae for six species of British butterflies (weighed 24 hours after larval–pupal ecdysis). ●, green-veined white, *Pieris napi*; ○ small white, *P. rapae*; ■, small tortoiseshell, *Aglais urticae*; ▲, large white, *Pieris brassicae*; △, brimstone, *Gonepteryx rhamni*; □, peacock, *Inachis io*.

Excluding the large white, over the range of average temperatures normally experienced by the developing larvae, body size is greater following development at lower temperatures. The offspring of the spring generation of large whites develop in Britain at an average temperature of about 20°C. The offspring of the summer butterflies, however, develop on average at a temperature of between 12·5° and 16°C. On this basis, spring large whites maximise the size of their offspring by flying toward lower temperatures whereas summer large whites maximise the size of their offspring by flying toward higher temperatures. A model that postulates that maximisation of the body size of the offspring is the critical selective pressure acting on the migration direction of spring and summer butterflies in temperate regions can therefore at present account for all observed directions. The precise advantage(s) of increased body size has not been identified but preliminary calculations have shown that increased fecundity may be important and by itself could outweigh the disadvantage of an increased generation interval when maximisation of body size involves flight toward lower temperatures.

[Compiled from Baker (1968a, 1969a,b) with some new analysis]

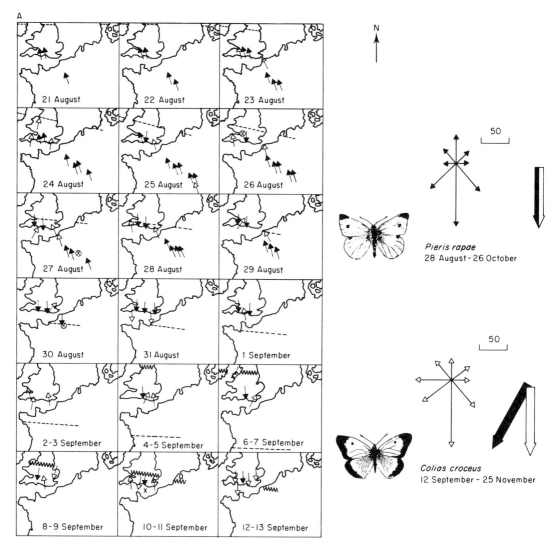

Fig. 19.12   Models of the autumn flight direction of temperate butterflies

A. Species that overwinter as larvae or pupae

The latitudinal temperature gradient seems likely to be the selective pressure that will exert the greatest influence on the autumn migration direction of species that are reproductive and for the offspring of which it is advantageous to attain a certain stage of development before winter. Individuals that migrate toward higher temperatures are likely to produce offspring with a higher probability of attaining the overwintering stage than individuals that fly in other directions.

Using this model, predictions have been made concerning the optimum date for a butterfly to change from the summer to the autumn direction (Baker 1968a) and the way that this date should vary latitudinally (Baker 1969a). In general, there should be on any given date in autumn a zone in which it is advantageous for the butterflies

concerned to change from the summer to the autumn direction. This zone may be wide or narrow but should shift to the south as autumn proceeds. It should shift earlier for species that overwinter as pupae than for species that overwinter as larvae.

The maps show the way that the predicted zones for the small white, *Pieris rapae* (–––), and clouded yellow, *Colias croceus* (⌇⌇⌇⌇) move south across western Europe, that of the small white, which overwinters as a pupa, advancing approximately two weeks earlier than that of the clouded yellow, which overwinters as a larva. The arrows (solid heads = *P. rapae*; open heads = *C. croceus*) show observed peak flight directions for these two species. ⊗ indicates no apparent peak direction for *P. rapae*; ×, the same for *C. croceus*.

The rose diagrams indicate observed migration directions for individuals (arrow length indicates the number of individuals flying in that direction) of these two species after the autumnal major change of direction in Britain. If

Inachis io
18 August – 10 November

Aglais urticae
18 August – 10 November

Cynthia cardui
28 August – 10 November

Vanessa atalanta
18 August – 25 November

the above model is correct, it follows from Fig. 19.11A(b) that the peak migration direction should be to the S (white arrow). The black arrows indicate the mean vectors calculated from the rose diagrams. In neither case are the two directions significantly different from S at the 5 per cent level.

B. Species that overwinter as reproductively diapausing adults in northwest Europe.

Two potential selective pressures that could act on the autumn migration direction of these species may be considered. The first concerns the severity of the winter experienced by the butterflies. The second concerns the length of the winter experienced by the butterflies (from the initiation of hibernation in autumn to emergence from

hibernation the following spring). From Fig. 19.11A(b) it follows that individuals of the four species shown that migrate in the direction of the dotted arrow will gain the greatest decrease in winter severity per unit distance whereas those that migrate in the direction of the dashed arrow will gain the greatest decrease in winter length per unit distance.

The solid arrow shown with the dotted and dashed arrows in the bottom half of each diagram indicates the mean vector for the species concerned as calculated from the illustrated observations of migration directions. In every case, observed mean migration direction supports the suggestion that it is the length and not the severity of the winter that is the critical selective pressure.

[*Compiled from data and diagrams in Baker (1968a, 1969a)*]

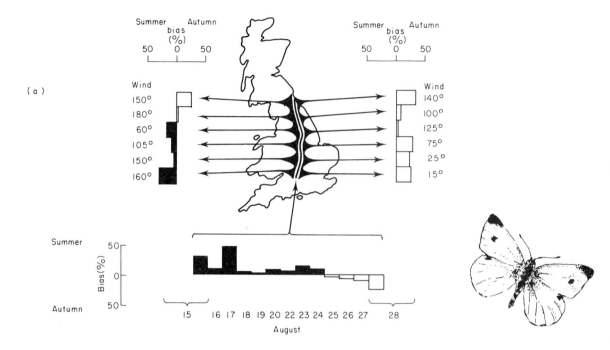

Fig. 19.13 Latitudinal and temporal patterns of the direction of migration and the environmental cues utilised by the small white butterfly, *Pieris rapae*, during the period of transition from summer to autumn peak direction in a single season (1973) in England

(a) Latitudinal and temporal behaviour patterns during the transition period.

Single-day transects were made along a 480 km N–S line (shown by solid line) at two stages in the transition period (i.e. 15 August and 28 August). Track directions (relative to the Sun's azimuth) of individuals crossing the road were recorded throughout each transect. Bias toward summer peak direction (159°) and autumn peak direction (0°) was calculated for 80 km lengths as indicated by the arrows. Bias toward the summer peak direction is shown by solid shading in the histograms. Bias toward the autumn peak direction is shown by open shading. A bias of 100 per cent would indicate a direction ratio of 100:0:0:0 whereas a bias of 0 per cent would indicate that the bias toward the summer peak direction was exactly equal to the bias toward the autumn peak direction. The figures in the wind columns show the estimated angular difference between the wind direction across the road (as determined by the direction of wind-blown thistle seeds) and the peak track direction of butterflies across the road. An angle of 180° would indicate a head-wind, an angle of 0° a tail wind, and an angle of 90° a cross-wind. Between 15 and 28 August counts were made of track direction at the southern end of the transect line as indicated by the arrow. Counts were made in a variety of situations, particularly of individuals crossing open fields over distances of not less than 100 m but also of individuals crossing roads. The majority of observations between these dates were made in the Vale of Pewsey but some were also made on the Marlborough

Downs and on Salisbury Plain. Observations were made in as many places as possible each day to minimise the effect of local topography and climate. Overcast skies prevented migration on 21 August. Number of individuals used in each calculation varies from 30 to 200.

On 15 August *P. rapae* was already showing bias toward the autumn peak direction in north-east England but 160 km to the south a bias toward the summer peak direction was still apparent and persisted throughout the remainder of the southward transect. A summer bias persisted at the southern end of the transect line until 25 August after which date a steadily increasing bias toward the autumn peak direction became established. On 28 August a bias toward the autumn peak direction was present throughout the length of the 480 km transect.

(b) Environmental cues utilised by *P. rapae* during the period of transition from summer to autumn peak direction.

During the accumulation of the data presented in (a) relevant environmental data were also collected. Dawn and dusk cloud cover in the Vale of Pewsey were recorded by the author. Cloud cover elsewhere and all maximum and minimum temperatures were obtained from the nearest recording centre. Day lengths (sunrise to sunset) were obtained by extrapolation from the *Smithsonian Meteorological Tables* (List 1951). Each dot on the graphs refers to a separate estimate of bias based on the track directions relative to the Sun's azimuth of not less than 30 individuals. A separate estimate was considered to have been made when the observer changed location or when at least an hour had elapsed since the previous estimate at the same place. In the graphs the horizontal dashed line indicates the position of balance between bias toward

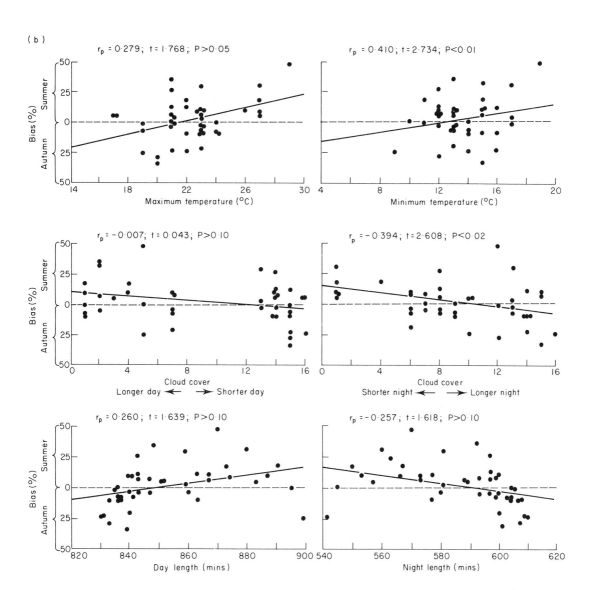

(b)

summer and autumn peak directions. The solid line is the calculated regression line of the data presented. The right-hand graphs show variation in bias in relation to the characteristics of the previous night. The left-hand graphs show variation in bias in relation to the characteristics of the previous day. Partial correlation coefficients (r_p) have been calculated for the variation in bias in relation to either temperature, cloud cover at dawn and dusk, or time between sunrise and sunset, with the other two variables partialled out. The top pair of graphs show the relation between bias and the maximum temperature of the previous day and between bias and the minimum temperature of the previous night. Partial correlation is significant only between bias and the minimum temperature of the previous night. The middle pair of graphs show the relation between bias and the sum of cloud cover (measured in oktas) at the dawn and dusk of the previous day and between bias and the sum of cloud cover at the dusk and dawn of the previous night (i.e. the relation between bias and the extent to which cloud either shortens the previous day or lengthens the previous night). Partial correlation is significant only between bias and the extent to which cloud lengthens the previous night. The bottom pair of graphs show the relation between bias and the interval between sunrise and sunset of the previous day and between bias and the interval between sunset and sunrise of the previous night. Although bias shows a positive partial correlation with day length and a negative partial correlation with night length, in neither case is the correlation significant at the $P = 0.05$ level.

The evidence indicates that the transition from the summer to autumn peak directions of removal migration of *P. rapae* is dependent on the receptivity of the adult butterfly to the length and temperature of the previous night such that individuals subjected to colder and longer dark periods are more likely to show the autumn rather than the summer peak direction than individuals subjected to warmer and shorter dark periods. However, during the transition period the interval between sunset and sunrise, which is presumably a major factor at other times of year, is less important than temperature and the modifying effect of the amount of cloud at dusk and dawn on the length of the dark period. If this were not so, of course, butterflies in more northern areas, which have shorter summer nights, would not start to fly south before butterflies in more southern areas.

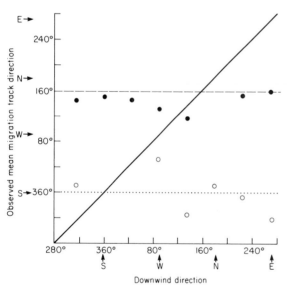

Fig. 19.14 Observed migration directions of the small white, *Pieris rapae*, in southern England (1965 and 1966) in relation to wind direction at the time of observation

————, relationship that would be expected on the basis of downwind displacement; – – – –, relationship that would be expected on the basis of the preferred mean compass direction on dates before 26 August; · · · · · ·, relationship that would be expected on the basis of the preferred mean compass direction after 27 August; •, observed mean migration directions on dates before 26 August; ○, observed mean migration directions on dates after 27 August.

The different track directions at the two times of year must be considered to be due to a real change in preferred compass direction and cannot be attributed in any sense to wind displacement.

So far, a detailed analysis of peak direction in relation to potential selective pressures is available only for temperate butterflies. The only tropical regions for which peak migration directions have been described are seasonally arid areas and although it is possible that temperature could be an important environmental gradient in these areas, it seems much more likely that the critical factor will be the seasonal distribution of rainfall (p. 425). Data are as yet insufficient to permit a detailed analysis but Fig. 19.15 tends to support this suggestion.

Finally in this section on butterfly migration let us consider the direction ratio component. In the seasonally arid regions of the tropics, the gradient of

habitat suitability is likely to be steep but non-linear, habitat suitability being low, but not necessarily zero, over wide areas, and then rising steeply near to the zone of maximum probability of rainfall (i.e. rainfall an appropriate length of time earlier— p. 674). In such a situation, it might be expected that the vast majority of individuals should fly in the direction of the zone of maximum habitat suitability. Most recorded observations of tropical butterflies in such regions give the impression of a direction ratio that approaches 100:0:0:0. However, recorded observations of butterfly migration in temperate regions often also give such an impression whereas it is unlikely that the direction

Fig. 19.15 The migration directions of tropical butterflies

Most data concerning the migration directions of tropical butterflies refer to the inhabitants of seasonally arid regions. In such areas it seems likely that seasonal variation in the spatial distribution of rainfall is the critical selective pressure on migration direction. In the cooler areas of sub-tropical regions, however, temperature may assume an importance approaching that suggested for temperate species.

Catopsilia florella in Africa

The maps show the peak migration directions of C. florella in Africa in relation to the spatial distribution of rainfall. Density of stippling reflects the amount of rainfall. In (a), migration direction refers to the period from December to February and the rainfall refers to the period from 1 November to 30 April. In (b), migration direction refers to the period from March to the end of May and the rainfall refers to the period from 1 May to 31 October. Even at this crude level, a broad correlation between migration direction and the direction of heavier rainfall is apparent. This correlation persists and improves when analysed on a monthly and local basis. It seems likely that the African population of this species consists of a number of demes, each of which perhaps has its own migration circuit.

The brown-veined white, Belenois aurota in southern Africa

This African species feeds as a larva on caper, Capparis spp., a bush that occurs in arid districts. A seasonally regular migration occurs in December/January in South Africa, Botswana, and Rhodesia. This migration is always to the E, towards an area of higher rainfall (see (a)).

Belenois aurota

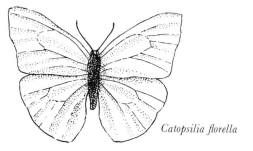

Catopsilia florella

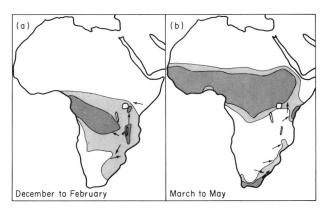

The common crow, Euploea core asela, in Sri Lanka

In Sri Lanka the common crow butterfly migrates from October to December during the NE monsoon season. Peak migration direction at this time is to the N. In March and April, during the SW monsoon, there is a second migration season, this time with a peak direction to the W or SW toward the area of heavy rainfall on the west and southwest coasts, the east coast remaining dry at this time.

The great southern white, Ascia monuste, in Argentina and Florida

In the grasslands of northeastern Argentina, the great southern white migrates with a peak direction to the S or SW during December and part of January. From mid-January until the end of March, however, peak direction is to the N. Similar directions are shown by Colias lesbia in the same region. Whether in this area it is rainfall or temperature or both that imposes selection on migration direction cannot be determined.

In Florida the great southern white migrates to the S from late autumn (November) until late spring or early summer (May/June) and to the N during summer and early autumn. The date of change of direction from S to N may vary from year to year by at least as much as from 17 May to 10 June. It seems likely that in this case temperature and not rainfall imposes the major selective pressure on migration direction.

[Compiled from data and information in Williams (1958), Bruggen (1961) and Hayward (1967). Rainfall from Fullard (1959)]

ratio of temperate butterflies ever approaches 100:0:0:0, as has been demonstrated by careful observation of the flight direction of *all* individuals (Baker 1969a). A student of mine, K.M. Adams, recently visited Botswana at a time when it was likely that a migration of the brown-veined white, *Belenois aurota*, would be observed and he was asked to pay particular attention to individuals migrating in directions other than in the main direction. On 1 January 1975 he counted 15 individuals flying W as compared to 720 flying E, with none flying in other directions. The direction ratio on this occasion, therefore, was 98:0:0:2, a value far nearer to 100:0:0:0 than has probably ever been encountered in temperate regions.

The explanation for those few *Belenois* that do not migrate in the peak direction could be that a few scattered suitable habitats are available at sites other than within the main zone of suitable habitats. It seems reasonable to suppose that the situation is free and that migration cost combines with deme density and habitat suitability (h_p—p. 70) in each place to give each individual the same potential reproductive success. Similarly, as discussed at length for birds (see p. 635), the autumn direction ratio of temperate butterflies that overwinter as adults, in combination with the frequency distribution of migration distance and the variation of both components with latitude, should interact to produce that latitudinal distribution of deme density that in combination with resource abundance leads, over the late autumn, winter, early spring period, to a free situation. In the case of the autumn migration of reproductive butterflies, such as the small white, *Pieris rapae*, the most applicable model is likely to be that based on the unpredictable variation in $\bar{E}$ (p. 71), for different directions, based on year-to-year probabilities of survival coupled with some other advantage acting in the opposite direction. Perhaps the larval offspring of autumn migrants to the W, N, and E gain an advantage through decreased deme density as long as they survive but suffer a greater probability that they will fail to reach the overwintering stage (Fig. 19.12). In some mild autumns, these individuals will gain the greater reproductive success. In cold autumns, the S-flying individuals gain the greater reproductive success. The proportion flying in each direction could thus reflect the probability of a mild or cold autumn.

In Chapter 12 it was demonstrated that a peak flight direction can evolve only if the geographical flight direction favoured by selection shows a regular

Fig. 19.16   The direction ratio of three west European pierids

The three species for which direction roses are illustrated have lifetime tracks that are probably of similar lengths, most individuals dying within perhaps 20–300 land km of their emergence site. Yet their direction ratios are different. The orange tip, *Anthocharis cardamines*, has a direction ratio that approximates to 25:25:25:25. The green-veined white, *Pieris napi*, has a peak migration direction but in Britain the bias toward this direction is slight, the direction ratio being only 36:23:23:18. The small white, *P. rapae*, has the same mean direction as *P. napi* but has a stronger bias, the direction ratio being 42:21:21:16. Analysis of this variation has shown that it is consistent with the model presented in Chapter 12.

The orange tip is univoltine, being on the wing only in spring (April to June). It cannot, therefore, be exposed to an environmental gradient of habitat suitability that shows a seasonal major change of direction. In such a situation the model predicts that 25:25:25:25 is the optimum direction ratio.

The green-veined and small whites are both multivoltine and should both be exposed to an environmental gradient of habitat suitability that shows a seasonal major change of direction. Both should therefore evolve a peak migration direction given a lifetime track of the length suggested above. By trial and error a computer model has been developed that both simulates the ecology of the small white and at the same time generates an optimum direction ratio of 42:21:21:16. The phenology of the green-veined white differs from that of the small white in Britain in that 30 per cent of the offspring of the spring butterflies enter an obligatory pupal diapause and, instead of producing further generations in the summer and autumn of the same year, aestivate and hibernate and produce butterflies the following spring. When this 30 per cent obligatory diapause is included in the computer model for the small white, the optimum direction ratio generated by the model changes from 42:21:21:16 to 36:23:23:18.

[*Rose diagrams from data in Baker (1969a). Details of the calculation of optimum direction ratio and the computer programs used are in Baker (1968d). Photos by R.R. Baker*]

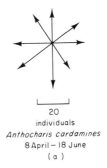

20
individuals
Anthocharis cardamines
8 April – 18 June
(a)

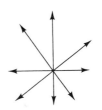

50
individuals
Pieris napi
1 March – 27 August
(b)

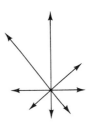

N

300
individuals
Pieris rapae
1 March – 27 August
(c)

major change. The degree of return favoured need not be great, however, as long as the difference between the two preferred directions is fairly large (e.g. greater than approximately 120°) and as long as the migration distance in the different peak directions by the same or succeeding generations is not too dissimilar. Figure 19.16 shows the direction ratio of the orange tip, *Anthocharis cardamines*, that has a migration distance that is probably little less than that of the small white, *Pieris rapae*. Female orange tips probably die between 20 and 200 km from their emergence site, though some males may move less than 1 km. It has also been demonstrated that each butterfly has an individual-specific non-compensated Sun orientation angle (Baker 1969a). However, because of its seasonal incidence, the species is subjected to selection only for migration in a single direction. In such circumstances, according to the direction ratio model (p. 86), a direction ratio of 25:25:25:25 should be favoured, an expectation that seems to be fulfilled. Evidence is also accumulating that a similar expectation is fulfilled for the white-letter hairstreak, *Strymonidia w-album*, though in this case migration distance is probably much less, the vast majority of individuals dying within perhaps 10 km of their natal site, within the range of local 'noise' (p. 654).

The expectation of the direction ratio model is also fulfilled in that all of those species that manifest a peak flight direction also manifest a seasonal major change of mean flight direction (e.g. Figs. 19.5–19.7 and 19.12).

In previous papers (Baker 1968d, 1969a) I have presented models, including the computer programs used in the application of these models, that have attempted to simulate in reasonable detail the movements of various species of British butterflies and that calculate the optimum direction ratio for that particular situation. The apparent ability of these models to account for observed inter-specific variation in direction ratio is quite marked, particularly so with respect to the difference (Fig. 19.16) between the direction ratios of the small white, *Pieris rapae*, and green-veined white, *P. napi* (Baker 1968d). In the present state of data shortage concerning the details of the ecology of even these butterflies, however, it is probably as well not to make too much of these models. Perhaps we are justified, however, in accepting the following conclusions: (1) the direction ratio of butterflies represents an evolutionary equilibrium situation with selection acting to favour those individuals that

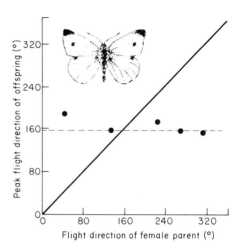

Fig. 19.17   The inheritance of mean migration direction and direction ratio in the small white, *Pieris rapae*

If selection has favoured butterfly lineages characterised by a direction ratio that is optimum for the species concerned, it must also have favoured a genetic mechanism whereby an individual produces offspring that manifest the optimum direction ratio irrespective of the preferred migration direction of the parent.

24 female small whites were captured during rectilinear migration on 15 June 1967, 3 for each of 8 migration directions relative to the Sun's azimuth. The offspring of these butterflies were reared at 30°C under continuous light. Virus infection killed the offspring of females with 3 of the 8 migration directions. When adult, the butterflies were given at least 2 days at 17 hours photoperiod and 20°C before being released in the middle of a large open field (different batches in different fields) between 8 and 15 July 1967.

The graph shows the observed mean migration direction (●) for the different batches as a function of the migration direction of the female parent. All angles are measured clockwise from the Sun's azimuth. The solid line indicates the relationship that would be expected if female parent passes on to its offspring a bias towards its own particular flight direction. The dashed line indicates the relationship that would be expected according to the thesis presented here.

Direction ratio relative to a mean migration direction of 159° varied for the different batches from 58:0:33:9 to 43:21:21:15. None, however, was significantly different at the 5 per cent level (χ^2 test) from the direction ratio of 42:21:21:16 that characterises the free-living population.

The available evidence supports the suggestion that: (1) selection has favoured a lineage of small whites with a direction ratio of 42:21:21:16 about a mean migration direction (in spring and summer) of 159°; and that (2) in order for this to evolve, selection has favoured a genetic mechanism whereby the offspring have a mean migration direction and direction ratio that is independent of the migration direction of the parents.

[Re-drawn from Baker (1969a), where full data are presented]

produce offspring with an optimum frequency distribution of migration directions; (2) except in the case of altitudinal migrants, a sufficiently long (greater than about 200 km) migration distance must evolve before selection acts to favour a direction ratio other than 25:25:25:25, the mean direction of which shows inter-deme consistency (p. 605); (3) a regular, usually seasonal, and major change of direction of the latitudinal gradient of habitat suitability must occur for selection to favour an innate direction ratio significantly different from 25:25:25:25; and (4) a peak flight direction is not, however, dependent on a high degree of return, but the higher the degree of return, the nearer the optimum direction ratio can approach 100:0:0:0.

Note that as in all equilibrium situations, the individual does not gain an advantage relative to other individuals as a result of its own behaviour (i.e. in this case, its own migration direction). Selection stabilises at that direction ratio that confers an equal advantage on all individuals. In a sense, of course, the individual gains an advantage in that if it flew in any other direction the equilibrium situation would be disrupted and it would suffer a disadvantage. Primarily, however, the advantage gained by the individual lies in producing offspring that show the optimum direction ratio. This implies, of course, that selection has favoured a genetic mechanism whereby offspring are produced with a direction ratio that is independent of the flight direction of the parent. Evidence that this is so, at least in the small white, is presented in Fig. 19.17.

19.4 Locusts

The term 'locust' is applied to a number of species of short-horned grasshoppers (Acrididae) which have developed a habit of mass migration. All locusts are tropical or sub-tropical, and their association with the seasonally arid parts of these regions, when compared with the world-wide distribution of acridids as a whole, strongly suggests that locust facies (e.g. the occurrence of *gregaria* and *solitaria* phases— p. 258—and migration by discrete swarms of *gregaria*) are adaptive specifically to this type of climatic regime.

The Rocky Mountain locust, *Melanoplus spretus*, was once extremely common in seasonally arid northwestern areas of the United States. The spread of agriculture, however, and disappearance of areas in which large aggregations of young stages ('hoppers') could pass unnoticed led to the rapid demise of the species towards the end of the nineteenth century. It has not been seen since (Williams 1958). The South American locust, *Schistocerca paranensis*, which is congeneric with the desert locust, *S. gregaria*, of Africa, is found throughout most of South and Central America as far north as southern Mexico. It also occurs on occasion on some of the West Indian Islands. The Australian plague locust, *Chortoicetes terminifera*, has only recently received intensive study, but is perhaps the first locust for which the movements of *gregaria* and *solitaria* have been evaluated in an integrated manner rather than as separate animals.

Most locusts, however, are found in Africa. The migratory locust, *Locusta migratoria*, also occurs elsewhere, different sub-species covering southern Europe, nearly all Africa, Malagasy, southern Asia, Indonesia, north Queensland, and New Zealand (Williams 1958). The desert locust, *Schistocerca gregaria*, the migration of which has perhaps been studied in more detail than that of any other animal species, is found primarily in and around the desert belt of Africa, extending south of the equator only in Kenya and Tanzania, but perhaps also occasionally in South Africa. It also occurs in Arabia, Iraq, Persia, Pakistan, and northern India. A swarm of the species has been seen in the Atlantic, midway between North America and Africa, 2250 km from the nearest possible site of origin on the latter continent. The Moroccan locust, *Dociostaurus maroccanus*, is found primarily in northwest Africa and in the hilly regions of other Mediterranean countries. Its total range of occasional occurrence extends from the Canary Islands in the west to the Aral Sea in the east. Finally, the brown locust, *Locustana pardalina*, and the red locust, *Nomadacris septemfasciata*, are found in the southern half of Africa.

The selective pressures on the inhabitants of seasonally arid areas have been discussed in relation to butterflies (p. 425), birds (p. 676), and, in some connections, locusts (pp. 246 and 258). Briefly, upon emergence, the probability that the most suitable habitats are elsewhere is extremely high. At any given time of year, the gross area in which rainfall has been, is, or will be most likely is relatively predictable. If the generation interval is less than a year, then this area is in a relatively predictable direction relative to the site of emergence. Selection will favour migration in that direction by means of the migration mechanism that incurs least migration cost. All locust migration can be viewed as an adaptation to this pattern of selection.

19.4.1 Migration mechanisms

19.4.1.1 Variation in migration threshold

Locusts that have developed under crowded, un-crowded, and intermediate conditions have the facies, respectively, of phases *gregaria*, *solitaria*, and *transiens* (p. 258). When locusts of any phase first become adult, their migration threshold is high but gradually lowers over the next few days as the cuticle hardens and flight becomes mechanically possible (p. 236). In most species, the migration threshold reaches its lowest level soon after the cuticle has hardened and remains at this level until the locusts encounter rain and other undefined conditions whereupon sexual maturation begins, the colour changes from pink to yellow (in *gregaria* of the desert, migratory, and red locusts), and the threshold begins to rise, slowly at first and then rapidly with the approach of copulation and oviposition (Chapter 17). The migration threshold falls again between each bout of copulation and oviposition but there is an overall decrease with age in both males and females (Michel 1969).

Phase *gregaria* migrate *en masse* during the day and phase *solitaria* migrate solitarily by night (p. 258). Diel variation in migration threshold can be modified, however, by variation in the threshold to other time-investment activities. Waloff (1972) observed two dense mature swarms of the desert locust, *Schistocerca gregaria*, phase *gregaria*, in northern Tanzania in late February that were accompanied by huge concentrations of predatory birds, primarily Abdim storks, *Sphenorhynchus abdimii*, and white storks, *Ciconia ciconia*. These birds were harassing the locusts by continual settling and taking-off within the moving swarm and seemed to be interfering with the feeding of the locusts which had empty guts and

Fig. 19.18 The migration of the desert locust, *Schistocerca gregaria*

The adult desert locust has a wing-span of 10–15 cm and a weight of 1–3 g. The eggs are laid in moist sand or soil at a depth of about 5–10 cm. Although eggs can be laid in areas in which rain fell no more recently than several weeks earlier, an optimum habitat for oviposition is one in which more than 20 mm of rain has very recently fallen and in which, therefore, there will be a supply of green vegetation for the resulting offspring. Sexual maturation is delayed, sometimes for up to six months, until the locusts have encountered the corresponding zone of seasonal rainfall. The egg stage lasts from 2 to 10 weeks, depending on soil temperature. The wingless nymph or 'hopper' stage lasts

from 5 to 6 weeks, at the end of which the insect moults for the last time to become a fully winged adult. Hoppers, if sufficiently concentrated, react gregariously to each other, assemble into dense groups, and march as coherent bands which themselves can migrate distances of 3–25 km before becoming adult. Such hoppers, which are yellow or orange in colour with black patterning, give rise to adults of phase *gregaria*. Hoppers that become isolated from their conspecifics assume a uniform green colour and give rise to adults of phase *solitaria*.

The air speed of a desert locust is of the order of 20 km/h soon after take-off, settling down to a cruising speed of 13–15 km/h. Migrating locusts alternate longer periods of flapping flight with shorter periods of gliding.

The maps show the seasonal distribution of desert locust swarms as determined during the period from May 1954 to May 1955. Arrows show the migration directions in the different areas during the period concerned and in the bottom right-hand map these arrows are collated to demonstrate the major migration circuits of the species. These migration circuits are quasi-regular though the precise routes and destinations of swarms on these circuits show year-to-year variation. The seasonal position of the Inter-Tropical Convergence Zone (ITCZ), a major rainfall area, is indicated by a heavy line but is only included for the months of January and July. Some details of the illustrated migration circuits are described in the text and only general features need be noted here.

All migration circuits result in locusts moving in a direction that is most likely to take them to the area containing the most suitable habitats for that particular season. Four major demes may be recognised, each with its own migration circuit: (1) the eastern Arabia, Iraq, Persia, Afghanistan, Pakistan, northern India deme; (2) the western Arabia, central and eastern North Africa deme; (3) the West and North West Africa deme; and (4) the East Africa and southwestern Arabia deme. Some interchange between these demes occurs, though the extent cannot yet be quantified. Note that once a locust swarm has encountered rain, and maturation has occurred, the locusts deposit offspring along the remainder of their lifetime track.

Although the different demes have their own particular migration direction at a particular season, certain broad inter-deme similarities are evident. In March and April nearly all locust tracks have a component to the N, though those in East Africa also orient toward high ground. In May and June, surviving swarms of mature locusts and swarms to the south of about latitude 15°N continue to orient to the N or to high ground but new swarms of immature locusts produced to the north of 15°N orient to the S. This process continues into July until all swarms are aggregated either in the vicinity of the ITCZ or on high ground (in East Africa). From October until February, locust tracks to the north of about latitude 20°N have a component to the N whereas those to the south of this latitude have a component to the S.

Such a pattern strongly implies that the preferred compass direction of locust swarms, and their migration thresholds to topography, are a function of photoperiod and/or temperature, with receptivity being greatest during the immature adult and early reproduction period of ontogeny.

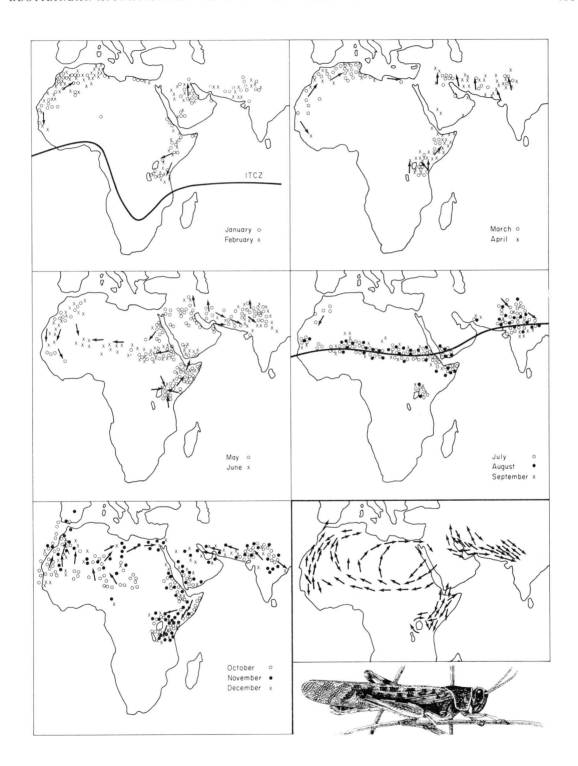

ITCZ

January ○
February ×

March ○
April ×

May ○
June ×

July ○
August ●
September ×

October ○
November ●
December ×

[*Data, but not interpretation, from Rainey (1963) and Rainey and Aspliden (in Rainey 1963)*]

greatly reduced fat bodies. On 27 and 28 February, in weather conditions that were favourable for continuation of migration, the locusts failed to initiate flight but remained settled and feeding actively. They also dropped into the undergrowth when disturbed instead of taking off. In the evening, however, some of the locusts were seen taking off from about 20 min after sunset, at first individually but then in small groups. These two swarms, therefore, which showed typical *gregaria* behaviour at first observation, for two days behaved like *transiens* or *solitaria*.

19.4.1.2 The annual migration cycle

There is strong evidence that in the two species for which relevant information is available, the desert locust, *Schistocerca gregaria*, and Australian plague locust, *Chortoicetes terminifera*, there is a more or less regular annual migration cycle that is adapted to the seasonal shift in zones of rainfall. As illustrated in Fig. 19.18, in some demes of the desert locust this annual cycle takes the form of a migration circuit, in others it is more or less a to-and-fro migration. In both cases, however, the migration is usually a re-migration, different stages being performed by different generations.

The red locust, *Nomadacris septemfasciata*, is univoltine (i.e. one generation per year) and breeds during the wet season in equatorial regions of Africa. After attaining the adult stage, locusts enter an imaginal diapause and pass the dry season in stands of long grass 2–3 m high. Up to 8 months may be spent in imaginal diapause and during this period little movement occurs. The most likely annual cycle in a univoltine inhabitant of seasonally arid areas is the pattern shown by some birds in such regions (p. 674). Here, at an appropriate season, the animal migrates to the area where rain has fallen or will fall (depending on the stage of the vegetational succession that follows rainfall to which the species is adapted) and then follows the direction of movement of the rainfall zone. Red locusts from northern Rhodesia follow the Inter-Tropical Convergence Zone northwards in June and July. Those from Uganda follow the same rainfall zone northwards in July and August (Rainey *et al.* 1957). Whether the same individuals then migrate back to the south with the returning rainfall zone does not appear to be known. If such a return does take place, perhaps it is performed by solitary reproducing adults migrating at night in a manner analogous to the Australian

plague locust (Roffey 1972a, Clark and Davies 1972).

Finally, it seems that the Rocky Mountain locust also performed a seasonally regular to-and-fro migration. According to Riley (in Williams 1958), writing in 1878, swarms of this locust from July to September migrated SE from breeding grounds in the Rocky Mountains to parts of Kansas, Missouri, and Texas. The following spring, in May and June, new swarms would migrate NW from these areas back to the Rockies.

19.4.1.3 The mechanics of migration

Locusts of phase *solitaria* migrate primarily at night and their migration mechanism is considered in Section 19.7 (p. 470) along with that of other nocturnal migrants. Phase *gregaria* perform the major part of their migration as discrete, dense, immature swarms that retain their identity for days or weeks. When these immature swarms encounter unsettled weather and rainfall and the maturation process begins they are liable frequently to disperse and converge and in strong winds in which flight is intermittent, either against or across the wind and at only 5–10 m above ground level, migration may well occur in diffuse formation. As maturation proceeds, the swarms become more diffuse and in the desert locust, *Schistocerca gregaria*, sexually mature swarms seem invariably to migrate as diffuse swarms (Waloff 1972). Examples of dense, cohesive swarms of sexually mature desert locusts are invariably small and often associated with large concentrations of predatory birds, particularly storks (see above). Swarms of the red locust, *Nomadacris septemfasciata*, also disperse at the approach of oviposition (Gunn 1960).

The migration mechanism of desert locusts migrating as dense discrete cohesive swarms is illustrated in Fig. 19.19 and shows a number of similarities to the migration mechanism of swarms of honey bees, *Apis mellifera* (p. 152). Both swarms are cohesive and in both the direction of cross-country displacement seems to be determined by the orientation of only some of the individuals that make up the swarm. In locusts, however, the swarm also feeds as it migrates and the speed of cross-country migration is a function partly of the ground speed of the oriented locusts and partly of the proportion of time that each individual locust spends on the ground as opposed to in the air (Waloff 1972). The migration mechanism of desert locusts migrating as diffuse

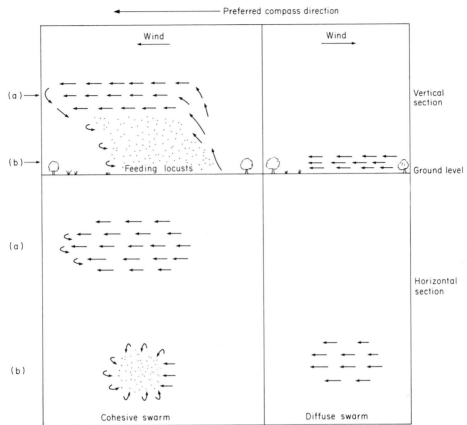

Fig. 19.19   The structure of cohesive and diffuse swarms of the desert locust, *Schistocerca gregaria*

The left-hand drawings illustrate the structure of a cohesive locust swarm and the right-hand drawings that of a diffuse locust swarm. Both sets of drawings are based on descriptions and data in Waloff (1972). An arrow indicates the mean direction of orientation of locusts in that part of the swarm (when the frequency distribution of orientation directions is significantly non-uniform at the 5 per cent level). Stippling indicates that no peak orientation direction is apparent.

The top pair of drawings presents a vertical section of these two types of locust swarms. Diffuse swarms fly relatively low with all locusts oriented in more or less the same direction at all heights and at all points within the swarm. The distance between individual locusts is relatively large and little swarm-cohesion behaviour is observed. Locusts may or may not settle beneath the swarm to feed. Cohesive swarms extend from ground level up to heights of several hundred or even thousand metres. Middle and high flying locusts are oriented more or less in the direction of travel (see p. 456). At all heights, locusts that emerge from the leading edge of the swarm reduce height and orient back into the swarm. Many settle on the ground to feed. Low-flying locusts show a uniform distribution of orientation and track directions, though 'streams' of locusts may temporarily develop. The direc-

tions of these streams, however, show no relationship to the direction of travel of the swarm as a whole. Grounded or low-flying locusts that emerge from the tail-end of the swarm take flight and/or orient back into the swarm. The result is that a cohesive swarm 'rolls' across country in the manner illustrated. The ground speed of the swarm is a function partly of the ground speed of the high-flying locusts, partly of the proportion of individuals that settle beneath the swarm to feed, and partly of the relative lengths of time that individual locusts spend on the ground and in the air. Because the locusts feed during migration, they may actually gain weight from the beginning of migration to the beginning of reproduction, despite having travelled 1000 km or more.

The lower drawings represent horizontal sections through the two types of swarm at the levels indicated by (a) and (b) in the top diagram. It can be seen that in a cohesive swarm, whenever a locust emerges from the edge, it immediately orients back into the main body of the swarm. In a diffuse swarm all locusts orient in the migration direction.

The factors that determine whether a given swarm should migrate in cohesive or diffuse formation are unknown. It is suggested in this book that, as illustrated, a cohesive formation is adopted whenever the downwind direction coincides with the preferred compass direction of the swarm whereas a diffuse formation is adopted whenever the downwind direction is opposed to the preferred compass direction.

swarms is quite different. On such occasions each individual locust orients in the appropriate direction and there is no indication of behaviour that would lead to swarm cohesion.

Cross-country displacement by desert locusts, phase *gregaria*, is thus achieved by orientation in a particular direction irrespective of whether the migrating swarm is dense, discrete, and cohesive or diffuse. The questions that remain are whether the locusts orient to the wind or in a particular compass direction or both, the way that this orientation can change with ontogeny and time of year, and the factors that favour migration on some occasions as a dense, cohesive swarm by day, on some occasions as an individual or small group within a diffuse 'swarm', also by day, and on some occasions as solitary individuals by night. These problems are as yet largely unresolved, but such evidence as is available may now be reviewed. All of this evidence concerns the desert locust.

Perhaps the most important problem is the extent to which locusts orient to the wind. It was suggested for butterflies that inhabit seasonally arid regions of the tropics that optimum strategy is to orient in the geographical direction that gives the route to the location of the zone of maximum habitat suitability with the minimum migration cost. When the wind has no vector in this preferred direction, optimum behaviour is to fly low. When the wind has a vector in this direction, optimum behaviour is to fly high above ground level. When the wind has a vector at right angles to the required track, optimum behaviour is to adjust the oriented heading by the amount necessary for the track direction to continue to be in the required direction (see p. 876 for bees). Such compensation, which in butterflies and locusts is unlikely to be complete (being that angle that gives the track with the maximum vector in the required direction per unit energy rather than simply the track with the maximum vector in the required direction), could be achieved by orienting secondarily (Fig. 19.2) to a succession of distant landmarks and/or by monitoring the angle at which the image of the substratum moves across the retina. In this connection it should be noted that even in the densest swarms, locusts are spaced sufficiently for at least those at 400 m above ground level to retain visual contact with distant landmarks and the substratum (Waloff 1972). The extent to which this strategy is used by locusts may now be considered.

Figure 19.20 presents an analysis of all records of the track directions of swarms of immature locusts

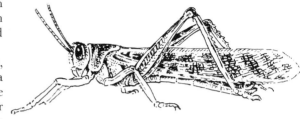

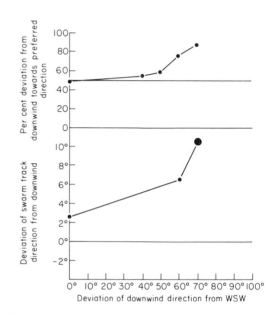

Fig. 19.20 A test of the hypothesis that cohesive swarms of immature desert locusts, *Schistocerca gregaria*, have a preferred compass direction

Current theses of the orientation of desert locusts state either that swarms are displaced downwind or that the high and middle flying locusts (see Fig. 19.19) orient downwind. On either of these hypotheses, which for present purposes may be termed downwind theses, the track direction of cohesive swarms of desert locusts should always be downwind irrespective of compass direction.

The thesis presented in this chapter is that locusts have a preferred compass direction that is location- and season-specific and that will maintain the locusts in the appropriate 'migration circuit' direction (see Fig. 19.18). On this thesis, as downwind direction deviates from the preferred compass direction, the locusts should adjust their orientation (probably to the Sun) in such a way that the swarm track deviates from the downwind direction in the direction that reduces or minimises the deviation from the preferred direction.

The graphs examine the thesis that desert locusts in Kenya in January and February have a preferred compass direction to the WSW. The upper graph categorises swarm track directions in terms of whether their deviation from the downwind direction is in the direction expected on the preferred direction thesis. The per cent that do so deviate is indicated. The lower graph indicates the mean angular deviation of observed track directions from the downwind direction (where the deviation is considered to be positive if in the direction predicted by the preferred compass direction thesis and negative if not). A small dot indicates that the mean deviation is not significantly different from downwind ($P = 0.05$); a large dot that the mean deviation is significant. These measures are plotted as a function of the minimum deviation (in the observations used in the calculation of that point) of the downwind direction from WSW. Hence, $0°$ indicates that all of the 52 available observations were used but $70°$ indicates that only those observations were used for which downwind direction deviated from WSW by $70°$ or more. In both graphs, a relationship that falls on the horizontal line supports a downward thesis, a relationship that is entirely above the horizontal line (except for the top graph at $0°$) and that shows a positive, though not necessarily linear, correlation supports the preferred compass direction thesis. Both graphs support this latter thesis.

It is concluded that as long as the downwind direction has a strong component toward the WSW, the locusts show little compensation for wind drift. The further the downwind component shifts from this direction, however, the more likely is the track direction of the swarm to be on the WSW side of the downwind direction. Finally, when the downwind direction deviates by $70°$ or more from WSW, the track direction is no longer downwind but is shifted by a significant amount ($a = 10.4°$; $n = 7$; $R = 6.94$; $X = 6.82$; $P < 0.05$) to the WSW side of the downwind direction.

[*Analysis based on all observations presented by Rainey (1963) and Waloff (1972)*]

migrating in Kenya in January and February in relatively settled weather presented by Rainey (1963) and Waloff (1972). Fifty-two records are available in all. The swarms that move across Kenya at this time of year originate in the Somali Republic, southeastern Ethiopia, and northeastern Kenya from eggs laid during the 'Short rains' season of October to December. The tracks of these swarms are more or less W to the north of the equator and more or less SW to the south of the equator. Figure 19.20 examines and supports the thesis that these swarms have a preferred track direction to the WSW (i.e. $247.5°$ measured clockwise from north) and that the effect of wind is to displace the locusts from this track.

There is, therefore, some indication that even in the most spectacular cross-country migrations of desert locusts, primary orientation is in a specific compass direction rather than relative to the wind. Waloff (1972), like Kennedy (1951), considered the possibility that the orientation of locusts flying at medium to higher levels (above 200 m) within cohesive swarms was taken from the Sun. Although

Waloff did not pursue this possibility, the above and following analyses suggest that such a mechanism is the most likely. Further support for this possibility is provided by Waloff's finding that the overall mean orientation of high-flying locusts was non-random ($P < 0.001$ in a Rayleigh's test) for both swarms for which comprehensive data were available. In both cases, however, the mean orientation deviated markedly both from the swarm track direction (deviations of $26°$ and $19°$) and from the downwind direction (deviations of $27°$ and $11°$).

When we come to consider the migrations of swarms in the vicinity of rainfall and/or topography that is likely to be associated with rainfall, such as high ground, the situation becomes more complex. It seems likely that an animal as dependent on local rainfall as the desert locust would, like other animals in similar environments (p. 95), have evolved the ability to perceive distant rainfall and to migrate towards it. Migration thresholds to topography also seem likely to have evolved where topographical features correlate with favourable conditions of temperature and rainfall.

Two examples may suffice to suggest that there is a real possibility that rain-in-sight may be an important factor in locust migration. Figure 18.3 shows the reversal of migration direction of a swarm of immature desert locusts in the Darfur province of Sudan in June. The 'migration circuit' direction in the region at this time of year is W, most swarms having originated in Arabia (Fig. 19.18). On both the 10 and 11 June, the mornings were fine to fair but Cumulonimbus developed each afternoon and local rain fell from the afternoon through the night. On the afternoon of 10 June, rain was recorded to the west of the swarm and the swarm on that day migrated 25 km in that direction. Overnight on the 11/12 of June, however, rain was recorded to the east of the swarm. During the course of the 11 June, the swarm moved about 65 km back to the east in the direction of the coming rain. These changes of direction, however, were associated with similar changes in the downwind direction, though on the morning of the 10 June there was some indication of a delay in the initiation of migration until the morning westerly winds gave way to winds from the east. On the morning of the 11 June, however, when the direction of movement reversed, there appears to have been no such delay. On the basis of this example, however, there is no means of determining whether the locusts changed their compass orientation or whether they were orienting downwind throughout. Waloff (1972), however, gives an example of a swarm of desert locusts migrating under conditions in which downwind orientation and orientation in either the 'migration circuit' direction or towards rain-in-sight would be in quite different directions. The observation was made in open thorn scrub on level ground in southeastern Ethiopia at a time of year (mid-October) when the 'migration circuit' direction is to the SW (Fig. 19.18). The swarm was low-flying, comparatively diffuse, and mature. Sampled females were near to ovipositing and may already have oviposited on a previous occasion. Conditions were overcast and humid and the wind was light and ranged from ESE to SE. Waloff remarks, however, that there was rain-in-sight to the SE and SW. The locusts appeared at 15.10 hours from the NW. There was no indication

of swarm cohesion, all locusts heading mainly ESE or SE (i.e. into or slightly across the wind) at all heights up to the top of the swarm (30–40 m above ground level) and throughout the 80 min required for the passage of the swarm past the observers. Although the track of the swarm was initially SE, the recorded tracks of the locusts while passing the observers were between S and SW. Whether the locusts were moving in the migration circuit direction or were migrating to rain-in-sight cannot, therefore, be determined. What is certain is that they were neither orienting nor being displaced downwind.

Figure 19.21 shows two examples of desert locusts being deflected from their previous track by an apparent attraction to high ground. There is also an indication from this figure that high ground raises

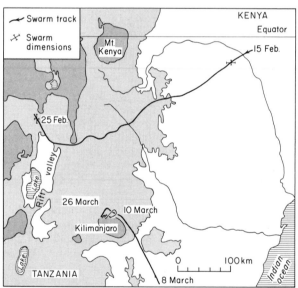

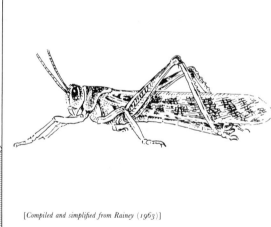

[Compiled and simplified from Rainey (1963)]

Fig. 19.21   The orientation and migration thresholds of swarms of the desert locust, *Schistocerca gregaria*, as a function of topography

The migration circuit direction of desert locusts in Kenya in January is WSW. This causes the swarms to intercept the rainfall associated with the Inter-Tropical Convergence Zone (ITCZ) in the vicinity of the Kenya highlands. These swarms show a marked reaction to high ground in that first they orient to it and secondly the proximity of high ground appears to raise their migration threshold to other variables. Contact with the ITCZ has the effect of changing the migration direction, whenever the migration threshold is exceeded, from WSW to NE. The relative involvement of photoperiod, temperature, rainfall, topography, and wind direction in this change is unknown. The nett effect of all these factors, however, to which may perhaps be added orientation to rain-in-sight,

is that young immature swarms produced in Somali, southeastern Ethiopia, and northeastern Kenya migrate to the WSW until they arrive in the potential high rainfall areas of either the Kenya highlands or the ITCZ. In the former case they await the arrival of the ITCZ. Swarms that encounter the ITCZ before they encounter high ground change their migration direction to N or NE, thus remaining in the vicinity of the ITCZ.

All of these effects can be seen in the diagram. The northernmost swarm migrated from its place of origin to the highlands in about two weeks. It then remained in the vicinity of the Rift Valley where it matured and oviposited on the March rains produced by the ITCZ. The southernmost swarm encountered the ITCZ at the end of February/beginning of March further to the south near the coast of Tanzania. It then migrated to the N until it encountered Mount Kilimanjaro on the slopes of which it remained from 10 until at least 26 March.

the migration threshold, at least at certain times of year. Perhaps the contribution of high ground to the composite migration threshold (and to migration direction) is a function of temperature and photoperiod. Whatever the physiological basis, however, it seems to be established that throughout the southeastern part of the desert locust range in Africa, swarms accumulate on high ground, particularly in Kenya in March, and again in Kenya and northern Somali from July to September during the hottest and dryest part of the year (Fig. 19.18). During this same period the swarms in northern Somali show delayed sexual maturation, suggesting that the entire physiology is influenced by indirect variables (e.g. temperature, photoperiod).

Turning now to swarms of mature desert locusts we find even stronger support for the suggestion that migration direction is determined primarily by orientation in an appropriate compass direction and that the locusts migrate high with favourable winds, low against head winds, and show compensation for cross winds. Mature cohesive swarms are seen only infrequently but when they do occur they show a structure similar to that illustrated in Fig. 19.19 except for a tendency for the major portion of airborne individuals to be oriented in the direction of travel (Waloff 1972). Individuals at the swarm edge also tend to vector the swarm track and the shortest route to the main body of the swarm rather than orienting directly into the swarm as do immature locusts. Both of the cohesive swarms of mature adults observed by Waloff were migrating in the 'migration circuit' direction (WSW and SW in northern Tanzania in February—Fig. 19.18) on days with a following wind, but they were also accompanied by huge concentrations of predatory birds (p. 452). It is difficult, therefore, to know the extent to which the usually diffuse structure of a mature swarm was modified in response to the presence of predatory birds and to what extent it was modified solely in accordance with the requirements of migration.

Most observations of mature swarms, however, involve diffuse formations with all individuals orienting, apparently independently, in a similar direction rather than to other locusts. One example of a mature swarm migrating across country against the wind but in the appropriate migration circuit direction has already been given, and because the point is so important further examples are given below.

Roffey (in Waloff 1972) observed a diffuse swarm of mature desert locusts, which had not yet oviposited, in the Northern Somali Republic on 20 March at a time when throughout East Africa the migration circuit direction is to the N. On this occasion there was a gradually strengthening headwind from the NE. Throughout the period of observation from 10.00 to 13.00 hours, the locusts were heading and migrating upwind, the major part of the swarm flying at less than 15 m above ground level.

A further observation in the Somali Peninsula was made by Waloff (1972) on 1 November at a time when the migration circuit direction is to the S (Fig. 19.18). The swarm observed may already have laid twice and according to the state of the ovaries of sampled females was passing through an interoviposition period. During the morning the winds were from the E and SE for up to a considerable height above the ground. Nevertheless, with locusts flying against the wind at heights no greater than 60 m by 10.00 hours but up to 300 m by 11.00 hours, air reconnaissance at 11.30 showed that the swarm had extended 4 km S and 5 km E of the observation site. Some time between 12.00 and 13.00 hours, however, a wind from the NW became established in the first few hundred metres above ground and locusts rose to heights of 460 m. Further air reconnaissance to the ESE showed that in this direction at least the locusts had spread for 30 km. Throughout the day, therefore, this swarm attempted to migrate to the SE, against the wind at low levels in the morning but with the wind at higher levels when a favourable wind became established.

Finally, Kennedy (1951) observed a diffuse swarm of mature desert locusts in central Persia in May at a time when the migration circuit direction is to the N. These locusts were flying low against NW winds.

Not only mature locusts form diffuse swarms. Waloff (1972) presents two observations of swarms of immature locusts that were also diffuse and on both occasions the tracks were into or across the wind. Unfortunately, on neither occasion was the swarm track recorded. On the first occasion, the swarm was observed in August in the Somali Republic at a time of year when the swarms are aggregated in the highlands and when relatively little movement takes place except of a local nature, the locusts in the region apparently being in a state of reproductive diapause. As far as the migration circuit is concerned, the observation was made towards the end of a period of movement to the N but before the beginning of a movement to the S (Fig. 19.18). The wind was strong from the SW and departure from the roosting site was delayed. The locusts oriented to the WNW and individual track directions were recorded as between NW and NE. Flights rose at times to 50 m. It seemed likely that the track direction of the swarm was to the N. The W–NW orientations of the locusts persisted even when the wind decreased towards sunset and backed to southerly.

The second observation was also made in the Somali Republic but this time in late September at about the time of onset of migration to the S. The locusts were flying intermittently in a low loose formation, below 5–10 m. The wind was strong (27–31 km/h) and persistent from the SSW and SW. The locusts were oriented between SW and SSE and recorded tracks of individual locusts varied between S and E. The direction of the swarm track was not recorded.

Examination of the records of locust orientations, tracks, and swarm formation in relation to wind direction presented by Rainey (1963) and Waloff (1972), including the examples presented in this section, shows a strong correlation, which presumably is not fortuitous, between swarm type and on the one hand wind direction and on the other hand state of maturity. The latter correlation is the least clear, but there seems to be a decrease in the manifestation of swarm-cohesion behaviour as the locusts mature and a tendency to fly in a diffuse formation with all locusts taking on the appropriate compass direction and reacting less strongly to each other. The other and stronger correlation, however, is between swarm structure and the relative direction of wind and the required direction of travel. It seems that in the vast majority of cases, a dense, discrete, cohesive swarm is associated with a downwind direction that has a strong vector in the required direction of travel (e.g. in the 'migration circuit' direction, the direction of rain-in-sight, or the direction of suitable topography). When the downwind direction has a vector against the required direction of travel, however, the swarm seems to adopt a diffuse formation with the individual locusts orienting more or less independently in the required direction and showing little if any swarm cohesion behaviour of the type illustrated in Fig. 19.19.

The diffuse formation approximates to that of all other insects that migrate and orient individually and requires little further discussion. The question of interest with respect to locusts concerns the advantage of migrating on occasion as a cohesive swarm. Apart from some social insects, such behaviour is unique to locusts. In social insects, such a swarm formation is adopted irrespective of wind direction, whereas in locusts it seems primarily to be adopted when the downwind direction coincides with the required direction of travel. Such cohesion is not an inevitable consequence of mass emergence as is shown by the adoption of a diffuse formation on other occasions. Let us take as an *a priori* assumption that having been aggregated as hoppers by environmental conditions, selection favours the aggregation being maintained, perhaps as an adaptation to reduce the risk of predation by birds (Waloff 1972) or other animals. When there is a head wind, the locusts fly low and there is less differential in ground speed within the swarm. The result is that by and large the swarm integrity can be maintained without the use of energy-consuming orientation to other locusts and consequent movement in directions other than the required direction of travel. When the wind is favourable, however, the efficiency of cross-country migration in terms of kilometres per unit energy is maximised by flying high above the ground at the altitude of the most favourable wind. At the same time there is an advantage in coming to the ground from time to time in order to feed. The result is a tremendous differential in ground speed between locusts at the top of the swarm (sometimes at a height of 2 km) and locusts on the ground. This, coupled with the possibility of variation in wind speed and direction with height and time seems likely soon to disperse the swarm over a large area were the locusts not to employ swarm-cohesion behaviour of the type illustrated in Fig. 19.19. Given, therefore, an advantage in: the maintenance of swarm integrity; flight at the height of the most favourable wind; and feeding during migration; a correlation between cohesive swarms and downwind migration and diffuse swarms and upwind migration would be expected. From the observations that I have so far examined, such a correlation does indeed seem to exist. There is a need, however, for a comprehensive analysis of all the observations that are available.

Because of the relationship between wind fields and major rainfall zones, immature locusts that have not yet encountered rain invariably migrate in dense cohesive swarms in a direction that varies only slightly from the downwind direction. Once rain is encountered, however, and maturation begins, or in regions and seasons in which the locusts prefer high ground, the relationship between wind direction and either the migration circuit direction or the direction of rain-in-sight/rain clouds or the direction of preferred topography breaks down. Although dense, cohesive swarms may still be adopted when the required direction coincides with a downwind component, on frequent occasions the required direction is upwind and the locusts adopt a diffuse swarm formation. The situation becomes complex when winds are favourable at one height but unfavourable at another.

19.4.2 Distance, direction, and direction ratio

Although some swarms of the desert locust may reproduce within a few hundred kilometres of their place of emergence, it seems likely that the majority of such locusts die within 1000 and 3000 km of their emergence sites. Similar distances are probably travelled by the migratory locust, *Locusta migratoria*, red locust, *Nomadacris septemfasciata*, and Australian plague locust, *Chortoicetes terminifera*, but the Moroccan locust, *Dociostaurus maroccanus*, probably migrates less far. Some extremely long-distance overland migrations have also been recorded, such as, for example, the desert locust swarm that flew over 5000 km from the Arabian peninsula to Mauretania in two months in 1950 and the swarm of the same species that flew over 3000 km from the Somali peninsula to the borders of Jordan and Iraq in less than 4 weeks in 1951–2 (Waloff 1959).

The major part of the cross-country distance is undoubtedly usually achieved during the pre-reproductive stage of ontogeny (Fig. 18.6) before rain is encountered and the maturation process begins. However, this is not always the case. In the Somali Peninsula immature swarms originate in June and July from eggs laid in March and April by mature swarms migrating NNE with the first rains across Tanzania, Kenya, southeastern Ethiopia, and Somali (Fig. 19.18). Those swarms that originate from eggs laid in the highlands of Kenya and Tanzania most often remain around their emergence site. Those originating further north migrate to the N during July and August and accumulate in the highlands of northern Somali. Towards the end of September these swarms, which are still immature, begin to migrate S (p. 453). With the arrival of the short rains in October, the swarms

mature and reproduce as they move southwards over the Somali Peninsula. Such swarms may migrate distances of perhaps 1000 km between successive bouts of oviposition. Rainey (1951), for example, records a swarm of sexually mature locusts which covered some 650 km between the Dusa Mareb area in Somali Republic and northeastern Kenya in 9 days from 18 to 27 October.

There is also no doubt that the major part of cross-country migration is achieved during periods that the locusts are migrating with favourable winds. Under such conditions, a migration rate of 130 km/day is not unusual, though on occasion swarms may achieve only 2 km/h (about 20 km/day) even when migrating with a wind of 9 km/h (Waloff 1972). Swarms migrating against the wind may achieve no more than 5 km/h (about 40–50 km/day) even if the individual locusts are continuously airborne (Waloff 1972). This is unlikely and an upwind migration rate of only 10 to 20 km/day seems more reasonable. Accurate measurements of the rate of upwind migration are not available, due largely to the difficulty of obtaining an accurate fix of position with a diffuse swarm.

In Persia in May, desert locusts migrate to the N, both against the wind (Kennedy 1951) and on temporary spells of warm southerly winds ahead of cold fronts in passing westerly depressions (B. Shaw 1965) and as elsewhere it is the latter migrations with the wind that make the major contribution to cross-country displacement. On the figures presented in the previous paragraph, however, the contribution of upwind migration to the total displacement cannot be considered to be negligible and may hasten arrival at regions of most suitable habitats by several hours, or even days, with obvious advantage to the animal. The advantage of upwind migration to rain-in-sight or to suitable topography is also clear.

As far as migration direction is concerned there is no difficulty in relating evolved migration circuits to the seasonal movements of major rainfall zones (Fig. 19.18), a fact pointed out in a now-classical paper by Rainey (1951). Further, as major rainfall zones, such as the Inter-Tropical Convergence Zone, are areas of convergence, such migration circuits are oriented for much of their length parallel to and down the wind flow in quasi-uniform wind fields. In the desert locust this is true for the spring (April to June) movements of swarms to the S across the Sahara and Arabia, to the SW and W from Arabia to the Sudan, Chad, Niger and Mali, and to the SE across Pakistan and India (Fig. 19.18). In all cases the migration is toward the rainfall associated with the Inter-Tropical Convergence Zone, as also is the simultaneous migration to the N from Kenya and Tanzania. The migration to the S across the Somali Peninsula from October to December is also toward the receding Inter-Tropical Convergence Zone and so also takes place in primarily quasi-uniform wind fields. Other sections of migration circuits are not, however, associated with quasi-uniform wind fields and over such sections the zone of maximum suitability may not always be downwind. The spring migration to the N across Persia is an example, as also is the autumn migration to the NE across the Sahara. Over such migration-circuit sections up-wind migration may make a significant contribution to the cross-country displacement, even though the greater part of the distance may occur on temporarily favourable winds during the passage of depressions. Having arrived in the rainfall zone, however, whereupon the relationship largely breaks down between wind direction and the local distribution of rainfall and topography and continuation of the migration-circuit direction, the swarms spend much more time migrating across or into the wind.

It has been demonstrated (p. 456) that the desert locust compensates for wind drift whenever the downwind direction deviates by more than about 60°–70° from the preferred compass direction. However, as long as the wind vector has a component in the direction of the preferred compass direction (i.e. as long as the deviation of the downwind direction from the preferred compass direction is less than 90°), the track direction of the swarm is to a large extent a function of wind direction. The result is that although migration circuits remain broadly similar from year to year, the precise destination of swarms performing a given circuit varies considerably from year to year and some areas may be devoid of locusts for several successive years.

This year-to-year variation in migration direction leads us to a consideration of direction ratio. Clearly, if swarms tolerate displacement by wind from their preferred track direction of up to about 60° (presumably because they select that orientation that gives the maximum component *per unit energy* in the required direction rather than simply the maximum component), the potential scatter of swarm directions that can be produced solely as a result of different swarms entering different wind regimes is as great as 120°. The directions of most swarms from a single area of origin seem to fall within such a

direction ratio and it may be unnecessary to postulate that different swarms have different preferred directions for any given place and time. However, a greater readiness of some individuals to depart from the site of adult emergence when the wind is in one direction and of other individuals to depart when the wind is in another direction could well result in different swarms being composed of individuals with different preferred directions and thus result in swarms with different preferred directions. It may be assumed that the mechanics of swarm cohesion (Fig. 19.19) impart to the swarm a preferred direction that is the mean of the preferred directions of its constituent individuals.

With most breeding in East Africa from October to December occurring well to the north and east (Fig. 19.18), mainly in the Somali Republic, it might be expected that swarms passing across northern Kenya in January and February would have a track more or less to the W whereas those passing across southern Kenya to Tanzania would have a track more or less to the SW. In fact, the mean track direction of swarms recorded by Rainey (1963) at this time and place can be calculated to be 283° (WNW) north of the equator and 220° (SW) south of the equator. However, the winds in the two regions at these times also differ in the same sense. It cannot, therefore, be determined to what extent the direction ratio results from swarms with different preferred directions and to what extent it results from wind drift of swarms all of which have a preferred direction to the WSW.

In most migration circuits of the desert locust in which there is a major change in direction at a particular time of year, the return migration seems likely to be a re-migration. In East Africa, however, the change seems likely on two occasions to cut across a generation. This happens firstly in late February/early March in southern Kenya and northern Tanzania when the migration direction changes from WSW to N or NE, and again in September in northern Somali when the direction changes from N to S or SW. At both times it is the same swarms that migrate in the different directions. Observations of swarms in northern Somali from August through September during the change-over period do not seem to have been made (but see p. 459). Observations have been made, however, on swarms in northern Tanzania during the change-over period in February/March (Fig. 19.21). In this case, the swarms migrate WSW to intercept the northward-moving Inter-Tropical Convergence Zone and then, having done so, the maturing swarms follow the Zone back to the north, though many settle permanently in the high rainfall area of the highlands. As yet there are insufficient data to evaluate the relative contributions of photoperiod, temperature, rainfall, wind direction, and topography to the physiological process involved in the change of direction. It seems likely that all of these factors are important, but it is too soon to determine whether the change of direction results simply from continuous downwind migration as suggested by Rainey (1963) or from a change in preferred compass direction and/or a changed reaction to topographic features.

19.4.3 The migration of an insect predatory on the desert locust

The following account of the migration of the hunting wasp, *Sphex aegyptius*, in East Africa is taken from Williams (1958). This solitary wasp, in common with many other hunting wasps, digs burrows into which are placed paralysed prey. The female then oviposits in the burrow which is then sealed. When the young larva hatches it has a ready supply of fresh but paralysed food upon which it can feed until ready to pupate. *Sphex* spp. are specialised to prey on grasshoppers and in the case of *S. aegyptius* in East Africa the prey is the desert locust, *Schistocerca gregaria*. It seems that the wasp accompanies the locust on its migration and attacks those individuals that settle underneath the swarm to feed (Fig. 19.19). The movement of wasps within the locust swarm is unknown but evidently behaviour exists that at intervals takes the wasps from the back of the locust swarm to the front. At any event, within 15 min of the first locusts settling on the ground under a swarm the wasps appear in numbers and are observed running about on patches of bare ground. Immediately on arrival the wasps begin to burrow and a little later begin to drag paralysed locusts along the ground and into their burrows (see p. 874). When the locusts settle to roost, the wasps continue to dig and stock burrows until dusk and then begin again the following morning after dawn. As soon as the tail edge of the locust swarm passes overhead, however, the wasps leave immediately, no matter what stage they have reached with their latest burrow. Presumably the wasps then fly through the swarm to the leading edge whereupon the process is

repeated. The same wasp species has probably been observed migrating within a locust swarm within a few metres above ground level, but whether they also fly at higher levels within the swarm is unknown.

19.5 Houseflies, blowflies, and hoverflies

The only remaining strong-flying diurnal insects for which data are available are the various species of two-winged flies (Diptera), particularly those houseflies, blowflies, and hoverflies belonging to the families Calliphoridae, Muscidae, and Syrphidae. By and large, the available data suggest that the migrations of these insects are extremely similar to those of butterflies. The following data, but not their interpretation, have been taken except where otherwise stated, from the review presented by Johnson (1969, pp. 382–397, 402–403).

Work on houseflies and blowflies in temperate regions has demonstrated, without exception, that the distance and direction of movement outwards

Fig. 19.22　A housefly, *Musca domestica*

[*Photo by E.J. Hudson (courtesy of Frank W. Lane)*]

from a release point and presumably, therefore, from any source, is independent of wind speed and direction, almost all migration taking place within a metre or so of the ground. Houseflies, *Musca domestica*, and the blowflies, *Lucilia sericata* and *Phormia regina*, released in a flat area, were recaptured at distances of up to 19, 6·5, and 13 km respectively with little or no detectable relationship to the direction of the wind. In other experiments, *P. regina* were considered to migrate at a faster rate than *Musca domestica* and some individuals were recaptured 45 km from the release point, 13 days after release.

Some flies are invariably recorded as flying in all directions from the release point, but the direction ratio may not always be 25:25:25:25. Often a peak flight direction is recorded which is sometimes downwind, sometimes across wind, and sometimes upwind. Hanec recorded that 79 per cent of *M. domestica* recaptured after release in farmland were upwind of the release point within 24 hours, 65 per cent being 3 km or more upwind. The compass bearing of peak flight directions has not been analysed in the manner described for butterflies. In any case, the situation may be confused over short distances by the existence of olfactory sources upwind from the release site which may attract or repel the flies.

Unfortunately, direct observation of houseflies and blowflies is so difficult that the nature of the lifetime track can be pieced together only from scattered observations. It seems likely, however, that at least for many temperate species, the lifetime track is of the linear range type illustrated for a butterfly in Fig. 18.1. Rectilinear flight over reasonable distances has been observed for a number of species. Thus the sun-loving *Lucilia sericata* was seen to enter, penetrate, and emerge from the other side of 90 m of woodland. *Calliphora vicina* and *C. vomitaria* have been seen to migrate through 50 m of woodland and out into the open at the other side and the hoverflies, *Syrphus vitripennis* and *S. ribesii*, have been observed flying through a gap between buildings in August at the rate of about 200/min.

As far as time investment in suitable habitats is concerned, up to 90 per cent of hungry female houseflies, *Musca domestica*, that enter houses have undeveloped ovaries, and up to 79 per cent of those that fly out have well-developed ovaries. In this species, feeding and oviposition take place in spatially separate habitat-types. Released screw-worm flies, *Callitroga hominivorax*, show maximum stay-time around stock ponds at which they remained, often for several days.

Alternation of time investment in suitable habitats with rectilinear migration thus seems to be established for a variety of temperate houseflies and blowflies and, for most, migration distance seems likely to be facultative. Little is known concerning the rectilinear migration, however, there being no indication whether, like butterflies, the preferred compass direction is an individual-specific characteristic (Fig. 19.2). The length and speed of some cross-country migrations, however, such as the 290 km in 2 weeks for female screw-worm flies, *C. hominivorax*, in the southern United States, and the 45 km in less than 13 days for *Phormia regina*, suggest that for some species at least each individual does have a preferred compass direction. Although there is no indication of the orientation mechanism used by flies, it seems clear that, as with butterflies, leading lines are important, both for shelter from the wind and for compensating for wind drift. Thus *Musca domestica* and *Lucilia* spp. preferred to migrate E and W along a road rather than cross a nearby range of hills. *Calliphora vomitaria*, a woodland species, preferred to migrate through a narrow gorge rather than over the surrounding hills and both *Protophormia terraenovae* and *Phormia regina* migrated along valleys. Even so, perhaps when no leading line made a small enough angle to the individual-specific preferred compass direction, individuals of all these species would cross open hillsides. *Calliphora vicina* and *Protophormia terraenovae* crossed hills that involved a climb of 460 and 365 m respectively and wooded slopes up to 150 m did not prevent *Phormia regina* from flying beyond them.

There seems no doubt of the similarity between the migration pattern of temperate houseflies and blowflies and that of temperate butterflies. Johnson (1969), however, makes out a case for the possibility that some temperate houseflies and blowflies may be displaced long distances by wind at altitudes of 300–1500 m above ground level. Such a possibility cannot be denied as readily for these flies as it can for butterflies, largely because there are no observations of long-distance displacement of the larger temperate Diptera against which Johnson's suggestion can be examined. High-altitude wind displacement is certainly unlikely to have been involved in the experimental data so far accumulated in temperate regions. Even the longest-distance displacement, 290 km in 2 weeks for female screw-worm flies, involves an average rate of only 20·5 km/day, a distance that could have been achieved by migrating within 3 m of the ground for, at the most, 2 hours/day and more probably for only 30 min to 1 hour/day.

There are no unequivocal examples of return or re-migration in these flies. Williams, Lack, and others have reported large numbers of the blowflies, *Cryptolucilia caesarion*, *Calliphora vicina*, and *C. vomitaria*, and the hover fly, *Syrphus balteatus*, flying fast within a metre of the ground, sometimes with, sometimes against the wind, but always to the S. The observations were made in autumn in a Pyrenean pass and could indicate that these flies perform re-migrations comparable to those of butterflies. In the absence of estimates of migration distance, however, such observations are difficult to interpret.

The screw-worm fly, *Callitroga hominivorax*, is observed throughout the winter in Texas, more or less south of the polar front. In the spring, the species is observed gradually further and further to the N, the rate of spread of such reports being of the order of 56 km/week as measured over an 800 km range. Eventually, the species is observed in Oklahoma, 2400 km to the N. Without marking–release/recapture data for much longer distances than are at present available it is impossible to evaluate the extent to which these reports refer to migration or emergence from hibernation.

A much better case for long-distance, seasonal migration has been made out by Hughes and Nicholas (1974) for the Australian bushfly, *Musca vetustissima*. Each spring, bushflies re-appear in the Canberra area having been unobserved all winter. The flies are relatively old and are suggested by Hughes to be derived ultimately from populations that have bred through the winter in central and southern Queensland. It is suggested that in some years flies reach Canberra from Queensland within the life-time of individual adults. In other years, however, it is suggested that flies from Queensland reach only as far as New South Wales and then it is the adults of the next generation which arrive in Canberra.

If the screw-worm fly and bushfly do disappear from large parts of their breeding range each year, it is possible that there are two periods in the year when selection favours an obligatory migration distance in these two species. One is in the autumn when habitat suitability is high a few hundred or a thousand kilometres or so towards lower latitudes. The other is in spring when all habitats toward higher latitudes are empty, or nearly so, and selection should favour an appropriate frequency distribution of obligatory migration distances (p. 84). At both of these times selection should favour flies taking advantage of favourable winds to cover the individual-specific obligatory distance with the minimum migration cost. Selection may also favour a delay in the onset of reproduction until this distance has been travelled. Hughes and Nicholas (1974) provide some evidence that bushflies take advantage of favourable winds in spring.

19.6 Small and weak-flying diurnal insects

Aspects of the migration of small, weak-flying insects such as aphids have been discussed elsewhere (pp. 228 and 242). It was concluded that as with many other animals a hierarchy of migration thresholds should evolve, each one adaptive to a particular inter-habitat distance. In the facultative initiation of migration, the lower, short-distance migration thresholds should be exceeded first but as higher and higher migration thresholds are exceeded, the insect should adopt mechanisms appropriate to longer and longer migration distances.

Most small and weak-flying insects are extremely limited in their still-air range, partly by their flying ability and partly by the limited amount of fuel that they can carry. Furthermore, their powers of flight are such that they have little ability to prevent themselves from being carried to high altitudes by convective winds or to prevent themselves from being blown into potentially dangerous situations, such as spider's webs, when flying within the vegetation layer. It is presumably because of these selective pressures that the lowest (short-distance) migration thresholds in the hierarchy receive a major contribution from wind speed (p. 255), and that when higher (long-distance) migration thresholds are exceeded, the insects employ a wind-displacement mechanism above tree height. It is this latter mechanism and the way it may be involved in re-migration that is of major concern in this section, though for reasons of perspective it is necessary also to bear in mind the shorter-distance migrations.

There are considerable problems in the interpretation of data for the migration of small insects. Marking–release/recapture experiments must inevitably grossly underestimate the migration distance of a large proportion of the released deme.

On the other hand, the process of capture and marking and then the release of often huge numbers of individuals within a very limited area must inevitably exceed the migration threshold of far more individuals than would otherwise have been the case. No examples are given here of the innumerable calculations of migration distance that exist in the literature but that are based on the observation of individuals at a certain distance from the nearest 'known' source. Such estimates always tend to exaggerate migration distance, sometimes grossly. I also prefer not to base conclusions on observed migrations across large bodies of water. Such observations, even for land birds, exaggerate migration distance over land and also the importance of wind in determining migration direction. The following few examples indicate the range of other data that are available. These examples, but not their interpretation, have been taken from Johnson (1969) except where otherwise stated.

The fruit fly, *Dacus oleae*, migrates up to 2 km within olive groves, but individuals also move out from groves in semi-mountainous places to adjacent plains. Flies have been captured 2·5 km from their release point after 24 hours and 4·3 km after 10 days. Experiments with *Drosophila pseudoobscura* have suggested that within 10 months after release about 95 per cent of the progeny of released flies would be within a circle of radius of 1·76 km, centred on the release point.

The mosquito, *Aedes cataphylla*, after emerging from the pupa, flies upwards at a steep angle of 30°–60° into the wind. Take-off into the wind gives initial maximum lift and is common to all flying animals. Having climbed, by flight, above the vegetation layer, the mosquitoes turn and fly downwind. In calm air they orient to the lowest point on the horizon. Wind-borne mosquitoes, upon reaching a valley, enter it and thereafter fly in various directions within the valley, sometimes upwind and sometimes downwind. It seems, therefore, that post-teneral behaviour in this species is adapted to take the mosquito to a valley, using the mechanism that imparts least migration cost under the particular conditions. One wonders what behaviour is shown by mosquitoes that, having taken off in calm air and having oriented to the lowest point on the horizon, then encounter a cross-wind.

The mosquito, *Culex tarsalis*, at least on its first three days after emergence, spirals upwards in the early evening to a height of 4–5 m above ground level. When wind speed exceeds 5–6·5 km/h, the mosquitoes fly downwind and are carried over trees and other obstacles, but when the wind speed is less, the mosquitoes are said to fly upwind. Downwind migrants were recaptured at distances of up to 8 km on the night of recapture and up to 26 km two nights later. Some individuals, however, remained near the release site and other authors found no indication of migration further than 9 km even after 2–3 weeks. The species has, however, been captured at heights of up to 600 m above ground level, though most migration is considered to occur at between 1·5 and 15 m. This type of variation is totally consistent with a hierarchy of migration thresholds (all of which vary with age), each threshold associated with migration of a given distance by a given mechanism, and thresholds of different levels being exceeded in different experiments.

As with these other weak-flying insects, an aphid for which the threshold for longer-distance migration has been exceeded flies vertically upwards, at least until it is above the vegetation level. As this is probably judged by the visual horizon, the experiments of Kennedy and Booth (1963a,b) (p. 231) in which the visual horizon was permanently above the insect, probably greatly extended the period of climbing flight compared with that under normal conditions. Having climbed above vegetation level (e.g. above tree-height), longer-distance migrants probably attempt to maintain themselves at this height when migration distance is facultative and examine each potentially suitable habitat encountered. The longer the preceding flight, the higher the aphid's threshold for migration away from the next habitat encountered (pp. 82 and 231).

Under conditions of strong convection, aphids migrating at above tree height, as well as some shorter-distance migrants that inadvertently stray out of shelter, are liable to be carried to much greater heights, perhaps up to several kilometres. It seems reasonable to assume that when these aphids return to near the ground, the only strategy available to them is to continue their migration as before and examine and perhaps settle in the next suitable habitat encountered. There is no reason for the place at which the aphids return to near the ground from high altitudes to be any more or less suitable than the place being traversed when the aphid was first taken aloft by rising air. The critical factors, therefore, as far as migration distance is concerned, are the time interval between being taken aloft and returning to within alighting distance of the ground, the number of times that an aphid is taken aloft during a given migration sequence, and the horizontal component of wind speed.

Figure 19.23 shows curves of abundance of aphids at 3 m and 300 m above ground level with time of day. Johnson (1969) suggests that these curves

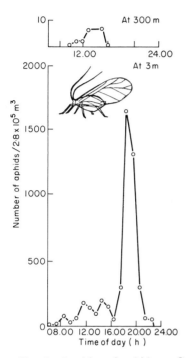

Fig. 19.23 Hourly densities of aphids at Cardington, England, on 8 August 1955

The diagram illustrates the height to which two waves of migrant aphids were carried during the course of a single day. Only aphid densities as measured by suction traps at heights of 3 m and 300 m above ground level are illustrated.

The first wave of aphids in the morning initiated migration into an atmosphere that was warm at the surface but unstable and buoyant. Many aphids were carried aloft to at least 300 m. At 15.00 hours the atmosphere cooled, became calmer and more stable, and cleared of aphids at least above 300 m.

The second wave of initiation of migration, involving more aphids than the first, began at about 16.00 hours. An inversion developed from the ground and the lower, cooler air became more stable. Few aphids were carried as high as 75 m and none were caught at 300 m.

Although such curves give an insight into the way that aphid migration is influenced by changes in atmosphere stability, they give no conclusive indication of the time spent by an individual aphid at the different heights above ground level.

[Simplified from Johnson (1969)]

indicate that migratory flights of aphids last 1–2 hours, the implication being that much of this time is spent at the higher altitudes. On this basis migration distance for these long-distance flights would vary from about 2 km on calm days to 30–100 km on windy days. However, I consider that these must be taken to be maximum estimates. The curves shown in Fig. 19.23 could equally well be generated by flights lasting any length of time less than 1–2 hours, even as short as a few minutes, but involving a different maximum height at different times of day. In this case, the contribution of high-altitude displacement to total migration distance is likely, for most individuals, to be less than a few to a few tens of kilometres. Certainly if aphids are aloft for an hour, they are likely to alight in the first suitable habitat encountered upon returning near to the ground. Experiments on individual black bean aphids, *Aphis fabae*, in tethered flight showed that after only 15 min flight, 93 per cent stayed for an hour on broad bean plants with which they were presented though only 47 per cent remained permanently. After an hour of flight, however, 100 per cent settled permanently and autolysed their flight muscles when presented with a broad bean plant. If presented with a very suitable broad bean plant (p. 231), the aphids settle after only 10 s flight.

Figure 19.24 illustrates the historical spread across North America and Europe of the Colorado beetle, *Leptinotarsa decemlineata*, from its sites of initial introduction. Despite presumably being wind displaced and despite on occasion being carried high up in the air, rate of spread was relatively slow, the maximum spread recorded being 350 km in a single year (i.e. probably less than 100 km/generation).

Figure 19.25 illustrates an example of a weak-flying insect in North America that is considered to migrate relatively long distances to the N each spring, and to re-migrate in the autumn. Johnson (1969) has pointed out that although there seems little doubt that migration occurs with the wind, the details of the migration mechanism are not well understood. In particular, the form of the lifetime track is unknown. However, as such insects settle out throughout the length of the migration track and as males spread along this track slower than females it seems likely that these insects have considerable control over the length of their lifetime track. In the case of the beet leafhopper in spring there is reasonably good evidence concerning the length of this track. In this species it seems that whereas some individuals probably migrate no further than a few

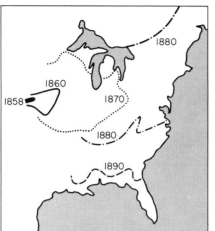
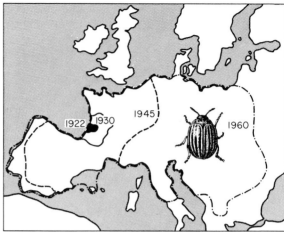

Fig. 19.24 The spread of the Colorado beetle, *Leptinotarsa decemlineata*, across North America and Europe

It is thought that the Colorado beetle originated in tropical Central America where it was associated with the tropical herb, *Solenum rostratum*. The seeds of this plant are thought to have been carried into Arizona, New Mexico, and Texas in the hair of the cattle and horses of Spanish caravans. Both the plant and the beetle become widespread on the eastern slopes of the Rocky Mountains. In 1845–50, settlers in the Mississippi Valley brought the potato, *Solanum tuberosum*, to the edge of the beetle's habitat, thus bridging the previously wide stretch of host-free country. Thereafter the beetle spread throughout North America. In 1922 it gained its first foothold in France, near Bordeaux.

Most spread across North America and Europe seems likely to have been by non-accidental, wind-aided migration, though presumably the trans-Atlantic migration was accidental. Across North America the maximum rate of spread was 2400 km, from Nebraska to Quebec, in the 15 years between 1859 and 1874. This gives an average rate of spread to the NE in the downwind direction of the prevailing winds of 160 km/year. The southward advance of the beetle down the Mississippi Valley system, against the prevailing wind, though possibly with temporary and/or local winds, proceeded at an average rate of only 25 km/year (800–960 km in 35 years). In Europe, the spread was slower, 35–40 years being required for the beetle to spread permanently over a range of about 2000 km (i.e. approximately 50 km/year). In some years, however, the advance seemed more rapid, perhaps up to 350 km. Such rates of spread do not, of course, necessarily indicate migration distance because longer-distance migrants may not achieve sufficient deme density for future generations to reproduce. Nevertheless, they do suggest that the major portion of each generation migrates relatively short distances, perhaps only tens of kilometres. Yet the beetles are often swept up to high altitudes, being found on mountains up to 2000 m, and other similar chrysomelid beetles have been caught at a height of 1500 m above ground level. It

must be concluded, therefore, either that even these individuals migrate only a few tens of kilometres or that such individuals represent a relatively minor part of the migrant population.

In the Northern Hemisphere, adult Colorado beetles of the late summer generation burrow into the soil and hibernate while sexually immature. They emerge in the following spring and migrate whenever conditions are such as to exceed the migration threshold. Larvae of other generations pupate in the soil and each generation migrates, primarily during the post-teneral period. In North America and central and southern Europe the longest flights seem to be made in the autumn by pre-hibernation immature adults. Although these adults also migrate the following spring, migration distance seems to be less. In northern Europe, the most obvious migrations are made in spring and early summer after emergence from hibernation. As long as feeding occurs, repeated migrations can be initiated for up to a month after hibernation. Summer generations do not migrate until after 10 days or so of feeding.

Colorado beetles are relatively weak fliers, achieving only 8 km/h in calm air, and there seems little doubt that for migrations greater than a few hundred metres they make use of downwind displacement at heights above the vegetation layer. Some individuals clearly have higher migration thresholds than others (Chapter 17), even during the post-teneral period, and fly briefly or not at all, producing a lifetime track that probably takes them no further than a few metres from their emergence site. Others initiate longer-distance migrations. Estimates of the maximum length of the lifetime track have varied from 150 to 300 km, but the implication of the observed rate of spread of the species is that for most individuals it is much less even than this. Where these longer lifetime tracks are produced it is likely that they are achieved by a succession of shorter-distance migrations, despite the beetles concerned being carried aloft to high altitudes.

[*Compiled from* Tower (*1906*) *and* Johnson (*1969*)]

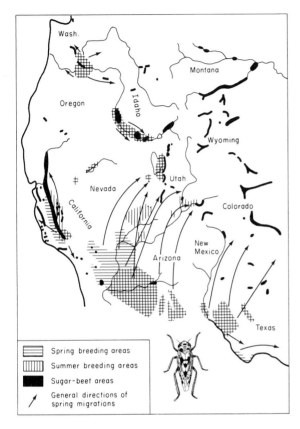

Fig. 19.25   The migrations of the beet leafhopper, *Circulifer tenellus*

In the United States, the beet leafhopper breeds on a variety of wild annual plants in arid and semi-arid areas where the rainfall is 25–30 cm annually. The map shows the major breeding areas to the west of the Continental Divide and along the Rio Grande in Texas and New Mexico. The solid black indicates areas of sugar beet cultivation that are affected by the leafhoppers. The arrows show the general directions of spring migrations but should not be considered to represent non-stop flights.

Leafhoppers settle out and examine plants at all points along their migration route. The migration distance may be obligatory during the pre-oviposition period of several days. It may also be obligatory for the immature adults produced in the autumn that do not become mature until the following spring after hibernation. If migration distance is obligatory in spring, however, there is a well-marked frequency distribution of migration distance. In males, in which migration distance in spring may be entirely facultative, there seems to be a frequency distribution of migration thresholds. In this case the densities of receptive females and competing males seem likely to be important habitat variables.

[*Modified from Johnson (1969)*]

kilometres and the majority of individuals die within 200 km of their birth-place, a small portion of the female population in spring may die in habitats as far as 700 km from their birth-place. As this spread in spring is taking the animals to previously unoccupied habitats it seems likely that such a frequency distribution of migration distances is an equilibrium situation, though whether it is achieved by facultative or obligatory (due, in this case, to selection for dispersal—p. 83) determination of migration distance, or a combination of both, is unknown.

In most cases of wind-displaced insects there are no data concerning the contribution of wind direction to the composite migration threshold. Whether the beet leafhopper (Fig. 19.25) only initiates migration when, in spring, the wind is from the south or whether it initiates migration independently of wind direction and is displaced to the north only when the wind happens to blow in this direction is unknown. Data are equivocal even in those areas in which a regular re-migration is known to occur.

The return migration and re-migration of the sunn pest, *Eurygaster integriceps*, from breeding sites on the plains to aestivation and hibernation sites at higher altitudes in Russia, the Middle East, and North Africa has already been described (p. 339). E.S. Brown (1965) has shown that this species, and another bug, *Aelia rostrata*, in Turkey have a peak flight direction in the study area that is to the SW when flying to the breeding sites and to the NE when flying to aestivation sites. Flight occurred most often when the downwind vector had a component in these directions. However, not all individuals flew in the peak direction, the track direction ratio approximating to 46:20:21:13 relative to the Sun and 60:14:14:22 relative to the wind (68:14:13:5 when wind speed was greater than 7 km/h) for *A. rostrata* leaving hibernation sites. The track direction ratio for *Eurygaster integriceps* leaving the breeding area approximated to 53:16:16:15 relative to the Sun and 50:20:19:11 relative to the wind. The prevailing winds at the different seasons are more or less opposite and most often are favourable for migration in the appropriate direction. It is not known whether the bugs only initiate migration when the wind is favourable. Nor is it known if the direction ratio reflects a frequency distribution of preferred compass directions within the deme or whether some individuals were not at the time initiating the seasonal migration. If the former, the implication is that, like butterflies and locusts, these bugs can migrate against the wind even at recorded wind speeds of 7 km/h. Migration distances between breeding and aestivo-hibernation sites are said to range from 20 to 200 km and marked bugs have been recaptured at distances of up to 20 km from the release point.

In late spring and early summer (May/June) the convergent ladybird, *Hippodamia convergens*, of California, soon after emerging from the pupa, migrates from the Californian plains to the aestivation and hibernation sites on the Sierra Nevada. It is assumed that the flight is with favourable winds though it seems likely that some visual orientation to mountains also occurs. This migration is convergent and large numbers aestivate and then hibernate *en masse*. The following spring (February/March) the ladybirds emerge from hibernation and, presumably with the favourable winds that now prevail from the opposite direction, migrate back to the plains. It is not known whether the beetles initiate migration only when the wind is favourable, though this seems likely. Having arrived on the plains, further but shorter-distance migrations may be performed if aphids, the main food, are scarce. The minimum distance of migration away from the mass hibernation site seems to be obligatorily determined, presumably as an adaptation to dispersal (p. 83). Ladybirds displaced from hibernation and released in their spring habitat-type nevertheless migrate long distances just as if they were still at their hibernation site (Williams 1958).

The optimum height for migration by weak-flying insects for which migration distance is facultative seems to be just above tallest vegetation height, thus giving the advantage of wind displacement without high predation risk yet maximising the opportunity to exploit any suitable habitat encountered. An insect using such a strategy, however, is inevitably often displaced to much greater heights, occasionally up to 4 km, by convective air currents. When such individuals return or are returned to ground level, they have to continue their downwind migration until they encounter a suitable habitat. These individuals will clearly travel longer distances than other individuals, the actual distance depending on how long the insect remains at high altitudes, wind speed, and inter-habitat distance. In an insect for which migration distance is facultative, optimum strategy seems to be to reduce this time at high altitudes to a minimum.

Melon flies, *Dacus cucurbitae*, released from an aircraft, drifted an average of 0·8 km for every 244 m

of altitude and for every 12·8 km/h of wind velocity. Thus it can be calculated that even flies carried to a height of 4 km should only migrate, in a wind of 10 km/h, 28 km further than flies that migrate downwind but just above the vegetation layer (assuming equal cross-country displacement before and after the moment of attainment of the height of 4 km, and assuming a constant wind speed at all heights). In a wind of 40 km/h the corresponding distance would be 31 km. These calculations should not, of course, be taken too seriously. It is instructive, however, that the released melon flies spent little time aloft, though their behaviour cannot necessarily be taken to represent that of free-living flies. Also, wind regimes in which the wind vector had a greater upward component would delay the flies' descent.

When selection has favoured an obligatory migration distance, however, there is less disadvantage (there may even be an advantage in the avoidance of predation by those birds, such as swallows and swifts, that specialise in feeding on the aerial plankton) in spending time at higher altitudes, at least until the optimum distance has been covered, whereupon the insect should return to just above the vegetation layer. Selection is most likely to favour an obligatory migration distance when the habitat suitability at the destination is a positive function of migration distance. This is most likely to apply to insects that are dispersing from a source but that are also entering an area where habitats are empty of other individuals. Examples are species such as the beet leafhopper (Fig. 19.25), the sunn pest, and the convergent ladybird. However, species such as the black bean aphid, *Aphis fabae*, for which deme density at the destination does not seem to be a function of migration distance would be expected to show a mechanism adaptive to a facultative determination of migration distance.

Even in weak-flying insects, therefore, the migration mechanism is likely to differ depending on whether migration distance is determined on a facultative or obligatory basis.

19.7 Night-flying insects

Figure 19.26 presents the results of a remarkable marking–release/recapture experiment carried out by Li *et al.* (1964) on the oriental armyworm moth, *Pseudaletia separata*, a noctuid, in China. This is another species that is thought not to overwinter in the northern part of its range. Once again, however, the possibility still exists that the entire range is occupied throughout the year but that seasonal patterns of aestivation and hibernation in as yet unidentified sites causes an apparent shift of the occupied range toward higher latitudes in spring and summer and toward lower latitudes in autumn and winter. There is however, no doubt from Fig. 19.26 that whether or not the entire range is occupied throughout the year, a proportion, and perhaps a large proportion, of the population also performs long-distance migrations of up to at least 1400 km. Furthermore, there is a clear indication that the peak migration direction is to the N throughout spring and summer. It is perhaps also significant that the only individual recaptured in autumn (September) had flown 800 km to the S.

The migration mechanism is unknown. The migration threshold seems to be lowest in moths 3–4 days old and in the laboratory moths can be made to

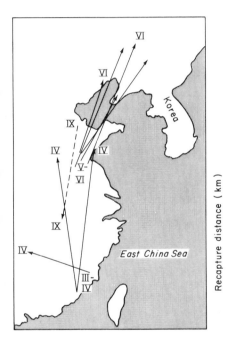

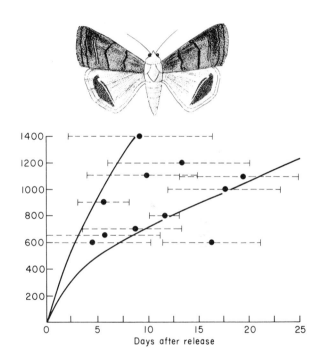

Fig. 19.26 The migration of the oriental armyworm moth, *Pseudaletia separata*, in China

The map shows the results of a marking–release/recapture experiment on the oriental armyworm moth. Arrows join the positions of release and recapture. Solid lines indicate spring and summer migrants; dashed line indicates an autumn migrant. Roman numerals indicate the months of release and recapture.

The graph shows variation in the distance of recapture with days after release. Because each batch of moths was released over a period of several days, the precise time interval between release and recapture cannot be determined. The solid dot refers to the time interval between the median release date of the appropriate batch of moths and recapture of the individual concerned. The dashed bar indicates the range of possible intervals between release and recapture for each individual. The two solid lines delimit most of the observed recapture rates.

There is a clear indication that in spring and summer successive generations of moths migrate to the N. Members of autumn generations may migrate predominantly to the S.

Migration rate seems to be greatest immediately after release and then to decrease. This may indicate an obligatory migration distance during the 3–4 days of the pre-reproductive period. The maximum migration rate that it is necessary to postulate to account for recapture distances seems to be about 200 km/night. Even allowing 1–2 hours for time investment in feeding and other activities, such a rate could be achieved by a moth flying only at night, even in the absence of a favourable wind. However, if migration distance is obligatory, it would be expected that the moths would fly at the height of the winds most favourable for migration in the required direction.

[*Raw data and map (simplified) from Li et al. (1964)*]

fly continuously for 36 hours at 17°C (Hwang and How 1966). The migration rates recorded in Fig. 19.26, however, are of insufficient magnitude for it to be necessary to postulate non-stop flight. Indeed, it is not even necessary to postulate that the moths concerned took advantage of favourable winds by flying at heights at which they could no longer monitor the area being traversed. Such a strategy seems likely, however, particularly during the pre-reproductive period. In this case the outstanding question seems to be whether, in spring and summer, moths preferentially migrate to the N and in autumn preferentially migrate to the S, taking advantage of favourable winds when they occur, as shown for some butterflies and locusts, or whether they migrate in any direction and delegate displacement solely to the prevailing wind, a strategy that has not yet been demonstrated for any animal. This question cannot be answered for the oriental armyworm moth from the data available. For at least two moths, however, it would seem that it is the former strategy that is adopted.

The bogong moth, *Agrotis infusa*, in southeastern Australia, is multivoltine and breeds in the lowland areas, particularly in New South Wales, where it is a serious pest of several crops. Some spring-emerging adults migrate to aestivate in huge numbers in caves and among rocks in the Australian Alps. In the autumn, these aestivated individuals return to the plains. Common (1954) observed unidirectional flights of bogong moths migrating to the aestivation sites in spring and just leaving aestivation sites in autumn. The moths showed a preferred compass direction to the S (i.e. towards the mountains) in spring and to the N (i.e. towards the plains) in autumn, independently of wind direction (Fig. 19.27).

An extremely instructive situation that has recently been subjected to analysis by Taylor *et al.* (1973) is provided by the silver-Y moth, *Plusia gamma*, which habitually migrates both during the day and night, at least for the first few days after emergence. In Britain, observations of diurnal migration over many years by a large number of observers have demonstrated convincingly that the silver-Y has a peak migration direction to the NNW in spring and summer and to the S in autumn (Williams 1958). In its diurnal migration, therefore, this moth behaves in a manner similar to many of the butterflies that occupy the same region. Like these butterflies (p. 438), the silver-Y shows a direction ratio that includes all possible flight directions, a point that has to be borne in mind in any analysis.

Taylor *et al.* (1973) have presented an analysis of 30 years of observation of flight direction that has been argued to demonstrate that although the silver-Y migrates like a butterfly during the day, track direction being independent of wind direction, during the night it migrates downwind, independently of compass direction. This is an extremely critical conclusion for, as rightly pointed out by the authors, if it is valid it casts doubts on the thesis of butterfly migration that I have presented earlier (Baker 1968a,b,d, 1969a) and that has been presented earlier in this chapter. Clearly if these moths migrate with an individual-specific preferred compass direction during the day, but at night fly, probably for longer distances, in some other direc-

Spring		Autumn	
Wind	Moths	Wind	Moths

Fig. 19.27 The migration mechanism of the bogong moth, *Agrotis infusa*, in the Australian Alps

Each spring, bogong moths migrate from breeding habitats on the plains of southeastern Australia to aestivation sites in the Australian Alps. The diagram illustrates the migration directions of the moths as recorded at dusk in the Australian Alps in spring and again in autumn when the moths leave their aestivation sites to migrate back to the plains to breed. Wind direction at the time of observation is also recorded. o indicates calm.

The data support the thesis that the moths have a preferred compass direction that in the study area is to the S in spring and to the N in autumn. There is no support for any form of downwind thesis.

[*Modified from Common (1954)*]

tion that varies according to the direction of the wind, then there is a strong implication that the maintenance of a given compass direction during the day is of little ecological consequence.

The arguments presented by Taylor *et al.* along these lines would be quite valid if it was found that by day the silver-Y migrates in a specific compass direction but by night migrates in any direction as long as it is downwind. It was something of an oversight, therefore, that these authors failed to test whether this was the case. As it was, their analysis totally ignored compass direction, apparently deliberately (their p. 753), and considered only the relationship between wind and moth-track direction, a methodological error that has also been made by other authors in relation to butterflies (p. 436). Taylor *et al.* also failed to distinguish between moths migrating over land and over water. Although the authors acknowledge that 'there is some slight correlation between downwind night-time flight and over-sea flight' in the records used in their analysis, the assertion that if this is so it implies visual orientation by ground, not celestial, cues is either unacceptable or irrelevant, there being no way that a silver-Y could migrate in a given compass direction solely by reference to the ground. In fact, the suggestion of a slight correlation between night-time flight and over-sea flight is something of an understatement. Using the same weighting adopted by the authors, 80 per cent of the night-time flights were over water but only 21 per cent of the day-time flights, and as all terrestrial flying animals, including landbirds, either allow themselves to drift, or are drifted, when crossing water, especially at night, or alternatively only start to migrate across water when the downwind direction more or less coincides with the preferred compass direction when one exists, then this 'slight correlation' cannot lightly be disregarded. It is clear, therefore, that some re-analysis of the data presented by Taylor *et al.* is necessary before the conclusions reached by these authors can be accepted as evidence against the thesis presented in this chapter. Critical re-analysis should examine the relationship between compass direction, moth-track direction, and wind direction, and should also separate data for flights over land and water by day and night. Such an analysis is presented in Fig. 19.28 using precisely the same data as analysed by Taylor *et al.*, including using the same weighting coefficients. It can be seen clearly that when the analytical precautions outlined above are observed there is every reason to accept that the silver-Y has

the same preferred compass direction during the night as during the day and that overland, in following winds and cross-winds, the moths compensate for the displacing effect of the wind, both by day and by night. Only when crossing water at night do the moths not compensate for wind-drift. The critical difference, therefore, is not between diurnal and nocturnal migration as suggested by Taylor *et al.*, but between migration over land and migration over water. To this extent, therefore, the migration strategy adopted by moths is the same as that of all other flying, terrestrial animals, including landbirds.

Interestingly, silver-Y's flying by day over land also appear to cease to compensate for wind-drift when wind direction deviates by more than 150° from the preferred compass direction, even though the moths still deviate significantly ($P < 0.05$) from flight in the downwind direction (Fig. 19.28). Similar behaviour is show by birds (p. 632) and it seems reasonable to assume that the same explanation is applicable to both. This is that at any one time, different individuals have different preferred compass directions and that wind speed *and direction* contribute to the individual's composite migration threshold. Thus, beyond a certain wind speed only those individuals the preferred compass direction of which coincides $\pm x°$ (where x probably approaches 90°) with the downwind direction initiate migration. In the case of the silver-Y, which would appear to have a wide scatter of individual-preferred compass directions around 360°, it might be expected on this basis that given a wind that deviates by 150°–180° from the peak preferred direction, the majority of individuals will settle or remain settled, leaving in the air only those individuals for which these winds are tail-winds. However, as individuals with these preferred compass directions are fewer in number than individuals with preferred directions up to, say, 90° different, but that may still find the wind direction sufficiently favourable for migration, the peak flight direction for all airborne individuals is likely to be significantly different from the downwind direction by about the amount illustrated in Fig. 19.28. Moths flying over water, however, cannot settle and optimum strategy for any individual that finds itself unable to compensate for wind drift from the preferred direction should be, as with birds, to drift with the wind and to expend only just enough energy to remain airborne at an appropriate height. Over water, therefore, particularly at night but also during the day, a track direction not

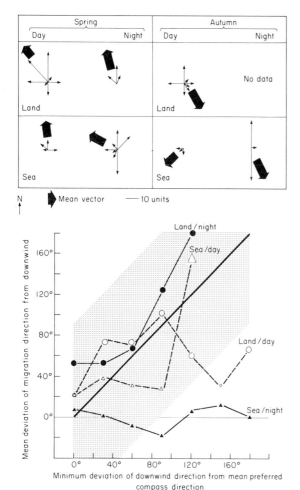

Fig. 19.28 The migration mechanism of the silver-Y moth, *Plusia gamma*, in Britain

The silver-Y is a plusiid moth that as a larva feeds on a wide variety of low-growing plants. It almost certainly occurs throughout its maximum breeding range at all times and in Britain at least hibernates as a young (first or second instar) larva. Nevertheless, circumstantial evidence suggests that the modal population density moves to the north in spring and summer and to the south in autumn and winter. The situation seems to be similar to that described for the monarch butterfly, *Danaus plexippus* (Fig. 19.5), red admiral, *Vanessa atalanta* (Fig. 19.6), and painted lady, *Cynthia cardui* (Fig. 19.7), and probably also the oriental armyworm moth, *Pseudaletia separata* (Fig. 19.26). For a few days after adult emergence in spring and summer and perhaps for longer in autumn, silver-Y moths are immature and make no time investment in reproductive activities. These periods of immaturity are probably associated, for a large proportion of the population, with an obligatory migration distance that is relatively

long. It is also associated with a period during which investment in roosting is much reduced and both migration and feeding occur by day as well as by night. Observations of diurnal migration directions over many years have clearly shown that in Britain the mean migration direction is to the NNW in spring and summer (March to August) and is to the S or SSW in autumn (September to November) (Williams 1958).

The diagram examines the hypothesis that these peak directions result from a direction ratio of individual-specific preferred compass directions that are the same both by day and by night. Any similarity between migration direction and downwind direction is due to a coincidence between the latter and the preferred compass direction. The data and units used are those presented by Taylor *et al.* (1973).

The arrows in the top part of the diagram break down these data into observed migration directions over land and sea by day and night in spring and summer (March to August) and autumn (September to November). Although data for each category are inevitably few, the indication is still that in all situations the mean migration direction is to the NNW in spring and summer and to the S or SSW in autumn.

The graph analyses the relationship between observed migration direction, wind direction, and mean preferred compass direction, for moths flying by day and night over land and sea. The x-axis indicates the minimum deviation of the downwind direction (for a given series of observations) from the mean preferred compass direction (NNW for spring and summer; S for autumn from 1 September until December). Thus 0° on the x-axis indicates that all observations are used in the calculation; 90° on the x-axis indicates that only those observations are used for which downwind direction deviates by 90° or more from the mean preferred compass direction. The y-axis is the mean deviation of observed migration directions from the downwind direction where a deviation toward the preferred compass direction is considered to be positive and a deviation away from the preferred compass direction is considered to be negative. Open symbols and dashed lines indicate diurnal observations. Solid symbols and continuous lines indicate nocturnal observations. Circles indicate migration over land. Triangles indicate migration over sea. Large symbols indicate that the mean deviation is significantly different from the downwind direction at the 5 per cent level; small symbols indicate that the mean deviation is not significantly different.

If moths are displaced downwind irrespective of compass direction, observed relationships should fall on the horizontal line. If moths have a preferred compass direction, the relationship should fall on the diagonal line (except perhaps at 0°) if compensation for wind drift from this compass direction is complete, or at least within the stippled area if compensation is not complete (p. 462). As the deviation of downwind direction from the mean preferred compass direction increases, the probability should increase that the mean migration direction will differ from the downwind direction by a significant amount.

The evidence supports the suggestion that silver-Y moths migrating over sea at night are displaced downwind. Moths migrating over land, whether by day or by night, and to some extent over sea by day, show more or less complete compensation for any tendency to be wind-drifted from their mean preferred compass direction. When the downwind direction deviates from the mean preferred compass direction by more than about 140°, although mean migration direction may continue to be significantly different from downwind, it is no longer in the mean preferred compass direction for the deme, nor even in this direction ±90°. This effect can be attributed to the alighting of those individuals the individual-specific preferred compass direction of which is such that the wind direction is unfavourable. In consequence, the only moths left airborne are those for which the wind direction is favourable. Whether or not the moths that remain airborne are those for which migration distance is facultative (see p. 433 for butterflies) cannot be determined at present.

[Photo by R.R. Baker]

significantly different from downwind is to be expected as shown in Fig. 19.28. Differential settling and take-off by different cohorts of moths, each with a different preferred compass direction, could also account for the observations made by Larsen (1949), again on the silver-Y, in Denmark.

There is now an impressive literature that demonstrates that moths and solitary locusts, as well as the vast majority of nocturnal weak-flying insects, migrate downwind. In this respect, nocturnal insects are identical in their migration strategy to many nocturnally migrant birds (p. 632). Only in birds, however, has the possibility been explored, and convincingly demonstrated, that this correlation results from the contribution of wind direction to the individual migration threshold and the presence in the population at any one time of individuals with different preferred compass directions. The example of birds, therefore, shows categorically that a demonstrated coincidence between track direction and downwind direction is not enough to prove that compass direction is an unimportant part of the migration strategy of individual insects. To do this, it is necessary to demonstrate that the same individual flies in different directions on different nights, but always downwind.

Brown (1970) studied the flight direction of night-flying insects, mainly moths and beetles, near Nairobi, Kenya, in April/June by observing insects passing through the beam of a vertically directed lamp. It was found that on calm nights, all flight directions were represented, but that as wind speed increased so too did the frequency of downwind flight. The published figures show 80 per cent of all insects flying downwind ±22.5° with only about 2 per cent of observed tracks deviating by 90° or more from the downwind direction. All but one of the observations were made with wind from the eastern quarter, so the peak direction was to the W, though there was an indication of a slight bias to the north side of west. One observation, however, lasting 15 min, was made during a westerly wind. Although the peak track direction reversed, being still downwind, there was no way of knowing whether it was the same individuals on the wing as on the nights with an easterly wind. Only 46 per cent of observed insects showed a downward track ±22.5° on this occasion with 13 per cent of observed tracks deviating by 90° or more from the downwind direction. This lowered bias toward downwind migration could have been a function of wind speed, though no evidence was presented that this was so, or it could have been a preferred compass direction effect within insect populations the majority of individuals of which have a preferred compass direction to the W, or slightly N of W (see Fig. 19.18 for locusts in the same region). In this region at this time and during the next few months the areas of higher rainfall lie in this direction, though perhaps significantly rainfall also increases to the E toward the coast (Fullard 1959). A dichotomy between individuals with preferred compass directions to the E and W might therefore be expected.

The only hope of settling the question of whether individual insects of non-swarming species such as moths and solitary locusts have a preferred compass direction or whether they simply orient downwind irrespective of compass direction seems to derive, as it did for birds, from the possibility of radar observation of individuals. Only by observing the migration track of a single individual during periods of shift in wind direction can the migration mechanism be unequivocally identified. Such evidence is not yet available. Schaefer (1972) describes the use of 3 cm radars in the Sahara, New South Wales, and the Sudan and claims that individual moths and grasshoppers can be detected at distances of 1·7 to 2·6 km. Insect concentrations can be detected at 70 km. At such a distance, however, there is no way of determining the extent to which individuals leave and join such concentrations. Detection at a distance of 2·6 km is not, of course, the same as being able to track an individual for 5 km. Nevertheless, the use of radar and telemetry in biological research is increasing in sophistication yearly and it is to be hoped that in the near future entomologists will have a tool, already available to ornithologists, that will permit resolution of the above questions.

Another question that could be resolved by radar some time in the future is the nature of downwind migration with respect to the frequency of examination of habitats encountered *en route* and the total duration of the migratory flight. As a result of the above study, Schaefer describes the characteristics of nocturnal migration of insects of body length exceeding 8 mm as follows: peak take-off occurs 55 min (40–70) after sunset; the insects climb at 0·3 m/s (0·6 maximum) to a median height of 400 m and a temperature-limited ceiling of 1·5 km (1·0–1·8); fly for 3 hours (1–12) at a ground speed of 30 km/h (10–80) travelling 90 km (30–500) per night, generally in a low level jet stream. The insects were frequently oriented, usually but not always downwind. It was found possible to identify species and sex of Acrididae and Noctuidae by wing-beat frequency. Roffey (1972b) obtained essentially similar results in Australia. The rapid increase in height soon after take-off was followed by 1–4 hours of gradual decline in the height of the highest insects. If nocturnal insects do have a preferred compass direction as suggested here, such a 'dusk ascent' could serve the function of vertical sampling by the insect to determine the height of wind with the most favourable characteristics of temperature, speed, and direction.

The estimates of migration distance per night should perhaps not be taken too literally at present and should be taken as estimates of the maximum possible on any given night. If individuals can be detected only at 2·6 km and if, as at present, *individuals* can be followed for distances much less than the 5 km that is theoretically possible from such a range, it is not yet possible to reach acceptable conclusions from radar concerning migration distance. For the moment, therefore, estimates of migration distance per night must rely on the indirect methods of back-tracking on the basis of synoptic weather situations and of marking–release/recapture.

Back-tracking, assuming continuous downwind displacement within a certain height range, up to the assumed time of arrival of a given cohort of insects in a given mass of air is extremely susceptible to the confounding effects of false correlations. In temperate areas, for example, night-flying insects are caught in numbers only on warm nights (p. 253). Such nights in the northern hemisphere are invariably characterised by winds from the south. Whether or not the insects have just arrived from the south, therefore, their date of first appearance is much more likely to coincide with winds from the south than from any other direction. Such an analysis has not, to my knowledge, been published, but it seems likely that back-tracking for the date of first appearance of moths that are known to have emerged locally would also, in the majority of cases, suggest an origin far to the south.

Large-scale back-tracking has therefore to be viewed with some reservation for it can rarely distinguish between conditions in the study area suitable for flight and conditions that would lead to immigration from 'reasonable' sources. Mikkola (1967), for example, analysed 100 observations of appearance of 52 different species of moths and butterflies in Finland between 1946 and 1966. It was found 'a bit difficult' to analyse the effects of warm days and airflows separately and it was also concluded that on some occasions a 'highway from more southern areas to Finland, formed by the warm air mass, is more important than aircurrent direction'. Thus, although there was a general correlation with winds from the S or SE and although most back-tracked trajectories indicated a source in the steppe area of south Russia, the relevance of this analysis to the real reason for the date of appearance of the various species in Finland cannot be considered to have been demonstrated. Mikkola acknowledges (his p. 96) the problems associated with an interpretation of such an analysis.

Interestingly, Mikkola found that two sphingids, the death's-head hawk moth, *Acherontia atropos*, and the convolvulus hawk moth, *Herse convolvuli*, both of which are large and extremely strong-flying, showed no correlation either with temperature or wind direction. The apparent lack of contribution from temperature to the migration threshold of hawk moths has already been noted (p. 253). Yet a preliminary analysis carried out by a student of mine, A. Brennan, demonstrated that the time interval in days between 1 May and the date of the first report of these species in Britain in any one year from 1950 to 1967 showed a significant positive correlation ($r = +0.795$; $n = 14$; $P < 0.05$) with the percentage of those days characterised by winds from the S or SE but no significant correlation ($r = 0.296$; $n = 14$; $P > 0.05$) with the percentage of those days characterised by winds from the S or SW. The effect of temperature can therefore be ignored and the indication is that these species only arrive in Britain with favourable winds (i.e. favourable in terms of direction). The combination of the results obtained by Mikkola and Brennan suggest that these species normally migrate with a preferred compass direction and at a speed that is relatively uninfluenced by wind direction and temperature but that either the moths attempt to cross large water bodies such as the English Channel only with a favourable wind or that they do not survive sea crossings unless the wind is favourable. Of these alternatives, the former seems the most likely for such powerful moths.

If moths and other night-flying insects do have a preferred compass direction, as seems to be the case for at least the silver-Y moth, *Plusia gamma* (Fig. 19.28), we had until very recently no idea of the cues by which it was determined. A field research programme that began in the summer of 1976, however, has started to provide answers to this problem (Sotthibandhu 1977 and *personal communication*).

Celestial cues have always seemed the most likely, especially as the response of most nocturnal insects to light traps suggests that light sources have some biological meaning to these animals. Sotthibandhu's research has the twin aims of identifying the nature of the light-trap response in moths and the cues used in their compass orientation. His apparatus consists of a semi-automatic portable flight-mill designed to make continuous records in the field of the orientation of tethered moths. So far, data are only available concerning the large yellow underwing moth, *Noctua pronuba*.

As far as the light-trap response is concerned, the surprising result has been obtained that large yellow underwings fail to orientate to a 125 W MV light source beyond a distance, which is a function of climate, of between 1.5 and 3.0 m. Using multiple regression analysis, Sotthibandhu has shown that the critical distance is a positive function of temperature and cloud cover.

In the absence of a light, moths nevertheless continue to show a significant mean vector (Rayleigh's test, p. 30). Lateral displacement has demonstrated that the moths are showing compass orientation and are not orientating to near- or middle-distance visual or olfactory landscape features. When the Moon is visible, the moths orient by it, though not necessarily at 0°. They do not, however, compensate for the shift in the Moon's azimuth, the mechanism being similar to that of some butterflies with respect to the Sun's azimuth (p. 416). The individual-specific angle to the Moon's azimuth is the same as the individual-specific angle to an artificial light source. In the absence of the Moon, the moths still show compass orientation as long as the stars are visible. Even so, preferred compass direction shifts clockwise at a rate of 16°/h, suggesting orientation to stars or star-patterns. In a dark room or with painted eyes in the field the moths are disoriented. In a planetarium moths captured from the field are similarly disoriented.

Fig. 19.29   Large yellow underwing moth, *Noctua pronuba*

19.8 Summary

The facultative and obligatory determination of linear range distance

1. When habitat suitability is not a function of migration distance, selection favours continuous and efficient monitoring of the environment and examination of each potentially suitable habitat encountered. The length of the linear range in such a situation is a function of the suitability of the succession of habitats encountered by the individual and is said to be facultative (p. 83).

2. When habitat suitability is a positive function of migration distance, selection favours individuals that do not invest time and energy monitoring the suitability of habitats encountered until the appropriate distance has been traversed. The length of the linear range in such a situation is said to be obligatory, though it invariably gives way eventually to a section of linear range the length of which is facultative (p. 83).

Some examples of facultative and obligatory migration distance

3. A facultative migration distance seems to characterise the lifetime track of lemmings that migrate only within their natal mountain slope (p. 413) and of most temperate butterflies at most times of year (p. 422), most temperate houseflies, blowflies, and hoverflies (p. 463), and probably most weak-flying temperate insects (p. 465).

4. An obligatory migration distance seems to characterise the lifetime track of lemmings that initiate migration away from their natal mountain slope (p. 413). It also seems to be found in the autumn generation of those temperate butterflies (p. 426), moths (p. 471), and possibly also some houseflies and blowflies (p. 465) as well as weaker-flying insects (p. 470) that migrate long distances toward lower latitudes for the winter. It may also characterise the early stages of the re-migration of these insects back to higher latitudes the following spring, especially if the habitats at these latitudes are unoccupied due to a failure of the species to hibernate in these regions. The inhabitants of seasonally arid areas, especially locusts (p. 451), and some butterflies (p. 425) and moths (p. 475) also seem likely to have an obligatory migration distance, at least during the pre-reproduction stage of ontogeny.

The migration mechanism

5. Lemmings for which migration distance is facultative produce a tortuous exploratory migration track that places the emphasis on maximum shelter from predators and strictly nocturnal activity. When migration distance is obligatory, time and energy expenditure seem to make a larger contribution to migration cost than does predation risk. The track becomes straight, even across open ground (though predators are avoided when encountered), and may continue during the day. Behavioural strategies have evolved that permit bodies of water and ice to be crossed with the minimum migration cost. Orientation seems to be up or down the mountain slopes as appropriate and then toward distant mountain slopes when crossing the valley floor (p. 413).

6. Butterflies for which migration distance is facultative take their primary orientation from the Sun's azimuth but also orient secondarily to leading lines. There is no compensation for the diel shift of the Sun. The orientation angle to the Sun is individual-specific (p. 416). Correction for wind-drift occurs by the use of leading lines for shelter and orientation and by the use of distant landmarks. Wind speed contributes to the composite migration threshold but wind direction does not (p. 419). This reaction to wind may well result from selection for mechanisms to maintain a constant ground speed. It is suggested that there exists a disadvantage, that is perhaps an inverse function of the air speed of the insect, that just balances the disadvantage that results from energy expenditure, which is a positive function of air speed (p. 421). The optimum height above ground level during migration is that height that maximises the area searched with maximum efficiency for a given resource (p. 415). This height is perhaps never very great (e.g. less than 10 m) and is an inverse function of wind speed at a given height (p. 421).

7. Butterflies for which migration distance is obligatory probably orient by the same mechanism except that they compensate for the diel shift of the Sun's azimuth (p. 425). When not searching for feeding or roosting habitats, when they behave as in item 6, the optimum height for migration is high above ground level if wind speed and direction is favourable for travel in the required direction but near the ground if wind speed and direction is unfavourable (p. 425).

8. Butterflies for which selection on migration distance has produced lifetime tracks longer than about 200 km seem to have been subjected to environmental gradients of habitat suitability (p. 435). When these gradients are in one direction at one time of year and in a more or less opposite direction at another time of year, selection favours a direction ratio with a peak migration direction that is also different at different times of year in the same sense (p. 448). Which direction ratio is shown by any given generation of a species is a proximate function of the photoperiod/temperature regime experienced by that generation, either as a larva or as an adult, depending on the species (p. 439). The individual gains no advantage from its own direction relative to other individuals, the advantage deriving from the pro-

duction of offspring with the optimum direction ratio (p. 451).

9. A swarm of the desert locust, *Schistocerca gregaria*, upon attaining flight maturity, has a preferred track direction that is characteristic for the place and season and that is part of a migration circuit for the deme concerned (p. 454). This migration circuit direction is achieved by flying in a dense cohesive swarm when the wind is favourable and in a low-flying diffuse swarm when the wind is unfavourable (p. 460). In dense cohesive swarms only the medium- and high-flying locusts at heights at which the wind is favourable adopt an orientation appropriate to the migration-circuit direction. Low-flying locusts and those medium- and high-flying locusts that pass beyond the leading edge of the swarm show orientations adaptive to swarm cohesion and feeding rather than to cross-country migration (p. 455). Dense cohesive swarms migrate across country in a downwind direction. However, if the downwind direction, while still retaining a component in the migration circuit direction, deviates by more than about 60°–70° from that direction, the locusts compensate to some extent for wind drift in a way that reduces the deviation of their track direction from the migration-circuit direction and that probably maximises the migration component in the required direction per unit energy (p. 462). In diffuse swarms, locusts at all points in the swarm orient independently primarily in the migration-circuit direction and thus move upwind (p. 459). Because of the relationship between wind fields and major rainfall zones, immature locusts that have not yet encountered rain most often migrate in dense cohesive swarms in a direction that varies only slightly from the downwind direction. Once rain is encountered, however, and maturation begins or in regions and seasons in which the locusts prefer high ground, the relationship between wind direction and either the migration-circuit direction, or the direction of rain-in-sight/rain-clouds, or the direction of preferred topography, breaks down (p. 458). Although dense cohesive swarms may still be adopted when the required direction coincides with a downwind component, on frequent occasions the required direction is upwind and the locusts adopt a diffuse swarm formation. The situation becomes complex when winds are favourable at one height but unfavourable at another.

10. Most temperate houseflies, blowflies, and hoverflies seem to possess the same form of lifetime track as butterflies (p. 463). At least two species, the screw-worm fly of North America and the Australian bushfly, may have been exposed to selection for an obligatory migration distance, at least during the pre-reproductive period in spring for individuals migrating to higher latitudes (p. 465). Insofar as data are available, the migration mechanism appears to be the same for stronger-flying flies as for butterflies (p. 464).

11. Weak-flying insects avoid wind displacement whenever possible during short-distance migration (up to a few tens of metres) (p. 465) but when the migration threshold for a longer-distance migration (perhaps more than about 100 m) is exceeded, these insects migrate downwind unless the wind is very light, when they may attempt to orient to a suitable distant destination (p. 466). In probably most temperate species, migration distance is facultative (p. 466). In this case the optimum mechanism is wind-assisted migration just above the height of the tallest vegetation (p. 466). Such insects may often be carried to great heights by convective air currents but optimum strategy is to reduce the time spent aloft to a minimum (p. 466). Such insects in areas where the inter-habitat distance is short may migrate no further than a very few tens of kilometres, even if carried to a height of 4 km in a moderate wind (10–40 km/h) (p. 470). Some weak-flying insects have been exposed to selection for an obligatory migration distance (p. 470). Such species may make less attempt to reduce the time spent at high altitudes (p. 470) until a certain time interval has passed and migration distance becomes facultative. When direction is important, as in seasonal re-migration, selection may favour weak-flying insects that initiate migration when the wind direction is favourable (p. 469) but few relevant data are available (p. 469).

12. The nocturnal strong-flying insects for which data are so far available seem to show a period of obligatory migration distance, at least during the pre-reproductive phase of ontogeny. In the only case for which critical data are available, the silver-Y moth in Britain, an individual-specific preferred compass direction is evident throughout the day/night cycle (p. 474). The moths compensate for wind displacement but, at least when wind is unfavourable for the peak direction, individuals flying in this direction seem to drop out of the air leaving airborne only those individuals for which the wind direction is favourable (p. 473). This could also explain visual and radar observations of other night-flying insects in tropical seasonally arid regions (p. 475). Whether distance is obligatory for all individuals or only for those migrating in the peak direction, as seems to be the case in butterflies (p. 434), is unknown. The large yellow underwing moth has been shown (p. 477) to use the Moon's azimuth for compass orientation, but not to compensate for its shift across the sky during the night. In the absence of the Moon, compass orientation is still shown, at least as long as the stars are visible. Moths in a dark room or in a planetarium and moths in the field but with their eyes painted over are all disoriented.

20

Aspects of the linear range: the return migrations of zooplankton and the animals of soil, sand, and leaf litter

20.1 Introduction

In Chapter 9 it was argued that adaptation to spatially separate habitat-types that varied out of phase with one another in their suitability to a particular animal was a necessary pre-adaptation for return migration to evolve in that animal. Such adaptation in itself, however, was considered insufficient to lead directly to the evolution of return migration rather than removal migration. Except, that is, for those species adapted to spatially separate habitats with a particular pattern of spatial distribution that led automatically to return migration, albeit with a low degree of return.

The particular spatial pattern concerned is that of separate but stratified habitat-types. Many examples of such stratification could be given, but those most relevant to this section are those produced by the vertical gradients of light, temperature, and food abundance in large water bodies, such as lakes and oceans, and the vertical gradients of temperature, humidity and food availability in soil and leaf litter. Some other situations are considered in Chapter 31. It will be noted that these stratifications are over a relatively short distance. Similar situations that differ only with respect to the distances involved are presented by the altitudinal zonation of temperature and vegetation on large-scale topographical features such as mountains and the similar but 'horizontal' and longer-distance latitudinal zonations of temperature, vegetation and day-length etc. The animals that have adapted to these latter stratified environments, however, tend to be larger and more mobile and on the whole to be animals for which

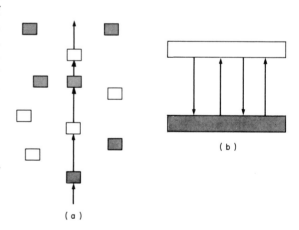

Fig. 20.1 The relationship between the linear range form of the lifetime track and return migration in a stratified environment

▢   type-A habitat
▢   type-B habitat

When type-A and type-B habitats are spatially scattered as in (a) in such a way that return migration is not favoured (p. 61), the linear range is the optimum form of the lifetime track. When type-A and type-B habitats are stratified as in (b), however, adaptation to these two habitat-types inevitably involves a return migration, though unless the requirements for the evolution of return migration discussed in Chapter 9 are also met, degree of return is unlikely to be high.

long-distance return migration is most likely to have crystallised out from the familiar area phenomenon. Alternatively, they tend to be animals such as butterflies and locusts, etc. (Chapter 19) for which degree of return is very low.

We are left, therefore, in this chapter, with those animals that have been exposed to selective pressures resulting from environmental gradients and stratification in water bodies and soil. These animals may conveniently be divided into the zooplankton and the animals of soil, sand, and leaf litter. Figure 20.1 indicates the relationship of the return migrations of these animals to the linear range.

20.2 Zooplankton

The members of the zooplankton of larger water bodies such as lakes and oceans live in an environ-

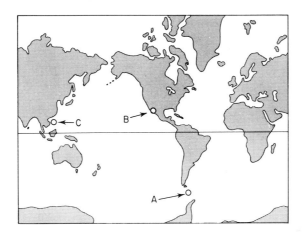

(a)

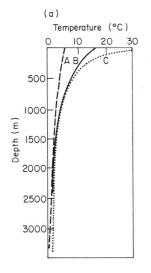

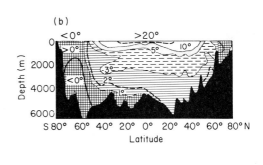

(b)

[*Compiled from Raymont (1963)*]

Fig. 20.2 Variation in sea temperature with depth and latitude

(a) Depth profile of temperature in three geographical situations

A, *B*, and *C* are respectively the sub-polar, sub-tropical, and tropical sites indicated on the map.

(b) Variation in temperature with depth and latitude in the Atlantic Ocean

The steepness of the temperature gradient is a function of both depth and latitude. The gradient is steeper in surface waters than in deeper waters and is steeper in the tropics than in polar regions

ment that is strongly stratified with respect to light intensity and also, often, temperature. In addition, there is a marked stratification of the abundance and species composition of the phytoplankton upon which the zooplankton is dependent (Figs. 20.2 and 20.3). The stratification changes in a quantitative and often qualitative way with time of day and also season (Fig. 20.4). Superimposed upon this variation there are the changes in stature and resource requirements of an animal as it ages. In view of these sources of spatial and temporal variation it is not surprising that many animals have become adapted to exploit different habitat-types at different stages of

ontogeny, time of day, and season. In a stratified environment, alternation between habitat-types involves alternation of direction of the rectilinear track which results in a to-and-fro vertical return migration, albeit with a low degree of return.

Vertical migration by zooplankton may have evolved in response to selection for non-calculated removal migration (Hardy 1958). Where this is so, the model developed in Chapters 8 and 16 can be applied directly. Alternatively, the vertical migration may be a return migration between a 'feeding habitat' and a 'non-feeding habitat' (or, more correctly, 'a vertical succession of non-feeding hab-

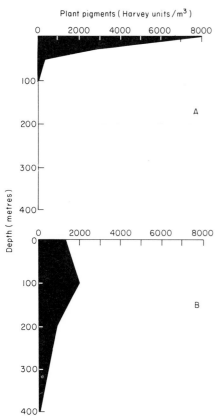

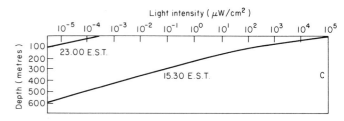

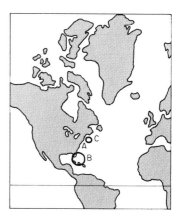

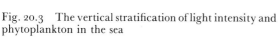

Fig. 20.3   The vertical stratification of light intensity and phytoplankton in the sea

Vertical stratification of light intensity

The right-hand graph shows decrease in light intensity with depth. Two series of readings are shown. Both were taken on 19 July, one series during the day at 15.30 hours EST, and the other series at night at 23.00 hours EST.

Vertical stratification of phytoplankton

The data on which the two left-hand graphs are based

were collected in May/June. In northern waters the major concentration of phytoplankton, measured as plant pigment units, is more or less at the surface and little, if any, healthy phytoplankton is found at depths greater than 100 m. In tropical waters the major concentration of phytoplankton is found at a depth of 100 m and some is still occurring and apparently healthy at depths exceeding 300 m.

[*Modified from Clarke and Wertheim (1956) and Riley (1939)*]

itats': where the former term is used in this discussion of vertical return migration by zooplankton it should be interpreted as shorthand for the latter term). In this case, the downward and upward migrations may be considered to be the unit removal migrations of which the return migration is composed as discussed in Chapter 7. The distinction between vertical migration as removal and return migration is illustrated in Fig. 20.5.

An example of the importance of distinguishing

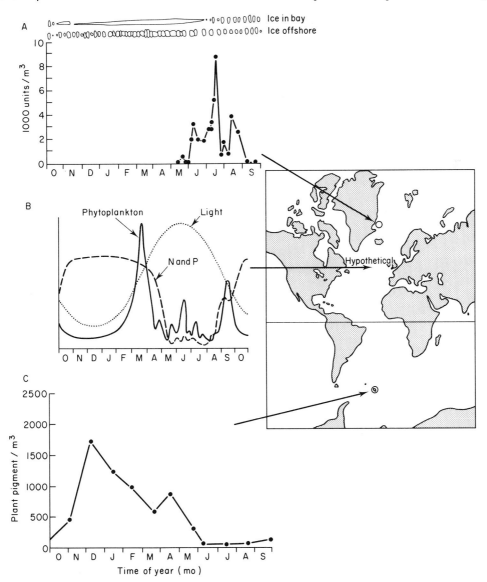

Fig. 20.4 Seasonal incidence of phytoplankton at different latitudes

A. Seasonal abundance of phytoplankton in Scoresby Sound, Greenland

The transition from loose pans of old and new ice to fast ice and back to scattered pans of old fjord ice and heavy polar floes in Rosenwinge Bay and the amount of packing by the offshore ice is indicated above the diagram.

B. Diagrammatic representation of seasonal cycles in light, nitrate and phosphate, and phytoplankton in a typical northern temperate sea

C. Seasonal variation in abundance of phytoplankton in the northern Antarctic region

[Compiled, with modifications, from Hart (1942), Digby (1953), Raymont (1963)]

between removal and return migration is provided by considering the results of J.E. Harris (1963) for the freshwater cladoceran, *Daphnia magna*, and for the marine copepod, *Calanus finmarchicus*. Harris demonstrated that these species, when kept under conditions of constant light or constant dark, would nevertheless continue to show vertical migration on a 24 hour basis in continuous dark and a 28 hour basis in continuous light. Animals collected during the winter months, however, showed no vertical

migration and remained at the bottom of the tank. These animals, therefore, performed movements similar to daily return migration (see Fig. 20.10), albeit over a limited vertical distance, and Harris interpreted his results from the viewpoint of the involvement of an endogenous cycle of activity in daily return migration.

The model developed in this section attempts to generate the ideal pattern of daily return migration and the ideal system of migration thresholds by

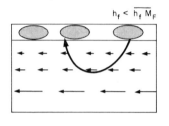

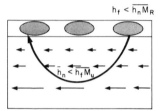

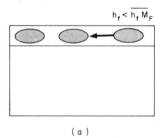

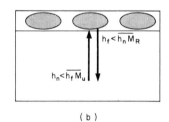

(a)　　　　　　　　　　　　　　(b)

Fig. 20.5　The distinction between vertical migration as removal and as return migration

In all diagrams the horizontal band indicates the vertical limits of the distribution of feeding habitats. The hatched areas indicate the patchiness of food availability in a horizontal plane. The arrows in the top two diagrams indicate current speed and direction relative to the surface layer. In the bottom two diagrams there is no differential speed with depth.

(a) Vertical migration as removal migration

In the upper diagram the water below the feeding layer moves at a different speed. Animals that drop into this zone are displaced horizontally. Migration thresholds should evolve that correspond to the situation in which the suitability of the feeding habitat occupied falls below $h_f M_F$ (where h_f is the suitability of feeding habitats and M_F is the migration factor for migration between feeding habitats). Ideally, upward migration should probably be initiated after a given time interval that gives maximum probability of rising to the surface in a new aggregation of food. This time interval should be a function of mean inter-aggregation distance and relative current speed.

In the lower diagram there is no differential current speed. In this case when h_f falls below $\overline{h_f M_F}$, removal

migration is probably achieved to greatest advantage by horizontal movement.

As food also has a vertical distribution, when h_f falls below $\overline{h_f M_F}$ (for the appropriate removal migration) at a particular depth, vertical removal migration within the feeding layer may also occur. In a free situation, selection should stabilise when the frequency distribution of multiple migration thresholds within the deme to depth, deme density, and food availability results in all individuals experiencing the same potential reproductive success (Chapter 17, p. 269).

(b) Vertical migration as return migration

Vertical migration as return migration can evolve whether or not different levels of the water mass move at different speeds. The return migration can be broken down into two removal migrations. The downward migration is initiated when h_f becomes less than $\overline{h_n M_R}$ where h_n is the suitability of the deep non-feeding habitat and M_R is the migration factor for the complete return migration. The upward migration is initiated when h_n becomes less than $\overline{h_f M_u}$ (where M_u is the migration factor for the upward migration).

All zooplankton perform removal migration, either by vertical or horizontal movement, or both. Some zooplankton, perhaps the majority, perform both removal and return migrations.

which this pattern may be achieved for each of a variety of combinations of environmental and morpho-physiological variables. The ideal system of migration thresholds (e.g. to light intensity, pressure and temperature) for daily return migration is considered to result from linkage to events in the sequence of collection and processing of food, not to an endogenous rhythm. The results obtained by Harris could therefore be taken to indicate a shortcoming in the model. Unless, that is, Harris was observing removal migration and not return migration. The experiments were carried out in tanks in the total absence of food. Inevitably, therefore, the animals would perform removal migration except during the winter months when habitat suitability is not a function of food availability. The question presents itself, therefore, of whether the observed endogenous rhythm could be adaptive for removal migration.

The critical factor for a removal migrant would be distance achieved through the different current speeds at different depths. Presumably there will be a removal migration distance that is the optimum compromise between probability of movement away from the area of low food availability and minimum time spent away from the feeding layer. Selection can only act on the animal to achieve this optimum distance by 'measurement' of the time interval between the initiation of movement to a lower depth and the initiation of movement back to the level most likely to support food organisms. Similarly, it is the time invested in search for food in a habitat of low suitability that is acted on by selection to produce optimum stay-time at the upper levels. In the case of removal migration, therefore, selection would be expected to have acted on the measurement of time (i.e. on an endogenous rhythm).

It is tentatively suggested, therefore, that by carrying out his experiments in the absence of food, Harris triggered the removal migration response in his animals. In this case, an endogenous rhythm might well have been expected. This does not necessarily mean, however, that an endogenous rhythm is involved in return migration.

Unfortunately, it is not at present possible to determine whether any particular observed vertical migration has resulted from environmental variables exceeding the migration thresholds for return or removal migration. The aim of this section on zooplankton is to develop a model that generates the ideal form and variation of return migration given the variables entered into the model. Observed

anomalies should then be examined to determine whether they indicate a shortcoming in the model for return migration or whether they are the result of the observed behaviour having been a removal migration as suggested for the example just presented.

20.2.1 Daily return migration: development and application of an adaptive model

Zooplankton perform daily return migrations not only in the oceans but also in lakes, fish-ponds, ditches and quiet rivers (Cushing 1951). In these fresh-water situations the migrants are mainly Crustacea, particularly copepods and cladocerans (Bainbridge 1961), but other animals, such as insect larvae, may also be involved. In such situations the one-way distance component of the migration may be measured in centimetres rather than metres. It is in the oceans, however, that daily return migration is most spectacular. Hardly a single major group of animals represented in the marine zooplankton does not include at least some species that show daily vertical return migration (Raymont 1963). It has been observed, for example, in many medusae, siphonophores (Fig. 20.11.A) and ctenophores, in sagittae and in pteropods, as well as in fish larvae, salps and appendicularians. Most information, however, is available for the various groups of Crustacea: euphausids (Fig. 20.9), amphipods, copepods (Fig. 20.10), ostracods, and decapods (Fig. 20.11C), as well as in meroplanktonic larval species. It occurs in bathypelagic species as well as those that live nearer the surface (Fig. 20.11). Waterman *et al.* (1939) demonstrated a daily vertical return migration by various decapods, mysids, euphausids, and amphipods living at depths down to at least 1000 m (Fig. 20.11C). In such species, most of which do not reach the surface, the one-way distance component of the migration exceeds 200 m and might be as great as 600 m, 800 m (Cushing 1951), or even 1000 m (Bainbridge 1961). At the other extreme, in Lake Mendota the one-way distance component may be as short as 0·5 m (Cushing 1951).

Not only the all-day members of the zooplankton may be considered to perform daily vertical return migration. Some animals which live essentially in the bottom sand and mud during the day leave the bottom for water immediately above them, or even reach the intermediate layers and swim in the plankton during the night. Mysids, cumaceans and some amphipods were the most typical animals which entered the night plankton over intertidal areas of north-east England (Colman and Segrove 1955), though some isopods may also leave the bottom at night (Raymont 1963). Many of these animals may enter the plankton for only a short time, and in interpreting these movements care must be taken to distinguish between return and removal migrations. If these animals are feeding in the plankton at night but roosting and perhaps feeding in the sand or mud by day, then the migration is a return migration similar to that of the other members of the plankton. If, on the other hand, these animals feed in the sand and mud by day and migrate at night to a new area of mud and sand then the migration is a removal migration as described in Fig. 20.5. Possibly both types of migrants are involved. Only the former type concerns us here, however, and this type need not be distinguished from other daily vertical return migrants that do not happen to reach the bottom during the day.

The general pattern of daily return migration by zooplankton is remarkably constant from polar regions to tropical regions, from the Pacific to the Atlantic, and from oceans to ditches. The components of migration, however, particularly distance and periodicity, vary with weather, latitude and season, with age and sex, and with species, race and generation.

The plan of the next few subsections is first to enter into the model developed below those selective pressures appropriate to female herbivorous zooplankton, secondly to add selective pressures appropriate to males of such species, thirdly to add selective pressures appropriate to carnivorous zooplankton, and finally to add selective pressures that are likely to be species-specific. These last selective pressures are necessary to generate ideal patterns of migration for individual species as opposed to broad ecological groups.

The model used in this section is based on the assumption (Chapter 9) that a vertical return migration can be analysed as two consecutive removal migrations, one downward from a feeding habitat to a non-feeding habitat and the other upward from a non-feeding habitat to a feeding habitat. According to Chapter 9, it follows that downward migration should be initiated when h_f falls below $\overline{h_n M_R}$ (where h_f is the suitability of the feeding habitat, h_n is the suitability of the non-feeding habitat, and M_R is the migration factor for the return migration). As each

depth level below the feeding layer may be considered to be a discrete non-feeding habitat, the non-feeding habitat considered by the model is at that depth for which, in the situation being considered, $h_n M_R$ is greatest. Upward migration should be initiated when h_n falls below $\overline{h_f M_u}$ (where M_u is the migration factor for the upward migration). In this case, h_n is the suitability of the non-feeding habitat occupied by the animal at the particular moment in time.

Factors contributing to variation in suitability of the feeding habitat: For female herbivorous marine zooplankton the feeding habitats are distributed in a more or less horizontal layer within the top 100 m of temperate seas and within the top 400 m of tropical seas (Fig. 20.3). In fresh water the depth of the feeding layer is in part a function of the turbidity of the water. The vertical distribution of feeding-habitat suitability is a function of the vertical distribution of those species of phytoplankton upon which the species is specialised to feed.

The suitability for all time-investment components of habitats within the feeding layer may be considered to be a function of efficiency of capture of food, efficiency of processing of food, and predation risk.

Fig. 20.6   The contribution of temperature to the suitability of food-processing habitats in zooplankton

Two animals are considered, the copepod, *Pseudocalanus minutus*, and the chaetognath, *Sagitta elegans*.

(a) Both species are larger when they develop at lower temperatures. (b) Larger individuals seem likely to have a greater fecundity. However, individuals that develop at lower temperatures also develop more slowly. In (c), the relative rates of increase are calculated (taking into account the opposing effects of increased fecundity and increased generation interval) for hypothetical lineages for which the mean temperature at the feeding depth is higher by the stated amount (°C) than the mean temperature at which food is processed. Rates of lineage increase/month can then be compared with that of non-migrants for which both feeding and food processing take place at the 'surface' temperature.

It is concluded that in thermally stratified waters those individuals are at an advantage that capture their food at one depth but process it at some other, greater, depth at which the temperature is lower.

[*Compiled from McLaren (1963) to which reference should be made for full details of the arguments and calculations involved in the production of these figures*]

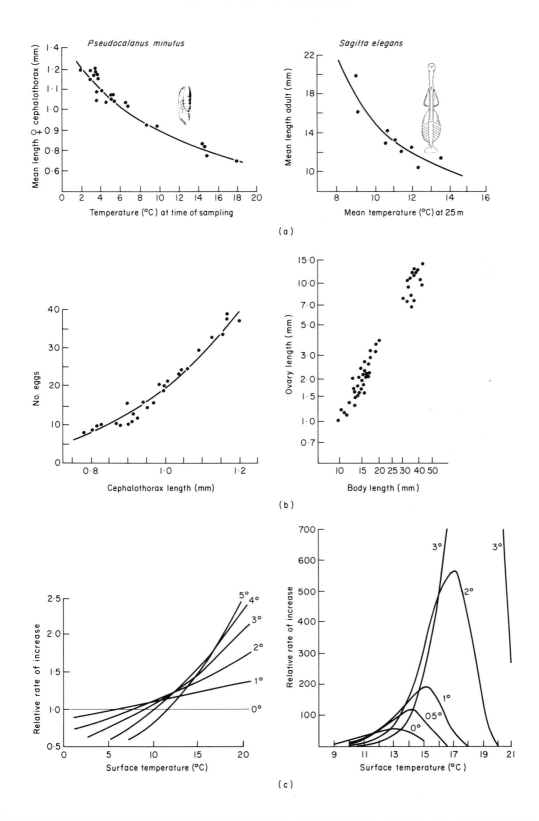

(a)

(b)

(c)

The contribution of the efficiency of capture of food (where efficiency is given by units of energy uptake per unit of energy expended) to habitat suitability should increase with time after the last feeding bout and should also be a negative function of the success of the last feeding bout. It follows, therefore, that immediately after a successful feeding bout there is no correlation between availability of food and habitat suitability. Habitats containing zero food are as suitable as habitats containing abundant food.

Efficiency of food processing seems to be a function of temperature (Fig. 20.6) such that at lower temperatures the processing of food (i.e. digestion and the utilisation of the energy released) is more efficient. In many water bodies, thermal stratification exists such that the feeding habitat near the surface is at a higher temperature than habitats at levels below the feeding layer (Fig. 20.2). The contribution of the efficiency of food processing to habitat suitability should be a function of the amount of food that has been captured and is awaiting processing. Consequently, as food capture proceeds, efficiency of food processing contributes to a decrease in suitability of the feeding habitat relative to deeper habitats.

Predation risk, assuming a constant array of predators, seems likely to be a function of light intensity (Bainbridge 1961, Fraser 1962), given that at least some predators make some use of the visual sense in finding and catching their prey. To some extent the higher predation risk at higher light intensities seems likely to be offset by the apparent ability of at least the larger zooplankton, such as the Antarctic krill, *Euphausia superba* (Fig. 20.9) to take evasive action more effectively at higher than at lower light intensities (Marr 1962). This effect, however, seems unlikely to destroy a general positive correlation between light intensity and predation risk. The contribution of predation risk to variation in habitat suitability seems likely to be constant. Consequently, the feeding-habitat layer, which in most areas at most seasons shows a marked daily variation in light intensity, should be less suitable during the hours of daylight than during the hours of darkness. Furthermore, the light gradient from the surface downwards means that during the day the feeding habitat is less suitable than habitats at lower depths.

Factors contributing to the variation in suitability of habitats below the feeding-habitat layer: The vertical distribution of non-feeding habitats is usually less restricted than the vertical distribution of the feeding habitats. This vertical distribution can be taken to be from the lower limit of the feeding layer down to the water-body bottom. However, these non-feeding habitats are not homogenous for suitability and in many places at certain times of year are characterised by gradients of, among other factors, temperature and light intensity. The factors that contribute to the variation in suitability of non-feeding habitats have in effect already been considered. It is necessary only to summarise the conclusions reached, but from a slightly different viewpoint.

Efficiency of capture of food makes little contribution to habitat suitability immediately after a successful feeding bout and there is no differential due to this factor between the suitability of habitats at these non-feeding levels and the habitats in the food layer. Increase in time after the previous feeding bout, decreased success at collecting food in the previous feeding bout, and increased requirements for energy uptake all decrease the suitability of habitats at lower levels relative to that of the habitats in, or nearer to, the food layer.

The contribution of efficiency of food processing to habitat suitability is a function of the amount of food captured ready to be processed. Efficiency of food processing is a function of temperature. Where temperature decreases from the feeding habitat downwards then as a feeding bout proceeds the gradient of increasing habitat suitability from the surface downwards due to this factor becomes more steep.

Predation risk is a function of light intensity. By night there is no gradient in habitat suitability due to predation risk. By day there is a gradient of increasing suitability from the surface downwards to the zone of darkness.

Factors contributing to variation in M_R, the migration factor for return migration: In most examples of return migration, little error is introduced by making the assumption that the migration factor for the outward journey (i.e. from A to B) is the same as the migration factor for the return journey (i.e. from B to A). This is not the case, however, for the vertical return migration of zooplankton. Some members of the zooplankton, of course, can maintain their level, or even rise, by means of substances such as gases or oils that maintain or change their body density at or to an appropriate value. In such cases, migration cost is a function of the energy demands of the density-regulation process. Many more members, however, have a body density that is greater than that of water. When individuals of such species are anaesthetised, they sink. Maintenance of level of such species, as well as movement upwards, requires expenditure of energy. Downward movement, on the other hand, could be achieved by passive sinking, though downward directed movement may also be employed. In general, therefore, downward migration involves less energy expenditure than station-keeping or upward migration.

It seems likely that nett energy expenditure contributes to a migration factor (M_d) for the downward migration from the feeding habitat layer that is greater than unity. On the other hand, the contribution of energy expenditure to the migration factor (M_u) for the upward migration back to the feeding-habitat layer is likely to result in a migration factor that is less than unity. It also seems likely that the migration factor for both journeys combined (i.e. M_R, which is given by $M_R = M_d M_u$) is also less than unity.

The contribution of nett energy expenditure to the variation in M_R is likely to be relatively constant. Nett energy expenditure is likely to be a function of migration distance, muscle loading, and body size. Increase in migration distance should be associated with an increase in the nett energy expenditure and hence a decrease in M_R. Increase in muscle loading is also likely to be associated with an increase in nett energy expenditure and hence a decrease in M_R. Thus animals that have just fed but which have not yet defaecated have a higher muscle loading than those that have defaecated. Hence M_d is higher for animals that have just fed and M_u is higher for animals that have just defaecated. Gravid females similarly have greater muscle loading than non-gravid females. Hence, M_d is greater for gravid females but M_u and probably also M_R is less. Finally, the influence of energy expended per unit distance on potential reproductive success is likely to be greater for smaller animals than larger animals (i.e. movement over a long distance is likely to reduce the potential reproductive success of a small animal with limited energy reserves more than that of a large animal with larger reserves).

The contribution of nett energy expenditure to variation in M_R is therefore likely to be to generate a negative correlation between M_R and each of distance and muscle loading, and a positive correlation between M_R and body size.

The integrated effect of factors affecting habitat suitability and migration factor on daily vertical migration by female herbivorous zooplankton: The combined action of factors affecting the efficiency of on the one hand food collection and on the other hand food processing seems to be as follows. When an animal begins a feeding bout, efficiency of food collection is at a premium and the most suitable habitats are those within the vertical distribution of feeding habitats. As more and more food is collected, however, and begins to enter the digestive tract, the contribution to habitat suitability of the efficiency of food collection decreases and the contribution of the efficiency of food processing increases. The result is that the gradient of habitat suitability (i.e. decreasing suitability with increasing depth) that existed when feeding began first disappears and then reverses such that when food processing is at a premium, habitat suitability increases with depth. If the suitability of the feeding habitat is h_f and that of a non-feeding habitat is h_n, then selection will favour downward migration, according to the migration equation (p. 43), if a level exists for which h_f is less than $\overline{h_n M_R}$ or alternatively for which h_f/h_n is less than $\overline{M_R}$. The probability of this situation arising is a function in part of the steepness of the gradient of increasing habitat suitability with depth and in part a function of body size. The gradient of increasing habitat suitability with depth is a function of the extent to which an animal can collect food at a faster rate than it can process food. The greater the abundance of food, therefore, the greater this gradient should be. The probability of h_f/h_n becoming less than $\overline{M_R}$ is therefore likely to be a function of food abundance and body size. Of the two, it might be expected that body size may have a higher coefficient of determination of $h_f/h_n M_R$ than food abundance, though quantitative proof is lacking. In general, therefore, larger animals should be more likely to initiate downward migration than smaller animals and any given animal should be more likely eventually to initiate downward migration when food is abundant than when it is scarce. In species and ontogenetic stages for which downward migration is advantageous, selection should favour linkage of the initiation of migration with some measure of amount of food collected, such as, perhaps, muscle loading or nett weight of collected but unprocessed food, in such a way that migration is initiated as soon as h_f becomes less than $\overline{h_n M_R}$ (Fig. 20.7).

For animals for which it is advantageous to drop

out of the feeding-habitat layer then, as food processing proceeds, the gradient of increasing habitat suitability with depth becomes less steep with time. At some stage during the sequence of food processing the gradient is reversed and habitat suitability decreases with depth down to the level at which food disappears and from there downwards remains constant. As food is processed and defaecation occurs, muscle loading decreases and the migration factor for any future upward migration increases. Upward migration should be initiated at the time that h_n becomes less than $\overline{h_f M_u}$. Selection should favour stay time at the lower depth being a function of the amount of food collected during the preceding feeding bout. This could be achieved by linkage of the initiation of migration with some time-correlated endogenous event such as the last food passing a certain stage of processing and/or accomplishment of a certain amount of defaecation.

If only efficiency of capture and processing of food and migration factor were to be entered into the model, a prediction would be made that species with smaller adults and the earlier ontogenetic stages of larger species would be more likely in all areas to remain within the vertical distribution of the feeding habitat. In waters with a steep thermal gradient a given range of adult size and ontogenetic stages would be likely to initiate migration down from the feeding level. In waters with a less steep thermal gradient, downward migration would be advantageous for only the larger species and stages within this range. The periodicity of return migration in such animals should be a function of the rates of food capture and processing. Linkage of certain events in the sequence of food collection and processing with the initiation of migration are likely to have been favoured by selection.

Although quantitative analysis does not seem to have been made, there is good qualitative evidence that the first part of this generated pattern shows good agreement with the observed behaviour of plankton. A notable proportion of the zooplankton remain in the same surface layers at all times of the day and night and have a much reduced or zero distance component to their daily vertical migration. In boreal seas, non-migratory plankton are species of small body size and the young stages of larger species (McLaren 1963). For example, the eggs and nauplii of *Calanus* show little daily vertical migration (Nicholls 1934). If the overwintering generation of copepodite V is excluded, however, Stages I–VI show a gradual increase in the distance component of their daily migration (Fig. 20.10).

The pattern generated of the relation between the size range of the migrants and the steepness of the thermal gradient also seems to show good agreement with observed behaviour. By contrast to temperate waters, tropical seas are characterised by a narrow seasonal range of temperature at the surface and a relatively deep isothermal layer of water through the year (McLaren 1963). Here, only the larger more mobile zooplankton, such as euphausids and chaetognaths show vertical migration.

At first sight, the Antarctic krill, *Euphausia superba* (Fig. 20.9) appears to be an exception to the pattern generated by the model. In this species, the young stages (body length less than 20 mm) show a gradually decreasing distance component to their daily vertical migration until, by the time they are adolescent (body length, 16–20 mm), they show little if any daily vertical migration (Marr 1962). The young adults and adults (body length greater than 20 mm) then show daily vertical migration once more (Fig. 20.9a). However, if steepness of thermal gradient is taken into account, it still seems possible that the species fits the pattern generated by the model. The pattern of exposure to thermal gradients by the different ages is extemely complex (Marr 1962). In general, however, it seems that the young stages may experience a surface temperature as high as $+1.99°C$ and consequently a thermal gradient of the type entered into the model. As they grow, however, these young stages experience a gradually decreasing surface temperature until, by the time adolescence is attained during spring, even the maximum surface temperatures experienced are sub-zero $(-0.01°C)$. Any gradient of temperature that may occur, therefore, is likely to be minimal and it is quite possible that despite their size, downward migration may no longer be advantageous. As they become adults, however, the krill experience increasing surface temperatures which may, in February, reach as high as $+3.99°C$. Ontogenetic variation in daily vertical migration in *E. superba*, therefore, could still be considered to be consistent with the ideal pattern generated by the model so far.

Not only body size but also muscle loading has been suggested (p. 489) to contribute to variance in M_u and hence to the optimum distance component of migration (Chapter 11). It is notable, therefore, that female *Calanus* have a shorter distance component when gravid than when non-gravid (Marshall and Orr 1960).

Let us now add to the model the contribution of predation risk, as a function of light intensity, to the variation in suitability of the feeding habitat relative to that of habitats at lower levels. This will have the major effect of integrating the vertical migration already generated by the model into a light/dark cycle. As a starting point, let us consider an animal that arrives at the feeding habitat and begins a feeding bout at dusk.

Throughout the hours of darkness, predation risk makes no contribution to variation in habitat suitability with depth. Cycles of food collection and food processing and, for larger animals in waters with a sufficiently steep thermal gradient, vertical return migration proceed as previously described. If all individuals start to feed at dusk, the deme will be concentrated at the feeding level. As individuals collect the threshold amount of food, they initiate downward migration. Individual variance in success at collecting food results in some individuals migrating earlier than others and some may be migrating back up again as others migrate down for the first time. Observed as a vertical distribution, this behaviour will be seen as a dispersal, some time after the dusk period of convergence at the feeding level. The dispersal may be followed by a second period of convergence at the feeding level. The critical factors are the optimum distance component and hence time required for the migration (Chapter 11), relative optimum stay times at the food processing and food collecting levels (Fig. 20.7), and length of night. Quite conceivably some species with a large distance component may have a single feeding period (p. 504) whereas species with a smaller distance component, if the night is sufficiently long, may have two or more feeding bouts per night. Where two feeding bouts are found, the intervening spread of individuals performing their return migration will appear as a 'midnight dispersal' if individual feeding success has a large variance and as a 'midnight sink' if the variance is less and the individuals migrate over a narrow time period. The occurrence of a midnight dispersal or sink is a well-known feature (Cushing 1951) of the vertical migration of some species (Fig. 20.10).

So far, then, entry of predation risk into the model has not in any way altered the pattern of migration generated earlier. This is because predation risk makes no contribution to variation in habitat suitability with depth during the hours of darkness. With the arrival of dawn, however, predation risk, as a function of light intensity, introduces a gradient of increase in habitat suitability with depth. For individuals coming to the end of a single or second feeding bout this gradient reinforces the gradient that should be developing in any case (in thermally stratified waters) due to the contribution of of food processing to variation in habitat suitability. For individuals at an earlier stage in a feeding bout the gradient due to predation risk will act to reverse the gradient due to the efficiency of food collection. The value of h_f will fall below $\overline{h_n M_R}$ later for these individuals than for the former individuals, and later for smaller individuals than larger individuals, and again not at all for individuals below a certain body size. Downward migration at about dawn should be found in animals down to a smaller body size in waters with a steep thermal gradient, as in temperate waters, which reinforces the gradient due to predation risk, than in waters with only a slight thermal gradient, as in tropical waters (below a certain depth—Fig. 20.2). Finally, an ideal system for the initiation of downward migration should include a migration threshold to light intensity as well as the triggering of migration beyond a certain muscle loading or nett weight of collected but unprocessed food. The relationship between these two factors should be a function of the relative gradients of habitat suitability due to predation risk and efficiency of food processing. In any case, migration should be initiated when the stage is reached that corresponds to the situation at which h_f becomes less than $\overline{h_n M_R}$ (Fig. 20.7).

The gradient of habitat suitability due to predation risk should increase toward mid-day as the Sun gets higher in the sky and then decrease toward dusk as the Sun gets lower in the sky. The gradient should not disappear, however, until dark. Early on in the day this gradient will be reinforced by the gradient of habitat suitability due to the efficiency of food processing (in thermally stratified waters). As food is processed, however, this reinforcement will grow less, eventually disappear, and finally be replaced by the reversed gradient due to the efficiency of food collection. Whether this reversed gradient becomes sufficiently pronounced to mask the gradient due to predation risk depends on a number of factors. Two of these are the time required to process the food collected and the time required to use up the energy thus released. Both of these factors are likely to be a function of temperature and for animals for which it is advantageous to migrate down from the feeding level the advantage of a lowered day-time temperature in slowing down the rate of food processing and energy utilisation acts to reinforce the advantage due to the greater efficiency of food processing. Both factors are also likely to be a function of the amount of food collected the previous night. The more food collected, the less likely is the gradient due to efficiency of food collection to mask the gradient due to predation risk. Amount of food collected is likely to be a function of food availability and length of the collecting period. When food is abundant, nights are relatively long, and days are relatively short, the chances are high that the gradient due to efficiency of food collection will not mask the gradient due to predation risk until the latter is decreasing in steepness toward dusk. When food is scarce, nights are relatively short, and days are relatively long, it seems likely that selection

will often favour individuals that, having experienced low food-collection success the previous night, perform a migration to the feeding level at some time during the following day. Here, the upward migration is again likely to be triggered by some endogenous event that is part of the food-processing sequence. Now, however, because of the added effect of predation risk on habitat suitability, h_n will not become less than $\overline{h_f M_u}$ until later on in this sequence. Selection is therefore likely to favour individuals for which time of initiation of upward migration is influenced by some factor that correlates with predation risk, such as, in this case, light intensity. It is possible to envisage that selection will act on perception of light intensity to be linked with the stage of food processing at which upward

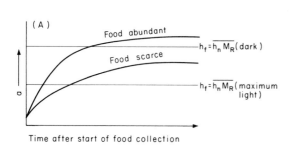

Fig. 20.7 Optimum interaction of environmental variables and endogenous events in the daily initiation of upward and downward migration by female herbivorous zooplankton: a thesis

h_f = suitability of the feeding habitat; h_n = suitability of a non-feeding habitat at a depth greater than the feeding habitat; M_R = migration factor for the entire return migration; M_u = migration factor for the upward migration.

A. The initiation of downward migration

a is the value of some variable that varies with the stage of the food collection sequence. Muscle loading or nett quantity of collected but unprocessed food are possibilities. The graph shows the variation in a with time after the start of food collection. The value of a increases more rapidly and to a higher level when food is abundant than when it is scarce.

Ideally, migration should be initiated when a reaches the value at which h_f becomes less than $\overline{h_n M_R}$. This value is a function of light intensity and temperature. Two values are indicated that correspond to the situations in which, respectively, $h_f = \overline{h_n M_R}$ in dark and in maximum light intensity. The coefficient of determination of this point

may be increased (p. 346) if temperature is included in the migration threshold as well as light intensity.

Ideally, downward migration should be initiated after a shorter period during the day than during the night. An animal at a given stage of food collection should show a downward-migration threshold to light intensity and temperature.

B. The initiation of upward migration

b is the value of some variable that varies with the stage of the food processing sequence. Muscle loading or amount of food in the intestine are possibilities. The graph shows the variation in b with time after the onset of food processing. The value of b starts and remains at a higher level when more food was collected than when less food was collected.

Ideally, migration should be initiated when b reaches the value at which h_n becomes less than $\overline{h_f M_u}$. This value is a function of light intensity, temperature, and depth (perceived as pressure). Ideally, upward migration should be initiated after a shorter period at lower light intensities and greater depth. The relationship to temperature is likely to be complex. An animal at a given stage in the food processing sequence should show a multiple upward-migration threshold to light intensity, temperature and depth.

migration is initiated, the greater the light intensity, the later in the food processing sequence upward migration should be initiated (Fig. 20.7).

This mechanism could possibly also accommodate a complication to this process that results from the fact that whether upward migration in daytime is advantageous is also likely to be a function of the steepness of the gradient of habitat suitability due to predation risk. The less steep the gradient, the greater the possibility of it being masked by the opposite gradient that develops with time after feeding due to efficiency of food collection. It is this factor that would tend to favour a synchronised ascent toward dusk as the gradient due to predation risk diminishes. There may also be a seasonal effect such that at times when the gradient is slight, as in winter and early spring in temperate latitudes, the gradient due to predation risk may be outweighed during daylight by the gradient due to efficiency of food collection. This may occur despite the relatively long night and high food abundance in spring (Fig. 20.4). Indeed, the high food abundance, by increasing the steepness of the gradient due to efficiency of food collection, may contribute to the higher probability that h_n will become less than $\overline{h_f M_u}$. The nett result of all these pressures may be a migration threshold as described (Fig. 20.7B) that results in daytime ascent at times when food is abundant and the gradient of predation risk is low (e.g. early spring for some species) and again when food is scarce and the nights are short (e.g. summer and early autumn for some species).

The existence of a daytime upward migration as a normal event for a proportion of the population, perhaps those with a low feeding success the previous night, is indicated by the upward 'tail' in Fig. 20.10 (a) C and D, and Fig. 20.10(b). The correlation with short nights and scarce food is also indicated by the presence of this 'tail' in summer, as in these figures, but not in January, as in Fig. 20.10(a)A. Most reports in the literature concerning daytime ascent inevitably concern large numbers swarming at the surface. However, a review of such observations by Cushing (1951) for the copepod, *Calanus finmarchicus*, indicates that daytime ascent occurs in early spring and again in late summer and early autumn. This seasonal incidence is quite consistent with the pattern generated by the model in the previous paragraph.

Each water depth can be considered for conceptual convenience as a discrete habitat for which a specific migration threshold for upward migration

can evolve. Quite possibly the combination of linkage of initiation with an endogenous event, part of the digestion process, that is influenced by light intensity can give a high coefficient of multiple determination for the variation with depth of the point at which h_n becomes less than $\overline{h_f M_u}$. It is likely, however, that inclusion of perception of pressure and perhaps temperature in the determination of the position in the digestion process at which upward migration is triggered may increase still further the coefficient of determination (p. 346). Such variables, however, probably usually contribute less to the determination than light intensity (Fig. 20.7B).

The pattern generated by the model for daytime descent is for the initiation of downward migration when h_f becomes less than $\overline{h_n M_R}$. If the conditions of light, pressure, and temperature (Fig. 20.7B) are above the composite threshold for upward migration at each depth encountered, the animal continues to descend. Descent ceases when a habitat is encountered for which the migration threshold for upward migration is exceeded, whereupon upward migration is initiated. Upward migration should continue until a habitat is encountered for which the migration threshold for upward migration is not exceeded, whereupon the process should be repeated. Bearing in mind that the migration threshold for upward migration is postulated to decrease with time after the last feeding bout (Fig. 20.7B), the optimum behaviour should be a fairly rapid descent around dawn to a daytime depth followed by a more gradual rise to the surface toward dusk. Unless, that is, due to a low feeding success the previous night the migration threshold for upward migration becomes sufficiently low, whereupon a daytime ascent right up to the food layer takes place.

It is clear from this generated pattern and Fig. 20.7B that the maximum daytime depth is a function of the depth to which light penetrates the water. This again is in agreement with observed variation. Thus *Calanus* are distributed higher in the water on foggy days and deeper on cloudless days and about twenty other species of zooplankton have been shown to be higher in the water on dull than on sunny days. In water covered with ice, zooplankton are deeper in the water under gaps in the ice than under the ice itself and higher in the water when snow is on the ice than when it is not (Cushing 1951). Similarly, when light intensity is reduced due to the surface being broken by wind and rain plankton may approach much nearer the surface. Unless, that is, the waves become heavy and the rain continues for long periods. Then it seems that the upper layers become physically uninhabitable and, particularly at night, the plankton show a deeper distribution.

These and other field observations (Cushing 1951), as well as laboratory experiments (Harris and Wolfe 1955), provide convincing evidence for the involvement of light intensity as a component of a composite migration threshold for upward migration and perhaps as the only habitat variable involved in the downward migration threshold, though conceivably temperature of the feeding layer may also be involved. The work of Moore (1950, 1952) suggests strongly that temperature may also be a component of a composite migration threshold for upward migration. Even when temperature and light intensity were both taken into account, however, Moore still encountered discrepancies in changes in the depth distribution of many species of zooplankton. This unexplained variance was considered to be explicable by the inclusion of a third variable that varies directly with depth. More recent work on the responses of marine animals to changes in hydrostatic pressure (Knight-Jones and Morgan 1966) makes it seem certain, as suggested by Moore, that this is the third component.

Experimental investigation of each of these three components on their own produces notoriously variable results, perhaps because, with the exception of Moore and Corwin (1956), few workers have viewed the three factors as constituting a composite migration threshold. As far as I am aware, also, no workers have considered that this composite migration threshold might be linked to an event in an endogenous sequence, which is the ideal pattern generated by the model developed here (Fig. 20.7B). This linkage means, of course, that the measured migration threshold to any one of the components of the multiple threshold for upward migration should be a function both of the time since the last feeding bout and the success of that feeding bout. Some support for this predicted linkage may perhaps be claimed from the observations of Clarke and Backus (1956). These workers noted that although animals forming a deep scattering layer followed the depth of constant light intensity during the middle part of the day (i.e. the migration threshold to light intensity remained constant), during the afternoon the layer moved upwards faster than the changing light intensities (i.e. the upward migration threshold decreased with time) and in fact reached illuminations in one case more than a hundred times

brighter than they had experienced at mid-day. As far as I am aware, also, experimenters have not distinguished between the upward migration threshold and the downward migration threshold. According to the model developed here, the upward migration threshold should be a composite threshold to light intensity, temperature and pressure, with the threshold linked to a variable associated with the food-processing sequence (Fig. 20.7B). The downward migration threshold, on the other hand, at least for herbivorous surface feeding species, should be either a simple threshold to light intensity or a composite threshold to light intensity and temperature. Which of these is ideal probably varies with the species, geographical location, and season. In either case, the threshold should be linked to a variable associated with the food-collection sequence (Fig. 20.7A). Finally, at the practical level, future experiments on return migration should be designed to avoid triggering the removal migration response, as may have occurred in the experiments of Harris (1963), for removal migrants and return migrants are likely to manifest different migration thresholds (Fig. 20.5). It is perhaps hardly necessary to stress also that different species, races, generations, ontogenetic stages, and sexes are likely to have different migration thresholds.

In summary, there seems to be convincing evidence that light intensity is involved in the downward migration threshold and that light intensity, temperature, and pressure are involved in some way in a composite upward migration threshold. The linkage of these thresholds with a variable event in the food collection and food processing sequences does not, however, seem to have been investigated.

Factors contributing to variation in habitat suitability for male herbivorous zooplankton: Males of herbivorous zooplankton should be subject to the same selective pressures as those entered into the model for females. In addition, probability of encounter with receptive females makes a major contribution to habitat suitability during the breeding season.

When females are equally receptive at all points in their daily vertical migration, probability of encounter with receptive females is unlikely to impose differential selection on males and females. Similarly, suppose that probability of encounter with receptive females was greatest at one particular point during the female migration. Although this may impose selection on males to arrive at this point slightly in advance of the females and to leave perhaps slightly after, on the whole the total array of selective pressures on males would act to favour a pattern of daily migration similar in all essential respects to that of the females. Except, that is, insofar as males and females may be of different body size and have different muscle loading.

The only situation that can readily be envisaged in which selection favours a slightly different pattern in males and females is as follows. Suppose that females during the period immediately preceding the onset of receptivity do not show daily vertical migration but are distributed permanently below the feeding level. According to the model this could occur if h_n was permanently greater than $\overline{h_f M_u}$ due to h_f being low as a result of food scarcity and due to h_n being relatively high due to a low temperature and a low light intensity. Such a situation is encountered during winter in temperate and subpolar latitudes (p. 498 and Fig. 20.4). Suppose now that immediately upon initiating migration back to the feeding level (i.e. when h_n does become less than $\overline{h_f M_u}$ as, say, in spring in temperate and subpolar latitudes) females become receptive. Such an occurrence seems quite likely as the daily migration becomes re-established due to the reappearance of an adequate supply of food which may also favour the onset of reproduction. This means that a gradient of probability of encounter with receptive females is established, with habitat suitability increasing with depth, as soon as receptive females start to appear. The result is selection on males to decrease stay-time in the feeding layer and to increase stay-time in the level at which females are becoming receptive. This ideal result would be manifest as a more even vertical distribution of males than migrating females at night and a male distribution with a peak at the level at which receptive females should appear during the day.

In the event that a reverse situation should occur (i.e. pre-receptive females being permanent residents in the feeding layer with vertical migration being initiated upon the onset of receptivity) males are favoured that increase stay-time in the feeding layer and decrease, perhaps to zero, stay-time in the non-feeding layers.

The influence of inclusion in the model of probability of encounter with receptive females in generating a sexual difference in vertical migration receives strong support from field observation for at least one species. In the marine copepod, *Calanus finmarchicus*, males have a much longer stay-time at lower levels than adult females when receptive

females are moulting from their Stage V copepodite overwintering stage at a deep and constant level (Fig. 20.10(a)A and B). When, as in Fig. 20.10(b), however, the Stage V shows daily vertical migration, the male pattern of migration is much nearer to that of the adult females. In this species males attach their spermatophores to newly moulted adult females. Indeed, Stage V copepodites have been found with spermatophores attached (Marshall *et al.* 1934). Usually, females are found with only one spermatophore attached, an observation which suggests that female receptivity is reduced after receiving a spermatophore. The situation is just that suggested to be likely to generate a sexual difference in return migration of the form observed at the different seasons in Fig. 20.10. Indeed, the data for vertical migration in Fig. 20.10(a)A and B were obtained at just about the date that, according to Marshall *et al.* (1934) females show the highest proportion of individuals carrying spermatophores and at which, therefore, any behaviour adaptive to selection due to probability of encounter with receptive females would be most advantageous.

Factors contributing to variation in habitat suitability for carnivorous zooplankton: Carnivorous zooplankton that are predatory on non-migratory zooplankton are subject to the same selective pressures as herbivores. Those carnivorous zooplankton that are predatory on migratory zooplankton are also subject to the same gradients of habitat suitability with respect to efficiency of food processing and predation risk. Now, however, selection is acting with respect to a feeding layer that itself shows a daily variation in depth. On previous arguments, selection should favour a migration pattern similar to that for herbivorous zooplankton but with respect to a level that itself is moving in a similar way. The result is selection for an increased distance component to the migration. This is opposed by selection that results from the migration factor. However, as, in general, predators are larger and more mobile than their prey, the migration factor also tends to be greater. The result is that selection should favour a migration with a greater distance component for carnivorous zooplankton than for their prey species. However, because of this greater distance component a complete return migration during the middle part of the night is perhaps unlikely but an upward migration to the daytime depth of the prey is perhaps more likely. Viewed as a distribution the ideal result is probably that the daytime distribution of the carnivore has its upper limit at the daytime distribution of the prey, but extends some distance below the distribution of the prey species. The same is probably ideal at night but unlike herbivores the lower limit to distribution is well above the daytime distribution level (Fig. 20.8C).

The generation of broad categories of migration: Figure 20.8 illustrates three broad categories of vertical migration generated by the model given the entry into it of efficiency of food collection, efficiency of food processing, predation risk, and migration factor.

The first category (Fig. 20.8A) consists of those groups for which $\overline{h_f/h_n}$ does not become less than $\overline{M_R}$. These are either young stages or adults of small body size for which M_R would be low. Reference has

already been made to the absence of daily vertical migration by such groups. Alternatively, they are species for which h_n is low. This is most likely to arise in waters with a shallow or non-existent gradient of temperature and/or predation risk. Both factors are absent during polar winters and the zooplankton exposed to such conditions show little except seasonal return migration (Bogorov 1946). During the remainder of the year a shallow gradient exists of light intensity and hence, presumably, predation risk. Even so, h_n is likely to be less than if both gradients exist. The result is that a relatively high value of $\overline{M_R}$ is equal to $\overline{h_f/h_n}$, and as optimum migration distance is a function of the value of M_R at this point of equilibrium (p. 678), selection should favour a reduced distance component. Even relatively large and mobile forms, such as the adults of

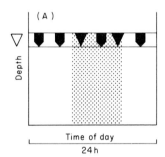
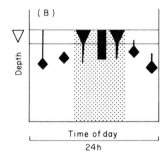
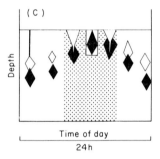

Fig. 20.8   Ideal patterns of daily variation in vertical distribution for three types of zooplankton

The stippled area indicates hours of darkness; horizontal lines in A and B and clear triangle at left indicate vertical distribution of food; solid black kites indicate vertical distribution of zooplankton under consideration.

A. Ideal pattern for zooplankton of small body size

In general, selection should favour animals of small body size remaining within the vertical distribution of the food species. Gradients of selection due to efficiency of food processing (in thermally stratified waters) and predation risk may lead to some vertical migration over short distances. The distance component for the migration should be a function of body size, steepness of gradients of habitat suitability, and vertical range of food species.

B. Ideal pattern for zooplankton of larger body size that feed on non-migratory food species

In general, selection should favour a deeper distribution during daylight. The 'tail' upward to the surface in the far left distribution results from selection favouring individuals with poor food collection success the previous night performing an upward migration to the feeding zone

some time during the day. Optimum stay-time in the feeding zone is shorter during the day than during the night (Fig. 20.7A). The 'tail' downward after dusk results partly from individuals that fed during the day initiating upward migration later than those that did not and partly from individuals which achieved rapid feeding success initiating downward migration. As more and more individuals initiate downward migration the vertical distribution becomes more even. Optimum stay time at lower levels is less at night than during the day (Fig. 20.7B). As more individuals migrate upwards for a second or third feeding bout, a second concentration in surface waters occurs, followed by initiation of downward migration at dawn.

C. Ideal pattern for zooplankton that feed on migratory food species

For species that live above the permanent zone of darkness, selection should favour migration as in B but relative to the mobile distribution of the prey species (clear kites). Species that arrive by day at the zone of darkness are no longer subject to selection for further downward migration due to predation risk and any gradient of increased advantage with depth decreases. In such species the daytime distribution is likely to be nearer to that of the prey species.

the Antarctic krill *Euphausia superba*, have a distance component of only about 50 m (Fig. 20.9). Where daily vertical migration in polar waters has a larger distance component, it is usually found to be occurring in thermally stratified waters (McLaren 1963). Most of the major genera of copepods, for example, including *Calanus*, *Pseudocalanus*, *Microcalanus*, and *Metridia*, in polar waters in late summer and autumn perform daily return migrations with a one-way distance component as great as 100–200 m (Bogorov 1946).

The second and third categories (Fig. 20.8B and C) consist of groups for which $\overline{h_f/h_n}$ does fall below $\overline{M_R}$. These are medium to large species and older stages of development (i.e. forms with a higher value for M_R) in waters with relatively steep gradients of temperature and predation risk (i.e. lower value of h_f/h_n, particularly during the day). The second category (Fig. 20.8B) consists of species that feed on non-migratory food. The generated pattern should be compared with Fig. 20.10 for the marine copepod, *Calanus finmarchicus*, which feeds near the surface on non-migratory phytoplankton. The third category (Fig. 20.8C) consists of forms that feed on species that themselves migrate. Here the generated pattern should be compared with Fig. 20.11C. In both cases, the generated patterns would seem to be acceptable descriptions of the observed patterns.

Factors contributing to species differences: Figure 20.8 indicates that differences can be expected between certain broad categories of zooplankton according to age, adult size, and feeding strategy. In addition, certain other factors are acting that are likely to produce species differences within each category.

Figure 20.5 indicated the way that the vertical distribution while in the feeding zone should result from the optimum frequency distribution of removal migration thresholds for the deme. These thresholds should typically be a multiple function of deme density and food availability (Chapter 16). There is probably a similar frequency distribution of composite removal migration thresholds to deme density and light intensity and perhaps also pressure and temperature while at the non-feeding depths. The result is that a deme is distributed in such a way that individuals have the same potential reproductive success (Chapter 17).

However, suppose that during the evolution of the frequency distribution of migration thresholds, some individuals become exposed to a different range of selective pressures sufficient to lead to speciation. If some of the characteristics involved in the speciation are concerned with the pattern of vertical migration, the result will be two or more species with different patterns of vertical migration that result from 'competition' between the species rather than solely from the general factors that have been entered into the model so far.

For example, Fig. 20.8B suggests that for some species there may during the night be two major periods of feeding activity involving the majority of individuals. These two periods are separated by a period during which individuals are more evenly distributed. This automatically imposes selection in favour of other individuals of the same or other species that concentrate their feeding during this middle period when competition is reduced. Individuals that concentrate their feeding in this middle period, however, are exposed to selection due to the gradient of efficiency of food processing and reduction in energy expenditure during earlier and later periods of the night. Such selection favours a deeper distribution than individuals that are feeding at these times. This may be achieved either by a later or slower dusk ascent and earlier or faster dawn descent from and to the same daytime depth or by a deeper daytime depth. Which of these represents the optimum adaptation depends on selection on migration distance (Chapter 11). The result is the evolution of several species showing species-characteristic periods of feeding activity and, in those species that feed near the surface, species-characteristic periods of appearance at the surface. Time of appearance at the surface is likely to show a multiple correlation with rate of climb, daytime depth, and stay-time at the deepest daytime depth.

Figure 20.11A illustrates a species for which surface activity shows a unimodal peak that occurs some time after midnight. Figure 20.12 shows even more convincingly the way that different species in the same area show a peak of surface activity at different times during the night. Whereas some may show a bimodal night surface activity pattern, which includes the midnight descent emphasised by Cushing (1951) and generated by the model here (Fig. 20.8B), if competition with other species is not included, other species show a unimodal pattern of surface activity as would be expected from the arguments of this section. There is also some indication of a correlation between daytime depth and time of arrival at the surface at night such that the later the time of arrival, the deeper the daytime depth (Cushing 1951).

It may be concluded, therefore, that a model based solely on efficiency of collection and processing of food, predation risk, migration factor, and, for males, probability of encounter with a receptive female, can generate the patterns of certain broad categories of vertical migration. In order to generate the optimum pattern of vertical return migration for a particular species, however, it is necessary to enter into the model such selective pressures as are likely to have been involved in the speciation sequence by which the species evolved. In general, such selective pressures are those that result from factors such as the particular food species and the nature of competition with other species.

20.2.2 Seasonal return migration

It is not always meaningful to separate seasonal and ontogenetic return migrations and the example given in this section, that of the copepod, *Pseudocalanus minutus*, in the Norwegian Sea, furnishes an example of ontogenetic return migration as well as of a seasonal return migration. In the previous section on daily return migration, certain seasonal and ontogenetic effects on the components of the daily pattern were noted. These effects should not be confused with the seasonal and ontogenetic return migrations described in this and the next section.

The copepod, *Pseudocalanus minutus*, in the Norwegian Sea spawns during the period of the diatom maximum in April and May (McLaren 1963). The young develop near to the surface but coincidental with their development to older stages they vacate the surface waters. By August the entire population is below 600 m, at which depth there is a sharp break in the temperature curve. The winter is thus passed at sub-zero temperatures, 7°–8°C cooler than the surface waters at any season. The animals overwinter largely as Stage V copepodites, moulting into adults during the upward migration that takes place in March and April. In brief, the species migrates down into colder waters at the end of the season of high bioproductivity (Fig. 20.4) and migrates upward again when next productivity is high. A similar seasonal return migration seems to be shown in northern waters by another copepod, *Calanus finmarchicus* (Fig. 20.10).

In both of these species it may be postulated that a time of year arrives at which h_f becomes less than $\overline{h_n M_R}$ but for some time after which h_n fails to become less than $\overline{h_f M_u}$. As this situation persists throughout

a period that food availability is low, it seems likely that it is the low value of h_f that is the critical factor. In addition, because of the temperature gradient, perhaps reinforced by the gradient of predation risk, nett loss of energy reserves is less at lower levels than at higher levels. The reverse gradient of habitat suitability that in daily migration develops each night and occasionally during the day, does not develop in the present situation until food starts to increase in abundance the following spring (Fig. 20.4). Although this seasonal migration fits the patterns generated by the model for daily migration insofar as it evokes no new types of selective pressures, it cannot be produced by assuming that these selective pressures favour the same migration thresholds. In daily migration, when little food is collected during a feeding bout, optimum strategy is to decrease stay time in non-feeding habitats and to increase stay time in feeding habitats (Fig. 20.7). In this case of seasonal migration, however, such strategy is clearly not optimum. It must be postulated, therefore, that adaptation to selection for seasonal migration imposes selection for a migration threshold separate to that for daily migration. The obvious candidate for the downward migration threshold is the indirect variable of the photoperiod/temperature regime (Chapter 16). It seems likely, therefore, that a migration threshold to indirect variables exists that results in initiation of a downward seasonal migration at the time of year at which h_f becomes less than $\overline{h_n M_R}$. It is difficult, however, to envisage any simple migration threshold that would result in upward migration at the appropriate season (at which h_n becomes less than $\overline{h_f M_u}$) and only at that season. The problems involved are those encountered by all overwintering animals and the ideal adaptation for plankton seems likely to be a diapause similar to that well known for terrestrial arthropods. We may postulate, therefore, that once the downward migration has been performed the animal is unreceptive to environmental variables except those involved in the maintenance of the optimum depth until the diapause has been broken by sufficient exposure to the appropriate temperature regime. Once the diapause has been broken an upward migration threshold should appear. The most appropriate threshold variables seem to be light intensity, temperature, and photoperiod. The ideal multiple migration threshold to these variables obviously varies with species, latitude, and optimum depth for passing the winter. Ideally, of course, the seasonal migration thresholds must over-ride the daily migration thresholds.

20.2.3 Ontogenetic vertical return migration

Only one example of ontogenetic return migration is presented here, that of the Antarctic krill, *Euphausia superba* (Fig. 20.9b), though as already indicated the seasonal migrations described in the previous section are also effectively ontogenetic migrations. In this example, however, we are concerned with ontogenetic stages that do not feed. Gradients of habitat suitability due to efficiency of food collection and processing are not therefore involved in the advantage of the migration though return to the level of food availability is undoubtedly a factor in the initiation of upward migration. Yet sufficient advantage must be postulated to outweigh the undoubtedly high migration cost. The value of $\overline{h_f/h_n}$ can only be less than $\overline{M_R}$ if h_n is much more suitable than h_f for reasons unconnected with food collection or processing. Quite possibly the gradient of predation risk on the eggs and young larvae may make some contribution to this. Perhaps also, in this case, it is the time spent in the deeper *warmer* water, thus speeding up development through the naupliar stages, that also makes a notable contribution.

20.2.4 Summary

Vertical migration as a return migration by zooplankton has been analysed by the application of an adaptive model that considers the return migration as two removal migrations, one a downward migration, the other an upward migration. On the basis of the model developed in Chapter 9 for return migration, the downward migration was assessed to become advantageous when $\overline{h_f/h_n}$ became less than $\overline{M_R}$. The upward migration was assumed to become advantageous when $\overline{h_n/h_f}$ became less than $\overline{M_u}$. In all, six selective pressures were entered into this model: efficiency of food collection; efficiency of food processing; predation risk; energy expended during migration; probability of encounter with receptive females; and competition with other species. It was found that these six selective pressures, when entered into the model, could account for a large proportion of the observed characteristics of vertical return migration. For daily migration, the model could account for the general diel characteristics as well as for seasonal, ontogenetic, sexual, and species differences. For seasonal and ontogenetic migration, the model could at least account for the general characteristics of the migrations of the few species described. The model also generated certain ideal characteristics for downward and upward migration thresholds. It is suggested that ideally the initiation of daily migration should be linked with events in, for downward migration, the food-collection sequence, and for upward migration, the food-processing sequence. The precise event, however, should also be a function of light intensity and perhaps temperature for downward migration and a multiple function of light intensity, pressure, and temperature for upward migration. A variety of observational and experimental data exist that provide circumstantial support for the suggestion that these ideal characteristics generated by the model have evolved. It is found, however, that the migration thresholds favoured by selection in relation to daily vertical return migration are non-adaptive for seasonal vertical return migration. It has to be postulated, therefore, that selection has favoured separate thresholds for daily and seasonal return migrations and that the migration thresholds for the latter migrations over-ride the thresholds for the former.

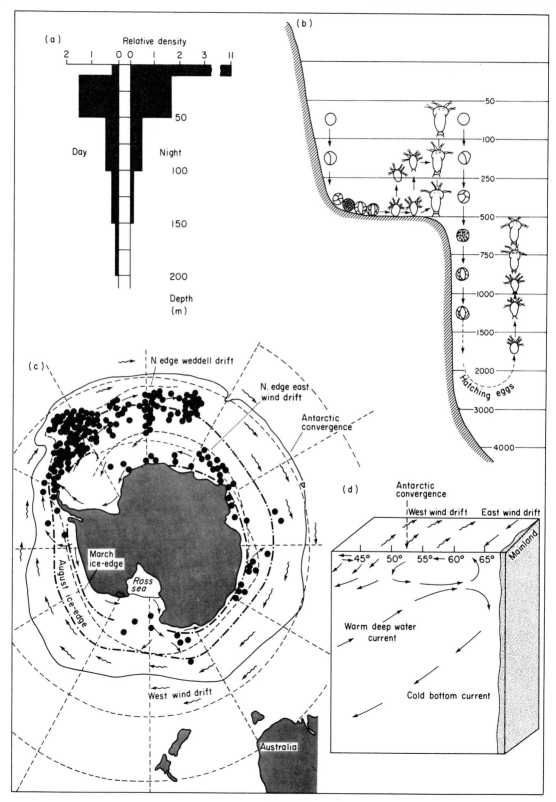

(a)

Relative density

Day

Night

Depth
(m)

(b)

Hatching eggs

(c)

N. edge weddell drift

N. edge east
wind drift

Antarctic
convergence

March
ice-edge

August ice-edge

Ross
sea

West wind drift

Australia

(d)

Antarctic
convergence

West wind drift   East wind drift

Mainland

Warm deep water
current

Cold bottom current

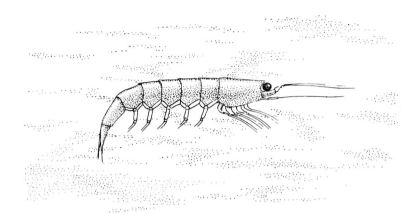

Fig. 20.9   Daily and ontogenetic return migration of the Antarctic krill, *Euphausia superba*

(a) Day and night vertical distribution of krill greater than 20 mm in length
(b) Ontogenetic return migration during early development
(c) Geographical distribution and some major features of the Antarctic environment
(d) Major surface and deep-water currents of the Southern Ocean

General ecology
The Antarctic krill, *Euphausia superba*, is confined almost exclusively to water south of the Antarctic convergence (c). As an adult (above), it lives predominantly in surface waters (a) and for most of its life in waters the temperature of which rarely rises above 0°C. The animal has been known to grow to a length of 65 mm, becoming full grown in its third year. The species lives in cohesive swarms that on the surface appear as patches varying from a few square metres to about 0·2 ha. The patches themselves tend to be clumped and in the Weddell drift may occupy areas of as much as 400 km². In such areas there are distances of about 0·5 km between patches. The patches are conspicuous by virtue of the pale yellow through ochre and brown to brick or blood red colouration. The individuals within the swarms are in a state of continuous movement with the result that the swarms are in a continuous state of flux. Each swarm consists of individuals of approximately similar age that possibly have remained together since soon after hatching from the egg. *E. superba* is a voracious herbivore, possibly feeding selectively on small spineless diatoms. Spawning is a relatively shallow-water phenomenon associated principally with the shelf or slope waters of the high latitudes of the continental land. A major spawning, lasting from November to March, seems to be associated with the far western reaches of the Weddell drift and another lasting from late January to March with the slope waters of continental Antarctica extending from the mouth of the Ross Sea westwards. The species is the staple food of a wide variety of vertebrates, particularly the crabeater seal, *Lobodon carcinophagus*, various penguins,

and of course the majority of southern hemisphere baleen whales (p. 746). Whales feed principally on adolescent and adult krill over 20 mm in length, though in spring and autumn earlier stages are taken ranging from 11 to 20 mm long. Seasonal incidence is such that the overwhelming mass of larvae are protected from surface predators by the ice-cover of the polar winter (c).

Vertical return migration
Diagram (b) shows the development of sinking eggs in shelf and oceanic water and demonstrates how hatching in the shallower conditions gives rise to nauplii and metanauplii unusually close to the surface. Depths are indicated in metres, though not on a linear scale. The nauplii and metanauplii are partly diagrammatic.

Available evidence suggests that eggs are produced in the surface layers and then sink, developing as they go. Hatching as a successful event is essentially a phenomenon of the open sea and takes place at great depths, possibly far below 2000 m if the water is deep enough, perhaps having been carried there by the sinking water that forms the Antarctic bottom current (d). From this northward- and eastward-spreading current the new-born krill climb towards the surface passing through their early developmental phases (second nauplius, metanauplius, and first calyptopis) as they do so. As they climb toward the surface the young krill spend some time, perhaps a month, in the warm deep current which in most parts of the Antarctic has a southerly component and which therefore plays an important part in maintaining the population within the normal limits of its geographical range. Later larval stages (notably the later furcilias) tend to be massed more or less permanently in the surface (50–0 m) layer. The distance component of such daily vertical movement as may occur within this layer decreases with age from the calyptopes stages onwards, only to re-appear once more in the older swarms of subadults and adults. Even so, the movement does not seem to extend to depths greatly in excess of 40–50 m (a), and furthermore involves only part of the surface population. There is no mass seasonal descent of either younger or older stages.

[*Compiled with modifications from Marr (1962). Diagram (d) from Slijper (1962)*]

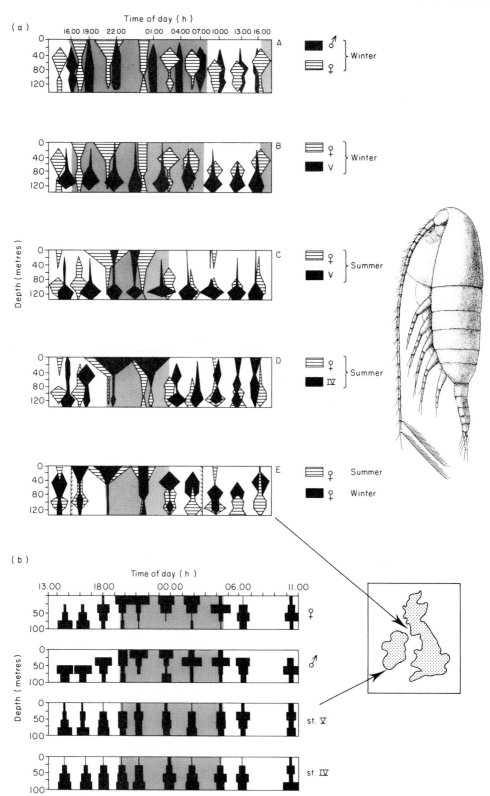

Fig. 20.10 Sexual, ontogenetic, seasonal, and geographical variation in the components of daily vertical return migration in the calanid copepod, *Calanus finmarchicus*

General ecology
Calanus finmarchicus is perhaps the best known of all marine copepods. It is found in vast numbers in the Arctic Ocean and in both the Pacific and Atlantic sectors of temperate seas of the northern hemisphere. It also occurs, though in lesser abundance, in various parts of tropical and subtropical waters. The species feeds only on phytoplankton, chiefly diatoms, but also peridinians and micro-algae. Food may be actively seized as well as filtered. Daily food requirements are about twice as great in summer as winter. As with all calanoid copepods, the species passes through six naupliar stages and six copepodite stages of which the last, Stage VI, is the adult. In Arctic waters *C. finmarchicus* has only one brood per year but individuals may live for up to two years. In temperate waters, however, there may be several generations in a year, and the time span from birth to death may be as short as two months in spring-born individuals, though up to eight months for middle to late summer-born individuals. In Arctic and temperate waters spawning begins with the spring outburst of phytoplankton (Fig. 20.4). In temperate waters, eggs are produced in early spring which give rise to individuals that spawn in late spring. These eggs in turn produce individuals that spawn in midsummer. These eggs, however, produce individuals that do not develop beyond Stage V in that year. These individuals overwinter, moulting to adults from December onwards at a time when deme density is reaching its lowest point. Successive generations of adults in temperate waters are of different body size, length being greatest in late spring and least in autumn. In polar and sub-polar waters and in places such as the fjords of Norway the species may perform a seasonal return migration, migrating into deeper layers (200–300 m) to overwinter, and swimming to the surface layers again the following spring.

Daily vertical return migration
Individuals of *C. finmarchicus* are heavier than water and constant upward swimming is necessary to maintain their level in the water. *Calanus* can swim upwards at a rate of some 15 m/h and downwards at over 100 m/h. The basic pattern of daily migration consists of a rise to the surface at night and a drop to deeper waters by day. There is some indication of dispersal at midnight followed by convergence and perhaps a rise before dawn.

In the diagrams the period of darkness is indicated by stippling except in (a)E, in which only summer darkness is indicated by stippling. The times of sunset and sunrise in winter are indicated by vertical lines with a hatched edge.

(a) Sexual, ontogenetic and seasonal variation in the Clyde area

The winter data refer to the night of 25/26 January. The summer data refer to the night of 11/12 July.

(b) Sexual and ontogenetic variation in summer in an area south of Ireland
The data refer to the night of 11/12 August

Sexual variation: In the Clyde area in winter males do not converge at the surface as do the females and in general are more dispersed throughout the day (A). Off Ireland in August, however, there is much less sexual difference though the males have perhaps a slightly smaller distance component to their migration and perhaps have a shorter stay time at the surface. The difference in sexual variation between (a) and (b) could be either seasonal or geographical but is consistent with the pattern generated by the model (p. 495).

Ontogenetic variation: In the Clyde area in winter the moult from Stage V to adult coincides with a marked change in the pattern of migration. Stage V individuals are distributed lower in the water than adults and, unlike the latter, the distribution changes only slightly with time of day (B). In summer, Stage V individuals are still distributed lower in the water than adults at night but during the day the two stages are found at more or less the same depth (C). The distribution of Stage V individuals also changes with time of day in a way similar to, but less pronounced than, that of adults. Stage IV individuals in July also show a daily change in vertical distribution similar to that of adults but in general are found rather higher in the water during the day (D). Off Ireland in August, however, daily change in distribution of Stage IV was slight and like Stage V individuals at the same place and time were distributed lower in the water than the adults. The differences between (a) D and (b) could be either geographical or indicate a change in behaviour between July and August. Excluding the seasonal migration of Stage V in autumn and spring the ontogenetic variation in daily return migration is consistent with the pattern generated by the model if the variables are entered of ontogenetic change in body size and lessened advantage for smaller stages to partition food collection and food processing into long time periods.

Seasonal variation: In the Clyde area in winter the distribution of adult females follows the conventional daily sequence described above. In summer, however, not only adults (E) but also Stage V(C) and even more so Stage IV(D) tend to return to the surface at some time during the day. This behaviour is also indicated for all groups except males in the Irish data and can perhaps be attributed to the limited time available for feeding during the short summer night (p. 493).

[*General ecology from Raymont (1963). Figure (a) compiled from Marshall and Orr (1955) and figure (b) modified from Farran (1947)*]

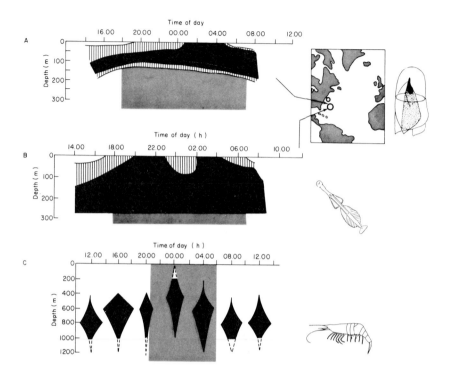

Fig. 20.11   Daily vertical return migrations of three species of marine zooplankton

The stippled area indicates the hours of darkness. In A and B the main concentration of animals is indicated by solid shading and lesser concentrations by hatching.

A. The siphonophore, *Eudoxoides spiralis*, off Bermuda
This species reaches the surface waters after midnight.

B. The chaetognath, *Sagitta bipunctata*, off Bermuda
Note the midnight descent of a large proportion of the deme studied.

C. The bathypelagic natant decapod crustacean, *Gennadas elegans*, off New Jersey

[A and B modified from Moore (1949). C from Waterman et al. (1939)]

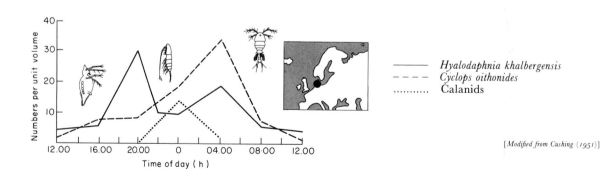

Fig. 20.12   Differences in times of appearance in surface waters

The figure shows the numbers of Entomostraca in surface tow nets in 24 h in the Grosser Plöner See.

Different species may appear at the surface at different times in the diel.

[Modified from Cushing (1951)]

20.3 Animals of soil, sand, and leaf litter

The habitat-types of soil, sand, and leaf litter are characterised by pronounced vertical gradients of environmental variables such as food abundance, predation risk, temperature, humidity, etc. Furthermore, these gradients are not constant but vary in steepness and sometimes sign with time of day and year.

Sampling problems in these habitat-types are acute and there are few data available on the movements of the animals that are adapted to them. The work of Gabbutt (1970) and Wood (1971) on pseudoscorpions, of Gerard (1967) and others on earthworms, and of Renaud-Debyser (1963) on the meiofauna of sandy beaches represent fine pioneer investigations but there are still a great many major gaps in our understanding of the vertical migrations of these and other animals that live in soil, sand and leaf litter. Perhaps the major gap in the available data is the absence of investigations relating to daily vertical return migration. Without this information it is extremely difficult to interpret seasonal variation in vertical distribution with any confidence.

Although it is known that the pseudoscorpions and earthworms discussed here live longer than a year, data on the longevity of individuals in the meiofauna of sandy shores are limited and it is possible that in this latter case we are dealing with re-migrations rather than return migrations. However, as the meiofauna of sand seem almost certain to show circadian and/or circatidal periodicities to their vertical return migrations, it is most probable that any seasonal effects noted are seasonal variations in one or more components of the daily or tidal return migrations.

In this brief consideration of the migrations of the inhabitants of soil, sand, and leaf litter, the definite seasonal return migrations are considered first followed by a consideration of seasonal and ontogenetic variation in possible short-period (daily or tidal) return migrations.

20.3.1 Seasonal return migration

Seasonal migrations of earthworms are closely associated with periods of inactivity (Fig. 20.13). 'Inactive' worms are not, in this context, simply worms that are not moving but are worms in a distinct physiological stage. Thus, according to Gerard (1967), inactive worms can be recognised during hand-sorting in the field by their empty gut and by their behaviour of coiling themselves within roughly spherical cells in the soil. Inactivity is of three types (Edwards and Lofty 1972): quiescence (in which the worm responds directly to adverse conditions and becomes active again as soon as conditions become favourable); facultative diapause (also a response to adverse conditions but which terminates after a certain time interval); and obligatory diapause (which occurs at a certain time or times of year independently of current environmental conditions).

All three types of inactivity are associated with a movement into deeper soil. Both inactivity and downward migration seem to be a response to the condition of the surface soil, particularly when it is dry or its temperature is low (near freezing) or high. At such times, temperature and soil-moisture depth profiles are such that habitat suitability increases with depth. Perhaps one of the critical ultimate factors is migration cost due to increased energy expenditure. Burrowing becomes difficult both when the soil is frozen and when the soil is dry. Both factors, therefore, would favour reduction in migration incidence, perhaps to zero (i.e. inactivity) when conditions became sufficiently adverse. The depth at which inactivity would be optimum would be in part a function of migration cost and in part a function of the energy involved in: reducing and then recovering from water loss (when the soil is dry); metabolism without energy intake (when the soil is hot); and production of anti-freeze (when the soil is cold). The arguments presented for zooplankton (p. 498) and inter-tidal organisms (p. 830) are probably also relevant here. The optimum depth for earthworms in response to each of these variables is likely to be deeper than the near-surface feeding depth. The nett result of all these pressures is that as soon as surface conditions become adverse (above-threshold), optimum strategy involves downward migration followed by a period of inactivity. In most cases the optimum threshold variable for downward migration is likely to be either a certain value for soil moisture or temperature at the feeding level. The optimum threshold variable for upward migration is likely to vary according to the depth at which inactivity occurs. Those nearer the surface probably perceive variables (soil moisture and temperature) that correlate well with events at the feeding level, and upward migration thresholds to these variables

Fig. 20.13 Seasonal variation in the vertical return migration of six species of lumbricid earthworms in Britain

General ecology

Five of the species considered here belong to the genus *Allolobophora* and one to the genus *Lumbricus*. The genus *Allolobophora* is found from Japan in the east, through northern India, Pakistan, and the Middle East, to North Africa, the Canary Islands, and Europe. It has also been introduced into North and South America, southwest Africa, New Zealand, and Australia. The genus *Lumbricus* has a more northerly distribution being found in Siberia, Europe, Iceland and North America as well as having been introduced throughout the world. The six species listed in the diagram are arranged more or less in the order of the level they occupy in the soil. Thus *Allolobophora chlorotica* in general is found nearer to the surface than the other species whereas *Lumbricus terrestris* during the day is found deeper than the other species. This order also approximates to the order of adult sizes of the species as indicated.

The three largest and deepest-living species, *L. terrestris*, *A. longa*, and *A. nocturna* form permanent burrow systems with smooth walls cemented together with mucus secretions and ejected soil. Burrows of *L. terrestris* and *A. nocturna* go down to a depth of 150–240 cm, being vertical for most of their depth but branching extensively near the surface. The species that live nearer the surface form extensive but temporary burrow networks. The burrows are formed by pushing through existing crevices and by the earthworms literally eating their way through the soil. The large, deep-burrowing species produce casts on the soil surface near the exits to their burrows, particularly in heavy compact soil where there are no spaces or crevices below the soil surface into which the faeces may be passed. Both dead and living organic matter present in the soil is digested as it passes through the earthworm's gut. *L. terrestris* can also feed on leaf and other plant material directly. This is obtained from the soil surface but is first pulled down into the mouth of the burrow to a depth of 2·5–7·5 cm, so forming a plug which may also help to prevent the entry of water and cold air into the burrow. Earthworms are hermaphrodite but not self-fertilising. *Allolobophora* spp. and *Lumbricus* spp. can only produce fertile eggs if they mate with another individual. *L. terrestris* mates on the surface but other species mate below ground. Cocoons containing the eggs, nutritive fluid, and spermatozoa are produced at intervals after copulation until all the stored seminal fluid has been used up. The horizontal hatched kites in the diagram for *A. chlorotica* give some idea of the seasonal incidence of cocoon production and indicates peaks in spring and autumn. The cocoons of *A. chlorotica* and *A. caliginosa* are produced at depths of 0–25 mm and 25–50 mm respectively. The period from hatching to maturity for *A. chlorotica* and *A. caliginosa* that hatch in July is 5–6 months. In captivity, *L. terrestris* has lived for 6 years and an *Allolobophora* sp. for 10.

The horizontal open bars indicate periods of inactivity, broken bars indicating that only part of the deme, usually those below the moist top 75 mm in summer, was inactive, but continuous bars indicating that all of the deme at all depths was inactive. Note that *L. terrestris* showed no extended periods of inactivity. For most species the inactivity is a response to adverse current conditions. It has been reported for *A. longa* and *A. nocturna*, however, that the inactivity is part of an obligatory diapause which may begin as early as May and not terminate until September or October. There is also a marked circadian rhythm to the activity pattern, *L. terrestris* being most active between 18.00 and 24.00 hours. Copulation and surface foraging in this species is also mostly confined to these hours of the diel.

Vertical return migration

The diagram shows the daytime vertical distribution (expressed as percentage in each layer for each species in each sample) of six species of earthworms under pasture in monthly soil samples from January to December, 1959, at Rothamsted, England. Samples were taken in 75 mm layers to a depth of 30 cm and sometimes deeper. The mean weekly temperature at 10 cm under grass and weekly rainfall from a nearby (1 km away) meteorological station are also indicated. The data refer to two categories of vertical return migration: seasonal and daily.

Seasonal: At any time during the year, *Allolobophora* spp. respond to very cold or very dry surface soils by migrating into deeper soil and becoming inactive. In most species the migration is a facultative response to habitat variables but in *A. longa* and *A. nocturna* the movement may be obligatory and a response to indirect variables.

Daily: In the case of *L. terrestris* extended periods of inactivity did not occur and the movements shown in the diagram should be interpreted more as seasonal variation in the distance component of daily return migration rather than as a seasonal return migration. The same probably applies to the *Allolobophora* species during periods that individuals were active. On this basis the results indicate that the distance component of daily return migration is long in January and February when surface temperatures are near freezing. Distance decreases most rapidly in March in smaller species and in the small and medium-sized individuals of *A. longa*, *A. nocturna*, and *L. terrestris*, the larger worms of these three species retaining a one-way distance component greater than 10 cm. By April, all sizes of all species have a one-way distance component less than 10 cm. From June to October when in the year of study the rainfall was mainly light and sporadic the distance component of those worms that remained active increased again, particularly in September when the soil was extremely dry (water content at 5 cm = 7·5 per cent). During September all worms except *L. terrestris* were inactive and, where indicated by arrows, were below the depths sampled (0–45 cm). A large proportion of the worms found in the top 7·5 cm of soil from June to October consisted of newly hatched individuals. In November and December the distance component for the migration again decreased for all worms.

Both the seasonal return migration and the seasonal and ontogenetic variation in the distance component of the daily return migration of the various species can be generated by entering the following factors into the return migration model (where these factors are gradients, habitat suitability due to the factor concerned is indicated by a + sign when suitability is greatest at the surface and a − sign when it is least): (1) a + gradient of food abundance; (2) a − gradient of predation risk that is

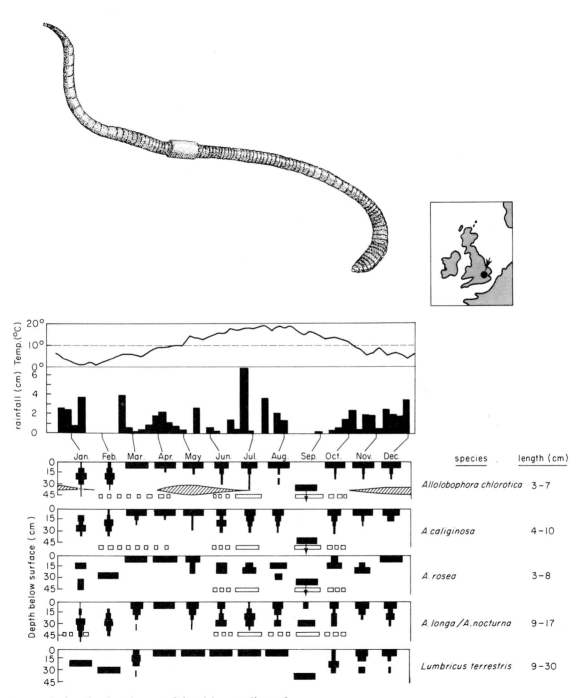

species	length (cm)
Allolobophora chlorotica	3–7
A. caliginosa	4–10
A. rosea	3–8
A. longa / A. nocturna	9–17
Lumbricus terrestris	9–30

steeper during the day than at night; (3) a gradient of temperature that is − in midsummer and midwinter but probably non-existent at other times; (4) a − gradient of soil moisture that is steepest in summer; and (5) migration cost due to energy expenditure as a direct function of body size.

[*General ecology from Savory (1955), Gerard (1967) and Edwards and Lofty (1972). Vertical distribution diagram modified from Gerard (1967)*]

probably evolve. Species that become inactive at deeper levels, however, probably receive little or no information concerning events at the surface. In such a situation, selection is likely to favour a diapause-like phenomenon with inactivity, once initiated, not ceasing until after the passage of a certain time-interval (where stay-time evolves as a function of the frequency distribution of periods of adverse conditions at the feeding layer experienced by the ancestors during the evolution of the behaviour). Finally, the obligatory downward migration in summer that seems to occur in *Allolobophora longa* and *A. nocturna* (Fig. 20.13) should, according to the arguments concerning obligatory migration (Chapter 16), evolve when the environmental events to which the migration is an adaptation are predictable. It is perhaps significant that of the deep-living species found in southern England it should happen to be those species, members of a southerly genus, that show obligatory downward migration in summer rather than the north temperate *Lumbricus terrestris* (Fig. 20.13).

Seasonal variation in vertical distribution of earthworms has been reported in other parts of the world. Thus members of the genus *Diplocardia* in the southeast United States moved from a depth of about 10 cm in early autumn (October) to about 40 cm in mid-winter (January) followed by a return to the surface in spring. There seems to be a temperature threshold for downward migration by this species of about 6°C (Dowdy 1944). In the northeast United States the peregrine megascolecid earthworm, *Pheretima hupeiensis*, originally indigenous to the temperate zones of northern China and Korea, was active near the surface (15–20 cm depth) in early spring (March) and early autumn (September). From late autumn through winter (November through February) the species was deeper than 55 cm (Grant 1956). In tropical Africa, Madge (1969) observed that two species of eudrilid worms moved down into deeper soil layers during the dry season.

Where information is available concerning other animals that live in soil beneath herbage, similar migrations seem to be performed to those of earthworms. As most of the animals concerned are insects or their larvae, however, that are feeding on the roots of plants, temperature seems to be the critical gradient rather than soil moisture. Thus young 'white grubs', subterranean larvae of scarab beetles, such as *Phyllophaga rugosa*, that are found throughout North America and which feed on the roots and other underground parts of plants, migrate downwards in the soil in early autumn. Winter is passed below the frost line and grubs have been taken 150 cm below the surface (Metcalf and Flint 1962). As the soil warms up in spring they migrate upward and are feeding a few centimetres beneath the surface by the time plant growth has started. Two winters are passed during larval development. In the spring following their second hibernation they pupate 15 cm or so beneath the surface and emerge as adults the following autumn. The adult beetles remain in the soil to hibernate and again migrate downward in early autumn. The distance is less than for a young larva, however, a characteristic that could be generated by the model by assuming greater migration cost for the adult due to the difficulty experienced by a large beetle in burrowing through soil by comparison with a small larva. The adults migrate to below the line of severe freezing rather than below the frost line. Downward migration during periods that the soil is hot and dry has also been described for subterranean insect larvae, such as the 'wireworm' larvae or elaterid or click beetles (e.g. the North American *Melanotus cribulosus*).

A special problem arises in relation to the interpretation of the downward winter migration of the pseudoscorpions illustrated in Figs. 20.14 and 20.15. The extraction technique most appropriate for these animals is that of the Tullgren funnel (Gabbutt 1970). Figure 20.14 shows that in winter a pseudoscorpion deme consists partly of animals that are extractable by the Tullgren funnel and partly of animals that are not. The problem of interpretation arises because it is not known whether these two groups represent, respectively, free-living and chambered, hibernating, individuals or whether all individuals are hibernating in chambers but that only some are extracted by the funnel. There is likely to be a similar reduction in efficiency of extraction of individuals in moulting chambers and females in breeding chambers (Gabbutt 1970). There is no way of knowing at present, unlike the situation for earthworms, to what extent the vertical distribution of free-living and chambered individuals may differ. Free-living individuals in winter could continue to perform daily migration whereas individuals in hibernating chambers presumably would not.

The downward movement in August and the absence of attainment of the most-surface distribution until June/July (Fig. 20.14) suggests that unlike earthworms the downward migration is not directly associated with entry into hibernation chambers, though it is impossible here to distinguish between the seasonal migration and seasonal variation in the components of any daily migration that may occur. Rather it seems that in autumn individuals either gradually migrate downwards or gradually lengthen the distance of their daily return migration or gradually abandon a daytime rise to the litter layer and then, when low temperatures arrive in October or November, construct a hibernation chamber at their normal daytime depth. In spring, when individuals leave their hibernation chamber, the reverse seems to occur. Individuals either come to live nearer and nearer the surface (a seasonal migration) or the distance component of daily return migration becomes shorter, or more and more individuals perform a midday upward migration.

A similar migration seems to be performed by the millipede, *Cylindroiulus punctatus*, in Britain (Blower and Gabbutt 1964, Banerjee 1967). Studies in oak, *Quercus robur* L., woodlands in southern England indicate that mating and oviposition occurs during spring and summer under the bark of oak logs. The first three developmental instars also are found in this position but the remainder of the immature period (fourth to seventh instars) is passed in the litter. Adults are found most commonly in the soil during the winter months but in spring they are found increasingly in the humus and then the litter layers and eventually under the bark of the logs themselves. In early autumn (September and October) at the end of the breeding season the reverse takes place. Once again, however, these observations are based solely on day-time samples. Consequently it is not known whether the migrations are strictly seasonal/ontogenetic or whether the effects observed represent variation in the components of a shorter-period return migration such as that described in the next section for the earthworm, *Lumbricus terrestris*. For example, adult *Cylindroiulus punctatus* are found in all layers (soil, humus, leaf litter, and logs) at all times of year. It is only the proportion in each layer that changes. This effect could result from one or any combination of the following: a partial seasonal return migration; a wide scatter of timing of upward and downward seasonal migration; seasonal variation in stay-time in each layer during a short-period vertical return migration; seasonal variation in the distance and timing components of a daily vertical return migration.

20.3.2 Short-period return migration

There seems little doubt that the meiofauna of sandy shores (Fig. 20.16) and at least the deeper-living earthworms (Fig. 20.13) perform short-period (daily and/or tidal as appropriate) return migrations. Many of the seasonal effects noted for these animals, therefore, are likely to represent variation in the components, particularly distance, migrant/non-migrant ratio, and periodicity, of daily return migration. This type of seasonal variation has already been considered for zooplankton and need not be considered here in detail.

A simple winter/summer comparison such as is available for the meiofauna (Fig. 20.16) allows little discussion of the variables that upon entry into the return migration model generate the most accurate fit to observed events. Renaud-Debyser (1963) considered that temperature was involved though whether the migration follows change of sign of temperature gradients or whether other factors are involved such as gradients of food abundance cannot be considered without monthly measurements. If temperature is a critical variable then, unlike plankton, the advantage lies with higher, not lower, temperatures.

Monthly measurements are available both for earthworms (Fig. 20.13) and pseudoscorpions (Fig. 20.14). Where there is no question of confusion of seasonal migration and seasonal variation in daily migration, as for *Lumbricus terrestris* which in the study area remained active throughout the year and which is known to be nocturnal, the distance component of the daily migration seems to be a function of the same variables as seasonal migration. Thus when the surface layers are cold or hot and dry, the distance component is greater than when the surface layers are of moderate temperature and moist. Similar effects can be distinguished for the *Allolobophora* spp. shown in Fig. 20.13 at times when the seasonal downward migration thresholds have not been exceeded. It is evident from the earthworm data that change of sign of the temperature gradient is not involved as it may be for some littoral and sub-littoral animals (p. 836). The observed pattern can be generated by assuming some pressure against increase in the distance component. Migration cost could be the critical factor especially as there is evidence that migration distance increases with

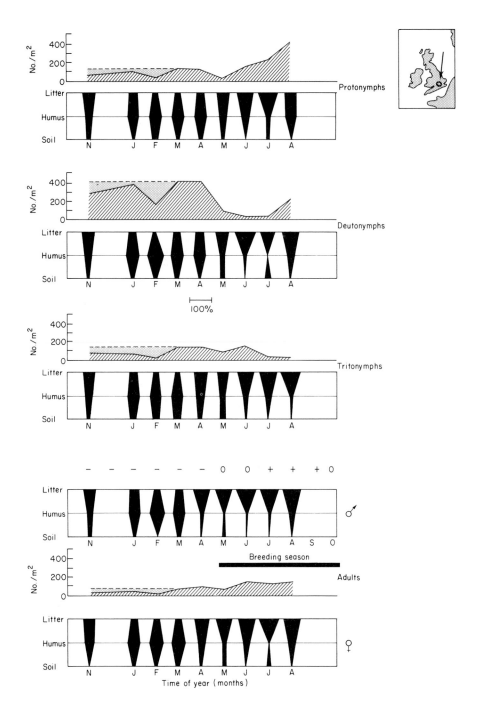

Fig. 20.14   Seasonal variation in the vertical distribution
of the pseudoscorpion, *Chthonius ischnocheles*

General ecology

Pseudoscorpions pass through three nymphal stages (pro-
tonymph, deutonymph, and tritonymph) between hatching from the egg and development to the adult stage. In
Oxfordshire, England, inseminated females of *Chthonius
ischnocheles* construct silken breeding chambers from May
to July. The female oviposits while in this chamber and
remains there until the young protonymphs have hatched.
Most protonymphs emerge from these breeding chambers
before winter but some do not emerge until spring

(March–April). Earlier emerging protonymphs may develop to the tritonymph stage before winter but the majority only develop to the deutonymph stage. Each moult takes place in a silken chamber constructed by the nymph. Protonymphs, deutonymphs, and tritonymphs overwinter in the ratio 1: 2: 1. Some, perhaps all, individuals spend part of the winter in a silken hibernation chamber. Individuals in silken chambers are extracted less efficiently from soil and litter samples than free-living individuals. In the diagrams of seasonal variation in deme density (number/m²) the stippled areas represent the difference between the total of extracted individuals (hatched area) and the total of individuals present but that are not extracted due to the effect of hibernation chambers (the difference is based on the assumption of limited overwinter development and mortality and is obtained by levelling off the March density). Overwintered nymphs reach maturity some time between March and September. Adults seem likely to live for at least two years. *Chthonius ischnocheles* is predatory, the main prey species apparently being collembolans, mites, and dipterous larvae.

Seasonal variation in vertical distribution
The kite diagrams show the vertical distribution at about midday of all stages of *C. ischnocheles* during nine months of the year at a site in Oxfordshire, England, in which the leaf litter was predominantly that of beech, *Fagus silvatica* L. Vertical distribution is expressed as percentage in each of three layers (leaf litter, humus, and soil) of the total individuals of each stage per unit area at the surface. Note that the vertical axis, which covers several centimetres, is not drawn to a linear scale and represents simply the three layers as indicated. Deme density for each stage (all adults are summated) is also indicated. The sign of the vertical gradient of temperature for each month is indicated above the diagram for adult males. A + sign indicates that the temperature decreases with depth, a − sign indicates that temperature increases with depth, and a ○ sign indicates

that there is less than 1°C difference between the litter layer and the soil. Temperature gradients refer to midday.

The seasonal variation in vertical distribution may indicate: (1) a seasonal return migration; (2) seasonal variation in the distance and/or timing components of a daily return migration. Where the migration is seasonal the data indicate that the optimum level for the species is nearer the surface in summer than in winter. Where, and if, the seasonal effect is the result of variation in the distance component of a daily return migration then, assuming the daily migration would involve an upward migration at night, the influence of season is for the distance component of daily migration to be greater in winter than in summer. Both the migrations could be generated by entering into the return migration model a + (i.e. habitat suitability greatest in the leaf litter) gradient of habitat suitability due to food abundance and a − gradient of habitat suitability due to temperature during the period November to March. This latter gradient is consistent with the period that the temperature gradient is − as expressed in the diagram. Daily return migration could be generated by entry of a diel periodicity of sign change in gradients of predation risk and/or temperature and/or humidity. If the seasonal effect is due to the occurrence of a midday as well as nocturnal upward migration in summer but not in winter (as in plankton—p. 492), day-length not temperature should be the critical factor. In fact, the peaks of occurrence at the surface seem to be distributed about June/July (i.e. shortest day-length) rather than August (highest temperature), a fact that tends to support the latter possibility. The situation cannot be resolved, however, until data are available concerning the absence, or nature, of daily return migration in this species.

[General ecology from Gabbutt (1969) and Wood (1971). Kite diagrams and temperature gradients drawn from data in Wood (1971) and Wood and Gabbutt (1978)]

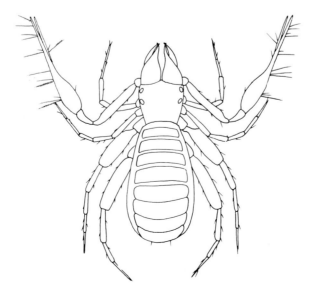

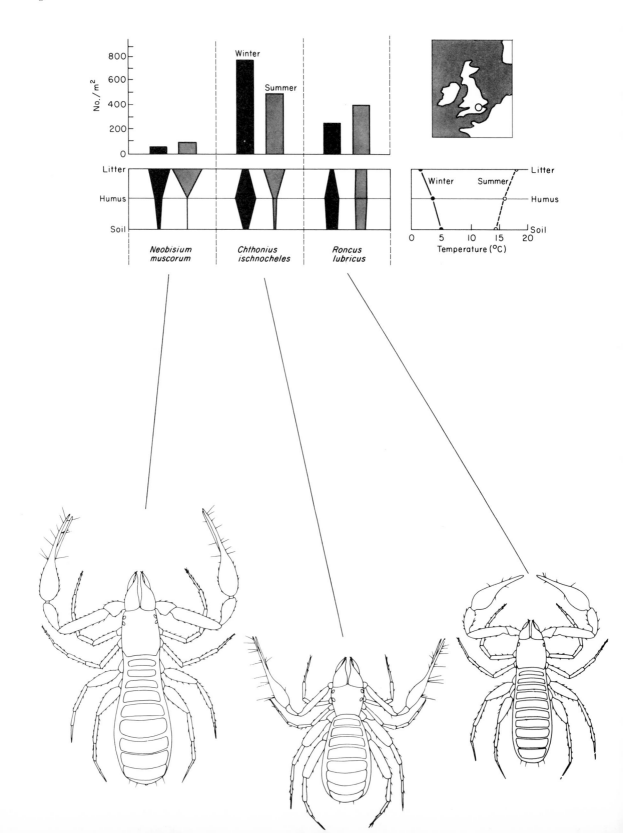

Fig. 20.15   Seasonal effects in the vertical distribution of three species of pseudoscorpions in leaf litter, humus and soil

General ecology
Of the three species considered, two are members of the family Neobisiidae (*Neobisium muscorum* and *Roncus lubricus*) and one is a member of the family Chthoniidae. Although all three species may be found in leaf litter, humus, and soil, *N. muscorum* when adult is primarily an inhabitant of the leaf litter layer and *R. lubricus*, at least when young, is primarily an inhabitant of the humus and soil. Adult size seems to be associated with the size of the spaces through which the animal moves, the largest species, *N. muscorum*, being the leaf-litter inhabitant. The life-history and feeding ecology of *Chthonius ischnocheles* has been described in the legend to Fig. 20.14. Female *N. muscorum* construct breeding chambers in April and emerge with protonymphs in June. These overwinter as deutonymphs, become adult the following September, and breed for the first time the following year after hibernation. Female *R. lubricus* construct breeding chambers in May/June and emerge with protonymphs in August. These either immediately give rise to deutonymphs or do not do so until the following August. After overwintering, these deutonymphs moult into tritonymphs the following May and these achieve adulthood in September the same year. Breeding commences the following spring after hibernation. Other aspects of the biology of these species, however, are similar to that of *C. ischnocheles*. Nymphs and adults of *R. lubricus* enlarge crevices or depressions in the soil and sit in them with their chelicerae and pedipalps oriented toward the opening.

Vertical return migration
The figure shows the vertical distribution at about midday in winter (February) and summer (July) at a site in Oxfordshire, England, at which the leaf litter was predominantly that of beech, *Fagus silvatica* L. The winter temperature in the litter, humus and soil was taken in late December and the summer temperature was taken in July. Both sets of readings were taken in ground covered by herbage. The vertical distribution is expressed as a percentage in each layer of the total individuals collected of each species. The deme density expressed as number of individuals/m² is indicated in the top half of the diagram. The animals were extracted by Tullgren funnels. Note that the vertical axis has not been drawn to a linear scale but refers simply to the three layers of the depth profile as indicated.
   All three species are found nearer to the surface at midday in summer than in winter. However, the vertical zonation of the three species relative to one another remains more or less constant with *N. muscorum* being found nearest the surface and *R. lubricus* being found furthest from the surface with *C. ischnocheles* being intermediate between the two. As described in the legend to Fig. 20.14, interpretation depends on the extent to which the seasonal effect observed is that of a seasonal migration or of seasonal variation in the distance, timing, and periodicity components of a daily return migration. Similarly, the between-species variation seems likely to be due to differences in the zonation of the three species but it

could be due to differences in the timing of daily vertical migration, different species arriving at the surface at different times of day (see Fig. 20.12).

General ecology from Gabbutt (1970) and Wood (1971). Migration and temperature diagrams drawn from data in Wood (1971) and Wood and Gabbutt (1978)]

body size. Thus the primary gradients (Fig. 20.13) of food availability and predation risk combined with migration cost generate a pattern of upward migration at night and limited downward migration by day with perhaps a food collection effect as postulated for zooplankton (p. 488). Only when an additional gradient is imposed, such as the energy involved in anti-freeze production in winter and the toleration of, and recovery from, water loss in summer, does the distance component increase. As soon as these additional pressures are relaxed, as in March/April, the distance component decreases again even though the sign of the temperature gradient has not changed. The gradients it is necessary to enter into the model in order to generate the observed vertical migration behaviour of earthworms are summarized in the legend to Fig. 20.13.
   The seasonal variation in pseudoscorpion vertical distribution also does not correlate with the change in sign of the temperature gradient. Thus the upward shift in the distribution in Fig. 20.14 starts in March/April while the temperature gradient is non-existent or negative and the downward shift in distribution starts in August when the gradient is still positive. Indeed, the maximum surface distribution seems to be centred on June/July which suggests a daylength effect rather than July/August or August as would be favoured by temperature gradients or their correlates. Without knowing the nature of any daily migration that may occur, the pseudoscorpion seasonal variation cannot be analysed further. However, the similarity to the seasonal variation in midday distribution of some zooplankton (e.g. Fig. 20.10) is striking and it could be that some selective pressures are common to both groups. The species differences indicated for the pseudoscorpions in Fig. 20.15 can also, as indicated in the legend, be interpreted in two ways in the absence of information concerning the nature of any daily return migration that may occur.

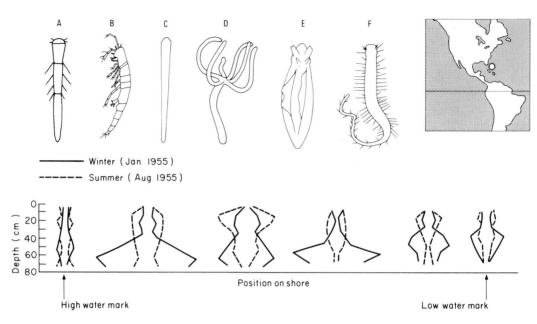

Fig. 20.16. Seasonal variation in the vertical distribution of the meiofauna of a sandy shore

General ecology

The meiofauna, or interstitial fauna, is defined as a group of animals the individuals of which move through the substratum without displacing the particles of which the substratum is composed. The meiofauna, therefore, is an assemblage of small animals that have become adapted to live in the interstices between particles of sand, mud, or soil. An interstitial fauna is found among the sand and mud of the sea, in the bottoms of freshwater lakes and rivers, and in the waterlogged subsoil on land. The animals illustrated here are members of a sandy shore meiofauna. Almost all animal phyla are represented in the meiofauna and some groups are practically confined to it. The animals illustrated are A, a polychaete; B, a harpacticoid copepod; C, a turbellarian; D, a hydrozoan coelenterate; E, an opisthobranch mollusc; and F, a gastrotriche. All of these are small and show the worm-like body shape characteristic of the meiofauna even though they are not all members of groups that normally include vermiform animals. Locomotion is often by ciliary gliding, though 'writhing' is also common. Crawling is less common, and even some animals that have legs, such as the harpacticoid copepods (B), move predominantly by writhing. The members of the meiofauna feed either on detritus, or on diatoms and other epiphytes that grow in a film surrounding each sand grain, or they are predatory on other members. The number of gametes is small, clutches of single eggs being common, and about 98 per cent of the meiofauna lack a pelagic phase in the life history. Parental care is common, particularly viviparity and the habit of incubating the eggs in a brood pouch. Copulation, rather than gamete liberation, is common, as also is hermaphroditism.

Seasonal variation in vertical distribution

The kite diagram shows the vertical distribution of the meiofauna (expressed as number of individuals of all species). Data are presented for six different heights above low water mark in both summer and winter at Archachon, Bahamas. Mean air temperature at sea level in this region varies from about 19°C in January to about 28°C in July. At all levels on the shore the meiofauna is distributed deeper in the sand in winter than in summer. It is not known whether these data refer to a seasonal return migration or to seasonal variation in the components of a daily or tidal vertical return migration, though the latter seems the more likely.

[*Compiled from Swedmark (1964). Kite diagram modified from Renaud-Debyser (1963)*]

21

Aspects of the limited-area range: the familiar-area hypothesis for the return migrations of vertebrates and some invertebrates

The familiar-area phenomenon has received attention in several of the preceding chapters. Chapters 14 and 15 were concerned with the familiar area in relation to the inherent advantage of calculated rather than non-calculated migration and the way that this advantage persists as long as it is not outweighed by the cost of the mechanisms whereby calculated migration is achieved. Chapter 18 was concerned with the way that by means of the social communication (in some species)–exploratory migration–calculated migration process an animal builds up a familiar area during its ontogeny. Consideration was also given to the way that a home range and/or territory may crystallise out from this familiar area and also to the way that the exploratory–calculated migration sequence can lead to type-H (total home range to total home range) removal migration.

The familiar-area hypothesis of return migration states that an animal that builds up a familiar area in the conventional way may be presented with habitats that are more suitable in one part of its familiar area at one time but more suitable in another part of its familiar area at another time. The result would be that the animal would spend part of its time in one part of its familiar area and part of its time in another part of its familiar area. Where the periodicity is seasonal, the animal migrates between seasonal home ranges; where ontogenetic, between ontogenetic home ranges; and so on.

Acceptance that the seasonal and ontogenetic return migrations of vertebrates and some invertebrates take place within a familiar area that has to be established by the conventional process has important repercussions in the interpretation and understanding of these movements. The remainder of this book examines the extent to which seasonal and ontogenetic (and in some cases shorter-period) return migrations of animals can be described by the familiar-area hypothesis. In the final two chapters the applicability of this thesis is reviewed and then, in the light of the conclusions reached, consideration is given to the orientation and navigation mechanisms involved in the execution of return migration.

22

Seasonal return migration within a familiar area by some terrestrial invertebrates

Clear examples of seasonal return migration within a familiar area in terrestrial invertebrates are scarce. There are, perhaps, two major reasons for this. The first is that most of the more conspicuous invertebrates, such as the insects, have evolved a short longevity and in many cases a linear lifetime range (p. 361). Consequently any familiar area that is formed is only temporary (p. 374). Secondly, many authors seem to feel, perhaps justifiably and perhaps not, that invertebrates lack the sensory and neural capacity to establish a familiar area, information concerning which must be stored within a spatial memory long enough for seasonal return migration to be performed. Nevertheless, two groups of terrestrial invertebrates include members that, at least as far as preliminary investigations can determine, appear to perform seasonal return migrations within a familiar area. These groups are gastropod molluscs and arachnids.

Fig. 22.1 The possibility of a seasonal/ontogenetic return migration within a familiar area in the wolf spider, *Lycosa lugubris*, in central Scotland

General ecology
Lycosids are a group of ground-dwelling spiders that are among the largest and most conspicuous members of the terrestrial invertebrate fauna of temperate regions. Adult females of European *Lycosa* are about 6 mm in length. Lycosids have relatively good eyesight, are good sprinters, and do not make use of silk to catch their prey. The female wraps the egg mass in a layer of silk to form an egg sac which is then carried, attached to her spinnerets.

The world distribution of *Lycosa lugubris* is very approximately as shown. The species is found predominantly in and around clearings in deciduous woodland and around woodland edges. In Scotland, the life history extends over about two years as indicated in the spiral diagram. Females normally produce two egg sacs during their life (E in the diagram), the spiderlings from the first sac emerging in July and those from the second emerging about September. Spiderlings overwinter in their third to

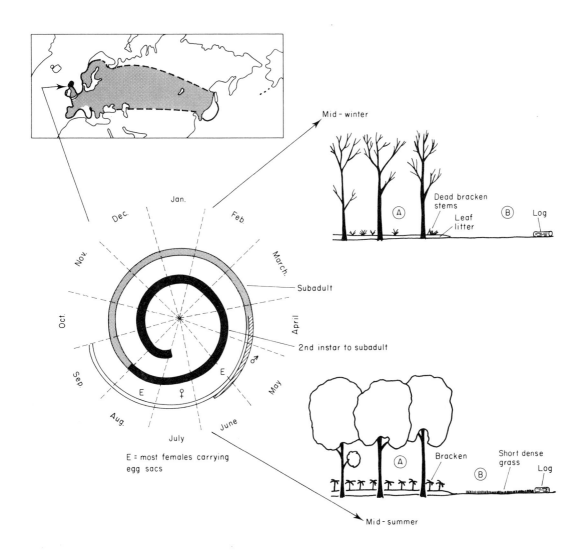

Mid-winter

Dead bracken stems
Leaf litter
Log
(A)   (B)

Jan.
Dec.   Feb
Nov   March.
Oct   April — Subadult
Sep — 2nd instar to subadult
Aug   May
July   June

Subadult

E = most females carrying egg sacs

Bracken
Short dense grass
Log
(A)   (B)

Mid-summer

fifth instar and then the following summer develop to the sub-adult (final non-reproductive) instar in which they over-winter. The following spring they moult to become adults. The first matings occur in April and most males have died by the end of June. Females carry the first egg sac for 5–6 weeks before tearing them open to allow the spiderlings to emerge. These are then carried on the female's back for up to a week before they disperse.

The highest values for density estimates in the study area for adult females, sub-adults, and small instars were 5·8, 10·3, and 22·4 spiders/m^2, respectively.

Seasonal/ontogenetic return migration
Spiderlings leave their mother's back in late summer or autumn while in clearings (B in the diagrams) and from then until winter gradually move into the wood where they overwinter in the leaf-litter. Perhaps a familiar area is

established during this first autumn. The following spring and summer seems to be spent in the wood (position-type A in the diagrams). Consequently, at the onset of their second winter, the young spiders, which are now sub-adults, are already in their overwintering home range. In spring, when the spiders moult into adults, they begin to extend their home range out of the overwintering area. Perhaps this is the stage of familiar-area extension or perhaps the movement takes place within the familiar area established during the first autumn. In late May, when the trees (oak, *Quercus petraea*, in the study area) break their leaves, the adult females accumulate in the clearings to sun their first egg sacs. After their first batch of spiderlings have dispersed, the females migrate back into the wood to feed before returning to the clearing to sun their second egg sacs.

[*Compiled, with modifications, from Edgar (1971)*]

Work on the ability of snails to negotiate T- and Y-mazes indicates some ability for spatial learning (Thorpe 1963), though one is forced to wonder whether such a laboratory-based approach gives an adequate picture of the ability of such animals to establish a long-term familiar area in the field where far more, and more appropriate, sensory cues are continuously available. Certainly, work on snails in the field, which, however, is at a tantalisingly early stage, suggests the possibility of a seasonal return migration within a fairly large familiar area. Edelstam and Palmer (1950) marked individuals of the European Roman or edible snail, *Helix pomatia*, in winter. During spring and summer it was found that the proportion of marked snails in the study garden gradually decreased but that toward winter the proportion increased once more. The most likely explanation seemed to be that marked individuals left the garden in spring and summer, their place being taken by unmarked snails, and that in autumn the snails gradually returned to their winter home range of the previous year. Displacement experiments, which unfortunately were not adequately controlled, suggested a high homing ability to the winter home range at displacement distances of up to 40 m, though the ability to orientate in the direction of the winter home range did not seem to disappear completely until somewhere between displacement distances of 150 and 2000 m. Furthermore, orientation to the winter range seemed to be most marked from late summer onwards. The evidence is strongly indicative, therefore, of a seasonal return migration between a winter home range and summer home range with a one-way distance component that for some individuals is perhaps as great as 40 m. These findings have subsequently in most respects been confirmed by Pollard (1975).

A similar ability to orientate in autumn to the hibernation site of the previous winter is indicated for the American snail, *Allogona profunda* (Blinn 1963). This species hibernates for six months of the year. In the spring, after hibernation, there seemed in the study area to be a directed movement to an area particularly rich in fragmented leaf mould. In this species, therefore, the annual migration circuit may consist of several seasonal home ranges.

In neither of these species has it been proved beyond doubt that the migration is performed within a familiar area. The critical test, if properly controlled, would be whether an individual, transferred to a part of its familiar area inappropriate for that time of year, returned to the appropriate part of its migration circuit. As described in Chapter 33, however, care has to be taken in this type of experiment not to exceed the threshold for type-C removal migration (p. 871). It is perhaps significant in this connection that if *Helix pomatia* is given 'adverse' conditions during the displacement experiment, such as being enclosed in a bag for several days, it no longer orientates back to its original circuit (Edelstam and Palmer 1950).

In the absence of a conclusive demonstration that the migration is performed within a familiar area, the possibility must remain that the snails merely settle in the first area encountered during random search that provides the resources appropriate for that time of year and that on occasion this behaviour results in some individuals performing the same migration circuit over several years. Nothing is yet known, of course, about the tracks of these snails during their movement from one home range to another. In most cases, also, nothing is known about the environmental features that cause one habitat to be more suitable as a summer or winter etc. home range than any other habitat. Nor, if the animals do live in a familiar area, is anything known about the way in which such a familiar area is established during ontogeny, though it may be assumed that it follows the type of sequence outlined in Chapter 18.

There is a similar lack of knowledge concerning the lifetime track of spiders. Papi and Tongiorgi (1963) have demonstrated the way that wolf spiders (Lycosidae) of the genus *Arctosa* use environmental clues to gain a sense of location (p. 878) which could well indicate that these spiders live within a familiar area. If the same is true of the wolf spider, *Lycosa lugubris*, then the seasonal behaviour described in Fig. 22.1 could well be achieved by a seasonal return migration within a familiar area. In this case, a reasonable assessment can be made of at least some of the factors that cause the different habitats to change in relative suitability as ontogeny and the seasons progress. In cases where the young are carried for a period on the mother's back after hatching as in *L. lugubris*, it is tempting to suggest that the young acquire an initial familiar area during this period. The experiments of Papi and Tongiorgi on *Arctosa* spp., however, suggest that this does not occur, familiar-area establishment beginning 2–3 days after the young have dispersed.

23

Seasonal and ontogenetic return migration by Amphibia

The vast majority of species of Amphibia pass their larval life in freshwater, either in a large, relatively permanent, water body, such as a stream, river, lake, or pond, or in some small and temporary puddle. Soon after metamorphosis from the larval stage, however, most Amphibia leave their natal water body and begin life on land. Some time later, when mature and at the onset of the breeding season, the amphibian returns to water to spawn. From then on, until death, the animal alternates between a terrestrial feeding period and an aquatic or semiaquatic breeding period.

It follows that, although there are exceptions, such as those species that are permanently aquatic, the majority of amphibians have two distinct phases to their lifetime track. The first phase consists of the migrations from hatching to maturity and the second phase consists of an annual, or occasionally biennial or longer-period, return migration. Both phases are return migrations between water and land. There is little information available concerning the degree of return to the natal site at the end of the first ontogenetic phase, but in the majority of species in which it has been investigated, the degree of return to the breeding site during the second phase from the age of first spawning onwards is high.

It also seems likely that in most species investigated so far the migration takes place within a familiar area. Attempts have been made to show that navigation to the breeding site is based on a single environmental clue. All such attempts have failed, however, and although in some species it seems clear that the familiar area is based to a large extent on olfactory clues, the only general conclusion is that amphibians establish a familiar area on the basis of all the environmental clues available to them. If some clues are removed, such as visual clues, then the animal continues to find its way about by making use of the clues that remain. No single

class of clues has yet been found to be indispensable. Five examples, two involving anurans and three involving urodeles, will suffice to show the type of evidence that is available concerning the return migrations of Amphibia.

Studies of the American toad, *Bufo americanus*, in London, Ontario, showed that the one-way distance component of the seasonal return migration varied from a few metres to 650 m (Oldham 1966). Spawning took place during a limited period in mid-May and degree of return to the breeding pool or river used the previous year seemed high. Degree of return to the natal site was not established, the time interval between metamorphosis and first breeding for males probably being about 2 or 3 years. Homing experiments during the breeding season demonstrated the existence of a familiar area, displaced toads orientating towards the 'home' breeding site at distances as great as 300 m, even though the presence of nearer, alternative breeding sites was known to the toad because of the breeding chorus emanating from them. Only if toads were displaced well beyond the limits of their familiar area (i.e. 2 km from their breeding site) did they show a tendency to orientate towards a breeding chorus emanating from a site other than from the home site. The environmental clues by which the toads built up their familiar area were probably auditory, olfactory, and visual. However, no single category was indispensable. No experiments were performed that showed at what stage of ontogeny the toads begin to establish their familiar area but it seems likely that the process begins from the moment the newly metamorphosed toad crawls out of its natal pool, or even earlier (p. 884).

The Pacific tree frog, *Hyla regilla*, males of which in West Oregon arrive at the breeding ponds in January or February, shows a high degree of return to the breeding home range during each perfor-

Fig. 23.1   Amphibia

[Photos by Jack Dermid (courtesy of Bruce Coleman Ltd.)]

The species shown are: Fowler's toad, *Bufo Fowleri* (top); the spring peeper, *Hyla crucifer* (bottom left); and the marbled salamander, *Ambystoma opacum* (bottom right).

mance of the migration circuit (Jameson 1957). Individuals released in a neighbouring pond, even if it contains a breeding chorus, still return to their home pond.

The western red-bellied newt, *Taricha rivularis*, breeds in the streams of the redwood belt of northwest California. In the study area of Twitty (1959), the newts migrate after breeding back up to the heavily wooded mountain slope to the south of the stream. There they spend the dry summer months underground. When the rains begin in the autumn or winter, the animals emerge again and forage on the forest floor, often high up on the mountainside. Division of the breeding stream into 50 m unit lengths along a 2·5 km stretch demonstrates a high degree of return of marked males to the unit occupied the previous year, a degree of return that remains high even after 6 years. Some removal migration occurs, up to a distance of at least 2·1 km, but, at least among those ever recaptured, is at a low incidence. Newts experimentally displaced up to 4 km showed some homing ability in subsequent years (Twitty *et al.*, 1964).

In the spotted salamander, *Ambystoma maculatum*, on Rhode Island, northeast United States, 76 per cent of marked individuals returned to their breeding pond two years after the initiation of the study (Whitford and Vinegar 1966). Displaced individuals oriented and returned to their home pond from a distance of 128 m.

The vermilion-spotted newt, *Notophthalmus viridescens*, shows a slightly different migration timing that seems to be associated with a greater propensity when adult for the adoption of a full-time aquatic existence (Hurlbert 1969). Typically, the life cycle entails the deposition of eggs in spring in a lake or pond, an aquatic larval stage of two to several months, and a terrestrial juvenile stage of two to four years. In New York State, maturing juveniles migrate back to water either in the autumn (August to December) or in the spring (April to May), the timing being much less precise than in most Amphibia so far studied. Circumstantial evidence suggests that degree of return to the natal site may not be high. Once in the water, adults may remain there more or less permanently or they may leave to spend the winter on land, migrating back to water in the spring. Migration thresholds to the migration-cost variables of rainfall and temperature exist, particularly in spring, migration being more likely after rain has fallen.

As is apparent from these studies, most infor-

mation available concerning the seasonal return migrations of Amphibia has been derived from studies carried out in the breeding home range where the animals are often concentrated and accessible. Relatively little information is available concerning animals in their feeding home range. Consequently, the variation in the distance component of the annual return migration between breeding home range and feeding home range is incompletely known except for those individuals with a short-distance component. One relevant series of observations, however, comes from the work of Stille (1952) on Fowler's toad, *Bufo fowleri*, around the southern beach of Lake Michigan. During the summer months, toads establish a home range which varies little with time, near the southern edge of the lake. The week's home range for each of all the toads in this region extends to the edge of the lake. The daytime home range where feeding occurs and where the toads dig themselves into a shelter during the hottest part of the day, may be as far as 213 m from the edge of the lake. About every fifth night during the summer the toads migrate to the edge of the lake and absorb water through their skin. The main point of interest for present purposes, however, is that the summer home ranges of five of the toads studied were about 1·5 km from the nearest known breeding ponds. Unfortunately, however, it was not proved that these individuals did perform a return migration over such a distance.

The evidence available for amphibia strongly supports the view that the seasonal return migration of each individual takes place between a breeding home range and a feeding home range (and perhaps other seasonal home ranges forming a migration circuit) which together constitute a part of the familiar area, the total size of which in all probability is much larger (p. 384). In particular, support is derived from the failure of males of the American toad, *B. americanus*, and the Pacific tree frog, *Hyla regilla*, to join a different breeding chorus when released nearby. The implication is that the threshold for type-C removal migration (p. 311) is high, partly for the reasons presented in Chapter 17 and partly because, as argued in Chapters 17 and 18, some time previously the exploratory–calculated migration sequence has led to the animal establishing the most suitable migration circuit available to it. Presumably some assessment of the cost of a homing migration is also involved (p. 912). As shown by Twitty *et al.* (1964) for the western red-bellied newt, *Taricha rivularis*, the proportion of animals that adopt the release point as their new breeding home range increases as the distance of displacement increases. Some males, displaced 2 km or so, adopted the release point as their new breeding home range for two years and then at a future breeding season migrated from their terrestrial home range back to their original breeding site, behaviour that strongly implies calculated migration within a familiar area of the type described in Chapter 14. Twitty *et al.* consider that their displaced male newts could not have been familiar with their release point prior to displacement. The experimental design of this work, however, gives little information concerning the terrestrial movements of these newts. Quite possibly the new breeding sites to which the newts are displaced are joined to the previously existing familiar area during the course of the terrestrial migrations. The full implications of the familiar-area hypothesis to the navigation of amphibia, however, are discussed in Chapter 33 rather than here.

Two more points are perhaps worth making with respect to the seasonal return migrations of Amphibia. First, the migrations, although obligatory (p. 738), manifest migration thresholds to migration-cost variables (p. 49). This is particularly true in connection with the migration to the breeding home range. The migration-cost variable involved is most often humidity, though temperature may also contribute (Noble 1931). Thus, in temperate regions the migrations of frogs, toads, newts, and salamanders are not initiated until after rain. In the tropics, it is the 'cooling thunderstorms of the wet season that bring forth the loudly-calling frogs' (Noble 1931).

Secondly, the degree of dispersal and convergence and the intra-deme variation in distance component seem likely to be an evolutionary response to the same type of selective pressures described for a bird on p. 652. Degree of overlap of the individual home ranges at the different seasons may fit the model presented in Fig. 18.25.

24

Seasonal and ontogenetic return migration by terrestrial reptiles

Displacement experiments, some of which are described elsewhere (pp. 195 and 885), on a wide variety of terrestrial and freshwater amphibious reptiles, are consistent with the view that such animals establish a familiar area, part of which becomes a home range (p. 384). Examples of seasonal return migration within a familiar area, however, are few.

Perhaps the most detailed example is that presented by Woodbury and Hardy (1940, 1948) for the American desert tortoise, *Gopherus agassizii*. During the months of November through February, these tortoises occupy communal winter dens that later work (Nichols 1953) has shown to be the result of cooperative digging by several individuals working either together or in relays. In March, however, the tortoises disperse, each individual migrating to a summer home range. This summer home range is used for breeding and/or feeding, depending on state of maturity, and may be occupied by either a solitary individual or by a mating pair. Feeding occurs at night. The size of the summer home range, which contains a hole or holes in which the tortoise may shelter during the heat of the day, rarely exceeds 4 ha, though one was recorded as being about 40 ha. In October and November the tortoises return to their winter home range.

Communal winter dens are also known for some snakes, such as the prairie rattlesnake, *Crotalus confluctus*, in Colorado for which at one site a collection of 153 snakes by one author (Klauber 1936) was described as 'just a few of the snakes present'. Marking experiments by Woodbury *et al.* (1951) on a similar den in Utah which was used by seven different species showed that there was a high degree of return to the cave concerned in successive winters. Only adults used the cave, however, and the way that juveniles incorporated the site into their familiar area was not discovered. The seasonal return migrations of the species concerned seemed to have a one-way distance component of anything up to 16 km, this distance being performed by a garter snake, *Thamnophis* sp, only 18 cm in length. In most cases, however, the components of snake return migrations are unknown, though one author (Neill, W.T. in Oliver 1955), using rather suspect reasoning, claimed that the one-way distance component for the canebrake rattlesnake, *Crotalus horridus*, in Georgia, USA could be as great as 30 km. This species also uses a winter den, to and from which a series of well-used radiating travelways may be seen. The distance between the hibernation site and summer home range of the adder, *Vipera berus*, in Dorset, England, is of the order of 0·6–1·9 km (Prestt 1971).

The adult females of the giant tortoise, *Testudo gigantea*, on the Galapagos Islands perform a return migration between a feeding home range in the cooler moist highlands and a nesting home range on the coastal lowlands (Orr 1970a). The eggs are laid in soil or sand. Nothing is known, however, concerning degree of return. In southeastern Iowa, snapping turtles, *Chelydra serpentina*, seem regularly in late summer (August) to leave the water bodies of their summer home range (Klimstra 1951). Individuals have been followed for distances of 600 m, but other details are not available and the movements need not necessarily have been part of a return migration. Davis, A.W. (in Netting 1936) records a spring (April) migration of the spotted turtle, *Clemmys guttata*, that seemed to be leading from possible hibernation sites in a more upland situation, across fairly flat terrain, consisting of fields and farm woodland, to a shallow swamp that would have offered suitable summer home ranges.

25

Seasonal return migration by terrestrial and freshwater-aquatic and amphibious mammals

There seems little doubt from the type of evidence presented in Chapters 15 and 18 that the vast majority, if not all, terrestrial mammals live within a familiar area that is established from the moment of birth onwards. The only potential exception seems to be the polar bear, *Ursus maritimus*, which may drift with the pack-ice of the Arctic Ocean making landfalls as and where it is deposited. Other evidence (p. 178), however, suggests that polar bears do establish a familiar area and that, if the occasion arises, they can navigate back to a point within that area.

All seasonal return migrations of terrestrial mammals can therefore be considered to take place within a familiar area. Consequently, the performance of a seasonal return migration by any particular individual or deme depends entirely on the existence within the familiar area (of optimum size—p. 378) of spatially separate habitats that vary in relative suitability with the season. Clearly such conditions are highly variable, not only from place to place, often being quite local, but also from year to year.

25.1 Small mammals

As far as small mammals are concerned a thesis that appears to be of general application is as follows. Young become familiar with at least part of the parental home range and then, possibly after a period of exploratory migration, perform a type-H removal migration (p. 308) which may or may not be calculated (p. 44) and which is most likely to occur before the animal begins to reproduce (p. 349). Depending upon the time of year at which type-H removal migration takes place, the animals

settle in a home range suitable for winter or summer or both (or wet or dry season, etc.). As the season progresses, the home range may decrease in suitability below the exploratory migration threshold (p. 64). If the situation is such that spatially separate habitat-types are most suitable at different seasons as in all the other cases mentioned in this section, the animal's familiar area then comes to include home ranges, each suitable for a particular season. Any subsequent extension of the familiar area arises only if one of the seasonal home ranges falls below the exploratory migration threshold for its type with the result that type-C removal migration may take place (p. 311).

Apart from the fact that the short period of parental care in many small mammals necessitates the young establishing themselves in a home range suitable for one season before they have experienced the conditions that prevail during other seasons, this thesis is no different to the familiar-area hypothesis presented for larger mammals in the major part of this chapter. As far as small mammals are concerned, however, there is one point that should be made. Many species are adapted to a high mortality rate due to predation and disease and characteristically start to reproduce at an early age, breed through much of the year and on average have a short longevity. Lemmings, for example, may breed at 3–4 weeks of age and may continue to breed throughout the year (Krebs 1973). Although in captivity lemmings may live as long as three years, the majority die much earlier and in the wild it seems likely to be unusual for lemmings to live more than one year. Consequently, much of the seasonal shift in distribution that occurs involves different generations. Not all of the migrants are return migrants, therefore, and as yet there are no estimates of the proportion of individuals of small mammal species that perform seasonal shifts in distribution that consists of return migrants.

Some examples of seasonal shifts in distribution by small mammals are described on p. 220 and in this section only one group is considered, those cricetid rodents of the genera *Lemmus* and *Dicrostonyx* known coloquially as lemmings. Most demes of lemmings seem to show a seasonal change of habitat-type (Marsden 1964). The brown lemming, *Lemmus trimucronatus*, at Barrow, Alaska, like all lemmings, spends the winter in burrows under the snow feeding on the vegetation. In this case, the lemmings occupy depressions on the tundra where more snow accumulates offering more protection from the predations of the Arctic fox, *Alopex lagopus*. When the snow melts and the water accumulates in these depressions the lemmings move for the summer to the relatively dry 'owl mounds' where they excavate burrows in the soil. In Siberia there is a period after the winter nests are flooded with spring meltwater during which the lemmings cannot dig new burrows in higher habitats because the soil is still frostbound. During this period the lemmings are forced to remain on the tundra surface, moving from one shelter to another. The varying lemming, *Dicrostonyx groenlandicus*, in the region of the Mackenzie River, Canada, migrates to well-drained slopes on higher ground in summer, inhabiting shallow burrow systems among the boulders. In the autumn it returns to the lower areas and the protection of the snow banks.

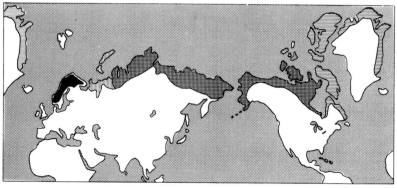

● *Lemmus lemmus* (Norwegian or Scandinavian lemming)
IIIIIII *L. trimucronatus* (Brown or Siberian lemming)
≡ *Dicrostonyx groenlandicus* (varying lemming)- photo

Fig. 25.1   The geographical distribution of lemmings

Note that only the varying lemming is found in Greenland, the high Arctic Islands, and the Ungava peninsula of Labrador. Only the Siberian lemming is found between the White Sea and the River Ob and perhaps also the northern Rockies. Only the Scandinavian lemming is found in Scandinavia.

The Scandinavian lemming, *Lemmus lemmus*, which unlike the other species is primarily a montane animal, is found throughout the mountain plateaux of Norway, Sweden and Finland. Although it occurs on the lower-level tundra of Finnmark and the Kola peninsula, throughout most of its range the preferred habitat-type is the subalpine zone of the Scandinavian mountains (i.e. above the coniferous forests, from the birch and willow scrub regions upwards nearly to the snow-line). A typical lemming habitat has marshy or tussock-covered ground, clothed with dwarf trees and shrubs (Marsden 1964). Within the subalpine zone, however, the lemmings perform a seasonal shift in habitat-type which may often be obscured by the complex changes in distribution brought about by the re-migrations described in Chapter 19 (p. 411). In summer, *L. lemmus* lives chiefly in the peatlands where mosses and sedges are abundant. In winter, however, the watery fens and muskeg-like thickets of the peatlands become unsuitable due to the freezing of the ground and the withering of the sedges and consequently there is a slight upward shift in altitude to the alpine zone where mosses are abundant and much herbage remains green under the snow.

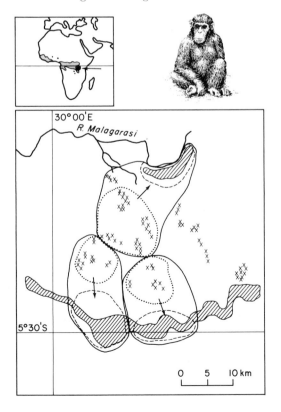

25.2 Large mammals

Before considering the migrations of various of the larger terrestrial mammals, an example of seasonal return migration by a large aquatic mammal may be included here that is so unique that it contributes little to our general understanding of mammalian migration but which I feel ought to be mentioned somewhere in a book such as this. This example is that of the manatee, *Trichechus manatus*. This permanently aquatic dugongid of the shallow tropical waters of America normally migrates between the sea and fresh water, though information concerning its movements are fragmentary (Layne 1965). In St. John's river on the east coast of Florida a deme exists that can no longer reach the sea due to human artefacts. The behaviour of the manatee in this river has been recorded in a film made by Jacques Cousteau. This deme migrates up and down St. John's river with a one-way distance component of about 300 km. The summer is spent upstream and the downstream migration begins in autumn, when the river temperature begins to fall, and ends at Blue Springs where the water is a constant 23°C. However, none of the aquatic vegetation on which this

Fig. 25.2   Wet season–dry season return migration by three large-sized groups of the chimpanzee, *Pan troglodytes*, in western Tanzania

General ecology
See Fig. 18.23 and p. 162. The area illustrated is one of the driest in which the species is found.

Seasonal return migration
The map shows the year's home range for three large-sized groups of chimpanzees. During the first half of the dry season, between June and August, the groups migrate to the open-land vegetation of the *Brachystegia* woodlands (hatched areas) and feed extensively on the beans of *B. bussei*. The dashed line indicates the extent of the home range of each group during this period. For the last half of the dry season (i.e. September and October) and for the first part (i.e. November to January) of the wet season, which lasts until May, the groups migrate to the dense riverine forests (× ×) in the inner parts of the hills in which they feed on the fruits of one or more species of *Landolphia*. The dotted line indicates the limits of the home range during this period (September to January). Arrows indicate the migration route. The solid line indicates the year's home range for each group, areas of which measured about 200–400 km². Deme density averages about 0·2 km²/individual.

[*Simplified from Kano (1971)*]

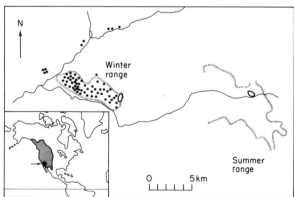

Fig. 25.3 Seasonal return migration of a herd of mule deer, *Odocoileus hemionus*, on the west slope of the Sierra Nevada in Tulare County, California

The herd consists of males, females and young of the year in September in the ratio 12:100:47 respectively. Females were trapped and marked either with a bell or with a radio transmitter in the summer range in July and August. The areas outlined by dashes represent the main winter and summer ranges. Dots show winter sightings of deer with bells. The small area of about 1 km diameter represents the home range of a single individual in summer (after trapping) and winter as determined by radio telemetry. The summer range is at an altitude of 1800–3000 m whereas the winter range is at an altitude of 450–1100 m. The actual ground distance between the two ranges is 15–30 km.

The autumn migration begins some time in October with the first significant snow fall and most individuals have left the summer range by about the end of October. The spring migration begins at the end of May. The radio-tagged female returned in the spring to the home range occupied the previous summer and in September was accompanied by a young of the year. This female accomplished the 15 km migration in the spring in less than three days.

[Figure modified from Schneegas and Franklin (1972). Photo by Dr. C.G. Hampson]

large cow-like mammal feeds is present in Blue Springs and the animals make daily return migrations between the Springs, where they roost, sleeping on the bottom, and the river where food is abundant. At the end of winter, the animals migrate back up the river. Presumably, this migration is as much within a familiar area as that of large terrestrial mammals. The environmental cues that the manatees use to establish their familiar area, however, must involve much fewer visual cues than that of terrestrial mammals because manatees are virtually blind.

Now, let us consider the seasonal return migrations of some of the larger terrestrial mammals, beginning with mainly herbivorous animals, such as ungulates, and then considering the migrations of some of the animals that prey upon them. Some examples of seasonal return migration by predominantly herbivorous terrestrial mammals are illustrated in Figs. 25.2–25.7. The animals concerned are: chimpanzee, *Pan troglodytes*, in western Tanzania (Fig. 25.2); mule deer, *Odocoileus hemionus*, in California (Fig. 25.3); moose, *Alces alces*, in Canada (Fig. 25.4); plains' zebra, *Equus quagga*, in the Ngorongoro crater (Fig. 25.16) and again, along with wildebeest, *Connochaetus taurinus*, Thomson's gazelle, *Gazella thom-*

soni, and eland, *Taurotragus oryx*, on the Serengeti Plains (Fig. 25.7); and the caribou or reindeer, *Rangifer tarandus*, in Canada, Alaska, and Scandinavia (Fig. 25.6).

The commonest situation in which such animals are likely to develop a seasonal return migration is on the slopes of hills and mountains, as in the case of the chimpanzee, mule deer, moose, and reindeer. This type of migration, which is always altitudinal, can develop at any latitude. Although the ungulate examples that are illustrated all involve temperate and subpolar species, similar examples are found in the tropics. Thus, the deme of the African elephant, *Loxodonta africana*, that occupies the slopes of Mount Kenya migrates to high altitudes during the dry season in January and February (Hill *et al.* 1953). In India, the gaur, *Bos gaurus*, lives in mixed herds of 2–40 ($\bar{x}=8$–11) between which frequent removal migration takes place. The species prefers to feed on green grass but also consumes coarse, dry grass, various herbs, and also the leaves, bark, and fruit of trees (Schaller 1967). A herd of these cattle has a familiar area that may be large enough to include a valley and surrounding hillsides between which it migrates from the cool season to the hot season.

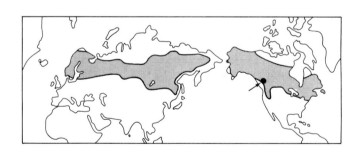

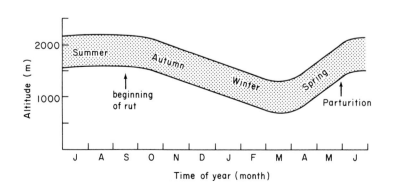

Fig. 25.4   The altitudinal migration of moose, *Alces alces*, in Wells Gray Park, British Columbia, Canada

The stippled area in the graph opposite represents the main altitudinal distribution of moose at different times of year. In the summer, the moose are found at high elevations among the coniferous forests up to the timber-line particularly where there are meadows among the trees and the ground is wet from seepage or where there are streams or lakes in which they often feed. By December they have moved down to the highest of the lowland streams on the valley floor and the remainder of the winter is occupied in a gradual descent down the gradient of a broad valley. All of the winter migration is through open, deciduous growth, rich in browse.

The critical snow depth beyond which moose find survival difficult but in which movement is not restricted is about 75 cm. Autumn migration is triggered by snow fall and with snow depths increasing faster at higher altitudes during the winter the moose appear to be pushed gradually downwards. The spring migration is conversely triggered by warm weather but a cold spell during the spring migration can cause a temporary reversal in direction.

[*Drawn from information in Edwards and Ritcey (1956). Photo by Wilford L. Miller (courtesy of Frank W. Lane)*]

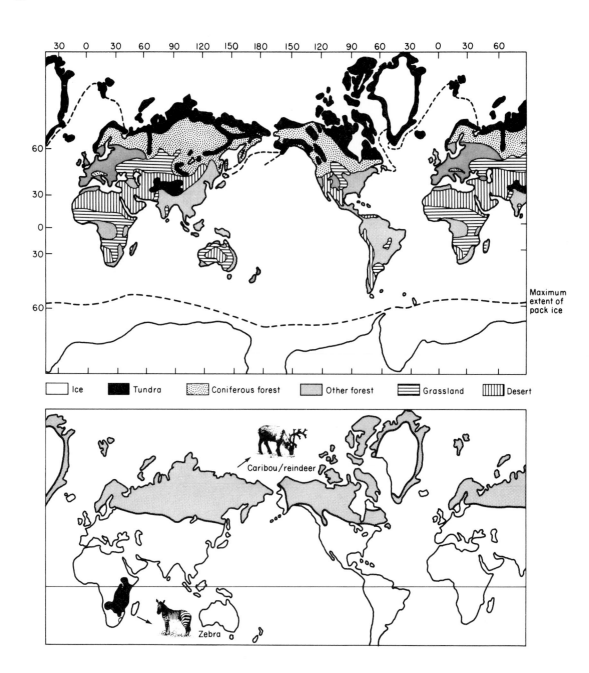

Fig. 25.5   The geographical distribution of seven species of ungulates with long-distance seasonal return migrations in relation to habitat-type

The top map on this page shows the major vegetation zones of the Earth.

The other two maps show the geographical distribution of seven species of ungulates, some demes of which perform or performed seasonal return migrations with a one-way distance component greater than about 100 km. The geographical distribution of the American bison is historical (early nineteenth century). The dashed line to the west of the present distribution of the saiga indicates a similar historical range.

Long distance components in the seasonal return migrations of ungulates evolve primarily on the tundra, grassland, and desert habitat-types.

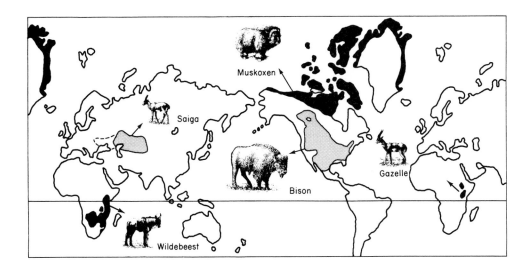

Seasonal return migrations that are not primarily altitudinal by contrast show marked latitudinal variation in their components, particularly in their distance component. Figure 25.5 shows the geographical distribution in relation to vegetation of some of the animals mentioned in this section. It can be clearly seen that those migrants that show the longest distance components, such as the caribou (Fig. 25.6), the ungulates of the African plains (Fig. 25.7), the saiga, *Saiga tatarica* (p. 536), the American bison, *Bison bison* (p. 554), and the muskox, *Ovibos moschatus* (p. 545), all occupy one of three habitat-types: grassland, desert, or tundra. They occur, therefore, in regions in which the availability of food is highly seasonal and erratic, either through drought or through snowfall or both. Such animals establish huge familiar areas (see p. 170 for a discussion of the size of the familiar area of the herds of barren-ground caribou on the Canadian tundra), a process that seems likely considerably to be aided by social communication (p. 201).

The influence of climate, acting through the spatial and temporal availability of food and water, is seen particularly clearly in and around the African plains and on the Eurasian steppes.

Most of the larger African mammals live in the grassland or in the zones of transition between grassland and primary forest (i.e. the different types of savanna). In these transitional zones, although the number of species is relatively high, large concentrations of animals are rare and any migra-

tion that takes place rarely involves seasonal home ranges with a low degree of overlap. The black rhinoceros, *Diceros bicornis*, is an example of an animal adapted to such a habitat-type (Schenkel and Schenkel-Hulliger 1969) as also is the giraffe, *Giraffa camelopardalis* (Guggisberg 1969). Even when there is some migration between wet- and dry-season home ranges, as in the white rhinoceros, *Ceratotherium simum*, the distance component is very short (Heppes 1958). In the grassland proper, however, the number of species is relatively small, concentrations of animals are temporarily high, and the seasonal migrations of herbivores are conspicuous and long-distance (Fig. 25.7).

On the Eurasian steppes occur antelopes, such as the dzeren, *Procapra gutturosa*, dzheiran, *Gazella subgutterosa*, and saiga, *Saiga tatarica*, and perissodactyls, such as the kulan, *Equus hemionus*, and wild horse, *E. przewalskii* (Formozov 1966). Other artiodactyls also occur, such as the European wild ox, *Bos primigenius*, and European bison, *Bison bonasus*. All of these ungulates perform seasonal return migrations, the best documented of which are perhaps those of the saiga (Bannikov *et al.* 1967). This antelope is an inhabitant of the steppe and forest-steppe habitat-types (Fig. 25.5). Except for brief periods during which parturition and copulation occur, the herds of saiga are more or less continuously on the move. In the western Caspian area, on the right bank of the Volga, a northward spring migration, performed by small groups as well

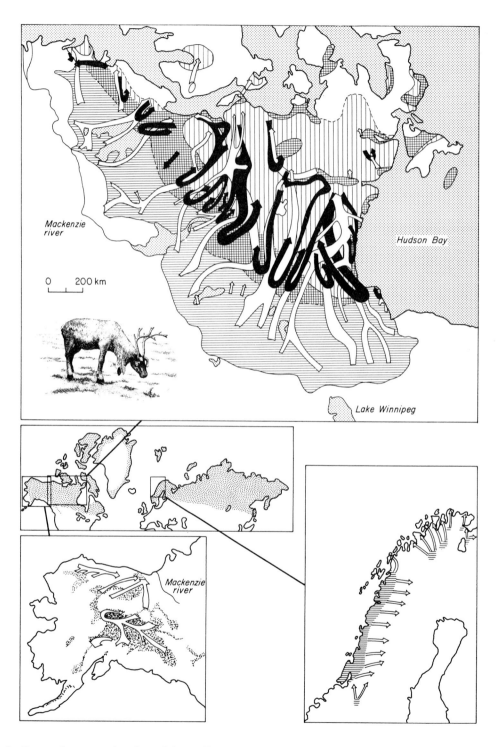

Fig. 25.6   Seasonal return migration of the caribou or
reindeer, *Rangifer tarandus*, in Canada, Alaska, and Scandinavia

General ecology
Rangifer tarandus has a circumpolar distribution as shown.
Old World forms are known as reindeer, New World forms

as caribou. The species is exclusively a grazer, eating no twigs or stems. A wide variety of plants are taken but, contrary to popular belief, the species does not live primarily on lichens. Even in winter, lichens constitute only up to 30 to 55 per cent of stomach contents, and in summer the proportion is much less.

The species occupies two major habitat-types, the coniferous forests and the tundra (see Fig. 25.5). Some demes live permanently in the forest and others live permanently on the tundra. The seasonal return migrations of these demes cover only a few, perhaps tens, of kilometres. The demes with the largest one-way distance components (up to 1000 km) to their seasonal return migrations are those living near the tundra-forest transition which alternate seasonally between the two habitat-types.

Females conceive for the first time at 1·5–3·5 years of age. Males are sexually mature at 1·5 years of age but achieve little success at copulation until 4 years of age. Longevity may be as great as 18–20 years.

Woodland demes

The southernmost demes of *R. tarandus* occupy mature spruce and other coniferous forests and remote alpine meadows. Some woodland demes may perform altitudinal migrations between alpine tundra ranges at high altitudes in summer and winter ranges lower down in the forest. Where the topography is less mountainous, demes may graze areas of bogs, low-lying grassy areas, or lichen-rich glades and move a few kilometres to a winter range where tree and ground lichens are available.

Tundra demes

The northernmost demes of *R. tarandus*, such as those that occupy the islands of the Canadian Arctic and those of Spitsbergen, spend the summer moving across the tundra where food is available. On Spitsbergen in the autumn the deer descend to the coasts where they feed on seaweed washed up by the tide. Winter range on the tundra, however, consists of windblown slopes, albeit where the windchill factor is low, where the wind keeps the slopes relatively free of snow and where, consequently, food remains accessible all winter. The migrations of some tundra demes are shown in the northernmost part of the top diagram.

Woodland–tundra transition demes

Throughout its range, those demes of *R. tarandus* that are near to the transition between tundra (either lowland or alpine) and woodland have evolved a seasonal return migration between a winter range in the forest and a summer range on the tundra. In the diagrams, the spring migration is shown by a white arrow and the first part of the autumn migration is shown by a black arrow. In the diagrams for Canada and Alaska, the relative number of animals involved is indicated by the width of the arrow. Where hatching is used, horizontal hatching shows the winter range and vertical hatching shows the summer range. Where stippling is used in the Alaska diagram, the relative density of the animals on their winter range is indicated by the density of the stippling. The diagrams show clearly the way that distance and direction vary from deme to deme. Apart from these components, however,

the migrations are essentially similar and the migrations of only one major deme, that of the barren-ground caribou of Canada (upper diagram) is described in detail to illustrate the annual cycle.

Winter: Most caribou spend the winter in the northern boreal forest. A few always, and many sometimes, winter on the tundra. Usually, adult males migrate furthest into the forest where snow tends to be deepest, and adult females and juveniles remain nearer the tree-line where snow is shallower. The herds converge prior to or during the spring migration as shown.

Spring: The caribou start the spring migration sometime between February and April and reach the tree-line on the way north in the first week of May. From the tree-line to the nearest calving grounds is 400–500 km. The first females arrive at the end of May or the beginning of June. Average speed is 25–50 km/day. During migration, the caribou tend to follow lines of least topographical resistance such as frozen lakes and rivers, open snow-free uplands and long narrow hills of soil and rock deposited by glaciers. Pregnant females lead the migration, the juveniles and adult males gradually lagging further and further behind. A migrating herd may be 300 km in length, thus often taking several weeks to pass through a given area on the migration route.

Parturition: Calving grounds are usually on rugged, inhospitable uplands still swept by spring storms. Calves can travel within a few hours of birth and start to graze during their first few weeks. Suckling ceases during the first week in July. After calving in June, large herds of several thousands of females, calves and yearlings descend in early July from the calving grounds to lower and greener pastures, perhaps 200 km away. Females show a high degree of return to the calving grounds after their first parturition year.

Summer: During early summer the caribou are often harassed by hordes of mosquitoes, warble flies, nose bot flies, and, in places, black flies, and much of the summer movement can be attributed to attempts to reduce insect harassment. Even so, many individuals become emaciated during this period (Fig. 25.12). In July, the herds start to migrate towards the forest. During August, however, the herds are small in size, never more than a thousand, and the vast majority of individuals are scattered across the tundra in small groups of three or less individuals. At the end of August, the animals converge once more.

Autumn: By late September, the herds, now fat and in good condition, arrive at and move along the tree-line or perhaps even back onto the tundra again, as shown. Autumn migration proceeds at up to 64 km/day. Copulation occurs in late October and early November, usually in the region of the tree-line. Males do not collect harems but instead attempt to copulate with and then guard receptive females against the attentions of other males.

[*Compiled from Murie (1935), Banfield (1954), Kelsall (1968), Ricard (1971), and Information Canada (1973)*]

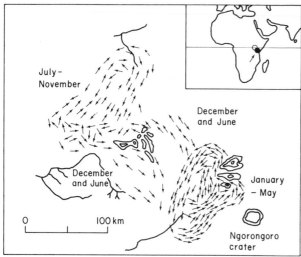

Partial Migrant

Eland *Taurotragus oryx*

Fig. 25.7 Long-distance seasonal return migrations of some African ungulates in the Serengeti region

The diagram shows the long-distance migrations performed by zebra, wildebeest, and Thomson's gazelle. The dry season (July to November) is spent in an area of open thorn woodland which in the wet season harbours mosquitoes and tsetse flies. In the dry season, however, tsetse flies have a more localised distribution in the region, being associated with water. In December, when the first storms appear, the herds migrate to the central plains of the Serengeti. During the rainy season (January to May) the herds migrate in a predominantly anticlockwise direction, following the rain and new vegetation. Movement is, however, entirely opportunistic and, particularly at the beginning of the wet season, is directed to any area where

rain can be seen or heard to be falling (p. 95) at distances of up to 100 km.

The species of long-distance migrants follow each other in a characteristic sequence during their seasonal return migrations. In the Serengeti region the succession is: first zebra, second wildebeest, and lastly Thomson's gazelle. Stomach analyses show that zebras, which feed on grass, have a high tolerance of stem material. They open up the herb layer by trampling and increasing the relative frequency of leaf by selection of stem. Wildebeest feed mainly on grass leaf and thus experience higher habitat suitability through following the zebra. The combined action of zebra and wildebeest exposes the dicotyledonous plant material on which the Thomson's gazelles feed.

Eland are partial migrants, some individuals remaining in the dry-season range all the year while the other half

Migrants

Zebra *Equus quagga*

Thomson's Gazelle *Gazella thomsoni*

Wildebeest *Connochaetus taurinus*

Non–Migrants

Impala *Aepyceros melampus*

Giraffe
Giraffa camelopardalis

Topi *Damaliscus korrigum*

Hartebeest
Alcelaphus caama

migrate with zebra herds. Their feeding ecology is intermediate between that of migrants and non-migrants. Animals that include browse in their diet, such as the impala, topi, hartebeest and giraffe, remain in the dry-season range of the long-distance migrants all year.

The zebra, wildebeest, and gazelle give birth in December and January. However, the impala, hartebeest, giraffe, and others lacking a seasonal return migration give birth in October during the second half of the dry season while accompanied by the seasonal migrants, behaviour that could be adaptive to a period of lower predation pressure at this time.

Studies of the mortality factors on these species in the Timbavati Private Nature Reserve in the lowveld region of the eastern Transvaal, South Africa, showed that seasonal migrants (zebra and wildebeest) suffered little mortality through starvation. Rather, mortality was due largely to predation, particularly by lions, *Panthera leo*, with juvenile and sub-adult, and particularly solitary males, being especially susceptible. Animals without seasonal migrations, such as the impala, giraffe, kudu (*Tragelaphus strepsiceros*), waterbuck (*Kobus ellipsiprymnus*), and warthog (*Phacochoerus aethiopicus*) all showed considerable mortality through starvation at the end of the dry season. In most species this mortality was greater than that due to predation.

[Compiled from information and diagrams in Grzimek and Grzimek (1960), Talbot and Talbot (1963), Gwynne and Bell (1968), and Hirst (1969). Photo by Norman Myers (courtesy of Bruce Coleman Ltd.)]

as herds of 60 000–100 000 moving in a 1·0–1·5 km wide column, begins at the end of March and beginning of April. Migration thresholds to migration-cost variables such as snow cover and storms have evolved in relation to this spring migration. The migration lasts about 14 days and by mid- or late-April the saiga have arrived at their parturition grounds. The maximum recorded distance between winter home range and parturition home range is 300–350 km. By mid-May, after parturition, the females and their young migrate in a direction between west and south to their summer feeding grounds, often being preceded by the males and barren females that had separated from the pregnant females before parturition. Again the distance between parturition and summer grounds may be 200–250 km. The middle of the summer is a period of continuous movement in search of food and water. The dryer the particular summer, the further west the saiga migrate to areas with more water. During wet years, little movement may occur. At the end of the summer, some time between the end of August and October, the saiga migrate eastwards, back to the Caspian lowland. Individuals differ considerably in the date at which they commence this migration away from the summer feeding grounds. The autumn and winter migrations show considerable year-to-year variability, depending on the weather. During this period, the animals search for fodder in the semidesert zone of the Caspian lowland. In autumn (October to November) when food is abundant, little movement occurs. At the end of November and beginning of December, with the arrival of snow, migrations are initiated that are often to the south. Although the migrations on the right bank of the Volga are predominantly north to south in direction, elsewhere this is not necessarily the case. The distance component of the migration also varies considerably, not only from place to place but also from year to year. In Kazakhstan, however, the animals may alternate between the arid steppe and desert zones (Fig. 25.5), in which case the winter and summer ranges may always be some distance apart.

Predatory terrestrial mammals show geographical variation in the components of seasonal return migration similar to that shown by the herbivorous mammals on which they prey. The majority, therefore, have a year's home range which shows little spatial separation into seasonal home ranges. Perhaps the commonest pattern of seasonal return migration involves seasonal variation in home-range

size (p. 390) rather than a spatial separation of seasonal home ranges. In most such species, the young are born into some sort of den or nest and for part of the year the pattern of movement within the home range is a return movement between the den and a part where food is available. When the prey species are migratory ungulates, such as caribou, *Rangifer tarandus*, for some demes of the wolf, *Canis lupus*, the distance between den and hunting ground may be 30 km or so (Murie 1944). A similar distance develops for some African hunting dogs, *Lycaon pictus*, towards the end of the dry season. A common pattern, then, is to produce young in a den at a time when food is abundant nearby. In the coyote, *Canis latrans*, the diameter of the home range from May to August during the period that the young are being fed may be as little as 5–7 km (Schwartz and Schwartz 1959). The young are then fed up until, at a time when food is becoming less available and the den to hunting site distance is becoming relatively large, the young become old and mobile enough to accompany the adults on their search for food. For the rest of the year the family group then moves nomadically within a home range which is much larger but which usually includes the breeding home range. This is the case in the coyote, which in winter has a home range the diameter of which may be as great as 40–50 km. The wolf in winter also has a large home range which often takes the form of a hunting circuit, approximately 200 km in circumference (S.P. Young 1964) that is completed every 10 days or so (Thompson 1952). In the African hunting dog, *Lycaon pictus*, the critical factor in breeding success is that the young can attain sufficient strength and mobility to accompany the adults on their migration to the area occupied by the migratory plains' animals before the distance between the den and the ungulates becomes too great (p. 214).

In the animals just described, three types of movement follow each other in a seasonal sequence. These are: short-distance movement between a den and a feeding place; long-distance movement between a den and a feeding place; and migration not based on a fixed site. This latter type may involve either nomadism, or a fixed circuit, or perhaps, when the prey is migratory, a simple following of the migrating prey herds. Predators of migratory ungulates have three alternative strategies available to them when the ungulates migrate away from the area occupied previously. Either they can switch to other prey species that are still present, or they can

Fig. 25.8   Hunting wolf, *Canis lupus*   [*Photo by Dr. C.G. Hampson*]

'commute' over an increasing distance between their home site and the current position of the ungulates, or they can migrate with or just behind the ungulate herd. All three types of adaptation can be found among terrestrial mammals, a variety of conditions determining which strategy is most advantageous in any one case.

Demes of the lion, *Panthera leo*, for example, feed on the migratory herds when they pass through their home range but at other times switch to alternative prey species (Schaller 1972). In eastern Canada, the micmac amerindians, *Homo sapiens*, spent the winter in the woods hunting moose (Fig. 25.4), caribou (Fig. 25.6) and smaller animals such as porcupines. In the spring, when the larger animals migrated to higher altitudes, the humans moved to the nearby coast and fed on bivalve molluscs and other sea foods (Orr 1970a).

In regions away from the coast, however, other demes of hunter–gatherer North American indians, such as those of the west, followed the game animals in their altitudinal migrations. On the plains, other amerindians followed the herds of bison (p. 554) on their seasonal return migrations. The spotted hyena, *Crocuta crocuta*, follows the herds of migrating ungulates across the Serengeti plains at the beginning of the dry season (Kruuk 1972). Some, however, particularly those females with relatively immobile young, stop short of the ungulates' dry season range (Fig. 25.7), establish dens on the plains near the edge of the thorn woodland, and from there commute to and from the ungulate herds. Other, more mobile, individuals remain with the herds throughout their dry-season migrations.

The habit of following the migrating herds of game animals is common among hunter–gatherer humans other than the amerindians. The bushmen of the Kalahari Desert, within the limits of their territories (p. 120), follow game animals as they migrate in search of grazing and water in the dry season. There is some evidence that such behaviour was well established in some demes of humans, especially in arid areas, during the Pleistocene (Howell 1970) at which time man was everywhere a hunter–gatherer. It is not surprising, therefore, that among those demes contiguous to the Middle East, in which continuous movement with the ungulate herds of the steppes (p. 531) was likely to be most pronounced, some humans developed a symbiotic relationship with their prey species. Marriage of this symbiosis with the practice of the cultivation of plants that in the Middle East was developed in

areas contiguous with the steppes at about the same time, about 10 000 years B.P., led to the development and spread of the agricultural practices that so revolutionised human ecology (p. 124).

Marriage of plant cultivation with a symbiotic relationship with ungulates could only take place in regions in which the ungulates had a sufficiently small distance component to their seasonal return migration to be consistent with the sedentary behaviour necessary to cultivate plants. In areas in which the distance component of the ungulate seasonal migration circuit was large, however, which also were areas (p. 530) in which plant cultivation was least reliable, such a marriage could not take place and in such areas human ecology evolved into, and remains, that of nomadic pastoralism (i.e. permanent association with ungulate herds during the execution of their seasonal migrations). Pastoral nomads are found, therefore, in parts of all of those habitat-types in which the distance component of ungulate

Fig. 25.9   The seasonal return migration of a deme of nomadic pastoralist humans, *Homo sapiens*, and an example of type-C removal migration

(a) Distribution of the Fulani pastoral nomads in relation to the African savanna

(b) Seasonal return migration

At the peak of the wet season (August/September) a number of family units, in this case 14, the males of which share common ancestry, are aggregated at the northern-most point of the migration circuit. There is standing water and, though continuous, cattle movement is limited. At the beginning of the dry season, which starts earlier in the north than the south, the southward migration begins. Standing water persists along the route until January at which time the group begins to disperse. Eventually, the group splits into family units which are age-graded male units (i.e. consisting of one reproductive male, one to several reproductive females and young, and perhaps post-reproductive males and females). Occasionally more than one family unit stays together as indicated by the numbers at the final dry season roosting and grazing area (bar). Family unit size varies from 2 to about 15 with an average of ten cattle per member. With the advent of the wet season at the beginning of April (in the south), the return migration starts until the groups are north of the main areas of tsetse risk. During the dry season the areas of tsetse risk (×) are localised.

(c) Type-C removal migration over a period of nearly 50 years

Straight lines join the extreme wet and dry season limits of

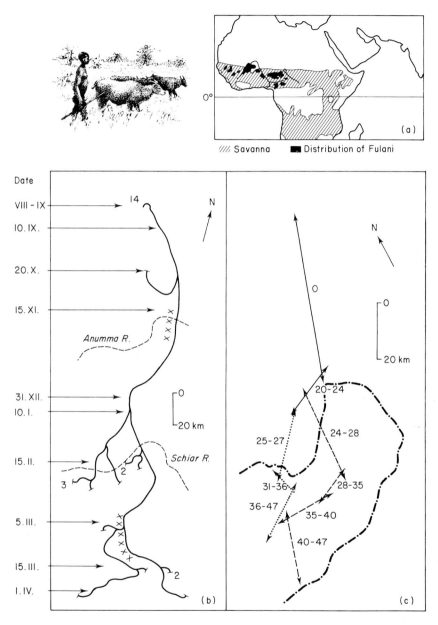

///. Savanna   ■■ Distribution of Fulani

a particular seasonal migration circuit. Solid lines, migration circuit of original group; dotted and dashed lines, migration circuits after dichotomy into two groups.

The sequence starts at 0 (which is taken to be 1880). Numbers by the lines indicate the period, in years after year 0, that a particular migration circuit was used. In this sequence, removal migration has occurred as a result of piecemeal change in the migration circuit each year. In particular there has been almost continuous extension of the southern limit of the dry season range and simultaneous retraction of the northern limit of the wet season range.

The heavy line indicates the edge of the 1300 m high Jos Plateau which is well watered and free of tsetse fly but which, prior to years 24–28, was barred to the pastoral nomads by an intervening group of sedentary, territorial humans. When the territorial defence ceased, the plateau was utilised. At first, the plateau was grazed only during the dry season, but when it was discovered that these dry season pastures, due to tsetse absence, were also suitable for wet season grazing, the southward drift that had already been apparent was further reinforced.

[*Compiled, with modifications, from Stenning (1957, 1959)*]

seasonal return migration is relatively large (i.e. mountain regions, tundra, deserts and grassland—p. 530).

The herds with which pastoral nomads are associated may be divided into two categories (Krader 1959). On the one hand they may consist of a single species of ungulate, such as the reindeer herds of the Lapps, Samoyeds, Tungus, Yakut, Chukchi, Koryak, Yukagir, Kamchadel, and Eskimo. Other single-species herds are those of the cattle raised by the Masai, Herero, Hottentot, and others of eastern and southern Africa and the camels raised by the Bedouin of the Arabian Peninsula. Alternatively, the herds may be composed of more than one species. Such 'complex' herds are found in association with the pastoralists that occupy the great arid grassland and desert belt that stretches from the Atlantic Ocean across North Africa, including the Sudan, to southwest Asia and beyond to interior Asia as far as Tibet and western Manchuria (Fig. 25.5). The complex herdsmen are invariably tent-dwellers, as are the Bedouin nomadic pastoralists.

The relationship between a pastoral nomad and his herd is clearly symbiotic. The herds supply the man with transport, milk, hides, wool and meat as well as dried dung for fires. In fact, the ungulate is killed only infrequently by the man, being much more used for the replenishable products: milk, blood, wool and dung. The man supplies the ungulate with protection from predators and assistance with parturition and suckling and in some instances with assistance in the search for suitable pastures.

As with the ungulates that live in the arid zone, the size of the familiar area of the pastoral nomads that live in this region may be huge. The average year's home range of a deme of the camel- and sheep-herding Bedouin of the Arabian Peninsula is about 55 000 km^2 with an average diameter of about 240 km (Dickson 1949). The size of the year's home range is determined by the relative abundance of water holes and by the presence of sufficient grazing for the camels and sheep. When rainfall is small, a larger proportion of the familiar area may be utilised than when rainfall is greater. During the summer, from June to mid-October, the various demes are concentrated around the water holes, for at this time of year the camels, sheep, and horses have to be watered daily. By mid-October, just before or at the arrival of the rains, the water holes are vacated and dispersal begins, though at first return to the water holes is frequent. The timing of the autumn migration is based on the visual perception of celestial cues. Preparation for the initiation of migration follows the first observation before the dawn of the star Canopus, which in Kuwait is visible soon after the 10 September. Preparation to leave the summer water-hole roosting site begins soon after and migration is often possible within a month. The final date of departure is to some extent governed by the arrival of rain, but by the end of October migration is initiated whether rain has arrived or not. From November to May a migration circuit is performed which remains similar from year to year but the details of which may vary according to the local distribution of rainfall. Decisions concerning the precise direction to be taken at each stage are made according to the directions of night lightning or the positions of cloud banks in the evening. Tents are shifted about every 10 days, partly for sanitary reasons and partly because the grass has been eaten in the vicinity. Normally, each shift involves a movement of about 16–20 km to an area which has been investigated previously by a male member of the deme. In summer, roosts are usually within a kilometre of water but in winter, when camels are not watered at all and sheep and horses drink only once in four days, roosts may be established 30–50 km from water.

In view of the size of the migration distance in the desert and other arid regions, the adoption of the animals being followed as a means of transport seems an inevitable step. Horse riding, for example, first appeared among the pastoral nomads of the Eurasian steppes about 4000 years B.P.

Nomadic pastoralism has occurred in Africa for 5000 years or more, but for 4000 of those years it was shown only by the caucasoid Beja on the western coast of the Red Sea (Murdock 1959). The Beja herds consisted of sheep, goats and cattle and in the north, for the past 1000 years, the camel also. It was not until 1000 years B.P. that nomadic pastoralism appeared elsewhere in Africa and then mainly in the area of the eastern horn surrounding the Beja. About 1300 years B.P. a hybrid group of humans, the Fulani, appeared in Senegal and spread eastward across the Sudan. The more-caucasoid of the Fulani adopted a predominantly nomadic pastoral economy, based largely on cattle, within the savanna zone of the Sudan. The seasonal return migration of a deme of the Fulani and the way in which type-C removal migration may lead to a considerable shift in migration circuit over a period of time are illustrated in Fig. 25.9.

The migration circuit shown in Fig. 25.9 is fairly typical of the migration circuits of the nomadic pastoralists of the arid zone. If we can judge from what is known of ungulate migration, then the migration routes taken by nomadic pastoralists and their herds approximate to those that would be followed by the same or similar ungulates in that region even in the absence of their human symbionts. This is not only true in the arid zone but also in mountain regions and on the tundra, as a few examples may illustrate. Everywhere, the direction and timing of movement is adaptive to the local variation in climate and food availability etc.

The pastoral nomads that winter in the vicinity of the Persian Gulf take their herds to the upland pastures of north and east of Shiraz in the summer. The autumn migration usually occurs in September or October, but timing is facultative. In years of drought, the migration can begin in mid-August.

The pastoral nomads that summer their sheep along the River Euphrates perform a seasonal return migration that takes them up to 320 km from their summer range. These people ride and transport their possessions on donkeys. In mid-October, at about or just before the arrival of the rains, a migration away from the Euphrates is initiated that terminates at the gravelly deserts some 320 km to the south. The return begins at the end of April or early May and is timed so that the people and their sheep arrive back at the Euphrates before the grazing on the intervening desert dries up (Dickson 1949).

The Lapps migrate with their reindeer between their winter home range among the sheltered forests and their summer home range along the coast of the Arctic Ocean (Fig. 25.6). At the present time, some Norwegian Lapps are the only people that still perform this migration, travelling some 400 km each spring and autumn from near the edge of the Finland border along a migration route that is some 400 years old. The northward trek starts at the end of April and the 400 km is covered in 10–12 days. The southward movement takes place at the end of September or beginning of October.

Having reviewed the range of species of terrestrial mammals that perform seasonal return migrations and described the general form of these migrations, let us now consider the application of the familiar-area hypothesis.

The red deer, *Cervus elaphus*, of Europe is found in winter in northern Scotland at altitudes from sea level to 520 m but in summer is found from 520 m to the tops of the hills (Darling 1937). These zones

Fig. 25.10   Reindeer, *Rangifer tarandus*

[*Photo courtesy of Asahi optical and Rank photographic*]

could, therefore, be described as winter and summer ranges. However, females also occupy the winter range in early summer (June) for parturition and both males and females enter the winter range in autumn (end of September to beginning of November) for the duration of the copulation season. The migration pattern is also highly flexible and the deer will migrate down to low altitudes even in summer if the weather is bad. In spring there is a daily altitudinal return migration from lower levels at night to higher levels by day. It seems likely, therefore, that the deer are familiar with the greater shelter from wind and storms that is available at lower altitudes and with the gradual succession of new food at increasing altitudes during the spring and summer and, perhaps, also with the greater protection from the attentions of blood-sucking and parasitic insects at higher altitudes in summer.

Given this familiarity, the deer then respond to prevailing conditions of climate and food availability etc. by migrating to the part of their familiar area that they assess (see p. 347) will offer optimum conditions given that particular situation.

This concept of an animal that is familiar with the resources offered by different parts of a given area and that has learned which habitat offers the optimum conditions at different times of year and under different conditions of climate and food availability, etc. is the central aspect of the familiar-area hypothesis for seasonal return migration. The general applicability of this hypothesis to the migrations of terrestrial mammals must await further investigation. In particular, there are two major questions that require answers. Firstly, how does an individual establish its familiar area during ontogeny and secondly, to what extent are migration thresholds to habitat, migration cost, and indirect variables inherited and to what extent are they learnt?

The problem of the ontogenetic establishment of a familiar area has already been considered at some length (Chapter 18). In most of the larger mammals considered in this section, however, social communication seems likely to be of considerable importance. First, in the majority of the species with a relatively long-distance annual return migration, the young mammal accompanies its mother for at least one performance of the migration circuit. The majority also live in social groups, often in large herds, for the major part of their lives. Even if an animal performs type-C removal migration (p. 311), therefore, it can immediately perform a suitable migration circuit by associating with the animals already present on that circuit. Only when it is necessary for a herd to initiate exploratory migration beyond the previously existing familiar area of all of its members is social communication unlikely to be involved.

The existence of migration thresholds to habitat variables is discussed for the caribou, *Rangifer tarandus*, in relation to snow depth on p. 212. Similar migration thresholds are encountered in all ungulates that experience snow. Thus, the white-tailed deer, *Odocoileus virginianus*, and the mule deer, *O. hemionus*, of America have a migration threshold to snow depth in autumn that in both cases is about 45 cm (A. Kramer 1971). One consequence of such migration thresholds is that an area that is visited one year when the migration threshold is exceeded in some other part of the range may not be visited

another year when snowfall is less. Such facultative year-to-year variability in migration circuit is one of the major characteristics of ungulate seasonal return migration, not only amongst those that encounter snow but also elsewhere. In addition to the other examples given in this section, the Indian elephant, *Elephas maximus*, in the Upper Perak region of Malaya shows such variability. Although the herds may take an established route to reach a certain place on a certain date they do not necessarily take the same route to reach the same place on the same date the following year (Mohamed Khan bin Momin 1967).

The existence of migration thresholds to indirect variables such as photoperiod and temperature and migration-cost variables such as wind and predators (see p. 265) does not seem to have been critically investigated. All of these migration thresholds could be learnt from other, older, individuals as seems likely to be the case for human pastoral nomads such as the Bedouin, and could also be the case for other animals that live in demes containing several generations. As yet, however, there is no critical evidence to indicate to what extent, if any, such learning occurs and to what extent such thresholds are inherited.

One particularly striking component of the migration threshold of at least some ungulates is herd size, the effect of which is most obvious from its influence on the distance component of the return migration. Thus, the barren-ground caribou, *Rangifer tarandus*, of the Canadian Arctic, typically migrates from summer grounds on the tundra to wintering grounds below the tree line (Fig. 25.6). If herd size falls below a certain level, however, the caribou of that herd no longer initiate migration to the tree-line but instead remain on the tundra, thus considerably reducing migration cost, and spend the winter on windward slopes where the snow cover remains sufficiently slight to permit continuous access to food. A similar migration threshold could well exist in the muskox, *Ovibos moschatus*. Early observers of the species in the 1820's concluded that the animals spent their summers on the Arctic islands and their winters on the mainland, undertaking lengthy migrations in the spring and autumn. Recent workers, however, have shown that the spatial separation of summer and winter ranges is now rarely greater than 80 km and may often be zero (Tener 1965). These conflicting reports could, of course, be attributed to differences in observer reliability, but it is also possible that they represent a real change in distance component brought about by a decrease in deme density.

Although the seasonal migrations of ungulates are often conspicuous because they involve a shift of range by whole herds, some important characteristics of the migration are obscured by considering the migrations only at this level. Often, the smaller social units of the herd (i.e. an individual, a single female plus offspring, a monogamous or polygamous family group, an all-male group, or a female band plus their offspring, depending on the species and time of year) are themselves migrating between one season's home range and another season's home range. This is shown clearly by the female mule deer, *Odocoileus hemionus*, in Fig. 25.3 and to some extent also by the female elk, *Cervus canadensis*, shown in Fig. 25.11. The social units of zebra, *Equus quagga*, in the Ngorongoro crater (Fig. 25.16) also are likely to have a wet season and dry season home range incorporating roosting and grazing sites. In this case, however, the situation is less clear because the wet and dry season home ranges show considerable spatial overlap. The distance component of seasonal return migration in the Ngorongoro crater is short because water and grazing are present all year. This is not the case on the Serengeti Plains, however (Fig. 25.7), where the distance component is much greater. Here, however, we have no information concerning the wet and dry season home ranges of individual family units.

In all cases in which an animal migrates as a herd but disperses into smaller social units at either end of the migration, it can perhaps be assumed that the herding behaviour during migration reduces migration cost. A number of factors could contribute to this reduction. One possibility is that there is less predation risk to the individual if in a herd than if solitary. Another is that by travelling in a herd the total familiar area of the group is likely to be much greater, thus giving greater flexibility to the migration route in the event of unusual contingency, such as the normal route being impassable. Finally, an individual displaced off the migration route through, for example, being chased by a predator, is less likely to become lost if other individuals are still present on the migration route. As already stressed (p. 378), familiar areas often consist of a number of smaller areas connected by narrow tracks, either side of which is outside of the familiar area.

Not all animals have just two home ranges (i.e. winter and summer, wet season and dry season, etc.) as did the mule deer in Fig. 25.3. Some may have a number of feeding home ranges, each one reaching peak suitability at a different time of year. In addition, many ungulates often have a parturition home range and a copulation home range. Males of Dall's sheep, *Ovis dalli*, and bighorn sheep, *O. canadensis*, have been found to have 6–7 seasons' home ranges. Females may have up to four (Geist 1971). These home ranges are often on different mountain ranges, necessitating the sheep crossing 5–6 km of low country (see p. 153) during which they travel quickly and in a compact group and during which they often fall prey to wolves, *Canis lupus* (Murie 1944).

Male mountain sheep establish their migration circuit within a familiar area built up by association in their younger years with older rams. Females establish their migration circuit by association with the ewe band (Geist 1971). Extension of the familiar area involves group-to-group removal migration (Chapters 15 and 17). The fact that migration circuits crystallise out of familiar areas established in this way and the infrequent advantage of type-C removal migration (Chapter 17) once an animal has established a migration circuit can lead to situations that appear odd to the human observer. Thus a male mountain sheep at the approach of the copulation season may leave females near which he is feeding and migrate 25 km or so to the copulation grounds on his own migration circuit. Another example is provided by the female elk, *Cervus canadensis*, in Fig. 25.11 that performed a 65 km spring migration that included the crossing of a high divide that was still snowbound to arrive at a parturition ground that seemingly could have been reached by a much more direct route of only 20 km. Of course, the history of this female was unknown but it seems likely that the route taken reflects the exploratory migration route by which that female first came to associate her winter home range and parturition and summer home range.

Most authors, when considering the migrations of ungulates, search for the 'causes' of the migration, often searching for a single habitat variable. A search for a 'cause' is not really appropriate in a study of migrations within a familiar area. A better approach seems to be on the one hand a consideration of the *calculated* migration thresholds (see p. 347) that are involved and on the other hand an evaluation of those features that result in different habitats reaching peak suitability at different times of year. This latter consideration must, of course, take into account the optimum seasonal incidence of the different ontogenetic events such as parturition and copulation. In only one case of which I am aware, however, has such an evaluation been attempted in anything other than an anecdotal way, the exception being the analysis of the winter distribution of caribou, *Rangifer tarandus*, in relation to snow depth that was described on p. 212. Nevertheless, some general observations are possible.

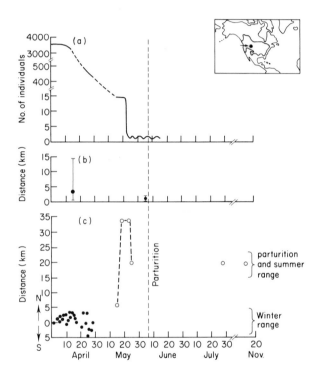

Fig. 25.11 Spring migration of a female elk, *Cervus canadensis*, at Jackson Hole, Wyoming

The elk or wapiti, *Cervus canadensis*, is the American counterpart of the Eurasian red deer, *C. elaphus*, and some authorities consider them to be the same species.

In the experiment illustrated a female elk was fitted with a collar containing a radio transmitter. The movement of this female was tracked during the period from 1 April to 1 May by an Earth-orbiting satellite and thereafter by ground-based tracking systems.

(a) Variation in degree of dispersal and convergence

During the winter the elk of Jackson Hole formed a loose winter herd of 3000–4000 animals. In April the number on the winter range reduced to 400–500. The tagged female performed the spring migration as a member of a group of 15 individuals. Just before parturition (vertical dashed line), however, the female alternated between a solitary existence and association with one other female.

(b) Variation in daily migration distance on winter and summer home ranges

The dot shows the mean, and the vertical line the range, of the straight-line distance between recorded positions on subsequent days while on the winter range and again during the period leading up to parturition.

On the winter range the female moved almost continuously, sometimes finishing the day as much as 15 km from the starting position. On the summer range, however, at least during the period immediately preceding parturition, the female either moved much less and/or had a higher degree of return and never shifted position more than 2 km on any single day.

(c) Shift in latitude

The dots indicate the position of the female expressed as kilometres north or south of the latitude occupied on 1 April. The solid dots are measured positions. The open dots are approximate positions.

Migration from the winter range to the summer range 20 km to the north was executed by a circuitous route of approximately 65 km (dotted line) that involved crossing a high divide that was still snowbound.

[*Drawn from data in Craighead* et al. (*1972*). *Photo of bull elk in autumn by Dr. C.G. Hampson*]

On the Arctic tundra, the major habitat variables seem to be snow depth (through its influence on the accessibility of food and ease of movement), shelter, the chilling effect of the wind, food availability, insect harassment, and the abundance of predators. The way that these factors combine to influence the succession of habitat-types in the case of the caribou are described in the legend to Fig. 25.6. In the muskox, *Ovibos moschatus*, which spends all year on the tundra (Tener 1965), the winter home range occupies windward slopes and hilltops where snow depth is only about 5 cm and the feeding areas can be scraped clear of snow by the front hoofs of the animals. Muskoxen eat a wide variety of plants, including woody species, flowering plants, grasses, sedges and mosses. Calving takes place from mid-April to mid-June, usually on the winter home range. Summer home ranges are largely centred around water sources, either rivers, ponds, or lakes, where willows, *Salix* spp, the preferred summer food, are abundant. In Canada and Greenland, copulation takes place in midsummer (July and August) on the summer home range.

Altitudinal variation in habitat suitability seems likely to result from altitudinal variation in temperature, rainfall, snow-depth, and wind and the effect that this may have on an animal's homeostatic mechanisms and on the seasonal abundance of food, water, and perhaps also insect pests and predators. On the arid grasslands and deserts of the world (Fig. 25.5) the distribution of rainfall and snowfall seems likely to be the factor of primary importance and the migration of animals such as the inhabitants of the African plains (Fig. 25.7) and of the American bison, *Bison bison* (p. 554), and saiga, *Saiga tatarica* (p. 536), are geared in any one area to exploit both the local succession of food and the erratic nature of rainfall (see p. 153).

One of the most conspicuous components of the seasonal return migrations of terrestrial mammals is that of the degree of dispersal and convergence at the different times of year. The observed variation in this component is consistent with the suggestion that selection has acted to give, throughout the year, an optimum group size and composition (p. 390), in a home range of optimum size for the time of year and local conditions (p. 388) with an optimum degree of overlap between the home ranges of neighbouring groups (p. 404). Apart from this, however, no other generalisation is possible because of the way the effect of a specific environmental variable differs from place to place and from time to time depending on the range and nature of other variables in combination with which it is acting.

Fig. 25.12   Caribou, *Rangifer tarandus*   [*Photo by Ernie Kuyt (courtesy of Canadian Wildlife Service)*]

Near the end of the fly season, exhausted and solitary caribou sometimes stand for hours on breeze-swept hill-tops. By autumn (late September) the herds are once more fat and in good condition.

At first sight, it might appear from a wide range of species that when water is the limiting resource, selection favours convergence upon the available water sources, followed, when water becomes less limiting, by dispersal, but that when food is the limiting resource, selection favours dispersal when food availability decreases and convergence when food availability increases. A few examples illustrate the extent to which this is applicable.

Fig. 25.13 Muskoxen, *Ovibos moschatus*, in defence formation

[Photo by Fred Bruemmer]

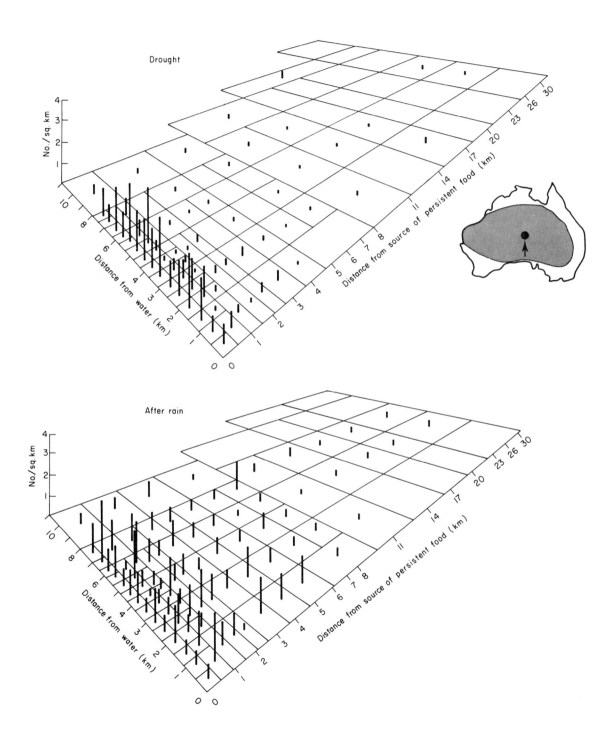

Fig. 25.14   The dispersive-convergent component of the
return migration of the red kangaroo, *Megaleia rufa*, in
central Australia

General ecology
The red kangaroo lives in an area in which rainfall is low
and unreliable and like many inhabitants of this region of

Australia has evolved an opportunistic rather than a regular, periodic breeding physiology. Males mature at 2·5 years of age and females at 3 years of age, though drought delays the onset of maturity by 6 months or so. During drought, females enter a period of anoestrus, 50 per cent entering such a state within 3–5 months of the beginning of a drought period. All mature females ovulate and conceive following rainfall, irrespective of the time of year. If good conditions persist for 8 months, all young-in-pouch survive. If drought conditions intervene, however, some of the young die. After 1·5–2·5 months of drought, 50 per cent of the young are likely to be dead.

The main resources required by the red kangaroo are green herbage and shady trees. The animal prefers sprouting grass and only drinks during severe drought.

Return migration
The diagram shows the distribution of a deme of the red kangaroo in one case during a drought and in the other case after rain. Distribution is expressed as distance from water and persistent food. There is a convergent migration during periods of drought and a dispersive migration following the onset of rain. Because of the irregular nature of the rainfall, the periodicity of the return migration is also irregular. Degree of return has not been investigated.

[*Compiled from Newsome (1965a, b, c)*]

Figure 25.14 shows the change in spatial distribution of a deme of the red kangaroo, *Megaleia rufa*, in central Australia as drought gives way to rain. The response to water shortage is convergence and the response to rain is dispersal. A similar migration is shown by the waterbuck, *Kobus ellipsiprymnus*, in Africa which establishes a home range along the borders of rivers. During the wet season, the home range is extended further away from the river than during the dry season (Herbert 1972). Water and its spatial relationship to suitable grazing is also the critical factor influencing the degree of dispersal and convergence in the African plains' animals (Fig. 25.7). Wildebeest, *Connochaetus taurinus*, during the wet season, are dispersed across the open plains, albeit in large herds. In the dry season, however, after migration into brush country, the aggregation that occurs around the limited permanent water and the influence of occasional local rain showers on the availability of temporary water and green grass, may produce local convergences of up to 200 000 animals (Talbot and Talbot 1963). The peak of parturition occurs in January and February when the plains are green and the herds dispersed. There is a higher calf mortality at higher herd densities due to separation from their mother. The African elephant, *Loxodonta africana*, in Uganda also migrates from wet season ranges on the grasslands to dry season ranges in the savanna woodlands and forests (Field 1971). Again the dispersive–convergent migration is from dispersed groups of 3–20 individuals during the wet season to temporary large herds of up to 200 (Carrington 1958) or even 1000 (Hill *et al.* 1953) individuals some time between the beginning of the dry season and the return of the rains. Finally, as far as large herbivorous mammals are concerned, in the hippopotamus, *Hippopotamus amphibius*, an animal that uses water bodies as diurnal roosting sites (p. 179) as well as for physiological purposes, there is once again a convergence on permanent water bodies during the dry season and a dispersal to short-lived inland wallows during the rains, particularly by the males (J.R. Clarke 1953, Field 1970).

The bushmen, *Homo sapiens*, of the Kalahari Desert, collect a wide variety of plants, not only for their nutritional value but also for their water content. Rain water is stored in ostrich egg-shells and calabashes and buried in cool earth or concealed in a tree until needed. Such supplies are rapidly exhausted in the dry season, at which time tsama melons, *Colocynthis citrullus*, and gemsbok cucumbers, *C. naudinionus*, and other water-rich species are eaten as well as the rumen-contents of buck. If all of these sources of water fail, bushmen make rapid calculated convergent migrations to the nearest of the permanent springs that are dotted about the Kalahari and establish a roost nearby until the drought breaks (Tobias 1964). During the period that the Hadza were studied by Woodburn (1968a,b) they too, in contrast to when studied by Tomita (1966) (Fig. 25.15) performed a dispersive migration during the wet season and a convergent migration during the dry season. Pastoral nomads of the desert, whether they herd camels, sheep or goats, converge for the summer and disperse for the winter. Convergence is a response to the scarcity of water during the summer. The camel- and sheep-herding and horse-riding Bedouin of Kuwait may form aggregations of as many as 3000 tents around a water hole during the summer. The sheep-herding,

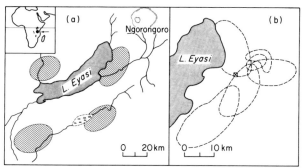

Fig. 25.15 Degree of dispersal and convergence in the seasonal return migration of the Hadza: a deme of hunter–gatherer humans, *Homo sapiens*

(a) Distribution of the Hadza

The hatched areas show the distribution of four large-sized groups into which the Hadza may be divided around Lake Eyasi, northern Tanzania.

(b) Seasonal home ranges of the northeastern large-sized group

The northeastern large-sized group at the time under consideration consisted of about 80 individuals. ×, main convergence site during the wet season (November through May).

Aggregation is at a maximum during the wet season when fruit is abundant. Fruit is collected mainly by the females and young. Hunting is performed in the morning and evening. When the prey is an ungulate the hunting unit is usually either one or two males. Bows, and arrows with poisonous tips, are used. Females may participate in the following of wounded prey and in the wet season females and juveniles hunt birds with sticks and stones. Dead prey is cut up and carried back to the roosting site, a temporary hut, to be cooked and eaten.

From May, at the end of the wet season, through August, honey is gathered as well as roots and stalks. During this period the group splits into increasingly small units, eventually into families, and disperses into the mountain area. Hunting becomes increasingly important and during September and October, when honey is no longer available, is virtually the only means of acquiring food.

[*After Tomita (1966)*]

donkey-riding people that converge on the Euphrates in summer, in late winter and early spring (February and March) cover the whole of the area to the south and west of Kuwait (Dickson 1949).

The zebra of the African plains behave in a way similar to the wildebeest described above (Fig. 25.7). Where water is available throughout the year, however, and food abundance and the distance of daily return migration between food and water become the critical factors, then the seasonal timing of the dispersive and convergent migrations changes. Such a situation is illustrated for the zebra, *Equus quagga*, in the Ngorongoro crater in Fig. 25.16. Here, the dispersive phase of the migration takes place during the dry season and the convergent phase during the wet season. In this case, however, as shown, the situation is not totally independent of the availability of water. In the chital or axis deer, *Axis axis*, of India, however, water does not seem to be a limiting factor. The species occupies secondary or open forests with a good understorey of grasses, forbs, and tender shoots but avoids the interior of extensive tall forests. It feeds mainly on green grass less than 10 cm high and also on the tips of coarse grass as well as on some browse. Typically, the species forms large groups when food is abundant but disperses into smaller groups when food is scarce (Schaller 1967).

The largest prides of lions, *Panthera leo*, are formed in regions where game is plentiful (Guggisberg 1961). Among hunter–gatherer humans, the bushmen of the Kalahari aggregate when food is plentiful during the rainy season (Murdock 1959) as do the Hadza (Fig. 25.15), and the Birhor of India (B.J. Williams 1968). The Yumbri of the monsoon forest of Northern Thailand perform a convergent migration once a year and for a period of about three months up to 100 individuals live in communal roosts. When food becomes scarce once more a dispersive migration is performed during which some sexual segregation takes place (G. Young 1962). Finally, during periods that the food supply is stable, large-sized groups of chimpanzees, *Pan troglodytes*, in savanna woodland congregate frequently and rarely disperse as solitary, sexual, or nursery groups for more than a day. During periods when the food supply is scarce, however, the group either shows rapid movement in a single unit or becomes extremely dispersed for perhaps a few weeks at a time (Izawa 1970).

These different patterns of dispersal and convergence depending on whether food or water is the limiting resource are shown even among congenerics living in different parts of the world. Among cattle, for example, the demes of semi-wild Camargue cattle, *Bos taurus*, of southern Europe, disperse during the winter when food is scarce but converge again in spring when food once more becomes abundant (Schloeth 1961). The banteng, *B. javanicus*, in Udjung Kulon on the western tip of Java, however, which experiences a dry season from August through December, forms its largest concentrations, up to 50 individuals, during this period (Hoogerwerf 1970).

Convincing though the correlation may appear from the examples presented, however, the critical factor is not whether the limiting resource is food or water but rather whether or not the limiting resource is evenly distributed or clumped. As it happens (and hence the above correlation), for most of the mammals described in the previous few paragraphs, when food is limiting it is spread fairly evenly but at a low density over a large area. When water is limiting, however, it occurs in local abundance in permanent water bodies such as rivers or lakes. In some situations, however, this is not the case and the correlation apparent from the previous paragraphs breaks down. On the Eurasian steppes, especially to the east of the Caspian Sea, the ungulates and humans of the arid zone experience severe winter cold which contrasts with the relatively mild winters and extremely hot summers of the other part of the arid zone to the south and southwest of the Caspian Sea (Fig. 25.5). In the latter zone, as already described for both ungulates and humans, the animals converge in the dry season and disperse in the wet season when water is limiting, as for the Bedouin, and converge in the wet season and disperse in the dry season when food is limiting, as for the Fulani (Fig. 25.9). In the cold part of the arid zone, however, water is plentiful in winter, albeit as frost, and food is scarce. On the previous correlation these conditions should favour dispersal. However, the distribution of food is that of local abundance wherever the ground has not been covered by snow. In this region, therefore, the antelopes and the kulan, *Equus hemionus*, and the human pastoral nomads, form their largest aggregations in winter (Krader 1959, Formozov 1966). Aggregation also protects the ungulates against winter storms. If the winter becomes severe, however, and food becomes non-localised, remaining scarce, as it does also in the summer, then a dispersive migration is performed.

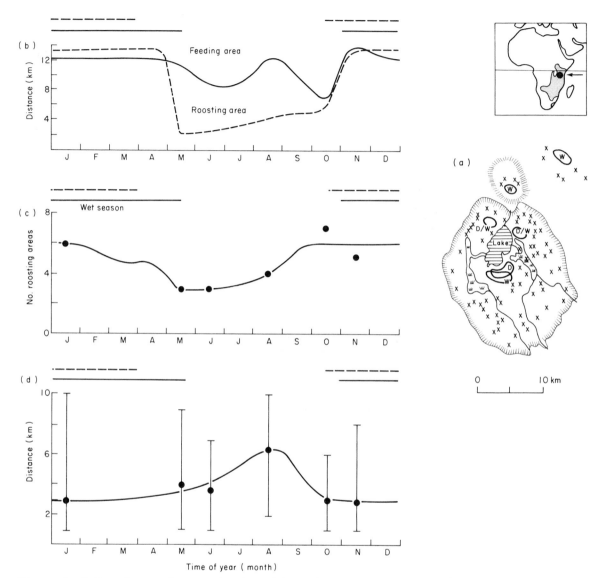

Fig. 25.16   Daily and seasonal return migrations of the plains' zebra, *Equus quagga*, in the Ngorongoro crater, Tanzania

The primary social units of the plains' zebra are a one-male bisexual group and an all-male group. On average a one-male bisexual group in the Ngorongoro crater consists of one adult male, three adult females and three young. Maximum group composition is one adult male, six adult females, and nine young. All-male groups contain 1–9 individuals. Average size for an all-male group, including solitaries, is 2·9 individuals. These primary social units are further organised into herds in that large numbers of units congregate in specific sleeping (roosting) and grazing grounds between which daily return migration takes place. The primary groups do not show territorial behaviour but move freely in large ranges which they share

with other primary groups. The roosts are open, dry places with short grass cover (top photo). At the end of the wet season (bottom photo) and beginning of the dry, grazing is in long grass areas in dry parts of the crater floor. At the end of the dry season, grazing is almost entirely in the dried-up swamps which still have green vegetation.

(a) Map of Ngorongoro crater to show the lake, streams and marshes. Crosses, grazing areas; open areas, roosting areas which are used either during the dry season (D), the wet season (W), or throughout both (D/W). Seasonal return migration can be considered in terms of the seasonal shift of roosting areas.

In (b), (c) and (d) the solid line above each figure indicates the wet season and the dashed line indicates the peak parturition period.

(b) Seasonal variation in the distance between the nearest point of the lake and the roosting and feeding areas. Dashed line, maximum distance from lake of the roosting areas; solid line, maximum distance of feeding areas from lake. Roosting areas are all close to the lake during most of the dry season but by the wet season some may be further away. In October, at the end of the dry season, grazing occurs close to the lake in the dried-up swamps.

(c) Dispersal/convergence component of seasonal migration: variation in the number of roosting areas (solid dots), and therefore herds, with season. The herds split and use an increasing number of roosting areas as the dry season progresses. As vegetation increases again during the wet season, the animals use fewer roosting areas once more.

(d) Variation in the distance component of daily return migration with season. Solid dot, mean distance between roosting and feeding areas; vertical bar, range of distances. Mean distance is more or less constant at 3 km during the wet season. Mean, minimum and maximum distances all increase from the beginning to the middle of the dry season. By the end of the dry season, however, all distances are shorter than at any other time of the year.

In summary, the plains zebra, *Equus quagga*, in the Ngorongoro crater has a daily return migration of a few kilometres between roosting and feeding areas. Throughout the dry season, roosting areas are near to the lake. As the dry season progresses, the distance between roosting and feeding areas increases. Half-way through the dry season the herds split and use more roosting areas that are near both to the lake and the feeding areas. As soon as the wet season starts, when food is still at a premium but water less so, the herds remain split and some factions move to roosting areas away from the lake. At the end of the wet season the distant herds return and join others in using the roosting areas near to the lake.

[*Drawn from data in Klingel (1969). Photos by H. Klingel*]

In a similar way it is the spatial distribution of prey in relation to the spatial distribution of areas suitable for establishing roosting sites that is the critical factor for predators. The spotted hyena, *Crocuta crocuta*, of the Serengeti, for example, which is predatory on the migratory ungulates, has dens scattered over the plains during the wet season when the prey are similarly scattered (Kruuk 1972). Only 13 per cent of the dens observed at this time were within 5 km of the woodland edge of the plains. During the dry season, however, when the plains' ungulates have migrated into the woodland savanna (Fig. 25.7), the hyenas perform a convergent migration to the edge of the plains, 72 per cent being within 5 km of the woodland edge. It is during this period that the hyenas divide into two categories, some becoming 'commuters' and others remaining with the migrating herds throughout their migration (p. 538).

Finally, the situation becomes even more complex when group size and efficiency is included. One example, the insulating effect of large herds of ungulates, has already been given. For predatory animals there is always the possibility that in some situations an increase in the size of the co-operative group may increase rate of prey capture per individual. Among humans, for example, the Netsilik, Iglulik, and Copper eskimoes form their largest groups in the winter (Damas 1968). Through the presence of sufficient males to wait at the majority of the breathing holes within a seal's winter home range (p. 711), the chances of catching that seal are greatly improved. Similarly, the Mbuti pygmies require large co-operative bands to catch their prey with nets, but at times when food is easily caught a dispersal migration takes place (Turnbull 1968).

The examples that have been given of the degree of dispersal and convergence in the seasonal return migrations of terrestrial mammals tend to support the thesis that this component can be attributed to the selective pressures acting on group size and composition, size of home range, and degree of overlap of neighbouring home ranges (p. 404). At the same time, they emphasise the range of environmental features that have to be taken into account and the way that the same variable acting within different combinations of other variables can favour quite different adaptations in the timing of the dispersive and convergent migrations.

One characteristic of all demes that perform a dispersive–convergent return migration is that some individuals travel further than others. It might be supposed that in most cases the situation is free (p. 652). However, the possibility that in some instances the situation is despotic cannot be ignored.

Associated with the question of degree of dispersal and convergence is that of sexual segregation. In many species, males and females segregate, often for long periods, during the annual migration cycle. Thus males and females of mountain sheep tend to occupy the same home range only during the copulation season (Geist 1971) and male barren-ground caribou migrate northwards in the spring later than the females (Banfield 1954). In most, if not all, cases sexual segregation can be attributed either to the different requirements or the different capabilities of the two sexes. Thus male sheep are not encumbered by young offspring and can reach feeding grounds not available to females. Pregnant female caribou and female saiga in spring both have to migrate to parturition grounds spatially separate from spring feeding grounds, produce their offspring, suckle them to a suitable stage, and then migrate to the feeding grounds in time not to miss the spring vegetation. There is evidence that it is the advantage of performing parturition etc. and yet still being able to arrive at the feeding grounds in good time that is the critical factor in determining the date, speed, and threshold of spring migration. This evidence is provided by the fact that barren females, at least in the saiga, often accompany the males rather than the pregnant females (Bannikov *et al.* 1967).

Finally, let us consider the direction component of seasonal return migration. Some authors, doubtless as a result of an ornithocentric approach to migration, have gone to great lengths to attribute a N–S migration axis to ungulates. It has long been considered, for example, that before their eradication during the nineteenth century the great herds of buffalo or, more accurately, bison, *Bison bison*, on the American prairies (Fig. 25.5), performed regular N–S seasonal return migrations between northern summer grounds and southern wintering grounds, often over distances of many hundreds of kilometres. Recent critical analysis (Roe 1972) of historical references, however, shows that nowhere were the migrations of bison regular and that only occasionally were they oriented along a N–S axis. Rather, the bison formed herds, albeit often large (perhaps on occasion as many as 4 million), each one of which performed a migration circuit adaptive to local conditions of the type characteristic of the other plains' ungulates already considered. Inevitably, of course, on occasion some of these circuits included a movement in spring that was more or less to the north and a movement in autumn that was more or less to the south. Only along the southern edge of the range, however, on the plains of Texas, does such a N–S axis seem to have been a noticeable component of the migration circuit. Parturition occurred in late spring and early summer (mid-April to end of June), copulation coming a month or so later in mid-summer to early autumn (July to September).

This absence of a N–S axis to the seasonal migration of the bison of the plains and the seasonal incidence of parturition and copulation deduced by Roe (1972) is fairly consistent with information deriving from surviving demes of the species such as those of the woods and plains of Alberta and the northwest territories of Canada studied by Soper (1941). Here, at about latitude 60°N, there is a dispersive westward migration in spring and a convergent eastward migration in autumn with a one-way distance component of between 8 and 240 km. The bison that have spent the winter on the plains start to migrate to the west in a methodical manner in April and May. Summer movements are to habitats providing a number of resources such as water, forage, wallows, and salt licks, and some

protection from harassment by flies. Copulation takes place in August and at the end of the month a leisurely migration begins toward the southeast. The soft and boggy plains of the region are not invaded for the winter, however, until the ground freezes. Much of the migration takes place along well-established trails.

The only ungulate that consistently performs a N–S migration is the barren-ground caribou (Fig. 25.6) and even in this case the movement is not strictly N–S but rather NE–SW. It is probably even more meaningful, however, to say that the migration is perpendicular to the tree-line (Banfield 1954). In other words, from their summer range on the tundra the caribou migrate to the tree-line by the shortest possible route. The migrations of other caribou and reindeer are similarly oriented (in a spatial sense rather than a behavioural sense) to topographical or vegetational features, such as coast lines, in such a way as to reduce migration cost to a minimum and to increase the habitat quotient (p. 78) to a maximum. As with all of the migrations considered in this section, therefore, the migration is adaptive to conditions within the familiar area, but in few cases are the latitudinal gradients within the familiar area sufficiently steep to produce a N–S migration axis (see p. 654).

26

Seasonal and ontogenetic return migration by bats

Table 26.1 summarises the classification, distribution, and feeding habits of the seventeen families of existing bats.

Despite the structural modifications that were involved in the evolution of flapping flight, the common ancestry of bats (order Chiroptera) and those insectivorous mammals that belong to the order Insectivora seems clear (J.Z. Young 1962). However, the earliest known fossil bat, *Icaronycteris index*, from an early Eocene (50 million years B.P.) Wyoming lake is not very different from modern bats and tells us little about the sequence of events that took place during the early evolution and differentiation of bats as a group (Jepsen 1970). By the middle Miocene (17 million years B.P.), bats existed that were substantially similar to some recent microchiropteran genera, such as *Rhinolophus*, *Hipposideros*, and *Myotis* (Tate 1946). However, not all mid-Miocene bats were so similar to recent forms. Contemporary with the above, bats of the vespertilionid line existed which still possessed the claw of the index finger. Among recent bats this feature is found only among the Megachiroptera which, as a group, retain more primitive characteristics than do the Microchiroptera.

From a presumed palaeotropical origin, bats have spread throughout the world from as far north as the Arctic circle to as far south, excluding Antarctica, as there is continental land. In the Eastern Hemisphere, the vespertilionid long-tailed bat, *Chalinolobus tuberculatus*, and the mystacopid short-tailed bat, *Mystacina tuberculata*, range south to Stewart Island, south of New Zealand, at 47°–48°S latitude (Dwyer 1962), whereas in the Western Hemisphere, the Chilean myotis, *Myotis chiloensis*, has been found at latitude 55°S on the north shore of Navarino Island, just south of Tierra del Fuego (Koopman 1967). Four species of *Myotis* are found in extreme

southern Alaska at 55°N (Hall and Kelson 1959) whereas in Scandinavia bats may range to 70°N (Ryberg 1947), The majority of bat families and species, however, in both the Old and New Worlds, occur in the tropical regions.

Bats, along with rodents and one species each of primates and carnivores, are the only placental mammals to have reached Australia prior to about 200 years B.P. Bats appear to have migrated from Asia from about the early Tertiary (60 million years B.P.) onwards (Simpson 1961). However, there has been no extensive radiation in Australia. New Guinea appears to have been the main centre for bat differentiation in the region, though mainly at the generic and species levels. Similarly, the species groups that are still found in New Guinea, the Philippines, and the Solomon Islands contributed many of the genera that are found on the more remote Pacific Islands such as the Carolines, Fijis, Hawaii, New Caledonia, and New Zealand (Tate 1946). Other genera are drawn from much wider-ranging intercontinental groups. Only one species, however, the hoary bat, *Lasiurus cinereus*, which is found on Hawaii, has reached Oceania from America.

26.1. Factors contributing to habitat suitability

Except where indicated to the contrary, the details of bat biology presented in this section are taken from G.M. Allen (1939) and Yalden and Morris (1975).

The food habits of the 17 families of bats are summarised in Table 26.1 and Fig. 26.1 shows the way that the number of species with different food habits are distributed according to latitude.

Throughout the Old World tropics, the bats that specialise in feeding on nectar, blossom, and fruit all belong to the only megachiropteran family, the Pteropidae (flying foxes or fruit bats). The niche left vacant in the neotropical region due to the failure of the Pteropidae to reach the New World has been filled by some sub-families of the microchiropteran leaf-nosed bats, the Phyllostomidae. The neotropical blood-sucking vampire bats, the Desmodidae

Table 26.1 Classification, distribution and food habits of bats

Family	Common name	Distribution	Food
Megachiroptera			
Pteropidae	fruit bats and flying foxes	*palaeotropical* Cyprus, Africa, India, Indonesia, Australia, Pacific Is. E. to Samoa	fruit, blossom
Microchiroptera			
Megadermatidae	big-eared bats, false vampires	*palaeotropical* Africa, India, Malaysia, Philippines, Australia	insects, fish, small bats, birds, mice, frogs, lizards
Rhinopomatidae	mouse-tailed bats	*palaeotropics and subtropics* Egypt, India, Burma, Sumatra	insects
Hipposideridae	horse-shoe bats	*palaeotropics and subtropics* Africa, E. to India, Malaysia, N. Australia, Solomon Is.	insects
Rhinolophidae	horse-shoe bats	*palaeotropics and temperate* Africa, India, S. China, Indonesia, Eurasia N. to England and Japan, Australia	insects
Nycteridae	hollow-faced bats	*palaeotropical* Central Africa, Malay peninsula, Java, Timor	insects
Mystacopidae	short-tailed bats	New Zealand	insects
Myzopodidae	sucker-footed bats	Malagasy	insects
Noctilionidae	hare-lipped bats	*neotropical* S. and Central America, West Indies	insects, crustaceans, fish
Natalidae	long-legged bats	*neotropical* America N. to Central Mexico, Caribbean	insects
Furipteridae	smoky bats	*neotropical* S. America	insects
Phyllostomidae	leaf-nosed bats	*neotropical* S. and Central America, Mexico, SW border USA, West Indies	insects, frogs, lizards, birds, mice, small bats, fruit, nectar, blossoms
Desmodidae	vampires	*neotropical* S. Brazil N. to S. Mexico	blood of birds and mammals
Thyropteridae	disc-winged bats	*neotropical* S. and Central America	insects
Emballonuridae	sheath-tailed bats	*world-wide* tropical, subtropical, incl. Pacific Is.	insects
Molossidae	free-tailed bats	*world-wide* tropical, subtropical, incl. Pacific Is.	insects
Vespertilionidae	vespertilionid bats	*world-wide* within limits of tree growth	insects

(Fig. 26.4), probably share common ancestry with the Phyllostomidae.

Although insectivorous bats are everywhere more numerous than bats with other food habits, they are rivalled in numbers in the tropics by the nectar, blossom, and fruit-feeding bats (Fig. 26.1) and many tropical plants (chiropterophiles) have become spec-ialised to be pollinated by bats (Vogel 1968). As far as frugivorous bats are concerned, any one species tends to feed on a wide variety of fruit and blossom, changing from one food to another as each plant species comes into season (Baker and Bird 1936).

The vast majority of bats restrict their time investment in feeding activity to the dark hours.

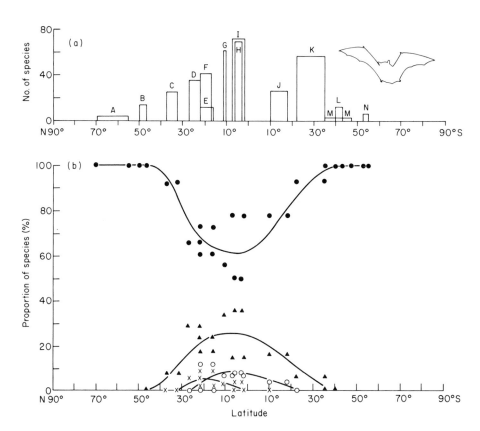

Fig. 26.1 Latitudinal variation in the number of species and food habits of bats

(a) Variation with latitude in the number of species of bats. Height of columns shows the number of species; width of columns shows the range of latitudes over which the sample was taken. Letters refer to the area in which the number of species is found: A, Alaska; B, Washington; C, Arizona; D, Tamaulipas; E, Veracruz >1200 m; F, Veracruz <200 m; G, Trinidad; H, Surinam; I, Malay Peninsula; J, Cape York Peninsula, Australia; K, South Africa; L, Tasmania; M, New Zealand; N, Tierra del Fuego. Number of species increases toward the equator.

(b) Variation with latitude of the food habits of bats. Each point refers to the percentage of species in the sample area with a particular food habit: solid dot, species that feed on insects; triangle, species that feed on blossom, nectar or fruit; open dot, species that feed on other vertebrates such as fish, frogs, lizards, birds, small rodents, or other bats; crosses, species that feed on the blood of birds and mammals. Lines have been drawn only to facilitate interpretation. Insectivorous species are more common at all latitudes than all other species together and at temperate latitudes represent the only feeding type. Other habits, particularly that of feeding on blossom, nectar or fruit, increase in species representatives toward the equa-tor. A graph of this type is inevitably biased by the choice of localities. For instance, sanguivorous bats do not disappear immediately south of the equator as suggested but are found in South America as far south as about 35°S. However, the figure serves to show general trends.

[Drawn from data in McNab (1971)]

Temperate and tropical insectivorous bats can fill their stomachs with food within about an hour of beginning to feed and in general they forage most actively shortly after sunset, being almost completely inactive during the latter part of the night (J.H. Brown 1968). In contrast, tropical fruit-eating species, and perhaps fish-eating species as well, tend to have their activity more evenly distributed throughout the dark hours, though some species have definite peaks later in the night when the insectivorous bats are less active.

As far as enemies are concerned, predatory birds, particularly owls and hawks, are probably the major predators. Owls may catch bats in flight or at roost at night and any bat that takes to the wing during the day, because of its relatively slow flight, has a high probability of falling prey to any hawk that may be in the vicinity (Mueller 1968). In addition, tree-roosting bat species may be taken from their day roost by hawks or eagles. For instance, wedge-tailed eagles, *Aquila audax*, and white-breasted sea eagles, *Haliaecstus leucogaster*, have been seen to swoop into a tree being used as a day-roost by the Australian grey-headed flying fox, *Pteropus poliocephalus* (Fig. 26.5), and to succeed in taking unsuspecting individuals. If the eagle fails to catch a bat with its first swoop, however, the now-alert and defensive bats can invariably ward off the predator (Nelson 1965b).

Bats may be preyed upon by other bats. The false vampire, *Macroderma gigas*, which is the only representative of the family Megadermatidae in Australia (Keast 1968b), preys extensively on smaller bats as well as on other small vertebrates (Allen 1939). A parallel evolution has taken place in the neotropical region among some subfamilies of leaf-nosed (phyllostomid) bats, and in particular *Phyllostomus hastatus*. Bats that prey upon other bats, however, appear to be absent from temperate regions (a possible exception being the hoary bat, *Lasiurus cinereus* (Fig. 26.12), which may take smaller species—Barbour and Davis 1969). Some Carnivora prey on bats if they encounter them, but, in North America, only skunks, *Mephitis mephitis*, opossums, *Didelphis marsupialis*, and raccoons, *Procyon lotor*, seem to enter caves specifically to search for bats. In some places, man seems to be an important predator, particularly of the larger fruit bats and flying foxes (Megachiroptera) (Allen 1939). Such predation occurs, for instance, in the Middle East, India, Indonesia, and Eastern Australia. In some parts of China and Africa, some members of the Microchiroptera are also eaten by man.

In Australia, pythons, *Morelia* spp., are often found to have fruit bat, *Pteropus* spp., remains in their intestine (Nelson 1965b) and, in North America, rat snakes, *Elaphe guttata* and *E. obsoleta*, enter the communal cave roosts of various species and climb up among and prey upon the roosting individuals. Other snakes, such as boas (in Cuba), colubrids (in India and Thailand), and mambas (in Africa) are also known to capture bats (Allen 1939).

Bats often scoop low over ponds and streams in order to drink or, in some species, to catch fish and crustaceans (Table 26.1) and while doing so have occasionally been known to be captured by large trout. Insects may also prey upon bats. In the tropics, migrating driver ants may invade a bat roost and although most adults usually escape, any young that fall from the roost or are otherwise left behind have little chance of surviving. Falling from the roost is a major hazard to young bats that cannot as yet fly. Although these fallen bats may be rescued by their mothers, those that remain on the ground are liable to be eaten by cockroaches. For example, any young of the southeastern myotis, *Myotis austroriparius*, in Florida that fall to the floor of the roost are eaten by the American cockroach, *Periplaneta americana* (Rice 1957). Finally, bats act as hosts to a wide range of ecto- and endo-parasites (Ryberg 1947).

Estimates of the nett effect of the above and other mortality factors, such as drowning, freezing, and starvation, etc. are unreliable (Fig. 17.21), but an annual mortality of the order of 40–50 per cent seems to be a reasonable estimate. Nevertheless, whereas it is probably true that the majority of bats live less than three or so years, some individuals can live much longer. For instance, an adult female of the little brown myotis, *Myotis lucifugus*, that was marked in a summer cave roost at Mashpee, Massachusetts, in 1936 was recaptured in a winter colony 270 km away in 1959 when the bat was at least 24 years old (Griffin and Hitchcock 1965).

Herreid (1964) found no evidence that the longevity of north temperate bat species was greater than that of tropical species.

Bats utilise a wide variety of environmental features as day-roosts but undoubtedly the single most widely used roosts are caves. Cave-roosting bats usually secrete themselves in the cracks and crevices of, or hang singly or in groups from, the walls and ceiling of the cave. However, this is not always the case. Daubenton's myotis, *M. daubentoni* (Fig. 26.22), in the underground limestone quarries of North Jutland, Denmark, also hibernates amongst, and often as far as 60 cm beneath the surface of, the layer of gravel that litters the floor of the cave (Roer and Egsbaek 1966).

The importance of caves as a roost varies widely among the different cave-roosting species. Some species may roost in caves throughout the year. Other species, although roosting in caves at one time of year, may use other sites at other times. In addition, some species only use caves in certain parts of their range. For example, the noctule, *Nyctalus noctula* (Fig. 26.13), hibernates in caves in Eastern Europe (Strelkov 1971). In Central Europe, however, the same species hibernates in buildings and trees (Eisentraut 1936) and in northwest Europe tree-cavities are used as roosts throughout the year (Sluiter and Heerdt 1966).

When not roosting in caves, temperate species use a variety of sites. The undersides of bridges, storm sewers, tombs, attics, eaves, cellars, hollow trees, cracks or crevices in rocks and trees, and even the burrows of some vertebrates, may all serve as roosts (G.M. Allen 1939). In the tropics many species in addition roost in such places as birds' nests, hollow bamboo stems, or the central rolled-up leaves of plants such as banana. Some species of leaf-nosed bats (Phyllostomidae) modify the palm frond in which they roost by cutting half-way through the frond so that its distal portion hangs down, thus providing shelter from sun and storms. Among the Megachiroptera, members of the genus *Pteropus*, the flying foxes, seasonally form large colonies in trees during the daylight hours. Other genera, however, may be solitary and may use caves, and the rousette bats, *Rousettus* spp., are all cave dwellers. It seems to be a more or less general rule among bats that tree-roosters are more brightly coloured than cave-roosters and a number of cryptic colour patterns seem to have evolved in the former type.

Species which roost in tree-cavities often find themselves in competition with other animals that use the cavities. For instance, the noctule, *Nyctalus noctula*, when it leaves hibernation in Holland at the end of March and beginning of April (Fig. 26.13), is restricted in its choice of roosting site in the summer area by the presence of nesting starlings, *Sturnus vulgaris*, in the tree-cavities. When the birds leave in May the noctules shift from their roosts into the more suitable roosts that had been occupied by the birds (Sluiter and Heerdt 1966).

The factors that influence the thermoregulation and hibernation of bats are discussed in Chapter 14 (p. 98).

Figure 26.2 shows the variation in reproductive season with latitude. In temperate latitudes there tends to be a distinct season for copulation and a distinct season for parturition and in general copulation occurs in the autumn and/or winter and spring and parturition occurs in the summer. In the tropics and subtropics, although each species tends to have its own limited season for reproduction, different species reproduce at different times of year. A single breeding season per year is usual but some species, such as the Australian large-footed myotis, *Myotis adversus*, in southeast Queensland, have more than one reproductive season per year. Hence, in this species copulation and ovulation occur from April to June and again from September to November and the females give birth in October and again in January (Dwyer 1970). Within a degree or so of the equator some species, such as the East African free-tailed bat, *Chaerephon hindei*, at Jinja, Uganda, at latitude 0°26′ N, reproduce throughout the year (Marshall and Corbet 1959). However, even in this case, although throughout the year at least 37 per cent of the females are pregnant at any one time, peaks of births still occur that coincide with the period of peak rainfall.

Copulation usually occurs while the female is hanging from a roost. In the red bat, *Lasiurus borealis*, however, copulation may be initiated in flight and may be in progress when the pair fall to the ground (Barbour and Davis 1969). Males of the western big-eared bat, *Plecotus townsendii*, maintain throughout hibernation a higher level of activity than females and copulation may occur, often while the female is still torpid, at any time during hibernation from September to February (Pearson *et al.* 1952).

The grey-headed flying fox, *Pteropus poliocephalus*, of Australia, in common with other *Pteropus* spp., establishes territories during the mating season (Nelson 1965b). Mate selection takes place before the setting up of the territories, which then function as a pre-copulatory guarding area (Parker 1970a). The males mark out the territories by rubbing branches with their scapular glands and also perform most of the fighting in defence of the territory. However, if the male is asleep, the female defends the territory initially. The noise then invariably wakes the male who takes over the defensive role (Nelson 1965b). Once the female has conceived, she leaves the territory, the sexes segregate, and the territories break up (Fig. 26.5). The territorial units are of two types (Nelson 1965b). One type consists of a male, a female, and the young bat that was born to that female four or five months earlier. The other type of unit consists of one male and either one or several females but without attendant young. By the time territories have been established, a gradient of density has developed within the summer camp in which, in *P. poliocephalus*, parturition and copulation take place. At the edge of the camp there is a density of about 1–5 bats per tree. At the centre, the density approaches 400 bats per tree.

Territorial behaviour has not been studied directly in other species. However, statistical analysis (Dwyer 1970) of temporal and spatial distribution patterns of the Australian large-footed myotis, *Myotis adversus*, suggests that this microchiropteran species, in southeast Queensland, shows male territorial behaviour during the mating seasons (April to June and September to November). Harem

formation also seems to take place, though with the females remaining 'loyal' to a particular spatial unit in the cave rather than to a particular male. Although the grey-headed flying fox, *Pteropus poliocephalus*, and the Australian large-footed myotis, *Myotis adversus*, appear to be the only species in which the phenomenon has been studied, it seems likely that territorial behaviour will be found to be a quite common phenomenon in bats. For instance, it will be surprising if the areas patrolled by solitary red bats, *Lasiurus borealis* (Barbour and Davis 1969), or the tree holes occupied in autumn by solitary male (though often accompanied by female) noctules, *Nyctalus noctula* (Sluiter and Heerdt 1966), are not territories, in any sense of the word (p. 408).

The gestation period of bats can vary from between 50 days, as in some pipistrelles, to 8 months,

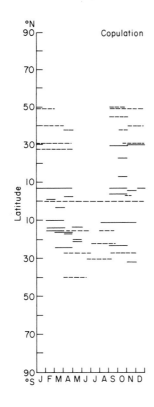

Fig. 26.2 Variation with latitude of the seasons for copulation, parturition and hibernation of bats

Each line represents the season for a particular species. Solid line, megachiropteran; dashed line, microchiropteran.

Most species have restricted breeding seasons, though in the tropics different species breed at different times of year.

Only a few tropical species breed throughout the year and even these usually have peaks of breeding activity.

It is difficult to delimit the beginning and end of hibernation due to the habit of bats of intermittent arousal and flight during winter and in some species the habit of movement from one hibernation roost to another. In general, however, length of the hibernation period increases at higher latitudes.

as in the common vampire, *Desmodus rotundus* (Orr 1970b). Copulation and ovulation may occur together or they may be separated by some months. In the latter case, the female stores the sperm in her reproductive tract until ovulation and consequently fertilisation take place. Female bats usually give birth to only one young at a time, though a few species give birth to two. Some species give birth to one young in one part of the range and two in another. In Europe, some species normally produce one young at a time in England but two at a time on the continent (Allen 1939). In North America, the big brown bat, *Eptesicus fuscus*, normally produces two young in the east and one in the west (Christian 1956). The only major exception to this general rule of a single or double birth is found in the tree bats of North America belonging to the genus *Lasiurus*. The red bat, *L. borealis*, is known to have up to four young at a time. Other North American *Lasiurus* spp. have more than two mammae, as does *L. borealis*, and are therefore also likely to produce more than one or two young at a time (Allen 1939).

Newborn bats weigh 15–30 per cent of the weight of the non-pregnant female. By the time the young bats of most species are 3–4 months old they are usually indistinguishable from the adult, though it is

rare for them to achieve sexual maturity before they are 1 year old and the males of some species may not become sexually mature until they are two years old (Orr 1970b). In the majority of microchiropterans there is effective sexual segregation of the adults during the period that the young are suckling. Females and dependent young occur either alone, as is usual among most American tree bats, *Lasiurus* spp., or with other females and their young, thus forming small to large 'maternity' or 'nursery' colonies.

26.2 Seasonal shift in home range

It is possible from the considerations of the previous section to identify a number of types of home range that are occupied by bats. These are primarily the home ranges used for: feeding; copulation; parturition; lactation; and hibernation.

A feeding home range consists of one or several day-roosts surrounded by, or adjacent to, an area containing a supply of food and water of the types that the species is adapted to exploit. The day-roosts must offer an appropriate thermal environment and

Fig. 26.3 A maternity colony of the big brown bat, *Eptesicus fuscus*

adequate protection from predators, at least during daylight hours. Copulation, parturition, and lactation home ranges must also contain these resources, though a copulation home range for a male must in addition support an appropriate supply of females, and a parturition/lactation home range for females must have a day-roost with a thermal environment that is optimum for the rapid growth of offspring. This thermal optimum may differ from, and is usually higher than, the thermal optimum for an adult-sized individual for which conservation of energy is the critical factor.

In addition to a day-roost, these various types of home range for many species also contain a suitable night-roost. Flying foxes, *Pteropus* spp., leave their day-roosts every evening, establish a small feeding territory each night, and return to their day-roost the following morning (Nelson 1965b). Fruit feeders, such as *Pteropus*, hang from a tree while they feed. Similarly, many of the slower-flying insectivorous species roost at some stage during the night while they manipulate or digest their food. Often specific night-roosts are used for this purpose. These night-roosts are usually separate from the day-roost and within, or near to, the feeding area. Some species, such as the faster-flying insectivorous species and others, such as the common vampire, *Desmodus rotundus* (Wimsatt 1969), do not use a night-roost.

A hibernation home range consists primarily of a combined night- and day-roost that is unlikely to be entered by predators and that offers optima of temperature and humidity. In some species, however, such as those discussed in Chapter 14 (p. 101) a hibernation home range consists of several such roosts, each offering slightly different conditions, surrounded by an area suitable for feeding during any period that the weather is favourable.

In the tropics, the same short-period home range is often suitable for all activities throughout the year and no seasonal shift takes place. The Yucatan free-tailed bat, *Tadarida yucatanica*, for example, uses the same day-roost, a cave, throughout the year (Villa-R. 1960). The common vampire, *Desmodus rotundus*, in Mexico may have one or several day-roosts within its home range but migrates between them on an irregular and probably opportunistic basis depending on the distribution of prey-species on any particular night. Although a daily shift in home range may occur, a slightly longer-term home range, such as, say, that occupied in a month (p. 384) shows no seasonal shift. There is a strong possibility,

however, that demes differ in the extent of any seasonal shift in home range. In the Mexican demes of *D. rotundus* studied so far, seasonal shift seems to be zero (Fig. 26.4). In Peru, however, the vampire is considered to be only seasonally resident in the Andean passes at altitudes of 2000–3000 m (Tuttle 1970). As with so many examples of migration within a familiar area, therefore, the pattern of migration of the vampire may be highly variable and depend entirely on the local temporal and spatial juxtaposition of resources to which each deme is subjected.

Where a seasonal shift in home range is found in tropical bats it is often one of two types. One type is found among frugivorous species and is related to the changing distribution and availability of fruit and to the phases of the bat's reproductive cycle. For example, males and females of the fruit bat, *Pteropus geddei*, on the islands of the New Hebrides in the Pacific Ocean at about 16°0′ S, 166°30′ E, after parturition in August and September, roost together, often near the shore, and subsist largely on breadfruit, *Artocarpus* sp. (Baker and Baker 1936). After copulation in February and March, the females leave the shore and feed on the flowers and fruits of various hardwood trees while the males remain at colonial roosts near the shore. By June, aggregations of females are found roosting inland while the males have dispersed and roost solitarily. Papau, banana (*Musa*), and the figs of *Ficus copiosa* are eaten throughout the year. Subtropical frugivorous species, such as the grey-headed flying fox, *Pteropus poliocephalus*, of Australia show a similar type of seasonal return migration (Fig. 26.5).

The other type of seasonal return migration shown in the tropics by bats, particularly insectivorous species such as the Mexican funnel-eared bat, *Natalus stramineus* (Fig. 26.6), takes place between wet and dry season home ranges.

The pattern of seasonal shifts of home range for which most information is available, however, is that shown by the vast majority of insectivorous temperate species and which in effect consists of movement between winter and summer home ranges. The optimum characteristics of winter and summer roosts are so different, it is not surprising that the same home range is rarely suitable at all seasons and for both sexes. In winter both males and females require a roost (or roosts—p. 101) which offers protection from predators, which is cool but not freezing, and possibly which contains or is near to a supply of water. In summer, males require a site which is near to an abundant supply of suitable food and water, is protected from predators during the day, and is warm enough to facilitate day-time digestion but not so warm that the daytime metabolic rate is unnecessarily high. Lactating females require a roost which is also near to an abundant supply of suitable food and water but which is protected from predators both day and night. In addition, due to the pressure on young temperate bats to accumulate as much body reserve as possible before the onset of winter (p. 99), females require the site to be as hot as possible during the day to facilitate the digestion of milk and development of the young but not so hot that much energy has to be expended in thermoregulation.

In many cases the distance between winter and summer roosts is such that movement between the two does not occur within a single night and consequently it is necessary to use transient roosts. Transient roosts may be used for just one day or may, as in the case of the lesser horseshoe bat, *Rhinolophus hipposideros*, in Czechoslovakia (Hanak *et al.* 1962), be occupied by an individual for several days or weeks. Examples of seasonal return migration which involve movement between winter and summer roosts and during which spring and autumn transient roosts are used are the American free-tailed bat, *Tadarida brasiliensis* (Fig. 26.18), the noctule, *Nyctalus noctula*, in Eurasia (Fig. 26.13), and Schreiber's bat, *Miniopterus schreibersi*, in Australia and Eurasia (Fig. 26.17).

The requirements for a site to be suitable as a transient roost are similar to those for male summer roosts though, of course, the roost need retain its suitability for only a short period. In view of this

similarity, and because an extended period of summer residence is not forced on males as it is to a large extent on females by the requirements of the young, it is perhaps not surprising that the males of some species have been found to be more or less nomadic (p. 27) throughout the spring, summer and autumn. This has been found, for instance, in the grey myotis, *Myotis grisescens* (Fig. 26.26), and the Indiana myotis, *M. sodalis* (Fig. 15.1). It seems likely that this habit of spring, summer, and autumn nomadism will be found to be widespread among male temperate insectivorous bats. Rate of change of day-roost may be less, however, in the autumn in those species which copulate away from the winter roost. For example, males of Schreiber's bat, *Miniopterus schreibersi*, in eastern Australia (Dwyer 1966) and the noctule, *Nyctalus noctula* (Sluiter and Heerdt 1966), establish themselves in autumn in roosts that are used by females as autumn transient roosts. Copulation apparently takes place in the day-roost. In *Miniopterus schreibersi*, the males reside in these copulation roosts, though leaving each night to forage, for much longer during the autumn than the females. In the red bat, *Lasiurus borealis*, (Barbour and Davis 1969), the male patrols an area around the day-roost, probably defends the area (if only by his presence—p. 408), and in autumn initiates copulation in flight with females that enter the area. Until investigated further, however, it is not possible to be sure that a male remains in one of these areas over several days or weeks. Possibly, as with some migrant insects the males of which have an almost identical courtship strategy (Baker 1972a), a male establishes a new territory every day or so along his migration route.

Fig. 26.4   Distribution, seasonal behaviour, and foraging distance of the common vampire, *Desmodus rotundus*

Vampires (Desmodidae) are confined to the mainland of South and Central America and Mexico where they occur up to within 280 km of the United States border. They are absent from all islands except those nearest to shore, such as Trinidad in the West Indies. Of the three species of vampire, only two are observed with any frequency. These are the common vampire, *Desmodus rotundus*, and the delightfully named hairy-legged vampire, *Diphylla ecaudata*.

The map shows the distribution of the common vampire which, partly because of its sanguivorous habits, can often

occupy areas that at first sight would appear most unsuitable for colonisation by bats. In Peru, for instance, *D. rotundus* is found at elevations of 3000 m with the consequence that the Andes, the lowest passes of which, in the north, are at elevations of just over 2000 m, have presented little barrier to the westward spread of the species. Furthermore, by subsisting on the blood of seabirds, *D. rotundus* has successfully colonised the western arid regions of Peru that are completely devoid of either fresh-water or vegetation.

At the present time, the distribution, feeding, and possibly also migration behaviour of *D. rotundus* is considerably influenced by the distribution and biology of man. Man arrived in the neotropical region between

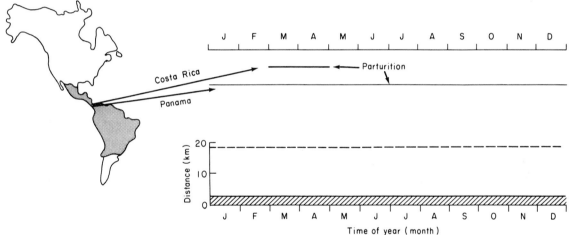

20 000 and 10 000 years B.P. The initiation of agriculture in the region about 3000 years B.P. and the subsequent increase in human deme density probably resulted in human blood becoming an important part of vampire economy. Finally, the advent and spread of pastoralism, particularly over the past 500 years, must have considerably modified vampire biology. Most present studies involve vampire groups that subsist largely on the blood of cattle.

Solid lines, parturition periods in Costa Rica and Panama as indicated; hatched area, observed daily foraging range; dashed line, distance from day roost from which homing has occurred during displacement experiments. Such homing could be the result of random search or the result of an ability to recognise home direction from outside the familiar area. It seems more likely, however, that individuals have a familiar area that is much larger than that within which the major part of foraging activity takes place (see p. 384).

D. rotundus roosts in caves and hollow trees in aggregations of 20–100 individuals. Individuals roosting in hollow trees in Costa Rica in an area where cattle are permanently near showed a more or less 100 per cent degree of return to the same day-roost every night over a period of 11 months. In Mexico, however, a group was found to have a home range containing multiple day-roosts between which individuals shift on a more or less nightly and perhaps opportunistic basis.

D. rotundus forages only once per night, usually in the early hours of the evening after dark. One to three hours are required to find prey, the blood of which, if it is a cow, is sucked for an average of about 9 minutes in the wet season and about 17 minutes in the dry season.

[*Compiled from data in Ryberg* (*1947*), *Wimsatt and Trapido* (*1952*), *Dalquest* (*1955*), *Hall and Kelson* (*1959*), *Cruxent* (*1968*), *Griffin* (*1968*), *Keast* (*1969*), *Wimsatt* (*1969*), *Tuttle* (*1970*) *and A.M. Young* (*1970*)]

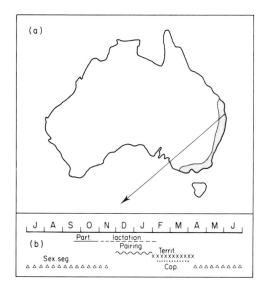

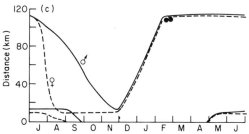

Fig. 26.5 Seasonal return migration of the grey-headed flying fox, *Pteropus poliocephalus*

(a) Map of Australia to show distribution (stippled area) of *P. poliocephalus*. Data presented in (b) and (c) are for the Brisbane area as indicated by the arrow.

(b) Seasonal behaviour. Solid line, parturition period; dashed line, lactation period; waved line, pairing (mate acquisition) period; crosses, territorial period; dotted line, copulation period; triangles, period of segregation of the sexes. *P. poliocephalus* feeds on blossom and fruit. Trees are used as day roosts. Aggregations of 5000–20 000 individuals are formed from August through March in the summer maternity roosts. Although both males and females have arrived at the maternity roost by October, the sexes remain segregated until December when pairing and then later territorial defence and copulation take place. The territorial unit consists in February, March and early April of either one male and one female with attendant young or one male and one to several females but without attendant young. After copulation the sexes segregate and disperse for the winter. The only major winter aggregations consist of immature (i.e. 6–10 months) individuals with which are associated a number of adults, mainly males, the function of which seems to be to protect the young from predators and from each other. The following summer, individuals about 16 months old which are not quite mature return to the site of their

previous winter camp during the period that the adults are copulating. Maturity is reached at about 18 months.

(c) Distance component. Details of (c) are speculative and based on general description. Furthermore, it is assumed that the only mark–release/recapture data available represent the maximum distance moved by the species during seasonal migration, an assumption that is unlikely to be true. A dot refers to the distance an individual has been recaptured from the maternity roost in which it was marked. The maximum and minimum distance limits are indicated for males (solid line) and females (dashed line). The maximum distance away from the maternity roost achieved during daily return migration is assumed to be 15–20 km. In July and August, females initiate convergence migration toward the maternity roost. Most females have arrived by September. Parturition begins soon after. The males start to arrive in September and all have arrived by December when pairing begins. After mate acquisition, some of the pairs leave the maternity roost. Territories may be established as far as 120 km from the maternity roost at which pairing took place. The majority of territories, however, are established at the maternity roost. Following copulation, the sexes segregate, groups disperse, and the maternity roost is vacated. Movement during the winter is assumed to be nomadic and to cover a wide area. The greatest dispersal and most movement occurs during winters that food is most scarce.

[*Compiled from data in Nelson (1965a, b)*]

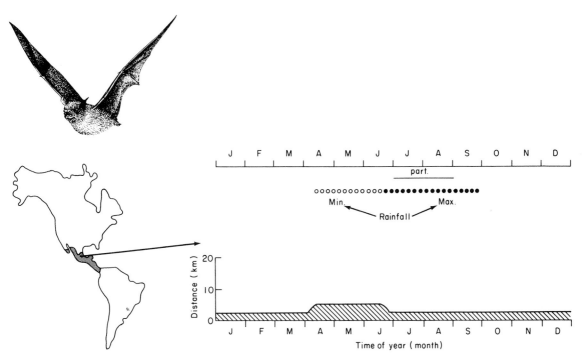

Fig. 26.6 Seasonal return migration of the Mexican funnel-eared bat, *Natalus stramineus mexicanus*

N. stramineus mexicanus feeds on insects and roosts in caves and mine shafts. This map shows the distribution (stippled area) of the subspecies. Return migration has been studied in Mexico as indicated.

Solid line, parturition period; open circles, period of minimum rainfall; solid circles, period of maximum rainfall; hatched area, distance within which individuals are likely to be found at roost or foraging as measured from the wet-season roost.

Males and females roost together throughout the year. During the dry season, some dispersal occurs and roosts are used within 2–3 km of the wet season roost.

[*Compiled from data in Hall and Kelson (1959) and Mitchell (1967)*]

26.3 Degree of return

The difficulty of distinguishing between mortality and removal migration, particularly pronounced in the study of bats (Fig. 17.21), renders impossible at present any precise statement concerning degree of return by bats. Where relevant data are available, however, they imply a degree of return to hibernation sites by males and females and to maternity sites by adult females that is at least as high as for most of those animals discussed in subsequent chapters that are also suggested to perform a seasonal return migration within a familiar area. The level and ontogenetic variation of the incidence of re-moval migration seems to be similar to that for type-C removal migration by other animals (p. 308).

Most investigations have been concerned in temperate regions with degree of return to hibernation or maternity roosts and in tropical regions with degree of return to the day-roost. Such studies do not necessarily indicate the degree of return to the season's home range as a whole. Nor have many studies examined degree of return to copulation or transient home ranges. The few that have been made, however, suggest a similarly high degree of return and hence the existence of a fairly regular annual migration circuit.

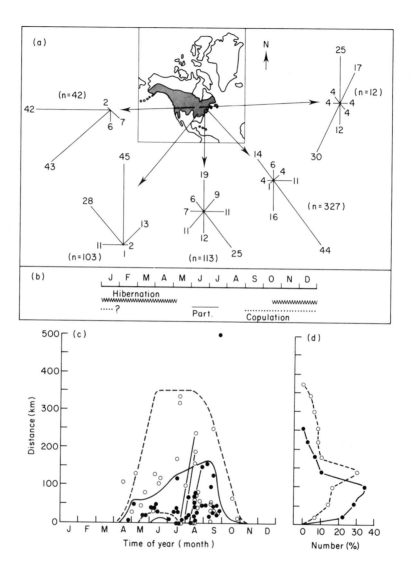

Fig. 26.7   Seasonal return migration of the little brown myotis, *Myotis lucifugus*

(a) Distribution and direction ratio. Stippled area, distribution. Direction ratios are shown for groups hibernating in five different areas as shown by arrows (i.e. Illinois, Kentucky, New York State, Vermont and Ontario). Length of line and number indicates the percentage of each group that was recaptured in the summer in each of eight directions from the hibernation site in which the bats were marked. *M. lucifugus* shows no consistent peak movement direction from the hibernation sites.

(b) Seasonal behaviour. Solid line, parturition period; waved line, hibernation period; dotted line, probable copulation period. Caves and mines are used as hibernation roosts by aggregations of up to 300 000 individuals. Females form summer maternity aggregations of 12–1200

individuals (average about 300) in the attics of buildings usually close to lakes and streams. Males are solitary in summer and roost under loose bark on trees and behind window shutters, etc. Each individual seems to visit its hibernation site for one day some time between July and September.

(c) and (d) Distance component of seasonal return migration. In (c), a solid dot (male) and open dot (female) refers to the distance an individual has been recaptured from the hibernation site in which it was marked. The maximum and minimum distance limits are indicated for males (solid line) and females (dashed line). Recaptures outside these lines are assumed to be removal migrants. Straight lines connect some recaptures in July, August and September with the date of marking during their summer visit to the hibernation site. In (d), the number of recaptures at different distances from the hibernation site

are expressed as a percentage of the total recaptures. Data are included in (d) that are not included in (c) because of unknown recapture date.

In summary, seasonal return migration of *M. lucifugus* consists of: (1) a post-hibernation movement away from the hibernation site during which females travel further than males; (2) a summer (post-parturition in females) return movement to the hibernation site for one day only;

(3) a subsequent pre-copulatory movement away from the hibernation site during which females again appear to move further than males but during which males seem to fly further than in the spring; and (4) an autumn return to the hibernation site at which most copulation seems to take place.

[*Compiled from data in Griffin (1940, 1945), Humphrey and Cope (1964), Davis and Hitchcock (1965), Barbour and Davis (1969), Fenton (1969) and Walley (1971). Photo by Roger W. Barbour*]

The males of Schreiber's bat, *Miniopterus schrei-bersi*, for example, in eastern Australia, show a high degree of return on an annual basis with respect to the autumn copulation site (Dwyer 1966). Degree of return of male and female bats with respect to other autumn and spring transient sites has, however, not been evaluated, though there is some indication for the American free-tailed bat, *Tadarida brasiliensis* (Constantine 1967) and Schreiber's bat, *Miniopterus schreibersi* (Dwyer 1966) that the same individuals may use the same transient sites each year.

In the grey-headed flying fox, *Pteropus poliocephalus*

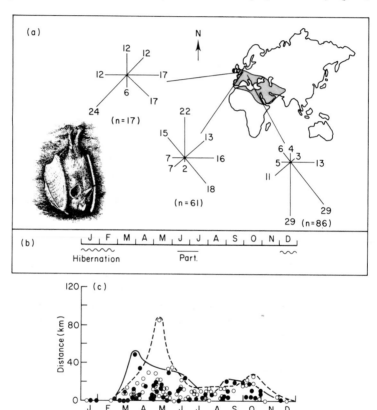

Fig. 26.8 Seasonal return migration of the lesser horseshoe bat, *Rhinolophus hipposideros*

(b) Seasonal behaviour. Solid line, parturition period; waved line, hibernation period. Caves are used as hibernation sites. Individuals roost solitarily. Females form maternity aggregations in buildings.

(a) Distribution and direction ratio. Stippled area shows the distribution of *R. hipposideros*. Spring direction ratio is shown for the species in three areas as indicated: (1) England; (2) Czechoslovakia, Austria and Hungary; and (3) the Netherlands and West Germany. Length of line and number show the percentage of individuals that were captured in each of eight directions relative to their hibernation roost. A peak spring direction to the NNE is apparent in the northern study areas on the European mainland and to the SSE in the southern study area. No peak spring direction is evident in England. In any case, the direction ratio and distance components (see (c)) are such that there is no seasonal shift of geographical range. This remains true even as far east as European USSR.

(c) Seasonal distance component. A solid dot (male) or open dot (female) refers to the distance between capture at a hibernation roost and subsequent or previous capture. Maximum distance limits are shown for males (solid line) and females (dashed line). Although the distances covered are small, there is an indication that both males and females disperse in the spring with those individuals that have migrated furthest returning to the hibernation area in mid-summer. A second, though smaller scale, dispersal may take place in the autumn before the final return to the hibernation area for winter.

[*Compiled from data in Bels (1952), Topal (1956), Hooper and Hooper (1956), Heerdt and Sluiter (1959), Felten and Klemmer (1960), Schmaus (1960), Hanak et al. (1962), Ellerman and Morrison-Scott (1966), Strelkov (1970), and Burton (1971)*]

(Fig. 26.5), and also the Gouldian flying fox, *P. gouldi* (Nelson 1965b), in eastern Australia, the pairing roost, which is not necessarily the copulation roost (Fig. 26.5), is also the maternity roost. Consequently, degree of return is probably high with respect to both the pairing and the maternity roost. This may not be the case, however, in the more inland species of eastern Australia, the little red flying fox, *P. scapulatus*, in which parturition occurs during May when the bats have dispersed and are nomadic (Nelson 1965b). A similar lack of data exists concerning degree of annual return to the daily, weekly, monthly, and season's home range of the males of those temperate, insectivorous microchiropterans that are also nomadic (p. 564).

The periodicity of return to the hibernation home range in most species is annual. In at least two species, however, the bats also briefly visit their hibernation home range during the summer, a return migration that on occasion apparently may have a two-way distance component of 500 (Fig. 26.7) or perhaps even 1000 km (Fig. 15.1). In the Indiana myotis, *Myotis sodalis* (Fig. 15.1), groups of bats visit during the summer the guano pile in the cave system that also contains the hibernation roost. These groups consist of 90 per cent males and stay in the cave for just one to two days. Hall (1962) has shown that virtually all the males that hibernate in the area make this visit at least once during the summer. The little brown myotis, *M. lucifugus* (Fig. 26.7), also returns briefly to its hibernation site during the summer (Fenton 1969). The visit starts in August, at which time the visitors are predominantly (62 per cent) young of the year. By September, when the adults are beginning to copulate, only 26 per cent of the visitors are young of the year. The average length of the visit is 61 ± 31 min. Each individual usually visits the cave only once. If the cave is visited a second time, this does not occur until September. Of individuals banded during hibernation, 3 per cent were caught while visiting the cave in the summer. Of individuals banded during their summer visit, 15 per cent were subsequently caught at that cave in hibernation. This behaviour is discussed in Chapter 15 (p. 160).

Although the habit of a summer visit to the hibernation home range has only been studied in these two species, inspection of the marking-release/recapture results presented in this chapter suggests the possibility that similar behaviour may be shown in Europe by both males and females of the lesser horseshoe bat, *Rhinolophus hipposideros* (Fig.

26.8), and by the males of the lesser mouse-eared myotis, *M. oxygnathus* (Fig. 26.9). Quite possibly, such behaviour is much more widespread than at present is suspected.

26.4 Distance

Information concerning bat return migration derives almost exclusively from marking-release/recapture experiments. Only for three species, the silver-haired bat, *Lasionycteris noctivagans* (Fig. 26.10), the red bat, *Lasiurus borealis* (Fig. 26.11), and the hoary bat, *L. cinereus* (Fig. 26.12), all of which are relatively solitary and roost in trees, has it been necessary to take recourse to the much less satisfactory method of the analysis of sightings.

The technique of the analysis of sightings is undoubtedly valid as a means of mapping the seasonal return migration of birds and also, with some minor reservations, that of cetaceans, pinnipeds, and fish. I have a deep mistrust, however (Baker 1972b), of the sighting method as a means of determining the seasonal migration patterns of animals, such as bats and insects, that differ by several orders of magnitude in their conspicuousness from season to season and from latitude to latitude. When seasonal and latitudinal differences in output and efficiency of observation on the part of research workers are added to these differences in conspicuousness it can be seen that the method has to be used with caution. It is, of course, a powerful method to demonstrate that a particular species is resident in an area. Its major weakness, however, is that absence of a sighting cannot be regarded as proof that an animal is absent from an area.

If a bat hibernates in some inaccessible place, such as high up in a tree, and does not fly from autumn until spring, the chances of discovery are slight. One account (R.T. Baker, *personal communication*), describes an incident concerning a hollow tree that was felled one summer in Wiltshire, England. The tree was felled, lopped, and cut up into short lengths using a mechanical saw. Eventually, an hour or so after the initial felling, while a section was being cut into still smaller lengths of only about 30 cm, two unidentified but adult bats were discovered, one of which immediately took flight. In winter, a solitary bat, perhaps even more reluctant to arouse itself, could easily be overlooked.

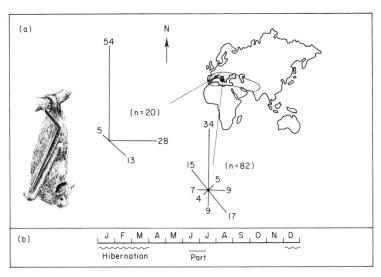

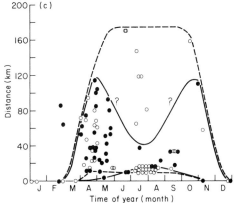

Fig. 26.9   Seasonal return migration of the lesser mouse-eared myotis, *Myotis oxygnathus*

(a) Distribution and direction ratio. Spring direction ratio is shown for *M. oxygnathus* in Austria and Hungary as indicated. Length of line and number show the percentage of individuals that were captured in each of eight directions relative to their hibernation roost. Peak spring direction is to the NE in Austria and to the N in Hungary. However, the direction ratio and distance component (see (c)) are such that there is no seasonal shift of geographical range. This remains true even as far east as European USSR.

(b) Seasonal behaviour. Solid line, parturition period; waved line, hibernation period. Caves are used as hibernation roosts and attics as summer roosts.

(c) Seasonal distance component. A solid dot (male) or open dot (female) refers to the distance between capture at a hibernation roost and subsequent or previous capture. The maximum and minimum distance limits are shown

for males (solid lines) and females (dashed lines). Females seem to travel further between winter and summer roosts than males though both males and females may occupy in July and August roosts that are less than 20 km from their hibernation roosts. The scatter of recapture distances shows a number of points. First, both males and females are widely scattered in spring and autumn. Second, males are found within about 40 km of their hibernation site in July and August. Finally, females, although found up to 180 km away, in July and August are found predominantly within 20 km of their hibernation roost. All of these results could be artefacts due to bias in collecting or to seasonal variation in the chance of recapture with distance. Alternatively, the results could be of biological significance. It is possible that males and females of *M. oxygnathus* disperse in spring, show a return movement to the hibernation area in summer, disperse again in autumn, and then return to the hibernation area for the winter.

[*Compiled from data in Topal (1956), Bauer and Steiner (1960), Hanak et al. (1962), Ellerman and Morrison-Scott (1966), and Strelkov (1970)*]

Felling seems likely to be the only means of accidental discovery. Yet tree-bats do not necessarily roost in those hollow and rotten trees that are the most likely to be felled. Systematic search for roosting tree-bats, especially solitary individuals, is also unlikely to be fruitful, especially during the winter. The more severe the winter, the less time workers are likely to spend searching, especially once the idea has become entrenched that such search is likely to be fruitless. Even when tree-bats are known to be present they are notoriously difficult to locate when at roost. Snyder (1902), for example, reports that a thorough search in summer of trees where hoary bats, *Lasiurus cinereus*, were almost certainly roosting failed to find any specimens. Yet the efficiency of search must be even less in winter. Hibernating bats, especially solitary individuals, would not leave conspicuous guano streaks outside their hibernation site as they do outside their summer sites. Nor can the occupancy or otherwise of a given site readily be determined simply by observation of emergence or entry as can be done any evening or morning during the summer. A great many trees would have to be climbed during the winter before an occupied site is encountered and even then the danger of not noticing the bat must be great. Furthermore, until a few individuals have been discovered, the most profitable places to search cannot be determined.

It is unfortunate that it is the solitary tree-roosting species that are at the same time the species most difficult to find and the species for which the advantage of overwintering at warmer latitudes seems to be the greatest. It is perhaps significant that the northern winter limits indicated in Figs. 26.10–26.12 are based almost entirely on observations of individuals seen flying during the winter months and not of individuals found in hibernation (e.g. Davis and Lidicker 1956, for the red bat, *Lasiurus borealis*). Neither can the observation that date of appearance in the spring is progressively later further north be taken as corroborative evidence of long-distance seasonal migration. Even if hibernation occurred throughout the summer range of the species, a precisely similar sequence might be expected if individuals emerged from hibernation later further north. Finally, the regular autumn appearance of *Lasiurus* spp. at Bermuda could just as easily reflect the seasonal incidence of exploratory migration (p. 158) as it could reflect a stage of a seasonal return migration.

The reality of the migrations that are apparent for tree bats from Figs. 26.10, 26.11, and 26.12 seems to be accepted, at least provisionally, by the majority of students of bats in North America (e.g. Barbour and Davis 1969, Orr 1970a). However, until there is corroborative evidence from marking–release/recapture experiments or from morphological analysis of bats captured in winter and summer in different regions as has been attempted for some butterflies for which similar problems apply (p. 430 and Baker 1972b), it cannot be accepted without reservation that these figures represent the true migration patterns of the species concerned. It should be stressed, of course, that there can be little doubt that these tree bats perform seasonal return migration. The major question mark concerns the components of the migration, particularly distance, direction, and direction ratio.

Having gone to some lengths to point out the shortcomings of the analysis of sightings as a technique for the evaluation of seasonal return migration, especially for tree bats, it has to be admitted that marking–release/recapture is also unsatisfactory in many respects. In particular there is the difficulty of distinguishing between return and removal migrants. Yet it is imperative that the distinction should be made. The difficulty of distinguishing between these two phenomena, which has also been discussed in connection with pinniped migration (p. 724), is particularly acute when recapture data are scarce. When recapture data are more numerous, however, as for many species of bats, a graphical method presents itself which enables a distinction to be made between removal and return migration with an at least acceptable degree of reliability. In most of the figures that accompany this chapter on bats, recapture distances are plotted against time of year. When recaptures are sufficiently numerous, the form of the seasonal return migration shows up clearly and it is possible to draw maximum distance lines that show this form around the bulk of the recaptures. Points outside these lines then have a high probability of representing removal migrants. A few of these points have been included in the figures of return migration (e.g. Figs. 26.22 and 26.28). The possibility remains, of course, that points outside these lines represent potential return migrants. The possibility is even stronger that some points inside these lines represent short-distance removal migrants. However, this graphical method of distinguishing between return and removal migrants seems to be the best that is available at present and it permits some analysis of the components of the two types of migration. The method used, however, should be borne in mind in evaluating any conclusions reached.

Other problems that exist in the determination of

migration distance are also functions of the technique of marking–release/recapture. It is noticeable, for example, that the fewer recaptures there are of a

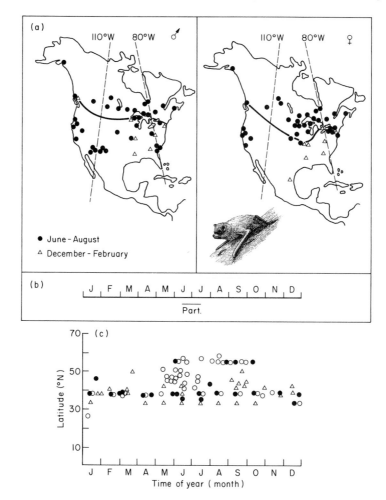

Fig. 26.10   Seasonal return migration of the silver-haired bat, *Lasionycteris noctivagans*

(a) Seasonal distribution of males and females. Dots, distribution from June through August; triangles, distribution from December through February; heavy line, northern limit of main body of observed population between December and February. In the autumn the observed population shifts southwards, though less in the west than in central and eastern North America.

(b) Seasonal behaviour. Solid line, parturition period. Hollow trees, buildings, and crevices in rocks and trees are used as hibernation sites by solitary individuals or small groups. Silica mines, but rarely limestone caves, may also be used. Hollow trees, birds' nests and the space behind loose bark on a tree are used as day-roosts in summer by solitary individuals. Females may form nursery colonies in

hollow trees but reliable information on this point is scarce. More open sites such as open sheds or piles of slabs or fenceposts may be used as spring and autumn transient roosts, especially when migrating through prairie country where roosts are scarce. In summer *L. noctivagans* is found in woods where it forages about ponds and streams.

(c) Latitudinal distribution with season in central North America (i.e. between longitude 80°W and 110°W). Solid dot, male; open dot, female; open triangle, sex unknown. There is a clear indication that the northward limit of the distribution shifts to the north in spring and to the south in autumn. The degree of confidence that can be placed in the reality of this apparent migration is discussed in the text (p. 571).

[*Compiled from data in Merriam (1888), Nero (1957a, b) Cowan and Guiget (1965), Barbour and Davis (1969), and Yates et al. (1976)*]

species, the shorter is the maximum distance recorded. In part, this is a straightforward reflection of chance in relation to the frequency distribution of migration distance among individuals but in part it also seems likely to be an artefact of the marking–release/recapture method, at least as applied to bats. The rarer or more difficult to find a species may be, the less the chance of 'casual' recovery by persons not actually connected with the experiment, and the density of experimenters inevitably decreases with distance from the experiment centre. This, coupled with the greater local knowledge on the part of the experimenters and consequently the greater efficiency of search near to the roost at which marking occurred, inevitably has the above result.

Bearing these problems in mind, however, it can perhaps be accepted that the longest seasonal return migrations by bats are performed by individuals of the American free-tailed bat, *Tadarida brasiliensis*, (Fig. 26.18) and the noctule, *Nyctalus noctula* (Fig. 26.13). The silver-haired bat, *Lasionycteris noctivagans* (Fig. 26.10), red bat, *Lasiurus borealis* (Fig. 26.11), and hoary bat, *L. cinereus* (Fig. 26.12) may perhaps also be included in this list. Among the *Myotis* spp., the longest seasonal return migrations are performed by the Indiana myotis, *M. sodalis* (Fig. 15.1), pond myotis, *M. dasycneme* (Fig. 26.21), and little brown myotis, *M. lucifugus* (Fig. 26.7).

Apparent sexual differences in the maximum migration distance in some cases may be attributed partly to the above artefacts associated with marking–release/recapture and partly to the additional artefact of a difference in the ease with which males and females are recaptured. In 4 of the 5 species in which male recaptures outnumber female recaptures at the appropriate time of year, male maximum distance is greater than female maximum distance, whereas in 11 of the 11 species in which female recaptures outnumber male recaptures, female maximum distance is greater than male maximum distance. If this factor is taken into consideration it seems that only for Geoffroy's myotis, *M. emarginatus* (Fig. 26.23), little brown myotis, *M. lucifugus* (Fig. 26.7), and Schreiber's bat, *Miniopterus schreibersi* (Fig. 26.17), and to a lesser extent for the American free-tailed bat, *Tadarida brasiliensis* (Fig. 26.18), and the noctule, *Nyctalus noctula* (Fig. 26.13), does there seem to be substantial evidence that the maximum straight-line distance attained from the winter roost by the two sexes is different. In all five of these cases it is the females that attain the greatest distance. Similarly, it seems that

for the lesser horseshoe bat, *Rhinolophus hipposideros* (Fig. 26.8), and the large mouse-eared myotis, *Myotis myotis* (Fig. 26.20), and to a lesser extent for the pond myotis, *M. dasycneme* (Fig. 26.21), there is substantial evidence that there is little difference in the maximum straight-line distance from the winter roost attained by the two sexes. In addition, but depending on the validity of the evidence on which the estimates of distance are made, females of the hoary bat, *Lasiurus cinereus* (Fig. 26.12), red bat, *L. borealis* (Fig. 26.11), and silver-haired bat, *Lasionycteris noctivagans* (Fig. 26.10), may also be considered to migrate further than the males.

Some authors have concluded that a sexual difference in migration distance may exist on the basis of data other than those derived from marking–release/recapture experiments. Strelkov (1971), for example, following his analysis of sightings of the bats of European USSR (p. 579), pointed out that the six species, one of which was the noctule, *Nyctalus noctula*, that appeared to perform long-distance seasonal return migrations, were also characterised by a predominance of females or a total absence of males in the northern part of their range in summer. Such a pattern of sightings could of course be generated by differences in the roosting behaviour of the two sexes in the northern part of the range. On the other hand, it could also be generated by a sexual difference in migration distance.

In the eastern pipistrelle, *Pipistrellus subflavus*, in North America, the percentage of males in hibernation caves increases with increase in latitude (W.H. Davis 1959). In southern Louisiana, the bats are active throughout the winter and males constitute 32 per cent of the overwintering colony (Jones and Pagels 1968). Further north (Davis 1959), the percentage of males increases to about 50 in southern Georgia, 70 in Missouri and northeast Alabama, 80 in west Virginia, and 90 in Ontario and Quebec. Davis (1959) has suggested that these figures indicate different migration distances between the sexes, but in the absence of other information they could equally well be taken to indicate sexual differences in the choice of hibernation site at different latitudes.

In those species that visit the winter home range at some time during the summer, the maximum distances attained from the winter site during the spring and autumn return migrations seem to be similar. There is perhaps some indication that it is the spring movement of female little brown myotis, *Myotis lucifugus* (Fig. 26.7), but the autumn move-

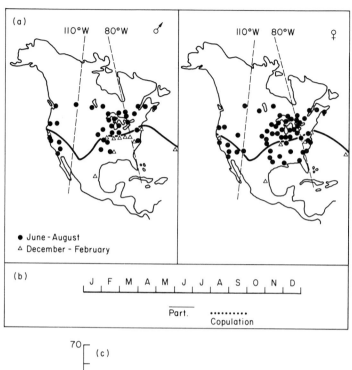

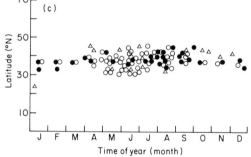

Fig. 26.11 Seasonal return migration of the red bat, *Lasiurus borealis*

(a) Seasonal distribution of males and females. Dots, distribution from June through August; triangles, distribution from December through February; heavy line, northern limit of main body of observed population between December and February. In the autumn the observed population shifts southwards, though less in the west than in central and eastern North America.

(b) Seasonal behaviour. Solid line, parturition period; dotted line, copulation period. Trees are probably used as hibernation sites. In summer, bats spend the daylight hours hanging in the foliage of trees, usually on the south or southwest side, in such a position that they are protected from view from all directions except below. *L. borealis* seems to establish individual feeding territories about the day roost in summer. Individuals roost solitarily, though

females roost with their 1–4 young. Copulation may be initiated in flight and may be in process when the pair land on the ground.

(c) Latitudinal distribution with season in central North America (i.e. between longitudes 80°W and 110°W). Solid dot, male; open dot, female; open triangle, sex unknown. There is a clear indication that the northward limit of the distribution shifts to the north in spring and to the south in autumn. The degree of confidence that can be placed in the reality of this apparent migration is discussed in the text (p. 571).

[Compiled from data in Grinnell (*1918*), Lewis (*1940*), McClure (*1942*), Stuewer (*1948*), Orr (*1950*), Davis and Lidicker (*1956*), Jennings (*1958*), Layne (*1958*), W.B. Davis (*1960*), Myers (*1960*), Jackson (*1961*), Elwell (*1962*), Downes (*1964*), Hoffmeister and Downes (*1964*), Wilson (*1965*), Glass (*1966*), Constantine (*1966*), Baker and Ward (*1967*), Barbour and Davis (*1969*), and Yates et al. (*1976*)]

[*Photos by Roger W. Barbour*]

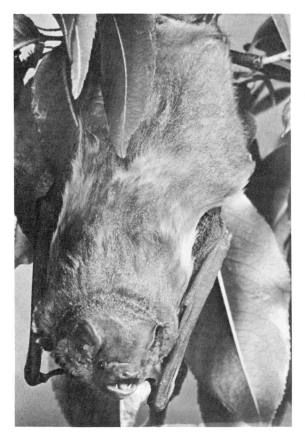

ment of male *M. lucifugus* and female Indiana myotis, *M. sodalis* (Fig. 15.1), that is further. However, the evidence is slight. To judge from the slope of the lines in Figs. 26.7 and 15.1, maximum rate of movement of spring and autumn outward migrations seem to be similar except possibly for male little brown myotis, *M. lucifugus*, which may migrate at a faster speed in autumn than spring.

The most useful information for the evaluation of the speed of movement of seasonal return migration would consist of two components (i.e. the speed of flight during migration from one day-roost to another and the frequency with which change of day-roost occurs). At best, marking–release/recapture or the graphical analysis of the type of figures presented in this chapter can give only an average rate of straight-line cross-country movement. Even if an individual is recaptured on consecutive days there is no way of knowing how much of the intervening period was actually spent on the seasonal return migration. Furthermore, the strong possibility exists that such an individual may well be

showing removal migration in response to the handling involved in marking, and the speed of removal migration may well be different from the speed of return migration.

Examples of maximum recorded speeds during what may be assumed to have been stages of a seasonal return migration are: 60 km/night for a female Indiana myotis, *M. sodalis*, during the outward phase of the autumn return migration; 81 km/night for a female little brown myotis, *M. lucifugus*, during the same phase of the same migration; 70 km/night for a female American free-tailed bat, *Tadarida brasiliensis*, in May. *T. brasiliensis* is reported (Davis *et al.* 1962) to fly during foraging at a normal speed of about 60 km/h but probably to be capable of reaching speeds of 90 km/h even in horizontal flight. It seems likely, therefore, that *T. brasiliensis* could perform even the maximum average rate of migration by devoting an average of about one hour per night to seasonal as opposed to daily return migration. In species that shift home range only in spring and autumn, the maximum

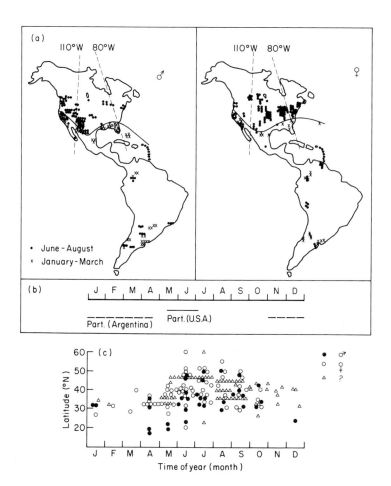

Fig. 26.12   Seasonal return migration of the hoary bat, *Lasiurus cinereus*

(a) Seasonal distribution of males and females. Dots, distribution from June through August; crosses, distribution from January through March; heavy line, northern limit of main body of observed population between January and March. In North America, males are concentrated in summer in the west while females are concentrated in the central and eastern states. In the autumn the observed population shifts southwards, though less in the west than in central and eastern North America. An apparent absence of females on the eastern mainland in January suggests that most pass the northern winter in the Caribbean or perhaps even in northern South America. Less information is available for South America, but there is no evidence of major seasonal latitudinal shifts south of the equator. There is some indication of an altitudinal migration in Chile and possibly an altitudinal or more probably no seasonal return migration in Colombia and Venezuela.

(b) Seasonal behaviour. Solid line, parturition period in the central North American states; dashed line, parturition period in Argentina. Females give birth on average to two young at a time. Throughout the year individuals roost solitarily or in groups of two or three amongst the foliage of trees. Copulation of northern hemisphere bats presumably takes place just before, during, or, more probably, just after the southernmost extension of range.

(c) Latitudinal distribution with season in northern hemisphere. Sightings of one or several individuals within the area between longitudes 80° W and 110° W (vertical dashed lines in (a)) are indicated by solid dot, male; open dot, female; triangle, sex unknown. There is a clear indication that the northward limit of the distribution shifts to the north in spring and to the south in autumn. The degree of confidence that can be placed in the reality of this apparent migration is discussed in the text (p. 571).

[*Compiled from data in Poole (1932), Sanborn and Crespo (1957), Findley and Jones (1964), Cowan and Guiget (1965), Constantine (1966), and Barbour and Davis (1969). Photo by Roger W. Barbour.*]

speed of migration in spring seems to be greater than that in autumn in male lesser horseshoe bats, *Rhinolophus hipposideros* (Fig. 26.8), and male and female Schreiber's bats, *Miniopterus schreibersi*, less than that in autumn in the American free-tailed bat, *Tadarida brasiliensis* (Fig. 26.18), and female large mouse-eared myotis, *M. myotis* (Fig. 26.20), and about the same as that in autumn in female lesser horseshoe bats, *Rhinolophus hipposideros*, male large mouse-eared myotis, *M. myotis*, female Geoffroy's myotis, *M. emarginatus* (Fig. 26.23), and male big brown bats, *Eptesicus fuscus* (Fig. 26.29).

So far, we have been concerned primarily with the maximum distance of seasonal return migration in the various species. Further problems present themselves when the minimum seasonal return migration distance is considered. In this case, however, the problem is not so much that of the difficulty of recapture of marked bats at short distances. Instead, the problem is that authors differ in the extent to which they consider short-distance recaptures to be worth reporting. For instance, Strelkov (1971) only reports noctules, *Nyctalus noctula*, that were re-

captured at distances greater than about 400 km from the release point.

The available data, however, show that some individuals of at least the greater horseshoe bat, *Rhinolophus ferrumequinum* (Fig. 26.15), lesser horseshoe bat, *R. hipposideros* (Fig. 26.8), Mediterranean horseshoe bat, *R. euryale* (Topal 1956); noctule, *Nyctalus noctula* (Fig. 26.13), large mouse-eared myotis, *M. myotis* (Fig. 26.20), male Indiana myotis, *M. sodalis* (Fig. 15.1), cave myotis, *M. velifer* (Barbour and Davis 1969), big brown bat, *Eptesicus fuscus* (Fig. 26.29), and at least male Schreiber's bat, *Miniopterus schreibersi* (Fig. 26.17) rarely move more than 2–3 km between their winter and summer roosts. Only for the American free-tailed bat, *Tadarida brasiliensis* (Fig. 26.18), and possibly the female pond myotis, *Myotis dasycneme* (Fig. 26.21), is there convincing evidence that the minimum one-way migration distance between the winter and summer roosts for at least some demes is greater than about 50 km. Only for the pond myotis, *M. dasycneme*, is there sufficient evidence to suggest that this distance is different for males and females and that in this case it is the females that travel further.

Some authors have tried to evaluate the minimum migration distance of bats by an analysis of sightings. Hoffmeister (1970), for example, has classified a number of North American bats according to their summer (April–October) and winter (November–March) status in the State of Arizona. This classification is based on individuals sighted or captured. If we assume that the shortest distance from the centre of Arizona to the border is about 250 km, this represents the minimum distance that must be travelled by any species that is reported to be present throughout the State in summer but absent or present in only the southeast corner in winter. Included in this category are the little brown myotis, *M. lucifugus*, cave myotis, *M. velifer*, and small-footed myotis, *M. leibii*, which on the basis of marking–release/recapture experiments in other States have been shown to have a minimum seasonal return migration distance that is considerably less. The American free-tailed bat, *Tadarida brasiliensis*, is also included in this category but in this case the conclusion is in accord with marking–release/recapture results. Similarly, the conclusion that the big brown bat, *Eptesicus fuscus*, and western big-eared bat, *Plecotus townsendii*, are present in Arizona in both summer and winter is also in accord with marking–release/recapture results. Strelkov (1970, 1971) has attempted a similar exercise for the bats of European USSR. The classification as residents of sixteen species, all of which hibernate to some extent in caves, is completely in accord with marking–release/recapture results for many of these species elsewhere in Europe. Six other species, however, which, perhaps significantly in view of the above discussion for tree bats, are all much less predominantly cave hibernators, are considered to be absent from north and central European USSR during the winter. The implication of this is that all six species show a minimum seasonal migration distance that is in excess of about 800 km. At least as far as the noctule, *Nyctalus noctula*, is concerned, which is one of these six species, this implication is at variance with marking–release/recapture results obtained elsewhere in Europe.

In both of these studies, it is possible that the discrepancies in minimum migration distances determined by an analysis of sightings on the one hand and marking–release/recapture experiments on the other are a function of geography rather than of the technique employed. It has been firmly established for the American free-tailed bat, *Tadarida brasiliensis* (Fig. 26.18; Cockrum 1969), that different demes migrate different distances. This is also true, though to a less spectacular extent, for the little brown myotis, *Myotis lucifugus* (Fig. 26.16). Consequently, it is quite possible that in the severe climate of European USSR, all individuals of the noctule, *Nyctalus noctula*, pass the winter more than 800 km to the southwest of their summer roost, whereas in milder western Europe, some individuals winter no more than 2–3 km from their summer roost. Similarly, in Arizona local conditions may favour a much longer minimum migration distance for the little brown myotis, *M. lucifugus*, small-footed myotis, *M. leibii*, and cave myotis, *M. velifer*, than in regions where marking–release/recapture experiments on these species have been carried out.

I have used a variety of different measures and transformations of body size and wing-length of bats, but can find no correlation between these variables and migration distance. This lack of correlation persists whether the analysis concerns all or just a single family. Even within the genus *Myotis*, the highest correlation coefficient obtained was 0·366, between maximum length of fore-arm and the square root of maximum migration distance, a value that is still far from being significant ($n = 11$, $P > 0\cdot10$).

Rather than size, the chief factor influencing inter-specific variation in maximum migration distance seems to be the type of specialisation for flight as a function of feeding habit. Specialisation to feed on different types of food in different situations has involved selection on, amongst other things, the type of flight and consequently on the gross morphology of the flight apparatus (Vaughan 1970). Such selection is likely to interact with selection on bat migration in that the type of flight that may be optimum for a particular type of feeding may or may not lend itself to exploratory or calculated migration over relatively long distances.

Frugivorous bats (Table 26.1) tend to have a strong direct flight but are not highly manoeuvrable. Some can also glide. Most of the pteropids, except the rousette bats, *Rousettus* spp., orient visually during flight and not by echo-location. Nectar feeders have evolved the ability to fly slowly near vegetation and to hover while their long, protrusible tongues seek out their food. None of the vegetarian bats are capable of turning and manoeuvring as rapidly as the insectivorous bats. Even among this latter group, however, the type of flight differs according to the particular type of insect prey and according to the situation and height at which the insect is caught.

The majority of insectivorous bats perform slow but highly manoeuvrable flight and remain on the wing continuously throughout the foraging period (Vaughan 1970). Insects are located in most cases by echo-location and many species of bats have large ears. Large ears are good for echo-location but introduce considerable drag during flight and are found only in the slower-flying forms. The free-tailed bats (Molossidae) as a group represent the bat-equivalent of the swallows (Hirundinidae) and swifts (Apodidae) amongst the birds. They have long, narrow wings and are adapted to feed fairly high above the ground by fast, enduring flight. When molossids dive, their wings are folded. In contrast to the Molossidae, which are the least manoeuvrable of the insectivorous bats, the horseshoe bats (Rhinolophidae) and some megadermatid, nycterid, phyllostomid, and vespertilionid bats have evolved a highly manoeuvrable 'flycatcher' habit. These bats forage close to the ground or near vegetation and perform slow, delicate flight. Unlike the other insectivorous bats, the 'flycatchers' do not remain continuously on the wing but retire to a night roost to eat their prey once it has been caught. Some species hang from the branches of trees and fly out at passing insects. Others, such as the greater horseshoe bat, *Rhinolophus ferrumequinum*, in Eurasia,

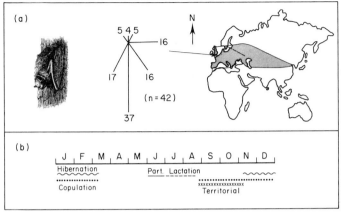

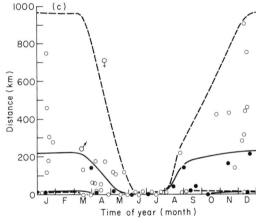

Fig. 26.13 Seasonal return migration of the noctule, *Nyctalus noctula*

(a) Distribution and direction ratio. Map shows gross world distribution (stippled area) of *N. noctula*. Dashed line, possible northern limit of winter distribution in Europe. Direction ratio is shown for bats that are in the Netherlands, as indicated, during the summer. Length of line and number indicate the percentage of individuals that were captured in each of eight directions relative to their summer roost. A peak autumn direction to the south is indicated. Preliminary results suggest a similar peak direction elsewhere in Europe. In European USSR it seems probable that the direction ratio approaches 100:0:0:0.

(b) Seasonal behaviour. Solid line, parturition period; dashed line, lactation period; waved line, hibernation period; dotted line, copulation period; crosses, territorial period. Buildings and hollow trees are used as hibernation roosts by aggregations of up to 200 individuals. Females form smaller maternity colonies in hollow trees with which are associated juveniles born the previous year and not yet mature. One or two young are born at a time (20 per cent twins in the Netherlands). In western Europe the young

can fly by mid-July. The females do not restrict themselves to a single maternity day roost. Instead, there is a maternity roost complex within a given area and females and their young change roosts within this area at irregular intervals. The sexes are segregated from March to August. In the autumn, individual males establish themselves in a hollow tree and the roost is then defended against other males. Copulation occurs with females that use the site as a transient autumn roost. Copulation also occurs during hibernation.

(c) Seasonal distance component based on data obtained in western Europe. A solid dot (male) or open dot (female) refers to the distance between capture at a summer roost and subsequent or previous capture. The maximum and minimum distance limits are indicated for males (solid line) and females (dashed line). Females travel further than males but both males and females may move as little as a few kilometres from summer to winter roosts. North and east of the dashed line in (a), greater distances of up to about 2000 km appear to be travelled by 100 per cent of the population.

[Compiled from data in Eisentraut (1936), Ryberg (1947), Bels (1952), Krzanowski (1960), Cranbrook and Barrett (1965), Sluiter and Heerdt (1966), and Strelkov (1971)] and the desert pallid bat, *Antrozous pallidus*, in the United States, have been seen to land on the ground and catch flightless insects.

As a result of their feeding specialisations, bats with long ears, such as *Plecotus* spp., and bats with a 'flycatcher' habit, such as most Rhinolophidae, probably have a high migration cost per unit distance, whereas many pteropids, vespertilionids, and molossids, probably have a low migration cost per unit distance. It is probably this factor more than any other that contributes to the inter-specific variation in maximum migration distance. Inspection of the figures shows clearly that in tem-

perate regions the shortest migration distances are found among species of *Plecotus* (Fig. 26.14) and *Rhinolophus* (Figs. 26.8 and 26.15), and the longest migration distances are found among the molossids (Fig. 26.18) and some vespertilionids, such as *Nyctalus noctula* and perhaps also the tree-bats of North America (Figs. 26.10–26.12). Nor is this correlation between maximum migration distance and feeding specialisation confined to the northern hemisphere. In South Africa, also, *Rhinolophus clivosus* and *Rhinolophus simulator* migrate only short distances, whereas Schreiber's bat, *Miniopterus schreibersi*, migrates relatively long distances (Laycock 1973).

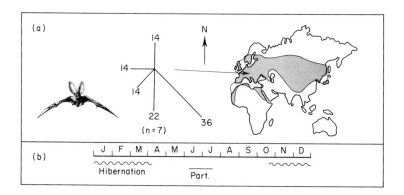

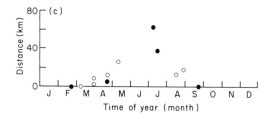

Fig. 26.14 Seasonal return migration of the long-eared bat, *Plecotus auritus*

(a) Distribution and direction ratio. Stippled area shows world distribution of *P. auritus*. Spring direction ratio is shown based on data for bats in the Netherlands, East Germany and Czechoslovakia. Length of line and number show the percentage of individuals that were captured in each of eight directions relative to their hibernation roost. Peak spring direction is to the S but it should be noted that the number of recaptures is small. In any case, the direction ratio is such that there is no seasonal shift of geographical range. This remains true even as far east as European USSR.

(b) Seasonal behaviour. Solid line, parturition period; waved line, hibernation period. Deep crevices in caves

and cellars are used as hibernation roosts by groups of individuals. Solitary individuals also hibernate in hollow trees. Hollow trees and buildings are used as summer roosts in which large numbers of females may form maternity colonies. *P. auritus* forages amongst trees and picks insects off leaves. Individuals often land on the ground in order to eat the prey thus caught.

(c) Seasonal distance component. A solid dot (male) or open dot (female) refers to the distance between capture at a hibernation roost and subsequent or previous capture. Data are insufficient for maximum and minimum distance lines to be drawn but there is some indication that males may have a greater maximum distance between summer and winter roosts than females.

[*Compiled from data in Ryberg (1947), Bels (1952), Felten and Klemmer (1960), Natushke (1960, Hanak et al. (1962), Strelkov (1970), and Burton (1971)*]

It should not be concluded, of course, that the possession by an individual of a flight apparatus that confers a low migration cost per unit distance inevitably means that the individual concerned will perform a long-distance migration. Intra-deme variation alone (e.g. in the noctule, *Nyctalus noctula*, between a few and several hundred kilometres) serves to emphasise this fact as also does geographical variation between demes. In the Molossidae, for example, the Yucatan free-tailed bat, *Tadarida* *yucatanica*, shows zero seasonal return migration distance, whereas the Arizona and New Mexico breeding demes of the American free-tailed bat, *T. brasiliensis* (Fig. 26.18), have a one-way distance component of over 1000 km. Even within this species, however, demes, such as those that breed in California, may have a near-zero distance component to their seasonal return migration.

Migration distance is discussed further in Section 26.7 (p. 591).

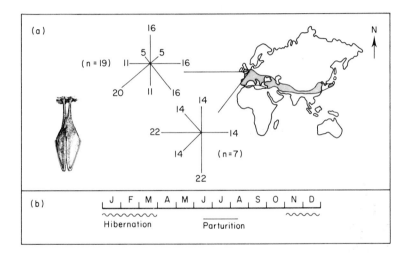

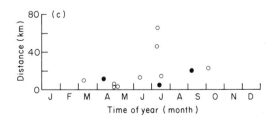

Fig. 26.15 Seasonal return migration of the greater horseshoe bat, *Rhinolophus ferrumequinum*

(a) Distribution and direction ratio. Stippled area shows the world distribution of *R. ferrumequinum*. Spring direction ratio is shown for the species in England and the Netherlands as indicated. Length of line and number show the percentage of individuals that were captured in each of eight directions relative to their hibernation roost. A peak spring direction appears to be absent in northwest Europe. Consequently there is no seasonal shift of geographical range. This remains true even as far east as European USSR.

(b) Seasonal behaviour. Solid line, parturition period; waved line, hibernation period. Caves are used as hibernation roosts (see p. 101). Maternity colonies are formed in caves and buildings by aggregations of up to 200 individuals. *R. ferrumequinum* lands on the ground in order to find prey and to drink.

(c) Seasonal distance component. A solid dot (male) or open dot (female) refers to the distance between capture at a hibernation roost and subsequent or previous capture.

[Compiled from data in Ryberg (1947), Bels (1952), Hooper and Hooper (1956), Topal (1956), Hanak et al. (1962), Hooper (1962), Ransome (1968), Strelkov (1970), and Burton (1971)]

26.5 Direction and direction ratio

Inspection of the direction ratios presented in Figs. 26.20–26.29 shows that many demes of bats manifest a peak direction. However, in the vast majority of cases this direction is neither constant from deme to deme nor does it vary in a constant way with latitude. This inter-deme variability is shown particularly clearly in Fig. 26.16 in relation to the little brown myotis, *Myotis lucifugus*, for which direction ratio has been determined for a number of different demes. The only reasonable conclusion, therefore, is that direction and direction ratio, in the same manner as distance, varies according to the local temporal and spatial juxtaposition of different resources and competitors. Any influence of latitudinal gradients of resources on migration direction seems to be negligible relative to the influence of local conditions. This deme-to-deme variation in distance and direction is typical of most animals that perform seasonal return migration within a familiar area.

The only species in which latitudinal effects, particularly temperature gradients, seem to influence migration direction more than the local juxtaposition of hibernation and feeding home ranges seem to be the American free-tailed bat, *Tadarida brasiliensis* (Fig. 26.18), the noctule, *Nyctalus noctula* (Figs. 26.13 and 26.19), and perhaps also the North American tree bats (Figs. 26.10–26.12). Significantly these are the species that have the greatest migration distance and hence experience the greatest influence of latitudinal gradients.

In the light of the discussion concerning the winter strategy of the greater horseshoe bat, *Rhinolophus ferrumequinum* (p. 101), it would seem that the critical gradient in winter results from the interaction of energy expended on thermoregulation and energy uptake as a function of number of nights on which feeding can occur. The figure for the noctule (Fig. 26.19) suggests that, as for some butterflies (Fig. 19.12), it is the shortness of the winter that is the critical factor rather than the severity of the winter. For species such as *Tadarida* and *Nyctalus* that, because of their flight mechanism, have the potential to establish a large familiar area

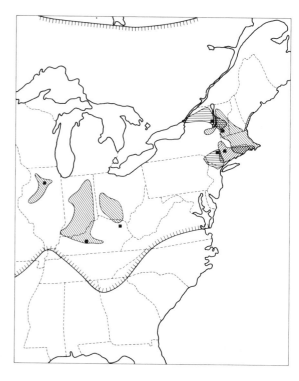

Fig. 26.16 Migration direction of the little brown myotis, *Myotis lucifugus*, between hibernation roost and summer roost

Squares, hibernation caves; hatched areas, areas within which bats banded at the nearest hibernation roost were recaptured during summer months; hatched line, approximate edge of range. Migration direction of this species seems to vary according to local environment rather than with latitude or season.

[*Compiled from data in Griffin (1940), Davis and Hitchcock (1965), Barbour and Davis (1969), Fenton (1969), and Walley (1971)*]

if, that is, local conditions frequently exceed the exploratory migration thresholds, then latitudinal gradients may well outweigh local effects.

The critical gradient in summer is likely to be a function of energy expended in thermoregulation and of the nett rate of energy uptake while feeding. Significantly, bats being nocturnal, this gradient seems to have acted on *Tadarida brasiliensis* and *Nyctalus noctula* to favour spring migration to the north, a fact which suggests that, contrary to the proposal for many birds (p. 660), time available for collecting food does not seem to make a major contribution to this gradient.

Experimental support for the suggestion that, despite the migration cost involved, a migration 1400 km or so to the south for winter may not be disadvantageous is provided by Twente (1956), though the validity of the premises involved may perhaps be questioned. Comparison was made in south central Kansas between the weight loss of the American free-tailed bat, *Tadarida brasiliensis*, and the cave myotis, *Myotis velifer*. During the period from October to April *T. brasiliensis* migrates south to Mexico, feeds throughout the winter, and returns to Kansas (Fig. 26.18). *M. velifer*, on the other hand, remains and hibernates in Kansas. Comparison of the two bats was assumed to be more or less valid as they are of similar size and weight. Twente found that, by April, *Tadarida brasiliensis* had lost less weight than *Myotis velifer*. Male *T. brasiliensis* had lost 17·2 per cent of their body weight as opposed to 26·8 per cent by *M. velifer*. Female *T. brasiliensis* had lost 21·2 per cent of their body weight as opposed to 29·9 per cent by female *M. velifer*.

26.6 Degree of dispersal and convergence

Most bats converge to roost and to hibernate and disperse to feed, a pattern that appears to be consistent with the area required to perform the two functions, the relative availability of suitable sites, and perhaps also the optimum group sizes, composition, and spacing for thermoregulation, protection from predators, and perhaps also social communication and the exchange of information.

The distance, direction, and direction-ratio components of seasonal return migration combine to produce a particular degree of dispersal and convergence for each deme. It has not yet been exam-

ined for any bat species, but it seems likely that the situation is free. That is, each member of the deme experiences the same potential reproductive success despite migrating different distances in different directions. If the effect of a presumably higher mortality risk through migrating up to 1400 km than through hibernation is taken into account, then the figures obtained by Twente (1956) for weight loss of the American free-tailed bat and the cave myotis, as presented at the end of the previous section, perhaps tend to support the existence of a free situation.

Where sexual segregation occurs, as in the majority of temperate species in summer, it can perhaps be attributed to the different roost requirements of the two sexes (p. 563).

26.7 Application of the familiar-area hypothesis to the seasonal return migrations of bats

Bats perceive a number of types of environmental cues that seem to permit the establishment of a familiar area. Perhaps the most useful cues are those perceived by means of echo-location, vision, and olfaction. Of these abilities, echo-location has received the most attention and vision has been the most maligned.

Echo-location (i.e. a mechanism by which an animal can locate an object or objects by producing sounds and assessing the characteristics of the resulting echoes) is highly developed in bats and is used for finding and catching food, particularly insects (Griffin *et al.* 1960), and for the avoidance of obstacles. Echo-location is used by all Microchiroptera but among the Megachiroptera it has evolved only in the genus *Rousettus*. The echo-location system of *Rousettus* spp. is based on audible sounds, unlike that of the Microchiroptera which is based on ultrasonic emissions (Griffin *et al.* 1958, Novick 1958). Recent work into the neurophysiological basis of echo-location in bats has been described by Suga (1972). Bat echo-location functions over a distance of about 100 m (Griffin *et al.* 1960).

Many of the early attempts to investigate the role of vision in echo-locating bats involved experiments carried out in normal room lighting. Under such conditions the bats were unable to utilise their visual sense which, not surprisingly, is adapted to use at low light intensities. Deafened individuals of the little brown myotis, *Myotis lucifugus*, can avoid obstacles in dim light, as long as the contrast is high, better than they can in normal room lighting or in total darkness (Bradbury and Nottebohm 1969). The bat eye is, in fact, well adapted for nocturnal vision and depth perception in that it has an all-rod retina, a large and spherical lens and reduced pupil mobility (Moehres and Kulzer 1956, Bradbury and Nottebohm 1969). Suthers (1966) tested the opto-motor responses to moving stripes at low light intensities of eight species of bats belonging to the families Emballonuridae, Phyllostomidae, Desmodidae, and Vespertilionidae. The species chosen included insectivorous, frugivorous, nectarivorous, carnivorous, and blood-sucking forms. Representatives of each feeding type and of all families except the vespertilionid species tested, the little brown myotis, *M. lucifugus*, showed response to a visual angle of at least $0.7°$. *M. lucifugus* showed a minimum visual angle of between $6.0°$ and $3.0°$. Experiments of this type, of course, measure only the threshold of movement to which the animal responds under experimental conditions, which is not necessarily the same as the threshold for perception. The possibility cannot be precluded that bats are capable of a visual performance under normal conditions that is better even than that demonstrated. In particular, from the viewpoint of navigation, it would be useful to know the visual performance of bats in relation to a point light source as well as to stripes.

The olfactory sense of at least the megachiropteran bats appears to be well developed. Rousette bats, *Rousettus* spp., can locate by olfaction 100 mg of banana substance and furthermore can distinguish between banana substance and banana essence (Moehres and Kulzer 1956). The microchiropteran large mouse-eared myotis, *M. myotis*, of Eurasia finds part of its food by hopping along the ground and searching for insects. Dung beetles appear to be found by olfaction (Burton 1971). Finally, in the cases in which it has been studied, individual recognition between mother and infant bats seems to be based on olfactory cues and probably also on auditory communication (Nelson 1965b, Turner *et al.* 1972).

There is some direct evidence that bats have a well-developed spatial memory. Neuweiler and Moehres (1967) have shown that if an obstruction is removed from the flight path of *Megaderma lyra*, the bats nevertheless continue to take avoiding action for a further six weeks. The previous position of the obstruction is not forgotten until nearly twelve weeks.

As with all animals, few data are available concerning the process of establishment of a familiar area during the period from birth to maturity. Movements of young pre-reproductive bats seem to have been studied in detail in only two species, one from the Megachiroptera and one from the Microchiroptera.

Juvenile grey-headed flying foxes, *Pteropus poliocephalus*, in eastern Australia segregate from the adults about three months after birth during the period that the adults are pairing (Fig. 26.5). During the following 2–3 months of late summer and early autumn, while the adjacent adults are territorial, juvenile group size increases as more and more juveniles leave the adult roosting area. These juvenile groups are attended by a few non-breeding adults, mostly males. When the adults disperse at the onset of winter, the juveniles move as a group, along with the adult male attendants, to a winter camp. The number of adults to be found with the juveniles in the winter aggregation is greater in years when food is abundant than in years when food is scarce. In spring, the juveniles, now called yearlings, return to the summer camp of their birth along with the converging adults. In autumn, when the adults establish territories, the yearlings return to the site they had used the previous winter. In their second winter, however, when the camp is invaded by the juveniles of the year, the yearlings disperse and presumably establish the remainder of the familiar area out of which will crystallise an adult seasonal migration circuit.

Probably the most detailed evaluation of the migrations of young bats is that of Dwyer (1966) for Schreiber's bat, *Miniopterus schreibersi*, in eastern Australia. Juvenile *M. schreibersi* leave the cave maternity colony 3–4 months after birth and for 2 months in the autumn are on the move from one transient roost to another (exploratory migration?) before settling down in a cave winter roost that often also contains adults. The following spring, about 3 months later, the young bats, now called yearlings, leave the winter roosts and again travel across country from one spring transient roost to another. During this spring movement both male and female yearlings may return to the maternity roost in which they were born. After this visit, which ends before

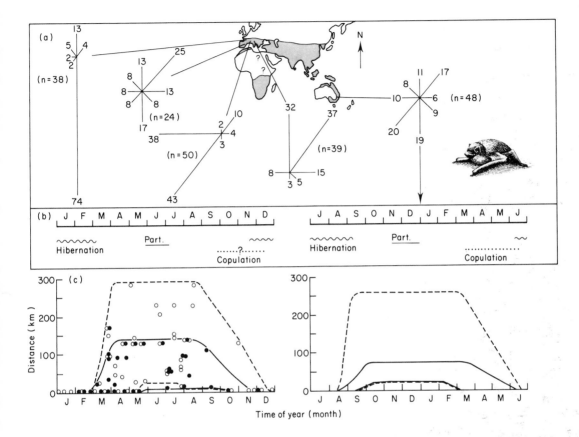

Fig. 26.17 Seasonal return migration of Schreiber's bat, *Miniopterus schreibersi*

(a) Distribution and direction ratio. Map shows the gross world distribution (stippled area) of *M. schreibersi*. Direction ratios are shown for groups hibernating in five different areas as shown by arrows (i.e. Czechoslovakia, northeast Spain, Germany, and Austria in Europe and northeast New South Wales in Australia). Length of line and number indicate the percentage of each group that was captured in each of eight directions from the hibernation sites in which the bats were found. Peak direction and direction ratio varies from area to area on a local rather than continental scale.

(b) Seasonal behaviour in Europe and Australia as indicated. Solid line, parturition period; dotted line, copulation period; waved line, hibernation period. Caves are used as hibernation roosts by both males and females. Not all males, however, hibernate communally in caves. The majority seem to winter in solitary, as yet undiscovered, sites. Females use caves as maternity roosts and large aggregations are formed. In Australia the migration of females from hibernation roost to maternity roost is a convergent migration. One or a few maternity roosts are used by females that for the rest of the year are spread over a wide area. Males also use caves as day-roosts during the summer and both males and females use caves

as spring and autumn transient roosts. Males use particular roosts during the autumn through which passes a steady stream of females with which copulation takes place. Juveniles during their first year after leaving the maternity roost are partially segregated from the adults to the extent that some roosts contain a majority of juveniles and a minority of adults.

(c) Distance component in Europe and Australia as indicated. A solid dot (male) or open dot (female) refers to the distance from its hibernation site that a marked individual has been captured. The maximum and minimum distance limits are indicated for males (solid line) and females (dashed line). The figure for Australia has been compiled from a description of the migration rather than from tabulated mark–release/recapture data.

M. schreibersi shows a pattern of seasonal return migration that with respect to the periodicity and distance components and the way these two components vary between the sexes is similar in Australia and Europe. Only the direction ratio and peak direction components show much variation. However, the variation appears to be local rather than continental.

[Compiled from data in Ellerman et al. (1953), Topal (1956), Bauer and Steiner (1960), Schnetter (1960), Hanak et al. (1962), Balcells (1964), Dwyer (1966, 1969), Hamilton-Smith (1966), and Purchase (1969)]

the adult females start to give birth, yearling bats, during the late spring and summer, visit the adult roosting sites. Some adopt these roosting sites and others do not. Instead, these latter yearlings return to, or travel to new, predominantly yearling roosts. In mid- to late-summer the yearlings start to arrive at winter roosts by way of autumn transient roosts. Some of the yearling females may be mated by adult males that use the autumn transient roosts as copulation roosts. The following spring the now mature females migrate to a maternity roost, usually the one in which they were born, and those that were mated give birth. Males also use adult summer, autumn, and copulation roosts. In this species, therefore, the adult seasonal migration pattern (Fig. 26.17) is established from the second winter onwards within a familiar area that apparently is established during the first year of life, a period that seems to be followed by a period of habitat assessment and calculated removal migration (p. 160).

Some clues concerning the pattern of familiar area establishment are also available for the Arizona and New Mexico breeding demes of the American free-tailed bat, *Tadarida brasiliensis* (Fig. 26.18). In the autumn, after weaning, the young and often also the adults seem to engage in a period of exploratory migration, for it is during this period that most records outside of the normal geographical range occur. This period of exploratory migration is followed in the Arizona and New Mexico breeding demes by the southward autumn migration. The following spring, after the adults have copulated, the bats return to the north. Constantine (1967) has suggested, however, on the basis of their absence in the north, that the young males, now yearlings, may stay in Mexico in all-male groups instead of joining the other age and sex classes on their migration to the normal summer home range as they do when adult. If this is correct, it could be that males gain a greater advantage through spending their first spring, summer, and autumn establishing a familiar area and performing habitat assessment within the region most suitable for copulation than through migrating to the thermally more suitable part of their familiar area to the north.

The role of social communication in the establishment of a familiar area by young bats is largely unknown. Certainly in most cases a strong bond exists between mother and offspring for the duration of lactation, though large young which cannot as yet fly are often left at the roost while the mother forages. Female big brown bats, *Eptesicus fuscus*,

leave even newly born young in the roosts when they go out to forage but are able, as are the females of many other species, to recognise and retrieve their own young if they fall onto the floor of the roosting site (Davis *et al.* 1968). In addition to the mother–offspring bond that exists during lactation, there is an indication that the young of the Australian large-footed myotis, *Myotis adversus*, associate with their mother for a short time after the end of the normal eight-week lactation period (Dwyer 1970), a behaviour pattern that if found in other species may have an important bearing on the mechanism of the establishment of a familiar area by the young bat.

It is tempting, for example, to suggest that the summer visit of little brown bats, *M. lucifugus*, to their copulation and hibernation roost (p. 571) is part of a process of familiar-area establishment and habitat assessment by young bats in association with their mothers. However, as, when the visit begins in August, 62 per cent of the bats may be young of the year, some at least cannot be associated with their mothers. Visiting males of this species and the Indiana myotis, *M. sodalis* (Fig. 15.1), seems likely to be engaged in some form of habitat assessment (p. 160) and habitat ranking (p. 379), perhaps in relation to hibernation or perhaps in relation to the copulation home range to be adopted in the coming autumn.

An exception to the usual mother–offspring bond is found in the American free-tailed bat, *Tadarida brasiliensis*. In Texas, the females of this species deposit the new-born young in large continuous colonies of often many thousands of individuals within the maternity cave. When the mother returns from the nocturnal feeding flight, she delivers her milk to the first two aggressive young encountered on landing in the nursery colony. The female does, however, usually return to the cave in which she gave birth. The young bats grow rapidly in the incubator-like climate of the caves and with the 'abundant milk supplied by the dairy herd of lactating mothers' (Davis *et al.* 1962).

The greatest development in parental and social care of young bats, at least as far as is known, is found among the megachiropteran fruit bats or flying foxes. From about three weeks after parturition onwards, female flying foxes, *Pteropus* spp., transport their young every evening to special, more foliaged, trees surrounding the summer camp. The following morning, just before dawn, the female returns to this roost, picks up her offspring, and transports it back to the defoliated day-roost. The female can recognise her own young, in this case, apparently, mainly by olfaction (Nelson 1965b). Another example of care for the young, in this case non-parental, is found in the grey-headed flying fox, *P. poliocephalus*, of Australia and takes place at the same time that many of the adults are establishing territories (Fig. 26.5). Coincident with territory formation by the adults, many of the young aggregate in juvenile groups. These juvenile groups, however, are attended by a number of adults, the majority of which are males (Nelson 1965b). These attendants not only drive away other adults and predators but also intervene in any fights that break out between the juveniles. Also it seems almost certain that these adults must in some way influence the pattern of familiar area formation of the juveniles with which they are associated.

Adult flying foxes groom one another in a manner reminiscent of, if not functionally similar to, the grooming of primates. When the sexes are segregated, from April to September, most of the grooming is homosexual. Heterosexual grooming takes place once males and females have paired and are establishing territories. Females also groom their young (Nelson 1965b).

There is evidence that groups of some species of bats may show considerable organisation while in flight. The American free-tailed bat, *Tadarida brasiliensis*, which in autumn may occur in large caves in Texas, New Mexico, and Arizona in aggregations of up to 20 million individuals, often leaves and returns to its roost in groups, thus suggesting that at times group flight is normal for the species (Davis *et al.* 1962). When foraging groups arrive back at the roost the following dawn they often first appear at altitudes of 2000–3000 m, from which height they dive down to the cave entrance. If more bats arrive at dawn than can enter the cave at once an orderly landing pattern is established. The groups, although maintaining their integrity, form a loose, wide-wheeling circle at about 2000 m altitude. One group at a time then peels off and dives down to the entrance. As one group dives the next in order moves in from outside the circle to take its place.

The results of some homing experiments on bats also suggest that migration may occur as a group (p. 892).

Although there is as yet, therefore, no evidence that social communication plays a part in the establishment of a familiar area by young bats, other observations suggest that eventually some evidence that this is so will be found.

The ontogenetic pattern of familiar area establishment that may exist in the American free-tailed bat, *Tadarida brasiliensis* (i.e. yearling males remaining in the winter/copulation part of the familiar area throughout their first summer), is broadly similar to that described for the minke whale, *Balaenoptera acutorostrata* (p. 768), and is presumably a reflection of a similar ontogenetic succession of exploratory and calculated migration thresholds. Presumably, also, this ontogenetic succession in thresholds has a largely inherited basis. As discussed in relation to pinnipeds (p. 736) and cetaceans (p. 767), even though bat migrations are suggested to take place within a familiar area, the migration thresholds to habitat, indirect, and migration-cost variables are likely to have a genetic basis. The initiation of breeding, hibernation, and copulation migrations, therefore, and perhaps also the initiation of summer visits to hibernation/copulation sites by adults, are likely to be obligatory, at least in areas where the seasonal changes in suitability of different habitats are predictable (p. 50) as is invariably the case in temperate regions. In some tropical regions, however, the seasonal changes may be less predictable and the migration may be facultative, especially when the critical habitat variables are availability of food and deme density. Thus the frugivorous sub-tropical grey-headed flying fox, *Pteropus poliocephalus* (Fig. 26.5), shows considerable year-to-year variation in the date of the dispersive autumn migration and in the proportion of individuals that take part (Nelson 1965a). In years when food remains abundant around the summer roost, the migration threshold of fewer adults are exceeded (p. 566) and long delays in the initiation of migration may occur, more individuals remaining aggregated with the juveniles during the winter. Similarly, the proportion of individuals of the Mexican funnel-eared bat, *Natalus stramineus* (Fig. 26.6), that initiate migration between wet and dry season home ranges may also be a function of the conditions that prevail in any particular year.

It seems likely that migration thresholds to the migration-cost variable of wind have evolved in bats as they have in birds (p. 632). However, data are scarce. Nevertheless, some support may be derived for this suggestion from the observation that the arrival of American free-tailed bats, *Tadarida brasiliensis*, in Texas in spring often coincides with winds from the south and that the departure in autumn often coincides with strong winds from the north (Davis *et al.* 1962).

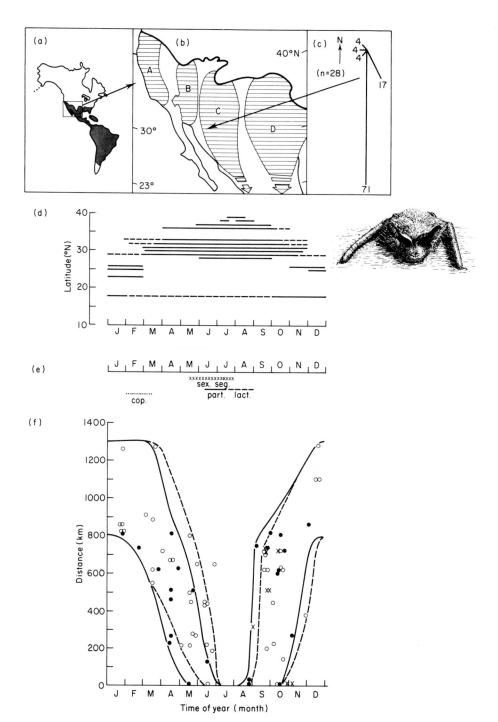

Fig. 26.18   Seasonal return migration of the American free-tailed bat, *Tadarida brasiliensis*

(a) Distribution. Stippled area, maximum distribution. Most work on migration has been carried out in the area marked off by a square. There is some indication of a slight latitudinal seasonal return migration in the southeastern United States and the possibility exists that some part of the mainland group passes the northern winter in the Caribbean.

(b) Geographical variation in the direction and distance components of migration within the subspecies *T. b. mexicana*. Area covered is that indicated by a square in (a). Heavy line indicates northernmost extent of summer range. Although the species is more or less continuously distributed throughout the area during the summer, four major demes, *A*, *B*, *C* and *D* may be recognised. Deme *A* has a much reduced distance component and a direction ratio that is probably more or less 25:25:25:25. Deme *B* has a slightly greater distance component and may have a peak autumn direction to the west. Demes *C* and *D* have a considerable seasonal distance component and a pronounced peak autumn direction to the south.

(c) Direction ratio of a group of individuals marked at summer roosts in Arizona as indicated. Length of line and number indicates the percentage of individuals that was recaptured in each of eight directions from the summer roosts at which the bats were marked. Direction ratios obtained throughout demes *C* and *D* give similar results.

(d) Latitudinal distribution with season of demes *C* and *D* as determined by observation. Solid line, common; dashed line, present but scarce. The main mass of individuals passes the northern winter at latitudes lower than about 27°N, though some individuals remain further north. As spring and summer progress, the species moves northwards but a few males, in particular immature individuals, are thought to remain in the south.

(e) Seasonal behaviour. Solid line, parturition period; dashed line, lactation period; dotted line, copulation period; crossed line, period of maximum segregation of the sexes with females occupying lower and warmer and males occupying higher and cooler altitudes. Individuals of deme *A* and also those occupying the southeast United States roost mainly in buildings. Males and females of demes *C* and *D* during spring and autumn migration use bridges, buildings, mines and caves as transient day-roosts. In summer, buildings predominate as day-roosts for males of these groups while females form maternity aggregations under bridges and in buildings, mines and caves. In some caves, especially in autumn, aggregations of up to 20 million individuals may form.

(f) Distance component of demes *C* and *D*. A solid dot (male), open dot (female) or cross (juvenile) refers to the distance an individual has been recaptured from the summer roost in which it was marked. The maximum and minimum distance limits are indicated for males (solid line) and females (dashed line). The October to April position of the maximum line for males is speculative. Males and females are probably similarly distributed during the copulation period even though northward migration has probably begun. After the copulation period, males appear to precede females in their northward migration. Similarly, males seem to initiate southward migration in the autumn before females. Some juveniles remain at their summer roosts after most adults have left.

[Compiled from data in Glass (*1958*, *1959*), Short et al. (*1960*), Davis et al. (*1962*), Villa-R. and Cockrum (*1962*), Constantine (*1967*), Barbour and Davis (*1969*), and Cockrum (*1969*)]

Although seasonal and ontogenetic variation in migration thresholds seem likely largely to be inherited, as with most of the animals discussed in subsequent chapters (except birds) this seems unlikely to be the case for the distance and direction components of migration. The arguments presented for pinnipeds (p. 736) and cetaceans (p. 767) are equally applicable to bats and need not be repeated here in detail.

The distance of seasonal return migration is likely to be a function of the size of the familiar area established by the young bat. This in turn is a function of the flying ability of the species (p. 580), the frequency with which exploratory migration is initiated, and the constancy of the direction of exploratory migration. The frequency with which exploratory migration is initiated is a function of the interaction of the inherited level of exploratory migration thresholds to habitat variables and the local spatial and temporal juxtaposition of conditions, including deme density. The constancy of the direction of exploratory migration (p. 654) for each individual is likely to be partly inherited (p. 610) and partly a response to local conditions of topography, resources, and deme density. In most species, however, either the individual or the deme as a whole is likely to initiate exploratory, though not necessarily calculated, migration in all directions (p. 654) except where modified through social communication of traditional directions. As with some cetaceans (p. 767) and birds (p. 605), however, there seems a strong possibility that in those demes or species with a long-distance migration with a strong bias in the peak direction and a peak direction that shows deme-to-deme consistency over a large geographical area, selection may have favoured an inheritable bias to the direction ratio of exploratory migration.

The major candidates for such a bias are the American free-tailed bat, *T. brasiliensis* (Fig. 26.18), the noctule, *Nyctalus noctula* (Figs. 26.13 and 26.19), and perhaps also the American tree bats (Figs. 26.10–26.12). In all cases it seems likely that selection has favoured a linkage between indirect variables and the direction, speed, and constancy of exploratory migration. Thus, in late summer and early autumn during establishment of a familiar area around the natal site, selection should not favour a directional bias to exploratory migration. Beyond a certain date, however, an inherited bias to the south may be advantageous (bearing in mind the discussion concerning social communication—

p. 588. A detailed discussion of the distance and other components of such a system is presented in relation to birds (p. 654). An inherited northward bias the following spring may or may not be advantageous for, once a familiar area has been established, the most advantageous migrations could be achieved solely by seasonal variation in migration thresholds. Except, that is, in the case of the males of the hoary bat, *Lasiurus cinereus* (Fig. 26.12). If the analysis of sightings for this species is valid, males are born predominantly in the central and eastern region of

North America. Then, in the autumn, they migrate to the south. The following spring, however, they do not return to the central region but instead take up a home range in the northern and western part of North America. Unless the exploratory migrations of the previous summer and autumn encompassed these northwestern areas, it presumably has to be assumed that the northwest bias to the spring migration of yearling male hoary bats is either inherited or results from social communication.

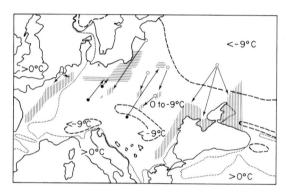

Fig. 26.19 Deme-to-deme variation of direction and direction ratio of the seasonal return migration of the noctule, *Nyctalus noctula*, in Europe

General ecology and migration
See Fig. 26.13. Noctules hibernate in caves in eastern Europe, buildings and trees in central Europe, and in tree-cavities in western Europe. In summer, tree-cavities are used as roosting sites throughout the range.

Direction and direction ratio
The large dots indicate roosts at which the bats were marked: ○, summer roosts; ●, winter roosts. ×, some other known winter roosts but at which no marking occurred. The hatching represents the area over which marked bats were recaptured: ≡, in summer; ||||, in winter. Arrows join roost of release to area in which the bats were recaptured.
According to the figures, the noctule in Europe shows a relatively slight deme-to-deme variation in migration direction and direction ratio with most individuals migrating in the autumn to the SE, S, or SW. Lines show midwinter (January) isotherms: ...., 0°C isotherm; ----, -9°C isotherm.
Neither autumn nor spring migration has a large dispersive or convergent component as far as the species as a whole is concerned, though summer and winter roosting groups do not retain their integrity through both seasons but rather break up and then re-combine with individuals from other groups to form a new roosting group. Migration direction and direction ratio, combined with

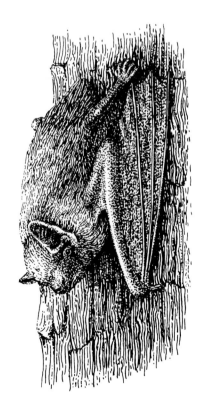

migration distance, tend to result in the individual experiencing a shorter winter and cooler summer than if seasonal return migration did not occur. Perhaps the critical factor during winter is the number of days on which temperature fails to rise sufficiently for feeding to be possible (p. 101).

[Modified from Strelkov (1971). Isotherms from Fullard (1959)]

26.8 Appendix: a graphical presentation of the results of marking–release/recapture experiments on bats

The literature on bats abounds with the results of marking–release/recapture experiments. At present, however, these results are scattered through the journals of many countries, with no attempt so far having been made to bring them together and collate them. A number of diagrams for various bat species are presented elsewhere in this book because they illustrate certain specific aspects of bat migration. This appendix represents an attempt to draw together the data for those other species that have received attention from bat ringers. In all cases the diagrams are based on raw ringing data as presented by the authors listed under each figure. This collation of results permits some degree of inter-specific comparison and forms the basis for many of the conclusions reached earlier in this chapter.

The single-species diagrams in this appendix are each divided into three sections.

Section (a) indicates the world distribution of the species concerned. Either the spring or autumn direction ratio (expressed as eight directions instead of the usual four) is indicated for each area for which mark–release/recapture experiments have been carried out. The convention adopted in the representation of the direction ratio is for the length of line and the number to indicate the percentage of individuals that were captured in each of eight directions relative to either their hibernation roost (if the spring direction ratio is shown) or their summer roost (if the autumn direction ratio is shown as in Fig. 26.28). In all cases north is to the top of the page.

Section (b) shows the seasonal behaviour. In all cases, where included, the solid line shows the parturition period, the waved line the hibernation period, and the dotted line the probable copulation period.

Section (c) shows the distance component of the seasonal return migration. A solid dot (male), open dot (female), and in some diagrams, a triangle (juvenile) or question mark (unknown age and/or sex) refers to the distance between capture at a hibernation roost (or summer roost in the case of the serotine, *Eptesicus serotinus*, in Fig. 26.28) and subsequent or previous capture. Where data appear adequate, the maximum and minimum distance limits are indicated for males (solid line) and females (dashed line). Dots outside these lines may refer to removal migrants.

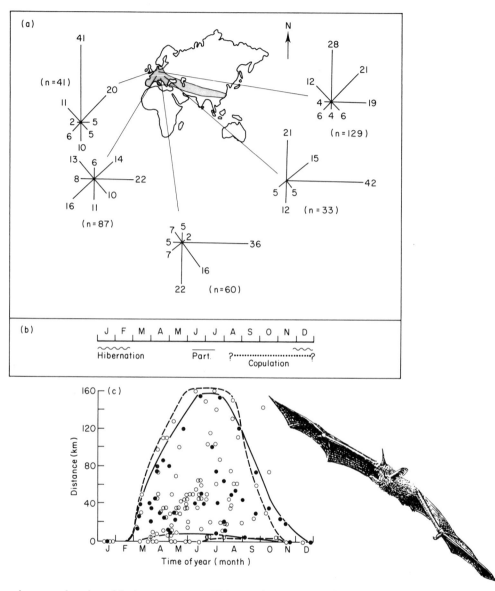

Fig. 26.20   Seasonal return migration of the large mouse-eared myotis, *Myotis myotis*

(a) Distribution and direction ratio. Spring direction ratio is shown for bats from five different areas of Europe as indicated (i.e. Netherlands, East Germany and northern West Germany, central and southern West Germany, Austria, and southern Poland). A peak spring direction to the NNE is apparent in the northern part of the species range. Further south and east, spring direction ratio becomes firstly more or less random, then shows a peak to the E and then to the ESE. In all places, however, whatever the peak direction, the direction ratio is such that there is no seasonal shift of geographic range. Individuals are found in all regions throughout the year.

This remains true even further to the east in European USSR

(b) Seasonal behaviour. Caves, mines and ventilator shafts are used as hibernation roosts. Females form maternity colonies in buildings. During the autumn, males establish solitary roosts in attics and copulate with females that use the site as a transient autumn roost. *M. myotis* obtains a large part of its food, such as spiders and dung beetles, on the ground. The bats hop over the ground, apparently finding dung beetles by smell.

(c) Seasonal distance component. There appears to be little difference in distance component between the sexes but males, which are territorial at the time, seem to show a slower autumn return than the females. There is some indication that in their first autumn, juveniles may move less far than adults from summer to hibernation roost.

[Compiled from data in Eisentraut (1936, 1960), Bels (1952), Felten and Klemmer (1960), Frank (1960), Gruber (1960), Hummitzsch (1960), Krzanowski (1960), Schmaus (1960), Hanak et al. (1962), Ellerman and Morrison-Scott (1966), Strelkov (1970), and Burton (1971)]

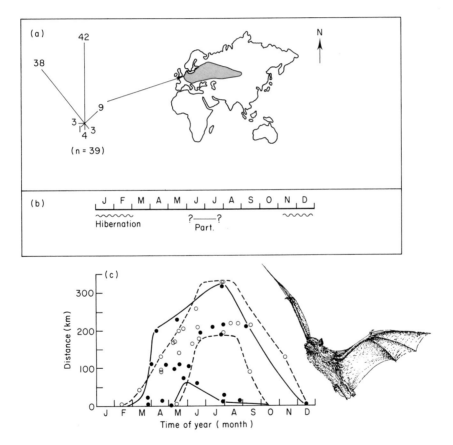

Fig. 26.21 Seasonal return migration of the pond myotis, *Myotis dasycneme*

(a) Distribution and direction ratio. Map shows world distribution (stippled area). Spring direction ratio is shown for *M. dasycneme* in the Netherlands as indicated. Peak spring direction is to the NNW in this region. However, the direction ratio is such that in the Netherlands there is no seasonal shift of geographical range. Individuals are found throughout the year. This remains true even further to the east in European USSR.

(b) Seasonal behaviour. Limestone caves are used as hibernation sites, though other situations are used as well

in European USSR. Attics of buildings are used as summer roosts. *M. dasycneme* is found near water in summer.

(c) Seasonal distance component. Males and females seem to travel similar maximum distances at about the same speed and time of year. However, although some males move only relatively short distances between winter and summer roosts, the data so far indicate that all females travel 200 km or more.

[Compiled from data in Ryberg (1947), Bels (1952), Heerdt and Sluiter (1953, 1954, 1956, 1957, 1958, 1959), Egsbaek and Jensen (1963), Strelkov (1970)]

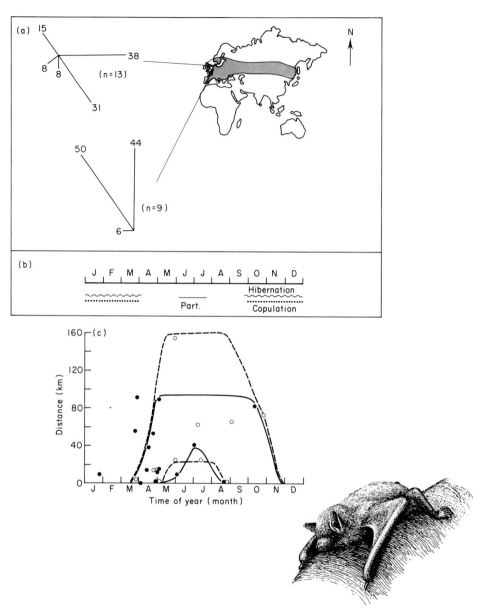

Fig. 26.22   Seasonal return migration of Daubenton's myotis, *Myotis daubentoni*

(a) Distribution and direction ratio. Map shows world distribution (stippled area). Spring direction ratio is shown for two different areas of Europe as indicated (i.e. Netherlands and Denmark). Peak direction is different in the two areas. In all places, however, the direction ratio and distance components of migration are such that there is no seasonal shift of geographic range. Individuals are found in all regions throughout the year. This remains true even further to the east in European USSR.

(b) Seasonal behaviour. Limestone caves are used as hibernation sites and individuals and small groups may be found buried amongst the gravel on the floor of the cave. Crevices in trees, buildings and caves are used as summer roosts with hollow trees being used particularly for maternity roosts. *M. daubentoni*, which is often known as the water bat, forages over water. As many as 200 individuals have been seen feeding over a single lake.

(c) Seasonal distance component.

[*Compiled from data in Ryberg* (*1947*), *Bels* (*1952*), *Heerdt and Sluiter* (*1959*), *Hanak et al.* (*1962*), *Egsbaek and Jensen* (*1963*), *Roer and Egsbaek* (*1966*), *Strelkov* (*1970*), *and Burton* (*1971*)]

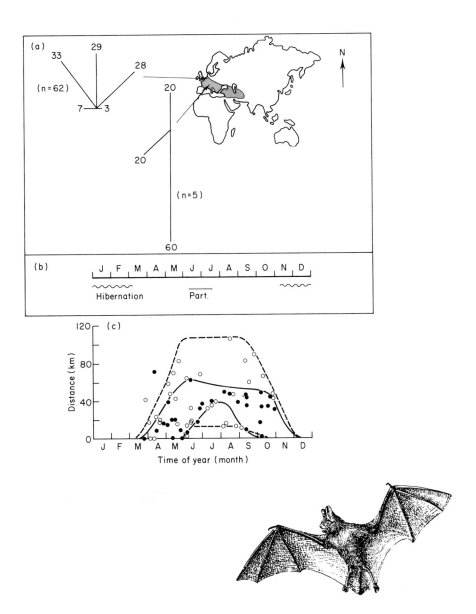

Fig. 26.23 Seasonal return migration of Geoffroy's myotis, *Myotis emarginatus*.

(a) Distribution and direction ratio. Spring direction ratio is shown for the species in two areas (i.e. the Netherlands and Austria) as indicated. Peak spring direction is to the N in the Netherlands and to the SSW in Austria. It should be noted, however, that the number of recaptures are few in the latter area.

(b) Seasonal behaviour. Caves and mines are used as hibernation roosts. Females form maternity aggregations of up to 100 individuals in caves and attics.

(c) Seasonal distance component. Females seem to travel further between winter and summer roosts than males, though some females may travel less than 20 km. One result of this minimum distance of travel is that despite the direction ratio found in the Netherlands, a seasonal shift of geographical range is unlikely to be apparent. This remains true even as far east as European USSR.

[*Compiled from data in Bels (1952), Heerdt and Sluiter (1959), Gruber (1960), Hanak et al. (1962), Ellerman and Morrison-Scott (1966), and Strelkov (1970)*]

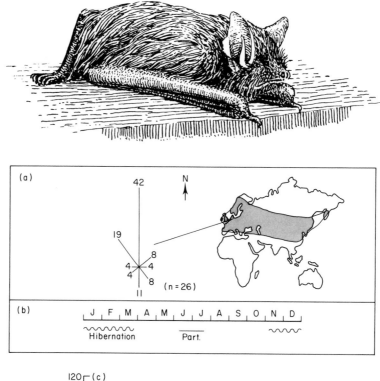

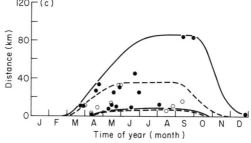

Fig. 26.24 Seasonal return migration of the whiskered myotis, *Myotis mystacinus*

(a) Distribution and direction ratio. Stippled area shows the world distribution of *M. mystacinus*. Spring direction ratio is shown based almost entirely on data obtained in the Netherlands as indicated. A peak spring direction to the N is indicated. However, the direction ratio is such that there is no seasonal shift of geographical range. Individuals are found throughout the year. This remains true even further to the east in European USSR.

(b) Seasonal behaviour. The majority of individuals use caves and mines as hibernation sites in which they may roost singly or in groups of up to 100. Males have solitary summer day-roosts in hollow trees. Females form maternity aggregations of 10–20 individuals in buildings. *M.*

mystacinus forages in summer around ponds and streams that are surrounded by woodland. The prey is not caught in the air but is picked off twigs etc.

(c) Seasonal distance component. Males seem to travel further between winter and summer roosts than females though some individuals of both sexes may move only a few kilometres. An individual of *M. mystacinus* has been captured in Bulgaria approximately 1950 km from its marking and release point in the USSR. There is no way of knowing whether this individual was performing a seasonal return migration or whether it was a removal migrant. However, the fact that *M. mystacinus* is present in the USSR throughout the year has already been noted.

[*Compiled from data in Ryberg (1947), Bels (1952), Heerdt and Sluiter (1959), Natushke (1960), Hanak et al. (1962), Krzanowski (1964), Strelkov (1970), and Burton (1971)*]

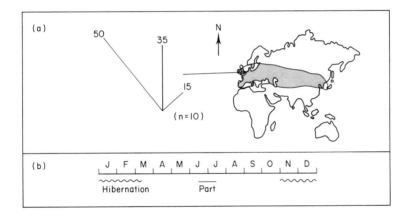

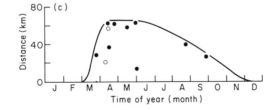

Fig. 26.25 Seasonal return migration of Natterer's myotis, *Myotis nattereri*

(a) Distribution and direction ratio. Stippled area shows the world distribution of *M. nattereri*. Spring direction ratio is shown for the species in the Netherlands as indicated. A peak spring direction to the NNW is indicated in this region. Although recapture data suggest that all individuals migrate between NE and NW, it should be noted that the number of recaptures is small.

(b) Seasonal behaviour. The majority of individuals use caves as hibernation sites in which they roost singly. Males

use crevices in trees, rocks and buildings as summer roosts. Females form maternity aggregations of 10–25 individuals in buildings. In summer, *M. nattereri* is found in woodland around fishponds.

(c) Seasonal distance component. The minimum distance travelled between winter and summer roosts is so small that even if the direction ratio shown in (a) is valid, a seasonal shift in geographical range is unlikely to be apparent. Even as far east as European USSR a seasonal shift does not occur.

[*Compiled from data in Ryberg* (*1947*), *Bels* (*1952*), *Heerdt and Sluiter* (*1959*), *Hanak* et al. (*1962*), *Strelkov* (*1970*), *Burton* (*1971*)]

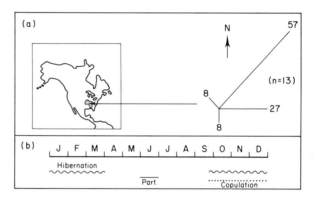

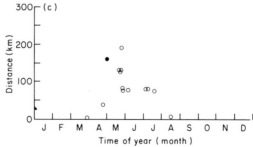

Fig. 26.26 Seasonal return migration of the grey myotis, *Myotis grisescens*

(a) Distribution and direction ratio. Stippled area, distribution. Direction ratio is shown for a group (data mainly for females) that hibernates in Kentucky as indicated. A peak post-hibernation direction of ENE is indicated.

(b) Seasonal behaviour. Caves are used as hibernation roosts and aggregations of as many as 100 000 individuals occur. Females form summer maternity aggregations of 100–250 000 individuals in caves with substantial streams. During the summer, males form nomadic groups of 100 or more individuals.

(c) Distance component of seasonal return migration. Maximum and minimum distance limits are not included due to the general lack of data for males and the lack of summer and autumn data for females.

[*Compiled from data in Hall and Wilson (1966) and Barbour and Davis (1969)*]

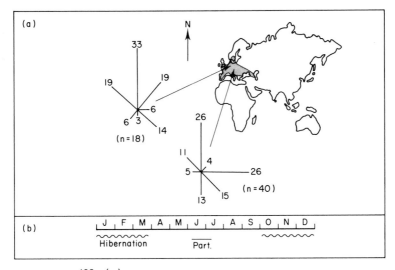

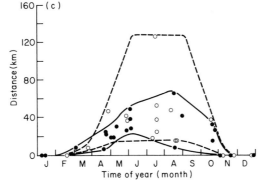

Fig. 26.27   Seasonal return migration of the barbastelle, *Barbastella barbastellus*

(a) Distribution and direction ratio. Stippled area shows world distribution of *B. barbastellus*. Spring direction ratio is shown for bats in two areas (i.e. the Netherlands, northern West Germany and East Germany; and southern West Germany and Austria) as indicated. Peak spring direction is to the N in the northern area and ENE in the southern area. However, the direction ratio and distance component (see (c)) are such that there is no seasonal shift of geographical range. This remains true even as far east as European USSR.

(b) Seasonal behaviour. Caves are used as hibernation sites in which bats roost either solitarily or in groups of up to 200 individuals. Crevices in trees and buildings and similar sites, such as behind shutters, are used as summer roosts.

(c) Seasonal distance component. The maximum distance between winter and summer roosts may be greater for females than for males.

[*Compiled from data in Ryberg (1947), Felten and Klemmer (1960), Frank (1960), Hoehl (1960), Hanak et al. (1962), Strelkov (1970), and Burton (1971)*]

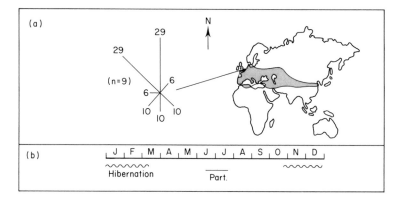

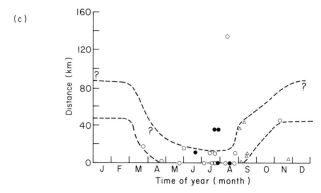

Fig. 26.28 Seasonal return migration of the serotine bat, *Eptesicus serotinus*

(a) Distribution and direction ratio. Stippled area shows world distribution of *E. serotinus*. Autumn direction ratio is shown based on data for bats near Oldenburg, West Germany as indicated. Peak autumn direction is to the NNW but it should be noted that the number of recaptures is small. In any case, the direction ratio is such that there is no seasonal shift of geographical range. This remains true even as far east as European USSR.

(b) Seasonal behaviour. Crevices in buildings, under the bark of trees and the inside of hollow trees are used as hibernation roosts. Females form maternity aggregations

of 10–30 individuals in buildings. The males are solitary during the summer. *E. serotinus* forages in woodland glades.

(c) Seasonal distance component. A solid dot (male), open dot (female) or triangle (juvenile) refers to the distance between capture at a summer roost and subsequent or previous capture. Maximum and minimum distance limits are shown for females (dashed lines). In this case, however, these lines are included solely as an aid to interpretation of the figure and are not intended to be a definitive biological statement. Recaptures outside of these lines could be removal migrations.

[*Compiled from data in Ryberg (1947), Havekost (1960), Natushke (1960), Hanak et al. (1962), Strelkov (1970), and Burton (1971).*]

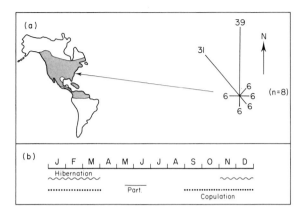

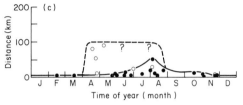

Fig. 26.29 Seasonal return migration of the big brown bat, *Eptesicus fuscus*

(a) Distribution and direction ratio. Stippled area, distribution. Direction ratio is shown for bats marked during hibernation in Kentucky and Indiana as indicated. A peak post-hibernation direction to the north is indicated, but the number of recaptures is small.

(b) Seasonal behaviour. All lines refer to Kentucky. In Nebraska, parturition occurs in late July. Caves and mines are used as hibernation roosts by single individuals and small groups of both sexes. In addition females may hibernate in buildings. In Indiana, individuals move frequently during the winter from one hibernation roost to another. Hibernation therefore occurs in a roost complex rather than a single roost. Females form summer maternity aggregations of between 20 and 300 individuals in buildings and hollow trees. Males are assumed to be mainly solitary and nomadic during spring and early summer and may continue to use caves and mines as day roosts.

(c) Distance component of seasonal return migration. The line for females from June to November is based firstly on the observation that maternity aggregations persist until August and secondly on the assumption that from the beginning of copulation males and females are the same distance from the hibernation-site complex.

In summary, seasonal return migration of *E. fuscus* may consist of: (1) a post-hibernation movement away from the hibernation site by all females, except perhaps some that hibernate in buildings, and most males. During this a peak direction may be apparent, and females move further than males even though female movement has terminated by May whereas male movement continues until August; and (2) a gradual return movement to the hibernation-roost complex by males from August onwards and a more rapid return movement by females upon the cessation of lactation.

[Compiled from data in Beer (1955), Christian (1956), Mumford (1958), Phillips (1966), Davis et al. (1968), Barbour and Davis (1969) and Goehring (1972)]

27

Seasonal and ontogenetic return migration by birds

27.1 Introduction

Two reptilean lineages are known to have adopted an essentially aerial way of life. These are the pterosaurs, such as *Pteranodon*, which became extinct towards the end of the Cretaceous (70 million years B.P.) (Bramwell 1971, Bramwell and Whitfield 1974), and the birds.

The bird lineage can be traced back from the present time to the Jurassic (165 million years B.P.), the position of the Jurassic fossil, *Archaeopteryx lithographica*, as on or near to the ancestry of all modern birds now being reasonably well settled (Yalden 1971a). The time of appearance of sustained flapping flight in this lineage, however, remains a subject for controversy. Current argument (Ostrom 1974) favours common ancestry of *Archaeopteryx* and coelurosaurian theropods among the archosaurian reptiles. It is also suggested: (1) that plumage in the form of contour feathers evolved initially in response to selection on the efficiency of thermoregulation (Ostrom 1973); (2) that enlargement of the primordial 'primary' and 'secondary' feathers enhanced the prey-catching function of the hands of a highly predaceous animal, transforming the forelimbs of the 'proto-*Archaeopteryx*' into natural insect nets (Ostrom 1974); and (3) that these adaptations were admirably pre-adaptive for the evolution of active, flapping flight. *A. lithographica*, which has a wing-span of about 58 cm (Yalden 1971a), was itself probably capable of some flapping flight even though such flight was probably not long sustained (Yalden 1971b). It is suggested (Ostrom 1974) that the species was a ground-dwelling, bipedal, cursorial predator, rather like the modern secretary bird, *Sagittarius*, and that the wings were used primarily in fluttering leaps while assaulting prey.

Only later in the bird lineage were the wings subsequently modified into stronger, flapping appendages, capable of supporting sustained flapping flight.

An alternative and still extant hypothesis of the origin of flapping flight in the bird lineage maintains that primitive ancestors, at the 'proto-*Archaeopteryx*' stage, were arboreal creatures that leaped from branch to branch or tree to tree. Selection for increased body-surface area that provided lift and thus permitted longer and longer leaps led first to the evolution of gliding and eventually to the evolution of flapping flight (Bock 1965). Whichever hypothesis is accepted, it seems likely (Yalden 1971a,b) that *Archaeopteryx* was capable of some flapping flight and that by 60 million years or so later (100 million years B.P.) birds existed that were capable of sustained flapping flight and that had colonised both land and sea.

There is, of course, no way that we can be certain when or how often the familiar-area phenomenon evolved in the vertebrate lineage but if, as suggested in this book, most, if not all, modern vertebrates live within a familiar area, then it seems highly probable that the archosaurian reptiles of the Triassic (200 million years B.P.) did so also. The large reptiles ('dinosaurs') that persisted from the late Triassic (200 million years B.P.) to the end of the Cretaceous (70 million years B.P.) may well have evolved seasonal return migrations within a familiar area, the distance and direction, etc., components of which may have been comparable with those of modern large terrestrial mammals (p. 526), such as ungulates (Ostrom 1972). As far as the bird lineage is concerned, however, it seems likely that the major part of the evolutionary sequence leading to flapping flight occurred among animals with a familiar area comparable in size with that of most modern reptiles (p. 195) and small mammals.

This remains true even if the arboreal hypothesis for the origin of flapping flight is accepted. Most gliding vertebrates, such as various squirrels, 'flying lemurs' (*Cynocephalus*), and marsupial gliders among the mammals and various lizards and at least one snake among the reptiles, have a home range that, despite the animals concerned being able to glide 100 m or so in the case of the mammals and 10–15 m or so in the case of the reptiles, is more or less similar in size to that of a comparable (p. 392) but non-gliding animal. For example, the marsupial greater glider, *Schoinobates volans*, which lives among the tall trees of the wet sclerophyll forests of Australia and which can glide for up to 120 m (Bergamini 1965), nevertheless has a year's home range of only about 1·2–2·4 ha (Tyndale-Biscoe 1973). This is comparable to the home range size of other similar, but non-gliding, phalangers such as the brush possum, *Trichosurus vulpecula*, which has a year's home range of about 3·01 ha for males and about 1·08 ha for females. To judge from the situation in mammals, therefore, it is the evolution of flapping flight, as in bats, that makes the major impact upon the distance and hence direction and direction ratio components of seasonal return migration within a familiar area.

Tentatively, therefore, we may suggest that until the evolution of flapping flight (some time between 200 and 100 million years B.P.) bird-ancestors lived within a familiar area comparable in size with that of modern small, flightless birds and 'terrestrial' and gliding reptiles and small mammals. Any seasonal return migration that occurred was likely to have had a short distance component, perhaps rarely more than a kilometre, though non-calculated and/or exploratory/calculated removal migrations may have occurred over much longer distances. With the evolution of flapping flight, however, there was likely to be a considerable increase in the distance of exploratory migration, an increase in the size of the familiar area, and consequently an increased possibility of a seasonal shift in home range and the evolution of a seasonal return migration.

In the discussion of bat migration it was suggested that an inherited bias to the distance, direction, and direction ratio components of exploratory migration was likely to evolve only when there was selection for relative inter-deme consistency in these components. Inter-deme consistency is likely to evolve only when the selective pressures acting upon these components are latitudinal rather than local. Most

temperate bats spend the winter in caves, the conditions inside which are relatively independent of latitudinal and seasonal variations. Consequently, the direction, direction ratio, and distance components of bat migration are most often a function of the local spatial juxtaposition of the areas providing the resources required during a year and caves suitable for hibernation. Only in those species that are the strongest fliers and that establish large familiar areas and in those species that make least use of caves as hibernation sites has latitudinal selection impinged sufficiently to produce long-distance seasonal return migrations with a strong bias to the direction ratio that is relatively consistent in direction from deme to deme. The most likely candidates for such a pattern were adjudged to be the American free-tailed bat, *Tadarida brasiliensis* (Fig. 26.18), the noctule, *Nyctalus noctula* (Fig. 26.13), and perhaps also the silver-haired bat, *Lasionycteris noctivagans* (Fig. 26.10), red bat, *Lasiurus borealis* (Fig. 26.11), and hoary bat, *L. cinereus* (Fig. 26.12). These species, therefore, should also have the highest probability among bats of an inherited bias to the direction ratio of exploratory migration. If, that is, it is assumed that during the evolution of the behaviour involved in familiar-area establishment, young bats have sufficiently often been in the position of having to establish a familiar area away from social communication with older individuals (see p. 767). This thesis seems equally applicable to birds.

From probably an early stage in their emergence as a distinct lineage, birds seem to have adopted a strategy of continual feeding without hibernation. Perhaps the only exception is Nuttall's poor-will, *Phalaenoptilus nuttallii*, (p. 664) which lives in the western and central parts of North America from British Columbia to Mexico (Howell and Bartholomew 1959). Taking the bird lineage as a whole, therefore, caves seem likely to have played no more than a small part in their ecology and have probably never been used for hibernation, though they are used by some species for nesting. From the beginning of their colonisation of temperate regions, therefore, many land birds seem likely to have been exposed to latitudinal selection on the distance, direction, and direction-ratio components of exploratory migration for the same reasons given for a minority of bats.

Sea birds, on the other hand, like pinnipeds (p. 706) and odontocete cetaceans (p. 755), are adapted to a habitat-type which perhaps imposes less in the way of latitudinal gradients of selection than the terrestrial habitat-type. There is, however, likely to be one major difference between the ecology of sea birds and that of these marine mammals. As shown in the chapter on their migration (p. 809), many species of marine fish in temperate and sub-polar regions adopt a greater depth at some stages of their seasonal migration, particularly during the winter months in temperate and sub-polar regions. Although this behaviour may not prevent these fish being preyed upon by pinnipeds and cetaceans, it may well reduce the suitability of high latitudes as a habitat-type for marine birds. The result could well be a latitudinal gradient of habitat suitability that could impose greater selection for deme-to-deme consistency in the distance, direction, and direction-ratio components of the seasonal return migration of sea birds than for deme-to-deme consistency in the same components of the seasonal return migrations of pinnipeds and odontocete cetaceans. Particularly, that is, with regard to the migration components of birds that are adapted to feed near the sea surface. The steepness of any seasonally recurring latitudinal gradient of habitat suitability may be less for birds that dive for their food.

Selection on degree of return by both sea and land birds have already been discussed (p. 321). In degree of return in birds seems to be comparable to that of other seasonal return migrants that have the opportunity when young for exploratory migration around the natal site and for assessment of breeding habitat suitability.

The hypothesis to be presented and evaluated in this chapter is that the seasonal return migrations of birds, like those of all other vertebrates, take place within a familiar area and hence, in this respect, are no different from the movement of birds and other vertebrates that do not perform a seasonal return migration.

27.2 Environmental cues available for the establishment of a familiar area

The establishment of a familiar area involves the perception of environmental structures and the storage of their spatial relationships within a spatial memory. Much current research on bird migration is concerned with a determination of the range of environmental cues that can be perceived by birds, though rather less research is concerned with the existence and potential of the spatial memory.

The vast majority of environmental cues likely to be used by a bird in the establishment of a familiar area are visual. Broadly speaking, visual cues can be divided into two categories: topographical (landmarks) and celestial. Sufficient evidence seems to be available (Matthews 1968) that not only do birds perceive topographical features and store the relevant information in a spatial memory but also they are capable of retrieving this information after an interval of several years (p. 893). The use of landmarks by homing pigeons in the vicinity of their lofts has been noticed by most workers on this species and the recognition of topographical features such as coastlines, major rivers, and hills along migration routes by many other birds also seems to be widely accepted, particularly in relation to waterfowl (Hochbaum 1955, Bellrose and Sieh 1960, Mathiasson 1963, Bellrose 1967b, 1972). As our knowledge stands at present there seem to be available far fewer 'topographical' cues that would enable a bird to establish a familiar area over large stretches of open ocean. There may, however, be visual cues available to a bird that are not readily apparent to a human. Even apart from obvious features such as coasts and islands, the sea may not be as featureless as at first it may appear. Perhaps patches of water of different colours and predictable cloud formations provide some features (Hochbaum 1955) as also may the direction of waves (Bellrose 1972) and wind and perhaps even the presence of certain other birds, etc., characteristic of a particular area at a particular time of year.

There appears to be an integration of topographical and celestial features within the spatial memory. The result is perhaps that the spatial position of the various topographical features with which the bird is familiar are memorised relative to the position of celestial features such as the Sun and stars and also, perhaps, the Moon or other, more subtle, features such as the axis of rotation of the night sky. Most experiments on the integration within a spatial memory of topographical and celestial features involve training the bird to go to a particular fixed position while exposed to a certain celestial pattern and then varying this pattern with the result that the bird goes to a different position in accordance with the new pattern of celestial variables.

As far as the Sun is concerned, birds have been

trained to go to a particular feeding site at a fixed time of day when the Sun is in a particular position in the sky (e.g. Kramer 1952). Experiments at different times of day demonstrate that the birds learn the way that the spatial relationship of the Sun and the feeding site changes throughout the day. In fact, it is easier to train a bird to compensate for the movement of the Sun across the sky than it is to train it not to compensate. The ability to compensate, of course, requires an endogenous 'clock' mechanism by which the bird may measure time of day. In keeping with most such biological clocks, the circadian clock of the bird requires a phase-setter or 'zeitgeber' which in this case is invariably the timing of the light/dark interphase. By exposing a bird to an experimental lighting regime in which the light/dark interphase is shifted by a certain time interval from that of the 'natural' day, the circadian clock of the bird can be re-phased. Subsequent exposure to the original experimental set-up under natural lighting conditions shows that the bird goes to a feeding or 'home' site an appropriate amount clockwise or anticlockwise of the target site, thus demonstrating not only the existence of an endogenous circadian clock mechanism in the bird, but also the zeitgeber role of the light/dark interphase, and confirming the integration within the bird's spatial memory of the feeding or home site and the position of the Sun (see also Chapter 33, p. 857).

Recent work has shown that starlings, *Sturnus vulgaris*, can be trained to distinguish between two geographical locations 200 km apart on an east–west axis. This remains true even if tested in a nearby 'unfamiliar' landscape, but not if the landscape is screened off (Cavé *et al.* 1974). The presence of both landscape and the Sun seems to be essential for the retention of this ability to discriminate the two locations, but according to Cavé *et al.* the landscape need not be familiar. These workers suggest that the critical factor is the presence of a horizon which would then enable the birds to measure the altitude, or rate of change of altitude, of the Sun. I feel, however, that in the interpretation of this and similar work we should be careful in our categorisation of areas as familiar or unfamiliar. In such circumstances it is quite possible that distant topographical features on the horizon can be stored in the spatial memory separately from nearer, more conspicuous, landscape features and that experimental displacement to a nearby area with different short-distance landscape features is not necessarily displacement to an unfamiliar area. The presence of the

Sun as well as landscape features on the horizon may be necessary either for the bird to recognise the direction of these features or simply to render them conspicuous. These experiments, therefore, admirably demonstrate the perception and integration of landscape and celestial features within a spatial memory but so far (only a progress report is available at the time of writing) do not demonstrate the particular features of landscape and Sun that are being perceived and memorised.

Experiments on the recognition of star patterns carried out in a planetarium by Wallraff (1972) on mallard, *Anas platyrhynchos*, and European teal, *A. crecca*, have shown that these birds can be trained to a particular direction under all positions of the rotating sphere at a constant latitude. The longest interval between training and testing was six months during which time other additional star patterns had been learned. Training was achieved just as readily under totally artificial stellar patterns never occurring in nature as under simulated natural patterns. It appears that birds can learn to distinguish between different patterns composed of more than a thousand star-like spots and have a pronounced capacity for memorising such patterns whether or not the patterns simulate natural skies (see also p. 609). No experiments were carried out with the Moon superimposed on the star pattern.

In general, therefore, it seems likely that most birds perceive a wealth of environmental cues through their visual sense, some of which may be integrated and stored within a spatial memory and permit the establishment of a familiar area. There are, however, at least two other environmental cues, perceived through senses other than vision, that can also be integrated into a spatial memory and that may become of particular importance under certain meteorological conditions, such as overcast skies and thick fog, when visual cues are reduced or nonexistent.

The first of these environmental cues is that of wind direction. A bird on the ground or on a tree has a fixed reference point from which wind direction can be assessed. A bird in flight, on the other hand, is part of the moving wind mass. Nevertheless, as pointed out by Nisbet (1955), near the surface of the Earth the air is anisotrophic as a result of friction with the Earth's surface features. These produce wave-like currents of air that are correlated with wind direction. Most birds fly within 6000 m of the Earth's surface which is well within the range of the turbulent wind pattern. Even in the absence of visual cues from the ground, therefore, a bird should be capable of detecting wind direction. Vleugel (1952, 1959), as a result of observation of chaffinches, *Fringilla coelebs*, in the Netherlands, concluded that the species could integrate other spatial information with wind direction in that they could maintain a straight course when all cues other than wind direction apparently had disappeared. Bellrose and Crompton (1970) placed groups of blue-winged teal, *Anas discors*, in circular cages at the top of a 30 m high tower. One group, in a wire cage, were exposed to the wind whereas another group, in a plexiglass cage, were protected from the wind. After several days the teal were released under overcast skies and over unfamiliar landscape. Both groups went downwind, but those previously prevented from perceiving the wind produced a markedly more tortuous track than those exposed to the wind.

The second of these additional environmental cues that could be integrated with others in the establishment of a familiar area, is that of the Earth's magnetic field. After many years of confusion and controversy (Merkel and Fromme 1958, Perdeck 1963, Merkel and Wiltschko 1965, Meyer and Lambe 1966) involving laboratory experiments on European robins, *Erithacus rubecula*, and homing pigeons, *Columba livia*, it now seems to be established not only that fluctuations in the Earth's magnetic field interferes with the directional selection of ring-billed gulls, *Larus delawarensis* (Southern 1969a, 1971a), and homing pigeons (Keeton 1972a, Walcott 1972, Keeton *et al.* 1974) but also that magnetic fields, both natural and induced, can provide a directional cue. So far the perception of direction from a magnetic field has been demonstrated for the European robin, *Erithacus rubecula* (Fig. 33.4), homing pigeon, *Columba livia* (Walcott and Green 1974), and two species of European warblers, the whitethroat, *Sylvia communis* (Wiltschko and Merkel 1971) and the garden warbler, *S. borin* (Wiltschko 1974). The limit of sensitivity to magnetic cues of the birds investigated so far is less than 70 (Keeton *et al.* 1974) or 50 gamma (Southern 1970), a sensitivity that approaches that of 10 gamma (less than a factor of 10^{-4} of the Earth's normal magnetic field) reported for honeybees, *Apis mellifera* (Lindauer and Martin 1972).

The mechanism of perception of this cue is unknown. Possibly the magnetic field could either mechanically displace some structure in a manner analogous to the displacement of otoliths by gravity or perhaps it could alter molecules or atoms in sensory cells in a way similar to the molecular alteration by light in photoreceptors (Keeton 1972a). In the experiments performed by Walcott and Green (1974) on homing pigeons, *Columba livia*, small Helmholtz coils were attached to the bird in such a way that they induced a magnetic field through the bird's head. As the induced magnetic field provided the bird concerned with directional cues, it seems likely that the sense organ is located in the head.

In conclusion, it seems that there are abundant environmental cues available to a bird for the establishment of a familiar area. It is suggested, therefore, that from the moment of hatching onwards, the topographical, celestial, meteorological, and magnetic cues perceived by the bird throughout its exploratory and other migrations are integrated within a spatial memory. The result is that from any point within the established familiar area the bird can migrate to any other point. The ways in which such navigation may occur, both while within the familiar area and following experimental displacement beyond the familiar area, are discussed in Chapter 33.

27.3 The migrations of young birds: establishment of a familiar area by exploratory migration and social communication

The general aspects of familiar-area formation by young birds have already been discussed (pp. 184 and 205) with particular reference to exploratory migration, habitat assessment, and calculated removal migration. This section, therefore, concentrates on the formation of a familiar area by the juveniles of species that perform seasonal return migrations.

In discussing the data available that are relevant to the familiar area hypothesis for bird migration, the following terminology will be used. Juvenile exploratory migration is sub-divided into postfledging, autumn, winter, and spring phases. In species that do not perform seasonal return migrations, the components of each of these phases are similar except for the migration thresholds to habitat variables that are involved. In all probability also, beyond a certain date in late autumn, most migration is calculated and not exploratory. In species that do perform seasonal return migrations, the autumn phase of the exploratory migration is considered to be the phase that takes the animal from its breeding home range to its eventual winter home range. The spring phase takes the animal back to its breeding home range. It is apparent that this terminology is strongly biased toward the description of migration by temperate land birds. This is because these are the birds for which most of the arguments are first developed. The application of the conclusions reached to sea birds and tropical land birds is discussed separately when appropriate.

As for all animals, (p. 378), little information is available for birds concerning the sequence by which environmental cues are committed to a spatial memory during the establishment of a familiar area. Where such information is available, it relates solely to the ontogenetic commitment of celestial cues to a spatial memory.

Emlen (1970, 1972) studied the commitment to memory of cues emanating from the night sky during the ontogeny of indigo buntings, *Passerina cyanea*. It was found that young birds are predisposed, particularly during their first summer of life before their first autumn migration, to perceive the apparent rotational motion of the night sky. Stars near the celestial axis have a smaller arc than stars near the celestial equator and by blacking out areas of the night sky in a planetarium it was found that young buntings concentrate their attention on areas of slow rotation (i.e. the circumpolar areas). Within this general area, individuals learn a variety of star groupings. Large individual variability existed in the star groupings that were learned. The advantage of concentrating on circumpolar areas seems to be that familiar star groupings are visible at all seasons and all times of night. Once stellar and rotational information have been integrated, it is possible to rely on the geometry of star configurations alone to locate the rotational axis and hence a reference direction, and indeed it was found experimentally that once the birds had performed this integration, celestial motion *per se* was no longer an important cue. Other workers have demonstrated that some nocturnal migrants can also orient by the Sun (St. Paul 1953, Shumakov 1965, 1967). Consequently, in

these species, an integration must also take place between stellar and solar information.

As suggested earlier (p. 607), while stellar configurations and the rotational axis are being learned they are likely at the same time to be integrated with a variety of other environmental cues within the spatial memory. Most authors, however, tend to minimise the importance of such integration and place the greatest emphasis on the importance of the celestial cues themselves. According to Bellrose (1972), for example, experiments by Serventy (1967) on the Tasmanian mutton bird, or short-tailed shearwater, *Puffinus tenuirostris* (Fig. 27.30), suggest that the nocturnal sky is the major cue by which these birds can locate their natal site. This species ceases to feed its chicks when the latter are about 83 days old. Fourteen days later the chicks make their first emergence from the burrow in which they were born in order to exercise their wings. Serventy transferred sixteen fledglings, calculated by Bellrose as being 85 days of age, from nesting islands off the Tasmanian coast to Fisher Island in the Bass Strait. Nineteen per cent of these birds over the next few years returned to Fisher Island as compared to 28 per cent of the fledglings born on Fisher Island. However, not one of fifty fledglings transferred to Fisher Island when they were 97 days of age returned in comparison to 36 per cent of the indigenous Fisher Island fledglings. It appears, therefore, that the period of sensitivity to locale imprinting in this species occurs during the fourteen-day desertion period before the young leave the island of their birth. During this period the young birds are active only at night and only at the mouths of their burrows. However, although perception and learning of parts of the night sky seems likely to occur, it seems highly improbable that the birds should perceive no other environmental cues at all during this period. Such cues can also be committed to the spatial memory, the celestial cues perhaps serving principally to fix the directional relationships of these other cues.

The post-fledging phase of exploratory migration of birds has already been discussed (pp. 190 and 331) but has received little critical study in the field. During marking–release/recapture experiments, although the inherited direction ratio of post-fledging exploratory migration might be expected to approach 25:25:25:25, it is usually only possible to distinguish between post-fledging exploratory migrants and autumn exploratory migrants in the case of those individuals in the former phase that explore in directions well outside the autumn migration directions. In Northern Hemisphere temperate regions these are those juveniles that are recaptured in early autumn to the north of the area in which they were banded, often as nestlings, a month or so earlier. For example, before the beginning of the autumn migration, some juveniles of the starling, *Sturnus vulgaris*, that are born in Switzerland, are recaptured to the north of their natal area. The extent and direction of this post-fledging exploratory migration depends on the time of hatching (Studer-Thiersch 1969), the main wintering area of the birds being in the region of the western Mediterranean (Fig. 27.1).

Beyond a certain date in autumn, therefore, these birds change their exploratory migration behaviour and begin to migrate to the SW. Displacement experiments on juveniles of starlings and other long-distance seasonal return migrants have demonstrated that the observed bias to the autumn migration is endogenously determined. Figures 27.5 and 27.1 show respectively the results of such displacement experiments with hooded crows, *Corvus corone cornix*, and starlings, *Sturnus vulgaris*. Similar experiments have been performed with the American crow, *Corvus brachyrhynchus* (Rowan 1946), the white stork, *Ciconia ciconia* (Schüz 1951, 1963), the chaffinch, *Fringilla*

coelebs (Perdeck 1958), and the blue-winged teal, *Anas discors* (Bellrose 1958b), in each case with the same result. Juveniles of all species flew in a direction after displacement that was parallel to the standard direction.

Evidence is also accumulating that not only the direction but also the distance of autumn migration in juveniles is endogenously determined. Two types of evidence are available that tend to support this hypothesis, one type coming from displacement experiments and the other from an apparent correlation between Zugunruhe (= migratory restlessness) and migration distance.

As an example of the first type of evidence, hooded crows, *Corvus corone cornix*, captured by Rüppell and Schüz (1948) half-way along their migration from a Baltic breeding area to winter quarters in northern Germany, were displaced to a locality beyond their normal winter quarters. These displaced young birds continued to migrate in their original direction and were recaptured at distances from the point of release not exceeding the distance from the point of initial capture and the southwestern boundaries of their normal winter quarters. Similar results have been obtained by Bellrose (1958b) for blue-winged teal, *Anas discors*, as well as by Perdeck (1958) for the starling (Fig. 27.1).

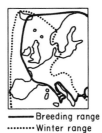

—— Breeding range
·········· Winter range

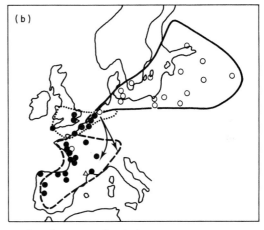

○ Breeding season (April to August)
△ Migration (March and September to November)
● Winter (December to February)

——Breeding range of controls
······Winter range of controls
——Range in 1st winter after displacement

Fig. 27.1 The behaviour of young starlings, *Sturnus vulgaris*, displaced experimentally from the Netherlands to Switzerland

General ecology
The starling is a blackish bird, glossed bronze-green and purple. The plumage is closely speckled in winter, particularly in the female. Juveniles are mouse-brown with a whitish throat. Flight is direct and rapid. The species feeds and roosts gregariously and in autumn and winter occurs in huge flocks that converge at dusk to form dense and noisy aggregations at night-roosts on city buildings, woods, or reed-beds (see p. 652). Nesting occurs in holes in trees, buildings, thatches, or even holes in the ground in barren areas. Through most of its range in Europe it is a partial seasonal return migrant (top map) but in north and east Europe it is a complete migrant.

(a). *Recapture of controls*
Starlings were captured and ringed in the Netherlands (arrow) in October and November. Young birds (approximately 6 months old) were divided into two groups. Controls were released at the point of capture. Subsequent recaptures are indicated according to the season of recapture.

(b). *Recapture of displaced birds*
The second group of young birds captured and ringed at the same time and place as the controls were transported by aeroplane to release points at Basle, Zurich, and Geneva as indicated by the arrows. The birds were transported in large bamboo cages wrapped in paper with a window cut at top-centre. They were fed during the journey and usually released within 24 hours of capture.

In the diagram the continuous and dotted lines enclose the breeding and winter range of the controls as shown in (a). Recaptures of displaced birds during their first winter are not shown but the dashed line encompasses the total area in which these recaptures were made. The recaptures shown, using the same convention as in diagram (a), refer to birds recaptured in seasons and years after their first winter.

The results indicate that after displacement young birds continue their migration as if they have not been displaced. The increased migration distance of displaced birds may perhaps be attributed to the influence of sea on the termination of the migration of the controls (see p. 696). The following summer the birds return with few exceptions to the same range as the controls, though whether each individual has returned to its natal area is unknown. In subsequent winters, again with a few exceptional individuals that migrate to the same area as the controls, the majority of individuals seem to return to the area that they occupied during their first winter after displacement.

The results are consistent with the thesis that the first autumn migration of young starlings is exploratory but with an endogenously determined bias to the direction ratio. It is also consistent with the suggestion that the seasonal return migration of starlings takes place within a familiar area established from the moment of hatching onwards.

[*Diagrams compiled from figures in Perdeck (1958). General ecology from Peterson et al. (1974). Photo by Eric Hosking*]

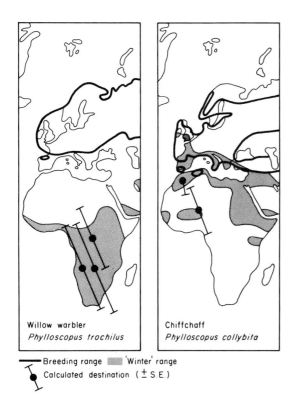

Willow warbler
Phylloscopus trochilus

Chiffchaff
Phylloscopus collybita

━ Breeding range ▒ 'Winter' range

◆ Calculated destination (± S.E.)

Fig. 27.2 Comparison of calculated and observed end-points of autumn migration for two species of *Phylloscopus*

General ecology

The willow warbler is the most abundant summer visitor to the northern half of Europe. It is a less arboreal species than the chiffchaff and has a greater preference for low vegetation. It nests on the ground in open bushy localities whereas the chiffchaff normally nests just above the ground in brambles and evergreens, etc. in light woods and bushy commons. In Africa, the willow warbler is highly adaptable, frequenting all sorts of country with trees at altitudes of up to 3500 m except for evergreen forest and the northern acacia steppe. The chiffchaff is found in various habitat-types from patches of acacia in Sénégal and mangroves in Gambia to willow thickets in the Inundation Zone. Both species moult in their winter quarters, the willow warbler after having moulted already in the Palaearctic.

Migration and Zugunruhe

The speed of autumn migration in the willow warbler appears to be at its greatest when the birds cross Europe and northern Africa (see Fig. 27.35), the largest numbers passing through the countries just south of the Sahara only about 4 weeks after peak migration occurs in Europe. Thereafter, migration appears to slow down. Rhodesia is not reached before the second half of October and the southern tip of Africa is not reached until the end of November.

Observation of caged individuals of this species during the same period in Germany indicated that the pattern of nocturnal Zugunruhe (= migratory restlessness) follows a similar temporal pattern. On the basis of this similarity, it has been suggested that the nocturnal Zugunruhe shown in cages is an expression of an endogenous temporal programme which determines the duration of, and temporal variations in, migratory activity. Using the formula $D = \mathcal{Z}d/z$ (where D is the total distance travelled when freeliving, z is the amount of Zugunruhe displayed by a caged bird in a given time interval as measured by an activity meter, d is the distance travelled by free-living conspecifics during the same time interval, and $\mathcal{Z}$ is the total amount of Zugunruhe displayed by a caged bird during the total autumn migration period) the end-point of migration has been calculated for groups of 20 and 24 individuals respectively of the willow warbler and the chiffchaff.

These calculated end-points are shown on the map in relation to the breeding and winter distributions of these two species. The agreement between the calculated destinations and the 'winter' range is taken to support the suggestion that the total migration distance of these two species is under the control of an endogenous programme.

[*Modified from Gwinner (1972). Distribution of chiffchaff modified according to Peterson* et al. *(1974) from which the breeding biology is also taken. Winter ecology from Moreau (1972)*]

An example of the second type of evidence is provided by comparison of the amount of Zugunruhe shown by the chiffchaff, *Phylloscopus collybita* relative to that shown by the congeneric but longer distance migrant, the willow warbler, *P. trochilus* (Fig. 27.2) and by the garden warbler, *Sylvia borin*, relative to that shown by the congeneric but shorter distance migrant, the blackcap, *S. atricapilla* (Gwinner 1972). Altogether, six species of *Sylvia* have now been compared with results that are consistent with a hypothesised correlation between Zugunruhe and migration distance (Berthold 1973). The Zugunruhe shown by three races of the white-crowned sparrow, *Zonotrichia leucophrys* (Gwinner, Farner and Mewaldt, in Gwinner 1972), also correlates roughly with the relative distances travelled during migration. However, unlike *Sylvia* spp., the temporal pattern of Zugunruhe in *Zonotrichia* is different from the pattern of actual migration. Autumn Zugunruhe in this case exceeds the time spent on migration.

It should not, of course, be assumed that migration distance and direction are rigidly determined by endogenous events. The observations that follow show clearly that this is not the case. Rather, it should probably be interpreted, particularly with regard to distance, that the endogenous mechanism determines only that the bird is in the physiological state that is optimum for migration. Whether or not the bird initiates migration at any particular moment in time depends on whether its migration threshold to migration-cost, and perhaps also habitat, variables is exceeded. Migration thresholds and endogenous programmes are discussed further in Section 27.5 (p. 626).

In many species, a factor that seems likely to have a major influence on the distance and direction of migration by juveniles is the presence or absence of individuals that have performed the migration previously. The flocks of some bird species, particularly shorebirds (Fig. 27.27) may be made up entirely of juveniles that migrate without the aid of adults. This is particularly true in the case of Arctic-nesting shorebirds in which the adults migrate south before the immatures, though the dunlin, or red-backed sandpiper, *Erolia alpina*, is an exception (Holmes 1966), all ages arriving together at the western North American wintering grounds. Like this latter species, most birds migrate in flocks made up of both juveniles and adults, in which case the former, in the establishment of a basic familiar area, can have recourse to the experience of the latter. This is especially evident in waterfowl (Hochbaum 1955, Bellrose and Crompton 1970) which, because of adult guides, not only use, at least initially, the same wintering home ranges as previous generations but also the same transient home ranges (see also p. 187). Among the several species of North American swans and geese (Figs. 27.40 and 27.41), flocks are composed of family units that follow the same routes as in previous years. Ducks do not travel as family units but, even so, because the juveniles accompany the adults, and because the latter often return to the same marshes that they visited the previous year, so the juveniles are indoctrinated into a series of water areas during migration.

It has also been shown, for at least one species, that when the endogenously determined direction and the socially communicated direction are in conflict, then it is the latter that prevails. When juvenile white storks, *Ciconia ciconia*, which had been experimentally displaced from East to West Germany, were released before the local adults had departed from the area of release, the juveniles became part of the flocks of the West German storks migrating to the SW (Fig. 27.3). When the native storks had departed, released East German juvenile storks adhered to their innate SE direction of migration (Schüz 1951).

Social communication during migration may be entirely by visual association, involving leading and following, or it may be acoustic. The nocturnal call-notes of migratory birds in particular seem likely to serve a communicatory function (Hamilton 1962b).

The data presented so far have demonstrated for birds what so far has not been demonstrated for any other vertebrate (see pp. 591, 736, and 767), namely that for a species that shows inter-deme consistency in the direction and distance of seasonal return migration, selection favours an inheritable bias to the distance, direction, and direction-ratio components of exploratory migration by the young animals. As long, that is, as there is a sufficiently high failure rate of familiar-area establishment through social communication. Although these data are consistent with the familiar-area hypothesis as presented for vertebrates in this book, it may quite correctly be objected that none of the evidence so far has demonstrated that the autumn migration of young birds is an exploratory migration as required by the familiar-area hypothesis.

In order to qualify as an exploratory migration the autumn migration must be a migration (p. 26): (1) beyond the limits of the pre-migration familiar

Fig. 27.3   The migration of the white stork, *Ciconia ciconia*

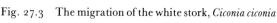

[*Re-drawn from Verheyen (1950) (modified from Rüppell)*] ·

area; (2) during which the ability to return to the pre-migration familiar area is retained; and (3) during which the area perceived is incorporated within the familiar area. There appear to be two other major alternatives to the categorisation of the autumn migration as an exploratory migration (Fig. 27.4). First, the migration could be a movement between a breeding familiar area and a (to be established) winter familiar area during which the bird orients to some single-coordinate, compass-direction-finding system such as the Sun, stars or magnetic field, etc. but during which the area over which the animal passes is not incorporated within the familiar area. On the other hand, the autumn migration could be considered as a series of movements to a number of intermediate 'goals' as suggested by Rabøl (1970). In this case the temporal

and spatial coordinates are assumed to be genetically programmed in the young bird such that the animal is due to be at a certain place at a certain time. Any error in fulfilling the programme elicits compensatory bi-coordinate navigation back to the appropriate goal. On such a view a familiar area of a sort exists but is encoded into the bird's genetic material, rather than learned. These three hypotheses of the nature of the autumn migration by juvenile birds may be termed, respectively, the 'exploratory migration', 'orientation', and 'programmed' hypotheses and are illustrated in Fig. 27.4. It is of paramount importance for the interpretation of homing experiments (p. 911) that a distinction should be made between these three alternatives.

The displacement–release experiments described

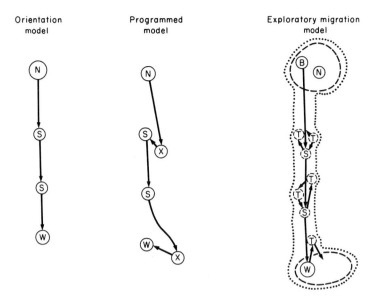

Fig. 27.4 The 'orientation', 'programmed', and 'exploratory migration' models of the autumn migration of young birds

In all three diagrams, N indicates the natal site, W the winter site, and S the place at which a migrant lands following completion of a migration unit.

The 'orientation' model
In this model a young bird leaves its natal or some other site within the breeding range of the species and initiates a migration unit in the endogenously determined standard autumn direction. During migration the bird orientates to some mono-coordinate compass-direction-finding system involving probably either the Sun, Moon, stars, wind, or magnetic field or any combination of these. After migrating for a certain length of time the bird lands, thus establishing position S. This process is repeated until the bird arrives in the wintering area. During each migration unit, mechanisms are employed to prevent or compensate for factors such as wind-drift. This model demands a considerable precision of orientation and timing on the part of the bird.

The 'programmed' model
In this model the coordinates of the position of each S and W are encoded within the birds genetic material. Each migration unit, therefore, involves bi-coordinate navigation to a pre-determined (by selection) target area that gradually shifts along the migration route as the migration season proceeds. If, for some reason, such as wind-drift, the bird fails to arrive at S but instead lands at X, at the first opportunity the bird migrates back to the appropriate position of the target area.

The 'exploratory migration' model
In this model the autumn migration is part of the process of establishment of a familiar area that is continuous from the moment of hatching onwards. In addition to the conventions used in the previous two models, the dashed line encloses that part of the familiar area for which detailed environmental information (e.g. position of feeding, roosting, and nesting sites, etc.) has been committed to the spatial memory. The dotted line encloses the total familiar area including that for which only gross environmental information (e.g. the positions of mountain ranges, coasts, hills, woods, drainage systems, etc.) has been committed to the spatial memory (see also Fig. 33.28).

The young bird establishes a familiar area around N, part of which (B) is assessed as a suitable potential breeding home range for the following year. When the first autumn migration unit is initiated in an endogenously determined direction (using the same mono-coordinate system for orientation as required by the orientation model), the gross familiar area is extended accordingly to enclose an area the limits of which are determined by the perceptual range of a bird flying in a relatively straight line. The position of S is determined partly by the endogenously determined duration of migration, partly by the amount of remaining energy reserves, and partly by perceptual encounter with a habitat-type that at the gross level is assessed to present a suitable transient home range. If S proves not to be suitable as a transient home range or if, in any case, exploration of other potential transient home ranges is advantageous for possible use in future years, the bird migrates back along its migration track to areas that at the gross level were assessed to be potentially suitable transient home ranges (T). After visiting one, several, or no T's, the bird initiates a further migration unit in the standard autumn direction.

This model, which can only fully be considered in association with seasonal variation in migration thresholds (p. 625), requires neither a high degree of orientation and timing on the part of the bird nor the involvement of bi-coordinate celestial navigation. The response to wind-drift is likely to be navigation back along the drift track to the pre-drift transient home range.

here (Fig. 27.1 and p. 610) argue against the programmed hypothesis but are consistent with both the exploratory migration and orientation hypotheses. Other types of displacement experiments, however, tend to argue against the orientation hypothesis while being consistent with the exploratory migration and programmed hypotheses.

First, let us consider 'natural' displacement by the wind. Such displacement, usually termed drift, is not common over land. Radar observations of night-to-night variations in the mean direction of migration in Northumberland, northeast England, in autumn (Evans 1972) suggested that both passerines (mainly sylviid warblers, chats, *Saxicola* spp., and thrushes, *Turdus* spp.) and waders compensate in flight for wind displacement. Flight was on preferred tracks rather than standard headings, a conclusion that is in agreement with the findings of Drury and Nisbet (1964), Nisbet and Drury (1967), and Bellrose (1967a) in America and Lack (1969) in western Europe. However, a direct consequence of compensation for wind drift is paradoxically that from time to time conditions occur in which compensation becomes physically impossible. If, on such occasions, it is not possible for the birds to land, they allow themselves to drift rather than expend energy trying to compensate under conditions in which compensation is not possible (Evans 1966, Pennycuick 1969). Evans (1966) has suggested that birds are likely to allow themselves to drift before the speed of the opposed wind reaches the cruising speed of the migrant.

Passerines moving in autumn from Scandinavia to Iberia do not normally pass over northeast England, though further south in East Anglia SSW movements are often seen by radar that are believed (Evans 1972) to be the western fringe of such a migration. Occasionally, however, under certain weather conditions, particularly with strong winds from the southeast, the western fringe drifts westwards and may then land in considerable numbers on the coasts of northeast England and eastern Scotland. Most of the birds are juveniles of redstarts, *Phoenicurus phoenicurus*, garden warblers, *Sylvia borin*, and pied flycatchers, *Ficedula hypoleuca*. Unlike crows, *Corvus corone* (Fig. 27.5), and starlings, *Sturnus vulgaris* (Fig. 27.1), displaced artificially, these young birds, when subsequently they continue their migration, do not fly parallel to their original route but compensate for their westward displacement and re-orient toward their original migration route. Evidence from banding recoveries, experiments in

orientation cages, and observation by radar of the direction of subsequent departure, all points to this conclusion. Furthermore, these wind-displaced birds rejoin their pre-drift migration route at some point to the north of their winter range (i.e. they are orientating to rejoin their migration route rather than orientating to their final destination). Whether these birds are orienting to intermediate rest and feeding points, the coordinates of which are known or inherited (Evans 1972), whether the re-oriented direction is derived by the bird vectoring its standard direction of migration, or whether they are returning to the transient home range occupied immediately preceding the drift migration, cannot be assessed on the basis of present information (see p. 918).

The second source of evidence concerning the nature of the autumn migration of young birds derives from experiments on night-migrating passerines, mainly willow warblers, *Phylloscopus trochilus*, garden warblers, *Sylvia borin*, and redstarts, *Phoenicurus phoenicurus*, trapped in autumn in Denmark by Rabøl (1970). Some of these birds were then tested in orientation cages at the place of capture and some were transported out or down their migration routes up to distances of about 500 km and tested for their directional preference in orientation cages the same night. The non-displaced birds continued to orient in their standard E or SE direction. Some groups of displaced birds also continued to orient in their standard direction. Others showed a major change of preferred direction that was consistent with navigation to an intermediate goal. However, a shift in direction from E to SE that was observed for willow warblers, *Phylloscopus trochilus*, captured, respectively, on 12–13 August and 18–23 August at Hanstholm, could have been due to the passage of different demes with different innate directions rather than due to a shift in intermediate target area for a single deme as suggested by Rabøl.

It can be seen, therefore, that displacement–release, wind-drift, and displacement–caging, 'experiments' on young birds give different results. It is always possible, of course, to fall back on the suggestion that the different species involved are employing different strategies. For the moment, however, let us attempt to reconcile these different results on the assumption that the different species involved are using the same strategy but that the different experimental procedures elicit different reactions. On this assumption the programmed hypothesis of Rabøl can immediately be rejected because of the results obtained by displacement–release experiments (Figs. 27.1 and 27.5, and p. 610) which show clearly that displaced and released juveniles eventually continue and complete their migration as if they have not been displaced. Yet birds that are displaced by the wind do not do this but instead orient back to their original migration route. This argues against the orientation hypothesis, at least in its simplest form, in that it shows that birds are not simply orienting to a mono-coordinate source. The difference between displacement–release and wind drift can only be attributed to the fact that in the second type of 'experiment' the bird can and does continuously perceive its total environment, as would be expected on the exploratory migration hypothesis, whereas in the former type of experiment the bird cannot perceive its environment during displacement and can only assume, upon release, that it has not been displaced. Incidentally, this difference between the two types of experiment argues against the existence of any form of inertial navigation whereby sense organs in the bird's body register and store information concerning every movement through space to which the bird is subjected during experimental displacement. However, both types of experimental evidence are consistent with the exploratory migration hypothesis. It remains, therefore, to reconcile the results of the displacement–caging experiments with those of the other types of experiment and to evaluate these results in relation to the exploratory migration hypothesis.

It is suggested here that the phenomenon being studied by Rabøl in his displacement–caging experiments is reversal rather than navigation to programmed and endogenously encoded intermediate goals. The relevance of reversal migration to the familiar-area hypothesis has been discussed on p. 188. It is suggested that one of the triggers for reversal by young birds is arrival at the end of a day or night's migration in an area assessed to be unsuitable. In this case the young bird may perform calculated migration back along the migration track to an area perceived during migration that is assessed to be more suitable. Observations of European robins, *Erithacus rubecula*, in Poland (Szulc-Olech 1965) perhaps provide a clue to the nature of autumn migration in young passerines, a question that is discussed in some detail in a later section (p. 642). Individuals arriving in the study area may either continue migration the following night or may remain for several days, feeding and increasing

weight. Territories are established, the situation is despotic, and the smaller individuals may be forced some distance away from the landing point before they succeed in establishing territories. Once a certain weight has been achieved, these individuals continue on their migration. The picture is therefore that of a number of birds landing in a certain area and then re-distributing themselves through the initiation of short-distance migrations in response to thresholds to habitat variables.

Reversal migrations are characterised by small numbers of individuals relative to those arriving and passing through in the standard direction. According to Parslow (1969), the total number of reversal migrants during a given autumn or spring is less than the number of standard migrants on a single night during the peak migration period. On any given night, however, most migrants may be travelling in the reverse direction (Evans 1972). This can be attributed to the fact that reversal migration, like the standard migration, has a clear migration threshold to the migration-cost variable of the wind (p. 632). Consequently, a wind that favours reversal migration depresses the standard migration. The lightness of reversal migration as recorded by radar could be attributed to a combination of low numbers and/or a short migration distance per individual, and perhaps also to much of reversal migration occurring at low altitudes below the level of radar reconnaissance.

The phenomenon of reversal, therefore, is consistent with the suggestion that birds arrive in an area during migration and, in association with the need to feed-up for a period before continuing migration, then re-distribute themselves within the area according to habitat suitability. The direction of migration can then be attributed partly to calculated migration within an area already perceived (p. 188), which in juveniles can only be back along the migration track, and partly to further exploratory migration for which thresholds exist to habitat and migration-cost variables (p. 632). Day-migrating birds such as hirundines also show reversal migration. Here, however, wind direction may be a habitat variable as well as a migration-cost variable, for hirundines (swallows and martins) feed while on the wing during migration. The same applies to swifts (Apodidae). When moving during windy periods, swifts and swallows move against the air current, a behaviour that presumably facilitates feeding upon the insects that constitute the aerial plankton (p. 465). At Ottenby Bird Station,

Sweden, observation in autumn (August–September) during periods of northerly winds showed that swifts and hirundines were moving predominantly to the north, not in the standard direction to the south. Of the birds observed, 90 per cent were house martins, *Delichon urbica*. Of 2000 of this species counted on one day when the wind was from the north, only 200 were moving to the south (Ramel 1960). Before and after the reversal migration, however, when the wind was from the south and west, the migration was in the standard direction to the south. Experiments showing that wildfowl in autumn also alternate migration in the standard direction with migration in other directions have been described in Chapter 15 (p. 188).

It is beginning to look, therefore, as though the classical view of birds migrating from their natal area in a fixed autumn direction in one or several long hops that take them to their winter quarters in as few migration stages as possible is incorrect. This point is taken up again in the section on migration distance (p. 677). The observations presented here, however, suggest rather that young birds in particular migrate much more gradually and alternate longer-distance movement in the standard direction with shorter-distance movements in other directions. These migrations in other directions seem to be partly calculated and partly exploratory and to be adaptive partly to the advantage of maximisation of feeding efficiency throughout the migration period and partly to the advantage of familiar-area establishment. Migration thresholds to habitat and migration-cost variables seem to have evolved (see Section 27.5, p. 644) such that the migration performed by a given individual on any particular day or night is partly a function of its energy reserves, partly a function of habitat suitability, and partly a function of migration-cost variables such as wind direction. Far from migrating from the natal area to the winter area in as few steps as possible, guided only by an orientation mechanism and ignoring the country over which it passes save for the purpose of correcting wind drift, therefore, the picture that emerges is that of an animal extending its familiar area from its natal area by a series of exploratory migrations during which the environmental cues emanating from broad tracts of country are integrated into a spatial memory along with other, particularly celestial, cues. Habitat assessment probably also takes place.

Returning now to the displacement–caging experiments of Rabøl (1970), it is suggested that the experimental procedure exceeded the reversal migration threshold of some of the birds. That these birds did not necessarily orient in precisely the opposite direction as found by Evans (1972) in radar studies may be attributed to the absence of the influence of wind direction on the bird while in the experimental cages. The variability of response of Rabøl's birds would be expected on the hypothesis developed here, for the response should be a function partly of the energy reserves of the birds, partly of their assessment of habitat suitability (an assessment that would be influenced by the amount of handling and length and conditions of imprisonment during the experiment), and partly of the proportion of standard and reversed migration units during the preceding nights and days.

Leaving hypotheses of the autumn migration for the moment, it can perhaps be accepted in the absence of any direct evidence, and no matter what hypothesis is favoured for the autumn migration, that upon cessation of the autumn migration the bird establishes a familiar area in its winter quarters out of which crystallises by the usual process (p. 384) a winter's home range. The following spring, endogenous events *per se* or endogenous events triggered in relation to migration thresholds to indirect variables (p. 647) trigger the spring migration. The question of interest here is the proportion of the migration units that make up the spring migration that are exploratory and the proportion that are calculated.

Much less evidence is available for the spring migration than for the autumn migration. Such evidence as is available, however, supports the suggestion that some exploratory migration occurs or at least that it can occur given an appropriate situation. The displaced hooded crows, *Corvus corone cornix* (Fig. 27.5), and starlings, *Sturnus vulgaris*, (Fig. 27.1), for example, continued migration beyond the displacement point northeastwards the following spring. This cannot have been achieved entirely by calculated migration. However, many of the starlings and some of the crows returned to their natal area. It seems, therefore, that although spring exploratory migration may be timed and directed by endogenous mechanisms as in autumn, where displaced individuals encounter the familiar area established during their post-fledging and autumn exploratory migrations prior to displacement they return by calculated migration to their natal area. Displacement in spring of 3000 juvenile (in their second calendar year) starlings from the

Netherlands to Zurich, Switzerland, showed that although nine eventually established a breeding home range back in the Netherlands, four established a breeding home range at the point of release (Perdeck and Speek 1974). Without knowing the spatial relationship of the release point and the familiar area of each individual bird, interpretation

of this result is not possible (see p. 912), but it suggests that exploratory migration still continues in spring. Finally, the observation that reversal migration occurs in spring to the same extent as in autumn (Parslow 1969) also suggests that there is little difference in the nature of spring and autumn migration in this respect.

Carrion crow

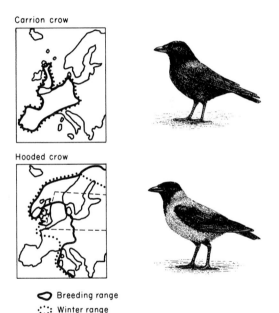

Hooded crow

⬤ Breeding range
⁙: Winter range

Fig. 27.5 Parallel shift of migration axis following experimental displacement of hooded crows, *Corvus corone cornix*

The maps show the winter and summer distributions in Europe of two sub-species of *Corvus corone*. Unlike the hooded crow, *C. c. cornix*, the carrion crow, *C. c. corone*, shows little seasonal shift in distribution. Both sub-species of *Corvus corone* are inhabitants of moors and cultivated country where some trees are present and are also found on sea shores and town parks. Usually they nest in trees but occasionally also on cliffs.

Hooded crows were captured at Rybatschi and transported to Flensburg, 750 km to the west as indicated by the arrow.

◼ normal breeding area of crows that migrate through Rybatschi.

⁙ normal wintering area of crows that migrate through Rybatschi.

≡ subsequent breeding area of displaced crows excluding some older (> 2 years old) individuals that returned to the normal breeding area.

The breeding area of the displaced individuals is consistent with the thesis that the direction of spring exploratory migration by young, first-year crows can be produced by an endogenously determined preferred compass direction.

[*Displacement data from Rüppell (1944). Distribution from Peterson et al. (1974)*]

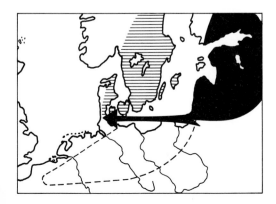

27.4 The seasonal return migrations of adults as calculated migrations within a familiar area

The conclusion reached in the preceding section was that the autumn and spring migrations of juvenile birds should be considered not as a rapid, uni-directional sequence of migration units oriented primarily to celestial cues, but rather as a sequence of exploratory migrations, the longer-distance stages being in the peak direction but alternating with other movements in other directions, the result being that the familiar area is gradually pushed further and further in the direction of the eventual wintering or breeding area. In species that do not mature until several years of age, the exploratory migration/habitat assessment process may continue over several years. In species that breed when only one year old, however, the major part of familiar-area establishment is likely to be completed in a year, though if habitat variables subsequently fall below the exploratory migration threshold then (by definition) further exploratory migration is initiated. The exploratory migration threshold, however, is likely to increase with age (Chapter 17, p. 321).

Let us now consider the migration of adults within the framework of this thesis. The first expectation is that, as a result of selection on degree of return (p. 321), individuals of most species (see pp. 323 and 658 for exceptions) should migrate at a particular time of year to the same part of the familiar area that was occupied at the same time the previous year. This expectation for birds seems first to have been expressed by Isakov (1949), albeit based on a view of the inflexible nature of bird behaviour. In this book, however, the expectation is an inevitable outcome of an application of the familiar-area hypothesis to the seasonal return migration of birds. Nevertheless, bird behaviour is viewed here as being highly flexible, type-C removal migration (p. 308) readily taking place if the calculated migration threshold is exceeded (p. 347). Selection on the pattern of ontogenetic variation of these migration thresholds, however, is suggested to have led to a characteristic ontogenetic variation in removal migration incidence which in adults of most species is manifest as a high degree of return (p. 321).

Evidence for a high degree of return to season's home ranges other than the breeding home range is now accumulating rapidly. Moreau (1969, 1972) reviewed the evidence for a high degree of return by palaearctic-breeding migrants (mainly passerines and waders) to Africa and additional new data are appearing yearly. Skead (1973), for example, reported that out of eight individuals of the red-backed shrike, *Lanius collurio*, banded on a farm in the Transvaal, two, both females, re-appeared on the same farm the following year, presumably, in the meantime, having migrated to and from their breeding home range in northern or central Europe or Asia (Fig. 27.46).

Not only is evidence available that demonstrates a high degree of return to breeding and 'winter' home ranges, but evidence is also accumulating that demonstrates, as for some bats (p. 567), a high degree of return to transient home ranges. Because of the short time periods involved, such evidence is much more difficult to obtain. Even so, nearly 1 per cent of whinchats, *Saxicola rubetra*, banded in transit through the oasis of Gabès in Tunisia, have been recaptured in subsequent years (Moreau 1972).

A special category of transient home range is the moulting home range, similar in function to that of pinnipeds (p. 710), adopted by many species of wildfowl. Three examples are perhaps sufficient to illustrate this widely known phenomenon. Common eider ducks, *Somateria mollissima*, and king eiders, *S. spectabilis* (Fig. 27.6), migrate to the west from breeding grounds around the Beaufort Sea to moulting grounds in the Chukchi Sea. Flocks passing Point Barrow, Alaska, are almost entirely males in late summer (July and August) though from 8 to 17 August the sex ratio shifts to a preponderance of females, young-of-the-year passing in September (Thompson and Person 1963). Both species winter in the ice-free parts of the Bering Sea. Emperor geese, *Anser canagicus*, that are one year old or more but which are still immature migrate in mid-summer (mid to late June) from West Alaska to St. Lawrence Island at the southern end of the Bering Straits, where they moult until the end of September (Blurton-Jones 1972). Finally, few common scoters, *Melanitta nigra*, remain to moult in Estonian waters but instead migrate westwards, passing through Estonia during late summer (July and August) (Jacoby and Jögi 1972). It seems likely that all of these moult migrations take place within a familiar area and are directed to moulting home ranges discovered by exploratory migration or, especially in the case of geese, social communication during the first one or two years of life.

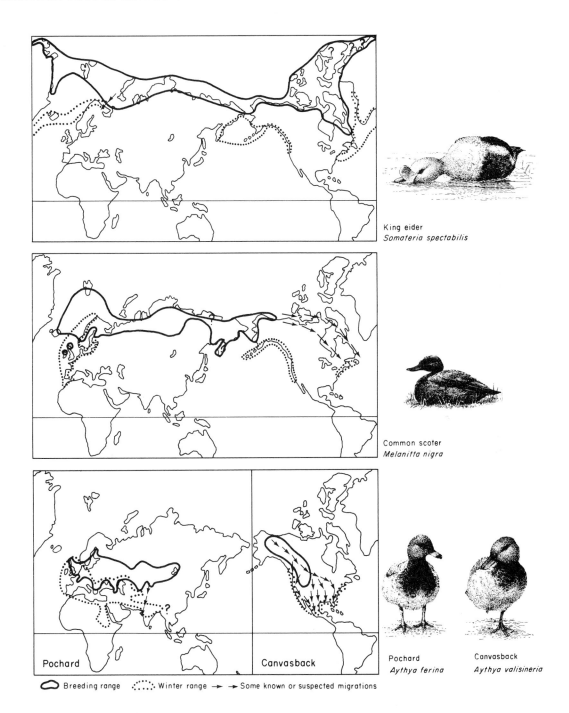

King eider
Somateria spectabilis

Common scoter
Melanitta nigra

Pochard
Canvasback

Pochard
Aythya ferina

Canvasback
Aythya valisineria

◯ Breeding range ⋯⋯ Winter range → Some known or suspected migrations

Fig. 27.6 Seasonal return migrations of four species of diving ducks (Aythyinae)

Diving ducks are found at sea and on estuaries, rivers, lakes, and marshes and feed by diving on a diet of small aquatic animals and plants. Maritime species feed predominantly on molluscs and crustaceans. The legs are placed closer to the tail than in surface species, with the result that diving ducks are more awkward on land. On the water, most require a pattering run across the surface for take-off.

[*Compiled from figures and descriptions in Bannerman (1958), Voous (1960), Peterson (1961), Dorst (1962), Robbins et al. (1966), Moreau (1972), and Peterson et al. (1974)*]

Having stressed the existence of a relatively high degree of return to all season's home ranges as would be expected from the familiar-area hypothesis of bird migration, it is perhaps necessary to stress also the flexibility of bird migration as achieved through selection in favour of type-C removal migration thresholds (p. 308). Any change in environmental variables, if they exceed the migration threshold, has been argued to trigger the exploratory-calculated migration sequence in a bird of any age. Not surprisingly (p. 620), however, studies are rare of degree of return to season's home ranges other than the breeding home range. Perhaps the most complete study so far is that by Shevaryova (1969) of wildfowl that breed in the region to the east of the Baltic. Here it seems that, at least for mallard, *Anas platyrhynchos*, shoveler, *A. clypeata*, wigeon, *A. penelope*, and garganey, *A. querquedula*, degree of return to the breeding home range is somewhat greater than degree of return to the moulting home range, winter home range, and transient home ranges. This finding is quite consistent with the selective pressures postulated to be acting on the incidence of type-C removal migration in birds (p. 321).

To judge from pinnipeds, cetaceans, and perhaps also marine fish, cephalopods, and marine reptiles, it seems to be a feature of adaptation to feed on pelagic marine organisms that the animal concerned establishes a relatively enormous familiar area. Sea-birds seem to be no exception to this general rule and indeed may even be taken to be the outstanding

example. Among the sea-birds that inhabit the Southern Ocean, for example, it seems highly probable that the young of many species perform circumpolar exploratory migration eastwards from their natal site in the direction of the prevailing westerly winds. Although not entirely convincing, analysis of the age of recoveries of juveniles of the royal albatross, *Diomedea epomophora*, banded at New Zealand, suggests that age may increase eastwards. Newly fledged birds appear, therefore, to perform a full circumpolar migration, including a rather longer pause in the region of the southwest Atlantic (Robertson and Kinsky 1972). The young of the giant petrel, *Macronectes giganteus*, may similarly perform a circumpolar exploratory migration (Tickell and Scotland 1961, Dorst 1962) as also may some individuals of the wandering albatross, *Diomedea exulans* (Fig. 27.8). The exploratory migrations of these species, particularly the latter, also take them northwards into the Pacific, Atlantic, and Indian Oceans and it seems likely that a major part of the Southern Ocean plus also parts of one or more of the other oceans may be encompassed within the familiar area thus formed by even a single individual.

In the case of the giant petrel, *Macronectes giganteus*, the adult home range is only a small fraction of the likely familiar area. In the albatross species, however, the adults seem to continue to exploit a major portion of the familiar area. Adults of the royal albatross, *Diomedea epomophora*, may perform a

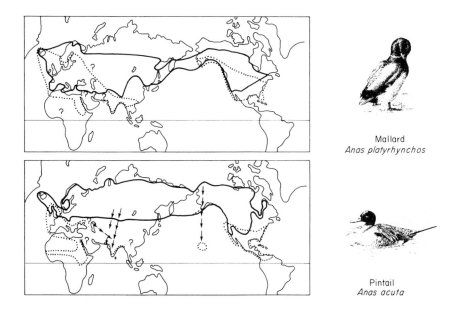

Mallard
Anas platyrhynchos

Pintail
Anas acuta

circumpolar migration similar to that performed when newly fledged (Robertson and Kinsky 1972). Analysis of banding and observations in the coastal waters of New South Wales, Australia, and the Tasman Sea and of reciprocal recoveries with South Georgia, Kerguelen, Marion, and Auckland Islands have demonstrated a high degree of return to the banding site in successive years by the wandering

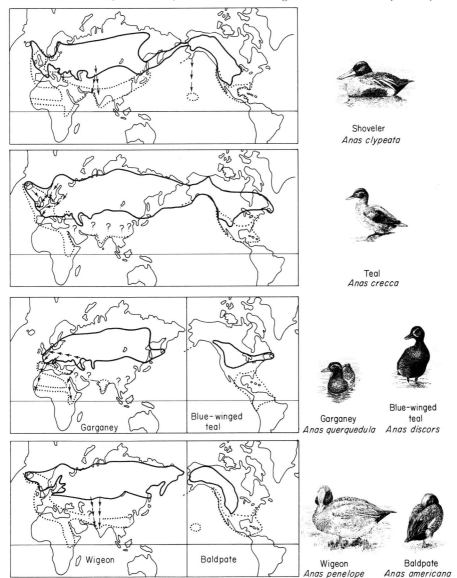

Fig. 27.7 The migrations of eight species of surface-feeding ducks (Anatinae)

⬭ breeding range;   ⋰⋰ winter range;
→ → some known or suspected autumn (post-moult) migrations.

Surface-feeding ducks are characteristic of shallow waters, creeks, rivers, ponds, and marshes and feed, by 'dabbling' and 'up-ending', on a diet of aquatic plants, grass, small aquatic animals, and insects. Occasionally they feed on land. Flight can be initiated, without running along on the water, by springing directly into the air, an ability that results from the central position of the legs. These ducks are also able to balance and walk reasonably well on land. Ground speed during migration may range from 60 to 100 km/h.

[Compiled from figures and descriptions in Bannerman (1958), Voous (1960), Peterson (1961), Dorst (1962), Robbins et al. (1966), Moreau (1972), Lister (1973), and Peterson et al. (1974). Where range description varies between authors, the most recent authority has been followed]

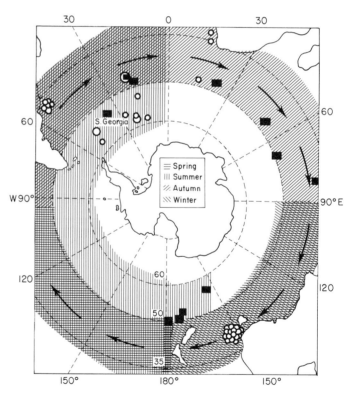

Fig. 27.8 The circumpolar migration of the wandering albatross, *Diomedea exulans*

Hatching indicates the main oceanic regions in which wandering albatross are observed in different seasons. Arrows show the direction of the prevailing winds. Breeding occurs (■) on South Georgia, Kerguelen, and other islands to the north of the Antarctic convergence. Young are fledged in spring/summer after having spent nearly a year on the nest. Albatross do not begin to reproduce until they are several years old and may live for up to 50 years.

There is a strong indication from the apparent population shift that many individuals encircle the Earth at the latitude of the Southern Ocean. Migration seems to be from W to E in the direction of the prevailing winds with a more or less annual periodicity. o, recaptures of birds marked on South Georgia.

[*Compiled from Dixon (1933) and Tickell (1968)*]

albatross, *D. exulans* (J.D. Gibson 1967). The results suggest a facility for navigation to remembered feeding areas between breeding seasons that is entirely consistent with calculated seasonal migration when adult within a familiar area established when young.

Further possible support for the view that adults migrate within a familiar area established by exploratory migration and/or social communication when young may be derived from the existence of otherwise bizarre migration routes similar to those encountered in some ungulates (p. 543). An example is provided by the migration routes of Finnish demes of common terns, *Sterna hirundo*, and Arctic terns, *S. paradisea* (Lemmetyinen 1968). All Finnish common terns migrate SW in autumn whereas only Arctic terns from the Gulf of Finland migrate to the SW (Fig. 27.45). Arctic terns that breed round the Gulf of Bothnia and on the southwestern archipelago of Finland migrate NW across Scandinavia to the coast of Norway before turning S. As pointed out by Lemmetyinen, these latter breeding demes of Arctic terns probably originated on the Norwegian coast some time towards the end of the nineteenth century. Using the terminology employed in this book, it seems that exploratory migrants from the Norwegian coast performed type-C removal migration to the Gulf of Bothnia at this time. Previously, these individuals had probably established a familiar area encompassing the coasts of Norway, Western Europe, West Africa, the waters of the Southern Ocean, and the Atlantic coast of South America (Fig. 27.45). The exploratory migration that led these individuals to the Gulf of Bothnia from Norway would have been an extension of this familiar area. If seasonal return migration occurs within a familiar area, these birds would adopt as the first stage of their post-nuptial seasonal migration the exploratory migration route by which they arrived in the area from their place of origin on the Norwegian coast. As it seems likely that for nearly the first year of life young Arctic terns rely to a large extent on social communication in the establishment of a basic familiar area, this route has persisted. On the familiar-area hypothesis it is likely to continue to persist until exploratory migrants from the Gulf of Bothnia happen to encounter their familiar area to the southwest of their breeding site. As it is, over the past two decades, it seems (Lemmetyinen 1968) that the existing migration route has been reinforced by an influx of further removal migrants from the Norwegian coast.

Further evidence for the familiar-area hypothesis is provided by the migration circuits eventually adopted by the starlings, *Sturnus vulgaris*, displaced when young from the Netherlands to Switzerland (Fig. 27.1). These young birds continued their autumn migration on a track parallel to their standard direction and, it may be assumed, performed their spring migration in the opposite direction, continuing to explore in this direction when the break in their familiar area was encountered that resulted from the experimental displacement. By whatever means (Chapter 33, p. 911), many of these young birds returned to breed in their natal area, rather than an area to the southeast. The following and subsequent autumns, however, the majority of these starlings did not migrate in the standard direction but instead returned to the home range that they established during their first winter. This result, which has been considered to be 'curious' (Dorst 1962, p. 331) within the framework of other hypotheses, can only readily be accommodated by a model that approximates to that of the familiar-area hypothesis. Displacement experiments on adult birds are also claimed in this book (p. 894) to provide evidence in support of the familiar-area hypothesis. However, these cannot easily be discussed without at the same time giving detailed consideration to the question of navigation in relation to the familiar-area hypothesis. Displacement experiments with adults are therefore discussed in Chapter 33 rather than here.

As a conclusion to this and the preceding sections, it is suggested here that enough supporting evidence is available for the familiar-area hypothesis to constitute the most valid model of the pattern of seasonal return migration by birds. Further discussion is presented in the chapter on navigation (p. 855). The remainder of this chapter, however, assumes the validity of the familiar-area hypothesis and discusses migration thresholds, degree of dispersal and convergence, distance, direction, and direction ratio accordingly.

27.5 Migration thresholds

It is pointed out on a variety of occasions (e.g. pp. 348, 736, and 768) in Part IIIB of this book that the physiological basis of the movements of animals within a familiar area rests with the phenomenon of the migration threshold. As far as seasonal return migration is concerned, more data are available for the seasonal variation in migration thresholds of birds than for any other group of animals. The following discussion is based to a considerable extent on the conclusions reached in Chapters 14–18.

Much of the seasonal return migration of birds is adaptive to predictable changes in the environment, particularly changes in climate and the availability of food. In consequence (p. 50), much of bird migration is obligatory, being based on variation in migration thresholds to habitat and/or migration-cost variables that are triggered either solely by endogenous events or by endogenous events that are themselves triggered when migration thresholds to indirect variables are exceeded.

A great deal remains to be discovered concerning the relative contribution of endogenous cycles of migration thresholds to indirect variables to the seasonal variation in migration thresholds to habitat and migration-cost variables. It has been demonstrated for the willow warbler, *Phylloscopus trochilus*, a trans-equatorial palaearctic migrant, however, that the annual sequence and timing of Zugunruhe and moult are controlled by an endogenous circannual rhythm that, even in birds given apparently constant environmental conditions, can persist for at least three cycles (Gwinner 1972). This does not mean, of course, that under normal conditions environmental variables do not take on a zeitgeber role but it does suggest that in this and perhaps some or many other species migration thresholds to indirect variables may not exist.

Let us take as a starting point for a consideration of seasonal variation in migration thresholds the fledging of young birds and the end of the breeding season in adults. Both of these events seem to trigger, modify, or are part of, an ordered programme of physiological change. This programme involves a species-specific sequence of moulting, variation in the utilisation and mobilisation of energy reserves, and variation in migration thresholds. The period immediately following fledging in young birds seems to be the period of ontogeny during which exploratory migration thresholds relative to breeding habitat variables are at their lowest (p. 331). For adults, the post-nuptial period seems to be the time of year that exploratory migration thresholds to breeding habitat variables are at their lowest (p. 191). However, the post-fledging and post-nuptial phases of migration have received little study (p. 190) and nothing is known of the factors contributing to the migration thresholds concerned.

The autumn sequence of variation in migration thresholds seems likely to be highly species-specific. Even within a single genus, the moult and autumn migration sequence can vary considerably. In the various species of *Empidonax*, for example, a genus of American tyrant flycatchers (Fig. 27.9), there is almost every variation between species (e.g. the Hammond flycatcher, *E. hammondii*), in which both juveniles and adults moult before initiating autumn migration and species (e.g. the gray flycatcher, *E. wrightii*, and the dusky flycatcher, *E. oberholseri*), in which both juveniles and adults initiate autumn migration before moulting. In the yellow-bellied flycatcher, *E. flaviventris*, and the least flycatcher, *E. minimus*, the juveniles moult before initiating autumn migration whereas the adults initiate autumn migration before moulting. Preliminary analysis of the situation suggests that the ultimate factors involved in producing this interspecific variation should be sought in the relative characteristics of the summer and winter habitat-types inhabited by the species (Johnson 1963). Taking temperate birds as a whole it is perhaps more common for moulting to be completed before the initiation of autumn migration (Dorst 1962).

In species and age classes in which moulting occurs before the initiation of autumn migration, it is likely that exploratory and calculated migration thresholds will increase considerably during the period of the moult of the flight feathers due to the ultimate effect on the efficiency of flight. Indeed, some wildfowl are incapable of flight for a short period during the moult of the flight feathers. During the fledging-to-moult or end of breeding-to-moult period there is likely to be a change in exploratory and calculated migration thresholds to habitat variables. At first these are likely to be those appropriate to breeding habitats but at some point they are likely to change to those appropriate to moulting habitats. These thresholds are likely predominantly to be exploratory in juveniles and calculated in adults. In the adults of many species, this change in calculated migration thresholds is unlikely to be manifest by a change in home range but as already described (p. 620), wildfowl tend to have separate breeding and moulting home ranges.

In many species there may be considerable overlap at the deme level between the moulting period and the period of autumn migration. This overlap is particularly marked at high latitudes, where the short summer means that breeding, moulting and migration is forced into a relatively short favourable period. Dolnik and Blyumental (1967) have suggested that in Russia at latitude 62°N the autumn migration of chaffinches, *Fringilla coelebs*, is initiated before the moult is completed. It has been shown, however, by Haukioja (1971) that although this is true at the level of the deme it is not true at the level of the individual except in relation to non-flight feathers. Each individual completes the post-nuptial moult of the flight feathers before initiating autumn migration.

It is clear from the results of Gwinner (1972) that at least for the willow warbler, *Phylloscopus trochilus*, the timing and duration of physiological changes, including those affecting migration thresholds, may be entirely under endogenous control and totally independent of indirect variables. In most cases, however, it seems likely that such endogenous rhythms can be modified or even re-phased by exogenous events. The onset of the post-nuptial moult in adult finches, for example, is in part a function of the date that breeding ends (Newton 1972). In the juveniles of two species of finch (i.e. the bullfinch, *Pyrrhula pyrrhula*, and chaffinch, *Fringilla coelebs*) for which relevant studies are available, the onset and duration of the post-juvenal moult is a function of date of hatching (Dolnik and Blyumental 1967, Newton 1972). Those young born late in the season also moult later, but more rapidly and at an earlier age, than young born early. In the chaffinch in the Baltic region the result is that although the date of onset of post-juvenal moult spans a period of about six weeks, the date of completion spans only a period of less than three weeks. Evidence is also available that although the timing of the post-nuptial moult and of the autumn migration is under endogenous control, the phasing of endogenous events in autumn is influenced by the photoperiod experienced six months or so earlier (Dolnik and Gaurilov 1972) between January and March. Photoperiod experienced in late spring, summer, and early autumn (mid-May to September), however, fails to influence the timing of onset of moult and migration, though it may influence moult duration.

Summarising so far: The sequence of physiological changes that in autumn affect the timing and duration of moulting are part of an endogenous programme that in at least one species has a free-running circannual periodicity. At certain times of year, however, indirect variables and exogenous events may influence the phasing of the programme. Part of the programme is a variation in exploratory migration thresholds from a low level immediately after fledging or breeding to a high value during the moult. Exploratory and calculated migration thresholds also change from those appropriate to breeding habitats to those appropriate to moulting habitats during the period leading up to the onset of moulting. Some time before or after moulting, depending on the species, and age-class, exploratory and/or calculated migration thresholds change to those appropriate to autumn or perhaps even winter habitats.

Associated with the change of migration thresholds to those appropriate to autumn or perhaps winter habitats are two further changes, one behavioural, the other physiological. The first concerns a change in the components of the migration initiated by the bird whenever its autumn migration threshold is exceeded. These components are discussed in Sections 27.7 (p. 654) and 27.8 (p. 677). The second concerns a change in metabolism that is marked by the deposition of fat in characteristic sites in the bird's body (King and Farner 1965).

Fat, which has an energy yield per unit weight of 9·2 kcal/g compared with only 4·2 for carbohydrate or protein and which has the added advantage of yielding water upon oxidation that is equal in weight to the fat used (Newton 1972), is the main energy source for autumn migration. The relationship between the amount of fat deposited and migration distance is discussed in Section 27.8. Here the main point of interest is the temporal relationship between fat deposition and other autumn events, particularly the variation in migration thresholds. It should be stressed, however, that the changes in migration thresholds and the deposition of fat are physiologically integrated, each one influencing the other and together constituting a distinct migration syndrome (p. 644).

We are now in a position to consider some of the migration thresholds involved in autumn migration by temperate birds. In species that moult before migrating, the change in migration thresholds for autumn migration (i.e. a migration that once initiated has the distance, direction, and other components that for the species concerned are standard for the migration from breeding to winter areas) seems likely to begin once the moult is coming to an end. Autumn migration thresholds seem likely to be high during the peak of moult and then gradually to decrease. Support for the suggestion that migration thresholds exist during the moult is provided by the observation that if habitat suitability deteriorates to a sufficient extent, migration is initiated. In the crossbill, *Loxia curvirostra*, for example, migration has usually ceased by the time the bird begins to moult. In some years, however, migrations may be initiated (see p. 656) while the moult is in progress (Newton 1972). In some migrant adults caught between June and September on Fair Isle in 1953, the moult of the flight feathers had apparently stopped, for the birds had some old feathers and some new but none in growth. Apparently, therefore, if migration is initiated during the moult, further moulting may be suppressed.

As moulting ends, it seems likely that autumn migration thresholds will fall, and continue to fall, throughout the period of the deposition of fat, reaching their lowest value when the energy reserves are optimum for the first migration unit (i.e. first 'stage' or first 'leg') of the autumn migration. Few birds initiate autumn migration until this time, but evidence can be presented that migration thresholds exist throughout the period of fat deposition but initially at such a high level that they are rarely exceeded. Many European warblers change from a diet of insects to a diet of soft fruit in autumn at a time when the availability of insects is decreasing and the soft fruit offers a rich source of suitable carbohydrates. However, soft fruit ripens earlier in southern than northern Europe and warblers toward the northern edge of their range, such as northeastern England, seem to be confronted with a later supply of suitable habitats than those further south. Autumn habitat suitability is likely to be a function of the availability of fruit, the deme density of warblers, and perhaps also temperature, and it seems likely that a composite autumn migration threshold to these variables will exist in warblers. It might be expected that this threshold will start high and then decrease as fat is deposited. Nevertheless, on the whole warblers leave northeastern England in advance of the deposition of fat (Evans 1966) suggesting that, although the migration threshold may be high, it is exceeded by the conditions experienced in the region. As is usual, when deme density is involved in the migration threshold (p. 268), however, some individuals remain and in Northumberland those blackcaps, *Sylvia atricapilla*, garden warblers, *S. borin*, and whitethroats, *S. communis*, that are still present in late September and October are feeding on the blackberries, *Rubus fruticosus*, and elderberries, *Sambucus niger*, that are then in season (Evans 1966).

It may be suggested, therefore, that during the period of fat deposition there is normally a decrease in migration threshold, though the rate of decrease need not be linear. Towards the end of this period, migration thresholds reach their lowest level. In caged birds, Zugunruhe appears.

The phenomenon of Zugunruhe, in which captive birds hop and flutter around their cages, is particularly conspicuous in nocturnal migrants which at the same time of the diel at other times of year would normally be roosting (Fig. 27.10). So much of migration research on birds makes use of this very convenient phenomenon that it is rather important to try to obtain a clear understanding of its meaning.

Fig. 27.9　Variation in the relative timing of moult and migration within a single genus: the tyrant flycatchers, *Empidonax* spp., of North America

The tyrant flycatchers (Tyrannidae) perch upright on exposed branches and fly out to catch passing insects. They are a New World family of about 365 species, the metropolis of which is the tropical lowlands of Mexico and central and northern South America.

The species of the *Empidonax* complex show considerable variation in the timing of autumn migration (AM) and moulting. Note that in the diagram AM (or SM for spring migration) does not indicate the calendar timing of the migration concerned, only the relationship of the migration to the moult. Horizontal lines indicate the span of time over which a particular moult has been recorded. Expanded portions indicate the periods over which all specimens examined of a given age were in moult. Broken lines represent periods for which data are incomplete.

It can be seen that moult may occur either before or after the autumn migration and that even within a single species adults and juveniles may show a different sequence. Even if moulting occurs after autumn migration, there may still be a pre-nuptial moult.

[Maps drawn from figures and descriptions in Peterson (1961), Robbins et al. (1966), and Orr (1970a). Moult data from Johnson (1963)]

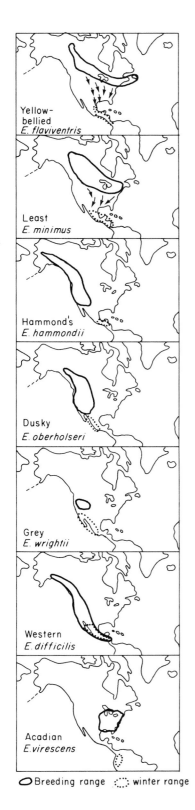

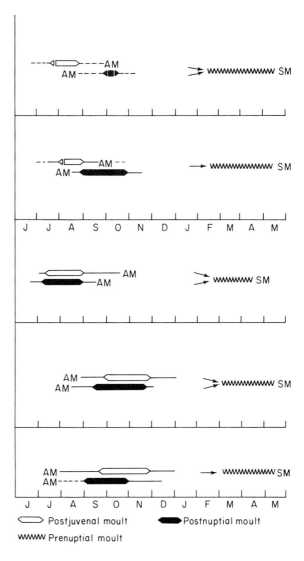

Yellow-bellied
E. flaviventris

Least
E. minimus

Hammond's
E. hammondii

Dusky
E. oberholseri

Grey
E. wrightii

Western
E. difficilis

Acadian
E. virescens

J J A S O N D J F M A M

▢ Postjuvenal moult   ■ Postnuptial moult

wwww Prenuptial moult

◯ Breeding range   ⋯: winter range

At present, however, the critical experiments do not seem to have been performed. The major alternatives would seem to be that Zugunruhe either: (1) reflects a low migration threshold on the part of the bird; or (2) reflects that the bird's migration threshold has been exceeded. It is also important to determine whether the migrations reflected by Zugunruhe are solely those of autumn and spring migrations or whether low post-fledging and post-nuptial exploratory migration thresholds and also low reversal migration thresholds are also reflected

by Zugunruhe. The exceptions to the general correlation between Zugunruhe and migration distance discovered by Gwinner (1972) and Berthold (1973) could perhaps be attributed to these other phenomena being reflected by Zugunruhe. Where demes do not perform autumn migration but nevertheless show Zugunruhe, as with the southern Pacific coastal race (*nuttalli*) of the white-crowned sparrow, *Zonotrichia leucophrys*, of North America, perhaps the Zugunruhe should be interpreted as a reflection of low post-fledging and post-nuptial exploratory mi-

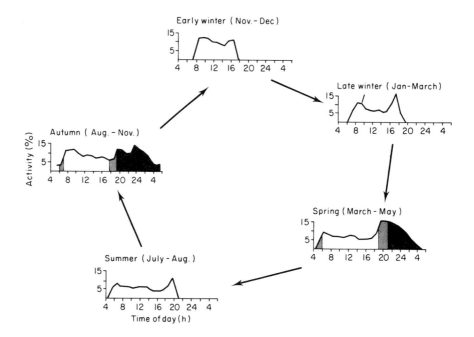

Fig. 27.10   Annual cycle of diel activity in the European robin, *Erithacus rubecula*

For migration map, see Fig. 33.4.

The diagram illustrates the diel incidence of activity of caged individuals of the robin during five different periods of the year. Diurnal activity is unshaded, nocturnal activity is black, dusk, and dawn activity is stippled.

Time investment in migration in both spring and autumn is at the expense of roosting rather than of any diurnal activity.

[*Modified from Merkel (1956)*]

gration thresholds (see p. 379) rather than as the 'atavistic remnant of ancestral migratory behaviour' suggested by Mewaldt *et al.* (1968).

As far as the behaviour shown by caged birds is concerned, the distinction between alternatives (1) and (2) above is not critical. If migration thresholds to habitat and migration-cost variables are low it seems inevitable that these thresholds will be exceeded in a caged bird. This does not necessarily mean, however, that these thresholds would be exceeded and the bird initiate migration if the bird were in its natural habitat, and it is perhaps significant that where correlations between Zugunruhe and migration distance break down, it is because the amount of Zugunruhe (in caged birds) is greater than expected from normal migration. A second possible reason for the existence of Zugunruhe in 'non-migratory' demes therefore presents itself. Conceivably, even these demes have migration thresholds beyond which migration is initiated. These thresholds may well be exceeded in captive birds but may be so high that normally, in a natural environment, they are rarely if ever exceeded.

Returning now to the question of the relationship between stage of fat deposition and level of migration threshold, it is clear that the physiological bases of the two processes are not linked in a simple manner. The usual timing of fat deposition and the initiation of migration suggest that migration thresholds start high early on in the fat deposition sequence and then decrease to their lowest level when the optimum amount of fat has been deposited. Failure to deposit fat, however, which presumably results from low habitat suitability, eventually leads to the initiation of migration without the deposition of fat as in European warblers in northeastern England. This, presumably, results from selection for the inclusion of time in the migration threshold (p. 51). Experiments on caged birds, such as bramblings, *Fringilla montifringilla* (Newton 1972), and white-crowned sparrows, *Zonotrichia leucophrys* (Mewaldt *et al.* 1968), also show that failure to deposit fat eventually leads to Zugunruhe.

Part of the physiological change associated with the change in migration threshold is a change in circadian rhythms of metabolism such that migration can be performed efficiently should it happen to be initiated. The diel incidence of migration varies from species to species. Diurnal migrants include pelicans, storks, birds of prey, turtledoves, bee-eaters, swifts, kingbirds, swallows, Corvidae,

some Turdidae (American robin, eastern bluebird), shrikes, and many Fringillidae (Dorst 1962) as well as starlings, red-headed woodpeckers, and certain waterfowl (Bellrose 1972). Nocturnal migrants include a number of waterbirds (Charadridae and Anatidae), cuckoos, numerous New World tyrannid flycatchers, most Turdidae, Old World warblers (Sylviidae) and flycatchers (Muscicapidae), vireos, New World warblers (Parulidae), American orioles (Icteridae), tanagers, and some insectivorous Emberizidae (buntings). Thus, among passerines there is a general rule that most seed-eating birds migrate by day and most insect-eating species, except swallows, migrate at night (Dorst 1962). A comprehensive model of diel incidence of autumn migration does not yet exist but a small number of selective pressures seem to account for the major part of the observed variation. Thus many of the largest birds conserve energy during migration by soaring and gliding (p. 689), activities that are possible only during that period of daylight hours that thermals are generated. Species that feed during flight, such as swallows and swifts, have the potential if they migrate during the day to integrate migration and feeding in such a way that migration can be performed without making inroads into other time-investment activities (p. 57). In most other species, however, selection seems to have favoured migration at the expense of time investment in roosting rather than at the expense of time investment in other activities, such as feeding, and perhaps also exploratory migration (in relation to assessment of habitat suitability as a transient home range), both of which are likely to be performed most efficiently in daylight. Perhaps other factors are also involved, such as reduced predation from birds of prey at night and possibly, also, more efficient navigation and habitat assessment by day. All of the suggested factors, however, imply that time of day should be regarded as migration-cost variable (p. 257). Superimposed on the gradual decrease in migration thresholds to all other habitat and migration-cost variables that it is suggested will occur during the period of fat deposition, therefore, will be a deme- or species-specific diel variation. The physiological events involved in this diel variation in migration thresholds are presumably based on a circadian endogenous rhythm.

All of the variation in migration thresholds so far described has been under endogenous control, either with or without a zeitgeber. Whether or not a given bird initiates migration on a given day, however, depends upon whether or not the migration thresholds to habitat and migration-cost variables as endogenously determined on that day are exceeded by the value of the environmental variables actually experienced on that day. The last decade has seen a great advance in the measurement of threshold variables, particularly migration-cost variables.

Perhaps the major work on migration thresholds to migration-cost variables is that of Nisbet and Drury (1968). Although this work was concerned with migration in spring, the thresholds to migration-cost variables at this time of year follow a pattern that is so similar in general form (though not necessarily in details) to that of autumn that there is no point in considering them separately. Nisbet and Drury, working in Massachussets, United States, followed by radar the nocturnal NE migration of songbirds over the mainland of New England and the E and NE migration of waterbirds over the Gulf of Maine. Records were made between late April and early June, 1959 to 1961, and the data obtained were then subjected to multivariate analysis with respect to environmental variables that in this book are classified as migration-cost variables. It was found that the density of migration correlated with high and rising temperature, low and falling pressure, low but rising humidity, and the onshore (SE) component of wind velocity. Songbirds also appeared to have a migration threshold to rain. Migration density correlated with conditions both at take-off and during the previous 24 hours and for waterbirds it was the change in conditions in the preceding 24 hours that was most important. Songbirds showed a migration threshold to the Moon, being less likely to migrate in the days preceding the full Moon except in cloudy weather. Quite possibly however, this is an artefact produced by birds flying at low altitude below radar reconnaissance whenever the Moon provides illumination.

These migration thresholds result in migration being strongest in the west-central part of high pressure systems but decreasing after the arrival of humid tropical air and in consequence result in the birds arriving at a transient or summer home range in suitable conditions. The migration density correlates with potential high temperatures, high clouds, and a lack of rain at the following day's destination but not with cloud cover and only weakly with wind direction. Computer simulation by Nisbet and Drury of the bird behaviour shows that these migration thresholds are close to the optima that would result in the birds being able to

predict the conditions en route and at the destination and to avoid arrival in rain, fog, and cold air (see p. 919).

These results have been confirmed in part by other authors working elsewhere (e.g. Evans 1972, Richardson 1974). Able (1973) showed that measures of wind direction, 24 hours change in temperature, and an index of synoptic weather could account for 54 per cent of night-to-night variation in the incidence of passerine migration. The major difference in migration thresholds to migration-cost variables between spring and autumn concerns temperature and wind direction. In autumn in the Northern Hemisphere the migration threshold to temperature is exceeded by a drop in temperature and by an increase in the northerly component of wind direction whereas in spring the threshold to temperature is exceeded by a rise in temperature and by an increase in the southerly component of the wind direction. Consequently, autumn migration tends to occur with the passage of a cold front and spring migration tends to occur with the passage of a warm front. Temperature, in this case, may be acting as a habitat variable rather than as a migration-cost variable (p. 265).

The contribution of wind direction to the composite migration threshold of an individual is grossly underestimated by multivariate analysis of variation in the density of migration. As first stressed by Nisbet and Drury (1967), the fact that mean track direction is a function of wind direction does not necessarily mean that birds are being drifted by the wind but rather that demes with different preferred tracks are involved. Alerstam and Ulfstrand (1974) in a radar study of the autumn migration of wood pigeons, Columba palumbus, in southern Scandinavia showed that the entire migrating deme had a migration threshold to the northerly component of the wind. However, individuals that initiated migration if the wind was NW may have been less likely to initiate migration if the wind was NE. Other individuals were more likely to initiate migration if the wind was NE than if it was NW. As wind direction changed, therefore, individuals and demes with different preferred directions made up a different proportion of the migrating population, with the result that the measured peak track direction was a function of wind direction even though each individual showed full compensation for any tendency to drift while over land. Migration over sea in particular was performed with a tail wind and there was a much greater tendency to drift while over water. Wherever possible, the pigeons flew parallel to the coast, though a little distance offshore. Until, that is, the direction of the coast made too great an angle to the required direction.

Presumably as a result of the danger of drift the migration thresholds of land-birds to migration-cost variables are influenced by whether they are about to migrate over land or water. Terrain acts, there-

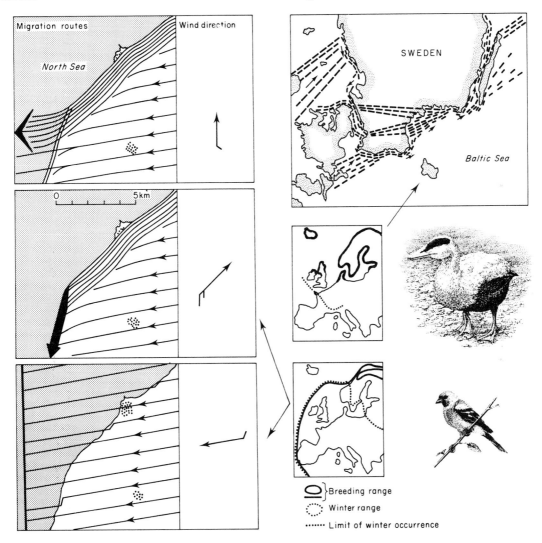

Fig. 27.11 Terrain as a threshold variable

The figures show the influence of terrain on the migration routes of a land bird, the chaffinch, *Fringilla coelebs*, and a maritime species, the common eider, *Somateria mollissima*.

In the case of the chaffinch in the Netherlands, levels of environmental variables that exceed the threshold for migration over land do not necessarily exceed the thresholds for migration over water. Evidence that the phenomenon is one of a raising of the level of the thresholds to all other component variables is provided by the observation that chaffinches fly higher and, as illustrated, migrate across the North Sea to Britain with following

easterly winds but turn along the Dutch coast with southwesterly ones.

The eider is a strongly maritime duck that also occurs along rocky coasts and locally inland around lakes or on river islands. It is mainly resident but in winter some may migrate to the English Channel, Switzerland, or southern France. During spring migration in southern Sweden, these migrants make detours to avoid flying over the mainland.

[*Compiled from Dorst (1962) and Newton (1972) after Tinbergen and Svärdson. Breeding and winter range maps and habitat requirements from Peterson et al. (1974)*]

fore, as a migration-cost variable. The critical factor seems to be the cost to the bird of landing or allowing drift if it were to encounter below-threshold values of other migration-cost variables during the coming migration. Hence land-birds have a higher migration threshold if about to cross water whereas water-birds have a higher migration threshold if about to cross land (Fig. 27.11).

Two other aspects of autumn migration thresholds to migration-cost variables can be observed. The first is that there is often an upper as well as a lower threshold to the tail-wind component. Most passerines, for example, have an upper migration threshold to this component of about 20 km/h. Gliders, such as eagles, hawks, vultures, and white storks, also have a low upper threshold to their with-wind component (Bellrose 1967a). Wind interferes with the thermals upon which these birds are dependent to minimise migration cost, and if they glide with the wind there is a danger of reaching stalling speed. Waterfowl, however, have a lower migration threshold that, at about 50 km/h, is higher than the general upper migration threshold of these other birds. This does not mean, of course, that these two categories of birds do not migrate at the same time. Not only do they migrate at different heights where wind speeds are different (p. 415) but also they both manifest the second aspect of migration thresholds to be discussed here. As expected for such a composite migration threshold as that shown by birds, time is an important component (p. 51). Thus, measured migration thresholds to wind speed and direction change with time since the last migration unit and birds have been observed to initiate migration against a relatively strong wind if they have been grounded by bad weather for several days (Bellrose 1967a).

Although the discussion so far has made considerable use of data collected for birds on migration, we are essentially still concerned with the initiation in the standard direction of the first unit of the autumn migration. This is a convenient point, therefore, to consider the extent to which habitat variables contribute to the observed inter- and intra-specific variation in migration threshold to migration-cost variables.

A number of bird species are partial migrants, a portion of a species (or deme) consisting of individuals that initiate seasonal return migration, the other portion consisting of individuals that retain the breeding home range throughout the year, showing little or no seasonal shift of home range. In

such a situation it is convenient to term the two groups, respectively, migrants and non-migrants. In other species a similar phenomenon exists but the 'migrants' also show considerable variation in distance and direction components. The variation in such components has rarely if ever been adequately evaluated and may show one, two, or several modes. In such cases it is usual, though dangerous, to describe the resident and short-distance migrants, especially those, in the Northern Hemisphere, that migrate in autumn to the W, N, or E, as 'non-migrants', and to describe the long-distance migrants, as long as they migrate with a more or less recognisable bias to the S, as 'migrants'. This convention, which seems to have been adopted either explicitly or implicitly by most previous authors, should be borne in mind in relation to the examples given below.

It is tempting, of course, to postulate a balanced, genetically determined, behavioural polymorphism consisting of two morphs, non-migrant and migrant. However, such a view is unsatisfactory for it is now well known that individuals may change their status from being migrant in one year to non-migrant the next and perhaps even back once more to being migrant. Such a situation has been reported in the United States in the blue jay, *Cyanocitta cristata* (Laskey 1958, Nunnely 1964), and the song sparrow, *Melospiza melodia*. In the latter species, one study (Nice 1937) in Ohio followed the change in status of seven individuals. Of these, one male was a migrant in two consecutive winters but was a non-migrant the following winter, and two males were non-migrants for two consecutive winters but migrants the following winter, re-appearing the following spring. No significant difference was found between migrants and non-migrants, either in wing-length, weight, or colouring. The most likely model, therefore, is not one that postulates migrant and non-migrant strains but rather one that postulates an intra-deme frequency distribution of migration thresholds determined in a way similar to that described in Chapters 17 and 18 (pp. 268 and 403). A number of studies now exist that suggest these migration thresholds are composite (p. 49), receiving contributions not only from the migration-cost variables already described but also from habitat variables.

It has long been accepted that some birds migrate in response to habitat variables, but such birds have usually been placed into a separate category and described as irruptives, especially if they have a variable maximum migration distance and a direction ratio that approaches 25:25:25:25. Some examples have already been presented (p. 208) and many other examples are available, including, for instance, the waxwing, *Bombycilla garrulus*. However, such a complete gradation of distance, direction, and direction-ratio components has now been discovered between 'irruptives' and 'migrants' among birds that clearly all include habitat variables as components of their migration threshold and the distinction no longer seems justifiable, even on the grounds of convenience. Three examples of species which include habitat variables within their composite migration threshold may be given.

A portion of the Finnish and Norwegian breeding demes of the brambling, *Fringilla montifringilla* (Fig. 27.21), migrates regularly each year to a relatively constant wintering area in western and central Europe, the portion from Finland wintering significantly further to the east than the portion from Norway (Eriksson 1970a). In years when the berry crop of rowan, *Sorbus aucuparia*, is particularly heavy, however, large numbers fail to initiate migration and instead spend the winter in Fennoscandia. Possibly the failure to initiate autumn migration releases more time for juveniles to continue exploratory migration and habitat assessment in relation to breeding habitats for, following wintering by large numbers in Fennoscandia, the breeding range extends further to the south than is usual.

Another study in Finland by Eriksson (1970b), in this case on the siskin, *Carduelis spinus*, has demonstrated a positive linear correlation between the number that fail to initiate autumn migration and the size of the seed crop of both birch and spruce. There is also a positive correlation with the maximum temperature during the migration season. Perhaps, in this case, temperature is also a habitat variable.

Finally, an analysis (Tyrväinen 1970) of the mass failure of the fieldfare, *Turdus pilaris*, to initiate migration from Finland in the winter of 1964–5 also suggests the contribution of habitat variables to the migration threshold. Once again, the major threshold variables seemed to be the size of the rowan berry crop, autumn temperature, and, in this case, also the date of the arrival of snow cover.

The involvement of habitat variables in the migration threshold renders it highly likely that deme density is one of the contributory components (p. 69). Consequently, the model to be applied is that presented on p. 269 which suggests that the

equilibrium position occurs when the distribution of migration thresholds is such that all individuals gain the same potential advantage. Given the perfect adaptation, no difference should be found in the reproductive success of migrants and non-migrants in any year (unless the situation is despotic, in which case migrants should always suffer a disadvantage relative to non-migrants). Lack (1943, 1944, 1954) implied a slightly different type of equilibrium, more comparable to the model outlined on p. 72, which is strictly applicable only if partial migration is due to the existence of two distinct genetic strains (i.e. migrant and non-migrant). On Lack's thesis, the equilibrium position results from the balance of advantages, the migrants surviving best in some years, the non-migrants surviving best in others. It will be difficult to determine which model is most appropriate for any given species for it is unrealistic in the former model to expect the perfect adaptation to have occurred. Equal reproductive success every year for all individuals is only as probable as the degree of correlation between the factors contributing to the migration threshold and the geographical variation in habitat suitability during the coming winter. Given a perfect correlation, 'migrants' and 'non-migrants' should gain the same advantage every year according to the first model (in species with a free, not a despotic, situation), whereas according to Lack's model there should be considerable year-to-year variation in the selective advantage gained by 'migrants' and 'non-migrants', though with both groups gaining the same average advantage over a period of years. Field studies of the relative success of migrants and non-migrants in partially migrant species are few and usually too incomplete for present purposes (e.g. Österlöf 1966 for the goldcrest, *Regulus regulus*, in Finland in which only about 10 per cent of the non-migrant portion survives until spring). However, from data spanning four years, Nice (1937) concluded that the survival of migrant and non-migrant portions of an Ohio-breeding deme of the song sparrow, *Melospiza melodia*, was similar for three of the four years but with a higher survival of migrants in the fourth year. This conclusion tends to support, for this species, the former of the two models presented here. The change of status of some individuals from migrant to non-migrant is also easier to reconcile with the first model than with that of Lack.

Both models would predict geographic, ontogenetic, and sexual variation in the migrant/non-migrant ratio. There is, however, a fundamental difference between the two in this respect. The latter model, that of Lack, has to predict that these differences are always genetic. The former model, which postulates a near-ideal frequency distribution of migration thresholds, can also accommodate a genetic cline of frequency distributions and hence of migrant/non-migrant ratios, but in addition allows some such variation to occur without the existence of genetic differences. In the discussion that follows, only this model is considered further.

It is suggested that in any one area a deme of birds exists that shows a characteristic frequency distribution of migration thresholds to which relative resource-holding power may or may not make a contribution depending on the species, deme density, and resource abundance (p. 404). In any one place, the migrant/non-migrant ratio will then change from year to year as described depending on the proportion of individuals the migration threshold of which is exceeded in each particular year. The same variation will occur from place to place, the migrant/non-migrant ratio being lower when both migration cost and habitat suitability are higher. As, in temperate regions, the gradient of habitat suitability *in winter* may be assumed to vary from a low level at high latitudes to a high level at low latitudes (see p. 659), it would be expected that the migration threshold of a greater proportion of individuals will be exceeded at high than low latitudes. This should still be the case whether the situation is free or despotic. It would be conceivable, therefore, for the same frequency distribution of migration thresholds to be adaptive for all demes no matter at what latitude they breed. However, the inherent advantage of obligatory rather than facultative migration when variation in habitat suitability is predictable (p. 50) will probably result in the evolution of a genetic cline of frequency distributions of migration thresholds. The direction of this cline should be such that the proportion of obligatory non-migrants decreases from low to high latitudes where the proportion of obligatory migrants increases from low to high latitudes. Even so, some portion of the population should remain as facultative migrants with a particular frequency distribution of migration thresholds. Within any given area, therefore, it is still conceivable that a panmictic deme should still show consistent area-to-area variation in the migrant/non-migrant ratio without there being any genetic differences between the groups occupying the different areas.

Numerous examples are available that demonstrate geographical variation in the migrant/ non-migrant ratio and in general they all conform to the model just outlined. There is, for example, a linear series of demes of the white-crowned sparrow, *Zonotrichia leucophrys*, that inhabits 1900 km of the Pacific coast of North America from about Santa Barbara, California, to just north of Comox on Vancouver Island, British Columbia (Mewaldt *et al.* 1968). The southern demes are 'non-migratory' and have been assigned to the race *nuttalli* whereas the northern demes are 'largely migratory' and belong to the race *pugetensis*. Here there is an undoubted genetic cline from north to south in the frequency distribution of migration thresholds and obligatory migrants and non-migrants. In Europe, a great many species that have a high migrant/non-migrant ratio on the mainland have a low migrant/non-migrant ratio in Britain. In general, migrant/ non-migrant ratio increases from southwestern to northeastern Europe, more or less at right angles to the mid-winter isotherms. In many finch species, the southernmost demes are completely non-migratory and the northernmost demes are completely migratory with the intermediate demes showing a cline of migrant/non-migrant ratio (Newton 1972). As another example, the great tit, *Parus major*, rarely migrates in England further than about 24 km from its natal home range whereas in European Russia it is 'mainly migratory' (Kluijver 1951). Some individual great tits are non-migrants, however, even at high latitudes. In some species, a south to north cline in the migrant/non-migrant ratio is evident even in Britain (Fig. 27.12), though whether or not the frequency distribution of migration thresholds shows a cline is unknown.

Within any given deme it is common for the frequency distribution of migration thresholds to be such that sexual and ontogenetic differences in migrant/non-migrant ratio exist. In passerines, particularly those in which male reproductive success is a function of the early occupation and defence of a suitable breeding territory, males show a lower migrant/non-migrant ratio than females and young (e.g. the brambling, *Fringilla montifringilla*— Eriksson 1970a). This presumably can be attributed to the same selective pressures discussed in Chapter 17 (p. 321).

As well as geographical, sexual, and ontogenetic differences in the frequency distribution of migration thresholds within a single species, there are also interspecific differences. These differences tend to be consistent even when the species concerned show geographic, sexual, and ontogenetic differ-

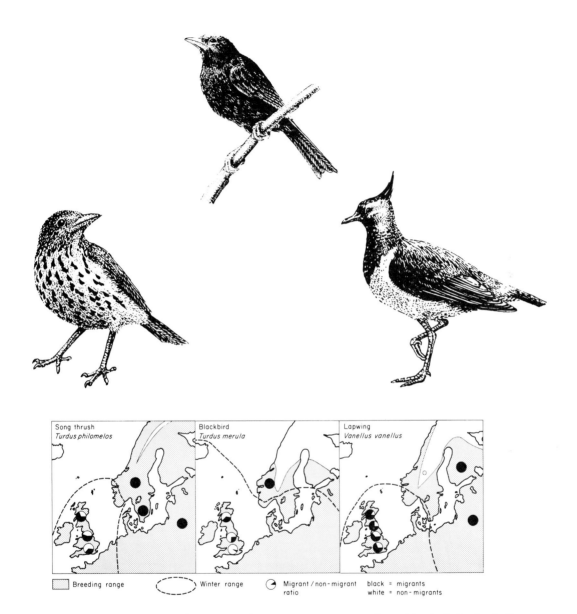

Fig. 27.12 Geographical and inter-specific differences in migrant/non-migrant ratio

The song thrush and blackbird are congeners with many ecological similarities. Both occur in woodlands, and gardens and parks surrounded by hedgerows. Both feed primarily on the ground on a variety of invertebrates, though they also take a variety of berries, and both nest in bushes and hedges. The blackbird may show a greater predilection for nesting in wood-piles and buildings. Both aggressively defend their home range, often against each other. As illustrated, the song thrush everywhere has the higher migrant/non-migrant ratio, except possibly in southwest Norway.

Migrants from Britain in this diagram are considered to be individuals that winter either in Ireland or in France or Spain. Migrant blackbirds from Scotland and northern England winter almost solely in Ireland.

The lapwing is a gregarious plover that in winter may form large straggling flocks. It is typically an inhabitant of farmland, marshes, and mudflats, breeding on arable land, moors, and marshes, etc. Eggs are laid on the ground. Food consists primarily of invertebrates collected from the ground.

In general in western Europe the migrant/non-migrant ratio is lowest in the south and west and highest in the north and east.

[*Compiled from Lack (1944) and Peterson et al. (1974)*]

ences. In Britain, for example, the blackbird, *Turdus merula*, shows near-zero seasonal return migration whereas the song thrush, *T. philomelos*, is a partial seasonal return migrant (Lack 1966), the migrant/non-migrant ratio increasing from south to north (Fig. 27.12), but always being higher in females than males. In the demes that breed around the Baltic, the blackbird is a partial migrant, with females having a higher migrant/non-migrant ratio than males, whereas the song thrush is virtually a complete migrant (Davydov and Keskpaik 1972).

This type of consistent species-difference leads us to a brief consideration of the selective pressures acting to produce interspecific variation in migrant/non-migrant ratio. Although certain broad generalisations are possible (Fig. 27.13), however, it is not always easy to appreciate the details of the selective pressures acting on individual species. In the case of the blackbird and song thrush just presented, for example, it is difficult to see just what differences there can be in the ecology of the two species that result in selection for a higher migrant/non-migrant ratio in the latter species. Both feed on a broadly similar range of invertebrates and berries with the song thrush having a greater predilection for feeding on snails. Yet for some reason the song thrush seems to experience a lower habitat suitability than the blackbird in northern latitudes. This possibility, suggested by an apparently higher winter mortality in the song thrush than the blackbird (Lack 1966) is presumably sufficient, if true, to generate the observed difference in the migrant/non-migrant ratio.

Restricting discussion now to general considerations, it appears that non-insectivorous species have a lower migrant/non-migrant ratio than insectivorous species except where the latter are adapted to specific categories of insects such as eggs, small larvae, or stages that secrete themselves inside buds or under bark and thus are available to a similar extent throughout the year. Analysis of long-distance migrant birds of North America (Fig. 27.13) shows a high correlation with vegetation and latitude but little direct variation with climate. The chief conclusion is that the critical factor is the magnitude of the change from summer to winter in

Fig. 27.13 Geographical variation in the migrant/non-migrant ratio in North America

The black sector of each circle shows the proportion of migrant individuals in the species-summated breeding populations of birds in various relatively undisturbed vegetation communities in North America. In this analysis a 'migrant' is classed as an individual that leaves the Nearctic Region (the area north of the Mexican border plus the Mexican highlands) for the northern winter.

Nearby censuses show similar migrant/non-migrant ratios except when the habitat-type changes rapidly. Any correlation with climate is not simple for the prairies with severe winters and warm summers have a lower migrant/non-migrant ratio than the west coast forests with mild winters and cool summers. The east coast is intermediate in climate but has an even higher migrant/non-migrant ratio.

The critical factor seems to be the magnitude of the change in predictable food supply from summer to winter. Where this is great, the migrant/non-migrant ratio is high. Thus prairies and deserts which have unpredictable and short-term fluctuations and also provide a large seed crop which persists into winter have a low migrant/non-migrant ratio. The highest migrant/non-migrant ratio is found in the deciduous forest of the northeast.

In addition, a slight latitudinal effect is also apparent, the migrant/non-migrant ratio being slightly greater at higher latitudes.

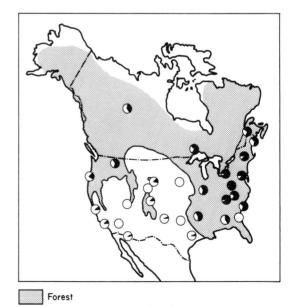

� Forest

◔ Black sector : proportion of breeding deme that migrates to the neotropics

[Re-drawn from MacArthur (1959)]

the predictable supply of food available to the birds. When summer food is predictably abundant but winter food is predictably scarce, the migrant/non-migrant ratio is high (MacArthur 1959).

Another contributory factor has been suggested by Haartman (1968). It is pointed out that one well-known advantage of not migrating is that there is invariably an opportunity to find a nesting territory and/or mate before the return of the migrants. This advantage, however, is counteracted by the disadvantage for many species of a higher winter mortality. Hole-nesting birds are likely to experience more competition for nesting sites yet at the same time perhaps have a lower winter mortality through roosting in tree cavities. These two pressures could well favour an earlier occupation of breeding sites and a lower migrant/non-migrant ratio in hole-nesting than in non hole-nesting species. Such a correlation with hole-nesting indeed seems to exist.

The migrant/non-migrant ratio is not the only component of migration that is likely to show inter-specific variation that is a function of habitat variables. The timing of the autumn migration period is also likely to be such a function. In this respect the influence of habitat variables is likely to be both ultimate and proximate. As an ultimate factor, seasonal variation in habitat suitability imposes selection upon the circannual endogenous programme of the species and favours the optimum seasonal timing and sequence of breeding and moulting, and of variation in migration thresholds as already discussed. In a proximate sense, selection is also likely to have favoured habitat variables contributing to the composite migration threshold of the species. The result is inter-specific and also year-to-year variation in migrant/non-migrant ratio and migration timing that is adaptive for each species or deme. The broad-winged hawk, *Buteo platypterus*, of America, for example, lives largely on reptiles and amphibians which hibernate early in the northern parts of its range. Consequently the autumn migration of this species is initiated early (M. Brown 1937). The sharp-shinned hawk, *Accipiter striatus*, in the same region, however, times its migration to coincide with that of the small passerines on which it preys. The prime example of such timing, however, is that of Eleonora's falcon, *Falco eleonorae* (Fig. 27.14) in which the entire annual programme is adapted to exploit the concentrated autumn passage of small passerines across the Mediterranean.

Ontogenetic variation may also occur in the timing of the initiation of autumn migration. To some extent this may result from selection for different amounts of time investment in exploratory migration by juveniles and adults but in the most conspicuous examples it seems to result from, or is at least associated with, selection for differences in the moulting season. In the yellow-bellied flycatcher, *Empidonax flaviventris*, and the least flycatcher, *E. minimus*, of North America, for example, the adults migrate to their winter home range before undergoing a protracted post-nuptial moult, whereas the juveniles undergo a post-juvenal moult in advance of initiating autumn migration (Johnson 1963). The result is that adults initiate autumn migration well in advance of the juveniles. In 1965, for example, the least flycatchers passing the Long Point Bird Observatory on the north shore of Lake Erie between 25 July and 2 August were all adults, whereas those passing from 13 August onwards were all juveniles (Hussell *et al.* 1967).

So far in this discussion of the selective pressures acting on migration thresholds we have considered the nature of intra- and inter-specific variation in the frequency distribution of migration thresholds and in those individuals, demes, and species that eventually achieve a low autumn migration threshold, the physiological changes have been outlined that lead to such a low threshold. The range and nature of environmental variables (habitat and migration cost) that contribute to this composite migration threshold have also been described. Now it is possible to consider the factors that influence the level of the migration threshold from the time the bird leaves its natal, post-fledging, breeding, or moulting home range to the time of arrival in the wintering area.

According to the familiar-area hypothesis, the adult bird terminates the first unit of its autumn migration in an area (= transient home range) with which it is familiar as a result of the exploratory migrations carried out during its first autumn migration as a young bird. Conceivably, further exploratory migration is performed each year as a preliminary to type-C removal migration (p. 308) if the transient home range is below the threshold suitability for such exploratory migration, but it seems unlikely that such exploratory migration is ever as extensive as during the bird's first autumn. Each unit of the total autumn migration of an adult bird thus seems likely to be calculated rather than exploratory. As with all calculated migration, however, it is conceivable that type-C removal migration may occur solely on the basis of social communication (p. 187).

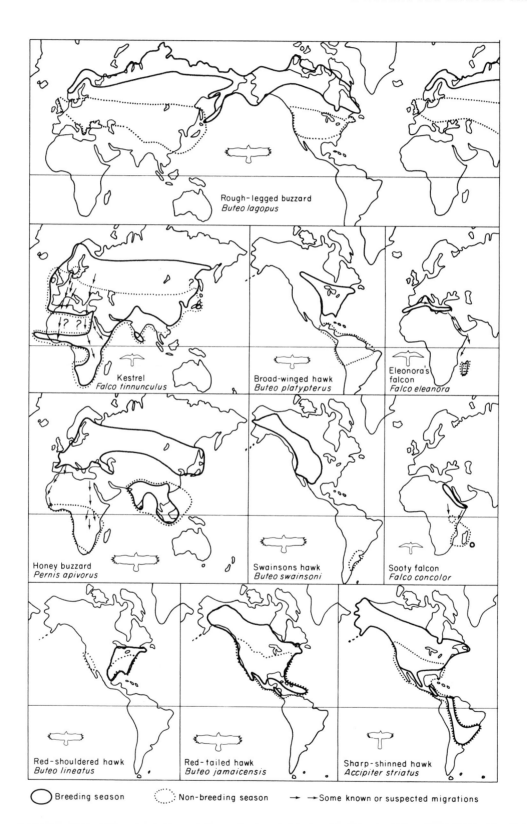

Rough-legged buzzard
Buteo lagopus

Kestrel
Falco tinnunculus

Broad-winged hawk
Buteo platypterus

Eleonora's
falcon
Falco eleanora

Honey buzzard
Pernis apivorus

Swainsons hawk
Buteo swainsoni

Sooty falcon
Falco concolor

Red-shouldered hawk
Buteo lineatus

Red-tailed hawk
Buteo jamaicensis

Sharp-shinned hawk
Accipiter striatus

⬭ Breeding season   ⬭ Non-breeding season   → Some known or suspected migrations

Fig. 27.14 Seasonal return migrations of ten species of diurnal birds of prey

The diurnal birds of prey the migrations of which are illustrated represent two families and three sub-families.

The buteos, buzzards, or buzzard hawks (Accipitridae: Buteoninae) are large, thick-set hawks that habitually soar in wide circles. As a group they feed principally on rats, mice, and rabbits and occasionally small birds and reptiles. However, Swainson's hawk also eats large insects, particularly grasshoppers, and the honey buzzard specialises on the larvae, pupae, and adults of social wasps and bees.

The accipiters or bird hawks (Accipitridae: Accipitrinae) are primarily woodland birds that soar less than buzzards and hunt among trees and thickets using a hedge-hopping technique. As a group they feed primarily on birds but also take small mammals.

The falcons (Falconidae: Falconinae) are streamlined birds of prey built for speed, not sustained soaring, though the kestrel (photo) is highly adapted for hovering. As a group they feed primarily on birds, rodents, and insects. The sooty falcon and Eleanora's falcon are peculiar in that they do not nest until after the summer solstice and are specialised to feed their young on the autumn passage migrants. During the breeding season the sooty falcon preys mainly on small passerines in the northern part of its range and on larger birds, such as hoopoes, *Upupa epops*, and orioles, *Oriolus* spp., in the southern Red Sea region. On passage through Kenya the sooty falcon takes queleas (Fig. 27.32). In Malagasy it eats mostly insects. Eleanora's falcon takes a variety of migrant species ranging in size from small warblers (*Phylloscopus* and *Sylvia*) to hoopoes. It has been estimated that during the nesting season (August through October) a pair of Eleanora's falcons plus their brood would need 1750 migrant birds averaging about 20 g in weight.

[Compiled from Voous (1960), Peterson (1961), Robins et al. (1966), Brown and Amadon (1968), Moreau (1972), and Peterson et al. (1974). Photo by Arthur Christiansen (courtesy of Frank W. Lane)]

It has been suggested (p. 618), however, that each unit of autumn migration by a young bird, if there is no social communication, consists of exploratory migration in the standard direction alternating with shorter-distance exploratory cum calculated migrations in other directions. The movement in the standard direction is considered to function largely as a major extension of the familiar area during which there is a minimum of habitat assessment except at the gross level of major habitat-types and optimum migration routes. The movements in other directions are considered to function less as extensions of the outer boundaries of the familiar area, though this may also occur, but more as calculated migrations to gross areas perceived during the standard migration. These calculated migrations should then be followed by more intensive exploratory migration and habitat assessment within the gross areas. These non-standard migrations, many of which may qualify as reversals (p. 617), also have thresholds to migration-cost and habitat variables. As far as thresholds to migration-cost variables are concerned, these seem from such information as is so far available to be similar to those for migration in the standard direction. Wind speed and direction in particular makes an important contribution and all studies of reversal, whether in spring or autumn, stress that reversal occurs either in calm conditions or with a tail-wind (e.g. Nisbet and Drury 1968, Parslow 1969, Evans 1972). If anything, the migration thresholds to wind speed and direction for these non-standard migrations (Fig. 27.15) seem to be even higher than those for migration in the standard direction in that these latter migrations may be initiated with light head-winds whereas this rarely if ever happens for non-standard migrations. Perhaps the critical difference is that time is a component of the standard-migration threshold (p. 51) whereas it may not be a component for non-standard migrations. No critical data are available concerning non-standard migration thresholds to habitat variables, though it has been suggested (p. 188) that these may make an important contribution to the composite non-standard migration threshold.

Most of the studies discussed earlier have demonstrated that migration thresholds to migration-cost variables at the time of take-off are a function of the conditions experienced during the previous 24 hours. To some extent, of course, these previous conditions are themselves migration-cost variables in that the accuracy of prediction of migration cost is maximised when conditions throughout the 24 hours contribute to the depth of the diel low of the composite migration threshold. To some extent also, however, these conditions are habitat variables. In this capacity they can act in two ways. On the one hand, they correlate with habitat suitability and contribute directly to the migration threshold. On the other hand, by influencing the energy reserves and other of the bird's physiological parameters, they can instigate a change in the migration threshold.

Observations on the blue-winged teal, *Anas discors*, for example, show that the birds feed more during the day but cease earlier in the evening on days preceding standard migrations than on days preceding non-migration nights (Owen 1968). This clearly demonstrates the way that the endogenous basis of diel variation in migration threshold influences the bird's physiology and behaviour at times other than at take-off and also emphasises the importance to a migrant of the feeding that occurs in the transient home range.

There are few critical studies of variation in migration threshold while a bird is in a transient home range, but such studies as are available suggest the following working hypothesis. When a bird, whether young or adult, arrives in a transient home range, its diel cycle of migration threshold fluctuates about a higher mean level than during the 24 hours preceding the migration just terminated (Fig. 27.16). If the transient home range has a sufficiently low suitability, the bird initiates further migration. A hierarchy of migration thresholds exists (p. 229) at any given time of day such that the most likely migrations to be initiated at any time of day are short-distance exploratory migrations by juveniles and exploratory or calculated migrations by adults, habitat suitability being assessed on the basis of migration thresholds appropriate to a transient home range. The hierarchy should, however, consist for different types of migrations of respective migration thresholds that fluctuate independently, both with time of day and with time after arrival in the transient home range. The short-distance exploratory and calculated migrations should follow the usual sequence (pp. 65 and 378), their direction being a function of migration-cost variables.

The highest migration threshold of the hierarchy upon arrival in the transient home range is that for the initiation of a further migration in the standard direction. The mean low level about which the autumn migration threshold should fluctuate on a diel basis should be a function of the energy reserves used up during the previous migration as well as a function of the stage reached in the endogenous programme for autumn migration. From a high mean level upon arrival in the transient home range, the migration threshold for the initiation of a further migration in the standard direction should decrease with time after arrival at a rate that is a function of the rate of acquisition of energy reserves. A low rate of acquisition causes a low rate of decrease in the mean level of diel fluctuation in autumn migration but at the same time reflects a low habitat suitability which is more likely to exceed any given autumn migration threshold. A high rate of acquisition of energy reserves causes a high rate of decrease in mean migration threshold but at the same time reflects a high habitat suitability which is therefore less likely to exceed any given autumn migration threshold.

Support for this thesis derives from a number of studies, three of which may be mentioned here. The first concerns the chaffinch, *Fringilla coelebs*, at Rossitten on the Baltic Coast (Dolnik and Blyumental 1967). Here the transient birds remain for a period of 2–4 days. Lean birds, however, stay for

longer than do fat ones and during their stay the former birds feed and replenish their fat. Not only does this study support that part of the thesis concerned with the influence of energy reserves on the rate of decrease in the autumn migration threshold, it also suggests that the various migration thresholds (for exploratory and calculated short-distance non-standard migrations and longer-distance standard migrations) indeed fluctuate independently and in a way that is a function of energy reserves. This follows from the observation that whereas lean birds stay longer and feed and presumably perform a variety of exploratory and calculated migrations associated with efficient feeding, fat birds may not feed at all during their short stay and may even lose weight. In lean birds, therefore, the feeding migration thresholds start lower than the autumn migration thresholds, whereas in fat birds the autumn migration threshold starts lower than feeding migration thresholds and presumably remains so unless unfavourable migration-cost variables prevent the bird from initiating autumn migration during the next 24 hours or so. Most chaffinches in the Baltic region migrate at a weight of about 23 g of which about 2 g is fat, but

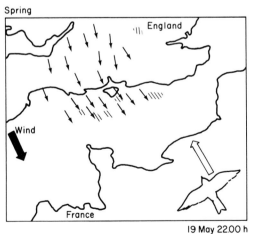

Spring

England

Wind

France

19 May 22.00 h

Autumn

England

Wind

France

9 September 21.00 h

Fig. 27.15 Two examples of 'reversal' migration by nocturnal migrants over southern England as detected by radar

Small arrows indicate the direction of bird migration. Large black arrows indicate wind direction. Large white arrows indicate the standard migration direction for the appropriate time of year (NNW in spring; SSE in

autumn). Hatched areas in the spring diagram indicate the position of rain showers.

Some reversal migration was recorded on one-third of all spring evenings. A migration threshold to wind direction is evident both for reversal and standard migration.

[*Modified from Parslow (1969)*]

some may weigh 26 g in which case about 5 g is fat. Lean birds in a transient home range can replenish their fat at a rate of about 0·5–1·0 g/day.

The second study suggests similar conclusions but is complicated by the fact that in the species concerned, the European robin, *Erithacus rubecula*, the composite migration threshold receives a major contribution from relative resource-holding power. The species migrates at night and feeds during the day (Fig. 27.10) and most individuals that arrive

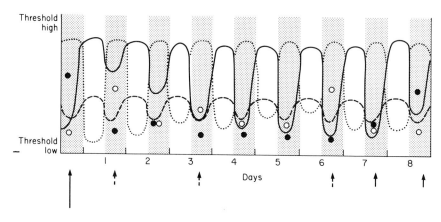

Fig. 27.16 Postulated pattern for a young bird of variation in the hierarchy of migration thresholds with time after arrival in a transient home range

The hierarchy is assumed to consist of thresholds for three migrations:

——————   threshold for a migration unit in the standard autumn direction;

– – – –   threshold for migration in non-standard autumn directions;

....   threshold for short-distance calculated and exploratory migrations associated with feeding.

● level on a particular night of environmental variables relative to the standard migration threshold;

○ level on a particular night of environmental variables relative to the non-standard threshold.

Arrows indicate the initiation of a standard (———▶) or non-standard (– – – –▶) migration. Length of arrow indicates migration distance.

Stippling indicates hours of darkness.

The figure refers to a species that migrates at night and feeds by day. A standard migration is a rectilinear exploratory migration in an endogenously determined direction during which gross features of the environment are committed to a spatial memory. Non-standard migrations are calculated migrations to gross areas perceived and assessed to be potentially suitable as a transient home range. These areas were first perceived during the previous standard migration unit. Detailed exploratory migration is then performed in these gross areas.

In the diagram, the fuel reserves, time since last migration, and time of day, components have been withdrawn from the composite migration threshold. The thresholds illustrated, therefore, which are composite for all other factors, become a function, though to varying degree, of these three variables (see p. 346). Standard and non-standard thresholds are assumed to be highest by day and lowest between dusk and midnight. Feeding migration thresholds are assumed to be lowest during the day. It is also assumed that the standard migration threshold (——————) is a function of fuel reserves and time since last (standard) migration. Non-standard migration thresholds (– – – –) are assumed to be independent of both these variables. Feeding migrations are assumed to be a function of fuel reserves but not of time since last standard or non-standard migration.

After a long standard migration, the standard migration threshold starts high (night 1) but decreases (nights 1 to 5) as fuel reserves are replenished and with time after last migration. Once optimum fuel reserves have been attained, further decrease in threshold (nights 5–7) is less and solely a function of time since last standard migration. A further standard migration is not initiated, however, until the level of environmental variables (●) on a particular night exceeds the standard migration threshold (nights 7 and 8).

Non-standard migration thresholds fluctuate about a constant level and are exceeded whenever the level of environmental variables (○) exceeds the non-standard migration threshold (nights 1, 3, and 6).

Feeding-migration thresholds should be lower on days following a standard or non-standard migration due to lower fuel reserves. As the standard migration threshold for a given night receives a contribution from environmental conditions during the preceding 24 hours, feeding migration thresholds may also be lower on days preceding the probable initiation of a standard migration.

The migrations initiated by a bird on any given day or night depend on which migration threshold of the hierarchy happens to be exceeded.

during autumn migration on the Baltic coast of Poland continue their migration the following evening (Szulc–Olech 1965). Birds with a relatively high resource-holding power, even if they are of above-average weight when they arrive, may stay for 12 or so days before initiating the next stage of their autumn migration. During this time, they establish and defend territories in the most suitable habitats. Although they may show no gain in weight during the first few days they later show a spectacular increase. Birds of relatively low resource-holding power that arrive with low energy reserves also stay for several days. During this time they are forced to perform considerable exploratory and calculated migration due to the presence of birds of higher resource-holding power in the otherwise most suitable habitats. Such birds remain in the low suitability habitats that they eventually occupy only for as long as is necessary to maintain a minimum recovery weight. These individuals thus initiate the next stage of their autumn migration at a lower weight than other individuals. In this species, therefore, although all birds may have the same composite migration threshold (p. 346) to other variables, different individuals have such different resource-holding powers and occupy habitats of such different suitabilities that measured migration thresholds to the different migration-cost variables will show considerable individual variation.

Finally, Berthold and Dorka (1969) examined the frequency distribution of migration dates in autumn for various bird species and found that whereas typical long-distance migrants showed a normal distribution of migration dates or showed a peak at the start of migration, short-distance migrants typically showed a peak at the end of the migration period. It was suggested by these authors that these results indicated a largely endogenous timing of migration for long-distance migrants and a greater influence of exogenous events on the timing of short-distance migrants. As we have seen, however, the timing of migration of all species involves exogenous factors as a result of migration thresholds to migration-cost variables. Many species also show migration thresholds to habitat variables. Perhaps the data presented by Berthold and Dorka can be re-interpreted as showing that the composite migration threshold of long-distance migrants consists almost entirely of migration-cost variables and time, whereas that of short-distance migrants, such as, for example, the robin, *Erithacus rubecula*, consists of a mixture of habitat and migration-cost variables and

includes time to a much less important extent. Inserting this effect into the thesis already presented, it could be suggested that for long-distance migrants the decrease in the mean level of the composite autumn migration threshold during the period after arrival in a transient home range is much more a function of time since arrival than it is for short-distance migrants (see also p. 696).

It seems likely from the results of Gwinner (1972) that in at least some species the pattern of change in the mean level of the autumn migration threshold varies during the period of the autumn migration in a way that is part of a circannual endogenous programme. Included in this programme is presumably an eventual sudden and permanent increase in migration threshold. The timing of this increase, which is probably not completely uninfluenced by the sequence and duration of environmental conditions encountered throughout the autumn migration, should be fixed by selection within the endogenous programme in such a way that the increase in migration thresholds occurs at about the time that the bird has arrived in the wintering area. However, because the endogenous effect should only be an increase in migration thresholds, if the bird is not within an appropriate area for wintering, environmental conditions exceed the migration threshold and further migration is initiated. It is not necessary, therefore, as has previously been assumed (e.g. Evans 1972), to postulate great precision in the determination of the total migration distance by the endogenous programme. This would be necessary on the assumption that it is the initiation and termination of migration that is under endogenous control, but is not necessary if, as suggested here, the endogenous programme only influences the temporal variation in migration thresholds.

Once again, however, it seems likely that a hierarchy of migration thresholds exists, one threshold being appropriate for continuation of autumn migration, another threshold being appropriate for shorter-distance exploratory migration in non-standard directions, and so on. The result should be that juveniles, upon arrival in a general area appropriate for wintering, no longer initiate exploratory migration in the standard autumn direction but instead perform exploratory and then calculated migration appropriate for the establishment of a winter home range. Similarly, adults behave either in a way similar to juveniles or, if they have arrived in the home range occupied the

Fig. 27.17   European robin, *Erithacus rubecula*

[*Photo by Ronald Thompson (courtesy of Frank W. Lane)*]

previous year, perform calculated migrations from the beginning. In both cases the migration thresholds set by the endogenous programme are those adaptive to winter or non-breeding habitats.

Migration thresholds seem likely to exist, however, throughout the winter, albeit, as far as long-distance exploratory or calculated migrations in the standard autumn direction are concerned, at a high level. That such thresholds exist, however, is supported by the observation that many birds, particularly those that feed on the ground or in still water and that winter in temperate regions initiate further migration, even in mid- or late-winter, if snow or frost prevents access to food. Such migrations, normally termed hard-weather movements, are often relatively long-distance and often in the same direction as the standard autumn migration. It seems not unreasonable to suggest that the autumn migration threshold persists, albeit at a high level, throughout the winter months. Even in the case of the mass occurrence of the fieldfare, *Turdus pilaris*, in the winter of 1964/5 in Finland that earlier (p. 635) was attributed to levels of migration-cost and habitat variables that did not exceed the autumn migration threshold, many individuals initiated migration in the winter when the previously abundant food supply became exhausted (Tyrväinen 1970). The correlation between probability of initiation of hard-weather migration and winter-feeding ecology is well shown by fringillid finches (Newton 1972). The species that initiate such migration following a sudden cold spell or fall of snow are mainly those that feed on or near the ground (e.g. the chaffinch, *Fringilla coelebs*, brambling, *F. montifringilla*, greenfinch, *Carduelis chloris*, goldfinch, *C. carduelis*, linnet, *Acanthis cannabina*, and twite, *A. flavirostris*).

The length of time that migration thresholds remain appropriate to the winter habitat varies considerably from species to species. Perhaps the minimum period is that shown by the Lapland-breeding deme of the twite, *A. flavirostris*. These finches initiate autumn migration in August and September and do not reach the south of their wintering range in Poland until December or January. Even then the majority stay for only a few weeks, perhaps only one, before starting back (Tomialojć 1967).

Except for the possibility of a relatively greater incidence of calculated rather than exploratory migrations in the case of young birds during the spring migration (p. 618), the pattern of variation in

migration threshold seems likely to be similar in spring and autumn. At some point in the spring either the endogenous circannual programme or an endogenous sequence initiated when the threshold to indirect variables is exceeded, promotes a species-specific sequence of pre-nuptial moulting, fat deposition, and change in migration thresholds to those appropriate to spring and/or breeding habitats. Many species do not include a moult in this sequence.

The available evidence suggests, at least in those species and demes that pass the winter on the same side of the equator as that on which they breed, that daylength is much more likely to be involved in the initiation or phasing of the endogenous sequence than in the autumn. Thus, in the chaffinch, *Fringilla coelebs*, which migrates entirely within the temperate zone of the Northern Hemisphere, long photoperiods between January and March induce vernal pre-migratory fattening, gonadal growth, and Zugunruhe (Dolnik and Gaurilov 1972). Bramblings, *F. montifringilla*, that are suddenly switched experimentally from photoperiods of 8 hours (appropriate to mid-winter) to 14·5 hours (appropriate to mid-spring) develop large fat stores and Zugunruhe within eight days while others kept on winter day lengths throughout the normal season of spring migration showed no such changes (Lofts and Marshall 1960). Similar results have been obtained for slate-coloured juncos, *Junco hyemalis*, and American crows, *Corvus brachyrhynchos* (Rowan 1925, 1926, 1929). Temperature may also influence the initiation and phasing of the endogenous sequence. Thus the white-crowned sparrow, *Zonotrichia leucophrys*, of America begins to deposit fat 10–12 days earlier in warm dry springs than in cold damp ones (King and Farner 1965). Presumably in this case temperature is acting as an indirect rather than as a habitat variable.

All of these species, however, remain within the Northern Hemisphere during migration and it is possible that among trans-equatorial migrants, such as the willow warbler, *Phylloscopus trochilus*, circannual endogenous programmes are much more general. It is possible, however, to conceive of the initiation and phasing of endogenous programmes by indirect variables even in a trans-equatorial migrant.

Where selection has acted on different demes of a species to winter in the same area but to breed at different latitudes (p. 696), then clearly it has also acted to produce inter-deme differences, either in endogenous programmes or in thresholds to indirect variables. Thus the goldfinches, *Carduelis carduelis*, present in winter in Iberia are made up from local, non-migratory individuals and individuals that breed in Britain. In spring, although exposed to identical conditions, one group initiates migration at about the time that the other begins to breed (Newton 1972). The same is true of the local and Scandinavian chaffinches, *Fringilla coelebs*, that winter in Britain, and of many other species throughout the world.

Similar inter-deme differences in migration thresholds are likely to apply to demes wintering in different areas. Thus, the honey buzzard, *Pernis apivorus*, crosses the Mediterranean northward in spring both in the region of the Bosphorus and in the region of Gibraltar (Fig. 27.14). Two separate demes seem to be involved. Those migrating to east and central Europe where the temperatures rise relatively early, cross the Bosphorus as early as late March and early April (Collman and Croxall 1967). Those migrating to Scandinavia where spring arrives relatively late cross Gibraltar no earlier than between the end of April and early May. Presumably these demes have different thresholds to indirect variables in response to which the vernal endogenous sequence is initiated.

Inter-specific differences exist in the timing of the vernal sequence. Most of this inter-specific variation can be attributed to selection for arrival in the breeding area as soon as conditions there become suitable for survival, which may be several weeks before conditions become suitable for nesting (Newton 1972). Inter-specific differences in the time of arrival of such conditions is sufficient to generate observed differences in timing of migration except where conditions for survival in transient home ranges impose modifying selective pressures.

Sexual differences in the initiation of spring migration can also be attributed to selection on time of arrival in the breeding area. Where the reproductive success of males is a function of the early establishment of breeding territories, selection for early arrival is greater on males than on females. In such species either the threshold to indirect variables that trigger or phase the spring endogenous sequence differs between the sexes such that males initiate migration earlier, or the migration thresholds differ such that those of males are lower. Alternatively, in many species, the same selective pressure has favoured a lower migrant/non-migrant ratio in males than females, and/or males migrate a

shorter distance than females. Any or all of these sexual differences may be found in a given species.

In species in which the young breed when one year old, selection acts similarly on all age groups with respect to date of arrival at the breeding range and hence on date of initiation of spring migration. In species in which breeding does not occur until two or more years of age, selection on the timing of spring migration is not as intense on immatures as on adults and the only advantage of return to the breeding range by the young derives from the advantage of exploratory migration and habitat assessment appropriate to a breeding home range (p. 184). In many such species, therefore, the immature birds either remain in the winter range throughout their first summer (e.g. the gannet, *Sula bassana*, in the North Atlantic—Thomson 1974), return only part of the way towards the breeding range (e.g. white stork, *Ciconia ciconia*—p. 190), or return to the breeding range much later than the adults (e.g. the immatures and sub-adults of the manx shearwater, *Puffinus puffinus*, in the North Atlantic—Fig. 33.26).

The migration thresholds to migration-cost variables shown by birds during the spring migration have already been described (p. 632) and in general, except for the different thresholds to temperature and wind direction, these thresholds seem likely to be similar in form, though not necessarily in level, to those involved in the autumn migration. The pattern of variation in migration threshold in relation to transient home ranges (Fig. 27.16) also seems likely to be similar. Delays due to unfavourable migration-cost variables are particularly conspicuous in birds that land in oases during passage across deserts, such as the Sahara (Erard and Larigauderie 1972). On the whole, however, variation in migration thresholds while in spring-transient home ranges has received even less attention than in autumn-transient home ranges. There is an indication, however, that variation in migration threshold with variation in energy reserves follows the same pattern as in autumn. The reed warbler, *Acrocephalus scirpaceus*, sedge warbler, *A. schoenobaenus*, garden warbler, *Sylvia borin*, whitethroat, *S. communis*, and yellow wagtail, *Motacilla flava*, for example, arrive in Britain carrying the same or slightly more fat than when they left Morocco (Ash 1969). Although other interpretations are possible, it seems likely that these birds, by virtue of the occupation of several transient home ranges en route, maintain a roughly constant energy reserve throughout this phase of the journey, each unit of the spring migration being initiated after the deposition of an appropriate amount of fat.

Significantly, however, these birds, in common with other trans-Saharan migrants, both in spring and in autumn, seem to lay down more fat immediately before crossing the Sahara than before initiating any other migration unit. Thus in birds such as wheatears, *Oenanthe oenanthe*, yellow wagtails, *Motacilla flava*, and sub-alpine warblers, *Sylvia cantillans*, in Nigeria in spring, there is an increase in weight by up to 30–40 per cent before initiating migration across the Sahara (Ward 1963). These and the other birds mentioned in the previous paragraph, however, have lost this weight by the time they reach Morocco (Ash 1969) and, as shown, do not regain it during the remainder of the spring migration. That this is not simply due to the lack of opportunity to replace fat used up during the long trans-Saharan flight is implied by the observation that many of these birds have already stopped to feed at other sites just to the south of Defilia, the Moroccan trapping site. More data are needed to check this point but the implications are important to the general thesis being developed here. It appears that the birds lay down more fat before initiating trans-Saharan flight than before initiating migration from any other transient home range (see also p. 694). This physiological activity could be encoded into the circannual endogenous programme. Alternatively, terrain being a migration-cost variable that contributes to the composite migration threshold (p. 633), perception of the desert by the birds during exploratory migration could well raise the threshold to other habitat and migration-cost variables, and time. The result would be that the birds stay longer in the transient home range at the desert edge (where such an edge exists) than at any other transient home range and lay down more fat. Eventually, however, the migration threshold lowers sufficiently for trans-Saharan migration to be initiated. Clearly the development of the Sahara desert on a bird's migration route will have imposed selection on the level and variation of the migration threshold for that species. Whether or not such selection has also led to a special phase of fat deposition as part of an endogenous programme remains to be determined.

One type of analysis that has been performed for a variety of species in both the Old and New Worlds and that provides a further clue as to variation in migration threshold is that of the production of isochronal lines showing the rate of northward

spread of the species (e.g. Fig. 27.18). If these lines are compared with the northward shift of an appropriate isotherm, a gross similarity is often found. Such coincidence may, of course, be purely fortuitous, but it seems not unreasonable to assume that the correlation is valid even though it is unlikely to indicate an actual migration threshold. In the case of the northward spread of the barn swallow, *Hirundo rustica*, through Europe, there is a broad correlation between the isochrone and the 9°C isotherm (Fig. 27.18). Date of passage of this isotherm through an area seems likely in a very general way to indicate the approximate date that insects become sufficiently abundant for swallow survival to become possible in that area. It is to be expected, therefore, that selection has favoured migration thresholds to habitat and migration-cost variables that permit the swallows to predict (p. 347) when such conditions have arrived at the next transient home range on the migration route. It is likely, however, indeed it is certain, that except by chance none of the thresholds to temperature, either as a habitat or migration-cost variable, has the value of 9°C.

To conclude this section it should be pointed out that although description of annual change in migration threshold has been strongly biased toward the terminology of temperate land-, shore-, and freshwater-birds, the same principles are applicable, with little or no modification, to sea-birds and tropical species. Terminology has also been biased in favour of temperate birds that migrate within a single hemisphere. It is, for instance, semantically incorrect to refer to the 'winter' range of a trans-equatorial migrant. Despite these descriptive shortcuts it seems likely that the adaptive seasonal pattern of variation in migration threshold presented here is applicable, at least in broad terms, to birds in general.

27.6 Degree of dispersal and convergence

The model applied throughout this book concerning the degree of dispersal and convergence as a component of migration within a familiar area is that of selection for an optimum degree of overlap of home ranges of optimum size occupied by a group of optimum size and composition. As these optima change with the season and with the range of

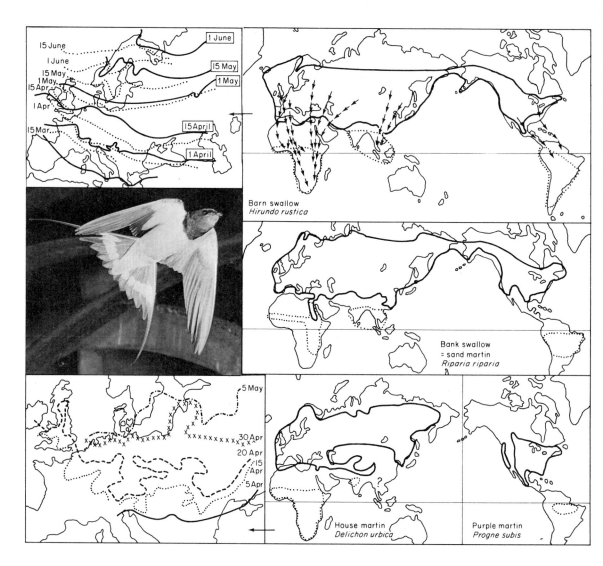

Fig. 27.18   Seasonal return migrations of swallows, martins, and swifts

◯ breeding range; ⋯⋯ non-breeding range;
→ → some known or suspected migrations.

Swallows and martins (Hirundinidae) are slim, streamlined birds with small feet, long-pointed wings, and short bills with very wide gapes. World-wide there are about 75 species. Swifts (Apodidae), although morphologically different, are similar in general appearance and behaviour to swallows and martins. Swifts, however, tend to have slimmer scythe-like wings and shorter tails. World-wide there are about 79 species. Both families feed on the insects of the aerial plankton, though different species feed at different heights and some, such as the barn swallow, take insects, including larvae, from plants or even the ground.

The house martin shows a marked change of behaviour between summer and 'winter', feeding at relatively low altitudes (above ground level) in its nesting range but feeding at high altitudes in its 'winter' non-breeding range.

The common swift of Europe does not normally mature during its first year and many yearlings remain south of the Sahara while the adults are breeding in north temperate regions.

The maps in the left-hand columns present isochronal lines for the species indicated and in the case of the barn swallow the isochrones (⋯⋯⋯⋯) are compared with the 9°C isotherm (——).

[Compiled from diagrams and descriptions in Voous (1960), Peterson (1961), Dorst (1962), Robbins et al. (1966), Orr (1970a), Moreau (1972), and Medway (1973). Photo of barn swallow by Eric Hosking]

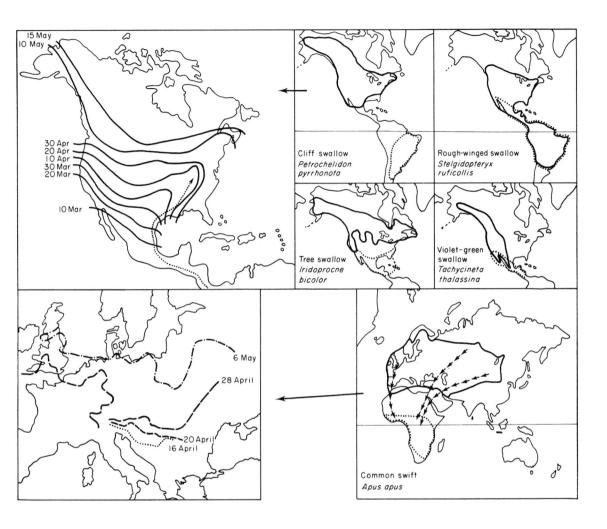

Cliff swallow
Petrochelidon
pyrrhonota

Rough-winged swallow
Stelgidopteryx
ruficollis

Tree swallow
Iridoprocne
bicolor

Violet-green
swallow
Tachycineta
thalassina

Common swift
Apus apus

habitat-types occupied, so a seasonal return migration is characterised by a certain degree of dispersal and convergence. In terms of other components of migration, selection on the degree of dispersal and convergence acts partly through selection on the direction ratio, partly through selection on the frequency distribution of migration distance, and partly through selection on the components of the migration threshold at different times of year. In any

one deme or species, selection may act through any one, two, or all three of these components.

Over a wide range of resource abundance and deme density the situation is likely to be free, all individuals gaining the same potential advantage with respect to distribution (p. 402). In terms of seasonal return migration, short-distance migrants must therefore experience a low habitat suitability as well as a low migration cost, whereas long-distance migrants experience a high migration cost but a high habitat suitability. In a free situation, deme density makes a major contribution to habitat suitability and hence also to the migration threshold. The stable equilibrium is achieved when the intra-deme frequency distribution of migration thresholds, migration distance, and migration direction (= direction ratio) in such that all members of the deme experience the same reproductive success.

Beyond a certain point in the interaction of resource abundance and deme density a despotic situation prevails and the reproductive success experienced by an individual or territorial unit becomes a function of its resource-holding power (p. 404). In terms of seasonal return migration, individuals with the highest resource-holding power migrate in the direction and for the distance (perhaps most often a minimum distance) that maximises reproductive success. Individuals with a lower resource-holding power migrate a shorter or (most often) longer distance, whichever offers the greater advantage, but still experience a lower reproductive success than those individuals with a greater resource-holding power (the situation is discussed in more detail for stream-living fish— p. 802). The influence of resource-holding power on the pattern of seasonal return migration has been indicated for the European robin, *Erithacus rubecula* (p. 645).

As far as I am aware, data that could test this model are not available for the seasonal return migration of any animal. This seems an appropriate place, however, to present data that do in effect permit some evaluation of the model for a bird, albeit in relation to daily rather than seasonal return migration. These data were collected by Hamilton *et al.* (1967) for the starling, *Sturnus vulgaris*, near San Francisco, California. In effect, the experiment was designed to test a model that in all but name was the same as the free-situation model presented in this book.

During winter and spring, starlings form communal night-roosts consisting of many thousands of individuals. Each morning, a dispersal migration is performed during which some individuals remain within a kilometre or so of the roost while other individuals fly to feeding areas up to 100 km away. In the evening, a convergent return migration is performed. The direction ratio of this migration is essentially 25:25:25:25, though it may be modified considerably by the local distribution of resources.

Individuals that migrate the shortest distance during this return migration maximise their foraging time, minimise their energy expenditure, but suffer greater intra-specific competition for food. Individuals that migrate the longest distance experience the opposite conditions. If the situation is free, as postulated by Hamilton *et al.* (1967), all individuals should experience the same energy balance unless other factors, such as predation, impose additional distance-correlated pressures. According to the model developed here (p. 635), if the situation is free the behaviour involved should result from selection on the frequency distribution of exploratory migration thresholds combined with ideal calculated migration thresholds (p. 91).

Hamilton *et al.* (1967) collected data for a variety of components of the daily return migration of the San Francisco starlings in March, the night-roost in question being among thick marsh vegetation. Assessments were made that were quantitative wherever possible of: potential foraging time; foraging success; and daily total energy expenditure (in kcal); for six different cohorts of birds. The cohorts were, respectively, birds with a one-way migration distance of 0–16, 16–32, 32–48, 48–64, 64–80, and 80–96 km. The data collected permitted the calculation of an index of the daily energy balance for individuals in each cohort. The values obtained were 3·10, 2·86, 2·86, 2·82, 2·66, and 2·52 for the 0–16, 16–32, 32–48, 48–64, 64–80, and 80–96 km cohorts respectively. There is a correlation between energy balance and distance of -0.9568 ($n = 6$; $P < 0.01$) on the basis of these figures, a result that tends to indicate that on the basis of energy balance alone, those birds that migrate a shorter distance gain a greater advantage than those that migrate further. If this conclusion is valid, it indicates that the situation is despotic rather than free. Hamilton *et al.*, however, consider that the calculated values for the energy balances are sufficiently similar and the direction of any experimental error is such that on the whole the results tend to support the existence of what, in this book, is termed a free situation. Whether the results in fact indicate that the communal night-roost of starlings gives rise during the day to a free or a despotic situation, this San Francisco study provides some insight into the way that selection may act on the intra-deme frequency distribution of migration thresholds, distances, and directions, and any other components that influence the degree of dispersal and convergence during seasonal return migration.

Fig. 27.19 Part of a flock of starlings, *Sturnus vulgaris*, at a night roost in winter

In winter, most major cities in temperate regions provide sites suitable for communal roosting by many thousands of starlings. Woods and marshy areas may also be used. Each day the starlings disperse from the roost, migrating initially in large flocks, some individuals travelling up to 100 km away from the night-roost. Towards dusk, the convergent return migration takes place. Analysis of the return migration and day-time distribution has failed to demonstrate unequivocally whether the situation is free or despotic.

[*Photo by S.C. Porter (courtesy of Bruce Coleman Ltd)*]

27.7 Migration direction and direction ratio

It was argued at the beginning of this chapter (p. 605) that birds, because they are mobile and have the potential to establish a large familiar area and because, unlike bats, they have adopted a strategy of year-round feeding activity, have been subjected from an early stage in the evolution of flapping flight to climatic selection on migration direction. In the absence of climatic selection, acting along either latitudinal or altitudinal gradients, selection tends to favour a seasonal return migration with a direction ratio that approximates to 25:25:25:25. This is because, on average, non-climatic variables, such as the distribution of roosting sites and, at least over relatively short distances of a few tens of kilometres or so, feeding habitats tend to be homogeneously distributed. Where a local bias does exist for a given habitat-type to be found in a given direction, either the normal exploratory–calculated migration process or social communication permits adoption by the deme of the optimum direction ratio. In this case, the exploitation of resources involves a local bias to the direction ratio that is characterised by considerable inter-deme variation and by the absence of inter-deme differences in genotype, at least with respect to direction ratio. Such a situation has already been encountered in terrestrial mammals and bats.

As already suggested (p. 644), birds are likely to possess a hierarchy of exploratory migration thresholds. In general in such hierarchies (p. 229), the lowest threshold relates to the shortest distance migration and the highest threshold relates to the longest distance migration. When environmental variables exceed a given threshold, only the appropriate migration is initiated. Relatively short-distance exploratory migrations of birds, such as those initiated in the immediate post-fledging period in land-birds, should for the reasons presented have an inherited direction ratio of 25:25:25:25. In other words, the direction taken up by the individual is endogenously determined but the direction taken up differs from individual to individual to produce the appropriate direction ratio.

Beyond a certain migration distance, however, perhaps, at a guess, in the region of 100–200 km for latitudinal migrants but less for altitudinal migrants, it seems likely that the gradient of climatic selection will show through the 'background noise' of local climatic and other effects. Birds initiating a mi-

gration that during the evolution of the thresholds associated with that migration has invariably led to cross-country displacement of 100–200 km or so elicit selection for a peak direction parallel to the gradient of increasing habitat suitability. The degree of bias to the direction ratio is a function of the steepness of the gradient and selection on dispersal and convergence (p. 649). The steeper the gradient and the less the selection for dispersal, the greater is the bias. Even with a bias, the direction of exploratory migration taken up by the young bird is determined endogenously, but now (by definition) the proportion of individuals having the same endogenously determined direction is greater for some directions and less for others then when there is no bias.

When optimum migration distance is short, selection for optimum degree of dispersal must act mainly on direction ratio and invariably adaptation seems likely to tend toward a direction ratio of 25:25:25:25 to minimise migration distance. When migration distance is longer, deme size remaining constant, a similar degree of dispersal can be obtained by selection on the frequency distribution of migration distance. Even if all individuals move in the same direction, the same spacing between individuals can be obtained given a long enough migration by some individuals. Selection favours such a situation if the advantage gained from long-distance migration in the peak direction is greater than that gained from short-distance migration in some other direction.

Application of the direction-ratio model developed on p. 85 to the seasonal return migration of birds can be seen therefore to generate the following migration pattern. If birds migrate within a familiar area, selection acts primarily on the direction ratio of the exploratory migration of young birds. A hierarchy of migration thresholds should evolve, each one relevant to a migration of a particular distance. Short-distance (app. < 200 km) exploratory migrations (where distance is the straight line distance between starting point and end-point) should have an inherited direction ratio of 25:25:25:25. Longer-distance migrations should have an inherited bias to the direction ratio in the direction of any gradient of increasing habitat suitability. The degree of bias is a function of the steepness of this gradient and of the distance of the migration. This model can now be evaluated against what is known concerning the direction and direction-ratio components of the seasonal return migration of birds.

Most studies of bird migration have concentrated on long-distance migrants and, as stressed by Bellrose (1972), there is now a real need for a study of the migration components of the less spectacular migratory species. This is particularly necessary for an evaluation of models of direction ratio. The limited information available, however, suggest that short-distance seasonal return migrations are indeed characterised by a direction ratio that approaches 25:25:25:25. The main problem in such studies is not only to establish the distance and direction ratio of the species concerned but also to establish that the migration is a return migration. The type-H (= total home range to total home range) removal migrations (p. 321) of birds such as the blue tit, *Parus caeruleus*, great tit, *P. major*, and blackbird, *Turdus*

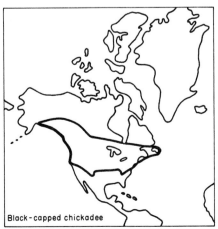

Fig. 27.20 Two members of the genus *Parus*

The photograph shows the great tit, *P. major*, which occurs throughout Europe and much of Asia. The map shows the distribution of the black-capped chickadee, *P. atricapillus*, in North America. The willow tit, *P. montanus*, of Europe and Asia is closely related to the chickadee and may be conspecific.

In common with most members of the genus, the great tit and black-capped chickadee are mainly 'resident' but in winter they form flocks (often mixed-species) and then may be nomadic within a relatively large home range. Both species feed on insect eggs, larvae, and pupae as well as on spiders, etc. and some seeds and both species in late summer and autumn when food is plentiful may hide it under bark or amongst lichen for use in winter. Such behaviour is only likely to evolve in an animal that occupies the same home range all year (see p. 209).

The great tit in northeast Europe and northern Asia and the black-capped chickadee in northern latitudes are both considered to be partial seasonal return migrants. The great tit may migrate several hundred kilometres and seems to migrate along a NE–SW axis. The chickadee, however, seems to migrate only about 100 km and to have a direction ratio that approaches 25:25:25:25.

[*Compiled from Lack (1966) and Lawrence (1973). Photo by Eric Hosking*]

merula, in Britain which show little if any seasonal return migration also seem, as would be expected (p. 86), to be characterised by a direction ratio that approaches 25:25:25:25. The selective pressures acting on the direction ratio of removal and return migration (Chapter 12), however, are different, a direction ratio of 25:25:25:25 always being favoured for the former, independently of migration distance, unless there is also selection for a major change of peak direction or unless the environment is linear (p. 783).

In at least one species, however, the necessary information has been obtained, albeit incompletely. In the black-capped chickadee, *Parus atricapillus*, of America (Fig. 27.20), migration distance seems to be no greater than about 200 km and enough cases of individuals captured in one area, then some distance away, then back at the original site, have been reported to suggest that the movement is a return migration. Most (up to 85 per cent) of the migrants are young birds (Bagg 1969) and for these the migration must surely be exploratory. The seasonality of the migration, however, does not seem to have been firmly established for, although the autumn migration, beginning as early as August, is fairly consistent, evidence relating to the spring timing of the return does not seem to be strong. Nor are estimates available for degree of return. Nevertheless, in this species we have the only situation that seems so far to have been described that approximates to a short-distance seasonal return migration. As far as direction ratio is concerned, Bagg (1969) clearly anticipates a peak direction on the N–S axis but the data available seem to indicate a direction ratio that approaches 25:25:25:25 as generated by the model. Conceivably the blue jay, *Cyanocitta cristata*, of North America behaves similarly (Bellrose 1972). Both the jay and chickadee have long been considered to be 'residents' and evidence that they may be partial, short-distance, seasonal return migrants has only recently been obtained.

The composite exploratory migration threshold of the black-capped chickadee evidently receives a major contribution from the size of the seed crop of certain conifers and hardwood trees upon which the species feeds during the winter (Bagg 1969). It seems, however, that a sufficiently wide variety of seed species are taken for a longer-distance migration not to have been favoured as it has been for those species that are specialised to feed on a much more limited range of tree seed species. It is these latter species, however, that provide the best examples of the next level of selection on migration distance and direction.

The pattern of temporal and spatial variation in the abundance of a tree seed crop has already been described (p. 208). Briefly, tree seeds are synchronously abundant at any one time over a relatively large area, but areas of abundance are separated by areas of scarcity. Also the areas of abundance change from year to year. The situation is of the type that inevitably produces a hierarchy of migration thresholds. Low tree seed abundance in the natal or nuptial home range could simply reflect extremely local shortage and should ideally first trigger short-distance exploratory migrations. However, after a period of short-distance exploratory migration, if the scarcity applies to a wider area, then optimum behaviour is likely to be a long-distance exploratory migration of the length necessary to leave the average-sized patch of scarcity for that tree-species. Such migration normally needs to be of several hundred kilometres. Beyond the lowest migration threshold, therefore, the exploratory migration initiated should be short-distance. Beyond the next or highest migration threshold, depending on how many form the hierarchy, a long-distance exploratory migration should be initiated. The former migration should have a direction ratio that approximates to 25:25:25:25. The latter, however, is likely to have been exposed to climatic selection and to have a direction ratio with some degree of bias.

The migrations of the crossbill, *Loxia curvirostra*, have been introduced elsewhere (p. 209) and are highly relevant here. The composite exploratory migration threshold of this species seems to receive a major contribution from the abundance of the spruce seed crop and from deme density. There is good evidence for at least two migration thresholds in the hierarchy, and as required by the model the migrations initiated every year in response to the lowest migration threshold are short distance, seem likely to have a direction ratio approximating to 25:25:25:25, and rarely lead to movement beyond the range occupied by the species during the previous 12 months. The migrations initiated on the less frequent occasions that the higher migration threshold is exceeded have a much longer distance component and accordingly have a strong bias to the direction ratio. Crossbills inhabiting Russian forests migrate in directions that are chiefly between W and SW, and those from Fenno-Scandia migrate chiefly between W and S (Newton 1972).

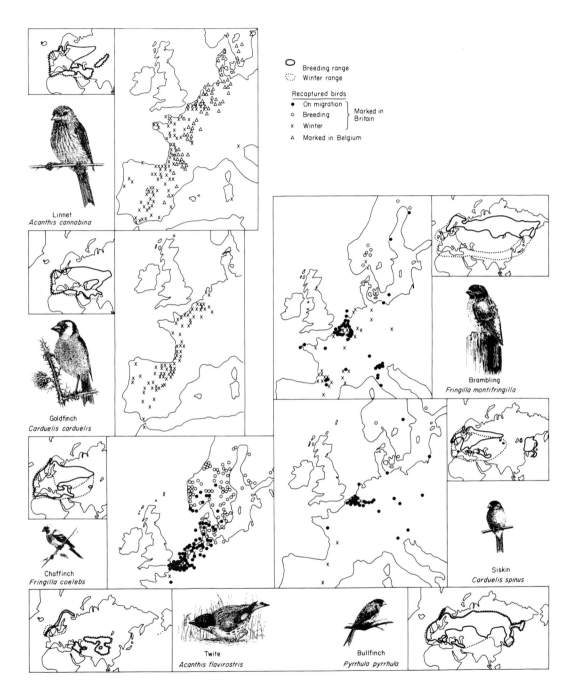

Fig. 27.21   The influence of feeding ecology on direction ratio: the migration of fringillid finches in western Europe

The left-hand column consists of finches that feed on the seeds of herbaceous plants. The right-hand column consists of finches that feed on the seeds of trees. The latitudinal gradient of increasing habitat suitability in winter (from NE to SW) is likely to be steeper for the former species than the latter species (p. 659). According to the direction-ratio model (Chapter 12), therefore, herbaceous seed feeders should show less scatter to their direction ratio than tree-seed feeders. Such a difference is clearly shown by the marking–release/recapture data presented in the diagrams.

[Compiled from Newton (1972) and Peterson et al. (1974)]

Nicely though the crossbill illustrates the point being made here, there is a complication that arises in the Eurasian deme of the species that is not found in other birds, including the North American deme of the crossbill, that in all other respects show a similar hierarchy of migration thresholds. For reasons discussed on p. 209, the Eurasian deme of the crossbill migrates just once each year, in mid-summer, and breeds before initiating a return migration. There is a problem, then, concerning the migration direction of young crossbills born in the southwestern part of the species range to parents that originated in and may return to areas 4000 km to the NE. What happens if the upper migration threshold of these young birds is exceeded? There are a number of possibilities, among which are: (1) the upper exploratory migration threshold is too high to be exceeded in the southwestern part of the range; (2) when the upper migration threshold is exceeded, the young birds initiate a migration which has a SW bias to the direction ratio as initially did that of their parents; (3) the young birds accompany the adults on the latters' return to the NE, inherited bias being superseded by social communication whenever the two conflict (p. 613); (4) direction ratio is influenced by indirect variables such as photoperiod and temperature in such a way that an upper migration threshold exceeded in midsummer when days are short and temperatures are high triggers a migration with a direction ratio biased to the NE, whereas an upper migration threshold exceeded in midsummer when days are long and temperatures are low triggers a migration with a direction ratio biased to the SW. Of these four alternatives the last seems to be the most adaptive strategy in the case of the Eurasian crossbill. Evidence, however, is lacking, though in at least one seasonal return migrant, the chaffinch, *Fringilla coelebs*, the innate directional preference has been found to be a function of day length (p. 674), albeit on an annual, not a latitudinal, basis.

A number of other finch species exist that are adapted to a winter diet consisting primarily of tree seeds. Examples are the siskin, *Carduelis spinus*, redpoll, *Acanthis flammea*, northern demes of the bullfinch, *Pyrrhula pyrrhula*, pine grosbeak, *Pinicola enucleator*, and brambling, *Fringilla montifringilla* (Newton 1972). The migration thresholds of some of these species have already been discussed (p. 635) and in most species the highest migration threshold of a portion of the population is exceeded every year. Some individuals always initiate a long-distance

seasonal return migration with a biased direction ratio, therefore, but the number varies greatly from year to year depending on conditions in the breeding area. Once again, however, although there is a bias to the direction ratio, the degree of bias is still not great. Furthermore, individuals show a high incidence of type-C removal migration as described in Chapter 17 (p. 323).

Finally, in the majority of birds that initiate long-distance seasonal return migration, there is a strong bias to the direction ratio of autumn exploratory migration and hence of seasonal return migration, a bias that in many species approaches 100:0:0:0. If we continue, for the moment, to take examples from European finches, such a direction ratio is characteristic of those species, such as the greenfinch, *Carduelis chloris*, goldfinch, *C. carduelis*, linnet, *Acanthis cannabina*, and chaffinch, *Fringilla coelebs*, that feed primarily on the seeds of herbaceous plants (Newton 1972). Direction ratios typical for this category are illustrated in Fig. 27.21.

Before extending this discussion to include other bird groups it is convenient at this point to consider this variation in relation to the model outlined at the beginning of this section. The increase in bias with increase in migration distance and the existence of a hierarchy of migration thresholds for each species, each threshold appropriate to a given migration distance, has already been considered. Tree- and herbaceous-seed-eating finches, however, respectively show a relatively low and high bias to their direction ratio despite performing migrations with a comparable distance component. According to the model, if migration distance is constant the degree of bias to the direction ratio should be a function of steepness of the gradient of selection. In order to estimate the steepness of this gradient it is necessary to identify the contributory variables.

A number of studies (e.g. West 1960, for the American tree sparrow, *Spizella arborea*; and Newton 1972, for the bullfinch, *Pyrrhula pyrrhula*) have been made that demonstrate that for an animal, such as a temperate bird, for which selection has apparently favoured a strategy of continual feeding throughout winter, the critical factors in the winter ecology are temperature, availability of food, and day length. At low temperatures the bird requires a greater food intake than at higher temperatures due to the increased demand of thermoregulation. Longer days and shorter nights allow more time for food acquisition and give a greater probability of an accumulation of food reserves that is sufficient for the bird to survive until the following morning. Habitat suitability is thus a function of ambient temperature, day length, and food abundance. The steepness of the gradient of selection on the direction of the autumn exploratory migration of the young of European finches is a function of the extent to which the gradients of day length, temperature, and food availability act in the same direction.

In the Northern Hemisphere, the gradient of day length acts on all diurnal species at all longitudes to favour migration to the south. The steepness of this gradient is greatest at intermediate latitudes. The gradient of temperature acts to favour migration in different directions at different longitudes (Fig. 27.22). For example, in west and northwest Europe the gradient of increasing temperature runs more or less from NE to SW. In central, southern, and eastern Europe, however, the gradient runs more or less from N to S. On the basis of these two factors alone, therefore, the steepest gradient of selection occurs in central, southern, and eastern Europe and runs from N to S. The gradient of selection in western and northwestern Europe is likely to be a little less steep and to run from NE to SW or SSW or N to S, depending on the relative importance of temperature and photoperiod. If we now consider the contribution of food abundance to these gradients it seems likely that for most species this factor reinforces the gradient of temperature. This is not the case, however, for most species of tree-seed eaters. For these, an area the size of Europe can be considered a mosaic of suitable and unsuitable habitats, imposing little in the way of a latitudinal selection gradient. A slight gradient may exist due to

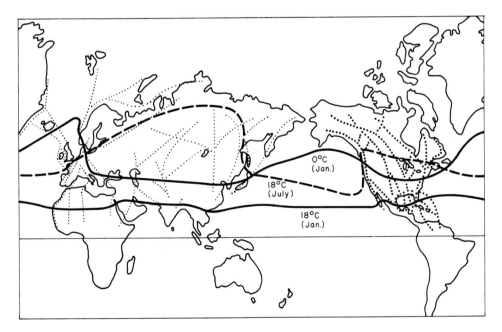

Fig. 27.22 The contribution of winter temperature to selection on the peak migration-axis direction of Northern Hemisphere birds

The dotted lines indicate the major migration axes of Northern Hemisphere migrant birds. The solid lines indicate the positions of the 0°C and 18°C winter (January) isotherms. The dashed line indicates the position of the 18°C isotherm in summer (July).

In critical areas, such as western Europe, in which winter and summer isotherms run in quite different compass directions, migration axes are more or less perpendicular to the winter isotherm. The conclusion that winter temperature is an important factor in selection on peak migration direction is supported elsewhere in Europe, Asia, and America. Because the isotherm angle (Fig. 19.11) is different at different latitudes, long-distance migrants often have a peak direction that is different from that of short-distance migrants. Hence, in western Europe long-distance migrants (see Fig. 27.18) have a N–S migration axis whereas short-distance migrants (see Fig. 27.21) have a NE–SW migration axis.

[*Migration axes compiled from Dorst (1962). Isotherms from Fullard (1959)*]

the time interval between major seed-production bouts by individual trees being shorter in warm temperate than cold temperate latitudes (p. 208). The direction of this gradient is likely to be some compromise between day length and temperature. Its contribution to the steepness of the gradient of selection is likely to be much less, however, than the gradient of abundance and availability of other foods, such as insects, other invertebrates, and herbaceous seeds, especially as the availability of these foods may be a function of other factors, such as snow cover. Consequently, the composite gradient (day length, temperature, and food abundance) of selection on the direction of autumn exploratory migration is likely to be less for tree-seed eaters than herbaceous-seed eaters. The greater bias to the direction ratio of the latter than the former is therefore quite consistent with the model developed here.

Figure 27.22 shows the major migration directions of Eurasian and North American land birds and it is concluded that on the whole these are consistent with the suggestion made here that they have evolved in response to a gradient of selection, the resultant of winter gradients of day length, temperature, and availability of food, that acts upon the direction of autumn exploratory migration. There is no indication, however, that a gradient of selection acting on the direction of spring migration is involved. This perhaps surprising conclusion requires further discussion.

A number of advantages of migration to higher latitudes in spring have been suggested. Lack (1954) has suggested that the increased foraging time that is gained as a result of the longer day length at higher latitudes is a critical selective pressure and Hussell (1972) has demonstrated a clear correlation between clutch size and breeding latitude for a number of Emberizidae (Fig. 27.23). In the Lapland longspur, *Calcarius lapponicus*, the correlation with latitude ceases beyond those latitudes at which there is a 24 hour day and the activity is no longer limited by day length. Surprisingly, however, as far as this thesis is concerned, this does not appear to be the case in the snow bunting, *Plectophenax nivalis*, which continues to increase clutch size even beyond this latitude and for which, therefore, at least one other factor must make an important contribution to clutch-size variation.

It has also been suggested that movement to higher latitudes removes the animal from too high temperatures at lower latitudes and that partly for

this reason and partly as a result of the longer foraging time available the spring migration leads to an improved energy balance. Experiments on the energy balance of birds that perform a spring migration are, however, of dubious usefulness as far as the evaluation of this particular situation is concerned. If a bird has become evolutionarily adapted to low latitudes in winter and high latitudes in summer, then it also seems likely that the physiology of that bird should also have become evolutionarily adapted to these latitudes. It should come as no surprise, therefore, and is of doubtful use in elucidating the selective pressures that led to the bird becoming adapted to an alternation between these latitudes, if it is discovered that the bird's energy balance is optimised by so alternating. The demonstration that resident neotropical finches stand to gain little if any advantage in energy balance by migration to the north in spring (Cox 1961), and that the dickcissel, *Spiza americana*, gains an improved energy balance through both its spring and autumn migrations (Zimmerman 1965), is not particularly surprising and in no way demonstrates that selection for an optimum energy balance as influenced by temperature was an important component of the selection that led to the evolution of residency or migration, respectively. On the other hand, the demonstration by West (1960) that the American tree sparrow, *Spizella arborea*, suffers a less favourable energy balance as a result of its spring migration to higher latitudes can be taken to indicate that selection for an improved energy balance as influenced by temperature and day length was not important in the evolution of the migration. It should be stressed, however, that all of the experimental data were obtained for caged birds, albeit under 'natural' conditions of day length and temperature. These birds were fed an artificial diet by the experimenter and were not, therefore, competing for food with local birds in a familiar area. Whatever the advantage of spring migration, it is likely, unless the critical factor is predation, to act through an improved energy balance. These experiments of West (1960) serve to show, however, that at least in the tree sparrow it is not the effect of day length and temperature on the energy balance that is the most critical.

Another observation that argues against the general application of a simple day-length hypothesis to the spring migration of birds concerns the migration of owls (Strigidae) and nightjars (Caprimulgidae). Many Holarctic owls show either zero or short-distance seasonal shifts in home range but a few perform longer-distance seasonal return migrations (Fig. 27.24). Similar migrations are performed by nightjars and poor-wills (Fig. 27.25). Where such migrations occur they follow the usual pattern of a higher latitude breeding home range and a lower latitude winter home range. Because these birds are nocturnal, their movements are in the opposite directions to those that would be expected to be favoured if day length made the sole or major contribution to the gradient of selection on spring and autumn migration. In the case of autumn migration, the implication is that the combined contribution of temperature and food abundance

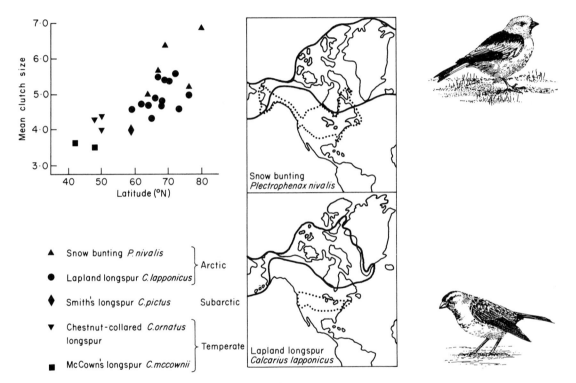

Fig. 27.23 Clutch size in relation to latitude in five species of Emberezidae

The maps show the seasonal distribution of two species of emberezids, the snow bunting and the Lapland longspur.

⬭ breeding distribution; ·::::· winter distribution.

Both species breed on the tundra and both are circumpolar, though only the distributions in North America are illustrated. While on the tundra, more than 90 per cent of the food consists of adult insects and more than 80 per cent consists of chironomids. Both species produce only a single brood.

North of about latitude 70°N, the activity of these birds is not restricted by day length. South of latitude 70°N, the clutch-size of both species correlates with latitude, suggesting that perhaps clutch-size is a function of the time available to collect food. This correlation is further

reinforced when the mean clutch size for three other longspurs, *Calcarius* spp. (that breed at sub-arctic and temperate latitudes) are included. It should be noted, however, that these species produce more than one clutch during the summer.

The suggestion that day length may be a critical factor in the determination of clutch size receives further support from the absence of correlation between clutch size and latitude in the Lapland longspur at latitudes north of about 70°N. In the snow bunting, however, the correlation seems to persist even at these latitudes. In this case some factor other than day length must be involved in the general correlation for at least this species. Until this factor can be identified (the interaction of deme density and food abundance seems the most likely candidate) and quantified, the support provided for the day-length hypothesis by this analysis cannot be evaluated.

[*Graph modified from Hussell (1972). Maps compiled from Peterson (1961) and Robins* et al. (*1966*)]

outweighs the negative contribution of day length. Even so, the result must be a composite gradient of selection that is less steep than similar gradients for diurnal species. Indeed owls as a group do seem to show a low bias to their direction ratio compared with diurnal birds of prey, a result that is consistent with the model presented for autumn migration. As far as spring migration is concerned, however, removal of day length as a contributory selective pressure means the removal of the only variable for which there is experimental support for its validity as a selective pressure on migration direction.

Fig. 27.24 Seasonal return migration of four species of owls

⬭ breeding range;

:⁚:⁚: non-breeding range;

→ → some known or suspected migrations.

Owls as a group tend to consist of birds that occupy the same season's home range all year. Where seasonal return migration does occur, however, even in nocturnal species, it is toward lower latitudes for the winter and toward higher latitudes for the summer. This suggests that day length makes less contribution to habitat suitability than other factors such as temperature and the availability of food.

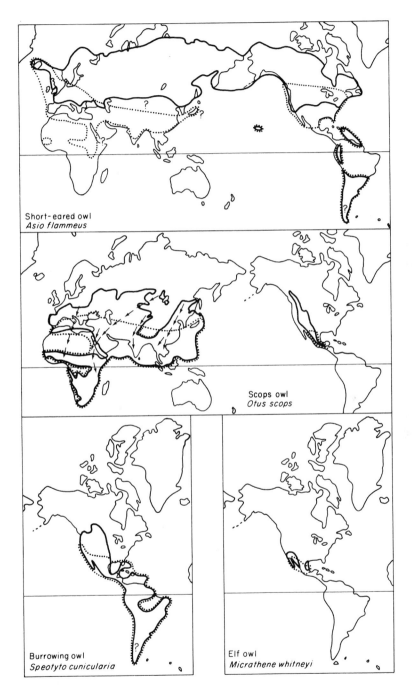

Short-eared owl
Asio flammeus

Scops owl
Otus scops

Burrowing owl
Speotyto cunicularia

Elf owl
Micrathene whitneyi

Of the species represented, the short-eared owl inhabits open country and feeds predominantly on voles, shrews, mice, rats, and small birds. The distribution of all members of the scops owl species-group (or superspecies) is illustrated. Most scops owls live in open woodland or the woodland of riversides or mountains. The diet consists mainly of insects but some small mammals and birds are included. The elf owl also feeds primarily on insects and is found in a variety of relatively open habitats. The burrowing owl roosts and nests in holes in the ground in open, tree-less grasslands.

[*Compiled from maps and descriptions in Voous (1960), Peterson (1961), Robbins et al. (1966), Moreau (1972), Burton (1973), and Peterson et al. (1974)*]

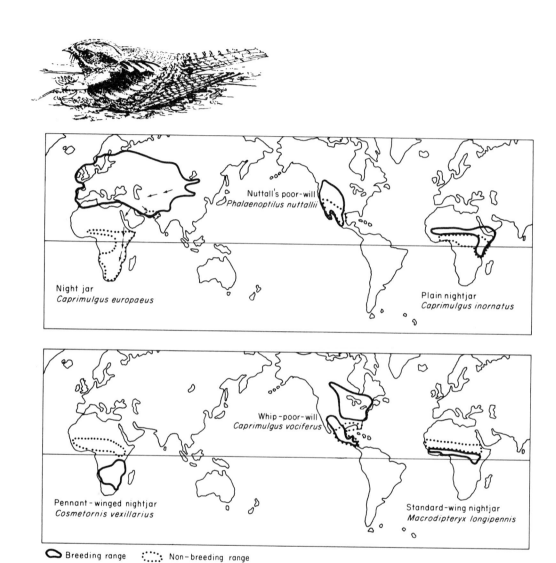

Night jar
Caprimulgus europaeus

Nuttall's poor-will
Phalaenoptilus nuttallii

Plain nightjar
Caprimulgus inornatus

Whip-poor-will
Caprimulgus vociferus

Pennant-winged nightjar
Cosmetornis vexillarius

Standard-wing nightjar
Macrodipteryx longipennis

◯ Breeding range   ⋯⋯ Non-breeding range

Fig. 27.25 Seasonal return migrations of six species of caprimulgids

The Caprimulgidae form a nearly world-wide family of nocturnal land-birds with ample tails, large eyes, short but broad bills, huge gapes, and very short legs. During the day they roost on a limb or on the ground, camouflaged by the 'dead-leaf' pattern of their plumage. The food consists primarily of nocturnal insects.

The northern temperate species illustrated (the poor-wills and the Eurasian nightjar) breed in temperate regions and migrate to warm-temperate or tropical regions for the northern winter in the same way as diurnal species, thus suggesting for the latter a relatively small contribution of day length to the gradient of habitat suitability that acts as a selective pressure on migration direction. Nuttall's poor-will (top centre) is perhaps unique among birds in that in the northern part of its winter range (in California and Arizona) it enters hibernation, behaviour that can perhaps in part be attributed to adaptation to nocturnal predation on insects.

The tropical species show a variety of patterns of seasonal incidence of breeding and migration. The plain nightjar (top right) breeds only during the wet season (July–September) but in the most arid sector of its yearly range. During the dry season (November–April) it moves to moister regions. The standard-wing nightjar (bottom right), which occupies approximately the same yearly range as the previous species, migrates in the same direction at the same time but breeds in the dry season (February to April) but in the most moist sector of its yearly range. The pennant-wing nightjar (bottom left) breeds (September–December) at the beginning of the wet season during the peak abundance of insects, especially termites. Once the brood is reared, the nightjars follow the rainy season northwards, thus coinciding in each place with the northward shift of peak insect abundance. The northernmost non-breeding range (illustrated) is occupied from February to August, during which moulting occurs.

[Compiled from Voous (1960), Dorst (1962) (after Chapin), and Moreau (1972) for Old World species and from Peterson (1961) and Robbins et al. (1966) for New World species]

One final variable may be mentioned that could serve as a selective pressure on spring migration direction. An analysis of the geographic size variation of non-migratory North American birds by James (1970) revealed a tendency for a general agreement with Bergmann's law, larger birds being found at lower temperatures. Humidity also contributed to the variation in size. It has been suggested for other animals, such as butterflies (Fig. 19.11) and zooplankton (Fig. 20.6) that the effect of decreased temperature on body size and fecundity has been a major selective pressure on the evolution of migration direction. Possibly a similar explanation

could be applied to birds, the advantage of migration to higher latitudes in spring being that the young are reared at lower temperatures and thus achieve a larger body size and hence a greater potential reproductive success. However, as the birds studied by James (1970) were non-migratory and as Dorst (1962) has pointed out that where Bergmann's law is applicable to migratory birds it is usually applicable during the winter and not during the breeding season, this apparent selective pressure seems likely either to be unreal or to make little if any contribution to selection on migration direction.

Figure 27.22 indicates no deviation of migration axes from those that could be attributed solely to the winter gradient of habitat suitability acting as a selective pressure on the direction of autumn migration. This is consistent with the existence of clear experimental support for such a gradient whereas neither observation nor experimentation has so far provided convincing support for a latitudinal gradient of habitat suitability in spring or summer. All that can be concluded for the spring migration direction is that it is the reverse of the autumn direction. Perhaps it is in this sense alone that the advantage of the direction of the spring migration should be sought. As soon as the winter gradient of habitat suitability wanes in spring and then disappears, the gradient of deme density along the autumn migration axis automatically creates a gradient of increasing habitat suitability in the opposite direction to that prevailing during the winter (inter-specific as well as intra-specific competition must be included in this model). Even in the total absence of contributions to habitat suitability from day length and temperature, therefore, deme density in combination with the winter gradient of selection imposes the annual oscillation of relative suitability of winter and summer habitat-types necessary for the generation of seasonal return migration (p. 62). Furthermore, a model that includes a spring gradient based solely or primarily on deme density generates a migration axis that is entirely a function of the gradient of selection acting on the direction of autumn migration. Consequently a pattern of migration is generated that is nearer to that observed than is the case if spring migration direction is considered to be adaptive to summer gradients of day length and temperature.

Before leaving the question of selection on the migration axis of land birds one further point is perhaps worth making. Because the gradients of habitat suitability for most species in western Europe run from NE to SW and because to the SW of southern Britain the sea offers habitats of only zero suitability, there is no gradient of selection acting on the birds breeding in this region. Migration to the SE would take the birds at right angles to the gradient of increasing habitat suitability and thus offers no advantage. Migration in a direction between SE and about SSW still offers a lower rate of increase in habitat suitability per unit distance than experienced by birds living on the European mainland to the SE of Britain that can migrate to the SW. It might be expected, therefore, that because of this lowered advantage per unit distance, British birds

would tend to have a lower migrant/non-migrant ratio than might otherwise be expected. Because of the general decrease in migrant/non-migrant ratio that exists in any case along the NE to SW axis across western Europe (p. 637), this possibility is difficult to evaluate.

Discussion has so far centred predominantly on the migration direction of land-birds. Extension of the model. As already suggested (p. 606), autumn or birds, however, continues to support the validity of the model. As already suggested (p. 000), autumn or post-fledging gradients are likely to be much less pronounced for sea-birds than land-birds, especially

as many sea-birds feed at night. It is in accordance with the model, therefore, that sea-birds in general show a lower bias (i.e. more scatter) to the direction ratio of exploratory migration than land-birds. Shore birds are, of course, a special case, for the linear nature of the habitat to which they are adapted inevitably favours a strong bias to the direction ratio. At temperate latitudes this bias may approach $100:0:0:0$, but in warm temperate and sub-tropical latitudes the bias may be nearer $50:0:0:50$ (Figs. 27.27 and 27.28).

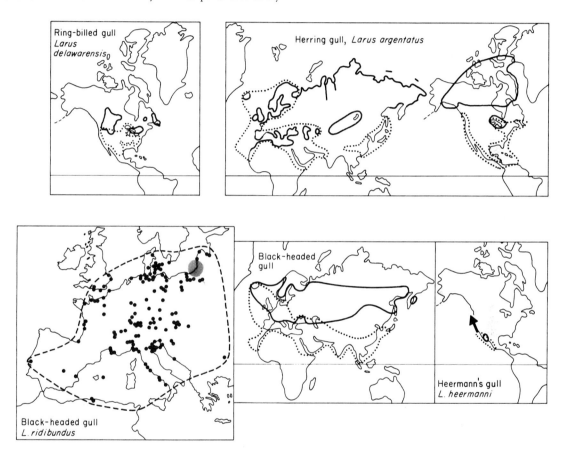

Fig. 27.26 Patterns of migration of gulls, *Larus* spp.

◯ breeding range; ⠐⠂⠄ winter range; ➜ summer migration.

Gulls are long-winged swimming birds that seldom dive and that are omnivorous, feeding on a variety of marine plants and animals as well as refuse and carrion. The group (Larinae) is nearly world-wide, favouring coasts, and consists of about 43 species.

The four species represented show the major features of gull migration. In North America, both the ring-billed gull and the herring gull manifest a seasonal return migration that is along a more or less N–S migration axis, though directed primarily toward and along coasts. The herring gull manifests a similar migration through most of Asia, but in Europe it is a partial migrant, with a declining migrant/non-migrant ratio along a NE–SW axis.

The black-headed gull occurs inland far more than the herring gull. It too is a partial migrant in Europe. Marking experiments at Rybatschi (large stippled circle) near the edge of the winter range has demonstrated a peak autumn direction to the SW (● = recaptured bird) but there is a considerable scatter to the direction ratio. Furthermore, some individuals (young exploratory migrants?) are highly nomadic during the winter months, travelling through large sections of the area enclosed by the dashed line.

Heermann's gull illustrates another type of return migration that achieves some expression in a number of species. In summer and early autumn these gulls migrate toward the higher latitude side of their breeding range, returning to lower latitudes in late autumn.

[Compiled from maps and descriptions in Voous (1961), Peterson (1961), Dorst (1962) (after Schüz and Weigold), Vaurie (1965), Robbins et al. (1966), and Peterson et al. (1974). Photo by J. Twiselton]

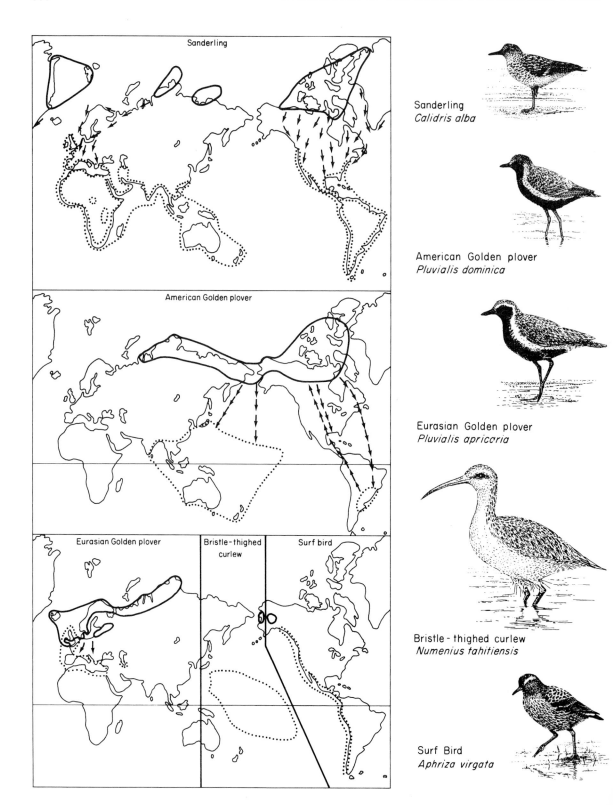

Sanderling
Calidris alba

American Golden plover
Pluvialis dominica

Eurasian Golden plover
Pluvialis apricaria

Bristle-thighed curlew
Numenius tahitiensis

Surf Bird
Aphriza virgata

Fig. 27.27 The migrations of some shore-birds and waders

⬭ breeding range; ⋰⋰ non-breeding range.

When birds are adapted to a linear habitat-type, such as the sea-shore, selection for optimum dispersal has less effect on the direction ratio and acts mainly on migration distance. Shore-birds and waders, particularly those adapted to continental coasts, thus manifest a wide range of migration distances.

[*Drawn from figures and descriptions in Voous (1960), Peterson (1961), Dorst (1962), Vaurie (1965), Robbins et al. (1966), Orr (1970a) and Moreau (1972)*]

Fig. 27.28 The return migration of the brown pelican, *Pelecanus occidentalis*

Pelicans are huge water-birds with long flat bills and great throat pouches that feed mainly on fish and crustaceans. World-wide there are about 6 species, being found in North and South America, Africa, southeast Europe, southern Asia, Indonesia, and Australia.

The brown pelican lives in salt bays and along oceanic coasts and nests colonially on the ground on islands. Its breeding range around the coasts of North and South America is shown by the black line. Migrants are also found in the area indicated by the dotted line. Such migrants seem certain to be exploratory migrants that eventually return to the breeding range. Neither the periodicity (at the individual level) nor the direction ratio seems to be known, but taking the range as a whole the latter seems likely to be 50:0:0:50.

[*Drawn from figures and descriptions in Peterson (1961), Robbins et al. (1966), and Orr (1970a). Photo by Paul Turner*]

Brown pelican
Pelecanus occidentalis

Sea-birds (Fig. 27.29) and penguins that live along the edge of the polar pack-ice are confronted with a latitudinal gradient similar to that described for pinnipeds. Each deme of penguins seems to have an autumn migration direction appropriate to the direction of the open sea (p. 671).

In the case of land-birds, selection on direction can in most cases be considered out of context of wind direction except insofar as wind speed and direction is a migration-cost variable to which appropriate migration thresholds have evolved (p. 632). Some species, however, seem to have been subjected to sufficiently predictable shifts in the direction of the prevailing wind for selection to have favoured slightly different migration directions in autumn and spring (e.g. Figs. 27.27 and 27.46). Sea-birds, however, are subjected to wind almost continuously. In fact, to judge from the migration circuits of species such as the short-tailed shear-water, *Puffinus tenuirostris* (Fig. 27.30), Wilson's petrel, *Oceanites oceanicus* (Fig. 27.31), and the various Southern Ocean species that appear to perform a circumpolar migration (p. 622; Fig. 27.8), the influence of the prevailing wind direction on migration cost seems to have made as much contribution to the optimum direction of migration as have gradients of availability of food and temperature, etc. Consequently, sea-birds have evolved 'looped' migration circuits to a far greater extent than land-birds. The same concept that is applied to fish (p. 809) (i.e. of demes and migration circuits crystallising out from less clear-cut situations in response to selection for a minimisation of migration cost and a maximum coincidence of arrival at a succession of areas at times of high local availability of food) seems to be of equal applicability to sea-birds.

Fig. 27.29   The seasonal return migrations of two Antarctic birds

The maps show the seasonal distribution of the adelie penguin, *Pygoscelis adeliae*, and snow petrel, *Pagodroma nivea*. The adelie breeds around the coasts of the Antarctic continent. The snow petrel breeds on the islands indicated by circles, though it also breeds at some localities along the continental shore and inland. Winter distribution (stippled) for both species and most other Antarctic animals is governed by the extent of the pack ice. Both of the species shown migrate N with the edge of the pack ice in autumn and winter. In spring (October), adelies, which feed on krill (Fig. 20.9), swim S (at least 650 km) to their Antarctic nesting sites.

Adelies displaced from their nesting sites and released on featureless snow–ice surfaces nevertheless walk in the direction that would take them to the sea from their home colony. Solid arrows refer to birds displaced from Cape Crozier (solid triangle). Open arrow refers to birds displaced from Mirnyy (open triangle) and released immediately. Birds from Mirnyy caged at the release site for several days adopt the same direction as birds from Cape Crozier. The initial difference, therefore, could be a clock-shift effect.

[*Maps drawn from descriptions in Orr (1970a) and Stonehouse (1972). Displacement–release data from Penny and Emlen (1967). Photos by D.G. Bone*]

In the case of loop migrations, many of which seem likely to have remained relatively unchanged for long periods of time, selection for an endogenous programme of migration thresholds seems likely. Perhaps selection has also favoured this programme including an endogenously determined sequence of preferred exploratory migration directions. Also, in many, but by no means all, sea birds, social communication is likely to make a major contribution to familiar-area establishment by the young bird.

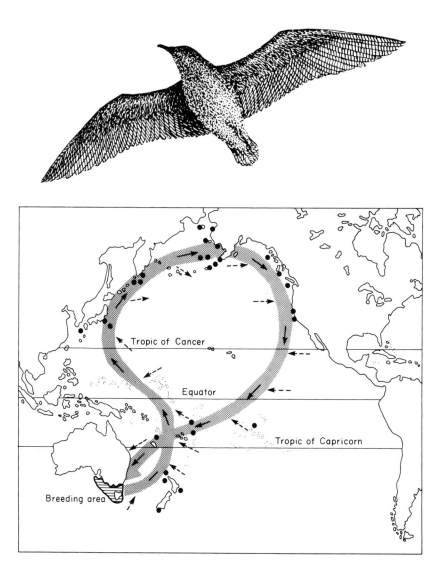

Fig. 27.30 The loop migration of the slender-billed shearwater (= Tasmanian mutton bird), *Puffinus tenu-irostris*

Arrows on the stippled loop indicate the direction of migration. Dashed arrows indicate prevailing wind direction. Solid dots indicate the positions of collected specimens. Young birds do not return to the breeding shores and islands until several years old, though they may be present at sea in the general area. Females breed for the first time when 5 years old, males when 7 years.

The loop would seem to give the maximum coincidence between arrival of shearwaters in an area and the local seasonal abundance of resources while at the same time minimising migration cost by migrating predominantly downwind.

[*Compiled from Dorst (1962) and Orr (1970a), after Serventy*]

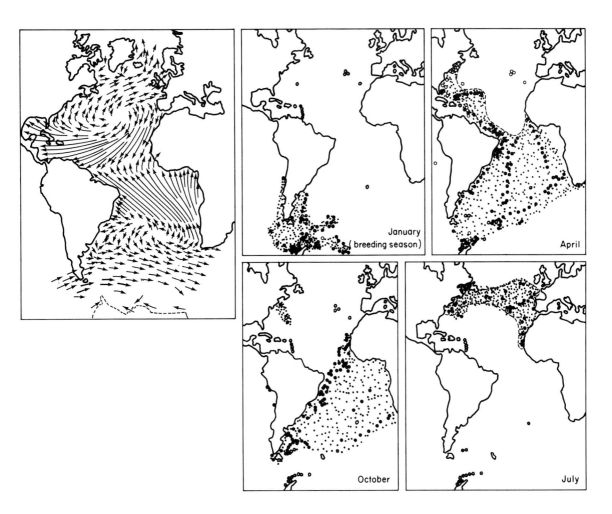

Fig. 27.31   The migration of Wilson's petrel, *Oceanites oceanicus*, in the Atlantic

The stippled area in each of the four outer maps shows the main area of distribution of Wilson's petrel at the time of year indicated. The left-hand map shows the prevailing winds in the Atlantic in August. As with the slender-billed shearwater, *Puffinus tenuirostris*, in the Pacific (Fig. 27.30), it seems likely that the migration circuit gives the maximum coincidence between arrival of petrels in an area and the local seasonal abundance of resources, while at the same time minimising migration cost by migrating predominantly downwind over long sections of the circuit.

[*Compiled from Dorst (1962) (after Roberts). Photo by Dr. I. Everson, British Antarctic Survey*]

Evidence for an annual change in the directional preference of a migrant bird has been provided by Perdeck and Clason (1974) for caged individuals of the chaffinch, *Fringilla coelebs*. It was found that the directional preference of caged birds turns through 360° in a clockwise direction during the course of a year such that in the migration months the preferred direction is the standard direction for the species. The birds were originally maintained under natural day/night conditions and tested for direction preference under a sunny sky. When tested in a solarium, however, with a fixed equinoxial Sun arc, the birds showed no change in directional preference from June to December. This preliminary result, therefore, suggests that a circannual endogenous programme is not involved and that directional preference is a response to changing day length.

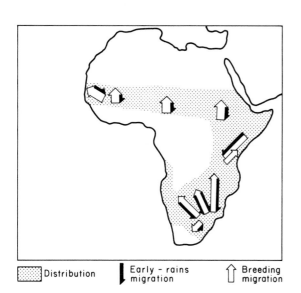

Distribution | Early - rains migration | Breeding migration

Fig. 27.32 Seasonal return migration of the red-billed quelea, *Quelea quelea*

The red-billed quelea is a member of the Ploceidae, a family of about 263 species such as the weaver birds and sparrows, including the house sparrow, *Passer domesticus*. The group is widespread in the Old World with most species in Africa.

The red-billed quelea is adapted to seasonally arid areas and feeds on seeds and insect larvae. During the dry season it feeds on seeds that are dry and lying dormant on the ground. As the dry season proceeds the birds converge on the more suitable habitats. When the rains start, the seeds in the dry season home range germinate and are no longer available as food to quelea. The flocks then perform an 'early rains' migration back along the track of the advancing rain front to an area where rain began a sufficient time previously for the plants to have grown and reached the green seed stage. These seeds, plus the larvae of insects feeding on the plants, provide the main source of food for the quelea. The return migration back to the dry season home range proceeds gradually and keeps pace with the movement of the green-seed belt along the track of the rain front. This is a breeding migration. Breeding, however, is opportunistic and occurs at any suitable area along the migration track. A part of the flock may settle to breed in one area while the rest of the flock moves on. After raising a brood, the birds continue migration and catch up with the rest of the flock, many of which are now breeding. The earlier breeders may join the later breeders and raise another brood or they may 'leapfrog' over them before doing so. Individual females may raise up to three broods during the course of a single breeding migration back to the dry season home range, each brood at a different point along the migration track.

[*Compiled from Ward (1971)*]

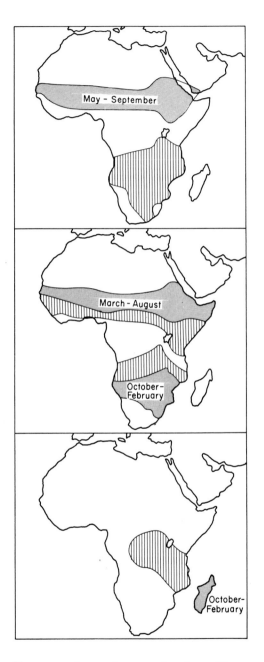

Abdim's stork
Sphenorhynchus abdimii

Carmine bee-eater
Merops nubicus

Southern carmine bee-eater
Merops nubicoides

Roller
Eurystomus glaucurus

Fig. 27.33 Seasonal return migrations of four species of tropical birds in Africa

☐ breeding range; ▥ non-breeding range.

In the tropics, most regular seasonal return migrations are synchronised with the movement of major rainfall zones, such as the Inter-Tropical Convergence Zone (ITCZ) (Fig. 19.18). Abdim's stork, for example, is a major predator of locusts (p. 452). Compare the breeding distri-bution of this species with the distribution of swarms of the desert locust, *Schistocerca gregaria* (Fig. 19.18), and the position of the ITCZ during the same period.

Note that the anti-tropical distribution of the two bee-eater species results in them moving N and S at the same time but that while one is in its breeding range, the other is in its non-breeding range.

[*Compiled from Dorst (1962) (after Chapin)*]

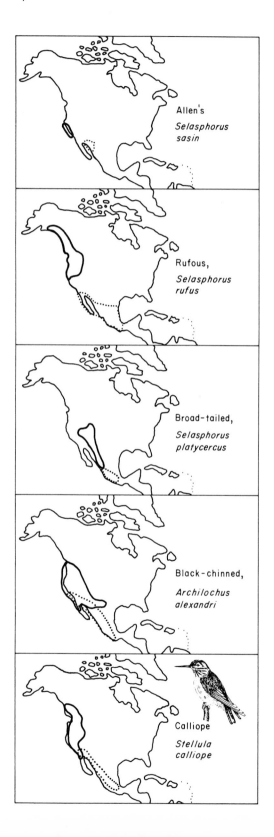

Fig. 27.34 Seasonal return migrations of five species of North American hummingbirds

⌒ breeding range; ⋰⋰⋰ winter range.

Hummingbirds (Trochilidae) are a predominantly tropical group. A few have colonised more-temperate regions of North America, but all that have done so manifest a seasonal return migration. The calliope hummingbird, which at 3 g is the smallest North American bird, has a one-way migration distance of up to 5000 km.

[*Drawn from figures and descriptions in Robbins* et al. (*1966*) *and Orr* (*1970a*)]

Quite possibly a similar, or perhaps endogenously programmed, change in directional preference is involved in the change in direction shown by most passerines leaving Britain in the autumn (Lack 1963, Evans 1972). The majority of small nocturnal migrants (warblers, chats, flycatchers, etc.) leave England on a track that is east of south yet many nevertheless reach western Iberia to the SSW of England. It is suggested (Evans 1972) that departure east of south takes maximum advantage of the west winds that predominate over Britain in autumn, whereas over southern France the westerly winds on average are less strong and in any case, flight is over land not over the sea. The most advantageous direction from Britain has been argued to be west of south (p. 666), but conceivably migration cost is reduced by crossing the North Sea, Dover Straits, or English Channel in a S or SSE direction. Perhaps similar considerations apply in the case of passerines that leave the eastern United States in the direction of Bermuda and then perform a loop toward the Carribbean Islands and northern South America (Williams *et al.* 1972).

Analysis of the migration patterns of birds inhabiting tropical latitudes, particularly those in arid or seasonally arid regions, is at an early stage (p. 324; Keast 1968a). The impression given is that of widespread nomadism and opportunistic breeding achieved by establishment of a large familiar area through exploratory migration and social communication. Many such birds move in flocks. The budgerigar, *Melopsittacus undulatus*, of Australia, for example, forms large flocks and is nomadic within the dry Australian interior (Serventy 1965). It is tempting to postulate a similarity between the migrations of birds living in such conditions and those of some ungulates, such as the wildebeest, *Connochaetus taurinus* (p. 95 and Fig. 25.7) which also live in a seasonally arid environment. From time to time, however, the budgerigar appears away from the interior on the coastal plains (Serventy 1965). Presumably, therefore, the hierarchy of migration thresholds of the species includes one that triggers a long-distance migration but which is only infrequently exceeded. Certainly it seems likely that part of the behavioural repertoire of birds like the budgerigar includes calculated migration thresholds that permit migration to distant areas where rain has recently fallen (p. 95).

Other tropical birds live in areas where the seasonal progression of rainfall is much more regular and in such cases regular shifts in seasonal home ranges may occur. Usually such species are in an area for which the seasonal timing and direction of rainfall vary on an annual basis in a relatively predictable way. Such species, of which perhaps the red-billed quelea, *Quelea quelea*, is the best worked example (Fig. 27.32), are likely to be subjected to selection for endogenous programmes, a hierarchy of migration thresholds, and an inherited bias to the exploratory migration direction ratio, that is comparable in almost every respect to that of temperate species, except perhaps in the case of quelea with respect to the length of the breeding season in any one place.

Finally, the observed migration direction of some demes of some species, such as the Alaska-breeding deme of the wheatear, *Oenanthe oenanthe* (Fig. 27.46), is such that it cannot be explained by the model developed here without the addition of one further element. This element is the post-Pleistocene history of the species and is discussed in Section 27.9 (p. 701).

27.7.1 Summary

The direction of all exploratory migration is endogenously determined in the individual level (p. 654) except where over-ridden by social communication (p. 613) or distance perception (p. 92). A hierchy of migration thresholds will evolve (p. 654), each level corresponding to migration of a specific distance (p. 229). The optimum direction ratio for short-distance exploratory migration approximates to 25:25:25:25 (p. 654) as does that for longer-distance exploratory migration if there is no gradient of increasing habitat suitability in a specific direction, The longer the migration distance, given such a gradient, the greater the optimum bias to the exploratory migration direction ratio (p. 654). The steeper the gradient of directional selection, the greater the bias to the direction ratio (p. 660). Winter gradients of day length, tempera-

ture, and food availability make the major contributions to the gradient of directional selection that acts on the autumn direction of exploratory migration in land-birds, but wind-direction may also be important when flight over the sea is favoured, and makes a particularly large contribution to selection on the track pattern of sea-birds (p. 672). The gradient of directional selection in winter, acting on autumn direction, makes the major contribution to the evolved directional axis of migration, particularly in land-birds. If a gradient of directional selection exists in spring that acts in a direction other than opposite to the gradient acting in autumn, it makes little contribution to the evolved directional axis of migration of land-birds (p. 659). Here, selection for return and for migration down a gradient of advantage imposed by deme density is suggested to be the major factor (p. 665). In sea birds for which a 'loop' migration circuit has evolved, 'autumn' and 'spring' tracks are equally the result of directional selection. There is some evidence that the endogenous determination of migration direction may require the stimulus of changing day length, at least in the case of birds that do not cross the equator (p. 674).

27.8 Migration distance

Selection on migration distance in any particular lineage has always to be considered in relation to migration direction, degree of dispersal and convergence, and a hierarchy of migration thresholds. Any new variant with a new migration distance that appears in a deme may be at a disadvantage if it is linked with one particular combination of migration direction and migration threshold but at an advantage if linked with another. The action of selection, therefore, is to link migration distance with an appropriate migration direction and migration threshold, the equilibrium position being achieved when the frequency distribution of all three characters results in the optimum degree of dispersal and convergence (p. 649). In a free situation, this occurs when all individuals have the same potential reproductive success. In a despotic situation it occurs when all individuals perform the migration that is optimum for each individual's particular resource-holding power.

There should be two factors that contribute to a correlation between the frequency distribution of migration distances within a deme and the direction ratio. First, as already described (p. 654), for any given deme size an increase in the main migration distance in the peak direction leads to selection, acting through optimum degree of dispersal, to favour more individuals migrating in that direction. Secondly, the steeper the gradient of increase in habitat suitability in the optimum direction, the greater the optimum bias in direction ratio (p. 660) and the greater the mean migration distance that is likely to be favoured (p. 654).

Turning now to the evolution of a hierarchy of migration thresholds, each one corresponding to a particular migration distance, the situation can be introduced by a further reference to the migration of the crossbill, *Loxia curvirostra*, in Europe as analysed by Newton (1972). Exploratory migration is initiated every year by at least some individuals of this species, but in most cases the exploratory migration initiated is relatively short distance. If we term the areas of hundreds or thousands of square kilometres that show synchronous abundance or scarcity of the spruce seed crop (p. 208) a 'gross area', it can be said that the lowest exploratory migration threshold is adaptive to local removal migration within a gross area. A higher exploratory migration threshold should exist, however, that is relevant to the suitability of the gross area itself. This threshold is likely to receive contributions from food abundance and deme density perceived during the course of the exploratory migrations initiated in response to the lower migration threshold. Once this upper migration threshold is exceeded, however, it is predictable that the bird will have to initiate a migration over at least hundreds of kilometres in order to leave its original gross area and have a chance of arrival at another gross area with an abundant spruce crop. To some extent, the distance the individual has to travel is a function of the number of other individuals also likely to arrive in the new area. Conceivably, therefore, there are several migration thresholds, each one corresponding to a migration of a particular distance. In the case of European common crossbills, the highest migration threshold seems to correspond to a migration with a one-way distance component of about 4000 km. This study emphasises the main factor that is involved in the evolution of migration thresholds and migration distance. For any given threshold it is predictable that a bird will have to migrate a certain minimum distance before it encounters a destination of above threshold suitability (i.e. migration distance is obligatory—p. 83). Let us now consider in a general way the factors that influence the end point of this migration and hence the total migration distance.

The total migration performed by a bird in shifting from a breeding home range to a winter home range can be broken down into a number of migration units of a particular distance performed by non-stop flights which alternate with periods of residence in a succession of transient home ranges. These home ranges are suggested (p. 615) to be formed by exploratory migration and/or social communication during the first migration of the young bird and to be reached by calculated migration in subsequent years when adult. Migration units can be described in terms of their distance, duration, and frequency.

According to the model developed in Chapter 11, a phase of a return migration, such as the autumn phase of the seasonal return migration of birds, terminates when h_1 no longer falls below $h_2 M^x$ for a migration unit in any direction. In those birds, particularly sea birds, that perform a migration circuit or are nomadic, this situation may never arise for stay-times longer than a few weeks. In most sea-birds it seems that after a relatively short stay-time in each transient home range, a further shift in home range becomes advantageous. This situation may also apply in some land-birds that perform a to-and-fro migration. In the twite, *Acanthis flavirostris*, for example (p. 647), that migrate between Finland and Poland, the birds spend no more than one or two weeks at their southernmost home range before, presumably, h_1 becomes less than $h_2 M^x$ (where, in this case, $x = 1$) for a migration unit to the north. Similarly, all those birds that may continue their autumn migration given hard weather may also experience a value of h_1 less than $h_2 M^x$ (where $x \simeq 2$) for migration in the autumn direction at any time during the winter. In many, perhaps most, migrant land-birds, however, a home range is eventually reached in autumn for which h_1 does not fall below $h_2 M^x$ until the following spring. This is the winter home range.

If H_w is the winter home range and H_n is the nuptial, natal, or moulting home range, and if M_R is the migration factor for the total return migration between the two, selection on the timing of departure from H_n and arrival at H_w and selection on the migration mechanisms should act to produce behaviour that maximises $h_w M_R / h_n$. Selection on the date of departure from H_n is likely primarily to be a function of selection on the optimum phasing of, and time investment in, breeding, moulting, post-fledging exploratory migration, etc. (p. 626). One additional factor that seems likely to be involved is the advantage of departure from H_n, as well as each of the earlier transient home ranges while the efficiency of energy acquisition in that home range is high. It is implied, therefore, that in this case the migration from H_n may be initiated before h_1 becomes less than $h_2 M^x$ $(x=2)$. This situation (initiation of a migration unit that is part of a long-distance return migration) possibly provides a second example (see p. 382) of a case for which the gross migration cannot necessarily be analysed solely by the advantage gained by each migration unit, a procedure that was based on the working hypothesis adopted in Chapters 7 and 18. Once again, there is the possibility that selection on the total migration may impose selection on the individual migration units in such a way that some units, in isolation, are not necessarily advantageous. In this case it would be selection for maximisation of $h_w M_R / h_n$ that imposed selection for the first few migration units to be initiated before h_1 was less than $h_2 M^x$ in order that more of the later migration units might be initiated while h_1 was less than $h_2 M^x$. It should be stressed that there is no critical evidence on this point, though many authors remark that autumn migration is initiated while the nuptial, natal, or moulting habitat appears still to be suitable (e.g. Newton 1972) and, by implication, at least as suitable as the first transient home range to be utilised.

However, not all species and demes need be considered to initiate migration from H_n at a time when the first migration unit is disadvantageous, particularly those that depart later in the autumn. For these birds, analysis of the total migration in terms of each migration unit is perhaps totally acceptable, each unit being initiated only when h_t becomes less than $h_{t+1} M^x$ (where H_t is the transient home range occupied and H_{t+1} is the next transient home range along the migration route). Even for those species or demes discussed in the previous paragraph, the later migration units as the birds approach the wintering area are likely only to be initiated when h_t becomes less than $h_{t+1} M^x$. Indeed, the advantage of initially 'disadvantageous' migration units early on in the autumn migration is likely to be a function of an increase in the possibility of advantageous migration units later on in the migration. If such birds delayed the start of the migration such that transient home ranges were of low suitability upon arrival, the birds would be unable to deposit sufficient fuel reserves for M to be sufficiently high for h_t to be less than $h_{t+1} M^x$. The result would be that the optimum strategy for the

bird would be to attempt to winter in an area much less suitable than would otherwise have been the case. Thus $h_w M_R / h_n$ would not be maximised.

The most likely adaptation to selection in favour of the initiation of the early migration units of long-distance migrants under conditions in which h_t is not less than $h_{t+1} M^x$ is a reduction in the contribution of habitat variables to the migration threshold and an increase in the contribution of time and/or indirect variables. The conclusion reached from observation of the departure dates of long- and short-distance migrants described earlier (p. 645), that habitat variables make much less contribution to the composite migration threshold of long distance migrants, provides some support for this suggestion.

So far in this consideration of selection for the maximisation of $h_w M_R / h_n$ we have been concerned with those adaptations that increase h_w / h_n. The major part of this section on migration distance, however, concerns those adaptations that act to increase M, predominantly through a reduction in migration cost, and of particular concern initially is a consideration of the optimum length, duration, and frequency of the migration units by which the total migration is achieved.

The critical factors in the determination of air distance (i.e. distance flown through the air as opposed to over the ground) for any given bird are the lift:drag ratio, which is also the ratio of weight to thrust, and the proportion of the body weight that consists of fuel (Pennycuick 1969). Any bird with a particular lift:drag ratio consumes a fixed proportion of weight per unit distance through the air independently of whether it is large or small. Thus, if a swan and a hummingbird happened to have the same lift:drag ratio and if the same percentage of the take-off weight of both birds was fuel, then they should have the same potential migration distance, measured as air-kilometres. That is not the same, of course, as having the same potential migration distance across the ground; large birds fly faster than small birds and hence are less affected by wind.

As a bird migrates, it uses up its fuel stores and its weight decreases. Pennycuick (1969) has pointed out that the result of this loss of weight is that the bird achieves a greater migration distance per unit weight of fuel towards the end of the flight than at the beginning. Thus a bird that began a migration unit with fuel reserves equal to the bird's fat-free weight, by the end of the flight is achieving 2·5 times as much distance per unit fat than it achieved at the beginning. If we apply this fact to a model that selection has acted to minimise migration cost then it follows that the most advantageous strategy is the minimisation of the fuel reserves, thus achieving maximum migration distance per unit of fat stored and hence also per unit of time investment in feeding. If only this selective pressure were involved, therefore, the favoured strategy would be a frequent alternation of feeding bouts and short migration units. Because of the circadian variation in migration cost (p. 631), this should mean the initiation of a short-distance migration every day (given favourable levels of other migration-cost variables) rather than long periods of residence alternating with long-distance migration.

Another selective pressure that seems likely to favour frequent but short migration units is that of minimum exposure to migration-cost variables. If we exclude relatively recent environmental hazards, such as radio masts and lighthouses, etc., that are encountered by a relatively small proportion of migrating birds, and excluding predators, such as Eleonora's falcon, *Falco eleonorae* (Fig. 27.14), which exert a migration cost that is probably independent of migration unit distance, the major relevant sources of high migration cost are changes in the weather to conditions that either displace birds over unfavourable terrain or force them to land under difficult conditions such as fog, strong winds, and heavy rain. Most authors in their discussion of bird migration invariably stress the spectacular dimensions of such mortality factors (e.g. Orr 1970a). On the whole, however, it probably gives a false impression of the migration cost experienced by migrating birds to labour such occurrences. In general, migration appears to be a highly adaptive phenomenon that leads to a mortality rate that is probably little if at all higher than that experienced by residents. Indeed, as argued in the discussion of partial migration (p. 635) and as demonstrated by Nice (1937) for the song sparrow, *Melospiza melodia*, selection should stabilise at the point where migrants and non-migrants achieve the same potential reproductive success. Juveniles of migrant birds may well experience a high mortality during autumn migration but then so too do juveniles that restrict their migrations to exploratory movements based on their natal area.

This de-emphasis of the hazards of migration should not be interpreted as implying that migration cost does not impose selection on migration mechanisms but rather that, on the whole, migration is a well-ordered and highly adaptive behavioural strat-

egy. Most of the time, therefore, the major component of migration cost seems likely to result from the level of efficiency of the acquisition and expenditure of energy. When mortality occurs through factors other than those resulting from changes in energy reserves (see p. 693), it seems likely that it results most often from changes in the weather. The low predictability for much of the area occupied by migrant land birds of the temporal and geographical changes in the weather is notorious, especially with respect to local conditions and disturbances. It is always possible, of course, perhaps even certain, that selection has led to birds attaining a greater standing as weather forecasters than humans, the migration thresholds on which selection seems likely to have acted already having been described. Even so, it is evident from observations of high bird mortality during storms, etc., that an accuracy of weather prediction of 100 per cent has not been achieved. If this is so, it seems likely that the longer and further a bird travels during a single migration unit, the more inaccurate its weather prediction becomes. This is not to suggest that the bird does not continue to monitor migration-cost variables during flight. However, as already shown (p. 632), maximisation of predictability is possible only if all weather changes during the preceding 24 hours are incorporated into the migration threshold as well as conditions at the moment of take-off. Yet the further a bird migrates from its take-off point, the less data there are available concerning longer-term weather changes. The result seems likely to be an increase in migration cost per unit distance with increase in distance from the take-off point. This selective pressure, therefore, seems to be another that favours frequent, short-distance, migration units rather than infrequent, long-distance migration units.

One final selective pressure also seems likely to favour frequent, short-distance migration units, though here we are concerned specifically with young birds for which the autumn migration is exploratory rather than calculated. Long-distance exploratory migration units can serve to fix gross topographical features such as hills, drainage systems, woodlands, and, nowadays, villages, towns, and cities, within the spatial memory and permit habitat assessment at the gross level (i.e. forest rather than moorland, etc.). It is doubtful, however, whether the assessment of habitat suitability for use as a transient home range can be carried out with acceptable reliability by a bird that does nothing but migrate over an area. This seems particularly likely to be true in the case of nocturnal migrants. Yet, because of the possibility that in future years the individual may be forced by unfavourable weather to settle and use the area concerned as a transient home range, selection seems likely to favour more rather than less habitat assessment by the young bird during its first migration. To some extent, the technique of reversal, as interpreted here (pp. 188 and 615) seems to offer the most efficient strategy. That is, the young bird migrates over a certain distance in the standard autumn direction, and performs habitat assessment at the gross level as migration proceeds. The next day or night the bird then migrates back to the areas assessed at the gross level to be the most suitable and there performs more detailed exploratory migration and habitat assessment.

The advantage of such detailed habitat assessment in sufficient areas along the migration route is suggested by an observation that at the same time demonstrates a selective pressure that would tend to oppose those already described (i.e. favour longer and less frequent migration units). In describing the weight changes of the European robin, *Erithacus rubecula* (p. 645), with time after arrival in a transient home range, it was stated that individuals with relatively high resource-holding power showed little or no weight increase during the first day or so after arrival but that thereafter the increase in weight was considerable. Obviously various interpretations are possible but the most likely seems to be that the efficiency of energy acquisition is a function of familiarity with the area. During the period of familiarisation, efficiency is low but beyond a certain point increases considerably. As it happens, the robin lives in a despotic situation and it may be that this is a significant determining factor. However, it seems not unreasonable to suggest that the same principle applies whether the situation is free or despotic: that efficiency of energy acquisition increases with time after arrival in a transient home range. If this is accepted, two further selective pressures seem likely. First, assuming that familiarisation with an area leads to a higher starting efficiency the following year, there is an advantage in performing more rather than less exploratory migration and habitat assessment during the ontogenetically first autumn migration. Second, there is an advantage in staying for a longer rather than a shorter time in each transient home range.

Turning now to such evidence as is available concerning the length, duration, and frequency of migration units, it is often considered that whereas diurnal migrants only migrate for a few hours during the day, nocturnal migrants migrate from dawn to dusk. Fringillid finches, for example, the vast majority of which are diurnal migrants, move primarily in the mornings, hardly at all in the afternoons, but may move again a little in the evenings before settling to roost before dark (Newton 1972). Consequently, these passerines migrate for only 3–4 hours/day and cover perhaps 100 to 200 km in each migration unit, behaviour that is consistent with the conclusions reached above. If, on the other hand, as suggested by Dorst (1962), nocturnal migrants are on the wing from dusk to dawn, there is a strong implication that for these species selection has favoured long-distance, and perhaps less frequent migration units. However, counts of the diel pattern of migration density, whether obtained visually (Lowery and Newman 1955) or by radar (Sutter 1957a,b) indicate that there is a peak during the middle part of the night, with a maximum between 23.00 and 24.00 hours. The existence of such a peak is not consistent with a general pattern of continuous migration from dusk to dawn but suggests a situation in which each individual migrates for only part of the night. Quite possibly, each species of nocturnal migrant shows a limited period of migration similar to that characteristic of diurnal migrants.

Marking–release/recapture results can give only average rates of movement, and without more precise measurement of the ground speed of migrating birds, such as that provided by Schnell (1965), and the length of time spent in migration during each 24 hours, such results are difficult to interpret. In Fig. 27.35, some marking–release/recapture results are presented for various warblers (Sylviidae) banded in Britain, particularly England. If these birds flew in long-distance but infrequent migration units it would be at least expected that birds recaptured within a few days of banding would be found at a distance that was either below or well above the average rate of cross-country progression. It might also be expected, as all the birds were banded within a relatively limited area, that there would be clusters of recapture distances, with relatively few intermediate recaptures, the distance between each cluster corresponding to the length of the migration unit. Although recaptures are few, such a situation can indeed be seen later on in their migration for those species that cross the Sahara, but there is no indication of such a situation over the European mainland. There is every indication,

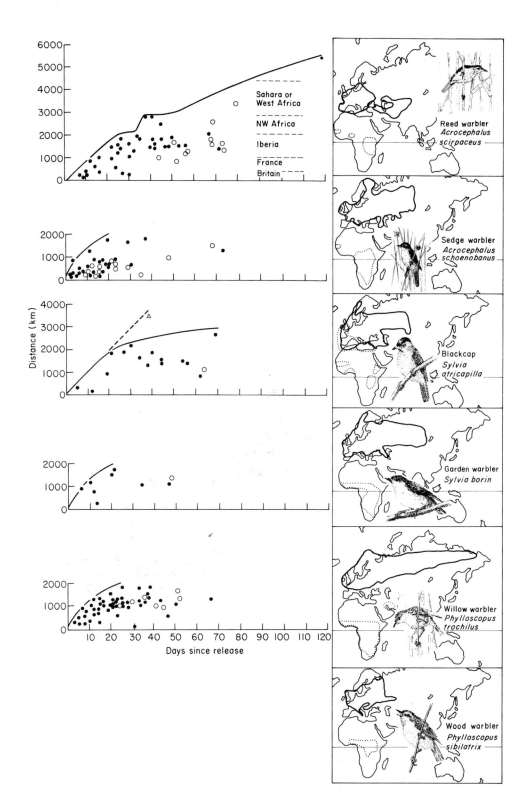

Fig. 27.35   Rate of migration of some Old World warblers

⬮ breeding range;   ⸭⸭⸭⸭⸭ non-breeding range.

The graphs show the distance between release and recapture (measured as a straight line between the two points) for birds recaptured in autumn or winter less than four months after marking and release in Britain. ○, birds marked in July; ●, birds marked August/September. Circles indicate that the birds were recaptured on the Britain–France–Iberia–northwest Africa–west Africa migration route. Four blackcaps were recaptured on a different route, three in Italy and one (△) in the Lebanon.

The dots form such a clear upper boundary (indicated by the solid line) that it is most unlikely that the distance of each non-stop migration unit while a bird is crossing France and Iberia exceeds about 300 km. The model that fits the data most easily is that of migration units at the more frequent, shorter-distance end of the available spectrum. The rather uneven scatter of points below the solid line is likely to be due to the intra-deme frequency distribution of migration rates and perhaps also to some birds (young?) alternating migration units in the standard direction with reversal or other migrations in non-standard directions.

There is an indication for both the reed warbler and the blackcap that the birds pause for some time in Iberia or northern Africa before beginning to cross the Sahara or migrating down the west African coast. This pause seems particularly apparent for the blackcap when rate of migration on the SW route from Britain is compared with the rate (dashed line) on the SE route. The situation is confused, however, by the lack of data and by the fact that birds ringed in July may not begin their autumn migration until several weeks after being released. The rate of migration of the wood warbler is discussed on p. 696.

Maps compiled from Voous (1960), Moreau (1972) and Peterson et al. (1974). Marking–release/recapture data from ringing reports in British Birds, 1950 to 1970 (extracted by M. Lewis)]

therefore, that for most of their migration these birds perform short but frequent migration units.

The final evidence in support of the predominance of selection for the minimisation of migration cost rather than the maximisation of distance flown during each migration unit relies on acceptance of the validity of the theory of the mechanics of bird migration presented by Pennycuick (1969). According to this theory, a bird of a given weight, fuel reserve, and lift:drag ratio that attempts to obtain maximum migration distance per unit of fuel has an extremely limited range of air-speeds available to it, though, because of fuel consumption during flight, optimum air-speed grad-

ually decreases during the course of a migration unit. Nevertheless, if selection acts to maximise migration distance, each species should have a characteristic air-speed that is optimum for a bird with that particular set of characteristics. Deviation from this air-speed is suggested by Pennycuick to occur in two situations. First, in some head-wind conditions a bird can only make progress over the ground by increasing its air-speed above the optimum to a level greater than the ground speed of the air. Secondly, if a bird is blown off-course and, because of the terrain, cannot settle, the most advantageous strategy may be to slow down the air-speed and delegate cross-country movement almost entirely to wind displacement. In this way, the bird uses less energy, can remain in the air longer, and thus has a greater chance of encountering land.

As far as comparison of maximum distance and minimum migration cost models is concerned, however, the critical test comes when a bird is migrating under tail-wind conditions. According to the familiar-area hypothesis, adult birds (excluding type-C removal migrants that during exploratory migrations should behave in the same way as juveniles) in autumn are performing calculated migration, moving from one familiar transient home range to another along a migration route that is also familiar. Migration is therefore over a relatively fixed distance, though selection probably favours sufficient exploratory migration when young to permit some flexibility of destination. Juvenile birds, on the other hand, are assumed to be performing exploratory migration during which longer-distance migration in a standard direction alternates with shorter-distance migration in the reverse or some other direction (p. 615). During the first type of movement, perception and habitat assessment utilises gross features alone. During the second type of movement, more detailed features are utilised. It seems likely that, as far as habitat assessment and the commitment of data to a spatial memory are concerned, there is an optimum compromise between the utilisation of gross and fine detail of the environment. Young birds as well as adults, therefore, are likely to be performing a relatively fixed-distance migration in the standard direction, though this is not to imply that there is no flexibility to the behaviour. If adults encounter highly favourable conditions during a particular migration unit, optimum strategy may be to miss one transient home range and fly on to the next, particularly, perhaps, if the stay at the previous home range had been prolonged due to unfavourable migration-cost variables. Young birds also, if they encounter highly favourable conditions, may gain greater advantage by slightly lengthening the distance of gross feature utilisation. Even so, a relatively limited range of migration-unit distances is likely to exist for both adults and young. These fixed-distance migrations should be completed, according to the selective pressures generated by a minimum-cost hypothesis, by a migration strategy that invokes the minimum migration cost for that distance. Returning now to the critical situation involving a tail-wind, if the minimum cost hypothesis is correct, selection should have favoured birds that reduce their air-speed (on the assumption, that is, that energy expended makes a greater contribution to migration cost than time in the air). A reduction in air-speed leads to reduced total energy expenditure during a fixed distance flight. If a maximum-distance model is more appropriate, selection should have favoured birds that retained their maximum range air-speed even with a tail-wind. Ground-speed, according to a maximum-distance model, should be maximum range air-speed plus wind-speed for all values of wind-speed. According to a minimum-cost model, increase in wind-speed should lead to a decrease in air-speed.

The advent of the use of radar in the study of bird migration has led to many measurements of the air- and ground-speed of migrating birds in varying wind-speeds and directions, and although these measurements vary in details they seem to provide unanimous support for a minimum-cost rather than a maximum-distance model. Some authors (e.g. Schnell 1965, Bellrose 1967a) have concluded from their results that birds maintain a constant ground-speed irrespective of wind-speed and direction, though Evans (1972) has questioned the validity of this conclusion. Bruderer and Steidinger (1972), using tracking radar by which single individuals may be followed, have also demonstrated that the ground-speed of a bird is not influenced in an additive way by the wind vector along the bird's track. In this case, however, it was found that approximately 67 per cent of the component is added to the bird's ground-speed.

Evidence drawn from a variety of different types of phenomena, therefore, all points to the validity of the conclusion that migration unit strategy has evolved in response to selection for the minimisation of migration cost. Evidence from the long-distance tracking of individuals, however, is equivocal (Fig. 27.36). Accepting this conclusion, however, exam-

ples of adaptive strategies that have so far been described are an air-speed that decreases with increase in the wind vector along the bird's track and migration units that are toward the shorter and more-frequent end of the possible spectrum. Other similar strategies may be described which may also be considered to be adaptive to selection for reduced migration cost. Most of these strategies were identified by Pennycuick (1969) in his comprehensive analysis of the mechanics of bird migration.

In terms of a reduction of migration cost there are at least two advantages to be gained by a bird that migrates at as high an altitude as possible. The first is that the optimum speed for minimum energy expenditure per unit distance increases with increase in altitude, thus giving a shorter flight time. The second is that as temperature decreases with increased altitude (approximately 2°C/300 m), less water reserve is used through evaporation during thermoregulation. The limit to the altitude at which migration can occur is set by the availability of oxygen which also decreases with increased altitude. At some height, the maximum rate of oxygen absorption is just enough to maintain the optimum air-speed. This is the optimum height, a level which may be identified by the birds climbing until they can no longer maintain the optimum air-speed and then descending until they just can. As weight decreases during flight, a reduced power is needed to maintain speed. Consequently, there is a reduced oxygen requirement and theoretically the optimum height for migration should increase.

In general, larger birds with faster air-speeds seem to migrate at greater heights (Bruderer and Steidinger 1972). In Europe, some small passerines, such as the meadow pipit, *Anthus pratensis*, often fly at altitudes less than 100 m above ground level, though others, as determined by radar, may reach 1500 and occasionally 4250 m. Waders, especially lapwings, *Vanellus vanellus* (Fig. 27.12), usually fly at 900–1800 m but may reach 3500 m (Dorst 1962). In the United States, most of the starlings, blackbirds (Icteridae), red-headed woodpeckers (*Melanerpes erythrocephalus*), blue jays (*Cyanocitta cristata*), and many swallows migrate at 15–45 m above the ground, below radar surveillance. Crows, gulls, and cormorants migrate at 30–300 m. White pelicans (*Pelecanus erythrorhynchos*), and hawks have been observed at 300–900 m, whereas snow geese, cranes, and whistling swans migrate most commonly at 600–1500 m, though they have been reported by pilots up to 2400 m, though rarely as high as 3600 m (Bellrose

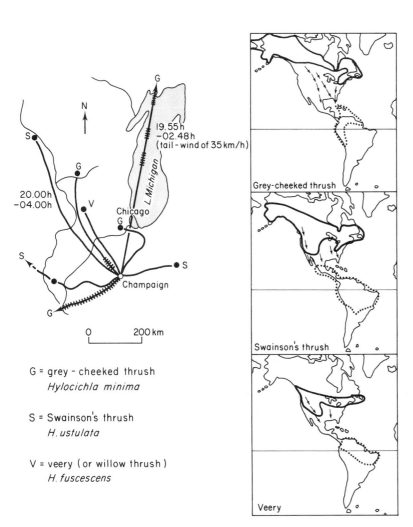

Fig. 27.36 Rate of migration of some New World thrushes

⬭ breeding range; ⫶⫶⫶⫶ non-breeding-range.

In the diagram, solid lines show the tracks of 8 individuals of 3 species of thrush engaged in nocturnal migration in spring. Tracks were monitored by attaching miniature (3 g) transmitters to the birds, releasing them (on different nights) at Champaign, and then following them by aeroplane. ●, indicates that the bird landed; ▲, indicates that the bird was still flying when contact was lost. Cross-hatching on the track indicates that the section concerned was performed under an overcast sky.

Data are as yet insufficient to draw general conclusions. However, even among the individuals investigated, a number of facets of bird migration as determined by other methods are apparent. Thus, at least two individuals seem to be executing a migration unit in the standard spring direction and at least two seem to be executing a reversal or other non-standard migration. Except when long leading lines (lake shore or river) and highly favourable winds are available, the migration units are toward the short end of the available spectrum, even for these relatively long-distance migrants. Finally, there is a conspicuous lack of precise linearity to the observed tracks that could be considered to be more consistent with the establishment of, or migration within, a familiar area than with any other model of migration.

[*Maps drawn from figures and descriptions in Peterson (1961) and Robbins et al. (1966). Tracking data from Cochran et al. (1967)*]

1972). Among nocturnal migrants, most small birds migrate at 100–1500 m. Waterfowl migrate at higher altitudes than smaller birds but lower than shore-birds, which at night migrate at consistently higher altitudes, as determined by radar at Havana, Illinois, than any other group of birds.

Selection on the altitude of migration, however, acting through the minimisation of migration cost, does not act solely on the compromise between minimum flight time, minimum water loss, and sufficient availability of oxygen. Recent radar studies (e.g. Bruderer and Steidinger 1972, Gauthreaux 1972), have shown that birds migrate most frequently at the height of favourable winds. Determination of this height by the bird could be achieved by the performance of a vertical transect. It could also be achieved by social means, either visually by day or acoustically by night. By day, a bird can compare its ground-speed relative to that of

other conspecifics above and below. By night, a bird could similarly compare its ground-speed relative to that of other conspecifics by monitoring the direction of successive call notes of another individual at a different altitude. All nocturnal migrants, which migrate as loose groups (about 100–300 m between individuals—Bruderer and Steidinger 1972), emit species-specific call notes that probably perform a variety of social functions (Hamilton 1962b) as well as permitting individuals to determine variation in the optimum altitude for migration.

Hamilton (1967) has also suggested that flock formation, particularly the inverted-V formation, characteristic of many species including most geese, some ducks, cormorants, certain herons, and cranes, and many gulls, increases the accuracy of orientation, the mean vector of a group being likely to be more accurate than that of an individual. This suggestion, however, was based on the concept of

long-distance precision orientation to astronomical clues and is not really consistent with the familiar-area hypothesis presented in this chapter. In any case, homing experiments on the relative performance of pigeons, *Columba livia*, released in small flocks and singly have shown that groups of four do not orient more accurately than single birds (Keeton 1970a). Even within the context of the familiar-area phenomenon, however, it is possible that migration cost is reduced for some species by migrating as a group rather than individually. Group migration may reduce migration cost in two ways. First, as suggested for ungulates (p. 543), if several individuals with similar but not identical familiar areas migrate together, there is a greater possibility of an advantageous contingency response and a consequent increase in flexibility, especially if the

group is familiar with more transient home ranges than each individual. Secondly, there is the possibility that group migration decreases the energy required per unit distance. Hamilton (1967) has argued against the second possibility by pointing out that birds that migrate with a V formation often retain this formation while resting or swimming on the water. It is quite possible, however, that the same principle is involved both while in the air and on water. That in the first place flight and in the second place swimming are performed with a reduced energy expenditure per unit distance when in a V formation. In addition, a V formation while resting on the water seems to give the optimum compromise between minimum spacing while still retaining sufficient wing-beat space for all to take to the wing at the same time if disturbed.

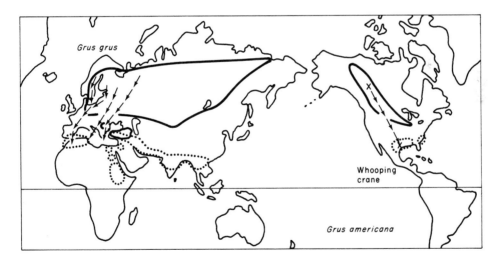

Fig. 27.37   The seasonal return migration of two species of crane

⬭ breeding range;   ⣉⣉⣉ winter range;
→ → some known or suspected migrations.

World-wide, there are about 14 species of cranes, found on all continents except South America. The maps show the seasonal distribution of two of these species, 'the' crane, *Grus grus*, of the Old World, and the whooping crane, *G. americana*, of the New World. The latter is a large, satiny-white bird with a long neck, long, dark pointed bill, and long, thin black legs. When erect, a large male stands more than 1·5 m tall.

The illustrated ranges for the whooping crane indicate the position in 1850 for the world population of this species of 1500 individuals. By 1941, the total population was down to 15. From that low point the population had climbed to a total of 59 in 1971. The two crosses on the map indicate the nesting and wintering areas of this population.

The 4000 km long migration of the whooping crane is executed in spring in units of 300–500 km, several days being passed in transient home ranges in Nebraska and Saskatchewan. The young bird migrates S in autumn with its parents but in spring it migrates N independently.

[*Compiled from Voous (1960), Dorst (1962), Vaurie (1965), Moreau (1972), and Goldman (1973). Photo of* Grus grus *by Arthur Christiansen (courtesy of Frank W. Lane)*]

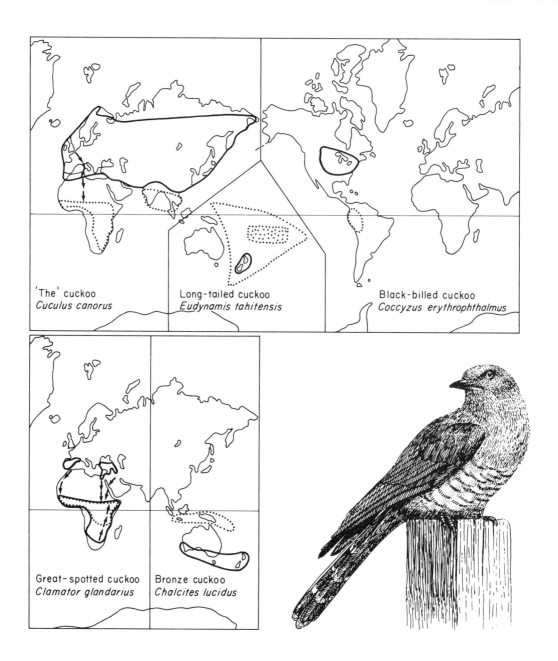

Fig. 27.38 The seasonal return migrations of five species of cuckoo

Cuckoos and other members of the Cuculidae (a family of about 128 species) are found in nearly all warm and temperate parts of the world. Many species do not build nests but are parasitic, laying their eggs in the nests of other bird species.

⬭ breeding range; ⬝⬝⬝ non-breeding range; →→ some known or suspected migrations.

The stippled rectangle within the general winter range of the long-tailed cuckoo indicates the major wintering area of the species.

Note that in the great-spotted cuckoo, a species parasitic primarily on corvids, northern and southern demes probably move N and S together and so may overlap little or not at all in equatorial regions.

[Compiled from Voous (1966), Peterson (1961), Dorst (1962), Vaurie (1965), Robbins et al. (1966), Moreau (1972), and Mead (1974)]

Lissaman and Shollenberger (1970) have calculated that any formation flight improves efficiency. Theoretically, 25 birds could increase their range per unit fuel by 70 per cent compared to a lone bird, the advantage being greater with a tail-wind. The V formation can be calculated to distribute the load evenly amongst the birds involved. Even the lead bird need not necessarily have a greater load. In fact, the optimum formation is not a precise V but rather a form somewhat intermediate between a V and a U. Optimum flight speed is slower when flying in formation than when solo.

The advantage of an increase in the number of familiar transient home ranges seems to be particularly great for birds that migrate over long total distances and indeed, as pointed out by Hamilton (1967) in support of the precision orientation hypothesis, a general correlation does seem to exist between group migration and longer-distance migration. Within the genus *Buteo*, for example, broad-winged hawks, *B. platypterus*, and Swainson's hawks, *B. swainsoni*, migrate relatively long distances and migrate in flocks whereas red-tailed hawks, *B. jamaicensis*, and red-shouldered hawks, *B. lineatus*, migrate shorter distances (Fig. 27.14) and migrate singly or in dispersed, malformed flocks (Bellrose 1972). I do not consider, however, that a correlation between flock formation and the required precision of navigation has been demonstrated. The major example presented by Hamilton (1967) concerned cuckoos. One species, the long-tailed cuckoo, *Eudynamis tahitensis*, breeds in New Zealand and migrates between 1000 and 3000 km to winter quarters on the high-volcanic and low-atoll Pacific Islands (Fig. 27.38). The required precision of orientation is considered to be high and the bird forms large flocks before and during migration. Most cuckoos, however, migrate solitarily, a fact which Hamilton attributes in the case of North American cuckoos to the relatively broad destination, the tropical regions of Latin America (Fig. 27.38). In view of the demonstration of the high degree of return to winter home ranges by individuals of many species of temperate-tropical migrants (p. 620), it does not necessarily follow, however, that the precision of navigation of these solitary cuckoos is in any way less than that of the flocking species.

Birds that migrate by soaring form loose, wheeling flocks in rising air currents which may be produced by wind deflected from a slope or by air that is heated by the ground. Storks, raptors, vultures, and also occasionally seagulls (when over land) use this migration mechanism. Generally 10–70 raptors may participate in a single thermal though there may be 250 or even more (Dorst 1962). Soaring birds do not generally begin to migrate before about 10.00 hours because they have to wait for upward currents which do not develop until the Sun has warmed the ground. Then the birds spiral upwards in each bubble of rising air and from the height thus gained glide to the next bubble forming along the migration route (Fig. 27.39).

Pennycuick attributes the evolution of migration by soaring and gliding to the reduction in the distance of each migration unit that otherwise should be experienced by large birds. Theoretically, birds with a fat-free weight greater than 1·5 kg should not be able to fly with a fuel load equal to the fat-free weight and so should experience a shorter flight range than smaller birds. The maximum load that can be carried by birds above this fat-free

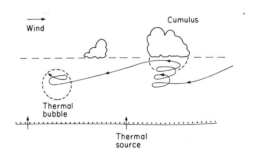

Fig. 27.39 The mechanism of migration by soaring and gliding

Once the Sun begins to heat the ground, various areas begin to act as thermal sources above which the air heats up more than over surrounding areas. The thermal bubble thus formed begins to rise. When such a bubble reaches the dew point, a cumulus cloud is formed, though many bubbles disperse before they reach the dew point, in which case there is no visible manifestation of the thermal.

Birds that migrate by soaring and gliding spiral round inside thermals, gaining height from the rising air, and then glide down until they encounter another bubble, when the process is repeated.

Apart from within thermals, air currents may have a vertical component in a number of other situations such as when wind is deflected upwards by the side of a hill or cliff, etc. Many birds also make use of these currents during migration.

[*Re-drawn from Dorst (1962) (after Forster)*]

weight gradually decreases with increased weight until it becomes zero at a fat-free weight of about 12 kg. It is implied that large migrants such as Accipitridae (approximately 2 kg) and Ciconiidae (3–4 kg) have evolved soaring flight in response to selection acting in response to this range reduction.

There is no indication that soaring flight increases the distance of each migration unit, however, and indeed, as pointed out by Pennycuick, some birds, such as geese and swans, which are among the largest migratory birds, nevertheless continue to migrate by flapping flight.

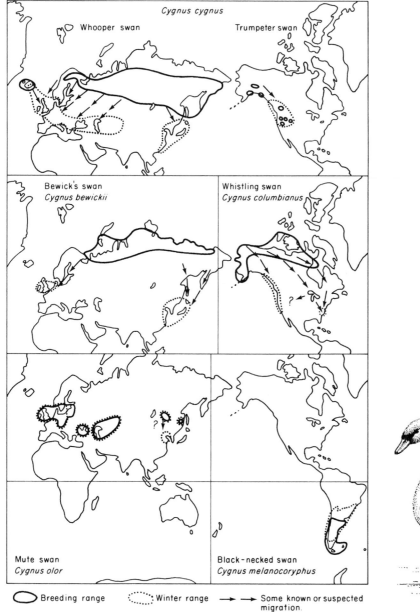

◯ Breeding range   ⣀⣀: Winter range   ⟶ ⟶ Some known or suspected migration.

The primary advantage of soaring flight seems to be that it involves minimum fuel consumption. Measurement of the energy cost of gliding flight in the herring gull, *Larus argentatus*, demonstrated an oxygen consumption that was only twice the resting level compared with an oxygen consumption during flapping flight that is seven times the resting level (Baudinette and Schmidt-Nielson 1974). Migration by soaring flight, however, need not necessarily result in a reduction in migration cost. Ground-speed is much reduced, with the result that the birds are in the air much longer per unit distance compared with birds that migrate by flapping flight. Conditions suitable for low-cost migration are also likely to be less frequent than for flapping flight and changes in the weather seem much more likely to cause a soaring bird to fail to reach a particular transient home range except by high-cost flapping flight, adaptation for efficient soaring probably reducing the efficiency of flapping flight. Adaptation

for soaring thus seems likely to increase the chances of a bird being stranded for a period in an area of low suitability. Whether any particular lineage becomes adapted to migrate by soaring or flapping, or both, seems likely, therefore, to be a function of on the one hand the relative contribution to migration cost of the probability of being stranded in an unsuitable area, and on the other hand of energy expenditure. If this is so, then evidently the relative contribution of these two factors is not solely a function of body weight. In the hawk and stork lineages, it would seem that energy expenditure made the major contribution to migration cost whereas in the geese and swan lineage(s) it seems that the probability of being stranded in an unsuitable area made the major contribution to migration cost. The reasons for this difference remain unsolved but may be associated in some way with the more-aquatic habits of the latter birds.

Any thesis based on the minimisation of migration

Fig. 27.40   Seasonal return migrations of five species of swans (Cygninae)

Swans, along with ducks and geese, are members of the Anatidae. Swans are larger and have longer necks than geese and are also more aquatic. They feed by immersing the head and neck, and sometimes by 'up-ending', on a diet that consists primarily of aquatic plants and seeds.

Swans migrate, often in a V formation at a speed of 50–80 km/h, and telemetry has shown that several hundred kilometres may be covered in a single night. As with geese, the young birds perform the first year's migration circuit with their parents. Evidently swans have a marked

capacity for long-term individual recognition. Observation of Bewick's swans has shown that in their second winter young swans arrive independently from their parents (presumably having spent the summer in independent exploratory migration and habitat assessment associated with attempting to find a suitable breeding site). If, however, both parents and second-year young arrive at the same winter home range, they reunite and move as a family, even if the parents are accompanied by young-of-the-year.

[*Compiled from Scott and The Wildfowl Trust (1972)*]

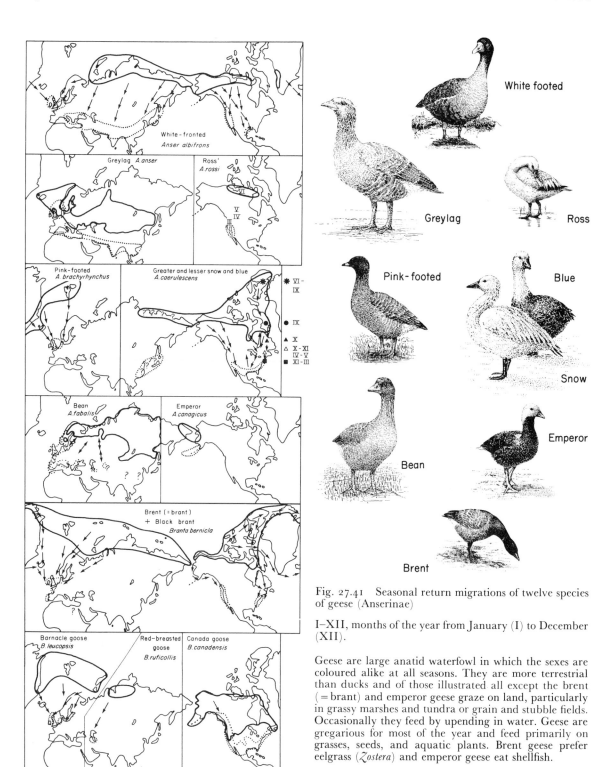

Fig. 27.41   Seasonal return migrations of twelve species of geese (Anserinae)

I–XII, months of the year from January (I) to December (XII).

Geese are large anatid waterfowl in which the sexes are coloured alike at all seasons. They are more terrestrial than ducks and of those illustrated all except the brent (= brant) and emperor geese graze on land, particularly in grassy marshes and tundra or grain and stubble fields. Occasionally they feed by upending in water. Geese are gregarious for most of the year and feed primarily on grasses, seeds, and aquatic plants. Brent geese prefer eelgrass (*Zostera*) and emperor geese eat shellfish.

[*Compiled from figures and descriptions in* Bannerman (*1957*), Voous (*1960*), Peterson (*1961*), Dorst (*1962*), Robbins et al. (*1966*), Orr (*1970a*), Heyland (*1973*), *and* Peterson et al. (*1974*)]

Barnacle   Red breasted

Canada

cost such as the one presented in this section makes the premise, if migration cost receives a significant contribution from energy expenditure, that energy used during migration cannot be replenished in the next transient home range without incurring some disadvantage. Furthermore, the assumption must be made that the greater the amount of energy used, the greater the disadvantage of replenishment. This premise is not an axiom. During the migration period, selection has favoured birds with a physiology geared to the rapid and efficient usage and replenishment of fuel reserves (Dorst 1962, Newton 1972). A bird that arrives in a transient home range with fuel reserves totally exhausted, by feeding with sufficient intensity throughout the remainder of the 24 hours could accumulate enough fuel reserves to sustain a further migration unit during the next appropriate phase of the diem. How, then, does the bird gain an advantage through using as little of its fuel reserves as possible in the performance of a migration of a given distance? The answer, if the premise is justified, can only be that the advantage lies in the reduced time investment in feeding that will be necessary in the next transient home range to be visited. Evidence that supports this conclusion may be derived from the observation that chaffinches, *Fringilla coelebs*, invest only just enough time in feeding in a transient home range as is necessary to attain what appears to be a threshold weight (acting through the relationship between fuel reserves and migration thresholds to migration cost variables—

p. 644). Individuals that arrive with a weight that is approximately threshold for the next migration unit (see below) may not feed at all (Dolnik and Blyumental 1967). Whether the advantage derives from the reduced time investment in feeding or from the increased time investment in other activities, such as roosting, cannot as yet be determined. Quite conceivably, however, a bird suffers a greater predation risk while feeding than when not feeding and so perhaps any increase in the proportion of time invested in non-feeding activities while in a transient home range could be advantageous.

The optimum amount of fat to be laid down in a transient home range before initiating the next migration unit may be divided into two components (Pennycuick 1969). One component represents the fuel required to complete the next migration unit. The other component represents a contingency fuel reserve in case a head-wind springs up on the way or in case the bird has to allow itself to drift to a previously unfamiliar area. Pennycuick suggests that for land-birds crossing the sea or a desert it may be advantageous to carry a reserve that is 100 per cent of the still air requirements in small birds and 50 per cent in large birds. A third component can be added to the two suggested by Pennycuick, particularly in the case of young birds. This is the fuel reserve necessary to perform exploratory non-standard migrations and generally to become sufficiently familiar with the transient home range for feeding to become efficient (p. 645).

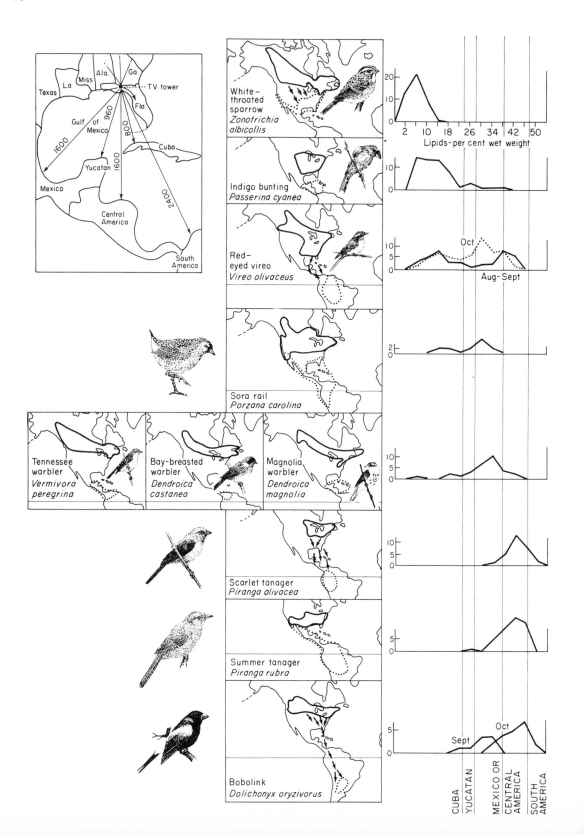

Fig. 27.42 The fat content of migrating birds as a function of the migration unit about to be performed

Every year a variety of nocturnal migrants are killed by flying against tall human artefacts such as radio masts and television towers. The diagram illustrates an analysis of the fat content of birds killed during autumn migration by striking a television tower situated on the north coast of the Gulf of Mexico (map, top left). The birds concerned belonged to a variety of species with a variety of destinations as indicated by the distribution maps (short distance migrants at the top).

⬭ breeding range; ⁖⁖⁖ non-breeding range; →→ some known or suspected migrations.

The right-hand column shows the number of individuals of each species or group with a particular fat content. The vertical lines indicate the range of destinations that are available by non-stop migration to a bird of a given fat content.

It can be seen that birds arriving at the tower usually have a fat-content appropriate to the migration unit about to be performed. There is therefore a clear inference that selection has acted to favour different fat mobilisation programmes in the different species. Without knowing how long the birds have been in the vicinity of the Gulf Coast, however, the analysis does not allow us to determine whether the programmes are entirely endogenous or whether some physiological feedback also occurs from, for example, perception of the sea. Certainly for all species except the two tanagers, individuals occur that have insufficient fat for any sea crossing and that ideally should remain on the Gulf Coast and feed up until they have laid down more fat. There is some indication for the bobolink and perhaps the red-eyed vireo that either the endogenous part of the fat mobilisation programme changes as autumn proceeds or different demes with different programmes (that perhaps reflect adaptation to different migration routes) pass by the tower in different months.

[Maps drawn from figures and descriptions in Odum (1960), Peterson (1961), and Robbins et al. (1966). Fat data from Odum (1960)]

The threshold value of fuel reserves for the initiation of a migration unit (p. 644) should be a function of the likely cost of that migration unit. According to the above suggestions, this cost is a function of distance, the probability of encountering a head wind yet being unable to land, and the cost of exploratory migration and the level of feeding efficiency at the next transient home range. Adult birds performing a calculated migration have information concerning all of these factors and physiological feedback from the memory to the migration threshold and fat deposition system seems likely.

Young birds, and those adults that are also performing exploratory migration, however, have less information available. It seems likely that the migration thresholds increase upon perception of unsuitable terrain such as sea or desert for land-birds and continental land for sea-birds (p. 633). Sudden increase in migration threshold should cause many individuals to cease migration upon perception of unfavourable terrain or even to perform calculated reversal migration to the last suitable habitat encountered, there to remain until the migration threshold for crossing such terrain is exceeded. The level of such thresholds is determined by selection. Whether there is a general system of thresholds for each species, one adaptive to 'sea', another for 'desert', etc., or whether selection has favoured a more sophisticated system with, in the case of North European birds, for example, one threshold appropriate to the Baltic, another to the Mediterranean, another to the Sahara Desert, etc. must await further investigation. If the latter system has evolved it seems likely that selection can only have acted to some extent for an endogenous programme of fat deposition for exploratory migrants. Calculated migrants, however, could achieve a highly adaptive strategy through selection on the appropriate physiological interaction between the deposition of fat and an endogenous programme, physiological feedback from a spatial memory, and topographical perception.

Whatever the mechanism, there seems little doubt that birds deposit fat of an amount adaptive to the migration unit about to be performed. In the Old World, passerines lay down more fat before crossing the Sahara than before any other migration, at least in Europe (p. 648). In the New World, captures of passerines arriving in autumn at the Gulf Coast of North America showed that those species that cross the Gulf to the Caribbean Islands, Central or South America were carrying more fat than those that winter along the Gulf Coast (Fig. 27.42). Different fat thresholds for different migration units coupled with the existence of contingency components to the fuel reserves can also account for the observation that some individuals of the chaffinch, *Fringilla coelebs*, arrive in transient home ranges south of the Baltic carrying more fat than the apparent threshold for the next migration unit. Such birds, which perhaps experienced low migration cost while crossing the Baltic, stay in their transient home range without feeding and may even lose weight between arrival and the initiation of the next migration unit

(Dolnik and Blyumental 1967). Other birds that perhaps experienced a higher migration cost, or previously occupied a less suitable transient home range and arrive with low fat reserves, feed up and replenish their reserves before initiating the next migration unit.

So far, the adaptations that evolve in response to maximisation of $h_w M_R/h_n$ have been discussed in two halves: those resulting from selection for a maximisation of h_w/h_n and those resulting from selection for a maximisation of M_R. However, because optimum adaptation to each may often differ, the observed adaptation is likely to be the optimum compromise between these two pressures. Thus selection on optimum length, duration, and frequency of the migration units of species that perform short-distance total migration may well favour dates of departure from H_n and arrival at H_w that permit maximum expression of selection for short-distance, frequent migration units. Selection on dates of departure from H_n and arrival at H_w by long-distance migrants may well shift the optimum length, duration and frequency of migration units away from the short-distance, frequent end of the spectrum, particularly if longer-distance migration units produce a higher average migration rate (km/day) than shorter-distance migration units. That longer-distance migration units produce a faster average migration rate seems likely even though an increase in distance brings about an increase in the energy requirements and hence an increase in the time investment in feeding and even perhaps total stay-time in each transient home range.

Support for this postulated compromise between selection acting through increase in h_w/h_n and selection acting through increase in M_R is provided by a comparison of different breeding demes of the chaffinch, *Fringilla coelebs*. Those that migrate from breeding grounds in northern Finland to winter grounds in France and Iberia migrate 3000 km in about 5 weeks, thus achieving an average rate of 90 km/day (Dolnik and Blyumental 1967). Those chaffinches that migrate from Britain to Iberia, however, a distance of only 1200 km, achieve an average rate of only 40 km/day. This does not necessarily prove that the former perform longer migration units than the latter as would be expected from the model, but it seems likely.

So far we have been concerned entirely with the frequency and duration of the migration units that comprise the autumn or post-nuptial migration. Little needs to be added for the same arguments to

be applicable to the spring or pre-nuptial migration except in relation to a comparison of the average rates of the two migrations. In spring, the advantage seems to lie with the first individuals to arrive in the breeding area (p. 648) after the date that conditions become just suitable enough for survival to be possible. Selection for the maximisation of $h_n M/h_w$ during the spring migration thus receives a much greater contribution from h_n/h_w due to the importance of the date of arrival at the breeding site than does selection for the maximisation of $h_w M_R/h_n$ during the autumn from h_w/h_n. The contribution from factors that decrease migration cost (i.e. increase M) are thus correspondingly less important. Depending on how much more critical is the date of arrival at the breeding site relative to the date of arrival at the winter site for any particular deme, the average rate of migration in spring is the same, a little, or much greater than the average rate of migration in autumn. In general, it seems that the average rate is greater in spring than autumn (Dorst 1962). In those demes of the wood warbler, *Phylloscopus sibilatrix* (Fig. 27.35), that breed in northern Germany, for example, the average rate of spring migration (approx. 180 km/day) has been estimated to be about twice the average rate of the autumn migration (Stresemann (in Dorst 1962)).

Finally, in this section on migration distance, let us consider the selective pressures that act on the total one-way migration distance in birds that perform a to-and-fro annual migration. According to the model (p. 82), and as already described in this chapter (p. 678), autumn migration terminates in that area for which h_1 remains greater than $h_2 M^x$ for migration in any direction until such time in spring that the first migration unit of the pre-nuptial migration becomes advantageous. Some evidence that supports this assertion derives from the observation that a great many migrations (e.g. Figs. 27.1, 27.7, and 27.21) terminate in areas from which the cost of further migration units would be higher than for previous migration units. Such areas are those that border stretches of sea (e.g. Mediterranean and Gulf of Mexico), desert (e.g. the Sahara), or other unsuitable habitat-types (e.g. the equatorial rain forest for many palaearctic visitors to Africa).

One particularly noticeable feature of the total migration distance of bird migration is the inter-deme variation that leads to 'leap-frogging'. Many, perhaps even the majority, of land-bird species show this phenomenon, two examples of which are shown in Fig. 27.43. Briefly, leap-frogging results from those demes that breed at higher latitudes (or, more accurately, at the most disadvantageous end of the winter gradient of habitat suitability that acts as a selective pressure on the direction of autumn migration—p. 658) having a longer total migration distance such that they winter at lower latitudes than those demes that breed at lower latitudes. In some species the disparity in total migration distance

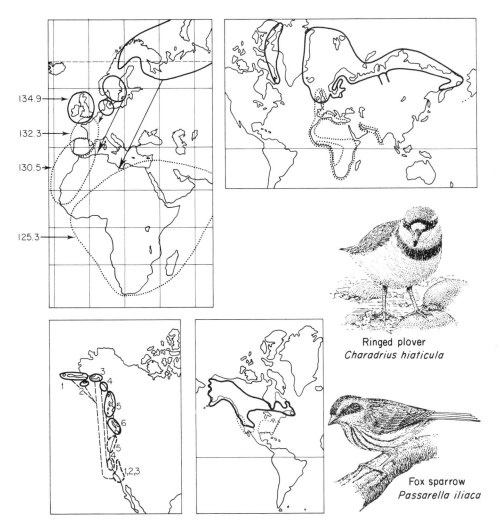

Ringed plover
Charadrius hiaticula

Fox sparrow
Passarella iliaca

Fig. 27.43 Inter-deme variation in migration distance: the leap-frog effect

◯ breeding range; ⋯⋯ non-breeding range

Demes of the two species illustrated provide clear examples of the phenomenon of leap-frogging, whereby the demes that breed at higher latitudes migrate to lower latitudes than demes that breed at intermediate latitudes. The right-hand diagrams show the seasonal population shift of the two species. The left-hand diagrams show the migrations of different demes within these populations.

Note that the British Isles deme of the ringed plover and the Vancouver Island, etc. deme (deme 6) of the fox sparrow show an extremely low migrant/non-migrant ratio and short or zero seasonal return migration distance and are more or less 'resident'.

The figures to the left of the 'winter' quarters of the ringed plover give the average wing length (mm) of the deme concerned. There is a clear cline of wing-length such that birds with shorter wings migrate further, overwinter in warmer areas, and breed in cooler areas than birds with longer wings.

[*Compiled from Voous (1960), Peterson (1961), Dorst (1962) (after Salomonsen and Swarth), Vaurie (1965), Robbins et al. (1966), and Moreau (1972)*]

is only such that the most migratory deme winters in the same area as the least migratory deme (Fig. 27.21).

Many authors (e.g. Dorst 1962) do not accept that a leap-frog pattern can be attributed to intra-specific competition. Others (e.g. Mayr and Meise 1930, Cox 1968) consider that intra-specific competition is the only factor that could lead to such a phenomenon. The latter is the view that is subscribed to here. It is considered that intra- and, to some extent, inter-specific competition has acted and still acts as a selective pressure on the endogenous determinants of migration distance. At the same time, as a factor that contributes to the composite migration threshold, it seems likely that deme density will also be involved at the proximate level in the determination of migration distance.

Let us consider the deme that breeds at the lowest latitude at which the species breeds in an area where it is possible for the entire deme to winter. Now consider a deme adjacent to this deme but at a slightly higher latitude and that consequently shows partial migration, some individuals being resident, others migrating to lower latitudes and encountering the sedentary deme. The migrant/non-migrant ratio, given a free situation, should give equal advantage to migrants and non-migrants (p. 635). Some migrants may settle among the sedentary deme. All migrants may do so if further migration to lower latitudes is disadvantageous. However, if further migration is not prohibitively disadvantageous, a free distribution is obtained by some, perhaps the majority, of the migrants by distributing themselves at lower latitudes than the sedentary deme. When we now consider the optimum strategy for a deme at even higher latitudes, it is clear that the greatest 'vacuum' for the formation of a free situation lies to the lower-latitude side of the wintering demes already described, though in order to maintain a free situation, some must also settle among the non-migrant portion of the partially migrant deme and some must also settle among the sedentary deme. Basically, therefore, the intra-specific competition component of the latitudinal gradient of winter habitat suitability (acting primarily through the availability of food and other resources) seems to be sufficient to generate the observed characteristics of inter-deme variation in migration distance. Where the lowest latitude breeding deme occupies an area that borders on an area of high migration cost, the migration of all demes is likely to terminate in this area and the free situation is built from this

area back along the standard migration axis. Where this is not the case and migration to lower latitudes is advantageous, the free situation is built in both directions from the latitude of the sedentary deme, again along the migration axis. In this case, the free situation towards higher latitudes results from latitudinal variation in the migrant/non-migrant ratio and towards lower latitudes from the leapfrog effect of migration distance. Finally, if there is no sedentary deme but it is assumed that the lowest latitude breeding deme can arrive first in the nearest suitable wintering latitude, then it is this deme that performs the shortest distance migration and the free situation builds from this deme but only further in the migration direction.

It should be noted that in neither winter nor summer should the habitat suitability experienced by all individuals be the same. Because some demes migrate further than others it would be expected that in a free situation the frequency distribution of migration distances should stabilise such that there is a differential in the habitat suitability experienced. This frequency distribution should be such that long-distance migrants experience winter and/or summer habitats that are sufficiently more suitable than those of short-distance migrants as to just balance the extra migration cost experienced. Where all migrants land in the same winter area and no leap-frogging occurs (e.g. the goldfinch, *Carduelis carduelis*—Fig. 27.21) this differential must come solely from the suitability of the breeding habitat. Where leap-frogging occurs (e.g. Fig. 27.43) it seems likely that the geographically most extreme non-breeding and breeding areas are slightly more suitable than intermediate areas.

One final aspect of the migration distance of landbirds that merits comment is that of the difference between birds that breed in the Northern and Southern Hemispheres. Every year, many landbirds that have bred in the temperate and sub-polar zones of the Northern Hemisphere cross the equator and spend the northern winter in the Southern Hemisphere. As far as I am aware, however, not a single land-bird that breeds in the Southern Hemisphere passes the southern winter on continental land in temperate or higher latitudes in the Northern Hemisphere. This difference is not, however, a question of migration direction. In both hemispheres, the selective pressures acting on direction (p. 658) have acted to favour migration towards higher latitudes for breeding and towards lower latitudes for winter. Instead, the difference is a question of migration distance, some birds from the Northern, but none from the Southern, Hemisphere continuing across the equator beyond lower latitudes and on towards higher latitudes in the opposite hemisphere.

The reason for this difference can undoubtedly be attributed to the difference in land area in the two hemispheres. Birds from the Northern Hemisphere, as they migrate S, are likely to be forced to converge on the E–W axis, whereas birds from the Southern Hemisphere, as they migrate N, are able to disperse on an E–W axis. Selection for any given degree of dispersal must therefore favour a greater migration distance for birds from the Northern than from the Southern Hemisphere (p. 654). If this thesis is correct, the opposite situation should be found in sea-birds because the area of sea is greater in the Southern than the Northern Hemisphere. To some extent, this expectation is fulfilled for, although some Northern Hemisphere birds, such as the Arctic tern (Fig. 27.45) are long-distance trans-equatorial migrants (perhaps due to the influence of association with coasts), the longest-distance, more-pelagic, migrants, such as the slender-billed shearwater (Fig. 27.30) and Wilson's petrel (Fig. 27.31) breed in the Southern Hemisphere.

It seems, therefore, that the differences in migration distance between birds that breed in Northern and Southern Hemispheres adds further emphasis to the importance of intra-specific competition (or, perhaps more accurately, selection for the optimum degree of dispersal or convergence) in the determination of the frequency distribution of migration distances.

27.9 The post-glacial history of bird migration

Although bird migrations of patterns essentially similar to those observed at the present day must have evolved at the same time and in association with the evolution and radiation of birds with flapping flight (p. 605), there seems little doubt that most modern migration routes are post-glacial in origin. At the height of the last glaciation, most major vegetational zones occupied lower latitudes than at the present (Fig. 27.44), and the vegetation of large areas of the tropics was quite different in character (see Moreau 1966 for Africa).

It would almost certainly be a mistake, however, to envisage that 20 000 years B.P. the tropical and sub-tropical regions were crowded with birds equivalent to those that at the present time occupy both the tropical and temperate regions. Instead, it seems probable that the glacial advances and retreats of the Pleistocene witnessed widespread

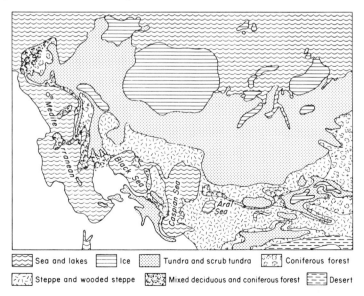

Sea and lakes   Ice   Tundra and scrub tundra   Coniferous forest

Steppe and wooded steppe   Mixed deciduous and coniferous forest   Desert

Fig. 27.44   Major vegetation zones of Europe and western Asia about 18 000 years B.P. at the height of the last glaciation

[Modified from Moreau (1972) (after Frenzel and Troll)]

changes in distribution patterns and migration routes and widespread extinction and reduction in the abundance of bird species (Mayr and Meise 1930). It is doubtful whether Africa or Central America at the peak of the last glaciation supported a density and diversity of birds any greater than that found in comparable warm temperate and subtropical areas at the present day.

Even at the peak of the last glaciation, seasonal return migrations must still have been performed. All of the selective pressures described in this chapter must have been in existence at that time and there is no reason to suppose that suitably adaptive migrations would not have occurred. Tundra-breeding species would still have found it advantageous to migrate to lower latitudes for winter as also would the inhabitants of the coniferous and deciduous forest zones, etc. The migration distance may have been much less then than now for many, though not necessarily all, of the lineages that have survived to the present day. Undoubtedly, however, some species must have performed trans-equatorial migrations due to the gradient of increased habitat suitability generated by gradients of day length, temperature, and the availablity of food (p. 658).

As the glaciers began to retreat towards the end of the last glaciation (approximately 15 000 years B.P.), the vegetational zones of temperate regions would have begun to shift to higher latitudes, particularly in the Northern Hemisphere, while the vegetational pattern of the tropics would have begun a gradual but eventually spectacular change, particularly in relation to the formation of areas of grassland and desert (Moreau 1966). We may envisage that through the process of exploratory and calculated migration, particularly by young birds, the breeding distribution of many lineages expanded to higher latitudes with the retreat of the ice. Whether the lower latitude edge also shifted to higher latitudes or whether speciation occurred, leading to high latitude/more migratory and lower latitude/less migratory (perhaps 'sedentary') species must undoubtedly have depended upon the specific array of selective pressures to which the lineage was subjected. In some lineages, also, the wintering range is likely to have shifted in the same direction as the breeding range, the total migration distance thus remaining relatively constant. In other lineages, the winter range is likely to have retained its original latitude or even to have shifted in the opposite direction to the breeding range, the latter particularly in the case of trans-equatorial migrants. In

either case, the total migration distance is likely to have increased. Which of these three options was adopted by any particular lineage should have been a function of the direction of shift of the habitat-types to which the lineage was adapted or pre-adapted.

In modern species, different demes breeding in different areas, migrating different distances, and often wintering in different areas, frequently manifest racial or sub-specific differences. During the retreat of the glaciers, many lineages must have formed several species as a result of the different conditions encountered by different demes during expansion of the range occupied by that lineage. Inter-specific differences in migration routes and distances can be envisaged to have crystallised out from an expanding range during the post-glacial period. The answer to the question 'why does species X migrate only from A to B whereas species Y migrates from A to D?' seems to be that it was the selective pressures encountered as a result of migration from A to B by a deme of the XY lineage that led to the formation of species X and the selective pressures encountered as a result of migration from A to D by another deme of the same lineage that led to the formation of species Y. The migrations of the European warblers of the genus *Sylvia* (Fig. 27.35) provide examples of the present position of such a process.

As a further example, but one that probably has nothing to do with the post-glacial period, glaciation probably having had relatively little influence on the migration routes of sea-birds, the lineage of the genus *Sterna* (terns) expanded over the Northern Hemisphere (as well as the Southern Hemisphere) to give rise to a number of species that breed at a variety of latitudes. One branch of the lineage that presumably speciated from higher latitude breeding demes performed trans-equatorial migrations down and up the coasts of the various continents. As with all shore- and sea-birds, migration routes were highly adaptive to the direction of the prevailing wind (p. 672) and also as with all shore-birds a characteristic frequency distribution of migration distances evolved, presumably in response to selection on the optimum degree of dispersal and convergence (pp. 654 and 667). Different demes with different migration distances must have been exposed to different selective pressures with the result that speciation occurred, each species consequently having a different migration distance. The longer-distance deme encountered the Southern Ocean.

Selection could have acted against individuals that entered this Ocean in which, as we have seen (p. 624), selection on sea-birds seems to favour circumpolar migration in the direction of the prevailing westerlies. However, this long-distance deme of *Sterna* evidently was not subjected to such absolutely prohibitive selection and gave rise to the Arctic tern, *S. paradisea*. It remains to be proved, but it seems highly likely that adaptation to the Southern Ocean by this lineage has also involved circumpolar migration as indicated (Fig. 27.45).

Two further points arise from a consideration of the post-glacial history of bird migration. The first concerns those lineages that during the peak of the last glaciation were confined to a single tropical population within a relatively circumscribed area.

It seems likely, for example, that the lineages of many widespread modern Palaearctic species were confined during the last glaciation to a range within tropical Africa. During the post-glacial period, as the breeding range of the lineage expanded, exploratory migrants must have encountered suitable wintering areas in other tropical regions and are now spread right across the Old World during both winter and summer (e.g. Fig. 27.46). In other species, however, it seems that exploratory migrants have so far failed to find suitable winter habitats anywhere in the tropics other than in Africa. For example, four sub-species of the wheatear, *Oenanthe oenanthe*, are recognised, two of which are resident in Africa, one of which nests in Northeastern Canada, Greenland, Iceland, Jan Mayen, and the Faeroes,

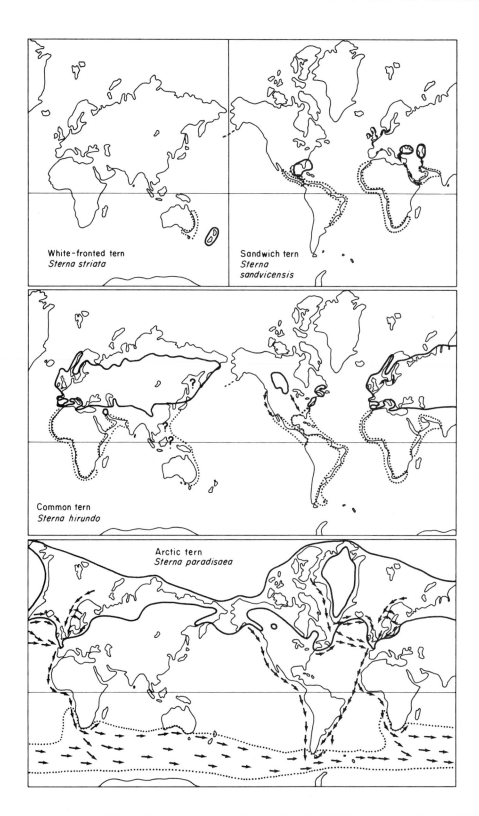

White-fronted tern
Sterna striata

Sandwich tern
*Sterna
sandvicensis*

Common tern
Sterna hirundo

Arctic tern
Sterna paradisaea

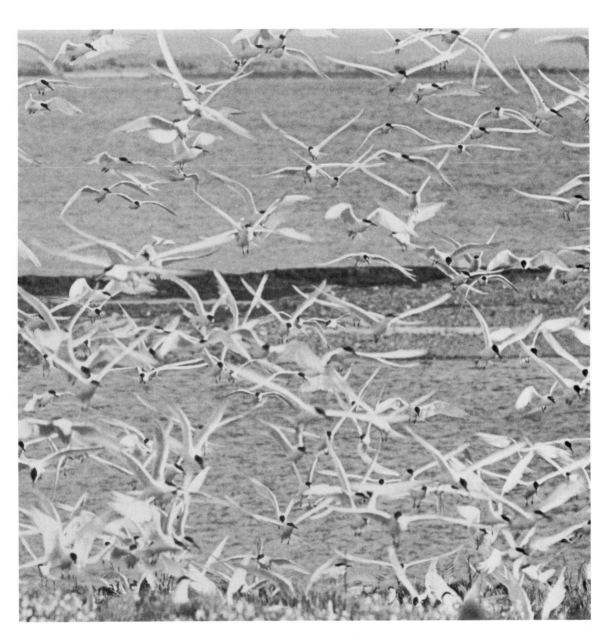

Fig. 27.45 Seasonal return migrations of four species of terns (Laridae: Sterninae)

⬭ breeding range; ⫶⫶⫶ non-breeding range; → → some known or suspected migrations.

Terns are an almost cosmopolitan group of about 39 species of graceful, streamlined, and manoeuvrable water birds. They often hover and plunge headfirst into the water but do not normally swim. Their diet consists primarily of small fish and large insects and, while at sea, various members of the zooplankton.

Of the species illustrated, the three Northern Hemisphere terns show a leap-frog pattern to their migrations, those breeding furthest from the equator in the Northern Hemisphere (i.e. Arctic tern) migrating furthest from the equator in the Southern Hemisphere. This pattern could be taken to suggest that the three species and their migration circuits crystallised out from a generally distributed common ancestor that, like so many species (p. 696), evolved a leap-frog pattern.

[*Drawn from descriptions and diagrams in Voous (1960), Peterson (1961), Bannerman (1962), Dorst (1962), Robbins et al. (1966), Lemmetyinen (1968), Peterson et al. (1974). Photos by C. Stephen Robbins*]

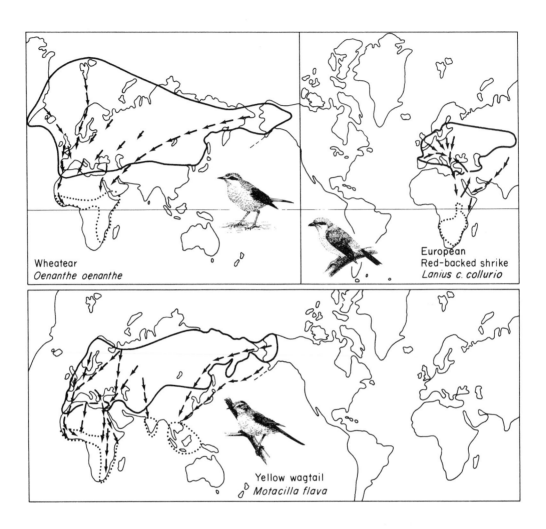

Fig. 27.46 The possible influence of post-Pleistocene history on the modern migration routes of three species of palaearctic birds

⬭ breeding range; ⟨⟨⟨⟩⟩⟩ non-breeding range; arrows, some known or suspected migrations → → autumn; ⟶ ⟶ spring).

Although other explanations are possible, let us assume that at the height of the last glaciation the lineages of the three illustrated species were entirely confined to Africa in winter and became adapted to particular habitat-types on that continent at that time. Given this assumption, modern migration routes are to a large extent explicable in terms of the post-Pleistocene history of these lineages (in conjunction with selection on migration direction— p. 658).

With the amelioration of the climate and the northward migration and expansion of the various vegetation zones (see Fig. 27.44), it seems likely that the breeding range expanded to the north and east across Europe and Asia. It seems that autumn exploratory migrants of the shrike and wheatear lineages failed to find suitable winter habitats in India and southeast Asia with the result that the only demes to survive were those with an autumn migration direction to the WSW and Africa. Although the advantage per unit distance of the autumn migration may be less for these demes than for demes that migrate at right angles to the winter isotherms (Fig. 27.22), the demes may still persist as long as the suitability of the breeding habitat is sufficiently great as to exceed migration cost. In the yellow wagtail lineage it seems that autumn exploratory migrants to India and southeast Asia succeeded in locating suitable habitats. Such individuals, because of reduced migration cost, were thus at an advantage relative to any individuals that, also breeding in northeast Asia or Alaska, migrated to Africa.

[*Wheatear and shrike maps modified from Dorst (1962). Yellow wagtail map drawn from descriptions and maps in Moreau (1972)*]

and the other of which nests across northern Europe and Asia to Alaska. In winter, all of these sub-species are confined to Africa (Fig. 27.46). This situation is perhaps the classic example of the influence of post-Pleistocene history on moden migration routes. On the familiar-area hypothesis, however, it is only fully explicable if it is assumed that all winter exploratory migrants have failed to encounter suitable habitats elsewhere in the tropics. The existence of such exploratory migrants is supported by the occasional winter occurrence of individuals in the eastern United States (Moreau 1972). As the winter range of this species is bare ground in savanna and steppe, however, the failure of exploratory migrants to encounter and establish themselves in suitable alternative areas may perhaps be accepted as an explanation of the persistence of ancestral routes in this species. Except possibly for certain areas of southeast India, there are few areas of the tropics other than Africa that seem to provide the winter conditions to which the species became adapted before or circa the beginning of the post-Pleistocene era.

The second and final point concerns the competition between tropical and Northern Hemisphere temperate birds that must inevitably occur during the northern winter. Every year, the neo- and palaeo-tropical regions receive during the northern winter a tremendous influx, respectively, of nearctic and palaearctic migrants (Orr 1970a, Moreau 1972). In Africa 185 species of Palaearctic migrants insinuate themselves every year into a resident avifauna of about 1480 species. After correction for the 409 African species that live in the evergreen forest that is not the prime winter habitat of any migrant species and after correction for the prevalence of superspecies in Africa, the comparable figures become 185 and 770. Thus, during the northern winter, about 20 per cent of the African non-forest avifauna consists at the species level (no data are available at the individual level) of immigrant birds (Moreau 1972). Inevitably a competitive element must exist in this and the comparable New World situation, even though as demonstrated by Moreau for Africa, in the majority of cases there is spatial and/or ecological separation between the resident and migrant species. Such separation can only have resulted from competitive exclusion during evolution and implies that during the northern summer, niches are available that are exploited with reduced efficiency by the resident tropical avifauna.

The situation is unlikely to be peculiar to modern times. Even during the last glaciation, seasonal return migration by some species but not others must have led to a similar situation. As the breeding range of migrants expanded during the glacial retreat, however, and as, presumably, the migrant population increased in number and perhaps variety, the equilibrium position must have changed until reaching the situation obs ved at the present time. The critical questions cc cern the influence of migrant/resident competitic on: (1) the population size of both residents and migrants; (2) the migration distance of migrants; and (3) the feeding and other adaptations of both groups.

It seems inevitable that the population size of residents and migrants is influenced by winter competition. However, survival in a competitive situation is likely to a large extent to be a function of the relative proportions of the two factions that enter into competition (as well as of their 'competitiveness'). On this basis, therefore, the critical factor is less a function of events while the two factions are in competition during the northern winter but rather a function of the relative numbers that are achieved while not in competition. In this case, the critical factor should therefore be the relative numbers that can be produced during the northern summer while the two are not in direct competition, for it is during this period that the number that subsequently enter into competition is determined. This number is itself, of course, in absolute terms, a function of the number that survived competition the previous winter. This effect, however, influences absolute numbers rather than relative numbers.

If, as seems likely, there is an element of interspecific competition, this element must be added to that of intra-specific competition (Cox 1968) in any consideration of the establishment of a free or even despotic situation. The existence and distribution of a species, resident or migrant, that offers interspecific competition will therefore exert an important selective pressure on the frequency distribution of migration distance of any species with which it is competing.

Finally, as indicated by Cox (1968) and Moreau (1972), interspecific competition might be expected to exert selection on the feeding and other adaptations of both migrants and residents. The result is an evolutionary modification of the migration thresholds involved in habitat assessment.

28

Seasonal and ontogenetic return migration by pinnipeds

28.1 Introduction

The pinniped nomenclature used in this section is that of J.E. King (1964, 1966, 1969) for all except the fur seals and harbour seals. For fur seals the nomenclature followed is that of Repenning *et al.* (1971) except that firstly a specific rather than subspecific distinction is retained between the South African fur seal, *Arctocephalus pusillus*, and the Australian fur seal, *A. doriferus*, and secondly the fur seals north and south of the Antarctic convergence are placed in a single species, the Kerguelen fur seal, *A. tropicalis*. For the harbour seal species-group the proposals of McLaren (1966) are followed with regard to the West Pacific harbour seal, *Phoca largha*, but not with regard to the proposed Kurile seal, *P. kurilensis*. These decisions, however, are to some extent based on convenience and should not be regarded as informed taxonomic comments.

There is some controversy over the origin of the different groups and among the more interesting speculations is the suggestion of a biphyletic origin for the phocoids (true seals) and the otarioids (eared seals, i.e. fur seals, sea lions, and walrus) (McLaren 1960b, King 1964). On this suggestion, the phocoids share common ancestry with otters and evolved in the freshwater lakes of Tertiary Asia (40 million years B.P.), moving into the sea during the Miocene (20 million years B.P.), and that the otarioids share common ancestry with canids and evolved during the Oligocene (40 million years B.P.), principally along the Pacific coast of North America. However, despite the strong possibility of a biphyletic origin for the pinnipeds, subsequent convergent evolution has resulted in the animals as they exist at the present time being so similar that it is convenient, as far as their migrations are concerned, to consider them together.

The geographical distribution of surviving pinnipeds, as well as the classification used here, is shown in Figs. 28.1 and 28.2. In general, the pinnipeds are found in the cooler ocean currents of the world. In either hemisphere, the 20°C summer isotherm, where it approaches the continental coasts, forms a reasonable pointer to the limits of where one might expect to find pinnipeds (King 1964). The only real exceptions are the monk seals, *Monachus* spp. The 25°C summer isotherm coincides more or less with the limits of distribution of the Mediterranean monk seal, *M. monachus*, and the Hawaiian monk seal, *M. schauinslandi*. The West Indian monk seal, *M. tropicalis*, however, is, or was, found in water the temperature of which may go up to 29°C in summer.

There are grounds for believing that the two major land-locked species, the Caspian seal, *Pusa caspica*, and Baikal seal, *P. sibirica*, arrived at their present locations in the late Miocene (8 million years B.P.) when what are now lakes were part of a large water mass that was connected to the Arctic basin (McLaren 1960a).

Fig. 28.1 Classification and distribution of nineteen surviving species of true seals (Phocoidea).

[*Classificaton based partly on that of King (1964) and partly on that of McLaren (1966). Distribution compiled from many sources.*

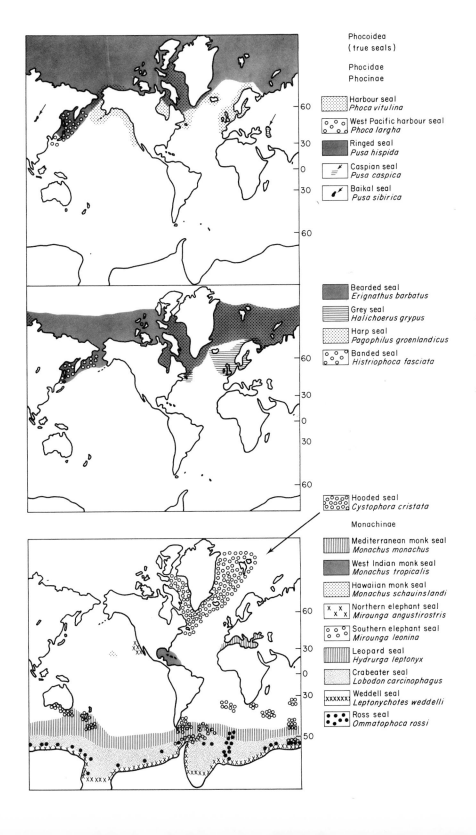

Phocoidea
(true seals)

Phocidae
Phocinae

Harbour seal
Phoca vitulina

West Pacific harbour seal
Phoca largha

Ringed seal
Pusa hispida

Caspian seal
Pusa caspica

Baikal seal
Pusa sibirica

Bearded seal
Erignathus barbatus

Grey seal
Halichoerus grypus

Harp seal
Pagophilus groenlandicus

Banded seal
Histriophoca fasciata

Hooded seal
Cystophora cristata

Monachinae

Mediterranean monk seal
Monachus monachus

West Indian monk seal
Monachus tropicalis

Hawaiian monk seal
Monachus schauinslandi

Northern elephant seal
Mirounga angustirostris

Southern elephant seal
Mirounga leonina

Leopard seal
Hydrurga leptonyx

Crabeater seal
Lobodon carcinophagus

Weddell seal
Leptonychotes weddelli

Ross seal
Ommatophoca rossi

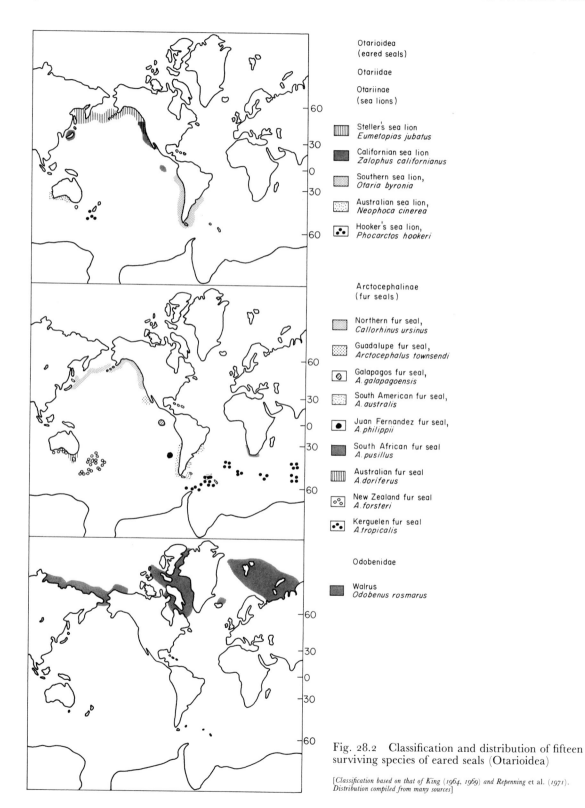

Fig. 28.2   Classification and distribution of fifteen
surviving species of eared seals (Otarioidea)

[*Classification based on that of King (1964, 1969) and Repenning et al. (1971).
Distribution compiled from many sources*]

At least two of the species the migrations of which are considered later were close to extinction at the beginning of the twentieth century as a result of large-scale predation by man. The Guadalupe fur seal, *Arctocephalus townsendi*, was thought to be extinct at the end of the nineteenth century, but in 1927 a few individuals were discovered on Guadalupe and this herd has since recovered dramatically (King 1964). The northern elephant seal, *Mirounga angustirostris*, was reduced by 1890 to a single herd of less than 100 individuals that survived on Guadalupe. It is from this herd that the numbers have built up and spread over their present range (Fig. 14.11).

28.2 Seasonal home ranges and behaviour

All pinnipeds are amphibious, alternating between periods on land and periods in the water during which virtually all feeding occurs. In all species, however, parturition occurs out of the water, either on mainland coasts, islands, sand-banks, ice-fields, or ice-floes (Table 28.1). All pinnipeds normally produce just a single young. Twins, although they occur, are rare. For any given deme, births are usually concentrated into a relatively short period of the year (Fig. 28.3), a period that is usually followed fairly quickly by the main period for copulation.

Copulation may occur either on land, as in most otarioids, or in the water, as in most phocoids. However, the division between copulation on land or water correlates less with phylogeny than with the existence or absence of a harem system (Bertram 1940). A majority of the otarioids, but a minority of the phocoids, exhibit a harem system during the breeding season (Table 28.1). The number of females that make up the harem of an individual male may vary from an average of three in the South American fur seal, *Arctocephalus australis*, to an average of 50 in the northern fur seal, *Callorhinus ursinus*. The number of females in the harem is a function of breeding site. The largest harems are formed on oceanic islands and the smallest on continental coasts. Where breeding occurs on ice, little or no harem formation takes place.

In most harem-forming species, the males do not leave the land to feed until all of the females have been inseminated. The males of species that do not have a harem system usually compete for the females in the water, and during the copulation season males

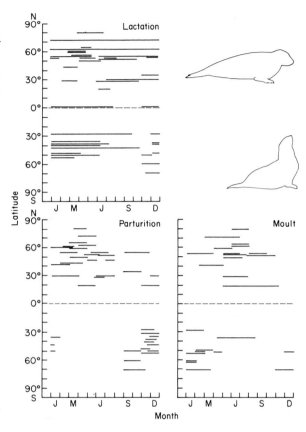

Fig. 28.3 Variation in season of parturition, lactation and moult of pinnipeds in relation to latitude

Each solid line represents a single deme or species. At higher latitudes moulting occurs at the warmest time of the year and parturition tends to occur in the spring. In temperate and sub-tropical latitudes, particularly in the northern hemisphere, there is much greater variation between species in time of year at which parturition and moulting occur, even though each deme gives birth or moults over only a limited period.

are frequently observed with wounds inflicted by rival males, especially in the region of the genitalia (Bertram 1940).

The hooded seal, *Cystophora cristata*, has evolved a slightly different breeding strategy. In this species the breeding unit consists of one male, one female, and a pup (conceived by the female the previous year). This unit is a territorial unit that occupies an ice floe which is then defended by both adults against occupation by other individuals (King 1964).

As well as leaving the water to give birth and, in many species, to copulate, pinnipeds also leave the water to suckle and to moult. The thermoregulatory problems that confront an amphibious homeotherm have led to the moult assuming a conspicuous part of the annual cycle. In general, moulting takes place out of the water at a time of year, usually shortly after the end of the breeding season, when heat conservation requirements are at a minimum (Fig. 28.3).

Exceptions to this general rule are some polar/sub-polar species, such as the Weddell seal, *Leptonychotes weddelli*, and the crabeater seal, *Lobodon carcinophagus*, which continue to feed at frequent intervals during the moulting period (Bertram 1940).

Although in most species breeding and moulting are the two major activities that take place on land, pinnipeds may also leave the water to avoid aquatic predators or to sleep. It is more usual, however, for

Table 28.1   Parturition sites and harem sizes of the different species of pinnipeds

	island	mainland coast	sandbank	fast-ice	ice-floe	Average size of harem (no. females/male)
Otariidae						
Otariinae (sea lions)						
Eumetopias jubatus	×					15
Zalophus californianus	×	×				4?
Otaria byronia		×				8
Neophoca cinerea		×				5
Phocarctos hookeri	×					12
Arctocephalinae (fur seals)						
Callorhinus ursinus	×					50
Arctocephalus townsendi	×					9
Arctocephalus australis	×	×				3
Arctocephalus pusillus	×	×				4?
Arctocephalus tropicalis	×					5
Arctocephalus doriferus		×				7
Arctocephalus forsteri	×	×				9
Odobenidae (walrus)						
Odobenus rosmarus					×	?
Phocidae (true seals)						
Phocinae						
Phoca largha					×	*
Phoca vitulina			×			*
Pusa hispida				×		*
Pusa caspica				×		*
Pusa sibirica				×		*
Erignathus barbatus					×	*
Halichoerus grypus	×	×			×	7 to *
Pagophilus groenlandicus				×		*
Histriophoca fasciata					×	*
Cystophora cristata					×	1
Monachinae						
Hydrurga leptonyx					×	*
Lobodon carcinophagus					×	*
Leptonychotes weddelli		×				*
Ommatophoca rossi					× ?	*?
Monachus monachus	×	×				?
Monachus tropicalis	×					?
Monachus schauinslandi	×					?
Mirounga leonina	×					30
Mirounga angustirostris	×					12

×   indicates type of parturition site
*   indicates that harems are not formed and that the males search for, and copulate with, the females in the water

pinnipeds to sleep while floating on the surface of the water (King 1964), and individuals of many species spend weeks or months at a time at sea without hauling-out onto land. Seals living under ice, however, cannot sleep for any length of time and are forced to maintain a fairly constant high level of activity. This is because pinnipeds, in common with all mammals, breath air. Consequently, any underwater activity, such as searching for food, has to be interrupted at intervals while the animal returns to the water surface to breathe. Through various respiratory and vascular mechanisms (King 1964), a pinniped can dive to depths of 15 m or so and remain submerged for a maximum of perhaps 20–30 min. Diving patterns, of course, vary between species and some, such as the northern elephant seal, *Mirounga angustirostris*, habitually feed at greater depths than others. Species that live in areas that are covered by ice during the winter, such as the ringed seal, *Pusa hispida*, bearded seal, *Erignathus barbatus*, Caspian seal, *P. caspica*, and Baikal seal, *P. sibirica*, in the northern hemisphere, and the Weddell seal, *Leptonychotes weddelli*, in the southern hemisphere, maintain contact with the atmosphere through air-holes in the ice that are either natural stress holes or have been kept open by the gnawing of the seals themselves. The possibility cannot be ruled out, however, that some use is made of air trapped beneath the ice (Bertram 1940). Some predators in the northern hemisphere, especially polar bears, *Ursus maritimus*, and man, regularly catch these species by waiting at the air holes.

From this brief consideration of pinniped behaviour it can be seen that it should be possible to recognise a variety of home ranges for this group. For both males and females a moulting home range can be recognised as can the feeding home ranges occupied at different seasons. For females, there are also breeding and lactation home ranges.

The breeding home range of females consists of the out-of-water parturition site, the copulation site, which may be either on land or ice or in the water, and the surrounding area of water within which any feeding that occurs during this period is carried out. The lactation home range consists of the out-of-water site on which suckling occurs and the surrounding area in which the female feeds. To judge from the time that a female may spend at sea during lactation, this home range may be large. In the northern fur seal, *Callorhinus ursinus*, for example, the first few trips to sea of the female after giving birth average about eight days in length, followed by about a day and a half on land suckling the pup

(Bartholomew and Hoel 1953). Lactating female New Zealand fur seals, *Arctocephalus forsteri*, in Australia, alternate between five days at sea and then a day and a half on land during the period immediately following parturition (Stirling 1971). Females of the South African fur seal, *A. pusillus*, which lactate for almost a whole year (Fig. 28.21) spend 2–3 continuous weeks of every month at sea and the remaining 1–2 weeks on shore feeding the young offspring (Rand 1959). In some species, however, such as the walrus, *Odobenus rosmarus* (Fig. 28.11), the young offspring, which is suckled for a year, accompanies its mother on a seasonal return migration. In this case, the lactation home range is also breeding and lactation home ranges.

For males there is a copulation home range which in those species that form harems (Table 28.1) is limited to a small area of land that is defended against other males. Little is known, however, concerning the copulation home range of males of species that copulate in the water.

Little is known, also, concerning feeding home ranges at different times of year. However, the winter home ranges of pagophilic species, living under the ice, must consist not only of suitable feeding grounds but also of a number of breathing holes.

At present we have little understanding of the environmental features that influence habitat suitability for a particular species at different stages of the annual cycle, nor therefore of the evolved migration thresholds. The evolution of the habit of breeding on oceanic islands, rocky shores, and ice was presumably an adaptation to the absence of terrestrial predators in the case of oceanic islands and their relative absence at the other sites. Presumably, also, any seasonal shift in feeding home range reflects the migration patterns of the prey species, some of which may be those species of fish described in Chapter 31 (p. 809). Evidence presented in the section on migration direction (p. 724) suggests that, either directly or indirectly, water temperature may also contribute to habitat suitability.

28.3 Degree of dispersal and convergence

Dispersal in pinnipeds is associated with feeding. Convergence is associated with parturition, copulation, moulting and roosting.

Pinniped species such as the northern fur seal, *Callorhinus ursinus* (Fig. 28.15), grey seal, *Halichoerus grypus* (Fig. 28.8), and harp seal, *Pagophilus groenlandicus* (Fig. 28.18), that are distributed at well-marked foci during the breeding season, disperse after breeding by individuals migrating different distances in different directions. Species such as the southern sea lion, *Otaria byronia*, which breed along continental coasts, and the crabeater seal, *Lobodon carcinophagus* (Fig. 28.4), and leopard seal, *Hydrurga* *leptonyx* (Fig. 28.5), which breed among or at the edge of the circumpolar band of Antarctic pack ice, disperse by different individuals moving different distances rather than in different directions. An estimate of the relative dispersal of pinniped species is given in Table 28.2. Also indicated in Table 28.2 is the relative gregariousness of the individuals while in the water during the non-breeding season. Some species, such as the leopard seal, *Hydrurga leptonyx*, and Ross seal, *Ommatophoca rossi*, are more or less

Table 28.2 Degree of dispersal and convergence during return migration and gregariousness during feeding of the different species of pinnipeds

Species	Degree of dispersal/convergence	Gregariousness
Otariidae		
Otariinae (sea lions)		
Eumetopias jubatus	×××	○
Zalophus californianus	××	?
Otaria byronia	×	○○○
Neophoca cinerea	×	?
Phocarctos hookeri	××	?
Arctocephalinae (fur seals)		
Callorhinus ursinus	××××	○
Arctocephalus townsendi	×× ?	○ ?
Arctocephalus australis	×	?
Arctocephalus pusillus	×	○○
Arctocephalus tropicalis	×××	○○
Arctocephalus doriferus	?	?
Arctocephalus forsteri	××	?
Odobenidae (walrus)		
Odobenus rosmarus	×	○○○
Phocidae (true seals)		
Phocinae		
Phoca largha	×× ?	○ ?
Phoca vitulina	××	○
Pusa hispida	×	○
Pusa caspica	×	?
Pusa sibirica	××	?
Erignathus barbatus	×	○
Halichoerus grypus	×××	○
Pagophilus groenlandicus	×××	○○
Histriophoca fasciata	×	?
Cystophora cristata	×××	○
Monachinae		
Hydrurga leptonyx	×××	○
Lobodon carcinophagus	××	○○○
Leptonychotes weddelli	×	○
Ommatophoca rossi	?	○ ?
Monachus monachus	?	?
Monachus tropicalis	?	?
Monachus schauinslandi	×	?
Mirounga leonina	××××	○ ?
Mirounga angustirostris	×××	○ ?

××××−×   decreasing degree of dispersal/convergence
○○○○−○   decreasing gregariousness when feeding (○ = solitary)
Degree of dispersal/convergence is estimated by a consideration of the maximum area occupied at any one time by all individuals of a species in relation to the minimum area occupied at any one time.

solitary throughout their lives (King 1964). Other species, such as the northern fur seal, *Callorhinus ursinus*, are intensely gregarious during the breeding season but more or less solitary while at sea, though some convergence does occur where there is local food abundance (Wilke and Kenyon 1954). The southern sea lion, *Otaria byronia*, the crabeater seal, *Lobodon carcinophagus*, and the walrus, *Odobenus rosmarus*, are more or less gregarious throughout the year (Hamilton 1934, Bertram 1940, King 1964).

Inspection of Table 28.2 in relation to Tables 28.1 and 28.3 suggests the possibility that species that feed predominantly on fish and cephalopods may disperse more than other species and may also be more inclined to have individual home ranges. On the other hand, species that feed on invertebrates such as krill or bottom-living molluscs and crustaceans may show less dispersal and be more likely to form social groups during the non-breeding season. Also, those species that breed on oceanic islands seem to show greater convergence during the migration to the breeding range, whereas coastal breeders and ice-breeders seem respectively to show less and less convergence during this phase of the migration.

Table 28.3 Feeding habits of the different species of pinnipeds

Species	crustaceans	bivalve molluscs, etc.	cephalopods	fish	penguins	seals
Otariidae						
Otariinae (sea lions)						
Eumetopias jubatus			×	×		
Zalophus californianus	×	×	××	×		
Otaria byronia	×		××	×	×	
Neophoca cinerea				×	×	
Phocarctos hookeri	×	×		×	×	
Arctocephalinae (fur seals)						
Callorhinus ursinus			×	××		
Arctocephalus townsendi			×	×		
Arctocephalus australis	×		×	×		
Arctocephalus pusillus	×		×	×		
Arctocephalus tropicalis	××		×	×		
Arctocephalus doriferus	×		×	×		
Arctocephalus forsteri	×		××	×	×	
Odobenidae (walrus)						
Odobenus rosmarus	×	××				×
Phocidae (**true seals**)						
Phocinae						
Phoca largha	×	×		××		
Phoca vitulina	×	×		××		
Pusa hispida	××		×	×		
Pusa caspica	×			××		
Pusa sibirica				××		
Erignathus barbatus	××	××	×	×		
Halichoerus grypus				××		
Pagophilus groenlandicus	×			××		
Histriophoca fasciata	×		××	××		
Cystophora cristata	×	×	××	××		
Monachinae						
Hydrurga leptonyx			×	×	××	×
Lobodon carcinophagus	××					
Leptonychotes weddelli	×		××	××		
Ommatophoca rossi	×		××	×		
Monachus monachus				××		
Monachus tropicalis				× ?		
Monachus schauinslandi			×	××		
Mirounga leonina			××	×		
Mirounga angustirostris			×	×		

× food source   ×× major food source

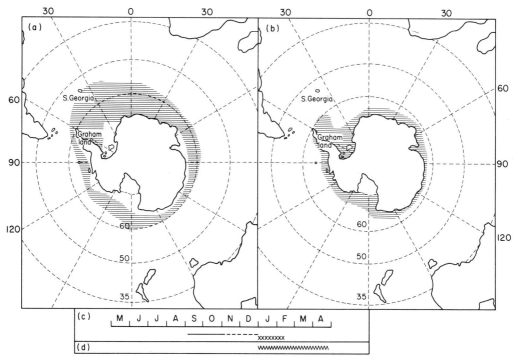

Fig. 28.4 Seasonal migration of the crabeater seal, *Lobodon carcinophagus*

Hatched area, distribution: (a) maximum distribution which occurs just before breeding, and (b) minimum distribution between January and April after the break up of the pack-ice around the Antarctic coasts. (c) Seasonal behaviour. Solid line, parturition period; dashed line, lactation period; crossed line, moulting period. Parturition occurs on the pack-ice before the spring break-up is really under way. (d) Seasonal incidence among the inshore waters of the west coast of Graham Land.

Seasonal return migration seems to consist of a post-reproductive southward movement to coastal waters followed by a northward movement at about May when the autumn extension of pack-ice begins. Maximum northward distribution probably occurs about August and is perhaps followed by a short southward movement into the main pack-ice for breeding. It is not known to what extent individuals of this species return to the same area to breed as was occupied the previous year

[*Drawn from information in Bertram (1940) and King (1964). Photo courtesy of Institute of Oceanographic Sciences*]

28.4 Degree of return

Information concerning degree of return in pinnipeds is only available in relation to the breeding site and even then only for a few species. Such information as is available is considered in Chapters 14 and 17 in relation to the incidence of removal migration. Although it appears that degree of return is invariably high, both to the natal site for first breeding and to the site used for breeding the previous year, some removal migration does occur, particularly during the pre-reproductive period of ontogeny.

Data are at present only available for species that breed on coasts or islands, little being available concerning the degree of return of those species that breed on ice.

28.5 Distance and the degree of overlap of seasonal home ranges

Inevitably, most information concerning pinnipeds relates to their out-of-water site or sites where they are most conspicuous and accessible to study. Delimitation of home ranges at sea have rarely been attempted and consequently degree of overlap of seasonal home ranges cannot be estimated except at the population level. It is at this level that the seasonal return migrations of pinnipeds are illustrated in the maps accompanying this section.

If spatial shifts in population had been the level at which the return migrations of animals other than birds and cetaceans had been investigated, few examples of seasonal return migration would have presented themselves, even though demes and individuals within the population occupy seasonal home ranges that show little or no overlap. As it happens, however, a number of species of pinnipeds do show a seasonal return migration even when analysed at the gross population level.

It is clear, however, from Chapters 26 and 31 that a great deal of information concerning seasonal return migration must be lost if consideration is restricted to the level of the population. Some conclusions can be reached from the marking–release/recapture data so far available, as shown for the northern fur seal, *Callorhinus ursinus* (Fig. 28.15), grey seal, *Halichoerus grypus* (Fig. 28.8), and South African fur seal, *Arctocephalus pusillus* (Fig. 28.21). The nature of pinniped migration is such (by contrast to the latitudinal migrations of whales— p. 745—and birds—p. 604, however, that many more recapture data are necessary, particularly of the same individual over a long period of time, before we can begin to piece together the structure of pinniped migration circuits.

One other level at which the seasonal return migration can be analysed is that of the breeding deme. Two measures are possible. One is the longest period of time that individuals are away from the out-of-water site that is part of the breeding home range (Table 28.4) and the other is the maximum distance away from the site achieved during the annual cycle by an individual that seems likely to be engaged on a return migration beginning from that site (Table 28.5)

In many species the average maximum period between successive visits can be estimated because all of the individuals leave the breeding home range at one time of year and do not re-appear until some time later. Some species, however, are characterised by individuals being present at the breeding area throughout the year (e.g. the grey seal, *Halichoerus grypus*, on the Farne Islands—Fig. 28.8). In situations such as this it is difficult to estimate the extent that the shore deme is in a state of flux. The only reliable method is that of the marking and observation of individuals. Marking–release/recapture experiments carried out by Rand (1959) on the South African fur seal, *Arctocephalus pusillus*, have shown clearly that although the shore is always occupied, each individual has a distinct feeding and resting cycle. The males, for example, outside of the breeding season, spend about 3 continuous weeks of every month at sea and then return to shore for the remaining week. Females, which lactate nearly all year (Fig. 28.21), spend 2–3 continuous weeks of every month at sea and the remaining 1–2 weeks on shore feeding the young offspring. For other species, however, in the absence of individual recognition by marking, it is not possible to determine whether a proportion of the population is away from the breeding area for long periods while the remainder goes to sea daily, or whether, as in *A. pusillus*, each individual alternates between days or weeks at sea and days or weeks at the breeding site. In Table 28.4 it is assumed that the latter situation is most common. If this is so, some estimate of the maximum time away can be obtained from a comparison of the minimum and maximum number of individuals on the shore. Hamilton (1939b), for example, observed that 5854 individuals of the southern sea lion, *Otaria byronia*, including 723 adult males, were present on shore at a breeding site on the Falkland Islands at the end of the breeding season in January. In August, however, the average count was 2545, including 183 adult males. If it is assumed, as seems reasonable from the descriptions of the behaviour of observed individuals on shore, that each individual spends about three consecutive days on shore in August, an estimate can be made of the length of time spent at sea (Table 28.4)

An estimate of the maximum period away from the breeding home range can be obtained for some species, such as the crabeater seal, *Lobodon carcino-*

phagus (Fig. 28.4) and the leopard seal, *Hydrurga leptonyx* (Fig. 28.5), not from observations of absence from the breeding home range but from observations of presence in non-breeding areas. In fact, for these two species any other method would be difficult as they breed in inaccessible areas among thick Antarctic pack ice.

Some species, such as the Australian sea lion, *Neophoca cinerea*, are considered to move very little from their breeding area (King 1964) but have not been sufficiently observed for any estimate to have been made of the maximum period between successive returns. Such species are given a nominal maximum period of one day in Table 28.4.

The maximum time away from the primary out-

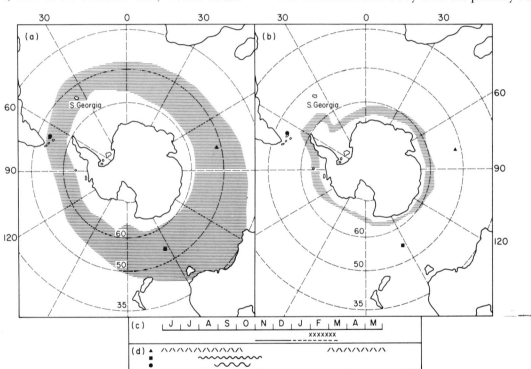

Fig. 28.5 Seasonal migration of the leopard seal, *Hydrurga leptonyx*

Hatched area, distribution: (a) maximum distribution between June and October, and (b) minimum (breeding) distribution between November and May; triangle, Heard Island; square, Macquarie Island; solid dot, Falkland Islands.

(c) Seasonal behaviour. Solid line, parturition period; dashed line, lactation period; crossed line, moulting period. Parturition occurs on the pack-ice, perhaps toward the outer fringes.

(d) Seasonal incidence on three islands indicated by

the symbols used in (a) and (b). Lines indicate period that animals are present. Heard Island is frequented by adults and immature individuals. Macquarie Island and the Falkland Islands are frequented almost entirely by young individuals. Males are mature at four years of age; females at three. It is not known to what extent individuals of this species return to the same area of pack-ice to breed as was occupied the previous year. Possibly circumpolar removal migration occurs as a result of a low degree of return. return.

[*Drawn from information in Hamilton (1939a) and King (1964). Photos by Dr. J.R. Beck and British Antarctic Survey*]

of-water site on a seasonal basis is shown by the walrus, *Odobenus rosmarus*, and hooded seal, *Cystophora cristata*, which may spend up to 9 months of each year away from the breeding home range. Both

of these species breed on ice. Of the species that breed on land, the female northern fur seal, *Callorhinus ursinus*, and the males of the Californian deme of the Californian sea lion, *Zalophus californianus*, show the maximum time away from their breeding home range.

Occasionally, from observed shifts in distribution of a population or deme, it is possible to fill in some of the details of the migration circuits of individuals during the period that they are away from their breeding home range. Individuals of the southern elephant seal, *Mirounga leonina*, with a breeding home range centred on South Georgia, for example, have a moulting home range centred on the South Orkneys (Fig. 28.16). Similarly, the hooded seals, *Cystophora cristata*, that breed on ice floes in March and April in the regions of Newfoundland and Jan Mayen, in June and July have a moulting home range in the region of the Denmark Strait between Iceland and Greenland (Fig. 28.17).

Table 28.5 compares the distance components of the return migrations of the different sexes, ages, and, in some cases, demes, of the different species of pinnipeds. The figures for adults represent the average maximum distance (measured as a straight line) from the breeding site that some members of the category normally achieve during the longest

return migration of the yearly cycle. The figures for juveniles refer to the average maximum distance from the birth site achieved during the period from birth to the first reproduction back at the birth site. These figures have been assessed in different ways for different species.

The most acceptable measure of distance is, of course, a measure obtained from marking–release/recapture experiments. Such experiments have provided usable data for the South African fur seal, *Arctocephalus pusillus* (Fig. 28.21), grey seal, *Halichoerus grypus* (Fig. 28.8), and northern fur seal, *Callorhinus ursinus* (Fig. 28.15). For all other species, however, some other means of estimating distance

Table 28.4 The periodicity component of the return migrations of different age and sex classes of the different species of pinnipeds

Species	Maximum period (days) away from the primary out-of-water site			Location of primary site (when data not for whole range)
	Adults ♂	♀	Young	
Otariidae				
Otariinae (sea lions)				
Eumetopias jubatus	200	100	100	
Zalophus californianus	270	1?	1?	California
	1?	1?	1?	Mexico
	1?	1?	1?	Galapagos Is.
Otaria byronia	12	7	7	Falkland Is.
Neophoca cinerea	1?	1?	1?	
Phocarctos hookeri	30?	30?	30?	
Arctocephalinae (fur seals)				
Callorhinus ursinus	220	250	300	
Arctocephalus townsendi	?	?	?	
Arctocephalus australis	1?	1?	1?	Uruguay
	150?	150?	150?	Falkland Is.
Arctocephalus pusillus	21	18	400?	
Arctocephalus tropicalis	120	120	400	
Arctocephalus doriferus	?	?	?	
Arctocephalus forsteri	90	90	90	New Zealand
	?	?	90?	Australia
Odobenidae (walrus)				
Odobenus rosmarus	270	270	270	
Phocidae (true seals)				
Phocinae				
Phoca largha	?	?	?	
Phoca vitulina	1?	1?	1?	
Pusa hispida	180	180	240	
Pusa caspica	180	180	180	
Pusa sibirica	210	210	210	
Erignathus barbatus	?	?	?	
Halichoerus grypus	70?	70?	700	Northern Britain
	70?	70?	?	West Atlantic
	1?	1?	?	SW Britain
Pagophilus groenlandicus	240	240	330	
Histriophoca fasciata	120?	120?	120?	
Cystophora cristata	270	270	270	
Monachinae				
Hydrurga leptonyx	240	240	700	pack-ice fringe
Lobodon carcinophagus	120	120	120	pack-ice
Leptonychotes weddelli	200	170	700	
Ommatophoca rossi	?	?	?	
Monachus monachus	?	?	?	
Monachus tropicalis	?	?	?	
Monachus schauinslandi	1?	1?	40?	
Mirounga leonina	180	180	270	
Mirounga angustirostris	90	120	90	Año Nuevo Is.

was necessary. In many cases the distance travelled can be estimated from a consideration of seasonal changes in distribution. In species that do not show seasonal changes in distribution, but which have recently expanded their geographical breeding range, such as the Kerguelen fur seal, *Arctocephalus* *tropicalis* (Fig. 28.22), the distance between the nearest original breeding area and the new breeding area can be taken as an indication of distance covered. It has to be stressed, however, that this measure may only be a measure of removal migration and may not indicate the distance from which

Table 28.5 Distance component of the return migrations of different sexes, age and geographical demes of the different species of pinnipeds

Species	Breeding area	Migration distance (km)		
		males	females	juveniles
Otariidae				
Otariinae (sea lions)				
Eumetopias jubatus	whole range	1800	120	120
Zalophus californianus	California	1000	40?	40?
	Mexico	40?	40?	40?
	Galapagos Is.	40?	40?	40?
Otaria byronia	Falkland Is.	200	200	200
Neophoca cinerea	whole range	40?	40?	40?
Phocarctos hookeri	,,	760	760	760
Arctocephalinae (fur seals)				
Callorhinus ursinus	,,	800	2000	4000?
Arctocephalus townsendi	Guadeloupe	680	?	?
Arctocephalus australis	Uruguay	100	100	100
	Falkland Is.	2000	2000	2000
Arctocephalus pusillus	whole range	800	800	1200
Arctocephalus tropicalis	,,	1000	1000	5000?
Arctocephalus doriferus	,,	?	?	?
Arctocephalus forsteri	New Zealand	630	630	630
Odobenidae (walrus)				
Odobenus rosmarus	Pacific	730	950	950
	Atlantic	50	50	50
Phocidae (true seals)				
Phocinae				
Phoca largha	whole range	100?	100?	100?
Phoca vitulina	Bering Sea	400?	400?	400?
	SE North Pacific	40?	40?	40?
	Atlantic	50	50	50
Pusa hispida	Bering Sea	400?	400?	400?
	Arctic Sea	8	8	15
Pusa caspica	whole range	830	830	830
Pusa sibirica	,,	300	300	300
Erignathus barbatus	,,	100?	100?	100?
Halichoerus grypus	Northern Britain	400	400	750
	Scilly Is.	40?	40?	?
Pagophilus groenlandicus	whole range	1900	1900	1900
Histriophoca fasciata	,,	300?	300?	300?
Cystophora cristata	,,	2000	2000	2000
Monachinae				
Hydrurga leptonyx	,,	1700	1700	1700
Lobodon carcinophagus	,,	1000	1000	1200
Leptonychotes weddelli	,,	40?	40?	?
Ommatophoca rossi	,,	?	?	?
Monachus monachus	,,	?	?	?
Monachus tropicalis	,,	?	?	?
Monachus schauinslandi	,,	40?	40?	40?
Mirounga leonina	,,	2000?	2000?	2000?
Mirounga angustirostris	,,	1400	260?	450

For males and females figures represent the average maximum distance (straight line) away from the breeding site achieved during the longest return migration of the yearly cycle. For juveniles figures refer to the average maximum distance from the birth site achieved during the period from birth to first reproduction back at the birth site.

an animal normally returns as part of its seasonal or ontogenetic return migration. However, as data are scarce, for present purposes such information is taken as being relevant to both return and removal migration. Some species, such as the Guadalupe fur seal, *A. townsendi*, are so rare and breed in such a localised area that any observation of an individual away from this area can be used to estimate distance travelled. Again, however, there is no guarantee that such individuals will eventually return to their breeding home range and are therefore performing a return migration. Other species have occasionally been observed at their feeding grounds. For distance, Hamilton (1934) observed the southern sea lion, *Otaria byronia*, feeding 200 km away from the nearest known breeding site, in this case on the Falkland Islands. In the absence of any other information, such a figure can be used as an estimate of the average maximum distance away from the breeding home range travelled by the species. Finally, there are some species for which little information is available but which are thought by various authors not to move very far from their breeding home range. These species, such as the Australian sea lion, *Neophoca cinerea*, ringed seal, *Pusa hispida*, Weddell seal, *Leptonychotes weddelli*, and the Scilly Isles deme of the grey seal, *Halichoerus grypus*, have all been given a nominal figure of 40 km in Table 28.5. This figure may well need considerable revision when the movements of the species concerned become known. For example, both the ringed seal, *Pusa hispida* (McLaren 1958, King 1964) and the harbour seal, *Phoca vitulina*, (Bigg 1969) are considered to move very little. Nevertheless, both of these species have been observed moving north through the Bering Strait in spring and early summer (Bailey and Hendee 1926, Burns 1970). Smith (1965) has suggested that 85 per cent of Weddell seals, *Leptonychotes weddelli*, migrate north during the southern winter. More recent evidence,

[Photo by D.G. Bone]

Fig. 28.6   Weddell seal, *Leptonychotes weddelli*

Although adults of this species probably move little during the course of a year, the young, after weaning, disappear for one to two years and seem likely to perform extensive exploratory migrations however (Stirling 1969), suggests that the adults at least move very little from their breeding area.

It is perhaps quite evident from the above that a great deal remains to be discovered about the distances that pinnipeds migrate. Table 28.5 has been compiled using such information as was available to me and I have no doubt that over the next few decades many alterations will be made to the figures presented. Nevertheless, for the moment, Table 28.5 provides sufficient information for a few general conclusions to be drawn.

In the majority of cases the average maximum distance travelled during return migration appears to be similar for all age groups and sexes. In some species, however, the distance component of return migration varies markedly from deme to deme, age to age, and/or sex to sex. Steller's sea lion, *Eumetopias jubatus* (Fig. 28.12), the Californian deme of the Californian sea lion, *Zalophus californianus* (Fig. 28.13), and the northern elephant seal, *Mirounga angustirostris* (Fig. 28.14), all have males that migrate much further than the females and young. In the case of the northern fur seal, *Callorhinus ursinus* (Fig. 28.15) and the Pacific race of the walrus, *Odobenus rosmarus* (Fig. 28.11), the females and young migrate further than the males. In the South African fur seal,

Arctocephalus pusillus (Rand 1959), Kerguelen fur seal, *A. tropicalis* (King 1964), grey seal, *Halichoerus grypus* (Hickling 1962), crabeater seal, *Lobodon carcinophagus* (Lindsey 1938), and the leopard seal, *Hydrurga leptonyx* (King 1964), the young pre-reproductive individuals travel much further than the adults (see also p. 106). Finally, the Californian sea lion, *Zalophus californianus*, and the South American fur seal, *Arctocephalus australis* (Fig. 28.23), and possibly also the grey seal, *Halichoerus grypus* (Davies 1957, Hickling 1962), and the harbour seal, *Phoca vitulina* (Bailey and Hendee 1926, Bigg 1969), show marked differences in the distance component of return migration in different parts of their range.

It seems likely that when more is known of the return migrations of other species, similar age, sexual, and geographical differences in the distance component will manifest themselves.

Where data are available for a breeding deme concerning both the maximum period away from the breeding site and migration distance, it is possible to examine the relationship between the two. It is found that the correlation between distance and time is highly significant ($P < 0.001$) for both adult males ($r = 0.668$; $n = 25$) and adult females ($r = 0.761$; $n = 25$). Such a correlation is not

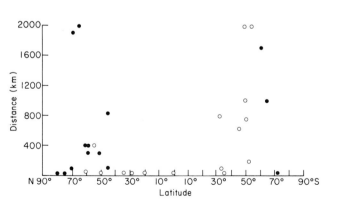

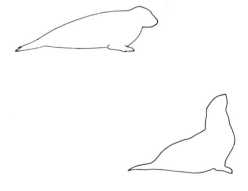

Fig. 28.7 Latitudinal variation in the one-way distance component of seasonal return migration in pinnipeds

The graph shows the maximum distance from the breeding site achieved by members of different demes of pinnipeds during the longest return migration of the annual cycle as a function of the mean latitude of the breeding sites for that breeding deme. Demes that show sexual segregation during seasonal return migration are excluded as also are those species for which the migration distance recorded in Table 28.5 seems highly likely to have referred to removal rather than return migrations.

Solid dots indicate demes that breed on sites that are reached by pack ice during its maximum annual extension. Open dots indicate demes the breeding sites of which are not reached by the ice.

There is a clear bimodal distribution, the greatest migration distance being found at cold-temperate and sub-polar latitudes by species that live close to the edge of pack ice. Minimum migration distances are found at warm-temperate and equatorial latitudes in demes that do not live near to the edge of the pack-ice and also at polar latitudes among demes that live well within the distribution of pack ice and that in winter establish a home range under the ice and maintain a number of air-holes.

inevitable for a return migration as is evident from the case of the green turtle, *Chelonia mydas*, in Fig. 30.1.

The meaning of this correlation is not clear, but perhaps a clue is provided by Fig. 28.7, which shows the latitudinal variation in maximum distance of seasonal return migration. The variation is clearly bi-modal, showing peaks in both the northern and southern hemispheres at sub-polar and high-temperate latitudes. These results are discussed below in conjunction with the similar results available for the direction and distance-ratio components.

Fig. 28.8   Seasonal and ontogenetic return migrations of the grey seal, *Halichoerus grypus*, in the east Atlantic

(a) Coasts edged in black, breeding distribution; stippled area, feeding distribution; arrows, some results of marking–release/recapture experiments on individuals marked soon after birth.

(b) Seasonal behaviour. Dotted line, territorial period (harems average about 7 females/territorial male); solid line, main parturition period (the young are born on rocky shores and births have been observed throughout the year); dashed line, lactation period; open circles, female moulting period; crosses, male moulting period.

The Baltic *H. grypus* group produces its young on drifting ice in February and March and has no harem formation. The west Atlantic *H. grypus* group (Fig. 28.1) produces its young in January and February.

(c) Counts of relative numbers on the Farne Islands (off eastern Britain) for each month of the year. Individuals are present throughout the year but with peaks during moulting and at the beginning of the parturition period.

(d) Direction ratio of young seals born on the Farne Islands as measured by marking–release/recapture experiments. Figures refer to the percentage of individuals moving in each of four direction sectors relative to the peak direction. Although this measured direction ratio seems reasonable in that the Farne Islands are near the southeast edge of the range of *H. grypus*, there is no way of knowing to what extent it is an artefact due to the relative probability of recapture in different directions due to the uneven distribution of human predators.

(e) Ontogenetic return migration: distance component during the first two years of life. Dots, recaptured individuals; hatched area, distances from the birth site within which individuals are likely to be found. Many young *H. grypus* leave their birth site immediately after weaning and within a month may be found at distances greater than 600 km. Thereafter the maximum distance from the birth sites decreases gradually. Whether this indicates a gradual return movement over a period of two

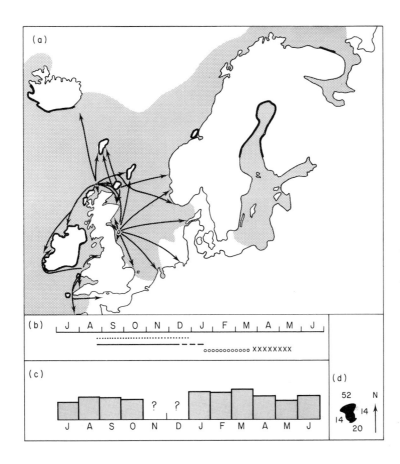

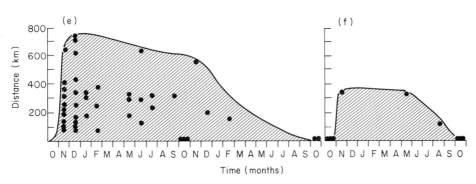

Time (months)

years or is an artefact due to the gradual decrease in the number of recaptures is unknown. Because recaptured seals are usually either dead or killed when recaptured there is little opportunity for tracing the movements of individuals. Consequently it may yet be proved that young *H. grypus* forage backward and forward between their birth site and the edge of their familiar area. At present, however, it is generally assumed that many young seals are extensively nomadic and away from their birth site for the first two years of life.

(f) As for (e) but for adult *H. grypus*. From (c) it can be inferred that the hatched area in this case represents the foraging range and that return to the breeding site is frequent. Data for adults are scarce and the gradual decrease in foraging range after May could be an artefact. Females give birth for the first time when three years old. Males are sexually mature at six or seven years.

[Compiled from Davies (1957), Hickling (1962), King (1964), Cameron (1970), Curry-Lindahl (1970), and Boyd and Campbell (1971). Photos by C. Stephen Robbins]

28.6 Direction and direction ratio

Little information is available on pinniped direction ratios, even for those species for which marking–release/recapture programmes have been carried out. Marking–release/recapture is, in fact, not a very satisfactory method of measuring the direction ratio of any animal unless steps are taken during experimental design to ensure that recapture sites are randomly distributed. This is rarely possible for pinnipeds where the majority of recaptures are the result of human sealing or fishing activities. For example, a large number of grey seals, *Halichoerus grypus*, that were marked soon after birth on the Farne Islands have been recaptured along the east coast of Scotland and northern England (Fig. 28.8) and a direction ratio of 54:14:14:20 about a peak direction of NNW can be calculated directly from the recaptures. However, these figures could be just as, if not more, indicative of the distribution of the

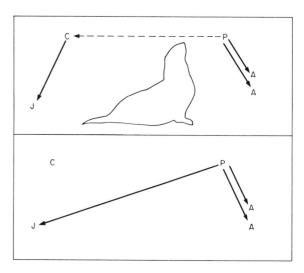

Fig. 28.9   The problem of determining the direction ratio of a return migration from the results of marking–release/recapture experiments

The figure is a diagrammatic representation of the movements of the northern fur seal, *Callorhinus ursinus*. C, Commander Islands; P, Pribilof Islands; solid arrows, outward stage of the return migration; dashed arrow, removal migration; J, recapture point in winter feeding quarters off Japan of an individual marked several years earlier at P; A, as for J but off the Pacific coast of North America. In the top diagram recapture J is irrelevant to the measure of the direction ratio of the outward stage of the return migration of Pribilof seals. In the lower diagram recapture J is relevant to the direction ratio. In practice, the two alternatives can be distinguished only if the same individual is observed at both ends of the return migration, or if many recaptures of different individuals are made at different stages in the migration. This latter method often enables the place of origin and the target for return to be determined by extrapolation (see Chapter 26).

fishing industry about the release point. In addition, those individuals that swim to the N or E have to travel further (i.e. to the coasts of Norway and Holland, etc.) before they are likely to be recaptured. If, as is possible, the probability of recapture decreases with distance, this factor also modifies the direction ratio as measured. On the other hand, it seems reasonable that the Farne Island grey seals should have a peak direction to the N and W because the Farne Islands are near the SE edge of the species' range.

Another problem in the measurement of direction ratio by marking–release/recapture is the difficulty

in distinguishing between return migration and removal migration. As an example, the case of the northern fur seal, *Callorhinus ursinus*, can be considered. It was believed for many years that the breeding demes on the Pribilof and Commander Islands (Fig. 28.15) were completely separate and it was not until individuals marked on the Pribilof Islands were recovered on the Commander Islands, and vice versa, that this was known not to be the case. Such recaptures, however, give information that in all probability is relevant only to removal migration, not return migration. Figure 28.9 shows the problem of distinguishing between return migration and removal migration on the basis of marking–release/recapture when an individual is recaptured once only. Although a direction ratio of 69:0:27:4 about SE (from the Pribilof Islands) can be calculated for the northern fur seal from the recapture data given by Kenyon and Wilke (1953), because most, if not all, of the recaptures in the west Pacific refer to removal migration, the true direction ratio for the return migration could just as easily approach 100:0:0:0 about SE.

Figure 28.10 summarises the direction components of pinniped return migrations, insofar as they are known. Many species, including all those for which maps (other than the distribution maps, Figs. 28.1 and 28.2) are not given, appear to show no peak migration direction. Among species in which

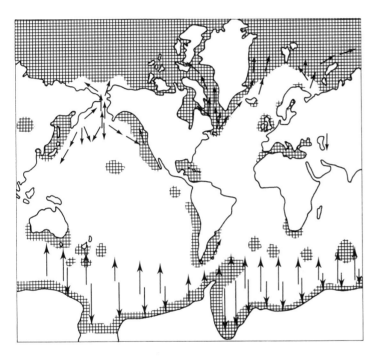

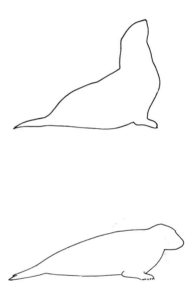

Fig. 28.10 The direction component of the return migration of the different species of pinnipeds

Arrows, peak post-breeding movement direction; hatched area, breeding distribution of species that do not show a peak movement direction.

The majority of species that show a peak movement direction are in some way associated with drifting ice of the open sea. These are the harp seal, *Pagophilus groenlandicus*, Caspian seal, *Pusa caspica*, leopard seal, *Hydrurga leptonyx*, crabeater seal *Lobodon carcinophagus*, and hooded seal *Cystophora cristata*. In addition, five species (i.e. walrus, *Odobenus rosmarus*, ringed seal, *Pusa hispida*, harbour seal, *Phoca vitulina*, bearded seal, *Erignathus barbatus*, and banded seal, *Histriophoca fasciata*), that elsewhere in their range do not show a peak movement direction, in the Bering Sea area show a peak movement direction that is also associated with drifting ice. All of these ten species, except the Caspian seal, *Pusa caspica*, show a peak direction that is toward the equator for the winter and toward the pole for the summer. *P. caspica* moves in the opposite directions.

Four other species (i.e. Steller's sea lion, *Eumetopias jubatus*, Californian sea lion, *Zalophus californianus*, northern fur seal, *Callorhinus ursinus*, and the northern elephant seal, *Mirounga angustirostris*), all of which are found in the north and northeast Pacific, also show a peak movement direction. These four species are characterised by geographical and/or sexual differences in either the direction or distance components of return migration. In *E. jubatus* and *C. ursinus* peak movement direction, where it occurs, is toward the equator for the winter and toward the pole for the summer. In *Z. californianus* and *M. angustirostris* peak movement direction, where it occurs, is to the polar side of the breeding area for both summer and winter.

All other pinniped species show no peak movement direction and occur at either tropical or temperate latitudes or, if associated with ice, are found coastally (bearded seal, *Erignathus barbatus*, walrus, *Odobenus rosmarus*, and Weddell seal, *Leptonychotes weddelli*) or coastally and in open water but not in the open sea (ringed seal, *Pusa hispida*) or in relatively land-locked areas such as the Sea of Okhotsk (banded seal, *Histriophoca fasciata*, and the West Pacific harbour seal, *Phoca largha*) and, as an extreme example, Lake Baikal (Baikal seal, *Pusa sibirica*).

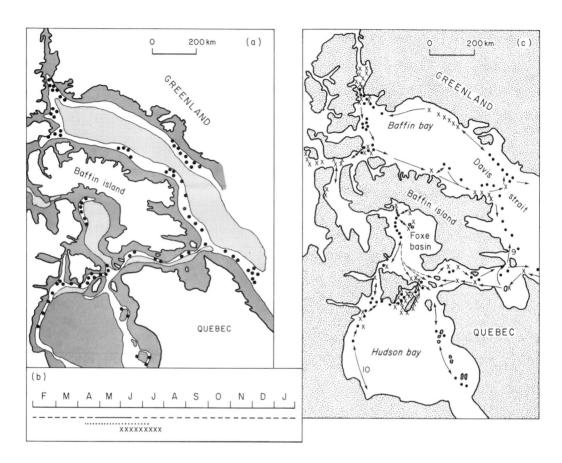

Fig. 28.11   Return and removal migrations of the walrus, *Odobenus rosmarus*

(a) Winter distribution of the walrus in the eastern Canadian Arctic in relation to the distribution of ice and open water. Dense stippling, fast ice (the shelf of ice that forms along a shore at the beginning of winter and gradually extends outwards as winter progresses); light stippling, pack ice (sheet of floating ice that moves with wind and current and which breaks up and re-joins and which for periods may join onto the fast ice). Within the distribution of fast and pack ice, open water leads occur. The position of open water is more or less constant from winter to winter but may vary according to temperature and wind direction etc. Walrus feed, mainly in the morning, on bottom-living bivalve molluscs at depths of up to 80–100 m and haul-out on land or ice for the remainder of the day. Walrus distribution therefore coincides with land or ice next to water < 100 m in depth. Solid dots show the winter distribution of the species.

(b) Seasonal behaviour. Solid line, parturition period (parturition is believed to occur on floating ice); dashed line, lactation period; dotted line, copulation period; crossed line, moulting period. Females give birth only every other year. Lactation occurs for a year or more and the young walrus accompanies the female for a further two

years or so after being weaned. Males and females mature when about six years old.

(c) Return and removal migrations in the eastern Canadian Arctic. Stippled areas, land masses. Walrus movement is related mainly to the changing position of the ice. Solid dots, winter distribution (February); crosses, summer distribution (August); numbers, month of year that walruses are present when occurrence is limited to a month or so; arrows, track directions; double-headed arrows, to-and-fro movements that in most cases are seasonal; single-headed arrows, one-way movement; dashed arrows, movements that do not occur every year. A great deal of walrus movement occurs by drifting on ice. Ice is vacated either when it takes the walrus over deep water or when it approaches land which may then be used for roosting instead of the ice. In areas such as Foxe Basin and northern Hudson Bay, the break-up of the ice in spring leads to a dispersal of the walrus from the localised open water occupied during the winter. Many individuals may transfer from ice to land during this dispersal for roosting and possibly also parturition and copulation. In the autumn, return migration occurs as a result of adherence to the edge of the developing fast and pack ice which leads eventually to concentration in the remaining areas of open water. In years with appropriate wind conditions, removal migration from Foxe Basin may occur

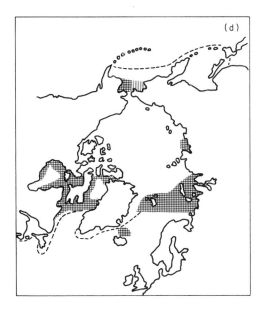

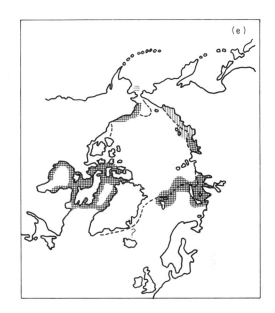

as a result of the ice-floes drifting southwards. Elsewhere conditions of wind and current may be such that removal migration may occur continuously. Hence in the Davis Strait/Baffin Bay area there appears to be a counter-clockwise migration circuit. Although essentially a removal migration, the wind and currents are such that many individuals or their offspring may return several years later to their original starting place. A similar migration circuit may exist between Quebec and Baffin Island in the Hudson Strait area.

(d) World distribution during the maximum extension of ice in February/March. Dashed line, average maximum limit of fast and pack ice; horizontal hatching, distribution of males; vertical hatching, distribution of females. In the Bering Sea, non-parturient females are found with the males in the east whereas parturient females (that do not copulate until the following year) are found in the west.

(e) World distribution at the time of minimum extension of ice in August/September. Dashed line, average minimum extent of ice; horizontal hatching, distribution of males; vertical hatching, distribution of females and young. Walruses pass through the Bering Strait northwards in May. Ice has gone from the Strait by the beginning of July. The return southward movement through the Strait occurs in November. The same behaviour that in the eastern Canadian Arctic produces a seasonal return migration of less than 100 km (one-way), in the Bering Strait area produces a seasonal return migration of up to 1000 km (one-way).

In summary, seasonal return migration in the walrus, *Odobenus rosmarus*, consists of drifting on ice in the spring and summer and a return at the edge of the advancing ice in the autumn and winter. In some areas, however, this strategy leads to removal, not return, migration. In yet other areas, it is a fortuitous result of topography that

removal migration may lead, after an interval of several years, to a return migration.

[*Compiled from Bailey and Hendee (1926), Degerbøl and Freuchen (1935), Mansfield (1958), Loughrey (1959), King (1964), and Burns (1970). Photo by Steve McCutcheon (courtesy of Frank W. Lane)*]

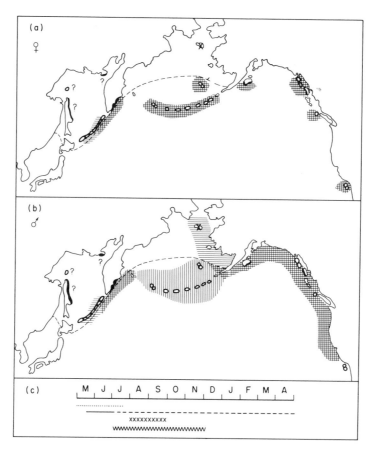

Fig. 28.12 Seasonal return migration of Steller's sea lion, *Eumetopias jubatus*

(a) Seasonal distribution of females and young. Solid black, breeding distribution (May–July); horizontal hatching, summer and autumn (July–November) distribution; vertical hatching, winter and spring (December–April) distribution; dashed line, average extreme limit of pack-ice (February/March). Ice has disappeared from the Bering Strait by the beginning of July.

(b) Seasonal distribution of males. Shading as in (a). In the Bering Sea, males move northward for summer and autumn and return to the vicinity of the breeding area for winter and spring. Along the American coast, males tend to move to the north in July and August and not to move south again until the following April. In the vicinity of the Kurile Islands, males and females move east after breeding and then north for the winter, males moving further north than females.

(c) Seasonal behaviour. Dotted line, territorial period: harems average 15 females/territorial male; solid line, parturition period: the young are born on rocky shores; dashed line, lactation period; crossed line, moulting period; zig-zag line, period during which *E. jubatus* is present on St. Lawrence Island (marked with a cross in (a) and (b)) in the north Bering Sea.

[*Compiled from Kenyon and Rice (1961), King (1964), Orr and Poulter (1965), Belkin (1966), Fiscus and Baines (1966), and Orr (1970a). Photo by Karl W. Kenyon (courtesy of Frank W. Lane)*]

both sexes have a peak migration direction, only in the Pacific deme of the walrus, *Odobenus rosmarus* (Fig. 28.11), and possibly also the Californian deme of the Californian sea lion, *Zalophus californianus* (Fig. 28.13), do males and females migrate in different directions. After travelling N through the Bering Straits, the females with dependent young of the walrus go NW along the northern coast of Siberia while the males and non-parturient females (Burns 1970) go NE along the northern coast of Alaska.

The Californian deme of the Californian sea lion and the Falkland Island deme of the South American fur seal, *Arctocephalus australis* (Fig. 28.23) are, as far as is known, the only demes of these two species that show a peak migration direction. Similarly, in the Bering Strait region, individuals of several species that elsewhere are not thought to have a peak direction move N in the spring and summer and S in the autumn (Fig. 28.10).

It is concluded in Fig. 28.10 that the existence of a peak migration direction is associated with the seasonal movement of pack ice or with sexual segregation. The question of sexual segregation during migration is discussed in the next section. Only the effects of the pack ice are considered here.

Several species of pinnipeds have become adapted to living for extended periods under fields of ice. In the Caspian seal, *Pusa caspica* (Fig. 28.20), and the Baikal seal, *P. sibirica* (Fig. 28.19) such adaptation was the only available strategy, but in the ringed seal, *P. hispida*, and Weddell seal, *Leptonychotes weddelli*, it was presumably the advantage of a lack of competition for food under the polar ice that led to their origin as separate species. Individuals of these latter two species throughout most of their geographical range remain within the same home range all year and simply take to the water to establish and maintain breathing holes for the duration of the winter period.

Species adapted to the edge of the pack ice, however, as well as those individuals of the ringed seal that occur near to the ice margin, migrate in spring and summer to areas exposed by the receding and disintegrating pack ice and then retreat in autumn as the ice once more advances. The clearest relationship between the pack ice and migration is that illustrated for the walrus, *Odobenus rosmarus*, in Fig. 28.11, but similar relationships exist for the other species mentioned in Fig. 28.10.

As well as the direct effect of the ice in preventing access to air, there are probably associated changes in food supply due to the lack of light penetration to the phytoplankton (Fig. 20.3), the influence of this on zooplankton (p. 498), and the consequent migration of many sub-polar fish (Figs. 31.12 and 31.13). Species that breed in the summer in areas close to or within the winter distribution of pack ice are therefore breeding in a home range that becomes unsuitable as a winter feeding home range relative to areas away from the edge of the ice. It seems likely that the nearer the breeding home range to the edge of the pack ice then the longer the period that the breeding home range is unsuitable as a feeding home range and the greater the distance (in a direction away from the ice) to a suitable feeding home range. This thesis could, perhaps, explain the bimodal distribution of latitudinal variation in migration distance (Fig. 28.7) and direction ratio (Fig. 28.10) and also the correlation between migration distance and the length of the maximum period away from the breeding home range (p. 721).

There is probably no compromise strategy that is adaptive to the conditions engendered by the winter pack ice. Either, as with the ringed seal and others, the animal remains in its summer home range and maintains a series of breathing holes, or it migrates away from the ice-edge in winter. Animals that attempt to feed under the ice without establishing a winter home range seem likely to run the risk of being unable to gain access to air at sufficiently frequent intervals.

These conclusions concerning the direct and indirect selective pressures generated by the advance and regression of the ice seem to be sufficient to explain the major part of the variation in migration distance and direction except for that variation associated with the sexual segregation of Steller's sea lion, *Eumetopias jubatus* (Fig. 28.12), northern fur seal, *Callorhinus ursinus* (Fig. 28.15), the Californian deme of the Californian sea lion, *Zalophus californianus* (Fig. 28.13), the northern elephant seal, *Mirounga angustirostris* (Fig. 28.14), and the Pacific deme of the walrus, *Odobenus rosmarus* (Fig. 28.11).

28.7 Sexual segregation

Bartholomew (1970) appears to be the first author to point out that the males and females of at least four of the five species just mentioned distribute themselves in such a way that the males occur in water of lower temperature than do the females. No other correlation is apparent for these species—certainly, at least, not with respect to distance or direction. In Steller's sea lion, *Eumetopias jubatus*, Californian sea lion, *Zalophus californianus*, northern elephant seal, *Mirounga angustirostris*, and the Pacific deme of the walrus, *Odobenus rosmarus*, the males migrate to the north during separation from the females whereas in the northern fur seal, *Callorhinus ursinus*, males migrate to the south. Females of Steller's sea lion and the northern elephant seal seem to remain around the breeding islands whereas females of the walrus migrate to the north and those of the northern fur seal and perhaps also those of the Californian deme of the Californian sea lion migrate to the south. Finally, in Steller's sea lion, the Californian sea lion, and the northern elephant seal, it is the males that migrate furthest whereas in the northern

fur seal and walrus it is the females that migrate the furthest.

The walrus is perhaps an exceptional species in relation to the other species in this group. In the Bering Sea area, segregation occurs during the copulation and parturition seasons. Non-parturient females associate with the males in the east of the region whereas parturient females aggregate in the west (Fig. 28.11). The effect of water currents on the migration of the animals, especially on the drift of the ice-floes, and the advantage of keeping over shallow water, results in the males and non-parturient females migrating northeastward through the Bering Strait and along the Arctic coast of Alaska while the females with young-of-the-year migrate northwestward along the Arctic coast of Asia. The critical question in the case of the walrus, therefore, concerns the variables that favour one part of the Bering Sea to be selected as the most suitable by females about to give birth and another part to be selected as the most suitable for copulation by males and non-parturient females. As a female walrus conceives only every other year, then one year she will occupy one part of the Bering Sea and the next year the other part. As yet, no analysis seems to be available that demonstrates why one area should be suitable for parturition whereas the other area is suitable for copulation.

In the other four species, however, the segregation occurs with respect to the feeding home range of males and females and is achieved by the two sexes

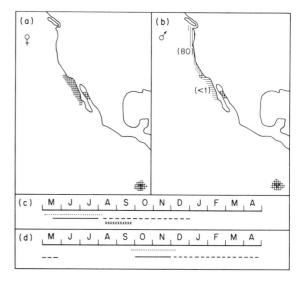

Fig. 28.13   Seasonal return migration of the Californian sea lion, *Zalophus californianus*, in the eastern Pacific

(a) Seasonal distribution of females. Horizontal hatching, breeding distribution; vertical hatching, maximum non-breeding distribution. There is some indication that the females that breed off the Californian and Baja Californian coast may extend slightly further to the south in winter (December–February). The females inhabiting the Gulf of California and Galapagos Islands do not seem to show any seasonal movement.

(b) Seasonal distribution of males. Conventions as in (a). At the end of July, after copulation, males leave their territories on the Californian and Baja Californian coastal islands and travel north. They are present in great numbers on the islands off the coast of central California in August and September and may be found there and further north from October to April. The numbers in brackets refer to the percentage of individuals present during the winter months that are adult males. In the northernmost figure the remaining 20 per cent are im-

mature individuals older than one year; in the southernmost figure the remaining 99 per cent are adult females and immature individuals. The males inhabiting the Gulf of California and the Galapagos Islands do not seem to show any seasonal movement.

(c) Seasonal behaviour off the coast of California and Baja California and in the Gulf of California. Dotted line, territorial period (Males maintain their territories partly by swimming backward and forward along a stretch of coast. Consequently territories are less well defined than in other species and the number of females/territorial male is more variable); solid line, parturition period (the young are born mainly on small offshore islands); dashed line, lactation period; crossed line, moulting period.

(d) Seasonal behaviour on the Galapagos Islands. Symbols as for (c).

[*Compiled from Bartholomew and Hubbs* (*1952*), *Bartholomew and Boolootian* (*1960*), *Orr and Poulter* (*1965*), *Fiscus and Baines* (*1966*), *Norris* (*1967*), *and Orr* (*1967*). *Photo by J. Twiselton*]

migrating different distances and/or in different directions. We are left, therefore, with Bartholomew's suggestion that temperature is the critical habitat variable to which the pinnipeds are responding in habitat assessment. Bartholomew further points out that males, being larger and therefore having a smaller surface area to volume ratio as well as having a thicker blubber layer, can thermoregulate better at lower temperatures than can females. Sexual segregation during seasonal return migration is thus taken to illustrate Bergmann's law that, in brief, larger animals live at lower temperatures, and thus to show the way that the evolution of one aspect of an animal, in this case

sexual dimorphism in size, can influence the course of some other aspect, such as migration distance and direction.

This is an attractive thesis but unfortunately it is not supported by the evidence available. Clearly Bergmann's law cannot be supported by the sexual segregation of pinnipeds and the sexual segregation be explained by Bergmann's law if the sexual segregation is the only evidence available that supports Bergmann's law. The argument is circular. What is necessary is first to demonstrate that Bergmann's law is applicable to pinnipeds, secondly to show that the distance and direction of sexual segregation is consistent with the general re-

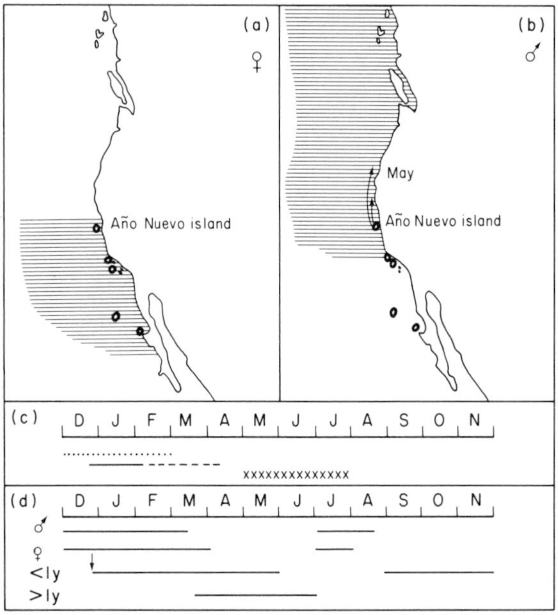

Fig. 28.14 Seasonal return migration of the northern elephant seal, *Mirounga angustirostris*

(a) Seasonal distribution of females. Solid black, breeding distribution; hatched area, maximum feeding distribution.

(b) Seasonal distribution of males. Conventions as for (a). Arrows and month, place and date of recapture of individuals marked on Año Nuevo Island.

Maximum feeding distribution in (a) and (b) along coast is based on sightings. Oceanic distribution is delimited arbitrarily due to lack of information. Quite possibly individuals feed much further out in the Pacific than indicated. The species feeds at depths of 100 m or more.

(c) Seasonal behaviour. Dotted line, territorial period (harems average about 12 females/territorial male); solid line, parturition period; dashed line, lactation period; crossed line, moulting period (young individuals moult earlier than adults).

(d) Seasonal occurrence on Año Nuevo Island. Adult males and females are present during the reproductive and moulting periods. Newly born (arrow) individuals stay until May, leave when the adults reappear to moult, return in September, leave in December when the adults return to breed, then return in March and leave in June after moulting.

It is not known whether males occupy the maximum feeding area both between March and July and between August and December.

[*Compiled from Bartholomew and Hubbs (1952, 1960), Bartholomew and Boolootian (1960), King (1964), Orr and Poulter (1965), Radford et al. (1965), and Orr (1970a)*]

lationship for pinnipeds between size and temperature, and finally to show that if sexual segregation were not to occur, that the size/temperature relationship would no longer be the same as the general relationship for pinnipeds.

I have calculated a great many correlation coefficients for pinnipeds using a variety of different measures and transformations for body size and temperature experienced at different times of year but have consistently failed to find a significant relationship. The nearest that such calculations came to demonstrating that larger pinnipeds occupied cooler waters was a correlation coefficient of -0.181 ($n = 55$; $P > 0.10$) for body length in centimetres against the maximum surface water temperature experienced at the time of year that the animals were furthest from their breeding site. When the five species of pinnipeds that show sexual segregation were added to the analysis, $r = -0.178$ ($n = 65$; $P > 0.10$). A similar analysis for just these five species yielded a value for r of -0.173 ($n = 10$; $P > 0.10$). The northern elephant seal could be considered to be anomalous in such a series in that it is exceptionally large, it is probably a relatively recent arrival in the warm temperate zone of the Northern Hemisphere, having perhaps arrived from the Southern Ocean during a Pleistocene glacial (Davies 1958), and in that it feeds at greater depths than the other species. Excluding this species from the analysis the relationship, for species that show sexual segregation, between body size and surface water temperature improves considerably ($r = -0.534$) but is still not significant, even at the 10 per cent level.

We are forced to conclude, therefore, that there is no indication from the figures available that pinnipeds as a group obey Bergmann's law. Nor is there any indication that species that show sexual segregation obey Bergmann's law. If this law is valid for pinnipeds at all then it is so only at the inter-sexual level of each of these four or five species, a conclusion that does not help to break the circularity implicit in Bartholomew's original thesis.

It still seems likely that temperature, or some correlate of temperature, is the variable that influences habitat suitability differently with respect to the males and females of these species. It seems, however, that the reasons for this influence of temperature on habitat suitability may have to be sought elsewhere than a relationship between body size and thermoregulation.

28.8 The application to pinnipeds of the familiar-area hypothesis

Evidence has already been presented in Chapter 14 (p. 106) that the movements of young pinnipeds are exploratory and involve assessment of habitat suitability and calculated removal migration, and consequently the establishment of a familiar area, seasonal home ranges, and perhaps also a migration circuit. The environmental clues used during the establishment of a familiar area encompassing large areas of sea as well as one or several terrestrial sites seem likely to be primarily olfactory and visual. It is possible, however, that some environmental clues suitable for the establishment of a familiar area may also be obtained through the echo-location mechanism possessed by at least some species.

In species, such as the walrus, *Odobenus rosmarus*, in which the young accompany the mother for some time, perhaps up to three years, it seems likely that considerable familiar-area establishment occurs during this period. In the case of the Pacific deme of the walrus, young individuals that accompany their mother for three years have the opportunity to establish a familiar area that includes not only the east and west sides of the Bering Sea but also the Arctic coasts of both Alaska and Asia (p. 727). Most species of pinnipeds, however, have young that appear to take up a solitary existence while in the sea (Table 28.2). How much communication occurs between apparently solitary individuals while in the water is unknown but it seems likely that the establishment of a familiar area by most pinnipeds is a solitary process achieved primarily by exploratory migration and owing little to social communication. Out of this familiar area, however, should crystallise a home range, seasonal home ranges, or a migration circuit, depending on the species. Because of the low initiation factor for such animals as pinnipeds (Chapter 17, p. 321), any migration circuit that develops is most likely to include a return for breeding purposes to the natal site.

One facet of the establishment of the familiar area is of particular interest in relation to pinnipeds, particularly those with a peak direction to their return migration (Fig. 28.10). This is whether or not there is also a peak direction to exploratory migration by the juveniles and, if so, whether this direction is inherited or learnt. A single example will serve to illustrate the situation and indicate the way that a

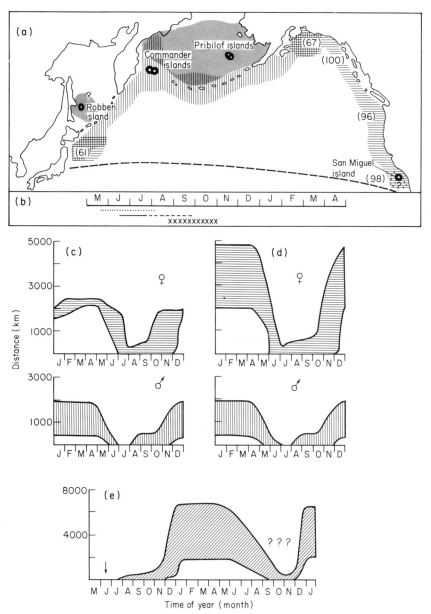

Fig. 28.15 Seasonal and ontogenetic return migrations
of the northern fur seal, *Callorhinus ursinus*

(a) Named islands edged by thick line, breeding areas;
stippled area, feeding distribution of males (August–
November), females (July to November) and immature
individuals (September–December); vertically hatched
area, feeding distribution of males (November–May);
horizontally hatched area, feeding distribution of females
(November–June); dashed line, approximate edge of
oceanic feeding distribution of nomadic immature individuals; figures in brackets, percentage of individuals in
area that are adult females.

Females migrating from the Pribilof Islands to the coast
of California migrate directly southeast and do not move
along the coast. The breeding colony on San Miguel Island
was established as recently as 1966 by individuals from
both the Pribilof and Commander Islands. The seasonal
return migration pattern of the individuals comprising
this colony will be of considerable theoretical interest.
Future observations should show to what extent the
direction, direction ratio, and distance components of the
return migration of *C. ursinus* are innate and to what extent
they are learned or a direct response to prevailing climatic
and geographical conditions.

(b) Seasonal behaviour. Dotted line, territorial period (harems consist on average of about 50 females/territorial male); solid line, parturition period; dashed line, lactation period; crossed line, moulting period.

(c) Distance component of the seasonal return migration of adult males and females that breed on the Commander Islands. Hatched area, distances from breeding site within which individuals are likely to be found.

(d) As for (c) but for adult males and females that breed on the Pribilof Islands.

(e) Ontogenetic return migration: distance component during the first 19 months of life (speculative in parts).

Hatched area, distances from the birth site within which individuals are likely to be found. After birth (arrow) in June the young *C. ursinus* stays on land through July, then enters the water to learn to forage until weaning occurs at the end of September. The young start to leave their birth site after the adults. All have left by the end of December. From January to June they have a solitary, nomadic, oceanic way of life. Yearlings do not haul-out at islands of their birth until the breeding season is over in September and some may not return until two years old. Females may be sexually mature when about three years old, males when about five or six.

[*Compiled from Kenyon and Wilke* (*1953*), *Wilke and Kenyon* (*1954*), *King* (*1964*), *Chugunkov and Prokhorov* (*1966*), *and Peterson* et al. (*1968*). *Photo courtesy of the National Marine Fisheries Service of the United States Department of Commerce*]

solution to this problem at least could be found for one species in the next few years.

The demes of the northern fur seal, *Callorhinus ursinus*, that breed on the Commander Islands and Pribilof Islands of the Bering Sea seem as likely, during the summer months, to search for food to the north of their breeding islands as to the south (Fig. 28.15). During the winter, however, the direction of the outward stage of the seasonal return migration is to the south for all individuals, though females migrate much further than males. Juveniles, during their winter exploratory migrations and establishment of a familiar area, are found all over the northern part of the North Pacific, but always to the south of their natal islands. The question is, therefore, whether the southerly bias to exploratory migration in winter is inherited or itself results from the exploratory–calculated migration process.

This latter process seems likely to function as follows. During the later stages of suckling and throughout weaning the young fur seal establishes a familiar area around its natal island-group, partly in company with its mother and partly solitarily, the latter particularly during the later stages of weaning. The familiar area extends in all directions from the natal island-group, both in the Bering Sea and to the south, eventually encompassing the Aleutians. Young northern fur seals haul-out on their natal islands for some time after the adults have left on their seasonal return migration, a fact that could imply that these islands are still being used as a base from which exploratory migrations are performed. At the same time as exploratory migrations are becoming long-distance and the familiar area is becoming sufficiently firmly established and large for frequent return to the natal site to become unnecessary, ice is spreading down from the north eventually to reach, or nearly to reach, the natal islands. Inevitably, therefore, the juvenile learns to concentrate on exploratory migration to the south of its existing familiar area rather than to the north. The following spring, potential breeding sites are visited while occupied by adults and compared for suitability relative to the natal site (Chapter 14, p. 110 and Chapter 17, p. 325). Later, the young fur seal hauls-out to moult on either the natal island-group or some other island within its now large familiar area. Over the next few years a migration circuit, assessed to be optimum by that individual, crystallises out of the familiar area, a migration circuit that involves migration to the south each autumn.

We are fortunate to be witnessing a natural experiment that will enable us to test this latter hypothesis against the inherited exploratory direction hypothesis. The new breeding deme of the northern fur seal on San Miguel Island off the coast of California (p. 108) is composed largely of removal migrants from the Bering Sea demes. If exploratory migration direction is inherited, these fur seals should establish migration circuits to the south from San Miguel or, alternatively, if the water to the south is too warm, as seems likely, to remain around the breeding islands all year. If the non-inherited hypothesis is correct, however, fur seals from San Miguel, following early exploratory migrations to the south and to the north of the natal site, should avoid further exploratory migration to the south and instead explore the region to the north which is likely to continue to be the same region of the north Pacific as individuals from the Bering Sea. One, perhaps probable, result is that males will perform long-distance seasonal return migration to the Aleutian Islands while the females perform a shorter-distance northward migration to the Pacific coast of North America and the Gulf of Alaska. Study of the San Miguel deme over the next few years should provide evaluations of these alternative hypotheses.

Although migration distance and direction may not be inherited in pinnipeds, there seems no doubt that some components of the migration have a large inheritable basis. The different predilections for solitary and gregarious feeding and breeding in the different species seem likely to be examples. The threshold values that cause one species to assess ice as the most suitable breeding habitat and another to assess a rocky shore as the most suitable breeding habitat also seem likely to be largely inherited, though some imprinting could take place during the period from birth to weaning. The migration thresholds to the habitat variables of water temperature and ice, etc., that cause different species to live in different geographical areas are probably also inheritable species-specific characteristics. As far as migration is concerned, however, perhaps the major inheritable component is that of seasonal timing, though as with ungulates it has not been proved that timing is not learnt from other individuals.

Let us consider as an example the seasonal timing of the migration to the breeding grounds. Presumably, the sequence of events is similar to the following. During the main feeding period, calculated-migration thresholds (p. 347) consist largely of habitat variables such as food availability,

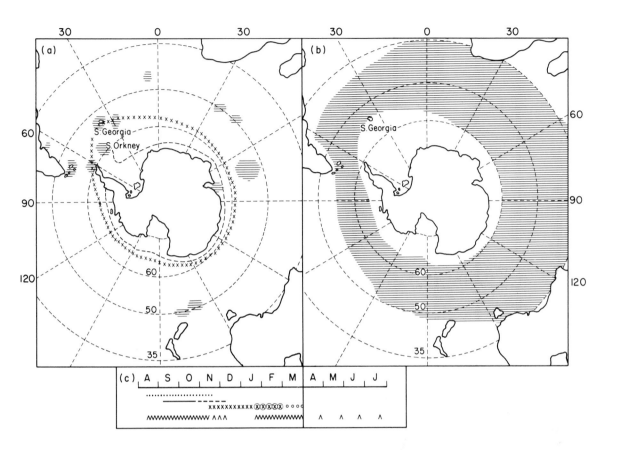

Fig. 28.16 Seasonal return migration of the southern elephant seal, *Mirounga leonina*

(a) Hatched areas, breeding and moulting distribution, late August to early April; dashed line, average minimum extent of pack-ice, February/March; crossed line, average maximum extent of pack-ice, August/September.

(b) Hatched area, maximum feeding distribution between April and July.

(c) Seasonal behaviour. Dotted line, territorial period; solid line, parturition period; dashed line, lactation period. Harems average about 30 females/territorial male. Parturition occurs on rocky shores of, in most cases, oceanic islands. Crossed line, moulting period of yearlings; crosses in circles, moulting period of females; line of open circles, moulting period of males; zig-zag line, period that adults are out of the water.

At the end of the territorial period in males and at the end of the lactation period in females, the adults are at sea until the moulting period. Animals that breed on South Georgia moult on the South Orkney Islands.

[*Compiled from Laws (1956) and King (1964). Photo courtesy of Institute of Oceanographic Sciences*]

predator abundance (the major aquatic predators of pinnipeds are sharks and killer whales, *Orcinus orca*), abundance of conspecifics, and temperature, etc. At some point toward the end of the feeding period, some mechanism, probably either a response to photoperiod/temperature and/or an annual endogenous rhythm triggers the sequence of physiological processes that lead up to the attainment at the appropriate season of a breeding physiology. In pregnant females, of course, this process inter-relates

with the stage of pregnancy. At some point in this physiological sequence, there is a change in calculated-migration thresholds such that those habitat variables associated with a suitable breeding habitat make the largest contribution. The result is the initiation of migration to that habitat that the animal assessed during familiar-area establishment to have the highest initiation factor (p. 77) relative to the home range that the animal has occupied for feeding. In other words, the initiation of migration is

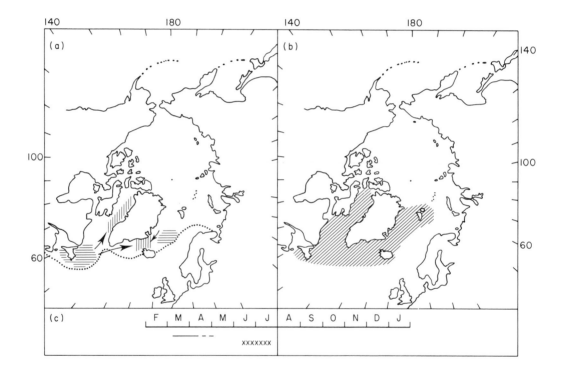

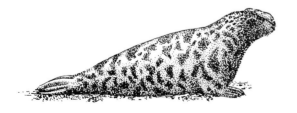

[*Drawn from information in King (1964)*]

Fig. 28.17   Seasonal return migration of the hooded seal, *Cystophora cristata*

(a) Horizontally hatched area, breeding distribution; vertically hatched area, moulting distribution; arrows, movement directions; dotted line, average extreme (February/March) limit of drift-ice.

(b) Diagonally hatched area, maximum feeding distribution between August and January.

(c) Seasonal behaviour. Solid line, parturition period; dashed line, lactation period; crossed line, moulting period. Parturition occurs on drifting ice-floes. There is no harem formation but territorial behaviour is shown in that a male and female pair defend their floe and the female's young offspring against other individuals.

likely to be obligatory (p. 50) even though the destination is determined by the previous experience of the animal.

It seems likely that a similar interaction of inheritable and learnt components are characteristic of all animals that perform seasonal breeding migrations within a familiar area. Similar systems probably exist in relation to moulting in pinnipeds and hibernation migrations in some other animals.

Where such migrations occur, the re-establishment of migration thresholds to habitat variables characteristic of the feeding period mean, in effect, that the feeding migration is also obligatory. Migrations between feeding home ranges suitable at different seasons, however, seem most likely to be primarily facultative and to contain little inheritable component except the species-specific feeding-period migration thresholds.

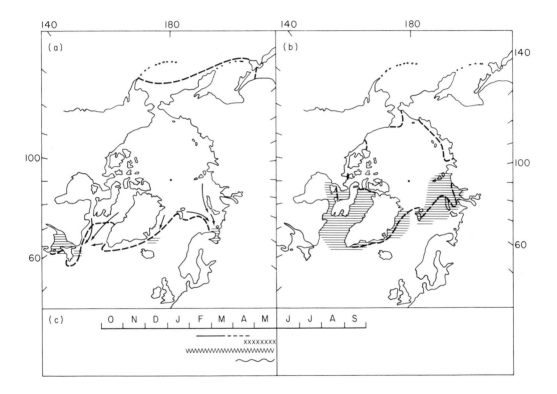

Fig. 28.18   Seasonal return migration of the harp seal, *Pagophilus groenlandicus*

(a) Hatched area, breeding distribution (January–May); dashed line, average maximum limit of pack-ice (February/March); arrows, convergence migration routes between October and January.

(b) Hatched area, maximum feeding distribution (June–September); dashed line, average minimum limit of pack-ice (August/September).

(c) Seasonal behaviour. Solid line, parturition period; dashed line, lactation period; crossed line, moulting period; zig-zag line, period females out of water on ice; wavy line, period males out of water on ice. There is no harem formation. Parturition occurs on fast-ice.

[*Drawn from information in Sivertson (1941)*]

28.9 Appendix: the seasonal return migrations of some pinnipeds

Figures 28.19–28.23 illustrate the seasonal return migrations of pinnipeds not illustrated in the main body of this chapter. Because the form of the migration varies so much from species to species it has been necessary to adopt different graphical conventions for different species in order to emphasise the major points. These conventions are described under each figure. For the seasonal behaviour, however, the convention has been adopted of: a line of small dots for the period of territorial behaviour in harem-forming species; a solid line for the parturition period; and a dashed line for the lactation period. Other lines refer to other seasonal behaviour components and are described in the appropriate figure.

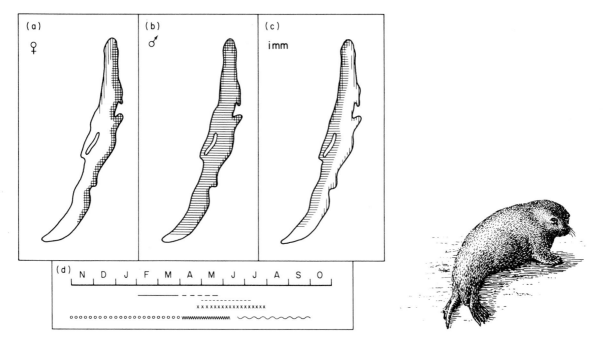

Fig. 28.19 Seasonal return migration of the Baikal seal, *Pusa sibirica*

(a), (b), and (c). Seasonal distribution within the freshwater Lake Baikal (length = 636 km) of (a) adult females, (b) adult males, and (c) immature individuals (both sexes mature when about four years old). Horizontal hatching, main distribution during winter when the seals are living under the ice; vertical hatching, main distribution of terrestrial roosts during summer (June to October).

(d) Seasonal behaviour. Solid line, parturition period (parturition occurs in lairs made by the female under snow drifts on the ice that covers the lake's surface but near an air-hole through which the female enters and leaves the water); dashed line, lactation period (the young seal stays in the lair while the female is feeding in the water); dotted line, copulation period; crossed line, moulting period; open circles, winter period during which the seals are in the water under the ice and breathing through air-holes kept open by the seals (the seals are in the water continuously except for the adult females which come onto the ice to give birth and suckle their young); zig-zag line, period during which seals frequently haul-out of the water onto the ice-surface; wavy line, summer period during which there is no ice-cover and during which the seals roost on shores and on rocks protruding above the lake's surface, in particular along the northern and southeastern coasts as shown in (a), (b), and (c).

[*Compiled from Ivanov (1938), Kozhov (1963), and King (1964)*]

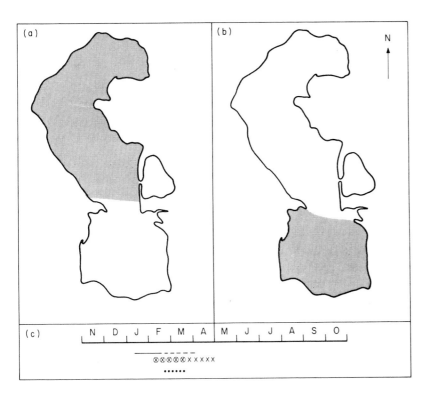

Fig. 28.20   Seasonal return migration of the Caspian seal, *Pusa caspica*

Stippled area, distribution: (a) November to April, and (b) May to October. This species is unique among pinnipeds in that it moves toward the equator for the summer and toward the pole for the winter. The water of the Caspian Sea is cooler in the southern regions in summer.

(c) Seasonal behaviour. Solid line, parturition period (the young are born on the ice); circled crosses, moulting of females; crosses, moulting of males; dotted line, copulation. There is no harem formation.

[*Drawn from information in King (1964)*]

Fig. 28.21 Ontogenetic and monthly return migration of the South African fur seal, *Arctocephalus pusillus*

(a) Hatched area, breeding and feeding distribution; arrow, a marking–release/recapture result which shows that individuals wander through a major part of the total range.

(b) Seasonal behaviour. Dotted line, territorial period (harems are of variable size but may average about four females/territorial male); solid line, parturition period (the young are born on rocky shores of the mainland or offshore islands); crosses, moulting period.

(c) Distance component of ontogenetic migration. Dots, recaptured individuals; hatched area, distances from birth site within which individuals may be found. For some months after birth (arrow) the young *A. pusillus* stays within a few kilometres of its birth site. After weaning, rapid migration (exploratory?) occurs and 1200 km may be covered in a few weeks. The first return to the birth site seems to take place when the animal is 2 years of age. Females copulate for the first time at this age.

(d) Distance component of the monthly return migrations of individuals older than 2 years. Dots, recaptured individuals; hatched area, distances from birth site within which individuals may be found. Males first hold a territory when about 6 years of age. Adults do not travel as far during migration as during their first 2 years and some individuals are always to be found at the primary roost. Females leave the roost for 2–3 weeks at a time and then spend 1–2 weeks at the roost feeding the young. Males leave the roost for 3 weeks at a time and upon return spend 1 week at the roost.

[Compiled from Rand (1959) and King (1964)]

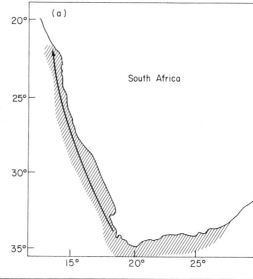

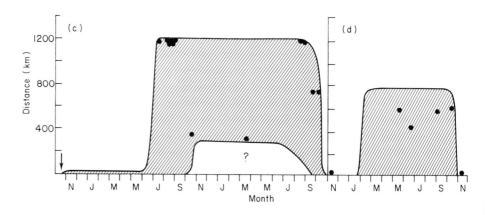

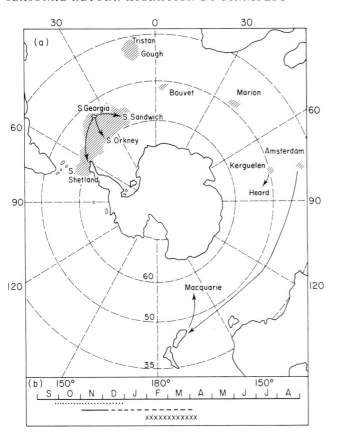

Fig. 28.22 Seasonal return migration and removal migration of the Kerguelen fur seal, *Arctocephalus tropicalis*

(a) Hatched area, breeding distribution (September to April); arrows, recent presumed removal migration.

(b) Seasonal behaviour. Crossed line, moulting period; dotted line, territorial period (harems average about 5 females/territorial male).

Seasonal return migration consists of post-moulting disperasal in April/May and pre-breeding convergence in September. Immature and non-breeding adults keep away from the breeding grounds between September and Feburary. Young individuals are though to be performing exploratory migration when away from the breeding area. The distances travelled can perhaps be estimated on the assumption that removal migration take place within the range of exploratory migration.

[*Drawn from information in King (1964). Photo courtesy of D.G. Bone*]

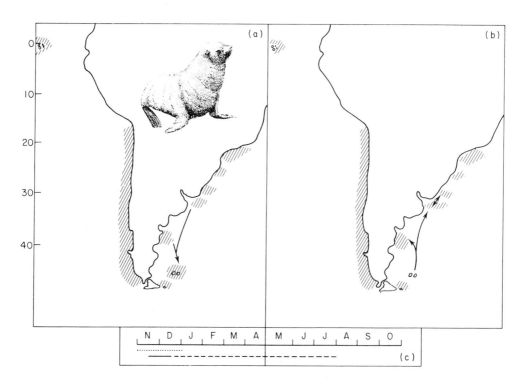

Fig. 28.23   The distribution and possible seasonal return migration of the South American fur seal, *Arctocephalus australis*

Hatched area, distribution: (a) November–April, and (b) May–October; Arrows, possible movements. Some movement may occur between the Islands off Uruguay, though these may be removal, as well as return, migrations. The *A. australis* that breed on the Falkland Islands are thought to migrate northwards for the winter, perhaps as far as Uruguay.

(c) Seasonal behaviour. Dotted line, territorial period (harems average about three females/male).

[*From information in King (1964). Photo courtesy of Institute of Oceanographic Sciences*]

29

Seasonal return migration by cetaceans

29.1 Introduction

Taxonomically, cetaceans are divided into two major groups, the Odontoceti and the Mysticeti. The odontocetes are toothed cetaceans and consist of all dolphins and porpoises and also a number of larger species such as the sperm whale, *Physeter catodon*, and killer whale, *Orcinus orca*. The mysticetes are the whalebone or baleen whales and consist of the large plankton- and fish-eating species that filter their food from the water through the fringed transverse rows of keratin, termed baleen, in their mouths.

The cetacean fossil record begins in the lower Eocene (50 million years B.P.) (Kellogg 1928), though the first unquestioned odontocetes and mysticetes are not found, respectively, until the upper Eocene (40 million years B.P.) and lower Miocene (25 million years B.P.) (True 1907, Davies 1963). The initial invasion(s) of the sea occurred early in the radiation of mammals, perhaps in the Palaeocene or even in the Cretaceous (65 million years B.P.) (Flower 1883) and may have been carried out by members of a group of animals that were the ancestors both of cetaceans and of artiodactyl ungulates (Slijper 1966). It has been suggested that odontocetes and mysticetes do not share a common aquatic ancestry but are the result of invasion of the sea by separate lineages. The mid-Eocene (45 million years B.P.) fossils *Protocetus* and *Pappocetus*, however, are sufficiently intermediate in form and early enough to have been directly ancestral to both odontocetes and mysticetes (Valen 1968).

The first cetaceans were probably tropical in origin and even mid-Eocene (45 million years B.P.) fossils occur only in warm water areas (i.e. Egypt, Nigeria, and Texas). The first cold-water forms

appeared in the mid-Tertiary (30 million years B.P.), perhaps in response to expansion of the cold-water environment that took place at that time (Davies 1963). Even so, only three genera (*Balaena*, *Delphinapterus*, and *Monodon*) have become adapted to breed in polar waters, none having done so in the Antarctic. Although many cetacean species are found in both the northern and southern hemispheres and in both the Atlantic and Pacific Oceans, many show an anti-tropical distribution, avoiding the equatorial warm-water zone (Davies 1963). This is particularly true of those species, such as most mysticetes, that feed primarily on plankton.

29.2 Factors affecting seasonal variation in habitat suitability

Unlike pinnipeds, cetaceans are permanently aquatic. The vast majority are marine, though a few odontocetes have entered large river systems, either as marine-to-freshwater migrants or as permanent residents. The finless black porpoise, *Neomeris phocaenoides*, for example, which is found mainly in coastal waters, especially lagoons and estuaries, from the Cape of Good Hope to Japan, in China travels up the Yangtse Kiang river beyond the Tung Ting Lake, almost 2000 km from the sea. It is perhaps significant, in view of the arguments presented concerning the anadromy/catadromy paradox (p. 843), that where odontocetes have crossed the river/sea boundary they have done so entirely within the tropics and sub-tropics (e.g. the Gangetic dolphin, *Platanista gangetica*, in the Ganges and

Indus; the Amazonian dolphin, *Inia geoffrensis*, in the Amazon; the La Plata dolphin, *Stenodelphis blainvillei*, in the Rio de la Plata; and the Chinese river dolphin, *Lipotes vexillifer*, in the Tung Ting Lake) (Slijper 1962).

Where data are available, it seems that marine cetaceans have a number of seasonal home ranges. As the majority of these are feeding home ranges, it seems appropriate first to consider some of the factors that influence the suitability of a habitat as a feeding home range. The first, and perhaps most important, point is that suitability is not solely a function of food density for, at least among mysticetes, different species are adapted to exploit food resources at different densities. For example, two major categories of feeding mechanisms, which have been termed 'swallowing' and 'skimming', may be recognised among baleen whales (Nemoto 1959). The main food species of swallowing whales are euphausids, though swarming fish and copepods may also be taken. Swallowing-type whales have generally thick and coarse baleen fringes that are suited to take macroplanktonic organisms in dense concentrations. Such whales open the mouth to take in large quantities of food and then turn on their side to swallow. Skimming types, however, such as the right whales, *Balaena mysticetus* and *Eubalaena glacialis*, and sei whale, *Balaenoptera borealis*, have fine baleen fringes and swim with considerable or moderate velocity below the surface of the sea with their jaws widely open while the sparsely distributed microplanktonic organisms are filtered and stay in the cavity of the mouth.

Most mysticetes feed by one or other of these two methods on some combination of planktonic crustaceans, fish, and cephalopods. The proportions of food-types taken by the individual species the migrations of which are illustrated are indicated by asterisks in the legends to the respective figures. The grey whale, *Eschrichtius gibbosus* (Fig. 29.2) is unusual, however, in that, or so it is believed, it ploughs into the soft sea floor at shallow depths (less than 27 m) and filters bottom-living invertebrates, particularly amphipods, from the mud and sand.

Figure 29.1 indicates the distribution of the fertile areas of the oceans. Although the factors influencing the standing crop at different latitudes are not fully understood, some general theories have emerged (Raymont 1963). For example, carbonic acid and oxygen are both more plentiful in cold than in warm waters whereas bacteria are more abundant in warmer waters. This latitudinal variation could contribute to the abundance near the poles of both the phytoplankton and the zooplankton that feeds upon it. On average, the Antarctic has a standing crop of plankton that is 10–20 times greater than that in the tropics (Slijper 1962).

Presumably as a result of this geographical variation in standing crop of plankton, all mysticetes that are specialised to feed on plankton spend the polar summer in polar regions. The precise distribution of home ranges, however, is a function of the distribution of plankton of the size and at the density upon which the particular whale species is adapted to feed. This in turn is a function of bottom topography and the positions of cyclones and the upwellings of deep water that these variables combine to produce (Beklemishev 1960, Chernyi 1966, Zaverin 1966). In the Antarctic, the major food species is the Antarctic krill, *Euphausia superba* (Fig. 20.9). Off New Zealand, the Falkland Islands, and Patagonia, whales feed also on 'lobster krill', larvae of crustaceans of the genus *Munida* (Slijper 1962). In the South Pacific, rorquals (*Balaenoptera* spp.) and the humpback, *Megaptera novaeangliae*, also add the euphausids, *Thysanoessa macrura* and *T. vicina*, to their diet. In the Arctic, which lacks the abundance of a single zooplankton species that characterises the Antarctic, the mysticete diet is more varied. Right whales (Balaenidae) and the blue whale, *Balaenoptera musculus*, feed predominantly on *Thysanoessa inermis* and *Meganyctiphanes norvegica* and pteropods, *Clio* and *Limacina*. The sei whale, *Balaenoptera borealis*, feeds on smaller crustaceans, such as the copepod, *Calanus finmarchicus* (Fig. 20.10), though in the north Pacific it also eats cephalopods.

Odontocetes prey predominantly on fish and cephalopods and usually swallow their prey whole. Most dolphins (Delphinidae) feed on small fish whereas most porpoises (Phocaenidae) also include cephalopods in their diet. The killer whale, *Orcinus orca*, preys on fish when young but when older feeds also on pinnipeds, porpoises, dolphins, and the larger cetaceans, including even the blue whale, *Balaenoptera musculus*, the largest living animal. Sea birds are also taken (Slijper 1962). Sperm whales, *Physeter catodon*, feed on abyssal cephalopods and fish (Tarasevich 1968), many of which are taken from mid-water at night when they are away from the sea bottom (Gaskin and Cawthorn 1967). From the incidence of sperm whales that entangle their jaws in submarine cables, however, it seems as though they sometimes feed by swimming along the ocean bottom with their jaws open (Caldwell *et al.* 1966).

Cetaceans, like all mammals, breathe air. From time to time, therefore, feeding and other activities under water have to be interrupted to swim to the surface, exhale ('blow'), and then take in more air. The sperm whale, *P. catodon*, is among the longest and deepest cetacean divers. There is a tendency within the species for time of dive to increase with size of the individual (Caldwell *et al.* 1966) and this relationship could well hold true for cetaceans as a whole. The maximum recorded dive for the sperm whale is 90 min and the shortest is 5 min. A general relationship for male sperm whales is that the animal 'blows' once for every minute of submersion. Females tend to blow more than once for each minute submerged. Male sperm whales dive to depths of up to 1500 m but females perhaps only dive to about 500 m.

By the end of the polar winter, because of the need for access to air, the pack ice cuts off all the normal concentrations of plankton and other food from the whales, a fact that itself could contribute towards the extraordinary abundance of the polar krill (Marr 1962). Some species of whale, such as the bowhead or Greenland right whale, *Balaena mysticetus* (Fig. 29.3), among the mysticetes, and the beluga or white whale, *Delphinapterus leucas*, and narwhal, *Monodon*

monoceros, among the odontocetes, have become adapted to living and breeding at the edge of the pack ice where they make use of holes in the ice in order to gain access to air. Occasionally, also, there are reports of other species being trapped by the developing ice (Taylor 1957). On the whole, however, cetaceans keep to the open sea.

Cetaceans, such as most odontocetes, that feed on fish and cephalopods which themselves migrate as the areas of production within the sea change with the seasons (p. 483), have in consequence a food supply that may be found not only within the fertile areas but also on migration in between. Cetaceans, such as most mysticetes, that feed primarily on plankton, however, when cut off from polar sources, have a limited food supply with a patchy distribution. It is most striking that rorquals (*Balaenoptera* spp.) and the humpback whale, *Megaptera novaeangliae*, are not evenly distributed over the oceans when away from polar regions but are concentrated in certain areas which, in general, coincide with areas of greatest biological productivity (Slijper *et al.* 1964). In the tropics and sub-tropics these areas are the Caribbean, the North African west coast, the Atlantic coast of South Africa, the Arabian Sea, Gulf of Aden, Bay of Bengal, Indonesian Archipelago,

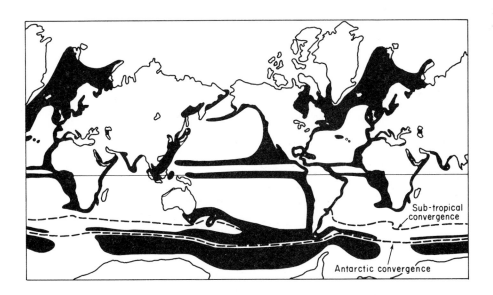

Fig. 29.1   Geographical distribution of the fertile areas of the oceans

   fertile areas; – – – – convergences

[Re-drawn from Walford (1958)]

and the African east coast between 30°S and 40°S (Fig. 29.1). For a long time, however, it was supposed that plankton-feeding mysticetes do not feed while in warm waters (Mackintosh 1942), a suggestion that gained its main support from the observation that the stomachs of whales taken at South African, Australian, and New Zealand land stations contain little or no food. To some extent the controversy continues as to whether or not plankton-feeding mysticetes feed while away from polar waters (Mackintosh 1965). However, the evidence of the distribution of such whales while in warm waters relative to the geographical variation in the productivity of the sea (Slijper *et al.* 1964), combined with the observation that in places whales do have food in their stomachs, seems to be fairly conclusive. At least for the humpback whale, *M. novaeangliae* (Chittleborough 1965), and probably for most other species there now seems little doubt that feeding occurs whenever food is available in sufficient quantity. Loss of weight during the period away from polar waters and the relatively small amounts of food found in the stomachs can be attributed to the scarcity of food rather than to the behaviour of the whale. The adherence of the grey whale, *Eschrichtius gibbosus*, to coastal and continental shelf regions throughout its migration circuit also provides strong circumstantial evidence that this bottom-feeding species feeds throughout the year (Pike 1962). It may be assumed also that all species that feed on fish and cephalopods also feed throughout the year.

Cetaceans are uniparous, producing just one young at a time (Slijper 1966). Twins, although they occur, are rare (about 1 per cent of all births). Suckling always takes place under water, though sometimes very close to the surface. When suckling, the mother moves very slowly and as the newly born young is suckled frequently (about every half-hour, day and night, in the bottlenose dolphin, *Tursiops truncatus*—McBride and Kritzler 1951), rate of movement is restricted at this stage. However, growth is rapid. The young of the blue whale, *Balaenoptera musculus*, for example, doubles its birth weight in seven days (Bernhardt 1961). By the age of six months, feeding frequency has decreased (seven times a day in the bottlenose dolphin) and weaning may have begun. Young humpback whales, *Megaptera novaeangliae*, for example, have been reported to feed on plankton during migration when a few months old (Dawbin 1966).

Cetaceans swim by moving their tail flukes

straight up and down. Assisted locomotion has been known to occur and may be widespread. Assistance is given in that a small individual swims by the side of a large individual in such a position that less energy is needed to maintain full speed, the larger individual supplying the necessary extra energy (Caldwell *et al.* 1966). This mechanism enables a young individual to maintain the speed of the mother or the group and is particularly advantageous in escape from predation by killer whales, *Orcinus orca*, or in the reduction of migration cost.

The restrictions on movement during the first few weeks of lactation and the danger of predation on the young by killer whales and perhaps large sharks has led the females of many species to establish

Fig. 29.2 Seasonal return migration of the grey whale, *Eschrichtius gibbosus*

General ecology
The grey whale is a mysticete that feeds primarily on bottom-living invertebrates (****) but will also take euphausids (*) and fish (*). It may plough the soft sea floor at depths less than 27 m and strain its food from the mud and sand. The species occurs either singly or in groups of 2–4 individuals.

The seasonal incidence of copulation (....), parturition (———), and lactation (– – –) in the eastern North Pacific is indicated above diagram (b). Gestation lasts 10–12 months.

Average maximum body length is 14·5 m for males and 15·5 m for females.

Seasonal return migration
(a) Geographical distribution in August (△) and February (○)

Dots show sightings at the appropriate month. Arrows show migration routes.

(b) Seasonal shift in latitude

The dots indicate latitudes at which grey whales have been sighted on different sides of the North Pacific during each month of the year.

There is a clear polar migration in summer and a tropical migration in winter. The southward migration is carried out predominantly within 5 km of the shore in the eastern Pacific. Part of the migration, however, also takes place some distance offshore, particularly during passage past the Channel Islands off Southern California. Nevertheless, all whales observed are migrating within the 1800 m isobath.

[Compiled from data in Gilmore (1960), Pike (1962), Slijper (1962), Slijper et al. (1964), Mackintosh (1965), Rice (1965), Berzin and Rovnin (1966), Nasu (1966), Hubbs and Hubbs (1967), and Nishiwaki and Kasuya (1970)]

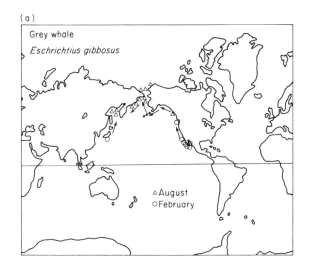

(a)

Grey whale
Eschrichtius gibbosus

△ August
○ February

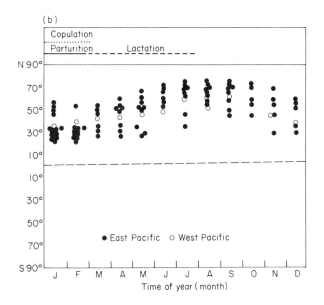

(b)

Copulation
Parturition ———— Lactation ————

● East Pacific   ○ West Pacific

Time of year (month)

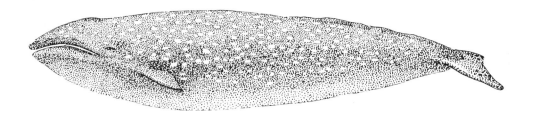

parturition and early-lactation home ranges in shallow water. The best known example is perhaps that of the grey whale, *Eschrichtius gibbosus* (Fig. 29.2), which every winter produces its young in the bays and lagoons of Baja California.

The gestation period of most cetaceans varies between 10 and 12 months, though it is about 16 months in sperm whales, *Physeter catodon*, and pilot whales, *Globicephala melaena*, both of which are odontocetes (Slijper 1966). In the smaller species the time interval between successive births is one year, females copulating soon after parturition while still lactating. In the larger species the pattern is more often a gestation period of 12 months or less followed by a lactation period of 12 months or less, followed soon after by copulation. In these species, therefore, the time interval between successive births is about 2 years. Finally, in the sperm whale, *Physeter catodon*, which has a 16 month gestation period and a 24 to 25 month lactation period (Ohsumi 1965), the time interval between successive births is usually 4–5 years. In all species, however, it seems that from time to time a female has a 'rest' year during which she is neither pregnant nor lactating. Little is known concerning the requirements of the copulation home range for males and females.

The role of water temperature as a habitat variable has yet to be thoroughly evaluated. Cetaceans occur at all water temperatures from $-2°C$ to $30°C$. Throughout this ambient range they maintain a body temperature of between $35°$ and $38°C$ (Kanwisher and Sundnes 1966). Heat loss to the water is decreased through the insulating effect of a thick layer of blubber and by cooling the outer layer of skin down to ambient temperature. Consequently, the peripheral structures are functionally poikilothermic. Countercurrent heat-exchange structures in the blood system reduce heat loss from the blood that has to circulate through the layers external to the blubber.

The problems of homeothermy change with change in body size, and as a large cetacean may be 10 000 times the size of a small cetacean it seems likely that the two extremes have quite different problems. In general, the larger cetaceans should find it easier to stay warm whereas the smaller species should find it easier to stay cool. Small species survive in cold water through a high metabolic rate and a thick blubber layer. Forty per cent of the body weight is blubber in the common or harbour porpoise, *Phocaena phocaena*, which is about 1·4 m long. Because of the effect of body size the problems of thermoregulation are even greater for the young of the smaller cetaceans. The resultant selection seems to have favoured the young being born at a relatively large body size. Young porpoises, for example, at birth are already 30 per cent of their adult weight (Slijper 1966).

A young large mysticete, on the other hand, is still a large animal and theoretically should be sufficiently insulated for even the coldest ocean (Kanwisher 1966). Indeed, keeping cool could be the major problem for large mysticetes, especially when adult, but there is no evidence that they overheat. The primary function of the blubber, however, in the large mysticetes seems unlikely to be thermoregulatory (Kanwisher and Sundnes 1966). If a porpoise can maintain thermal balance with a blubber layer 2 cm thick, all else being equal except for size, a large whale should need even less. Consequently, the 15–20 cm thick blubber layer on a large whale cannot be considered as insulation. Furthermore, for reasons of size and movement, the whale has more than twice the metabolic heat flux of the porpoise. It should therefore need only half the insulation. Instead, it has 10 times as much. Insofar as the blubber is concerned, therefore, the whale is many times over-insulated and a primary thermal function seems unlikely. One possible alternative is subsistence in areas where food is scarce or absent. For example, half of a whale's blubber could maintain the animal at its basal rate for 4–6 months (Parry 1949). Hydrostatic buoyancy is also a possible function.

Consideration of the thermoregulatory ability of cetaceans therefore provides little indication of the selective pressures, if any, that are generated by water temperature. Given that cetaceans can thermoregulate in any water within their range, therefore, the critical factor becomes the energy expended in thermoregulation rather than the danger of failing to thermoregulate. From the previous discussion it seems likely that young and small cetaceans expend less energy on thermoregulation in warmer waters whereas large cetaceans expend less energy on thermoregulation in cooler waters. To some extent, circumstantial evidence exists that supports this suggestion. Thus, the newly weaned calves of the minke whale, *Balaenoptera acutorostrata* (Fig. 29.10), which at an adult length of 9·1 m is the smallest but one of extant mysticetes, but which is still a large animal, remain in summer in tropical waters while the adults travel along the Norwegian coast as far as the Barents Sea (Slijper 1962).

Further, figures given by Nishiwaki (1966) for the sperm whale, *Physeter catodon*, in the North Pacific, indicate larger individuals penetrate further north than smaller individuals. A similar correlation between increase in size with age and extension of the migration circuit into colder and colder waters is also found in the blue whale, *Balaenoptera musculus* (Fig. 29.5), finback whale, *B. physalus* (Fig. 33.13), and sei whale, *B. borealis* (Fig. 15.12).

In conclusion, therefore, it seems that the major habitat variables for cetaceans are food, access to air, conditions suitable for copulation, an absence of predators, and optimum water temperature. With the progression of ontogeny, the seasons, and, for females, the reproductive cycle, a change takes place in the relative suitability of different geographical regions and different habitat-types. The way that this temporal and spatial fluctuation in habitat suitability has generated different patterns of seasonal change in home range can now be considered.

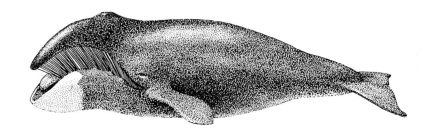

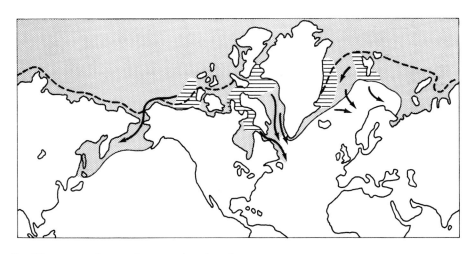

Fig. 29.3 Possible pattern of seasonal return migration of the bowhead, *Balaena mysticetus*

General ecology
The bowhead, or Greenland right whale, is a mysticete that is confined more or less to the Arctic Ocean. The species feeds primarily on smaller species of plankton and employs a skimming type of feeding behaviour. It occurs either singly or in a family group of a male, female, and calf. When food is abundant, however, larger schools may be formed. Average maximum length is about 15·5 m. Copulation takes place in late winter and early spring (February and March) while on the winter grounds and is followed by a gestation period of 9–10 months and a lactation period of 12 months.

Seasonal return migration
The map shows the estimated summer feeding grounds (horizontal hatching) and the migration routes (arrows) to the winter range. Stippling indicates the area covered by pack ice in winter and the dashed line shows the extent of the pack ice in summer.

The bowhead spends a large part of its life at the edge of the Arctic pack ice in areas where the ice is sufficiently broken to allow access to the air. Most of the migration is related to the seasonal advance and regression of the ice.

[*Compiled, with modifications, from data and diagrams in Slijper (1962) and Vibe (1967)*]

29.3 Seasonal shifts in home range

Most of the information available concerning seasonal change in home range has been obtained by a consideration of seasonal shifts in population of the type presented in the figures accompanying this section. As also indicated in the figures, where marking–release/recapture experiments have been carried out they have so far only served to emphasise the conclusions that can be drawn from the population shifts. Except, that is, insofar as they have also supplied information concerning degree of return (Section 29.4). Unfortunately, this means that conclusions concerning the migration circuits of cetaceans can really only be drawn for those species that show a latitudinal shift in distribution. Once we try to look at seasonal shifts in home range at the deme or individual level, the data become extremely fragmentary. As it happens, however, the migrations of the larger plankton-feeding species do appear to be strongly latitudinal and to be performed by the vast majority of the population. Were it not for this fortuitous event, our knowledge of cetacean migrations would be virtually non-existent, as indeed is the case for those species that do not show latitudinal shifts, such as the killer whale, *Orcinus orca* (Fig. 29.7).

Where latitudinal shifts of entire populations occur, they are found primarily in large, plankton-feeding mysticetes. The general pattern is for the whales to be feeding at polar latitudes in the summer when the rich aggregations of plankton are uncovered by the pack ice. As the ice re-forms and advances in winter, however, the whales migrate away from the poles. Most species migrate as far as tropical waters where, during the polar winter, they give birth and copulate. In most cases, as they are by and large distributed in the more fertile areas of tropical waters, some feeding probably also occurs. The bowhead, *Balaena mysticetus* (Fig. 29.3), however, seems to maintain a position near to the edge of the Arctic ice throughout the year (Nishiwaki 1972). In the spring, off the Arctic coast of Alaska, for example, the northward movement begins as soon as the ice opens enough to allow the whales to pass, holes in the ice being used as blowholes (Bailey and Hendee 1926). The first appearance at Cape Prince of Wales generally occurs at the end of April, a few days before the first appearance further north near Point Barrow.

Most of the species that show latitudinal shifts in

seasonal home range are found (in both the northern and southern hemispheres) in each of the Pacific, Atlantic, and Indian Oceans. The only exception is the grey whale, *Eschrichtius gibbosus* (Fig. 29.2), which is the sole survivor of an archaic group (Davies 1963) and occurs only in the North Pacific. The result for most species of being distributed in both hemispheres is that while the northern hemisphere deme is breeding in the tropics, the southern hemisphere deme is feeding in the Antarctic, and while the northern hemisphere deme is feeding in the Arctic, the southern hemisphere deme is breeding in the tropics. Most classical accounts stress, therefore, that the northern and southern hemisphere demes are separate and do not meet at any point on their respective migration circuits, though it is suggested (Davies 1963) that considerable contact and mixing occurred during the Pleistocene whenever the equatorial waters became cooler. Inspection of the figures suggests that complete separation may well exist for the black right whale, *Eubalaena glacialis* (Fig. 29.4), and the finback whale, *Balaenoptera physalus* (Fig. 33.13). Indeed, the southern hemisphere deme of the right whale has in the past been considered to be a separate species, the southern right whale, *Eubalaena australis* (see Slijper 1962). The impression of lack of contact between the southern and northern demes is further emphasised by the results of marking–release/recapture experiments that are presented.

Fig. 29.4   Seasonal return migration of the black right whale, *Eubalaena glacialis*

General ecology
The black right whale is a mysticete that feeds primarily on copepods (**) and euphausids (*). It employs a skimming type of feeding behaviour.

Before its depletion due to human predation, the species used to occur in large schools. Now, however, it usually occurs singly or in pairs or at any event in groups of less than 5 individuals.

The seasonal incidence of copulation (....), parturition (———), and lactation (– – –) in the northern and southern hemispheres is indicated, respectively, above and below diagram (b). Gestation lasts for 11–12 months and is followed by a lactation period of up to 12 months at the end of which copulation takes place.

Sexual maturity is attained at between 14·5 and 15·5 m for males and between 15·0 and 16·0 m for females.

Seasonal return migration
(a) Geographical distribution in August (△) and February (○)

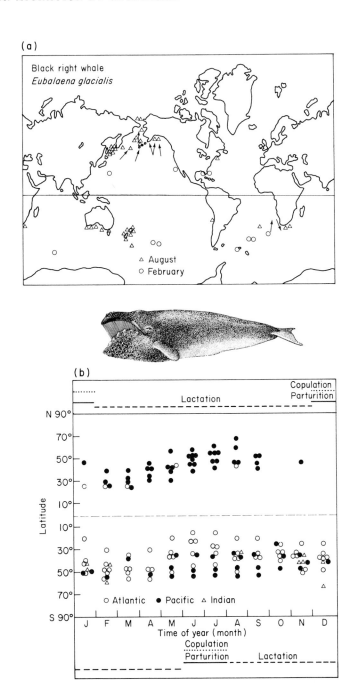

(a)

Black right whale
Eubalaena glacialis

△ August
○ February

(b)

Copulation
Parturition
Lactation

Time of year (month)
Copulation
Parturition
Lactation

○ Atlantic ● Pacific △ Indian

Dots show sightings at the appropriate month. Arrows show possible migration routes.

(b) Seasonal shift in latitude

The dots indicate latitudes at which right whales have been sighted in different oceans during each month of the year.

There is a clear polar migration in summer and a tropical migration in winter. There is also a clear separation of the demes of the northern and southern hemispheres.

[Compiled from data in Chittleborough (1956), Omura (1958), Moore and Clark (1963), Slijper et al. (1964), R. Clarke (1965), Layne (1965), Berzin and Rovnin (1966), Gaskin (1968b), Neave and Wright (1968), Rice and Fiscus (1968), Omura et al. (1969), and Best (1970b)]

(a)

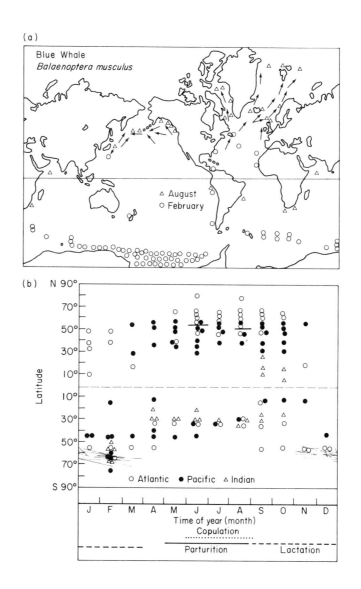

(b)

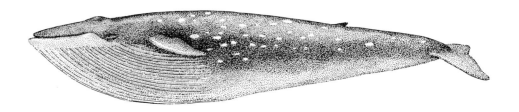

Fig. 29.5   Seasonal return migration of the blue whale, *Balaenoptera musculus*

General ecology

The blue whale, which is the largest living animal, is a mysticete that feeds primarily on euphausids and employs a swallowing type of feeding behaviour.

The species is relatively solitary, being found either singly or in male/female pairs, perhaps with a calf. Larger groups may be formed while in tropical waters.

The seasonal incidence of copulation (....), parturition (———), and lactation (– – –) in the southern hemisphere is indicated under diagram (b). Gestation lasts for 10–12 months and is followed by 7 months of lactation.

Sexual maturity is attained at 4·5 years in males and 5·0 years in females at body lengths of, respectively, 23 m and 24 m. Average maximum body length is 24 m for males and 26 m for females.

Seasonal return migration
(a) Geographical distribution in August (△) and February (○)

Arrows show possible migration routes

(b) Seasonal shift in latitude

The dots indicate the latitudes at which blue whales have been sighted in different oceans during each month of the year. The lines join the latitudes and times of release and recapture of marked individuals.

There is a clear indication of a polar migration in summer and a tropical migration in winter. Throughout most of the year there is a clear separation between northern and southern hemisphere demes but in September there seems to be some overlap between the two in the Indian Ocean.

In the southern hemisphere, immature individuals occur further from the poles than adults and they also spend more time near the coasts when in tropical waters. As an individual ages, therefore, it contracts its home range away from coastal regions and extends its familiar area into colder and colder waters.

[*Compiled from data in Chittleborough (1953), S.G. Brown (1954, 1956, 1957, 1958, 1959), Slijper (1962), Omura and Ohsumi (1964), Slijper et al. (1964), Bannister and Gambell (1965), Berzin and Rovnin (1966), Ichihara (1966), Jonsgård (1966), Mackintosh (1966), Nishiwaki (1966, 1967), Gaskin (1968b), Neave and Wright (1968), and Small (1971)*]

Further inspection of the figures for the blue whale, *Balaenoptera musculus* (Fig. 29.5), sei whale, *B. borealis* (Fig. 15.12), minke whale, *B. acutorostrata* (Fig. 29.10), and humpback whale, *Megaptera novaeangliae* (Fig. 29.9), especially the latter, however, suggests that in these species at least there may be some contact between the two demes.

In fact, the figure for the humpback whales looks very little different from the figure for the sperm whale, *Physeter catodon* (Fig. 29.8), even though the two are classically described as showing two distinct migration patterns (Orr 1970a). The sperm whale is usually described as having a tropical distribution, that of the females being restricted to the zone between 40°N and 40°S, while the larger males migrate much further towards the poles during the polar summer. However, at least in the North Pacific, the northern limit for females is very variable. In years when the water further north rises to an above-average temperature, the females are found much further to the north, even around the Aleutian Islands (Nishiwaki 1966). This is not to say, however, that there are not quantitative differences in the migration patterns between the two species. The sperm whale is relatively more common in the tropics than the humpback and only non-harem males of the sperm migrate to high-temperate and sub-polar latitudes, whereas all age and sex classes of the humpback do so. Finally, it is known from marking–release/recapture data (Fig. 29.8) that sperm whales migrate between the northern hemisphere and southern hemisphere whereas this has not yet been proved for the humpback except insofar as it seems likely that southern Indian Ocean and South Pacific demes have seasonal migration circuits that may extend across the equator.

The sperm whale is a fish- and cephalopod-eating odontocete. In the single example of a mysticete that feeds primarily on fish, Bryde's whale, *Balaenoptera edeni* (Fig. 15.13), the species also has a tropical distribution. In this case, however, there is only a slight indication of a seasonal latitudinal shift in distribution, with the result that if there is any seasonal shift in home range by demes or individuals it has not yet been recorded in any detail. Other fish- and cephalopod-feeding species, all odontocetes, such as the pilot whales, *Globicephala* spp., (Fig. 29.6) are similarly characterised by a small or zero latitudinal shift in distribution with season. The extreme case is the killer whale, *Orcinus orca* (Fig. 29.7), which appears to occupy all of the world's oceans throughout the year. Any information concerning the migrations of these species, therefore, is derived solely from observation of appearance and disappearance of the species at different points within the species' range.

The odontocetes of the Arctic Ocean, such as the beluga, *Delphinapterus leucas*, and narwhal, *Monodon monoceros*, both of which measure about 4·5 m in length, both of which inhabit coastal regions and feed on bottom-living cephalopods, decapod crustaceans, and fish, and both of which are preyed upon

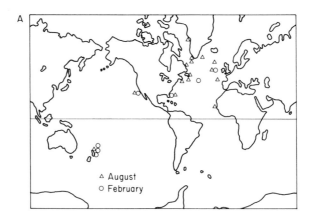

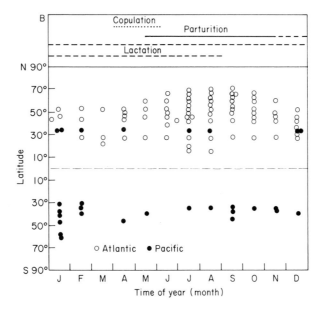

Fig. 29.6 Possible seasonal return migration of pilot whales of the genus *Globicephala*

General ecology
Pilot whales are odontocetes of the genus *Globicephala*. They occur in all the major oceans but are represented by different species in different areas. *G. melaena* occurs in the northern part of the North Atlantic (north of 34°N), including the Mediterranean. It is probably the same species that occurs in all waters south of 30°S in the Atlantic and on the Pacific coast of South America. *G. macrorhyncha* occurs in the southern part of the north Atlantic (south of 39°N) and along the American coast is found from Virginia to the West Indies. *G. scammoni* occurs in the North Pacific.

Pilot whales feed mainly on cephalopods and fish. They occur in schools of 2–100 ($\bar{x} = 20$) individuals when feeding and of 2–1000 or more ($\bar{x} = 85$) when not feeding. Some schools are all male and others contain about three times as many females as males.

The seasonal incidence of copulation (....), parturition (———), and lactation (– – –) in the northern hemisphere is indicated above diagram B. Gestation lasts 16 months and is followed by 22 months of lactation.

Males become sexually mature when 12 years old and females when 6–7 years old. Average maximum length is 6·17 m for males and 5·11 m for females.

Seasonal return migration
A. Geographical distribution in August (△) and February (○)

B. Seasonal shift of latitude

The dots indicate the latitudes at which pilot whales have been sighted in different oceans during each month of the year.

Through most of the range, demes probably perform migration circuits that do not have a N–S axis. In both the North Atlantic and the South Pacific, however, there is an indication that individuals at the highest latitudes migrate toward the poles during the polar summer and toward the equator in the polar winter.

[*Compiled from data in Brown and Norris (1956), Sergeant and Fisher (1957), Brown (1961), Slijper (1962), Layne (1965), Neave and Wright (1968), and Gaskin (1968b)*]

by killer whales (Slijper 1962), migrate with the advance and regression of the pack ice in the same manner as the bowhead (Fig. 29.3). Of the three species (bowhead, narwhal, and beluga), it is the beluga that migrates furthest to the south in winter (Vibe 1967, Slijper 1967). The beluga, for example, is seen near Point Barrow, Alaska, in October and remains there until forced south by the ice (Bailey and Hendee 1926). The following spring the species starts northwards through the Bering Straits at the

end of April, usually a little before the bowhead. The same species (beluga) is found in the Barents and Kara Seas at all seasons. At any one time about 30 per cent of the deme is found in the coastal zone with 56 per cent being found more than 100 km from the shore (Golenchenko 1963). In winter in the White Sea when the temperature falls and coastal ice floes begin to form, the whales are no longer found in the bay but instead aggregate in the White Sea basin, up to 500 individuals being present from February to April. To what extent this is a single deme showing high flux, performing inshore–offshore return migrations, and to what extent it is composed of a number of separate demes, some occupying coastal waters and some living further out at sea, appears to be unknown.

The Pacific white-sided dolphin, *Lagenorhynchus obliquidens*, for example, has a seasonal inshore–offshore return migration off the coast of California (Brown and Norris 1956). In winter and spring the species moves inshore to feed on anchovies, *Engraulis mordax*. In summer and autumn the anchovies move into water that is too shallow for the dolphins which instead move offshore to feed on another schooling fish, the saury, *Cololabis saira*. Dall's porpoise, *Phocoenoides dalli*, shows a similar migration in the same area. However, the common dolphin, *Delphinus delphis*, which at an adult length of about 2·1 m is a smaller species, stays inshore and feeds on anchovies all year.

The pilot whale, *Globicephala melaena* (Fig. 29.6), which off Newfoundland feeds on short-finned squid, *Illex illecebrosus*, and cod, *Gadus morhua* (Fig. 31.13), moves into bays in summer, as does the common porpoise, *Phocaena phocaena*, in the Bay of Fundy, Nova Scotia (Neave and Wright 1968). In this area the porpoises are following the shoals of herring, *Clupea harengus* (Fig. 31.12), but copulation also takes place in the bay. The same species of porpoise occurs in the Baltic but only in summer. As the ice forms the species migrates and spends the winter in the Skagerrak and Kattegat around Denmark (Wølk 1969). Finally, the killer whale *Orcinus orca* (Fig. 29.7), is considered by some authors to follow the walrus, *Odobenus rosmarus* (Fig. 28.11), on its migration from the Bering Sea to the Arctic Ocean (Bailey and Hendee 1926) and to follow rorquals on their seasonal return migration off Eastern Canada (Sergeant and Fisher 1957).

It seems likely, therefore, that demes of most odontocetes, and also, perhaps, Bryde's whale, *Balaenoptera edeni*, have annual migration circuits

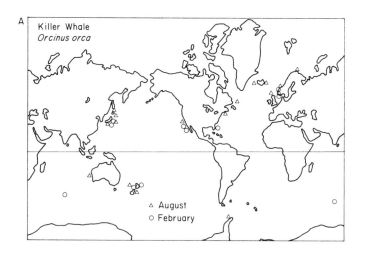

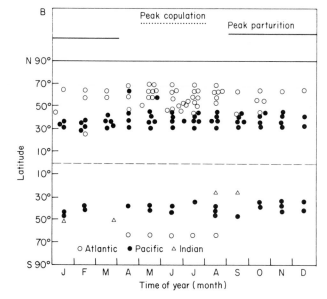

Fig. 29.7 Geographical distribution of the killer whale, *Orcinus orca*

General ecology

The killer whale is an odontocete and feeds primarily on fish and cephalopods when young and also on pinnipeds and other cetaceans when larger. Even the blue whale, *Balaenoptera musculus*, falls prey to schools of killer whales occasionally.

The species often occurs singly and also in large schools of up to 200 or so individuals, which usually contain both sexes but occasionally are of a single sex. Copulation and parturition seasons are not known and may occur throughout the year. Maximum incidence of copulation in the northern hemisphere may be May through July and of parturition may be September through March as shown above diagram B. The gestation period may be as long as 16 months. Average maximum length of males is 9 m and of females 6·5 m.

A. Geographical distribution in August (Δ) and February (o)

B. Latitudinal distribution throughout the year

The dots indicate the latitudes at which killer whales have been sighted in different oceans during each month of the year.

Despite the apparent gap in tropical regions in the diagram, killer whales occur throughout the world's oceans throughout the year. There is little indication of a seasonal shift in latitude. Nevertheless, each deme seems likely to be highly migratory. In the North Pacific a deme or demes of killer whales may follow the walrus, *Odobenus rosmarus*, on its seasonal return migration (Fig. 28.11) and off Eastern Canada a deme or demes of killer whales may follow the larger cetaceans, particularly the rorquals, on their seasonal return migration. However, because distance and direction is likely to vary from deme to deme, a latitudinal shift in distribution of the species is unlikely to be observed.

[*Compiled from data in Bailey and Hendree (1926), Jonsgard and Øynes (1952), Chittleborough (1953), Paulian (1953), Brown and Norris (1956), Sergeant and Fisher (1957), Taylor (1957), Nishiwaki and Handa (1958), Gaskin (1968b), Neave and Wright (1968), Rice (1968), Jonsgård and Lyshoel (1970)*]

based on the seasonal succession and migration circuits of their prey species. Perhaps, also, the home ranges adopted during parturition and copulation are based on habitat variables other than the availability of food. However, because the successions and migration circuits of prey species vary so much from area to area the distance and direction components of the most advantageous migration circuit vary considerably from deme to deme. In many cases, the distance component is probably just

as great as for the plankton-feeding species. Because of deme-to-deme variability, however, latitudinal shifts of the entire population do not occur. The situation is very similar to that already described for terrestrial mammals. Each deme performs a conspicuous migration circuit with little overlap of the home ranges adopted at different seasons yet the geographical range of the species as a whole shows little, if any, seasonal change.

Only in the case of the plankton-feeding mysticetes, therefore, is it possible to piece the data together to produce with any confidence the migration circuits of particular demes. One point in particular seems clear. Mysticetes not only have a polar feeding home range and a tropical parturition/copulation/feeding home range (except, that is, for the Arctic bowhead and the tropical Bryde's whale), but they also have a number of home ranges between the two. Two types of evidence point to this being the case. The first type of evidence comes from the disparity between observed speed of locomotion during migration and the rate of shift of the population. Thus, most large cetaceans migrate at speeds greater than 5 km/h. The normal speed of the sperm whale, *Physeter catodon*, is about 4–7 km/h, though 20 km/h can be reached for short periods (Caldwell *et al.* 1966), particularly if under stress. The average speed of the northward shift of southern hemisphere mysticetes has been estimated to vary from about 15° of latitude per month for humpbacks, *Megaptera novaeangliae*, to about 40° per month for finbacks, *Balaenoptera physalus*. The rate of movement of humpbacks is thus less than 2 km/h, a figure that would be difficult to reconcile with the much greater cruising speed of the species (about 7 km/h) were it not for the observation that the whales engage in a great deal of 'random' movement and circling, and casual feeding, during the migration (Dawbin 1966). That this 'random' movement is in fact movement within a home range is supported by the second type of evidence which, unfortunately, consists of a single observation. An individual humpback whale, *Megaptera novaeangliae*, with certain recognisable morphological characteristics was observed by Schevill and Backus (1960) to traverse repeatedly, with a daily periodicity, an area of about 50 km² off the coast of Portland, Maine, eastern United States. This individual remained within this area for a period of at least 10 days from 29 October through 7 November during the period of southward migration (Fig. 29.9).

It seems, therefore, from the scant evidence

available, that mysticetes have a succession of seasonal home ranges that form a migration circuit stretching, in most species, from polar latitudes in summer to tropical latitudes in winter. Just what the habitat variables may be that favour such a migration circuit, however, have still not been totally evaluated. The abundance of the standing-crop of plankton in polar waters during the summer seems likely to be an important variable. The fact that within so many species, however, there is a segregation by size whereby larger individuals occupy colder waters (p. 750) seems to suggest that temperature is also an important variable. As the polar ice advances and cuts off access to the major plankton fields there is again a clear advantage in migrating to less extreme latitudes. Nevertheless, some individuals (p. 765) and species migrate before others, suggesting that some other factors are involved, such as perhaps temperature (as a habitat variable—p. 265) or the possible advantage to a male of arriving early in its copulation home range. Also it should not be necessary to migrate all the way to tropical waters simply because of the lack of access to the polar plankton due to the advance of the ice. If access to food is the critical factor, food remains available at shorter distances from the polar ice than the tropics (Fig. 29.1). To judge from the coincidence between mysticete distribution and food abundance in the tropics and subtropics (Slijper *et al.* 1964), food availability must make some contribution to habitat suitability, but it cannot be the only contributory factor.

The only other habitat variable that seems likely to contribute in a major way to habitat suitability, however, is temperature. If anything, to judge from previous arguments (p. 750), adult mysticetes seem likely either to have to expend energy, or reduce activity and hence also energy uptake, in order to keep cool in tropical waters. The only reasonable suggestion, therefore, is that young mysticetes have to expend less energy to keep warm in tropical waters than in sub-polar or temperate waters. It is suggested, therefore, that the advantage gained by adult female mysticetes through spending the winter in tropical waters is gained through the reduced energy demands for thermoregulation on the part of the offspring and that this advantage is greater than the disadvantage that results from the increased expenditure of energy by the mother in order to keep cool. As selective pressures resulting from seasonal effects and the physiological potential of females in relation to gestation period generate, for most

species, an optimum copulation season that is also during the winter, then males also find it advantageous to migrate to tropical waters. This question is explored further in relation to sexual segregation.

29.4 Degree of return

There are few or no data available concerning the degree of return of odontocetes. However, analyses of marking–release/recapture results (p. 318) and serological studies (Fujino 1962) have shown that mysticetes consist of demes smaller than simply North Pacific, South Atlantic, Indian Ocean, etc. Within each deme there is a high degree of return, even though, inevitably, some deme-to-deme removal migration by individuals does occur (p. 318). Perhaps the three best known demes are those of the humpback whale, *Megaptera novaeangliae*, in the southwest Pacific and Indian Ocean that are considered in Chapter 17 and that are also indicated in Fig. 29.9. Within the limits of these demes, degree of return is high, both with regard to the tropical and to the polar region. Similar demes seem to exist for all mysticetes for which marking–release/recapture results are available, such as the blue whale, *Balaenoptera musculus* (Fig. 29.5), and the finback whale, *B. physalus* (Fig. 33.13) (Brown 1954).

Although degree of return may be high, however, seasonal home ranges are sufficiently flexible to accommodate year-to-year and moment-to-moment variation in habitat variables. Shifts in the distribution of plankton due to water currents or due to changes in the position of water fronts between warm and cold waters invariably leads to a corresponding shift in home range by the whales (Nemoto 1959).

As far as degree of return is concerned, however, it is important to distinguish between the degree of return of the deme and the degree of return of the individual. It was suggested by Dawbin (1959) that, within the deme of humpback whales that breeds along the east coast of Australia and the southwest Pacific Islands, degree of return of individuals may be extremely high in that each individual returns to the same coast or island group to breed that it occupied the previous year. However, Chittleborough (1965) has analysed recapture data for this region and concluded that although this may be true in some years, in other years individuals seem to return at random to the different breeding sites within the area occupied by the deme while in tropical waters. Chittleborough's conclusion, if it is accepted, has far-reaching consequences in the understanding of whale migration. It seems in particular that the winter and summer home ranges are occupied by the entire deme. This is likely to be adaptive only if mechanisms exist, such as some combination of social communication and random mixing, through which the deme can deploy itself to exploit available resources in the most efficient way. The situation may not be unlike that suggested as a possibility for the barren-ground caribou, *Rangifer tarandus*, on the Canadian tundra (p. 171) or the wildebeest, *Connochaetus taurinus*, on the Serengeti Plain of Africa (p. 95). It was suggested for the caribou that the entire Canadian tundra is within the familiar area of the herd as a whole and that by a process of convergence and dispersal, mixing and communication, the population divided and deployed itself (through intra-deme variation in migration thresholds—p. 268) in such a way as to take maximum advantage of the conditions available. Perhaps each deme of humpback whales should also be regarded as such a unit (see also p. 767) rather than as a herd of the type formed by the mule deer, *Odocoileus hemionus* (Fig. 25.3), in which each individual has its own winter and summer home range.

29.5 Degree of dispersal and convergence

Visible aggregations of cetaceans are usually termed schools and may either result from convergent migration upon areas of abundant food or upon copulation grounds or they may result from aggregation for the purposes of migration or as a result of selection through predation, etc. on optimum group size while in a particular type of home range. Schools of mysticetes are of a size that is characteristic for the species (Nemoto 1964) and may consist of any age and sex combination.

The finback whale, *Balaenoptera physalus* (Fig. 33.13), forms schools of 300 or even 1000 individuals in areas where food is abundant (Slijper 1962). However, a school of 10–15 individuals is more common and even this contains recognisable subgroups of 2–3 individuals. In the feeding area the mother and calf usually form a social unit separate from other units though during migration they may

have been part of a larger school (Nemoto 1964).

In accordance with the principle established during a consideration of the dispersal and convergence of terrestrial mammals (p. 547), cetaceans invariably seem to converge where food is abundant and disperse where food is scarce. This is the case, for example, with the schools formed by the common dolphin, *Delphinus delphis*, bottlenose dolphin, *Tursiops truncatus*, false killer whale, *Pseudorca crassidens*, and white-sided dolphins, *Lagenorhynchus* spp. In these species very large schools of up to 1000 individuals of mixed age and sex are formed where food is plentiful. Where food is scarce, however, the group disperses as a number of smaller schools (Slijper 1962). Among the mysticetes, the bowhead, *Balaena mysticetus* (Fig. 29.3), and blue whale, *Balaenoptera musculus* (Fig. 29.5), tend to be solitary or to live in family groups, each consisting of a male, female, and young offspring. Where food is abundant, however, the bowhead may show a convergent migration. The blue whale, on the other hand, tends to be solitary during the main feeding season but to form larger groups in winter during the parturition and copulation season in tropical waters.

There is some evidence that convergence and dispersal of the sperm whale, *Physeter catodon* (Fig. 29.8), while in the Antarctic has a lunar periodicity such that there is a strong tendency to collect into schools at about full and new moon (Holm and Jonsgård 1959). Under the influence of a waxing moon, sperm whales in the Antarctic begin to collect into schools and to migrate to the banks. This convergence reaches a maximum during the full-moon phase. Concentrations are rarely found during a waning moon but at new-moon phase concentrations occur both near banks and in deep water. Presumably this behaviour is related in some way to the behaviour of the prey species.

29.6 Sexual segregation

Most of the examples of convergence and dispersal presented in the previous section concern sexually mixed groups of cetaceans. Fragmentary evidence exists, however, that sexual segregation is common at least in some species. The common dolphin, *Delphinus delphis*, for example, lives in sexually segregated schools most of year, the two sexes occupying the same home range only during the copulation season. The grey whale, *Eschrichtius*

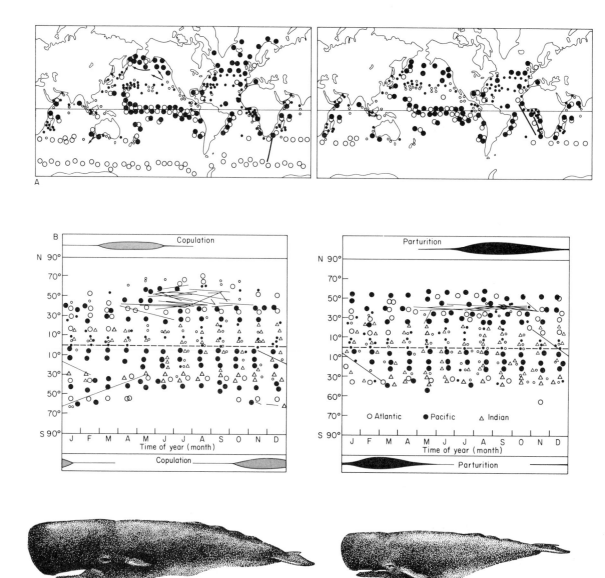

Fig. 29.8   Seasonal return migration of males and females
of the sperm whale, *Physeter catodon*

General ecology
See Fig. 17.19.

The sperm whale is an odontocete and feeds mainly on
cephalopods in water deeper than 100 m. The average
maximum length is 16·0 m for males and 11·0 m for
females. Males reach sexual maturity at 26 years of age
and females at 9 years of age. Males between 25 years and
40 years of age are usually segregated from females until
they become the leader of a harem which may eventually
contain up to 16 females.

The seasonal incidence of copulation in both the
northern and southern hemispheres is shown, respectively,
above and below diagram B for males. The seasonal
incidence of parturition in both hemispheres is shown
above and below diagram B for females. Gestation lasts 16–
17 months and is followed by 24–25 months of lactation.
This in turn may be followed by 7 months of anoestrus.

Seasonal return migration
A. Geographical distribution in August (●) and February
(○) for males and females

Large dots are sightings in which the sex of the individual

was identified. Small dots are sightings in which sex was not recorded but which are within the latitudes delimited by the large dots. Straight lines join positions of release and recapture of marked individuals. Arrows indicate possible migration routes.

B. Seasonal shift of latitude by males and females

The dots indicate the latitudes at which sperm whales have been sighted in different oceans during each month of the year. The convention with regard to dot size is as in A. The lines join the latitudes and times of release and recapture of marked individuals.

The sperm whale has a tropical distribution from which some males and females migrate towards the poles for the polar summer. Males migrate further than females but occasional large females in the North Pacific may migrate as far as the Aleutian Islands. Within the tropical and warm temperate zone, demes may exist that have distinct migration circuits that cannot be traced by gross plots of latitudinal position. Thus it has been suggested that a deme exists that winters in the region of the Cape Verde Islands and summers to the west of the English Channel.

[Compiled from data in Boschma (1938), Clarke (1956), Brown (1956, 1958), Jonsgård (1960), Slijper (1962), Omura and Ohsumi (1964), Slijper et al. (1964), Clarke et al. (1964, 1968), Ohsumi (1965, 1966), Berzin and Rovnin (1966), Caldwell et al. (1966), Nishiwaki (1966, 1967), Bannister (1968), Best (1968, 1969a,b), Gaskin (1968a,b), Roe (1969), and Berzin (1972)]

water and at greater depths than females. Males that have successfully gained a harem are inevitably found within the range of distribution of the females, despite the presumably greater energy expenditure in thermoregulation, as also are younger and smaller males. Solitary adult males greater than 12 m in length, however, migrate into colder water than females, larger individuals penetrating to colder waters. Sexual segregation in the sperm whale, therefore, can perhaps be attributed to sexual differences in the cost of thermoregulation in cold water (but see p. 730 for pinnipeds). These differences in turn can be attributed to the polygamous breeding strategy of the species.

The examples of segregation by size and sex presented earlier (p. 750) can perhaps also be attributed to differences in the energy cost of thermoregulation. The best worked example of sexual segregation, however, is that of the humpback whale, *Megaptera novaeangliae*, during migration which is described and discussed in Fig. 29.9.

29.7 Application to cetaceans of the familiar-area hypothesis for seasonal return migration

gibbosus, also seems to show sexual segregation for part of the year. Much of this segregation can perhaps be attributed to differences in mobility of males on the one hand and females with young on the other and hence to differences in the types and abundance of food available to the two sexes. The situation may be similar to that described for terrestrial mammals (p. 554).

The sperm whale, *Physeter catodon* (Fig. 29.8), shows perhaps the greatest sexual segregation of any cetacean. Males, after a long period of apparent solitude (Chapter 17, p. 301), may eventually become the possessor of harems containing up to 16 mature females (Ohsumi 1966). Associated with this polygamous behaviour is a marked sexual dimorphism, males attaining an average maximum length of 16 m whereas females attain only 11 m. Other factors that seem likely to be associated with this dimorphism are differences in the energy cost of thermoregulation (p. 750) and differences in the depth at which feeding occurs and the type of food taken (Tarasevich 1968), males feeding in colder

With the possible exception of the minke whale, *Balaenoptera acutorostrata* (Fig. 29.10), most young cetaceans associate with their mothers for at least as long as would be necessary to complete one, and usually both, legs of the year's migration circuit. This circuit could then serve as a base from which, as a result of exploratory migration and perhaps also social communication (p. 767), an adult-sized familiar area could be established. It seems likely from the peak of deme-to-deme removal migration that occurs during the pre-reproductive stage of ontogeny (p. 318) that the exploratory migrations performed during this stage may take the individual outside the geographical range of its natal deme. As a result of the exploratory–calculated migration process, the individual may then either join the appropriate deme (p. 183) and take on the appropriate migration circuit or else establish its own individual migration circuit, depending on the nature of seasonal return migration in that particular species.

A variety of environmental clues exist that could be used to establish a familiar area. Thermal cues seem to be one possibility, a suggestion that perhaps receives some support from the way that shifts in water fronts are followed by both mysticetes and odontocetes. There is some indication also that whales can perceive the direction of the wind (Zaverin 1966), which over much of the world's oceans has a relatively consistent direction, and also that they can perceive salinity gradients and their direction (Chernyi 1966). The question of the extent to which waterborne chemicals can be perceived, however, has yet to be fully evaluted. Mysticetes possess an olfactory organ and two olfactory lobes and, although these organs are small, do possess a sense of smell (Purves 1966a). Odontocetes, however, are unique amongst mammals in that apparently they lack completely the olfactory sense (J.Z. Young 1962). Insofar, that is, as after birth they have neither olfactory nerves nor olfactory lobes of the brain (Tomilin 1955) although they are found in the embryo (Kellogg 1938). However, pits are found at the root of the tongue which could permit a

Fig. 29.9 Seasonal return migration of the humpback whale, *Megaptera novaengliae*

General ecology
The humpback whale is a mysticete and feeds on euphausids (****) and fish (*). It employs a swallowing type of feeding behaviour. In the Antarctic, feeding occurs every 3–4 hours after 05.00 hours but not at all during the evening or night.

The species is fairly solitary, normally being found either singly, in pairs of any sexual combination, or in trios consisting of two adults and one young. These groups may at times converge to form schools of 10–20, or occasionally up to 100 individuals. The seasonal incidence of copulation, parturition, and lactation in the southern hemisphere is shown below diagram (b). Females are polyoestrous during the copulation season. Gestation lasts for 10–12 months and is followed by 10–11 months lactation which in turn is probably followed by a 'resting' year during which which the female is neither pregnant nor lactating. Parturition occurs in costal waters at temperatures of 25°C or so.

Sexual maturation occurs when 4 years old at body lengths of 11·6 m in males and 11·9 m in females. Average maximum body length is 13·1 m for males and 13·7 m for females.

Seasonal return migration
(a) Geographical distribution in August ($\triangle$) and February ($\circ$)

Dots show sightings at the appropriate month. Arrows show observed migration routes. The areas marked out in the southwest Pacific and southeast Indian Ocean represent areas occupied by three distinct demes. Marking–release/recapture experiments show that the individuals that occupy each of these areas may show random mixing within each seasonal home range, and have a low but characteristic pattern of deme-to-deme removal migration. The situation is analysed in Fig. 17.25.

(b) Seasonal shift in latitude
The dots indicate the latitudes at which humpbacks have been sighted in different oceans at each month of the year. The lines join the latitudes and times of release and recapture of marked individuals.

There is a clear indication of a polar migration in summer and a tropical migration in winter. There is no clear separation into northern and southern hemisphere demes.

The top right diagram indicates the percentage of individuals moving north and south as observed off Point Cloates, West Australia, in 1952. Note that the change of direction is gradual, thus emphasising the continuous movement that occurs while in the winter home range.

(c) Age and sex segregation during migration
The diagram indicates the sequence of different age and sex classes during the tropical and polar migrations measured in terms of days after the beginning of the migration during which a particular category is observed to pass a given point on the migration route. Day 0 is the day on which the first migrant is observed. The sequence is such that the length of time spent in polar waters by the different categories is as indicated in the table. Effectively the same sequence has been observed off each of New Zealand, West Australia, East Australia, Malagasy, Durban, and Zaïre in the southern hemisphere and off Japan in the northern hemisphere.

The pattern can be generated by the model developed for stay-time (Chapter 11, p. 82) if it is supposed that habitat suitability is a direct function for all classes of the availability of food, an inverse function of water temperature for the sex and age classes with the largest body size, and a direct function of water temperature for the smaller age classes, and if it is supposed that the advantage gained by lactating females is greater if their young offspring rather than they themselves experience the least cost of thermoregulation. Hence, lactating females are last to arrive and the first to leave the polar feeding grounds. Immature individuals have the next shortest time in cold but food-rich waters and leave for the tropics shortly after the lactating females. Finally, males, resting females, and pregnant females which, as listed, are in increasing order of body size and in increasing order of food requirements, spend, respectively, increasingly long at the polar feeding grounds.

[Compiled from data in Brown (1954, 1958, 1959, 1962), Chittleborough (1953, 1958, 1965), Slijper (1962), Omura and Ohsumi (1964), Slijper et al. (1964), Bannister and Gambell (1965), Berzin and Rovnin (1966), Dawbin (1966), Mackintosh (1966), Nasu (1966), Nishiwaki (1966, 1967), Best (1967), Bannister (1968), and Neave and Wright (1968)]

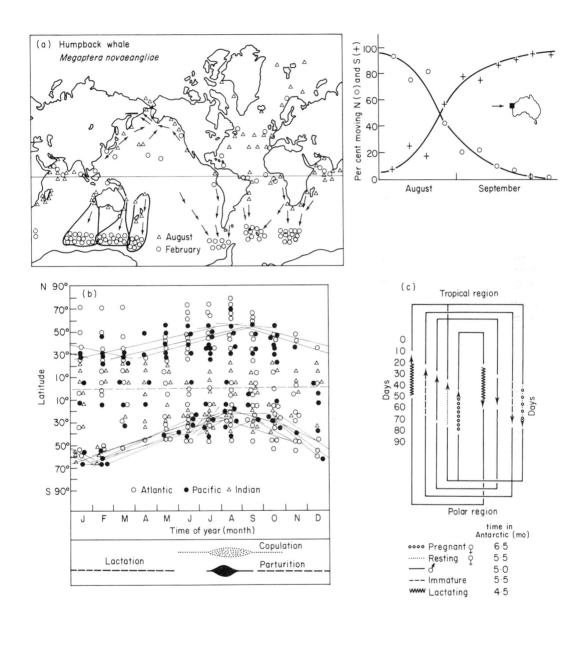

(a) Humpback whale
Megaptera novaeangliae

△ August
○ February

Per cent moving N(○) and S(+)

August   September

(b)

N 90°
70°
50°
30°
10°
0°
10°
30°
50°
70°
S 90°

Latitude

○ Atlantic   ● Pacific   △ Indian

J F M A M J J A S O N D
Time of year (month)

Lactation

Copulation

Parturition

(c) Tropical region

Days   Days

Polar region

time in
Antarctic (mo)

○○○○ Pregnant ♀   6·5
.... Resting ♀   5·5
——— ♂   5·0
– – – Immature   5·5
wwww Lactating   4·5

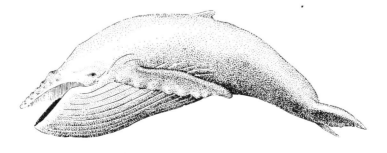

waterborne chemical sense and could conceivably be carried to the unreduced olfactory region of the cerebral cortex (Yablokov 1961).

Other information concerning the underwater environment that could be used to build up a familiar area could be obtained through the mechanism of echo-location. This mechanism is widespread and perhaps even universal among odontocetes (Norris 1968). On the other hand, there is no evidence that any mysticete actively echo-locates. In odontocetes, echo-location is accomplished by trains of brief (0·7–24·0 msec) broadband clicks that are projected forward in highly directional, structured sound fields. The head is moved around during echo-location, thus moving the projected sound beam. The entire process of sound generation in odontocetes is perhaps related to the existence of structures in the nasal diverticula (Norris 1968). Sound reception may occur in the external auditory meatus (Purves 1966b), or alternatively the fat-filled lower jaw may act as a wave guide, passively transmitting sound to the tympanic bulla to which the fat body is attached (Norris 1968).

A review of the work, chiefly on porpoises, that shows the peripheral auditory and central neurophysiological specialisations favourable for the analysis of echo-locating clicks and their echoes is given by Bullock and Ridgway (1972). Much of the environmental information that can be derived from echo-location is limited with respect to distance (see also p. 891 for homing experiments on echo-locating bats). However, the potential of the situation has not yet been evaluated, and it has been suggested (Norris and Harvey 1972) that the spermaceti organ of the forehead of the sperm whale, *Physeter catodon*, is a device for the production of *long-range* echo-location sounds useful to the sperm whale in its deep sea habitat in which food must be located at considerable distances in open water. In this case, any contribution of information derived through the echo-location mechanism to the establishment of a familiar area could be greatly increased.

The visual sense can provide information concerning underwater, terrestrial, and celestial structures that could also be used in the establishment of a familiar area. Experiments on the visual capacity of a captive bottlenose dolphin, *Tursiops truncatus* (Kellogg and Rice 1966), demonstrate the ability to distinguish between various pairs of patterns when the patterns were observed under water. Ripples on the surface, however, distort the image and raise problems in water-to-air vision. Even so, discrimi-

nation between patterns was found under such conditions as long as the forms had been previously differentiated underwater.

Sperm whales, *Physeter catodon*, and pilot whales, *Globicephala* spp., are known to 'stand' vertically with their heads out of the water in order to inspect visually surface objects, such as boats (Caldwell *et al.* 1966) and such behaviour could be used to perceive coastal topography. This behaviour is also particularly well shown by the grey whale, *Eschrichtius gibbosus*, which, while migrating along the coast of North America, periodically thrusts its head out of water (Pike 1962). It is perhaps significant in this connection that the southward migration of this species off the west coast of Baja California in January/February has been observed to continue at night under bright moonlight but to be fully suspended during the absence of moonlight (Hubbs and Hubbs 1967). Of course, in this case the position of the moon, either in its own right or relative to the orientation of the coastline, could also have been perceived and integrated with the familiar area within a spatial memory (p. 865).

As far as the perception of other celestial cues is concerned a variety of anecdotal observations have been made. For example, groups of 50–100 individuals of the common dolphin, *Delphinus delphis*, in the Mediterranean have been observed swimming very fast in a straight line consistently in the direction of the Sun (Busnel and Dziedzic 1966). The behaviour was always noticed between 15.00 and 17.00 hours with good weather, a glassy sea, and clear water. No sound signals, not even echo-locating clicks, were emitted on these occasions, during which the animals were swimming very close to the surface.

Apparently healthy blind sperm whales, *Physeter catodon*, have been captured as also has a supposedly deaf individual. These reports would suggest that a variety of mechanisms are employed during the movement through an area searching for and catching food and that failure of one mechanism does not necessarily result in the loss of both of these abilities. It seems likely, however, that under normal conditions echo-location is the most important food-finding and catching mechanism. It is, of course, not known whether the ability to move around a familiar area was affected in these whales.

One other aspect of familiar-area establishment that deserves attention is the possibility of the involvement of social communication by acoustic means. Experiments on bottlenosed dolphins,

Tursiops truncatus have domonstrated that relatively complex instructions can be passed from one individual to another solely by acoustic means (Bastian 1967). Many odontocetes live in tightly organised groups, the integrative signals of which are sounds. Whistle-like cries (for communication?) and echolocation clicks (for measuring relative position and finding food?) are often emitted simultaneously by the same individual implying the ability to use two domains of sound at once (Lilly and Miller 1961, Evans and Prescott 1962).

In addition to the ability to communicate, cetaceans show a well-known range of altruistic behaviour. For example, individuals of many, if not all, odontocete genera show standing-by behaviour when a conspecific young or adult has been injured (Caldwell and Caldwell 1966). As well as standing-by, the uninjured individual may position itself between the injured animal and the attacker, push the individual away, or even attack the attacker. Among the Delphinidae, the supporting of an injured animal at the surface has also been reported. Females are more likely to remain with distressed or injured females and young than with distressed or injured males. Males rarely remain with distressed males but may remain with distressed adult females of the same or another species. Mysticetes also stand-by the conspecific young or adult and attempt to free the young by, for example, breaking a harpoon line. Females rarely stay with injured adult males if danger threatens, though males of some genera remain with and support adult females. Such care behaviour may be directed, not only to individuals with which the individual has previously associated but also to individuals that are encountered for the first time.

An aspect of social communication that is of a particular interest in relation to the establishment of a familiar area and other aspects of seasonal return migration is the distance over which communication can occur. Zoologists are only just beginning to discover the potentials of the cetacean communication system. It is known, however, that the 20 c/s sounds produced by finback whales, *Balaenoptera physalus*, can be received under water many kilometres from source (p. 881) and that sperm whales, *Physeter catodon*, are capable of communicating alarm over distances at least as great as 10–11 km (Caldwell *et al.* 1966).

The area covered by a group of cetaceans that nevertheless retain acoustic contact with one another may therefore be large. Thus young male sperm whales between 10·5 and 14·0 m in length may form groups of up to 40 individuals. The individual members of such a group may be spread over an area of 5 × 10 km yet nevertheless all move together. It seems certain that we have not yet discovered the limits of the ability of cetaceans to communicate underwater and one wonders just how many individuals extending over how large an area can nevertheless continue to communicate with one another, either directly or by relays. Knowledge of the limits of this ability is of clear importance in understanding the nature of the familiar-area phenomenon in whales (p. 881). If, as suggested earlier, (p. 761), demes of the type identified for the humpback whale, *Megaptera novaeangliae*, do in effect constitute a social group comparable to an ungulate herd or perhaps a large-sized group of chimpanzees or hunter–gatherer humans, then the familiar area may be very much that of the deme rather than the individual.

As far as the inherited and learnt components of the seasonal return migrations are concerned, it seems likely that the situation of most of the fish- and cephalopod-feeding species of cetaceans are the same as that suggested for pinnipeds (p. 736). That is the distance and direction components are largely learned but the seasonal change in migration thresholds are largely inherited and obligatory, triggered either by thresholds to indirect variables or by endogenous events. In the plankton-feeding mysticetes, however, the situation may be slightly different. Here the optimum migration direction is more or less constant from deme to deme, ocean to ocean, and hemisphere to hemisphere, and is likely to have been so for tens of millions of years. If social communication is as widespread as has been suggested, of course, few individuals can have found themselves having to continue exploratory migration when young completely out of contact with other members of the species. In this case, selection for an inherited basis for an appropriate direction to exploratory migration may have been insufficient to cause such a genotype to spread through the population. On the other hand, because there is little disadvantage in an inheritable bias to the direction of exploratory migration, young calves would only have had to lose contact with their mothers (through disease or predation) and other members of their natal deme at a relatively low incidence during the past few millions of years for an inherited bias to spread though the population. Of course, once an individual has completed a migration circuit with its mother and established a basic familiar area with an appropriate directional axis, the advantage of an inherited bias disappears. The possibly unusual behaviour of minke whales, *Balaenoptera acutorostrata* (Fig. 29.10), during the first two years of life perhaps supports the thesis of an inherited bias to the direction of exploratory migration but could also be achieved by social communication in conjunction with ontogenetic variation in migration thresholds (i.e. habitat suitability being a direct function of water temperature during the first summer, an inverse function during the second winter, and thereafter showing seasonal variation).

The nature and direction of seasonal change in migration thresholds to habitat variables follows naturally from the consideration of seasonal change in habitat suitability (p. 747) and the model for migration thresholds derived in Chapter 8 (p. 44) and evaluated in Chapter 16 (p. 265) and need not be discussed further here. Migration thresholds to indirect variables and the way these trigger the seasonal change in migration thresholds to habitat variables probably follow the same sequence as that suggested for pinnipeds (p. 736). Figure 29.11 suggests that at least as far as the northward migration of female humpback whales is concerned, photoperiod may act as an indirect variable independently of temperature. If latitudinal variation were to be taken into account, however, it would probably be found that an ideal adaptation can only be achieved by a combination of photoperiod and temperature (see p. 445).

It is argued for fish (p. 852) and turtles (p. 851) and again for birds (p. 632) that, even for migration within a familiar area, selection would favour the adoption of migration routes and the evolution of migration thresholds to migration-cost variables such that migration cost was reduced to a minimum. In the animals just mentioned, selection seems to have favoured migration thresholds to the speed and direction of air and water currents and the adoption of migration circuits that wherever possible take maximum advantage of these currents. The migration routes of the large cetaceans, however, such as the rorquals and humpbacks, show no relationship with the major ocean currents except where such currents also show a great productivity as do, for example, the Gulf Stream and the currents in the northern part of the Indian Ocean (Slijper *et al.* 1964, Dawbin 1966). Perhaps current speed and direction is negligible as a migration-cost variable to an animal as powerful as a whale.

Fig. 29.10   The seasonal return migration of the minke whale, *Balaenoptera acutorostrata*

General ecology
The minke, or little piked, whale is the smallest member of its genus. It is a mysticete and feeds on euphausids (**), fish (**), and copepods (*). The feeding method is that of 'swallowing'. The seasonal incidence of copulation, parturition, and lactation in the northern hemisphere is indicated above diagram B. The species occurs in schools of less than 100 individuals and usually only 10–20. The schools are mixed with respect to age and sex. The gestation period is about 10 months.

Sexual maturation occurs when 7 years old at a length of 7·1 m for males and 7·9 m for females. Average maximum length is 8·3 m for males and 8·8 m for females. Average maximum age is 18–22 years for both sexes with a potential longevity that is less than 50 years. For females there is an interval of 1 or sometimes 2 years between successive births.

Seasonal return migration
A. Geographical distribution in August (△) and February (○)

B. Seasonal shift of latitude
The dots indicate the latitude at which minke whales were sighted in different oceans for each month of the year. Although there is a clear indication of a polar migration in summer and a tropical migration in winter, the separation into northern and southern hemisphere demes is not complete.

Establishment of a familiar area?
There is some indication that the individuals from the northern hemisphere that remain in tropical and subtropical waters in summer (July to September) are primarily young, newly weaned calves. If this is so then temperate and sub-polar regions are not encompassed within the familiar area of this species until its second summer. The following winter, however, there is some indication for the North Atlantic deme that the young (now 2 years old) whales do not return to the tropics. This is based on the observation that two-year-olds are not found stranded on British beaches yet they are caught in early spring (April) in the Barents Sea. Thereafter, young minke whales seem to perform a seasonal return migration, but only gradually, as they grow in size, extend into colder and colder waters. At latitude 60°S, 80 per cent of minke whales are males and are mainly larger individuals. Females still with young remain in the warmer water to the north. In the northern hemisphere, sexual segregation occurs during the spring northward migration along the Norwegian coast, the pregnant females migrating first.

[*Compiled from data in Chittleborough (1953), Moore and Palmer (1955), Slijper (1962), Sergeant (1963), Slijper et al. (1964), Kasuya and Ichihara (1965), Bourne (1966), Jonsgård (1966), Struhsaker (1967), Gaskin (1968b), and Ohsumi et al. (1970)*]

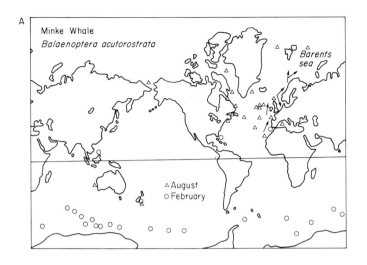

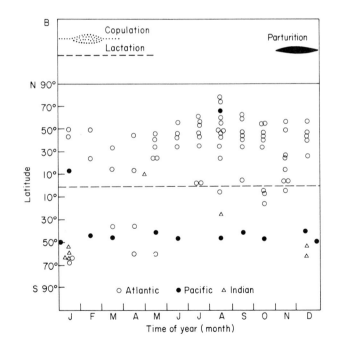

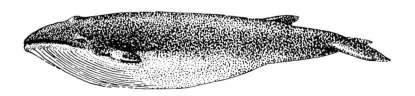

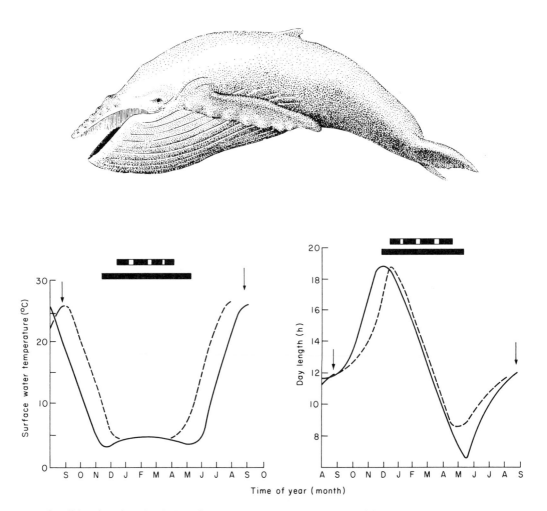

Fig. 29.11 Possible migration threshold to indirect variables for the northward migration of the southern hemisphere female humpback whale *Megaptera novaeangliae*

General ecology and seasonal return migration
See Fig. 29.9.

Migration threshold
In the two diagrams, the dashed line indicates the conditions experienced by a lactating female and her calf and the solid line indicates the conditions experienced by a pregnant female. The thick black lines along the top of each graph indicate the period that the whales are in Antarctic waters. Again the dashed line refers to lactating females and the solid line to pregnant females. Vertical arrows indicate parturition.

The left-hand diagram indicates the surface-water temperature conditions experienced during a year and the right-hand diagram indicates the daylength experienced.

Because temperature changes vary little during the period preceding the date of initiation of northward migration, it seems likely that selection has favoured a migration threshold to photoperiod. However, lactating and pregnant females initiate northward migration at different dates. This could be achieved by one or other of two adaptations. If the migration threshold is the same for both categories of females, the time lag between the migration threshold being exceeded and the initiation of migration could be a function of physiological state. Alternatively, the time lag could be the same for both categories of females and it could be the migration threshold that is a function of physiological state.

[*Modified from Dawbin (1966)*]

30

Seasonal and ontogenetic return migration by marine reptiles

There are five genera of sea turtles (Cheloniidae). These are: the green turtles, *Chelonia*; the logger-heads, *Caretta*; the hawksbills, *Eretmochelys*; the rid-leys, *Lepidochelys*; and the leatherbacks, *Dermochelys*; (Carr 1968). Of these, the green turtles are herbivor-ous, feeding in areas where there is abundant submerged vegetation and where there are small, deep, rocky potholes to roost at night (Carr 1952). The hawksbill turtles tend to be omnivorous with a slight preference for animal food such as ascidians, molluscs, and crustaceans. The remaining genera are carnivorous.

Along with the terrestrial tortoises and freshwater/amphibious terrapins (in British termi-nology), the sea turtles constitute the reptilian order Chelonia. The order is an old one, with repre-sentatives in Permian sediments at least 250 million years old (J.Z. Young 1962). Marine turtles of the *Chelonia* type were in existence 100 million years ago and there is evidence that this particular lineage had by then already adopted the herbivorous habit (Carr and Coleman 1974).

Information concerning the distribution and mi-gration of the various turtle species is not easily obtained (Carr 1968) and the only reasonably detailed information on migration refers to three nesting demes of the green turtle, *Chelonia mydas* (Fig. 30.1).

It seems clear that the arguments presented in relation to the migrations of marine fish (p. 809) which in turn are based on the arguments presented for fish that breed in freshwater (p. 786) and for stream-living invertebrates (p. 780) are equally applicable to marine turtles. Thus demes seem to have evolved that share a migration circuit linking nesting grounds (sandy beaches), nursery areas (as yet undiscovered), and feeding grounds. The form of

the migration circuit is not yet certain. It seems likely that young turtles go downcurrent from their hatching site to their nursery area, though where in relation to this circuit the young hatchlings of all five genera spend their first year of life remains a mystery (Carr 1968). Some time later, when mature, both males and females migrate to breeding grounds and thereafter perform a return migration between these and the feeding grounds. Migration to the feeding grounds seems invariably to be with the current. Migration to the nesting grounds may be performed, it is assumed, by one of the following means: swimming against the prevailing current (p. 813); swimming with a counter-current (p. 809); alternat-ing between orientation against a current that provides appropriate olfactory clues and swimming with a counter-current at another depth (p. 814); and/or using celestial navigation (Carr 1972, Carr and Coleman 1974). Turtles, of course, unlike fish, have to swim to the surface at intervals in order to breathe. As yet, there are no data that support any one of these alternatives rather than any other. It is assumed that males and females, once mature, complete the migration circuit with a high degree of return, at least to the nesting site, and with a certain periodicity. So far, however, this has only been demonstrated for female green turtles that nest on Tortuguero beach, Costa Rica, and on Ascension Island, and in both cases the periodicity is found to be either two or three years (Fig. 30.1).

One particular problem in a consideration of the evolution of sea turtle migration concerns those demes that breed on remote oceanic islands but that have a migration circuit with a one-way distance component of several thousand kilometres. The size of the distance component is not unique to turtles among marine organisms, for the migration circuits

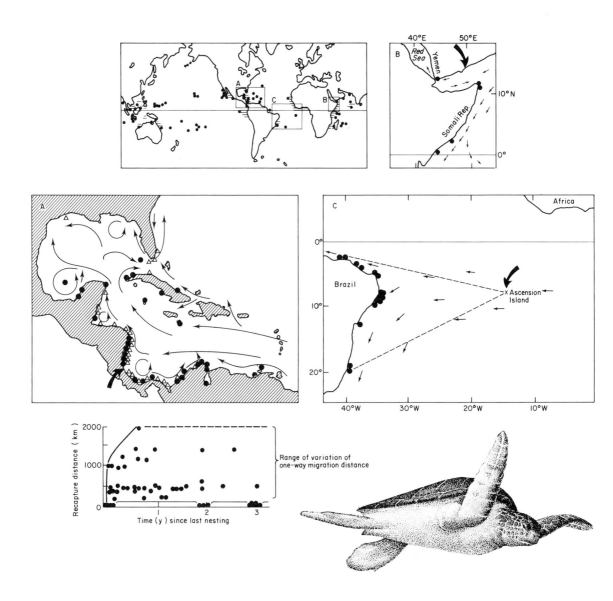

Fig. 30.1   The migration circuits of three nesting demes of the green turtle, *Chelonia mydas*

General ecology

Green turtles are amphibious, nesting and occasionally, in the Pacific, basking on land and spending the rest of their lives in the sea. Nesting occurs on particular sandy beaches throughout the tropical and sub-tropical regions of the world's oceans where the average temperature of the surface water during the coldest month of the year is above 20°C. The timing and duration of the nesting season varies from place to place. In some places it is restricted to 3–4 months of the year and in others it may run throughout the year, though with one or two peaks of incidence. Mating occurs in the water, just off the nesting grounds, and the males do not come out onto land. Females lay eggs every two or three years and the matings seem likely to fertilise the batch of eggs laid two or three years later rather than the present batch. During the time the female spends around the nesting area (1–3 months) she goes ashore 3–7 times at intervals of about 12 days and on each occasion lays about 100 eggs. The eggs are deposited in a hollow dug in the sand and then covered over. On hatching, about 57 days later, the young turtles head directly to the sea and enter the water. During this short migration, the young turtles are subject to heavy predation from terrestrial carnivores, birds, and fish, including sharks.

During their first year of life, young green turtles disappear and little is known of their habits and migrations except that they are mainly carnivorous, feeding on small marine invertebrates. By the time they weigh 4·5 kg, however, they have already become herbivorous and are feeding on the main adult foods, a few species of marine spermatophytic plants. The feeding habitat-type is any shallow water area containing abundant submerged vegetation, including the foodplants, and with small but deep rocky potholes in which the turtle can roost at night. Green turtles start to breed, perhaps a decade or so after hatching, when they weigh about 60 kg but may grow to weigh more than 225 kg. Potential longevity is probably considerable, to judge from other chelonids, and may perhaps be 100–200 years.

The green turtle lifetime migration circuit: general form
The green turtle lifetime migration circuit may be divided into two phases which may or may not prove to be similar. The first phase consists of the migrations performed by the young turtles during the period from hatching to first maturity. Hardly anything is known of these migrations. There is no doubt that the young turtles, upon hatching, immediately take to the sea and migrate away from their natal beach. It is assumed, but has not been proved, that they then drift with the current for a period. Whether this period is a short one, perhaps a few weeks, during which no feeding occurs but at the end of which the young turtle arrives in nursery areas or whether for a year they live a pelagic existence, feeding as they go on members of the zooplankton, is unknown, though perhaps the latter is more likely. Even once a herbivorous diet has been adopted, the turtles do not seem to remain long in one place. For example, there is a seasonally regular arrival of young green turtles, varying in weight from 2 to nearly 60 kg, off the west coast of peninsula Florida, near Cedar Key. These turtles arrive in April, perhaps from the south, and leave in November. The migrations of young green turtles could well be exploratory and possible examples of homing if displaced from these temporary areas supports the suggestion (p. 851) that spatial learning is being performed and a familiar area is being established.

Whatever the significance of the migrations of immature turtles, at maturity an appropriate beach is selected for nesting or copulation and the second phase of the lifetime migration circuit begins. Whether mature males and females about to breed for the first time have a high degree of return to their natal beach is unknown, though analogy to animals with a similar ecology, such as pinnipeds, sea-birds, and anadromous and marine fish, suggests that there is a high degree of return but that some removal migration (calculated?) will take place. Once a nesting beach has been used, however, and a home range has been established there, females have been shown to have a high degree of return. Tagged females from the Ascension Island and Tortuguero nesting demes have so far provided no example of removal migration to another comparable nesting deme, though some beach-to-beach removal migration may occur (p. 194). Once mature, therefore, females perform a regular migration circuit with a period of 2–3 years. After nesting, the females migrate with the surface current until they arrive at a feeding ground which may be a permanent feeding home range. Then, two or three years later, by a route and mechanism as yet unknown, they return to their nesting home range. The

migration circuits and degree of return of males, which because they do not come ashore do not lend themselves to marking or recapture, are totally unknown.

The migrations of three nesting demes of females: Tortuguero, the Yemen, and Ascension Island
The world map shows the distribution of recorded nesting grounds (solid dots) and major feeding grounds (horizontal shading) of the green turtle. Not all of the nesting grounds are still in use. Once mature, and perhaps even from birth, female (and perhaps also male) green turtles are members of nesting demes between which removal migration, which must surely occur, is infrequent. Each deme does not seem to have a distinct migration circuit. Rather, each individual migrates from its nesting ground to its feeding ground. Whether this is a nesting home range to feeding home range migration circuit from which there is little type-C removal migration (p. 308) is unknown.

Tortuguero: Diagram A shows the recoveries of females tagged at Tortuguero, Costa Rica, a 35 km stretch of nesting beach (indicated by the large black arrow) in the western Caribbean. Solid dots indicate turtle recaptures. Triangles indicate recovery sites of drift bottles also released off Tortuguero. Arrows show surface currents for July. The coincidence between the recoveries of bottles and turtles strongly suggests that currents are involved in the nesting site to feeding site migration. The extent to which females can control which part of which current they enter and thus perhaps arrive at the feeding home range occupied previously is unknown. The graph beneath shows the distance of recovery of marked turtles plotted against the time since last breeding. On Tortuguero, most breeding takes place between July and October. Approximately a third of females return to Tortuguero after an interval of 2 years and the remainder return after an interval of 3 years. There is no indication of an increase in recapture distance with time after last breeding, a fact that could indicate a nesting home range to feeding home range return migration or at least that females settle in one of the first suitable feeding habitats that they encounter.

Yemen: The best nesting beaches around the Gulf of Aden are those on the coast of the south Yemen. Diagram B shows the prevailing surface ocean currents for January (small arrows) and the recovery sites (solid dots) of tagged female turtles released from Mukalta (large black arrow) in November.

Ascension Island: Ascension is a small (8 km across) island on the mid-Atlantic ridge. Several beaches on the island are used for nesting and some beach-to-beach removal migration takes place (p. 194). Diagram C shows the recoveries of tagged female turtles (solid dots) in relation to the prevailing surface currents. All recoveries have been made 2000 km or so to the west on the Brazilian coast. Return to Ascension Island again occurs at intervals of two or three years. Nesting occurs from February to June.

[*Compiled from data, descriptions, and diagrams in Carr and Ogren (1960), Carr and Hirth (1962), Carr (1965, 1967, 1968, 1972), and Hirth and Carr (1970)*]

Fig. 30.2   The green turtle, *Chelonia mydas*

[*Photos by Paul Turner*]

of some marine fish (e.g. Figs. 31.12 and 31.15) involve similar distances, as do those of pinnipeds (p. 707) and sea-birds (p. 670). It is, however, the origin of a habit that involves such precise navigation that is of particular interest to students of turtles.

One possibility is that oceanic islands are discovered either by accidental or exploratory migrants and that in the latter case calculated breeding site to breeding site removal migration takes place. Such an event has been observed in recent years in the northern fur seal, *Callorhinus ursinus*, which has begun to breed on an island off the coast of California, several thousand kilometres from its nearest breeding site (p. 112). An explanation based on such a sequence of events is also all that is available for most animals that perform long-distance migration circuits but which use Pacific Islands as the breeding area. There seems no reason why such an explanation should not also be acceptable for all demes of the green turtle. Especially as the species is found in the tropical and sub-tropical regions of all the oceans of the world (Fig. 30.1) and as individuals, which may be either accidental or exploratory migrants, frequently appear outside of their normal geographical range.

Despite this, an attractive alternative hypothesis has recently been put forward in relation to the Ascension Island–Brazil migration circuit deme of the green turtle by Carr and Coleman (1974). This hypothesis is summarised in Fig. 30.3. The basic thesis is that an ancestral deme of the green turtle that inhabited the seas between 'North America' and northwestern Gondwanaland about 100 million years ago began to breed on the volcanic islands formed on the midoceanic ridge associated with the early formation of the northern South Atlantic about 80 million years ago. As the ocean expanded, successive volcanic islands disappeared beneath the sea while new ones appeared and the mainland-to-island distance gradually increased. This green turtle lineage, however, discovered and adopted as a breeding site each new island as it appeared, eventually ending up with the migration circuit with a one-way distance component of 2000 km that is observed today.

The main points of difference between this 'continental-drift hypothesis' and the generally applicable 'exploratory–calculated migration hypothesis' are first that the continental-drift hypothesis requires an unbroken and more or less inbred lineage from the deme that occupied the first mid-oceanic volcanic island some 80 million years ago to

the deme breeding on Ascension Island at the present day, and second that Carr and Coleman suggest the inclusion within the orientation repertoire of the deme of an inherited orientation to the position of the rising Sun during migration to Ascension Island. This axis of orientation is considered to have remained constant during the evolution of the deme, and as each volcanic island becomes unsuitable the next one is discovered by increasing migration distance along this axis.

The criticisms of these two points that distinguish the continental-drift thesis from the more general thesis are twofold. First, if the Ascension Island deme was as old and inbred as required by the thesis then, especially in the light of the unique migration circuit of the deme, it might be expected that the deme would have evolved other, perhaps morphological, differences from other demes. Such differences do not seem to exist (Carr 1968). Secondly, any species that successfully spreads through all the world's oceans seems likely to have had as a preadaptation a navigation system that is generally adaptive and not specific to a particular migration circuit. Thus, the instinctive orientation to the position of the rising Sun during the breeding migration is adaptive more or less only to the Ascension Island deme. It seems much more likely that the green turtle, throughout its range, shows a general navigation system that is more or less adaptive into no matter which deme the turtle is born. Such a navigation system is shown by salmon (p. 880) and probably by the vast majority of migrants (Chapter 33).

There is no doubt that continental drift must have had profound effects on the optimum migration circuits of marine animals, such as fish and turtles, that lived through the major changes brought about by drift such as the formation of the Atlantic Ocean. However, most animals that make use of oceanic islands as part of a migration circuit have evolved a behavioural repertoire that seems capable of permitting the colonisation and exploitation of any more suitable island with which they come into contact. There seems no reason to suppose that turtles are any different in this respect and it seems unnecessary to invoke continental drift to account for the origin of specific demes such as the Ascension Island deme of the green turtle. Furthermore, the points raised by the continental drift thesis are not supported by the above data which are, however, consistent with the general thesis which was developed in Chapters 8 (p. 44) and 14 (p. 91).

The thesis that the evolution of catadromous behaviour in the North Atlantic eel(s) (Fig. 31.18) is also the result of the formation of the Atlantic Ocean by continental drift can also be discounted. All members of the genus *Anguilla* are catadromous, even those in the presumed area of origin of the genus in the Indo-Pacific region where continental drift is unlikely to have been important as far as catadromous behaviour is concerned. Again, however, although not involved in the origin of catadromous behaviour, the widening of the North Atlantic must have imposed considerable selection on the form of the migration circuits in North Atlantic eels, though not only of catadromous species (Fig. 31.17), especially on the distance component of the migration.

Little is known concerning the migrations of the other marine turtles. The main problem, apart from that of identification, arises because there is as yet no way of demonstrating whether turtles that are encountered far outside their breeding ranges are accidental migrants, exploratory migrants, or

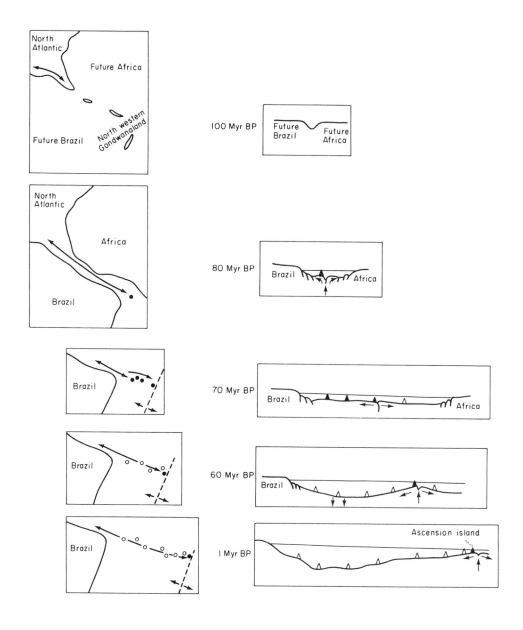

Fig. 30.3   The continental drift hypothesis for the origin of the Ascension-nesting deme of the green turtle, *Chelonia mydas*, as propounded by Carr and Coleman

Marine turtles of *Chelonia* type inhabited the seas between 'North America' and northwestern Gondwanaland by 100 million years B.P. Fossil evidence suggests that the herbivorous habit had already evolved and that there was a spatial separation of feeding and breeding grounds. It is postulated that these turtles included the ancestors of the modern *Chelonia mydas* (fossil *Chelonia* having been found in sediments deposited about 20 million years B.P.), and that mammalian egg predators were already imposing selection for nesting on islands.

The opening of the equatorial Atlantic began with rift valley formation about 110 million years ago or earlier. Sporadic but progressive ocean flooding 90–80 million years ago led finally to the formation of a single Atlantic Ocean about 80 million years ago. Volcanic piles are a frequent feature of mid-oceanic ridges and some may grow sufficiently to emerge as islands. In the diagrams, volcanic piles that emerge as islands are shown as solid dots in the left-hand diagrams and as solid cones in the right-hand diagrams. As sea-floor spreading continues, the volcanic piles become inactive and are carried outwards, away from the mid-oceanic ridge, and downwards, becoming submerged sea-mounts (open circles and cones in the diagrams). The direction of sea-floor spread away from the mid-oceanic ridge is indicated by the small arrows. Sporadically, new volcanoes appear on the ridge.

The large arrows in the left-hand diagrams indicate the axis of the feeding ground–nesting ground return migra-

tion of the turtle deme. At first, the axis was along the northern coast of the future Brazil. This axis persisted through the formation of the Atlantic Ocean and led to the discovery and incorporation of the first offshore islands into the migration circuit. As these islands became submerged and/or unsuitable, the turtles swam further on the same or nearly the same course and encountered the newly formed islands. The one-way distance of the return migration thus gradually increased until at the present day it is about 2000 km. 70 million years ago the distance must have been about 300 km. The present-day Ascension Island is less than 7 million years old. What was probably the penultimate island is now a sea-mount 15 km away and is submerged to a depth of 1500 m.

The main postulates of the thesis are:

1. There has been genetic continuity between the turtle deme that nested on the first mid-oceanic island in the newly formed equatorial Atlantic about 80 million years B.P. and the Ascension Island deme of the present day. Removal migration into the deme from other demes has been negligible.

2. The deme is characterised by an inherited tendency to swim a particular travel path, roughly WNW–ESE, perhaps by using the rising Sun as a beacon to stay on latitude. Final navigation to the nesting island is achieved when this course intersects a plume of sensory clues that arise from the island and to which the turtles were imprinted as hatchlings. Suitable current systems have probably existed from an early date.

[*Compiled, with some extrapolation, from Carr and Coleman (1974)*]

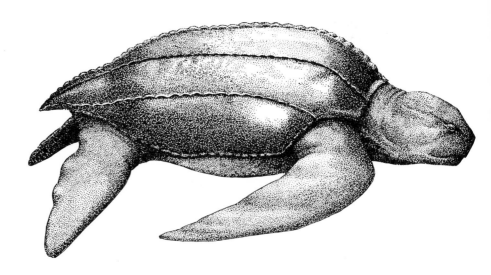

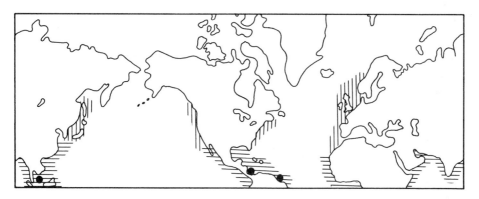

Fig. 30.4 The approximate breeding and non-breeding distribution of the leatherback turtle, *Dermochelys coriacea*, in the Northern Hemisphere

General ecology

The general ecology of sea turtles is poorly known. Like all sea turtles, of which it is the largest, the leatherback breeds in the warmer parts of the world's oceans, females hauling out onto sandy shores to lay and bury their eggs. In the diagram, the approximate range of the breeding area is indicated by horizontal hatching. The black dots indicate the three areas in the northern hemisphere where leatherbacks assemble to breed. There is a fourth known site on the Tongaland coast of Zululand in Africa. Elsewhere, the leatherback nests solitarily over a great extent of tropical shore. In the Caribbean, the nesting season of the leatherback is mainly March and April, three months or so earlier than that of other species. Except that leatherbacks are carnivorous, their feeding ecology is unknown. Large jellyfish have been found in leatherback stomachs and it has been suggested on the basis of the array of flexible, backward-projecting spines that line the oesophagus that leatherbacks may feed heavily on pelagic jellyfish.

Migration

Hatchling leatherbacks disappear, as do those of all sea turtles, for the first year of their life. Mature leatherbacks are sighted most frequently in the temperate zone, well outside of the breeding area, as indicated in the diagram by vertical hatching. Some of the sightings show such regularity that there is a strong implication of the existence of a migration circuit. On the coasts of Honshu Island, Japan, well outside the breeding area, leatherbacks show a bimodal curve of incidence. One peak is during the summer, from June to September, and is on the Pacific coast, and the other peak is in winter, in January and February, and is on the Japan Sea coast. In Casco Bay, Maine, United States, leatherbacks arrive regularly and stay several weeks as they do also in the coastal waters of New England and Nova Scotia. It appears, therefore, as though *Dermochelys* has a migration circuit that takes it regularly into temperate waters, returning by unknown routes to mate and nest in the tropics. The occurrence of leatherbacks off the British Isles and Norway strongly implies that part of that particular migration circuit is performed on the currents of the Gulf Stream and North Atlantic Drift.

[*Compiled from descriptions in Carr and Ogren (1959) and Carr (1968)*]

individuals engaged on a regularly performed migration circuit. Leatherback turtles, *Dermochelys*, for example, appear regularly in parts of the North Pacific and North Atlantic, often with such regularity that there is a strong implication of a migration circuit (Fig. 30.4). If such migrants are not accidental migrants, then there seems no doubt that ocean currents are of major importance to their lifetime track as they are for marine fish and for at least

the nesting site to feeding site migration of green turtles.

Although this section is concerned with marine turtles, it seems worth pointing out that although little is known concerning their movements, there must be a strong possibility that marine snakes, which also lay their eggs on land, also perform long-distance migration circuits (Fig. 30.5).

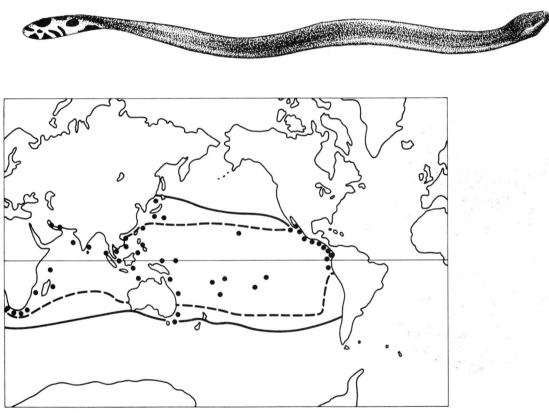

Fig. 30.5   The possibility of a long-distance migration circuit in the yellow-bellied sea snake, *Pelamis platurus*

The sea snakes (Hydrophidae) form a characteristic part of the fauna of the Indo-Pacific region. The family contains 15 genera and more than 60 species but very little is known concerning their ecology and migrations except that they are amphibious, coming ashore to lay their eggs and spending the rest of their lives in the sea.

The yellow-bellied sea snake is a weak swimmer that drifts with the current and feeds on small pelagic fish and other animals. It has a poisonous bite and its aposematic colouration results in its being ignored by most predatory fish. Research on the temperature physiology of the species suggests that its distribution may be limited by the 18°C water temperature isotherm.

The diagram shows the reported sightings of the snake (solid dots) in relation to the position of the 18°C isotherm in winter (dashed line) and summer (continuous line). The northernmost and southernmost sightings that are between these two lines seem likely to have been made in summer. Unfortunately, this cannot be confirmed from the literature, but if it is so it seems that each summer at least some yellow-bellied sea snakes are carried considerable distances beyond the limits of their winter range. Possibly this species has long-distance migration circuits, based on ocean currents, similar to the circuits of marine fish and turtles.

[*Modified from Graham* et al. (*1971*)]

31

The return migrations of fish and invertebrates that live in the sea or in running freshwater

31.1 Introduction

This chapter is concerned primarily with the daily, tidal, lunar-monthly, seasonal, and ontogenetic return migrations of fish. It is convenient, however, in the same context, to discuss the migrations of a variety of invertebrates that live in the same habitat-types. The chapter begins with a consideration of the removal and return migrations of stream-living invertebrates and uses this as an introduction to the migrations of fish that breed in running freshwater. Next, some consideration is given to the migrations of fish that breed in the sea. Those species that are primarily open-sea fish are considered first, along with a brief note on cephalopods, followed by a consideration of the migrations of littoral and sublittoral fish and invertebrates. After a brief discussion of the anadromy-catadromy paradox, consideration is given to the applicability of the familiar-area hypothesis to the range of migrations that have been discussed.

31.2 Stream-living invertebrates

A plankton net suspended in any type of running water catches considerable numbers of invertebrates that are normally considered to be benthic (bottom living). Such downstream migration has been recorded in water courses of all sizes, from small trout streams in Sweden to large rivers such as the Nile, Missouri and the Amur in much warmer climates.

All kinds of invertebrates may be involved and various species of annelid worms (Oligochaeta), crustaceans (Amphipoda and Isopoda), insects (Ephemeroptera, Plecoptera, Odonata, Trichoptera, Hemiptera, Diptera, and Coleoptera), mites (Hydrocarina) and molluscs have been collected like this (Hynes 1970).

Data that unquestionably identify this downstream drift as part of a downstream–upstream return migration are scarce. Where such data are available the return migration is ontogenetic. However, the most convenient way of considering upstream–downstream migration in invertebrates is not to do so in terms of its periodicity but first to consider the downstream movement, then to consider the selective pressures that act as a result of this downstream movement, and finally, in the light of these selective pressures, to consider such data as are available concerning upstream migration.

31.2.1 Downstream migration

Waters (1972) recognises three categories of downstream migration by stream-living invertebrates in which the major energy source for the migration is that provided by water currents rather than by the animal's own locomotory mechanism (= drift, in limnological terminology). These categories are: (1) catastrophic drift (in which there is physical disturbance due to flooding, drought, high temperature, anchor ice, or pollution); (2) constant drift (in which low numbers of all species participate at all times); and (3) behavioural drift (which is distinguished from the other two categories in that it is characterised by a diel periodicity).

Although these categories may be useful to a limnologist, they tend to cloud the issue as far as drift as a migration is concerned. For example, even catastrophic drift as a category is not homogenous from the standpoint of the selective pressures involved. Thus the scouring action associated with flooding and anchor ice is likely to qualify a large number of the downstream migrants as accidental migrants (p. 24). Although some individuals may chance to gain an advantage from the migration, therefore, it is not necessary to postulate that on average the migration is an advantage to the individuals concerned. When drought, high temperature, or pollution is involved, however, the migration is clearly non-accidental and has been subjected to all the selective pressures postulated to be associated with the facultative initiation of removal migration.

Some individuals performing both constant and behavioural drift could also be considered accidental migrants for which no advantage need be postulated for the migration. Any number of events, such as substrate movement, or dislodgement by larger but non-predatory animals (either vertebrate or invertebrate), could from time to time overcome an individual's normal station-keeping mechanisms, causing it to be carried a short distance by the current. Such events, where they occur at a low-level frequency, could contribute to the animals that constitute the constant drift. For such events to produce 'behavioural' drift, it is only necessary to postulate that the animals are at greater risk to accidental migration at some times in the diel than at others. It is significant, for example, that peaks of appearance in the drift for a particular species tend to coincide with peaks of feeding activity (Waters 1972). Thus the mayfly larvae of the genus *Baetis*, all stages of the amphipod crustaceans of the genus *Gammarus*, and indeed most other stream-living invertebrates are night-active, showing more drift during the hours of darkness than during the day. Some others, mainly the larvae of caddis flies (Trichoptera) are day-active. When animals such as these are at roost they are usually under stones and are relatively safe from accidental migration. When they forage, however, they have to come above stones and increase their frequency of crawling or even swimming. The risk of accidental migration is thus correspondingly increased.

Both constant and behavioural drift, on the other hand, probably also include non-accidental facultative removal migrants. The only difference from

similar migrants in the catastrophic drift is likely to be that such obvious physical factors as drought, temperature, and pollution are not the major contributors to habitat suitability and thus migration threshold. The availability of food and shelter per individual seems to be the most likely alternative. Even a simple comparison of bottom density with drift density sometimes produces a good positive correlation (Pearson and Kramer 1972) though this is not always the case (Waters 1972).

This consideration has perhaps been sufficient to demonstrate that the critical distinction in the present context is not between catastrophic, constant, and behavioural drift but rather between accidental and non-accidental migration. Where the migration is accidental it is unnecessary to seek an advantage for the migration. Where the migration is non-accidental, however, in which case the initiation of the migration is likely to be facultative, an advantage has to be sought. First, however, it is necessary to consider briefly what proportions of drift migrants are likely to fall into each of these categories.

Of all animals living on the stream bottom, those caught most abundantly in the drift (Hynes 1970, Waters 1972) are the larvae of mayflies (Ephemeroptera), blackflies (Simuliidae), caddis flies (Trichoptera), stone flies (Plecoptera), and midges (Chironomidae) as well as amphipod crustaceans of the genus *Gammarus* and isopods of the genus *Asellus*. These are all animals that crawl or swim, at least occasionally, and are thus perhaps most prone to accidental migration. Burrowing forms, such as oligochaete worms, and heavy creatures, such as snails and those caddis fly larvae that live in stony cases, that perhaps are least prone to accidental migration, are infrequent members of the drift. On the other hand, other crawling/swimming animals such as larval and adult beetles and hydrocarine mites are also relatively uncommon in the drift.

The faunal composition of the drift, therefore, perhaps tends to suggest that accidental migration is common but that not all drift migration can be attributed to this cause. Other characteristics of drift migration in fact tend to suggest that at least a notable proportion is non-accidental. Perhaps the most important characteristics are firstly that there is a marked ontogenetic variation in migration incidence and secondly that despite the large number of individuals collected in drift nets, there is normally no change in density of the deme immediately upstream, even though the distance drifted by

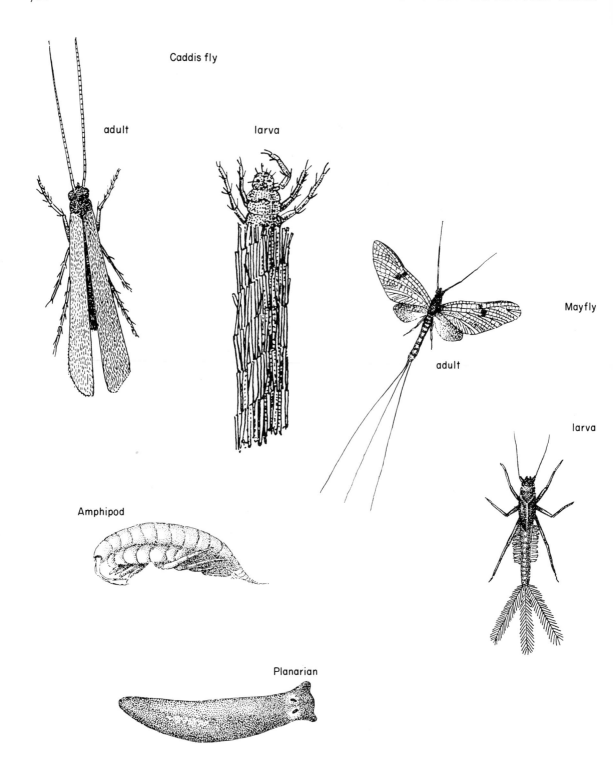

Fig. 31.1   Some members of the downstream drift

a single individual in a night's migration is probably only of the order of 50–60 m (Hynes 1970). As far as ontogenetic variation is concerned in most species there is greater drift during later and larger stages, though in some species the greatest drift occurs in the younger stages (Waters 1972). Although both of these ontogenetic patterns of variation in migration incidence are also commonly found among terrestrial forms (p. 342), thus suggesting a non-accidental migration, they could also be interpreted as reflecting ontogenetic variation in proneness to accidental migration. The observation that deme density of the stream bottom remains constant even when there are large numbers of animals in the drift is, however, strongly suggestive of the equilibrium situation described in Chapter 17, and when taken in conjunction with the existence of diel periodicity and ontogenetic variation it strongly suggests that a great deal of downstream migration has to be treated as non-accidental. The conclusions concerning the equilibrium situation stems from the work of Waters (1964, 1965) on the amphipod crustacean, *Gammarus pseudolimnaeus*, and the larva of the mayfly, *Baetis vagans*, in Valley Creek, Minnesota, United States. If an area of creek is cleared of inhabitants it is rapidly recolonised by downstream migrants. Yet once a certain density has been achieved, it remains relatively stable. Rate of recolonisation is directly related to rate of drift until this stable density is approached. It seems likely that the situation is free (p. 402) and that the frequency distribution of migration thresholds is such that all habitats become of equal suitability (p. 269). Such a situation can evolve only in relation to non-accidental migration. It should also be noted here that except where the removal migration is a response to drought, high temperatures, or pollution, it is invariably partial. Headwaters and upstream areas remain populated even when upstream migration does not seem to occur at a sufficient rate. Such partial migration is typical of free situations in which deme density makes a major contribution to habitat suitability (p. 69).

Having concluded that downstream drift consists of both accidental and non-accidental migrants, it is appropriate to consider the selection that acts on both types of migrant with respect to upstream migration. In order to appreciate this selection it is important to realise that selection has favoured species zonation along a river length from source to mouth just as much as it has favoured zonation on the sea-shore (p. 823) and in the soil (p. 513) (Hynes 1970). Individuals of most species, therefore, cannot drift downstream indefinitely without suffering a disadvantage.

Let us first consider an accidental migrant. Here it is unnecessary to postulate that the migration is advantageous. It is clear also that the major action of accidental migration is to impose stringent selection in favour of efficient body shape, attachment mechanisms, and habitat assessment. Nevertheless, once accidental migration has taken place and as long as it is not inevitable that the individual will die, some strategies will still be more advantageous than others. The most advantageous strategy will obviously be upstream migration during periods that migration is optimum behaviour. Indeed, for many species the danger of accidental migration could well be sufficiently great to impose selection for upstream migration at all times that migration is optimum behaviour, not only once an individual has experienced accidental migration.

The situation is slightly different with respect to non-accidental migration. Because we are dealing with an environmental force acting in only one direction (i.e. downstream) we are dealing with a migration that if found in all individuals results in a direction ratio of 100:0 (in this section, because of the linear nature of the habitat-type, only a two-item direction ratio will be used, i.e. 100:0:0:0 becomes 100:0 and 25:25:25:25 becomes 50:50). Essentially, therefore, the problem is one of optimum direction ratio, a model for which was developed in Chapter 12. According to this model, a peak downstream migration direction can evolve only if at some other time selection favours a major change of direction. This major change of direction must be such that all individuals perform a return migration, albeit without a high degree of return.

31.2.1.1 Conclusions

In conclusion, therefore, downstream migration through drifting is a common feature of stream-living invertebrates. This migration is performed by both accidental and non-accidental migrants. Accidental migrants suffer a disadvantage from the migration. Selection resulting from the danger and occurrence of accidental migration is likely to favour upstream migration, whenever non-accidental migration is optimum behaviour, either throughout life or after accidental migration. Non-accidental migrants, even if they are only a proportion of the deme, must on average experience an advantage from the migration, albeit an advantage equal to that experienced by the non-migrants (p. 69). Such migration can only be advantageous if there is also selection for some pattern of upstream mig.ation. If there is selection for a return migration to the natal habitat by all migrants then all downstream migrants migrate upstream again before reproducing. The direction ratio for the migratory portion of the deme, however, with regard to post-hatching (from the egg) direction ratio may be any value between 100:0 and 50:50, inclusive. Where the migration cost of an upstream migration is less than the cumulative saving in migration cost during all previous downstream migrations, however, the evolved post-hatching direction ratio is likely to approach 100:0 in a downstream direction. Even if there is no selection for return to the natal habitat selection can still favour a peak downstream migration during most of the year. As long, that is, as at other times of the year or at certain ontogenetic stages the upstream vector for the deme (p. 87) is greater than the downstream vector for the deme. Then a direction ratio will evolve that is intermediate between 50:50 and 100:0. The only other alternative, according to the direction ratio model, is for the centre of gravity of each deme to remain stationary as a result of equilibrium between upstream and downstream vectors (where the vectors are a function of number of migrants and mean migration distance in each direction—see also p. 30).

These expectations of the model in relation to upstream migration by drift invertebrates can now be compared with field observations of upstream migration.

31.2.2 Upstream migration

The idea that downstream drift by insect larvae was part of a return migration seems first to have been suggested by Müller (1954). The postulated advantage of the migration was that it permitted the larvae to colonise all suitable habitats, the adults then flying back upstream. This colonisation cycle suggested by Müller thus corresponds to the situation described at the end of the previous section in which there is selection for an accurate return migration by each individual.

There seems little doubt that many insects do perform a return migration of this type. Thus the limnephilid caddis fly, *Oligophlebodes sigma*, in the Logan River, Utah, United States, feeds diurnally as a larva on algae and diatoms (Pearson and Kramer 1972). The species is single-brooded in this area and produces adults in late summer and autumn (August–November). These adults drink but do not feed, and each individual lives for about 5 days. Males form mating swarms in more or less the area in which they emerge. From 120 to 30 min before sunset, adults leave the mating swarms in the male:female ratio of 0·4:1 and begin to migrate upstream. All of the migrating females are gravid. As long as the wind speed is less than 4–5 km/h (if

blowing upstream) or 2–3 km/h (if blowing downstream) the adults migrate upstream at a flight speed in calm air of 2·3–4·3 km/h ($\bar{x} = 3\cdot1$ km/h). Most fly 1–4 m above the stream but a few fly as high as 10–12 m. Only occasionally do individuals leave the stream and then usually to detour shrubbery etc. It is estimated that an individual may migrate upstream some 2–3 km during its adult life.

Roos (1957) placed nets to catch insects flying both up and down a river in central Sweden. He found that flight was predominantly upstream among the caddis flies (Trichoptera), mayflies (Ephemeroptera), stoneflies (Plecoptera) and female blackflies (*Simulium spp.*), especially among individuals that were ready to oviposit (see also p. 353). Visual observations of some caddis flies showed that they may move long distances upstream, even passing over small lakes.

It seems to have been established, therefore, that an upstream–downstream return migration has evolved in many stream-living insects. The following factors seem likely, upon entry into the return migration model, to be capable of generating the observed pattern. First, it is possible that for some species, within their zone in the river, there is a gradient of increasing habitat suitability for the larvae from the upstream end to the downstream end. This gradient may result from downstream changes in any or all of: availability of food; availability of shelter; temperature; and risk of accidental migration. Predation risk seems likely to change in the opposite direction. Even in the absence of such a gradient of habitat suitability, however, downstream migration is likely to be favoured at each facultative removal migration due to migration cost. For downstream movement to evolve as a general characteristic of the migrant portion of a deme, however, selection must also act to favour an upstream return migration by the downstream migrants before they reproduce. Such selection could result from an upstream gradient of habitat suitability for oviposition within the species zone due to the influence on the offspring of the risk of accidental migration and the possibility that facultative removal migration may become advantageous. Such selection could also, perhaps with an even higher probability, result from the fact that if an individual has survived to adulthood despite the migrations, both accidental and non-accidental, that it has performed during development, it is a likely correlate that the habitat into which that individual was laid will be at least of average suitability as an oviposition habitat, and perhaps even of above average suitability. Streams are relatively stable habitats with a relatively slow rate of change of suitability (p. 273). Furthermore, stream-living insects, like stream-living fish, but unlike birds (p. 608), usually have little opportunity for exploratory migration, either as young or when adult. That is, exploratory migration tends to have a high migration cost. On the whole, therefore, $h_0 M/h_e$ (where h_0 is the suitability of the site in which oviposition eventually occurs, h_e is the suitability for oviposition of the adult emergence site, and M is the migration factor for migration between these two sites) is maximised by migrating directly upstream to the site into which the individual was itself deposited as an egg.

For an insect that as an adult lives out of the water, the cost of upstream migration suddenly decreases upon attainment of the adult state and emergence from the water. Selection therefore tends to favour a pre-reproductive upstream migration by the females. For males, however, the critical gradient is probability of encounter with receptive females (p. 784), a gradient that is far less likely always to favour upstream migration.

Even an insect that cannot fly, as long as it leaves the water, is likely to have a much reduced upstream migration cost per unit distance after it has left the water than before. Thomas (1966) observed the emergence of the stone fly, *Capnia atra*, from a stream onto snow-covered ground in northern Sweden. At first these stone flies walked at right angles to the stream to the edge of a wood some 50 m away. They continued on the same course for a further 50–100 m and the majority then turned through 90° and, even though the stream was no longer visible, headed in an upstream direction. This is not an isolated case (Hynes 1970) and it may be that the behaviour is quite general for winter stone flies in areas where the temperature is too low for flight.

Insects that as larvae have performed downstream migration may well have passed several junctions of tributaries, not all of which are equally suitable for oviposition. Selection is likely to favour any behaviour that favours a female, not only flying upstream, but flying upstream to the natal site in the natal tributary. Whether such selection has resulted in appropriate navigation mechanisms in stream-living insects as it has in fish is unknown.

Not only insects seem to have evolved a return migration of this type. The common European flatworm, *Crenobia alpina*, is a cold-water species that lives under stones in rapid water. Beauchamp (1933, 1935, 1937) demonstrated that when individuals of this species are hungry or when they have bred they migrate downstream. Arrival in warm water or the onset of sexual maturity, on the other hand, triggers an upstream migration. Presumably an animal as flat as *C. alpina* has the potential to migrate upstream by a route that is continuously under and/or in the shelter of stones and is thus protected to a large extent from the strongest effects of the downstream water current.

Some of the larger fresh-water invertebrates, such as the decapod Crustacea, perform an upstream–downstream return migration (Hynes 1970). Thus crayfish, such as *Pacificastacus klamathensis*, in Oregon, United States, make regular upstream–downstream migrations as does the mitten-crab, *Eriocheir sinensis*. This latter species regularly travels 750 km up the Yangtse Kiang in China and has been found 1400 km from the sea. Since its introduction into Europe it has been collected as far inland as Czechoslovakia. This species, however, breeds in brackish water and it is the young crabs that travel upstream. The return migrations of stream-living decapod crustaceans can hardly, therefore, be attributed to the selection generated by the water current and should probably be considered in the same terms as littoral and sub-littoral species (p. 830) rather than here.

There is little doubt that a great many stream-living invertebrates have evolved a return migration with a high degree of return that is performed by that portion of the deme that at some time performs downstream migration. Perhaps such return migration will eventually be found to be much more widespread than has so far been demonstrated. There seems little doubt, however, that some species do not perform a return migration of this type. For example, the same study (Pearson and Kramer 1972) that demonstrated such a convincing return migration by females of the caddis-fly, *Oligophlebodes sigma*, found no evidence of such a return migration in the same place by the mayfly, *Baetis bicaudatus*, the larvae of which also formed an important part of the drift.

According to the model developed in this book, one of the other alternatives presented on p. 784 must apply to these other species. Unfortunately, all of these other explanations require much more detailed information on the relative incidence of accidental and non-accidental migration, migrant/non-migrant ratio, direction ratio, and frequency distribution of migration distances. Such data are extremely difficult to obtain in the stream environment and none are available. Waters (1972) has shown that upstream movement does not seem to balance downstream movement, at least at the times of year studied. However, only two of the alternatives generated by the model require a continuous balance between upstream and downstream migration. Indeed, if accidental migration is common for a species, it is unnecessary to postulate that upstream migration ever balances downstream migration. The only pattern for non-accidental migration, however, for which a peak migration direction downstream is optimum but that does not require a balance between upstream and downstream migration nevertheless requires a particular relationship between them. That is that for a short period at some time of year upstream migration is greater than downstream migration.

None of these alternatives can yet be tested except insofar as, even in the case of accidental migration, they all postulate that upstream migration will occur. To this extent, there is supporting evidence. Thus, when the amphipod, *Gammarus pulex*, was introduced into a small stream on the Isle of Man in the Irish Sea, it appeared in numbers in the headwaters 3 km upstream within four years of becoming established. Circumstantial evidence of this type suggests that upstream movement occurs in a wide variety of stream-living invertebrates (e.g. small and large snails, flatworms, amphipods, and the larvae of mayflies and caddis flies) (Hynes 1970).

31.2.2.1 Conclusions

The downstream water current results in the downstream migration of a wide variety of stream-living invertebrates, either as a result of accidental migration or as a result of downstream bias to each non-accidental removal migration. In both cases, the migration is usually partial. A bias towards downstream non-accidental migration can evolve only if there is also selection for certain characteristics of migration distance and direction ratio. In many animals selection is for a return migration with a high degree of return, and in these cases the direction ratio usually seems to approach 100:0. Where the animal is an insect, selection favours the upstream migration being concentrated into the pre-oviposition period of the female adult. This is also found to be the case in a flatworm and the pattern may be of widespread occurrence. Such return migration is most likely to evolve if the cost of upstream migration is less than the cumulative gain due to the earlier migrations being downstream rather than in any

other direction. Where this is not the case a return migration, even a partial migration, with a direction ratio of 100:0 cannot evolve. The model for direction ratio generates two other alternatives plus the alternatives associated with accidental migration. All of these alternatives demand the existence of some degree of upstream migration. Data suggest that upstream migration is widespread but are not sufficiently detailed to distinguish between the possible alternatives generated by the model. Nevertheless, the characteristics of downstream drift and associated behaviour by stream invertebrates as an upstream–downstream return migration and also where the phenomenon is apparently divorced from return migration seems as far as is known to be quite consistent with the migration model developed in this book.

31.3 Fish that breed in running water

Most of the observed characteristics of the migrations of fish that breed in running water fit quite satisfactorily within the patterns generated by the model developed in this book. Furthermore, some field data exist that support the reality of the selective pressures that it seems necessary to include in the model to generate these patterns.

31.3.1 Some examples of downstream–upstream migration in fish

Perhaps the commonest pattern of downstream–upstream return migration in stream-living fish is ontogenetic and consists of a downstream migration when young followed some time later when adult by an upstream migration to spawn. In the majority of cases this migration takes place entirely within freshwater. Many of the species concerned spend most of their lives in lakes but migrate out into streams, usually but not always inlet streams, to spawn. In the brown trout, *Salmo trutta*, of Europe the young fish live in the nursery stream for one to four years before moving into the main river or lake (Mills 1971). Adult trout become ripe and spawn in autumn and winter (October–February). Upstream migration precedes spawning, males tending to migrate in advance of females, with migration taking place throughout the day in turbid water but only at night in clear water. Spent fish migrate back downstream during winter and early spring and often take up the same feeding territory that they occupied the previous year (Hynes 1970). In Africa many varied fishes move out of lakes to breed in rivers. These include the cyprinids, *Barbus*, *Varicorhinus*, *Barilius*, and *Labeo*, the catfish, *Clarias*, and the characid tiger-fish, *Hydrocyon* (Hynes 1970). Similarly, throughout the world, a great many types of river fish move upstream to spawn. These include sturgeons and paddlefish (Chondostrei), garpikes (Holostei), shads (Clupeidae), suckers (Catastomidae), barbels, minnows, and other Cyprinidae, catfish (Siluroidei), and some darters (Etheostominae). Even apparently sluggish species, such as the carp, *Cyprinus* sp. have been observed to move up the Des Moines River in Iowa and to jump over lowhead dams.

Although the majority of examples of upstream–downstream return migrations concern movements entirely within freshwater, in some species the downstream migration continues into brackish estuarine or coastal sea water (Fig. 31.5) or even out into the open sea (Fig. 31.2). The fish concerned are then described as anadromous if they return to freshwater to spawn. Perhaps the best known examples of anadromous fish are salmon (e.g. the Atlantic salmon, *Salmo salar*–Fig. 31.2) and lampreys (e.g. *Lampetra fluviatilis*—Fig. 31.3). However, the anadromous habit is by no means confined to these groups and is shown by, among others, (Hynes 1970): the sturgeons, *Acipenser* spp. (Fig. 31.4) and *Huso huso*, of eastern Europe, specimens of which may weigh up to a tonne; the shads (Clupeidae), such as *Alosa* and similar genera in many temperate parts of the world; various other herring-like fish or Clupeiformes such as *Clupeonella* in the Danube and *Harengula* in the Volga; some whitefish or coregonids such as *Stenodus* and *Coregonus* in the USSR and the houting, *C. oxyrhinchus*, in the southeastern part of the North Sea (Mills 1971); the cyprinid roach, *Rutilus rutilus caspicus*, in the Caspian Sea and Volga; and the three-spined stickleback, *Gasterosteus aculeatus*, in northern Europe (Baggerman 1956).

Even among the salmon and lampreys, however, demes occur that do not reach the sea during their downstream migration, but instead feed and mature in lakes or in the lower reaches of the drainage system in which they occur. Such demes of normally anadromous species are often described as land-locked. To zoologists that are not specialists in fish migration, however, this term upon casual encounter can be misleading, for it implies that the deme concerned is denied access to the sea by some land barrier. Although this is sometimes the case, it is not generally so and most such demes have perfectly good but unused access to the sea. The absence of the sea-going habit in such demes must therefore be considered to be adaptive except perhaps where it is found in demes recently introduced by man, in which case it could be a founder effect, resulting from the genotypes of the original introductions. The majority of Atlantic salmon, *Salmo salar*, introduced into the rivers running into Lake Te Anau, South Island, New Zealand, descend only to the lake and after feeding there return to the head rivers to spawn (Mills 1971). Their passage to the sea is not barred as the River Waiau flows out from this lake. The explanation for the absence of anadromy in this deme is unknown. It could be a founder effect as just described or alternatively it may result from low deme density in the lake (see p. 802). It is known that the timing and distance components of salmon migration have evolved in the form of optimum response curves that are sensitive to environmental variables rather than being entirely under genetic control (Mills 1971).

The occurrence of such land-locked forms in species that are normally anadromous is only one aspect of the wide variation in distance component that is characteristic of all species that spawn in freshwater and perform an ontogenetic downstream–upstream return migration. This variation is discussed in Section 31.3.5 (p. 802).

The migrations of species that spawn many times over a period of several years become both ontogenetic and seasonal after the first spawning. The brook trout, *Salvelinus fontinalis*, a native of eastern North America, shows considerable variation in migration distance. Individuals that migrate only within freshwater migrate upstream during the autumn (September–November) spawning season and then downstream to deeper water for the colder months of winter. The Danube salmon, *Hucho hucho* occurs in the Danube river basin, preferring the upper reaches of the tributaries of this large river in the German Federal Republic, Austria, Czechoslovakia, Poland, Hungary, Rumania, Bulgaria, Yugoslavia, and Russia. It is a large fish that can grow up to 2 m in length and is a fierce predator of other fishes. The Danube salmon grows rapidly and may live for up to 20 years. It spawns in spring (March/April) in the upper courses of highland rivers. After spawning it migrates downstream and in autumn and winter seeks the deeper waters where other fish spend the cold season in large numbers (Mills 1971). Migration into deeper water for the winter is characteristic of many temperate species

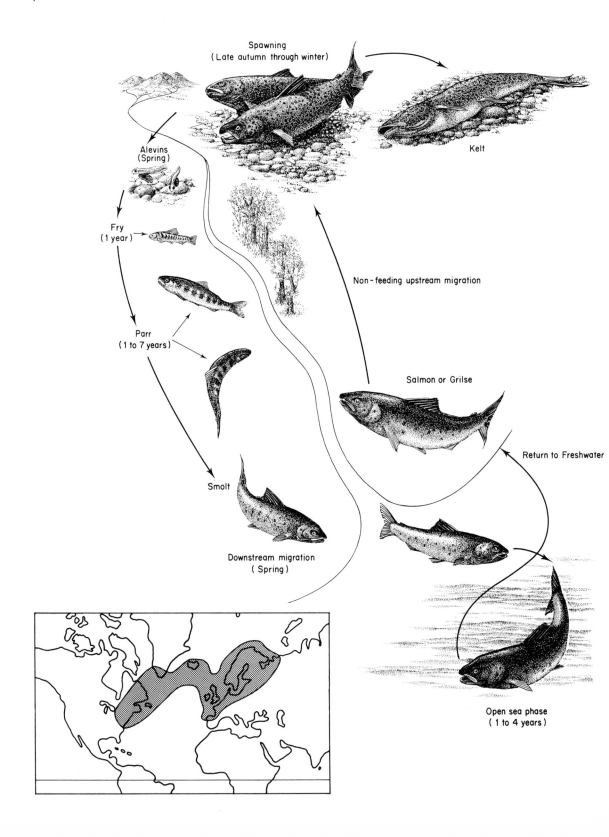

Spawning
(Late autumn through winter)

Kelt

Alevins
(Spring)

Fry
(1 year)

Parr
(1 to 7 years)

Non-feeding upstream migration

Salmon or Grilse

Return to Freshwater

Smolt

Downstream migration
(Spring)

Open sea phase
(1 to 4 years)

Fig. 31.2 Ontogenetic return migration and life history of the Atlantic salmon, *Salmo salar*

The map shows the approximate geographical range of the Atlantic salmon.

Most Atlantic salmon spawn in late autumn or winter, usually in the upper reaches of a river system where gravel exists that is suitable for the construction of the spawning pit or 'redd'. The fertilised eggs, which may be under 15–30 cm of gravel, require 70–160 days to hatch, depending on water temperature. The resulting alevins (approximately 1 cm long), which each possess a large yolk sac attached to the under surface, move about in the gravel and do not emerge until 3–4 weeks later when the yolk sac is nearly absorbed and they are ready to feed as fry. The fish remain in the fry stage for about a year during which mortality is high due to starvation, predation, and competition for space. At the end of this period the fish are known as parr.

In England the growth of the fry and parr takes place only during spring, summer, and autumn (April through October) but does not occur during the winter months. A wide variety of invertebrates are used as food. Young salmon establish a distinct home range to which they attempt to return if displaced. Some individuals have returned from displacement distances as great as 215 m, both upstream and downstream. It is thought that the situation is despotic and that fry that migrate downstream are individuals of low resource-holding power. Parr also migrate downstream, particularly in spring. This movement may also reflect a despotic situation but may also be due to individuals that have outgrown their previous territory and that are searching for a new territory.

On both sides of the Atlantic, parr that reach a length of about 10 cm towards the end of one growing season are likely to become smolts at the next season (spring) of smolt downstream migration. The time taken for a parr to grow to a length of about 10 cm varies with the geographical location of the river. In the Hampshire Avon, England, over 90 per cent of smolts are yearlings, whereas in northern Scandinavia parr stay in the rivers for 7–8 years. A varying proportion of male parr attain sexual maturity before leaving the river and may appear on the spawning grounds in large numbers following an upstream autumn migration (see p. 807).

In the spring following attainment of appropriate length, transformation to the smolt and downstream migration occurs. In the spring, during the early (April and early May) part of the smolt migration, the movement is during the hours of darkness, but later on (late May and early June) movement occurs during the middle part of the day. It is likely that smolt drop downstream tail first, though in a lake they almost certainly swim actively downstream. Migration speed downstream varies between 0·5 and 2·0 km/day. It is likely that smolts need some time to become acclimatised to saline conditions and they may remain in the estuary, where one exists, for a short time before entering the sea.

Little is known of the movements of smolts at sea. Marked smolts from northern Swedish rivers are recaptured as adults in the southern Baltic. Smolts from the British Isles may be found off Greenland during their first year. Older fish from these islands, Canada, and Norway are also frequently caught off Greenland (see also Fig. 31.9) but at least some of the salmon from Sweden and perhaps the major part from Norway and the Soviet Union may grow up in the eastern part of the North Atlantic. The sea food of salmon consists chiefly of fish such as herring, *Clupea harengus*, capelin, *Mallotus villosus*, and sand eels, *Ammodytes lancea marinus*, and large zooplankton, particularly euphausids and amphipods. Salmon are eaten by various fish and also seals, particularly the grey seal, *Halichoerus grypus*.

Atlantic salmon return to freshwater after a marine phase of one to four years. Fish returning after only one year are known as grilse whereas fish that have spent longer at sea are known as salmon. The time of year that salmon and grilse return to freshwater varies from river to river and may be at any season. Some fish may enter a river up to 12 months before spawning. There seems to be an upstream migration threshold to the migration-cost variables of water level and, in spring, temperature. Migration is more likely to be initiated during spates than during dry weather and above water temperatures of 6·5 °C. The average rate of upstream movement has been estimated at 10–20 km/day.

The salmon does not feed during upstream migration and may lose 25 per cent of its body weight. There is also redistribution of material, the scales, for example, being considerably absorbed and the calcium thus released being used in growth of the bones of the jaws and teeth, particularly in the male. On average, spawning salmon are approximately 70 cm long. There is, however, a relationship between the length of the salmon and the number of eggs produced. The average expression of this relationship for various Scottish rivers is given by the equation

$$\log_{10} N = 2\cdot3345 \, \log_{10} L - 0\cdot582$$

(where N = number of eggs and L = body length in cm).

After spawning, both male and female fish are known as kelts. The death rate after spawning is high, especially among males. Between 20 and 36 per cent ($\bar{x} = 26$ per cent) of kelts survive to perform downstream migration. Upon reaching the sea, the kelt may either remain within the influence of the river or it may perform long-distance migration to the open sea similar to that performed during the first visit to the sea. The period spent in the sea between spawnings varies from about 4 to 18 months. On average, 3–6 per cent of fish that spawn in a river have spawned previously, though in some short-course rivers in western Scotland and eastern Canada the proportion may be as high as 34 per cent. A salmon has been captured that was 13 years old and that had spawned four times.

Degree of return to the natal river system of those that survive to return to freshwater is extremely high (greater than 90 per cent) as both salmon and kelt. Degree of return to the natal tributary is somewhat less, but still high (approximately 80 per cent).

[Compiled from Tchernavin (1939) and Mills (1971)]

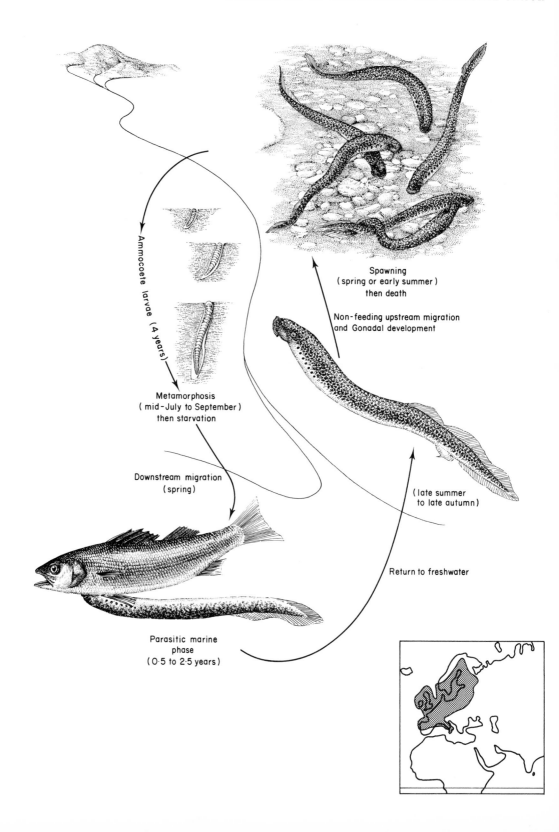

Ammocoete larvae (4 years)

Spawning
(spring or early summer)
then death

Non-feeding upstream migration
and Gonadal development

Metamorphosis
(mid-July to September)
then starvation

Downstream migration
(spring)

(late summer
to late autumn)

Return to freshwater

Parasitic marine
phase
(0·5 to 2·5 years)

Fig. 31.3 Ontogenetic return migration and life history of the anadromous river lamprey, *Lampetra fluviatilis*, and comparison with its paired species, the brook lamprey, *L. planeri*

The river lamprey

Lampetra fluviatilis spawns in the western European watersheds of both the Atlantic and Mediterranean (see map). Upon hatching from the egg the species, like all lampreys, begins a protracted freshwater larval stage during which it is termed an ammocoete. During this period, which lasts about 4 years, the ammocoete lies concealed in the silt deposits of rivers and streams, straining off from the water the micro-organisms on which it feeds. As with many filter-feeders, ammocoetes may be found in demes of extremely high density in suitable areas. Emergence from the burrows for removal migration occurs only during the hours of darkness during which ammocoetes may become members of the downstream drift (p. 780). The arguments presented with respect to invertebrate members of the drift may also be applied to ammocoetes. In streams with a high or logarithmic profile, older larvae become increasingly predominant in downstream regions, indicating that downstream removal migration occurs steadily throughout larval life. Predation, particularly by fish, may be heavy during the early larval stage.

At the end of the larval stage metamorphosis begins in late summer and autumn and for several months the lamprey does not feed. There is some loss of weight during this period but little or no reduction in length, which is approximately 10 cm. The following spring, downstream migration to the sea occurs, and is followed by the parasitic marine phase lasting for 6–30 months. The main food during this period is the blood, other body fluids, and cytolysed tissue of teleost fish. On the whole, only healthy fish are attacked and the lamprey leaves once its host/prey is dying or dead, some 76–220 h after the initial attack depending on the relative sizes of lamprey and host.

The river lamprey grows to a length of 20–40 cm, depending on how long it remains in the sea. Movements at sea are unknown but it is assumed that the species does not travel far into the open sea. In the Mediterranean it has been caught within 15 km of the coast at depths of 15–50 m. In any event, anadromous lampreys seem to concentrate in coastal regions during the summer preceding their upstream migration. During this period they may feed heavily on other coastal fish or on other anadromous fish, such as salmon, that pass through the coastal waters during their own migration. Feeding ceases once the upstream migration begins. Time of entry into freshwater appears to be a function of the distance up-river that the individual will spawn. Usually the run begins in late summer and early autumn and continues until spring with a peak in late autumn. Migration incidence is a function of temperature during this period and also shows a marked daily variation, reaching a peak about two hours after nightfall. The average rate of upstream migration varies from approximately 1 to 3 km/night depending on the speed of the current. Where the current is strong, migrants move along the stream edge in shallow water. Obstacles are surmounted by short bursts of activity punctuated by periods of attachment by means of their oral sucker. Such evidence as is available suggests that degree of return to the natal site is low by comparison with other anadromous fish such as salmon. Males tend to arrive at the spawning area first. Spawning, however, may not commence for several weeks until water temperatures reach 10–11°C. 'Nests' are built by small groups of variable sex ratio but which are often polygamous. The male attaches himself to the females as shown and the efficiency of fertilisation seems to be a function of the proximity of the cloacal apertures during spawning.

The brook lamprey

Lampetra planeri shares the geographical range of the river lamprey, *L. fluviatilis* and together the two species are referred to as 'paired species'. This term is applied to the pairs of closely related and morphologically similar lampreys of which one is a non-parasitic or brook lamprey and the other a parasitic species similar to the form from which it is generally considered that both species have been derived. Throughout the world, ten parasitic species may be paired with non-parasitic species. In addition there are also two brook lampreys (*Okkelbergia aepyptera* in east and south-east United States and *Lampetra zanandreai* of the Po basin in Italy) for which there are apparently no extant parasitic counterparts and which may be relict species which were formerly more widely distributed.

Lampetra planeri shows the typical facies of a brook lamprey. There is an extension of the period of larval life relative to the river lamprey with which it is paired (i.e. in this case 6+ years as compared with 4+ years) with the result that body size is greater at the time of metamorphosis (total body length up to 17 cm as compared to 10 cm). After metamorphosis, however, the brook lamprey neither feeds nor migrates to the sea and immediately begins gonad development. There is an inevitable reduction in potential and absolute fecundity in comparison with the parasitic species (see also Fig. 31.8), energy being made available for oocyte maturation by a massive destruction of other oocytes. After metamorphosis, the young adult brook lamprey retains the burrowing habit of the ammocoete, perhaps even to within a few weeks before spawning. Upstream migration begins only a short time before the onset of spawning and the distances travelled are often less than 2 km. Although *L. fluviatilis* and *L. planeri* have been observed spawning in the same areas, no examples of interspecific pairing are known. On the whole, however, *L. planeri* tends to spawn in small streams and upland tributaries whereas *L. fluviatilis* tends to spawn in the lower reaches.

[*Compiled from Hubbs and Potter (1971) and Hardisty and Potter (1971a,b,c)*]

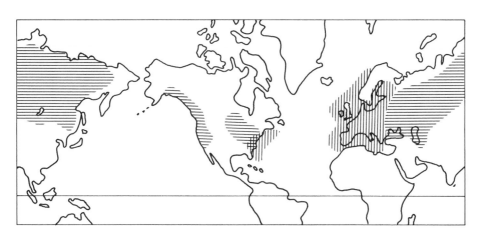

Fig. 31.4 Geographical distribution of sturgeons of the genus *Acipenser*

Sturgeons are members of the Chondrostei, an ancient group of fish that had a wide distribution in Palaeozoic time (300 million years B.P.). Fossil representatives of the genus *Acipenser* have been recorded from England and France since the Eocene (50 million years B.P.) and in Virginia, United States since the Miocene (25 million years B.P.). Altogether, about 16 species of sturgeons (Acipenseridae) are known at the present time and all perform a downstream–upstream return migration. Several species are also either totally anadromous or at least contain anadromous members. The adults are large fish, 2–3 m in length an in some species reaching as much as 7 m and weighing a tonne.

The distribution of the common sturgeon, *A. sturio*, is indicated by vertical hatching. It is a bathypelagic and bathydemersal fish of the north Atlantic and Mediterranean which spawns in the rivers of North America and Europe. It is seldom captured at sea and its marine distribution and migrations are little known, but in spring and early summer (March–July) at the time of the spawning migration it enters the rivers of continental Europe in shoals and in small numbers also the rivers of Britain, Denmark, Norway and Sweden. Upstream migration occurs mainly at night and in large rivers such as the Rhine, Elbe, and Vistula may be relatively long distance. In other rivers the migration may extend only just beyond brackish water. The female is escorted up-river by a number of males which are smaller than the female. The eggs (i.e. caviar) are laid in a depression dug in the river bottom. The young migrate to the sea in the same summer that they hatch.

A number of sturgeons occupy the Black and Caspian Seas and ascend the rivers to spawn. Sturgeons entering the Danube may have a one-way migration distance of 2000 km.

[Compiled from Meek (1916), Hynes (1970) and Ricard (1971)]

such as darters (Etheostominae), European and many other minnows (Cyprinidae), trout (Salmoninae), and also the sterlet, *Acipenser ruthenus*, in the Volga (Hynes 1970). Similarly in dry climates there is often a marked downstream migration during the dry season when the headwaters and swamps begin to dry out, followed by a return, some time later, when conditions improve.

Some fish that are found at some seasons or at some ontogenetic stages in streams and rivers are nevertheless essentially marine and breed in the sea. The migrations of such fish, among which are included eels (Anguillidae) and southern trout or jollytails (Galaxiidae), are discussed in Section 31.4 along with the migrations of full-time marine species.

This section has served to introduce the nature and incidence of the downstream–upstream return migration of fish that breed in running water. The following sections are concerned respectively with the selective pressures that could have produced these return migrations and with the selective pressures that could have acted on some of the components of these return migrations. Further details of the observed variation in such components as distance and timing are presented in the appropriate sections.

31.3.2 Selection for downstream–upstream return migration

Accidental downstream migration, particularly by young fish and particularly during spates, must occur from time to time as it does with stream invertebrates. However, studies of fish for which a home range has been demonstrated (Gerking 1959) have shown that a wide variety of species are capable of maintaining their position, even during very high spates. Nevertheless, the risk of accidental downstream migration must always be present. Where accidental migration occurs, the selection for subsequent upstream migration discussed in relation to stream invertebrates (p. 783) seems to apply. In an animal as mobile as a fish, however, there seems no reason why accidental migrants should not be capable of returning to their original position, if such return were to be favoured by selection. Such selection is perhaps most likely to act if the individual had invested in a home range. In such cases selection is likely to favour a return migration unless

the act of displacement exceeds the home range to home range removal migration threshold of the individuals concerned (p. 871). Displacement experiments on fish have demonstrated marked ability to return to their original home range, especially if the displacement is in a downstream direction (Hynes 1970).

Apart from the selection for return associated with accidental migration, a variety of features of the running water habitat-type also combine in a way that seems likely to impose selection for return migration. The first combination that can be considered is that also suggested for stream-living invertebrates in the preceding section: unidirectional facultative removal migration coupled with selection for reproduction in the natal site (p. 784). Here, however, there is some evidence to support the suggestion that, at least as far as migration cost is a function of energy expenditure, selection always favours removal migration to be in a downstream direction (within the constraints of selection on direction ratio—p. 85).

Bottom-dwelling fish species maintain their position in streams and rivers by physical contact with the substratum, and some species may even burrow (Hynes 1970). Swimming species may also at times go down to the bottom. At other times, however, swimming species maintain their position by swimming against the current and by sensory reference to fixed objects. In clear water when sufficient light is available reference is usually by vision, though in permanently muddy rivers and perhaps in the dark the lateral-line system seems likely to be used. Permanent swimming into the current to maintain position involves considerable energy expenditure. To a large extent this energy expenditure is reduced by seeking shelter in the 'dead' water behind obstacles and other parts of the stream. Even strong swimmers such as trout, *Salmo trutta*, do this for much of the time, and the shortage of suitable obstacles makes a major contribution to the appearance of territorial behaviour in some fishes. Despite sheltering behaviour, however, swimming fish must expend energy to maintain their position, and studies in central Germany (Tesch and Albrecht 1961) have indicated that the brown trout, *S. trutta*, grows better in slower flowing streams than in very fast ones and better in natural streams with varied depths and shelter than in straightened artificial channels. Growth rate is also decreased during floods and spates. There is a strong implication, therefore, that even this streamlined fish expends much energy in

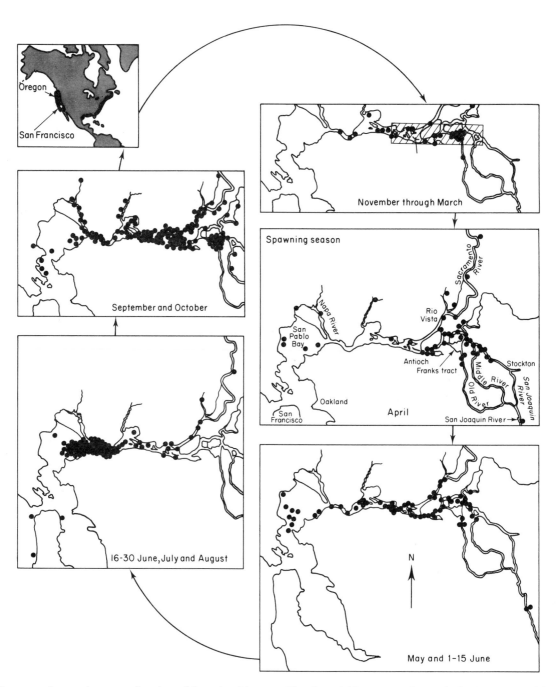

Fig. 31.5   Seasonal return migration of the striped bass, *Roccus saxatilis*, in the Sacramento-San Joaquin River, California

General ecology

The striped bass was introduced to the Pacific coast of North America from the Atlantic coast in 1879 and 1882. The original introductions were into the Sacramento River but within 20 years the species had spread north into Oregon. By 1906 the species was being caught nearly 1000 km north of the San Francisco Bay in the mouth of the Columbia River. Occasional individuals have also been captured hundreds of kilometres to the south of San Francisco. There is some evidence that the incidence of such removal migration shows a pre-reproductive peak (Chapter 17). Striped bass spawn at water temperatures

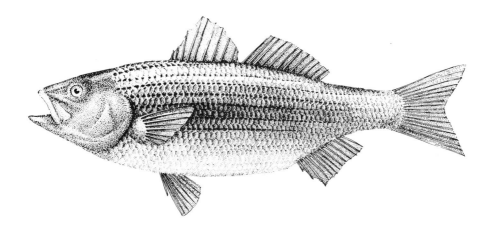

greater than 15·5°C. In the Sacramento–San Joaquin delta, spawning reaches a peak in mid-spring (April). The food is mainly shrimps and anchovies, *Engraulis mordax*, in summer and autumn and small fish in winter.

Seasonal return migration
the hatched area in the top right diagram indicates the area within which fish were marked. All marking was carried out between November and May. Solid dots indicate the position of subsequent recaptures at different times of the year.

During winter (November through March) the fish are concentrated in the river delta but as the height of the spawning season approaches (April) the fish spread out over the delta to spawn and some ascend the tributary rivers. After spawning the fish move downriver and the young fish use the estuary as a nursery area. During the summer (June through August), almost all adults are in salt and brackish water. On the Atlantic coast a higher proportion of adults enter the sea and there is some indication of a northward movement along the coast in late spring and summer and a southward movement in autumn. As yet there is no indication of such a marked movement along the Pacific coast, though individuals are often caught south of San Francisco Bay for a brief period in summer. In autumn (September and October) most of the fish pass back up into the delta again.

[*Compiled, with modifications, from Calhoun (1952)*]

fighting the current. As far as removal migration is concerned, therefore, it follows that the migration factor for upstream migration is low. The migration factor for downstream migration, on the other hand, is high, perhaps even greater than unity, depending on the extent to which migration cost is a function of energy expenditure. Evidence that other factors are involved, however, is provided by the observation that at both the smolt and kelt ontogenetic stages (Fig. 31.2) salmon often migrate downstream while still facing upstream (Hoar 1951, Jones 1959), though in a lake they swim actively in a downstream direction. Energy is thus expended to reduce the rate of downstream displacement. Possibly, given a sensory field that approaches 360°, escape from predators and habitat assessment are more efficient if speed of downstream movement is reduced and the animal is facing upstream.

The advantage of return to the natal site for reproduction has been suggested previously to be in part a function of the opportunity for, and cost of, exploratory migration (p. 785) (as well as a function of rate of change of habitat suitability). Given a habitat-type, such as a stream, that is characterised by a slow rate of change of habitat suitability, those species of fish with greater opportunity for exploratory migration are likely to receive less advantage from automatic return to the natal site when the time for reproduction approaches. As far as migrant fish are concerned, if selection on return to the natal site is as suggested, those species that migrate away from the natal site when young and for which energy expenditure and time availability are at a premium immediately before reproduction are most likely to gain maximum advantage by migrating directly to their natal site in order to

Table 31.1 Degree of return to the natal site as a function of the opportunity for exploratory migration in some salmonids†

Species	Opportunity for exploratory migration	Degree of return to natal site
pink salmon *Oncorhynchus gorbuscha*	*	**** (90%)
steelhead trout *Salmo gairdneri gairdneri*	**	**** (90%)
Atlantic salmon *Salmo salar*	**	*** (80%)
sea trout *Salmo trutta*	***	** (?)
brown trout *Salmo trutta*	****	* (?)

† From data in Harden Jones (1968) and Mills (1971)
* to **** = increasing manifestation of variable
    ($X\%$) = approximate per cent return to natal tributary of individuals that survive to spawn

reproduce. In other words, those species with the longest distance component to their ontogenetic return migration should show the highest degree of return to their natal site. Table 31.1 presents information for several species of salmonids which tends to support this suggestion and therefore supports also the suggestion that selection for return migration acts in the manner discussed above.

Unidirectional removal migration coupled with selection for return to the natal site provides a basis for the considerations of this section. The next point to be considered is the range of events that can lead to removal migration being advantageous in the first place. Removal migration thresholds seem to have evolved to biotic factors such as deme density, food abundance, and access to shelter from the current. As usual (p. 268), facultative removal migrations in relation to these factors seem to manifest a characteristic frequency distribution of migration thresholds such that invariably some individuals remain where they are whereas others initiate migration. The situation, unlike that suggested for invertebrates, may not be free. In rivers and streams, non-breeding territories are common, particularly among the salmonids. Shelter from the current seems to be a critical resource (Vibert 1962) and it has been observed that territorial behaviour may cease if current speed is reduced (Kalleberg 1958). Young brook trout, *Salvelinus fontinalis*, for example, are territorial in shallow rapids but tend to form shoals in pools and backwaters (Keenleyside 1962). Thus, in terms of the territory model (p. 404) energy expended in territory and defence is advantageous only when the current is strong and the energy saved

through being able to shelter behind an obstacle is high. As current speed decreases, the advantage of shelter also decreases, until at some point it is no longer greater than the disadvantage that results from energy that would be expended in territorial defence. At this point the position shifts to that in which alternative strategies, such as shoaling, are more advantageous than territorial defence. The distribution of fish such as trout, therefore, at least when young, fluctuates between free and despotic situations. The models developed in Chapters 8, 16, 17, and 18, suggest that the migration threshold consists of availability of food and shelter, deme density, and relative resource-holding power, the contribution of the latter being a function of current speed.

Experimental release of 2- and 3-year-old cutthroat trout, *Salmo clarkii*, into Gorge Creek, Alberta, was followed by considerable downstream movement of the introduced individuals. Downstream movement was accompanied by a loss of weight and often death. This poor performance was attributed to territorial interaction with the original residents (Miller 1958), it being usual in a despotic situation for intruders to show less resource-holding power than territory holders (Parker 1974b). Some young Atlantic salmon, *S. salar*, at the fry and parr stages (Fig. 31.2), while other individuals remain upstream, appear further and further downstream long before the initiation of the obligatory downstream migration characteristic of the later (smolt) stage of most demes of this species. This early downstream migration is usually considered to be performed by those fry and parr of low resource-holding power. Similar downstream migration is performed by the smaller fry of the brown trout, *Salmo trutta*, while larger fry of the same age remain upstream (Mills 1971).

Downstream migration thresholds also seem to have evolved in response to physical factors such as pollution, drought and extremes of temperature. Thus in some climates, with a pronounced dry season, headwaters and swamps may dry out. Although some fish, such as lung fishes, have evolved the habit of burrowing down into the mud to aestivate (Hynes 1970), the majority initiate downstream migration. Similarly, in cold areas such as the River Amur in Siberia, a number of species, such as the bighead, *Hypophthalmichthys molotrix*, a cyprinid, have more or less complete hibernation, moving into deep pools, aggregating on the river bed, and becoming enveloped in slime (Nikolsky 1963). More generally, however, not only in the Amur but throughout the north temperate zone, species move into deeper, and thus warmer, water for the winter. In many cases, such as brook trout, *Salvelinus fontinalis*, movement to deeper water is movement downstream.

Selection for return migration in these cases does not necessarily result from the advantage of reproducing in the natal site and is likely to act on fish whether or not they are mature. In all of the cases described in the previous paragraph, as conditions upstream recover, a gradient of increasing habitat suitability upstream is established due to the absence of fish from these reaches. Where the change in sign of the upstream–downstream gradient of habitat suitability is seasonal, quite possibly upstream migration is initiated either after a certain stay-time downstream or in response to indirect threshold variables such as photoperiod/temperature. Young rainbow trout, *Salmo gairdneri*, have been shown experimentally to respond to long days by moving downstream and to short days by moving upstream (Northcote 1958). Alternatively, or additionally, upstream migration may be facultative, the threshold variable(s) being some factor(s) carried by the water from further upstream, variation in which correlates with changes in upstream conditions. The upstream migration threshold should correspond to the situation in which $h_d = h_u M_u$ (where $h_d =$ the suitability of the downstream habitat, $h_u =$ the suitability of the upstream habitat, and $M_u =$ the migration factor for upstream migration).

In the majority of cases, upstream migrations seem to be initiated in response to rising water (Hynes 1970), not only in the case of seasonal migrants but also in the case of ontogenetic migrants such as salmon. Not only does rising water indicate an increase in suitability of upstream habitats, particularly following downstream migration in response to drought or high or low temperature, it probably also correlates with a decrease in migration cost. This decrease seems likely to result from change in energy expenditure and predation risk. Rising water correlates with increased width and depth of the streams through which migration is to occur and hence a corresponding decrease in impassable obstacles such as shallows or narrow, shallow, and swift currents. Increased width of streams spreads out the discharge and provides zones of slower water which are nevertheless deep enough to swim through (Hynes 1970). In rapids, even the strongest swimmers such as trout remain close to the bed where the current is slowest. Falls, however, necessitate jumping and many upstream migrants can jump very effectively. Atlantic salmon, *Salmo salar*, can clear 3·3 m (Jones 1959) and small falls and beaver dams are no obstacles to trout (Gard 1961). Leaping fish always take off from the crest of a standing wave where there is an upward component to the turbulence (Stuart 1962). The tail continues

Fig. 31.6 Atlantic salmon, *Salmo salar*

[*Photo by Ronald Thompson (courtesy of Frank W. Lane)*]

to vibrate as they move through the air so that as soon as they land on or above the crest of the fall, at which their aim is visual, they are swimming and hence move rapidly upstream away from the lip of the fall. The stimulus to leap is the impact of the water on the pool and larger fish respond to greater impacts.

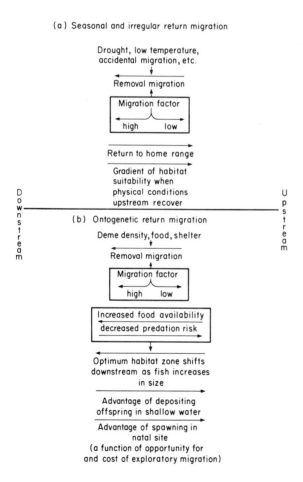

(a) Seasonal and irregular return migration

Fig. 31.7 Summary of the selective pressures acting to produce downstream–upstream return migration with: (a) an irregular or seasonal periodicity; and (b) an ontogenetic periodicity

Horizontal arrows show the direction of movement favoured by the particular selective pressure. Boxes isolate pairs of selective pressures or groups of related effects. Vertical arrows indicate cause and result.

Upstream migration is also influenced by temperature. There seems to be an upstream migration threshold to temperature of about 19°C in the king salmon, *Oncorhynchus tshawytscha*, in the San Joaquin Delta, California (Hallock *et al.* 1970). Most of the salmon observed failed to initiate upstream migration from the delta until the water temperature dropped below this level. Whether, in this case, temperature is a migration cost variable or a habitat variable that correlates with the suitability of the spawning habitat (Chapter 16) is unknown.

Finally, in this section, two other selective pressures may be described that also seem likely to contribute to the evolution of return migration. The first is that there is likely to be a gradient of increasing availability of food from upstream to downstream, a gradient that has been particularly well demonstrated for plankton (Greenberg 1964).

The second selective pressure seems likely to result from a gradient of predation risk. Towards the source of a stream there is a limit to the size of fish that can live in such shallow waters. Apart from the physical problems associated with movement in very shallow water, food is likely to be in short supply and there is the problem of exposure to terrestrial and avian predators. Any fish that is small enough to occupy such a niche, however, gains the advantage of absence of predation from larger fish. Selection due to lessened predation by other fish, therefore, seems likely to favour a given fish living in water as shallow as is consistent with its size, the limit being set by the point at which the advantage gained through decreased predation by other fish is balanced by the disadvantage due to increased problems of movement, finding food, and exposure to terrestrial and avian predators. As, in general, drainage systems increase in size and depth from source to mouth, this selective pressure should result in smaller fish occupying a position further upstream than larger fish. This expectation is generally supported by data on the source-to-mouth size distribution of fish (Hynes 1970) and provides the gradient of selection that seems to have generated the downstream movement of young salmon into larger and deeper stretches of water as the fish themselves increase in size (Mills 1971).

Not only does this gradient favour downstream migration of young fish as they grow, it also favours the very young fish occupying an upstream position in the first place. As it may be assumed that the cost of upstream migration is less for an adult than for a young hatchling it follows that this selective gradient imposes selection for a pre-spawning upstream migration by adults. As long, that is, as the disadvantages of the upstream migration cost for adults is not greater than the advantage they gain as a result of depositing their young in an upstream position.

This gradient seems to be sufficient, upon entry into the return migration model, to account for the fact (Hynes 1970) that almost without exception the direction of any pre-spawning migration performed by fish that spawn in running freshwater is upstream rather than downstream. Where the latter does occur, the movement is still from deeper water to shallower water as when lake-dwelling fish near the lake outlet stream migrate to this stream to spawn instead of to the inlet stream, though the latter migration is perhaps more common.

It can be seen, therefore, that running water as a habitat-type is characterised by a number of selective pressures that seem likely to favour the evolution of return migration. The pressures discussed in this section are summarised in Fig. 31.7.

31.3.3 Migrant/non-migrant ratio

Whether the migration concerned is a removal migration or a return migration, the model developed in Chapter 17 generates a characteristic frequency distribution of migration thresholds. When the animals concerned live in a free situation, the migrant/non-migrant ratio should stabilise when migrants and non-migrants experience the same potential reproductive success. When the animals concerned live in a despotic situation, however, the non-migrants should experience a higher reproductive success than the migrants and the migrant/non-migrant ratio is a function of the distribution of the critical resource(s).

As far as ontogenetic return migration is concerned, the selective pressures summarised in Fig. 31.7 seem to have led in the vast majority of fish that breed in running water to 100 per cent of the deme performing a return migration. The variable component of the migration seems to be distance rather than the migrant/non-migrant ratio. Where this is not so and individuals, such as of various grayling, *Thymallus* spp., reproduce in the area in which they developed to maturity (Mills 1971), the behaviour can be attributed to short-distance variations in river depth providing optimum conditions for feeding and reproducing within the non-spawning home range of a single individual.

If a time interval shorter than the period from birth to maturity is taken, however, some behaviour is encountered that is best described by the migrant/non-migrant ratio. Thus the young stages of many salmonids, although they gradually move down-river as they grow, at each stage in this migration establish themselves within a territory for a period ranging from days to months. The downstream migration that does occur during the 1–8 years or more that a young Atlantic salmon, *Salmo salar*, spends in freshwater before migrating to the sea (Fig. 31.2) is not really part of the ontogenetic return migration and some individuals may migrate very little during this period. The situation seems to be despotic and we may assume that the advantage lies with those individuals that move least during this period. Presumably, during the evolution of the habit, individuals that initiated downstream migration up to a certain ontogenetic stage after hatching were at a disadvantage to those that did not. Beyond this stage, however, those that initiated downstream removal migration were at an advantage. The action of selection would be to favour a downstream migration threshold to which relative resource-holding power made a major contribution (p. 403) up to the appropriate ontogenetic stage and then linkage of obligatory initiation of downstream migration with the attainment of that stage (Chapters 8 and 16). The selective pressures acting on the timing of migration are discussed in Section 31.3.6.

31.3.4 Direction and direction ratio

The linear nature of the running-water environment inevitably means that any selection on dispersal acts primarily on the frequency distribution of migration distance rather than on direction ratio. The result is that selection on direction ratio tends to favour either 100:0 or 50:50, depending, respectively, on whether the migration is or is not part of a return migration.

As far as return migration is concerned, the selective pressures summarised in Fig. 31.7 are constant with respect to direction. Seasonal, irregular, or accidental migration involves downstream migration when conditions become adverse followed some time later by an upstream migration. Ontogenetic return migration involves downstream migration when young followed by upstream migration to reproduce. Where reproduction continues seasonally over several years, inter-spawning return

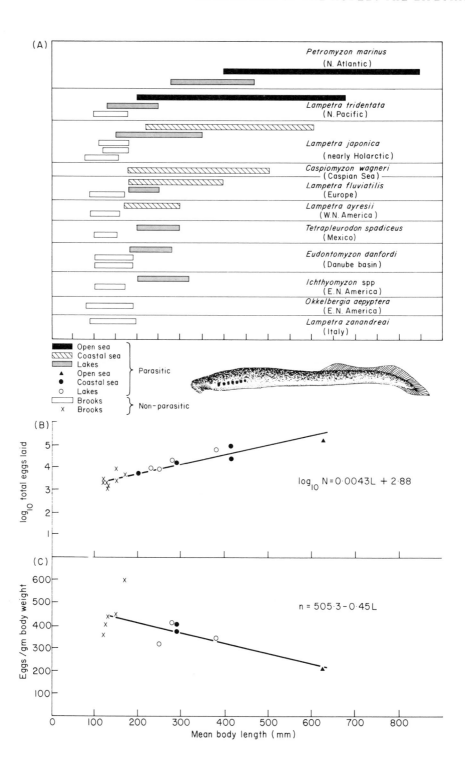

Fig. 31.8   Data relating to selection on migration distance in lampreys

The model
The model for migration distance (Chapter 11) suggests that selection should favour the migration distance for which $h_d M_R$ is maximum (where h_d is the suitability of the furthest destination reached by an individual and M_R is the migration factor for the complete return migration between that destination and the habitat of origin). In a despotic situation, individuals with the highest resource-holding power settle in the destination(s) for which $h_d M_R$ is maximum, other individuals migrating shorter or longer distances. In a free situation, selection stabilises when the frequency distribution of migration distances in a deme is such that all individuals experience the same value of $h_d M_R$.

The model applied to stream-breeding fish
The situation within a drainage system may be either despotic or free. In lampreys, which as young are filter-feeders, the situation is probably free. In salmonids, which are predatory, the situation is more likely to be despotic but may be free. Once in the sea, the situation for stream-breeding fish is likely to be free. In both despotic and free situations, the advantage of longer-distance migrations is generated by gradients of availability of resources, especially food and sites for sheltering from the current, and is hence in part a function of deme density. Within a drainage system, lakes and the river/sea boundary are likely to impose evolutionary brakes on selection for further increase in the maximum migration distance shown by some members of a deme. In those lineages for which selection generates adaptation to marine conditions and the ability to navigate back to the natal river, the possibility exists that $h_d M_R$ for migration to the sea is greater than $h_d M_R$ for any point in the drainage system. In which case, whether the situation in the drainage system is free or despotic, selection favours 100 per cent of the deme migrating to the sea. Alternatively, if $h_d M_R$ for migration to the sea is greater than $h_d M_R$ for any point in the drainage system except lakes, selection favours that distribution of deme proportions that migrate only to the lake and the sea respectively that render $h_d M_R$ the same for migration to the lake(s) and to the sea. In almost all cases, therefore, except where a despotic situation still exists, it should be found that $h_d M_R$ is the same for all migration distances found in a deme. For this situation to exist, those individuals that migrate furthest, thus expending more energy and, presumably, experiencing a higher mortality risk (i.e. lower M_R) should encounter a more suitable habitat (i.e. higher h_d) that gives greater nett energy uptake. The greater energy input should just balance the combined effects of greater energy output during migration and greater mortality risk. We would expect, therefore, that longer-distance migrants should attain a greater body size and potential fecundity (due to greater energy input), should have a lower efficiency of conversion into gametes of energy reserves accumulated in the feeding habitat (due to greater energy expenditure in migration), and finally should have a higher absolute fecundity that just balances the effect of higher total mortality risk.

Evaluation of the model on the basis of data for lampreys
A. Range of body size of species and demes with different migration distances

Demes of parasitic lampreys that migrate to the open sea, coastal waters and lakes, and demes of non-parasitic lampreys that stay in streams and rivers form a sequence of decreasing migration distance and decreasing body size. In the diagram, the species name refers to the parasitic species of the species pair (see Fig. 31.3) or species group where such a parasitic species exists. The open bars then refer to the non-parasitic species that is or are considered to share common ancestry with the parasitic species. Body size refers to length in millimetres of individuals just about to spawn.

B. Regression of total eggs laid against body size with an indication of migration distance

The regression (where N = total eggs laid and L = body length in mm) is a highly significant fit to the data ($F_{1,15}$ = 122·6; $P < 0·001$) with a coefficient of determination of 87 per cent. The coefficient of determination is even higher (R = 90 per cent) if $\log_{10} N$ is regressed on $\log_{10} L$. This regression is in agreement with the expectation of the model in that longer-distance migrants have a higher fecundity than shorter-distance migrants. Whether this fecundity is just that required to balance the increased mortality risk in the determination of the value of $h_d M_R$ is unknown.

C. Regression of eggs laid/g of body weight against body length with an indication of migration distance

The regression (where n is the number of eggs/g of body weight) is significant ($F_{1,9} = 8·22$; $P < 0·05$) and has a coefficient of determination of 48 per cent. This regression is in agreement with the expectation of the model in that it suggests that longer-distance migrants have a lower efficiency of conversion of energy reserves to gamete production. Whether this is the result of the energy expenditure component of migration cost as required by the model is unknown.

Given a relationship between body size and migration distance as indicated in A, all of the expectations of the model are manifest by lampreys.

[*Raw data from Hubbs and Potter (1971) and Hardisty (1971)*]

migration involves downstream migration by large post-spawning fish to feed-up followed by upstream migration to spawn. Because in all these migrations selection favours either a high degree of return or an upstream migration distance that on average is the same as the downstream migration distance, and because, in view of the linearity of the environment, selection on dispersal does not act on direction ratio, optimum direction ratio is 100:0.

31.3.5 Distance

According to the model developed in Chapter 11, the one-way migration distance for a return migration is fixed by the point at which $h_d M_R$ is maximal (where, in the present case, h_d is the suitability of the downstream habitat and M_R is the migration factor for the upstream–downstream return migration). In a free situation, selection favours a frequency distribution of migration distances such that all individuals experience the same advantage from the migration as a result of those individuals that migrate further (i.e. lower M_R) encountering more suitable habitats (i.e. higher h_d). In a despotic situation, individuals with a higher resource-holding power establish themselves at the distances at which $h_d M_R$ is greatest while individuals with a lower resource-holding power perform longer and/or shorter distance migrations to habitats for which $h_d M_R$ is less. Some individuals may even be forced to migrate such distances that M_R and/or h_d approach zero, as in the experiment on cutthroat trout, *Salmo clarkii*, described on p. 796.

To my knowledge, analysis of the frequency distribution of migration distance for fish in the same way as for a bird (p. 652) is not possible with the data available. However, data available for lampreys (Fig. 31.8) are consistent with the patterns generated by the model developed in Chapter 17 and it is assumed here that the migration distances of fish (see Table 31.2) have evolved in response to the selective pressures included in that model.

Superimposed on the gradual increase in habitat suitability from source to mouth along the length of a drainage system there may be areas of marked increase in suitability due to the existence of two particular habitat-types. These are lakes and the sea. Let us first consider the influence of lakes on the optimum migration distance of fish moving down from further upstream. If deme density is low, the lake is likely to offer the maximum value for $h_d M_R$ for all individuals. Fish migrating beyond the lake are likely to experience lower values for both h_d and M_R. Fish terminating migration before they reach the lake experience a higher value for M_R but a lower value for h_d. Not unless deme density in the lake is high enough to lower habitat suitability to near that of the inlet and outlet streams, therefore, does selection favour fish that settle anywhere but in the lake. The result is that lakes, where they occur, serve as a form of evolutionary brake on selection for longer-distance migration. Only if deme density increases to the level at which lakes are no longer more suitable for feeding than the streams and rivers that lead into and drain from them should selection favour fish that migrate beyond the lake system during their downstream migration. It is perhaps significant, therefore, that where land-locked demes of otherwise anadromous species occur they invariably are found to terminate downstream migration in a lake.

The freshwater/sea boundary probably also acts as an evolutionary brake on selection for an increase in the distance component of migration. Individuals that reach the sea during downstream migration inevitably are those members of the deme that have the greatest migration distance and hence the smallest migration factor. In a despotic situation they are also the individuals with the lowest resource-holding power. In a free situation such long-distance migrants persist only if there is an increase in habitat suitability that has the potential to compensate for the low migration factor. If we are dealing with fish the immediate ancestry of which was in freshwater, as seems likely for the salmonids (Tchernavin 1939), the physiological problems associated with the initial entry (in an evolutionary sense) into brackish and saline water most undoubtedly have contributed to a low habitat suitability. To this extent, therefore, the sea formed and forms an evolutionary brake on further increase in migration distance, especially where the river runs directly into the sea with little or no estuary. Only if some other factor contributes to an increase in habitat suitability through the river/sea transition that is greater than on the one hand the contribution of physiological problems to a decrease in habitat suitability and on the other hand the ever-decreasing migration factor due to increased migration distance, can adaptive extension of downstream migration into the sea take place. Food availability per unit volume and total food availability seem to be two such factors.

As long as selection acts to favour return migration to the natal site to spawn, entry into the sea imposes even further selection on the frequency distribution of migration distance. The further from the mouth of the natal river that a fish (or a shoal of fish) may swim following downstream migration, the more it is likely to increase its nearest-neighbour distance and hence increase h_d. Apart from the contribution to a decrease in M_R of the increased energy expenditure involved, however, the further from the river mouth that a fish swims the less likely it is to be able to return to its natal site. This leads to further reduction in M_R. If M_R decreases more than h_d increases, selection favours behaviour patterns that maintain the fish in sensory contact with the water from the natal river. Only if the increase in h_d is greater than decrease in M_R with distance from the river mouth, due perhaps to a marked increase in food availability or due to the influence of further

increase in deme density near the river, is selection imposed for increasingly sophisticated navigation mechanisms.

It is possible, however, that beyond a certain stage of sophistication of the navigation mechanism, at the point perhaps at which a fish becomes capable of returning to its natal river from any distance and any position (see Chapter 33), h_d increases per unit distance at a faster rate than M_R decreases. This could occur when the vast food resources of the sea suddenly become available. If the result is that $h_d M_R$ is higher for fish that enter the open sea than $h_d M_R$ for fish that migrate only within freshwater or to the sea around the mouth of the natal river, then individuals that migrate to the open sea will increase in number relative to those that migrate within freshwater. If $h_d M_R$ for the open sea continues to be much greater

Table 31.2 Variation in the range of migration distance in some species of salmonids†

Species	Range‡ of migration distance within species						Region
	open sea	coastal sea	rivers	streams	lakes	streams	
pink salmon *Oncorhynchus gorbuscha*	—	—	*				N. America
chum salmon *Oncorhynchus keta*	—	—	*				N. America
coho or silver salmon *Oncorhynchus kisutch*	—	—		*			N. America
chinook or king salmon *Oncorhynchus tshawytscha*	—	—	—	*			N. America
sockeye and kokanee salmon *Oncorhynchus nerka*	—	—		* L	* —	*	N. America
Japanese salmon *Oncorhynchus masu*	—	—	—	*			Japan
Atlantic salmon *Salmo salar*	—	—	—	*	—	*	N. Atlantic
steelhead, rainbow, and Kamloops trout *Salmo gairdneri*	—	—	* —	*	—	*	N. America
sea and brown trout *Salmo trutta*		—	* —	* —	—	* —	Europe
cutthroat trout *Salmo clarkii*		—	* —	* —	—	* —	N. America
brook trout *Salvelinus fontinalis*		—	* —	* —	—	* —	N. America
Arctic char *Salvelinus alpinus*		—	* —	L	* —		circumpolar
Windermere char *Salvelinus willughbii*					* —	*	England
European grayling *Thymallus thymallus*		—	* —	* —	* —	* —	Europe
Danube salmon *Hucho hucho*		—	* —				Europe

† Data from Tchernavin (1939), Hoar (1958), Orr (1970a), and Mills (1971).
‡ Two types of downstream–upstream migration are indicated: (a) the stream–river–sea–river–stream type; and (b) the stream–lake–stream type
— indicates that some individuals of the species complete the major part of their growth in the habitat-type
* indicates that spawning occurs in the habitat-type
L indicates that within an essentially stream–river–sea–river–stream migration, some individuals also spawn in lakes

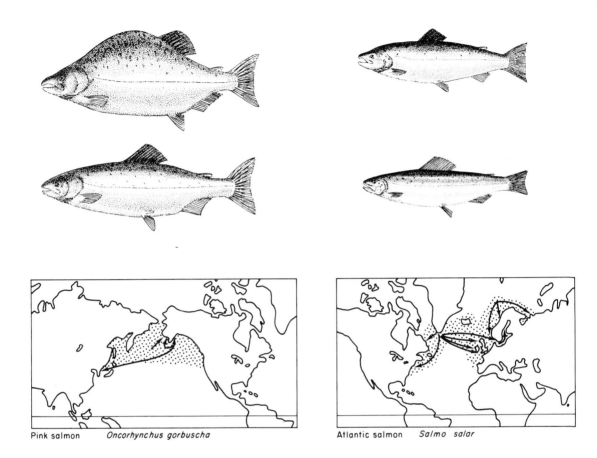

Pink salmon   *Oncorhynchus gorbuscha*

Atlantic salmon   *Salmo salar*

Fig. 31.9   Migration distance of two species of salmonids within the sea

The majority of species of salmonids exhibit a range of migration distances such that within a deme the shorter-distance migrants remain permanently within freshwater whereas the longer-distance migrants enter the sea but remain within the influence of the natal drainage system (Table 31.2). A few species of salmonids, however, have become adapted to the open sea and have evolved the ability to return to the natal drainage system from distances as great as at least 4000 km and perhaps even 8000 km. The maps show the results of some marking–release/recapture experiments on two species of salmonids and indicate the range of distances travelled by the species concerned while at sea. The maximum migration distance shown (approximately 7800 km) is that of a pink salmon that was marked in the Gulf of Alaska and recovered in the Sea of Japan, opposite Korea.

Arrows have been drawn from the open sea capture site to either the natal stream or the stream in which the individual performed a spawning run. Arrows are not intended to imply a migration route. The stippled area indicates the range of the species during the open sea phase.

[*Data from Tchernavin (1939), Hasler (1966), Harden Jones (1968), and Mills (1971)*]

Fig. 31.10   Distribution at sea in relation to ocean currents of different demes of four species of salmonids

The solid, dashed, and dotted lines in the top three diagrams mark the limits of distribution at sea of three demes for each of three species of Pacific salmon. The demes indicated are those fish that spawn in the rivers of Asia (solid line), Alaska (dashed line), and the remainder of the west coast of North America (dotted line). In the bottom diagram, the circles and crosses mark similar limits

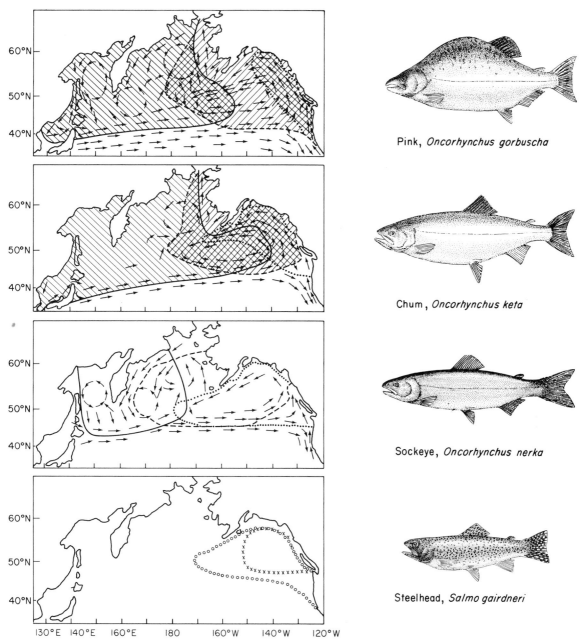

Pink, *Oncorhynchus gorbuscha*

Chum, *Oncorhynchus keta*

Sockeye, *Oncorhynchus nerka*

Steelhead, *Salmo gairdneri*

for two demes of the anadromous sub-species of the rainbow trout, *Salmo gairdneri*. This anadromous sub-species is usually referred to as the steelhead trout. The two demes indicated are those fish that spawn in the rivers of British Columbia (crosses) and Washington, Oregon, and California (open dots). The black edging shows the range of coastal distribution of the species. In the top three diagrams the surface ocean currents are indicated. These are shown in greatest detail in the top diagram. In the other two diagrams, only those currents that seem most relevant to the distributions are shown.

The distribution of salmon at sea seems to be consistent with the thesis that is presented in Section 31.4.1 for full-time marine fish. At sea, migration circuits consist of migration routes linking geographically separate spawning areas and winter and summer feeding areas. These circuits have evolved to a large extent in response to selection for migration direction that results from the unidirectional component of ocean currents.

[*Distribution maps modified from Hart (1973). Ocean currents from Harden Jones (1968). The latter author also explores further the relationship between salmon migration and ocean currents*]

than that for freshwater, those that migrate within freshwater will eventually disappear and the result will be a species that, upon attainment of the appropriate stage, performs an obligatory migration to the sea. Table 31.2 shows that only one species has reached this situation in the Atlantic, though in the Pacific the six species of *Oncorhynchus* consist almost entirely of individuals that migrate to the open sea. Even in these species, however, there appears to be considerable variation in migration distance while in the sea (Figs. 31.9 and 31.10).

Figure 31.31 indicates that there is strong latitudinal variation in the proportion of a deme that migrates to the sea. The maximum expression of the sea-going habit is found in north temperate and subarctic latitudes and again, though to a lesser extent, in south temperate latitudes. There is minimum expression of the habit in tropical regions. It seems from Fig. 31.31 that selection on fish to increase migration distance beyond the river/sea boundary and on into the open sea is greatest in temperate and sub-polar latitudes and least in tropical and sub-tropical latitudes. This latitudinal variation is discussed further in relation to the anadromy/catadromy paradox.

So far, this section on migration distance has been concerned primarily with distance in a qualitative rather than a quantitative sense. Such an approach has obvious descriptive advantages and is probably adequate to permit various broad conclusions to be reached. Yet rivers do vary in length and seas do vary in size. Obviously absolute migration distance is an important component of the selective pressures discussed in this section. In general, however, variation in absolute migration distance raises no new problems and is as in accord with the model discussed here as is variation in migration to and from qualitatively recognisable areas of a drainage system. Nevertheless, there is an inherent interest in the absolute distances moved by animals, especially when they are as great as those covered by migratory fish. Even the distances covered within freshwater may be quite spectacular. Some Pacific salmon, *Oncorhynchus* spp., for example, may swim 4000 km from first entry into the estuary of a drainage system to the spawning grounds (Hasler 1966). Among lampreys (Fig. 31.8), *Caspiomyzon wagneri* goes 2000–4000 km up the Volga. Sturgeons may swim 2000 km up the Danube and white salmon, *Stenodus leucicthys*, may swim 2800 km up the Volga, a journey that requires an entire year to complete (Hynes 1970). In this last species, the fat content drops from 21 per

cent at the beginning of the journey to 2 per cent at the end.

Inevitably, maximum migration distance varies according to the length of the river and the distance from the sea of suitable spawning grounds. Thus in the USSR, the omul, *Coregonus autumnalis*, goes 1600 km up the River Lena but only 500 km up the River Jana (Hynes 1970).

31.3.6 Timing

In temperate regions, fish that perform downstream–upstream migration and which spawn in freshwater tend to spawn either in autumn or in spring with a tendency for stream-spawners to spawn in autumn and for lake spawners to spawn in spring. Doubtless the timing of spawning of a particular species in a particular stream or lake is adaptive to seasonal cycles in availability of food for the offspring. In streams, minimum biomass tends to occur in late spring and early summer (Hynes 1970) whereas in lakes minimum biomass tends to occur in mid-winter with maxima in spring and autumn.

The timing of downstream migration by the young fish could probably also be attributed to seasonal cycles of food availability and the attainment of a certain size and strength suited to the downstream destination. Thus young salmon and trout gradually move downstream as they grow. The smolt transformation of species of *Salmo* and *Oncorhynchus* (Hoar 1958), however, does not take place until a certain size has been attained (Fig. 31.2). Once such a size has been achieved, however, migration to the lake or sea then occurs in spring at a time when biomass in the stream is decreasing and hence the advantage of stream-to-lake or stream-to-sea migration is at a maximum. Thus salmon smolt migrate to the sea in late spring and early summer (April–June—Mills 1971) as do lampreys (Fig. 31.3) and striped bass, *Roccus saxatilis* (Fig. 31.5).

Part of the speciation process in the genus *Oncorhynchus* (Table 31.2) seems to have involved not only adaptation to different niches within the stream (Hoar 1958) but also adaptation to drainage systems with and without lakes by an ancestor that had already evolved the habit of 100 per cent migration to the sea. Suppose that we accept the slightly more probable alternative that such a fish evolved in drainage systems without lakes. Upon entry into drainage systems that included lakes, those individuals during their young (fry and parr) stages that migrate downstream before the obligatory migration encounter the high habitat suitability of lakes (p. 802). If such individuals are at an advantage to those fry and parr that do not migrate until the obligatory migration, selection is likely to favour two obligatory downstream migrations, one from the spawning site to the lake and the second from the lake to the sea. In the absence of lakes downstream of the spawning site, however, selection continues to favour delay of an obligatory downstream migration until an appropriate size and strength has been attained. The result seems likely to be selection for linkage on the one hand between spawning well upstream beyond any lake system and two ontogenetically obligatory downstream migrations by the young salmon and on the other hand between spawning either in systems without lakes or on the downstream side of lakes. Such a situation could well lead to speciation. In the genus *Oncorhynchus*, three species, the silver or coho salmon, *O. kisutch*, the king or chinook salmon, *O. tshawytscha*, and the sockeye salmon, *O. nerka*, spawn well upstream (Orr 1970a) and the latter species in particular seems to be especially adapted to drainage systems containing lakes (Hoar 1958). The young sockeye fry migrate downstream soon after hatching until they enter the lakes in which they develop to the smolt stage. Two other species, however, the chum salmon, *O. keta*, and the pink salmon, *O. gorbuscha*, spawn in the lower reaches of drainage systems within about 200 km of the sea. The fry of these species only feed in freshwater for a few weeks before initiating downstream migration to the sea (Orr 1970a).

Selection on the time interval between downstream migration as a young fish and the upstream spawning migration seems likely to consist of a number of selective pressures. We may assume that increase in this time interval increases body size, increases number of gametes, decreases the influence of upstream migration cost on reduction in number of gametes, and for males increases the ability to gain sole reproductive access to one or several females. All of these selective pressures act to favour an increase in the time interval between downstream and upstream migration. In addition, however, increase in the time interval is likely to increase mortality risk and to increase the generation interval (Cole 1954), both of which act to favour a decrease in the time interval. We may expect that the frequency distribution of time intervals that occurs in any one species (Fig. 31.2) reflects the optimum compromise between these opposing selective pressures.

An intriguing polymorphism has evolved in the male of the Atlantic salmon, *Salmo salar*. Here, as in most salmonids, the female digs or 'cuts' a pit in the sand and gravel in the bottom of the stream by turning on her side and creating a negative pressure that lifts stones which are then carried downstream by the current. The negative pressure is created by vigorous flapping of the tail. Female and male then spawn in the pit or 'redd' thus formed after which the female moves forward and digs again, thus burying the newly laid eggs. The size and shape of the hole formed by this digging is such that neither male nor female salmon can get their vents right down to the bottom. Nevertheless, some of the eggs fall there and are much less likely to be fertilised by the sperm from a fully grown male than are the rest of the eggs. The existence of this group of often unfertilised eggs seems to have imposed selection on the range of variation of the time interval between downstream and upstream migration of male salmon (Jones 1959). Some males become sexually mature and migrate upstream to the spawning grounds while still parr (Fig. 31.2). These little males can swim right down into the redd and ejaculate close to the eggs, thus fertilising these bottom eggs and perhaps some of the others as well. It seems likely that the proportion of males with this short migration interval is such that their reproductive success is the same as those males which spend a year or more at sea, run a greater mortality risk, have a longer generation interval, but grow to a larger size.

As far as the timing of the sea-to-coast migration and entry into the river is concerned we might expect that selection will favour those individuals that link the behaviour with appropriate environmental clues that correlate with season such that, with the minimum upstream migration cost, they arrive at the spawning site at the optimum time for spawning. As rivers vary, not only with respect to optimum time for spawning, but also with respect to distance from mouth to source, it is to be expected that the demes inhabiting different rivers will show different timing. Thus the Atlantic salmon, *Salmo salar*, usually spawns between November and February. Even within the British Isles, however, there is considerable variation in the time that the salmon enter the various rivers. Thus the rivers Tweed, Tay, Findham, Spey, Invernessshire Garry and Ness have runs of fish in early spring while others, such as those on the west coast of Scotland, may have very few spring fish and the first large runs only start to enter the rivers in summer in June and July (Mills 1971).

In the Tweed, there is also a large run of autumn fish which continues until December. Salmon pass through the tidal reaches of the rivers Taw and Torridge in Devonshire in early winter (December). There is an early run of the same species in St. John River, New Brunswick, United States that occurs almost a full year before spawning.

Finally, the selective pressures acting on the repetition of upstream–downstream return migration may be considered. As a fish spawns it is in effect partitioning its energy resources between reproduction and continued survival. Increase in the energy put into gamete production and reproductive behaviour increases the immediate realisation of potential reproductive success but decreases the energy available to increase the probability of survival and future realisation of potential reproductive success. If the probability of surviving until the next spawning season is zero, the optimum strategy should be to invest every available quantum of energy into present reproduction. If the probability of surviving to the next spawning season is high, optimum strategy should be to invest as much energy into present reproduction as is compatible with minimum reduction in the probability of surviving to the next spawning season. In short, selection should favour that partitioning of energy between present reproduction and continued survival that maximises reproductive success where optimum partitioning is likely to be a function of probability of survival to the next spawning season.

Variation in the partitioning of energy reserves in salmonids seems to fit this model. Thus long-distance freshwater to open sea return migrants that would be likely to have a low probability of survival to a further spawning invest all or nearly all of their energy reserves in their first spawning and consequently die without returning to the sea. All members of the genus *Oncorhynchus*, the Pacific salmon, die after their first spawning and about 74 per cent of the individuals of the Atlantic salmon, *Salmo salar*, do so. The massive Danube salmon, *Hucho hucho*, on the other hand, which clearly has a high probability of survival, may continue to spawn for 20 years or so.

Associated with this question is that of feeding during upstream migration. A large fish adapted to feeding in the sea, if it attempted to feed during upstream migration could well expend more energy searching for food in freshwater than it obtains in the form of food captured. Upstream migration cost may well be minimised by not investing time and energy in searching for, capturing, and digesting food but instead by concentrating effort on efficient ascent. Smaller fish, adapted in any case to feeding on freshwater organisms, may well reduce migration cost by feeding during migration. Pacific salmon, *Oncorhynchus* spp. and the Atlantic salmon, *Salmo salar*, as well as all anadromous lampreys apparently do not feed during upstream migration despite the fact that, in the case of salmon, they still take bait offered by an angler. Sea trout and brown trout, *S. trutta*, and brook trout (Fig. 31.11) on the other hand, which migrate less far and continue to spawn for several years, continue to feed during upstream migration (Mills 1971) as do the vast majority of fish that migrate totally within freshwater.

Fig. 31.11   Brook trout, *Salvelinus fontinalis*

[Photo by W.T. Davidson (courtesy of Frank W. Lane)]

31.4 Fish that breed in the sea (with a note on cephalopods)

31.4.1 Fish that migrate entirely within the sea

Despite the immense amount of research that has been undertaken into the movements of marine fish, there is little information available that is suitable for the type of evaluation being given in this book.

Not all marine fish perform long-distance onto-genetic/seasonal return migrations, but the majority probably do so (Nikolsky 1963). Most of the fish that do not, such as many coral fish of the families Pomacentridae, Siganidae, Apogonidae, and others, probably spend most of their lives within a limited home range and perform type-H removal migration (p. 308). Others, particularly coastal species, such as gobies, may perform short-distance inshore–offshore seasonal or ontogenetic return migrations (p. 830) with varying degree of return. This section, how-ever, is concerned with those longer-distance return migrations of pelagic fish such as herring, *Clupea harengus* (Fig. 31.12), cod, *Gadus morhua* (Fig. 31.13), and tuna, *Thunnus* spp. (Figs. 31.14 and 31.15) and others, such as the bottom-living plaice, *Pleuronectes platessa* (Fig. 31.16).

Conceptually, it is most convenient to picture marine fish as performing a migration circuit rather than a to-and-fro migration. A migration circuit need not be circular and may be one of any of a wide variety of possible configurations. Essentially, how-ever, a migration circuit for a marine fish involves all or some of the following: a spawning area; nursery area; feeding area(s); summer area; winter area; assembly area; and so on, joined together by migration routes and visited in more or less a strict sequence that for the deme concerned remains constant, or at least changes only slowly, from year to year and generation to generation. The popu-lation of such a species can usually be divided into a number of demes where each deme has its own migration circuit. The migration circuits of neigh-bouring demes may overlap to any degree from not at all to considerably in both a spatial and a temporal sense (p. 812). Inevitably, also, there may be removal migration (type-C, p. 308) of individuals from one deme to another such as that

described for the Atlantic herring, *Clupea harengus*, on p. 197.

For most species, the environmental features are unknown that cause one part of an area to be more suitable for spawning etc. than other parts and at present we can only accept that habitats do vary in suitability and that fish tend to exploit those that are the most suitable. Given this, however, the action of selection seems likely to have been to favour those individuals that arrived at the appropriate season in the most suitable area for the optimum behaviour. Individuals that happen to experience the best sequence of habitats with the best timing of arrival and departure and the best timing of ontogenetic and seasonal physiological events, such as spawning, will increase in number relative to other individuals. No matter whether one takes a spatially randomly distributed group of individuals as an arbitrary evolutionary starting point or whether one considers a steady influx of removal or exploratory migrants into a new area (as seems likely to have been the case for the cod, *Gadus morhua*, off the west coast of Greenland—Fig. 31.13), the result should be the rapid establishment and growth of a deme that performs an adaptive migration circuit.

Most of the arguments concerning the evolution of downstream–upstream return migration by stream-living fish seem also to be relevant to marine fish. It seems likely, therefore, that at least some marine fish have evolved a downcurrent–upcurrent return migration similar to that of stream-living fish. In the sea, however, current systems have other properties in addition to those they share with the current systems of streams and rivers. In particular, most areas of the World's oceans are characterised by the presence of gyrals or counter-current systems. In the former case, currents go round in a circle. In the latter case, although the current at one depth at one place may be going permanently in a given direc-tion, some distance to the side of that current or at a different depth directly above or below that current there may be another going in exactly the opposite direction. Consequently, the possibility presents itself to marine fish, that does not present itself to stream or river fish, of performing a return migration that throughout its length is in the same direction as the current.

Any fish that spends a notable proportion of its time swimming is liable to displacement by ocean currents. In order to maintain position relative to the sea bottom, an individual has to expend time and energy swimming against the current (p. 793) or by

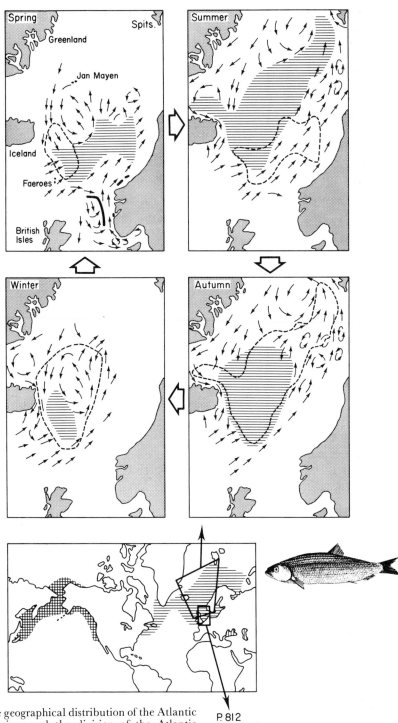

Fig. 31.12 The geographical distribution of the Atlantic and Pacific herrings and the division of the Atlantic herring, *Clupea harengus*, into a number of demes, each with its own migration circuit

P. 812

General ecology

The world map shows the distribution of the Pacific herring *C. pallasi* (cross hatching) and the Atlantic herring, *C. harengus* (horizontal hatching). The Atlantic herring may grow to 40 cm and has a dark blue back with silvery side and belly. It feeds mainly on plankton, particularly copepods, euphausids, and mysids. Both Pacific and Atlantic herring have demersal eggs and pelagic larvae, but whereas the Pacific herring spawns only in spring and comes right inshore to oviposit on seaweed at depths between tide marks, the Atlantic herring may spawn at any time of year and does so on shingle or gravel beds in depths from 40 to 200 m, sometimes on banks far from the shore. Herring show a daily vertical return migration that involves a rise to the surface at sunset, a midnight sinking, and a dawn rise, followed by a return to deeper water during the day (see p. 496).

The division of Clupea harengus into demes

Clupea harengus may be divided into demes at a number of levels. First, there is a division into demes of the northwest and northeast Atlantic. The northeast Atlantic deme may then be further divided into six major demes between which there is little removal migration. These demes are the Atlanto-Scandian, Hiberno-Caledonian, English Channel, North Sea, Skagerrak, Kattegat, Sound and Belts, and the Baltic. To this list may be added a number of demes that occupy estuaries or fjords, or are otherwise local. Within each of these major demes a number of smaller demes may be recognised, the members of which spawn in the same area at the same time and share the same or similar migration circuit. Removal migration between these spawning demes is likely to be greater than between the major demes but does not seem to be sufficiently great to invalidate the recognition of the deme as a biological unit. The migration circuits of four spawning demes are illustrated. One of these, the Norwegian spring-spawning deme, is a member of the Atlanto-Scandian deme, and the other three, the northern North Sea summer-, central North Sea autumn-, and Southern Bight winter-spawning demes, are members of the North Sea deme. The latter three illustrate the way that the migration circuits of spawning demes may overlap in space and time.

The annual migration circuit of the Norwegian spring-spawning deme

The upper sequence of four diagrams shows the winter, spring, summer and autumn distribution of mature herring in the Norwegian Sea. The solid black shading in the spring diagram indicates the spawning areas. The horizontal shading shows the approximate distribution of herring during the season indicated and the dashed line indicates the distribution of herring during the previous season. The fish have therefore moved from the position indicated by the dashed line to the position indicated by the shading. The arrows show the surface water currents in the relevant part of the Norwegian Sea.

In late winter and early spring, herring arrive off the Norwegian coast approximately mid-way between the northern and southern spawning areas. The coast is reached about 5–6 weeks before the fish begin to spawn.

Spawning commences in February and continues to early April. The distribution shown in the spring diagram probably consists of post-spawning as well as pre-spawning fish. The horizontal hatching is not continued onto the spawning areas to avoid obscuring the ocean currents. After spawning the herring appear to move northwest and west, perhaps along the western branches of the Atlantic current, and enter the cyclonic current system of the Norwegian Sea. Selection for such a migration circuit seems to arise from the seasonal changes in the production and distribution of food in this area. Biological spring spreads north and northwest (Fig. 20.4) across the Norwegian Sea along the line of the branches of the Atlantic current. Possibly, therefore, the herring migrate from region to region with the current, keeping pace with the shift of plankton production, and feed on the pre-spawning concentrations of *Calanus finmarchicus* (see Fig. 20.10) and euphausids, and reach the polar front at the time of biological spring (Fig. 20.4A) when *C. hyperboreus* is abundant in the cold water. During the summer the herring are in the 30–40 m below the surface of the sea and the shoals do not make any extensive daily vertical movements (see p. 496). Most fish stay above the thermocline on the warm side of the transition area between warm Atlantic and cold Arctic waters, though, if food is particularly abundant there, the herring may enter the polar water. Not only the distribution shift but also tagging experiments have shown that a migration to the southwest takes place from summer to winter, beginning in August/September, and in general following the course of the East Icelandic current. During this migration the shoals appear to make extensive daily vertical return migrations from about 300 to 400 m depth during the day to near the surface at night. Before their arrival in Norwegian coastal waters in January, the herring shoals cross the North Atlantic current that flows to the northeast. Possibly this is achieved at depth within the East Icelandic current which north and east of the Faeroes may dive under the North Atlantic drift. Norwegian spring-spawning herring are recovered at Norway year after year and it seems likely that the deme performs an annual migration circuit. It is by no means certain, however, that this is the case for all spawning demes.

Although Norwegian spring-spawning herring may perform an annual migration circuit, the young do not enter this circuit immediately upon hatching. Many of the young herring spend their first year in the fjord or coastal waters between the spawning areas and the coast of northern Norway, the site depending on the distance of drift during their first months of life. Others develop in more open waters. Immature herring perform an inshore–offshore return migration (see p. 837), migrating inshore for the summer and offshore for the winter. This return migration is repeated each year until the immature fish reach a certain critical length, which may decrease with age. Then they leave the coastal waters and do not return until they are ready to spawn, one or two years later. Immature herring are found all over the Norwegian Sea, particularly to the southeast of Jan Mayen. Recruitment to the spawning shoals may take place any winter from the fourth to seventh year of life. The available evidence supports the hypothesis that herring spawn in the same season as that in which they were born and that spawning takes place at least in the natal area and possibly also on the natal ground, though there is no evidence relating to whether spawning occurs at the natal bed. In general, therefore, it seems as though the young recruit to the same spawning and migration circuit deme as that of which their parents were a member.

The annual migration circuit of three spawning demes of herring in the North Sea
The bottom three diagrams illustrate the migration circuits of three demes of mature herring in the North Sea. These demes are: (a) the northern North Sea deme, which spawns in summer and autumn; (b) the central North Sea deme, which spawns in autumn; and (c) the Southern Bight deme, which spawns in winter. In the diagrams, the solid black indicates the spawning areas and the horizontal hatching indicates the wintering areas. In (c), the deme winters and spawns in the Southern Bight. The dotted outline in the middle of the North Sea shows the approximate position of the Dogger sandbank. The long arrows indicate probable migration routes, though in (c) there is only a little evidence that the migration proceeds northwards on the west side of the Dogger Bank rather than on the east side. Short arrows indicate surface ocean currents.

There is considerable spatial and temporal overlap of the three migration circuits and it is certain that all three share a common feeding area in the northern part of the North Sea. Nevertheless, although deme to deme removal migration does occur, the available evidence suggests that the majority of herring return to spawn with the group to which they recruited when mature. Furthermore, other evidence, although as yet not conclusive, tends to support the hypothesis that the majority of herring recruit to the deme that spawns on their natal ground.

Newly hatched herring are found in all the inshore waters surrounding the North Sea, particularly close to the German and Danish coasts and in the Scottish firths to which positions they are carried by currents. Metamorphosis occurs at a length of 3–5 cm and the young fish move offshore to deeper water when 9–10 cm long, particularly during the spring of their first or second year. These young herring become most numerous in the southern North Sea, particularly in the area south and east of the Dogger Bank. In the autumn, the fish spread out east, north, and west and by the following spring most are north of the Dogger Bank, migrating further and further to the north as spring and summer progress and eventually are spread out all over the middle North Sea. Those that do not recruit to spawning groups during their third or fourth years probably feed in the western part of the middle and northern North Sea and winter more towards the east.

In general, the migration circuits are based on current systems but there are exceptions. In (c), for example, herring migrating south down the east coast of the Southern Bight must be moving against the current.

[Compiled, with modifications, from Meek (1916) and Harden Jones (1968). Reference should be made to the latter author for a critical review of the evidence on which the various conclusions described above are based]

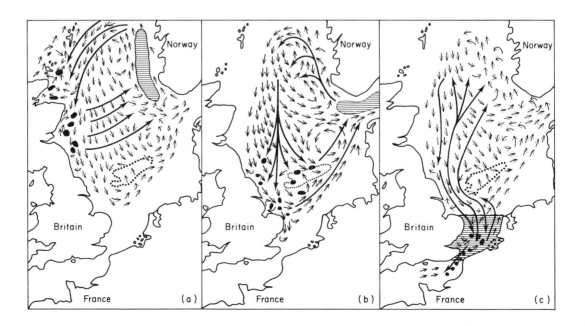

some other mechanism such as exploitation in the vertical plane of a counter-current system. Selection for such station-keeping is likely to be most pronounced during periods that the fish are in some way exploiting a particular area of sea bottom, such as when pelagic fish that lay demersal (i.e. bottom-living) eggs are on the spawning ground. When selection favours migration, however, individuals that migrate in the same direction as the current at the depth at which habitat suitability is optimum are likely to be at an advantage for the same reasons given for stream-living animals (pp. 783 and 798). This depth is likely to be near the surface for young and small fish that feed on plankton, bearing in mind selection for daily vertical migration (p. 496), and increasingly deep for older and larger fish and, of course, for fish adapted to life on the sea-bottom. It is tempting, therefore, to postulate that in the early stages of evolution or establishment of a migration circuit, those individuals are favoured that perform, solely by passive drift within a gyral or counter-current system, the circuit that is nearest to the optimum. However, as long as the migration circuit thus evolved in any evolutionary sequence is less advantageous than a modification of that circuit, selection will continue to act for change. Any individual that by virtue of swimming at a slightly different depth at a slightly different speed or in a slightly different direction or that maintains station in an area for a time that is slightly shorter or longer, and as a consequence achieves a slightly more advantageous migration circuit is favoured by selection. The result would be a deme of fish performing a migration circuit based largely on drift with the current but with modifications of timing achieved by one, some, or all of the behavioural mechanisms of depth regulation, variation in swimming speed, variation in swimming direction relative to the current, and variation in the time spent in station-keeping.

A major part of the complex of selective pressures acting on the behaviour by which a migration circuit is achieved is selection for arrival at the appropriate time at the different areas on the migration circuit. This can be achieved either by an endogenous clock that times each stage of the migration as suggested for some birds (p. 627) or by sensory recognition of an appropriate area followed by an optimum period of station-keeping. There appear to be no experiments on the fish described in this section (Figs. 31.12–31.16) or with comparable species that identify the sensory mechanisms that are involved in recognition of suitable areas. Harden Jones (1968) points out that given an annual synchronisation of physiological changes and an annual migration circuit, a deme of fish could spawn in the same spawning area year after year and generation after generation without any recognition on the part of the individuals that the area was either the natal area or the area used the previous year. The evolved mechanism could be instead that the fish spawn in the first area with habitat variables that are above threshold for spawning that the fish encounter after they become ready to spawn. A distinction has to be drawn, therefore, between recognition of an area suitable for spawning (or any other type of area as appropriate) and recognition of an area as either the natal area or the area in which the behaviour was performed previously. Both mechanisms would permit a degree of return that would be adaptive to the selective pressures discussed earlier (p. 809) but there is no evidence for long-distance marine migrants that indicates for any particular deme which mechanism has evolved. In either case, however, it seems likely that the relevant environmental variables are perceived by one or any combination of the acoustic–lateralis system, and the organs of sight and olfaction. As reviewed by Hasler (1966) and Harden Jones (1968), all of these sense organs have been shown to be involved in the recognition by fish of their home range.

In salmon, the perception of olfactory clues appears to be part of the navigation process by which the fish return to their natal tributary from the coast (Chapter 33). Briefly, upon detection while in the sea of the odour of the natal drainage system, a salmon swims against the current, thus entering the mouth of this drainage system. Similar behaviour may also be adaptive for marine fish. Selection for downcurrent migration by young fish followed by selection for return to the natal site for spawning could well favour a downcurrent–upcurrent return migration in marine fish for the same reasons presented for stream animals (pp. 785 and 798). By migrating downcurrent and thus remaining in sensory (olfactory) contact with the home site, to the olfactory emanations of which imprinting has taken place, the fish may return to this site at any time by migrating against the current containing the appropriate cues.

In the preceding paragraphs it has been argued that the selective influence of oceanic currents which at any given depth in a given area act predominantly in one direction will be to favour a migration circuit that is either a downcurrent–upcurrent return migration relative to the spawning area or a return migration which is predominantly downcurrent but is based on a gyral or on a counter-current system. Conceivably the migration circuit as a whole may consist of sub-circuits of both types as is suggested to be the case for those salmon that migrate to the open sea (Fig. 31.10). Of the two types of circuit, downcurrent–upcurrent migration seems most likely to evolve when there is a gradient of increasing feeding-habitat suitability with distance downcurrent from the spawning area, as described for stream-breeding fish (p. 802), especially in areas in which suitable current gyrals or counter-current systems are lacking. Where such systems are not lacking, however, a migration circuit based only on

downcurrent movement seems the more likely. It is, however, possible to envisage selection that favours a compromise between these two mechanisms. Suppose that a counter-current system exists but that animals using only this system have a low degree of return. Suppose, also, that there is strong selection for a high degree of return and that the natal site gives off olfactory cues to which the fish were imprinted soon after hatching as in salmon (p. 850). In this case it is possible that the optimum behaviour for the animal is to first locate water containing the olfactory cues given off from the natal site, orientate upcurrent, sink or rise into the counter-current, and then alternate between the current containing olfactory information (to maintain accurate orientation) and the counter-current (to reduce migration cost).

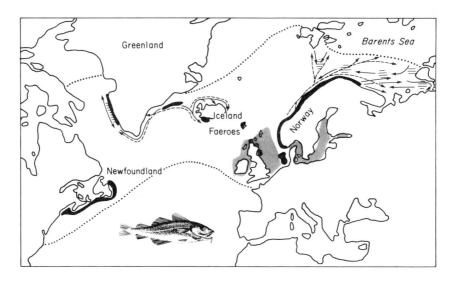

Fig. 31.13   Migration circuits of three demes of the cod, *Gadus morhua*

General ecology

Cod spawn from late winter to early summer (February to June), and occasionally also in the autumn, in water with a temperature usually from 3 to 6°C, though occasionally lower. Spawning takes place in midwater or near the sea bottom. Each female produces from 0·5 to 15·0 million eggs and extends egg laying over several days. Fertilised eggs just float in sea water and become evenly distributed

throughout the top 100 m of the water column. At 5–6°C the larvae hatch in about a fortnight and for the first 3–6 months swim about in mid-water. During this period they feed on members of the zooplankton, particularly copepods. At the end of this period they take to the sea bottom, even at depths of 200–400 m. Here they feed on bottom living animals, particularly crustaceans. Possibly there is a seasonal return migration to shallow water in summer and deeper water in winter. As they grow, the cod begin to feed on fishes which become the main food when adult. The diet of the adult cod is basically small pelagic fish, particularly herring, *Clupea harengus*, capelin, *Mallotus villosus*, haddock, *Gadus aeglefinus*, and young cod, though other pelagic organisms, particularly crustaceans, are also taken. When they are approaching maturity, the young cod join the mature fish on their migration circuit and there is some evidence that a complete migration circuit may be performed before the young cod themselves spawn. The age and size at which cod first spawn differs from deme to deme. In the North Sea, spawning may occur at 5 years of age and a length of 70 cm. The Arcto-Norwegian cod, on the other hand, first matures between 6 and 14 years of age. Once they have matured, cod seem to spawn every year until they die, which may not be until the age of 20 years or more.

In the diagram, the approximate distribution limits of the cod are indicated by dotted lines. Within this world population of the species, several geographical demes may be recognised between which there is little removal migration. The Newfoundland cod may be considered to be such a deme, as also may the cod at the Faeroes, in the North Sea, and in the Baltic. The cod captured at different times in the Barents Sea and along the Norwegian coast also seem to belong to a single deme. On the other hand, the relationship between the cod of Greenland and Iceland is complex. In the diagram the solid black shading indicates the approximate positions of the major cod spawning areas. Stippling indicates areas in which, although cod spawn, they do so only in small numbers. Horizontal lines indicate wintering areas. Continuous arrows indicate migrations of mature fish whereas dashed arrows indicate the drift migrations of eggs and larvae from the spawning grounds to the nursery grounds.

The cod demes of Greenland and Iceland

Along the west coast of Greenland, there are a number of well-established demes of cod that spawn in the fjords and that appear to have migration circuits entirely within their local fjord and perhaps also just extending into coastal waters. It seems unlikely that at the beginning of the twentieth century cod spawned with any success on the west Greenland banks. From 1917 onwards, however, with the advent of a warm period in the north Atlantic which has led to a northward shift in the geographical distribution of many boreal marine animals, there has been a great increase in the abundance of cod at west Greenland. The majority of the initial removal migrants seem to have been 5-year olds from the 1912 year-class. These removal migrants almost certainly came from the Iceland-spawning deme rather than from the local fjord fish. At the present time a proportion of the eggs and larvae from Iceland spawnings drift with the Irminger current to southwest Greenland as shown. When mature, many, if not the vast majority, of these fish migrate back to Iceland to spawn. Many of the eggs and larvae from spawnings off southwest Iceland also drift with the current around to the north and east coasts of Iceland, returning when mature to spawn off southwest Iceland. Whether these two portions of the eggs and larvae from Icelandic spawning constitute a single or several demes is unknown, but certainly fish that have performed the Iceland–Greenland–Iceland return migration have at a later date been captured off the north and east coasts of Iceland.

By the 1960's, the west Greenland cod, excluding the fjord demes, seemed to consist of two, at least incipient, demes. The southernmost fish seem to be members of the Iceland-spawning deme whereas the northernmost fish seem to be members of a newly formed West Greenland deme with its own migration circuit. This circuit consists of an annual north–south return migration with the northernmost position being reached in August and the southernmost position being reached at the end of spawning in May. During the northward migration from May to August the recently spawned adults are likely to be accompanied by their drifting eggs and larvae. Over the past 50 years or so, therefore, it seems likely that we have witnessed along the west coast of Greenland the early stages of the formation of a new deme of fish with a new migration circuit. If the development of this migration circuit is observed in the years to come, we may learn a great deal about the interaction of selection and learning that is involved in this process.

Little is known of the migration circuit of the cod that have been discovered recently to spawn along the east coast of Greenland. Quite possibly this deme is also the result of removal migration from the Iceland-spawning deme.

The cod deme of the Barents Sea and Norwegian coast

Eggs and larvae from spawnings along the Norwegian coast drift with the current into the Barents Sea where the young fish descend to deeper water in nursery areas. After spawning, the spent cod leave the coastal banks at the beginning of April and travel north at about 30 km/day. The northern and eastern limits of the feeding area are occupied from August to October after which the cod return to the south and west. The cod concentrate in wintering grounds in the Barents Sea as shown but do not remain long and gradually move back toward the spawning areas, perhaps by moving into deeper water and entering the Spitsbergen and North Cape currents. Average speed during the migration to the spawning grounds, at which the cod arrive in February and March, is about 11 km/day. Degree of return to feeding areas in subsequent years does not appear to be high and the migration circuit for the deme may literally be from the Norwegian coast to the Barents Sea, though it seems likely that there is some sub-division into more restricted spawning demes.

Migration in relation to currents

There seems little doubt that the movements of eggs and larvae of all demes are based on surface currents. Once they take to deeper water in the nursery area, however, the young fish seem to enter a period of station-keeping. Once they begin to feed on shoaling fish, the young cod seem to migrate with the prey shoals and hence predominantly with the current. When migrating back to the spawning ground, the cod are believed to migrate either near the sea bottom or, at least in deeper water, in midwater. Cod have been trawled at depths of 460 m. After spawning, the migration depth is unknown but some observations suggest that it may be much nearer the surface than at other times. Quite possibly, therefore, the migration circuits of cod are based on surface current–deeper counter current systems.

[*Complied from Ekman (1953), Zenkevitch (1963) and Harden Jones (1968). Reference should be made to the last author for a critical review of the evidence on which the various conclusions presented above are based*]

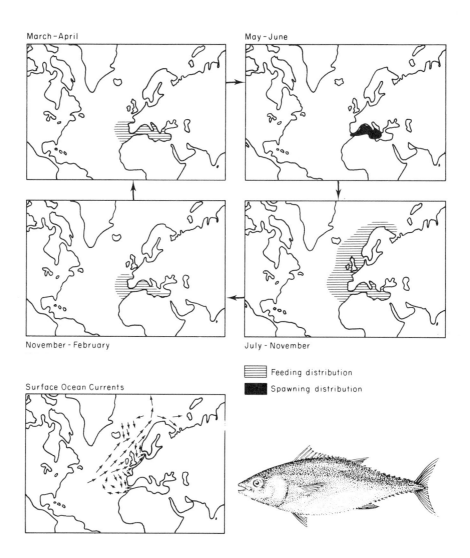

Fig. 31.14 Annual migration circuits of the Mediterranean-spawning deme of the tunny, *Thunnus thynnus*

General ecology

The tunny, or bluefin tuna, is an imposing fish that weighs 140–450 kg and may reach a length of 4 m. On average, however, a tunny of 14 years is about 2·5 m in length. The fish has a dark blue back and silver flanks flecked with gold or opalescent spots on the fins. It feeds mainly on pelagic schooling fish of small size, particularly sardines, *Sardinia pilchardus*, anchovy, *Engraulis encrasicholus*, mackerel, *Scomber scombrus*, and in the northern part of its range on young herrings, *Clupea harengus*. It also feeds on larger fish, such as cod, *Gadus morhua*, and coastal species, such as eels, as well as animals from other groups such as large cephalopods and members of the macroplankton. It spawns for the first time when 3 years old.

Annual migration circuit

The deme illustrated spawns in May and June in the area (dark shading) bounded by Tunisia, Sicily, Sardinia, and the Mediterranean coast of Spain. Other demes spawn in the Black Sea, the Atlantic just outside the Straits of Gibraltar, around Bermuda, and in various areas in the Pacific. After spawning, part of the Mediterranean-spawning deme has a migration circuit that is totally within the Mediterranean. The largest portion of the deme, however, leaves the Mediterranean, passing Gibraltar often in shoals containing up to 10 000 fish, and migrates northwards. Possibly the migration is based on the surface ocean currents shown bottom left. Until about 1920, the northern limit of the migration circuit seemed to be in the region of the English Channel. Now, however, the circuit extends as far as Norway, some individuals travelling up to 5000 km in the space of a month. In winter (November–February), there is a southward movement, the fish occupying deeper and deeper water, and in places the route is perhaps based on deeper counter-currents. In spring there is a slight expansion of the range as the fish return to the surface, an expansion that may be based on current gyrals such as that in the Atlantic off the west coast of Spain and Portugal. As the spawning season approaches, the tunny enter the Mediterranean in the current that passes continuously eastwards in the middle of the Straits of Gibraltar.

[*Compiled from Sella (1930) and Ricard (1971). Surface ocean currents from Harden Jones (1968)*]

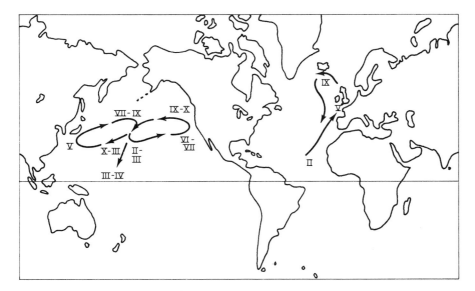

Fig. 31.15   Annual migration circuits of two demes of the albacore, *Thunnus alalunga*

General ecology
The albacore is a tuna and has a feeding ecology that is broadly similar to that of the tunny described in Fig. 31.14. The albacore spawns in tropical waters in both the Atlantic and Pacific Oceans.

Annual migration circuits
It seems likely that the whole of the North Pacific is occupied by a deme the members of which share an ontogenetic migration circuit. Three annual migration circuits are shown for this North Pacific deme. The numerals at the different points on each circuit indicate the months that that part of the circuit is occupied. The eastern circuit is performed primarily by albacore up to 3 years of age. In the fourth year, however, the fish enter the western (Japanese) circuit. With the onset of maturity at 6 years of age the albacore migrate southwards to breed in tropical waters.

In the North Atlantic, post-spawning adults and any 1-year-old young that have not yet left tropical waters migrate northwards in February. The northward migration shown in the diagram involves the fish remaining

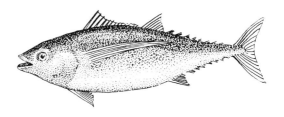

within water the temperature of which at a depth of 45 m is above or equal to 14°C. In the autumn, the albacore migrate southwards from Iceland but at a depth that gradually increases until, by winter, they have reached depths of 250–350 m in tropical waters.

[*Compiled from descriptions and diagrams in Ricard (1971)*]

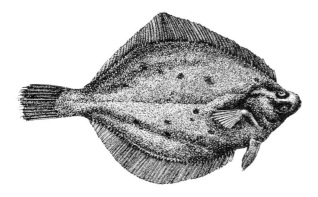

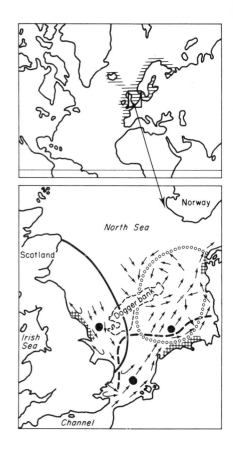

Fig. 31.16   Ontogenetic and seasonal return migration of the plaice, *Pleuronectes platessa*

General ecology

The plaice is a flatfish that lies on its 'left' side on sandy ground at depths down to 120 m. The geographical distribution is indicated by horizontal hatching in the top map. The eyed side is a greenish-brown with irregular darker blotches and white, orange, and red spots, the latter two with a blue halo. The blind side is white. After metamorphosis from planktonic eggs and larvae, the fish feed on benthic marine organisms, particularly lamellibranchs but also polychaete worms, crustaceans, and others. Plaice spawn from November to May. Maturity is reached at 3–4 years and individuals may live for 20 years or so.

Several demes can be recognised between which there is unlikely to be a high incidence of removal migration. The plaice of the Murman coast, Iceland, the Faeroes, and Baltic are all likely to be such demes. Around the coasts of Britain, removal migration may be more frequent and some certainly occurs between the demes inhabiting the Irish Sea, English Channel, and North Sea. In the North Sea there appear to be four spawning demes, three of which are considered below and the other of which spawns off the east coast of Scotland.

In the southern North Sea, plaice eggs can be found over a wide area during the spawning season but there are major concentrations, indicated by the large solid black dots in the bottom map, near Flamborough Head (left dot), central Southern Bight (bottom dot), and German Bight (right dot). Peak egg production in these three areas occurs, respectively, in March, January, and February. Plaice spawn in relatively warm water ranging from 4 to 10°C and a female appears to require 2–3 weeks to lay her 16 000–350 000 eggs.

Ontogenetic and seasonal return migration

Eggs and larvae are pelagic and drift with the current. Eggs hatch in 2–3 weeks. Metamorphosis starts about 5 weeks after hatching and is completed by the seventh or eighth week. The young plaice, now about 10–14 mm long, then take to the bottom. Current systems are indicated in the bottom diagram by small arrows and off Flamborough tend to keep the eggs and larvae near to the coast. From the Southern Bight and German Bight areas,

eggs and larvae will be carried to the northeast. Rate of migration from the Southern Bight is about 3.8 km/day with the result that the total migration distance during the pelagic phase is about 260 km. At the end of this time the young fish take to the bottom and move inshore to the nursery grounds. The approximate nursery areas of the three demes illustrated are indicated by hatched shading. Up to their third year, plaice are found at depths from 0 to 20 m. Plaice from 3–5 years of age are found in depths from 20 to 60 m, and larger and older fish in still deeper water. At maturity, the plaice seems likely to migrate back to its natal area. After the first spawning, however, the plaice do not necessarily return to deeper water off the nursery area but instead disperse over a wider feeding area which may overlap with the feeding areas of other spawning demes. The solid and dashed lines, and row of open circles, indicate, respectively, the limits of subsequent recaptures of mature plaice tagged on the Flamborough, Southern Bight, and German Bight spawning grounds. Recaptures of these fish in the first spawning season after liberation show that there is a high degree of return of mature plaice from their feeding areas to the spawning area used previously. How consistent the individual's annual migration circuit may be in the years after the onset of maturity is unknown.

[*Compiled from Bagenal (1973) and Harden Jones (1968). Reference should be made to the latter author for a critical review of the evidence on which the various conclusions presented above are based*]

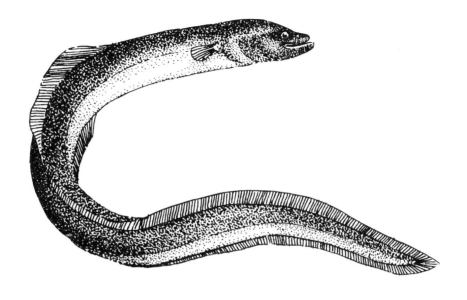

Fig. 31.17 The geographical distribution and life-histories of two genera of anguillid eels

The two maps show the coastal distribution of adults of species of *Conger* and *Anguilla* as indicated. In the top diagram, the cross-hatching shows the distribution of *Conger vulgaris* whereas the horizontal hatching shows the distribution of the other species of the genus.

Anguillids are characteristically marine and tropical and, whereas some of the species occur in moderate depths, the majority occupy deep-sea habitat-types. After hatching, anguillid larvae are leaf-shaped and known as leptocephali. Eggs and leptocephali, even of deep water species, are pelagic and drift with the current. In deep-sea species, metamorphosis occurs in the open sea and the young soon descend to the bottom. In some species, including those of the genera *Conger* and *Anguilla*, the leptocephali metamorphose over continental waters and the nursery, and growth and development, areas are not far from land. Two Indonesian eels, *Ophichthys boro* and *Muraena polyuranodon*, and all members of the genus *Anguilla* enter brackish water and freshwater to pass the major part of growth and development.

In most cases the spawning areas of eels are unknown. The Atlantic conger eel, *Conger vulgaris*, seems to have at least three spawning areas: in the Mediterranean, East Atlantic, and West Atlantic. Mediterranean eels of the catadromous East Atlantic *Anguilla* sp., on the other hand, do not spawn in that sea and may migrate to spawn in the Sargasso Sea along with other East Atlantic eels. This and other postulated migration circuits for *Anguilla* spp. are illustrated in Fig. 31.18.

[*Compiled from Meek (1916) and Bertin (1956)*]

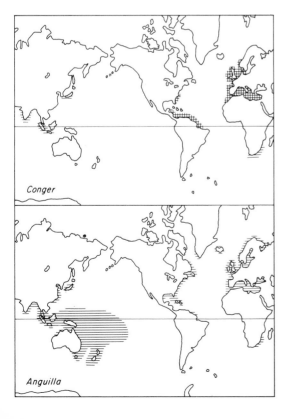

Conger

Anguilla

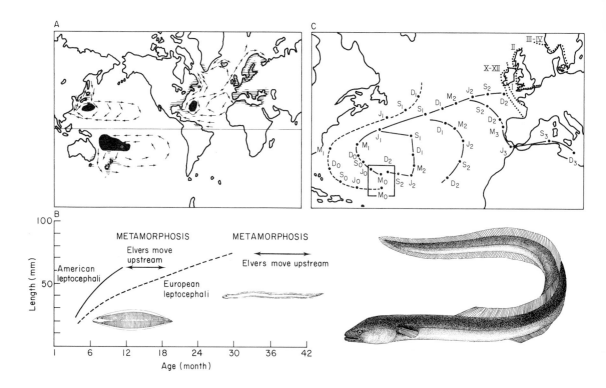

Fig. 31.18 Some postulated ontogenetic migration circuits of eels of the genus *Anguilla*

All members of the genus *Anguilla* are catadromous, spawning in the sea but undergoing the major part of their growth and development in brackish water or freshwater. 14 of the 15 or 16 species of *Anguilla* are found in the Indo-Pacific region and a tropical origin of the genus in this area seems likely. Fossil representatives of the genus have been found in Europe since the upper Eocene (40 million years B.P.) and the distribution is such as could arise from the separation of the large Mediterranean area from the Indian Ocean in the lower Miocene (25 million years B.P.).

All *Anguilla* spp. probably spawn in warm saline water at depths of 400–700 m but over very deep water. The eggs are pelagic and develop into leaf-like leptocephali larvae (B) which then drift with the current towards the coast. The period of larval life varies considerably from species to species but terminates when the leptocephali metamorphose into elvers (B) which swim up estuaries and rivers into freshwater where the eels grow. When the adults mature several years later, they migrate back to the sea and are believed to return to their natal area to spawn once and then die. Most, if not all, species are thought to consist of a single spawning deme with a broad migration circuit based on surface ocean current systems that carry the larvae from the spawning area to suitable coastal nursery areas. Nothing is known of the migration of adults from their feeding areas back to the spawning areas,

though it is assumed that migration occurs at some depth. Whether the migration circuit is based on migration entirely with the current within counter-current systems or whether the adults migrate against or across the current at the depth at which they migrate is unknown.

Three examples of postulated migration circuits are presented in diagram A. Postulated spawning areas are shown by solid black shading and growth and development areas are shown by horizontal hatching. Arrows show surface ocean currents. The North Pacific spawning area is used by a single species, *A. japonica*, whereas the South Pacific spawning area that is illustrated is used by two species: *A. australis*, a short-finned species that while in freshwater lives in the larger rivers and lakes; and *A. dieffenbachi*, a long-finned species that lives in smaller waters. Whether the North Atlantic spawning area is used by one or two species has for decades been the subject of a continuing controversy.

The Atlantic-spawning deme
Although eggs have not been taken, the distribution of the smallest larvae (less than 10 mm) has led to general agreement that the spawning of all Atlantic *Anguilla* takes place from early spring (February) until well into summer at depths of 400–700 m in the general area of the Sargasso Sea. It should be noted, however, that the spawning area as currently delimited cannot readily account for the presence of *Anguilla* on the north coast of South America (see currents in map A).

The pelagic eggs then float up to the surface where the larvae hatch out and feed. When young, the larvae may feed on diatoms but the food of the older larvae is not known. There is some evidence that the older larvae perform a daily vertical return migration (p. 496).

There is also a general agreement that American and European *Anguilla* derive from different regions of the spawning area. In diagram C, the spawning area is indicated by an open square. If the movement of two blocks of surface water from the north and south sections of this area are plotted, the tracks obtained are as indicated. The dashed line shows the movement of the southern block and the continuous line shows the movement of the northern block. The letters M, J, S, D by the side of the dots along these lines indicate, respectively, that the dot is the position reached in March, June, September, and December. The suffixed number indicates years from start, i.e. 0 = 0–1 year; 1 = 1–2 years; 2 = 2–3 years; and so on. The tracks show that American eels are more likely to be derived from the southern part of the spawning area whereas European eels are more likely to be derived from the northern part of the spawning area. However, larvae arrive at the American coast during their first or second year of life whereas larvae do not arrive at European coasts until their third or fourth year of life. Metamorphosis to elvers and migration upstream must therefore occur earlier in the American than the European eels. Diagram B shows that larval growth and metamorphosis is consistent with the different times of arrival at the coast and that differences in growth rate between the potential American and potential European eels are evident from an early age. So, too, are morphological differences, American eels having 103–111 vertebrae ($\bar{x} \pm$ S.D. = 107·2 $\pm$ 1·2) and European eels having 110–119 (114·7 $\pm$ 1·3). 'American' eels are found along west Atlantic coasts from the Guianas of South America in the south to southwest Greenland in the north and 'European' eels are found along east Atlantic, Mediterranean, Black Sea, and Red Sea coasts from Iceland in the north to north Africa in the south.

Metamorphosing eels arrive in coastal waters as small transparent eels referred to as glass eels. Time of arrival at various points off European coasts are indicated by dotted lines, the Roman numerals indicating the month of arrival. On the whole, the time of arrival is consistent with the illustrated calculated rate of drift, though it is not yet known whether the larvae entering the Mediterranean are really in their fourth year as required by the hypothesis nor whether the larvae that metamorphose near the Azores are really only two years old. If both of these predictions of the hypothesis are correct it is difficult to escape the conclusion that metamorphosis can be triggered and/or delayed by cues given off by different parts of the European continent. Unless, that is, the spawning area is subdivided into spawning grounds smaller than simply north and south with each geographical feeding deme (e.g. Mediterranean deme, Azores deme, etc.) also retaining its integrity as a spawning deme upon arrival at the Sargasso Sea. In which case, if removal migration between spawning demes was of a sufficiently low incidence, age at metamorphosis could be an inherited characteristic of each deme. This latter possibility receives some support from recent serological studies (de Ligny and Pantelouris 1973)

which seem to demonstrate consistent genetic differences, not only between American and European eels, but also, though to a lesser extent, between different geographical demes of the European eel.

Elvers can detect freshwater input into sea water and respond by orientating and swimming against the current, but pause before entering rivers until they become more pigmented. The young eels swim upstream, often in great numbers, keeping close to either bank of the river. Obstacles, such as falls, are surmounted by crawling like small wet worms beside the water. The elvers may go far upstream, migrating predominantly at night. Overland migration through wet grass at any stage of freshwater life may lead to eels being found in ponds cut off from running water (p. 196). During their life in freshwater, eels feed on a wide range of vertebrate and invertebrate animals as well, perhaps, as aquatic plants and other vegetable foods.

The time spent in freshwater depends on the rate of growth. Female European eels initiate downstream migration when 54–60 cm long and males when about 40 cm. The time required to attain these lengths may vary from 9–19 years for females and 7–12 years for males. After the eel has attained the critical length it undergoes a series of physiological and physical changes which culminate in the transformation of the freshwater yellow eel into the sea-going silver eel with enlarging eyes and with a retinal pigment characteristic of deep-sea fish. It is as a silver eel that the European eel migrates downstream to the sea in the autumn and early winter (mainly August in Sweden, September in Denmark and eastern Britain, October in Ireland, and November in southwest England).

Agreement among zoologists concerning the North Atlantic eel ceases with the entry of the silver eel into the sea. Two conflicting hypotheses are extant to which, for completeness, may be added a third. The classical thesis of Schmidt (1922) is that American and European eels are distinct species, known, respectively, as *A. rostrata* and *A. anguilla*. Mature eels from both sides of the Atlantic are considered to migrate back to the Sargasso Sea to spawn, *A. anguilla* taking up a more northerly distribution and *A. rostrata* taking up a more southerly distribution. Incorporating the recent serological evidence into this hypothesis, we would postulate that the different feeding demes of the European eel also occupy slightly different spawning areas as also must the South American deme of the American eel. On this hypothesis, the differences in rate of growth, age at metamorphosis, and number of vertebrae are considered to be inheritable species- and deme-specific characteristics. The alternative extant hypothesis is that of Tucker (1959). On this hypothesis, the European eels perish without leaving continental waters or at least before arriving to spawn at the Sargasso Sea. The entire population of North Atlantic *Anguilla* is spawned by the American freshwater deme. The differences between American and European eels are considered to be environmentally determined and to be the result of differences in the temperature conditions encountered during the ascent of eggs and larvae from the spawning depth to the surface. The vertical temperature gradient is steeper in the southern part of the spawning area than the northern part (see Fig. 20.2). Finally, an almost equally probable third hypothesis could be advanced that it is the American eels that fail to return to the

Sargasso Sea and that the entire North Atlantic population is spawned by the European freshwater deme.

From an evolutionary viewpoint, Schmidt's hypothesis is the most likely. If eels that spawn in one region of the Sargasso Sea consistently fail to produce offspring in succeeding generations, selection would surely have acted against the behaviour that led to eels being present in that part of the sea when they begin to spawn. For this reason, I am inclined to accept Schmidt's hypothesis rather than either of the other alternatives. As yet, however, the crucial evidence is not available. No fully grown mature eel, either American or European, has ever been captured at sea away from the continental shelf.

[*Compiled from Meek (1916), Bertin (1956), Harden Jones (1968), and Hynes (1970)*]

Harden Jones (1968) has reviewed the evidence relating to the migration mechanisms of marine fish. Most of the data so far available suggest that a migration circuit based on continuous downcurrent migration, particularly in gyrals, is widespread. As yet, however, there are no data that can be considered to demonstrate beyond dispute the existence of such a migration circuit. The migrations of adults of the spring-spawning deme of Norwegian herring (Fig. 31.12) are generally in the direction of the current and the deme appears to be contained within the Norwegian Sea gyral. Some of the spent fish, however, leave the Norwegian coast spawning grounds on a limb of the North Atlantic drift and reach major feeding grounds in the Jan Mayen area. They return on the East Icelandic current. For continuous downcurrent migration, however, it is necessary to postulate that north and east of the Faeroes this current and the herring dive under the North Atlantic drift, thus returning the herring towards Norwegian coastal waters. Most of the demes of herring with migration circuits in the North Sea also migrate within gyrals. On the eastern side of the Southern Bight, however, herring move south despite the northeast flow of the residual current. The migrations of Arcto-Norwegian and West Greenland demes of cod, *Gadus morhua*, could also be all downcurrent, though more data on currents and depth of migration are required to prove the point conclusively (Fig. 31.13). The changes in the depth at which different stages of the migration circuits are performed by tuna (Figs. 31.14 and 31.15) and perhaps also cod (Fig. 31.13) and plaice (Fig. 31.16) could perhaps indicate that counter-current systems are involved in the migrations of these species.

31.4.2 Cephalopods

It is perhaps a reasonable assumption that the more-active cephalopods in any given region will perform migrations that are similar to those performed by fish that live at approximately the same depths in the same region. Thus short-finned squid, *Ilex illecebrosus*, make seasonal movements in the Newfoundland fishing area (Squires 1957) that seem likely to be part of a migration circuit. Some tropical and sub-tropical species in the North Pacific migrate north to the Kurile Islands, travelling 1500–2000 km in 1–2 months (Akimushkin 1963), a migration that, if it is a limb of a migration circuit, is reminiscent of that of some fish, such as the albacore (Fig. 31.15) that also breed in tropical regions. Data are insufficient, however, to examine the assumption of a similarity between fish and cephalopod migration in any detail.

31.4.3 Fish that migrate to and from freshwater

Fish that spend a large part of their lives in freshwater but which migrate to the sea to breed are described as catadromous. Such migrants, with the exception of the eels of the genus *Anguilla*, are less well known than the anadromous migrants described in Section 31.3.

The southern trouts or jollytails, *Galaxias* spp., (see Fig. 31.32 for distribution), migrate downstream to breed in tidal marshes. The larvae then go to sea for a while before returning to grow in the rivers. *Galaxias attenuatus* of Australia, New Zealand, and South America is perhaps the best described of these species (Burnet 1965). Some species of *Galaxias*, however, such as *G. vulgaris* of New Zealand and *G. findlayi* of the Australian Alps remain permanently in freshwater. The southern continents also support a number of other catadromous species, notably the smelts (Retropinnidae) and graylings (Haplochitonidae), which have life histories similar to that of *G. attenuatus* (K.R. Allen 1956). Other catadromous species are the Japanese salmonid, the ayu, *Plecoglossus altivalis* (Kawanabe 1958), and the climbing goby, *Awaous guamensis*, of Hawaii which, before the introduction of the rainbow trout, *Salmo gairdneri*, was the only stream fish inhabiting the island (Needham and Welsh 1953). Without doubt, however, the best known group of catadromous fish and the group in which catadromous behaviour is most developed is that of the eels of the genus

Anguilla. Anguilla spp. perform migrations that in most respects are similar to those of other members of the eel family (Fig. 31.17). The only difference is that instead of growing and developing in the open sea or in marine coastal waters they enter freshwater for this phase of their life history.

Most, if not all, catadromous fish, except those of the genus *Anguilla*, are essentially coastal species that have retained a pelagic larval phase but that enter freshwater during their inshore–offshore seasonal/ontogenetic return migration (see Fig. 31.24). *Anguilla* spp., on the other hand, as mature adults have the facies of abyssal fish (enlarged eyes and a retinal pigment characteristic of deep-sea fish) that are characteristic of the majority of the eel family and are found where drainage to the sea is adjacent to deep water (Bertin 1956, Carlisle and Denton 1959). The life-history and migrations of the North Atlantic and other *Anguilla* spp. are illustrated in Fig. 31.18. While at sea, the migrations of these and other eels (Fig. 31.17) are quite consistent with the arguments put forward for other marine species in Section 31.4.1. The only additional component of the migration is that of the entry into freshwater.

There is a paradox here. In Section 31.3 of this chapter, arguments were presented to account for the almost universal pattern of downstream–upstream ontogenetic return migration of stream-breeding fish. Part of the hypothesis presented was the advantage of the sea as a feeding habitat for species that have the potential to evolve the physiology of a marine fish. Yet in this section we encounter fish that, although of marine ancestry, nevertheless enter freshwater to feed. This paradox, despite being so apparent, nevertheless seems previously to have escaped an evolutionary consideration. However, as such a consideration involves reference to several different sections from this chapter, it seems more appropriate to separate the paradox from any single group of animals and to discuss it in a separate section (p. 843).

31.5 Littoral and sub-littoral animals

This section is concerned with the inshore–offshore and/or shallow water to deeper water return migrations of a variety of marine animals. Although examples are taken from a range of animal groups (polychaete worms, molluscs, crustaceans, echinoderms, and fish) most information is available for the various crustacean groups and fish. The migrations of Crustacea have been reviewed by Bainbridge (1961) and Allen (1966) and those of fish by Gibson (1969). The physical characteristics of the littoral and sub-littoral zones are documented by Sverdrup *et al.* (1946), Raymont (1963), J.R. Lewis (1964), and Eltringham (1971). Except where stated to the contrary the information presented in this section has been taken from these sources.

31.5.1 Daily and tidal return migrations

The juxtaposition of sea and land results in gradients of selective pressures due to physical and biotic factors between essentially terrestrial conditions on the one hand and deeper water marine conditions off the continental shelf on the other. The result is the striking zonation of plant and animal species along the onshore–offshore axis that is so much a feature of both rocky shores (Lewis 1964) and sandy shores (Eltringham 1971). Onshore–offshore gradients of periods of exposure to air and water, gradients of temperature, light intensity, and salinity, and gradients of the size and nature of organic particles deposited on the shore all contribute to selection for zonation either directly or indirectly through the zonation of species that are the prey or predators of other species.

Specialisation to exploit a particular food source in a way that is related to the state of the tide inevitably means that an animal is exploiting a mobile food source. Any selection to maximise the feeding time, therefore, imposes selection on an animal to move in a way related to the tidal cycle. Against this, however, must be set migration cost due to energy expenditure and to the danger of being unable to maintain or regain position on the shore. This problem is particularly acute on rocky shores and here migration cost seems to be so great that few animals have evolved upshore–downshore movements in phase with the tide, even though activity rhythms inevitably show a circatidal and/or circadian periodicity. On rocky shores efficient attachment to the substratum and limited movement are at a premium. On sandy shores, however, the danger of being unable to maintain or regain position on the shore is less acute and by making use of the inshore–offshore components of tidal currents, energy expenditure may also be reduced. Migration cost seems likely to be less on sandy shores and many animals living on such shores show a tidal return migration. Even so, few remain with the tide during the whole of its tidal cycle, especially those tides of highest amplitude (Fig. 31.21). Instead, such animals tend to burrow into the sand or mud for a period and await the return of the tide or wait for the

tide to uncover a particular zone in which they are specialised to feed.

Superimposed upon the selection related to the tidal cycle is further selection, probably acting through predation, linked to the light/dark cycle. Hence, many of the animals described here are only active during a limited period of the diel, most being nocturnal.

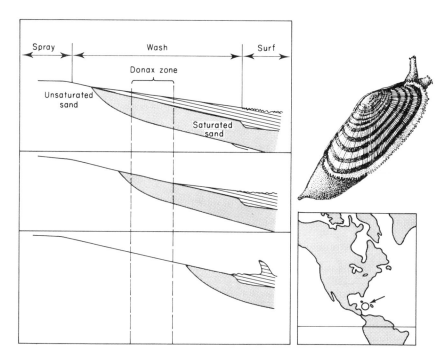

Fig. 31.19   The tidal migration of the small clam, *Donax denticulatus*, on the sandy shores of the West Indies

General ecology

The clam, *Donax denticulatus*, occurs in the saturated wash zone of sandy beaches throughout the West Indies. The species may occur in very large numbers and many open siphons can often be seen at the sand surface drawing in organic detritus from the backwash following each wave. The wash zone in which the species lives is a sloping area of the beach up which the surf runs. After each wave, some water flows back down the beach leaving relatively dry sand in an upper unsaturated zone. The duration of saturation is minimal at the top of the beach and maximal at the bottom. *Donax* requires 4 s to burrow into the sand after being disturbed by wave action. It can, however, only burrow in sand saturated with water. Low down the beach there is insufficient time between the backwash of one wave and the breaking of the next for the animal to burrow. Further up the beach the sand is saturated for too

short an interval. *Donax* is therefore adapted to a zone on the shore as indicated in which organic detritus is deposited and in which the animal can maintain its position with maximum efficiency.

Tidal migration

The zone that is optimum for *Donax* shifts up and down the beach with the tidal cycle. As the tide advances, animals lower down the shore emerge from the sand when the sand–water mixture becomes liquid by pushing down with the foot. These animals are then carried up the shore by the next wave. They then burrow. As the tide recedes and the sand partially dries out the animals emerge from the sand upon re-immersion in water by an incoming wave and are thus carried back down the shore as the wave recedes. In this way, *Donax* moves up and down the shore maintaining its position relative to the tide.

[*After Trueman (1971)*]

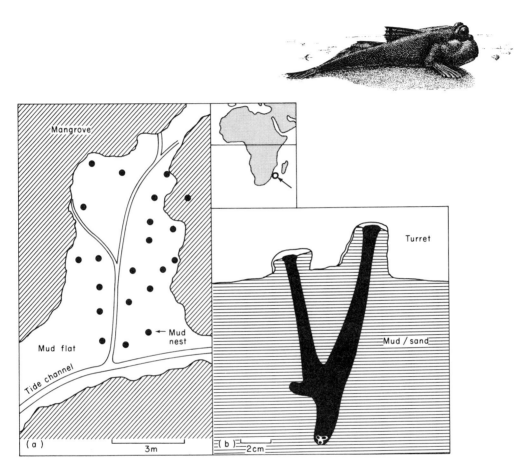

Fig. 31.20 Daily/tidal return migration of the mud-skipper, *Periophthalmus sobrinus*, on the tidal sand flats of Inhaca Island, Moçambique

General ecology
On Inhaca Island the mud-skipper is found in close association with the dense mangrove borders of the tidal mud/sand flats. The fish occurs along tidal channels that connect the open area of the high spring-tide zone with the low-tide flats. The banks of the tidal channels are lined with the arching buttress roots of the mangle, *Rhizophora mucronata*, the fish being most numerous among the pneumatophores of *Avicennia marina* in places where the tidal channels open out into the flats. At low tide mud-skippers are found at mid-tide level foraging within about 30 cm of the edge of the gentle flow of water in the tide channel (a) that drains from the mangrove area to the sand flat. The food consists mainly of small animals, often only a millimetre or so in length, that are detected chiefly if not entirely by sight. The animals are over-dispersed while feeding, each individual maintaining a feeding territory around itself. Nights are spent in the underground nest (b). Turreted nests such as the one illustrated often contain a pair of individuals, whereas nests without turrets are occupied by small single individuals. The nests (represented by solid black dots in (a) are over-dispersed on a given sand/mud flat. The species is probably ovoviviparous, producing an aquatic larva.

Daily/tidal migration
In winter, the mud-skippers enter their nests if the tide floods during the day. In summer, however, as the tide advances, individuals move from the mid-tide areas along the tidal channels and distribute themselves in the high-tide zone. Mud-skippers are commonly observed at high-tide during summer afternoons skipping from mangrove branch to bank. As the tide ebbs or as darkness approaches the fish return to their mud nests on the mud/sand flat, whether or not the nests are under water. Marking experiments do not seem to have been carried out and the degree of return is unknown. There is a strong possibility, however, that the fish are in fact moving within a familiar area between a low-tide home range centred on their mud nest and a high-tide home range.

[*Compiled with modification from Stebbins and Kalk (1961)*]

Finally, twice a day as the tide floods and covers large areas of shore that are uncovered at low water, a whole new food source becomes available with a 12·4 h periodicity, both on the shore and in tidal creeks and estuaries. Many animals have become adapted to exploit these food sources at high tide but to exploit offshore food sources at low tide, adaptations that inevitably involve an inshore–offshore return migration linked to the tidal cycle and often also to the light/dark cycle.

The examples presented below can all, therefore, be generated by entering into the return migration model: (1) a gradient, based on the availability of food, that changes sign with the tidal cycle; and (2) gradients of predation risk based on periods of exposure to marine predators (i.e. an upshore increase in habitat suitability) and terrestrial predators (i.e. a downshore increase in suitability) both of which are more pronounced during the hours of daylight than at night.

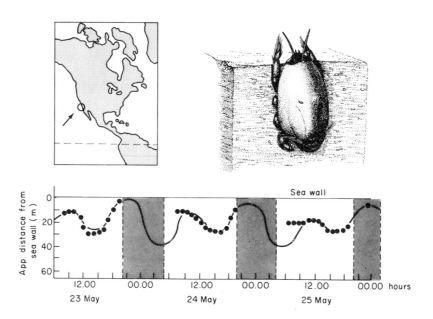

Fig. 31.21   The tidal migration of the mole crab, *Emerita analoga*, on the coast of California

General ecology
Aggregations of the adults and juveniles of the mole crab are a conspicuous feature of the exposed sandy beaches of the west coast of the United States and Baja California. Females release their newly hatched young into the sea as first-stage zoea and after a number of moults in the plankton the last zoea moults to a megalopa. The megalopa then swims or drifts onto the beach. As soon as they land the young crabs form dense aggregations, perhaps up to 22 000 individuals/m². The youngest post-megalopa individuals are at the top of the wash zone just below the surface of the sand. Within any aggregation, young individuals are further up the beach than older ones. By the time young females are sexually mature they are usually living in water of depths between 15 and 60 cm. Males are smaller and although remaining in the same aggregation as females are found higher up the beach. The animals feed by burying themselves in the sand with just the feeding apparatus clear as shown. Particles left behind

or filtered out of breaking waves are collected by the long antennae and pushed toward the oral apparatus. The zonation of individuals according to size is probably a function of particle size at different points in the breaking waves.

Tidal migration
The solid line in the diagram indicates the approximate position of the edge of the tide over a period of three days. The solid dots indicate the observed position of an aggregation of mole crabs over the same period. The stippled areas indicate hours of darkness. Hours are given as Pacific Standard Time. The aggregation moves up and down the beach at the edge of the water without deviating from the tidal cycle except during the lowest tides. When the tide recedes toward its lowest point for the day the crabs stop at the level of the maximum of the next high tide. This leaves the aggregation on the beach above the water's edge until the tide rises once more.

[Modified from Efford (1965)]

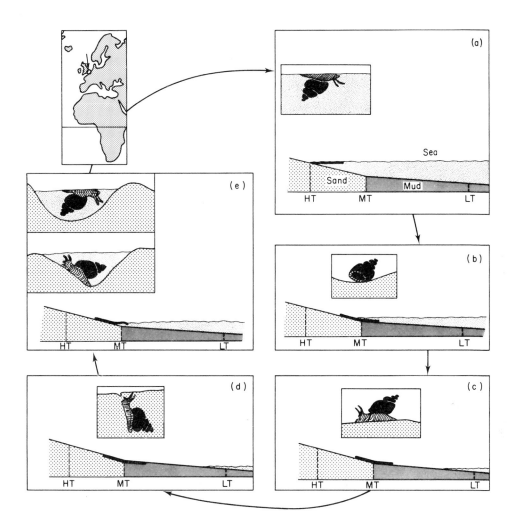

Fig. 31.22 The tidal return migration of the laver spire shell snail, *Hydrobia ulvae*

General ecology
Hydrobia ulvae is a small (height of shell, 6 mm) marine prosobranch mollusc that may be found, often at high densities (up to 60 000/m^2), on muddy sand and muddy flats, as well as in estuaries and salt marshes. Salinities down to 1·5 g/kg are tolerated. It is often found in association with the sea lettuce, *Ulva* spp. Geographically, the range extends from the Mediterranean northwards to Scandinavia, including the Baltic Sea. On the shore it is found usually to be centred upon mid-tide level, but maximum density coincides with the change from coarse deposits to fine silty mud. Although essentially inter-tidal, the species extends down to depths of 35 m. In Britain, the species breeds in late spring and summer. Eggs are attached to the shells of other individuals of the same species. There is a planktonic veliger stage which may be short-lived or even suppressed. In Britain, many maritime

ducks feed extensively on *Hydrobia*, particularly on the ebb tide when the mollusc is on or just below the surface of the mud (diagrams c and d). The shelduck, *Tadorna tadorna*, for example, feeds overwhelmingly on *Hydrobia* and on little else. A single bird may have up to 3000 specimens in its stomach. Pintail, *Anas acuta*, teal, *A. crecca*, shoveler, *A. clypeata*, and goldeneye, *Bucephala clangula*, are all major predators.

Tidal return migration
The figure shows the tidal return migration and associated behaviour of *Hydrobia ulvae*, as observed in the Thames estuary on the north Kent coast near Whitstable, England, in an area where the species occurred at a density of about 10 000/m^2. The solid black band indicates the distribution of *Hydrobia* at various points in the tidal cycle. The behaviour at each stage is indicated in the inset. When the tide is at the high water mark (HT), as in (a), the *Hydrobia* are floating upside down at the edge of the tide. The animals are supported by a mucus raft which

they secrete. This raft also traps plankton which is then eaten by the snail. As the tide ebbs, the snails are carried back down the beach. As they pass over the area of transition from coarse deposits to fine silty mud, often at about mid-tide level (MT), they drop from their mucus raft down onto the substratum (b). Once stranded by the tide the snails crawl around on the surface of the mud (c) feeding on the layer of mud-dwelling naviculoid diatoms. As the mud begins to dry out the snails burrow (d) below the mud-surface, though feeding may continue. Finally, as the incoming tide deposits the first water between the mud ripples (e), the snails emerge from their burrows, climb to the water surface, turn upside down on the mucus raft that they secrete, and are carried up the shore on the incoming tide.

[*General ecology from Fretter and Graham (1962) and Olney (1965a,b). Figure compiled from diagrams and descriptions in Newell (1962)*]

Tidal migrants may be divided into three classes. The first class consists of animals that move up and down the shore with the tide, thus maintaining a more or less constant position relative to the water's edge. The second class consists of animals that roost high up on the shore by burrowing in sand or mud or by hiding under rocks and weed and move down the shore to exploit areas that have been uncovered by the receding tide. The third and final class consists of

essentially sub-littoral animals that move into the littoral zone, tidal creeks, and estuaries to exploit the rich food sources available there when covered by the tide.

The first class, that of animals that move up and down with the tide maintaining a more or less constant position relative to the water's edge, consists of animals from a wide variety of taxonomic groups. Few, however, perform this migration continually, though an example of an animal that does so is provided by the bivalve mollusc, *Donax denticulatus* (Fig. 31.19). Littoral fish may also move up and down with the tide. Thus, female blennies, *Blennius pavo*, move up the shore as the tide advances, retiring again as it recedes. This following of the water line has also been described for the Periophthalmidae of Africa (Fig. 31.20) and India and also for the Red Sea blenny, *Alticus saliens*. In the majority of cases, however, the animals concerned do not follow the water line throughout its movement from low water to high water and back again. Either the movement is performed only at certain times of day or the animals follow the tide to a certain position on the shore and no further. There the animals wait until the tide, as it ebbs or floods, overtakes them once more. Thus at low water, two of the common shore fish of California, *Clinocottus*

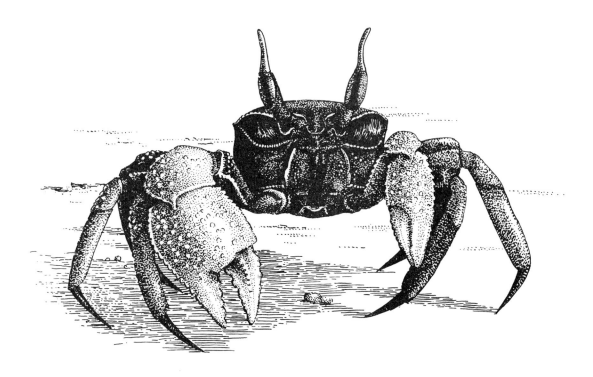

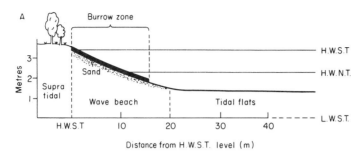

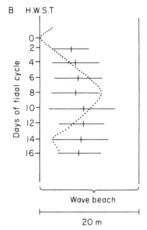

Fig. 31.23 Daily/tidal and semi-lunar return migration of the ghost crab, *Ocypode ceratophthalmus*, on Inhaca Island, Moçambique

General ecology

Crabs of the genus *Ocypode*, known as ghost, racing, or sand crabs, are found characteristically on the sandy shores of the tropics. At Inhaca, adult *O. ceratophthalmus* spend the day buried in the sand of the wave beach (A) and emerge and become active at night during periods when they are not covered by the tide. Before dawn or as the tide comes in the crabs burrow into the sand once more. Unlike the adults, the juveniles are active during the day as well as at night.

Daily/tidal return migration

Adults of *Ocypode ceratophthalmus* emerge from their burrows in the wave beach (A) at dusk or when uncovered by the ebbing tide, whichever is the later. Depending on the phasing of darkness and the period uncovered by the tide the crabs feed either on the wave beach or on the tidal flats (maximum width approximately 700 m) if these are uncovered. As the tide comes in once more or as dawn approaches, whichever is the sooner, the crabs return to the wave beach to burrow. Diagram A shows a profile of the study beach on Inhaca Island. The solid black shows

the zone in which *Ocypode* burrows as indicated. The positions of high water spring tides (HWST), high water neap tides (HWNT) and low water spring tides (LWST) are also indicated.

Semi-lunar return migration

Diagram B indicates that the crabs do not necessarily return to the level on the wave beach at which they had burrowed the previous day and that the relative distance components of the seaward and shoreward migrations vary with the tidal cycle. The horizontal bars show the approximate distribution of *Ocypode* burrows on the wave beach in A on alternate days. The vertical bar indicates the mean position. The dotted line indicates the variation in the extent of the tidal cover. The mean positions on days 2, 10, and 16 are significantly different ($P<0.025$). The change in distribution is consistent with the suggestion that the relative distance components of the seaward and shoreward migrations are a function of the duration of the tidal cover. When duration of tidal cover is decreasing, the seaward migration distance is greater than the shoreward migration distance. When the duration of tidal cover is increasing, the reverse is true. This behaviour decreases the migration cost of the daily/tidal return migration.

[*Compiled from Barras (1963)*]

analis, and *Girella nigricans*, rest in pools above the tide line and then actively migrate upshore as the depth of water increases. Both of these species perform the migration by day and night but another species of *Girella*, *G. punctata*, performs the migration only during the day. Another example of an animal that migrates upshore and downshore with the tide but not all the way is the mole crab, *Emerita analoga* (Fig. 31.21). The amphipod, *Corophium volutator*, on British beaches reacts to the incoming tide through perception of the increase in hydrostatic pressure by leaving its burrow but not swimming (Morgan 1965). The animals are thus carried shorewards by the incoming tide. As the tide begins to ebb, the animals react to the decrease in hydrostatic pressure by beginning to swim. In this way they avoid being deposited too high up on the beach. Instead they are carried back by the tide and settle in their original burrowing zone. A similar behaviour pattern is shown by the pycnogonid sea spider, *Nymphon gracile*, which also swims most when the pressure decreases as the tide ebbs (Morgan *et al.* 1964).

Analysis of laboratory responses of the opossum-shrimps or mysids, *Schistomysis spiritus*, *Siriella armata*, *Praunus flexuosa*, and *P. neglectus*, to changes in hydrostatic pressure in different lighting conditions suggests that when the tide comes in the animals rise to the surface through their light reactions by day or by their gravity reactions at night and are carried up the shore. As the tide recedes the animals swim away from the light by day or downwards by night and presumably are returned to their original position on the shore (A.L. Rice 1961). The different reactions of *Corophium* and *Nymphon*, on the one hand, which maintain contact with the substratum during the flood tide and these mysids, on the other hand, which maintain contact with the substratum during the ebb tide may perhaps be attributed to adaptation to food available at different stages in the tidal cycle. Caution should be used, however, in interpreting these laboratory analyses as return migrations for most were carried out in conditions likely to trigger the removal migration rather than the return migration response as already described in relation to zooplankton (p. 485).

Finally, larger males of the shore crab, *Carcinus maenas*, along British coasts move in and out of the shore each day (Edwards 1958). Smaller males, on the other hand, often remain on the shore for extended periods whereas females show a more gradual upshore–downshore movement that has a periodicity that is longer than daily or tidal.

The laver spire shell snail, *Hydrobia ulvae* (Fig. 31.22) shows a movement that is intermediate between this first class of tidal migrants that move up and down with the tide and the second class that exploit areas left uncovered by the receding tide. A more typical example of the second class is the ghost crab, *Ocypode ceratophthalmus* (Fig. 31.23) which in addition shows a semi-lunar variation in the distance components of the tidal migration such that migration cost is reduced. The beach flea, *Talorchestia longicornis*, a North American amphipod, comes out of its burrows after dark to feed (Smallwood 1903). It migrates down to the tide line as the water recedes and migrates back up the beach as the tide comes in, burrowing when overtaken by the tide. If dawn precedes being immersed by the tide at the burrowing zone the animals migrate rapidly back up the beach and burrow within about 10 min. Another amphipod, the sand hopper, *Talitrus saltator*, behaves similarly as do *Orchestoidea corniculata* (Enright 1963) and the sand beach isopod, *Tylos punctatus*, in California. This latter species is nocturnal and by day is buried at the high-tide line (Hamner *et al.* 1968). It feeds on brown seaweed and follows the receding tide to its foodplants, returning to the high tide line before dawn.

Finally, the third class of animals that show tidal or daily inshore–offshore return migration are those essentially sub-littoral species that migrate inshore to exploit the rich food sources that are accessible only at high tide. Many sub-littoral fish move in to the littoral zone at high water to feed and some, such as the Atlantic silverside, *Menidia menidia*, and the mummichog, *Fundulus heteroclitus*, on the beaches of Long Island, New York, may move up tidal creeks to feed or to spawn and to avoid incoming predators, migrating downstream again on the ebbing tide (Butner and Brattstrom 1960). In areas where tidal movements are less marked, other fish may perform similar movements but on a diel basis. Hence, in Hawaii the blenny, *Entomacrodus marmoratus*, migrates up the shore at night. Similar behaviour seems to be shown by *Blennius pavo* in the virtually tideless Mediterranean.

31.5.2 Seasonal and ontogenetic return migrations

The seasonal and ontogenetic return migrations of sub-littoral and littoral animals are notable for the lack of generalisations that can be made concerning

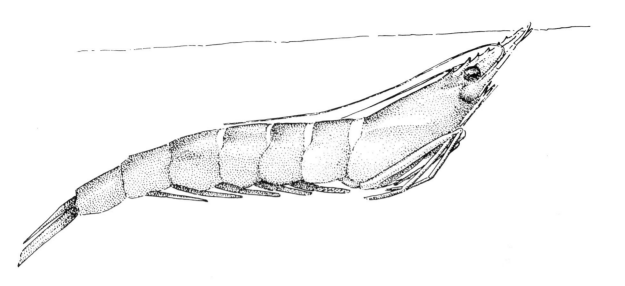

Fig. 31.24 The ontogenetic return migration of the greentail prawn, *Metapenaeus mastersii*, in the Brisbane River, Australia

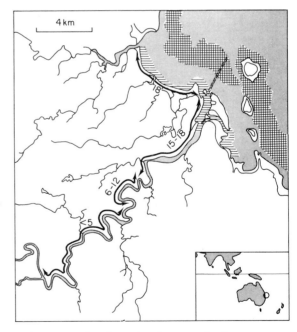

General ecology

In the Brisbane River, greentail prawns spawn from early spring to late autumn (October–May). In common with other penaeids, but in contrast with other decapod crustaceans, females do not carry their eggs externally for extended periods. The duration of planktonic life is probably of the order of two weeks. Age at sexual maturity is estimated as 12–15 months at 16 mm carapace length for males and 20 mm for females (equivalent to total lengths of 76 and 95 mm respectively). Prawns larger than 25 mm carapace length are rare and it is presumed that most individuals die after a full season of spawning. The species is nocturnal when adult, remaining buried at other times of the day, but appears during daylight hours when young. There is a reduction in activity during the winter period, activity being resumed in September. The species is not found as an adult in water deeper than about 10 m.

Ontogenetic return migration

The diagram shows the modal frequencies of carapace lengths (in mm) along the length of the Brisbane River and also the areas in which fertilisation (horizontal hatching) and spawning (cross hatching) take place. At the end of the larval period, as a member of the marine plankton the post-larvae (1·5–2 mm carapace length; 5–7 mm overall) enter the river and penetrate upstream for 40–80 km. When a sheltered shallow locality with abundant algal cover is encountered, the postlarvae become demersal (bottom living) and remain there until the end of the post-larval period (5 mm carapace; 25 mm overall). The majority start to move downstream at this stage, but some, mostly females, remain near the nursery grounds and almost reach adult size there. Most males and females 8–10 mm in carapace length (3–5 months old) occupy a region 15–30 km from the river mouth. At a carapace length of about 16 mm (7–9 months old) the prawns are within 15 km of the river mouth. The males are now sexually mature. Depending on the season at which maturity is attained, reproductive activity may commence either immediately or after the winter period of inactivity. Females are impregnated in the lower reaches of the river (horizontal hatching) and then migrate into water 6–10 m deep (cross hatching) to spawn.

[*Compiled, with modifications and some extrapolation, from Dall (1958)*]

their timing and direction. When such a situation is encountered it suggests that the gradients of selective pressures acting on the migration pattern are species-specific. This in turn suggests that these gradients are those of predation risk and food availability rather than gradients of physical factors such as temperature, light intensity, salinity, and day-length. Thus selection for one species to migrate in one direction at a particular time imposes selection on the predators of that species to move in the same direction at the same time but on the prey of that species to move in the opposite direction at that time. Temperature and light intensity effects, on the other hand, as argued for zooplankton, tend to favour, within broad limits, a similar migration pattern for a wide range of species. When a variety of migration patterns are encountered, as here for sub-littoral animals, sense cannot be made of the migrations until the general ecology of each species is known in some detail, a state that unfortunately has not yet been reached for these animals. The approach used in this section is first to consider those migrations that show least variability in timing and direction, and to leave until later those migrations that show the greatest variability.

The dominant group of natant decapod Crustacea in tropical and sub-tropical waters is the Penaeidea. Almost all species of this group that occur in shallow water migrate into deeper water at certain times of the year. Observations of such migrations have been reported from India, Malaya, Hong Kong, Australia (Fig. 31.24), the Mediterranean, Japan, and America.

Although these migrations are both seasonal and ontogenetic, there seems little doubt that they are primarily ontogenetic, being performed only by mature individuals during the breeding season, and are seasonal only as a result of selection on the seasonal incidence of ontogenetic events such as spawning. This reproductive return migration is in contrast to the primarily seasonal return migrations of the dominant natant decapod group in temperate waters, the Caridea. In this latter group all ages perform the migration which often takes place in the middle of the breeding season.

The reproductive return migration of penaeids is often superimposed on a general birth-to-death ontogenetic return migration. Unlike other decapod crustaceans, the penaeid sheds its eggs free into the water where they become pelagic. In general, this event and fertilisation takes place in deeper water than the remainder of the post-metamorphosis life history. After a planktonic larval phase, the newly metamorphosed juveniles appear inshore (Fig. 31.24). Most penaeid species extend into brackish water to salinities as low as 4 g/kg and frequently it is the juveniles that live in the least saline water. As they age they gradually move into increasingly saline areas (the 'birth-to-death' ontogenetic return migration), eventually when adult moving into deeper water to spawn. After spawning, which may last six months and which may involve the production of more than one egg batch, the adults move back inshore until the following year (the 'reproductive' ontogenetic return migration). This latter type of migration usually involves both sexes. However, in those few cases, such as the Indian species, *Metapenaeus brevicornis*, and the greentail prawn, *M. mastersii*, of Australia (Fig. 31.24) in which fertilisation takes place in relatively shallow water, only the females migrate. Migration distance appears to be a function of age, and perhaps body size. Most species live 2–3 years and on their second offshore migration the older penaeids migrate to deeper water than they did in their first breeding season.

Although a large number of species may live within a confined area, different species seem to occupy different substrata and use different food species. Associated with these differences are species differences in time of breeding and hence of migration. Most penaeids are nocturnal foragers, many remaining buried in the substratum by day and leaving the bottom at night to carry out a daily vertical migration similar to that performed by bathypelagic forms (Fig. 20.11C).

The distance component of the reproductive return migration shows considerable inter-specific variation. Some species, such as the Australian greentail prawn, *M. mastersii* (Fig. 31.24) may show a limited distance component to their reproductive return migration, release apparently taking place in only 6–10 m of water. This is unusual for the group, and most authors report long-distance return migrations for the species studied. Thus *Penaeus orientalis* in the Yellow Sea may have a one-way distance component of 700 km and marked *P. setiferus* off the Atlantic coast of North America have travelled as much as 600 km from their point of release. Although it is difficult to distinguish between removal and return migration by marking experiments, the return migration of this latter species appears not necessarily to be inshore–offshore. Off the mouth of the Mississippi the movement is inshore–offshore as usual for the group but off the Atlantic coast of

America observed movements are parallel to the coast.

The Caridea replace the Penaeidea as the dominant natant decapod group in temperate regions and unlike the latter group do not release their eggs into the water. Instead, females retain their eggs on the abdomen until the larvae hatch. The Caridea also show much more variation in the timing and direction of their onshore–offshore migration. The difference, however, seems to be latitudinal rather than taxonomic, for where the migrations of sub-tropical or warm-temperate carideans have been investigated they show certain similarities to those of the penaeids already described. Thus, the shrimps, *Crangon franciscorum* and *C. nigricauda*, on the coast of California also move as adults out of the estuaries as the spawning season approaches and release their larvae in deeper water of higher salinity. The spiny lobster, *Panulirus interruptus*, is a member of yet another decapod group, the Palinura. This species is one of the largest and most familiar crustaceans in southern California and Mexico (Lindberg 1955). Females lay their eggs in shallow inshore water in spring and summer, migrate offshore to water deeper than 13 m to release the larvae, and then migrate back inshore for the remainder of the summer. Similar behaviour is shown by this species and the congeneric *P. argus* off southern Florida. In Bermuda, the latter species migrates about 13 km out across a lagoon to the edge of the reefs.

Let us assume that the reproductive return migrations described so far represent adaptations to similar sets of selective pressures. Only one general feature emerges: the migration of females to deeper water to release their eggs or larvae. It is the release of eggs or larvae into the plankton that appears to be critical. Other reproductive events such as mating and egg laying, where the latter is a separate event from egg or larval release, may occur either inshore or offshore. Males only migrate with the females if mating and egg or larval release both occur at the same site. This suggests that the critical gradient does not involve factors that influence general adult physiology. It also suggests that the absence of such migration by younger pre-reproductive stages is not the result of an inverse correlation between migration cost and size. Again the absence of the migration by these younger stages may be taken to suggest that the critical gradient does not involve factors such as temperature that influence general physiology. Females migrate to deeper water to release their eggs or larvae whether the larvae, upon

metamorphosis, settle inshore, as in the Australian greentail prawn, *Metapenaeus mastersii* (Fig. 31.24), or in deep water offshore, as in the spiny lobster, *Panulirus interruptus* (Lindberg 1955). As all of these larvae have a planktonic phase we may perhaps conclude that the critical gradient is in some way related to change in the advantage of planktonic life with distance offshore. Increasingly clean suspension of phytoplankton is one possibility. Increasing depth of the water column in which vertical migration may take place is another. Although migration cost can be ruled out as the critical factor leading to the absence of a similar return migration by young individuals, decrease in migration cost per unit distance with increase in body size may well account for the increased migration distance on consecutive spawnings reported for most penaeids that live more than one year.

Although females of these species migrate offshore to release their eggs or larvae and although males may also perform the migration if copulation takes place only a short time before release of offspring or if the female produces several broods of eggs while offshore, most of the adult life of these species is spent inshore. These periods are essentially feeding periods and it seems likely that a gradient of increasing habitat suitability inshore at these times results from factors associated with food collection and processing. On the whole, it is likely that more nutrients are available inshore than offshore.

The suggestion is therefore that when feeding is the optimum behaviour, there is a gradient of increasing habitat suitability from offshore habitats to inshore habitats. When the release of larvae is optimum behaviour, however, the sign of the gradient of habitat suitability is reversed. Entry of this reversal of sign into a model for return migration seems to be sufficient to generate the observed pattern of inshore–offshore migration of adults of tropical or sub-tropical species that occur in shallow water.

No such generalisation can be made concerning the birth-to-death ontogenetic return migration of these species. Thus, penaeids and the two Californian carideans so far described settle as post-metamorphosis larvae inshore in shallow estuarine water and gradually migrate into deeper, more saline water as they approach maturity. Most Palinura, however, settle in deeper water and gradually move into shallower water as they approach maturity. Spiny lobsters, *Panulirus interruptus*, for example, do not arrive inshore until their second to sixth years. For most species migration to shallower water involves an inshore movement, but off Bermuda the direction is offshore, but from the deeper lagoon to the shallower reefs. Nor is the direction of this stage of ontogenetic return migration a function of estuarine and marine conditions. In temperate regions, for example, at depths of 200 m or more, a number of caridean species of the genus *Pandalus* are found that are zoned with respect to temperature, depth, and substratum. Larvae and newly metamorphosed juveniles of these species settle out from the plankton at the shallower end of the zone occupied

by their particular species. As they age they gradually migrate into deeper and deeper water within their species-zone. The migration is not continuous, however, and seems to occur in stages at yearly intervals. Males seem to move into deeper water after they have fertilised the females and females seem to migrate some time later when they have released the larvae. However, not all carideans living in moderate depths seem to perform such an ontogenetic return migration. There is no evidence of such a migration, for example, in the crangonid, *Pontophilus spinosus*.

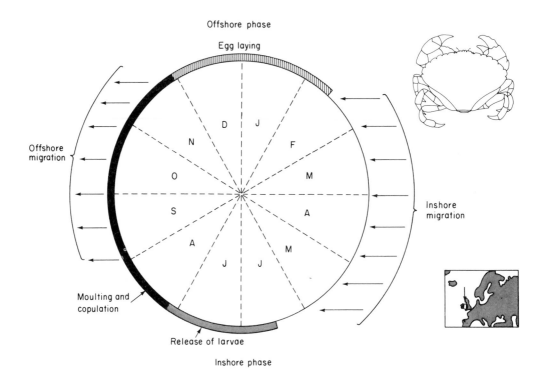

Fig. 31.25   Seasonal/ontogenetic return migration of the edible crab, *Cancer pagurus*, on British shores

General ecology
Cancer pagurus is a large (carapace up to 25 cm across) inhabitant of particularly rocky shores on which it is found at middle and lower shore levels. While the tide is out the crab is found mainly under stones and seaweed. After moulting and copulation the female lays eggs which are then carried for six months or so, not being released until they hatch into larvae.

Seasonal/ontogenetic return migration
The diagram indicates the integration of migration with

other behaviour in the annual cycle of those individuals that migrate. During the offshore phase of the cycle the crabs are found at depths of 40–60 m. The offshore migration is partial, only being performed, after copulation, by some females. This may be taken to indicate that the gradient of increasing habitat suitability offshore in winter and the change of sign in spring acts primarily on the eggs being carried by the females rather than on the adults themselves. The partial nature of the migration can then be generated by entering this gradient (due to temperature?) and a deme density effect into the model presented in Fig. 8.10 (p. 56).

[*Compiled from Pearson (1908) and Allen (1966)*]

In order to generate such a pattern it is necessary to postulate a complex variable or assemblage of variables that form gradients of selective pressure such that: (1) in estuaries, smaller forms are at an advantage to larger forms up-river and larger forms are at an advantage to smaller forms nearer the coast; (2) in shallow water (less than about 200 m) in the sea, larger forms are at an advantage inshore and smaller forms are at an advantage offshore; and (3) in moderately deep water (greater than about 200 m), smaller forms are at an advantage in shallower water and larger forms are at an advantage in deeper water. No single environmental factor springs readily to mind that could generate this sequence of

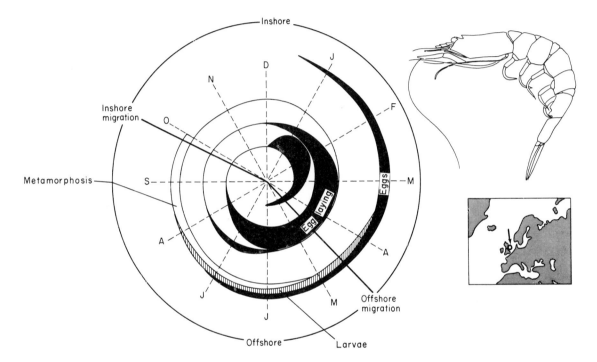

Fig. 31.26   The seasonal/ontogenetic return migration of the shrimp, *Crangon allmani*, off the coast of northeast England

General ecology
Crangon allmani is an eastern Atlantic boreal member of the Caridea, the dominant natant decapod group in temperate regions. The species occurs from the White Sea in the north to the northern part of the Bay of Biscay in the south. It has been recorded from Iceland and is present throughout the North Sea and Kattegat, being found on all substrata except rock at depths of 20–200 m. Feeding occurs by searching in the top few millimetres of the substratum. The most commonly taken food items off northeast England are living crustaceans and annelids, particularly *Nephthys*. Molluscs, foraminiferans, and ophiuroids are also taken. Significantly more worms are taken when offshore than inshore. Female *C. allmani*, in common with other carideans, retain the eggs on the abdomen after laying and fertilisation, not releasing them until the larvae hatch. The life span is probably $2–3\frac{1}{2}$ years.

Life history and migration
The diagram shows the integration of ontogenetic events and the seasonal inshore–offshore migration. Off northeast England the offshore phase is spent in some 80–100 m of water, 15–25 km off the coast. The majority of all year age-classes except perhaps the oldest (2–3 years age-class), which reproduce earliest, migrate inshore in late September and early October, perhaps in part by utilising the inshore component of the flood tide. Although the majority of females produce a brood of eggs and fertilisation takes place while still inshore, the major part of larval release takes place during or after the offshore migration (especially as the earliest larvae to be released are produced by the oldest females which tend to remain offshore). Both males and females migrate and a second brood of eggs may be produced and fertilised during the offshore phase. Larval development and metamorphosis takes place offshore. The inshore migration coincides with the temperature maximum of offshore deep water and the offshore migration coincides with the temperature minimum of inshore water.

[Re-drawn, with some modification, from Allen (1960)]

changes in sign of gradients of habitat suitability relative to age/size along a transect from brackish to moderately deep sea-water. Gradients of salinity, turbidity, and temperature could combine to generate some parts of the sequence. In addition, day length varies along this transect, being shorter with increased turbidity and increased depth. For nocturnal offshore species, therefore, increase in the period of activity with depth reinforces any advantage of decreasing temperature but detracts from any advantage of increasing temperature. The opposite, of course, applies to diurnal species and it could be instructive in the future to examine species differences in migration direction relative to diel activity incidence. Competition between different year classes may also contribute to the gradient of habitat suitability once the direction of the gradient has been established through some other factor.

In addition to the inshore–offshore mid-summer return migration of the spiny lobster, *Panulirus interruptus* (p. 833), on the Californian coast (Lindberg 1955), mature individuals of this species show an offshore movement in autumn (October) and an inshore movement in spring. Copulation occurs during this offshore phase, particularly from November through March, males depositing a spermatophore ventrally on the sternum of the female. Young and immature spiny lobsters do not take part in this migration. This either suggests that the migration is associated with gradients of habitat suitability that are irrelevant to immatures, such as suitability for mating, or that some seasonal reversal of gradient sign that takes place is of insufficient magnitude to overcome the higher cost of a seasonal inshore–offshore migration for smaller, younger individuals. The migration of this species is performed by walking.

Further problems are encountered in relation to the seasonal and ontogenetic return migrations of some other reptant decapod crustaceans. Thus the commercially exploited edible crabs of Europe and America, *Cancer pagurus* (Fig. 31.25) and *C. magister*, and the American blue crab, *Callinectes sapidus*, all of which occur in shallow water, migrate inshore to release their larvae. The migrant/non-migrant ratio is least in *Cancer* and the migration may only involve some of the females.

The spring and autumn migrations of the spiny lobster in California in waters the mean surface temperatures of which rarely fall below 12°C and of the edible crab in the Irish Sea (Fig. 31.25) may or may not be a seasonal rather than a reproductive

ontogenetic migration. Other species, however, particularly in temperate waters, show clear seasonal migrations that may even cut across breeding seasons. In European waters, for example, the shrimp, *Crangon vulgaris*, has a prolonged breeding season from March to November which covers both the inshore and offshore phases of the migration. There does, however, seem to be a depth effect. The congener, *C. allmani*, lives in deeper water but adjacent to *C. vulgaris*, the area of overlap occurring in water about 20 m deep. The migrations of the two species are such that when *C. vulgaris* migrates inshore, *C. allmani* migrates offshore and vice versa. The migration of *C. allmani* is illustrated in Fig. 31.26 and in many ways is similar to the migrations just considered that are performed by adult tropical and sub-tropical sub-littoral decapods. As argued in the legend to Fig. 31.26, by far the major amount of larval release occurs in offshore water, whereas the major feeding phase occurs in inshore water. The migration of *C. vulgaris*, on the other hand, is closely connected with the seasonal changes in sign of inshore temperature gradients with depth and is such that the species always migrates toward warmer water. In temperate regions, the temperature of coastal waters decreases with depth in summer but increases with depth (down to a certain level) in winter.

This change in sign of the temperature gradient at shallow depths in coastal waters coupled with variation with latitude and depth in the steepness of the gradient, when entered in a return migration model, can account for a large proportion of observed variation in inshore–offshore migration of shallow-water species. Note that according to such a model the timing of the migration is a response to the change in sign of the environmental gradient and that the distance component of the migration should be a function of the steepness of the gradient, the change in steepness (in this case with depth) of the gradient, and the change (again with depth) in migration cost per unit distance. The migration should terminate at the place where the advantage resulting from the (temperature?) gradient becomes less than the disadvantage of further migration. In the present case, if the temperature gradient is the critical factor or is correlated with the critical factor(s), the higher the inshore temperature in winter, the less steep the critical gradient. Above a certain winter temperature, therefore, a zero migration distance is favoured. Below that temperature, an increasingly large migration distance is favoured until the point is reached below which further decrease in surface temperature does not change the depth at which the temperature gradient is outweighed by the migration cost. Bearing these expectations in mind, a number of examples may be presented that suggest that temperature, or some variable or variables that correlate with temperature, is the variable that forms the critical gradient.

Analysis of the numbers of the shanny, *Blennius pholis*, a littoral fish, in rock pools at low tide throughout the year shows a clear correlation with temperature (Gibson 1967) and other circumstantial evidence also suggests that there is a downshore migration in winter (Fig. 31.27). A similar migration seems to be performed by many other littoral fish (Gibson 1969). There is, however, latitudinal variation in the components of the migration that is consistent with the model. In the Gullmarsfjord in Sweden, for example, the sandy-shore living goby fish, *Pomatoschistus minutus*, migrates offshore in winter. At Penpoul in Brittany, however, the distance component of the migration is much reduced, probably to zero. Similarly, the distance and migrant/non-migrant ratio components of the winter offshore migration of an ecologically similar congener, *P. microps*, has been found to be a function

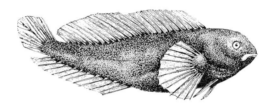

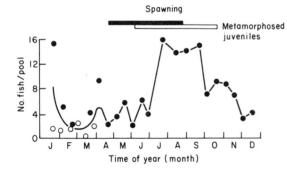

Fig. 31.27 Seasonal return migration of the shanny, *Blennius pholis*, on the west coast of Anglesey

General ecology

The shanny is a yellow or greenish-coloured fish blotched with dark patches such that it blends well with its usual background of irregular rock or weed. In summer, it is never found as low on the shore as the *Laminaria* zone and occurs mainly in the *Fucus* and *Himanthalia* zones (i.e. approximately from just above the high-water neap level to just below the low-water neap level). Like many shore fishes it has no scales and the body is soft and slimy. It is found in shallow rock pools or under stones and is an omnivore and a grazer, moving round biting off barnacles and feeding on slow-moving organisms such as annelids and molluscs as well as some rapidly moving Crustacea. The fish spawns in spring and summer when numbers of bright amber-coloured eggs are laid in shelter under rocks or in crevices. The male changes colour at this time to a sooty black except for the prominent white lips and stands guard over the eggs until the young hatch. These then spend about two months in the coastal plankton before metamorphosing into the adult form.

Seasonal return migration

The graph shows the seasonal variation in the mean number of *B. pholis* of all age groups per pool on the west coast of Anglesey in 1964 (solid circles) and 1965 (open circles). The line from January through March is the mean for the two years. In general, fish are less abundant in the winter months. The number of fish in the inter-tidal zone at low tide is correlated with temperature. A movement downshore in winter seems likely and is supported by the collection of a few shanny from the *Laminaria* zone in November and December but at no other times and by the presence in the shanny stomach in December and January of a mollusc that occurs almost exclusively on *Laminaria* fronds. There is some indication from marking experiments that the shanny establishes a temporary home range. Although only 20 per cent of the deme contained by a pool stayed there for more than 2 weeks, over half returned at least once after an absence. Whether seasonal home ranges exist that are part of a larger familiar area remains unknown.

[*General ecology from Yonge (1949) and Gibson (1967). Seasonal return migration from Gibson (1967)*]

of winter inshore sea temperature. In areas where winter sea temperatures fall below 5°C an offshore migration is performed. Where the minimum sea temperature is above 7°C, the distance component is zero. In those areas with intermediate temperatures the migrant/non-migrant ratio and distance components of the offshore migration varies according to the minimum temperature each year.

Similar variation is found in the distance component of the seasonal return migration of the palaemonid prawn, *Palaemon serratus*, in different parts of its geographical range. Off Plymouth, southwest England, where the minimum inshore surface temperature is about 8°C, the species tends to leave the estuaries in winter, but the offshore movement is slight and specimens can be found between tidemarks over most of the year. Further north in Scandinavian waters, however, where the minimum inshore temperature may be −1°C, the species over-winters in deep water and migrates

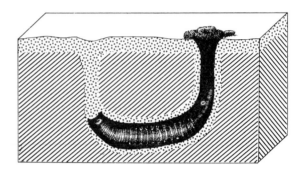

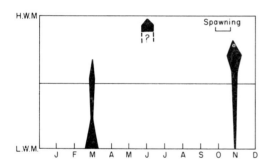

Fig. 31.28   Seasonal return migration of the lug worm, *Arenicola marina*, on the coast of northeast England

drawn into the burrow by the tail. *A. marina* is a northern species that spawns in the autumn as indicated, though spring spawnings have also been reported.

Seasonal migration
During the winter, *Arenicola* is exposed at low tides to sub-freezing temperatures. In exceptional winters up to 90 per cent of a deme may be killed by low temperatures. The probability of exposure to lethal temperatures is a function of the period for which the animal is uncovered by the tide. This gradient is most pronounced, however, for this boreal species during the period of minimum air temperatures from December through March. Another gradient acting in the same direction but throughout the year is that due to energy loss through the mechanisms for surviving asphyxiation while uncovered by the tide. A third gradient that probably acts in the opposite direction is that of predation risk, which is likely to be greater when covered by the tide. The nett result of these three gradients is a change of sign in the gradient of habitat suitability from summer to winter such that a distribution higher up the shore from April to November than from December to March is favoured. The kite diagram shows the distribution of numbers of lug worms at three times during the year. The observations for March and November are fully comparable and were made in Cullercoats Bay. The observation for June is part of a series made at Budle Bay and is not strictly comparable though it does indicate a higher distribution in mid-summer (June) even than in late autumn (November). The migration mechanism is unknown but could involve swimming (Seymour 1972), surface crawling, burrowing, or displacement by waves.

General ecology
The lug worm lives on sandy beaches characterised by enough mud to provide detritus for food but with sufficient sand to prevent its respiratory organs from being clogged. Its presence on a beach is apparent from the small depressions on the sand that are formed at the head end of the U-shaped burrow and by the earthworm-like casts that mark the position of the tail end of the burrows. The worm spends most of its time in the horizontal gallery lined with mucus. The vertical tail shaft is used only when the worm backs up it to defaecate at the entrance. The head shaft which is also vertical is invariably filled with sand but its position is apparent from its yellow colour (= stippled sand in diagram) which contrasts strongly with the sulphide layer of black sand (= hatched sand in diagram) in which it is normally located. The worm eats the sand in the head shaft. As this sand is consumed, more falls down from the surface. Water for respiratory purposes is drawn in through the tail shaft, passes over the worm, and then up the head shaft. When the tide is out, water movement is impossible but some oxygen is obtained from bubbles of air

[*General ecology from Wells (1945) and Eltringham (1971). Seasonal kite diagram compiled from Brady (1943)*]

inshore in the spring. During June and July, however, the animals perform a second offshore migration to deeper water where the eggs hatch and the larvae are released. The newly metamorphosed juveniles migrate immediately to shallow water to be followed a little later by the adults. The species then remains inshore until initiating the offshore winter migration. This mid-summer offshore migration seems likely to be similar to that described for the spiny lobster, *Panulirus interruptus* (i.e. associated with selection acting on larval release) though it may be interpreted as a response to high (greater than 20°C) inshore temperatures.

Some other upshore–downshore migrations can probably also be attributed to change in sign of the temperature gradient. Thus, the common shore crab, *Carcinus maenas*, on the south coast of England, makes its first appearance in the littoral zone in early April. In winter, However, it migrates offshore into deeper and permanent water (Edwards 1958). Boreal inter-tidal species, such as the shore crab, probably experience an added component to the temperature gradient to which they are exposed, a component, however, that acts with even greater intensity on 'full-time' littoral species such as the lug worm, *Arenicola marina* (Fig. 31.28). This component is that of exposure to air of a potentially lethal temperature. Period of exposure to such air is a function of height above low water spring tides. The other gradients that seem likely to be involved are outlined in the legend to Fig. 31.28. Downshore migration in winter, such as that found in the lug worm, will only evolve, however, if the ratio of efficiency of protection from freezing due to: (a) shorter exposure to air temperature; and (b) production of a given amount of 'anti-freeze'; is greater than the ratio of the cost of: (a) migration; and (b) production of anti-freeze.

The previous few paragraphs have been concerned with shallow-water temperate animals, the seasonal return migrations of which have been explicable almost entirely in terms of the seasonal change of sign of the temperature gradient in these waters. The absence of seasonal, as opposed to ontogenetic (where the two can be distinguished), return migration in tropical and moderately deep (200 m plus) temperate waters in which the change in sign of the temperature gradient with season does not occur also lends testimony to the importance of this environmental gradient in the evolution of seasonal return migration. Nevertheless, even for shallow-water boreal species, temperature gradients cannot always generate the observed migration pattern on their own.

For some species it is necessary to enter some other factor into the model in addition to temperature. An example of such a factor is the variation in distribution of the plants and animals on which a species may feed. Hence, the Aesop prawn, *Pandalus montagui*, in shallow water off English coasts feeds to a large extent on the small, tubiculous polychaete, *Sabellaria*. Off the southeast coast of England, *Sabellaria* is found over a wide depth range and here *Pandalus montagui* lives in 20–40 m of water during the winter but migrates inshore in spring and summer, often into depths of less than 4 m. Other demes of the prawn 450 km to the north, however, off northeast England where there is little *Sabellaria* present in shallow water are found predominantly at depths of 50–60 m, where *Sabellaria* is more common. The variable component of the migration, however, does not appear to be distance but the migrant/non-migrant ratio. Some *Pandalus montagui* migrate inshore in summer even off northeast England. Quite possibly the situation can be described by that part of the model described on p. 71 which suggests that selection on the temporal and frequency distributions of migration thresholds for migration from habitat-type A to habitat-type B will stabilise when A and B are of equal suitability throughout the year or when one habitat-type remains the most suitable despite supporting 100 per cent of the deme.

Some other species, however, that might be expected to show a migration that is explicable at least in part in terms of the seasonal change in sign of the temperature gradient nevertheless seem to have evolved a return migration that is independent of temperature as the gradient variable. This does not mean, however, that an environmental temperature event is not important in the behaviour that is integrated with the return migration. Thus, the sea urchin, *Echinus esculentus* (Fig. 31.29), migrates inshore two or more months after the temperature gradient sign has changed, spawns inshore at about the time that the surface temperatures of inshore waters pass their minimum level, and then gradually migrates offshore once more. This offshore migration effectively spans both the spring and autumn changes in sign of the temperature gradient. In no way, therefore, can the seasonal/ontogenetic return migration of *E. esculentus* be considered an adaptive response to temperature gradients. Some other species-specific suggestions are made in the legend to Fig. 31.29.

In many ways the seasonal return migration of the king or Kamchatka crab, *Paralithodes camtschatica*, a large (carapace up to 227 × 283 mm) species found in the North Pacific, may represent more of an adaptive response to temperature than the previous species, even though superficially the migrations appear to be very similar. Like the other crabs referred to in this section the king crab migrates inshore to breed. The inshore migration commences in December, the younger animals migrating first and the males migrating before the females. The animals do not, however, arrive inshore until about April (Powell and Nickerson 1965). Mating and egg laying occur in April and May in water as shallow as 12–16 m. The eggs are not released and are carried for about 11 months. Consequently the larvae are also released from a shallow-water position. Marked adults have been observed to have a one-way

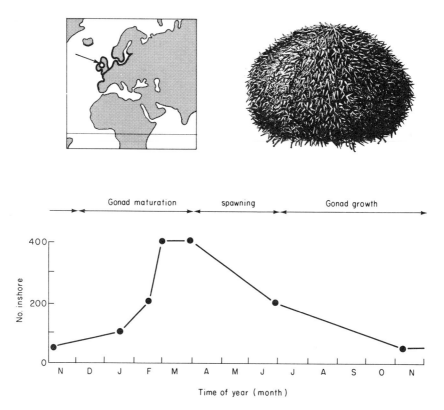

Fig. 31.29   Seasonal/ontogenetic return migration of the sea urchin, *Echinus esculentus*

General ecology
The sea urchin, *E. esculentus*, is found approximately from Iceland in the north to Finland in the east and Portugal in the south as indicated by the thick black coastal line in the map. As an adult, it feeds predominantly on polyzoans, *Laminaria*, *Membranipora*, and balanid barnacles. The time interval from settling out from the plankton to first spawning as an adult is approximately four years.

Seasonal/ontogenetic return migration
The graph shows counts of *E. esculentus* inshore on a breakwater at different times of year at Port Erin, Isle of Man. The major annual reproductive events are also indicated. When not inshore the species is found in the laminarian zone at depths down to about 40 m. Except that maximum numbers inshore coincide with minimum inshore water temperatures, the migration does not seem to correlate with changing sign of gradients of either temperature or day-length. Gradients of predation risk and efficiency of fertilisation seem the most likely, upon entry into the return migration model, to generate the observed migration components. Spawning more or less coincides with the vernal bloom of phytoplankton in inshore waters.

[*Compiled from Elmhirst (1922) and Stott (1931)*]

distance component of up to 100 km and it seems likely that the young larvae are displaced a similar distance before they metamorphose and become bottom living.

Like the sea urchin, *Echinus esculentus*, the king crab starts its inshore migration at a time unrelated to any change in sign of the temperature gradient. Unlike the sea urchin, however, the king crab does not arrive inshore until the inshore waters have started to warm up relative to offshore waters. Perhaps the initiation of inshore migration in December can be attributed to the time required for an animal to walk 100 km or so from its offshore stations and yet still arrive inshore by April. Alternatively, certain species-specific gradients based on predation risk and availability of food may be involved, similar to those postulated for *E. esculentus* (Fig. 31.29).

Gradients of predation risk certainly seem likely to be involved in the migrations of some sub-littoral fish. In the north of their range the lumpsucker, *Cyclopterus lumpus*, and the viviparous blenny, *Zoarces viviparus*, stay inshore in summer and migrate offshore in winter (Zenkevitch 1963). In the south of their range in Britain, however, the migration is reversed. Here the fish spends the summer in deeper water offshore and migrates inshore in winter to spawn, the lumpsucker laying attached eggs near low water mark (Jenkins 1936) and the viviparous blenny bearing live young which are released into the littoral zone (Yonge 1949). These southern demes therefore reproduce inshore at a time when many potential predators of both the adults and their larvae have performed a seasonal migration offshore. Further north, it may be supposed that the amplitude of the change of sign of the temperature gradient is sufficiently great to outweigh any other gradients that may be acting and thus generate the typical seasonal migration response of an offshore movement for winter and an inshore movement in summer.

In both the lumpsucker and viviparous blenny, of course, it is possible that southern demes are also showing adaptive response to temperature gradients but in the opposite direction (i.e. migration toward a lower temperature) to the northern demes. A gradient of predation risk in winter as suggested may either not be involved or may be additional to that of temperature. In some cases, however, the migration can scarcely be considered to be an adaptive response to anything other than opposing gradients of food availability and predation risk on offspring.

The clearest example is that of the Californian grunion, *Leuresthes tenuis* (Fig. 31.30). After hatching, the fish, as both larva and adult, lives pelagically offshore, presumably in the zone optimum for obtaining food and avoiding predation. Spawning, however, takes place inshore on sandy beaches in moist sand out of the water. The spawning and migration behaviour is so clearly adapted to ensure that the eggs develop in a position uncovered by the tide that it can really only be interpreted as an adaptive response to avoid predation on the eggs by marine predators. The closely related species, *Hubbsiella sardina*, behaves similarly to *L. tenuis* except that whereas the latter spawns at night, *H. sardina* spawns during the day.

The last few examples of return migration have involved timings and directions that it seems are only explicable in terms of more or less species-specific gradients of predation risk and food availability. A number of other inter-tidal animals do not perform an inshore–offshore migration even though one would imagine that they are mobile enough to have a relatively low migration cost, at least as far as energy expenditure is concerned. For example, although the spider crab, *Pisa tetraodon*, performs an inshore–offshore migration, many other spider crabs apparently do not, even though some of the species concerned are known to be able to swim. Similarly, in other brachyuran groups, such as the Xanthidae, not all species perform such a migration, even in the case of inter-tidal species. Any model that attempts to generate observed patterns of return migration in other littoral and sub-littoral species should also eventually be capable of including the species-specific circumstances that fail to generate a return migration in these species.

As a final example of an ontogenetic return migration that requires species-specific gradients to be entered into the model in order to generate observed behaviour, the case may be presented of the Florida land crab, *Cardisoma guanhumi*. This species may be found as much as 8 km from the sea. The eggs are produced while inland but the females migrate in sharp lunar and semi-lunar peaks preceding the new and full moons between June and November to the sea to release the hatching larvae. Here, the female remains a single night at the edge of the sea and periodically enters the water to release young by fanning the abdomen. Young crabs appear on land some 5 months later. This migration can only be generated by the model by entering into it gradients, from the inter-tidal area landwards, of

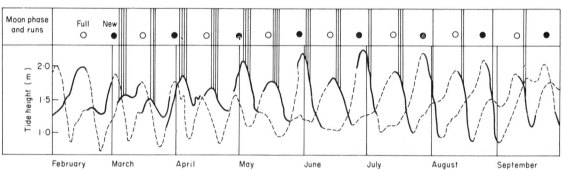

Fig. 31.30 Ontogenetic and semi-lunar return migrations of the grunion, *Leuresthes tenuis*

General ecology

The grunion is a small (15 cm) silvery atherinid fish found only along the coast of southern California and northern Baja California. Maturity is reached when 1 year old and individuals may live for up to 4 years. Most of its life is spent pelagically within a few kilometres of the shore line in water 5–15 m deep.

Return migration

Eggs are laid in moist sand out of the water on specific nights from February to September. The bottom diagram shows the timing of grunion spawning runs in relation to the tide and the phase of the moon in 1946. In the diagram the curves show the variation in height of high tides 24·8 h apart. The two tides of each day thus yield two series of curves. Solid lines indicate high tides occurring in darkness, dotted lines indicate high tides occurring in daylight. Nights on which a grunion run took place in 1946 are indicated by a vertical line (one line for one run) extending up into the part of the diagram indicating the phase of the moon. Observations were made on Scripps Beach, La Jolla, California. Variation in the number of individuals involved in each run is not indicated.

The grunion spawns only on 3 or 4 nights following each full or new moon. A spawning run is heralded by a few lone fish, usually males, that swim in with a wave at or soon after high tide. By swimming against the outflowing water the fish strand themselves on the beach until another wave

washes over them. Females arrive and spawning starts about 20 minutes later, the peak of activity, which lasts for about 30 min to 1 h, being reached about an hour after the first fish appeared. During a 'good' run thousands of shimmering silver fish may be on the beach at one time. Finally, when the tide has dropped 30 cm or so the run suddenly stops. The females swim onto the beach accompanied by 1 or more (up to 8) males. Each female swims as far up on the sand as possible and digs herself in as the wave recedes. As she deposits her eggs (1000–3000) in the sand, the males curve round her as in the drawing and discharge their milt onto the sand near the female. Digging in and egg laying takes about 30 s. The spent female frees herself from the sand and returns to the sea with the next wave to reach her. Large females produce about 3000 eggs every two weeks and spawn four to eight times during the season on consecutive runs.

The eggs are deposited about 5 cm below the surface by the females and are buried deeper by sand deposited by the outgoing tide, eventually being covered by 20–40 cm of sand. Here the eggs remain, in moist sand but uncovered by the sea, for about 10 days until the next series of high tides erodes the beach and washes the eggs out of the sand. 2–3 min after the eggs are freed from the sand the young grunion hatch and are washed out to sea. If the eggs are not reached and freed by the following series of high tides they can remain in the sand for another two weeks, and still hatch if washed free by the following series of high tides.

[*Compiled, with modifications, from Walker (1952)*]

food availability and predation risk that could act on a marine species with the pre-adaptations for the colonisation of the terrestrial environment for all except larval life.

Perhaps the major gap in our understanding of the ultimate factors involved in the seasonal/onto-genetic return migrations of littoral and sub-littoral marine animals is not so much a lack of gradients that, upon entry into the model, can generate observed migrations but rather a complete absence of demonstration of the mechanisms by which such gradients may act as selective pressures. For example, a great deal of observed behaviour can be generated as shown by entering temperature gradients into the model and postulating that it is an advantage to migrate the optimum distance in the direction of the higher temperature. However, the danger of exposure to inevitably lethal temperatures is present in only a small proportion of these cases. The advantage of migration toward warmer temperatures thus remains to be evaluated. Other gradients of advantage, such as that suggested for the release of larvae in deeper water, are similarly not understood. Gradients of predation risk and food availability that were suggested for some species, although readily understood, have in no case been demonstrated in the field.

No attempt has been made in this section to generate curves of migration thresholds. The reader should note, therefore, that all discussion in this section has been concerned with ultimate, not proximate factors.

31.6 The anadromy/catadromy paradox

Surprisingly, the evolutionary paradox of the existence in the rivers of some parts of the world of two groups of fish, one of which breeds in freshwater and migrates to the sea to feed and the other of which breeds in the sea and migrates to freshwater to feed, has received little detailed consideration. Yet clearly any model that attempts to generate optimum migration behaviour must, if it is to be acceptable, be capable of generating these differences in migration direction.

In Section 31.3 of this chapter, the selective pressures entered into the migration model seemed to be adequate to generate most of the observed

components of the migrations of fish that breed in freshwater. These selective pressures on their own, however, cannot accommodate the catadromous pattern of migration. Similarly, the selective pressures entered into the model in Section 31.4.1 seem to be adequate to generate the observed migration patterns, as far as they are known, of fish that breed in the sea. No selective pressures have yet been entered into the model, however, that are of sufficient magnitude to overcome the effect of the selective pressures entered into the model in Section 31.3 and thus generate the entry of sea-breeding fish into freshwater to feed. If such selective pressures that are specific to the animals that show catadromous behaviour can be found, they would be sufficient to resolve the anadromy/catadromy paradox.

It is important, however, to be clear concerning the points of difference between the behaviour of catadromous and anadromous fish while in fresh-water. In fact, there is only one major difference: young fish are not deposited in an upstream position as eggs but arrive there as a result of their own behaviour. Apart from this, catadromous fish, and crustaceans, behave in accordance with the selective pressures outlined for anadromous fish in Section 31.3 (p. 798). This includes, except for eels, a gradual downstream movement with increased size (Fig. 31.24). In eels this particular selective pressure, discussed on p. 799, has probably acted with less intensity than on other fish as a result of eels having been preadapted by virtue of their shape, which presumably evolved in the depths of tropical seas, for emergence onto land to feed. Upstream migration by young fish rather than by their parents will remain advantageous as long as the advantage of spawning into the sea is greater than the disadvantage due to increased migration cost. The only major difference characteristic of catadromous migration, therefore, rests, within the framework of selection for return migration, in migrating from sea to freshwater to pass the growth phase of ontogeny.

Our search for the selective pressures that have produced the anadromy/catadromy paradox can therefore concern itself solely with the advantage to some fish in crossing the river/sea boundary in one direction to feed and to other fish in crossing the boundary in the other direction. One possibility derives from selective pressures that could result from the specialised array of parasites and predators that an animal acquires as a result of the passage of time in a given habitat-type. Thus a fish with immediate freshwater ancestry is likely to have

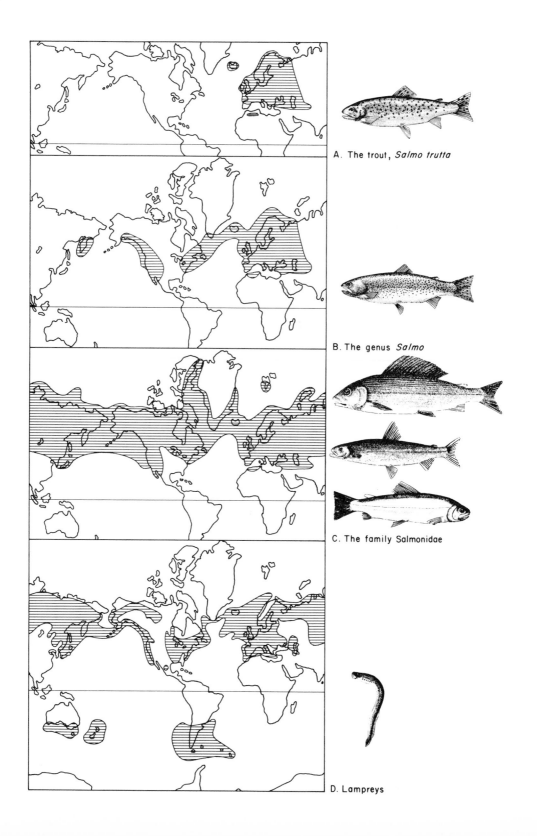

A. The trout, *Salmo trutta*

B. The genus *Salmo*

C. The family Salmonidae

D. Lampreys

Fig. 31.31 Latitudinal variation in the migration distance of fish that breed in freshwater

A. Geographical range of the trout, *Salmo trutta*, excluding recent introductions

Two forms of *S. trutta* are recognised. These are the sea trout, which is anadromous, and the brown trout, which performs downstream–upstream return migration but entirely within freshwater. As shown in the map, sea trout are found from the White Sea, Scandinavia, and Iceland in the north to the Bay of Biscay in the south. Brown trout are found in all the rivers in which sea trout spawn and are also found in rivers further to the south that enter the western Mediterranean and the northern part of the eastern Mediterranean. Brown trout are also found in Corsica and Sardinia. However, there are no sea trout in the Mediterranean.

B. Geographical range of the genus *Salmo*, excluding recent introductions

The genus *Salmo* includes the rainbow trout, *S. gairdneri*, of the Pacific coast of America and the Atlantic salmon, *S. salar*. In fish from the rivers leading both into the Pacific and the Atlantic the maximum expression of anadromy is found in the northern part of the geographical range of the genus. Furthermore, anadromous fish from the more southern part of the range tend to migrate to the open sea well to the north in order to feed (Fig. 31.9)

C. Geographical range of the family Salmonidae, excluding recent introductions

As well as the salmon and trout of the genus *Salmo*, the family Salmonidae includes the char, *Salvelinus* spp. (e.g. the Arctic char, *S. alpinus*, and the brook trout, *S. fontinalis*, of eastern N. America); huchen, *Hucho* spp. (e.g. the Danube salmon, *H. hucho*); Pacific salmon, *Oncorhynchus* spp.; whitefish, *Coregonus* spp.; white salmon, *Stenodus* spp.; and grayling, *Thymallus* spp. Not only is this family, which is one of the two major groups to contain anadromous fish, confined to temperate and sub-polar regions, but also, within the family, there is a trend toward less anadromous behaviour in the more southerly parts of the range. This trend is also shown within individual species in addition to the trout described in A. The Arctic char, *Salvelinus alpinus*, for example, which is circumpolar, is frequently anadromous in the northern part of the range but in the south anadromy is less frequent and the species tends to live in deep lakes and is known as the Alpine char.

D. Geographical range of lampreys

The lampreys form the other major group of anadromous fish. The geographical distribution of the group is temperate and sub-polar and shows the same trend as described for salmonids. In particular, the expression of anadromous behaviour is most marked in the cooler parts of the range.

Conclusion
The absence of a major group of anadromous fish in tropical and sub-tropical latitudes and the trend within whole groups as well as single species of anadromous fish for there to be a greater expression of anadromous behaviour in the cooler parts of the range suggests that one or more of the

selective pressures acting on migration distance shows latitudinal variation. The nature of this variation is such that selection for migration beyond the river/sea boundary is greater at higher latitudes.

[*Compiled from Tchernavin (1939), Frost and Brown (1967), Hubbs and Potter (1971) and Mills (1971)*]

acquired a specialised array of parasites and predators that are adapted to attack the animal while in freshwater. As long as the array of generalised parasites and predators that the animal contacts if it enters the sea do not impose a greater mortality risk than the specialist enemies in freshwater, selection could well act to favour those individuals that enter the sea. This pressure can be added to those entered into the model in Section 31.3. A similar pressure could apply to species with marine ancestry that with sufficient frequency contact freshwater or estuaries. It is possible, therefore, that whether an animal's ancestry is freshwater or marine, contact with the other habitat-type during return migration could impose selection to extend the distance component of the migration such that the animal spends some time in the other habitat-type. Intra-specific competition would exert a similar influence.

Although this pressure could by itself be sufficient to resolve the anadromy/catadromy paradox, there seem to be other factors that contribute to the variation of the phenomenon. Figures 31.31 and 31.32 explore the geographical variation in the expression of anadromous and catadromous behaviour, not only in fish but also in the decapod crustaceans described in Section 31.5.2. It is concluded that anadromous behaviour shows maximum expression in temperate and subpolar regions, particularly in the northern hemisphere, whereas catadromous behaviour shows maximum expression in tropical regions.

Other data also provide support for the view that the evolution of anadromy and catadromy is in part a function of latitude and/or climate. Figure 31.31 was concerned solely with two groups of anadromous fish, salmonids and lampreys. Where anadromous behaviour occurs in fish of other groups, however, it seems to show the same trends. Thus, the three-spined stickleback, *Gasterosteus aculeatus*, has a very wide range, being found on the coasts and in the rivers of the Arctic and temperate regions of the northern hemisphere, extending as far north as

Greenland, Alaska, and Kamchatka, and as far south as Japan, California, New Jersey, and Spain. In northern regions it is essentially a marine fish; in the British Isles it is equally common on the coasts and in the rivers; and in Spain and Italy it is almost entirely confined to freshwater (Norman 1931).

If we consider reports of fish that are essentially marine but that are found frequently to enter freshwater, albeit irregularly in most cases, we find a predominantly tropical and warm temperate bias similar to that shown by fish that are more strongly catadromous. Thus sharks (Chondrichthyes), gobies (Gobiidae), and pipefishes (Belonidae) have been recorded in the lower parts of the Zambesi, and sting-rays (Chondrichthyes) have been recorded far upstream in the rivers of Nigeria and South America (Hynes 1970). Menhaden, *Brevoortia* sp., anchovy, *Anchoa* sp., needle fish (Belonidae), and atherines are

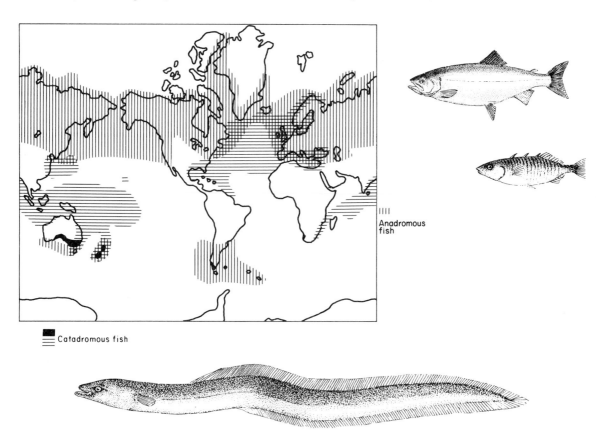

|||| Anadromous fish

▬ Catadromous fish

Fig. 31.32 Geographical distribution of anadromous and catadromous fish

Anadromous fish, as the term is used here, are fish that perform a return migration between freshwater and the sea with spawning taking place only in freshwater. Examples are salmon (*Salmo* and *Oncorhynchus* spp.), char (*Salvelinus* spp.), sticklebacks (*Gasterosteus* spp.), sturgeons (*Acipenser* and *Huso* spp.), and lampreys (*Petromyzon* and *Lampetra* spp.). See also p. 787).

Catadromous fish are species that perform a return migration between freshwater and the sea with spawning taking place only in the sea. Examples are eels (*Anguilla* spp.), southern trout (*Galaxia* spp.) and others (p. 822). In the diagram, the distribution of galaxiids is indicated by solid black shading (Australia, New Zealand, South America, and Africa). The distribution of other catadromous fish is indicated by horizontal shading.

Although the distributions of the two types of migrants overlap, anadromous behaviour seems to achieve maximum expression in temperate and sub-polar latitudes, whereas catadromous behaviour seems to achieve maximum expression in tropical and warm temperate latitudes. This impression is further reinforced if decapod crustacea are also taken into account. In this group, the maximum expression of catadromous behaviour is found in the Penaeidea, a tropical group (see p. 832).

[Compiled from Meek (1916) and Figs. 31.18 and 31.31)]

reported to enter freshwater in Virginia, United States (Hynes 1970) and the young of the flounder, *Platichthys flesus*, of Europe are often found far up rivers in freshwater, returning to the estuaries and the sea when they mature (Bagenal 1973).

Finally, we may consider the effect of climatic change on the distribution of anadromous fish. According to Netboy (1968), the distribution of the Atlantic salmon, *Salmo salar*, changed considerably in the 15 000 years or so between the peak of the last glaciation and the present day. As the ice-cap receded and sea temperatures in southern Europe increased it seems that the salmon disappeared from the Mediterranean and the southern part of the North Atlantic. Those demes that survived, such as perhaps the Dalmatian salmon, *S. obtusirostris*, of Yugoslavia (Mills 1971), which resemble a salmon parr, only seemed to do so by the restriction of migration entirely to freshwater. A similar adaptation seems to have taken place in the trout, *S. trutta* (Fig. 31.31).

While on the subject of the influence of the Pleistocene ice ages on fish migration it is perhaps worth digressing for a moment to note that previous authors have considered that the climatic changes that took place would have had a marked effect on the distance component of the downstream–upstream return migration of fish that breed in freshwater. Interestingly, however, students of different fish groups have reached quite the opposite conclusions as to what this effect may be. Thus students of salmonids have felt that the evolution of anadromous behaviour could be attributed directly to the progression and regression of the polar ice (Tchernavin 1939, Netboy 1968, Orr 1970a). The argument was that as the polar ice melted, the salinity of the sea became sufficiently reduced to be colonised by salmonids following downstream migration and that as the salinity later increased the fish gradually adapted. It is difficult to see, however, the advantage of this thesis over that of adaptation to marine conditions through the estuary habitat-type as a result of selection for gradual increase in migration distance as suggested in Section 31.3. Students of lampreys, on the other hand (e.g. Zhukov 1968, Hardisty and Potter 1971c), attribute the formation of non-anadromous, non-parasitic species (i.e. the brook lampreys—Figs. 31.3 and 31.8) to the effects of the glacial epoch. Here the argument is that demes of anadromous, parasitic lampreys became isolated in lakes, cut off from the sea, and thus lost the anadromous habit which then was not regained

when contact with the sea was re-established. The new species was genetically isolated from the ancestral anadromous species when next the two came into contact due to differences in body size and the inefficiency of any attempts at cross-breeding. There are, however, difficulties with this thesis. In particular, isolation in lakes hardly seems likely to lead to selection against parasitic feeding, lake lampreys invariably retaining the parasitic habit. The loss of parasitic feeding really only seems likely to have evolved in streams where suitable host species are scarce. It is difficult to see how glaciation could lead to isolation in streams. On the other hand, except that brook lampreys are given specific status and have evolved from an animal that was a filter-feeder when young, the paired-species phenomenon of lampreys falls well within the range of variability of migration distance that is such a marked characteristic of the majority of fish that breed in freshwater. As this variation is explicable (p. 802) without invoking the action of an ice age, and as the effect of the ice age, if it is to be invoked, must be postulated to have led to the evolution of anadromy in demes of one fish group and the loss of anadromy in another fish group, it seems best at present to accept a single model for variation in migration distance that does not include the ice age. This is especially so as in any case glaciation cannot be universally invoked for the loss of anadromy in lampreys (Hardisty and Potter 1971c). There are lamprey species, such as *Ichthyomyzon gagei*, *I. hubbsi*, and *Okkelbergia aepyptera*, the present distribution of which lies to the south of the glacial ice margins. For the moment, therefore, it is considered here that variation in migration distance would have evolved without the involvement of the ice age, even though inevitably such major climatic fluctuations must considerably have influenced the geographical distribution of the species with which we are concerned.

Having surveyed the evidence for latitudinal and climatic effects in the evolution of anadromous and catadromous behaviour, it is important now to isolate the critical questions concerning these effects. First, for example, the latitudinal effect does not influence the sign of the gradients of selection acting on the downstream–upstream return migration of fish that breed in freshwater. Throughout the world the general pattern of these migrations is the same (p. 787). Nor, probably, does the latitudinal effect influence the pattern of migration while at sea of those species that migrate beyond coastal waters. Thus both catadromous eels (Fig. 31.18) and anadromous salmon (Fig. 31.10) seem likely to have evolved migration circuits in the sea that are similar to those evolved by full-time marine fish, migration circuits that result in the fish occupying the areas most suitable for feeding at the appropriate time of year. We are left, therefore, with the conclusion that the critical site of action of the latitudinal effect is at the river/sea boundary.

The first selective pressure described in this section (p. 845) implies that throughout the world, wherever the return migrations of full-time marine and full-time freshwater fish bring them into contact with the river/sea boundary, there is always some residual selection for this boundary to be crossed in both directions. Latitudinal variation implies, however, that in tropical, sub-tropical, and warm-temperate regions the selection for fish to cross from sea to river is reinforced and/or the selection for fish to cross from river to sea is reduced. In cold-temperate and sub-polar regions, however, the selection for fish to cross from sea to river is reduced and/or the selection for fish to cross from river to sea is reinforced. In temperate regions, however, selection may continue to act in both directions. Also, of course, the selective pressures involved in specific cases may be sufficiently great to reduce or even reverse the general trends. Thus the absence of a flat, bottom-living fish in the freshwaters of the north temperate zone would increase the sea-to-river gradient for marine coastal flatfish and could account for the development of catadromous behaviour in temperate flatfish such as the flounder, *Platichthys flesus*, in Europe. Similar arguments could account for the northward spread without loss of the catadromous habit by the *Anguilla* eels of the North Atlantic (Fig. 31.18) from the Indo-Pacific region in which they seem likely to have originated.

The hypothesis of residual selection throughout the world for fish to cross the river/sea boundary in both directions, to which is added further uni-directional selection, the sign of the gradient of which shows latitudinal variation, seems sufficient to account for a large proportion of the characteristics of the anadromy/catadromy paradox. It remains, therefore, to consider the environmental variable(s) that could form this unidirectional gradient of selection across the river/sea boundary. Here it has to be admitted that suggestions are scarce, though temperature, or some correlate of temperature, seems to be the obvious variable.

Such latitudinal variation in the sign of the temperature gradient across the river/sea boundary

as occurs is not consistent with a hypothesis based on that as the critical variable. We are left, therefore, with absolute temperature as the critical variable that shows the best correlation with observed latitudinal variation in anadromy and catadromy. As absolute temperature in itself seems unlikely to act as the critical selective pressure, it is necessary to postulate that there is a gradient of some other factor, such as efficiency of osmoregulation or availability of food, the sign of which across the river/sea boundary correlates with absolute temperature, whether or not there is a causal relationship between the two. Such a gradient, in order to generate observed patterns of migration, would have to contribute to an increase in relative suitability of the sea as a habitat-type for growth and development at low temperatures and an increase in the relative suitability of freshwater as a habitat-type at high temperatures.

The critical gradient to be entered into the model, therefore, must await future identification. Apart from that gap, however, the past three sections have attempted the development of a comprehensive model of the evolution of fish migration. The model also emphasises the dynamic nature of the evolution of fish migration. Thus many fish lineages that contact the river/sea boundary as a result of their return migrations are likely to be subjected to selection for extension of the migration across this boundary. Marine fish are more likely to cross the boundary in tropical regions and to become catadromous, whereas freshwater fish are more likely to cross the boundary in cold temperate and sub-polar regions. Where there is strong selection for a high degree of return to the natal habitat-type, as opposed to simply the natal site (p. 784), the species are likely to remain catadromous or anadromous as they spread. Where there is only selection for a low degree of return, catadromous species may lose contact with the sea altogether and become freshwater species and anadromous species may lose contact with freshwater altogether and become marine species. The frequency with which such changes must have occurred in the past is evident from the partitioning of many taxonomic groups of fish into marine and freshwater species (Norman 1931, Hynes 1970). Quite conceivably, some lineages repeated the process several times during their evolution, passing over the years through marine–catadromous–freshwater–anadromous–marine sequences one or several times. Questions over whether a group is ancestrally marine or freshwater (e.g. Tchernavin

1939, and Jones 1959 in relation to salmonids) are therefore only meaningful with respect to a particular time interval. Thus the salmonid lineage, when it shared common ancestry with herring-like fish, was probably marine but passed through a catadromous phase before its members became freshwater fish, which is the stage that was of interest to Tchernavin (1939). It then seems to have become partly freshwater and partly anadromous as with most present species. Even now, however, some species, such as the pink salmon, *Oncorhynchus gorbuscha*, which occasionally spawns in estuarine waters, could be in the process of becoming marine. It goes without saying, of course, from the arguments concerning migration distance (p. 802) that the formation within anadromous species of demes that migrate entirely within freshwater and within catadromous species of demes that migrate entirely within the sea is an even shorter evolutionary step than the evolution, respectively, of marine and freshwater demes.

31.7 Application to fish of the familiar-area hypothesis for return migration

Figure 31.33 shows the result of an experiment carried out on the white bass, *Roccus chrysops*. The evidence strongly suggests that these lake-dwelling fish establish a familiar area which may even encompass the entire lake. The seasonal return migrations that occur between the main body of the lake and the spawning area in a small part of the lake seem, therefore, to take place within a familiar area in just the same way as those of most other vertebrates considered in Part IIIb. The way in which the familiar area is established and the range of sensory cues used by the fish during this process are unknown as also, to a large extent, are the factors that cause such a limited area of the lake to be suitable for spawning.

The seasonal and ontogenetic return migrations of fish have now been described under three separate headings in this chapter and it is perhaps desirable at this point to inter-relate these three sections. In this section a single example has been presented of a seasonal/ontogenetic return migration by a lake-dwelling fish that clearly seems to establish a

familiar area within which the migration takes place.

In the previous section (p. 837), we were concerned primarily with marine coastal fish species that perform inshore–offshore return migrations. In at least some of these species there seems a strong possibility that the migration was in fact taking place within a familiar area. Certainly, examples are known of coastal species for which individual fish have an inshore and an offshore home range between which they alternate (Gibson 1969). In the case of the mud skipper (Fig. 31.20), for example, it seems highly probable that the migration takes place entirely within a familiar area. The spatial memory of some coastal fish is well documented and has been shown to achieve a high degree of sophistication in some species. Perhaps the best-known example is that of the goby, *Bathygobius soporator* (Aronson 1951, 1971). The species lives in tidal pools of the coasts of the western Atlantic and it seems that the fish is capable of building up within a spatial memory a fairly detailed picture of its surroundings. Most familiar-area establishment seems to take place at high tide while the fish is swimming over the rocks. The system is sufficiently sophisticated, however, for the fish to be able, at low tide when the rock pools are cut off from one another, to jump from pool to pool without any significant risk of finding itself on dry land. This is so even though the fish are unable to see the neighbouring pools before leaping. Homing experiments carried out in Bermuda have demonstrated that the species can use this spatial memory to return to a particular point in its familiar area when displaced to some other point within that area (Beebe 1931). Until demonstrated to the contrary, therefore, there seems a strong possibility that the inshore–offshore return migrations of fish take place within a familiar area.

In Section 31.3 (p. 786), we were concerned primarily with fish exposed to water currents that favoured migration in a single direction. The major part of that discussion did not concern the familiar-area phenomenon, for most of the present literature prefers an interpretation of fish migration based primarily on displacement by currents and/or up-current orientation. In either case only a minimum sense of location is attributed to the fish. This view has already been questioned (p. 775) and where evidence is available, as for goldfish, *Carassius auratus* (Kleerekoper *et al.* 1974), and salmon, the view is supported that migration takes place within a familiar area. Hypotheses concerning the navigation process in salmon are given in Chapter 33. The most attractive of the extant theses in the light of the data available is the sequential hypothesis which suggests that as the young salmon fry or smolt migrates downstream it learns the sequence of olfactory cues that it encounters. Whether these olfactory imprints are stored in a spatial memory or simply in a sequential memory has not yet been investigated. It is suggested here, however, that a spatial memory is the system that would be favoured by selection, for it would permit return to the natal site even if some of the clues were changed or disappeared while the salmon is at sea. On the other hand, the sequential system would be simpler. In either case the salmon should have some sense of location and if we were concerned only with migration within running water both systems seem to be potential adaptations, depending on whether the advantage lies with greater reliability or greater simplicity.

Let us suppose for a moment that salmon establish a familiar area that encompasses the whole stream from the salmon's natal site to the sea. What sensory cues could be used in the formation of this familiar area? Inevitably, most considerations of fish place the greatest emphasis on the olfactory sense, though I can see no reason why the remainder of the sensory repertoire of a stream-living fish should not also be involved in the formation of the familiar area as indeed it seems to be in the case of lake and coastal fish (p. 880). In short, it seems quite possible that as a salmon fry or smolt moves downstream it establishes a familiar area utilising all the sensory information available to it as described in the section on Amphibia (p. 884). In this case, the downstream migration is in effect a linear exploratory migration. Extending this concept further, I can see no reason why, when the salmon enters the sea, it cannot be considered to build up a large familiar area based not only on olfactory cues but also on visual,

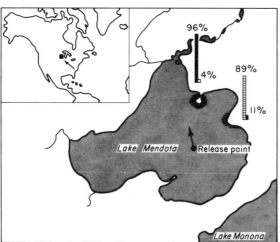

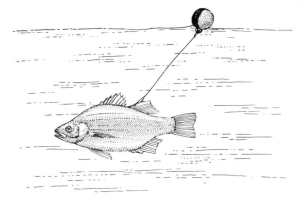

Fig. 31.33 Seasonal return migration of the white bass, *Roccus chrysops*, in Lake Mendota

The white bass, *Roccus chrysops*, has two spawning grounds in Lake Mendota. In the diagram, these two grounds are represented respectively by solid black shading and horizontal hatching. The distance between the two grounds is 1·6 km. The total shoreline of the lake is 32·4 km and its surface area is 39·4 km².

White bass spawn in Lake Mendota in late May and early June and spend the rest of the year in the open water. The histograms show the result of a displacement experiment on males collected during the spawning season from both spawning grounds. These were then released at midday in the centre of the lake as shown. Of those fish recaptured back at each spawning ground, the percentages that had returned to the spawning ground at which they were originally captured were 96 and 89 respectively, thus suggesting that the release point was within the familiar area of at least some of the released fish. Unfortunately, the full significance of the experiment cannot be evaluted as only about 15 per cent of the fish tagged were ever recaptured.

An attempt was made to follow the tracks of the displaced fish by attaching a small float by a nylon line. The arrow drawn from the release point shows the mean bearing of the released fish one hour after release.

[*Compiled from data and diagrams in Hasler* et al. (*1958*)]

auditory, thermal, and any other type of cue that is available. Thus, as the salmon moves away from shore, it does not enter a sensory void but instead continues to add information to the sequential or spatial memory that first began to function as a fry in the natal stream. If such a view is possible for the movement of salmon at sea, then it is also possible for other marine fish, as well as for other marine animals such as sea snakes, turtles, pinnipeds and whales.

One likely fact that at first sight seems to contradict this familiar-area hypothesis for the migrations of marine fish is that some demes arrive back at the natal site *with* a current that they seem unlikely to have encountered as an adult. Harden Jones (1968) uses this likely fact as an argument against the general application of the olfactory hypothesis to all marine fish. He argues that if a fish migrates back to a natal site by swimming with a current it cannot be swimming in water that is carrying olfactory cues from that site to the fish. However, not only is this not an argument against the familiar-area hypothesis, but the answer to it provides an insight into the way that marine fish could conceivably build up such large familiar areas as would be required to encompass the observed seasonal return migrations. When a fish hatches from its egg it is being carried in a body of water that not only carries sensory cues emanating from the natal site but is also carrying sensory cues picked up from locations up-current from the natal site. Sufficient cues, perhaps, for the fish, if it encounters the current at some distance up-current from the natal site, to be able to recognise that this is the current that passes through that site.

It seems likely that as currents flow on their course they gain some characteristics, mainly thermal and olfactory, and lose others. Where currents juxtapose, certain relationships are set up, including perhaps changes in depth which it seems likely a fish can measure. In addition, there is the possibility for fish that approach the surface, of perceiving relationships between characteristics of the water and other, perhaps celestial, cues. In short, no two points in space in the ocean need be identical (Aubert 1971, Aubert *et al.* MS, Aubert in Dowling 1975). It is now only necessary to postulate that fishes and other marine vertebrates can perceive these differences to the same extent that terrestrial vertebrates can perceive differences between two points in space in their habitat-type, and that these differences can be stored in a spatial memory, and a familiar-area basis for the migrations of marine vertebrates becomes a distinct possibility. It is then only necessary to add

the ability to recognise not only the region of the familiar area from which currents encountered during exploratory migration are emanating but also, as described in the previous paragraph, the region of the familiar area to which currents encountered are going, and it seems that the familiar-area hypothesis could account for all of the known characteristics of migrations of marine fish and reptiles.

The familiar-area hypothesis, when applied to marine fish and reptiles, would postulate that from the moment of birth onwards, the animals build up a familiar area in the same way as described for terrestrial animals (p. 378) but based in this case on the cues available to an animal living in the sea. The accumulated information is stored in a spatial memory that is such as to permit the animal to migrate from any one point in that familiar area to any other point as the situation demands or favours. Quite probably, also, the fish is able to learn which route gives the least migration cost. It is this perception of migration-cost variables that has led to the migration circuits of so many fish being predominantly with the current. The familiar-area hypothesis, however, permits sufficient behavioural flexibility for a deme, finding itself in a familiar area in which no suitable current exists, such as the Southern Bight deme of the herring, *Clupea harengus* (Fig. 31.12), still to be able to develop an adaptive migration circuit even though elsewhere other individuals on their spawning migration are migrating predominantly with the current.

On any other hypothesis, such inter-deme differences would have to be attributed to long periods of genetic isolation. However, deme-to-deme removal migration does occur (p. 196) and the case of the West Greenland cod, *Gadus morhua* (Fig. 31.13) serves to emphasise that many new demes must have been formed, particularly since the last glaciation, and that in most cases the initial removal migrants take up a migration circuit that is quite different in all but the most general of respects from that of the parent deme and yet nevertheless seems to be adaptive to local conditions. The behaviour of the immature cod of joining a spawning group and performing a complete 'trial' migration circuit before themselves beginning to spawn is also strongly reminiscent of the type of exploratory migration behaviour often associated with the familiar-area phenomenon (Chapter 15).

The association of the migration circuits of many demes of fish with current gyrals and counter-current systems discussed in Section 31.4.1 is still valid and in no way provides evidence against the familiar-area hypothesis. Most animals that are subject to a migration-cost variable (p. 49) evolve behaviour patterns that reduce migration cost to a minimum. Birds, for example, tend to migrate when the wind is favourable (p. 632). This does not mean that the birds migrate by wind displacement nor that they orientate to the wind during migration (though wind provides one of the many environmental cues that are included in a bird's navigational repertoire—p. 607). Rather, it shows that birds, like most animals, have evolved migration thresholds to migration-cost variables. The only basic difference between a bird being subjected to a current of air and a fish being subjected to a current of water is that for birds the strength and direction of the wind may change from day to day. Consequently, the initial phase of an adaptive response to

an unfavourable wind is simply to wait until it is favourable. For fish, current strength and direction are relatively constant, and the only available strategy if a current is unfavourable is to migrate to a place or depth where the current is favourable.

The hypothesis presented here, therefore, is that marine, catadromous, and anadromous fish establish a familiar area on the basis of the storage within a spatial memory of some of the olfactory and other environmental cues to which the fish is subjected from the moment of hatching onwards. As the fish builds up and moves around within this familiar area it is capable of learning the routes of minimum migration cost from one point in that familiar area to another. The establishment of the familiar area and the learning of routes is facilitated by the nature of current systems which may be recognisable as leading to a point in the familiar area as well as coming from a point in the familiar area.

32

The extent of the applicability of the familiar-area hypothesis of return migration

The familiar-area hypothesis of return migration presented in Chapter 21 and in various sections of each of Chapters 22–31 seems to be of general applicability to the return migrations of vertebrates. Indeed, for some groups, such as amphibians, terrestrial reptiles, and the larger mammals, it is difficult to conceive of any other model that could accommodate observed forms of the lifetime track. Paradoxically, however, the best evidence in support of the familiar-area hypothesis is found for birds. Avian migration is so spectacular, takes place over such long distances, and in many cases is so linear, that at first sight it might seem unlikely that it can be accommodated within a model based on the familiar-area thesis. Nevertheless, it is claimed here, on the basis of the evidence presented in Chapter 27, that it can be so accommodated. The migrations of bats can also readily be fitted within a familiar-area model, though less supporting data are available than for birds. Finally, it is beginning to look as though even some terrestrial snails and spiders may perform a seasonal return migration within a familiar area.

All terrestrial vertebrates and some invertebrates, therefore, seem to spend the major part of their lives within a familiar area, whether or not they perform a seasonal or ontogenetic return migration. The major problem in the application of the familiar-area hypothesis concerns aquatic animals. To a large extent, however, this problem seems likely to be conceptual rather than real. Man is a terrestrial animal, lacking the senses of animals that are adapted primarily to life under water. As a result he tends to imagine that fish and other aquatic organisms, even pinnipeds and cetaceans, live vir-

tually within a sensory void. Past attempts to construct migration circuits based on animals at the mercy of the currents with no sense of location except perhaps a tenuous olfactory thread leading back to some limited home site, bears testimony to this tendency. Such a view of the perceptual world of an animal such as a fish seems most unlikely to be valid and, as already suggested on p. 851, it seems much more likely that fish, and certainly pinnipeds and cetaceans and probably marine reptiles as well, live in a familiar area that is similiar in all respects except the environmental cues on which it is based to that of terrestrial animals.

If this is true for fish, then there seems no reason why it should not also be true for at least the larger aquatic invertebrates, such as cephalopods and decapod crustaceans. Evidence is presented in Chapter 33 that for one species at least, a spiny lobster (p. 879), return migrations take place within a familiar area.

It seems unlikely, however, that all return migrations take place within a familiar area. Zooplankton, for example (p. 481), must surely perform their return migrations solely by reference to the vertical plane, taking no account of horizontal displacements. Most inter-tidal migrants, such as beach fleas (p. 830), small molluscs (p. 827), clams (p. 824), and perhaps even crabs (p. 826), also seem likely to migrate up and down the beach without reference to their location along the length of the beach. Perhaps, also, some onshore–offshore migrants, even among fish, are concerned only with their position on this axis and take little account of along-the-beach displacements. In all of these cases the animals have a lifetime track that is of the modified linear range type illustrated in Fig. 20.1, and any extra-perceptual familiar area, if the phenomenon exists at all for these species, is temporary.

To some extent, the division between animals that perform return migrations within a familiar area and animals that perform return migrations as part of a modified linear range is taxonomic. This is inevitable in that a familiar area larger than the instantaneous sensory range demands a central nervous system capable of accommodating a spatial memory. Indeed, the advantage of living in increasingly large familiar areas due to the inherent advantage of calculated rather than non-calculated migration (p. 44) could well have been a major selective pressure on the vertebrate central nervous system. Especially as it seems likely that large familiar areas in water can be developed relatively easily (p. 869) but that on land even small familiar areas demand a relatively large central nervous system. The division of animals with respect to the two forms of return migration is not, however, wholly phylogenetic. The larger earthworms that develop large burrow systems (Fig. 20.13), for example, seem likely to live within a relatively sophisticated familiar area, whereas some inshore–offshore fish migrants may not.

The critical factor, superimposed upon phylogenetic considerations, seems to be the degree of return favoured by selection (Chapter 9). When there is no selection for a high degree of return, the optimum form of the lifetime track is some modification of the linear range, executed primarily by orientation mechanisms. When selection does favour a high degree of return, however, the optimum form of the lifetime track seems likely to be a limited-area that is executed primarily by navigation mechanisms. This conclusion, and the mechanisms involved in the different forms of return migration, are discussed further in the next chapter.

33

Orientation and navigation mechanisms in return migration

33.1 Introduction

In the previous chapters, two major categories of return migration have been described, those characterised by a low degree of return (e.g. those performed by zooplankton; many littoral and sub-littoral animals; soil animals; and perhaps some of those invertebrates that inhabit running freshwater) and those characterised by a high degree of return (e.g. those performed by most vertebrates and at least some invertebrates). If we were to confine our analysis strictly to the degree-of-return component, then we would encounter an almost complete spectrum from a near-zero level to a highly precise return over long distances (p. 7). As it is, however, at the mechanistic level return migrations appear to fall conveniently into two major categories: those that involve a familiar area and those that do not.

Return migrations out of context of the familiar-area phenomenon seem to have evolved solely in response to a layering of habitat-types such that at intervals a gradient of habitat suitability reverses. The result is selection for removal migration in one direction along the gradient at one time and in the reverse direction along the gradient at another time. There is no selection for destination, however, as long as it is in the optimum layer, and there is certainly no selection for compensation for any lateral displacement while in a particular layer (i.e. selection does not act against a low degree of return). In such animals, therefore, selection on the mechanism involved favours that system that takes the animal to the optimum layer by the shortest route. In most, perhaps all, cases this can be achieved by some form of orientation (i.e. movement in a particular direction—Fig. 33.1) to an environmental gradient, either directly, or indirectly by response to a token stimulus that happens to orient the animal along the gradient.

Return migrations within the context of the familiar area involve the animal, by definition, in a retention of the ability to return to particular points in that area. Where selection has favoured a familiar area (Chapter 18), therefore, it has at the same time favoured the evolution of navigation mechanisms (i.e. mechanisms that enable the animal to move to a particular point in space). This is so whether the animal performs a seasonal or ontogenetic return migration within its familiar area or not. When such

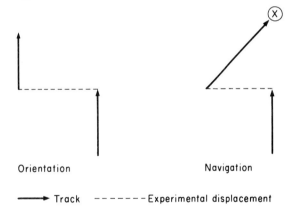

Fig. 33.1 Lateral displacement as a means of distinguishing between orientation and navigation

Orientation is the mechanism involved in the recognition and maintenance of direction. *Navigation* is the mechanism involved in the identification of the direction of a given point in space.

When an animal employing orientation is displaced laterally, it continues in its original direction as if it had not been displaced. When an animal employing navigation in order to reach point X is displaced laterally, it compensates for the displacement and continues to move toward X. Having determined the direction of point X by navigation, however, the mechanism employed by the animal is likely to be an orientation mechanism until some endogenous or external event, such as displacement, causes it once more to employ its navigation mechanism.

migrations evolve, they simply elaborate an already existing ability. As defined here, all animals that live within a familiar area can navigate within that familiar area. The question that presents itself later is whether some animals (i.e. bats and birds) can also navigate from outside their familiar area.

The critical test of whether an animal is orienting or navigating is performed by displacing that animal laterally as indicated in Fig. 33.1. However, even animals that live within a familiar area orient as well as navigate. Thus as already discussed at some length, when an animal performs exploratory migration, it does so by moving in a particular direction rather than by moving to a particular point in space, though the latter undoubtedly also occurs, particularly within the instantaneous perceptual range. Movement in a given direction is particularly the case in young birds of species that perform long-distance seasonal return migrations (p. 611). Often, also, an animal determines the direction of a particular resource by means of its navigation mechanism, but once it has determined the appropriate direction, that direction may be taken up and maintained by orientation.

This chapter, therefore, can be divided into three, albeit unequal, parts. The first part (Section 33.2), deals with the orientation mechanisms employed by animals that perform return migrations solely by orientation. The second part (Section 33.3) deals with the orientation mechanisms employed during exploratory migration. The third and longest part (Section 33.4) deals with the navigation and orientation mechanisms of those animals that perform return migrations within a familiar area.

33.2 Return migration solely by orientation

The orientation mechanisms involved in the vertical return migrations of zooplankton have in effect already been discussed (p. 492) and need not be elaborated here. Briefly, vertically migrant zooplankton have a composite migration threshold that receives contributions from light intensity, temperature, and hydrostatic pressure. This threshold may vary with the stage of food processing that the animal has reached. If the threshold is exceeded in one direction, the animal swims upwards until the threshold is no longer exceeded. If it is exceeded in the other direction, then the animal swims down-

wards. The direction of 'up' and 'down' may be determined by light intensity, gravity, or perhaps pressure gradients.

Experiments on the orientation of littoral animals have so far demonstrated two major categories of orientation mechanisms: the Moon and Sun compass and the beach-slope orientation.

Enright (1972) studied three species of talitrid amphipods: *Talitrus saltator*, *Orchestoidea corniculata*, and *Talorchestia martensii*. When deprived of all obvious orienting stimuli except a view of the Moon and stars, these amphipods usually, though not always, show non-random orientations that are consistent between replicate experiments. Further, by reflecting the Moon from a mirror, major deviations in the oriented direction of the amphipods can be achieved, indicating that lunar position is the main, if not the only, significant orienting stimulus, at least under experimental conditions.

The non-random directions are usually such (i.e. $\pm 60°$ and often $\pm 30°$) that they would be seaward on the beach of origin. The angle of orientation relative to the Moon changes more or less in compensation as the Moon changes position during the night, though some experimental series show much more precise compensation than others. There is also approximate appropriate compensation for the changes in day of the lunar cycle. Laboratory maintenance of the animals under constant conditions results in a deterioration of the capacity for seaward orientation.

On the basis of this evidence Enright suggests that talitrid amphipod crustaceans have an endogenous rhythm with a period of about 25 h on which is based the animal's ability to compensate for changes in direction of the Moon. The rhythm can persist for several days without exposure to a *zeitgeber*, which may be either moonlight or some factor associated with the tides. In the Californian sandhopper, *Orchestia*, the ability to compensate for the Moon's movement across the sky seems to be dependent on the animal being exposed to both sunset and moonrise on the night of observation. Failing this, the amphipod maintains a constant angle to the Moon's azimuth (Hoffmann 1965).

Species that are both diurnal and nocturnal have both a Moon and a Sun compass. Where investigated, the preferred direction is innate and differs from deme to deme according to the seaward direction on the home beach (Papi and Pardi 1963). The sand-beach isopod, *Tylos punctatus*, of Baja California, however, seems to perform its upshore–

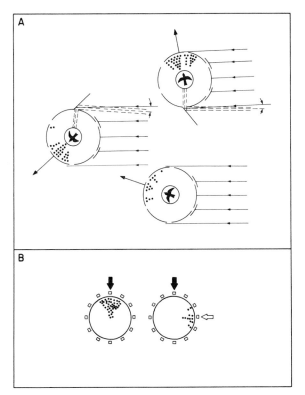

Fig. 33.2 The use of the Sun as a compass

A. In the experiment illustrated, starlings, *Sturnus vulgaris*, were placed in a circular cage in the centre of a circular room. When exposed to direct sunlight (the room possessing six windows), the starlings showed preferential compass orientation to the NW. When the apparent position of the Sun was shifted by a combination of shutters and mirrors, the starlings changed their compass orientation by an amount equal to the apparent shift of the Sun.

B. If the Sun is to be used as a compass, it is necessary for the animal concerned to learn to take into account the diel shift of the Sun's azimuth. In the bottom diagram, starlings were trained under a sunlit sky to feed from a pot in a fixed compass direction as indicated by the black arrow. When, as in the right-hand figure, the birds are exposed for a period to an artificial day/night regime that is 6 h behind the normal day, and then tested as before, they feed from the pot in a compass direction 90° clockwise from that to which they were trained. This experiment demonstrates: (1) that the birds have access to a Sun compass; (2) that they compensate for the diel shift of the Sun; and (3) that the timing of the light/dark interphase acts as a zeitgeber (= phase setter) for the clock used in such compensation.

[*Compiled from Matthews (1955) (after Hoffmann) and Dorst (1962) (after Kramer).*]

downshore daily return migration (p. 830) solely by reference to sand moisture and the slope of the beach (Hamner *et al.* 1968). When the substrate is wet, the isopod moves up-slope and when the substrate is dry it moves down-slope. The animal seems to be able to detect a slope as small as 1°. It is assumed, but has not been tested, that the animal reverses its direction when fed.

33.3 Exploratory migration by orientation

There is some indication that many animals, such as bats (Dwyer 1969) and terrestrial mammals (e.g. red deer, *Cervus elaphus*, p. 169), may orient primarily, if not solely, to topographical features such as drainage systems, mountain ranges, and valleys during exploratory migration, and it seems likely that the same may be true with respect to these or even smaller-scale features for a great many terrestrial mammals. Aquatic animals may well allow themselves to drift with the current during the outward phase of exploratory migration and then return to the pre-migration familiar area by migrating against the current (p. 813) or by completing a circuit (p. 809).

The main concern of this section, however, is with animals that orient their exploratory migrations in an endogenously determined compass direction. The main example is that of birds (p. 611) though it seems likely that the same will apply to individuals of bats, pinnipeds, and cetaceans. Note that an endogenously determined compass direction for exploratory migration at the individual level is not the same as an endogenously determined bias to the direction ratio of exploratory migration at the level of the deme (p. 654).

It is now apparent that at least three environmental cues are available to animals during exploratory migration by reference to which compass direction may be determined. These are the Sun (as a compass), the night sky, and the geomagnetic field. Most of the critical examples concern birds.

Displacement–release and displacement–orientation experiments have shown that young birds during their autumn and often also their spring migrations continue to perform their migration in the standard compass direction for the deme without compensating for lateral displacements. They are, therefore, performing these migrations by orientation, not navigation, though in spring some individuals do show a navigational ability (p. 618). On the familiar-area model (see below), it seems likely also that if an adult is deprived of all information of its familiar area in an orientation-cage experiment then it too orientates in the standard compass direction rather than compensating for displacement.

Experiments demonstrating the use of the Sun, the night sky, and the geomagnetic field as a compass are illustrated in Figs. 33.2, 33.3, and 33.4 respectively.

Clock-shifting (p. 607) of birds that use the Sun as a compass results in the birds changing their compass orientation by an appropriate amount (Fig. 33.2). However, Matthews (1963a) has shown that clock shifting of mallard, *Anas platyrhynchos*, affected a change only in their diurnal orientation but not in

their nocturnal orientation. Night directions of migrating mallard thus are obtained by reference to star patterns rather than the azimuth position of individual stars. This is consistent with the findings of Emlen (1972) that birds use the night sky as a compass by locating the axis of rotation firstly by observation of star movement (or by reference to the geomagnetic field (Wiltschko and Wiltschko 1976)) and then later by reference to star patterns, presumably by a method akin to the 'boy-scout' method of locating Polaris (in the Northern Hemisphere) by extrapolating from two stars in Ursa Major. Such a method is independent of the movement of stars with time and consequently does not involve a clock mechanism.

Compass directions obtained from the geomagnetic field are also virtually independent of time of day (see p. 877). Wiltschko (1972) has shown that European robins, *Erithacus rubecula* (Fig. 33.4), can

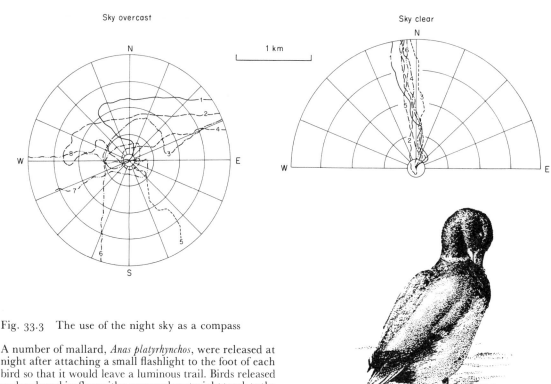

Fig. 33.3   The use of the night sky as a compass

A number of mallard, *Anas platyrhynchos*, were released at night after attaching a small flashlight to the foot of each bird so that it would leave a luminous trail. Birds released under clear skies flew with a more or less straight track to the N after release. Birds released under overcast skies flew in a wide variety of directions, often with a relatively tortuous track. It seems that although mallard have access to a compass when the night sky is clear, they do not have such access when the night sky is overcast.

[*Re-drawn from Bellrose (1958a)*]

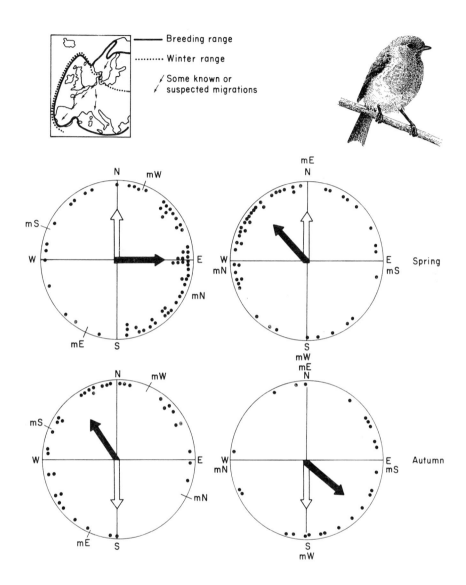

Fig. 33.4   The effect of an experimental manipulation of the magnetic field on the orientation direction of the European robin, *Erithacus rubecula*

The map shows the breeding and winter distribution of the robin in Europe. Through much of its range it is a partial migrant.

The experiments were performed in circular orientation cages during the periods of spring and autumn migration. Each solid dot indicates the mean direction of hop of a caged individual during migratory restlessness. The solid arrow shows the mean preferred direction and the open arrow shows the direction of migration of free-living conspecifics. N, S, E, and W are, respectively, geographical north, south, east and west whereas mN, mS, mE, and mW are experimentally produced magnetic north, south, east and west.

Manipulation of the magnetic field results in appropriate changes in preferred direction. Further, response of the birds to the same magnetic field is more or less opposite in direction in autumn and spring.

[*Modified from Wiltschko (1972). Map from Peterson et al. (1974)*]

perceive the axial direction of the field lines but not their polarity. In order to determine direction, therefore, some other cue is necessary and it is suggested by Wiltschko that this is achieved by the birds measuring the angle between gravity and magnetic field direction.

The robin is a nocturnal migrant, but use of the geomagnetic field as a means of determining compass direction may not be confined to nocturnal migrants. Southern (1969a,b,1971a) has shown that chicks of ring-billed gulls, *Larus delawarensis*, can orient in the direction appropriate for autumn migration (SE) equally well under clear and overcast skies but that selection of this direction is disrupted during magnetic storms of moderate severity (40 gamma and above).

One final type of orientation that may be mentioned appears to be relatively specific but may well be found to be of much greater applicability. The first stage of exploratory migration by young sea turtles after hatching involves an almost frenzied movement to the sea from their hatching site on a sandy beach (p. 773). Predation is severe during this movement and a reduction in migration time is at a premium. Displacement experiments have demonstrated that a compass direction is not involved and that the young turtles are orienting directly to the nearest sea (Carr and Ogren 1960). That young turtles can do this whether or not the sea is visible has been demonstrated many times and after a variety of experiments it has been concluded that the turtles are responding to some characteristic difference between sky over sea and sky over land. It seems that the sea-sky will draw the turtles away from all other cues. The only exception seems to occur when an area of unobstructed foreground attracts the turtles away from a more extensive background. Thus the turtles will enter a lagoon only 3 m distant if the sea is 100 m away in the opposite direction over a bush-strewn beach (Carr and Ogren 1960). After many years of experimentation (e.g. Carr 1968, 1972), the sea-sky hypothesis still stands, though the critical characteristics remain unidentified. Quite possibly many other animals, such as land and sea birds, that occasionally encounter or even exploit the other habitat-type can also distinguish between sky over sea and sky over land and can thus find their way from a distance to one or other habitat-types, even though initially they are not directly visible.

Finally, on the subject of orientation as opposed to navigation, it is important to realise that the homing mechanism involves two distinct processes: direction determination and direction maintenance. Only the former of these two processes always involves navigation, for direction maintenance can be achieved solely by orientation. Thus, having determined which is the compass direction that will lead the animal to the required resource, then clearly the remainder of the migration could be achieved solely by orientation. Obviously the migration is more efficient if the animal monitors its position from time to time to check that the required resource is still in the direction of travel. Such position checks need not be frequent, however, and experiments on homing pigeons, *Columba livia*, seem to have demonstrated (p. 897) that some individuals may travel 80 km before they check that they are moving in the appropriate direction. Observation of animals during migration, therefore, may give little if any information concerning the navigation mechanism. The critical period is either that preceding the moment that the animal first takes up a particular direction or else (Fig. 33.28) that preceding final arrival at the destination. Any change in direction that takes place during migration, however, may also be instructive as it implies that the animal has just checked its direction.

An animal that begins a migration with sufficient cues available for navigation may well lose these cues during the course of the migration. As long as at least one cue is available that permits continued orientation, however, the migration can continue. Thus, a bird may begin a migration with all cues available to it. During migration it is therefore flying at a certain angle to the magnetic field, wind direction, Sun's azimuth or axis of rotation of the night sky. Bellrose (1972) has also suggested that oceanic birds are able to maintain a particular track on overcast days by keeping a reference between the angle of the wind and the waves. Even when the wind changes direction, depending on the magnitude of the waves, the waves continue in a set direction for some time before the wind gradually forces a change in the direction of their roll. One by one these cues may disappear as the bird encounters cloud, fog, dead calm, or magnetic anomalies. However, the bird could switch from one cue to another and back again as conditions change and thus as long as one cue is always available, migration can continue. It seems likely, however, that the probability of drift from the required track increases with time and that if conditions that permit the direction of the required resource to be checked do not recur after a certain period the migration may be terminated until such conditions do recur. Conditions suitable for navigation, therefore, can be added to the list given on p. 632 of migration cost variables that contribute to the migration threshold.

33.4 Return migration by navigation and orientation

33.4.1 The least-navigation hypothesis

It will be evident from the preceding and subsequent sections that there is an abundance of navigational cues available to any given animal. These may be derived by sight, smell, touch, taste, echo-location, hearing, etc. and may consist of topographical features, olfactory sources, thermal cues, pressure cues, humidity, gravity, wind direction, the Sun, stars, Moon, magnetic field, and so on. The question is as much, therefore, what should not be used to solve a given navigational problem as it is what should be used. This has recently been stressed by workers on a wide variety of different animals (e.g. Ferguson (1971) for amphibians; Keeton (1972a) for birds), all of whom stress the hierarchical organisation of an animal's navigational repertoire and the fact that in solving any given problem a large proportion of this repertoire is redundant. As one set of cues is removed from the hierarchy by, for example, blinding or deafening the animal or releasing it under heavy cloud, the animal invariably switches to the next set of cues in the hierarchy. In order to identify the optimum strategy for an animal in a given set of circumstances, this section advances a 'least-navigation' hypothesis.

The basis of this hypothesis is that the greater the proportion of the available time an animal spends monitoring its environment for navigational cues, the less time that is available for other activities such as, for example, assessing habitat suitability, perceiving new resources, and maintaining a 'watch' for predators. In other words, the more navigation involved in the performance of a given migration, the greater the migration cost. Selection should always have favoured, therefore, the navigational repertoire of an animal being organised in such a way that for any given sensory input the animal uses that combination of cues that permits migration with a minimum time investment in navigation and orientation.

33.4.2 Non-map models

Models of navigation may be divided into two general categories. The first category consists of models that suggest an animal can arrive at the location of a required resource without reference to any form of map. Such models may be termed non-map models and may be contrasted with the second category which consists of models that suggest an animal has access to some form of map by means of which it can determine its current position relative to that of the required resource. It should be stressed, however, that the navigation models as presented here are not necessarily mutually exclusive. Even if an animal normally uses a familiar-area map (Section 33.4.3.5), for example, it does not necessarily mean that it does not have access, if released outside of its familiar area, to a Sun-arc map if the Sun is visible or a geomagnetic map if it is not. Alternatively, an animal released outside of its familiar area but which does not have access to either of the latter maps may well employ random search or downcurrent movement. However, some models cannot really be considered separately. Thus if an animal has sensory contact with the location of its required resources then it is automatically making use of a familiar-area map.

33.4.2.1 Random-search models

Even if animals used only random search, some members of a group could arrive by chance at the intended destination. This applies whether the randomness concerns only the starting direction (i.e. all individuals producing a rectilinear track but with the group as a whole showing a direction ratio of 25:25:25:25) or to the track of each individual (i.e. a random sequence of track directions) or both. Because chance can be expressed mathematically it is possible to make predictions concerning the proportion of individuals that will arrive at the destination and the speed at which they will do so given certain premises concerning the rate of change of direction, speed of movement, distance between starting point and destination, size of target area, and so on.

Perhaps the two most elegant attempts to analyse the potential of random or near-random search strategies are those of Wilkinson (1952) and Saila and Shappy (1963). Although these models were concerned primarily with birds and salmon, respectively, they are of general application to all animals given an appropriate modification of parameters. If random-search models are applied solely to the percentage arrival at the destination and the variation in time taken from departure to arrival from various distances, they show remarkably similar results to those actually observed. To this extent, therefore, they have performed a valuable service in that they have caused experimental zoologists to exercise much greater care in the interpretation of their results than probably would otherwise have been the case. However, all random-search models must inevitably generate certain characteristics to the spatial scattering of the searching animals and by and large these characteristics are inconsistent with the vast majority of experimental results. The first characteristic is that the animals depart from the release point with a mean vector that is not significantly different from zero. Although this occurs under certain conditions it is far from universal, homeward orientation being invariable under other conditions. At least in these latter conditions, therefore, random search is not applicable even though it may often mimic the speed and success of return. The other major characteristic is that the low percentage return that characterises, for example, the return of salmon to their natal stream, has to be attributed to other individuals being lost rather than to mortality or to type-H or type-C removal migration. In the case of salmon, for example, Saila and Shappy (1963) predict a 20 per cent return to the natal river, a figure that is quite consistent with field data. However, they also have to predict that 40 per cent are lost at sea and 40 per cent are lost in coastal searching. As pointed out by Hasler (1966), however, adult sexually mature salmon do not continue to be caught in the open ocean after their appearance in streams. A 40 per cent type-C removal migration to streams other than the natal stream is also not consistent with field data. In fact, even to generate this situation, it was necessary to include in the random-search model a slight (but 'negligible') homeward orientation. If even a slight homeward orientation is possible, however, we would have expected selection to act on the orientation process to produce a homing strategy that was no longer recognisable as random search.

The possibility that any given navigational problem has been solved by the animals concerned employing random search has, of course, always been borne in mind, particularly when the initial directions of a group are either not recorded or show a zero mean vector. As a general model for animal navigation, however, random search need not be considered further in this chapter.

33.4.2.2 Sensory-contact models

Sensory-contact models have been employed predominantly to account for the navigation of animals that make use of olfaction in their recognition of location. Certain fallacies have been inherent in sensory-contact models in the past. However, because the correction of these fallacies involves an integration of sensory-contact models with the familiar-area map model (Section 33.4.3.5), the potential of homing by sensory contact need not be discussed separately here.

33.4.2.3 Downcurrent models

Downcurrent models seek to explain the return migration of animals to a given location, such as a spawning or nesting site, by postulating either that they allow themselves to drift with wind or water currents or alternatively that they orient downcurrent such that the animal's speed through the air or water becomes added to that of the air or water. Although at first sight such a model might seem unpromising as a description of the migration of any animal, it appears in fact to be of surprisingly widespread applicability.

The evolutionary basis for this seems to be that selection for the minimisation of migration cost is so strong that migration circuits based throughout their length on downcurrent vectors of air and water currents have crystallised out on many occasions from more homogenous situations. This process has already been discussed in relation to marine fish (p. 809) and sea birds (p. 670). As pointed out in relation to both these groups, it is unlikely that the migration consists solely of passive drift. Rather a variety of migration thresholds will evolve such that, in combination with alternating periods of station-keeping and downcurrent drift or orientation, the migration circuit will give the maximum coincidence between the animal's arrival at, and departure from, each location and the local peak of resource abundance.

Emlen (1971) has extended the concept of downcurrent orientation to the migration of land-birds. The way that the evolved migration thresholds to migration-cost variables of land-birds confer on the bird the ability to predict weather conditions suitable for the next migration unit has already been discussed (p. 680). Without expounding this particular point, Emlen has argued that if birds can predict that the wind direction during the coming migration period (i.e. the next diel low of the migration threshold—p. 644) will be in the appropriate direction, then they can initiate the next migration unit even in the total absence of any other navigational cues. By riding the wind the bird can still migrate in the appropriate direction.

In a sense, therefore, downcurrent models do involve a form of navigation in that at a particular moment in time an animal requires to go to a specific location and achieves this by going in a specific direction at a specific angle to an environmental feature. However, if an animal that uses downcurrent vectors were to be displaced laterally, it could only navigate in the appropriate direction by using some mechanism other than downcurrent orientation in order to determine that direction. Downcurrent models are only applicable, therefore, in animals that either are not normally exposed to the danger of lateral displacement or that also use some other navigational system. As few animals come into the former category, it may be assumed that animals using downcurrent orientation do so as part of a much more flexible navigational system. An attempt to integrate downcurrent orientation with the familiar-area map model is presented in Section 33.4.3.5.

33.4.2.4 Inertial navigation

Inertial navigation postulates that an animal's body registers every twist and turn of an outward journey and from the information thus available can vector the mean direction of displacement. A system based on inertial navigation is incorporated into the navigational system of modern aircraft. Attempts have been made to disrupt such a mechanism, if it exists, in pigeons by rotating them on turntables during displacement. No reduction of homing ability has been observed following such treatment (Matthews 1968) though it could be argued that the mechanism is so perfect that it can accommodate such manipulations (but see p. 617).

The inertial navigation model has gained little support among zoologists though it was resurrected by Barlow (1964, 1971) on the sensory basis of the vestibular apparatus. However, Wallraff (1965) showed that cutting the semi-circular canals of pigeons did not impair their homing ability.

33.4.3 Map models

An animal can construct a map that is suitable for navigation in one of two ways. Either it can build up a mosaic of past experiences in some form of spatial memory or it can take measurements from one, two,

or several environmental variables that interact in such a way as to form a grid on the Earth's surface. Two coordinates are the minimum requirement for such a grid map to be used for navigational reference. Ideally a grid map should be such that each point on the Earth's surface is unique. Most animals, however, could probably manage as long as similar points were sufficiently far apart on the Earth's surface.

The past-experience map is the type generated by the familiar-area model developed throughout Part IIIb of this book and is elaborated in Section 33.4.3.5. The coordinates for a grid map could be derived from environmental phenomena such as the Sun's arc, stars, and the geomagnetic field, etc. Theoretically, there seems no reason why a familiar-area map should not be superimposed on one or several grid maps, the animal making reference to whichever map is available in a given situation. Indeed, all proponents of the existence of grid maps include within their model the superimposition of some form of familiar area on the grid in the vicinity of the home location. If a grid-map model is accepted, therefore, the main point at issue is the distance from the home location at which the animal ceases to use the familiar-area map and switches to the grid map. However, it is argued in the remainder of this chapter that whereas there are abundant experimental data in support of the existence of a familiar-area map, there are none as yet that support the existence of a grid map for any animal except man.

33.4.3.1 The Sun-arc map

The suggestion that birds navigate during migration by reference to a grid map derived from coordinates obtained from components of the Sun's arc can be attributed to Matthews (1955). This suggestion, and variants thereof (e.g. Pennycuick 1960), was the trigger for a period of intense experimentation on bird navigation and was the guide to most mainstream ornithological thinking for nearly 20 years. Details of the Sun-arc hypothesis and its variants may be obtained from Matthews (1968). Basically, the suggestion is that a bird can determine its position on the Earth's surface relative to the location of a required resource by measuring the altitude of the Sun's arc and how far along the arc the Sun has travelled, and then by comparing these measures with those that would be expected if the bird were already at the location of the required resource. The precise measurements likely to be made by the bird are open to debate. Thus Matthews (1955, 1968) considers that the culmination point of the Sun arc at noon is the basis for comparison between the current and home site. The bird is suggested to observe the movement of the Sun along its arc for a period of perhaps one to two minutes and then to 'extrapolate and guestimate' the culmination point of that arc. This is the position of the Sun at noon at the bird's current location. If the culmination point is higher than at the home location, home direction has a vector away from the Sun, if lower, home direction has a vector towards the Sun. If the culmination point is reached sooner than at the home site, the bird should move with a vector that adds to the movement of the Sun; if reached too late, the vector should be subtracted from the movement of the Sun. According to Pennycuick (1960) a more suitable measure would be obtained without extrapolating the Sun's arc to its culmination point. If only the vertical component of the Sun's movement is registered, then an appropriate grid could still be obtained by measuring the Sun's altitude and its rate of change of altitude.

It is with a tinge of regret that the next paragraph is written for, like so many zoologists, I have found the Sun-arc hypothesis, throughout its development, refinement, and evaluation, a constant source of academic stimulation. It is perhaps still a little premature to pronounce the total demise of the Sun-arc hypothesis, but there now seems little likelihood that it will be found to be applicable to the navigation of birds or, in that case, of any other animal except man.

Experiments that should have provided support for the Sun-arc hypothesis, were it valid, have been disappointing and inconsistent from the beginning (Matthews 1955, 1968) and during the last few years the evidence against the hypothesis has steadily been mounting (e.g. Keeton 1970c, Walcott and Michener 1971, Schmidt-Koenig 1972, Alexander and Keeton 1974). All attempts to manipulate the birds' measurement of the Sun's arc by clock-shift experiments, by altering the Sun's apparent altitude with mirrors, and by disrupting the birds' ability to measure time have failed to demonstrate that the Sun's arc or altitude has biological significance to the bird. There was an apparent reprieve when Whiten (1972) in an operant training study with homing pigeons appeared to demonstrate that a bird could perceive changes in the Sun's altitude, at least when tethered and given an artificial horizon. However, caged birds have proved themselves so receptive to training and so alert to the slightest visual or thermal, etc., irregularity in their environment (Howland 1973) and Whiten's experiments involve so few birds and are so open to interpretation as rapid retraining that, in the face of contradictory field data, they cannot as yet be taken to demonstrate that unrestrained birds in a natural environment perceive changes in the Sun's altitude.

33.4.3.2 The night-sky map

There is no doubt (p. 609) that birds can use the night sky to determine compass direction, an ability first demonstrated by Sauer and Sauer (1955) for Old World warblers of the genus *Sylvia*, and by Bellrose (1958a) for mallard, *Anas platyrhynchos* (Fig. 33.3). Sauer (1957, 1963) also claimed to have demonstrated that several bird species were able to use astral cues to form a grid map. This work, however, has received careful yet devastating criticism from other authors (Wallraff 1960, Matthews 1968) and since that time many studies have shown only star compass orientation (Hamilton 1962a, 1966, Mewaldt *et al.* 1964, Emlen 1967, 1972). Furthermore, the demonstration by Matthews (1963a) that the compass orientation of mallard, *Anas platyrhynchos*, was unaffected during clock-shift experiments and by Emlen (1972) that indigo buntings, *Passerina cyanea*, use learned star patterns, not the azimuth of individual stars, to locate the axis of rotation of the night sky, make it unlikely, if their findings are generally true, that bicoordinate stellar navigation is to be found in birds. Most recent authors (e.g. Bellrose 1972), therefore, with the exception of Ricard (1971) who apparently was not aware of the arguments against the possibility that birds derive a grid map from the stars, accept that birds do not have access to a night-sky grid map unless it is based on the Moon. The latter is a possibility but is unlikely in view of the vast amount of nocturnal migration that is performed during those times of the night during the lunar month that the Moon is beneath the horizon. As, of all animals, birds seem to have the greatest use for a night-sky grid map, to accept that they do not is to suggest that no other animal group uses such a map, either. Again this is consistent with what, so far, is known concerning other animal groups. Although various animals use a night-sky compass, none have given any indication of using a night-sky map.

In a sense, also, to accept that a grid map based on the night sky does not exist is also to accept a reduced possibility that a similar grid map exists but is based on the Sun. Even if, as suggested by Matthews (1968), nocturnal migrants determine their appropriate initial orientation from the setting Sun, it still has either to be accepted that nocturnal migrants have access to a map that is not based on celestial features or else it has to be accepted that nocturnal migrants can manage quite well without access to a map at any time during their migration. In either case, any argument that the Sun-arc map is the major map available to diurnal migrants is considerably weakened.

33.4.3.3 The geomagnetic map

The original suggestion that the Earth's magnetic field could provide a navigational grid seems to be attributable to Viguier (1882). He suggested that birds could detect and measure the three components of the field, its intensity, inclination (the angle which a magnetic compass makes with the horizontal), and declination (the angle between magnetic and geographic north). The inclination and intensity isolines generally cross one another at oblique angles, thus making a relatively poor grid for navigation, and the inclusion of declination isolines in the grid seems to be essential for the model to be plausible. It seems unnecessary to take the step proposed by Yeagley (1947) that the Earth's field was detected not directly but indirectly as a result of the flying bird acting as a linear conductor moving through the lines of force of the field. In any case, demonstrations of the use of a magnetic compass (p. 859) have involved birds that are caged as well as free-flying.

Theoretically, therefore, it seems that a geomagnetic map is a possibility, just as a Sun-arc map is a possibility. Practically, it seems that the geomagnetic field provides only a compass bearing, just as the Sun and stars appear to provide only a compass bearing. Even as a compass, however, it has been noted (Keeton 1972a) that birds are apparently reluctant to make use of the magnetic field. This certainly applies if celestial cues are present and has even been noted to be the rule when they are not, homing pigeons, for example, often preferring to settle if they encounter cloud rather than switch to a magnetic compas (p. 906). There are also abundant examples of birds failing to locate home direction when they had access to neither a Sun compass nor familiar landmarks and when, if they had access to a geomagnetic map, homeward orientation should have been possible. As yet, therefore, there is no

evidence, as far as I am aware, that birds, or any other animal (except man), have access to a geomagnetic map, but there is abundant evidence that birds fail to refer to such a map when such reference seems advantageous. This is not the same, of course, as having evidence against the existence of an ability to refer to a geomagnetic map and perhaps zoologists are simply in need of the discovery of the appropriate experimental technique to demonstrate such an ability. In the meantime, however, it has to be accepted that the available evidence does not support the suggestion that a geomagnetic map is used in animal navigation.

33.4.3.4 Other maps

It would of course be possible for an animal to construct a grid map by obtaining different coordinates from different geophysical cues. Thus one coordinate could be obtained from the Sun's arc and one from the geomagnetic field or one could be obtained from the night sky and one from the geomagnetic field. In the light of what has been said in the previous sections such a mixed grid seems unlikely and is totally devoid of experimental support. Other geophysical cues have also been suggested, such as measures derived from Coriolis and other forces produced by the Earth's rotation. The lack of plausibility of such suggestions at both the theoretical and practical levels has been convincingly demonstrated by Matthews (1955, 1968) and need not be considered further here.

33.4.3.5 The familiar-area map

In the previous few chapters it has been argued that the vast majority of vertebrates and at least some invertebrates spend their entire lives within a familiar area that is larger than their instantaneous perceptual range. Such a familiar area can only be established by the commitment of perceived environmental features to a spatial memory. From the moment of birth (or, more accurately, first perception of the environment), therefore, each individual builds up within its central nervous system a 'map' of its environment. The nature of this map as it is envisaged here for animals living in different environments with different sensory equipment is described in detail in Figs. 33.5–33.7. The text of this section is in effect a general discussion of some of the points to emerge from these figures.

In order to function as a map, the familiar area has to be built up in such a way that at any given point the animal can recognise its current position

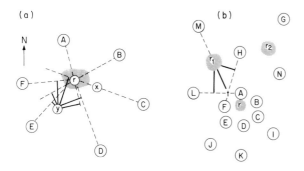

Fig. 33.5   The familiar-area model of navigation for an animal with a primarily visual map

Two situations are illustrated, (a) being relevant to a relatively immobile animal (e.g. a rodent or a 'resident' bird with a small foraging area) and (b) being relevant to a relatively mobile animal (e.g. a bat or a bird with a large foraging area).

A–N are gross environmental features, such as mountains, hills, woodland, moorland, large rivers, lakes, the sea, etc. on land, or islands, coastlines, patches of different-coloured water, submerged topographic features, etc. at sea. r, r_1, and r_2 are areas containing resources for the animal concerned.

The area enclosed by features A–N plus a large area beyond is mapped within the animal's spatial memory solely on the basis of the spatial relationships of gross features. The stippled area is that portion of the familiar area that is mapped by the spatial relationships of detailed features, such as individual trees, bushes, tracks, ponds, sounds, smells, etc., both relative to each other and relative to gross features as well as relative to compass-determining cues such as the Sun, stars, and magnetic field.

Navigation from the apparent compass shift of familiar features
As an animal moves or is moved within its familiar area there is an apparent shift in the compass direction of structures A–N (or at least those with which the animal has sensory contact) as perceived by the animal. Thus an animal moved from x to y observes, among other things, a compass shift of A, F, D, and C but a negligible shift of E and B. From all or even only a part of the information available the animal can readily compute its position and the appropriate direction in which to locate x or r, even if neither of these can be seen from position y.

In order to move to the location of a required resource (r) from its current position it is necessary for the animal to have learned on a previous visit to r the compass directions of gross features when at that location. These directions are indicated in diagram (a) by the dashed lines. The animal could make use of this memorised information in one of two ways. Either it could compute the direction (solid lines drawn from y) that would restore by the shortest track each environmental cue to the compass direction appropriate to the location of r and then move in the mean vector of these directions or it could extrapolate the required compass directions and move toward their crossover point. The latter method may conceivably give a

marginally more accurate fix on the required location but the former may be the easier to compute. Put in simple terms either mechanism involves migration to the N side of features that appear further N than at the required location, to the E of those features that appear further to the E, and so on, and directly towards those features that are shifted 180° from their compass bearings at the required location. The question of variation in the accuracy of homeward orientation with distance from home is discussed in Fig. 33.23.

In animals that experience situations such as that illustrated in diagram (b) an additional component is necessary for navigation. It is necessary for animals to learn the compass direction of gross features that pertain at each of a sequence of target areas. The most advantageous placing of each target area to reduce the number required to a minimum is at the point that one set of gross features passes out of perceptual range or take on an unfamiliar shape through being viewed from a different angle (as would an escarpment, for example) and at which another set comes into perceptual range. For an animal that requires to move between r and r_1, therefore, t seems to be a suitable site for a target area if features A and F become unrecognisable at this point. If they do not, the target area could be much further from A and F. At point t, if performing exploratory migration outward from r (the location of r_1 being so far undiscovered) features A–F pass from view or take on new shapes (which could also be learned eventually) and features L, H, and M may come into perceptual range. Having learned the compass bearings of L, H, and M from t during the first exploratory migration from r the animal can now return to r from anywhere within the area that L, H, and M are in sensory range, by first locating the compass direction of t as described in the previous paragraph. The advantage of this method is that an animal can perform exploratory migration anywhere within the area bounded by L, H, M, A, and F without learning detailed landmarks and without memorising the route travelled. Apart from not leaving the area within which at least one of these landscape features is within sensory range, the animal need take no further stock of its position until it requires to return to r, a process quite comparable to that required of an animal during displacement–release experiments. Once the animal has established that resources exist at r_1 and has determined that future migration between r and r_1 will be advantageous, the compass direction that leads from r_1 to r can be learnt and navigation is not required on future occasions. If r_1, t, and r do not lie on a straight line, the straight-line compass bearing between r_1 and r could be determined by vectoring the direction and distance from r_1 to t and t to r. The position of t may then have only transient significance in the exploratory migration– calculated migration sequence but may be retained in the spatial memory for use during further exploratory migration.

Other uses of a landscape map in navigation
If some assessment of the apparent size of landscape features and their apparent angular separation relative to that pertaining at the required location were also taken into account the accuracy of orientation could perhaps be increased. Such factors, however, require more con-

tinuous monitoring of the environment during navigation. Although they may be used when a compass is not available, therefore, they are less likely to be used on other occasions. The same applies to those other mechanisms of navigation by landmarks such as the retracing of previously followed tracks and the process of 'steeplechasing' from one known landmark to another.

This model does not demand a high degree of navigational accuracy on the part of the animal, nor does it require a mass of detail concerning the environment to be stored within the spatial memory. Only in the final stages of the navigation to a particular location need detailed features be utilised.

and can extrapolate from that position the direction of a track leading eventually to some other point within that familiar area. This track is then taken whenever the animal is motivated to migrate to that point according to its resource requirements and migration thresholds of the moment. This track need not necessarily be the straight line between the two points and may indeed be the track followed on previous occasions. Clearly, however, the ability to vector or extrapolate the straightline direction from previous tracks is often likely to be an advantage for many animals.

Broadly speaking, familiar-area maps can be divided into two categories. On the one hand we have those that are based primarily on olfactory cues and on the other we have those that are based primarily on visual or echolocation cues. The former type of map is likely to be possessed by animals such as fish (p. 849), marine and terrestrial reptiles (p. 851) and amphibians (p. 883) and the latter type by pinnipeds (p. 733), cetaceans (p. 763), bats (p. 585) and, it is argued below, birds (p. 606). Immediately, however, it is important to stress that these two categories have been constructed solely for descriptive convenience. It has been argued many times in the previous chapters that the familiar-area map is the integrated result of all the cues available to the animal concerned. Thus amphibians integrate olfactory and visual cues in the formation of their map and pinnipeds, bats, and perhaps some seabirds (Matthews 1968), may use some olfactory cues in their predominantly visual map. Many ungulates and other terrestrial mammals, such as canids, may well give relatively equal weight to olfactory and visual cues. Acoustic (as opposed to echo-location)

cues may be used by a wide variety of animals, though it seems unlikely that any animal will build a map solely on acoustic cues unless it is forced to do so through being blinded and rendered anosmic. Thus blind but otherwise apparently healthy pinnipeds and cetaceans have been reported (p. 766) as have apparently blind/deaf cetaceans. The possibility of such individuals being maintained by social means cannot be ruled out, however, in animals such cetaceans which show a high degree of reciprocal altruism (p. 767).

As man is an example of an animal that lives within a familiar area based primarily on visual cues but also making some use of olfactory and acoustic cues (and with the potential to use only these latter cues plus some others if blinded), the nature of a visually based familiar-area map should be well within the range of human experience, though the same cannot be said for maps with a predominantly olfactory basis. I shall resist the temptation, however, to develop the vision-based familiar-area map model by analogy to the mechanisms used by man in everyday movements, as by and large these do not seem to have been subjected to experimental evaluation in the field.

Perhaps the most important point in both the visual and olfactory-based map is the distinction between the use of gross and detailed features. Gross features are those such as mountains, valleys, moorland, woodland, cities, large rivers, lakes, coasts, sea, etc. that permit the instantaneous formation of an extremely large familiar-area map by any animal that has perceptual access to such features.

For most terrestrial animals this gross familiar area may be established early on in ontogeny and all subsequent movements are unlikely greatly to change the animal's position relative to these gross features. For such animals, mountain range A is always on the opposite horizon to hill D and all of the animal's movements can be related to these and similar features. Detailed features can then be integrated into this map for reference during short-distance movements. Thus, if large tree x is on the same side of the animal's horizon as hill D the animal is one side of its home range, and if it is on the same side as mountain range A the animal is the other side of its home range. Further information is obtained by perceiving how far to the right or left of hill D or mountain range A tree x appears to be. Further flexibility is added if the compass direction of A and B and other gross features is included on the map by reference to their position relative to geophysical

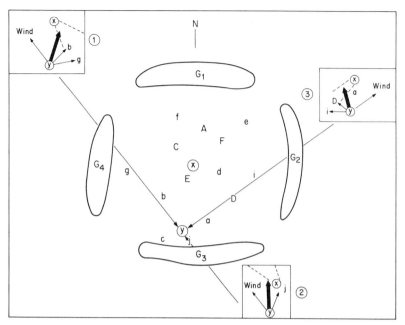

Fig. 33.6   The familiar area model of navigation for a terrestrial animal with a primarily olfactory map

In the main diagram, G_1–G_4 are gross features, such as hills, cities, woodlands, rivers, etc., the olfactory signatures of which are carried to an animal at x on winds from a wide range of geographical directions. A–F and a–j are more localised olfactory sources, each of which contributes to the olfactory signature of a much narrower range of wind directions. x is the animal's home site and y is a position attained through exploratory migration or experimental displacement. The insets represent the solution to three navigational problems. It is assumed that the animal can determine wind direction, either directly or indirectly (e.g. by observation of cloud movement), and can transpose this direction to a compass bearing by reference to a compass based on the Sun, stars, Moon, or geomagnetic field.

It is postulated that an animal, while living in and around x, experiences a variety of wind directions, each one with a different olfactory signature, and that the animal learns to associate a given olfactory signature with wind from a given compass direction. Only as many olfactory components (A–F and a–j) are learned as are necessary to recognise a given signature as representing wind of a given compass direction ($\pm$ an angle that gives the optimum compromise between the efficiency of direction determination and the disadvantage of having to commit too many olfactory cues to a spatial memory). The number of olfactory sources shown seems to approach optimum.

Suppose that at the end of a bout of exploratory migration during which the route taken was not monitored, the animal arrives at y and then requires to join up y with x as part of the familiar area. A number of directional cues are available to the animal, no matter what the wind direction, that would permit the animal to find its way back to x. One vector can be taken from wind direction. If the air current at y possesses a signature that is similar to a signature that may be experienced at x but with additional but familiar components, home direction has an upwind vector. If the signature has no additional components, then home direction has a downwind vector. In addition, home direction has a vector in each of the compass directions that is downwind from x on occasions that the signature has each of the detailed components currently perceived at y. Home direction is then given by the mean vector. This mean vector always intersects the downwind plume from x (shown by dashed lines in the insets), assuming that x is also an olfactory source.

In inset 1, the wind direction is NW and the olfactory signature at y is G_4gb. The nearest signature to this experienced at x is G_4g or G_4b. There is an additional component, therefore, and home direction has an upwind vector. This vector and the vectors due to the downwind compass direction for each of g and b are indicated by light arrows. The mean vector is shown by the heavy arrow which intersects the plume of olfactory cues downwind from x.

In inset 2, the wind direction is SE and the olfactory signature at y is G_3j. The nearest signature to this experienced at x is G_3jE. There are no additional components at y, therefore, and home direction has a downwind vector. The vector due to j is shown and gives a mean vector that again intersects the downwind plume from x.

In inset 3, the wind direction is NE and the olfactory signature at y is G_2iDa. The nearest similar signature experienced at x is G_2id. There are, therefore, additional components at y and the home direction has an upwind component. The vectors due to i, D, and a are shown and give a mean vector that again intersects the downwind plume from x.

The more olfactory cues the compass directions of which

are learned before the animal leaves x, the nearer to x does the home track intersect the downwind plume. Similarly, the nearer the animal comes to x, the more accurately home position can be determined.

As shown, the initial homeward orientation always intersects the downward plume from x and the animal need not necessarily monitor its position again but simply maintain a constant compass bearing (or angle to the wind) perhaps associated with interim visual fixes on distant, or even near, landmarks, until it perceives the olfactory cues from x when the final stages of homing can occur upwind. However, the animal can home by a shorter route if at intervals it monitors home direction from its new position. As the animal moves in the direction of the heavy arrow, the olfactory signature of the air

changes. If each time a familiar component is added to the signature, the animal monitors its new position, it gains an increasingly accurate fix on x and arrives sufficiently close eventually to use visual, acoustic, or short-distance olfactory cues in the final stages of homing.

This model does not require the animal to learn the compass directions of cues from every point visited, only those (e.g. x and y) between which the animal performs frequent return migrations. Once the compass direction of a given resource has been learned, however, there is no need for it to be determined by navigation on every subsequent visit. The model also does not require the animal to memorise its tracks in order that it may retrace its steps, though in the absence of a compass such a strategy could perhaps be adopted.

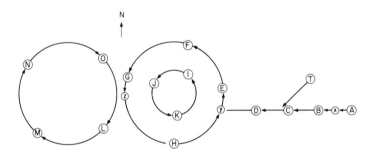

Fig. 33.7 The familiar-area model of navigation for an aquatic animal with a primarily olfactory/gustatory map

Current direction is usually much more predictable in aquatic systems than in terrestrial systems and it may not be necessary to include a compass in the map to be described. There seems little doubt, however, that a compass would increase the efficiency of use of the map.

In the diagram, A–O and T are stationary olfactory sources and x, y and z are locations important to the animal as well as being olfactory sources.

Consider an animal born at x that performs exploratory migration down the linear current and joins the gyral at y. As long as y can be re-located, x can also be re-located by swimming against the current for as long as any combination of D, C, T, B, x, and A (but not A or T alone) contribute to the olfactory signature of the current in which the animal finds itself. If the signature of A or T alone is encountered, the animal should swim downcurrent until re-locating a signature that also contains D, C, B, or x. It is not necessary to postulate that the animal learns the strict sequence $AxBCD$, though this may occur.

Having arrived at y from x an exploratory migrant receives two pieces of information. First, y is on a gyral with a signature $HEFG$. Second, y is located at a specific point on that gyral where the contribution of each of these components has a characteristic relationship (in this case

H probably makes the greatest contribution and E the least on the assumption that concentration decreases with distance from source). y may also be learned as that point on the $HEFG$ gyral at which the current flows to the north. Suppose on its first exploratory migration from y the animal migrates at an angle anticlockwise of the downcurrent direction and encounters water with a JIK signature. Suppose that the animal continued through the JIK gyral. At z it would re-encounter the $HEFG$ gyral. Sufficient information is immediately available for the animal to determine that if it remains with the gyral it will return to y, even though it has never visited z before. Because of the existence of gyrals, marine animals seem to have a considerable navigational advantage relative to those terrestrial animals that also use an olfactory familiar-area map.

By learning the spatial relationships of different gyrals (i.e. movement at an angle anticlockwise of downcurrent while in the $HEFG$ gyral always takes the animal into the JIK gyral, etc.) and by learning the particular contributions of different olfactory sources that identify a specific location (e.g. z is just downcurrent of the point where G makes its maximum contribution to the $HEFG$ gyral's signature and only from z does movement at an angle clockwise of the downcurrent direction—or west—enter the animal into the $MNOL$ gyral) the animal can readily establish a huge familiar area.

features such as the Sun, Moon, stars, or magnetic field. Then, under conditions of fog or if the distant horizon is obscured by trees or a hedge, etc., position can still be determined by reference to the position of detailed, close, features relative to the direction of the particular geophysical feature. Clearly other factors could be involved, such as a continuous monitoring by the animal of its own movements and through even more sophisticated recognition of detailed features. For example, tree x is likely to be a different shape when viewed from different directions.

It would, of course, be possible to build up a familiar-area map using only detailed features. The advantage of the use of gross and geophysical features as well as detailed features, however, is that it reduces by many orders of magnitude the amount of information that has to be committed to a spatial memory and hence, presumably, reduces considerably the proportion of the central nervous system that has to be set aside for familiar-area formation. It seems likely, also, that the use of gross features decreases considerably the danger of an animal becoming 'lost', a danger that must be great if the familiar area is based solely on detailed information. Any animal forced to lose sensory contact with its detailed familiar area, through being chased or carried by a predator, for example, should still be able to obtain a gross fix of position by reference to its position relative to gross features. This fix is most accurate if the compass direction of gross features is measured and compared with the compass direction of the same features from the home range.

Gross and geophysical features become particularly important in the case of highly mobile animals such as pinnipeds, cetaceans, some terrestrial animals, bats, and birds. The demands on the central nervous system would be excessive if these animals were forced to establish a familiar area solely on the basis of detailed features. Unlike the animals just discussed, however, animals such as bats and birds do not necessarily live their lives with mountain range A always to the north and hill D always to the south. If the animal flies over the mountain range A, then both A and D are to the south and hill D is almost certainly obscured by mountain range A. Nevertheless, by using these features to build up a map, a bird or bat confronted with a mountain range to the south and no sign of hill D could recognise its position to be to the north of the resources situated in the valley between A and

D. It would only be necessary for a bird or bat to learn the spatial position and relative compass directions of a few gross features of this type for it to be capable of recognizing its position at least in gross terms from any point within a huge area.

The integration of geophysical features into the familiar area has so far been suggested to function both as a means of recognising position when gross features are obscured and as a means of fixing position on the map by reference to the compass direction of other features, both gross and detailed, compared with their compass direction at the location of the required resource. Thus, if hill D is due south and city C is due east, the animal is in the immediate vicinity of resource R. However, if hill D is SE but city C is still E then the animal is to the W of resource R, whereas if hill D is SE and city C is ENE, then the animal is to the SW of resource R. The accuracy of this fix is a function of the number of gross features that have been incorporated into the familiar-area map, the proportion of the horizon in which suitable features are located, and the past experience of the degree of compass shift of a feature on the map for a given movement in a given direction. The ability of the animal to determine its position and vector the direction of resource R should increase with experience.

There is a likely third function of the integration of compass direction cues within the familiar-area map. Thus, an animal may be familiar with several locations at which, say, two river tributaries join. However, these tributaries are unlikely to have precisely the same compass direction in all places. By learning the compass direction of these tributaries in each place an animal could increase the chances of recognising which junction it had encountered even if no other cues were available.

So far, we have assumed that the same sense is involved in the recognition of gross and detailed features. This need not necessarily be the case. Some senses, such as vision and olfaction, are particularly suited to perception of gross features at long distances, given suitable intervening conditions (i.e. no fog and some wind, respectively). Other senses, however, such as echolocation and touch are suited to the perception of detailed features at short distances. Where an animal uses a combination of senses in map establishment, therefore, it seems likely that one sense is used for mapping gross features and another sense for mapping detailed features. Vision and olfaction, of course, are suited to perception of detailed features at short distances as well as to perception of gross features at long distances.

The concept of a familiar-area map based on olfactory cues is rather alien to man, but there seems no reason why this should not function in much the same way as any other type of familiar-area map. The concept of the olfaction-based familiar-area map is developed in Figs. 33.6 and 33.7. These figures also illustrate the fallaciousness of the assumption that because an animal uses olfaction it can only migrate to a resource if it is in olfactory contact with that resource and that consequently such migration can only be performed against a current of air or water.

33.4.4 Experimental evaluation of the least-navigation–familiar-area map model

33.4.4.1 Displacement experiments: their design and interpretation

The major experimental technique in the study of animal navigation is that of displacement. Observation of migrating animals may lead to some conclusions concerning orientation and its maintenance but rarely can it provide information concerning navigation. Only by lateral displacement can orientation and navigation be distinguished (Fig. 33.1).

Classically, experiments on navigation have been displacement–release or 'homing' experiments in which a group of animals are transported from their capture location to a new location where they are released. At first, the main aim of such experiments was to record the proportion that returned to their place of capture (= 'homed') and the speed with which they did so. The realisation of the potential of random search in generating observed homing performances if only return speed and proportion are recorded, however, led to the recording of the direction from the release point that different individuals vanished from sight (= the vanishing point) (see Matthews 1955, 1968, for a discussion of the experimental precautions necessary). The scatter of vanishing-point directions can then be analysed for the existence or absence of homeward orientation. A similar analysis can be undertaken for the 'recovery directions' of individuals recaptured some distance from the release point but not yet at 'home'.

Other authors have preferred to use a displacement–orientation experimental technique. In such experiments, animals are displaced as before, but then their directional preference has been examined by exposing them to a variety of conditions at the displacement point while they are restrained in an orientation cage (i.e. a circular cage with some facility for recording, either by observation or automatically, that part of the cage preferred by the experimental animal). The design of such orientation experiments and the analysis of the results obtained have been critically examined by Wallraff (1960) and, more recently, in a fine paper by Howland (1973). The apparent demonstration of bicoordinate stellar navigation (p. 864) was later found to be invalid due to the technique employed and the analysis and interpretation of experiments carried out in orientation cages.

A number of difficulties arise during the execution of displacement experiments. In particular, the true test of navigation is whether the animal can find its way to the location of the required resource. The experimenter, however, can only guess where this location might be, and usually, in fact always, assumes that it is the site at which the animal was originally captured. Whether or not this is a valid assumption depends on whether the animal's threshold for removal migration from that site has been exceeded since the time it was captured there. Thus an animal captured at a transient home range may, by the time of release, require the resources offered by its summer home range. The test of navigation in this case is not whether the animal returns to its site of capture but whether it finds its way to its summer home range. Another animal may be familiar with a number of winter home ranges (e.g. some bats, p. 101) and may well have a low threshold for range-to-range removal migration. If this threshold is exceeded by the displacement process, the animal may elect to go to an alternative winter home range rather than the range of capture. Alternatively, another animal may assess the area surrounding the release point to be more suitable than the place of capture and thus remain near the release point rather than returning to the place of capture. Finally, the threshold for removal migration may be higher at some times of year than others. Thus homing pigeons home better from July to September than they do from February to March. Gronau and Schmidt-Koenig (1970) suggest that this variation implies that landmarks are unimportant in pigeon homing. It seems more likely, however, that the variation in no way reflects variation in navigational *ability*, no matter on what that ability is based, but rather reflects a removal migration threshold that is higher (i.e. the 'motivation to return' to the loft is greater) from July to September than from February to March. In all animals so far investigated, homing is better to the breeding home range (i.e. copulation home range in males; oviposition or parturition home range in females) than to any other season's home range. Again this does not need to imply that the animals' navigational *ability* is greater when searching for the breeding home range. It is much more acceptable as a reflection of a type-C removal migration threshold that is higher for breeding home range to breeding home range removal migration than for home range to home range removal migration at any other season (pp. 321 and 622). Some indication of the migration threshold exceeded during displacement can be obtained from the nature of the results. Thus, if per cent returns are lower, the removal migration threshold of more

birds has been exceeded. If returns are slower, only the exploratory migration threshold has been exceeded.

It can be seen, therefore, that interpretation of homing performance in any given experiment is very much dependent on the experimenter making the correct guess concerning the animal's intended destination(s). We may perhaps accept that when individuals of a species that normally shows a high degree of return to the breeding home range are displaced from that home range, the majority will attempt to return to that home range. Even then, however, as Matthews (1968) has shown for the manx shearwater, *Puffinus puffinus*, a great deal depends on the stage of the breeding season that has been reached. Individuals displaced early on in the breeding season show a much higher per cent return than individuals displaced toward the end of the breeding season. Again there is no suggestion that navigational ability varies during the breeding season. At all times the birds probably show a

Fig. 33.8   Male and female mallard, *Anas platyrhynchos*

[*Photo by Edgar T. Jones*]

constant ability to migrate to the required destination. It is just that beyond a certain date, this destination is less and less likely to be the breeding home range from which the bird was displaced.

A particularly disconcerting variable in displacement experiments was discovered after a number of misleading experiments on birds had been performed. It was discovered, initially in relation to mallards, *Anas platyrhynchos*, and then later for other birds, including homing pigeons, *Columba livia*, that each particular roosting deme had a preferred compass direction that was adopted in displacement–release experiments. This orientation has been termed 'nonsense' orientation by Matthews (1968) and, despite its semantic shortcomings, the term is sufficiently appropriate to have gained widespread acceptance. The term is not meant to imply, of course, that the orientation direction is not significant to the birds, only that it seems nonsensical to the experimenter.

The mallard captured by Bellrose (1958a) during the autumn migration through Illinois and that oriented to the NNW upon release but were recaptured in the standard direction to the S were perhaps not showing nonsense orientation but reversal (see p. 188). There is a clear temptation, having created a nonsense orientation niche in zoological terminology, to use it as a dumping ground for observed preferred directions that cannot as yet be interpreted. Care will have to be taken in future that this category is not used too freely in the convenient dismissal of phenomena that in reality have a variety of cause and effect relationships. In the original, and valid, opening of a nonsense orientation category, the resident mallard of Slimbridge, southwest England, showed a preferred orientation to the NNW from a variety of release sites in a variety of directions from their home site (Matthews 1968). The orientation was maintained, however, for only 20 minutes and about 16 km before it gave way to random scatter. The mallard from London parks have a nonsense orientation to the SE. Those from Sweden also orient to the SE.

Whatever the eventual evaluation of the significance of nonsense orientation (see p. 896), insofar as it is a single phenomenon, there is no doubt that in the past it has caused considerable confusion, particularly in homing experiments on birds. Birds released in a given direction showed strong homeward orientation only for it to be discovered years later that birds from that particular roost always show a predilection for orientation in that direction

no matter from what direction relative to home they are released. In recent years, therefore, it has become standard practice in homing experiments on birds always to release the birds from both sides of the home site (Matthews 1968), a practice that should be adopted in future homing experiments on any animal.

33.4.4.2 Navigation by insects

Support for the least-navigation–familiar-area map model can be derived from a wide variety of field and laboratory experiments on insects. In this section I have selected three examples that provide particularly clear support for this model.

The first example is that of *Ammophila pubescens* (Fig. 33.9), a hunting wasp that paralyses caterpillars that are often too heavy to be carried in flight. Having prepared a nest hole, therefore, in which she will store the paralysed prey upon which she will eventually oviposit, the female wasp hunts and immobilises a suitable caterpillar which is then dragged over the ground, often for 100 m or more, back to the nest. The area has been explored previously from the air and perhaps also by foot and on occasion the wasp may leave the prey during the journey back to the nest and perform what can only be a short navigational flight. Figure 33.9 provides strong circumstantial evidence that these wasps assess the compass direction of gross features around the nest site and that, having determined home direction, orientation is maintained by Sun compass orientation as long as such a compass is available. As the female draws nearer and nearer to the nest hole, she refers to finer and finer landscape details. In other hunting wasps, the final location of the nest hole has been found to be based on the configuration of detailed features (such as stones and fir cones, etc.) around the hole (Thorpe 1963). In every sense, therefore, the behaviour of these hunting wasps seems to be consistent with the familiar-area map model presented in Fig. 33.5.

By far the most convincing support for the familiar-area map and least-navigation models that can be presented for any animal, however, derives from the immense amount of work that has been performed on the honey bee, *Apis mellifera*. The following summary of conclusions from this work insofar as it relates to an evaluation of the least-navigation–familiar-area model is taken from Frisch (1967) unless stated to the contrary.

Young bees that have never before left the hive show no navigational ability even from distances as short as 20 m, unless by chance a bee arrives sufficiently close to the hive to recognise it by olfaction as the home site. The first object for inspection during the first exploratory migration is the hive. After examining the hive by flying in front

of it for a few seconds, the young bee flies off in a straight line away from the hive. Five minutes or so later she returns and hovers in front of the hive for a few minutes before she settles. During this flight it is assumed that by a combination of Sun compass and

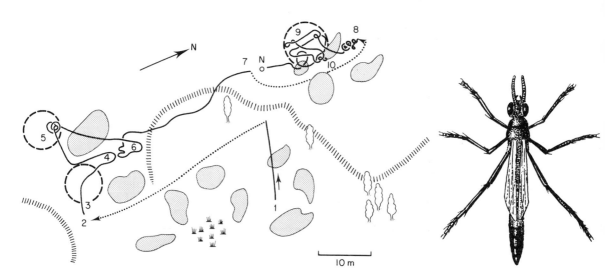

Fig. 33.9   Navigation by the hunting wasp, *Ammophila pubescens*

Ammophila hunts and paralyses the caterpillars of butterflies and moths that are too heavy to carry in flight and that therefore have to be carried back to the nesting hole by walking. In advance of this the wasp has performed exploratory migration over the area by flight. The illustrated experiment may be used to evaluate the thesis that the navigation mechanism consists of learning the compass directions of a variety of topographical features relative to the nest hole, the direction of which from some other position can then be determined from the compass directions of these same features from the current position (Fig. 33.5). Once the nest hole direction is determined, the animal moves in that direction by orientation.

In the diagram, the wasp's track is shown by a heavy line. A dotted line indicates experimental displacement during which the wasp and the caterpillar were placed in a box. The sky was clear throughout the experiment. The experimental area was a slight depression in gravelly heathland on which grew scattered plants of *Erica* and *Calluna* plus some denser patches of the same species (indicated by the stippled areas). The locations of small (about 1–2 m high) birch trees, *Betula*, are indicated by the tree symbols. The dashed circles indicate areas of confusion if the location of the nest hole (N) had been learned as SSW of a dense patch of heath, and NNE of an area without heath.

When first encountered (position 1), the wasp was moving to the W in a straight line toward the nest.

Evidence that the direction of movement was being maintained by orientation, probably to the Sun, was provided when, upon displacement to position 2, the wasp started to move in its original compass direction. Navigation was apparently not employed until position 3, at which point the animal made three flights. Following these, the insect set off in the home direction. Appearance of a dense patch of heath to the SW of the track (position 4), which would be inconsistent with the home site, led the animal to attempt to attain an appropriate compass position relative to the heath patch. The wasp thus moved to, and searched within (position 5), the dashed circle. The second time that the insect reached a position (6) NE of a patch of heath it began to repeat the previous manoeuvre before setting off finally in the home direction determined at position 3. Having nearly reached the nest (7), the wasp was displaced to position 8 whereupon it searched first nearby and then (9) in a further area of confusion. At position 10 a further flight was made which, after a further search in the area of confusion, led the wasp to locate the nest hole.

Although other features, such as the birch trees, may also have been used, though to a lesser extent, the observed track is most consistent with the thesis that in this case the wasp had learned the location of its nest hole in terms of the compass directions of the presence and absence of dense patches of heath as viewed on foot from that site.

[Diagram modified from Thorpe (1963)]

distant landmarks the bee establishes the position of the hive relative to the surrounding landmarks. Even after the outward phase of the first exploratory migration the bee can return to the hive from a distance of several hundred metres and after several exploratory migrations she can return from several, but less than 8, kilometres.

During the first few days of exploratory migration the bee cannot compensate for the Sun's shift in azimuth with time of day. This has to be learned and to do so the bee notes the Sun's shift during the day relative to familiar landmarks. The Sun's shift has only to be observed during part of the day for the bee to be able to extrapolate the Sun's position at any time of day. Bees transported from a place in the Northern Hemisphere, at which the Sun moves clockwise across the sky, to a place in the Southern Hemisphere, where the Sun moves anticlockwise, were at first disorientated. After a minimum experience of 500 flights over 5 afternoons, however, the bees were found to have an accurate compass based on the newly learned pattern of shift in the Sun's azimuth (Lindauer 1967b).

From the moment that bees have developed an accurate Sun compass, all information seems to be entered into the spatial memory along with the relevant compass information. The way that information concerning feeding sites is communicated from one bee to another by means of the bee-dance has been described in Fig. 14.21 (p. 147). A bee leaving the hive on the basis of information derived from another bee sets off in the appropriate compass direction as derived from the Sun's azimuth (the Sun's position in the sky being determined either directly or, if the Sun is not visible, from the pattern of polarised light). Compass direction can be determined as long as the Sun is further than $2°$ from the zenith (=directly overhead) position and as long as the cloud cover, if continuous, is not beyond a certain thickness. Perception by bees of the Sun through cloud cover is possible only in the ultra-violet region (300–400 mμ) and can occur long after the cloud has become too thick for humans to perceive the Sun's position.

Having left the hive on its first foraging trip in the appropriate compass direction the bee pays some attention to distant landmarks in the maintenance of this direction. The outward flight ceases when the bee has expended the amount of energy appropriate to that communicated (taking account of that involved in necessary detours to avoid buildings, etc.) by the dancing bee from which the location of

the food source was obtained. The final stage in the location of the food source is achieved by olfaction, based both on the smell of the food source and on pheromones deposited at the site by the dancing bee that communicated the information.

A bee that is performing its first foraging flight determines the home direction by navigation, individuals displaced laterally returning directly to the hive. If experienced bees are displaced laterally, however, they attempt to return on a track parallel to their outward track. In other words, during its first foraging trip the bee monitors its surroundings to determine the home direction. Having learnt this home direction, however, the bee automatically takes up this direction by reference to the Sun compass without investing time in navigation. This finding is identical to that found in birds and provides strong support for the least-navigation model. Further support for the least-navigation model is provided by the observation that if the flight path between the hive and a food source which the bee has visited previously takes the bee parallel to a continuous leading line (e.g. a forest margin, shore-line, or road), as long as it is not too far to the side, then the bee refers to the leading line rather than to the Sun compass. Presumably less time investment in navigation, particularly in compensating for wind drift, is necessary to follow a leading line than to orient to the Sun, even if a distant landmark is also employed. Bees displaced to the location of a similar leading line but which leads in a different compass direction from the original follow the leading line rather than orienting in the original compass direction. On the other hand, a succession of isolated trees or other landmarks is not adopted and bees with flight paths that pass a succession of such landmarks do not fly from one to another (= 'steeplechasing') but instead employ the Sun compass.

Having arrived at a food source for the first time, the bee evidently immediately registers the home direction, for if displaced (along with the food source) while feeding it subsequently indicates in its dance the direction of the source from the hive as it was immediately upon arrival at the source. Detour experiments suggest that the bee vectors the direction and distance of each leg that it flies during which the Sun is visible. If the Sun is invisible during the last leg the bee subsequently conveys in its dance the direction of the last target area before the Sun disappeared. Such vectoring may well occur when the outward exploratory flight is simple, but when it is tortuous, such as when a scout is searching for a new nesting site (p. 150), such vectoring seems unlikely. It may be suggested here that perhaps hive direction, either from the food source or potential nesting site or from the last target area for which a compass was available, is determined by the method outlined in Fig. 33.5. Support for this suggestion is provided by bees exposed to a crosswind on the way to the feeding site and that compensate by adopting an appropriately different compass heading from the Sun such that they head obliquely across the wind. Such bees subsequently announce the true compass direction of the food source, not the compass heading adopted by the bee. The suggestion is, therefore, that on their first foraging trip to a new food source or during exploratory migration by scouts, the outward journey is monitored only in areas at which familiar landmarks disappear. These then become target areas (Fig. 33.5). Immediately upon arrival at

a required resource its direction relative to the hive is determined by navigation. It is this direction or the direction of the last target area that was monitored before the compass disappeared that is announced in the subsequent dance. The route followed by the bee once it leaves the required resource, even if it is more direct (as can be manipulated experimentally—Fig. 33.10), has no further influence on the direction announced, except under certain experimental conditions (Frisch 1967, p. 169).

Bees that receive information from bees exploiting a food source the other side of a detour initially locate the source by flying over the detour in the appropriate compass direction. Subsequently, however, they locate the (shorter) detour route. This could be achieved by the reverse procedure to that described in Fig. 33.5. Thus, some point or target area that is visible from both the hive and the food

Fig. 33.10   Some aspects of the navigation of honeybees

A. When a honeybee that arrives at a feeding dish is displaced laterally (along with the feeding dish in a closed box), its behaviour upon attempting to return to the hive is a function of whether or not it has made the return flight between the dish (without displacement) and the hive on previous occasions. An experienced forager to that feeding site adopts the compass direction that it has learned will return it to the hive from that feeding site. Not until it has flown an appropriate distance without arriving at the hive does it employ a navigation mechanism to locate the hive position. A novice forager to that site, however, employs its navigation mechanism before attempting to return to the hive. Bees do not, therefore, simply reverse their outward orientation in their return to the hive. Instead, on at least the first occasion, they employ navigation and thereafter adopt the location-associated compass orientation.

B. In this experiment, involving the Indian honeybee, *Apis indica*, the bees reached the feeding dish by flying 3 m in the open and then (a) 3 m through a glass passageway. The announced direction (dashed arrow) upon return to the hive by a reversed track was the direct air-line between the hive and the feeding site. When, as in (b), the glass passageway was darkened so that neither the position of the Sun nor, therefore, the hive could be determined from the feeding site, the bees announced the direction of the end of the passageway which thus served as a target area (Fig. 33.5). This direction continued to be announced even if the bees were allowed to return to the hive by the direct route as shown.

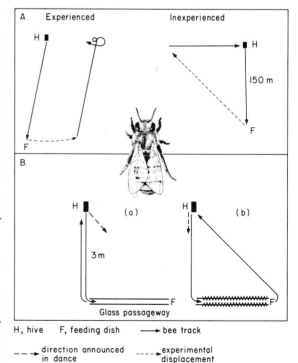

H, hive   F, feeding dish   ———▶ bee track

- - -▶ direction announced   - - - -▶ experimental
          in dance   displacement

[*Compiled from Frisch (1967) (after Lindauer and Otto)*]

source is recognised as such and at some stage on either an outward or return journey the bee navigates first to the target area.

There is mounting evidence that the Earth's magnetic field can be perceived by bees. The observations of Lindauer and Martin (1972) seem to show that the Earth's magnetic field can disturb the orientation of foragers that have recently returned to the hive during the execution of their wagging dances. During the dance the bees transpose the angle between Sun and foraging site into the gravity field (Fig. 14.21). Bees dancing parallel to the direction of the magnetic field lines orient themselves without error and exclusively in respect to gravity. In all other directions, the bees show an error of orientation (*Missweisung*) that is a function of the variations in the magnetic field. The error, which may be as much as 10° or so, depends in part on the orientation of the comb and in part on the diel fluctuations of the magnetic field. The diel variation of *Missweisung* follows closely a function of the curve of magnetic variation. The bees are sensitive to variations of the magnetic field in the range from $0-3 \times 10^{-3}$ oersted. Experimental compensation of the Earth's magnetic field results in the dancing bees accurately transposing the angle between goal and Sun into the gravity field.

These results, although showing that bees can perceive the Earth's magnetic field, provide no evidence of orientation. However, Oehmke (in Lindauer and Martin 1972) has provided such evidence. Daughter swarms were taken from bees that for generations had built their combs in a N–S direction and were placed in an empty cylindrical box with a central entrance in the bottom. Neither gravity nor incident light could provide orientation cues. Yet all of the colonies, without exception, built their combs in the same direction ($\pm 2°$ S.D.) as they had done in the hives from which they originated. Further, if the declination of the magnetic field is changed artificially, the direction of the combs is changed in the same sense.

It is perhaps not correct to think of the *Missweisung* as an error on the part of the performing bee (Frisch 1967). Presumably the recruit, while following the performer, is subjected to the same disrupting influence due to the magnetic field. In transposing back from the gravity field to the angle between Sun and goal, therefore, the effect of the magnetic field is automatically taken into account and the recruit flies in the 'correct' direction. It is perhaps because of this that there has been no

selection to compensate for the disrupting effect of variation of the Earth's magnetic field.

Even though it seems clear that bees should now be added to the list of various birds, beetles, flies, snails, planarians and protozoans that seem to be sensitive to magnetism (Adler 1970) and, like some birds, have access to a magnetic compass, there is as yet no evidence that bees make use of this compass in navigation. When the Sun is at the zenith or when it is too cloudy for a Sun compass to be accessible, the bees cease foraging until the Sun moves or reappears and do not take recourse to a magnetic compass. As with most birds, therefore, we have to assume that selection for least navigation acts against the use of a magnetic compass if any other alternative strategy is available.

Experiments on the foraging behaviour of the North African desert ant, *Cataglyphis bicolor*, which is a predatory and solitary hunter, have demonstrated (Wehner 1972) mechanisms of orientation and navigation that are similar in many respects to those of bees (see also p. 147). This ant, which forages at distances up to 100–200 m from the nest entrance, uses a time-compensated Sun compass, orientation to which is based on perception of the pattern of polarised light, and returns to the nest by reversing the outgoing direction. It does not make use of time-compensated orientation to the wind even though in the area of study the shifts in wind direction correspond with certain hours of the day. Although the ants orient frequently to topographical features, Sun compass angles are monitored throughout the complete return migration. The Sun compass direction of a specific feeding place from the nest is learned, as with bees, during the outward trip or upon first arrival at the feeding place, not during the preceding return.

Work on the navigation of insects, therefore, seems strongly to support the suggestion that navigation is achieved by monitoring the compass direction of known landmarks as viewed from one or several sites at which resources are located. Having associated a given site with a given compass direction from another site, least navigation favours return through learned location associated orientation, not navigation. The hierarchy of cues for orientation is by leading lines, wherever possible, followed by location-associated Sun compass orientation. There is no evidence for the use of a magnetic compass or for steeplechasing and if such mechanisms are employed they are at the bottom of the navigational hierarchy.

33.4.4.3 Navigation by other invertebrates

Perhaps the most widely known example of navigation by a non-insect invertebrate is that of various species of limpets (p. 199) that home, toward the end of each high tide, to a smooth, clean, slightly depressed area on the rock surface (Galbraith 1965). Some, such as the owl limpet, *Lottia gigantea*, of California and Mexico, defend the foraging area about this roosting site by prising and pushing off smaller individuals (Stimson 1973). It seems likely that navigation is achieved by committing tactile perception of the rock surface to a spatial memory on the outward journey (Galbraith 1965), though the possibility that a trail is laid that is resistant to scrubbing and detergent but not to chiselling of the rock surface cannot totally be eliminated. Either way, navigation occurs through the establishment of a familiar area.

The possibility that the land snail, *Helix pomatia*, shows a high degree of return to a winter home range has already been discussed (p. 518). Poorly controlled experiments suggested a mean deviation from the home direction of the direction of recoveries of less than 30° (mean deviation should be 90° if orientation is random) at distances of under 40 m. Homeward orientation apparently disappeared completely at release distances of between 150 and 2000 m (Edelstam and Palmer 1950). Homeward orientation did not occur if the snails were given very adverse treatment during displacement or if they were released at the edge of a blocking area (i.e. a wide path or a building). Presumably under both conditions the removal migration threshold was exceeded. The observation that orientation toward the winter home range was not found until middle to late summer is also consistent with the seasonal variation in threshold for migration to the winter home range if the snails performed a seasonal return migration. If navigation does occur, the only conceivable model that could be applicable is one approximating to that shown in Fig. 33.6 based on a primarily olfactory map.

Some species of wolf spiders (Lycosidae) of the genus *Arctosa* live on river banks or on sandy beaches. If such spiders are displaced from the bank onto the water the spiders can orient towards the bank. Each deme has its own preferred direction of orientation corresponding to the inland direction of the bank it inhabits (Papi and Tongiorgi 1963). When the sky is

covered the spiders orient visually to landmarks but when the sky is clear they use a compass based on the Sun and/or the pattern of polarised light. There is compensation for the Sun's movement across the sky during the day. Like all lycosids, the female lays her eggs in a cocoon which is carried attached to her spinnerets and after the young hatch they are carried on the female's back for a few days before they disperse. Under normal conditions the young spiders learn during the first few days after leaving the mother's back the appropriate direction to take to reach the bank when placed on water. The accuracy of orientation then improves with age. The learning process does not take place on the water and Papi and Tongiorgi suggest that the spiders extrapolate the appropriate direction from the slope of the bank. It seems possible, also, that the spiders can perceive the water visually from the bank and having learned the compass direction of water from the bank can reverse the direction when they find themselves on water. Even once learned, the preferred direction is not permanent and if transferred to a new location the spiders can learn the directional relationship between bank and water appropriate to the new location. Clearly, therefore, these spiders establish some form of familiar area during ontogeny and although degree of return has not been established the interaction between landmarks and compass and the strategy employed at any given moment as so far evaluated is quite consistent with the least-navigation–familiar-area map model.

When the young spider is a few days old it shows either random orientation or moves directly away from the Sun's azimuth, but if it is removed from a normal directional learning situation before it has adopted a preferred compass orientation an orientation to the N develops and by 30 days of age prevails over all others. The preference for N is found in all demes and is independent of the bank occupied by the mother. Papi and Tongiorgi consider this preferred direction to be innate, albeit of low inheritance, and most authors seem content to cast the phenomenon into the limbo occupied by all preferred directions once they are termed a nonsense orientation. It seems most unlikely, however, that the factors that result in a preferred direction to the N for an *Arctosa* will be found to be the same as the factors that result in an apparent preferred direction for a given deme of mallard (p. 873). For example, in the absence of all other cues, a spider given a clear view of the Sun seems slightly more likely to encounter a bank if it moves away from the Sun than towards it because the Sun is more likely to be obscured if it is over the nearest bank than if it is over the furthest. Support for this suggestion is given by the observation that until the spider has learned to compensate for the Sun's diel shift in azimuth it is most likely to orient away from the Sun's azimuth at all times of day. A spider that receives no other cues, therefore, once its clock mechanism begins to operate, seems more likely to adopt a N orientation than any other. It is perhaps not valid, therefore, to conclude that N orientation is innate. The mechanism of learning may give directional training to the N in the absence of other cues: having moved away from the Sun's azimuth the spiders eventually find themselves returned to a favourable situation, even though it may not have been achieved by their orientation.

The only major study on the navigation of sublittoral animals seems to be that of Herrnkind and McLean (1971) on the spiny lobster, *Panulirus argus*. Lobsters released in open sand areas moved offshore to shelter apparently by orienting into the current or surge. Since waves generally travel landward near shore, wave surge offers a more dependable directional indicator to offshore than does bottom slope which is masked by topographic irregularities. However, the ability to maintain directionality under a wide variety of conditions suggests that orientation is maintained by the use of multiple cues. No single guidepost, even wave surge, could account for all observed orientations. However, as 'homing' behaviour in this species, both spontaneous and after experimental displacement, even over distances of hundreds of metres (Fig. 33.11), suggests that learning plays an important role, it seems likely that much of the migration of this species takes place within a familiar area. As long-distance migration takes place *en masse* with thousands of lobsters joining in long, single-file columns, it seems possible that the familiar area within which seasonal return migration (p. 833) takes place is established through social communication. Whether the danger of not being able to establish a year's home range through social communication has been sufficiently great for recently mature lobsters to have an endogenous bias to the direction ratio of exploratory migration does not seem to have been determined.

33.4.4.4 Navigation by fish

There are as yet insufficient data on fish navigation to permit an evaluation of the least-navigation–familiar-area map model with respect to this group.

Displacement experiments on fish have in the main been restricted to fish with a limited home

Fig. 33.11 Homing tracks of three spiny lobsters, *Panulirus argus*, in Great Lameshur Bay, St. John, Virgin Islands

Three lobsters (1, 2, 3) were each captured in a den (rectangle), labelled with a sonic tag, released separately at R, about 200 m from the capture point, and then tracked by hydrophone. Solid lines indicate the homing track. Typically, lobsters moved in the general direction of the capture site until they reached the reef edge where they withdrew into a crevice or otherwise took shelter until dark. After dark, movement was resumed and by the following morning (circle) the lobsters were near to their place of capture.

Adoption of the homeward direction was rapid after release and as the release point was probably within the nightly home range of the lobsters it seems likely that some form of spatial memory was involved. Note that although the lobsters took shelter in the reef, initial orientation after release was nearer to the homeward direction than to the direction that would give the shortest route to the reef.

Water depth ranged from about 20 m at the release point to about 10 m at the denning sites.

[*Modified from Herrnkind and McLean (1971)*]

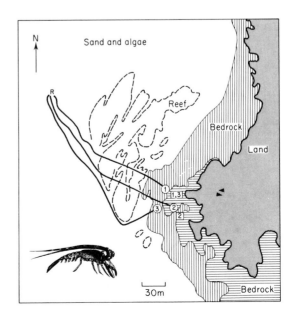

range (p. 849, Fig. 31.33) and have shown the same characteristics as displacement experiments on other animals. Most experiments involving the blocking of one or other sensory systems have shown that some fish recognise their home range using primarily olfactory cues (e.g. longear sunfish, *Lepomis megalotis*—Gunning 1959) and others recognise their home range using primarily visual cues (e.g. green sunfish, *L. cyanellus*, and European minnow, *Phoxinus laevis*—Hasler 1966). As pointed out by Harden Jones (1968), however, these experiments show only the senses used in recognising the home site having arrived there and give no information on the mechanism by which the fish recognise the home direction from their release site. The recent, but not yet abundant, evidence from various tracking techniques (Stasko 1971) indicates that homeward orientation is neither very precise nor very direct as might be expected if the fish were employing a familiar-area map based on primarily olfactory cues (Fig. 33.7), even if visual, thermal, electromagnetic, pressure, and other cues were also employed in map formation. Homeward orientation would particularly be unlikely to be precise and direct if the fish were making maximum use of favourable currents to return them home as illustrated in Fig. 33.7. There is still a great need for measurement of current speed and direction at the same depth as the fish.

Adult salmon which develop from eggs displaced from one stream to another return to the stream in which they grew up and not to the stream in which they were laid as eggs (Harden Jones 1968). Prespawning sockeye salmon, *Oncorhynchus nerka*, displaced from one tributary of Brook's Lake, Alaska, to three other tributaries showed a very high degree of return to the tributary of capture (Hartman and Raleigh 1964). Such experiments demonstrate clearly that some learning, whether sequential or spatial (p. 850) occurs during the downstream migration by the fry and/or parr. The latter experiment tends to favour a spatial rather than sequential memory but the evidence is not conclusive. Either way, the situation could be described by the model illustrated in Fig. 33.7.

The coincidence of the migration circuits of many marine fish with ocean currents was stressed by Harden Jones (1968) and has been elaborated in Chapter 31. Royce (in Hasler 1967, pp. 19–20) has pointed out that those Pacific salmon from streams in British Columbia and southeast Alaska that reach a point 3000 km away, some 800 km south of the Commander Islands (Fig. 31.10), would need to make a number (albeit few) of decisions in order to arrive back at the home coast after orienting for most of the time downcurrent. Such decisions are compatible with the map hypothesis presented in Fig. 33.7.

A compass is not an essential component of the model presented in Fig. 33.7, though it would seem that if a fish could monitor, say, the direction of a current in compass terms, the resulting map would be much more flexible. Waterman (1972) has argued that the scattering of sunlight in water could produce Sun-dependent polarisation patterns that, from photographic evidence, would provide a Sun compass for fish down to depths of at least 200 m. A magnetic compass also seems possible (McCleave *et al.* 1971) though there is as yet no evidence that fish have access to such a compass.

There is no doubt that perception of, and orientation to, a current is possible if visual cues are also available. There is as yet, however, no direct evidence that fish can detect and orient to water currents at depths where visual cues are not available. However, there are so many possible ways in which a fish could obtain information concerning current direction (Harden Jones 1968) and such an ability would be of such clear advantage to the animal, that there seems little doubt that fish will eventually be found to manifest such an ability.

33.4.4.5 Navigation by pinnipeds and cetaceans

The cues available for pinnipeds and cetaceans to establish a familiar-area map that could be used in navigation, and circumstantial evidence that they may do so, have already been discussed in Chapters 28 and 29. To my knowledge, the critical displacement experiments have not been performed on these animals and there is no evidence either way concerning the applicability of the familiar-area map and least-navigation models developed in this chapter. The observation of H.D. Fisher (in Sergeant 1965) that adult and immature harp seals, *Pagophilus groenlandicus*, (Fig. 28.18) which migrate mainly at night, were observed on 1 April migrating to the N off Labrador on a cloudy night, neither proves nor disproves the existence of celestial navigation or a celestial compass. Presumably it will be found that pinnipeds, like all animals, have a variety of direction finding *and direction maintenance* cues available to them and that as one set is removed they switch to another (p. 861), perhaps in accordance with the least-navigation hypothesis. Thus, these harp seals may well have determined migration direction by reference to a celestial compass and/or a familiar-area map at some other time and/or at some other place. When deprived by cloud of further celestial information, however, the animals could have maintained direction by switching to the use of a, perhaps familiar, leading line, by maintaining a fixed angle to the current, or by referring to a magnetic compass.

Some evidence that bottom topography may be used as a leading line or as part of a familiar-area map is provided by radio-tracking experiments on delphinids (Fig. 33.12). However, perhaps the most stimulating speculations concerning cetacean navigation involve the possible role that might be played by social communication. Thus Payne and Webb (1971) stress that use of the term 'herd' for whales may be misleading because it has in the past always been based on a visual assessment by the observer. The powers of communication of whales, however, could be so great that individuals, apparently alone and separated from other individuals by tens of kilometres or more of ocean may nevertheless be in acoustic contact (see also p. 767). The 20 Hz sounds made by finback whales, *Balaenoptera physalus*, even in modern deep-ocean, ambient noise conditions in areas of moderate shipping traffic, can be calculated to be audible, at a signal-to-noise ratio of 0 dB, out to ranges of 72 km when the sounds propagate by spherical losses and to ranges of

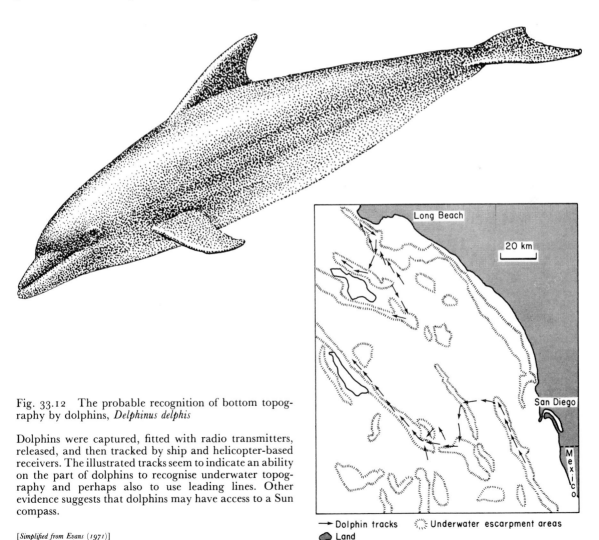

Fig. 33.12 The probable recognition of bottom topography by dolphins, *Delphinus delphis*

Dolphins were captured, fitted with radio transmitters, released, and then tracked by ship and helicopter-based receivers. The illustrated tracks seem to indicate an ability on the part of dolphins to recognise underwater topography and perhaps also to use leading lines. Other evidence suggests that dolphins may have access to a Sun compass.

[*Simplified from Evans (1971)*]

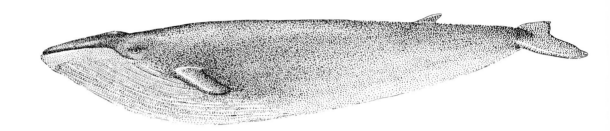

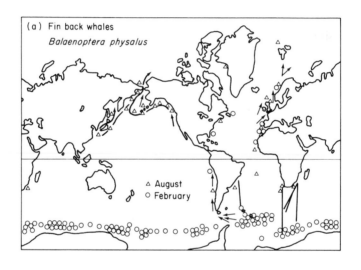

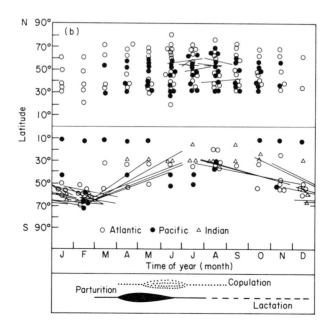

Fig. 33.13 Seasonal return migration of the finback whale, *Balaenoptera physalus*

General ecology

The finback whale is a mysticete and feeds on euphausids (****), copepods, (**), and fish (*). It employs a swallowing type of feeding behaviour (p. 746). The finback seems to be adapted to exploit areas containing euphausids at average density.

The species occurs both singly and in groups of usually 10–15 individuals, though schools of 300 or even 1000 are not unknown. These schools are often accompanied by dolphins. Pairs of finbacks are usually male/female pairs or female plus calf.

The seasonal incidence of copulation (....), parturition (———), and lactation (– – –) in the southern hemisphere is indicated below diagram (b). Gestation lasts for 10–12 months. The preferred temperature in winter (April–October) off South Africa is 16°C.

Sexual maturity is attained at an average length of 19·2 m for males and 20·3 m for females. Average maximum length is 20·7 m for males and 21·0 m for females.

Seasonal return migration

(a) Geographical distribution in August (△) and February (○)

Dots show sightings in the appropriate month. Arrows show possible migration routes. Straight lines join points of release and recapture of marked individuals.

(b) Seasonal shift of latitude

The dots indicate the latitudes at which finbacks have been sighted in different oceans during each month of the year. The lines join the latitudes and times of release and recapture of marked individuals.

There is a clear indication of a polar migration in summer and a tropical migration in winter. There is also an indication, however, that in both the Pacific and the Atlantic some individuals remain at tropical latitudes throughout the year. Off South Africa, all finbacks are migrating north from May to November.

There seems to be a clear separation between northern and southern hemisphere demes.

In the southern hemisphere, immature individuals occur further from the poles than adults and they also spend more time near the coasts when in tropical waters. As an individual ages, therefore, it contracts its home range away from coastal regions and extends its familiar area into colder and colder waters

[*Compiled from data in S. G. Brown (1954, 1958, 1959, 1962), Nemoto (1959), Slijper (1962), Omura and Ohsumi (1964), Slijper et al. (1964), Bannister and Gambell (1965), Berzin and Rovnin (1966), Jonsgård (1966), Mackintosh (1966), Nasu (1966), Nishiwaki (1966, 1967), Best (1967), Gaskin (1968b), and Neave and Wright (1968)*]

840 km when propagated via the deep-sound channel. Ranges of 1120 km represent an upper limit that is approached but not attained. Presumably, however, the 20 Hz sounds evolved for a set of background-noise conditions that no longer exist due to the traffic of ships. In remote areas the maximum transmission range can be calculated to be 224 km by spherical propagation and 5600 km by cylindrical propagation in the deep-sound channel. Indeed, other conditions can be envisaged that would extend this range even further. The characteristics of the 20 Hz frequency also encourages an interpretation of its function as long-range communication. It would be free of disruption by storms, being just below storm-generated noise. It loses almost no energy by reflection from the bottom. It has remarkably low attenuation with distance (3 dB in 8960 km) and is in the best octave for long-range propagation under polar ocean conditions.

In their paper, Payne and Webb show a reticence that is appropriate to such an astounding possibility. Their arguments are sufficiently convincing, however, for the possibility to be taken seriously that all finback whales in, say, the Atlantic Ocean are in communication with one another, either directly or in relays. Examination of Fig. 33.13 shows that finback whales are scattered throughout the World's oceans at all times of year and that the seasonal return migration is in effect a seasonal shift in modal distribution rather than a clear-cut to-and-fro population shift such as appears to be the case in the black right whale, *Eubalaena glacialis* (Fig. 29.4). The information content of the 20 Hz signals would not need to be very great for the signals to be used both as a means of comparing deme density and habitat suitability in widely separate parts of a given ocean and as a means of navigation and orientation, all entirely by social communication. The scatter of individuals throughout the ocean and the different responses of different individuals would still, of course, reflect the optimum frequency of migration thresholds as in any other system (Chapter 17).

It is to be hoped that the attempts that surely must be made to determine the existence or otherwise of this astounding possibility identified by Payne and Webb will not be thwarted by the premature extinction of the great whales.

33.4.4.6 Navigation by amphibians

Insofar as they are known, the navigational mechanisms employed by amphibians fall well within the framework of the least-navigation–familiar-area map model. Except where otherwise stated, the experimental data presented below, but not necessarily their interpretation, have been taken from the review by Ferguson (1971).

Almost all experiments to date show that if amphibians build up a familiar area, they do so on the basis of a wide variety of environmental cues. It seems likely, also, that all of the information used is entered into the spatial memory in relation to the compass direction of a given feature from a given location. Reports are available that the Sun, stars, and stars and Moon, are used as compass reference points by various frogs, toads, and salamanders, and that in the case of the Sun the animals compensate for its movement across the sky. The apparent lack of access to a day-time compass if subjected to conditions (e.g. constant dark, temperature, and humidity) that fail to provide a zeitgeber for the clock mechanism and the lack of access to either a day-time or night-time compass during periods of cloud suggests that if amphibians do have access to a magnetic compass then they are reluctant to make use of it. Even 'blind' amphibians have access to a Sun compass by means of the perception of light by extra-optic photoreceptors, apparently usually located in or under the optic tectum, mid-brain, or pineal region.

A commonly evaluated landscape feature that is committed to the spatial memory in terms of compass direction is that of the land–water interface or perhaps even the compass direction of the steepest gradient from shore to deep water. This information is learned by amphibian larvae soon after hatching and is utilised whenever the animal's resource requirements lie at a location on this gradient that is different from the location currently occupied. Detailed studies have been made in America on Fowler's toads, *Bufo fowleri*, narrow-mouthed toads, *Gastrophryne carolinensis*, bronze frogs, *Rana clamitans*, and bull frogs, *R. catesbeiana* (McKeown in Ferguson 1971). In narrow-mouthed toads the compass direction of this gradient has been learned within 20 h of hatching. Displaced newly hatched individuals orient shoreward to shallow water, but somewhat older individuals orient offshore toward deeper water. Individuals approaching metamorphosis orient shorewards. In all cases, orientation fails under complete cloud cover and is not shown at all in larvae reared in darkness without access to a celestial compass.

Proof that the compass directions of visual landmarks are learned seems to be lacking, though

observations such as the capture 14 times during a 2 year period of the same individual of Fowler's toad under the same streetlight and the familiarity of this and other toads with escape shelters in the vicinity indicates that at least the spatial relationships relative to each other of visual landmarks are committed to a spatial memory. Evidence is available, however, that acoustic cues are committed to the spatial memory with respect to their compass direction from a given location. Chorus frogs, *Pseudacris triseriata*, caged for several days near a breeding chorus of conspecifics, then moved in light-tight containers and released in an arena, selected a compass course that would have taken them to the chorus from the original site, but not from the site of testing.

Amphibians displaced in light-tight containers behave initially upon release as if they have not been displaced, though later they may attempt to return to their original location. Amphibians displaced in open containers, however, orient upon release in the compass direction opposite to the direction of displacement. If only part of the displacement is in

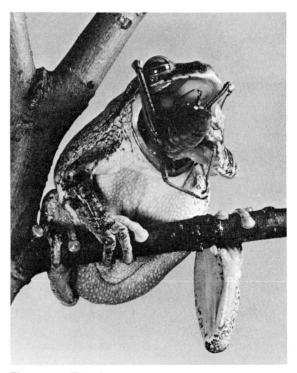

Fig. 33.14   Tree frog

open containers, orientation upon release is in the compass direction opposite to the direction of displacement during that leg and takes no account of the displacement direction for the leg during which the Sun was not visible. Presumably such behaviour is advantageous in animals liable to, say, downstream displacement or liable to be carried some distance by predators but then escaping. The mechanism involved seems likely to be to monitor the direction of shift of close, not necessarily familiar, landmarks and then adopt the same compass direction upon release. Thus if trees appear to be moving from E to W during displacement the amphibian should migrate W upon release. Alternatively, some inertial mechanism may indicate direction of shift, the compass direction of which could then be determined by reference to the Sun. Amphibians that do not show the ability to monitor the compass direction of shift in open containers (e.g. southern cricket frogs, *Acris gryllus*) and individuals displaced in light-tight containers, may nevertheless eventually show homing ability after release.

Frogs and toads caged on an unfamiliar shore show evidence of being able to orient to their new surroundings within 2 h and the process is nearly complete in 24–48 h. The role of a compass in this process was shown by southern cricket frogs which failed to orient if not allowed to see the daytime sky. Having been given visual access to this sky, however, the frogs could orient to the new shore in both day and night tests.

Many experiments designed to test the possibility that olfactory cues are used in navigation failed to take the precaution of displacing the animals in light-tight containers and so cannot be considered as critical experiments. Where proper precautions are taken, however, evidence is obtained that amphibians make use of olfactory cues. Blinded Californian toads, *Bufo boreas*, homed to breeding sites from 200 m whereas anosmic toads appeared to be disoriented. Neither blindness nor anosmia prevented homing of Mexican toads, *B. valliceps*, from 237 m, but both together did so. The tendency of homing salamanders to raise their heads above the body axis is also suggestive of the sampling of air-borne odours. Ferguson (1971) appears to condone the suggestion that the unreliability of wind direction and the existence of homing in the presence of winds that, because of direction, could not have carried 'home odours', argues against the use of olfactory cues. Great weight has also often been given to the observation that breeding ponds that have been

filled in and covered with soil, thus preventing the emission of home odours, are nevertheless visited by toads for up to five years. None of these arguments or observations, however, contradict the possible existence of olfaction-based maps of the type postulated in Fig. 33.6.

One final example may be presented that is strongly indicative of an important role for distance perception, either visual or olfactory, in the navigational mechanism of amphibians. The red-cheeked salamander, *Plethodon jordani*, is an American species found in the southern Appalachian highlands. It lives at high elevations in moist woodland habitats away from aquatic media, is lungless, and depends largely on skin moisture for cutaneous gaseous exchange. The species is active on the forest floor at night and spends the day under cover, usually in burrow systems. An adult has a total average length of 11·5 cm. When displaced at night from their roosting site for distances of up to about 300 m, individuals show highly oriented, direction-independent homing, the success of which is in part dependent on the distance of release (Madison 1972). Vision is not essential for successful homing, though it is possible that vision is used if unimpaired. A functioning olfactory system, however, seems to be essential, even though homing is not continuously against the wind. The most relevant factor in these experiments, however, was that there was an increased incidence of climbing on vegetation and trees by displaced individuals. This behaviour presumably facilitates the detection of airborne odours and/or more distant visual landmarks.

33.4.4.7 Navigation by reptiles

The critical experiments that permit some evaluation of the familiar-area map and least-navigation models do not seem to have been performed on reptiles.

The most relevant series of experiments with which I am acquainted are those on the box turtle, *Terrapene c. carolina*. Although the following data are taken from Gould (1957), their interpretation is my own except where indicated.

Displacement–release experiments have demonstrated an ability to home from distances of up to 1·3 km. Spontaneous movements of 774 m have been observed, though at any single season the animal occupies a home range of only about 100 m in diameter. The size of the familiar area is unknown but is likely to be considerably greater. Of 43 box turtles captured and released at distances of 0.2–2.7 km, only 22 showed homeward orientation. Not all of these 22, however, were necessarily orienting to home because on subsequent releases they adopted the same compass direction as on the previous occasion, even though it was no longer the home direction. We may conclude, therefore, that even in the complete absence of true clues concerning home direction, a box turtle adopts a particular compass direction. Perhaps the direction selected is the one that from the most similar situation in the animal's familiar area would return the individual to the appropriate home range. In any event, after capture the box turtles were maintained in a shallow pit and subsequent release experiments showed that some individuals at least had adopted this pit as 'home'. After displacement–release and being tested for homeward orientation the turtles were returned to this pit. It is reasonable to suppose, therefore, that the turtles associated their last compass orientation with successful homing. Whatever the explanation, subsequent release experiments, especially at distances greater than about 700 m and at all long-distance (288 and 640 km) releases, showed that the turtles tended to adopt the compass direction that they had adopted on the previous occasion, independently of home direction. Significantly, the only releases from different directions that consistently showed homeward orientation were at short-distance release sites where the home was between 240 and 700 m, and, in one instance, 1 km, from the release site. One individual was not recaptured after showing homeward orientation and was recovered at home, 512 m away, 10 days later. The compass used on all occasions seemed to be a Sun compass. No orientation occurred in cloudy weather.

In the box turtle, therefore, as in all the animals so far considered, the implication is strongly that navigation is achieved by some interaction of land-marks and a compass. Whether this interaction is as suggested in Fig. 33.5, as it seems to be in insects, other invertebrates, and amphibians, cannot as yet be determined. Again as in the other animals so far considered there is a strong indication that the animal drops the determination of homeward direction and relies entirely on a compass as quickly as possible in the solution of any given navigational problem, in a way that is consistent with the least navigation model.

As far as other reptiles are concerned, little progress appears to have been made in an evaluation of navigational strategy, even in the case of the sea turtles that have otherwise received most herpetological attention. Extant hypotheses of sea turtle navigation as well as the familiar-area model have already been described (pp. 775 and 851) as also has the sea-finding behaviour of newly hatched sea turtles (p. 860). Experimental data, however, as yet offer no possibility of the evaluation of these theses. Long-range (185 km) radio-tracking has been employed on females of the green sea turtle, *Chelonia mydas*, but only preliminary results have so far been obtained (Baldwin 1972). Such results obtained during homing experiments suggest firstly that, at least in the vicinity of the breeding grounds at Tortuguero, Costa Rica (Fig. 30.1), olfactory cues emanating from the home site need not be used during homing. Note, however, that this does not necessarily mean the absence of an olfaction-based map (Fig. 33.7). Furthermore, although based on only two observations, some indication was obtained that in water depths in excess of 180 m, sea turtles may have access to a magnetic compass. This was concluded when a turtle with a magnet fixed to its plastron in such a way that the magnet was free to pivot showed a lack of orientation whereas when the magnet was fixed so that it could not pivot the oriented path was 90° different from the home direction.

33.4.4.8 Navigation by terrestrial mammals

Insufficient data are available concerning the navigation of mammals to permit a critical evaluation of the familiar-area map and least-navigation models for these animals. A few intriguing observations are available, however, which suggest that these models may be applicable.

From time to time a report appears in the popular literature of some terrestrial mammal that has found its way home after having been displaced some spectacular distance, often 1000 km or so. We are not forced, of course, to take these instances into account in considering animal navigation. The 'experiments' were not controlled and we have no idea how many animals are similarly displaced each year, attempt to 'home', but fail. Those that succeed may well do so by the lucky chance that must befall some individuals of any group of animals forced to employ a random-search strategy. It is perhaps worth a passing reference, however, that, to the best of my knowledge, the animals concerned (dogs, bison, deer, etc.) are invariably species with a powerfully developed sense of smell. Of all the models presented in Section 33.4.3.5, that shown in Fig. 33.6 that is based on an olfactory map seems to be the most capable of permitting extremely long-distance homing in an animal with a sufficiently acute olfactory sense. Horses, for example, are considered to be able to home over short-distances by olfactory means (Hafez *et al.* 1962). It may well be worth employing long-distance displacement–orientation experiments on, say, dogs, to examine this possibility.

The one-humped dromedary of Arabia is reputed to have a well-developed capacity to return to its herd from distances of up to 50 km and in exceptional cases from distances as great as 240 km (Dickson 1949). By repute, therefore, the camel has the ability to navigate over long distances though again it has not been proved experimentally that the observed instances of homing cannot be explained by random search. If navigational ability exists, the mechanisms involved are unknown. It is perhaps worth noting, however, that the camel, adapted as it is to a habitat in which food, salt (in the form of salt-rich plants), and water are sparsely distributed, is likely to have a large familiar area. In those living symbiotically with man, this area averages about 55 000 km² with an average radius of about 240 km.

Critical experiments on human navigation also seem not to have been performed. The majority of people who read this book will have spent their entire lives within a familiar area that was established in the first place by social communication. Information is available not only through word of mouth from other individuals but also through maps, charts, road signs, films, etc. Physical entry into a new area is invariably preceded by, or occurs simultaneously with, a process of familiarisation with the area through social communication. This is least navigation. Indeed, the involvement of social communication in the navigational mechanism of any animal may be considered to be an adaptation to selection for least navigation. I would suggest also that the various elements of the familiar-area map based on primarily visual cues (Fig. 33.5) and also the other elements of the least-navigation model (e.g. use of leading lines, a minimum monitoring of outward tracks, the cessation of navigation once the direction of one resource from another has been learned, etc.) are within the experience of most of us. In the absence of critical experimentation, however, there is little to be gained from trying to develop this argument solely on the basis of subjective experience.

Some specialist humans make use in their everyday lives of socially communicated grid maps based on the Sun arc and geomagnetic fields and may well be the only animals that do so. This navigational aid, however, should perhaps be viewed as part of the wide range of socially communicated and traditional abilities developed by man. The majority of humans, for the major period of the evolution of the species, appear to have used a navigational system based on landmarks and a Sun, star, and Moon compass that is little different from that described in Fig. 33.5. There is no evidence that man has access to a geomagnetic compass without the use of instruments (Beischer 1971).

The homing performances of grizzly bears, *Ursus arctos*, from 45 km (Cole 1971) and of red foxes, *Vulpes vulpes*, from 56 km in 12 days (Phillips and Mech 1970) and of a wide variety of other carnivores have been described in Chapter 15. All such observations and experiments suffer from the inability to distinguish between navigation and random search and from a lack of knowledge of the size of the familiar area, as opposed to the year's home range, of the animals concerned. Long-range (158 km) radio tracking of polar bears, *Ursus maritimus*, has begun, but so far only provisional results are available (Baldwin 1972).

Brush rabbits, *Sylvilagus bachmani*, in Oregon were found to home from distances of up to 165 m but not further and it was suggested that beyond this distance familiar visual landmarks disappeared (Chapman 1971). Muskrats, *Ondatra zibethica*, released in the same or a different stream showed a 50–57 per cent return from 0·5 to 2·0 km, 31 per cent return from 3 km, and a 15 per cent return from 4 km. These rodents were not deterred by obstacles such as mills or power plants but the navigational mechanism cannot certainly be identified, though it seems likely to have involved visual landmarks (Mallach 1972). Some evidence that a compass as well as landmarks may also be involved in rodent navigation was provided by Bovet (1968) who trained deer mice, *Peromyscus maniculatus*, to home from distances of up to 500 m. Thereafter, even when released on featureless snow, they oriented immediately in the home direction. The experimental technique cannot, however, provide critical data.

The reason for the lack of critical experiments on the navigation of terrestrial mammals in the field is difficult to determine. It is to be hoped, however, that the advent of telemetry and the resulting

Fig. 33.15   The navigation of bears

The grizzly bear, *Ursus arctos horribilis*, and polar bear, *U. maritimus*, are thought to have evolved within the last 20 000 years and to stem from a common ancestor, the Pleistocene bear, *U. etruscus*. Despite this relatively recent common origin the two species show different patterns of lifetime track, though the difference may be less than once thought.

The grizzly is a solitary animal with a home range of 7–25 km². The familiar area, however, is much larger. Individuals perform exploratory migrations of up to 110 km from their pre-existing home range and suitable habitats (such as rubbish tips) encountered during such migrations may well be visited at regular or irregular intervals, even though they may be 80 km from the usual home range. Grizzlies, therefore, have the ability to navigate over such long-distances despite living-most of their lives within a much smaller home range in which well-worn tracks are often utilised.

The polar bear is thought to have a much larger home range and perhaps also, though not necessarily, familiar area than the grizzly. Within its home range the polar bear is thought to be nomadic, though this may simply reflect insufficient data to demonstrate a movement pattern. It seems unlikely that polar bears circumnavigate the pole on drifting pack ice. Rather, reports of the ability to navigate over distances of 100 km or so suggest the existence of a familiar area.

[*After Harington (1970) and Mundy (1973). Grizzly photo by Gary R. Jones. Polar bear photo by Fred Bruemmer.*]

possibility of monitoring initial orientation after release will lead to a rapid increase in the experimental field data available for this group.

33.4.4.9 Navigation by bats

The potential for bats to establish a familiar area and the cues available are discussed in Chapter 26. This section, therefore, is concerned solely with a consideration of displacement experiments on bats and an evaluation of the extent to which they support the least-navigation–familiar-area map model. In an excellent paper, Davis (1966) critically reviewed displacement–release experiments on bats up to that time and concluded that with two exceptions all previous releases had been within the familiar area of at least some of the bats concerned. Consequently, although these experiments were instructive in evaluating the nature of the familiar area in bats, they were of little help in attempts to determine whether bats used any form of grid map during homing.

This section on bat navigation is in effect divided into two parts. The first part surveys displacement–release experiments on bats wholly within their familiar area. The second part considers those very few experiments in which it is possible that the bats were released outside their familiar area. First, however, some discussion of experimental technique seems to be necessary.

Displacement–release experiments on bats are subject to the same shortcomings as those on other animals (p. 871) and are additionally hampered through being by necessity performed at night. Until the advent of telemetry (Fig. 33.16), displacement–release experiments produced only data pertaining to homing performance. The absence of data for orientation upon release was a serious drawback due to the difficulty of dismissing random-search models.

A further difficulty derives from the seasonal, sexual, and ontogenic variation in the threshold for type-C removal migration (p. 872) (= variation in 'locality-loyalty'; = variation in 'motivation to return'). As for most other animals this variation was confused with variation in the *ability* to navigate. Thus young bats have a lower removal migration threshold than adults (Chapter 17) and males have a lower threshold than females with respect to the maternity roost but not with respect to the copulation roost. A few examples will serve to illustrate this variation in removal migration threshold and the ease with which it may be confused with variation in navigational ability.

In the pallid desert bat, *Antrozous pallidus*, in southern Arizona, the maternity roosts are occupied from March to October by the adult females. The young bats are born, suckled, and weaned in these roosts from May to September. During the first part of the summer, from March to July, very few adult males are found in these maternity roosts, the majority using scattered summer roosts. In August there is an influx of adult males into the maternity roosts at the time that the juveniles are beginning to leave. Copulation occurs in September and October and the adults then transfer to the winter roosts. Homing experiments carried out on males and females collected from the maternity roosts from March to July show a much higher degree of return by the females. Homing experiments carried out on males and females collected at the maternity roosts from August onwards, after the influx of males at the beginning of the mating season, show no significant difference in degree of return between males and females (Davis 1966).

Sexual differences in removal migration thresholds are to be expected with respect to maternity and copulation roosts but are perhaps less likely with respect to hibernation roosts. Thus there is no difference in the homing performance of males and females to winter roosts in the little brown myotis, *Myotis lucifugus*, and Indiana myotis, *M. sodalis* (Mueller 1965, Hassell 1963). In some species, however, males and females have different requirements for hibernation (e.g. Schreiber's bat, *Miniopterus schreibersi*—Dwyer 1966) and in such cases sexual differences in homing to a particular type of winter roost might be expected. The threshold for removal migration might be expected to be lowest with respect to transient roosts and certainly American free-tailed bats, *Tadarida brasiliensis*, displaced from transient roosts in Mexico, show a relatively low degree of return (Davis 1966).

Occasionally, sexual differences in migration threshold combine with other factors to produce unexpected results, as in the homing experiments carried out by Gunier and Elder (1971) on the grey myotis, *Myotis grisescens*, in central Missouri. Both males and females were displaced from maternity colonies and were released, some at 60 km from the home roost and others at 120 km. The latter may, as suggested by Gunier and Elder, have been outside the familiar area but it is by no means certain (see Fig. 26.26). Females, with presumably high removal migration thresholds with respect to the maternity colony, showed a 27 per cent return from the shorter distance and a 24 per cent return from the longer distance. Males, on the other hand, showed a 1 per cent return from the shorter distance yet a 38 per cent return from the longer distance. The critical factor seems to have been that there were no suitable roosting sites between the more-distant release and the home site. The latter could well, therefore, have been the nearest available roosting site and males, with perhaps a greater flight speed, were less subject to mortality risk than females over this distance (see below). At the shorter distance, however, the lower removal migration threshold of males resulted in the majority going to other roosting sites with which they were familiar (Gunier and Elder 1971).

The low returns following displacement are characteristic of the majority of such experiments on bats and it seems likely that the combination of experimental handling and its interaction with mortality risk factors during homing is the major factor involved. Indeed, the combination of these factors often makes it extremely difficult to determine what is being measured by any given experiment. Thus a large number of experiments have attempted to identify the relative roles of vision and echo-location in the navigational strategy of bats. Mueller (1968) has pointed out, however, that the role of vision being measured by homing experiments may not necessarily be directly related to the navigation mechanism. As far as reduced homing by blinded bats is concerned, it is possible that the only aspect of vision being tested is the ability to distinguish between night and day. The ease with which bats are caught in daylight by predatory birds undoubtedly leads to higher mortality on the part of bats that cannot determine the appropriate time to roost. An alternative explanation to the use of the visual sense in navigation could also be advanced to account for the slower rate of return normally found for blinded bats. Blinded bats fly much lower than normal bats, presumably as a result of the need to keep within echo-location reach of the ground. 'Normal' bats fly much higher, are unimpeded by topographical obstacles, can maintain a straighter track, and as a consequence can home faster. From longer distances these two factors could reinforce one another. By homing more slowly a blinded bat may fail to arrive back at its roost before dawn, may therefore fly during daylight, and as a consequence may also suffer higher mortality. It is possible, therefore, that the experiments which have set out to evaluate the importance of vision in bat homing have not been evaluating the importance of vision in

navigation at all. Possibly all that has been tested so far is the ability of bats to distinguish between night and day and to perceive the visual horizon that enables them to fly at an altitude sufficiently great to avoid being impeded by topographical features.

This artefact could well account for the major finding concerning the relative importance of vision and echo-location in homing. That is that echo-location is important at short distances but that vision becomes increasingly important with increasing distance. This is also, of course, the conclusion that would make most biological sense if the least-navigation–familiar-area map model were valid. Thus, vision, even in the dark (p. 386), is best suited for the perception of gross features at long distances whereas echo-location, with its limited range (p. 585), is best suited to the perception in the dark of detailed features at short distances.

Male southeastern myotis, *Myotis austroriparius*, released in daylight up to 230 m from their roost in October showed that the visual sense was of major importance in their attempted return (Layne 1967). Of the control 'normal' bats released, 33 per cent returned. Of the experimental bats, 12·5 per cent of those the ears of which had been plugged and none of those whose eyes had been covered returned.

Releases of the little brown myotis, *M. lucifugus*, and Keen's myotis, *M. keenii*, at a distance of 50 km suggested that the sense of hearing and echo-location was important in homing from this distance, which for *M. lucifugus*, at least, is well within the familiar area (Fig. 26.7). None of the 32 bats released with impaired hearing returned to the roost, even though half had normal vision. Homing performance was similar for blinded and vision-unimpaired bats (Stones and Branick 1969).

Releases of the carnivorous leaf-nosed bat, *Phyllostomus hastatus*, showed no difference in homing ability between blindfolded and normal bats over a distance of 13 km, but over distances of 24 to 32 km none of the blindfolded bats returned as compared to 50 per cent of sham-blindfolded bats (Williams *et al.* 1966). In this instance, therefore, differences in homing ability cannot be attributed to the irritating effect of the mask.

Blinded Indiana myotis, *Myotis sodalis*, showed much reduced homing performance in comparison with normal controls (Barbour *et al.* 1966). Whereas 65 per cent of normal bats returned from 200 km and 28·6 per cent and 67·4 per cent returned from two releases at 320 km, only one blinded individual returned (5 per cent of one release group from

320 km), and this individual was subsequently found to have had the use of partial vision. In other releases in which blinded males were released at the same time and from the same place as the controls, blinded individuals returned in some numbers but not to the same extent as the controls (Davis and Barbour 1970).

There is some indication that, at least in some species, migration is not an individual phenomenon but that groups of individuals stay together. For example, of 26 female little brown myotis, *M. lucifugus*, that were banded in one roost in September of one year, 11 were found together in a single maternity colony the following June, yet none of the nearby maternity colonies contained banded individuals (Gifford and Griffin 1960). Cooperation between individuals during movement and exchange between individuals of information concerning routes and directions, either before or during movement, cannot entirely be ruled out on present

evidence. The tendency for group cohesion during seasonal return migrations and during homing experiments has been noted (Davis 1966). Such group cohesion could explain some of the homing returns of blinded bats when these are released along with other, sighted, individuals.

There seems no doubt from the experiments so far described that a bat's familiar area is based on both visual and echo-location cues. In view of Mueller's (1968) criticism of homing experiments with blinded bats it cannot be taken as proved that visual cues increase in importance relative to echo-location cues with increase in distance from the roost, even though this seems likely. Certainly the only major experiment so far performed in which bats were tracked during their homeward flights following displacement–release (Fig. 33.16) provides support for the suggestion that homeward orientation from long distances (for the species) is based initially on visual cues. It also seems likely that the American

Fig. 33.16   Radio tracking of the carnivorous leaf-nosed bat, *Phyllostomus hastatus*, released at varying distances from roosts in the limestone caves of the northern mountain range of Trinidad

Φ, average percentage and range (in different experiments) of released individuals that return to the home roost the same night during normal (i.e. not radio-monitored) homing experiments. Vertical dashed line indicates the limit of the normal foraging area, but not necessarily the familiar area, as determined by radio-tracking. Top diagram shows the results of radio-monitored homing experiments. Wavy line indicates period of 'random' movement after release. Black segment shows the angular scatter of homing orientations and the number shows the percentage of bats that had returned to their home roost after the time indicated.

The visual acuity of this species is such that the mountain range in which the home roosts were located seems likely to be clearly visible from 10 to 20 km, visible with difficulty from 30 km, and not at all from 60 km. Radio tracking suggests that in this case failure to home is due either to the inability to locate home direction or to the removal-migration threshold being exceeded and not due to mortality between release point and home roost.

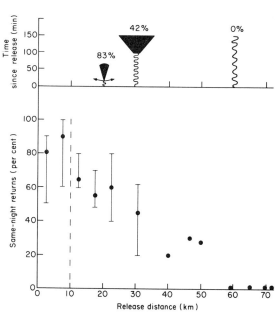

[Compiled from Williams and Williams (1970)]

free-tailed bat, *Tadarida brasiliensis*, travels too high during migration and homing for echo-location to be usable (Davis *et al.* 1962) and that in this species the familiar area is likely to be established primarily on visual cues. The same must also be true for those fruit bats or flying foxes (Megachiroptera) that lack an echo-location mechanism, though olfactory cues could also be used in these species.

Although bats, after bees and birds, are the group for which some of the most-successful progress has been made in evaluating the navigatory mechanisms, it is apparent that it will be some time before sufficient data are available to examine the detailed applicability of the familiar area map and least navigation models as presented in Fig. 33.5 and p. 861. Thus there is as yet no evidence relating to the use or lack of use of a compass, no analysis of the effect of cloud, little analysis of homeward orientation, no analysis of recoveries *en route* from the release site, and so on. Such analyses, most of which will be possible given an advance in telemetric techniques, must be awaited with interest. There is some indication from the movements of Schreiber's bat, *Miniopterus schreibersi*, in Australia, that rivers, hills, and coastlines may be used in navigation (Dwyer 1969), though whether such features function as gross features (Fig. 33.5) or as leading lines (pp. 875 and 899) or both cannot as yet be determined.

In some instances, information that has been gathered over the years has led to some knowledge of the likely familiar area of individuals from a particular roost. This knowledge enabled Hassell and Harvey (1965) to release females of the Indiana myotis, *Myotis sodalis* (Fig. 15.1), that had been captured in October in their hibernation roost, equal distances (320 km) from the home roost, but some from within the probable familiar area and some from outside. Furthermore, the releases were arranged such that if the bats followed waterway systems they would not return to their home roost. It was found that 67 per cent of those released from within their presumed familiar area but only 28 per cent of those released outside of their presumed familiar area returned home. In the absence of information on initial orientation or on interim recoveries it cannot be determined whether this 28 per cent located the familiar area by random search or by navigation. Even if by the latter, there is no way of knowing whether the navigation was achieved by distant landmarks as with *Phyllostomus hastatus*, and as suggested for some birds (p. 902), or

whether it would be necessary to postulate some form of grid map.

Similar problems surround the two experiments that Davis (1966) considered to provide evidence for a grid map in bats. One of these experiments concerns a single female pallid bat, *Antrozous pallidus*, that returned from six different release points in different directions and at distances of between 78 and 110 km from the home roost (Davis and Cockrum 1962). It is considered by the authors concerned that all of these release points were outside the female's familiar area. They can certainly be accepted as outside the year's home range for this individual. However, as already discussed for other animals (e.g. pp. 736 and 889), the year's home range is not the same as the familiar area. The latter is established during all of the exploratory migrations performed by the individual from birth onwards. The year's home range then crystallises out from this area and may represent a very small proportion of the animal's familiar area. Birds can evidently remember parts of their familiar area that they have not visited for up to 8 years (p. 606) and it might be expected that bats are at least as accomplished in this respect if not more so. Without knowing the lifetime track of the bat concerned, therefore, we cannot be certain that these releases were outside of her familiar area. As shown for birds, also, familiar-area navigation can operate for many tens of kilometres from outside of the area covered by the lifetime track if the landscape features employed are sufficiently general, conspicuous, and recognisable.

Similar problems surround the other experiment cited by Davis (1966) as evidence of a grid map. Of 36 female big brown bats, *Eptesicus fuscus*, released 400 km away from their home roost, 85 per cent returned at an average speed of up to 10 km/h (Cope *et al.* 1961). It may certainly be accepted that the release point was outside the year's home range (see Fig. 26.29) but it is by no means certain that it was outside the familiar area. Even if it were outside the familiar area, the result is so unique among homing experiments on bats that it cannot be accepted as evidence of a grid map unless the other possibilities can be excluded, which they cannot. One of these possibilities is that the release happened to be in the direction from which the bats could home as a result of false navigation to some environmental feature or features similar to those with which the bats were familiar within their familiar area (p. 896). Perhaps on this occasion the false home direction happened to be so similar to the true direction that the bats eventually encountered their familiar area (which in female *E. fuscus* has a diameter of at least 100–200 km—Fig. 26.29). Repetition of the experiment, taking the precautions that are now standard practice in homing experiments on birds, must be awaited with interest.

33.4.4.10 Navigation by birds that do not perform seasonal return migrations

The critical factor in the interpretation of homing experiments on birds that do not perform seasonal return migrations is to determine the extent of their familiar area. It is often tacitly assumed that this is relatively small and is similar in size to the adult's home range. As argued many times in this book, however, the familiar area is usually many times the size of the adult home range, often by several orders of magnitude, as a result of distance perception and the post-fledging exploratory migrations carried out by the bird when young. To judge from ringing recoveries, these migrations may take the young bird distances up to 100–200 km from its natal site in the case of a small passerine, several hundred kilometres or more in larger, more mobile, land birds, such as raptors, and perhaps several thousand kilometres in pelagic sea birds. Great caution has to be exercised, therefore, in categorising any particular release point as beyond a bird's familiar area.

A few homing experiments have been carried out on non-domesticated 'resident' birds, such as house sparrows, *Passer domesticus*. So far, no returns have been achieved from distances greater than 15 km (Matthews 1968). This could well indicate that this is the limit of the familiar area of these birds, though it could equally well indicate that beyond this distance the type-H removal migration threshold is exceeded (p. 871). Even random search should have given returns from this distance if the birds had attempted to home. However, whether or not 15 km is the limit of their familiar area, the implication is that small, 'resident', birds with a familiar area that presumably is smaller than that of any of the other birds discussed in this and the next section, use no form of navigation other than a type referable to the familiar-area map model.

The vast majority of experiments on the navigational mechanisms of 'resident' birds, however, have been carried out on homing pigeons, *Columba livia*, the domesticated version of the rock dove (Fig.

33.17) and it is with this voluminous work that the remainder of this section is concerned. Before evaluating in terms of navigation some of the multitudinous displacement–release experiments on pigeons, however, one alternative mechanism has to be considered. It has been shown for bees and box turtles that at the first opportunity the animal ceases to employ navigation in its attempt to return home and instead adopts an appropriate compass direction. This compass direction is either the one last employed (successfully) following displacement–release (i.e. a 'previous displacement-associated' compass direction) or the one that the animal has learnt will take it from one specific location to another (i.e. a 'location-associated' compass direction). Only at the first visit to a particular location, therefore, does the animal necessarily employ navigation. Subsequently, the mechanism used is compass orientation, or even leading-line orientation (p. 875). This behaviour is entirely consistent with the least navigation model and seems to be as likely to occur in birds as in any other animal. Examination of experimental data seems to confirm this expectation.

The first expectation, that pigeons associate a particular home compass direction with the process of being displaced and released if orientation in that direction was previously successful in returning the pigeon to its loft, has been known for many years, and pigeon races are still based on the one-direction principle. All birds taking part in a given race are those that have been trained in the appropriate direction. Usually the technique employed is that of releasing the bird at ever-increasing distances from the loft in a constant direction. This is the technique employed by Matthews (1968) in preparing his birds for critical experimentation, with the maximum distance training release being at about 128 km. German and American workers, however, have tended to adopt a minimal training method by releasing their birds at distances of only a few kilometres at various directions from the loft. Wall-raff (1974) has shown that even a single previous release in sunny weather can influence the direction taken on subsequent release.

The second expectation, that pigeons, having visited a site previously during exploratory migration and having learned the compass direction of home, adopt the appropriate compass direction to take them home with no further reference to their surroundings until they perceive the home site, also seems to derive support from experimental data.

Fig. 33.17   The biology of *Columba livia*

The solid line indicates the present distribution of the rock dove, the 'wild' form of *C. livia*. The dashed line indicates the approximate (very) distribution of the feral form of the species.

Columba livia probably evolved in arid or semi-arid and nearly tree-less regions and the wild form is found in areas where cliffs, gorges, potholes with caves, hollows, or sheltered ledges are available for nesting and roosting and are next to open country suitable for feeding. The rock dove is primarily a seed eater, though small snails and other molluscs are also eaten, and feeds primarily on bare ground or where the vegetation is short or scattered. The species is gregarious, forming pairs, parties, or small flocks that forage together, and large numbers may aggregate where food is abundant. They are a common prey of falcons and parties stay close together when crossing open ground and fly close in to cliffs or hillsides when these are available. Some seasonal return migration occurs in the northern and Saharan parts of the range, but in most places more or less the same home range is occupied throughout the year. The rock dove is said to be a regular passage migrant through Fair Isle.

Feral pigeons (i.e. birds or their offspring that have passed through a period of domestication by man) live primarily in towns and cities, roosting and nesting on artefacts that most resemble the original habitat-type, though in Australia some demes nest in hollows in gum trees, *Eucalyptus* spp., in open country. In mild temperate regions, such as Britain, feral pigeons may nest at all times of year but most pairs stop breeding during the autumn (August/September) moult. Some may breed again in late autumn (October/November) after moulting, others may not breed again until late winter or early spring (February/March). Daily return migrations occur from roosting and nesting sites in town and city centres to feeding sites at distances ranging from nearby to well out into the surrounding country.

[Drawn from maps and descriptions in Goodwin (1967)]

Using birds the clocks of which had been time-shifted by 6 h (p. 857), Graue (1963) found that when the pigeons were released 1–2 km from home in a position in which they had a direct view of the loft, they flew directly home. If, however, they were released at the same distance but at a site from which the loft was not directly visible, they flew at an angle to the left or right of the home direction in accordance with the shift in their clock. Graue (1963) and Matthews (1968) interpreted this as demonstrating that pigeons learn the area surrounding the loft and from the resulting 'aerial photograph' type of map can determine whether or not they are W or N of the loft, etc., and then home by Sun compass. I would prefer a slightly different emphasis to this interpretation: that from a variety of locations about the loft the birds have learned the compass direction of the loft and perhaps other resources. Upon finding themselves in that location, the birds automatically adopt the compass direction that they have learned will take them home, and consequently make no reference to their navigational mechanism. With either emphasis, however, the essential fact is that the birds have learned to associate a given compass direction with a given set of environmental features and the only navigation involved is for the bird to reach a decision concerning in which of the familiar locations it has arrived.

In my view, such a response could well account for the phenomenon of nonsense orientation, the characteristic feature of which is that the birds take up the same compass direction no matter where they are released relative to the home site. On the above basis, this would imply that at each release the birds 'think' they are in the same location which in turn implies that the release site always has the same characteristics. This might at first seem unlikely until the experimental requirements are examined. In order to observe the direction of the vanishing point of a released bird, whether a pigeon or a mallard (p. 873), especially if 16 × 40 binoculars are to be used and useful (Matthews 1955, 1968), it is essential that the immediate, and to some extent the less immediate, environment is relatively open. Not only open but equally open in all directions. It helps, also, if the background is sky rather than dark hills so a fairly flat and featureless area has to be chosen (Matthews 1968). Of necessity, therefore, release sites have some topographic characteristics in common, no matter how different the distant landmarks might be. If in the vicinity of the home, therefore, conditions similar to a standard release site lie

predominantly in one direction, then the majority of birds from a given roost will come to associate finding themselves in that type of location with having home lying in a particular compass direction.

Matthews (1963b) has attempted to determine whether the nonsense orientation of mallard, *Anas platyrhynchos*, is learned or innate by raising birds from hatching to flightworthiness in aviaries and then displacing and releasing them to determine whether any preferred direction was the same as the parent deme from which they were taken. Matthews tentatively concluded that, contrary to the interpretation of the previous paragraph, the direction of nonsense orientation is innate. There is some doubt, however, that the results justify such an interpretation. The birds used were raised from eggs taken from two demes. One deme, from Slimbridge, has a nonsense orientation to the NNW. The other, from London, has a nonsense orientation to the SE. Application of circular statistics to the data (using those grouped as the numbers flying toward the NW, NE, SE, and SW) shows that both groups of aviary-raised birds had a mean vector to the NNE, though only for the Slimbridge birds was the length of the mean vector significantly different from zero. A Rayleigh test shows that there was no significant difference ($P > 0.05$) between the directions of the mean vectors of those two groups but even more important (in view of the near-zero mean vector for the aviary-reared London birds) there is a highly significant difference ($F_{1, 1371} = 31.702$; $P < 0.01$) between the direction of the mean vector of the aviary-raised Slimbridge birds and the present deme from which they were taken. For neither the Slimbridge nor London birds, therefore, is there any evidence that they are born with the nonsense orientation of their parents. The difference in length of mean vector between the London and Slimbridge aviary-raised birds may perhaps be attributed to their different experience through being raised in separate aviaries, no matter how close these may have been.

Birds showing nonsense orientation, having taken up their compass direction, may maintain that direction for some time and distance (e.g. 20 min, 16 km; p. 873) before monitoring their new position and eventually, in some cases, heading in the correct home direction. The same behaviour is shown by displaced bees (p. 876) and it might be expected, therefore, that the same would apply to pigeons that adopt a compass direction without first monitoring their true position. Again there is evidence that this is the case. Michener and Walcott (1967) have shown that radio-tracked pigeons trained to home from a position 56 km west of their loft will fly east upon release and maintain that direction for from 8 to 80 km even when released at sites 48 km N or 32 km S of their loft. Only after this distance did the birds turn and migrate in the true home direction. This experiment of Michener and Walcott suggests that pigeons adopting the learned compass direction upon release may not begin to monitor their environment for other clues until 8 to 80 km (app. 10 min to 1·5 h) after release. When they did so, however, they turned and flew in the general direction of the loft, often with an accuracy of ± 2°. Michener and Walcott (1967) and Matthews (1968) dismiss the possibility that landmarks were involved in this direction determination. Their basis for this, however, is that the birds were over terrain that previously had not been traversed. However, as a previous traverse is not part of the familiar-area model presented here, the experiment could equally well be taken to support this model as any other.

Great care has therefore to be taken in the design, execution, and analysis of displacement–release experiments. One of the merits of the least-navigation–familiar-area map model, however, is that it accommodates the incredible variation found in the response of individual pigeons in this type of experiment without having to fall back solely on variation in navigational ability. All models, of course, can accommodate some variation because the analysis assumes that all birds are striving to return to the same resource location (i.e. the home loft) whereas many of them may elect to perform removal or exploratory migration (p. 872). Apart from this cause of variability, however, the Sun-arc map and magnetic-field map models must assume either that not all individuals have access to these maps or that they vary in their ability to read them and calculate the home direction. On the least-navigation–familiar-area map model, however, any variability over and above that due to removal or exploratory migration can be attributed to differences in experience. Already the model gives the pigeon upon release three alternative strategies, and additional strategies will be apparent from later considerations. These three strategies are : (1) to adopt, without navigation, the compass direction that was successful in returning the bird home last time it was displaced and released (i.e. a previous displacement-associated compass direction); (2) to adopt the compass direction that it has learned will return it

home from the same (or a similar) location which the bird has encountered during the course of exploratory migration (i.e. a location-associated compass direction); or (3) to employ its full navigational repertoire by monitoring the compass direction of familiar gross landscape features and from their apparent compass shift to assess home direction (i.e. navigation). Which of these three alternatives a given individual employs depends partly on its inherited predilection (= relative threshold) for each strategy and partly on the influence that recent experience has had on these thresholds.

There seems to be no shortage of experimental support for this view of the variability of initial orientation following displacement–release. Thus Wallraff (1974) has shown that a single experience of displacement–release procedure is sufficient to increase the proportion of individuals employing a previous displacement-associated compass direction strategy. That individuals vary in their selection of location-associated compass direction and navigation strategies is indicated in that, assuming nonsense orientation to be a form of location-associated compass direction strategy, when pigeons are released in such a direction from the home loft that both strategies give the same or similar compass direction, a greater homeward component to the scatter of vanishing-point directions is evident than if the two strategies give conflicting compass directions (Matthews 1968). Evidence that past experience can influence the proportion selecting each of all three strategies is also available. Thus, both Matthews (1968) and Michener and Walcott (1967) trained their experimental pigeons in a constant direction. When Michener and Walcott released their pigeons in a new direction, however, the birds selected either a previous displacement-associated or location-associated (falsely—see p. 899) compass direction strategy, whereas when Matthews released his pigeons in a new direction the majority selected a navigation strategy. The only apparent difference between the two groups was that Matthews' birds were trained from different locations and distances on the training line whereas Michener and Walcott's birds were trained from the same location as well, obviously, as from the same direction. Matthews' birds, therefore, in effect had a major choice between a previous displacement–associated compass direction and navigation strategies, whereas Michener and Walcott's birds had an equal choice between all three strategies. Tentatively, therefore, we may conclude that the predilection of pigeons for

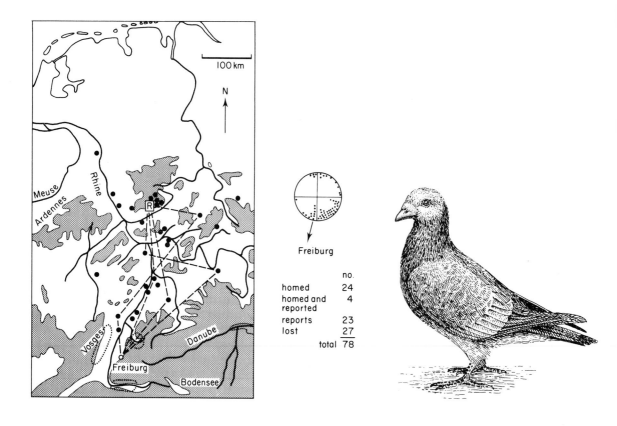

	no.
homed	24
homed and reported	4
reports	23
lost	27
total	78

Fig. 33.18 The navigation of homing pigeons by reference to a familiar-area map: pigeons from Freiburg

The diagram has been produced by superimposing a map (with no topographical features) illustrating the results of a release–recapture experiment on pigeons on a map showing the major rivers (heavy lines) and areas of high ground (stippling indicates areas more than 365 m above sea level) in the region concerned. Because of slight scale differences between the two maps, some distortion of recapture points may inadvertently have occurred, but this distortion is unlikely to be so great as to invalidate the general conclusions.

Seventy-eight pigeons from a loft at Freiburg were released at R. The fates of these pigeons are indicated in the table. The locations of the 27 individuals that were reported at places other than Freiburg are indicated by solid dots. Where an individual was reported on more than one occasion or was reported and homed, the reports are joined by dashed lines (n.b. these lines should not be interpreted as tracks).

The association of these Freiburg pigeons with rivers is clear. On the familiar-area map hypothesis it might be expected that Freiburg pigeons would learn that their loft was to the east of a large river in a town on the western edge of high ground. The Rhine also seems likely to have been adopted as a leading line, pigeons breaking away to the east when clues are perceived that they are level with

Freiburg. The individual that was first reported near to the Rhine and then 130 km to the east may well have broken away from the Rhine prematurely in response to some cue reminiscent of the home situation. Having failed to locate the home loft this individual may then have doubled back on its tracks and re-located the Rhine rather than migrating along the direct (dashed line) route to home.

The dotted circles and oblongs indicate areas within about 100 km of the loft with which the pigeons may be familiar and which share certain characteristics with the area of release (e.g. high ground to the north and west of a river). A pigeon that adopts a location-associated compass direction would therefore migrate SSW, SE, or N, depending on which of these areas the release site is assumed to be. Such a model is entirely consistent with the observed and illustrated scatter of vanishing-point directions. A model in which the pigeons have determined home direction (arrow) by the time they vanish from sight is much less consistent with the observed scatter.

Note that only 36 per cent of the released pigeons successfully returned to the home loft.

[Pigeon data from Matthews (1968) (after Pratt and Wallraff). Rivers and topography from Fullard (1959)]

the three different strategies is that they are most likely to adopt a location-associated compass direction, if given sufficiently similar conditions, and least likely to adopt a previous displacement-associated compass direction. This is perhaps as would be expected from the least-navigation–familiar-area map model and is consistent with what has been found for other animals.

In bees, an additional strategy was available that was also consistent with the least-navigation–familiar-area map model. When a leading line coincided with or was parallel and close to a flight path between a given location and home, the bees adopted the leading line rather than the compass direction when, experimentally, the two were put in conflict. There is evidence that birds may show a similar predilection, thus suggesting similar selection for least navigation in both groups. Southern (1971b) reports that if ring-billed gulls, *Larus delawarensis*, or herring gulls, *L. argentatus*, are taken from Lake Huron or from a colony adjacent to the lake and are transported to a release site close to one of the other great lakes that also has an extensive shoreline, the birds spend their time following the shore-line until the homing response wanes. Figure 33.18 illustrates a displacement–release experiment with homing pigeons that gives every indication that the birds followed leading lines that would have been appropriate near to the home loft and that, as it happened, for those birds that made the correct decision when presented with a potentially confusing situation, did lead many of the birds back to the home loft.

It now appears, therefore, that a released pigeon has four alternative strategies available to it. The order of preference of these four strategies seems to be: (1) a familiar leading line; (2) a location-associated compass direction; (3) navigation; and (4) a previous displacement-associated compass direction. This order of preference seems to be entirely consistent with the least-navigation model, the strategy chosen on any given release being determined by the recent experience of the individual concerned and the gross characteristics of the release site. On this basis, therefore, it follows that the analysis of results in terms of navigation may be extremely misleading if conditions are at all likely to lead the birds to adopt a leading-line or location-associated compass direction. If results relevant to an evaluation of the mechanisms involved in navigation are to be obtained, therefore, it is imperative that the birds are released from a site that they have never

visited before during exploratory migration. In view of the relative lack of navigational monitoring of country traversed during exploratory and other migrations, however, unless the area becomes biologically meaningful to the individuals in terms of providing a resource or as a target area (Fig. 33.5), it is probably less important that the bird has never before traversed the location. Even if the release site has not been visited before, however, the birds may (erroneously) employ a location-associated compass direction if the site is sufficiently similar to a site that they have visited before. In this case, the proportion employing such a compass direction is likely to be a function of the degree of similarity between the two locations. Because the crucial importance of this similarity has not previously been recognised, the best we can do at present in an attempt to evaluate the navigational mechanisms employed by pigeons is to examine those experiments in which there seems a reasonable chance the birds have never before physically visited the release site.

Experiments on young birds that have had the opportunity to perform exploratory migration in the vicinity of the home loft but which have received no training of any kind and which have never before been displaced by man, show marked homeward orientation upon release. Matthews (1968) released such birds 80 km to the SSW, 96 km to the N, and 125 km to the SSE of the home loft and found that of the 54 birds involved, 52 per cent produced vanishing points within 45° of the home direction. Wallraff (1967) released similar birds at distances of 85 and 165 km from the home loft with similar results. As far as the familiar-area map model is concerned, it can be calculated that even if the birds had previously flown no further than 20 km from their home loft in the direction of the release point and if they had never achieved a greater height above ground level than 60 m, then even in the longest-distance release the most conspicuous feature of the horizon need only be higher than about 120 m above the surrounding country to be used as a gross feature (Fig. 33.5). Given constant altitude above sea level the distance of the horizon in kilometres is given by $3 \cdot 83 \sqrt{h}$ where h is the height of the observer above ground in metres. In the above experiments, therefore, even a modest hill or the features thereon would be sufficient to provide such a reference point. Alternatively home direction could have been given by some rather more subtle landscape feature as argued in Figs. 33.19 and 33.20.

Figure 33.19 shows the results of a further experimental series that seems to demonstrate navigation. The interpretation of these results in terms of the familiar-area model is given in the legend. It certainly seems unlikely that the results given so far as demonstrating navigation can be attributed to the adoption of a wrong location-associated compass direction because of a chance similarity between the release site and a site with which the birds are familiar. It is also unlikely that the birds have visited the release sites before and are homing simply by adopting a valid location-associated compass direction. To remove these possibilities, however, and to determine with a reasonable degree of certainty the limits of a bird's experience before release, it is necessary to restrain a bird in some way and to limit its experience during the entire period from birth to release. At this point, therefore, it is convenient to consider a series of experiments in which the birds were given no opportunity to establish a familiar area either by exploratory migration or by obtaining a fix of the compass direction of distant landmarks before displacement and release. On the familiar-area map model such birds should not be able to recognise their position nor therefore be able to determine home direction. On the Sun-arc and magnetic field models, however, they should be able to recognise position and determine home direction, even though they may have been unable to find the precise position of the loft due to a lack of familiarity with its appearance from outside.

Kramer (1957, 1959) and Wallraff (1966) confined pigeons in an aviary (3 m high and 10 × 6 m at the base) from birth and prevented them from seeing landscape features either by keeping the aviary in a bomb crater or by giving the aviary opaque sides. When released 150 km to the south of the home loft, no birds returned but even more significantly 68 recoveries elsewhere gave no indication of a homeward component (Fig. 33.20). Other birds similarly caged but given a view of the distant horizon gave a 7·2 per cent return ($n = 125$) with other recoveries giving a homeward bias. Birds with a single gap in the opaque walls permitting a total view of 140° of the horizon (but not simultaneously) showed zero returns ($n = 165$). Others with five gaps, each 1/15th of the perimeter, so that the complete horizon was visible but not simultaneously from any single position, gave a 5·8 per cent return ($n = 52$).

These findings are not at all consistent with a Sun-arc or geomagnetic field map model but are entirely consistent with the familiar-area map model. Further, they strongly support the suggestion that the

critical feature of the familiar-area model is that the animal should learn its home position relative to the spatial relationships of gross features on its horizon, and that it is extremely important that all 360° of the horizon should be visible (Fig. 33.5). The importance of the Sun as a compass in this process is illustrated by the further experiment in which the aviary was given a roof such that only 4° of sky above the horizon was visible and in which a view of the rising and setting Sun was denied the birds by shutters. Although none homed when released there was an indication of a homeward orientation of recoveries if the birds were given a natural daylength, thus suggesting that some compass information could be obtained either from observation of

Fig. 33.19   The navigation of homing pigeons by reference to a familiar-area map: pigeons from Cambridge, England

Stippled areas are more than 100 m above sea level.

The length of the lines radiating from the three large circles indicates the number of pigeons with a vanishing-point direction in the given direction. The shortest line on each circle refers to a single individual. The same individuals were used at each release. All were experienced homers to the Cambridge loft before the experiment began.

The results are consistent with the model that the pigeons locate the home direction by monitoring the compass shift of high ground relative to the compass distribution of such ground around the home loft.

Pigeons released 205 km NW of the loft may or may not perceive higher ground to the E and W of the release point but none to the S. In all of the comparable situations around the home site, return is executed by flight in a direction somewhere between E and SW until just before reaching high ground to the SE.

Pigeons released 126 km WSW of the loft are likely to perceive higher ground in all directions except perhaps to the E and NE. In comparable situations around the home site, home direction lies between NE and SE.

Pigeons released 110 km SSE of the loft may perceive higher ground in all directions but perhaps the important factor is that the higher ground is oriented along an E–W axis. In comparable situations around the home site, home direction lies between W and NE. Having flown sufficiently far north, however, for the high ground (North Downs) to lie to the S, further high ground appears to the N, thus causing the pigeons to continue to assess home direction as lying to the N.

[Pigeon data from Matthews (1968). Map and topography simplified from Fullard (1959)]

shadows (see McDonald 1972) or from the brightness of different parts of the visible sky at different times of day. Whether sufficient information concerning the Sun's movement had been available for the birds to mature their Sun compass mechanism was not determined. A more detailed evaluation of parts of this experiment in relation to the familiar-area map model is given in Fig. 33.20.

This experiment, therefore, provides strong evidence that both landmarks and a compass are required for pigeons to be able to navigate with any accuracy. It does not, however, indicate con-

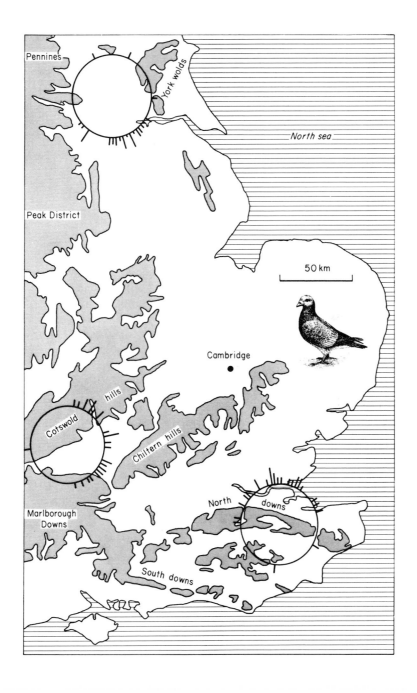

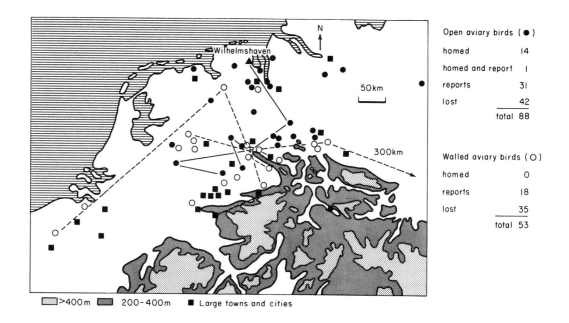

Open aviary birds (●)	
homed	14
homed and report	1
reports	31
lost	42
total	88

Walled aviary birds (○)	
homed	0
reports	18
lost	35
total	53

▨ >400 m   ▨ 200–400 m   ■ Large towns and cities

Fig. 33.20   The navigation of homing pigeons by reference to a familiar-area map: the effect of reduced pre-displacement experience

Pigeons at a loft just to the south of Wilhelmshaven were in effect divided into two groups. Both groups were retained in an aviary from fledging onwards and were thus prevented from performing exploratory migration around the loft. One of the groups was maintained in an open aviary and had visual access both to the Sun and to the horizon. The other group was maintained in a walled aviary such that although the Sun was visible the horizon was not. Both groups of birds were then released (separately) at R with the results illustrated. Lines (———, open aviary birds; – – – –, walled aviary birds) are shown only when a bird was reported more than once.

There is a clear difference in performance of the two groups, the walled aviary group showing no indication of being able to identify home direction even in the most general of ways. However, even in the open aviary birds, homing performance is not high (17 per cent) and the major difference between the two groups appears to be a greater tendency on the part of the open aviary birds to orient away from high ground. Almost as many open aviary birds failed as succeeded to locate the loft even after having arrived in the north. Those that failed nevertheless seem to show an affinity on the one hand for large towns and cities and on the other hand for the coast, though both effects could be artefacts.

The data are consistent with the following model. Birds in an open aviary, but not those in a walled aviary, establish a familiar area by visual perception of the landscape. The cues registered are: total absence of high ground in all directions; the presence of buildings and in particular a city skyline more or less due N; the proximity of the sea to the E and NE (or perhaps just the presence of

factors associated with the sea irrespective of direction, such as the sound and sight of gulls). Upon release at R, the birds are presented with a flat landscape only to the N, E, and W. Orientation predominantly in this sector brings some individuals to the cities and coastline of the north. Search of the south side of cities near to the sea leads some individuals to detailed landmarks that were visible from the loft.

[Data from Kramer (1959). Topography simplified from Fullard (1959)]

clusively the mechanism by which the birds make use of these cues, though the suggestion made in Fig. 33.5 is perhaps not entirely unreasonable. Some evaluation of the mechanism is possible, however, by means of clock-shift experiments.

Figure 33.21 indicates the effect of clock-shift experiments on the mean vector of vanishing-point directions on the basis of the compass-shift mechanism postulated in Fig. 33.5. It can be seen that clock-shifts should have a similar effect on both the

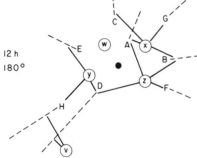

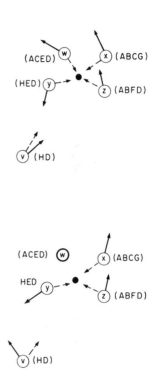

Fig. 33.21 The influence of a shifted clock on the vectored home direction of a pigeon that navigates on the basis of a familiar-area map

The left-hand pair of figures show the navigational process for pigeons the clocks of which have been retarded by 6 hours (i.e. real time 15.00 hours; pigeon time, 09.00 hours) in the top diagram, and 12 hours (i.e. real time, 21.00 hours; pigeon time, 09.00 hours) in the bottom diagram. The right hand pair of figures show the home direction that would be vectored by clock-shifted birds (solid arrows) relative to the true home direction (dashed arrow) from five different release points.

A–H are gross landscape features, the compass bearings of which from the loft (•) have been learned by the pigeons according to the familiar area model while their internal clocks were in phase with real time. w–z are release points of clock-shifted birds within their familiar area. The dashed lines indicate the tract of country from anywhere on which the relevant gross feature is on the same compass bearing as from the loft as assessed by the clock-shifted bird. Note that unlike the situation for a bird with an internal clock in phase with local time (see Fig. 33.5), these dashed lines no longer converge on the home loft but

are shifted clockwise through 90° (top diagram) or 180° (bottom diagram). The solid lines (shown only for x in the top diagram and for v and x, y, z in the bottom diagram) indicate the shortest route to restore each gross feature to its appropriate compass bearing. It is assumed that only those features indicated in brackets by each release site in the right-hand figures are visible, recognised, and/or used from each release site.

Normally the mean vector of the solid lines gives the home direction. In a clock-shifted bird, however, as shown in the right-hand figures, this is not the case. In general, vectored home direction deviates from the true home direction by an amount similar to the shift in the compass. This deviation is not as precise, however, as found in birds or other animals using simple Sun compass orientation. Navigating birds using the mechanism outlined in the familiar-area map model should show much less consistent shifts in orientation and a much greater scatter. Indeed, for w in the bottom diagram a mean vector is unlikely to be apparent. The lack of convergence of the dashed lines must be potentially confusing and is likely to add to the scatter of vanishing-point directions of clock-shifted birds.

All of these effects are observed in clock-shift experiments on pigeons released in a site from which normally they would be able to locate home direction.

vanishing point of birds employing a navigation strategy and on the vanishing point of birds employing a location-associated or previous displacement-associated Sun compass orientation. There are a number of differences, however, the derivations of which are illustrated in Fig. 33.21. The major differences are that birds attempting to navigate after their clock has been shifted should show a greater scatter in vanishing-point direction than birds using simply a Sun compass. Secondly, although the angular difference between the vectored home direction and the true home direction should be similar to the deviation that would be expected if only the birds' ability to adopt a given compass direction were affected (Fig. 33.2), it is not necessarily identical. As it happens, both of these effects seem to be apparent from the clock-shift experiments that have been performed.

Clock-shift experiments on homing birds have experienced difficulty from the beginning (Matthews 1955), except where only a compass reaction is used. The latter happens, for instance, when the birds elect to use a location-associated compass direction rather than navigation, either because they are familiar with the release site or because the release site has a relatively high degree of similarity with a familiar site (p. 899). Figure 33.21 perhaps accounts for the disappointing results obtained by this type of experiment. Clock-shift experiments, although they would 'make sense' to a bird navigating by a Sun-arc grid map, and have no effect on the navigational ability of a bird using a geomagnetic grid map, an aerial photograph landmark map, or a learned sequence landmark map, makes nonsense of the information available to a bird that navigates by monitoring the compass shift of familiar landmarks. As shown in Fig. 33.21, instead of different landmarks giving the same home direction, they instead give widely different home directions. The situation would be confusing and may well cause the bird to switch to some other mechanism. It may, for example, increase the proportion that adopt a location-associated compass direction, adopting the compass direction (shifted according to the clock-shift) that would be appropriate from the most similar familiar location.

Even if all birds persist in attempting to vector home direction, however, a wider scatter is to be expected because of the widely different vectors given by different landmarks. Nevertheless, as shown in Fig. 33.21, a mean vector is available from most locations and should emerge from the scatter of

vanishing points. Although this mean vector is shifted from the true home direction by an amount that is by and large similar to the shift that would be expected if an ordinary compass response was being affected (Fig. 33.2), the difference is nowhere near as precise, and deviations of 90° more or less than expected on a simple compass effect may occasionally occur, depending on the precise arrangement of landmarks relative to the release point.

All authors, with the possible exception of Matthews (1968), now seem to agree that the only effect obtained by clock-shift experiments on pigeons attempting to navigate is very similar to the effect obtained by clock-shift experiments on pigeons (or any animal) attempting to use a Sun compass (e.g. Walcott and Michener 1967, 1971; Schmidt-Koenig 1972; Wallraff and Graue 1973). All of these authors, however, have reported a much wider scatter of vanishing points with clock-shifted navigating pigeons. In some cases (Walcott and Michener 1971) the scatter was so great that with clock-shifts of 6 h or more there was no longer a significant mean vector. Other authors (Schmidt-Koenig 1972) have found a mean vector at most clock-shifts but one that was not precisely what would be expected if the birds were simply orienting by a Sun compass (as opposed to using a Sun compass to monitor apparent landmark shifts as required by the familiar-area model). Thus the mean vector was displaced by 72° and 93° instead of the 90° expected from a 6 h clock-shift and 168° instead of the 180° expected with a 12 h clock-shift. Other authors have reported mean vectors 33–121° greater than expected (Walcott and Michener 1967). Always, however, there is an increase in scatter such that shifts even as great as 30° cannot always be detected statistically (Schmidt-Koenig 1972).

None of these experiments are totally consistent with the suggestion that pigeons have access to a Sun-arc grid map and most are totally against such a suggestion. However, the fact that clock-shifts influence navigation at all argues against the preferential use of a geomagnetic grid map, but in favour of the Sun being involved at some level in the navigational process. All of the characteristics of these clock-shift experiments seem to be consistent with the suggestion that the role of the Sun compass in navigation is to monitor the apparent shift of landscape features.

It should also be noted that the fact that clock-shifts influence the direction of the vanishing point also argues against the use of an 'aerial photograph' type of landmark map, the general use of leading lines, and learned sequences of landmarks, in the determination of initial orientation. As long as the clock-shift does not cause the mean vector to be as great as 90° different from the true home direction and as long as the bird monitors its position with reasonable frequency, it will eventually spiral in to the home site. Most physical returns to the home site following clock-shift experiments, however, seem to occur on subsequent days when perhaps the birds' clock has been re-phased to local time.

As, to my knowledge, this is the first formulation of a model approximating to the least-navigation–familiar-area map model, then obviously experiments have not yet been performed that have been designed specifically to test the major predictions of the model. It is suggested here, however, that the above experiments and those described in the next section on migratory birds do support the thesis that when birds elect to use navigation they do so by

vectoring home direction from the apparent compass shift of familiar landmarks. These do not have to be a recognisable mountain, forest, or river, etc. and the reader is referred to Figs. 33.19 and 33.20 for an indication of the subtle range of landscape features that could be used in navigation.

Having established as far as seems possible at present the serious contention of the least-navigation–familiar-area map as a model of bird navigation, it is now necessary to examine the other variants of the model that seem likely to be used by the bird when those high up on the least-navigation hierarchy are not available. That is when neither leading lines nor a Sun compass are accessible.

Pigeons trained under sunny skies or simply allowed to perform their own exploratory migrations around the loft are sometimes influenced in their homing performance by heavy cloud and sometimes not. The scatter of vanishing points when birds are released under heavy overcast is invariably

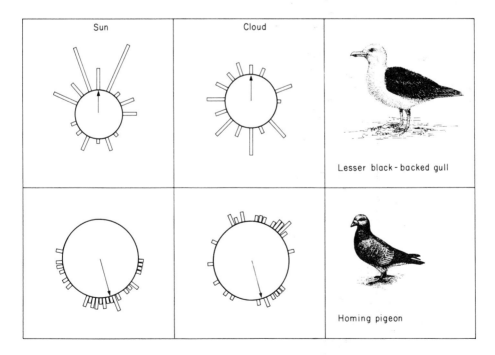

Fig. 33.22   The effect of cloud on bird navigation

The diagram shows two comparisons of vanishing point scatters obtained under conditions of sunshine and cloud. In all cases the shortest column refers to one individual and the arrow indicates home direction.

Without access to the Sun, birds either cannot or are reluctant to attempt to locate home direction. If the birds have access to some compass or map other than that involving perception of the Sun, they are reluctant to make reference to it.

[Re-drawn from Matthews (1968) (partly after Wallraff)]

random (Fig. 33.22) and radio-tracked pigeons have been found often to land if released in overcast conditions and not to fly until the Sun reappears (Walcott and Michener 1967). Furthermore, if they are released in sunny conditions but encounter Sun-obscuring cloud on the way many again land and wait until the cloud has passed. Even so some continue and home despite being released under cloudy conditions and some releases under cloud show no reduction in homing performance (Matthews 1968).

Assuming that the birds have access either to an alternative compass or to an alternative map, or both, but that all alternatives require 'more navigation' than Sun compass orientation following assessment of compass shift of familiar landmarks, then the above variability is consistent with the least-navigation model. Birds released in heavy overcast and thus prevented from using least-navigation methods (except leading lines, which are rarely appropriate) are faced with a choice of strategy. Either to wait until the preferred navigational mechanisms become available or to switch to alternative mechanisms which impose (presumably—p. 861) a higher migration cost. The possibility of using a geomagnetic compass is discussed below. For the moment, let us consider the possibility of using an alternative landmark map such as an 'aerial photograph' type, a learned sequence of landmarks, or an angular separation method (i.e. flying towards a pair of landmarks that have a smaller angular separation than at the home site or that show left-to-right inversion and away from pairs of landmarks that have a large angular separation). There seems little doubt that birds have access to such maps, at least in the immediate vicinity of their home site, due to their continued activity, but at shorter distances from home (Hamilton *et al.* 1967), in cloudy weather, even in birds that have not been trained in overcast conditions (see below). Even so, as we have seen (p. 896), pigeons prefer to use a location-associated Sun compass direction, if the Sun is visible, at distances from the loft as short as 1 km. We may assume, therefore, that the use of these alternative landscape maps is less favoured by selection for least navigation than maps based on landscape features and a Sun compass, but that such maps are established in the central nervous system and are used when a Sun compass is not available. It seems reasonable to assume that these alternative maps require more time investment in navigation on the part of the bird and may even be less accurate.

Certainly the birds in the experiments described in Figs. 33.19 and 33.20 could not have determined home direction with any accuracy without using a compass as well as landscape features, even though homing would still eventually be possible. We may perhaps also assume that birds do not have an infinite capacity for memorising landscape details and that, apart from in the vicinity of resources of biological importance, such as the loft, only gross and general landscape patterns would be committed to a spatial memory in the formation of these alternative landscape maps. Presumably whether a bird elects to wait for a Sun compass to become available or whether it employs an alternative landscape strategy depends on how 'urgent' it is to return to the loft, the probability of a Sun compass becoming available in the near future, and the degree of familiarity with the area. Such a model seems to be sufficient to account for observed variations in pigeon behaviour when the birds are released under overcast skies and a Sun compass is not available. Perhaps birds that home the same day as released despite having their clocks shifted (p. 905) also do so by switching to an alternative landscape map.

Another alternative, of course, is to switch to a different compass that is unaffected by cloud and to continue to use the mechanism of measuring the apparent compass shift of familiar landmarks. Results suggesting that pigeons may switch to a magnetic compass have not yet presented a clear picture of the factors involved. At Ithaca, N.Y., if homing pigeons are released without previous experience under overcast skies they go random (Keeton 1972b). If, however, the birds have been given a series of 4·8–16 km training flights under overcast skies, then even if released at 160 or 240 km they can orient under overcast. Keeton has performed a number of experiments on the effect of carrying magnets on the orientation of experienced and first-release pigeons trained and released under overcast skies (Keeton 1971, 1972a). The results obtained were variable, yet nevertheless seemed to show that under certain conditions the homing ability of pigeons was affected if the magnetic field surrounding the bird was disrupted. First-release birds seemed more prone than experienced birds to have their homeward orientation disrupted by bar magnets, an effect that persisted even if the birds were released under Sun. This could suggest that perhaps information concerning the magnetic compass is processed during exploratory migration, even under sunny conditions. It is difficult to reconcile this observation, however, with the observation that pigeons have to be trained under overcast conditions before they will adopt a magnetic compass. The experimental reversal of homeward orientation of such birds when exposed to a reversed magnetic field through their head has already been described (p. 608).

There seems to be little doubt that birds have access to a magnetic compass and when placed in a situation (e.g. in orientation cages in an enclosed room or under cloudy skies with no landmarks) in which no other information is available to them, they can be forced to make use of this information. The work with pigeons has also shown that they can be trained to make use of a geomagnetic compass in their determination of home direction. All of the available evidence, however, shows that birds allowed normal perceptual access to their environment have a marked reluctance to take recourse to a magnetic compass, often preferring to wait for Sun compass information to become available again. On the least-navigation hierarchy, therefore, navigation using a geomagnetic compass (even for determining the apparent compass shift of landmarks), is towards the bottom, perhaps not even as high as navigation by the alternative landscape maps described above, at least at distances less than 32 km from the loft (Keeton 1972a).

There is one final characteristic of displacement–release experiments that a model of pigeon navigation must be capable of accommodating if it is to be accepted as a valid description of the phenomenon. Three different authors, Matthews (1955), Schmidt-Koenig (1966), and Keeton (1970b) have now examined variation in homeward orientation with distance between the release point and home. Matthews and Schmidt-Koenig found that the homeward component was least from intermediate distances whereas Keeton found that it was greatest. No other extant model seems to be capable of accommodating this variation but as shown in Fig. 33.23 such variation may readily be accommodated by the familiar-area model.

The time required for homeward orientation to emerge after a pigeon has been released varies considerably but seems to be about 1·5 min (Matthews 1968) if navigation is employed but less if it is not. As far as the familiar-area model is concerned a finite time is required after release to locate recognisable landmarks, to determine their compass direction, and to vector the home direction. It is, of course, impossible to be sure, but a period of 1–2 min

to perform such an operation seems reasonable, though presumably the time required is a function of the ease with which landmarks can be recognised as being those visible from some other location.

In the past, authors have been virtually unanimous in dismissing a central role for landscape features in the navigation mechanisms of homing pigeons. However, the arguments presented by these authors have been aimed at a dismissal of maps based on the concepts of an 'aerial photograph' type and on learned sequences of landmarks. As we have seen, however, such strategies come very low in the least-navigation hierarchy of landscape-based strategies. Dismissal of the possibility that pigeons are using these strategies in any given experimental series does not dismiss the central role of landscape as postulated by the familiar-area model.

Michener and Walcott (1967) and Matthews (1968) concluded that the radio-tracking results obtained by the former authors dismissed the use of landmarks by pigeons when navigating from distances even as short as 56 km. Using 10 pigeons, Michener and Walcott made repeated releases 56 km WNW of a loft in Massachussets. No two tracks out of 36 flights covered exactly the same route. It was suggested that this demonstrated that the birds were not flying precisely from one known landmark to another. It does not, however, show that the birds had not taken an approximate directional bearing from a familiar landmark and then flown in that general direction having delegated interim orientation in the determined direction to some other mechanism such as a Sun compass. Indeed, as the birds were on a repeated-release schedule from a given location, they would automatically take up the approximate location-and previous displacement-associated compass direction without reference to any map (p. 897). That this is so is demonstrated by the fact that when the birds were released in a different direction from the loft as already described (p. 897) they took up their trained compass direction. The accuracy with which this compass direction is taken up is unlikely to be sufficiently great for every homing track to be precisely the same as the individual took on the previous occasion. It was also noted during the repeat releases that the birds homed and oriented equally well whether visibility was 8 or 80 km. Michener and Walcott and Matthews argue that this demonstrates the birds were not using distant landmarks. Again this is not necessarily so, although, as it happens, the birds probably were not

using distant landmarks on the repeat releases
because they automatically took up the location-and
previous displacement-associated compass direction
on each occasion. Even if this were not the case,
however, the observation does not demonstrate that
the pigeons were not determining direction from
distant landmarks. A pigeon at a release point 56 km
from the loft can, if visibility is reduced to 8 km, see
features 48 km from its loft. Such objects are 28 km

away from the nearest previous position of the
pigeon if, in its flights about the loft, the pigeon
never travelled more than 20 km from its loft. A
pigeon, even flying at a height as low as 60 m, has a
horizon at a distance of 30 km (given a constant
altitude of the Earth's surface above sea level). As
this distance is greater than the 28 km separating the
pigeon on this earlier occasion from the furthest
(8 km) landmark visible to the bird from the release

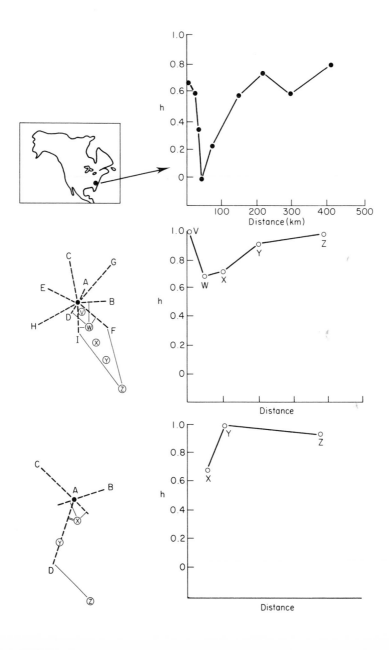

Fig. 33.23   Variation with distance of release from the loft in the homeward component of the vanishing-point orientation of pigeons: application of the familiar-area model

The homeward component, h, of a given scatter of vanishing points is given by: $h = r \cos (a - B)$, where r is the length of the mean vector, a is the direction of the mean vector, and B is the home direction. For a description of the calculation of the mean vector see p. 30.

Experimental data for variation with release distance in the homeward orientation of homing pigeons, *Columba livia*, have now been obtained by three authors (Matthews 1955, Schmidt-Koenig 1966, Keeton 1970b). In two cases the homeward component was greatest from short and long distances (top graph) and in the other case the homeward component was greatest from intermediate distances. If all of these results are accepted, any model of pigeon navigation must be capable of accommodating such gross variation in the influence of distance.

In the middle diagram, A–I are gross features, the compass bearings of which from the loft ($\bullet$) have been learned by the pigeons according to the familiar-area model. v–z are release points following displacement. It is assumed that the pigeons have visited location v previously and home by adopting a location-associated compass direction which would give a homeward component of 1·00. In the top diagram the shortest release distance was 3 km, so again most of the birds should have homed by adoption of a location-associated compass direction rather than navigation. The dashed lines in the middle diagram

indicate the tract of country from anywhere on which the relevant gross feature is on the same compass bearing as from the loft. The solid lines (shown only for w and z) indicate the shortest route to restore each gross feature to its appropriate compass bearing. Home direction is given by the mean vector of these lines. It is assumed that only *BDFI* are visible from w and x, and only *FI* from y and z. The scatter of vanishing points is considered to be a function of the scatter of solid lines on the assumption that each individual pigeon monitors home direction from only a few of the gross features available to it and/or monitors home direction with an accuracy that is a function of the scatter of the solid lines. Variation with distance in the homeward component is thus given by variation in the homeward component of the solid lines. The variation is similar to that shown in the top diagram and is due to the variation in accuracy of determination of the homeward component depending on whether the bird can adopt a location-associated compass direction or whether, if it employs navigation, it is surrounded by, at the edge of, or outside the circle of familiar gross landscape features.

It will be noted that the homeward component is consistently greater in the middle diagram than in the top diagram. This can be attributed to the fact that in the model it is assumed that all birds attempt to return directly to the loft. In a real experiment, however, this is rarely, if ever, the case, with the result that the measured homeward component relative to the loft is always lower than the true homeward component relative to the intended destination of each individual pigeon. While considering variation in motivation to return directly to the loft, it should be noted that this variation is itself unlikely to be constant with distance of release and quite conceivably those released at intermediate distances are often more prone at first to initiate exploratory or other non-return migrations and not to attempt to return directly home. The top diagram is therefore likely to be a product of the variation with distance in navigational accuracy (middle diagram) and variation with distance in motivation to return.

In the bottom diagram a different situation is shown that happens to give a greater homeward component from intermediate distances on the basis of navigational accuracy because release point (y) and the loft happen to be on a direct line between the two gross landscape features available to the birds. It is assumed that the birds have not previously visited x and that *ABCD* are visible from x, *AD* from y, and only *D* from z.

Neither the middle nor lower model is intended to replicate the experiments carried out by the other authors. Nevertheless they indicate that the familiar-area model has the potential to accommodate observed variation in homeward orientation with release distance. The same cannot be said for other models.

Conclusion

Most often, pigeons should show least homeward orientation from intermediate distances. It is possible, however, to conceive of landscape configurations in which this is not the case. Distance effects should not only be interpreted on the basis of variation in navigational ability but should also take into consideration variation with distance in motivation to return.

site, then theoretically the landmark need be no height above the Earth's surface to be available to the bird for navigation. Any landmark, even trees or houses, etc., could be used. In no way, therefore, do these experiments of Michener and Walcott provide evidence against the familiar-area model. Indeed, as it was observed that some birds were actually misled by topographical features reminiscent of an area nearer to the birds' loft, they may almost be taken to provide support for the model.

Schmidt-Koenig and Schlichte (1972) performed a series of experiments that were designed specifically to test the role of visual cues in pigeon navigation. The birds were conditioned to wearing clear contact lenses and then the evening before displacement–release the clear lenses were replaced by frosted glass lenses that impaired vision. The birds were then released N, S, E, and W of the loft at distances of 15 km and 130 km. Homeward orientation upon release and homing performance was then compared with that of controls wearing clear contact lenses from the same sites. For all but one release site there was no difference in the vanishing-point orientation and homing performance of birds that returned home, though more experimental birds than controls failed to return, a large number having been captured by hawks. In the exceptional case (from 130 km N), the experimentals showed better homeward orientation than the controls!

Clearly the critical factor in the interpretation of these results is an assessment of the degree of visual impairment produced by the lenses. The authors performed a series of not altogether appropriate experiments to test this. Birds were trained to a red pole 5 cm wide and to a mountain ridge 1 km away from the loft. It was found that the latter could no longer be recognised 'as soon as the frosted lenses were applied' and that the red pole could be recognised at a distance of 2 m but not at 6 m. There is no indication, however, that birds that homed successfully with frosted lenses were tested for visual acuity. It is possible, even likely, that the birds' vision was considerably impaired *immediately* upon application of the frosted lenses but that it improved dramatically during the course of the following 12 h or so as the eye accommodated. Also, the visual acuity necessary to perceive a red pole 150 cm high from 6 m (with no indication of whether or not it was silhouetted against the sky) is quite different from that required to monitor gross landscape configurations on the horizon. The authors did demonstrate that the birds still had access to a Sun compass, so this component of the navigation mechanism proposed here was unaffected. The final, and to me, convincing proof that the birds' vision was not so seriously impaired that they could not monitor the apparent compass shifts of landmark features was the fact (that emerged partly from the authors' descriptions and clearly from a film of the experiment shown in the BBC *Horizon* television series) that the birds could see sufficiently well to locate trees and roofs on which to land, albeit clumsily. Many could also locate and enter the loft. The authors' assertion that this indicates that senses other than vision are involved in loft location cannot seriously be considered in the absence of much more convincing evidence that the birds' vision was as impaired as the authors assume. All in all, therefore, I feel that these experiments by Schmidt-Koenig and Schlichte have failed to demonstrate that vision is not involved in pigeon navigation.

In all of the experiments so far described in which pigeons have shown a significant homeward component to the scatter of their vanishing points and in which this can clearly be attributed to a navigation mechanism other than a learned location-associated compass direction, it has been possible to postulate that the birds had access to sufficient environmental cues for them validly to use a familiar-area map (e.g. Fig. 33.20). Some experiments have been performed, however, in which the birds were clearly outside of their familiar area and yet apparently showed good homeward orientation. Perhaps the most spectacular of such experiments is that of Wallraff and Graue (1973) in which the pigeons concerned were displaced 7000 km across the Atlantic from Munich, Germany, to Toledo, Ohio, and *vice versa*. The authors claim that a homeward component towards the transatlantic home is apparent from both the scatter of vanishing points and the scatter of recoveries. Before transatlantic displacement the birds were released 190–320 km E and W of their home loft. The orientation shown during these cisatlantic releases then served as a control for the transatlantic releases.

Let us consider for a moment the likely behaviour of a bird subjected to these events if the only map available to it is a familiar-area map. It is suggested later (p. 921) that, contrary to current opinion, displacement–release experiments present a bird, or any other animal that normally lives within a familiar area, with a navigational problem with which it is familiar. Normally, however, a bird that requires to return to a specific location from a previously unfamiliar site needs to do so only from a point that is no further than the edge of its familiar-area map as outlined in Fig. 33.5. Note that the edge of this map may be tens or hundreds of kilometres from the nearest previous point reached by the bird on its lifetime track. Given both of these assumptions, it follows that a bird asked to solve a given navigation problem after being displaced and released can only begin by 'assuming' that it is at some point on its pre-existing familiar-area map. Whether or not the bird is actually within the boundary of that map it should begin a search for familiar landscape features by which to determine home direction. Enough examples have been presented of birds being misled in unfamiliar locations by landmarks reminiscent of familiar locations for it to be accepted that in such a situation a bird accepts as familiar that landscape feature or features that show the greatest similarity to features on the familiar-area map. If sufficient birds assess the same features as most similar to familiar landscape features, then a given group of released birds will show a non-random scatter about a compass direction that would take the birds home if the features really were the familiar features on the birds' familiar area map. This pseudo-home direction is unlikely to be in precisely the true home direction, of course, but the chances of a broad similarity are really quite high.

If we now return to a consideration of the transatlantic pigeon releases by Wallraff and Graue (1973) we find a number of discrepancies in the results obtained that suggest that the birds could well have been determining a pseudo-home direction rather than a true home direction. First, by no means all of the homeward components, particularly of the recoveries, were significant. Second, the length of the mean vector was by no means always significantly different from zero. Third, most of the angles of the mean vectors are significantly different from the true home direction, no matter whether the latter is taken as a Great Circle Route (i.e. shortest global route) or as a Rhumb Line (i.e. constant compass direction). This last point is important, but more important still is that there was a pattern to the form of the error relative to the true home direction. First, both the cisatlantic and transatlantic releases of birds from both America and Germany showed a preferred direction to the south of the true home direction in Germany and to the north of the true home direction in Ohio. This is just as would be expected if there were some landscape configuration (such as, perhaps, high

ground generally to the south of both Munich and Toledo) that was used by birds from both lofts as a gross feature but which was in a slightly different compass direction relative to one loft than relative to the other. There is little point in pursuing this possibility further here because the details of the lofts and release sites are not available. Nor are the results of the transatlantic releases to the east and west of the local loft given separately. There is, however, one other factor that adds considerably to the possibility that Wallraff and Graue were observing a pseudo-home direction rather than a true home direction.

If the pigeons were responding to the degree of similarity between the release site and a region within their familiar area, the relationship between pseudo-home direction and true home direction should remain consistent only as long as the same transatlantic release site was used and only as long as each group of birds had similar topographical experience. In fact, this was not the case. The German birds used in 1967 and 1968 were from a loft 20 km ESE of Munich whereas those used in 1969 were from a loft 30 km SW of Munich. German birds were released in 1967 at Bowling Green, 30 km S of Toledo, but in 1968 and 1969 were released along with the inexperienced American controls 250 km to the E and W of Bowling Green. If the birds were adopting a pseudo-home direction rather than a true home direction we might expect some difference between the directions adopted by the 1967/68 and 1969 German birds and particularly the 1967 transatlantic-released German birds and the 1968/69 birds. As it happens, there do indeed appear to be differences in the directions adopted by these birds.

Clearly, before it can be accepted that pigeons can locate the home direction following transatlantic, or any long-distance, displacement–release, this experiment needs to be repeated with care being given that pseudofamiliar features do not happen to give a pseudo-home direction that approximates to the true home direction.

33.4.4.11 Navigation by birds that perform long-distance seasonal return migrations

Having shown that the least-navigation–familiar-area map model has the potential to accommodate most, if not all, of the data relating to the navigation of homing pigeons, *Columba livia*, we may now turn to a consideration of whether the same model can accommodate such data as relate to navigation by seasonally migrant birds. The procedure followed is first to examine displacement–release experiments on such birds, secondly to examine data collected on birds in the process of migrating, and thirdly to attempt a synthesis of the solution of the navigational problems encountered by migrant birds from birth to the establishment of a year's migration circuit. First, then, let us consider the results of a few displacement–orientation and displacement–release experiments on migrant birds.

Rüppell and Schein (1941) kept young starlings, *Sturnus vulgaris*, in aviaries from fledging until they began to nest the following spring and then released them at a distance of 115 km. None returned. A second batch of birds was captured and caged after they had performed a seasonal return migration and thus established a large familiar area. When they started to nest they were displaced and released from the same spot as the previous birds. This group showed a high degree of return.

Sargent (1962), working on bank swallows (=sand martins), *Riparia riparia*, in Wisconsin, displaced birds for distances of up to 160 km from their nesting site. Initial orientation after release had a significant homeward component at distances of 2–40 km and none at all at distances of 80–160 km. Between 40 and 80 km the homeward component seemed apparent but was not significant. Birds similarly displaced but tested in circular orientation cages on a hill or roof-top so that distant landmarks could be seen similarly showed a significant homeward component at distances of up to 40 km but not beyond. Yet other birds similarly tested but in cages with opaque sides so that landmarks could not be seen showed no homeward orientation at 2 to 16 km from the nesting site.

Perdeck (1967) tested the orientation of black-headed gulls, *Larus ridibundus*, in an orientation cage and found that there was no homeward orientation in heavy overcast. As long as both Sun and landmarks were available, however, homeward orientation to the colony was found at a distance of 25 km but not at 70–150 km.

The apparent attenuation of homing ability at distances of 25–80 km in birds that migrate hundreds or thousands of kilometres need not necessarily indicate that the birds do not recognise their location at the greater distances, only that they cease to attempt to return from these distances. It seems likely that the further a bird is released from its breeding site, the greater the chance that its removal migration threshold will be exceeded. Not only will the experimental handling and period of detention be greater but also, if the bird does recognise its location, it will also recognise how long and how much energy will be required to return home and perhaps also whether it can return the same day. It is perhaps significant that in species such as the Manx shearwater, *Puffinus puffinus*, and Laysan albatross, *Diomedea immutabilis*, that normally leave their young for long periods while they forage away from the breeding area, the attenuation of homing performance with release distance is much less (Figs. 33.26 and 33.27). Other evidence that the influence of distance on the removal migration threshold may be the critical factor rather than the ability to recognise location is provided by the observation that although occasionally birds show better homing performance from release sites along their migration track, most often they do not (Matthews 1968). Great caution has to be exercised, therefore, before interpreting the absence of homing ability or orientation from a particular release site as an inability to recognise location.

Taken together, the three experiments described above and those illustrated in Figs. 33.24–33.27 provide considerable support for the applicability of the familiar-area map model to the movements of migrant birds in and around their breeding home range. Because of the experimental problem of an attenuation of the homing response with distance that is identified in the preceding paragraph, however, experiments on migrant birds displaced from their breeding home range can tell us little of the navigation mechanisms employed during migration. To obtain such data it is necessary to perform displacement–release experiments on birds that are either on migration or that perform a seasonal return migration between displacement–release and recapture. Here again there are problems. In particular it is difficult for the experimenter to guess the bird's intended destination after release. Consequently it is difficult to assess the mechanisms employed by the bird in attempting to reach that destination.

Autumn displacement–release experiments on the almost exclusively nocturnal migrant, the blue-winged teal, *Anas discors* (Bellrose 1958b, 1967b, 1972) have shown that adults, like juveniles, maintain the standard autumn direction of the species (Fig. 27.7) instead of homing to the expected winter area. This result may be attributed to the fact that wildfowl have been shown to perform considerable latitudinal exploratory migrations during their autumn migration (p. 188). It is quite possible, even likely, therefore, that these adult teal were released within their familiar area. Because wildfowl also have a large familiar area at the winter end of their range, these birds could well have elected to migrate to the nearest winter area which would be in the standard autumn direction rather than to compensate for the displacement. Adults and juveniles, having performed parallel migrations, may well have done so using different methods.

Another aspect of this problem is that it now appears that birds do not migrate solely in their standard direction but instead alternate movement in this direction with other movements in other directions (p. 618). Displaced and released migrant birds, therefore, have a much greater variety of potential local destinations available to them than might otherwise have been expected. This renders extremely difficult an interpretation of displacement–release experiments of the type performed by Rabøl (1970) and described on p. 616.

Figures 33.24–33.27 illustrate a few displacement–release experiments on migrant birds. An interpretation of each experiment on the basis of the familiar-area map model is offered in the legends. It is tentatively concluded that all displacement–

Fig. 33.24 The behaviour of adult starlings, *Sturnus vulgaris*, following displacement–release

All figures relate to birds marked at The Hague (Δ) in autumn (October/November) during the period that large numbers are migrating along the Dutch coast. Details of the displacement–release experiment and the results obtained with juvenile starlings have been given in Fig. 27.1.

The solid line shows the breeding range of starlings that migrate through The Hague in autumn and the dashed line encloses the winter range. Some birds, however, are also captured in a much wider area as indicated by the dotted line. Whether the relatively few recaptures outside as compared to within the dashed line results from few individuals performing this lateral migration or whether it results from each individual spending relatively little time outside of the dashed area, or both, is unknown. However, the fact that most (82 per cent) captures outside the dashed line occur in autumn (October/November) rather than winter (December–February) implies that the migration is brief and exploratory and that the birds return to settle for the winter within the dashed line. Such exploratory migration is most likely to be performed during the bird's first autumn when juvenile. On this model, the

dotted line could be taken to represent the familiar area of many of the adult starlings passing along the Dutch coast in autumn.

Adults displaced (arrow) from The Hague to Switzerland initially (○) migrate along the standard migration axis, with some reversal, and only later (●) do they seem to recognise their position and migrate back to regain their winter home range. Such behaviour is much more consistent with navigation by a familiar area map than by any other mechanism. Perhaps the topographical cues used are similar to those suggested for homing pigeons in Figs. 33.18, 33.19, and 33.20.

Conclusion

Release–recapture experiments on starlings (Fig. 27.1 and here) are consistent with the suggestion that the first autumn migration of the young starling is exploratory, and that during the course of it migration units in the standard autumn direction alternate with migration units in non-standard directions, and that subsequent migrations when adult take place within the familiar area thus established.

[*Release–recapture data from Perdeck (1958)*]

release experiments on migrant birds so far performed can be accommodated within the familiar-area map model developed in this chapter.

One final expectation of the familiar-area model that can be considered is that birds with larger familiar areas should perform better (e.g. greater per cent return, greater speed of return) in homing experiments than birds with smaller familiar areas. This expectation has been fulfilled in a variety of field studies. Thus, during the homing experiments carried out by Mewaldt (1963, 1964) on the white-crowned sparrow, *Zonotrichia leucophrys*, a 1·06 per cent return was obtained for the race *pugetensis* ($n=658$) as compared with 5·23 per cent ($n=325$) return for the race *gambelii* which has a much longer distance component to its return migration and hence a larger familiar area (Fig. 33.25). Over 75 per cent of herring gulls, *Larus argentatus*, displaced from a colony in New England homed at a speed greater than 100 km/day but only 6 per cent came into this homing category when displaced from a Scottish colony (Matthews 1968). The New England gulls perform a seasonal return migration (Fig. 27.26) and thus are likely to have a much larger familiar area than those from Scotland, despite the relatively long-distance exploratory migrations that are likely to be performed by both groups of birds. The influence of a large familiar area on the results of homing experiments is not only that it increases the chances of a bird being released within its familiar area but also that even if it is not released in that area, the probability is much greater that the familiar area will be encountered by random search. It certainly seems

adopted the area as their winter home range for future years.

The winter following each displacement, many of the displaced birds re-appeared at San José, presumably after having performed a seasonal return migration. The areas delimited by dashes and crosses indicate the range of destinations of displaced birds the following summer if it is assumed that they perform a spring migration from the release site that is the same in terms of distance and direction as that performed from San José. In fact, it would be expected that birds that failed to locate habitat-types suitable for breeding would continue spring migration until they did so (p. 645). The actual destinations on the basis of this model would therefore be concentrated in the northern and western parts of the areas indicated. In both experiments, execution of a normal, though perhaps extended, spring migration would cause some individuals of both *Z. atricapilla* and *Z. leucophrys gambelii* to intersect their familiar area towards its northern end and thus result in them re-appearing in San José the following winter. *Z. l. pugetensis*, which indeed showed fewer returns than the others from Baton Rouge and no returns from Laurel, would fail to intersect a familiar area from either release point unless the spring migration was greatly extended. From Baton Rouge, however, individuals would just encounter high ground (in the region of Iowa/South Dakota/Nebraska). The familiar area map of *Z. l. pugetensis* should involve, upon encountering high ground, the recognition that the migration route/breeding range is encountered by flying W until the coastal plain or coast is reached. This response should be sufficient for some individuals to locate their true familiar area in this instance.

As far as *Z. atricapilla* and *Z. l. gambelii* are concerned, those individuals most likely to return from Baton Rouge are those with a breeding home range furthest to the north and west. These are also the individuals most likely to return from Laurel and it is perhaps significant that individuals that returned to San José from Baton Rouge and that were then displaced to Laurel showed a much higher proportion of returns (4 out of 8 for *Z. l. gambelii*; 2 out of 7 for *Z. atricapilla*; but 0 out of 7 for *Z. l. pugetensis*) than those displaced for the first time (6 out of 231 for *Z. l. gambelii*; 9 out of 142 for *Z. atricapilla*; and 0 out of 287 for *Z. l. pugetensis*).

261 pre-migratory *Z. l. gambelii* and *Z. atricapilla* released in Seoul, Korea, in April failed to return to their Californian winter range the following autumn.

Conclusion

The results of release–recapture experiments on *Zonotrichia* sparrows are consistent with the suggestion that normal spring orientation causes some birds to encounter their familiar area. This support is reinforced by the fact that a golden-crowned sparrow that had been released at Laurel in winter was captured the following spring in the region of the Great Lakes (■). It is unnecessary to postulate that the birds can recognise home direction from their point of release.

Fig. 33.25 The behaviour of sparrows, *Zonotrichia* spp., following displacement–release

The lower maps show the breeding range (⬯) and winter range (⠿) of the golden-crowned sparrow and two races of the white-crowned sparrow. These ranges are reproduced on the large map which also shows (stippled) areas higher than 400 m above sea level.

Sparrows were captured and marked in their winter range at San Jose and then displaced 2900 km to Baton Rouge on the Gulf Coast. In a later year, further sparrows were captured, including some that had returned from Baton Rouge, and displaced 3860 km to Laurel, Maryland.

Adult sparrows disappeared from the vicinity of the release site but many young remained and some even

[*Data from Mewaldt (1964) and Mewaldt et al. (1973). Maps compiled from Fullard (1959), Peterson (1961), Robbins et al. (1966), and Mewaldt (1964)*]

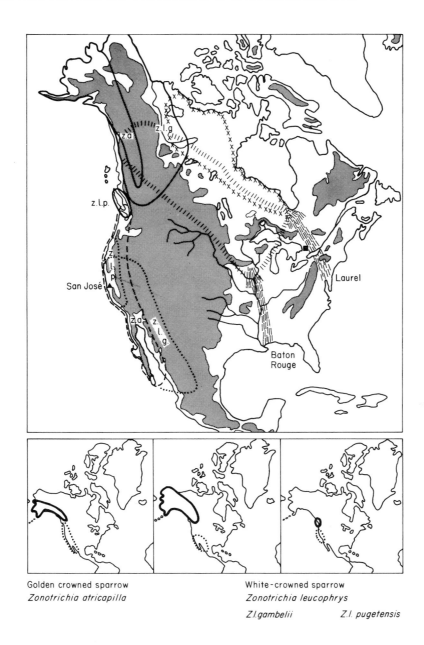

Golden crowned sparrow
Zonotrichia atricapilla

White-crowned sparrow
Zonotrichia leucophrys

Z.l.gambelii   *Z.l. pugetensis*

Fig. 33.26 The navigation of Manx shearwaters, *Puffinus puffinus*, by reference to a familiar-area map

The continuous line surrounds the breeding islands of two demes of the Manx shearwater. As far as the Atlantic deme is concerned, the months from September to February are spent off the east coast of South America as indicated by the dotted line. Young birds do not return to the breeding grounds during their first independent summer but instead explore large areas of the North Atlantic, especially along the Atlantic coast of North America. In subsequent years they return to the breeding grounds. Date of return is earlier each year until at 5 years of age they begin to breed.

Because of the exploratory migrations performed when young, Manx shearwaters are likely to have a huge familiar area (within the area enclosed by the dashed line). Consequently, even spectacular homing performances such as the one illustrated by an arrow are likely to be executed by reference to a familiar-area map.

[Figure and table compiled from Matthews (1968), Orr (1970a), and Mead (1974). Photo by Eric Hosking]

Return of young to breeding grounds in northern summer

Age
1 None (x)
2 Few (June / July)
3 May / June
4 April
5 Breed

significant that the most spectacular homing performances have been achieved with birds such as the Manx shearwater, *Puffinus puffinus* (Matthews 1968), Laysan albatross, *Diomedea immutabilis* (Kenyon and Rice 1958), and Leach's petrel, *Oceanodroma leucorhoa* (Billings 1968), which have an extremely large familiar area. Laysan albatrosses have homed from distances of up to 6629 km at a speed of 510 km/day. Comparable figures for Leach's petrel have been 4795 km at 349 km/day.

Although most of the data so far available on bird navigation have been obtained from displacement–release and displacement–orientation experiments, some interesting, though preliminary, results have been obtained with unmolested birds performing their normal migrations. Such results have been obtained since the advent of tracking radar which, because it produces a pencil beam, has conferred on the observer the ability to follow a single target.

Griffin (1972) followed birds for 3–4 km at altitudes that were judged on the basis of surface observations, radiosonde data, and aeroplane pilots' reports to be occupied by opaque cloud so that the birds could see neither the sky nor the land. All of the birds tracked continued in the appropriate direction and some maintained a straight and level path. Others climbed and descended and yet others produced non-linear flight paths (zig-zags, curves, reverse turns, and on one occasion a series of loops). However, these birds had established their flight direction before entering the stratus cloud. It could be assumed that having taken up the appropriate direction, the required course relative to wind direction could have been determined. Mainten-

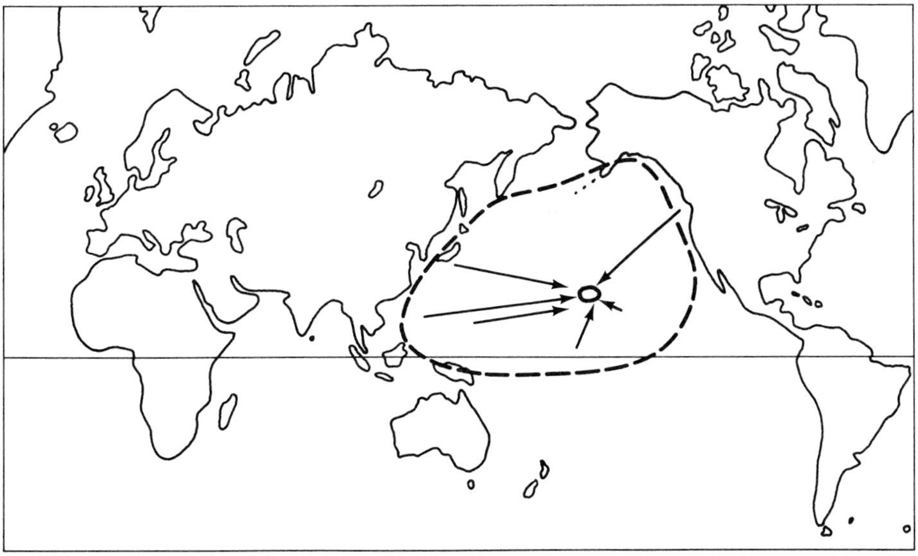

Fig. 33.27 The navigation of the Laysan albatross, *Diomedea immutabilis*, by reference to a familiar-area map

Displacement–release experiments on the Laysan albatross have produced some spectacular homing performances to the small breeding range ( ) as illustrated. However, individuals of this species wander widely within the area enclosed by the dashed line and it is quite possible for an individual's familiar area to encompass a large proportion of the species' range. Even the homing performances illustrated, therefore, could be executed by reference to a familiar-area map.

[*Compiled from Peterson (1961) and Matthews (1968)*]

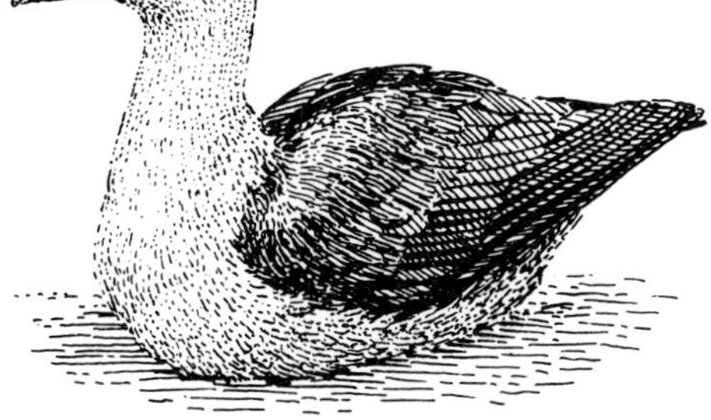

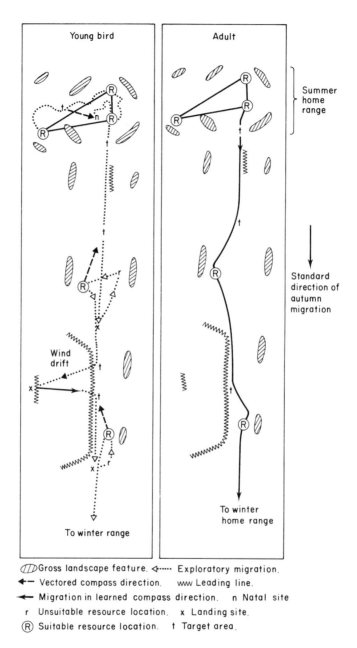

Fig. 33.28 A model of bird migration and navigation from birth to maturity

Migration by a young bird
During the period from hatching to fledging the young bird learns the compass bearings of various gross landscape features from the natal site (*n*). Just as Emlen (1972) demonstrated that different individuals learned different star patterns within the night sky (p. 609), so too does it seem likely that different individuals, even among siblings, select different topographical features to commit to a

spatial memory. After fledging, the bird begins, either alone or in company with its parent(s) or siblings, to perform exploratory migrations within this basic familiar-area map, the boundaries of which are set by the distance of the gross landscape features originally selected. During exploratory migration (dotted lines) the country traversed need not be monitored. Each time an area of potential biological importance (= resource location, *R*) is located, the direction of other areas of importance from that site is located by monitoring the apparent compass shift of the selected landscape features relative to their compass direction from the other site or sites. After the first one or two journeys between these different resource locations, the bird on all future occasions performs the journey simply by adopting the appropriate compass direction (solid arrows). If leading lines are available, these are learned and adopted in preference to compass bearings. Some learning of landmark sequences and of topographical details also occurs to an extent sufficient to permit access to resources on cloudy days but such maps are not used unless leading lines and a Sun compass are not available. All compass directions may also be related to a magnetic compass, recourse to which is rarely made. If the bird performs exploratory migration beyond the limit of its original familiar area map, it learns the compass bearings of landscape features from a target area (t) or succession of target areas, each one being placed at about the boundary of the previously existing familiar-area map (see Fig. 33.5b). For many ('resident') birds, this process is the extent of the exploratory–calculated migration sequence. In other ('migratory') birds, however, the process continues until a year's home range has been established.

In migratory birds, by the time of initiation of the autumn migration the bird has become familiar with a variety of resource locations scattered through a large area around or to one side of its natal site and has the ability to move from any one resource location to any other with a minimum of navigation (either by the use of leading lines or by adoption of a location-associated compass direction). The autumn migration is in effect a continuation of the process already described. Adopting an endogenously determined compass direction (by reference to the Sun, stars, wind, and/or geomagnetic field as appropriate) the bird migrates at an optimum height and for an optimum period (pp. 684 and 680), monitoring landscape features as it goes, partly for the purpose of preventing wind drift and partly as part of a process of locating potential transient home ranges. Features monitored, however, need not be committed to a spatial or any other memory except in relation to the establishment of target areas (see below). Leading lines and distant landmarks may be included in this category.

Target areas may not often be established. A bird flying at 2000 m has a potential horizon at a distance of about 170 km in all directions. If there was no restriction on visibility, a bird at this altitude could migrate 1000 km for the establishment of only 3 target areas if it is assumed that such an area is formed when the bird reaches the edge of its pre-existing familiar-area map. As it is, particularly with birds that migrate at lower altitudes and at night, rather more target areas will be necessary, but even so one or two such areas for each migration unit (x to x) seem to be sufficient for most species. This task presumably would not be too great, especially for a memory that has evolved in response to selection for such an ability.

It is suggested in Chapter 27 (p. 615) that in young birds migration units in the standard direction (down the page) alternate with migrations back along the migration track to areas (r and R) that are potential transient home ranges (= resource locations) and that are perceived during the preceding standard migration. If one or several of such locations are found to be suitable (R), the compass directions of surrounding gross landscape features are learned as also is the compass direction of the preceding

target area (dashed arrow) and the area becomes a transient home range for the following autumn. This process is repeated until the bird arrives at its winter quarters (p. 645) whereupon a procedure similar to that described relative to the natal site is performed, thus establishing a winter home range. The spring migration is a repeat of the autumn migration and may or may not make use of target areas learned during the autumn (p. 618). Where this is not the case, however, the familiar area established around the natal site in the preceding summer and autumn has to be large enough to be encountered by the migrant on its return the following spring.

Migration by an experienced adult

By the end of its first year the bird is familiar with a number of resource locations in its summer home range, and a sequence of target areas and transient home ranges between the summer and winter home ranges. For each location the compass bearings of a few gross landscape features are memorised, as also is the compass bearing that leads the bird from one location to another. In all future years, therefore, it is only necessary for the bird to adopt the location-associated compass bearing.

This implies that we should not look to the beginning of each migration unit for evidence relating to the navigational mechanisms employed by migrating birds but to the end of each migration unit when they navigate to the transient home range. At the initiation of a migration unit the bird need only adopt, perhaps crudely, the location associated compass direction. This permits an explanation of the frequent observation that birds may take off on the next stage of their migration in the appropriate direction even in heavy overcast (P. R. Evans 1972). It would be sufficient during the preceding 24 h whenever conditions were suitable to locate near and distant landmarks that lie in the appropriate compass direction. Quite possibly, also, the accuracy of prediction and/or observation from the ground (of cloud direction?) of wind direction is sufficiently high (p. 680) to permit the birds to move in the appropriate learned (or endogenous in the case of young birds) compass direction. Once started in that direction the birds could then adopt alternative compass-maintaining mechanisms (p. 860) such as flying at a fixed angle to the wind or by reference to a magnetic compass. Not until the bird nears the transient home range does it need to switch from orientation to navigation by monitoring the compass direction of familiar landmarks and determining the direction of the next transient home range. As long as the previous orientation was not too inaccurate, such a combination of orientation and navigation should permit arrival at the intended destination. Such a model is consistent with the otherwise odd observation that bird migration thresholds to migration-cost variables seem adaptive more to a prediction of conditions suitable for navigation at the destination (p. 632) than at the place of departure.

Wind drift

Normally a bird shows full compensation for wind drift, thus maintaining migration along a track with a fixed compass bearing (p. 632). Compensation for wind drift is possible only if reference can be made to topographical features. Land birds that lose sensory contact with land allow themselves to drift to conserve energy if they cannot settle (p. 683). A bird subjected to wind drift or that enters a zone where it will be exposed to the danger of wind drift should perhaps establish a target area at that point. Where the bird cannot land but reduces air speed to increase possible time in flight (p. 683) it should monitor the compass direction of displacement (p. 616). Upon arrival at a location where it is possible to land it should then, as soon as the threshold for such a migration is exceeded, migrate back along the direction of displacement until it encounters the target area learned when drift began. However, where the target area is part of a leading line, such as a coastline, as is almost invariably the case for a land bird, it is perhaps not necessary to migrate back to the precise area at which drift began. A vectored compass direction between the back-drift direction and the standard direction would normally be sufficient to re-locate the leading line and the resulting gap in the familiar area could be filled in during the return migration the following spring. In an adult bird, this is a particularly advantageous strategy, for a vectored direction will intersect an already familiar area.

ance of this angle relative to the wind would have been sufficient to account for the persistence of the appropriate direction upon entry into the opaque cloud. This mechanism could also have accounted for the non-linear flight paths though these could also have reflected attempts by the bird to find areas where perception of land or sky was possible.

Williams *et al.* (1972) also provide data relevant to migration under conditions of total overcast (with cloud cover determined by ground observations, radar, and satellite photographs). It was found that nocturnal migrants could maintain their track direction even when over the sea, though within sight of land, despite the absence of stars. One bird, migrating initially at an altitude of 2100–2400 m in a strong cross wind, initiated a marked decrease in altitude. This bird maintained a constant track direction but in order to do so, because of changing wind speed and direction with altitude, had to change its course by more than 120°. Yet when the bird encountered islands and the shore its track shifted. Two other birds were tracked during periods that cloud was both high and low and during which it was assumed that the birds could perceive neither land nor sky. These birds maintained their course (=compass heading) but not their track when confronted by shifting winds. Apparently they could not compensate for wind drift without reference to visual cues.

There is a strong implication from these results that even at night migrating birds monitor landmarks and the maintenance or change in track direction may be strongly linked with this process. This does not, of course, prove that landmarks were involved in the initial determination of migration direction. Similarly, these observations do not prove that the night sky, wind direction, or the Earth's magnetic field are involved in this initial determination of migration direction. They do suggest, however, that once a bird has set off in a particular compass direction it is aware of the angle it is making to a variety of directional variables, such as the above, and that sudden removal of perceptual access to one or several of these variables results simply in the bird maintaining direction by making use of those that remain, at least for a limited, as yet undefined, period.

By way of a summary of this and the previous section, we may now draw on the major conclusions reached to construct a model of the way in which from birth onwards a migrant bird establishes a year's home range. This model is presented in Fig. 33.28.

33.5 General conclusions

Although the orientation mechanisms involved in the return migrations of animals with an apparently low degree of return (e.g. zooplankton, littoral animals, soil animals, etc.) are of considerable physiological interest in terms of the mechanisms involved in, say, compensating for the movement of the Sun and Moon, they represent a relatively simple mechanism for the performance of return migration. Of considerably greater interest, however, is the conclusion that, as far as I am aware, has not been reached before, that many other animals from bees to birds employ wherever possible a similar technique in the performance of often spectacular return migrations with a high degree of return. It appears that as soon as possible after becoming acquainted with a location the animal learns the compass direction that will take it to other resource location(s) with which it is already familiar and thereafter uses only compass orientation to take it between these locations. Not until it has been moving for an appropriate period/distance and yet has failed to arrive at the required location does the animal monitor its position and employ navigation.

In other words, the animal uses navigation to determine the home direction on only the first visit (or perhaps the first and a few subsequent visits) to a particular site. Any displacement–release experiment that releases an animal into an area which the animal has visited before, or even if the animal only 'thinks' it has visited the place before (i.e. there are sufficient gross similarities between the release site and the site with which the animal is really familiar to trigger the same responses as at this latter site), will initially trigger only a compass orientation and not navigation. Associated with this is the ease with which animals (e.g. box turtles and birds) are trained to associate a given compass direction upon release with the displacement–release process itself, independently of the location or characteristics of the release site. All of this, plus the behaviour of following topographical leading lines (but not landmark sequences) in preference to the compass cue as long as the leading line coincides or is parallel with the animal's home track, is absolutely consistent with the least-navigation hypothesis. Indeed, the apparent absence of navigation in animals for which selection for a high degree of return is weak or zero can be attributed to the even stronger selection against navigation that results from selection for least navigation.

Many, indeed most, authors (e.g. Matthews 1968, Ferguson 1971) are of the opinion that displacement–release experiments require an animal to solve a navigational problem of a type that it has not been required to solve during the course of the evolution of the navigational mechanism. I disagree. I feel the demands on an animal that attempted to monitor the entire course of its exploratory migrations would be too great and in any case unnecessary and would be selected against by the advantage of least navigation. It seems much more likely that before leaving a given site the animal learns the location of that site within a familiar-area map (Figs. 33.5–33.7) and thereafter fails to monitor its exploratory migration track. Not until it has located an appropriate resource, therefore, does the animal attempt to locate the direction of the previously occupied site. In effect, therefore, the animal has to solve precisely the type of navigational problem that it is required to solve by displacement–release. The only difference is that normally the animal's exploratory migration would be limited to the familiar area map established from its original site or, if it requires to leave this area, upon doing so it will establish a second map which overlaps the first

based on an intermediate location. This intermediate location may have no biological significance to the animal other than as a target area (Fig. 33.5) and is likely to be situated at approximately the limit of the first map. During displacement–release, if they are displaced beyond the limits of their original map, animals do not normally have the opportunity to learn target areas.

It seems clear that few, if any, animals have a familiar-area map based on a single sense, and it seems as though those derived from the different senses are superimposed on one another. Thus most microchiropteran bats appear to have an 'echo-location map' superimposed upon a larger 'visual map'. Amphibians appear to have an acoustic map superimposed upon a visual map superimposed upon a much larger olfactory map. To what extent these are integrated within the spatial memory and to what extent they are stored separately is unknown. Certainly, however, the removal of one map rarely prevents the animal from making full use of one or all of the other maps available to it. All maps, however, seem to have one factor in common. They may well be entered into the spatial memory as a map in the aerial photograph sense and some memorised sequences of landmarks (visual, olfactory, thermal, etc.) may also be included. In addition, however, sites of biological importance to the animal (e.g. roosting sites, feeding sites, etc.) are entered into the spatial memory in association with the compass direction from that site of various sensory landscape features and, eventually, in association with the compass direction of other biologically meaningful sites. Furthermore, least navigation has dictated that wherever possible it is this latter type of information that is used, the other landscape maps being used only when compass information is not available. This view of biologically meaningful sites linked together by memorised leading lines or compass directions is totally consistent with the latticework nature of most familiar areas (p. 378).

Just as the maps are based on a variety of senses, so too do many animals seem to have access to more than one compass. However, the reluctance of most animals to make use of a magnetic compass even if they have access to one suggests that the, as yet unknown, sensory mechanisms involved render such a compass disadvantageous in terms of least navigation. The compass directions of sensory landmarks from each of the important sites are probably used only for exploratory migration and on those few occasions that the animal may be displaced by wind, water currents, or predators (from which they subsequently escape). Once the compass directions of the various resource sites relative to one another have been linked, the major part of an animal's movements can be performed solely by orientation, not navigation. Alternatively, and probably favoured even more by selection for least navigation, orientation may be based on a network of leading lines (where these happen to be appropriate) and/or tracks (where the animal's movement imparts some change to the area traversed as it does in molluscs and terrestrial mammals, etc., but not in bats, birds, or fish). Only when the animal performs further exploratory migration is navigation once more required.

It is suggested here that the least-navigation–familiar-area map model accommodates such a large proportion of the experimental data on animal navigation and migration and is not actively discredited by any, that an advantage would be gained from its adoption as a working model. Not only does it accommodate the major portion of experimental data but it also unites the navigational mechanisms of all animals within a single model. In consequence, it simplifies to the point of extinction the theoretical problem (but which would not have been insoluble) of accounting for the evolution of complex navi-

gational mechanisms by animals that perform long-distance seasonal return migrations. The mechanisms used are the same as those used by more sedentary species. It also resolves the paradox that the ability to navigate was best known, best studied, and equally well marked in a 'resident' animal, the homing pigeon, rather than in those animals that migrate over large portions of the globe. It now appears that the navigational mechanisms are virtually identical, at least in principle, whether the animal spends its entire life within a few square metres or whether it wanders over an entire ocean.

One final point is perhaps worth stressing. If the familiar-area–least-navigation model is accepted, much more care will be necessary in future displacement experiments to exclude the possibility of false navigation. On this model an animal can only assume after displacement that it is still within its familiar-area map and navigate accordingly. In so doing it can only navigate relative to that feature(s) that most resembles a feature(s) in the familiar area. Alternatively, it can take recourse to the compass direction adopted on some previous occasion that resulted in a successful return. Either way, non-random orientation may be found, perhaps by chance, in a direction that is broadly similar to the home direction, even though the animal has no idea in which direction the home site is really located.

References and author index

The numbers in brackets after each item indicate the page or pages on which reference to that item is made. Bold type indicates that reference is made in a legend to a figure or in a table.

Able, K.P. (1973) The role of weather variables and flight direction in determining the magnitude of nocturnal bird migration. *Ecology* **54**, 1031–41. (632)

Ables, E.D. (1969) Home range studies of red foxes (*Vulpes vulpes*). *J. Mammal.* **50**, 108–20. (174, 393)

Adams, K.M. (*pers. comm.*) (425, 448)

Adler, H.E. (1970) Ontogeny and phylogeny of orientation. *In:* Aronson, L.R., Tobach, E., Lehrman, D.S., and Rosenblatt, J.S. (eds.) *Development and evolution of behaviour: essays in memory of T.C. Schneirla.* Freeman, San Francisco. pp. 303–36. (877)

Akimushkin, I.I. (1963) *Cephalopoda in the seas of the USSR* (In Russian). Izdatel'stuo AN SSSR. (822)

Aldrich-Blake, F.P.G. (1970) Problems of social structure in forest monkeys. *In:* Crook, J.H. (ed.) *Social behaviour in birds and mammals.* Academic Press. London. pp. 79–101. (117, 162, 164, 297)

Alerstam, T. and Ulfstrand, S. (1974) A radar study of the autumn migration of wood pigeons, *Columba palumbus*, in southern Scandinavia. *Ibis* **116**, 522–42. (632)

Alexander, J. and Keeton, W.T. (1974) Clockshifting effect on initial orientation of pigeons. *Auk* **91**, 370–4. (864)

Allee, W.C., Emerson, A.E., Park, O., Park, T. and Schmidt, K.P. (1949) *Principles of animal ecology.* W.B. Saunders. Philadelphia. (70)

Allen, G.M. (1939) *Bats.* Harvard University Press. (556, 559, 560, 562)

Allen, J.A. (1960) On the biology of *Crangon allmani* Kinahan in Northumberland waters. *J. mar. biol. Ass. U.K.* **39**, 481–508. (**835**)

Allen, J.A. (1966) The rhythms and population dynamics of decapod Crustacea. *Oceanogr. Mar. Biol. Ann. Rev.* **4**, 247–65. (823, **834**)

Allen, J.H., Buck, R.C., and Wink, A.T. (1955) *Pulling up stakes and breaking apron strings: a study of mobility among Pennsylvania's rural youth.* W.B. Saunders. Philadelphia. (135, 138)

Allen, K.R. (1956) The geography of New Zealand's freshwater fish. *N.Z. Sci. Rev.* **14**, 3–9. (822)

Anderson, S. and Banerji, D. (1962) Report on a study of migration in four taluks of Bangalore district. *Population review* (Madras) **6**, 69–77. (134, 135, 137, 138)

Antonius, O. (1937) Über Herdenbildung und Paarungseigentumlichkeiten der Einhufer. *Z. Tierpsychol.* **1**, 259–89. (165–6)

Arnol'di, K.V. (1955) The hibernation of *Eurygaster integriceps* in the mountains of the Kuban, from investigations in 1949–53. *In:* Fedotov, D.M. (ed.) *The noxious little tortoise, Eurygaster integriceps Put.: Work of the Laboratory of Invertebrate Morphology in connection with the expedition to the province of Krasnodar in 1949–1953*, Vol. 3, pp. 171–237 (Akad. Nauk SSSR, Moscow) (*Rev. appl. Ent.* (A) **46**, 10). (241)

Aronson, L.R. (1951) Orientation and jumping behaviour in the gobiid fish *Bathygobius soporator.* *Am. Mus. Novitates* **1486**, 1–22. (849)

Aronson, L.R. (1971) Further studies on orientation and jumping behaviour in the gobiid fish, *Bathygobius soporator.* *Ann. N.Y. Acad. Sci.* **188**, 359–77. (849)

Ash, J.S. (1969) Spring weights of trans-Saharan migrants in Morocco. *Ibis* **111**, 1–10. (648–9)

Ashmole, N.P. (1963) The regulation of numbers of tropical oceanic birds. *Ibis* **103b**, 458–73. (111)

Aubert, M. (1971) Télémédiateurs chimiques et équilibre bioligique océanique. 1. Théorie générale. *Rev. Intern. Océanogr. Méd.* **21**, 5–15. (851)

Aubert, M. (in Dowling 1975). (851)

Aubert, M., Pesando, D., and Gauthier, M.J. (MS) Interspecific relationships between marine microorganisms. (851)

Bagenal, T.B. (1973) *Identification of British fishes.* Hulton Educational Publications, Amersham, England. (**818**, 847)

Bagg, A.M. (1969) A summary of the fall migration seasons, 1968, with special attention to the movements of black-capped chickadees. *Audubon Field Notes* **23**, 4–12. (656)

Baggerman, B. (1956) An experimental study on the timing of breeding and migration in the three-spined stickleback (*Gasterosteus aculeatus* L.). *Arch. Néerl. Zool.* **12**, 105–317. (787)

Bailey, A.M. and Hendee, R.W. (1926) Notes on the mammals of Northwestern Alaska. *J. Mammal.* **7**, 9–28. (720–1, **727**, 752, 757, **759**)

Bainbridge, R. (1961) Migrations. *In:* Waterman, T.H. (ed) *The physiology of Crustacea. Vol. II. Sense organs, integration and behaviour.* Academic Press, New York. pp. 431–63. (485, 488, 823)

Baker, J.R. and Baker, Z. (1936) The seasons in a tropical rainforest (New Hebrides). Part 3. Pteropidae. *J. Linn. Soc. Lond.* **40**, 123–41. (563)

Baker, J.R. and Bird, T.F. (1936) The seasons in a tropical rain forest (New Hebrides). Part 4. Insectivorous bats (Vespertilionidae and Rhinolophidae). *J. Linn. Soc. Lond.* **40**, 143–61. (558)

Baker, P.T. and Weiner, J.S. (eds.) (1966) *The biology of human adaptability.* Clarendon Press, Oxford. (129, 133)

Baker, R.J. and Ward, C.M. (1967) Distribution of bats in southeastern Arkansas. *J. Mammal.* **48**, 130–2. (576)

Baker, R.R. (1968a) A possible method of evolution of the migratory habit in butterflies. *Phil. Trans. R. Soc.* B, **253**, 309–41. (188, **335**, 362, 416–17, 435, 438–9, **440–3**, 472)

Baker, R.R. (1968b) Sun orientation during migration in some British butterflies. *Proc. R. ent Soc. Lond.* A, **143**, 89–95. (416–17, 433, **434**, 472)

Baker, R.R. (1968c) The relevance to bird migration and navigation of a theory of the evolution of the migratory habit in butterflies. *Ibis* **110**, 411–12. (63)

Baker, R.R. (1968d) The evolution of the migratory habit in British butterflies. Unpublished Ph.D. thesis, Bristol University. (86, **87**, **448–9**, 450, 472)

Baker, R.R. (1969a) The evolution of the migratory habit in butterflies. *J. Anim. Ecol.* **38**, 703–46. (9, 20, 32, 47, 86, **87**, 249, **335**, 352, 362, 415, 424, **428–32**, **435**, 436, 438–9, **440–3**, **448–9**, 450, **450**, 472)

Baker, R.R. (1969b) Die Entwicklung des Wanderverhaltens bei Schmetterlingen. *Umschau (Frkf./M)* **69**, 626–7. (**440–1**)

Baker, R.R. (1970) Bird predation as a selective pressure on the immature stages of the cabbage butterflies, *Pieris rapae* and *P. brassicae. J. Zool., Lond.* **162**, 43–59. (234, 243, **335**)

Baker, R.R. (1971) Comments at a meeting. *Proc. R. ent. Soc. Lond.* **36**, 33–6. (414, 421)

Baker, R.R. (1972a) Territorial behaviour of the nymphalid butterflies, *Aglais urticae* (L.) and *Inachis io* (L.). *J. Anim. Ecol.* **41**, 453–69. (57, 68, 197–9, 237, **369**, 374, 403, 406, 408, 560, 564)

Baker, R.R. (1972b) The geographical origin of the British spring individuals of the butterflies *Vanessa atalanta* (L.) and *V. cardui* (L.). *J. Ent.* A **46**, 185–96. (426, **428–32**, 571, 573)

Baker, R.R. and Parker, G.A. (1973) The origin and evolution of sexual reproduction up to the evolution of the male-female phenomenon. *Acta Biotheoretica.* **22**, 49–77. (57)

Baker, R.R., Gill, S.E. and Mackenzie, C.E.C.P. (1977) Wheel activity of small mammals as a specific indicator of exploratory migration. Exhibit at the Society for Experimental Biology Conference, Manchester, 29–31 March 1977. (379)

Baker, R.T. (*pers. comm.*). (571)

Balcells, E. (1964) Ergebnisse der Fledermausberingung in Nordspanien. *Bonner Zool. Beitr.* **15**, 36–44. (**587**)

Baldwin, H.A. (1972) Long-range radio tracking of sea turtles and polar bear—instrumentation and preliminary results. *In:* Galler, S.R., Schmidt-Koenig, K., Jacobs, G.J. and Belleville, R.E. (eds.) *Animal orientation and navigation.* Scientific and Technical Information Office, National Aeronautics and Space Administration; Washington, D.C. pp. 19–37. (178, 886–7)

Banerjee, B. (1967) Seasonal changes in the distribution of the millipede, *Cylindroiulus punctatus*

(Leach) in decaying logs and soil. *J. Anim. Ecol.* **36**, 171–7. (509)

Banfield, A.W.F. (1953) The range of individual timber wolves (*Canis lupus*). *J. Mammal.* **34**, 389–90. (174, **311**)

Banfield, A.W.F. (1954) *Preliminary investigation of the barren ground caribou*. Canadian Wildlife Service, Wildlife Management Bulletin, ser. 1, no. 10A, 79 pp., and 10B, 112 pp. (98, 166, 170, **533**, 554–5)

Bannerman, D.A. (1957) *The birds of the British Isles. VI Ciconiidae, Phoenicopteridae, Ardeidae, Anatidae (part)*. Oliver and Boyd, London. (**692**)

Bannerman, D.A. (1958) *The birds of the British Isles. VII. Anatidae (conclusion)*. Oliver and Boyd, London. (**621, 623**)

Bannerman, D.A. (1962) *The birds of the British Isles. XI. Glareolidae, Otididae, Burhinidae, Gruidae, Laridae*. Oliver and Boyd, London. (**703**)

Bannikov, A.G., Zhirnov, L.V., Lededeva, L.S. and Fandeev, A.A. (1967) *Biology of the saiga*. Translated from Russian: Israel program for Scientific Translations, Jerusalem. (531, 536, 554)

Bannister, J.L. (1968) An aerial survey for sperm whales off the coast of Western Australia. *Aust. J. Mar. Freshwater Res.* **19**, 31–51. (**301, 763, 764**)

Bannister, J.L. and Gambell, R. (1965) The succession and abundance of fin, sei, and sperm whales off Durban. *Norsk Hvalfangsttid.* **54**, 45–60. (**180, 755, 764, 883**)

Barbour, R.W. and Davis, W.H. (1969) *Bats of America*. Univ. Kentucky Press, Lexington. (**159**, 559, 560, 561, 564, 573, **574, 576, 578**, 579, **584, 591, 600, 603**)

Barbour, R.W., Davis, W.H. and Hassell, M.D. (1966) The need of vision in homing by *Myotis sodalis*. *J. Mammal.* **47**, 356–7. (891)

Barlow, J.S. (1964) Inertial navigation as a basis for animal navigation. *J. Theoret. Biol.* **6**, 76–117. (863)

Barlow, J.S. (1971) Comments at a meeting. *Ann. N.Y. Acad. Sci.* **188**, 331–58. (863)

Barnett, S.A. (1971) *The human species: a biology of man*. (5th ed., revised). Harper and Row, London. (113, 114, 117, 119, 124, 125, 133, 134)

Barnicot, N.A., Mukherjee, D.P., Woodburn, J.C. and Bennett, F.J. (1972) Dermatoglyphics of the Hadza of Tanzania. *Human Biol.* **44**, 621–48. (127, 128)

Barrass, R. (1963) The burrows of *Ocypode ceratoph-thalmus* (Pallas) (Crustacea, Ocypodidae) on a

tidal wave beach at Inhaca Island, Moçambique. *J. Anim. Ecol.* **32**, 73–85. (**829**)

Bartholomew, G.A. (1970) A model for the evolution of pinniped polygyny. *Evolution* **24**, 546–59. (730–3)

Bartholomew, G.A. and Boolootian, R.A. (1960) Numbers and population structure of the pinnipeds on the California Channel Islands. *J. Mammal.* **41**, 366–75. (**731, 732**)

Bartholomew, G.A. and Hoel, P.G. (1953) Reproductive behavior of the Alaska fur seal, *Callorhinus ursinus*. *J. Mammal.* **34**, 417–36. (711)

Bartholomew, G.A. and Hubbs, C.L. (1952) Winter population of pinnipeds about Guadalupe, San Benito, and Cedros Islands, Baja California. *J. Mammal.* **33**, 160–71. (**731, 732**)

Bartholomew, G.A. and Hubbs, C.L. (1960) Population growth and seasonal movements of the Northern Elephant Seal, *Mirounga angustirostris*. *Mammalia, Paris.* **24**, 313–24. (**732**)

Bastian, J. (1967) The transmission of arbitrary environmental information between bottlenose dolphins. *In: Cours d'Été O.T.A.N. sur les systèmes sonars animaux: Biologie et bionique. 26 Septembre–3 Octobre, 1966. Frascati, Italie. Tome II.* Laboratoire de Physiologie Acoustique, Paris. (184, 767)

Bateman, A.J. (1950) Is gene dispersion normal? *Heredity* **4**, 353–63. (138)

Batschelet, E. (1965) *Statistical methods for the analysis of problems in animal orientation and certain biological rhythms*. Amer. Inst. Biol. Sci., Washington, D.C. (30)

Bauder, W.W. (1959) Analysis of trends in population, population characteristics and community life in southern Iowa. *Seminar on adjustment and its problems in southern Iowa, CAEA report 4*. Iowa State University College of Agriculture. pp. 113–37. (135)

Baudinette, R.V. and Schmidt-Nielsen, K. (1974) Energy cost of gliding flight in herring gulls. *Nature, Lond.* **248**, 83–4. (691)

Bauer, K. and Steiner, H. (1960) Beringungsergebnisse an der Langflügelfledermaus (*Miniopterus schreibersi*) in Österreich. *Bonner Zool. Beitr.* **11**, 36–53. (**572, 587**)

Beall, G. (1946) Seasonal variation in sex proportion and wing length in the migrant butterfly, *Danaus plexippus* L. *Trans. R. ent. Soc. Lond.* **97**, 337–53. (426, **427**)

Beall, G. and Williams, C.B. (1945) Geographical variation in the wing length of *Danaus plexippus*

(Lep. Rhopalocera). *Proc. R. ent. Soc. Lond.* A **20**, 65–76. (**427**)

Beauchamp, R.S.A. (1933) Rheotaxis in *Planaria alpina. J. exp. Biol.* **10**, 113–29. (785)

Beauchamp, R.S.A. (1935) The rate of movement of *Planaria alpina. J. exp. Biol.* **12**, 271–85. (785)

Beauchamp, R.S.A. (1937) Rate of movement and rheotaxis in *Planaria alpina. J. exp. Biol.* **14**, 104–16. (785)

Beebe, W. (1931) Notes on the gill-finned goby, *Bathygobius soporator* (Cuvier and Valenciennes). *Zoologica N.Y.* **12**, 55–66. (849)

Beer, J.R. (1955) Survival and movement of banded big brown bats. *J. Mammal.* **36**, 242–8. (**603**)

Beer, J.R. and Meyer, R.K. (1951) Seasonal changes in the endocrine organs and behavior patterns of the muskrat. *J. Mammal.* **32**, 173–91. (**310**)

Beijer, G. (1969) Modern patterns of international migratory movements. *In:* Jackson, J.A. (ed.) *Migration.* Cambridge University Press, pp. 11–59. (139)

Beischer, D.E. (1971) The null magnetic field as reference for the study of geomagnetic directional effects in animals and man. *Ann. N.Y. Acad. Sci.* **188**, 324–30. (887)

Beklemishev, C.W. (1960) Southern atmospheric cyclones and the whale feeding grounds in the Antarctic. *Nature, Lond.* **187**, 530–1. (746)

Belkin, A.N. (1966) (Summer distribution, stocks, hunting propects, and some features of the biology of Steller sea lions living on the Kurile Islands) (in Russian). *Izv Tikhookean Nauch-Issled Inst. Ryb. Khoz. Okeanogr.* **58**, 69–95. (**729**)

Bellrose, F.C. (1958a) Celestial orientation by wild mallards. *Bird-banding* **29**, 75–90. (**858**, 864, 873)

Bellrose, F.C. (1958b) The orientation of displaced waterfowl in migration. *Wilson Bull.* **70**, 22–40. (610, 912)

Bellrose, F.C. (1967a) Radar in orientation research. *Proc. XIV Int. Ornith. Cong.* 281–309. (616, 634, 684)

Bellrose, F.C. (1976b) Orientation in waterfowl migration. *In: Storm, R.M.* (ed.) *Animal orientation and navigation.* 27th Ann. Biol. Colloq., Oregon State University Press, Cornwallis. pp. 73–98. (606, 912)

Bellrose, F.C. (1972) Possible steps in the evolutionary development of bird navigation. *In:* Galler, S.R., Schmidt-Koenig, K., Jacobs, G.J. and Belleville, R.E. (eds.) *Animal orientation and*

navigation. Scientific and Technical Information Office, National Aeronautics and Space Administration, Washington, D.C. pp. 223–57. (190, 201, **322**, 324, 606, 609, 631, 655–6, 684, 689, 860, 864, 912)

Bellrose, F.C. and Crompton, R.D. (1970) Migration behavior of mallards and black ducks as determined from banding. *Ill. Nat. Hist. Surv. Bull.* **30**, 168–234. (608, 613)

Bellrose, F.C. and Sieh, J.G. (1960) Massed waterfowl flights in the Mississippi flyway, 1956 and 1957. *Wilson Bull.* **72**, 29–59. (606)

Bels, L. (1952) Fifteen years of bat-banding in the Netherlands. *Publties natuurh. Genoot. Limburg Ser. V.* **1952**, 1–99. (**570**, **581**, **582**, **583**, **595**, **596**, **597**, **598**, **599**)

Bergamini, D. (1965) *The land and wildlife of Australasia.* Life Nature Library, Time Life Books, Amsterdam. (119, 122, 605)

Bergström, U. (1967) Observations on Norwegian lemmings, *Lemmus lemmus* (L.) in the autumn of 1963 and spring of 1964. *Arkiv för Zoologi* **20**, 321–63. (**310**, 413–14)

Bernhardt, F.W. (1961) Correlation between growth-rate of the suckling and percentage of total calories from protein in the milk. *Nature, Lond.* **191**, 358–60. (748)

Bernstein, I.S. (1968) The lutong of Kuala Selangor. *Behaviour* **32**, 1–16. (**295**, 297)

Berthold, P. (1973) Relationships between migratory restlessness and migration distance in six *Sylvia* species. *Ibis* **115**, 594–9. (613, 630)

Berthold, P. and Dorka, V. (1969) Vergleich und Deutung von jahreszeitlichen Wegzugs-Zugmustern ausgeprägter und weniger ausgeprägter Zugvögel. *Vogelwarte* **25**, 121–9. (645)

Bertin, L. (1956) *Eels, a biological study.* Cleaver-Hume, London. (196, **819**, **822**, 823)

Bertram, G.C.L. (1940) The biology of the weddell and crabeater seals: with a study of the comparative behaviour of the Pinnipedia. *British Graham Land Expedition, 1934–37. Scientific Reports.* **1**, 1–139. (709–11, 713, **714**)

Berzin, A.A. (1972) *The sperm whale.* Translated from Russian: Israel program for Scientific Translations, Jerusalem. (**763**)

Berzin, A.A. and Rovnin, A.A. (1966) (The distribution and migration of whales in the north-eastern part of the Pacific Ocean and in the Bering Sea and the Sea of Chukotsk) (In Russian). *Izv Tikhookean Nauch-Issled Inst. Ryb. Khoz. Okeanogr.* **58**, 179–207. (**748**, **753**, **763**, **764**, 883)

Beshers, J.M. and Nishiura, E.N. (1961) A theory of internal migration differentials. *Social Forces* **39**, 214–18. (134)

Best, P.B. (1967) Distribution and feeding habits of baleen whales off the Cape province. *S. Afr. Div. of Sea Fish. Invest. Rep.* **57**, 1–44. (**180, 182, 764, 883**)

Best, P. B. (1968) The sperm whale (*Physeter catodon*) off the west coast of South Africa. 2. Reproduction in the female. *S. Afr. Div. Sea Fish. Invest. Rep.* **66**, 1–32. (**763**)

Best, P.B. (1969a) The sperm whale (*Physeter catodon*) off the west coast of South Africa. 3. Reproduction in the male. *S. Afr. Div. Sea Fish. Invest. Rep.* **72**, 1–20. (**301, 763**)

Best, P.B. (1969b) The sperm whale (*Physeter catodon*) off the west coast of South Africa. 4. Distribution and movements. *S. Afr. Div. Sea Fish. Invest. Rep.* **78**, 1–12. (**301, 763**)

Best, P.B. (1970a) The sperm whale (*Physeter catodon*) off the west coast of South Africa. 5. Age, growth and mortality. *S. Afr. Div. Sea Fish. Invest. Rep.* **79**, 1–27. (**301**)

Best, P.B. (1970b) Exploitation and recovery of right whales, *Eubalaena australis*, off the Cape Province. *S. Afr. Div. Sea Fish. Invest. Rep.* **80**, 1–20. (**753**)

Beutler, R. (1930) Biologisch-chemische Untersuchungen am Nektar von Immenblumen. *Z. vergl. Physiol.* **12**, 72–176. (**335**, 352)

Bigg, M.A. (1969) The harbour seal in British Columbia. *Bull. Fish. Res. Bd Canada* **172**, 1–33. (720–1)

Billings, S.M. (1968) Homing in Leach's petrel. *Auk* **85**, 36–43. (917)

Blair, W.F. (1958) Effects of x-irradiation on a natural population of the deer-mouse (*Peromyscus maniculatus*). *Ecology* **39**, 113–18. (**310**)

Blair, W.F. (1960) *The rusty lizard, a population study.* Univ. Texas Press, Austin. (**309**)

Bleakly, J.W. (1961) *The aborigines of Australia. Their history, their habits, their assimilation.* Jacoranda Press, Brisbane. (117, 125)

Blinn, W.C. (1963) Ecology of the land snails *Mesodon* and *Allogona*. *Ecology* **44**, 498–505. (376, 518)

Blower, J.G. and Fairhurst, C.P. (1968) Notes on the life-history and ecology of *Tachypodoiulus niger* (Diplopoda, Iulidae) in Britain. *J. Zool., Lond.* **156**, 257–71. (343)

Blower, J.G. and Gabbutt, P.D. (1964) Studies on the millipedes of a Devon oak wood. *Proc. zool. Soc. Lond.* **143**, 143–76. (509)

Blurton-Jones, N.G. (1972) Moult migration of Emperor Geese. *Wildfowl* **23**, 92–3. (620)

Bobek, H. (1962) The main stages in socio-economic evolution from a geographical point of view. *In:* Wagner, P.L. and Mikesell, M.W. (eds.) *Readings in cultural geography.* University of Chicago Press, Chicago. pp. 218–47. (133, 134)

Bock, W.J. (1965) The role of adaptive mechanisms in the origin of higher levels of organization. *Syst. Zool.* **14**, 272–87. (604)

Bodenheimer, F.S. (1944) Studies in the ecology and control of the Moroccan locust (*Dociostaurus maroccanus*) in 'Iraq. 1. Results of a mission of the 'Iraq Department of Agriculture to N. 'Iraq in spring 1943. *Bull. Dir.-gen. Agric. Iraq.* No. 29, 121 pp. (240)

Bogorov, B.G. (1946) Peculiarities of diurnal vertical migration of zooplankton in Polar Seas. *J. Mar. Res.* **6**, 25–32. (496–7)

Boschma, H. (1938) On the teeth and some other particulars of the sperm whale (*Physeter macrocephalus* L.). *Temminckia* **3**, 151–278. (**763**)

Bourgeois-Pichat, J. (1951) Evolution générale de la population française depuis le XVIIIème siècle. *Population* **6**, 635–62. (145)

Bourne, A.G. (1966) A study of the minke whale (*Balaenoptera acutorostrata*). *Oryx J. Fauna Preserv. Soc.* **8**, 229–32. (**768**)

Bovet, J. (1968) Trails of deer mice (*Peromyscus maniculatus*) traveling on the snow while homing. *J. Mammal.* **49**, 713–25. (887)

Bowden, J. and Johnson, C.G. (in Johnson 1971). (439)

Boyce, A.J., Küchemann, C.F. and Harrison, G.A. (1967) Neighbourhood knowledge and the distribution of marriage distances. *Ann. Hum. Genet.* **30**, 335–8. (137–8)

Boyd, J.M. and Campbell, R.N. (1971) The grey seal (*Halichoerus grypus*) at North Rona, 1959 to 1968. *J. Zool., Lond.* **164**, 469–512. (**723**)

Bradbury, J.W. and Nottebohm, F. (1969) The use of vision by the little brown bat, *Myotis lucifugus*, under controlled conditions. *An. Behav.* **17**, 480–5. (586)

Brady, F. (1943) The distribution of the fauna of some intertidal sands and muds on the Northumberland coast. *J. Anim. Ecol.* **12**, 24–41. (**838**)

Bramwell, C.D. (1971) Aerodynamics of *Pteranodon. Biol. J. Linn. Soc.* **3**, 313–28. (604)

Bramwell, C.D. and Whitfield, G.R. (1974)

Biomechanics of *Pteranodon*. *Phil. Trans. R. Soc.* (B) **267**, 503–81. (604)

Brennan, A. (pers. comm.). (477)

Bristowe, W.S. (1939) *The comity of spiders.* London. (235)

British Ornithologists' Union (1971) *The status of birds in Britain and Ireland.* Blackwell, Oxford. (**206, 207**)

Broadbent, L. (1949) Factors affecting the activity of alatae of the aphids, *Myzus persicae* (Sulzer) and *Brevicoryne brassicae* (L.). *Ann. appl. Biol.* **36**, 40–62. (260)

Brookfield, H.C. (1968) New directions in the study of agricultural systems in tropical areas. *In:* Drake, E.T. (ed.) *Evolution and environment: a symposium presented on the occasion of the one hundredth anniversary of the foundation of Peabody Museum of Natural History at Yale University.* Yale University Press, New Haven and London. (**381**)

Brookfield, H.C. and Brown, P. (1963) *Struggle for land: Agriculture and group territories among the chimbu of the New Guinea highlands.* Oxford University Press, Melbourne. (125–6, 130–1)

Brooks, F.T. (1953) *Plant diseases.* (2nd ed.) Oxford University Press, London. (139, **141**)

Brown, C.H. (1960) Personal and Social characteristics associated with migrant status among adult males from rural Pennsylvania. Unpublished doctoral dissertation, Pennsylvania State University. (137)

Brown, D.H. and Norris, K.S. (1956) Observations of captive and wild Cetaceans. *J. Mammal.* **37**, 311–26. (**757, 759**)

Brown, E.S. (1962) Researches on the ecology and biology of *Eurygaster integriceps* Put. (Hemiptera, Scutelleridae) in Middle East countries, with special reference to the overwintering period. *Bull. ent. Res.* **53**, 445–514. (241)

Brown, E.S. (1963) Report on research on the sunn pest (*Eurygaster integriceps* Put.) and other wheat pentatomids in Middle East countries, 1958–1961. *Misc. (Print) Dep. tech. Co-oper. London.* No. 4, 1–46. (339)

Brown, E.S. (1965) Notes on the migration and direction of flight of *Eurygaster* and *Aelia* species (Hemiptera, Pentatomoidea) and their possible bearing on invasions of cereal crops. *J. Anim. Ecol.* **34**, 93–107. (241, 469)

Brown, E.S. (1970) Nocturnal insect flight direction in relation to the wind. *Proc. R. ent. Soc. Lond.* A, **45**, 39–43. (475)

Brown, J.H. (1968) Activity patterns of some neotropical bats. *J. Mammal.* **49**, 754–7. (559)

Brown, L. and Amadon, D. (1968) *Eagles, hawks, and falcons of the world* (2 vols). Country Life Books, Hamlyn, Middlesex. (**641**)

Brown, L.B. (1957) Applicants for assisted migration from the United Kingdom to New Zealand. *Population Studies* **11**, 86–91. (134, 136)

Brown, L.B. (1960) English migrants to New Zealand: the decision to move. *Human relations* **13**, 167–74. (136)

Brown, L.E. (1962) Home range in small mammal communities. *In:* Glass, B. (ed.) *Survey of biological progress 4.* Academic Press, New York. pp. 131–79. (220, 387)

Brown, M. (1937) Three seasons at Hawk Mountain Sanctuary. *Emergency Conserv. Com. Publ.* **61**, 10 pp. (639)

Brown, S.G. (1954) Dispersal in blue and fin whales. *Discovery Repts.* **26**, 355–84. (**755, 760, 764, 883**)

Brown, S.G. (1956) Whale marks recently recovered. *Norsk Hvalfangsttid.* **12**, 661–4. (**755, 763**)

Brown, S.G. (1957) Whale marks recovered during the antarctic whaling season 1956–7. *Norsk Hvalfangsttid.* **10**, 555–9. (**755**)

Brown, S.G. (1958) Whale marks recovered during the antarctic whaling season 1957–8. *Norsk Hvalfangsttid.* **10**, 503–7. (**755, 763, 764, 883**)

Brown, S.G. (1959) Whale marks recovered in the antarctic seasons 1955–56 and 1958–59 and in South Africa 1958 and 1959. *Norsk Hvalfangsttid.* **12**, 609–16. (**755, 764**, 883)

Brown, S.G. (1961) Observations on pilot whales in the North Atlantic. *Norsk Hvalfangsttid.* **50**, 225–54. (**757**)

Brown, S.G. (1962) A note on migration in fin whales. *Norsk Hvalfangsttid.* **1**, 13–16. (**764**, 883)

Bruderer, B. and Steidinger, P. (1972) Methods of quantitative and qualitative analysis of bird migration with a tracking radar. *In:* Galler, S.R., Schmidt-Koenig, K., Jacobs, G.J. and Belleville, R.E. (eds.) *Animal orientation and navigation.* Scientific and Technical Information Office, National Aeronautics and Space Administration, Washington, D.C. pp. 151–167. (684, 686)

Bruggen, A.C. van (1961) Observation on the migration of *Belenois aurota* (F) in the Transvaal (Lepidoptera: Pieridae). *J. ent. Soc. S. Afr.* **24**, 222–3. (**447**)

Brunner, E. de S. (1957) Population research. *In: The growth of a science: a half-century of rural sociological research in the U.S.* Harper, New York. pp. 42–63. (134, 138)

Buchbinder, G. and Clark, P. (1971) The Maring people of the Bismarck ranges of New Guinea: some physical and genetic characteristics. *Hum. Biol. Oceania* **1**, 121–33. (130)

Buckner, C.H. (1966) Populations and ecological relationships of shrews in tamarack bogs of southeastern Manitoba. *J. Mammal.* **47**, 181–94. (**309**)

Bullock, T.H. and Ridgway, S.H. (1972) Neurophysiological findings relevant to echolocation in marine animals. *In*: Galler, S.R., Schmidt-Koenig, K., Jacobs, G.J. and Belleville, R.E. (eds.) *Animal orientation and navigation.* Scientific and Technical Information Office, National Aeronautics and Space Administration, Washington, D.C. pp. 373–95. (766)

Bureau of Commercial Fisheries (1969) Fur seal investigations 1966. *U.S. Fish. Wildl. serv. Spec. Sci. Rep. Fish. 584.* 1–123. (110)

Burnet, A.M.R. (1965) Observations on the spawning migrations of *Galaxias attenuatus*. *N.Z. Jl Sci. Technol.* **8**, 79–87. (822)

Burns, J.J. (1970) Remarks on the distribution and natural history of pagophilic pinnipeds in the Bering and Chukchi seas. *J. Mammal.* **51**, 445–54. (720, **727**, 729)

Burrell, H. (1927) *The platypus.* Angus and Robertson, Sydney. (178, 179–80)

Burt, W.H. (1940) Territorial behaviour and populations of some small mammals in southern Michigan. *Misc. Publ. Mus. Zool., Univ. Michigan* no. 45, 1–58. (96, **310**)

Burton, J.A. (ed.) (1973) *Owls of the world: their evolution, structure and ecology.* Eurobook Ltd, Netherlands. (**663**)

Burton, M. (1971) *The observer's book of wild animals.* (Rev. ed.). Warne, London. (**570**, **582**, **583**, 586, **595**, **596**, **598**, **599**, **601**, **602**)

Busnel, R-G., and Dziedzic, A. (1966) Acoustic signals of the pilot whale, *Globicephala melaena* and of the porpoises, *Delphinus delphis* and *Phocoena phocoena*. *In*: Norris, K.S. (ed.) *Whales, dolphins and porpoises.* Univ. of California Press, Berkeley and Los Angeles. pp 607–46. (766)

Butner, A and Brattstrom, B.H. (1960) Local movement in *Menidia* and *Fundulus*. *Copeia* **1960**, 139–41. (830)

Cagle, F.R. (in Oliver 1955). (194)

Caldwell, D.K., Caldwell, M.C. and Rice, D.W.

(1966) Behavior of the sperm whale, *Physeter catodon* L. *In*: Norris, K.S. (ed.) *Whales, dolphins and porpoises.* Univ. of California Press, Berkeley and Los Angeles. pp. 677–717. (**301**, 746–8, 759, **763**, 766–7)

Caldwell, M.C. and Caldwell, D.K. (1966) Epimeletic (care-giving) behavior in Cetacea. *In*: Norris, K.S. (ed.) *Whales, dolphins and porpoises.* Univ. of California Press, Berkeley and Los Angeles. pp. 755–89. (767)

Caldwell, R.L. and Rankin, M.A. (in Dingle 1972). (352)

Calhoun, A.J. (1952) Annual migrations of California Striped Bass. *Calif. Fish and Game* **38**, 391–403. (196, **309**, **795**)

Calhoun, J.B. (1963) *The ecology and sociology of the Norway rat.* U.S. Dept. Health, Educ., Welfare. Public Health Service. (**310**)

Cameron, A.W. (1970) Seasonal movements and diurnal activity rhythms of the grey seal, *Halichoerus grypus*. *J. Zool., Lond.* **161**, 15–23. (**723**)

Cappieri, M. (1970) The racial homogeneity of the Andamanese: Part I. *Mankind Quart.* **10**, 199–212. (118, 120, **286**)

Carlisle, D.B. and Denton, E.J. (1959) On the metamorphosis of the visual pigments of *Anguilla anguilla* (L.). *J. mar. biol. Ass. U.K.* **38**, 97–102. (823)

Carr, A. (1952) *Handbook of turtles. The turtles of the United States, Canada, and Baja California.* Cornell Univ. Press. Ithaca, New York. (194, 771)

Carr, A. (1965) The navigation of the Green Turtle. *Sci. Amer.* **212**, 78–86. (**773**)

Carr, A. (1967) Adaptive aspects of the scheduled travel of *Chelonia*. *In: Storm, R.M. (ed.) Animal orientation and navigation.* Proc. 27th Annual Biol. Coll., Oregon State Univ. Press, Corvallis. pp. 35–55. (**773**)

Carr, A. (1968) *The turtle: a natural history of sea turtles.* Cassell, London. (771, **773**, 775, **778**, 860)

Carr, A. (1972) The case for long-range chemoreceptive piloting in *Chelonia*. *In*: Galler, S.R., Schmidt-Koening, K., Jacobs, G.J. and Belleville, R.E. (eds.) *Animal orientation and navigation.* Scientific and Technical Information Office, National Aeronautics and Space Administration, Washington, D.C. pp. 469–83. (193, **194**, 771, **773**, 860)

Carr, A. and Coleman, P.J. (1974) Seafloor spreading theory and the odyssey of the green turtle. *Nature, Lond.* **249**, 128–30. (771, 775, **777**)

Carr, A. and Hirth, H. (1962) The ecology and migrations of sea turtles. 5. Comparative features of isolated green turtle colonies. *Amer. Mus. Novitates* no. 2091, 1–42. (194, **773**)

Carr, A. and Ogren, L. (1959) The ecology and migrations of sea turtles. 3. *Dermochelys* in Costa Rica. *Amer. Mus. Novitates* **1958**, 1–29. (**778**)

Carr, A. and Ogren, L. (1960) The ecology and migrations of sea turtles. 4. The green turtle in the Caribbean Sea. *Bull. Amer. Mus. Nat. His.* **121**, 1–48. (**773**, 860)

Carr, N. (1962) *Return to the wild*. Fontana Books, London. (**295**)

Carrington, R. (1958) *Elephants*. Penguin Books, Harmondsworth. (550)

Carr-Saunders, A.M. (1922) *The population problem*. Oxford University Press, Oxford. (118)

Caspers, H. (1961) Beobachtungen über hebensraum und Schwärmperiodizitat des Palolowurmes *Eunice viridis* (Polychaeta: Eunicidae). 1. Die tages- und mondzeitliche Konstanz des Schwärmens; Bibliographie. *Int. Res. Ges. Hydrobiol.* **46**, 175–83. (235)

Catchpole, C.K. (1974) Habitat selection and breeding success in the reed warbler (*Acrocephalus scirpaceus*). *J. Anim. Ecol.* **43**, 363–80. (378)

Cavé, A.J., Bol, C. and Speek, G. (1974) Experiments on discrimination by the starling between geographical locations. *Progress Report 1973, Institute of Ecological Research, Royal Netherlands Academy of Arts and Sciences.* p. 82. (607)

Central Statistical Office (1972a) *Abstract of regional statistics 1972 No. 8.* HMSO, London. (**143**)

Central Statistical Office (1972b) *Annual abstract of statistics 1972 No. 109.* HMSO, London. (**143**, **284**)

Central Statistical Office (1973) *Abstract of regional statistics 1973 No. 9.* HMSO, London. (**143**, 144)

Chalmers, N.R. (1968) Group composition, ecology, and daily activities of free living mangabeys in Uganda. *Folia Primat.* **8**, 247–62. (**295**, 307)

Chapin, J.P. (in Dorst 1962). (**665**, **675**)

Chapman, J.A. (1971) Orientation and homing of the brush rabbit (*Sylvilagus bachmani*). *J. Mammal.* **52**, 686–99. (887)

Charles-Dominique, P. (1971) Eco-ethologie des prosimiens du Gabon. *Biol. Gabonica* **7**, 121–228. (92)

Chernyi, E.I. (1966) (The distribution of baleen whales in the Antarctic as a function of hydrological conditions) (In Russian). *Izv Tikhookean*

Nauch-Issled Inst. Ryb. Khoz. Okeanogr. **58**, 223–8. (746, 764)

Chittleborough, R.G. (1953) Aerial observations on the Humpback Whale, *Megaptera nodosa* (Bonnaterre), with notes on other species. *Australian J. Marine and Freshw. Res.* **4**, 219–26. (**755**, **759**, **764**, 768)

Chittleborough, R.G. (1956) Southern right whale in Australian waters. *J. Mammal.* **37**, 456–7. (**753**)

Chittleborough, R.G. (1958) The breeding cycle of the female humpback whale, *Megaptera nodosa* (Bonnaterre). *Aust. J. Mar. Freshwater Res.* **9**, 1–18. (**764**)

Chittleborough, R.G. (1959) Australian marking of humpback whales. *Norsk Hvalfangsttid.* **48**, 47–55. (183, **318**, 329)

Chittleborough, R.G. (1965) Dynamics of two populations of the humpback whale, *Megaptera novaeangliae* (Borowski). *Aust. J. mar. Freshwat. Res.* **16**, 33–128. (183, **318**, 329, 748, 760, **764**)

Christian, J.J. (1956) The natural history of a summer aggregation of the big brown bat, *Eptesicus fuscus fuscus. Amer. Midl. Natur.* **55**, 66–95. (562, **603**)

Chugunkov, D.I. (1966) (Distribution of herds of fur seals between Bering and Mednyi Islands (Commander Islands)) (In Russian). *Izv Tikhookean Nauch-Issled Inst. Ryb. Khoz. Okeanogr.* **58**, 15–21. (105, **106**, 110)

Chugunkov, D.I. and Prokhorov, V.G. (1966) (New information about the overwintering of fur seals in the Bering Sea) (In Russian). *Izv Tikhookean Nauch-Issled Inst. Ryb. Khoz. Okeanogr.* **58**, 233–4. (**735**)

Churcher, C.S. (1973) *Red fox.* Canadian Wildlife Service, Hinterland Who's Who. Information Canada, Ottawa. (**173**, 174)

Clark, D.P. and Davies, R.A.H. (1972) Migration and plague dynamics of Australian plague locust. *Ann. Rept. C.S.I.R.O. Div. Ent.* CSIRO, Canberra. pp. 108–9. (454)

Clark, J.D. (1962) The spread of food production in sub-Saharan Africa. *J. African History* **3**, 211–28. (125)

Clark, L.B. and Hess, W.N. (1940) Swarming of the Atlantic palolo worm, *Leodice fucata* (Ehlers). *Pap. Tortugas Lab.* **33**, 21–70. (235)

Clarke, C.M.H. (1971) Liberations and dispersal of red deer in northern South Island districts. *N.Z. J. For. Sci.* **1**, 194–207. (169)

Clarke,   G.L.   and   Backus,   R.H.   (1956)

Measurements of light penetration in relation to vertical migration and records of luminescence of deep-sea animals. *Deep Sea Res.* **4**, 1–14. (494)

Clarke, G.L. and Wertheim, G.K. (1956) Measurements of illumination at great depths and at night in the Atlantic Ocean by means of a new bathyphotometer. *Deep Sea Res.* **3**, 189–205. (**482**)

Clarke, J.R. (1953) The hippopotamus in Gambia, West Africa. *J. Mammal.* **34**, 299–315. (179, 213, 550)

Clarke, R. (1956) Sperm whales of the Azores. *Discovery Rep.* **28**, 237–98. (763)

Clarke, R. (1965) Southern right whales on the coast of Chile. *Norsk Hvalfangsttid.* **54**, 121–8. (753)

Clarke, R. and Aguayo, L.A. (1965) Bryde's whale in the Southeast Pacific. *Norsk Hvalfangsttid.* **54**, 141–8. (**182**)

Clarke, R., Aguayo, L.A. and Paliza, G. (1964) Progress report on sperm whale research in the south-east Pacific Ocean. *Norsk Hvalfangsttid.* **53**, 297–302. (**763**)

Clarke, R., Aguayo, L.A. and Paliza, O. (1968) Sperm whales of the Southeast Pacific. *Hvalradets Skr.* **51**, 5–80. (**301**, **763**)

Cochran, W.W. (1972) Long-distance tracking of birds. *In:* Galler, S.R., Schmidt-Koenig, K., Jacobs, G.J. and Belleville, R.E. (eds.) *Animal orientation and navigation.* Scientific and Technical Information Office, National Aeronautics and Space Administration, Washington, D.C. pp. 39–59. (191)

Cochran, W.W., Montgomery, G.G. and Graber, R.R. (1967) Migratory flights of *Hylocichla* Thrushes in spring: a telemetry study. *Living bird* **6**, 213–25. (**685**)

Cockbain, A.J. (1961) Fuel utilization and duration of tethered flight in *Aphis fabae* Scop. *J. exp. Biol.* **38**, 163–74. (262, **263**)

Cockrum, E.L. (1969) Migration in the guano bat, *Tadarida brasiliensis. Univ. Kans. Mus. Natur. Hist. Misc. Public.* **51**, 303–36. (580, **591**)

Cohen, D. (1967) Optimization of seasonal migratory behaviour. *Am. Nat.* **101**, 1–17. (73)

Cole, G.F. (1971) Preservation and management of grizzly bears (*Ursus arctos*) in Yellowstone National Park. *Bio. Science.* **21**, 858–64. (887)

Cole, L.C. (1954) The population consequences of life history phenomena. *Quart. Rev. Biol.* **29**, 103–37. (37, 76, 807)

Collman, J.R. and Croxall, J.P. (1967) Spring migration at the Bosphorus. *Ibis* **109**, 359–72. (648)

Colman, J.S. and Segrove, F. (1955) Tidal plankton over Stoupe Beck Sands, Robin Hood's Bay. *J. Anim. Ecol.* **24**, 445–62. (486)

Common, I.F.B. (1954) A study of the ecology of the adult bogong moth, *Agrotis infusa* (Boisd.) (Lepidoptera: Noctuidae), with special reference to its behaviour during migration and aestivation. *Aust. J. Zool.* **2**, 223–63. (**257**, **472**)

Constantine, D.G. (1966) Ecological observations on lasiurine bats in Iowa. *J. Mammal.* **47**, 34–41. (**576**, **578**)

Constantine, D.G. (1967) Activity patterns of the Mexican Free-tailed bat. *Univ. New Mexico Pub. Biol.* **7**, 8–79. (570, 588, **591**)

Coon, C.S. (1965) *The living races of man.* Knopf, New York. (117, 121, 125, 126, 130)

Cope, J.B., Koontz, K. and Churchwell, E. (1961) Notes on homing of two species of bats, *Myotis lucifugus* and *Eptesicus fuscus. Proc. Indiana Acad. Sci.* **70**, 270–4. (893)

Corbet, G.B. (1966) *The terrestrial mammals of Western Europe.* G.T. Fowlis and Co. Ltd, London. (**387**, **392**, **398**)

Corbet, P.S. (1962) *A biology of Dragonflies.* Witherby, London. (198)

Coulson, J.C. (1966) The influence of the pair-bond and age on the breeding biology of the kittiwake gull, *Rissa tridactyla. J. Anim. Ecol.* **35**, 269–79. (121, 187, 324)

Coulson, J.C. (1971) Competition for breeding sites causing segregation and reduced young production in colonial animals. *Proc. Adv. Study Inst. Dynamics Numbers Popul. (Oosterbeek)* 257–68. (187, **322**, 324)

Coulson, J.C. and Horobin, J.M. (1972) The annual re-occupation of breeding sites by the fulmar. *Ibis* **114**, 30–42. (187)

Cousteau, J. (film). (526)

Cowan, H.I. (1961) *British emigration to British North America: the first hundred years.* (rev. and enlarged ed.). University of Toronto Press. (139, **140–1**)

Cowan, I. McT. and Guiguet, C.J. (1965) The mammals of British Columbia. *British Columbia Provincial Mus.* **11**, 1–141. (**574**, **578**)

Cox, G.W. (1961) The relation of energy requirements of tropical finches to distribution and migration. *Ecology* **42**, 253–66. (660)

Cox, G.W. (1968) The role of competition in the evolution of migration. *Evolution* **22**, 180–92. (698, 705)

Craighead, F.C., Craighead, J.J., Cote, C.E., and Buechner, H.K. (1972) Satellite and ground radiotracking of elk. *In:* Galler, S.R., Schmidt-Koenig, K., Jacobs, G.J. and Belleville, R.E. (eds.) *Animal orientation and navigation.* Scientific and Technical Information Office, National Aeronautics and Space Administration, Washington, D.C. pp. 99–111. (**544**)

Cranbrook, Earl of, and Barrett, H.G. (1965) Observations on noctule bats (*Nyctalus noctula*) captured while feeding. *Proc. zool. Soc. Lond.* **144**, 1–24. (**581**)

Crofton, H.D. (1948) The ecology of immature phases of Trichostronglye nematodes. I. The vertical distribution of infective larvae of *Trichostrongylus retortaeformis* in relation to their habitat. *Parasitology* **39**, 17–25. (18, 235)

Crook, J.H. (1970) The socio-ecology of primates. *In:* Crook, J.H. (ed.) *Social behaviour in birds and mammals.* Academic Press, London. pp. 103–66. (92, 164, 304, 390)

Cruxent, J.M. (1968) Theses for meditation on the origin and dispersion of man in South America. *In: Biomedical challenges presented by the American Indian.* Pan American Health Organization, Scientific Publication No. 165. pp. 11–16. (114, **565**)

Csordas, S.E. (1962) The Kerguelen fur seal on Macquarie Island. *Vict. Nat.* **79**, 226–9. (**106**)

Curry-Lindahl, K. (1962) The irruption of the Norway lemming in Sweden during 1960. *J. Mammal.* **3**, 171–84. (413)

Curry-Lindahl, K. (1970) Breeding biology of the Baltic grey seal, (*Halichoerus grypus*). *Zool. Gart.* **38**, 16–29. (**723**)

Cushing, D.H. (1951) The vertical migrations of planktonic crustacea. *Biol. Rev.* **26**, 158–92. (485, 491, 493–4, 497, **504**)

Dall, W. (1958) Observations on the biology of the greentail prawn, *Metapenaeus mastersii* (Haswell) (Crustacea Decapoda: Penaeidae). *Aust. J. mar. Freshwat. Res.* **9**, 111–34. (**831**)

Dalquest, W.W. (1955) Natural history of the vampire bats of eastern Mexico. *Am. Midl. Nat.* **53**, 79–87. (**565**)

Damas, D. (1968) The diversity of Eskimo societies. *In:* Lee, R.B. and DeVore, I. (eds.) *Man the hunter.* Aldine, Chicago. pp. 111–17. (554)

Darling, F. (1937) *A herd of red deer.* Oxford University Press, London. (541)

Dart, R.A. (1964) The ecology of South African man-apes. *In:* Davis, D.H.S. (ed.) *Ecological studies in Southern Africa.* Dr. W. Junk, The Hague. pp. 49–66. (115)

Davenport, D. and Dimock, V.R. (in Evans 1971). (199)

Davey, J.T. (1959) The African migratory locust (*Locusta migratoria migratorioides* Rch. and Frm., Orth.) in the Central Niger Delta. Part 2. The ecology of *Locusta* in the semi-arid lands and seasonal movements of populations. *Locusta* No. 7, 180 pp. (225, 258, 263)

David, W.A.L. and Gardiner, B.O.C. (1961) The mating behaviour of *Pieris brassicae* (L.) in a laboratory culture. *Bull. ent. Res.* **52**, 263–80. (**334**)

David, W.A.L. and Gardiner, B.O.C. (1962) Oviposition and the hatching of the eggs of *Pieris brassicae* (L.) in a laboratory culture. *Bull. ent. Res.* **53**, 91–109. (333, **335**, 336)

Davidson, J. (1927) The biological and ecological aspect of migration in aphids. *Sci. Prog., Lond.* **21**, 641–58; **22**, 57–69. (231)

Davies, J.L. (1957) The geography of the gray seal. *J. Mammal.* **38**, 297–310. (721, **723**)

Davies, J.L. (1958) Pleistocene geography and the distribution of the northern pinnipeds. *Ecology* **39**, 97–113. (733)

Davies, J.L. (1963) The antitropical factor in cetacean speciation. *Evolution* **17**, 107–16. (745, 752)

Davies, L. (1957) A study of the age of females of *Simulium ornatum* Mg. (Diptera) attracted to cattle. *Bull. ent. Res.* **48**, 535–52. (352)

Davis, A.W. (in Netting 1936). (523)

Davis, J. and Williams, L. (1957) Irruptions of the Clark Nutcracker in California. *Condor* **59**, 297–307. (208–9)

Davis, J. and Williams, L. (1964) The 1961 irruption of the Clark's nutcracker in California. *Wilson Bull.* **76**, 10–18. (208–9)

Davis, R. (1966) Homing performance and homing ability in bats. *Ecol. Monogr.* **36**, 201–37. (158, 890, 892, 893–4)

Davis, R. and Cockrum, E.L. (1962) Repeated homing exhibited by a female pallid bat. *Science* **137**, 341–2. (893)

Davis, R.B., Herreid, C.F. II, and Short, H.L. (1962) Mexican free-tailed bats in Texas. *Ecol. Monogr.* **32**, 311–46. (577, 588–9, **591**, 893)

Davis, W.B. (1960) *The mammals of Texas.* Game and Fish commission, Austin. (**576**)

Davis, W.H. (1959) Disproportionate sex ratios in hibernating bats. *J. Mammal.* **40**, 16–19. (575)

Davis, W.H. (1966) Population dynamics of the bat, *Pipistrellus subflavus*. *J. Mammal.* **47**, 383–96. (**312**)

Davis, W.H. (1970) Hibernation: ecology and physiological ecology. *In:* Wimsatt, W.A. (ed.) *Biology of bats.* Vol. I. Academic Press, London. pp. 266–300. (99)

Davis, W.H. and Barbour, R.W. (1970) Homing in blinded bats (*Myotis sodalis*). *J. Mammal.* **51**, 182–4. (892)

Davis, W.H., Barbour, R.W. and Hassell, M.D. (1968) Colonial behaviour of *Eptesicus fuscus*. *J. Mammal.* **49**, 44–50. (588, **603**)

Davis, W.H. and Hitchcock, H.B. (1965) Biology and migrations of the bat, *Myotis lucifugus*, in New England. *J. Mammal.* **46**, 296–313. (**569**, **584**)

Davis, W.H. and Lidicker, W.Z. (1956) Winter range of the red bat, *Lasiurus borealis*. *J. Mammal.* **37**, 280–1. (573, **576**)

Davydov, A. and Keskpaik, J. (1972) (The winter energy metabolism in the song thrush and blackbird) (In Russian with English summary). *Communs Baltic Common Study Bird Migr.* **7**, 164–70. (638)

Dawbin, W.H. (1959) New Zealand and South Pacific whale marking and recoveries to the end of 1958. *Norsk Hvalfangsttid.* **5**, 213–38. (183, 760)

Dawbin, W.H. (1964) Movements of humpback whales marked in the south-west Pacific Ocean, 1952 to 1962. *Norsk Hvalfangsttid.* **3**, 68–78. (183, **318**)

Dawbin, W.H. (1966) The seasonal migratory cycle of humpback whales (Balaenopteridae). *In:* Norris, K.S. (ed.) *Whales, dolphins, and porpoises.* University of California Press, Berkeley and Los Angeles. pp. 145–70. (748, 759, 764, 768, 770)

Degerbøl, M. and Freuchen, P. (1935) Mammals. Report of the 5th Thule Expedition, 1921–24. *The Danish expedition to Arctic North America in charge of Knud Rasmussen Ph.D.* **2**, no. 4–5, 278 pp. (161, 223, **727**)

Delsman, H.C. (1922) Op de Java Zee. *Trop. Natuur.* **11**, 17. (195)

DeVore, I. and Hall, K.R.L. (1965) Baboon ecology. *In:* DeVore, I. (ed.) *Primate behaviour: field studies of monkeys and apes.* Holt, Rinehart and Winston, New York. (**295**)

Dickinson, G.C. (1958) The nature of rural population movement: an analysis of seven Yorkshire Parishes based on electoral returns from 1931–1954. *Yorkshire Bulletin of Economic and Social Research.* **10**, 95–108. (137)

Dickson, H.R.P. (1949) *The Arab of the desert.* George Allen and Unwin, Ltd, London. (153, 540, 541, 551, 887)

Digby, P.S.B. (1953) Plankton production in Scoresby Sound, East Greenland. *J. Anim. Ecol.* **22**, 289–322. (**483**)

Dingle, H. (1965) The relation between age and flight activity in the milkweed bug, *Oncopeltus*. *J. exp. Biol.* **42**, 269–83. (**340**, 351)

Dingle, H. (1968a) Life history and population consequences of density, photo-period, and temperature in a migrant insect, the milkweed bug, *Oncopeltus*. *Am. Nat.* **102**, 149–63. (225)

Dingle, H. (1968b) The influence of environment and heredity on flight activity in the milkweed bug, *Oncopeltus*. *J. exp. Biol.* **48**, 175–84. (225)

Dingle, H. (1972) Migration strategies of insects. *Science* **175**, 1327–35. (18, 37, 75, 76)

Dixon, A.F.G. (1969) Population dynamics of the sycamore aphid, *Drepanosiphum platanoides* (Schr.) (Hemiptera: Aphididae): migratory and trivial flight activity. *J. Anim. Ecol.* **38**, 585–606. (46, 198–9, 225, 228, 229–30, 246)

Dixon, A.F.G., Burns, D.M. and Wangboonkong, S. (1968) Migration in aphids: response to current adversity. *Nature, Lond.* **220**, 1337–8. (231)

Dixon, C.C. (1933) Some observations on the albatrosses and other birds of the Southern Oceans. *Trans. roy. Canadian Inst.* **19**, 117–39. (**624**)

Dolnik, V.R. and Blyumental, T.I. (1967) Autumnal premigratory and migratory periods in the chaffinch (*Fringilla coelebs coelebs*) and some other temperate-zone passerine birds. *Condor* **69**, 435–68. (189, 627, 643, 693, 696)

Dolnik, V.R. and Gaurilov, V.M. (1972) (Photoperiodic control of annual cycles in the Chaffinch, a temperate-zone migrant) (In Russian, with English summary). *Zool. Zh.* **51**, 1685–96. (627, 647)

Dorsett, D.A. (1962) Preparation for flight by hawkmoths. *J. exp. Biol.* **39**, 579–88. (252, 253)

Dorst, J. (1962) *The migrations of birds.* Heinemann, London. (3, 6, 13, 20, 32, **185**, **206**, 209, 262, **621**, 622, **623**, 625–6, 631, **633**, **650**, **659**, **665**, **667**, **668**, **672**, **673**, **675**, 681, 684, **687**, **688**, 689, **692**, 693, 696, 698, **703**, **704**, **857**)

Dowdy, W.W. (1944) Influence of temperature on vertical migration. *Ecology of invertebrates inhabiting different soil types* **25**, 449–60. (508)

Dowling, K. (1975) Tunny troubles. *Sunday Times* 20 April. (851)

Downes, W.L. (1964) Unusual roosting behavior in red bats. *J. Mammal.* **45**, 143–4. (**576**)

Downing, R.L., McGinnes, B.S., Petcher, R.L. & Sandt, J.L. (1969) Seasonal changes in movements of white-tailed deer. *In: White-tailed deer in the southern forest habitat (Symposium).* Southern Forest Expt. Sta., Nacogdoches, Texas. pp 19–24. (**311**)

Doxiades, C.A. (1970) Man's movement and his settlements. *Ekistics* **29**, 1–26. (119, 126, 139)

Drabble, P. (1973) Life with a stoat, and a weasel. *The Field* **243**, 1508–9. (378)

Drury, W.H. and Nisbet, I.C.T. (1964) Radar studies of orientation of songbird migrants in southeastern New England. *Bird-Banding* **35**, 69–119. (616)

Dunnet, G.M. (1962) A population study of the quokka, *Setonix brachyurus.* 2. Habitat, movements, breeding and growth. *C.S.I.R.O. Wildl. Res.* **7**, 13–32. (**309**, 313)

Duval, D.M. (1972) A record of slug movements in late summer. *J. Conch.* **27**, 505–8. (376)

Dwyer, P.D. (1962) Studies on the two New Zealand bats. *Zool. Publ. Victoria Univ. Wellington* **28**, 1–28. (556)

Dwyer, P.D. (1966) The population pattern of *Miniopterus schreibersii* (Chiroptera) in northeastern New South Wales. *Aust. J. Zool.* **14**, 1073–1137. (158, 160, **309**, 312, 564, 570, 586, **587**, 890)

Dwyer, P.D. (1969) Population ranges of *Miniopterus schreibersii* (Chiroptera) in south-eastern Australia. *Aust. J. Zool.* **17**, 665–86. (**587**, 857, 893)

Dwyer, P.D. (1970) Social organization in the bat, *Myotis adversus. Science* **168**, 1006–8. (560, 561, 588)

Easterlin, R.A. (1969) Towards a socioeconomic theory of fertility: a survey of recent research on economic factors in American fertility. *In: Behrman, S.J., Corsa, L., and Freedman, R. (eds.) Fertility and family planning: a world view.* University of Michigan Press, Ann Arbor. pp. 127–56. (144)

Eberhardt, L.L. (1970) Correlation, regression and density dependence. *Ecology* **51**, 306–10. (268)

Edelstam, C. and Palmer, C. (1950) Homing behaviour in gastropodes. *Oikos* **2**, 259–70. (376, 518, 878)

Edgar, W.D. (1971) The life-cycle, abundance and seasonal movement of the wolf spider, *Lycosa (Pardosa) lugubris*, in central Scotland. *J. Anim. Ecol.* **40**, 303–22. (**517**)

Edwards, C.A. and Lofty, J.R. (1972) *Biology of earthworms.* Chapman and Hall, London. (505, **507**)

Edwards, D.K. (1962) Laboratory determinations of the daily flight times of separate sexes of some moths in naturally changing light. *Can. J. Zool.* **40**, 511–30. (337)

Edwards, R.L. (1958) Movements of individual members in a population of the shore crab, *Carcinus maenas* L., in the littoral zone. *J. Anim. Ecol.* **27**, 37–45. (830, 839)

Edwards, R.Y. and Ritcey, R.W. (1956) The migrations of a moose herd. *J. Mammal.* **37**, 486–94. (**529**)

Efford, I.E. (1965) Aggregation in the sand crab, *Emerita analoga* (Stimpson). *J. Anim. Ecol.* **34**, 63–75. (**826**)

Egsbaek, W. and Jensen, B. (1963) Results of bat banding in Denmark. *Vidensk, Medd. Dansk. Naturh. Foren.* **25**, 269–96. (**595**, **596**)

Eisenberg, J.F., Muckenhirn, N.A. and Rudran, R. (1972) The relation between ecology and social structure in primates. *Science* **176**, 863–74. (117, 118, 131, 162)

Eisenstadt, S.N. (1955) *The absorption of immigrants: a comparative study based mainly on the Jewish community in Palestine and the State of Israel.* The Free Press, Glencoe, Illinois. (20)

Eisentraut, M. (1936) Ergebnisse der Fledermausberingung nach dreijähriger Versuchzeit. *Zeitsch. Morph. Okol. Tiere* **31**, 1–26. (560, **581**, 595)

Eisentraut, M. (1960) Die Wanderwege der in der Mark Brandenburg beringten Mausohren. *Bonner Zool. Beitr.* **11**, 112–23. (**595**)

Ekman, S. (1953) *Zoogeography of the sea.* Sidgwick and Jackson, London. (**815**)

Ellefson, J.O. (1968) Territorial behaviour in the common white-handed gibbon, *Hylobates lar* Linn. *In:* Jay, P. (ed.) *Primates: studies in adaptation and variability.* Holt, Rinehart and Winston, New York. (**296**)

Ellerman, J.R. and Morrison-Scott, T.C.S. (1966) *Checklist of Palaearctic and Indian mammals 1758–1946.* (2nd ed.) British Museum (Natural History), London. (**570, 572, 595, 597**)

Ellerman, J.R., Morrison-Scott, T.C.S. and Hayman, R.W. (1953) *Southern African mammals*

1758 to 1951: a reclassification. British Museum, (Natural History), London. (**587**)

Elmhirst, R. (1922) Habits of *Echinus esculentus*. *Nature, Lond*. **110**, 667. (**840**)

Eltringham, S.K. (1971) *Life in mud and sand*. English Universities Press, London. (823, **838**)

Elwell, A.S. (1962) Blue jay preys on young bats. *J. Mammal*. **43**, 434. (**576**)

Emlen, S.T. (1967) Migratory orientation in the indigo bunting (*Passerina cyanea*). Part I. Evidence for the use of celestial cues. *Auk* **84**, 309–42. (864)

Emlen, S.T. (1970) Celestial rotation: its importance in the development of migratory orientation. *Science* **170**, 1198–1201. (609)

Emlen, S.T. (1971) Comments at a meeting. *Ann. N.Y. Acad. Sci*. **188**, 331–58. (862)

Emlen, S.T. (1972) The ontogenetic development of orientation capabilities. *In:* Galler, S.R., Schmidt-Koenig, K., Jacobs, G.J. and Belleville, R.E. (eds.) *Animal orientation and navigation*. Scientific and Technical Information Office, National Aeronautics and Space Administration, Washington, D.C. pp. 191–210. (609, 858, 864, **918**)

Enright, J.T. (1963) Endogenous tidal and lunar rhythms. *Proc. XVI Int. Cong. Zool*. **4**, 355–9. (830)

Enright, J.T. (1972) When the beachhopper looks at the Moon: the Moon-compass hypothesis. *In:* Galler, S.R., Schmidt-Koenig, K., Jacobs, G.J. and Belleville, R.E. (eds.) *Animal orientation and navigation*. Scientific and Technical Information Office, National Aeronautics and Space Administration, Washington, D.C. pp. 523–55. (856)

Erard, C. and Larigauderie, F. (1972) Observations sur la migration prénuptiale dans l'ouest de la Lybie (Tripolitaine et plus particulièrement Fezzan). *Oiseau Revue fr. Orn*. **42**, 81–169; 253–84. (648)

Eriksson, K. (1970a) Wintering and autumn migration ecology of the brambling (*Fringilla montifringilla*). *Sterna* **9**, 77–90. (635–6)

Eriksson, K. (1970b) The autumn migration and wintering ecology of the Siskin, *Carduelis spinus*. *Ornis fenn*. **47**, 52–68. (635)

Evans, P.R. (1966) Migration and orientation of passerine night migrants in northeast England. *J. Zool., Lond*. **150**, 319–69. (616, 628)

Evans, P.R. (1972) Information on bird navigation obtained by British long-range radars. *In:* Galler,

S.R., Schmidt-Koenig, K., Jacobs, G.J. and Belleville, R.E. (eds.) *Animal orientation and navigation*. Scientific and Technical Information Office, National Aeronautics and Space Administration, Washington, D.C. pp. 139–49. (188, 189, 616, 617, 618, 632, 642, 645, 676, 684, **919**)

Evans, S.M. (1971) Behavior in polychaetes. *Quart. Rev. Biol*. **46**, 379–405. (199, 235)

Evans, W.E. (1971) Orientation behavior of delphinids: radio telemetric studies. *Ann. N.Y. Acad. Sci*. **188**, 142–60. (**881**)

Evans, W.W. and Prescott, J.H. (1962) Observations of the sound production capabilities of the bottlenose porpoise: a study of whistles and clicks. *Zoologica* **47**, 121–8. (767)

Ewer, R.F. (1968) *Ethology of mammals*. Logos Press, London. (378, 400)

Fairhurst, C. (1974) The adaptive significance of variations in the life cycles of schizophylline millipedes. *Symp. zool. Soc. Lond*. **No. 32**, 575–87. (343)

Fairley, J.S. (1969) Tagging studies of the red fox (*Vulpes vulpes* (L.)) in north-east Ireland. *J. Zool., Lond*. **159**, 527–32. (**173**)

Fairley, J.S. (1970) More results from tagging studies of foxes. *Ir. Nat. J*. **16**, 392–3. (**173**)

Farran, G.P. (1947) Vertical distribution of plankton (*Sagitta, Calanus* and *Metridea*) off the south coast of Ireland. *Proc. Roy. Irish. Acad*. **51B**, 121–36. (**503**)

Felten, H. and Klemmer, K. (1960) Fledermaus-Beringung im Rhein-Main-Lahn-Gebiet 1950–1959. *Bonner Zool. Beitr*. **11**, 166–88. (**570, 582, 595, 601**)

Fenton, M.B. (1969) Summer activity of *Myotis lucifugus* (Chiroptera: Vespertilionidae) at hibernacula in Ontario and Quebec. *Can. J. Zool*. **47**, 597–602. (160, **569**, 571, **584**)

Ferguson, D.E. (1971) The sensory basis of orientation in amphibians. *Ann. N.Y. Acad. Sci*. **188**, 30–6. (861, 884–5, 921)

Ferguson-Lees, I.J. (1968) Serins breeding in southern England. *Br. Birds* **61**, 87–8. (192)

Field, C.R. (1970) A study of the feeding habits of the hippopotamus (*Hippopotamus amphibius* Linn.) in the Queen Elizabeth National Park, Uganda, with some management implications. *Zoologica Africana* **5**, 71–86. (179, 550)

Field, C.R. (1971) Elephant ecology in the Queen Elizabeth National Park, Uganda. *East Afr. Wildl. J*. **9**, 99–123. (550)

Findley, J.S. and Jones, C. (1964) Seasonal distribution of the Hoary bat. *J. Mammal.* **45**, 461–70. (**578**)

Fischer, E. (1955) Insectenkost beim Menschen. *Z. Ethn.* **80**, 1–37. (114)

Fiscus, C.H. and Baines, G.A. (1966) Food and feeding behavior of Steller and California sea lions. *J. Mammal.* **47**, 195–200. (**729, 731**)

Fisher, H.D. (in Sergeant 1965). (880)

Fisher, R.A. (1930) *The genetical theory of natural selection.* Oxford University Press. (37, 75)

Fleming, T. H. (1971) Population ecology of three species of neotropical rodents. *Misc. Publ. Mus. Zool. Univ. Mich.* **143**, 5–77. (**310**)

Flower, W.H. (1883) On the characters and divisions of the family Delphinidae. *Proc. Zool. Soc. London* **1883**, 466–513. (745)

Flowerdew, J.R. (1974) Field and laboratory experiments on the social behaviour and population dynamics of the wood mouse (*Apodemus sylvaticus*). *J. Anim. Ecol.* **43**, 499–512. (**310**)

Folk, G.E. (1940) Shift of population among hibernating bats. *J. Mammal.* **21**, 306–15. (98)

Formozov, A.N. (1966) Adaptive modifications of behavior in mammals of the Eurasian steppes. *J. Mammal.* **47**, 208–23. (531, 551)

Formozov, A.N. (in Dorst 1962). (208–9)

Forster, G.H. (in Dorst 1962). (**689**)

Frädrich, H. (1965) Zur Biologie und Ethologie des Warzenschweines (*Phacochoerus aethiopicus* Pallas), unter Berücksichtigung des Verhalten anderer Suiden. *Z. Tierpsychol.* **22**, 328–93. (213)

Frank, H. (1960) Beobachtungen an Fledermäusen in Höhlen der Schwäbischen Alb unter besonderer Berücksichtigung der Mopsfledermaus (*Barbastella barbastellus*). *Bonner Zool. Beitr.* **11**, 143–9. (**595, 601**)

Fraser, J. (1962) *Nature adrift: the story of marine plankton.* Fowlis, London. (488)

French, R.A. (1964) Migration records, 1962. *Entomologist* **97**, 121–8. (426)

Frenzel, B. and Troll, C. (in Moreau 1972). (**699**)

Fretter, V. and Graham, A. (1962) *British prosobranch molluscs: their functional anatomy and ecology.* Ray Society, London. (**827**)

Fretwell, S.D. (1972) *Populations in a seasonal environment.* Princeton University, New Jersey. (39, 68, 69, 192, 211, 402, **405**)

Friedman, H. (1969) The violent Sun. *In: The amazing world of nature.* The Reader's Digest Association Ltd, London. pp. 299–301. (52)

Frisch, K. von (1954) *Sprechende Tänze im Bienenvolk.* Festrede in der Bayer. Akad. Wiss. (**147**)

Frisch, K. von (1967) *The dance language and orientation of bees.* Oxford University Press, London. (146, **147**, 425, 873–7)

Frost, W.E. and Brown, M.E. (1967) *The trout.* Collins, London. (**845**)

Fryer, J.C.F. (1937) Recent caterpillar plagues in Great Britain. *Nature, Lond.* **140**, 94–5. (232)

Fujino, K. (1962) Blood types of some species of antarctic whales. *Am. Nat.* **96**, 205–10. (760)

Fullard, H. (ed.) (1959) *Philips' new school atlas* (48th ed.). George Philip and Son Ltd, The London Geographical Institute, London. (**447**, 475, **592, 659, 898, 900, 902, 914**)

Gabbut, P.D. (1969) Life-histories of some British pseudoscorpions inhabiting leaf litter. *In:* Sheals, J.G. (ed.) *The soil ecosystem.* Systematics Association Publication No. 8. pp. 229–35. (**511**)

Gabbutt, P.D. (1970) Sampling problems and the validity of life history analyses of pseudoscorpions. *J. nat. Hist.* **4**, 1–15. (505, 508, **513**)

Galbraith, R. (1965) Homing behavior in the limpets *Acmaea digitalis* and *Lottia gigantea. Am. Midl. Nat.* **74**, 245–6. (199, 878)

Gard, R. (1961) Effects of beaver on trout in Sagehen Creek, California. *J. Wildl. Mgmt.* **25**, 221–42. (797)

Gardner, P.M. (1968) Comments at a meeting. *In:* Lee, R.B. and DeVore, I. (eds.) *Man the hunter.* Aldine, Chicago. p. 341. (119)

Gardner, R. and Heider, K.G. (1969) *Gardens of war: life and death in the New Guinea stone age.* Andre Deutsch, London. (125, 130–2)

Gartlan, J.S. and Brain, C.K. (1968) Ecology and social variability in *Cercopithecus aethiops* and *C. mitis. In:* Jay, P. (ed.) *Primates: studies in adaptation and variability.* Holt, Rinehart and Winston, New York. (162, **295**, 299)

Gaskin, D. E. (1968a) Analysis of sightings and catches of sperm whales (*Physeter catodon* L.) in the Cook Strait area of New Zealand in 1963–4. *N.Z.J. Mar. Freshwat. Res.* **2**, 260–72. (**301, 703**)

Gaskin, D.E. (1968b) The New Zealand Cetacea. *N.Z. Fish. Res. Lab. Bull.* **n.s. 1**, 1–92. (**180, 182, 753, 755, 757, 759, 763, 768, 883**)

Gaskin, D.E. and Cawthorn, M.W. (1967) Diet and feeding habits of the sperm whale (*Physeter catodon* L.) in the Cook Strait region of New Zealand. *N.Z.J. Mar. Freshwat. Res.* **1**, 156–79. (746)

Gauthreaux, S.A. (1972) Flight directions of passerine migrants in daylight and darkness: a radar and direct visual study. *In:* Galler, S.R., Schmidt-Koenig, K., Jacobs, G.J. and Belleville, R.E. (eds.) *Animal orientation and navigation.* Scientific and Technical Information Office, National Aeronautics and Space Administration, Washington, D.C. pp. 129–37. (686)

Gaymer, R. (1968) The Indian Ocean giant tortoise, *Testudo gigantea*, on Aldabra. *J. Zool., Lond.* **154**, 341–63. (195)

Geist, V. (1971) *Mountain sheep: a study in behaviour and evolution.* University of Chicago Press, Chicago and London. (**311**, 312, 320, 543, 554)

General Register Office (1967) *Sample census 1966 Great Britain: summary tables.* HMSO, London. (**288**)

General Register Office (1968) *Sample census 1966 England and Wales: migration summary tables, Part I.* HMSO, London. (135, **288**)

Gerard, B.M. (1967) Factors affecting earthworms in pastures. *J. Anim. Ecol.* **36**, 235–52. (505, **507**)

Gerking,, S.D. (1959) The restricted movements of fish populations. *Biol. Rev.* **34**, 221–42. (196, 793)

Gibson, J.D. (1967) The wandering albatross (*Diomedea exulans*): results of banding and observations in New South Wales coastal waters and the Tasman Sea. *Notornis* **14**, 47–57. (625)

Gibson, R.N. (1967) Studies on the movements of littoral fish. *J. Anim. Ecol.* **36**, 215–34. (837)

Gibson, R.N. (1969) The biology and behaviour of littoral fish. *Oceanogr. Mar. Biol. Ann. Rev.* **7**, 367–410. (823, 837, 849)

Gifford, C.E. and Griffin, D.R. (1960) Notes on homing and migratory behavior of bats. *Ecology* **41**, 378–81. (892)

Gilbert, L.E. and Singer, M.C. (1973) Dispersal and gene flow in a butterfly species. *Am. Nat.* **107**, 58–72. (225)

Gilbert, L.E. and Singer, M.C. (1975). Butterfly ecology. *Ann. Rev. Ecol. and Syst.* **6**, 120–52. (86, 424).

Gilmore, R.M. (1960) Census and migration of the California gray whale. *Norsk Hvalfangsttid.* **49**, 409–31. (**748**)

Glass, B.P. (1958) Returns of Mexican free-tail bats banded in Oklahoma. *J. Mammal.* **39**, 435–7. (**591**)

Glass, B.P. (1959) Additional returns from free-tailed bats banded in Oklahoma. *J. Mammal.* **40**, 542–5. (**591**)

Glass, B.P. (1966) Some notes on reproduction in the red bat, *Lasiurus borealis. Proc. Oklahoma Acad. Sci.* **46**, 40–1. (**576**)

Glass, D.V. (1969) Fertility trends in Europe since the second world war. *In:* Behrman, S.J., Corsa, L. and Freedman, R. (eds.) *Fertility and family planning: a world view.* University of Michigan Press, Ann Arbor. pp. 25–74. (144, 145, **290**)

Godfrey, G.K. (1957) Observations of the movements of moles (*Talpa europaea* L.) after weaning. *Proc. zool. Soc. Lond.* **128**, 287–95. (**309**)

Godfrey, G. and Crowcroft, P. (1960) *The life of the mole.* Museum Press, London. (218, **309**)

Goehring, H.H. (1972) Twenty-year study of *Eptesicus fuscus* in Minnesota. *J. Mammal.* **53**, 201–7. (**603**)

Goldman, L.C. (1973) *Whooping Crane.* Canadian Wildlife Service, Hinterland Who's Who. Information Canada, Ottawa. (687)

Golenchenko, A.P. (1963) (A few facts about the commercial accumulations of white whales in the White, Barents, and Kara seas. *In: Problems in the utilization of commercial resources in the White Sea and the internal reservoirs of Karelia*) (In Russian). Akad. Nauk SSSR, Moscow-Leningrad **1**, 152–6. *From: Ref. Zh. Biol.* (1965) No. 61161. (757)

Goodbody, I.M. (1952) The post-fledging dispersal of juvenile tit mice. *Br. Birds.* **45**, 279–85. (190, **322**)

Goodwin, D. (1967) *Pigeons and doves of the world.* Trustees of the British Museum (Natural History), London. (895)

Gould, E. (1957) Orientation in box turtles, *Terrapene c. carolina* (Linnaeus). *Biol. Bull.* **112**, 336–48. (885–6)

Graham, J.B., Rubinoff, I. and Hecht, M.K. (1971) Temperature physiology of the sea snake, *Pelamis platurus:* an index of its colonization potential in the Atlantic Ocean. *Proc. Nat. Acad. Sci. USA* **68**, 1360–3. (**779**)

Grant, W.C. (1956) An ecological study of the peregrine earthworm, *Pheretima hupeiensis*, in the Eastern United States. *Ecology* **37**, 648–58. (508)

Graue, L.C. (1963) The effect of phase shifts in the day-night cycle on pigeon homing at distances of less than one mile. *Ohio J. Sci.* **63**, 214–17. (896)

Green, G.W. (1962) Flight and dispersal of the European pine shoot moth, *Rhyacionia buoliana* (Schiff.). 1. Factors affecting flight, and the flight potential of females. *Can. Ent.* **94**, 282–99. (**259**, 337)

Greenberg, A.E. (1964) Plankton of the Sacramento River. *Ecology* **45**, 40–9. (798)

Greenwood, P.J. and Harvey, P.H. (1976) The adaptive significance of variation in breeding area fidelity of the blackbird (*Turdus merula* L.). *J. Anim. Ecol.* **45**, 887–98. (**323**)

Griffin, D.R. (1940) Migrations of New England bats. *Bull. Mus. Comp. Zool.* **86**, 217–46. (**569, 584**)

Griffin, D.R. (1945) Travels of banded cave bats. *J. Mammal.* **26**, 15–23. (**569**)

Griffin, D.R. (1972) Nocturnal bird migration in opaque clouds. *In:* Galler, S.R., Schmidt-Koenig, K., Jacobs, G.J. and Belleville, R.E. (eds.) *Animal orientation and navigation.* Scientific and Technical Information Office, National Aeronautics and Space Administration, Washington, D.C. pp. 169–88. (917)

Griffin, D.R. and Hitchcock, H.B. (1965) Probable 24-year longevity record for *Myotis lucifugus. J. Mammal.* **46**, 332. (559)

Griffin, D.R., Novick, A. and Kornfield, M. (1958) The sensitivity of echolocation in the fruit bat, *Rousettus. Biol. Bull.* **115**, 107–13. (585)

Griffin, D.R., Webster, F.A. and Michael, C.R. (1960) The echolocation of flying insects by bats. *An. Behav.* **8**, 141–54. (585)

Griffin, J.B. (1968) The origin and dispersion of American Indians in North America. *In: Biomedical challenges presented by the American Indian.* Pan American Health Organization Scientific Publication No. 165. pp. 2–10. (114, **565**)

Grinnell, H. W. (1918) A synopsis of the bats of California. *Univ. California Publ. Zool.* **17**, 223–404. (**576**)

Gronau, J. and Schmidt-Koenig, K. (1970) Annual fluctuation in pigeon homing. *Nature, Lond.* **226**, 87–8. (871)

Gruber, J. (1960) Vier Jahre Fledermausberingung in Eberschwang, Ob.-Österreich (1956–9). *Bonner Zool. Beitr.* **11**, 33–5. (**595, 597**)

Grzimek, M. and Grzimek, B. (1960) A study of the game of the Serengeti Plains. *Zeit. für Saugetierkunde* **25**. (95, **535**)

Guggisberg, C.A.W. (1961) *Simba.* Collins, London. (553)

Guggisberg, C.A.W. (1969) *Giraffes.* Golden Press, New York. (531)

Gugler, J. (1969) On the theory of rural-urban migration: the case of Subsaharan Africa. *In:* Jackson, J.A. (ed.) *Migration.* Cambridge University Press. pp. 134–55. (133, 134, 135, 136, 137)

Gunier, W. and Elder, W.H. (1971) Experimental homing of gray bats to a maternity colony in a Missouri barn. *Am. Midl. Nat.* **86**, 502–6. (890–1)

Gunn, D.L. (1960) The biological background of locust control. *A. Rev. Ent.* **5**, 279–300. (454)

Gunn, D.L., Perry, F.C., Seymour, W.G., Telford, T.M., Wright, E.N. and Yeo, D. (1945) Mass departure of locust swarms in relation to temperature. *Nature, Lond.* **156**, 628. (252, 254)

Gunn, D.L., Perry, F.C., Seymour, W.G., Telford, T.M., Wright, E.N. and Yeo, D. (1948) Behaviour of the desert locust (*Schistocerca gregaria* Forskål) in Kenya in relation to aircraft spraying. *Anti-Locust Bull.* No. 3, 70 pp. (262)

Gunning, G.E. (1959) The sensory basis for homing in the longear sunfish, *Lepomis megalotis megalotis* (Rafinesque). *Invest. Indiana Lakes Streams* **5**, 103–30. (880)

Gwinner, E.G. (1972) Endogenous timing factors in bird migration. *In:* Galler, S.R., Schmidt-Koenig, K., Jacobs, G.J. and Belleville, R.E. (eds.) *Animal orientation and navigation.* Scientific and Technical Information Office, National Aeronautics and Space Administration, Washington, D.C. pp. 321–38. (**612**, 613, 626–7, 630, 645)

Gwinner, E.G., Farner, D.S. and Mewaldt, L.R. (in Gwinner 1972). (613)

Gwynne, M.D. and Bell, R.H.V. (1968) Selection of vegetation components by grazing ungulates in the Serengeti National Park. *Nature, Lond.* **220**, 390–3. (**535**)

Haartman, L. von (1949) Der Trauerfliegenschnäpper. *Acta Zool. Fenn.* **56**, 1–104. (275)

Haartman, L. von (1960) The Ortstreue of the pied flycatcher. *Proc. XII Intern. Ornith. Congr.* 266–73. (275, **276**)

Haartman, L. von (1968) The evolution of resident versus migratory habit in birds: some considerations. *Ornis fenn.* **45**, 1–7. (639)

Haeger, J.S. (1960) Behaviour preceding migration in the salt-marsh mosquito, *Aedes taeniorhynchus. Mosquito News* **20**, 136–47. (**237**)

Hafez, E.S.E., Williams, M. and Wierzbowski, S. (1962) The behaviour of horses. *In:* Hafez, E.S.E. (ed.) *The behaviour of domestic animals.* Baillière, Tindall and Cox, London. pp. 370–96. (887)

Haine, E. (1955) Aphid take-off in controlled wind speeds. *Nature, Lond.* **175**, 474. (255, 262)

Hall, E.R. and Kelson, K.R. (1959) *The mammals of North America.* 2 vols. Ronald Press, New York. (556, **565, 567**)

Hall, J.S. (1962) A life history and taxonomic study of the Indiana bat, *Myotis sodalis. Reading Pub. Mus. Art Gal., Sci. Pub.* **12**, 68 pp. (99, 101, 154, **159**, 160, 571)

Hall, J.S. and Wilson, N. (1966) Seasonal populations and movements of the gray bat in the Kentucky area. *Amer. Midl. Natur.* **75**, 317–24. (**600**)

Hall, K.R.L. (1965) Behaviour and ecology of the wild patas monkey. *J. Zool., Lond.* **148**, 15–87. (**386**)

Hallock, R.J., Elwell, R.F. and Fry, D.H. (1970) Migrations of adult king salmon, *Oncorhynchus tshawytscha*, in the San Joaquin Delta as demonstrated by the use of sonic tags. *State of California, Resources Agency, Dept. Fish. Game, Fish. Bull.* **151**, 1–92. (798)

Hamilton, J.E. (1934) The southern sea lion, *Otaria byronia* (de Blainville). *Discovery Reports* **8**, 269–318. (713, 720)

Hamilton, J.E. (1939a) The leopard seal, *Hydrurga leptonyx* (de Blainville). *Discovery Reports* **18**, 239–64. (**716**)

Hamilton, J.E. (1939b) A second report on the southern sea lion, *Otaria byronia* (de Blainville). *Discovery Reports* **19**, 121–64. (716)

Hamilton, W.J. III (1962a) Does the bobolink navigate? *Wilson Bull.* **74**, 357–66. (864)

Hamilton, W.J. III (1962b) Evidence concerning the function of nocturnal call notes of migratory birds. *Condor* **64**, 390–401. (613, 686)

Hamilton, W.J. III (1966) Analysis of bird navigation experiments. *In:* Watt, K.E.F. (ed.) *Systems analysis in ecology.* New York. pp. 147–78. (189)

Hamilton, W.J. III (1967) Social aspects of bird orientation mechanisms. *In:* Storm, R.M. (ed.) *Animal orientation and navigation.* Proc. 27th Annual Biol. Coll., Oregon State Univ. Press, Corvallis. (686–7, 689)

Hamilton, W.J. III, Gilbert, W.M., Heppner, F.H. and Planck, R.J. (1967) Starling roost dispersal and a hypothetical mechanism regulating rhythmical animal movement to and from dispersal centers. *Ecology* **48**, 825–33. (68, 652, 906)

Hamilton-Smith, E. (1966) The geographical distribution of Australian cave-dwelling Chiroptera. *Int. J. Speleol.* **2**, 91–104. (**587**)

Hamner, W.M., Smyth, M. and Mulford, E.D. (1968) Orientation of the sand-beach isopod, *Tylos punctatus. Anim. Behav.* **16**, 405–9. (830, 857)

Hanak, V., Gaisler, J. and Figala, J. (1962) Results of bat-banding in Czechoslovakia 1948–60. *Acta Universitas Carolinae-Biologica* **1962**, 9–87. (312, 564, **570, 572, 582, 583, 587**, 595, **596, 597, 598, 599, 601, 602**)

Hanec, W. (in Johnson 1969). (464)

Harden Jones, F.R. (1968) *Fish Migration.* Arnold, London. (196, **796, 804, 805, 812**, 813, **815, 817, 818, 822**, 851, 880)

Hardisty, M.W. (1971) Gonadogenesis, sex differentiation and gametogenesis. *In:* Hardisty, M.W. and Potter, I.C. (eds.) *The biology of lampreys.* Vol 1. Academic Press, London. pp. 295–360. (**801**)

Hardisty, M.W. and Potter, I.C. (1971a) The behaviour, ecology and growth of larval lampreys. *In:* Hardisty, M.W. and Potter, I.C. (eds.) *The biology of lampreys.* Vol. 1. Academic Press, London. pp. 85–126. (**791**)

Hardisty, M.W. and Potter, I.C. (1971b) The general biology of adult lampreys. *In:* Hardisty, M.W. and Potter, I.C. (eds.) *The biology of lampreys.* Vol. 1. Academic Press, London. pp. 127–206. (**791**)

Hardisty, M.W. and Potter, I.C. (1971c) Paired species. *In:* Hardisty, M.W. and Potter, I.C. (eds.) *The biology of lampreys.* Vol. 1. Academic Press, London. pp. 249–78. (**791**, 847)

Hardy, A.C. (1958) *The open sea. I. The world of plankton.* Collins, London. (482)

Harington, C.R. (1970) *Polar bear.* Canadian Wildlife Service, Hinterland Who's Who. Queen's Printer for Canada, Ottawa. (**177, 889**)

Harris, J.E. (1963) The role of endogenous rhythms in vertical migration. *J. mar. biol. Ass. U.K.* **43**, 153–66. (484, 494)

Harris, J.E. and Wolfe, U.K. (1955) A laboratory study of vertical migration. *Proc. Roy. Soc.* B **144**, 329–54. (494)

Harris, M.P. (1972) Inter-island movements of manx shearwaters. *Bird Study* **19**, 167–72. (**322**)

Harris, V.A. (1964) *The life of the rainbow lizard.* Hutchinson Tropical Monographs. Hutchinson, London. (195, **295**)

Harris, W.V. and Sands, W.A. (1965) The social organization of termite colonies. *Symp. zool. Soc. Lond.* **No. 14**, 113–31. (236)

Hart, J.L. (1973) *Pacific fishes of Canada. Fish. Rs. Bd.*

Canada. *Bull. 180.* Information Canada, Ottawa. (**805**)

Hart, T.J. (1942) Phytoplankton periodicity in Antarctic surface waters. *Discovery Reports* **21**, 263–348. (**483**)

Hartman, W.L. and Raleigh, R.F. (1964) Tributary homing of sockeye salmon at Brooks and Karluk Lakes, Alaska. *J. Fish. Res. Bd Canada* **21**, 485–504. (880)

Hasler, A.D. (1966) *Underwater guideposts.* Univ. Wisconsin Press, Madison. (**804**, 806, 813, 862, 880)

Hasler, A.D. (1967) *In:* Storm, R.M. (ed.) *Animal orientation and navigation.* Proc. 27th Annual Biol. Coll., Oregon State Univ. Press, Corvallis. (880)

Hasler, A.D., Horrall, R.M., Wisby, W.J. and Braemer, W. (1958) Sun-orientation and homing in fishes. *Limnol. Oceanogr.* **3**, 353–61. (**850**)

Hassell, M.D. (1963) A study of homing in the Indiana bat, *Myotis sodalis. Tran. Kentucky Acad. Sci.* **24**, 1–4. (890)

Hassell, M.D. and Harvey, M.J. (1965) Differential homing in *Myotis sodalis. Am. Midl. Nat.* **74**, 501–3. (893)

Hatler, D.F. (1972) Food habits of black bears in interior Alaska. *Can. Field Nat.* **86**, 17–31. (175, **314**)

Haufe, W.O. (1954) The effects of atmospheric pressure on the flight responses of *Aedes aegypti* (L.). *Bull. ent. Res.* **45**, 507–26. (260)

Haufe, W.O. (1962) Response of *Aedes aegypti* (L.) to graded light stimuli. *Can. J. Zool.* **40**, 53–64. (261)

Haukioja, E. (1971) Summer schedule of some subarctic passerine birds with reference to post-nuptial moult. *Rep. Kevo Subarctic Res. Stat.* **7**, 60–9. (627)

Havekost, H. (1960) Die Beringung der Breitflügelfledermaus (*Eptesicus serotinus* Schreber) im Oldenburger Land. *Bonner Zool. Beitr.* **11**, 222–33. (**602**)

Hawkins, R.E. and Klimstra, W.D. (1970) A preliminary study of the social organization of white-tailed deer. *J. Wildl. Mgmt.* **34**, 407–19. (**311**)

Hayward, K.J. (1953) Migration of butterflies in Argentina during the spring and summer of 1951–52. *Proc. R. ent. Soc. Lond.* A **28**, 69–73. (425)

Hayward, K.J. (1967) Notes on Argentina butterfly migration, 1965–66. *Entomologist* **100**, 29–34. (**447**)

Healey, M.C. (1967) Aggression and self-regulation of population size in deermice. *Ecology* **48**, 377–92. (**310**)

Heape, W. (1931) *Emigration, migration and nomadism.* Cambridge University Press, Cambridge. (15, 27)

Heathcote, G.D. and Cockbain, A.J. (1966) Aphids from mangold clamps and their importance as vectors of beet viruses. *Ann. appl. Biol.* **57**, 321–36. (**261**)

Heerdt, P.F. van and Sluiter, J.W. (1953) The results of bat banding in the Netherlands in 1952 and 1953. *Natuurhist. Maandblad* **42**, 101–4. (**595**)

Heerdt, P.F. van and Sluiter, J.W. (1954) The results of bat banding in the Netherlands in 1954. *Natuurhist. Maandblad* **43**, 85–8. (**595**)

Heerdt, P.F. van and Sluiter, J.W. (1956) The results of bat banding in the Netherlands in 1955. *Natuurhist. Maandblad* **45**, 62–4. (**595**)

Heerdt, P.F. van and Sluiter, J.W. (1957) The results of bat banding in the Netherlands in 1956. *Natuurhist. Maandblad* **46**, 13–15. (**595**)

Heerdt, P. F. van and Sluiter, J.W. (1958) The results of bat banding in the Netherlands in 1957. *Natuurhist. Maandblad* **47**, 38–41. (**595**)

Heerdt, P. F. van and Sluiter, J.W. (1959) The results of bat banding in the Netherlands in 1958. *Natuurhist. Maandblad* **48**, 96–8. (**570, 595, 596, 597, 598, 599**)

Henson, W.R. (1962) Convective transportation of *Choristoneura fumiferana* (Clem.). *Proc. XI Int. Congr. Ent.* (Vienna 1960) **3**, 44–6. (337)

Heppes, J.B. (1958) The white rhinoceros in Uganda. *Afr. wildl.* **12**, 273–80. (531)

Herbert, H.J. (1972) *The population dynamics of the waterbuck, Kobus ellipsiprymnus (Ogilby, 1833) in the Sabi-Sand Wildtuin.* Verlag Paul Parey, Hamburg. (549)

Herreid, C.F. II (1964) Bat longevity and metabolic rate. *Expt. gerontology* **1**, 1–9. (559)

Herrnkind, W.F. and McLean, R. (1971) Field studies of homing, mass emigration and orientation in the spiny lobster, *Panulirus argus. Ann. N.Y. Acad. Sci.* **188**, 359–77. (**879**)

Heyland, J.D. (1973) *Greater snow goose.* Canadian Wildlife Service, Hinterland Who's Who. Information Canada, Ottawa. (**692**)

Hickling, G. (1962) *Grey seals and the Farne Islands.* Routledge and Kegan Paul, London. (721, **723**)

Higgins, L.G. and Riley, N.D. (1973) *A field guide to the butterflies of Britain and Europe* (2nd ed.). Collins, London. (**428–30**)

Hill, W.C.O., Barker, W., Stockley, C.H., Pitman, C.R.S., Offermann, P.B., Rushby, G.G. and Gowers, W. (1953) *The elephant in East Central Africa. A monograph.* Rowland Ward Ltd, London. (528, 550)

Hirst, S.M. (1969) Populations in a transvaal lowveld nature reserve. *Zoologica Africana* **4**, 199–230. (**535**)

Hirth, H. and Carr, A. (1970) The green turtle in the Gulf of Aden and the Seychelles Islands. *Verh. Kon. Neder. Akad. Wettensh. Nat.* **58**, 1–44. (193, **773**)

Hitt, H.L. (1956) Population movements in the southern United States. *The Scientific Monthly* **82**, 241–6. (137)

Hoar, W.S. (1951) The behaviour of chum, pink and coho salmon in relation to their seaward migration. *J. Fish. Res. Bd Canada* **8**, 241–63. (795)

Hoar, W.S. (1958) The evolution of migratory behaviour among juvenile salmon of the genus *Oncorhynchus J. Fish. Res. Bd Canada* **15**, 391–428. (**803**, 806)

Hochbaum, H.A. (1955) *Travels and traditions of waterfowl.* University of Minnesota Press, Minneapolis. (187–8, 321, 325, 327, 378, 606, 613)

Hock, R.J. (1951) The metabolic rates and body temperatures of bats. *Biol. Bull.* **101**, 289–99. (99)

Hoel, E. (1960) Beringungsergebnisse in einem Winterquartier der Mopsfledermäuse (*Barbastella barbastellus* Schreb.) in Fulda. *Bonner Zool. Beitr.* **11**, 192–7. (**601**)

Hoffmann, K. (1965) Clock-mechanisms in celestial orientation of animals. *In:* Aschoff, J. (ed.) *Circadian clocks.* North Holland, Amsterdam. pp. 426–41. (856)

Hoffmann, K. (in Matthews 1955). (857)

Hoffmeister, D.F. (1970) The seasonal distribution of bats in Arizona: a case for improving mammalian range maps. *Southwest Natur.* **15**, 11–22. (579)

Hoffmeister, D.F. and Downes, W.L. (1964) Blue jays as predators of red bats. *Southwest Natur.* **9**, 102–9. (**576**)

Holm, J.L. and Jonsgård, Å. (1959) Occurrence of the sperm whale in the Antarctic and the possible influence of the moon. *Norsk Hvalfangsttid.* **No. 4**. 161–182. (761)

Holmes, R.T. (1966) Breeding ecology and annual cycle adaptations of the red-backed sandpiper (*Calidris alpina*) in Northern Alaska. *Condor* **68**, 3–46. (613)

Hoogerwerf, A. (1970) *Udjung Kulon. The land of the last Javan rhinoceros.* E.J. Brill, Leiden. (166, 551)

Hooper, J.H.D. (1962) *Animals of Britain. 2. Horseshoe bats* (ed. Harrison Matthews, L.). The Sunday Times, North Western Printers, Stockport, England. (**583**)

Hooper, J.H.D. and Hooper, W.M. (1956) Habits and movements of cave-dwelling bats in Devonshire. *Proc. zool. Soc. Lond.* **127**, 1–26. (**570, 583**)

Hornocker, M.G. (1970) An analysis of mountain lion predation upon mule deer and elk in the Idaho primitive area. *Wildl. Monographs* **21**, 1–39. (174, **310**)

Hotz, M.B. (1955) A study of cohort migration in the U.S.: 1870–1950. Unpublished doctoral dissertation, Washington University. (135)

Howard, W.E. (1949) Dispersal, amount of inbreeding, and longevity in a local population of prairie deermice on the George Reserve, Southern Michigan. *Univ. Mich. Contrib. Lab. Vert. Biol.* **43**, 1–52. (349)

Howard, W.E. (1960) Innate and environmental dispersal of individual vertebrates. *Am. Midl. Nat.* **63**, 152–62. (37)

Howell, F.C. (1970) *Early man* (revised ed.). Time-Life International, Netherlands. (114, 116, 119, 125, 133, 538)

Howell, T.R. and Bartholomew, G.A. (1959) Further experiments on torpidity in the poor-will. *Condor* **61**, 180–5. (605)

Howland, H.C. (1973) Orientation of European robins *to* Kramer cages: eliminating possible sources of error and bias in Kramer cage studies. *Z. für Tierpsychologie* **33**, 295–312. (864, 871)

Hubbs, C.L. and Hubbs, L.C. (1967) Gray whale censuses by airplane in Mexico. *Calif. Fish Game* **53**, 23–27. (**748**, 766)

Hubbs, C.L. and Potter, I.C. (1971) Distribution, phylogeny and taxonomy. *In:* Hardisty, M.W. and Potter, I.C. (eds.) *The biology of lampreys*, Vol. 1. Academic Press, London. pp. 1–66. (**791, 801, 845**)

Hughes, R.D. and Nicholas, W.L. (1974) The spring migration of the bushfly (*Musca vetustissima* Walk.): evidence of displacement provided by natural population markers including parasitism. *J. Anim. Ecol.* **43**, 411–28. (465)

Hummitzsch, E. (1960) Fledermausberingungen in

Leipzig und Umgebung. *Bonner Zool. Beitr.* **11**, 99–104. (**595**)

Humphrey, S. and Cope, J.B. (1964) Movements of *Myotis lucifugus lucifugus* from a colony in Boone County, Indiana. *Proc. Indiana Acad. Sci.* **72**, 268–71. (**569**)

Hurlbert, S.H. (1969) The breeding migrations and interhabitat wandering of the vermilion-spotted newt, *Notophthalmus viridescens* (Rafinesque). *Ecol. Monog.* **39**, 465–88. (521)

Hussell, D.J.T. (1972) Factors affecting clutch-size in Arctic passerines. *Ecol. Mon.* **42**, 317–64. (660, **661**)

Hussell, D.J.T., Davis, T. and Montgomerie, R.D. (1967) Differential fall migration of adult and immature least flycatchers. *Bird-Banding* **38**, 61–6. (639)

Hwang, Guan-Huei and How, Wu-Wei (1966) (Studies on the flight of the Army-worm moth (*Leucania separata* Walker). I. Flight duration and wingbeat frequency.) (In Chinese, English summary.) *Acta ent. sin.* **15**, 96–104 (*Rev. appl. Ent. A* **55**, 88–9). (472)

Hynes, H.B.N. (1970) *The ecology of running waters.* Liverpool University Press, Liverpool. (780–3, 785–7, **792**, 793, 797–9, 806, **822**, 846–8).

Ichihara, T. (1966) The pygmy blue whale, *Balaenoptera musculus brevicauda*, a new subspecies from the Antarctic. *In:* Norris, K.S. (ed.) *Whales, dolphins and porpoises.* Univ. of California Press, Berkeley and Los Angeles. pp. 79–113. (**755**)

Imanishi, K. (1961) (The origin of human family—a primatological approach) (In Japanese with English summary). *Jap. J. Ethnol.* **25**, 119–30. (118)

Imms, A.D. (1964) *A general textbook of Entomology* (9th ed. Revised by Richards, O.W. and Davies, R.G.). Methuen, London. (98, 240)

Information Canada (1973) *Caribou.* Canadian Wildlife Service, Hinterland Who's Who, Ottawa. (**533**)

Isakov, J.A. (1949) (The problem of elementary populations in birds) (In Russian). *13 vest. Akad. Nauk SSSR Moskva-Leningrad. Biol. Ser.* **1**, 54–70. (620)

Itani, J. (1963) Vocal communication of the wild Japanese monkey. *Primates* **4**, 11–66. (162)

Itani, J. (1966) (Social organization of chimpanzees) (In Japanese). *Shizen* **21**, 17–30. (162)

Itani, J. (1967) (From the societies of non-human primates to human society) (In Japanese). *Kagaku* **37**, 170–4. (162)

Itani, J. and Suzuki, A. (1967) The social unit of chimpanzees. *Primates* **8**, 355–81. (117, 118, 162, **295**, 297)

Itani, J., Tokuda, K., Furuya, Y., Kano, K. and Shin, Y. (1963) The social construction of natural troop of Japanese monkeys in Takasakiyama. *Primates* **4**, 1–42. (216–17, **281**, 282)

Ivanov, T.M. (1938) (On the Baikal sea dog *Phoca sibirica* Gmelin, its biology and fishing) (In Russian with English summary). *Bull. Inst. Scient. Univ. Irkoutsk* **8**, 5–119. (**740**)

Izawa, K. (1970) Unit groups of chimpanzees and their nomadism in the savanna woodland. *Primates* **11**, 1–46. (162, 400, 551)

Jackson, H.H.T. (1961) *The mammals of Wisconsin.* Univ. Wisconsin Press, Madison. (**576**)

Jacobs, M.E. (1955) Studies on territorialism and sexual selection in dragonflies. *Ecology* **36**, 566–86. (198)

Jacoby, V.E. and Jögi, A. (1972) (The moult migration of the common scoter in the light of radar and visual observational data) (In Russian with English summary). *Communs Baltic Commn Study Bird Migr.* **7**, 118–39. (620)

James, F.C. (1970) Geographic size variation in birds and its relationship to climate. *Ecology* **51**, 365–90. (665)

Jameson, D.L. (1957) Population structure and homing responses in the Pacific Tree Frog. *Copeia* **1957**, 221–8. (196, 521)

Jansen, C. (1969) Some sociological aspects of migration. *In:* Jackson, J.A. (ed.) *Migration* Cambridge University Press. pp. 60–73. (137)

Jenkins, D., Watson, A. and Miller, G.R. (1963) Population studies on red grouse, *Lagopus lagopus scoticus* (Lath.) in north-east Scotland. *J. Anim. Ecol.* **32**, 317–76. (192–3)

Jenkins, J.T. (1936) *The fishes of the British Isles.* Warne, London. (841)

Jennings, W.L. (1958) The ecological distribution of bats in Florida. Unpublished. Ph.D. Thesis. Univ. Florida, Gainesville. (**576**)

Jepson, G.L. (1970) Bat origins and evolution. *In:* Wimsatt, W.A. (ed.) *Biology of bats.* Vol. 1. 1–64. Academic Press, London. (556)

Jewell, P.A. (1966) The concept of home range in mammals. *Symp. zool. Soc. Lond.* **18**, 85–109. (378, 384, **387–8**, 389, 401)

Johnson, B. (1954) Effect of flight on behaviour of *Aphis fabae* Scop. *Nature, Lond.* **173**, 831. (231)

Johnson, B. (1958) Factors affecting the locomotor and settling responses of alate aphids. *Anim. Behav.* **6**, 9–26. (231, 236)

Johnson, B. (1965) Wing polymorphism in aphids. II. Interaction between aphids. *Entomologia exp. appl.* **8**, 49–64. (229)

Johnson, B. and Birks, P. R. (1960) Studies on wing polymorphism in aphids. I. The development process involved in the production of the different forms. *Entomologia exp. appl.* **3**, 327–39. (228)

Johnson, C.G. (1969) *Migration and dispersal of insects by flight.* Methuen, London. (5, 7, 15, 17, 18, 20, 24, 31, 74, 76, 199, 232, 235, 236, 244, 247, 253, 257, 338, 351–2, 426, **436**, 463–6, **467–9**)

Johnson, C.G. (1971) Comments at a meeting. *Proc. R. ent. Soc. Lond.* **36**, 33–6. (421, 439)

Johnson, N.K. (1963) Comparative molt cycles in the tyrannid genus *Empidonax. Proc. XIII Int. Orn. Congr.* 870–83. (626, **628**, 639)

Johnston, R.F. (1961) Population movements of birds. *Condor* **63**, 386–9. (37, 190)

Jolly, A. (1966) *Lemur behaviour: a Madagascar field study.* Univ. Chicago Press, Chicago. (161, **394**)

Jones, C. and Pagels, J. (1968) Notes on a population of *Pipistrellus subflavus* in southern Louisiana. *J. Mammal.* **49**, 134–9. (575)

Jones, J.W. (1959) *The salmon.* Collins, London. (795, 797, 807, 849)

Jones, R. (1973) Emerging picture of Pleistocene Australians. *Nature, Lond.* **246**, 278–81. (114)

Jonkel, C.J. and Cowan, I. McT. (1971) The black bear in the spruce-fir forest. *Wildl. Monogr.* **27**, 5–57. (73, 175, **314**, 329, 331, 353, 397)

Jonsgård, Å. (1960) On the stocks of sperm whales (*Physeter catodon*) in the Antarctic. *Norsk Hvalfangsttid.* **49**, 289–99. (763)

Jonsgård, Å. (1966) The distribution of Balaenopteridae in the North Atlantic Ocean. *In:* Norris, K.S. (ed.) *Whales, dolphins and porpoises.* Univ. California Press, Berkeley and Los Angeles, pp. 114–24. (**180, 182, 755, 768, 883**)

Jonsgård, Å. and Lyshoel, P.B. (1970) A contribution to the knowledge of the biology of the killer whale, *Orcinus orca* (L.). *Nytt Mag. Zool. Oslo.* **18**, 41–8. (**759**)

Jonsgård, Å. and Øynes, P. (1952) Om Bottlenosen og Spekhoggeren. *Fauna* no. 1. (**759**)

Jordan, J.S. (1971) Dispersal period in a population of eastern fox squirrels (*Sciurus niger*). *U.S. For. Ser. Res. Pap. NE* **216**, 1–8. (**310**)

Kalleberg, H. (1958) Observations in a stream tank of territoriality and competition in juvenile salmon and trout (*Salmo salar* L. and *S. trutta* L.). *Rep. Inst. Freshwat. Res. Drottningholm* **39**, 55–98. (796)

Kano, T. (1971) The chimpanzee of Filibanga, Western Tanzania. *Primates* **12**, 229–46. (**526**)

Kanwisher, J. (1966) Comments at a meeting. *In:* Norris, K.S. (ed.) *Whales, dolphins and porpoises.* Univ. California Press, Berkeley and Los Angeles. pp. 407–9. (750)

Kanwisher, J. and Sundnes, G. (1966) Thermal regulation in cetaceans. *In:* Norris, K.S. (ed.) *Whales, dolphins and porpoises.* Univ. California Press, Berkeley and Los Angeles. pp. 397–407. (750)

Kasahara, Y. (1958) The influx and exodus of migrants among the 47 prefectures in Japan, 1920–35. Unpublished doctoral dissertation, University of Michigan. (134)

Kasuya, T. and Ichihara, T. (1965) Some information on minke whales from the Antarctic. *Sci. Rep. Whales Res. Inst.* **19**, 37–43. (**768**)

Kaufmann, J.H. (1962) Ecology and social behavior of the coati, *Nasua narica*, on Barro Colorado Island, Panama. *Univ. Calif. Publ. Zool.* **60**, 95–222. (**401**)

Kaufmann, T. (1965a) Meaning of flight in West African Firefly, *Luciola discicollis* Cast. *Proc. XII Int. Congr. Ent.* (London 1964) 299–300. (353)

Kaufmann, T. (1965b) Ecological and biological studies in the West African Firefly, *Luciola discicollis* (Coleoptera: Lampyridae). *Ann. ent. Soc. Am.* **58**, 414–26. (353)

Kawanabe, H. (1958) On the significance of the social structure for the mode of density effect in a salmon-like fish, 'Ayu', *Plecoglossus altivelis* Temminck and Schlegel. *Mem. Coll. Sci. Engng Kyoto imp. Univ.* **B25**, 171–80. (822)

Keast, A. (1968a) Seasonal movements in the Australian honeyeaters (Meliphagidae) and their ecological significance. *The Emu* **67**, 159–209. (324, 399, 676)

Keast, A. (1968b) Evolution of mammals on southern continents. IV. Australian mammals: Zoogeography and evolution. *Quart. Rev. Biol.* **43**, 373–408. (559)

Keast, A. (1969) Evolution of mammals on southern continents. VII. Comparisons of the contemporary mammalian faunas of the southern continents. *Quart. Rev. Biol.* **44**, 121–67. (**565**)

Keenleyside, M.H.A. (1962) Skin-diving observations of Atlantic salmon and brook trout in the Miramichi River, New Brunswick. *J. Fish. Res. Bd. Can.* **16**, 625–34. (796)

Keeton, W.T. (1970a) Comparative orientation and homing performances of single pigeons and small flocks. *Auk* **87**, 797–9. (687)

Keeton, W.T. (1970b) 'Distance effect' in pigeon orientation: an evaluation. *Biol. Bull.* **139**, 510–19. (907, **909**)

Keeton, W.T. (1970c) Do pigeons determine latitudinal displacement from the sun's altitude? *Nature, Lond.* **227**, 626–7. (864)

Keeton, W.T. (1971) Comments at a meeting. *Ann. N.Y. Acad. Sci.* **188**, 331–58. (906)

Keeton, W.T. (1972a) Effect of magnets on pigeon homing. *In:* Galler, S.R., Schmidt-Koenig, K., Jacobs, G.J. and Belleville, R.E. (eds.) *Animal orientation and navigation.* Scientific and Technical Information Office, National Aeronautics and Space Administration, Washington, D.C. pp. 579–94. (608, 861, 865, 906–7)

Keeton, W.T. (1972b) Comments at a meeting. *In:* Galler, S.R., Schmidt-Koenig, K., Jacobs, G.J. and Belleville, R.E. (eds.) *Animal orientation and navigation.* Scientific and Technical Information Office, National Aeronautics and Space Administration, Washington, D.C. p. 281. (906)

Keeton, W.T., Larkin, T.S. and Windsor, D.M. (1974) Normal fluctuations in the Earth's magnetic field influence pigeon orientation. *J. comp. Physiol. A.* **95**, 95–103. (608)

Kellogg, R. (1928) The history of whales: their adaptation to life in the water. *Quart. Rev. Biol.* **3**, 174–208. (745)

Kellogg, R. (1938) Adaptations of structure to function in whales. *Publ. Carnegie Inst.* **501**, 649–82. (764)

Kellogg, W.N. and Rice, C.E. (1966) Visual discrimination and problem solving in a bottlenose dolphin. *In:* Norris, K.S. (ed.) *Whales, dolphins and porpoises.* Univ. California Press, Berkeley and Los Angeles. pp. 731–54. (766)

Kelsall, J.P. (1968) *The migratory Barren-ground Caribou of Canada.* Canadian Wildl. Serv. Monogr. Ser. 3, Canadian Wildl. Serv., Ottawa. (**171**, 211–12, **533**)

Kendeigh, S.C. (1941) Territorial and mating behavior of the house wren. *Ill. Biol. Monogr.* **18**, 1–120. (**322**, 325)

Kennedy, J.S. (1951) The migration of the desert locust (*Schistocerca gregaria* Forsk.). I. The behaviour of swarms. II. A theory of long-range migration. *Phil. Trans. R. Soc.* (B) **235**, 163–290. (236, 420, 457, 459, 461)

Kennedy, J.S. (1956) Phase transformation in locust biology. *Biol. Rev.* **31**, 349–70. (263)

Kennedy, J.S. (1958) The experimental analysis of aphid behaviour and its bearing on current theories of instinct. *Proc. X Int. Congr. Ent.* (Montreal, 1956) **2**, 397–404. (18)

Kennedy, J.S. (1961) A turning point in the study of insect migration. *Nature, Lond.* **189**, 785–91. (18, 20, 244)

Kennedy, J.S. and Booth, C.O. (1963a) Free flights of aphids in the laboratory. *J. exp. Biol.* **40**, 67–85. (231, 466)

Kennedy, J.S. and Booth, C.O. (1963b) Coordination of successive activities in an aphid. The effect of flight on the settling responses. *J. exp. Biol.* **40**, 351–69. (231, 466)

Kennedy, J.S. and Stroyan, H.L.G. (1959) Biology of aphids. *A. Rev. Ent.* **4**, 139–60. (18)

Kenyon, K.W. (1969) The sea otter (*Enhydra lutris*) in the Eastern Pacific Ocean. *North Am. Fauna* **68**, 1–352. (178, 222–3, **311**)

Kenyon, K.W. and Rice, D.W. (1958) Homing of Laysan albatrosses. *Condor* **60**, 3–6. (917)

Kenyon, K.W. and Rice, D.W. (1961) Abundance and distribution of the Steller sea lion. *J. Mammal.* **42**, 223–34. (**729**)

Kenyon, K.W. and Wilke, F. (1953) Migration of the Northern Fur Seal, *Callorhinus ursinus. J. Mammal.* **34**, 86–98. (**105**, 725)

King, J.A. (1955) Social behavior, social organization and population dynamics in a black-tailed prairiedog town in the Black Hills of South Dakota. *Contr. Lab. Vert. Biol. Univ. Mich.* **67**, 1–123. (102–4, 124, 213, 329, **331**, 406, **407**)

King, J.E. (1964) *Seals of the world.* British Museum (Natural History), London. (9, **106, 107,** 110 706, **708,** 709, 711, 713, **714, 716,** 720–1, **723, 727, 729, 732, 735, 737, 738, 740, 741, 742, 743, 744**)

King, J.E. (1966) Relationships of the hooded and elephant seals. *J. Zool., Lond.* **148**, 385–98. (706)

King, J.E. (1969) The identity of the fur seals of Australia. *Aust. J. Zool.* **17**, 841–53. (706, **708**)

King, J.R. and Farner, D.S. (1965) Studies of fat deposition in migratory birds. *Ann. N.Y. Acad. Sci.* **131**, 422–40. (627, 647)

Kirk, D. (1969) Natality in the developing countries: recent trends and prospects. *In:* Behrman, S.J., Corsa, L., and Freedman, R. (eds.) *Fertility and family planning: a world view.* University of Michigan Press, Ann Arbor. pp. 75–98. (145)

Klassen, W. and Hocking, B. (1964) The influence of a deep river valley system on the dispersal of *Aedes* mosquitos. *Bull. ent. Res.* **55**, 289–304. (352)

Klauber, L.M. (1936) A statistical study of rattlesnakes. *Occ. Papers, San Diego Soc. N.H.* **1**, 1–24. (523)

Kleber, E. (1935). Hat das Zeitgedächtnis der Bienen biologische Bedeutung? *Z. vergl. Physiol.* **22**, 221–62. (**335**, 352)

Kleerekoper, H., Matis, J., Gensler, P. and Maynard, P. (1974) Exploratory behaviour of goldfish, *Carassius auratus. Anim. Behav.* **22**, 124–32. (196, 849)

Klimstra, W.D. (1951) Notes on late summer snapping turtle movements. *Herpetologica, San Diego* **7**, 140. (523)

Klingel, H. (1969) The social organization and population ecology of the plains zebra, *Equus quagga. Zoologica Africana* **4**, 249–63. (165, 166–7, **303**, 304, 553)

Kluijver, H.N. (1935) Waarnemingen over de levenswijze den Spreeuw (*Sturnus v. vulgaris* L.) met behulp van gerigde individuen. *Ardea* **24**, 133–66. (191)

Kluijver, H.N. (1951) The population ecology of the Great Tit, *Parus m. major* L. *Ardea* **39**, 1–135. (211, **322**, 636)

Kluijver, H.N. and Tinbergen, L. (1953) Territory and regulation of density in titmice. *Arch. Neerl. Zool.* **10**, 265–89. (**405**)

Knight-Jones, E.W. (1951) Gregariousness and some other aspects of the settling behaviour of *Spirorbis. J. Mar. Biol. Ass. U.K.* **30**, 201–22. (199)

Knight-Jones, E.W. and Morgan, E. (1966) Responses of marine animals to changes in hydrostatic pressure. *Oceanogr. Mar. Biol. Ann. Rev.* **4**, 267–99. (494)

Kock, L.L. de and Robinson, A.E. (1966) Observations on a lemming movement in Jämtland, Sweden, in autumn 1963. *J. Mammal.* **47**, 490–9. (**310**)

Koerwitz, F.L. and Preuss, K.P. (1964) Migratory potential of the army cutworm. *J. Kans. ent. Soc.* **37**, 234–9. (240)

Koford, C.B. (1957) The vicuña and the puna. *Ecol. Monogr.* **27**, 153–219. (166, **295**, 307, 318)

Kolenosky, G. (1970) *Black bear.* Canadian Wildlife

Service, Hinterland Who's Who. Queen's Printer for Canada, Ottawa. (176, **314**)

Koopman, K.F. (1967) The southernmost bats. *J. Mammal.* **48**, 487–8. (556)

Kozhov, M.M. (1963) *Lake Baikal and its life.* Dr. W. Junk, The Hague. (**740**)

Krader, L. (1959) The ecology of nomadic pastoralism. *UNESCO Int. Social Sci. J.* **11**, 499–510. (540, 551)

Kramer, A. (1971) Notes on the ecology of the mule and white-tailed deer in the Cypress Hills, Alberta. *Can. Field Natur.* **85**, 141–5. (168, 542)

Kramer, G. (1952) Experiments on bird orientation. *Ibis.* **94**, 265–85. (607)

Kramer, G. (1957) Experiments on bird orientation and their interpretation. *Ibis* **99**, 196–227. (900)

Kramer, G. (1959) Recent experiments on bird orientation. *Ibis* **101**, 399–416. (900, **902**)

Kramer, G. (in Dorst 1962). (**857**)

Krebs, C.J. (1964) The lemming cycle at Baker Lake, Northwest Territories, during 1959–62. *Arctic Inst. N. Amer. Tech. Paper* **15**, 1–104. (161, **310**)

Krebs, C.J. (1973) *Lemming.* Canadian Wildlife Service, Hinterland Who's Who. Information Canada, Ottawa. (524–6)

Krebs, J.R. (1971) Territory and breeding density in the great tit, *Parus major* L. *Ecology* **52**, 2–22. (211, **405**)

Krebs, J.R., Ryan, J.C. and Charnov, E.L. (1974) Hunting by expectation or optimal foraging. A study of patch use by chickadees. *Anim. Behav.* **22**, 953–64. (58)

Krogh, A. and Zeuthen, E. (1941) The mechanism of flight preparation in some insects. *J. exp. Biol.* **18**, 1–10. (253)

Kruuk, H. (1972) *The spotted hyena: a study of predation and social behavior.* University of Chicago Press Ltd, London. (538, 554)

Krzanowski, A. (1960) Investigations of flights of Polish bats, mainly *Myotis myotis* (Borkhausen, 1797). *Acta theriol.* **4**, 175–84. (**581, 595**)

Krzanowski, A. (1964) Three long flights by bats. *J. Mammal.* **45**, 152. (**598**)

Kühme, W. (1965) Freilandstudien zur Soziologie des Hyänenhundes (*Lycaon pictus lupinus* Thomas 1902). *Z. Tierpsychol.* **22**, 495–541. (174, 215)

Kuznets, S. (1969) Economic aspects of fertility trends in the less developed countries. *In:* Behrman, S.J., Corsa, L. and Freedman, R. (eds.) *Fertility and family planning: a world view.* University of Michigan Press, Ann Arbor. pp. 157–79. (144)

Lack, D. (1943) The problem of partial migration. *Br. Birds* **37**, 122–30. (635)

Lack, D. (1944) The problem of partial migration. *Br. Birds* **37**, 143–50. (635, 637)

Lack, D. (1954) *The natural regulation of animal numbers*. Oxford University Press. (76, 118, 635, 660)

Lack, D. (1963) Migration across the southern North Sea studied by radar. Part 4. Autumn. *Ibis* **105**, 1–54. (676)

Lack, D. (1966) *Population studies of birds*. Clarendon Press, Oxford. (111, 160, 185, 190–1, 192, 208, 211, 277, 321, **322**, 638, **655**)

Lack, D. (1969) Drift migration: a correction. *Ibis* **111**, 253–5. (616)

Landin, J. and Stark, E. (1973) On flight thresholds for temperature and wind velocity, 24-hour flight periodicity and migration of the water beetle *Helophorus brevipalpis* Bedel (Col. Hydrophilidae). *Zoon* Suppl. **1**, 105–14. (246, **253**, 255, **256**)

Larsen, E.B. (1949) Studies in the activity of insects. III. Activity and migration of *Plusia gamma*. *Biol. Meddr.* **21**, 1–32. (475)

Larsen, T. (1968) Ecological investigations on the polar bear in Svalbard. *Norsk Polarinst. Arbok 1966.* 92–8. (177–8)

Lasker, G.W. and Kaplan, B.A. (1964) The coefficient of breeding isolation: population size, migration rates and the possibilities for random genetic drift in six human communities in Northern Peru. *Hum. Biol.* **36**, 327–38. (138)

Laskey, A.R. (1958) Blue jays at Nashville, Tennessee: movements, nesting, age. *Bird-banding*. **29**, 211–18. (634)

Laughlin, W.S. (1966) Genetical and anthropological characteristics of Arctic populations. *In:* Baker, P.T. and Weiner, J.S. (eds.) *The biology of human adaptability*. Clarendon Press, Oxford. pp. 469–96. (119)

Lawick, H. van (film). (214–15)

Lawick-Goodall, J. van (1968) The behaviour of free-living chimpanzees in the Gombe Stream Reserve. *Anim. Behav. Monog.* **1**, 165–311. (114, 115)

Lawrence, L. de K. (1973) *Black-capped chickadee*. Canadian Wildlife Service, Hinterland Who's Who. Information Canada, Ottawa. (**655**)

Laws, R.M. (1956) The elephant seal (*Mirounga leonina* Linn.). II. General, social and repro-

ductive behaviour. *Falkland Is. Depend. Surv. Sci. Repts.* **13**, 1–88. (**737**)

Laws, R.M. and Clough, G. (1966) Observations on reproduction in the hippopotamus, *Hippopotamus amphibius* linn. *Symp. zool. Soc. Lond.* **No. 15**, 117–40. (179)

Laycock, P. (1973) Distribution and abundance of bats in the Natal midlands (Mammalia: Chiroptera). *Ann. Transvaal Mus.* **28**, 207–27. (582)

Layne, J.N. (1958) Notes on mammals of southern Illinois. *Am. Midl. Nat.* **60**, 219–54. (**576**)

Layne, J.N. (1965) Observations on marine mammals in Florida waters. *Bull. Florida State Mus.* **9**, 131–81. (180, 526, 753, **757**)

Layne, J.N. (1967) Evidence for the use of vision in diurnal orientation of the bat, *Myotis austrori-parius. Anim. Behav.* **15**, 409–15. (891)

Lee, R.B. and DeVore, I. (1968) Problems in the study of hunters and gatherers. *In:* Lee, R.B. and DeVore, I. (eds.) *Man the hunter.* Aldine, Chicago. pp. 3–12. (**115**)

Lees, A.D. (1961) Clonal polymorphism in aphids. *In: Insect Polymorphism. Symp. R. ent. Soc. Lond.* **No. 1**, 68–79. (228)

Lees, A.D. (1966) The control of polymorphism in aphids. *In: Advances in insect physiology* **3**, 207–77. Academic Press, New York. (263)

Lemmetyinen, R. (1968) The migration routes of Finnish Common and Arctic terns (*Sterna hirundo* and *S. paradisaea*) in Scandinavia. *Ornis fenn.* **45**, 114–24. (625, **703**)

Lent, P.C. (1971) Muskox (*Ovibos moschatus*) management controversies in North America. *Biol. Conserv.* **3**, 255–63. (169)

Lewis, J.B. (1940) Mammals of Amelia County, Virginia. *J. Mammal.* **21**, 422–8. (**576**)

Lewis, J.R. (1964) *The ecology of rocky shores*. English Universities Press, London. (823)

Lewis, M. (pers. comm.) (**683**)

Lewis, T. (1961) Records of Thysanoptera at Silwood Park with notes on their biology. *Proc. R. ent. Soc. Lond.* A **36**, 89–95. (248)

Lewis, T. and Taylor, L.R. (1964) Diurnal periodicity of flight by insects. *Trans. R. ent. Soc. Lond.* **116**, 393–479. (258)

Leyhausen, P. (1964) The communal organisation of solitary animals. *Symp. zool. Soc. Lond.* **No. 14**, 249–63. (218, 317)

Li, Kuang-Po, Wong, Hong-Hsiang and Woo, Wan-Sei (1964) (Route of the seasonal migration of the Oriental Armyworm Moth in the eastern part of China as indicated by a three-year result of releasing and recapturing marked moths) (In Chinese with English Summary). *Acta phytophyl. Sin.* **3**, 101–10. (*From: Rev. appl. Ent. A* **53**, 391). (470, **471**)

Lidicker, W.Z. (1962) Emigration as a possible mechanism permitting the regulation of population density below carrying capacity. *Am. Nat.* **96**, 29–33. (37)

de Ligny, W. and Pantelouris, E. M. (1973) Origin of the European Eel. *Nature, Lond.* **246**, 518–19. (**821**)

Lilly, J.C. and Miller, A. (1961) Sounds emitted by the bottlenose dolphin. *Science* **133**, 1689–93. (767)

Lin, N. and Michener, C.D. (1972) Evolution of sociality in insects. *Quart. Rev. Biol.* **47**, 131–59. (202)

Lind, H. (1969) Internal migration in Britain. *In:* Jackson, J.A. (ed.) *Migration.* Cambridge University Press. pp. 74–98. (136–7, 139, 141)

Lindauer, M. (1955) Schwarmbienen auf Wohnungssuche. *Z. vergl. Physiol.* **37**, 263–324. (150–2)

Lindauer, M. (1967a) Recent advances in bee communication and orientation. *Ann. Rev. Entomol.* **12**, 439–70. (146, 148)

Lindauer, M. (1967b) *Communication among social bees.* Harvard University Press, Cambridge, Massachusetts. (146, 147, 148, 150, 875)

Lindauer, M. (in Frisch 1967). (**876**)

Lindauer, M. and Martin, H. (1972) Magnetic effect on dancing bees. *In:* Galler, S.R., Schmidt-Koenig, K., Jacobs, G.J. and Belleville, R.E. (eds.) *Animal orientation and navigation.* Scientific and Technical Information Office, National Aeronautics and Space Administration, Washington, D.C. pp. 559–67. (608, 877)

Lindberg, R.G. (1955) Growth, population dynamics, and field behavior in the spiny lobster, *Panulirus interruptus* (Randall). *Univ. Calif. Publs Zool.* **59**, 157–246. (833, 836)

Lindroth, C.H. (1949) *Die fennoskandischen Carabidae. Ein Tiergeographische Studie*, Vol. III. Vetensk. Samh. Handl., Goteborgs. (**274**, 275)

Lindsey, A.A. (1938) Notes on the crabeater seal. *J. Mammal.* **19**, 456–61. (721)

Lissaman, P.B.S. and Shollenberger, C.A. (1970) Formation flight of birds. *Science* **168**, 1003–5. (689)

List, R.J. (1951) *Smithsonian meteorological tables.* Smithsonian Institution, Washington. (444)

Lister, R. (1973) *Mallard.* Canadian Wildlife Service. Hinterland Who's Who. Information Canada, Ottawa. (**623**)

Lockie, J.D. (1966) Territory in small carnivores. *Symp. zool. Soc. Lond.* **No. 18**, 143–65. (**310**, 350, **398**, 403, **405**)

Lofts, B. and Marshall, A.J. (1960) The experimental regulation of *Zugunruhe* and the sexual cycle in the brambling *Fringilla montifringilla. Ibis* **102**, 209–14. (647)

Löhrl, H. (1959) Zur Frage des Zeitpunktes Einer Pargung auf die Heimatregion beim Haldbandschnapper (*Ficedula albicollis*). *J. für Ornithologie* **100**, 132–40. (190)

Long, R.C. (1973) *American robin.* Canadian Wildlife Service, Hinterland Who's Who. Information Canada, Ottawa. (7)

Loughrey, A.G. (1959) Preliminary investigation of the Atlantic walrus *Odobenus rosmarus rosmarus* (Linnaeus). *Canada Wildlife Service, Wildlife Mgmt Bull. ser. 1* **No. 14**, 123 pp. (**727**)

Lowery, G.H. and Newman, R.J. (1955) Direct studies of nocturnal bird migration. *In: Recent studies in avian biology. Univ. Ill. Bull.* (*Urbana*). pp. 238–63. (681)

Lyman, C.P. (1970) Thermoregulation and metabolism in bats. *In:* Wimsatt, W.A. (ed.) *Biology of bats.* Vol. I. Academic Press, London. pp. 301–30. (98)

MacArthur, R.H. (1959) On the breeding distribution of North American migrants. *Auk* **76**, 218–25. (**638**, 639)

MacArthur, R.H. and Wilson, E.O. (1967) *The theory of Island Biogeography.* Monographs in Population Biology 1. Princeton Univ. Press, Princeton, N.J. (75, 76)

McBride, A.F. and Kritzler, H. (1951) Observations on pregnancy, parturition and postnatal behavior in the bottlenose dolphin. *J. Mammal.* **32**, 251–66. (748)

McCarley, H. (1966) Annual cycle, population dynamics, and adaptive behaviour of *Citellus tridecemlineatus. J. Mammal.* **47**, 294–316. (**309**)

McCleave, J.D., Rommel, S.A. and Cathcart, C.L. (1971) Weak electric and magnetic fields in fish orientation. *Ann. N.Y. Acad. Sci.* **188**, 270–82. (880)

McClure, H.E. (1942) Summer activities of bats (genus *Lasiurus*) in Iowa. *J. Mammal.* **23**, 430–4. (**576**)

McDonald, D.L. (1972) Some aspects of the use of

visual cues in directional training of homing pigeons. *In:* Galler, S.R., Schmidt-Koenig, K., Jacobs, G. and Belleville, R.E. (eds.) *Animal orientation and navigation.* Scientific and Technical Information Office, National Aeronautics and Space Administration, Washington, D.C. pp. 293–304. (901)

Mackenzie, N. (1954) Low wages and large families. *New Statesman and Nation* 26 June, p. 823. (136, 144)

McKeown, J.P. (in Ferguson 1971). (884)

MacKinnon, J. (1974) *In search of the red ape.* London, Collins. (397)

Mackintosh, N.A. (1942) The southern stocks of whale-bone whales. *Discovery Reports.* **22**, 197–300. (748)

Mackintosh, N.A. (1965) *The stocks of whales.* Fishing News, London. (748)

Mackintosh, N.A. (1966) The distribution of southern blue and fin whales. *In:* Norris, K.S. (ed.) *Whales, dolphins and porpoises.* University of California Press, Berkeley and Los Angeles. pp. 125–44. (**755, 764, 883**)

McLaren, I.A. (1958) The biology of the ringed seal (*Phoca hispida* Schreber) in the eastern Canadian Arctic. *Fish. Res. Bd Canada, Bull.* **118**, 1–97. (720)

McLaren, I.A. (1960a) On the origin of the Caspian and Baikal seals and the palaeoclimatological implication. *Am. J. Sci.* **258**, 47–65. (706)

McLaren, I.A. (1960b) Are the Pinnipedia biphyletic? *Systematic Zoology* **9**, 18–28. (706)

McLaren, I.A. (1963) Effects of temperature on growth of zooplankton, and the adaptive value of vertical migration. *J. Fish. Res. Bd. Canada.* **20**, 685–727. (**486**, 490, 497–8)

McLaren, I.A. (1966) Taxonomy of harbor seals of the western North Pacific and evolution of certain other hair seals. *J. Mammal.* **47**, 466–73. (706)

McNab, B.K. (1963) Bioenergetics and the determination of home range size. *Am. Nat.* **97**, 133–40. (**389**, 390, **392–4**, 395)

McNab, B.K. (1971) The structure of tropical bat faunas. *Ecology* **52**, 352–8. (**558**)

Madge, D.S. (1969) Field and laboratory studies on the activities of two species of tropical earthworms. *Pedobiologia* **9**, 188–214. (508)

Madison, D.M. (1972) Homing orientation in salamanders: a mechanism involving chemical cues. *In:* Galler, S.R., Schmidt-Koenig, K., Jacobs, G.J. and Belleville, R.E. (eds.) *Animal orientation and navigation.* Scientific and Technical

Information Office, National Aeronautics and Space Administration, Washington, D.C. pp. 485–98. (885)

Maguire, B. (1963) The passive dispersal of small aquatic organisms and their colonization of isolated bodies of water. *Ecol. Mon.* **33**, 161–85. (25)

Malcolm, L.A., Booth, P.B. and Cavalli-Sforza, L.L. (1971) Intermarriage patterns and blood group gene frequencies of the Bundi people of the New Guinea highlands. *Hum. Biol.* (131, 132)

Mallach, N. (1971) Markeirungsversuche zur Analyse des Aktionsraums und der Ortsbewegungen des Bisams (*Ondatra zibethica* L.). *Anz. Schaedlingskd Pflanzenschutz* **44**, 129–36. (**310**)

Mallach, N. (1972) Translokationsversuche mit Bisamratten (*Ondatra zibethica* L.) *Anz. Schaedlingskd Pflanzenschutz* **45**, 40–4. (887)

Manders, N. (1904) Some breeding experiments on *Catopsilia pyranthe* and notes on the migration of butterflies in Ceylon. *Trans. ent. Soc. Lond.* **1904**, 701–6. (236)

Mangalam, J.J. (1968) *Human migration: a guide to migration literature in English 1955–1962.* University of Kentucky Press, Lexington. (20)

Mansfield, A.W. (1958) The biology of the Atlantic walrus, *Odobenus rosmarus rosmarus* (Linnaeus), in the eastern Canadian arctic. *Fish. Res. Bd. Canada. Report No. 653 in Manuscript Report Series (Biol).* (**727**)

Margalef, R. (1963) On certain unifying principles in ecology. *Am. Nat.* **97**, 357–74. (73)

Marine Mammal Biological Laboratory (1970) Fur seal investigations 1967. *U.S. Fish. Wildl. Serv. Spec. Sci. Rep. Fish.* **597**. 104 pp. (**106**, 110)

Marr, J. (1962) The natural history and geography of the Antarctic krill (*Euphausia superba* Dana). *Discovery Reports* **32**, 33–464. (488, 490, **501**, 747)

Marsden, W. (1964) *The lemming year.* Chatto and Windus, London. (413–4, 525–6)

Marshall, A.J. and Corbet, P.S. (1959) The breeding biology of equatorial vertebrates: reproduction of the bat *Chaerephon hindei* Thomas at latitude 0° 26′ N. *Proc. zool. Soc. Lond.* **132**, 607–16. (560)

Marshall, S.M., Nicholls, A.G. and Orr, A.P. (1934) On the biology of *Calanus finmarchicus*. V. Seasonal distribution, size, weight, and chemical composition in Loch Striven in 1933, and their relation to the phytoplankton. *J. mar. biol. Ass. U.K.* **19**, 793–827. (495)

Marshall, S.M. and Orr, A.P. (1955) *The biology of a*

marine copepod: *Calanus finmarchicus* (*Gunnerus*). Oliver and Boyd, Edinburgh and London. (**503**)

Marshall, S.M. and Orr, A.P. (1960) On the biology of *Calanus finmarchicus*. XI. Observations on vertical migration, especially in female *Calanus*. *J. mar. biol. Ass. U.K.* **39**, 135–47. (490)

Martin, J.A. (1955) The impact of industrialisation upon agriculture: a study of off-farm migration and agricultural development in Weakly County, Tennessee. Unpublished doctoral dissertation, University of Minnesota. (137)

Martin, R.D. (1968) Reproduction and ontogeny in tree shrews (*Tupaia belangeri*) with reference to their general behaviour and taxonomic relationships. *Z. Tierpsychol.* **25**, 409–532. (**309, 394**)

Mason, W.A. (1966) Social organization of the South American monkey, *Callicebus moloch:* a preliminary report. *Tulane Stud. Zool.* **13**, 23–8. (**295, 297**)

Mathiasson, S. (1963) Visible diurnal migration in the Sudan. *Proc. XIII Intern. Ornithol. Congr. (Ithaca, N.Y.).* pp. 430–5. (606)

Matthews, G.V.T. (1955) *Bird navigation* (1st ed.). Cambridge University Press. (**857**, 863–5, **871**, 896, 904, 907, **909**)

Matthews, G.V.T. (1963a) The astronomical basis of 'nonsense' orientation. *Proc. XIII Intern. Ornithol. Congr. (Ithaca, N.Y.)*, 315–29. (858, 864)

Matthews, G.V.T. (1963b) 'Nonsense' orientation as a population variant. *Ibis* **105**, 185–97. (896)

Matthews, G.V.T. (1968) *Bird navigation* (2nd ed.). Cambridge University Press. (201, 606, 863–5, 867, 871, 894, 896, 897, **898**, 899, **900**, 904, **905**, 906–7, 912, 914, **916**, 917, 921)

Mayer, W.V. (1953) A preliminary study of the Barrow ground squirrel, *Citellus parryi barrowensis*. *J. Mammal.* **34**, 334–45. (161)

Maynard Smith, J. (1954) Birds as aeroplanes. *New Biol.* **14**, 64–81. (252)

Maynard Smith, J. (1974) The theory of games and the evolution of animal conflicts. *J. theor. Biol.* **47**, 209–22. (403)

Maynard Smith, J. and Price, G.R. (1973) The logic of animal conflict. *Nature, Lond.* **246**, 15–18. (403)

Mayr, E. (in Newton 1972). (**191**)

Mayr, E. and Meise, W. (1930) Theoretisches zur Geschichte des Vogelzuges. *Der Vogelzug (Berlin)* **1**, 149–72. (698–700)

Mead, C. (1974) *Bird ringing: British Trust for Ornithology Guide number sixteen*. B.T.O., Tring, England. (**688, 916**)

Medway, Lord (1973) A ringing study of migratory barn swallows in West Malaysia. *Ibis* **115**, 60–86. (**650**)

Meek, A. (1916) *The migrations of fish*. Edward Arnold, London. (**792, 812, 819, 822, 846**)

Merkel, F.W. (1956) Untersuchungen über tages- und jahres-periodische Aktivitätsänderungen bei gekäfigten Zugvögeln. *Z. Tierpsychol.* **13**, 278–301. (**630**)

Merkel, F.W. and Fromme, H.G. (1958) Untersuchungen über das Orientierungvermogen Nachtlich Ziehender Rotkehlchen (*Erithacus rubecula*). *Naturwiss.* **45**, 499–500. (608)

Merkel, F.W. and Wiltschko, W. (1965) Magnetismus und Richtungsfinden Zugunruhiger Rotkehlchen (*Erithacus rubecula*). *Vogelwarte* **23**, 71–7. (608)

Merriam, C.H. (1888) Do any Canadian bats migrate? Evidence in the affirmative. *Trans. Royal Soc. Canada 1887 Sec.* **4**, 85–7. (**574**)

Metcalf, C.L. and Flint, W.P. (1962) *Destructive and useful insects: their habits and control* (4th ed. Revised by Metcalf, R.L.). McGraw-Hill, New York. (508)

Meteorological Office (1939) *The weather map*. M.O. 225i, HMSO, London. (256, **419**)

Meteorological Office (1954–1967) *Daily weather report of the British Meteorological Office*. M.O., Bracknell. (436, **437–8**)

Mewaldt, L.R. (1963) Californian 'crowned' sparrows return from Louisiana. *Western Bird Bander* **38**, 1–4. (914)

Mewaldt, L.R. (1964) California sparrows return from displacement to Maryland. *Science* **146**, 941–2. (**914**)

Mewaldt, L.R., Cowley, L.T. and Pyong-Oh Won (1973) California sparrows fail to return from displacement to Korea. *Auk* **90**, 857–61. (**914**)

Mewaldt, L.R., Kibby, S.S. and Morton, M.L. (1968) Comparative biology of Pacific coastal White-crowned sparrows. *Condor* **70**, 14–30. (631, 636)

Mewaldt, L.R., Morton, M.L. and Brown, I.L. (1964) Orientation of migratory restlessness in *Zonotrichia*. *Condor* **66**, 377–417. (864)

Meyer, M.E. and Lambe, D.R. (1966) Sensitivity of the pigeon to changes in the magnetic field. *Psychon. Sci.* **5**, 349–50. (608)

Meyer, W. (1955) Arbeitsteilung im Bienen-schwarm. *Naturwissenschaften* **42**, 350. (150)

Michel, R. (1969) Étude experimentale des variations de la tendance au vol au cours du vieillissement chez le criquet pélerin *Schistocerca gregaria* (Forsk.). *Rev. Comp. Anim.* **3**, 46–65. (240, 452)

Michener, M.C. and Walcott, C. (1967) Homing of single pigeons—an analysis of tracks. *J. exp. Biol.* **47**, 99–131. (897, 907)

Mikkola, K. (1967) Immigrations of Lepidoptera, recorded in Finland in the years 1946–1966, in relation to aircurrents. *Ann. ent. fenn.* **33**, 65–99. (476–7)

Miller, R.B. (1958) The role of competition in the mortality of hatchery trout. *J. Fish. Res. Bd. Can.* **15**, 27–45. (796)

Mills, D. (1971) *Salmon and trout: a resource, its ecology, conservation and management.* Oliver and Boyd, Edinburgh. (786–7, **789**, **796**, 797–9, **803**, **804**, 806–8, **845**, 847)

Mitchell, G.C. (1967) Population study of the funnel-eared bat (*Natalus stramineus*) in Sonora. *Southwest Natur.* **12**, 172–5. (**567**)

Moehres, F.P. and Kulzer, E. (1956) Über die Orientierung der Flughunde (Chiroptera-Pteropodidae). *Zeitschrift fur vergleichende Physiologie* **38**, 1–29. (586)

Moericke, V. (1941) *Zur Lebensweise der Pfirsichlaus (Myzodes persicae Sulz.) auf der Kartoffel.* Doct. Thesis Univ. Bonn. (231)

Moericke, V. (1955) Über die Lebensgewohnheiten der geflügelten Blattläuse (Aphidina) unter besonderer Berücksuchtigung des Verhaltens beim Landen. *Z. angew. Ent.* **37**, 29–91. (231)

Mohamed Khan bin Momin, K. (1967) Movements of a herd of elephants in Upper Perak. *Malay. Nat. J.* **20**, 18–23. (542)

Mohamed Khan bin Momin, K. (1969) Population and distribution studies of Perak elephants. *Malay. Nat. J.* **23**, 7–14. (166)

Moor, P.P. de and Steffens, F.E. (1972) The movements of vervet monkeys (*Cercopithecus aethiops*) within their ranges as revealed by radio-tracking. *J. Anim. Ecol.* **41**, 677–87. (162, **391**)

Moore, H.B. (1949) The zooplankton of the upper waters of the Bermuda area of the North Atlantic. *Bull. Bingham Oceanogr. Coll.* **12**, 1–97. (**504**)

Moore, H.B. (1950) The relation between the scattering layer and the Euphausiacea. *Biol. Bull.* **99**, 181–212. (494)

Moore, H.B. (1952) Physical factors affecting the distribution of euphausids in the North Atlantic. *Bull. Mar. Sci. Gulf and Caribb.* **1**, 278–305. (494)

Moore, H.B. and Corwin, E.G. (1956) The effects of temperature, illumination and pressure on the vertical distribution of zooplankton. *Bull. Mar. Sci. Gulf and Caribb.* **6**, 273–87. (494)

Moore, J.C. and Clark, E. (1963) Discovery of right whales in the Gulf of Mexico. *Science* **141**, 269. (753)

Moore, J.C. and Palmer, R.S. (1955) More piked whales from southern North Atlantic. *J. Mammal.* **36**, 429–33. (768)

Moreau, R.E. (1966) *The bird faunas of Africa and its islands.* Academic Press, New York and London. (699–700)

Moreau, R.E. (1969) The occurrence in winter quarters (Orstreue) of trans-Saharan migrants. *Bird Study* **16**, 108–10. (620)

Moreau, R.E. (1972) *The Palaearctic-African bird migration systems.* Academic Press, London and New York. (**7**, **276**, **612**, 620, **621**, **623**, **641**, **650**, **663**, **665**, **668**, **683**, **687**, **688**, **697**, **699**, **704**, 705)

Morgan, E. (1965) The activity of the amphipod *Corophium volutator* (Pallas) and its possible relationship to changes in hydrostatic pressure associated with the tides. *J. Anim. Ecol.* **34**, 731–46. (830)

Morgan, E., Nelson-Smith, A. and Knight-Jones, E.W. (1964) Responses of *Nymphon gracile* (Pycnogonida) to pressure cycles of tidal frequency. *J. exp. Biol.* **41**, 825–36. (830)

Mueller, H.C. (1965) Homing and distance-orientation in bats. *Z. Tierpsychol.* **23**, 403–21. (890)

Mueller, H.C. (1968) The role of vision in vespertilionid bats. *Am. Midl. Nat.* **79**, 524–5. (559, 891–2)

Müller, H.J. (1953) Der Blattlaus-Befallsflug im Bereich eines Ackerbohnen- und eines Kartoffel-Bestandes. *Beitr. Ent.* **3**, 229–58. (231)

Müller, K. (1954) Investigations on the organic drift in North Swedish streams. *Rep. Inst. Freshwater Res. Drottningholm* **35**, 133–48. (784)

Mumford, R.E. (1958) Population turnover in wintering bats in Indiana. *J. Mammal.* **39**, 253–61. (98, **603**)

Mundy, K. (1973) *Grizzly Bear.* Canadian Wildlife Service, Hinterland Who's Who. Information Canada, Ottawa. (**889**)

Murdock, G.P. (1959) *Africa: its peoples and their culture history.* McGraw-Hill, New York. (117, 118, 121, 123, 124, 125, 128–9, 540, 551)

Murdock, G.P. and Wilson, S.F. (1972) Settlement patterns and community organization: Cross-cultural codes 3. *Ethnology* **11**, 254–95. (138)

Murie, A. (1940) Ecology of the coyote in the Yellowstone. *Fauna Nat. Parks U.S. No. 4*. U.S. Dept. Interior, Nat. Parks Serv., Washington, D.C. 206 pp. (98)

Murie, A. (1944) The wolves of Mount McKinley. *Fauna Nat. Parks U.S. No. 5*. U.S. Dept. Interior, Nat. Parks Serv., Washington, D.C. 238 pp. (116, 153, **311**, 536, 543)

Murie, O.J. (1935) *Alaska-Yukon Caribou*. U.S. Dept. Agric., N. Amer. Fauna 54. 93 pp. (**533**)

Murray, B.G. (1967) Dispersal in vertebrates. *Ecology* **48**, 975–8. (37)

Musgrave, M.E. (1926) Some habits of mountain lions in Arizona. *J. Mammal.* **7**, 282–5. (98, 174)

Myers, K. and Poole, W.E. (1959) A study of the biology of the wild rabbit, *Oryctolagus cuniculus* (L.) in confined populations. I. The effects of density on home range and the formation of breeding groups. *C.S.I.R.O. Wild. Res.* **4**, 14–26. (104)

Myers, K. and Poole, W.E. (1961) A study of the biology of the wild rabbit, *Oryctolagus cuniculus* (L.), in confined populations. II. The effects of season and population increase on behaviour. *C.S.I.R.O. Wildl. Res.* **6**, 1–41. (104, 205, 220, **310**)

Myers, R.F. (1960) *Lasiurus* from Missouri caves. *J. Mammal.* **41**, 114–17. (**576**)

Napier, J.R. and Napier, P.H. (1967) *A handbook of living primates*. Academic Press, London. (**394**)

Nasu, K. (1966) Fishery oceanographic study on the baleen whaling grounds. *Sci. Rep. Whales Res. Inst.* **20**, 157–200. (**180, 748, 764, 883**)

Nasu, K. and Masaki, Y. (1970) Some biological parameters for stock assessment of the Antarctic sei whale. *Sci. Rep. Whales Res. Inst.* **22**, 63–74. (**180**)

Natushke, G. (1960) Ergebnisse der Fledermaus-beringung und biologische Beobachtungen an Fledermäusen in der Oberlausitz. *Bonner Zool. Beitr.* **11**, 77–98. (**582, 598, 602**)

Neave, D.J. and Wright, B.S. (1968) Seasonal migrations of the harbor porpoise (*Phocaena phocaena*) and other Cetacea in the Bay of Fundy. *J. Mammal.* **49**, 259–64. (**180, 753, 755, 757, 759, 764, 883**)

Needham, P.R. and Welsh, J.P. (1953) Rainbow trout (*Salmo gairdneri* Richardson) in the Hawaiian Islands. *J. Wildl. Mgmt.* **17**, 233–55. (822)

Negus, N., Gould, E. and Chipman, R.K. (1961) Ecology of the rice rat (*Oryzomys palustris*) on Breton Island, Gulf of Mexico, with a critique of the social stress theory. *Tulane Stud. Zool.* **8**, 95–123. (**310**)

Neill, W.T. (in Oliver 1955). (523)

Nelson, J.E. (1965a) Movements of Australian flying foxes (Pteropodidae: Megachiroptera). *Aust. J. Zool.* **13**, 53–73. (**566**, 589)

Nelson, J.E. (1965b) Behaviour of Australian Pteropodidae (Megachiroptera). *Anim. Behav.* **13**, 544–56. (559, 560, 563, **566**, 571, 586, 589)

Nemoto, T. (1959) Food of baleen whales with reference to whale movements. *Sci. Rep. Whales Res. Inst.* **14**, 149–290. (**180**, 746, 760, **883**)

Nemoto, T. (1964) School of baleen whales in the feeding areas. *Sci. Rep. Whales Res. Inst.* **18**, 89–110. (761)

Nero, R.W. (1957a) Saskatchewan silver-haired bat records. *Blue Jay* **15**, 38–46. (**574**)

Nero, R.W. (1957b) New silver-haired bat records. *Blue Jay* **15**, 121. (**574**)

Netboy, A. (1968) *The Atlantic Salmon: a vanishing species?* Faber and Faber, London. (847)

Netting, M.G. (1936) Hibernation and migration of the spotted turtle, *Clemmys guttata* (Schneider). *Copeia, Ann Arbor* **1936**, 112. (523)

Neuweiler, G. and Moehres, F.P. (1967) Die Rolle des Ortsgedächtnisses bei der Orientierung der Grossblatt-Fledermaus *Megaderma lyra*. *Z. Vergl. Physiol.* **57**, 147–71. (586)

Newell, P.F. (1966) The nocturnal behaviour of slugs. *Med. and biol. Illust.* part 16, 146–59. (**374-5**)

Newell, R.C. (1962) Behavioural aspects of the ecology of *Peringia* (=*Hydrobia*) *ulvae* (Pennant) (Gastropoda, Prosobranchia). *Proc. zool. Soc. Lond.* **138**, 49–75. (**827**)

Newsome, A.E. (1965a) The abundance of red kangaroos *Megaleia rufa* (Desmarest) in Central Australia. *Aust. J. zool.* **13**, 269–87. (**549**)

Newsome, A.E. (1965b) The distribution of red kangaroos, *Megaleia rufa* (Desmarest) about sources of persistent food and water in central Australia. *Aust. J. Zool.* **13**, 289–99. (**549**)

Newsome, A.E. (1965c) Reproduction in natural populations of the red kangaroo, *Megaleia rufa* (Desmarest) in central Australia. *Aust. J. Zool.* **13**, 735–59. (**549**)

Newton, I. (1972) *Finches*. Collins, London. (189,

191, 208, **209**, **210**, 211, 262, **323**, 627, 628, 631, **633**, 636, 647–8, 656, **657**, 658, 678–9, 681, 693)

Nice, M.M. (1937) Studies in the life history of the song sparrow. 1. A population study of the song sparrow. *Trans. Linn. Soc. New York.* **4**, 1–247. (190, **323**, 634–5, 680)

Nicholls, A.G. (1934) On the biology of *Calanus finmarchicus*. III. Vertical distribution and diurnal migration in the Clyde Sea area. *J. mar. biol. Ass. U.K.* **19**, 139–64. (490)

Nichols, U.G. (1953) Habits of the desert tortoise, *Gopherus agassizii, Herpetologica* **9**, 65–9. (523)

Nielsen, E.T. (1961) On the habits of the migratory butterfly, *Ascia monuste* L. *Biol. Meddr.* **23**, 1–81. (18, 259, 415, **417**, **419**, 421, 423)

Nikolsky, G.V. (1963) *The ecology of fishes.* Academic Press, New York. (797)

Nisbet, I.C.T. (1955) Atmospheric turbulence in bird flight. *Br. Birds* **48**, 557–9. (607)

Nisbet, I.C.T. and Drury, W.H. (1967) Orientation of spring migrants studied by radar. *Bird-banding* **38**, 173–86. (616, 632)

Nisbet, I.C.T. and Drury, W.H. (1968) Short-term effects of weather on bird migration: a field study using multivariate statistics. *Anim. Behav.* **16**, 496–530. (632, 642)

Nishida, T. (1966) A sociological study of solitary male monkeys. *Primates* **7**, 141–204. (162, 218, **281**, 282, 308, 329)

Nishida, T. (1968) The social group of wild chimpanzees in the Mahali Mountains. *Primates* **9**, 167–224. (117, 118, 161, 162, **295**, 297, 307)

Nishiwaki, M. (1966) Distribution and migration of the larger cetaceans in the north Pacific as shown by Japanese whaling results. *In:* Norris, K.S. (ed.) *Whales, dolphins and porpoises.* Univ. of California Press, Berkeley and Los Angeles. pp. 171–91. (**180, 182,** 751, **755, 763, 764, 883**)

Nishiwaki, M. (1967) *Distribution and migration of marine mammals in the north Pacific area.* No. 1. 64 pp. Ocean Research Institute, University of Tokyo, Nakano, Tokyo, Japan. (180, **182, 222, 301, 755, 763, 764, 883**)

Nishiwaki, M. (1972) General biology. *In:* Ridgway, S.H. (ed.) *Mammals of the sea: biology and medicine.* Charles C. Thomas, Springfield, Illinois. pp. 3–204. (752)

Nishiwaki, M. and Handa, C. (1958) Killer whales caught in the coastal waters off Japan for recent 10 years. *Sci. Rep. Whales Res. Inst.* **13**, 85–96. (**759**)

Nishiwaki, M. and Kasuya, T. (1970) Recent record of gray whale in the adjacent waters of Japan and a consideration on its migration. *Sci. Rep. Whales Res. Inst.* **22**, 29–37. (748)

Noble, G.K. (1931) *The biology of the Amphibia.* McGraw-Hill, New York. (522)

Norman, J.R. (1931) *A history of fishes.* Ernest Benn, Ltd, London. (846, 848)

Norris, K.S. (1967) Some observations on the migration and orientation of marine mammals (Pinnipeds, cetaceans). *In: Animal orientation and navigation.* Proc. 27th Ann. Biol. Colloquium, 6–7 May 1966, Corvallis Oregon. Oregon State University Press, Corvallis. pp. 101–25. (**731**)

Norris, K.S. (1968) The evolution of acoustic mechanisms in odontocete cetaceans. *In:* Drake, E.T. (ed.) *Evolution and environment: a symposium presented on the occasion of the 100th anniversary of the foundation of Peabody Museum of Natural History at Yale University.* Yale University Press, New Haven, Conn. pp. 297–324. (766)

Norris, K.S. and Harvey, G.W. (1972) A theory for the function of the spermaceti organ of the sperm whale (*Physeter catodon* L.). *In:* Galler, S.R., Schmidt-Koenig, K., Jacobs, G.J. and Belleville, R.E. (eds.) *Animal orientation and navigation.* Scientific and Technical Information Office, National Aeronautics and Space Administration, Washington D.C. pp. 397–417. (766)

Norris, M.J. (1959) Reproduction in the red locust (*Nomadacris septemfasciata* Serville) in the laboratory. *Anti-Locust Bull.* **No. 36**, 46 pp. (262)

Norris, M.J. (1962) Group effects on the activity and behaviour of adult males of the desert locust (*Schistocerca gregaria* Forsk.) in relation to sexual maturation. *Anim. Behav.* **10**, 275–91. (262)

Norris, M.J. (1964) Environmental control of sexual maturation in insects. *Insect Reproduction Symp., R. ent. Soc. Lond.* **No. 2**, 56–65. (262)

North Pacific Fur Seal Commission (1969) *Report on investigations from 1964 to 1966.* Washington D.C. 161 pp. (110)

Northcote, T.G. (1958) Effect of photoperiodism on response of juvenile trout to water currents. *Nature, Lond.* **181**, 1283–4. (797)

Novick, A. (1958) Orientation in palaeotropical bats. II. Megachiroptera. *J. exp. Biol.* **137**, 443–59. (585)

Nunneley, S.A. (1964) Analysis of banding records of local populations of blue jays and redpolls at Granby, Massachusetts. *Bird-Banding* **35**, 8–22. (634)

Nutting, W.L. (1966) Colonizing flights and associated activities of termites. 1. The desert damp-wood termite, *Paraneotermes simplicicornis* (Kolotermitidae). *Psyche, Camb.* **73**, 131–49. (240)

Odum, E.P. (1960) Lipid deposition in nocturnal migrant birds. *Proc. XII Internat. Ornith. Congr.* 563–76. (**695**)

Odum, E.P. and Kuenzler, E.J. (1955) Measurement of territory and home range size in birds. *Auk* **72**, 128–37. (321)

Oehmke (in Lindauer and Martin 1972). (877)

O'Farrell, M.J. and Bradley, W.G. (1970) Activity patterns of bats over a desert spring. *J. Mammal.* **51**, 18–26. (98)

Office of Population Censuses and Surveys (1971) *The Registrar General's statistical review of England and Wales for the year 1969. Part II. Tables, population.* HMSO, London. (**289–90, 292**)

Ohsumi, S. (1965) Reproduction of the sperm whale in the north-west Pacific. *Sci. Rep. Whales Res. Inst.* **19**, 1–35. (**301**, 750, **763**)

Ohsumi, S. (1966) Sexual segregation of the sperm whale in the North Pacific. *Sci. Rep. Whales Res. Inst.* **20**, 1–16. (**301, 763**)

Ohsumi, S., Masaki, Y. and Kawamura, A. (1970) Stock of the Antarctic minke whale. *Sci. Rep. Whales. Res. Inst.* **22**, 75–125. (**768**)

Oldham, R.S. (1966) Spring movements in the American toad, *Bufo americanus. Canad. J. Zool.* **44**, 63–100. (519)

Oliver, J.A. (1955) *The natural history of North American amphibians and reptiles.* Van Nostrand, Princeton, N.J. (194, 195, 523)

Olney, P.J.S. (1965a) The food and feeding habits of shelduck, *Tadorna tadorna. Ibis* **107**, 527–32. (**827**)

Olney, P.J.S. (1965b) The autumn and winter feeding biology of certain sympatric ducks. *In: Trans. VI Congr. Int. Union Game biol.* The Nature Conservancy, London. pp. 309–20. (**827**)

Omura, H. (1958) North Pacific right whale. *Sci. Rep. Whales Res. Inst.* **13**, 1–52. (753)

Omura, H. (1959) Bryde's whale from the coast of Japan. *Sci. Rep. Whales Res. Inst.* **14**, 1–33. (**182**)

Omura, H. and Ohsumi, S. (1964) A review of Japanese whale marking in the North Pacific to the end of 1962 with some information on marking in the Antarctic. *Norsk Hvalfangsttid.* **53**, 90–112. (**180, 182, 755, 763, 764, 883**)

Omura, H., Ohsumi, S., Nemoto, T., Nasu, K. and Kasuya, T. (1969) Black right whales in the North Pacific. *Sci. Rep. Whales Res. Inst.* **21**, 1–78. (**753**)

Orr, R.T. (1950) Notes on the seasonal occurrence of red bats in San Francisco. *J. Mammal.* **31**, 457–8. (**576**)

Orr, R.T. (1967) The Galapagos sea lion. *J. Mammal.* **48**, 62–9. (731)

Orr, R.T. (1970a) *Animals in migration.* Collier MacMillan, London. (**3, 7**, 185, 192, 195, 196, 523, 538, 573, **628, 650, 668, 669, 671, 672, 676,** 680, **692,** 705, **729, 732,** 755, **803,** 807, 847, **916**)

Orr, R.T. (1970b) Development: prenatal and postnatal. *In:* Wimsatt, W.A. (ed.) *Biology of bats.* Vol. I. Academic Press, London. pp. 217–32. (562)

Orr, R.T. and Poulter, T.C. (1965) The pinniped population of Año Nuevo Island, California. *Proc. Calif. Acad. Sci.* **32**, 377–404. (111, **729, 731, 732**)

Osborne, R.H. (1956) Internal migration in England and Wales, 1951. *Advancement of Science* **12**, 424–34. (139)

Österlöf, S. (1966) (The migration of the Goldcrest (*Regulus regulus*).) (In Swedish, English summary). *Var. Fagelvarld* **25**, 49–56. (635)

Ostrom, J.H. (1972) Were some dinosaurs gregarious? *Palaeogeog. Palaeoclim. Palaeoecol.* **11**, 287–301. (604)

Ostrom, J.H. (1973) The ancestry of birds. *Nature, Lond.* **242**, 136. (604)

Ostrom, J.H. (1974) *Archaeopteryx* and the origin of flight. *Q. Rev. Biol.* **49**, 27–47. (604)

Otto, F. (in Frisch 1967). (**876**)

Owen, D.F. (1971) *Tropical butterflies.* Clarendon Press, Oxford. (426)

Owen, R.B. (1968) Premigratory behavior and orientation in blue-winged teal, *Anas discors. Auk,* **85**, 617–32. (642)

Packard, R.L. (1955) Release, dispersal and reproduction of fallow deer in Nebraska. *J. Mammal.* **36**, 471–3. (169)

Pages, E. (1970) Sur l'écologie et les adaptations de l'Oryctérope et des Pangolins sympatriques du Gabon. *Biol. Gabonica* **6**, 27–92. (**309, 311**)

Pajunen, V.I. (1962) Studies on the population ecology of *Leucorrhinia dubia* v.d. Lind. (Odon., Libellulidae). *Ann. zool. Soc. zool.-bot. fenn. Vanamo.* **24**, 1–79. (257)

Pajunen, V.I. and Jansson, A. (1969) Dispersal of the rock pool corixids *Arctocorisa carinata* (Sahlb.) and *Callicorixa producta* (Reut.) (Heteroptera, Corixidae). *Ann. Zool. Fenn.* **6**, 391–427. (246)

Papi, F. and Pardi, L. (1963) On the lunar orientation of sandhoppers (Amphipoda, Talitridae). *Biol. Bull.* **124**, 97–105. (856)

Papi, F. and Tongiorgi, P. (1963) Innate and learned components in the astronomical orientation of wolf spiders. *Ergebnisse der Biologie* **26**, 260–80. (518, 878–9)

Parker, G.A. (1970a) Sperm competition and its evolutionary consequences in the insects. *Biol. Rev.* **45**, 528–68. (57, 339, 341, 560)

Parker, G.A. (1970b) The reproductive behaviour and the nature of sexual selection in *Scatophaga stercoraria* L. (Diptera: Scatophagidae). I. Diurnal and seasonal changes in population density around the site of mating and oviposition. *J. Anim. Ecol.* **39**, 185–204. (95, 224)

Parker, G.A. (1970c) The reproductive behaviour and the nature of sexual selection in *Scatophaga stercoraria* L. (Diptera: Scatophagidae). II. The fertilization rate and the spatial and temporal relationships of each sex around the site of mating and oviposition. *J. Anim. Ecol.* **39,** 205–28. (**53,** 58, 68, 95, 224, 269–70, **270–2,** 356)

Parker, G.A. (1970d) Sperm competition and its evolutionary effect on copula duration in the fly, *Scatophaga stercoraria*. *J. Insect Physiol.* **16**, 1301–28. (224)

Parker, G.A. (1970e) The reproductive behaviour and the nature of sexual selection in *Scatophaga stercoraria* L. (Diptera: Scatophagidae). IV. Epigamic recognition and competition between males for the possession of females. *Behaviour* **37**, 113–39. (224)

Parker, G.A. (1970f) The reproductive behaviour and the nature of sexual selection in *Scatophaga stercoraria* L. (Diptera: Scatophagidae). V. The female's behaviour at the oviposition site. *Behaviour* **37**, 140–68. (224)

Parker, G.A. (1970g) The reproductive behaviour and the nature of sexual selection in *Scatophaga stercoraria* L. (Diptera: Scatophagidae). VII. The origin and evolution of the passive phase. *Evolution* **24**, 791–805. (224)

Parker, G.A. (1971) The reproductive behaviour and the nature of sexual selection in *Scatophaga stercoraria* L. (Diptera: Scatophagidae). VI. The adaptive significance of emigration from the oviposition site during the phase of genital contact. *J. Anim. Ecol.* **40**, 215–33. (224, **226–7**)

Parker, G.A. (1972a) Reproductive behaviour of *Sepsis cynipsea* (L.) (Diptera: Sepsidae). I. A preliminary analysis of the reproductive strategy and its associated behaviour patterns. *Behaviour* **41**, 172–206. (232–3)

Parker, G.A. (1972b) Reproductive behaviour of *Sepsis cynipsea* (L.) (Diptera: Sepsidae). II. The significance of the precopulatory passive phase and emigration. *Behaviour* **41**, 242–50. (232–3)

Parker, G.A. (1974a) Courtship persistence and female-guarding as male time investment strategies. *Behaviour* **48**, 156–84. (38, 57, 58, 121, 327)

Parker, G.A. (1974b) Assessment strategy and the evolution of fighting behaviour. *J. theor. Biol.* **47**, 223–43. (68, 402, 796)

Parker, G.A. (1974c) The reproductive behaviour and the nature of sexual selection in *Scatophaga stercoraria* L. (Diptera: Scatophagidae). IX. Spatial distribution of fertilization rates and evolution of male search strategy within the reproductive area. *Evolution* **28**, 93–108. (58, 224)

Parker, G.A. (pers. comm.) (95, 408)

Parker, G.A. and Stuart, R.A. (1976) Animal behavior as a strategy optimizer: evolution of resource assessment strategies and optimal emigration thresholds. *Amer. Natur.* **110**, 1055–76. (vi)

Parker, G.A., Baker, R.R. and Smith, V.G.F. (1972) The origin and evolution of gamete dimorphism and the male-female phenomenon. *J. Theor. Biol.* **36**, 529–53. (57)

Parks, G.H. (1973) *Evening grosbeak*. Canadian Wildlife Service, Hinterland Who's Who. Information Canada, Ottawa. (24)

Parman, D.C. (1926) Migration of the long-beaked butterfly, *Libythea bachmani* Kirtland (Lepid: Libytheidae). *Ent. News* **37**, 101–6. (425)

Parry, D.A. (1949) The structure of whale blubber and its thermal properties. *Q. J. Microbiol. Sci.* **90**, 13–26. (750)

Parslow, J.L.F. (1969) The migration of passerine night migrants across the English Channel studied by radar. *Ibis* **111**, 48–79. (617, 619, 642, **643**)

Paulian, P. (1953) Pinnepèdes, cétacés, oiseaux des Iles Kerguelen et Amsterdam. Mission Kerguelen 1951. *Mém. Inst. Scient. de Madagascar, Sér. A.* **8**, 111–234. (**759**)

Payne, R. and Webb, D. (1971) Orientation by means of long-range acoustic signaling in baleen whales. *Ann. N.Y. Acad. Sci.* **188**, 110–41. (881, 883)

Pearson, J. (1908) *Cancer.* Liverp. Mar. Biol. Cttee Memoir XVI. (**834**)

Pearson, K. and Blakeman, J. (1906) Mathematical

contributions to the theory of evolution. XV. A mathematical theory of random migration. *Drap. C. Res. Mem. Biom. Ser.* **3**, 54 pp. (15)

Pearson, O.P., Koford, M.R. and Pearson, A.K. (1952) Reproduction of the lump-nosed bat (*Corynorhinus rafinesquei*) in California. *J. Mammal.* **33**, 273–320. (560)

Pearson, W.D. and Kramer, R.H. (1972) Drift and production of two aquatic insects in a mountain stream. *Ecol. Mon.* **42**, 365–85. (781, 784–5)

Penny, R.L. and Emlem, J.T. (1967) Further experiments on distance navigation in the Adelie Penguin. *Ibis* **109**, 99–109. (**671**)

Pennycuick, C.J. (1960) The physical basis of astronavigation in birds: theoretical considerations. *J. exp. Biol.* **37**, 573–93. (863–4)

Pennycuick, C.J. (1969) The mechanics of bird migration. *Ibis* **111**, 525–56. (616, 679, 683–4, 689, 693)

Perdeck, A.C. (1958) Two types of orientation in migrating starlings *Sturnus vulgaris* L. and chaffinches *Fringilla coelebs* as revealed by displacement experiments. *Ardea* **46**, 1–37. (610, **611, 913**)

Perdeck, A.C. (1963) Does navigation without visual cues exist in robins? *Ardea* **51**, 91–104. (608)

Perdeck, A.C. (1967) Oriëntatievermogen van Kokmeeuven. *Limosa* **40**, 158–60. (912)

Perdeck, A.C. and Clason, C. (1974) Spontaneous migration activity of chaffinches in the Kramer cage. *Progress Report 1973, Institute of Ecological Research, Royal Netherlands Academy of Arts and Sciences* 81–2. (674)

Perdeck, A.C. and Speek, B.J. (1974) Displacement of juvenile starlings during spring migration. *Progress Report 1973, Institute of Ecological Research, Royal Netherlands Academy of Arts and Sciences.* 80–1. (619)

Perry, R. (1966) *The world of the polar bear.* Univ. Washington Press. (177–8, 218)

Peterson, R.L. (1966) *The mammals of eastern Canada.* Oxford, New York. (**392**, 393)

Peterson, R.S., LeBoeuf, B.J. and DeLong, R.L. (1968) Fur seals from the Bering Sea breeding in California. *Nature, Lond.* **219**, 899–901. (**106**, 108, 110, **112, 735**)

Peterson, R.T. (1961) *A field guide to western birds.* The Riverside Press, Cambridge. (**3, 621, 623, 628, 641, 650, 661, 663, 665, 667, 668, 669, 685, 688, 692, 695, 697, 703, 914, 917**)

Peterson, R.T., Mountford, G. and Hollom, P.A.D.

(1974) *A field guide to the birds of Britain and Europe* (3rd ed.). Collins, London. (**611, 612, 619, 621, 623, 633, 637, 641, 657, 663, 667, 683, 692, 703, 859**)

Phillips, G.L. (1966) Ecology of the big brown bat (Chiroptera: Vespertilionidae) in northeastern Kansas. *Am. Midl. Nat.* **75**, 168–98. (**603**)

Phillips, R.L. and Mech. L.D. (1970) Homing behavior of a red fox. *J. Mammal.* **51**, 621. (887)

Pike, G.C. (1962) Migration and feeding of the gray whale (*Eschrichtius gibbosus*). *J. Fish. Res. Bd. Can.* **19**, 815–38. (748, 766)

Pollard, E. (1975) Aspects of the ecology of *Helix pomatia* L. *J. Anim. Ecol.* **44**, 305–30. (518)

Poole, E.L. (1932) A survey of the mammals of Berks County, Pennsylvania. *Reading Public Mus. and Art Gallery Bull. No. 13* 1–74. (**578**)

Popov, G.B. (1954) Notes on the behaviour of swarms of the desert locust (*Schistocerca gregaria* Forskål) during oviposition in Iran. *Trans. R. ent. Soc. Lond.* **105**, 65–77. (262)

Powell, G.C. and Nickerson, R.B. (1965) Reproduction of king crabs, *Paralithodes camtschatica* (Tilesius). *J. Fish. Res. Bd. Can.* **22**, 101–11. (840)

Pratt, J.G. and Wallraff, H.G. (in Matthews 1968). (**898**)

Prestt, I. (1971) An ecological study of the viper, *Vipera berus*, in southern Britain. *J. Zool., Lond.* **164**, 373–418. (523)

Price, C. (1969) The study of assimilation. *In:* Jackson, J.A. (ed.) *Migration.* Cambridge University Press. pp. 181–237. (136)

Proctor, V.W. and Malone, C.A. (1965) Passive dispersal of small aquatic organisms via the intestinal tracts of birds. *Ecology* **46**, 728–9. (25)

Prokopy, R.J. and Gyrisco, G.G. (1965) Summer migration of the alfalfa weevil, *Hypera postica* (Coleoptera: Curculionidae). *Ann Ent. Soc. Am.* **58**, 630–41. (225, 273)

Pruitt, W.O. (1959) Snow as a factor in the winter ecology of the barren-ground caribou (*Rangifer arcticus*). *Arctic* **12**, No. 3. (212–13)

Purchase, D. (1969) Fifth annual report on bat-banding in Australia. *Tech. Pap. Div. Wildl. Res. CSIRO Aust.* **15**, 3–14. (587)

Purves, P.E. (1966a) Comments at a meeting. *In:* Norris, K.S. (ed.) *Whales, dolphins and porpoises.* Univ. of California Press, Berkeley and Los Angeles. p. 253. (764)

Purves, P.E. (1966b) Anatomy and physiology of the outer and middle ear in cetaceans. *In:* Norris,

K.S. (ed.) *Whales, dolphins and porpoises*. Univ. of California Press, Berkeley and Los Angeles. pp. 320–80. (766)

Rabøl, J. (1970) Displacement and phaseshift experiments with night-migrating passerines. *Ornis Scand.* **1**, 27–43. (188, 614, 616–18, 913)

Radford, K.W., Orr, R.T. and Hubbs, C.L. (1965) Reestablishment of the northern elephant seal (*Mirounga angustirostris*) off Central California. *Proc. Calif. Acad. Sci.* **31**, 601–12. (**107**, 111, **732**)

Rainey, R.C. (1951) Weather and the movements of locust swarms: a new hypothesis. *Nature, Lond.* **168**, 1057–60. (461)

Rainey, R.C. (1963) Meteorology and the migration of desert locusts. *Tech. Notes Wld met. Org.* **No. 54**, 115 pp. (240, **365, 452–3, 456–8,** 460, 462–3)

Rainey, R.C. and Aspliden, C.I.H. (in Rainey 1963). (**452–3**)

Rainey, R.C., Waloff, Z. and Burnett, G.F. (1957) The behaviour of the red locust (*Nomadacris septemfasciata* Serville) in relation to the topography, meteorology, and vegetation of the Rukwa Rift Valley, Tanganyika. *Anti-locust Bull.* **26**, 1–86. (241, 256, 454)

Ramel, C. (1960) The influence of the wind on the migration of swallows. *Proc. XII Internat. Ornith. Congr.* 626–30. (617)

Rand, R.W. (1959) The Cape fur seal (*Arctocephalus pusillus*). Distribution, abundance and feeding habits off the South Western coast of the Cape Province. *Union S. Africa Dept. Commerce and Industries Div. Fisheries Invest. Rep.* **No. 34**. 75 pp. (**106**, 711, 715, 721, **742**)

Ransome, R.D. (1968) The distribution of the greater horseshoe bat, *Rhinolophus ferrum-equinum*, during hibernation, in relation to environmental factors. *J. Zool., Lond.* **154**, 77–112. (98, 99, 101, 160, **583**)

Raymont, J.E.G. (1963) *Plankton and productivity in the oceans*. Pergamon Press, Oxford. (**481, 483,** 485–6, **503,** 746, 823)

Registrar General (1894) *Fifty-fifth annual report of the registrar-general of births, deaths, and marriages in England (1892)*. HMSO, London. (144, 145)

Registrar General for Scotland (1971) *Annual report of the Registrar General for Scotland 1969. Part II. Population and vital statistics. No. 115*. HMSO, Edinburgh. (**290, 292**)

Renaud-Debyser, J. (1963) Recherches écologiques sur la faune interstitielle des sables (Bassin d'Arcachon, Ile de Bimini, Bahamas). *Vie et Milieu*, suppl. **15**. 1–157. (505, 509, **514**)

Renfrew, J.M. (1973) *Palaeoethnobotany: the prehistoric food plants of the Near East and Europe*. Methuen, London. (124)

Repenning, C.A., Peterson, R.S. and Hubbs, C.L. (1971) Contributions to the systematics of the southern fur seals, with particular reference to the Juan Fernandez and Guadalupe species. *In:* Burt, W.H. (ed.) *Antarctic Pinnipedia*. Ant. Res. Series 18. pp. 1–34. (706, **708**)

Ricard, M. (1971) *The mystery of animal migration*. Paladin, London. (**533, 792, 817,** 864)

Rice, A.L. (1961) The responses of certain mysids to changes of hydrostatic pressure. *J. exp. Biol.* **38**, 391–401. (830)

Rice, D.W. (1957) Life history and ecology of *Myotis austroriparius* in Florida. *J. Mammal.* **38**, 15–32. (559)

Rice, D.W. (1965) Offshore southward migration of gray whales off southern California. *J. Mammal.* **46**, 504–5. (**748**)

Rice, D.W. (1968) Stomach contents and feeding behaviour of killer whales in the Eastern North Pacific. *Norsk. Hvalfangsttid.* **57**, 35–8. (759)

Rice, D.W. and Fiscus, C.H. (1968) Right whales in the southeastern North Pacific. *Norsk Hvalfangsttid.* **57**, 105–7. (**753**)

Richards, O.W. (1940) The biology of the small white butterfly (*Pieris rapae*), with special reference to the factors controlling its abundance. *J. Anim. Ecol.* **9**, 243–88. (233)

Richardson, A. (1959) Some psychosocial aspects of British emigration to Australia. *British Journal of Sociology.* **10**, 327–37. (136)

Richardson, W.J. (1974) Spring migration over Puerto Rico and the western North Atlantic: a radar study. *Ibis* **116**, 172–93. (632)

Richmond, A.H. (1969) Sociology of migration in industrial and post-industrial societies. *In:* Jackson, J.A. (ed.) *Migration*. Cambridge University Press. pp. 238–81. (136, 137)

Riley, C.V. (in Williams 1958). (454)

Riley, G.A. (1939) Plankton studies. II. The western North Atlantic, May–June, 1939. *J. Mar. Res.* **2**, 145–62. (**482**)

Robbins, C.S., Bruun, B., Zim, H.S. and Singer, A. (1966) *A guide to field identification: Birds of North America*. Golden Press, New York. (**621, 623, 628, 641, 650, 661, 663, 665, 667, 668, 669, 676, 685, 688, 692, 695, 697, 703, 914**)

Roberts, B. (in Dorst 1962). (**673**)

Robertson, C.J.R. and Kinsky, F.C. (1972) The dispersal movements of the Royal Albatross, *Diomedea epomophora. Notornis* **19**, 289–301. (622–3)

Robinson, J.T. (1961) Australopithecines and the origin of man. *Ann. Rep. Smithsonian Inst. 1961.* (114)

Rodman, P.S. (1973) Population composition and adaptive organization among orang-utans of the Kutai Reserve. *In:* Michael, R.P. and Crook, J.H. (eds.) *Comparative ecology and behaviour of primates.* Academic Press, London and New York. pp. 171–209. (161, **309**)

Roe, F.G. (1972) *The North American buffalo: a critical study of the species in its wild state.* David and Charles, Newton Abbot. (554–5)

Roe, H.S.J. (1969) The food and feeding habits of the sperm whales (*Physeter catodon* L.) taken off the west coast of Iceland. *J. Cons. int. Explor. Mer.* **33**, 93–102. (**301, 763**)

Roer, H. (1959) Über Flug- und Wandergewohnheiten von *Pieris brassicae* L. *Z. angew Ent.* **44**, 272–309. (**435**, 439)

Roer, H. (1961a) Zur Kenntnis der Populationsdynamik und des Migrations-verhaltens von *Vanessa atalanta* L. im paläarktischen Raum. *Beitr. Ent.* **11**, 594–613. (423, 426, **435**, 436, **437–8**, 439)

Roer, H. (1961b) Ergebnisse mehrjähriger Markierungsversuche zur Erforschung der Flug- und Wandergewohnheiten europäischer Schmetterlinge. *Zool. Anz.* **167**, 456–63. (421, 423, **435**, 436, **437–8**, 439)

Roer, H. (1962) Experimentelle Untersuchungen zum Migrationsverhalten des Kleinen Fuchs (*Aglais urticae* L.). *Beitr. Ent.* **12**, 528–54. (423, **435**, 436, **437–8**)

Roer, H. (1968) Weitere Untersuchungen über die Auswirkungen der Witterung auf Richtung und Distanz der Flüge des Kleinen Fuchses (*Aglais urticae* L.) (Lep. Nymphalidae) im Rheinland. *Decheniana* **120**, 313–34. (415, **418–19**, **422–3**, **435**, 436, **437–8**, 439)

Roer, H. (1969) Zur Biologie des Tagpfauenauges, *Inachis io* L. (Lep. Nymphalidae), unter besonderer Berücksichtigung der Wanderungen im mitteleuropäischen Raum. *Zool. Anz.* **183**, 177–94. (423, **435**, 436, **437–8**, 439)

Roer, H. (1970) Untersuchungen zum Migrationsverhalten des Trauermantels (*Nymphalis antiopa* L.) (Lep. Nymphalidae). *Z. angew. Ent.* **4**, 388–96. (249, 423, **435**, 436, **437–8**, 439)

Roer, H. and Egsbaek, W. (1966) Zur Biologie einer skandinavischen Population der Wesserfleder-mäuse (*Myotis daubentoni*) (Chiroptera). *Z. Säugetierk* **31**, 440–53. (560, **596**)

Roffey, J. (1972a) Migration and dispersal in the Australian plague locust. *Abstracts, 14th Int. Cong. Ent. (Canberra)* p. 155. (454)

Roffey, J. (1972b) Radar studies of flight activity of locusts and other insects. *Ann. Rept, CSIRO Div. Ent., Canberra* 111–13. (476)

Roffey, J. (in Waloff 1972). (459)

Roos, T. (1957) Studies on upstream migration in adult stream-dwelling insects. *Rep. Inst. Freshwat. Res., Drottningholm* **38**, 167–93. (784)

Roppel, A.Y., Johnson, A.M. and Chapman, D.G. (1965) Fur seal investigations, Pribilof Islands, Alaska, 1963. *U.S. Fish. and Wild. Spec. Sci. Rep.—Fish. No. 497.* 1–60. Washington, D.C. (110)

Rotenberg, M. (1972) Theory of population transport. *J. theor. Biol.* **37**, 291–305. (37)

Rowan, W. (1925) Relation of light to bird migration and development changes, *Nature, Lond.* **115**, 494–5. (647)

Rowan, W. (1926) On photoperiodism, reproductive periodicity, and the annual migrations of birds and certain fishes. *Proc. Boston. Soc. Nat. Hist.* **38**, 147–89. (647)

Rowan, W. (1929) Experiments in bird migration. I. Manipulation of the reproductive cycle: seasonal histological changes in the gonads. *Proc. Boston Sco. Nat. Hist.* **39**, 151–208. (647)

Rowan, W. (1946) Experiments in bird migration. *Trans. roy. Soc. Can.* (3) 40, **5**, 123–35. (610)

Royce, W.F. (in Hasler 1967). (880)

Rüppell, W. (1944) Versuche über Heimfinden ziehender Nebelkrähen nach Verfrachtung. *J. Orn.* **92**, 106–32. (**619**)

Rüppell, W. (in Verheyen 1950). (**614**)

Rüppell, W. and Schein, W. (1941) Über das Heimfinden freilebender Stare bei Verfrachtung nach einjähriger Freiheitsentziehung am Heimatort. *Vogelzug* **12**, 49–56. (912)

Rüppell, W. and Schüz, E. (1948) Ergebnis der Verfrachtung von Nebelkrähen (*Corvus corone cornix*) während des Wegzuges. *Vogelwarte* **15**, 30–6. (610)

Ryberg, O. (1947) *Studies on bats and bat parasites.* Svensk Natur, Stockholm. (556, 559, **565**, **581**, **582**, **583**, **595**, **596**, **598**, **599**, **601**, **602**)

Saila, S.B. and Shappy, R.A. (1963) Random movement and orientation in salmon migration. *J. Cons. perm. int. Explor. Mer.* **28**, 153–66. (861–2)

St. Paul, U.V. (1953) Nachweis der Sonnerorientierung bei Nächtlich Ziehenden Vögeln. *Behaviour* **6**, 1–7. (609)

Salomonsen, F. (in Dorst 1962). (**697**)

Salzano, F. M. (1971) Demographic and genetic inter-relationships among the Cayapo Indians of Brazil. *Soc. Biol.* **18**, 148–57. (137)

Sanborn, C.C. and Crespo, J.A. (1957) El Murcielago Blanqizco (*Lasiurus cinereus*) y sus subespecies. *Biol. Mus. Argentino de Ciencias Nat. 'Bernardino Rivadavia'* **No. 4**, 1–13. (**578**)

Sargeant, A.B. and Warner, D.W. (1972) Movements and denning habits of a badger. *J. Mammal.* **53**, 207–10. (390)

Sargent, T.D. (1962) A study of homing in the bank swallow (*Riparia riparia*). *Auk* **79**, 234–46. (912)

Sauer, E.G.F. (1957) Die Sternenorientierung nächtlich ziehender Grasmücken (*Sylvia atricapilla, borin* und *curruca*). *Z. Tierpsychol.* **14**, 29–70. (864)

Sauer, E.G.F. (1963) Migration habits of golden plovers. *Proc. XIII Intern. Ornithol. Congr. (Ithaca, N.Y.).* pp. 454–67. (864)

Sauer, F. and Sauer, E.G.F. (1955) Zur Frage Nachtlichen Zugorientierung von Grasmücken. *Revue swisse de Zoologie* **62**, 250–9. (864)

Savory, T. H. (1955) *The world of small animals.* University of London Press, London. (**507**)

Savory, T.H. (1964) *Spiders and other Arachnids.* The English Universities Press, London. (199)

Schaefer, G.W. (1972) Radar studies of the flight activity and aerial transportation of insects. *Abstracts, 14th Int. Congr. ent. (Canberra)* pp. 154–5. (476)

Schaller, G.B. (1963) *The mountain gorilla.* University of Chicago Press. (92–94, 117, 162, **164, 298,** 299, 305, 307, 381, **385,** 396)

Schaller, G.B. (1965) The behavior of the mountain gorilla. *In:* DeVore, I. (ed.) *Primate behavior: field studies of monkeys and apes.* Holt, Rinehart and Winston, New York. pp. 324–67. (**298, 385**)

Schaller, G.B. (1967) *The deer and the tiger: a study of wildlife in India.* Univ. Chicago Press. (218, **310,** 313, 528, 551)

Schaller, G.B. (1972) *The Serengeti lion: a study of predator-prey relations.* Univ. Chicago Press. (**295, 310–11,** 538)

Schenkel, R. (1966) Play, exploration and territoriality in the wild lion. *Symp. zool. Soc. Lond.* **No. 18,** 11–22. (378)

Schenkel, R. and Schenkel-Hulliger, L. (1969) *Ecology and behaviour of the black rhinoceros (Diceros bicornis L.): a field study.* Verlag Paul Parey, Hamburg. (531)

Schevill, W.E. and Backus, R.H. (1960) Daily patrol of a *Megaptera. J. Mammal.* **41**, 279–81. (183, 759)

Schloeth, R. (1961) Das soziallében des Camargue Rindes. *Z. Tierpsychol.* **18**, 574–627. (551)

Schmaus, M. (1960) Fledermausberingung im Hunsrück. *Bonner Zool. Beitr.* **11**, 198–203. (**570, 595**)

Schmidt, J. (1922) The breeding places of the eel. *Phil. Trans. R. Soc.* B, **211**, 179–208. (**821**)

Schmidt-Koenig, K. (1966) Über die Entfernung als Parameter bei der Anfangsorientierung der Brieftaube. *Z. vergl. Physiol.* **52**, 33–5. (907, **909**)

Schmidt-Koenig, K. (1972) New experiments on the effect of clock shifts on homing in pigeons. *In:* Galler, S.R., Schmidt-Koenig, K., Jacobs, G.J. and Belleville, R.E. (eds.) *Animal orientation and navigation.* Scientific and Technical Information Office, National Aeronautics and Space Administration, Washington D.C. pp. 275–86. (864, 904)

Schmidt-Koenig, K. and Schlichte, H.J. (1972) Homing in pigeons with impaired vision. *Proc. Nat. Acad. Sci. U.S.A.* **69**, 2446–7. (910)

Schneegas, E.R. and Franklin, G.W. (1972) The Mineral King deer herd. *Calif. Fish Game* **58**, 133–40. (**527**)

Schneirla, T.C. (1971) *Army ants: a study in social organization.* W.H. Freeman, San Francisco. (75, 147, 149–50, **345**)

Schnell, G.D. (1965) Recording the flight-speed of birds by doppler radar. *The Living Bird.* **4**, 79–87. (681, 684)

Schnetter, W. (1960) Beringungsergebnisse an der Langflügelfledermaus (*Miniopterus schreibersi* Kuhl) im Kaiserstuhl. *Bonner Zool. Beitr.* **11**, 150–65. (**587**)

Schoener, T.W. (1971) Theory of feeding strategies. *Ann. Rev. Ecol. and Syst.* **2**, 369–404. (58)

Schüz, E. (1951) Uberlick uber die Orientierungsversuche der Vogelwarte Rossitten (jetzt: Vogelwarte Radolfzell). *Proc. XIII Intern. Ornithol. Congr. (Upsulla, Sweden)* pp. 249–68. (610, 613)

Schüz, E. (1963) On the northwestern migration

divide of the white stork. *Proc. XIII Intern. Ornithol. Congr. (Ithaca, N.Y.)* pp. 475–80. (610)

Schüz, E. and Weigold, H. (in Dorst 1962). (667)

Schwartz, C.W. and Schwartz, E.R. (1959) *The wild animals of Missouri*. Univ. Missouri Press, Columbia. (98, 101–2, 161, 174, 178, 179, **310**, 536)

Scott, P. and The Wildfowl Trust (1972) *The swans*. Michael Joseph, London. (691)

Sella, M. (1930) Distribution and migrations of the tuna, studied by the method of hooks etc. *Rev. Hydrobiol.* **24**. (**817**)

Sergeant, D.E. (1963) Minke whales, *Balaenoptera acutorostrata* Lacépède of the western North Atlantic. *J. Fish. Res. Bd Canada* **20**, 1489–1504. (**768**)

Sergeant, D.E. (1965) Migrations of harp seals, *Pagophilus groenlandicus* (Erxleben) in the northwest Atlantic. *J. Fish. Res. Bd Canada* **22**, 433–64. (880)

Sergeant, D.E. and Fisher, H.D. (1957) The smaller Cetacea of eastern Canadian waters. *J. Fish. Res. Bd. Canada* **14**, 83–115. (**757, 759**)

Serventy, D.L. (1967) Aspects of the population ecology of the short-tailed shearwater (*Puffinus tenuirostris*). *Proc. XIV Intern. Ornithol. Congr. (Oxford)* 165–90. (609)

Serventy, D.L. (in Dorst 1962). (**672**)

Serventy, V. (1965) Budgerygah. *Pacific Discovery* **18**, 23–25. (676–7)

Seymour, M.K. (1972) Swimming in *Arenicola marina* (L.). *Comp. Biochem. Physiol.* **41A**, 285–8. (**838**)

Shaw, B. (1965) Depressions and associated Desert Locust swarm movements in the Middle East. *In:* 'Meteorology and the Desert Locust'. *Tech. Notes Wld met. Org.* **No. 69**, 194–8. (461)

Shaw, M.J.P. (1970a) Effects of population density on alienicolae of *Aphis fabae* Scop. I. The effect of crowding on the production of alatae in the laboratory. *Ann. appl. Biol.* **65**, 186–96. (231, 355)

Shaw, M.J.P. (1970b) Effects of population density on alienicolae of *Aphis fabae* Scop. II. The effects of crowding on the expression of migratory urge among alatae in the laboratory. *Ann. appl. Biol.* **65**, 197–204. (231, 355)

Sheldon, W.G. (1950) Denning habits and home range of red foxes in New York State. *J. Wildl. Mgmt.* **14**, 33–42. (**173, 311**)

Sheldon, W.G. (1953) Returns on banded red and gray foxes in New York State. *J. Mammal.* **34**, 125. (**173**)

Shelford, V.E. (in Dorst 1962). (**206**)

Shevaryova, T. (1969) (Stability and changes of the nesting, moulting and hibernation places of waterfowl) (In Russian with English Summary). *Communs. Baltic Comm. Study Bird Migr.* **6**, 13–38. (622)

Short, H.L., Davis, R.B. and Herreid, C.F. II (1960) Movements of the Mexican free-tailed bat in Texas. *Southwest. Naturalist* **5**, 208–16. (**591**)

Shull, A.F. (1938) Time of determination and time of differentiation of aphid wings. *Am. Nat.* **72**, 170–9. (228)

Shumakov, M.E. (1965) (Preliminary results of the investigation of migrational orientation of passerine birds by the round-cage method) (In Russian). *Bionica* 371–8. (605)

Shumakov, M.E. (1967) (An investigation of the migratory orientation of passerine birds) (In Russian). *Biol. Ser., Vestnik Leningradskogo Universitata* 106–18. (605)

Simpson, G.G. (1961) Historical zoogeography of Australian mammals. *Evolution* **15**, 431–46. (556)

Sivertson, E. (1941) On the biology of the Harp seal, *Phoca groenlandica* Erxl. *Hvalradets skrifter* **No. 26**. 166 pp. (**739**)

Skead, D.M. (1973) Red-backed shrikes returning to the same wintering grounds. *Ostrich* **44**, 81. (620)

Skertchly, S.B.J. (1879) Butterfly swarms. *Nature, Lond.* **20**, 266. (236)

Slijper, E.J. (1962) *Whales*. Hutchinson, London. (**180, 182, 301, 501,** 746, **748,** 750, **751,** 752, **755, 757,** 761, **763, 764, 768,** 883)

Slijper, E.J. (1966) Functional morphology of the reproductive system in Cetacea. *In:* Norris, K.S. (ed.) *Whales, dolphins and porpoises*. Univ. Calif. Press, Berkeley and Los Angeles. pp. 277–319. (745, 748, 750)

Slijper, E.J. (1967) White whales in Netherlands waters (*Delphinapterus leucas*). *Z. Saugetierkunde* **32**, 86–9. (757)

Slijper, E.J., Utrecht, W.L. van and Naaktgeboren, C. (1964) Remarks on the distribution and migration of whales, based on observations from Netherlands ships. *Bijdr. Dierk.* **34**, 1–93. (**180,** 747–8, **753, 755, 760, 763, 764,** 768, **883**)

Sluiter, J.W. and Heerdt, P.F. van (1966) Seasonal habits of the noctule bat (*Nyctalus noctula*). *Arch. nèerl. zool.* **16**, 423–39. (99, 560, 561, 564, **581**)

Small, G.L. (1971) *The blue whale*. Columbia University Press, New York and London. (**755**)

Smallwood, M.E. (1903) The beach flea: *Talorchestia longicornis*. *Cold Spring Harbor Monographs* **1**, 1–27. (830)

Smith, M.S.R. (1965) Seasonal movements of the Weddell seal in McMurdo Sound, Antarctica. *J. Wildl. Mgmt.* **29**, 464–70. (720)

Snyder, W.E. (1902) A list with brief notes of the mammals of Dodge Co., Wisconsin. *Bull. Wisconsin Nat. Hist. Soc.* **2**, 113–26. (573)

Soper, J.D. (1941) History, range and home life of the northern bison. *Ecol. Monographs* **11**, 347–412. (555)

Sotthibandhu, S. (1977) A portable, semi-automatic flight mill for the study of moth orientation in the field. Exhibit at the Society for Experimental Biology Conference, Manchester, 29–31 March 1977. (477)

Sotthibandhu, S. (pers. comm.) (477)

Southern, H.N. (ed.) (1964) *The handbook of British mammals*. Blackwell, Oxford. (104, 115, 174, **310**)

Southern, W.E. (1969a) Orientation behavior of gull chicks. *Condor* **71**, 418–25. (608, 860)

Southern, W.E. (1969b) Sky conditions in relation to ring-billed and herring gull orientation. *Ill. State Acad. Sci.* **62**, 342–9. (860)

Southern, W.E. (1970) Orientation in gulls: effect of distance, direction of release and wind. *The Living Bird* **9**, 75–95. (608)

Southern, W.E. (1971a) Gull orientation by magnetic cues: a hypothesis revisited. *Ann. N.Y. Acad. Sci.* **188**, 295–311. (608, 860)

Southern, W.E. (1971b) Comments at a meeting. *Ann. N.Y. Acad. Sci.* **188**, 337. (899)

Southwood, T.R.E. (1962) Migration of terrestrial arthropods in relation to habitat. *Biol. Rev.* **37**, 171–214. (15, 31, 50, 54, 230, 244–5, 246, **247**, 248, 249, 250, 337, 342–3)

Southwood, T.R.E., Jepson, W.F. and Van Emden, H.F. (1961) Studies on the behaviour of *Oscinella frit* L. (Diptera) adults of the panicle generation. *Entomologia exp. appl.* **4**, 196–210. (241)

Sowls, L.K. (1955) *Prairie ducks, a study of their behavior, ecology and management*. Stackpole, Harrisburg. (187, **322**, 325)

Spuhler, J.N. and Clark, P.J. (1961) Migration into the human breeding population of Ann Arbor, Michigan, 1900–1950. *Human Biology* **33**, 223–36. (**135**)

Squires, H.J. (1957) Squid, *Illex illecebrosus* (LeSueur), in the Newfoundland fishing area. *J. Fish. Res. Bd Canada* **14**, 693–728. (200, 822)

Stasko, A.B. (1971) Review of field studies on fish orientation. *Ann. N.Y. Acad. Sci.* **188**, 12–29. (880)

Stebbins, R.C. and Kalk, M. (1961) Observations on the natural history of the mud-skipper, *Periophthalmus sobrinus*. *Copeia* **1961**, 18–27. (**825**)

Steiniger, F. (1950) Beiträge zur Soziologie und sonstigen Biologie der Wanderratte. *Z. Tierpsychol.* **7**, 356–79. (161)

Stenning, D.J. (1957) Transhumance, migratory drift, migration; patterns of pastoral Fulani nomadism. *J. Royal Anthropological Inst. Gt. Britain and Ireland* **87**, 57–73. (**539**)

Stenning, D.J. (1959) *Savannah nomads: a study of the Wodaabe pastoral fulani of western Bornu Province, Northern Region, Nigeria*. Oxford University Press, London. (**539**)

Stevenson-Hamilton, J. (1947) *Wild life in South Africa*. Cassell, London. (174, 194)

Stickel, L.F. (1960) *Peromyscus* ranges at high and low population densities. *J. Mammal.* **41**, 433–41. (**405**)

Stille, W.T. (1952) The nocturnal amphibian fauna of the southern Lake Michigan beach. *Ecology* **33**, 149–62. (521)

Stimson, J. (1973) The role of territory in the ecology of the intertidal limpet, *Lottia gigantea* (Gray). *Ecology* **54**, 1020–30. (199, 878)

Stirling, I. (1969) Ecology of the Weddell seal in McMurdo Sound, Antarctica. *Ecology* **50**, 573–86. (721)

Stirling, I. (1971) Studies on the behaviour of the South Australian fur seal, *Arctocephalus forsteri* (Lesson): II. Adult females and pups. *Aust. J. Zool.* **19**, 267–73. (711)

Stoddart, D.M. (1970) Individual range, dispersion and dispersal in a population of water voles (*Arvicola terrestris* L.). *J. Anim. Ecol.* **39**, 403–25. (329)

Stonehouse, B. (1972) *Animals of the Antarctic: the ecology of the far south*. Eurobook, Netherlands. (**671**)

Stones, R.C. and Branick, L.P. (1969) Use of hearing in homing by two species of *Myotis* bats. *J. Mammal.* **50**, 157–60. (891)

Stones, R.C. and Wiebers, J.E. (1966) Body weight and temperature regulation of *Myotis lucifugus* at a low temperature of 10°C. *J. Mammal.* **47**, 520–1. (99)

Stott, F.C. (1931) The spawning of *Echinus esculentus*, and some changes in gonad composition. *J. exp. Biol.* **8**, 133–50. (**840**)

Strelkov, P.P. (1970) (Sedentary and migratory species of bats (Chiroptera) in the European part of the USSR: 1) (In Russian with English Summary). *Byull. Mosk. o-va ispyt. prir. otd. Biol.* **75**, 38–52. (**570, 572,** 579, **582, 583, 595, 596, 597, 598, 599, 601, 602**)

Strelkov, P.P. (1971) (Sedentary and migratory species of bats (Chiroptera) in the European part of the USSR: 2) (In Russian with English summary). *Byull. Mosk. o-va ispyt. prir. otd Biol.* **76**, 5–21. (560, 575, 579, **581, 592**)

Strelkov, P.P. (1972) (Sedentary and migratory species of bats (Chiroptera) in the European part of the USSR: 3) (In Russian with English Summary). *Byull. Mosk. o-va ispyt prir otd Biol.* **77**, 27–31. (99)

Stresemann, E. (in Dorst 1962) (696)

Struhsaker, P. (1967) An occurrence of the minke whale, *Balaenoptera acutorostrata*, near the Northern Bahama Islands. *J. Mammal.* **48**, 483. (**768**)

Stuart, T.A. (1962) The leaping behaviour of salmon and trout at falls and obstructions. *Freshwat. Salm. Fish. Res.* **28**, 46 pp. (797)

Studer-Thiersch, A. (1969) Das Zugverhalten schweizerischer Stare *Sturnus vulgaris* nach Ringfunden. *Orn. Beob.* **66**, 105–44. (610)

Stuewer, F.W. (1948) A record of red bats mating. *J. Mammal.* **29**, 180–1. (**576**)

Suga, N. (1972) Neurophysiological analysis of echolocation in bats. *In:* Galler, S.R., Schmidt-Koenig, K., Jacobs, G.J. and Belleville, R.E. (eds.) *Animal orientation and navigation.* Scientific and Technical Information Office, National Aeronautics and Space Administration. Washington D.C. pp. 341–53. (585)

Sugiyama, Y. (1960) On the division of a natural troop of Japanese monkeys at Takasakiyama. *Primates* **2**, 109–44. (216)

Sugiyama, Y. (1968) Social organization of chimpanzees in the Budongo forest, Uganda. *Primates* **9**, 225–58. (162, **295**)

Suthers, R. (1966) Optomotor responses by echolocating bats. *Science, N.Y.* **152**, 1102–1104. (586)

Sutter, E. (1957a) Radar als Hilfsmittel der Vogelzugforschung. *Orn. Beob.* **54**, 70–96. (681)

Sutter, E. (1957b) Radar-Beobachtungen über der Verlauf des nächtlichen Vogelzuges. *Rev. Suisse Zool.* **64**, 294–303. (681)

Sutter, J. (1969) The effect of birth limitation on genetic composition of populations. *In:* Behrman, S.J., Corsa, L. and Freedman, R. (eds.) *Fertility and family planning: a world view.* University of Michigan Press, Ann Arbor. pp. 213–51. (137)

Sutton, G.M. and Hamilton, W.J. (1932) The exploration of Southampton Island, Hudson Bay. II. Zoology (i). The mammals of Southampton Island. *Mem. Carnegie Museum* **12**, 1–111. (208)

Svärdson, G. (in Dorst 1962). (**633**)

Sverdrup, H. U., Johnson, M.W. and Fleming, R.H. (1946) *The oceans: their physics, chemistry and general biology.* Prentice-Hall, New York. (823)

Swarth, H.S. (in Dorst 1962). (**697**)

Swedlund, A.C. (1972) Observations on the concept of neighbourhood knowledge and the distribution of marriage distances. *Ann. Hum. Genet.* **35**, 327–30. (138)

Swedmark, B. (1964) The interstitial fauna of marine sand. *Biol. Rev.* **39**, 1–42. (**514**)

Sweeney, J.R., Marchinton, R.L. and Sweeney, J.M. (1971) Responses of radio-monitored white-tailed deer chased by hunting dogs. *J. Wildl. Mgmt.* **35**, 707–16. (169, **311**)

Szathmary, E.J.E. and Reed, T.E. (1972) Caucasian admixture in two Ojibwa Indian communities in Ontario. *Human Biol.* **44**, 655–71. (127, 128)

Szulc-Olech, B. (1965) The resting period of migrant robins on autumn passage. *Bird Study* **22**, 1–7. (617, 645)

Talbot, L. and Talbot M. (1963) The wildebeest in Western Masailand, East Africa. *Wildlife Monographs* No. 12. (95, 155, 166, 408, **535,** 549)

Talbot, M. (1964) Nest structure and flights of the ant *Formica obscuriventris* Mayr. *Anim. Behav.* **12**, 154–8. (236)

Tanner, J.T. (1966) Effects of population density on growth rates of animal populations. *Ecology* **47**, 733–45. (268)

Tarasevich, M.N. (1968) (Food connections of sperm whales in the northern Pacific) (In Russian with English summary). *Zool. Zh.* **47**, 595–601. (746, 763)

Tate, G.H.H. (1946) Geographical distribution of the bats in the Australian Archipelago. *Am. Mus. Novit.* **1323**, 1–21. (556)

Taylor, L.R. (1957) Temperature relation of teneral development and behaviour in *Aphis fabae* Scop. *J. exp. Biol.* **34**, 189–208. (**257**)

Taylor, L.R. (1963) Analysis of the effect of temperature on insects in flight. *J. Anim. Ecol.* **32**, 99–117. (253, 254, **257**)

Taylor, L.R., French, R.A. and Macaulay, E.D.M. (1973) Low-altitude migration and diurnal flight periodicity; the importance of *Plusia gamma* L. (Lepidoptera: Plusiidae). *J. Anim. Ecol.* **42**, 751–60. (472–3, **474**)

Taylor, R.H.F. (1957) An unusual record of three species of whale being restricted to pools in Antarctic sea-ice. *Proc. zool. Soc. London.* **129**, 325–31. (747, **759**)

Tchernavin, V. (1939) The origin of salmon. *Salmon and Trout Mag.* **95**, 1–21. (789, 802, **803, 804, 845**, 847, 848)

Tener, J.S. (1954) Facts about Canadian musk-oxen. *Trans. 19th N. Amer. Wildl. Conf., Mgmt. Inst., Washington D.C.* pp. 504–10. (166)

Tener, J.S. (1965) *Muskoxen in Canada: a biological and taxonomic review.* Queen's printer, Ottawa. (542, 545)

Tesch, F.W. and Albrecht, M.L. (1961) Über den Einfluss verschiedener Umweltfaktoren auf Wachstum und Bestand der Bachforelle (*Salmo trutta fario* L.) in Mittelgebirgsgewässer. *Verh. int. Verein. theor. angew. Limnol.* **14**, 763–8. (793)

Thomas, D.S. (1959) Age and economic differentials in internal migration in the United States: structure and distance. *In: International Population Conference, Vienna, 1959.* International Union for the Scientific Study of Population, Vienna. pp. 714–21. (**288**)

Thomas, E. (1966) Orientierung der Imagines von *Capnia atra* Morton (Plecoptera). *Oikos* **17**, 278–80. (**785**)

Thompson, D.Q. (1952) Travel, range, and food habits of timber wolves in Wisconsin. *J. Mammal.* **33**, 429–42. (174, 390, 393, 536)

Thompson, D.Q. and Person, R.A. (1963) The Eider pass at Point Barrow, Alaska. *J. Wildl. Mgmt.* **27**, 348–56. (620)

Thomson, A.L. (1974) The migration of the Gannet: a reassessment of British and Irish ringing data. *British Birds* **67**, 89–103. (648)

Thomson, A.L. (in Dorst 1962). (**185**)

Thorpe, W.H. (1950) A note on detour experiments with *Ammophila pubescens* Curt. *Behaviour* **13**, 257–63. (374)

Thorpe, W.H. (1963) *Learning and instinct in animals* (2nd ed.). Methuen, London. (518, 873, **874**)

Tickell, W.L.N. (1968) The biology of the great albatrosses, *Diomedea exulans* and *Diomedea epomophora. In:* Austin, O.L. (ed.) *Antarctic bird studies.* Antarctic Research Series, Vol. 12, American Geophysical Union, Washington. pp. 1–55. (**624**)

Tickell, W.L.N. and Scotland, C.D. (1961) Recoveries of ringed giant petrels *Macronectes giganteus. Ibis* **103a**, 260–6. (622)

Tilden, J.W. (1962) General characteristics of the movements of *Vanessa cardui* (L.). *J. Res. Lepidoptera* **1**, 43–9. (258)

Tillyard, R.J. (1917) *The biology of dragonflies.* Cambridge University Press. (236)

Tinbergen, L. (in Dorst 1962). (**633**)

Tindale, N.B. (1953) Tribal and intertribal marriage among Australian aborigines. *Human Biol.* **25**, 160–90. (118, 120, 122, 123, 126, 132)

Tinkle, D.W. (1967a) Home range, density, dynamics and structure of a Texas population of the lizard, *Uta stansburiana. In:* Milstead, W.W. (ed.) *Lizard ecology: a symposium.* University of Missouri Press, Columbia. pp. 5–29. (195, **309, 405**)

Tinkle, D.W. (1967b) The life and demography of the side-blotched lizard, *Uta stansburiana. Misc. Pub. Mus. Zool. Univ. Mich.* No. **132**, 1–182. (96, 195, **309, 405**)

Toba, H., Paschke, J.D. and Friedman, S. (1967) Crowding as the primary factor in the production of the agamic form of *Therioaphis maculata* (Homoptera: Aphididae). *J. Insect Physiol.* **13**, 381–96. (229)

Tobias, P.V. (1957) Bushmen of the Kalahari. *Man.* **57**, 33–40. (129)

Tobias, P.V. (1961) Physique of a desert folk. *Nat. Hist.* **70**, 17–25. (129)

Tobias, P.V. (1964) Bushman hunter–gatherers: a study in human ecology. *In:* Davis, D.H.S. (ed.) *Ecological studies in Southern Africa.* Dr. W. Junk, The Hague. pp. 67–86. (117, 118, 120, 122, 126, 132, 550)

Tobias, P.V. (1973) Implications of the new age estimates of the early South African hominids. *Nature, Lond.* **246**, 79–83. (**113**)

Tomialojć, L. (1967) (The twite, *Carduelis flavirostris* (L.) in Poland and adjacent territories) (In Polish, English summary). *Acta orn., Warsz.* **10**, 109–56. (647)

Tomilin, A.G. (1955) (On the behaviour and sonic signalling of whales) (In Russian). *Trudy Inst. Okeanol., Akad. Nauk. SSSR* **18**, 28–47. (764)

Tomita, K. (1966) The sources of food for the Hadzapi tribe: the life of a hunting tribe in East Africa. *Kyoto Univ. African Studies* **1**, 157–71. (114, 117, 118, 125, **550**)

Tomlinson, R.E., Wight, H.M. and Baskett, T.S.

(1960) Migrational homing, local movement, and mortality of mourning doves in Missouri. *Trans. 25th N. Amer. Wildl. Conf.*, Wildl. Mgmt. Inst., Washington D.C. pp. 253–67. (**322**)

Topal, G. (1956) The movements of bats in Hungary. *Ann. Hist. Natur. Mus. Nat. Hung.* **N.S. 7**, 477–88. (**570, 572,** 579, **583, 587**)

Tower, W.L. (1906) An investigation of evolution in chrysomelid beetles of the genus *Leptinotarsa. Publs Carnegie Instn* No. **48**, 1–320. (351, **468**)

Townsend, J.E. (1953) Beaver ecology in western Manitoba with special reference to movements. *J. Mammal.* **34**, 459–79. (**310**)

Trivers, R.L. (1971) The evolution of reciprocal altruism. *Quart. Rev. Biol.* **46**, 35–57. (202)

Trivers, R.L. (1973) Parental investment and sexual selection. *In:* Campbell, B. (ed.) *Sexual selection and the descent of man 1871–1971.* Aldine, Chicago. (57, 58, 118, 121, 144, 285, 304, 327)

True, F.W. (1907) Remarks on the type of the fossil cetacean *Agorophius pygmaeus* (Müller). *Smithson. Inst. Spec. Publ.* no. **1694**. 1–8. (745)

Trueman, E.R. (1971) The control of burrowing and the migratory behaviour of *Donax denticulatus* (Bivalvia: Tellinacea). *J. Zool., Lond.* **165**, 453–69. (**824**)

Tucker, D.W. (1959) A new solution to the Atlantic eel problem. *Nature, Lond.* **183**, 495–501. (**821**)

Tulloch, R.J. (1968) Snowy owls breeding in Shetland in 1967. *British Birds* **61**, 119–32. (**207**)

Tulloch, R.J. (1969) Snowy owls breeding in Shetland. *British Birds* **62**, 33–6. (**207**)

Turnbull, C.M. (1968) The importance of flux in two hunting societies. *In:* Lee, R.B. and DeVore, I. (eds.) *Man the hunter.* Aldine, Chicago. pp. 132–7. (554)

Turner, D., Shaughnessy, A. and Gould, E. (1972) Individual recognition between mother and infant bats (*Myotis*). *In:* Galler, S.R., Schmidt-Koenig, K., Jacobs, G.J. and Belleville, R.E. (eds.) *Animal orientation and navigation* Scientific and Technical Information Office, National Aeronautics and Space Administration, Washington D.C. pp. 365–71. (586)

Turner, J.R.G. (1971) Experiments on the demography of tropical butterflies. II. Longevity and home range behaviour in *Heliconius erato. Biotropica* **3**, 21–31. (198, 374, 424)

Tuttle, M.D. (1970) Distribution and zoogeography of Peruvian bats, with comments on natural history. *Univ. Kansas Sci. Bull.* **49**, 45–86. (563, **565**)

Twente, J.W. (1955) Some aspects of habitat selection and other behavior of cavern-dwelling bats. *Ecology* **36**, 706–32. (99)

Twente, J.W. (1956) Ecological observations on a colony of *Tadarida mexicana. J. Mammal.* **37**, 42–7. (585)

Twitty, V. (1959) Migration and speciation in newts. *Science* **130**, 1735–43. (521)

Twitty, V., Grant, D. and Anderson, O. (1964) Long distance homing in the newt *Taricha rivularis. Proc. Nat. Acad. Sci.* **51**, 51–8. (196, 521)

Tyndale-Biscoe, H. (1973) *Life of marsupials.* Edward Arnold, London. (114, **309**, 605)

Tyrväinen, H. (1970) The mass occurrence of the fieldfare (*Turdus pilaris* L.) in the winter of 1964–5 in Finland. *Ann. Zool. Fenn.* **7**, 349–57. (635, 647)

Udvardy, M.D.F. (1969) *Dynamic zoogeography, with special reference to land animals.* Van Nostrand Reinhold, New York. (274–5)

Urquhart, F.A. (1960) *The monarch butterfly.* University of Toronto Press. (241, 337, 423, **424, 426–8,** 433)

Vachon, M. (1947) Nouvelles remarques à propos de la phorésie des Pseudoscorpions. *Bull. Mus. Hist. nat. Paris* **19**, 84–7. (246)

Valen, L. van (1968) Monophyly or diphyly in the origin of whales. *Evolution* **22**, 37–41. (745)

Vaughan, T.A. (1970) Flight patterns and aerodynamics. *In:* Wimsatt, W.A. (ed.) *Biology of bats.* Vol. I. Academic Press, London. pp. 195–216. (580)

Vaughan, T.A. (1972) *Mammalogy.* W.B. Saunders, London. (124, 387)

Vaurie, C. (1965) *The birds of the Palearctic fauna; A systematic reference, non-passeriformes.* H.F. and G. Witherby, London. (**667, 668, 687, 688, 697**)

Verheyen, R. (1950) La Cigogne blanche dans son quartier d'hiver. *Gerfaut* **40**, 1–17. (**614**)

Verheyen, R. (1954) *Monographie éthologique de l'hippopotame, Hippopotamus amphibius Linné.* Exploration du Parc National Albert, Inst. parcs nat. Congo Belge. Brussels. (179)

Vibe, C. (1967) Arctic animals in relation to climatic fluctuations. *Medd. Gronland* **170**, 7–227. (**751**, 757)

Vibert, R. (1962) Quelques conséquences du courant d'eau et des champs électriques sur le comportement des poissons. *Schweiz. Z. Hydrol.* **24**, 436–43. (796)

Viguier, C. (1882) Le sens d'orientation et ses organes chez les animaux et chez l'homme. *Rev. Phil.* **14**, 1–36. (865)

Villa-R., B. (1960) *Tadarida yucatanica* in Tamaulipas. *J. Mammal.* **41**, 314–19. (563)

Villa-R., B. and Cockrum, E.L. (1962) Migration in the Guano bat *Tadarida brasiliensis mexicana* (Saussure). *J. Mammal.* **43**, 43–64. (591)

Vleugel, D.A. (1952) Über die Bedeutung des Windes für die Orientierung Ziehender Buchfinken (*Fringilla coelebs*). *Orn. Beob.* **49**, 45–53. (608)

Vleugel, D.A. (1959) Über die Wahrscheinlichste Methode der Wind-Orientierung Ziehender, Buchfinken (*Fringilla coelebs*). *Orn. Fennica* **36**, 78–88. (608)

Vogel, S. (1968) Chiropterophilie in der neotropischen flora: Neue Mitteilungen 1. *Flora Abt. B. Morphol. Geobot. Jena* **157**, 562–602. (558)

Voous, K.H. (1960) *Atlas of European birds.* Nelson, Elsevier. (**621, 623, 637, 641, 650, 663, 665, 667, 668, 687, 688, 692, 697, 703**)

Wada, S. (1969) Territorial expansion of the Iraqw: land tenure and the locality group. *Kyoto Univ. Afr. Stud.* **4**, 115–32. (127)

Walcott, C. (1972) The navigation of homing pigeons: do they use sun navigation? *In:* Galler, S.R., Schmidt-Koenig, K., Jacobs, G.J. and Belleville, R.E. (eds.) *Animal orientation and navigation.* Scientific and Technical Information Office, National Aeronautics and Space Administration, Washington D.C. pp. 283–92. (608)

Walcott, C. and Green, R.P. (1974) Orientation of homing pigeons altered by a change in the direction of an applied magnetic field. *Science,* **184**, 180–2. (608)

Walcott, C. and Michener, M.C. (1967) Analysis of tracks of single homing pigeons. *Proc. XIV Int. Orn. Cong. Oxford* 311–29. (904, 906)

Walcott, C. and Michener, M.C. (1971) Sun navigation in homing pigeons—attempts to shift sun coordinates. *J. exp. Biol.* **54**, 291–316. (864, 904)

Walford, L. A. (1958) *Living resources of the sea.* Ronald Press, New York. (**747**)

Walker, B.W. (1952) A guide to the grunion. *Calif. Fish Game* **38**, 409–20. (**842**)

Walley, H.D. (1971) Movements of *Myotis lucifugus lucifugus* from a colony in LaSalle County, Illinois. *Trans. Ill. State Acad. Sci.* **63**, 409–14. (**569, 584**)

Wallraff, H.G. (1960) Does celestial navigation exist in animals? *Cold Springs Harbor symposia on quantitative biology: biological clocks* **25**, 451–61. (864, 871)

Wallraff, H.G. (1965) Über das Heimfindevermögen von Brieftauben mit durchtrennten Bogengängen. *Z. Vergleich. Physiol.* **50**, 313–30. (863)

Wallraff, H.G. (1966) Über die Heimfindeleistungen von Brieftauben nach Haltung in verschiedenartig abgeschimten Volieren. *Z. Vergl. Physiol.* **52**, 215–59. (900)

Wallraff, H.G. (1967) The present status of our knowledge about pigeon homing. *Proc. XIV Intern. Ornithol. Congr. (Oxford)* 331–58. (899)

Wallraff, H.G. (1972) An approach toward an analysis of the pattern recognition involved in the stellar orientation of birds. *In:* Galler, S.R., Schmidt-Koenig, K., Jacobs, G.J. and Belleville, R.E. (eds.) *Animal orientation and navigation.* Scientific and Technical Information Office, National Aeronautics and Space Administration, Washington D.C. pp. 211–22. (607)

Wallraff, H.G. (1974) The effect of directional experience on initial orientation in pigeons. *Auk* **91**, 24–34. (894, 897)

Wallraff, H.G. (in Matthews 1968). (**905**)

Wallraff, H.G. and Graue, L.G. (1973) Orientation of pigeons after transatlantic displacement. *Behaviour* **44**, 1–35. (904, 910–11)

Waloff, Z. (1959) Notes on some geographical aspects of the desert locust problem. In Rep. of the FAO Panel of Experts on the Strategy of Desert Locust Plague Control. Document No. FAO/59/6/4737. pp. 23–6. (461)

Waloff, Z. (1962) Flight activity of different phases of the desert locust in relation to plague dynamics. *In:* Physiologie, Comportement et Écologie des Acridiens en Rapport avec la Phase. pp. 201–16. *Colloques int. Cent. natn. Rech. scient. Paris* No. **114**. (258)

Waloff, Z. (1963) Field studies on solitary and *transiens* desert locusts in the Red Sea area. *Anti-Locust Bull.* **40**, 93 pp. (258)

Waloff, Z. (1972) Orientation of flying locusts, *Schistocerca gregaria* (Forsk.), in migrating swarms. *Bull. ent. Res.* **62**, 1–72. (19, 240, 452, 454, **455–7**, 458–61)

Waloff, Z. and Rainey, R.C. (1951) Field studies on factors affecting the displacement of desert locust swarms in eastern Africa. *Anti-Locust Bull.* **9**, 50 pp. (257, 258, 259)

Walz, O.C. (1960) Migration of Ranch County

youth. *South-western Social Science Quarterly.* **41** (Supp.), 283–9. (137, 138)

Ward, P. (1963) Lipid levels in birds preparing to cross Sahara. *Ibis* **105**, 109–11. (649)

Ward, P. (1971) The migration patterns of *Quelea quelea* in Africa. *Ibis* **113**, 275–97. (674)

Ward, R.G. (1961) Internal migration in Fiji. *Journal of the Polynesian Society.* **70**, 257–71. (127)

Warnecke, G. and Harz, K. (1964) Zur Kenntnis der Populationsdynamik und des Migrationsverhaltens von *Vanessa atalanta* L. im paläarktischen Raum. *Beitr. ent.* **14**, 155–8. (426)

Waterman, T.H. (1972) Visual direction finding by fishes. *In:* Galler, S.R., Schmidt-Koenig, K., Jacobs, G.J. and Belleville, R.E. (eds.) *Animal orientation and navigation.* Scientific and Technical Information Office, National Aeronautics and Space Administration, Washington D.C. pp. 437–56. (880)

Waterman, T.H., Nunnemacher, R.F., Chace, F.A. and Clarke, G.L. (1939) Diurnal vertical migrations of deep-water plankton. *Biol. Bull.* **76**, 256–79. (485, **504**)

Waters, T.F. (1964) Recolonisation of denuded stream bottom areas by drift. *Trans. Am. Fish. Soc.* **93**, 311–15. (783).

Waters, T.F. (1965) Interpretation of invertebrate drift in streams. *Ecology* **46**, 327–34. (783)

Waters, T.F. (1972) The drift of stream insects. *Ann. Rev. Ent.* **17**, 253–72. (780–3, 786)

Watson, A. and Miller, G.R. (1971) Territory size and aggression in a fluctuating red grouse population. *J. Anim. Ecol.* **40**, 367–84. (**405**)

Watts, C.H.S. (1970) Long distance movement of bank voles and woodmice. *J. Zool., Lond.* **161**, 241–56. (349)

Wehner, R. (1972) Visual orientation performances of desert ants (*Cataglyphis bicolor*) toward astromenotactic direction and horizon landmarks. *In:* Galler, S.R., Schmidt-Koenig, K., Jacobs, G.J., and Belleville, R.E. (eds.) *Animal orientation and navigation.* Scientific and Technical Information Office, National Aeronautics and Space Administration, Washington D.C. pp. 421–36. (877)

Weiner, J.S. (1971) *Man's natural history.* Weidenfeld and Nicolson, London. (118, 124, 125)

Weis-Fogh, T. (1952) Fat combustion and metabolic rate of flying locusts (*Schistocerca gregaria* Forskål). *Phil. Trans. R. Soc.* (B) **237**, 1–36. (262)

Wells, G.P. (1945) The mode of life of *Arenicola marina. J. mar. biol. Ass. U.K.* **26**, 170–207. (**838**)

Werth, I. (1947) The tendency of blackbird and song-thrush to breed in their birthplaces. *British Birds* **40**, 328–30. (**323**)

West, G.C. (1960) Seasonal variation in the energy balance of the tree sparrow in relation to migration. *Auk* **77**, 306–29. (658, 660)

Whiten, A. (1972) Operant study of sun altitude and pigeon navigation. *Nature, Lond.* **237**, 405–6. (864)

Whitford, W.G. and Vinegar, A. (1966) Homing, survivorship and overwintering of larvae in spotted salamanders, *Ambystoma maculatum. Copeia* **1966**, 515–19. (196, 521)

Wilke, F. and Kenyon, K.W. (1954) Migration and food of the northern fur seal. *Trans. 19th N. Amer. Wildl. Conf., Wildl. Mgmt. Inst., Washington D.C.* pp. 430–40. (713, **735**)

Wilkinson, D.H. (1952) The random element in bird 'navigation'. *J. exp. Biol.* **29**, 532–60. (861)

Williams, B.J. (1968) The Birhor of India and some comments on band organization. *In:* Lee, R.B. and DeVore, I. (eds.) *Man the hunter.* Aldine, Chicago. pp. 126–31. (551)

Williams, C.B. (1930) *The migration of butterflies.* Oliver and Boyd, Edinburgh. (20, 257, 426)

Williams, C.B. (1958) *Insect Migration.* Collins, London. (7, 18, 20, 24, 75, 243, 258, 421, 426, **447**, 451, 454, 463, 470, 472, **474**)

Williams, C.B. (1961) Studies in the effect of weather conditions on the activity and abundance of insect populations. *Phil. Trans. R. Soc.* (B) **244**, 331–78. (256)

Williams, C.B. (1962) Studies on blackflies (Diptera: Simuliidae) taken in a light trap in Scotland. III. The relation of night activity and abundance to weather conditions. *Trans. R. ent. Soc. Lond.* **114**, 28–47. (256)

Williams, C.B., Cockbill, G.F., Gibbs, M.E. and Downes, J.A. (1942) Studies in the migration of Lepidoptera. *Trans. R. ent. Soc. Lond.* **92**, 101–283. (433)

Williams, T.C. and Williams, J.M. (1970) Radio tracking of homing and feeding flights of a neotropical bat, *Phyllostomus hastatus. Anim. Behav.* **18**, 302–9. (**892**)

Williams, T.C., Williams, J.M. and Griffin, D.R. (1966) Visual orientation in homing bats. *Science* **152**, 677. (891)

Williams, T.C., Williams, J.M., Teal, J.M. and Kanwisher, J.W. (1972) Tracking radar studies of bird migration. *In:* Galler, S.R., Schmidt-Koenig, K., Jacobs, G.J. and Belleville, R.E. (eds.) *Animal orientation and navigation.* Scientific and Technical Information Office, National Aeronautics and Space Administration, Washington, D.C. pp. 115–28. (676, 920)

Wilson, A.P. (1968) *Social behavior of free-ranging rhesus monkeys with an emphasis on aggression.* Doctoral thesis, University of California, Berkeley. (162)

Wilson, E.O. (1962a) Chemical communication among workers of the fire ant, *Solenopsis saevissima* (Fr. Smith). 1. The organization of mass-foraging. *Anim. Behav.* **10**, 134–47. (149)

Wilson, E.O. (1962b) Chemical communication among workers of the fire ant, *Solenopsis saevissima* (Fr. Smith). 3. The experimental induction of social responses. *Anim. Behav.* **10**, 159–64. (149)

Wilson, E.O. (1965) Chemical communication in the social insects. *Science* **149**, 1064–72. (146, 147)

Wilson, N. (1965) Red bats attracted to insect light traps. *J. Mammal.* **46**, 704–5. (**576**)

Wiltschko, W. (1972) The influence of magnetic total intensity and inclination on directions preferred by migrating European robins (*Erithacus rubecula*). *In:* Galler, S.R., Schmidt-Koenig, K., Jacobs, G.J. and Belleville, R.E. (eds.) Scientific and Technical Information Office, National Aeronautics and Space Administration, Washington D.C. pp. 569–78. (858, **859**)

Wiltschko, W. (1974) Der Magnetkompass der Gartengrasmücke (*Sylvia borin*). *J. Ornithol.* **115**, 1–7. (608)

Wiltschko, W. and Merkel, F.W. (1971) Zugorientierung von Dorngrasmücken (*Sylvia communis*) im Erdmagnetfeld. *Vogelwarte* **26**, 245–9. (608)

Wiltschko, W. and Wiltschko, R. (1976) Interrelation of magnetic compass and star orientation in night-migrating birds. *J. Comp. Physiol.* **109**, 91–9. (858)

Wimsatt, W.A. (1969) Transient behavior, nocturnal activity patterns and feeding efficiency of vampire bats (*Desmodus rotundus*) under natural conditions. *J. Mammal.* **50**, 233–44. (563, **565**)

Wimsatt, W.A. and Trapido, H. (1952) Reproduction and the female reproductive cycle in the tropical American vampire bat, *Desmodus rotundus murinus. Amer. J. Anat.* **91**, 415. (**565**)

Wølk, K. (1969) (Migratory character of the Baltic population of the porpoise, *Phocaena phocaena* (L.))

(In Polish, English Summary). *Prezegl. Zool.* **13**, 349–51. (757)

Wood, P.A. (1971) Studies on laboratory and field populations of three British pseudoscorpions, with particular reference to their gonadial cycles. Unpublished Ph.D. thesis, University of Manchester, England. (505, **511, 513**)

Wood, P.A. and Gabbutt, P.D. (1978) Seasonal vertical distribution of pseudoscorpions in beech litter. *Bull. Br. arachnol. Soc.* **4** (in press). (**511, 513**)

Woodburn, J. (1968a) An introduction to Hadza ecology. *In:* Lee, R.B. and DeVore, I. (eds.) *Man the hunter.* Aldine, Chicago. pp. 49–55. (550)

Woodburn, J. (1968b) Stability and flexibility in Hadza residential groupings. *In:* Lee, R.B. and DeVore, I. (eds.) *Man the hunter.* Aldine, Chicago. pp. 103–10. (550)

Woodbury, A.M. and Hardy, R. (1940) The dens and behaviour of the desert tortoise. *Science* **92**, 529. (523)

Woodbury, A.M. and Hardy, R. (1948) Studies of the desert tortoise, *Gopherus agassizii. Ecol. Monographs* **18**, 145–200. (523)

Woodbury, A.M., Vetas, B., Julian, G., Glissmeyer, H.R., Hyrend, F.L., Call, A., Smart, E.W. and Sanders, R.T. (1951) Symposium: a snake den in Tooele County, Utah. *Herpetologica* **7**, 1–52. (523)

Wynne-Edwards, V.C. (1962) *Animal dispersion in relation to social behaviour.* Hafner, New York. (76, 160, 230)

Yablokov, A.V. (1961) (The 'sense of smell' of marine mammals) (In Russian). *Trudy Sovesh. Iktiol. Komiss., Akad. Nauk SSSR* **12**, 87–93. (766)

Yalden, D.W. (1971a) The flying ability of *Archaeopteryx. Ibis* **113**, 349–56. (604)

Yalden, D.W. (1971b) *Archaeopteryx* again. *Nature, Lond.* **234**, 478–9. (604)

Yalden, D.W. and Morris, P.A. (1975) *The lives of bats.* David and Charles, Newton Abbott. (556)

Yamada, M. (1963) A study of blood-relationship in the natural society of the Japanese macaque: an analysis of co-feeding, grooming, and playmate relationships in Minoo-B-troop. *Primates*, **4**, 43–65. (**279**)

Yates, T.L., Schmidly, D.J. and Culbertson, K.L. (1976) Silver-haired bat in Mexico. *J. Mammal.* **57**, 205. (**574, 576**)

Yeagley, H.L. (1947) A preliminary study of a physical basis of bird navigation. *J. Appl. Phys.* **18**, 1035–63. (865)

Yengoyan, A.A. (1968) Demographic and ecological influences on aboriginal Australian marriage sections. *In:* Lee, R.B. and DeVore, I. (eds.) *Man the hunter.* Aldine, Chicago. pp. 185–99. (**286**)

Yerger, R.W. (1953) Home range, territoriality, and populations of the chipmunk in central New York. *J. Mammal.* **34**, 448–58. (**309**)

Yonge, C.M. (1949) *The sea shore.* Collins, London. (**837**, 841)

Yoshiba, K. (1968) Local and intertroop variability in ecology and social behavior of common Indian langurs. *In:* Jay, P. (ed.) *Primates: studies in adaptation and variability.* Holt, Rinehart and Winston, New York. (164, **295**)

Young, A.M. (1970) Foraging of vampire bats (*Desmodus rotundus*), in Atlantic wet lowland Costa Rica. *Rev. Biol. Trop.* **18**, 73–88. (**565**)

Young, E.C. (1963) *Studies on flight muscle polymorphism in Corixidae (Hemiptera-Heteroptera).* Ph.D. Thesis, London University. (241, 246)

Young, G. (1962) *The hill tribes of northern Thailand: a socio-ethnological report* (2nd ed.). Siam Society, Bangkok. (117, 118, 120, 125, 126, 551)

Young, J.Z. (1962) *The life of vertebrates* (2nd ed.) Clarendon Press, Oxford. (556, 764, 771)

Young, J.Z. (1971) *An introduction to study of man.* Oxford University Press, New York. (**284, 286, 290**)

Young, S.P. (1964) *The wolves of North America. Part 1. Their history, life habits, economic status and control.* Dover publ. Inc., New York. (116, 174, 536)

Zaverin, Yu. P. (1966) (The effect of hydrometeorological conditions on commercial concentrations of whales in the Antarctic) (In Russian). *Izv. Tikhookean Nauch-Issled Inst. Ryb. Khoz. Okeanogr.* **58**, 209–21. (746, 764)

Zenkevitch, L. (1963) *Biology of the seas of the U.S.S.R.* George Allen and Unwin Ltd, London. (**815**, 841)

Zhukov, P.I. (1968) (Routes of penetration of Ponto-Caspian fauna, Ichthyofauna into rivers of the Baltic Sea basin) (In Russian). **47**, 1417–19. (847)

Zimmerman, J.L. (1965) Bioenergetics of the dickcissel, *Spiza americana. Physiol. Zool.* **38**, 370–89. (660)

Geographical index

Subject index

Taxonomic Index

References are almost entirely to migrants. Foodplants, etc. are not itemised. Prey species and/or predators are only indicated when the predator/prey interaction is relevant to an understanding of the lifetime track of the animal concerned (or a close relative).

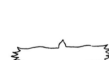

Sus spp (domesticated
  pigs)   124, 130–2
Infraorder Ancodonta
Hippopotamus amphibius
  (hippo)   178–9, 213, 550
Suborder Tylopoda (camels)
Camelus dromedarius (one-humped
  dromedary/domestic
  camel)   19, 133–4, 540,
  550, 887
Vicugna vicugna (vicuña)   166
  295, 306–7, 318
Suborder Ruminantia
  Family Cervidae   886
  Alces alces (moose)   12, 14,
    18, 528–9, 538
  Axis axis (chital/axis
    deer)   551
  Cervus canadensis (elk)   12,
    18, 543–5
  C. elaphus (red deer)   169,
    541–2, 544, 857
  Dama dama (fallow
    deer)   171
  Odocoileus hemionus (mule
    deer)   527–8, 542–3
  O. virginianus (white-tailed
    deer)   168–9, 311, 542
  Rangifer tarandus
    (caribou/reindeer)   14,
    18, 98, 166, 170–1,
    211–3, 219, 266, 528,
    530, 532–3, 536, 540–2,
    544, 546, 554–5

Family Giraffidae
Giraffa camelopardalis
  (giraffe)   531, 535
Family Bovidae
  Subfamily
    Antelopinae   551
  Aepyceros melampus
    (impala)   535
  Alcelaphus buselaphus
    (hartebeest)   535
  Connochaetes taurinus
    (wildebeest)   95–6,
    153–5, 166, 201, 408,
    528, 531, 534–5, 549,
    551
  Damaliscus korrigum
    (topi)   534–5

Gazella spp   95
G. subgutturosa
  (dzheiran)   531
G. thomsoni (Thomson's
  gazelle)   528, 531,
  534–5
Kobus ellipsiprymnus
  (waterbuck)   535,
  549
Procapra gutturosa
  (dzeren)   531
Saiga tatarica
  (saiga)   531, 536,
  546, 554
Taurotragus oryx
  (eland)   95, 528, 535
Tragelaphus strepsiceros
  (greater kudu)   535
Subfamily Caprinae
Capra spp (goats)   124,
  540, 550
Ovibos moschatus
  (muskox)   166, 169,
  531, 542–3, 545,
  547
Ovis spp (sheep)   18,
  124, 133, 234, 312,
  540–1, 550
O. canadensis (bighorn
  sheep)   311, 320, 543,
  554
O. dalli (Dall's
  sheep)   153, 311, 320,
  543, 554
Subfamily Bovinae
Bison bison (American
  bison/'buffalo')   531,
  538, 546, 554–5
B. bonasus (European
  bison)   531, 886
Bos spp (cattle)   124,
  133, 538–40
B. gaurus (gaur)   528
B. javanicus
  (banteng)   166, 551
B. primigenius (European
  wild ox)   531
B. taurus (feral cattle)   551

British Library Cataloguing in Publication Data

Baker, R R
  The evolutionary ecology of animal migration.
  1. Animal migration   2. Evolution
  I. Title
  591.5'2   QL754

  ISBN 0–340–19409–X